Phytotechnologies

Remediation of Environmental Contaminants

Phytotechnologies

Remediation of Environmental Contaminants

Edited by

Naser A. Anjum • Maria E. Pereira
Iqbal Ahmad • Armando C. Duarte
Shahid Umar • Nafees A. Khan

CRC Press is an imprint of the
Taylor & Francis Group, an **informa** business

CRC Press
Taylor & Francis Group
6000 Broken Sound Parkway NW, Suite 300
Boca Raton, FL 33487-2742

First issued in paperback 2022

ISBN-13: 978-1-439-87518-6 (hbk)
ISBN-13: 978-1-03-234026-5 (pbk)
DOI: 10.1201/b12954

Publisher's Note

The publisher has gone to great lengths to ensure the quality of this reprint but points out that some imperfections in the original copies may be apparent.

Library of Congress Cataloging-in-Publication Data

Phytotechnologies : remediation of environmental contaminants / editors, Naser A. Anjum ... [et al.].
p. cm.
Includes bibliographical references and index.
ISBN 978-1-4398-7518-6 (hardback)
1. Phytoremediation. I. Anjum, Naser A.

TD192.75.P489 2012
628.4--dc23 2012025720

Visit the Taylor & Francis Web site at
http://www.taylorandfrancis.com

and the CRC Press Web site at
http://www.crcpress.com

Contents

Foreword

Environmental pollution can be considered as an inevitable evil of human evolution-led immense scientific and technological progress, which has now become one of the most critical challenges facing the world today. Although the pollution level of the biosphere is rapidly going from bad to worse, there is still a dearth of sustainable strategies to resolve varied devastating environmental issues. From this important perspective, a variety of plant- and associated microbe-based technologies—collectively termed phytotechnologies—are now widely accepted as a nature-driven mighty biogeochemical process for cleaning environmental compartments contaminated with varied pollutants. In fact, the term itself describes the application of science and engineering to examine problems and provide solutions involving plants and their associates; thus, it is aimed at providing beneficial vital roles within both societal and natural systems to improve environmental and human health.

Moreover, it is an emerging technology that has the potential to treat a wide range of contaminants at lower cost than traditional technologies. This technology uses various types of plants and plant products to degrade, extract, contain, or immobilize contaminants in soil and water. Phytotechnology has been used for remediation of chlorinated solvents, metals, explosives and propellants, pesticides, polycyclic aromatic hydrocarbons, radionuclides, and petroleum hydrocarbon compounds. While phytotechnologies generally are applied *in situ*, *ex situ* applications (e.g., hydroponics systems) are possible. Typical organic contaminants, such as petroleum hydrocarbons, gas condensates, crude oil, chlorinated compounds, pesticides, and explosive compounds, can be addressed using plant-based methods. Phytotechnologies also can be applied to typical inorganic contaminants, such as heavy metals, metalloids, radioactive materials, and salts. Additionally, phytotechnologies provide numerous advantages for supporting sustainable water quality sanitation and environmental conservation, but the studies and application of phytotechnologies for water sanitation and conservation are still limited, particularly in Asian countries. Hence, more effort should be undertaken to explore scientific information and the use of this technology. Furthermore, some phytotechnology applications involving plants for housing, food, forage, and sources of medicine can create employment. This is particularly important in developing countries.

Researchers Drs. Naser A. Anjum, Maria E. Pereira, Iqbal Ahmad, Armando C. Duarte (CESAM-Centre for Environmental and Marine Studies and the Department of Chemistry, University of Aveiro, Portugal), Dr. Shahid Umar (Hamdard University, New Delhi, India), and Dr. Nafees A. Khan (Aligarh Muslim University, Aligarh, India) have done a timely admirable job of assembling a wealth of information on this sustainable environmental contaminants remediation technology in a single volume. As the current volume has successfully provided a common platform to a broad range of experts including environmental engineers, environmental microbiologists, chemical scientists, and plant physiologists/molecular biologists working with a common aim of sustainable solutions to varied environmental issues, I fervently believe that this volume will be a meaningful addition to the existing body of knowledge that is essential to develop good management practices in this field and for sure will also enlighten readers of various disciplines and at various levels, thus bridging theoretical knowledge to application.

M. N. V. Prasad
Professor, Recipient of Pitamber Pant National Environment Fellowship of the Ministry of Environment and Forests, Government of India
Department of Plant Sciences, University of Hyderabad
Andhra Prudesh, India

Foreword

Foreword II

Soils represent nonrenewable resources of the earth and are the basis of our agricultural production as well as our settlements. The ongoing contamination and subsequent degradation of soils by metals, metalloids, other inorganic pollutants, and organic xenobiotics from industrial production, household emissions, spills, and other anthropogenic activities are increasing processes that negatively impact our quality of life. We ought to be concerned about soil contamination, the origin of the degradation of other valuable core resources such as ground and surface waters, and food production. This negatively affects as well biodiversity and ecosystem services, which are also challenged by climate changes.

Worldwide, around 52 million hectares, representing more than 16% of the total land area, are impacted by some level of soil contamination. In Europe alone, there may be up to 3 million contaminated sites, as estimated by the European Environment Agency (EEA 2007). Only a small fraction of these areas have been risk assessed and cleaned up over the last 30 years. The most heavily contaminated areas are found near industrialized regions, but many contaminated areas also exist around major cities (EEA 2007). The EEA quotes further that new EU member states bring at least additional 250,000 polluted sites that urgently require our attention. For the United States, neither a comprehensive listing of the number of contaminated sites nor an analysis of the degree of contamination and potential risks at given sites is available. Nevertheless, 1,289 polluted Superfund Sites are listed in the National Priorities List (2011). Moreover, instead of decreasing, we can estimate that the number of polluted sites will increase by more than 50% in the next decade, due to more pollution on the one hand, and stricter legislation with better identification tools on the other (Mench et al. 2009). In rapidly developing countries such as Brazil, Russia, India, China, and South Africa, environmental legislation is often patchy and its application too lax, thus leading to severe impacts on soils and related ecosystem compartments with dramatic consequences on water quality, food safety, and human health. Emerging contaminants such as rare earths, pharmaceuticals, and personal care products will reinforce the need for highly efficient analytical and speciation tools, ecotoxicological tests, and bioremediation technologies based on the use of plants and microorganisms.

Given that these numbers of contaminated sites are correct or at least not completely wrong, we have to develop strategies for both preserving clean soils and remediating contaminated ones. As the world population is rapidly increasing and land degradation is progressing, the whole land bank, including agricultural, marginal, and remediated land, is requested to produce food and plant-based feedstock. Without the implementation of efficient remediation options, contaminated sites represent a potential or proven risk for human and/or animal health. In addition, a major nondietary exposure pathway for many inorganic and organic contaminants is via incidental ingestion of contaminated soil and inhalation of dust, notably accumulated in houses, often originating from eroded contaminated soils and tailings. They cannot be classified as waste nor be disposed in (not available) dump sites. On the contrary, we have to understand that even polluted soils are valuable resources that need to be reclaimed. Hence, one main goal must be to improve their matrix quality and functions, including biodiversity, at affordable costs (Mench et al. 2010).

Metals (Cd, Hg, Pb, Zn, Cu, Sn, etc.) and metalloids (Se, As, Ge, Sb, Te, etc.) are important environmental pollutants generally subsumed as trace elements. Many of them are potentially toxic even at very low concentrations in exposure pathways; for essential trace elements, toxicity occurs when their chemical species in a living organism exceed the range needed for biological action. Pollution of the biosphere with trace elements has accelerated dramatically since the beginning of the industrial revolution (Padmavathiamma and Li 2007), when ore melting, steelwork, and the use of metal(loid)s in tools and chemicals became favorable. Today, the primary sources of soil pollution by trace elements are the burning of fossil fuels, mining and smelting of metalliferous ores, and recycling of municipal wastes. The increasing production of various electronic equipment and their disposal contribute to enhancing anthropogenic dispersion of trace elements. In addition, fertilizers, pesticides, pig slurries, sewage, etc. often do contain trace elements in significant concentrations (Wei and Zhou 2008).

Increasing industrialization has provided two novel sources of foreign compounds: first, by the invention and use of agrochemicals for the protection of crops and the control of pests and weeds; second, by the emission of organic xenobiotics in the process of chemical production of goods and the use of synthetic chemicals. Those man-made chemicals represent a threat to our environment as they are emitted without any control. The carefree use of chemicals in the past has led to the accumulation of numerous pollutants in various ecosystems and soils, at and near former production and disposal sites, and also in urban areas. Many of them do represent a threat to human health. Besides, pollution from nonpoint sources leads to low but widespread diffuse pollution with numerous recalcitrant xenobiotics.

Phytoremediation, as the word itself suggests (from the Greek φυτό, meaning *plant*), is a green, plant-based way for (1) removing unwanted compounds from environmental compartments, (2) decreasing their presence and/or changing their speciation in the pollutant linkages, or (3) degrading organic xenobiotics. It is based on the inability of plants to move away from a pollutant source, and on their capacity to adsorb chemicals from the surrounding media. As adsorption is usually followed by uptake and transport in the plant, powerful sequestration and detoxification mechanisms exist that can be exploited to extract, render less mobile, or degrade unwanted substances from the environment. These features of plants and their associated microorganisms have been intensively studied in the framework of the European COST Actions 837 and 859 (http://w3.gre.ac.uk/cost859/). There has been less emphasis on a multidisciplinary approach and on long-term changes in the (bio)-remediated soil–plant system. Beneficial effects of interactions between plants and microorganisms (mycorrhizae, rhizospheric, and endophytic bacteria) have been recognized and better understood. They can contribute to the nutrient supply of their plant hosts, which is important for optimal plant growth in normal conditions, but also for improving plant survival in hostile environments and partial alleviation of contaminant stress. Hence, the scene is set for a new generation of phytoremediation options that can use the full spectrum of plant-based tools to remediate the environment. For instance, the Greenland project (European Commission FP7-KBBE-266124, http://www.greenland-project.eu/) will bring gentle remediation options (phytoremediation, *in situ* stabilization) into practical application to remediate trace element-contaminated soils at low cost and without significant negative effects for the environment.

The term *phytotechnologies* refers to the application of plant-based techniques and engineering to provide sustainable solutions based on the use of plants and their associated microorganisms (Mench et al. 2009). A variety of such techniques and strategies are directly based on the confrontation of living plants with a polluted environment, in many cases leading to contaminant degradation, removal (through accumulation or dissipation), or immobilization.

Compared with conventional "dig-and-dump" methods of soil remediation, the use of plants provides several striking advantages. It is cheap, and after planting, only marginal costs apply for harvesting and field management. One past option was to incinerate the harvested biomass, ashes from metal hyperaccumulators being recycled by smelters in some success stories. No additional carbon dioxide was released into atmosphere beyond what was originally assimilated by plants. However, in response to the EU Commission's support, the biobased industry in Europe has burgeoned, with production activities increasing and the organization of biomass-based product chains. Sustainable biomass production for plant-based feedstock on contaminated soils can be a part of a remediation strategy assisting in contaminant containment, stabilization, degradation, or removal. The choice of plant species depends upon the end use, (local) bioconversion option of interest, e.g. combustion, gasification, pyrolysis, fermentation, hydrothermal oxidation, and mechanical extraction of oils, and the energy form or feedstock required. Some plant species are amenable to nearly all of the potential conversion technologies while others such as wood and cereal crops are only suitable for a few of these. Growth rate, ease of management, harvesting, and the inherent properties of the biomass source determine both the choice of conversion process and any subsequent processing difficulties that may arise. Hence, plant-based environmental remediation techniques are efficient tools, offering environmentally friendly solutions for the rehabilitation of contaminated sites and water, the improvement of food safety, and the carbon sequestration to reduce global warming. Developing renewable energy sources will additionally contribute to sustainable land use management (Vangronsveld et al. 2009).

Worldwide, emerging phytoremediation industries consist of a small but growing number of successful companies forming discrete categories. Among them dedicated phytoremediation companies are most prominent. Their primary remediation technology is the use of plants and associated microorganisms predominantly for wastewater phytoremediation in constructed wetlands. A number of large to midsize consulting firms have developed an expertise in phytoremediation and channel their customers' demands into feasible projects performed by engineers and gardeners. The number of such companies has grown encouragingly in the last few years. Also some of the chemical industries conduct research and field remediation for internal needs, and a large number of academic, government, and other nonprofit research groups conduct research to develop new technologies (Pilon-Smits 2005). Until now, numerous commercial projects have been undertaken in several countries around the world and became visible during the 7th Phytotech conference in Parma, Italy, 2011. Smaller, but emerging, markets exist in developing countries, particularly in large portions of Asia.

Nowadays, we distinguish among *phytoextraction* (the use of pollutant-accumulating plants to remove metals or organics from soil by concentrating them in harvestable parts), *hytodegradation* (the use of plants and associated microorganisms to degrade organic pollutants), *phytovolatilization* (the use of plants to volatilize pollutants or metabolites), *rhizofiltration* (the use of plant roots to adsorb and absorb pollutants, mainly metal(loid)s, from water and aqueous waste streams), *evapotranspiration* (ET) cover systems (so-called pump and tree; the use of plants, notably trees, to evaporate water and to extract pollutants from the soil), *(aided) phytostabilization* (the use of plants with and without soil conditioners to reduce the bioavailability of pollutants in the environment), and *hydraulic control* (the control of the water table and the soil field capacity by plant canopies (Trapp 2007).

Laboratory experiments are widely used to investigate various plant and microbial responses toward pollutants under identical growth conditions. Whether such results are applicable to events in the field is highly debatable, as spatial variability of soil factors, total concentrations of contaminants, texture and depth of soil layers, nutrient and water availability, problems with the homogeneous input of amendments, differences in rooting depth and density, variability of soil moisture, impacts of pests, pathogens, and herbivores can affect both treatment efficiency and plant responses in the field are often not assessed (Friesl et al. 2006). The molecular and biochemical mechanisms of plant adaptation to real stress conditions are still poorly understood and signaling pathways involved remain unclear (Schröder 2001). However, adaptation to environmental changes is crucial for plant growth and survival. When growing energy plants on polluted land, they may suffer from various (a)biotic stresses such as drought, heat, pathogens, nutrient deficiency, and pollutant or metal toxicity. Hence, currently cultivated crops bred for productivity in high input cropping systems will prove unsuitable to do this job. In such a context, COST Actions 837 and 859 have therefore identified promising candidates for successful phytoremediation.

With some of the plants identified there, current research has developed best fit linear regression models for predicting As, Cd, and Pb bioavailability. To assess the efficiency of phytotechnologies in reducing risk from soil–plant–herbivore transfer of pollutants, several factors can be used to express element or compound accumulation in plants. Once in direct contact to the root, any transfer from soil or water to the plant is diffusion driven for compounds with a lipophilicity close to that of the respective plant root. Root uptake and transport of organic xenobiotics have been reviewed by a number of authors (Schröder and Collins 2002). New assessment methods and improved *in situ* containment technologies that directly address contaminant mobility and bioavailability have also enhanced our ability to effectively manage sediments. Organic matter is an important component regulating trace element dynamics in both terrestrial and aquatic ecosystems. The transport and bioavailability of trace elements are strongly influenced by the formation of soluble and insoluble complexes with organic matter.

It has long been known from applied botany that mixed stands develop unique properties enhancing the performance of the whole production system. Consequently, and to exploit plant metabolism for the decontamination of a site or a water body, it is of high interest to plant vegetation above and around a given pollution with species that combine low transpiration and high production and release of root exudates for bacterial stimulation, followed and secured by a belt of deep rooting species with high transpiration rates to take up what is available, and a mixed canopy of perennial plants for the general protection of the site. The correct choice of plants has to be assisted by experts, and computer-aided assessment methods of pollution distribution as well as novel decision tools based on the chemical ecology of plants have to be utilized to guarantee success. In any case, this kind of plantation must be adapted to site specific characteristics. The latter approach takes into account that the normal physiology of plants includes the synthesis of complex secondary compounds that might resemble structures of organic pollutants. Enzymes in the metabolic pathway of these natural compounds are good candidates for the detoxification of structurally similar xenobiotics. Such an approach has been followed by Schwitzguebel et al. (2002, 2008) for the phytoremediation of sulfonated anthraquinones. Additionally in this context, Collins et al. (2006) has demonstrated that modern molecular tools contribute to the search for promising species with the desired metabolic traits.

The current book attempts to summarize the knowledge available on phytoremediation. The first chapters set the scene for the technologies, the pollutants are described, and differences between metal(loid) remediation and the removal of organic pollutants are categorized. In depth studies exemplify the possibilities. The second set of chapters is exclusively dedicated to the genus *Brassica* and its applications in phytoremediation. This is a well-chosen topic, as the Brassicaceae contain some of the most important plant species available in phytoremediation. They are closely related to the model plant, *Arabidopsis thaliana*. They can also provide various oilseeds and plant-based feedstock, even woody crops and perennial grasses should not be neglected. The third compilation of chapters is concerned with case studies concentrating on phytoremediation aided by bacteria. It becomes clear that the key to enhance rhizosphere degradation is to find the right combination of plant and microorganism for an environmental system.

Many internationally renowned actors have contributed to this comprehensive volume, providing the readers with an up-to-date information on phytotechnologies to treat contaminated sites. The present book will be useful for students and teachers, scientists and engineers, policy makers, industrials, and stakeholders, demonstrating the possibilities of this sustainable strategy to remediate our impacted environment and mitigate the effects of many pollutants.

We should ensure that future generations are not deprived of the goods and resources we were able to use and that are exhaustible. Any intended land use of posttreated sites should be taken into consideration in any remedial undertaking. We cannot just use technological advancement without ensuring clean air and water, green landscapes, and healthy soils for the entire society. Finally, it is necessary to include as many stakeholders as possible for the successful implementation of phytoremediation and the assessment of its underlying economics, in order to convince policy makers and to enhance public support. Here, the present comprehensive book on phytotechnologies is a valuable tool to inform the interested reader, and to set the scene for the breakthrough of this sustainable and green technology.

Peter Schröder
Research Unit Microbe Plant Interactions, Helmholtz Zentrum Muenchen, German Research Center for Environmental Health, Neuherberg, Germany

Michel Mench
UMR BIOGECO Cestas, France
and
University of Bordeaux 1, Talence, France

Jean-Paul Schwitzguébel
Laboratory for Environmental Biotechnology, Lausanne, Switzerland

REFERENCES

Collins, C., M. Fryer, and A. Grosso. (2006). Plant uptake of non-ionic organic chemicals. *Environmental Science & Technology, 40*, 45–52.

EEA (European Environment Agency). (2007). Progress in management of contaminated sites. CSI 015, DK-1050 Copenhagen, K, Denmark.

Friesl, W., J. Friedl, K. Platzer, O. Horak, and M. H. Gerzabek. (2006). Remediation of contaminated agricultural soils near a former Pb/Zn smelter in Austria: Batch, pot and field experiments. *Environmental Pollution, 144*, 40–50.

Mench, M., N. Lepp, V. Bert, J. P. Schwitzguébel, S. W. Gawronski, P. Schröder, and J. Vangronsveld. (2010). Successes and limitations of phytotechnologies at field scale: Outcomes, assessment and outlook from COST Action 859. *Journal of Soils and Sediments, 10*, 1039–1070.

Mench, M., J. P. Schwitzguebel, P. Schroder, V. Bert, S. Gawronski, and S. Gupta. (2009). Assessment of successful experiments and limitations of phytotechnologies: Contaminant uptake, detoxification and sequestration, and consequences for food safety. *Environmental Science and Pollution Research, 16*, 876–900.

Padmavathiamma, P. K., and L. Y. Li. (2007). Phytoremediation technology: Hyper accumulation of metals in plants. *Water, Air and Soil Pollution, 184*, 105–126.

Pilon-Smits, E. (2005). Phytoremediation. *Annual Review of Plant Biology, 56*, 15–39.

Schröder, P. (2001). The role of glutathione and glutathione S transferases in plant reaction and adaptation to xenobiotics. In D. Grill (Ed.), *Significance of Glutathione to Plant Adaptation to the Environment* (155–183). Netherlands: Kluwer.

Schröder, P., and C. Collins. (2002). Conjugating enzymes involved in xenobiotic metabolism of organic xenobiotics in plants. *International Journal of Phytoremediation, 4*, 247–265.

Schwitzguebel, J. P., S. Aubert, W. Grosse, and F. Laturnus. (2002). Sulphonated aromatic pollutants—limits of microbial degradability and potential of phytoremediation. *Environmental Science and Pollution Research, 9*, 62–72.

Schwitzguébel, J. P., S. Braillard, V. Page, and S. Aubert. (2008). Accumulation and transformation of sulfonated aromatic compounds by higher plants—toward the phytotreatment of wastewater from dye and textile industries. In N.A. Khan, S. Umar, S. Singh. (Eds.), *Sulfur Assimilation and Abiotic Stress in Plants* (335–354). Berlin: Springer.

Trapp, S. (2007). Fruit tree model for uptake of organic compounds from soil and air. *SAR QSAR Environmental Research, 18*, 367–387.

Vangronsveld, J., R. Herzig, N. Weyens, J. Boulet, K. Adriaensen, A. Ruttens, T. Thewys, A. Vassilev, E. Meers, E. Nehnevajova, D. van der Lelie, and M. Mench. (2009). Phytoremediation of contaminated soils and groundwater: Lessons from the field. *Environmental Science and Pollution Research, 16*, 765–794.

Wei, S., and Q. Zhou. (2008). Trace elements in agro-ecosystems. In M.N.V. Prasad. (Ed.), *Trace Elements as Contaminants and Nutrients—Consequences in Ecosystems and Human Health* (55–80). New Jersey: Wiley.

Preface

During the last few decades, the exploration of sustainable solutions to myriad rapidly mounting global environmental issues has been the major subject of environmental pollution research. In this perspective, the strategic use of the natural and inherent traits of plants and their associated microbes to exclude, accumulate, immobilize, metabolize, or degrade various environmental contaminants—collectively called *phytotechnology*—significantly contributes to the final outcome of various environmental contaminants and efficiently and sustainably decontaminate the biosphere from unwanted hazardous compounds. In fact, phytotechnologies are essentially a form of ecological engineering that capitalizes on the naturally occurring relationships among plants, microorganisms, and environment. Moreover, as phytotechnologies use human initiative to enhance natural plant- and associated microorganism-assisted solutions of various environmental problems, they represent a technology that is intermediate between engineering and natural attenuation (McCutcheon and Schnoor 2003; ITRC 2001, 2009; Prasad et al. 2010).

Although a plethora of publications focused on plants and associated microbe-based remediation of various environmental contaminants has been exponentially growing in the last decades, available research reports and findings from different arenas are largely disorganized and are not critically cross-talked and/or integrated. Therefore, this book is an effort to provide a common platform to environmental engineers, environmental microbiologists, chemical scientists, and plant physiologists/molecular biologists who are working with a common aim of finding sustainable solutions to various environmental issues. This book addresses in a single volume the major aspects of phytotechnology that are crucial for understanding the ecosystem approaches that help achieve the sustainable development objectives.

Contributed by an international team of authors, this book, aside from assessing the current state of the science and application of *phytotechnologies* around the globe, presents an overview of major environmental contaminants, contaminated sites, and the significant role of plants including *Brassica* and vetiver grass species for the remediation of various environmental contaminants, and also an exhaustive exploration of potential strategies for enhancing various environmental contaminant remediation, and critically discusses the major physiological, biochemical, and genetic–molecular mechanisms responsible for plant tolerance/adaptation to important environmental contaminants. Thus, the outcome of the present book may provide a conceptual overview of ecosystem approaches and phytotechnologies and their cumulative significance in relation to various environmental problems and potential solutions. Moreover, as the present treatise ensures a good equilibrium between theory and practice without compromising the basic conceptual framework of the concerned subject covering a wide range of important topics under a common umbrella of phytotechnologies, this volume will be a useful asset to students, researchers, practitioners, policy makers specializing in the areas of soil-sediment pollution, environmental chemistry/microbiology/plant physiology/molecular biology, sustainable development, ecology, soil biology, and related disciplines.

We are thankful to the contributors for their interest, significant contributions, and cooperation that made the present volume possible. Thanks are also due to all the well-wishers, teachers, seniors, research students, colleagues, and our families. Without their unending moral support, motivation, endurance, and encouragement, the grueling task would have never been accomplished. We also extend our appreciation to Hilary Rowe, Amy Blalock, Jennifer Derima, and their team at CRC Press/Taylor & Francis Publications Group for their exceptional kind support, which made our efforts successful.

Last, but not least, the financial support to our research from the Foundation for Science & Technology (FCT), Portugal, the Aveiro University Research Institute/Centre for Environmental and Marine Studies (CESAM), University Grants Commission (UGC) and Department of Science and Technology (DST), New Delhi, India, Indian Potash Research Institute (IPRI), Gurgaon, India, and International Potash Institute (IPI), Switzerland, are gratefully acknowledged.

REFERENCES

Interstate Technology and Regulatory Council (ITRC). 2001. *Phytotechnology Technical and Regulatory Guidance Document*; [cited 2001 January]. Available from www.itrcweb.org/Documents/PHYTO-2.pdf.

Interstate Technology and Regulatory Council (ITRC). 2009. *Phytotechnology Technical and Regulatory Guidance and Decision Trees, Revised*. PHYTO-3. Washington, D.C.: Interstate Technology & Regulatory Council, Phytotechnologies Team. www.itrcweb.org.

McCutcheon, S.C., and J.L. Schnoor. 2003. *Phytoremediation: transformation and control of contaminants*. New Jersey: John Wiley & Sons, Inc.

Prasad, M.N.V., H. Freitas, S. Fraenzle, S. Wuenschmann, and B. Markert. 2010. Knowledge explosion in phytotechnologies for environmental solutions. *Environmental Pollution* 158: 18–23.

Contributors

Iqbal Ahmad
Centre for Environmental and Marine Studies (CESAM)
Department of Chemistry
and
Department of Biology
University of Aveiro
Aveiro, Portugal

Rashid Ahmad
Department of Crop Physiology
University of Agriculture
Faisalabad, Pakistan

Eki T. Aisien
College of Education
Ekiadolor-Benin, Nigeria

Felix A. Aisien
Department of Chemical and Environmental Engineering
University of Benin
Benin City, Nigeria

Naser A. Anjum
Centre for Environmental and Marine Studies (CESAM)
Department of Chemistry
University of Aveiro
Aveiro, Portugal

Iztok Arčon
University of Nova Gorica
Nova Gorica, Slovenia
and
Jožef Stefan Institute
Ljubljana, Slovenia

Muhammad Yasin Ashraf
Nuclear Institute for Agriculture and Biology (NIAB)
Faisalabad, Pakistan

Nazila Azhar
Institute of Horticulture
University of Agriculture
Faisalabad, Pakistan

Meri Barbafieri
Institute of Ecosystem Study
National Research Council
Pisa, Italy

M. Caçador
Centro de Oceanografia
Instituto de Oceanografia
Lisboa, Portugal

Cristina S. C. Calheiros
CBQF/Escola Superior de Biotecnologia
Universidade Católica Portuguesa
Porto, Portugal

Lixiang Cao
Department of Biochemistry
Sun Yat-sen University
Guangzhou, China

Marisol Castrillo
Departmento Biologia de Organismos
Universidad Simon Bolivar
Caracas, Venezuela

Paula M. L. Castro
CBQF/Escola Superior de Biotecnologia
Universidade Católica Portuguesa
Porto, Portugal

Bodhisatwa Chaudhuri
Post Graduate Department of Biotechnology
St. Xavier's College
Kolkata, West Bengal, India

Luu Thai Danh
College of Agriculture and Applied Biology
University of Can Tho
Can Tho City, Vietnam

Kaushik Das
Post Graduate Department of Biotechnology
St. Xavier's College
Kolkata, West Bengal, India

Katerina Demnerova
Department of Biochemistry and Microbiology
Institute of Chemical Technology Prague
Prague, Czech Republic

Andrea De Sousa
Postgrado Cs. Biologicas
Universidad Simon Bolivar
Caracas, Venezuela

Sharon Doty
College of the Environment
School of Environmental and Forest Sciences
University of Washington
Seattle, Washington

Armando C. Duarte
Centre for Environmental and Marine Studies (CESAM)
Department of Chemistry
University of Aveiro
Aveiro, Portugal

B. Duarte
Centro de Oceanografia
Instituto de Oceanografia
Lisboa, Portugal

Andrew Agbontalor Erakhrumen
Department of Forest Resources Management
University of Ibadan
Ibadan, Nigeria

Kevin Falk
Saskatoon Research Center
Agriculture and Agri-Food Canada
Saskatoon, Saskatchewan, Canada

Guido Fellet
Dipartimento di Scienze Agrarie e Ambientali
Università di Udine
Udine, Italy

Jan Fiser
Department of Biochemistry and Microbiology
Institute of Chemical Technology Prague
Prague, Czech Republic

Neil R. Foster
Supercritical Fluids Research Group
The University of New South Wales
Sydney, New South Wales, Australia

J. Freitas
Centro de Oceanografia
Instituto de Oceanografia
Lisboa, Portugal

Masayuki Fujita
Laboratory of Plant Stress Responses
Department of Applied Biological Science
Kagawa University
Kagawa, Japan

Elisa Gamalero
Dipartimento di Scienze dell'Ambiente e della Vita
Università del Piemonte Orientale
Alessandria, Italy

Jorge L. Gardea-Torresdey
Department of Chemistry
The University of Texas at El Paso
El Paso, Texas

Sarvajeet S. Gill
Plant Molecular Biology Group
International Centre for Genetic Engineering and Biotechnology
Aruna Asaf Ali Marg
New Delhi, India

Bernard R. Glick
Department of Biology
University of Waterloo
Waterloo, Ontario, Canada

Margaret Y. Gruber
Saskatoon Research Center
Agriculture and Agri-Food Canada
Saskatoon, Saskatchewan, Canada

Mirza Hasanuzzaman
Laboratory of Plant Stress Responses
Department of Applied Biological Science
Kagawa University
Kagawa, Japan
and
Department of Agronomy
Sher-e-Bangla Agricultural University
Dhaka, Bangladesh

Ismael Hernández-Valencia
Instituto de Zoología y Ecología Tropical
Universidad Central de Venezuela
Caracas, Venezuela

Carmen Infante
Instituto de Ciencias de la Tierra
Universidad Central de Venezuela
Caracas, Venezuela

Erik J. Joner
Bioforsk Soil & Environment
Aas, Norway

Rashid Kaveh
Department of Civil and Environmental Engineering
Temple University
Philadelphia, Pennsylvania

Nafees A. Khan
Department of Botany
Faculty of Life Sciences
Aligarh Muslim University
Aligarh, Uttar Pradesh, India

Zareen Khan
College of the Environment
School of Environmental and Forest Sciences
University of Washington
Seattle, Washington

Z. H. Khan
Department of Botany
Christ Church College
Kanpur, India

Špela Koren
Department of Biology
University of Ljubljana
Ljubljana, Slovenia

Pavel Kotrba
Department of Biochemistry and Microbiology
Institute of Chemical Technology Prague
Prague, Czech Republic

Peter Kump
Jožef Stefan Institute
Ljubljana, Slovenia

Nand Lal
Department of Life Sciences
Chhatrapati Shahu Ji Maharaj University
Kanpur, India

Xiang Li
Saskatoon Research Center
Agriculture and Agri-Food Canada
Saskatoon, Saskatchewan, Canada
and
Department of Biochemistry and Biomedical Sciences
McMaster University
Hamilton, Ontario, Canada

Liliana López
Instituto de Ciencias de la Tierra
Universidad Central de Venezuela
Caracas, Venezuela

Petra Lovecká
Department of Biochemistry and Microbiology
Institute of Chemical Technology Prague
Prague, Czech Republic

Tomas Macek
Department of Biochemistry and Microbiology
Institute of Chemical Technology Prague
and
Institute of Organic Chemistry and Biochemistry
Czech Academy of Sciences
IOCB & ICT Prague Joint Laboratory
Prague, Czech Republic

Martina Mackova
Department of Biochemistry and Microbiology
Institute of Chemical Technology Prague
and
Institute of Organic Chemistry and Biochemistry
Czech Academy of Sciences
IOCB & ICT Prague Joint Laboratory
Prague, Czech Republic

Khalid Mahmood
Nuclear Institute for Agriculture and Biology (NIAB)
Faisalabad, Pakistan

Raffaella Mammucari
Supercritical Fluids Research Group
The University of New South Wales
Sydney, New South Wales, Australia

Luca Marchiol
Dipartimento di Scienze Agrarie e Ambientali
Università di Udine
Udine, Italy

Lucie Musilova
Department of Biochemistry and Microbiology
Institute of Chemical Technology Prague
Prague, Czech Republic

Jitka Najmanova
Department of Biochemistry and Microbiology
Institute of Chemical Technology Prague
Prague, Czech Republic

Marijan Nečemer
Jožef Stefan Institute
Ljubljana, Slovenia

Martina Novakova
Department of Biochemistry and Microbiology
Institute of Chemical Technology Prague
Prague, Czech Republic

K. Nüsslein
Department of Microbiology
University of Massachusetts
Amherst, Massachusetts

Chris O. Nwoko
Department of Environmental Technology
Federal University of Technology
Owerri, Nigeria

Innocent O. Oboh
University of Uyo
Uyo, Nigeria

Francesca Pedron
Institute of Ecosystem Study
National Research Council
Pisa, Italy

Primož Pelicon
Jožef Stefan Institute
Ljubljana, Slovenia

Jose R. Peralta-Videa
Department of Chemistry
The University of Texas at El Paso
El Paso, Texas

Maria E. Pereira
Centre for Environmental and Marine Studies (CESAM)
Department of Chemistry
University of Aveiro
Aveiro, Portugal

Beatriz Pernia
Postgrado Cs. Biologicas
Universidad Simon Bolivar
Caracas, Venezuela

Sreeparna Pradhan
Post Graduate Department of Biotechnology
St. Xavier's College
Kolkata, West Bengal, India

Yuan Pu
Supercritical Fluids Research Group
The University of New South Wales
Sydney, New South Wales, Australia

António O. S. S. Rangel
CBQF/Escola Superior de Biotecnologia
Universidade Católica Portuguesa
Porto, Portugal

Marjana Regvar
Department of Biology
University of Ljubljana
Ljubljana, Slovenia

Rosa Reyes
Departmento Biologia de Organismos
Universidad Simon Bolivar
Caracas, Venezuela

Aryadeep Roychoudhury
Post Graduate Department of Biotechnology
St. Xavier's College
Kolkata, West Bengal, India

Ganapathi Sridevi
Department of Plant Biotechnology
Madurai Kamaraj University
Madurai, Tamil Nadu, India

L. M. Stout
Department of Microbiology
University of Massachusetts
Amherst, Massachusetts

Michal Strejcek
Department of Biochemistry and Microbiology
Institute of Chemical Technology Prague
Prague, Czech Republic

Rouzbeh Tehrani
Department of Civil and Environmental Engineering
Temple University
Philadelphia, Pennsylvania

Eva Tejklová
Plant Biotechnology Department
AGRITEC Plant Research Ltd.
Sumperk, Czech Republic

Palaniswamy Thangavel
Department of Environmental Science
Periyar University
Salem, Tamil Nadu, India

Marcia Toro
Instituto de Zoología y Ecología Tropical
Universidad Central de Venezuela
Caracas, Venezuela

Paul Truong
The Vetiver Network International
and
Veticon Consulting
Brisbane, Queensland, Australia

Ondrej Uhlik
Department of Biochemistry and Microbiology
Institute of Chemical Technology Prague
and
Institute of Organic Chemistry and Biochemistry
Czech Academy of Sciences
IOCB & ICT Prague Joint Laboratory
Prague, Czech Republic

Shahid Umar
Department of Botany
Faculty of Science
Hamdard University
New Delhi, India

Benoit van Aken
Department of Civil and Environmental Engineering
Temple University
Philadelphia, Pennsylvania

Mary Varkey
Department of Botany
Christ Church College
Kanpur, India

Primož Vavpetič
Jožef Stefan Institute
Ljubljana, Slovenia

Jitka Viktorova
Department of Biochemistry and Microbiology
Institute of Chemical Technology Prague
Prague, Czech Republic

Katarina Vogel-Mikuš
Department of Biology
University of Ljubljana
Ljubljana, Slovenia

Miroslava Vrbová
Plant Biotechnology Department
AGRITEC Plant Research Ltd.
Sumperk, Czech Republic

Ejaz Ahmad Waraich
Department of Crop Physiology
University of Agriculture
Faisalabad, Pakistan

Neil Westcott
Saskatoon Research Center
Agriculture and Agri-Food Canada
Saskatoon, Saskatchewan, Canada

Liu Xiaona
School of Land Science and Technology
China University of Geosciences
Beijing, China

Giuseppe Zerbi
Dipartimento di Scienze Agrarie e Ambientali
Università di Udine
Udine, Italy

Zhao Zhongqiu
School of Land Science and Technology
China University of Geosciences
and
Key Laboratory of Land Consolidation and Rehabilitation
Ministry of Land and Resources
Beijing, China

1 Introduction

Naser A. Anjum, Iqbal Ahmad, Armando C. Duarte, Shahid Umar, Nafees A. Khan, and Maria E. Pereira

CONTENTS

1.1 GENERAL CONSIDERATIONS

With the development of the Industrial Age and the rapid rise in world population over the last century, societies have allowed unchecked release of large amounts of varied contaminants into different environmental compartments, thus posing significant consequences for human health, biodiversity, and ecosystem stability (Gerhardt et al. 2009; Nordberg 2009). Therefore, during the last few decades, global exploration of sustainable solutions to a myriad of rapidly mounting environmental issues globally has been the major subject of environmental pollution research and of that regarding potential solutions. From this perspective, the strategic use of plants and their associated microbes to exclude, accumulate, immobilize, metabolize, or degrade varied environmental contaminants—collectively called *phytotechnology*—is contributing significantly to the fate of varied environmental contaminants and to efficiently and sustainably decontaminating the biosphere from unwanted hazardous compounds. Although the use of plants and associated microbes for contaminant remediation dates back to the Roman Empire, the concept of phytoremediation was born in the 1980s out of the extraordinary ability that some plant species display in accumulating high quantities of toxic metals in their tissues or organs (Jez 2011; Maestri and Marmiroli 2011). Over the years, the term *phytoremediation* began being used in the scientific literature, starting in 1993 (Cunningham and Berti 1993; Raskin et al. 1994; Salt et al. 1995). In subsequent years, the definition evolved into *phytotechnologies* (Interstate Technology and Regulatory Council 2001), meaning a wide range of technologies applied successfully to remediate pollutants through a number of significant strategies, such as stabilization; volatilization; metabolism, including degradation at the level of the rhizosphere; accumulation; and sequestration (McCutcheon and Schnoor 2003; Maestri and Marmiroli 2011) (Figure 1.1). Moreover, *phytotechnologies*, in fact, are essentially a form of ecological engineering, capitalizing on naturally occurring relationships among plants, microorganisms, and their environment. Additionally, as *phytotechnologies* employ human initiative to enhance natural plant- and associated microorganism-assisted solutions to varied environmental problems, they represent a technology that is intermediate between engineering and natural attenuation (McCutcheon and Schnoor 2003; Interstate Technology and Regulatory Council 2009; Prasad et al. 2010).

It is worth mentioning here that although there exists a plethora of publications focused on plant- and associated microbe-based remediation of varied environmental contaminants, which has been growing exponentially in the last decades, the available research reports and findings from different

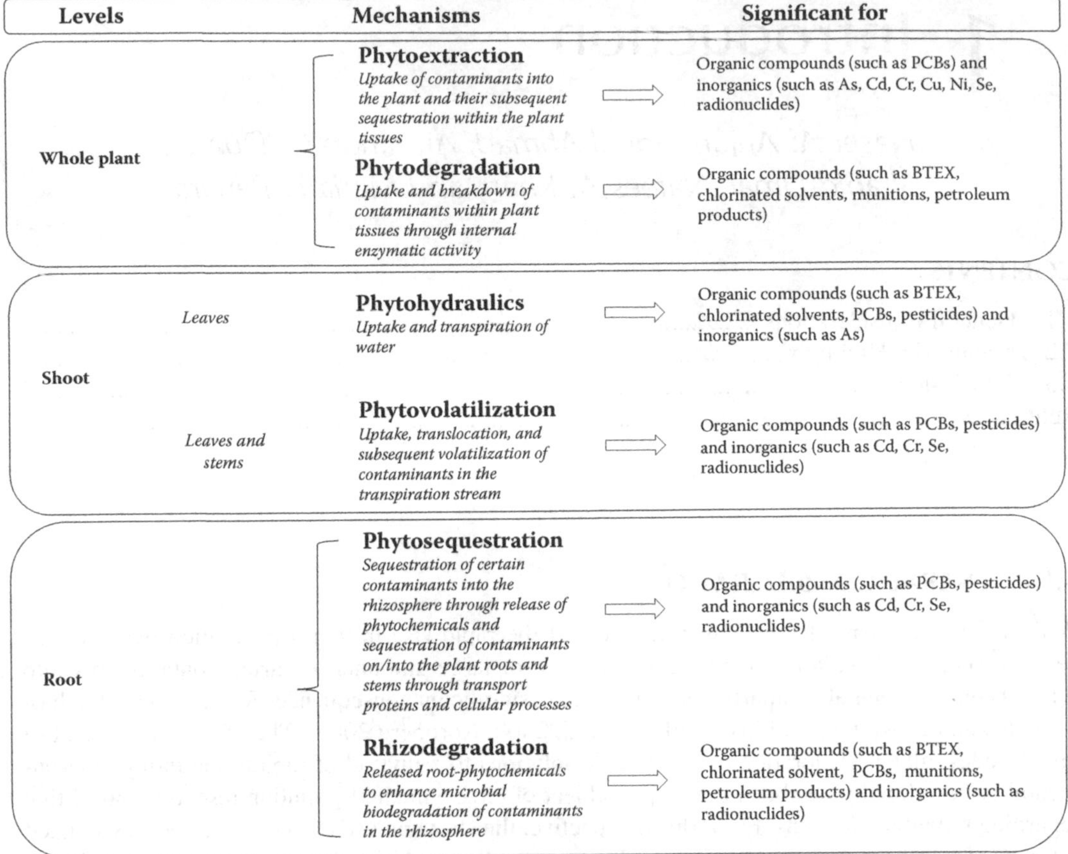

Levels
Mechanisms
Significant for
Whole plant
Phytoextraction
Uptake of contaminants into the plant and their subsequent sequestration within the plant tissues
Organic compounds (such as PCBs) and inorganics (such as As, Cd, Cr, Cu, Ni, Se, radionuclides)
Phytodegradation
Uptake and breakdown of contaminants within plant tissues through internal enzymatic activity
Organic compounds (such as BTEX, chlorinated solvents, munitions, petroleum products)
Shoot
Leaves
Phytohydraulics
Uptake and transpiration of water
Organic compounds (such as BTEX, chlorinated solvents, PCBs, pesticides) and inorganics (such as As)
Leaves and stems
Phytovolatilization
Uptake, translocation, and subsequent volatilization of contaminants in the transpiration stream
Organic compounds (such as PCBs, pesticides) and inorganics (such as Cd, Cr, Se, radionuclides)
Root
Phytosequestration
Sequestration of certain contaminants into the rhizosphere through release of phytochemicals and sequestration of contaminants on/into the plant roots and stems through transport proteins and cellular processes
Organic compounds (such as PCBs, pesticides) and inorganics (such as Cd, Cr, Se, radionuclides)
Rhizodegradation
Released root-phytochemicals to enhance microbial biodegradation of contaminants in the rhizosphere
Organic compounds (such as BTEX, chlorinated solvent, PCBs, munitions, petroleum products) and inorganics (such as radionuclides)

FIGURE 1.1 Schematic representation of major phytotechnology mechanisms and their significance for varied contaminants' cleanup. BTEX, benzene, toluene, ethylbenzene, and xylenes; PAH, polycyclic aromatic hydrocarbon; PCB, polychlorinated biphenyl; RDX, cyclotrimethylenetrinitramine; TNT, trinitrotoluene; chlorinated compounds, e.g., PCE, perchloroethylene (tetrachloroethylene); TCE, trichloroethylene; DCE, dichloroethylene; VC, vinyl chloride; TCA, trichloroethane; TCAA, trichloroacetic acid; PCP, pentachlorophenol; PCB, polychlorinated biphenyls; munition, e.g., TNT, RDX, and similar compounds; radionuclides, e.g., ^{137}Ce, ^{239}Pu, ^{90}Sr, $^{234/238}$U, tritium. (After McCutcheon, S. C., and J. L. Schnoor, *Phytoremediation: Transformation and Control of Contaminants*, John Wiley & Sons, Inc., New Jersey, 2003; Tsao, D. T., Overview of phytotechnologies, In *Advances in Biochemical Engineering/Biotechnology, Vol. 78*, T. Scheper (Ed.), Springer Verlag, 2003; Interstate Technology and Regulatory Council, Phytotechnology technical and regulatory guidance document, January, 2001, retrieved from www.itrcweb.org/Documents/PHYTO-2.pdf; Interstate Technology and Regulatory Council, Phytotechnology Technical and Regulatory Guidance and Decision Trees, Revised, PHYTO-3, Interstate Technology and Regulatory Council, Phytotechnologies Team, Washington, DC, 2009, retrieved from www.itrcweb.org; Prasad, M. N. V., H. Freitas, S. Fraenzle, S. Wuenschmann, and B. Markert, *Environmental Pollution*, 158, 18–23, 2010; Conesa, H. M., M. W. H. Evangelou, B. H. Robinson, and R. Schulin, *The Scientific World Journal*, 2012, DOI: 10.1100/2012/173829.)

arenas are largely disorganized and not critically cross-linked and/or integrated. Therefore, this book is an effort to provide a common platform for environmental engineers, environmental microbiologists, plant physiologists, and molecular biologists working with a common aim of sustainable solutions to varied environmental issues, and to address, in a single volume, major aspects of phytotechnology that are significant for understanding the importance of ecosystem approaches in helping to achieve sustainable development objectives.

1.2 MANUSCRIPT HIGHLIGHTS

To present a clear concept of the current book, chapters have been divided into three major sections. Section I (A–C), comprising Chapters 2–13, as a whole, deals with introductory aspects of contaminants, contaminated sites, and the significance of genus *Brassica* and vetiver grass species for the remediation of varied environmental contaminants. Encompassing Chapters 14–23, Section II presents an exhaustive exploration of potential strategies (including molecular-genetic methods) for enhancing varied environmental contaminants' remediation; whereas, Chapters 24–28, grouped into Section III, present an overview of major physiological, biochemical, and genetic-molecular mechanisms responsible for plant tolerance and adaptation to environmental contaminants.

Taking into account the themes of individual chapters, Hasanuzzaman and Fujita review the current status of major heavy metals in the environment in Chapter 2. Besides discussing heavy metals' phytotoxicity, this chapter also introduces and explores potential heavy metals remediation strategies. Aisen et al. (Chapter 3), Calheiros et al. (Chapter 4), and Infante et al. (Chapter 5) throw light on the phytotechnologies significant for the remediation of inorganic environmental contaminants, petroleum hydrocarbons, and tannery wastewater; whereas, Nwoko (Chapter 6) discusses the fate and transport issues associated with varied contaminants and contaminant by-products in phytotechnology. Chapters 7–11, contributed respectively by Anjum et al., Ashraf et al., Fellet et al., Li et al., and Roychoudhury et al., cumulatively explore the variability of contaminants' (including trace metals) accumulation in major brassica species and discuss the significance of genus *Brassica* in environmental contaminant–remediation strategies. The role of vetiver grass, *Chrysopogon zizanioides*, in the remediation of soils contaminated with heavy metals, metalloids, and radioactive materials is discussed by Danh et al. in Chapter 12. Several strategies' significance for enhancing plants' contaminant remediation potential are thoroughly discussed in Chapters 13–23. In this context, Joner (Chapter 13), Varkey et al. (Chapter 14), Zhongqiu and Xiaona (Chapter 15), Cacador et al. (Chapter 16), Stout and Nüsslein (Chapter 17), Lixiang Cao (Chapter 18), Gamalero and Glick (Chapter 19), Barbafieri et al. (Chapter 20), Uhlik et al. (Chapter 21), Khan and Doty (Chapter 22), and Macek et al. (Chapter 23) critically discuss the important role of a number of microbes; chemicals, chelates, and organic acids; plant growth regulators; and plant-transgenic technology in enhancing the potential of different plant species for remediating varied environmental contaminants. Significant contributions from Castrillo et al. (Chapter 24), Vogel-Mikuš et al. (Chapter 25), Thangavel et al. (Chapter 26), Erakhrumen (Chapter 27), and Van Aken et al. (Chapter 28) explore various analytical tools and physiological, biochemical, and genetic-molecular vital mechanisms significant for plants' tolerance and/or adaptation to varied environmental contaminants, including trace metals-metalloids, pharmaceuticals, and other important emerging contaminants.

It is evident from the above discussion that the current reference book is an effort to provide a common platform for environmental engineers, environmental microbiologists, plant physiologists, and molecular biologists working with a common aim of sustainable solutions to varied environmental issues and to address major aspects of *phytotechnology* significant for understanding the importance of ecosystem approaches in helping to achieve sustainable development objectives. Although there exist occasional overlaps of information between chapters, the significance of the manuscript as a whole reflects central and multiple aspects of plants and their associated microbe-based technologies important to the remediation of varied environmental contaminants.

1.3 CONCLUSIONS

In summary, the deliberations set out in the current volume have provided a conceptual overview of phytotechnologies and their importance in relation to various environmental problems and potential solutions. Although the chapters did not address all aspects of phytotechnology, but it is expected that the deliberations will make a major contribution to future studies aimed at understanding the importance of ecosystem approaches in helping to achieve sustainable development objectives.

REFERENCES

Conesa, H. M., M. W. H. Evangelou, B. H. Robinson, and R. Schulin. (2012). A critical view of current state of phytotechnologies to remediate soils: Still a promising tool? *The Scientific World Journal.* DOI: 10.1100/2012/173829

Cunningham, S. D., and W. R. Berti. (1993). The remediation of contaminated soils with green plants: An overview. *In Vitro Cellular and Developmental Biology: Plant*, *29*, 207–212.

Gerhardt, K. E., X. D. Huang, B. R. Glick, and B. M. Greenberg. (2009). Phytoremediation and rhizoremediation of organic soil contaminants: Potential and challenges. *Plant Science*, *176*, 20–30.

Interstate Technology and Regulatory Council. (2001). Phytotechnology technical and regulatory guidance document. Retrieved from www.itrcweb.org/Documents/PHYTO-2.pdf.

Interstate Technology and Regulatory Council. (2009). Phytotechnology Technical and Regulatory Guidance and Decision Trees, Revised. PHYTO-3. Washington, DC: Interstate Technology and Regulatory Council, Phytotechnologies Team. Retrieved from www.itrcweb.org.

Jez, J. M. (2011). Toward protein engineering for phytoremediation: Possibilities and challenges. *International Journal of Phytoremediation*, *13*, 77–89.

Maestri, E., and N. Marmiroli. (2011). Transgenic plants for phytoremediation. *International Journal of Phytoremediation*, *13*, 264–279.

McCutcheon, S. C., and J. L. Schnoor. (2003). *Phytoremediation: Transformation and control of contaminants.* New Jersey: John Wiley & Sons, Inc.

Nordberg, G. F. (2009). Historical perspectives on cadmium toxicology. *Toxicology and Applied Pharmacology*, *238*, 192–200.

Prasad, M. N. V., H. Freitas, S. Fraenzle, S. Wuenschmann, and B. Markert. (2010). Knowledge explosion in phytotechnologies for environmental solutions. *Environmental Pollution, 158*, 18–23.

Raskin, I., P. B. A. N. Kumar, S. Dushenkov, and D. E. Salt. (1994). Bioconcentration of heavy metals by plants. *Current Opinion in Biotechnology*, *5*, 285–290.

Salt, D. E., M. Blaylock, P. B. A. N. Kumar, V. Dushenkov, B. D. Ensley, I. Chet, and I. Raskin. (1995). Phytoremediation: A novel strategy for the removal of toxic metals from the environment using plants. *Biotechnology*, *13*, 468–474.

Tsao, D. T. (2003). Overview of phytotechnologies. In T. Scheper (Ed.), *Advances in biochemical engineering/biotechnology. Vol. 78*, (1–50). Springer Verlag.

Section I

Contaminants, Contaminated Sites, and Remediation

2 Heavy Metals in the Environment

Current Status, Toxic Effects on Plants and Phytoremediation

Mirza Hasanuzzaman and Masayuki Fujita

CONTENTS

2.1 INTRODUCTION

The rapid increase in population together with fast industrialization causes serious environmental problems, including the production and release of considerable amounts of toxic metals in the environment (Sarma 2011). Most of the toxic metals are heavy metals (HMs), which are ascribed to transition metals with atomic masses over 20 and having a specific gravity of above 5 g cm^{-3} or more. However, in biology, "heavy" refers to a series of metals and metalloids that can be toxic to both plants and animals even at very low concentrations (Rascio and Navari-Izzo 2011). Although HMs are thought to be synonymous with toxic metals, lighter metals, such as aluminum (Al), also have toxicity, and not all HMs are particularly toxic. Some are essential, such as iron (Fe), copper (Cu), zinc (Zn), and molybdenum (Mo). Therefore, the definition may also include trace elements when considered in abnormally high, toxic doses. A difference is that there is no beneficial dose for a toxic metal with a biological role. HMs are ubiquitous environmental contaminants in industrialized civilizations throughout the world. Because of rising environmental pollution in industrial areas, toxicity of various HMs for living organisms has become a matter of utmost global concern (Dubey 2011). Over the last few decades, we have witnessed a dramatic, troublesome increase in HM contamination in the environment globally. It would appear that humans are the only ones to be blamed because anthropogenic activities are the main source of the pollution that is causing the contamination (Gratão et al. 2005; Azevedo and Azevedo 2006). Extreme levels of metals in the water and soil may come up because of a range of activities, such as mining; metal industries; road traffic; power stations; burning of fossil fuels; crop production; animal rearing, including wastewater; use of agrochemicals; waste disposal; and so on (Dubey 2011). Contamination of soil with metals leads to losses in agricultural yield and are a threat to the health of wildlife and humans (Sharma and Dubey 2007; Sharma and Dietz 2008).

After excessive uptake by plants, HMs may participate in some physiological and biochemical reactions, which destroy the normal growth of the plant by disturbing absorption, translocation, or synthesis processes. More seriously, they may combine with a huge molecule, such as a nucleic acid, protein, or enzyme, or may substitute special functional elements in a protein or enzyme so as to induce a series of turbulence of metabolism. Therefore, the growth and procreation of the plant is prohibited, and the plant died (Wei and Zhou 2008). Once taken up by the plant, HMs interact with different cell components and disturb normal metabolic processes. However, making a generalization about the effect of HMs on plants is difficult because of the multidimensional variations in parameters under different concentrations, types of HMs, duration of exposure, target organs of plants, plant age, etc. The most obvious plant reactions under HM toxicity are the inhibition of growth rate, chlorosis, necrosis, leaf rolling, altered stomatal action, decreased water potential, efflux of cations, alterations in membrane functions, inhibition of photosynthesis, altered metabolism, altered activities of several key enzymes, etc. (Sharma and Dubey 2007; Dubey 2011). One of the obvious effects of HM toxicity at the cellular level of plants is the production of excessive reactive oxygen species (ROS), such as superoxide (O_2^-), hydroxyl radical (OH·), singlet oxygen ($^1O_2$), and hydrogen peroxide (H_2O_2) (Yadav 2010), which causes membrane lipid peroxidation, protein oxidation, enzyme inhibition, and damage to nucleic acids and subsequent cell death (Gill and Tuteja 2010; Gill et al. 2011c; Anjum et al. 2012).

Considering the increasing trend of HM contamination in the environment and their negative impact on plants and other organisms, ways of mitigating HM consequence is now a burning issue. Plant-based bioremediation or phytoremediation technologies have recently been well accepted as strategies to clean up metal-contaminated soils and water by employing green plants and their associated microbes *in situ* (Sadowsky 1999). Diversified and multidisciplinary research works have clearly established that a number of plant species acquire the genetic potential to remove, degrade, metabolize, or immobilize different HMs. However, in spite of having tremendous potential, phytoremediation is yet to become a commercial technology. Thus, exploring the basic plant mechanisms and the effect of agronomic practices on plant-soil-metal interactions would allow practitioners to optimize phytoremediation by manipulating the process to site-specific conditions.

This review attempts a comprehensive account of recent developments in the research on the current status and the sources of toxic HMs and their effects on humans and plants. We have also provided an overview of current facts and figures on different phytoremediation technologies and their potential role in mitigating the contamination of toxic metals in the environment.

2.2 CONCEPTS OF METALS, HEAVY METALS, AND TOXIC METALS

Our global environment now consists of numerous natural and artificial metals. Metals have played a major role in industrial development and technological advances. Most metals are not destroyed; indeed, they are accumulating at an accelerated pace because of the ever-growing demands of modern civilization. The term "metals" refers to elements with very good electrical conductance (this property declines with decreasing temperature) and that exhibit an electrical resistance that is proportional to the absolute temperature (Lyubenova and Schröder 2010). There are approximately 67 elements that may be termed "heavy metals" as they exhibit metallic properties (Figure 2.1).

From a chemical point of view, the term "heavy metal" is strictly ascribed to transition metals with atomic mass over 20 and with a specific gravity of 5 g cm^{-3} or more. In biology, "heavy" refers to a series of metals and metalloids that can be toxic to both plants and animals even at very low concentrations (Rascio and Navari-Izzo 2011). The other metals are referred to as light metals (<4.5 g cm^{-3}) (Lyubenova and Schröder 2010). The specific gravity of water is 1 at 4°C. Simply stated, specific gravity is a measure of the density of a given amount of a solid substance when it is compared to an equal amount of water. Some well-known HMs with a specific gravity of 5 or more times that of water are arsenic (As), 5.7; cadmium (Cd), 8.65; iron (Fe), 7.9; lead (Pb), 11.34; and mercury (Hg), 13.54 (Lide 1992) (Figure 2.1). In fact, some HMs, known as "trace metals," for example, Fe, Cu, Zn, Mo, nickel (Ni), and cobalt (Co), are essential for the growth and metabolism of organisms at low concentrations, and microorganisms possess mechanisms of varying specificity for their intracellular accumulation from the external environment. To the contrary, many other HMs, such as Pb, tin (St), Cd, Al, and Hg, have no essential biological function but can still be accumulated in biomass and are freely transferred from one organism to another through the food chain.

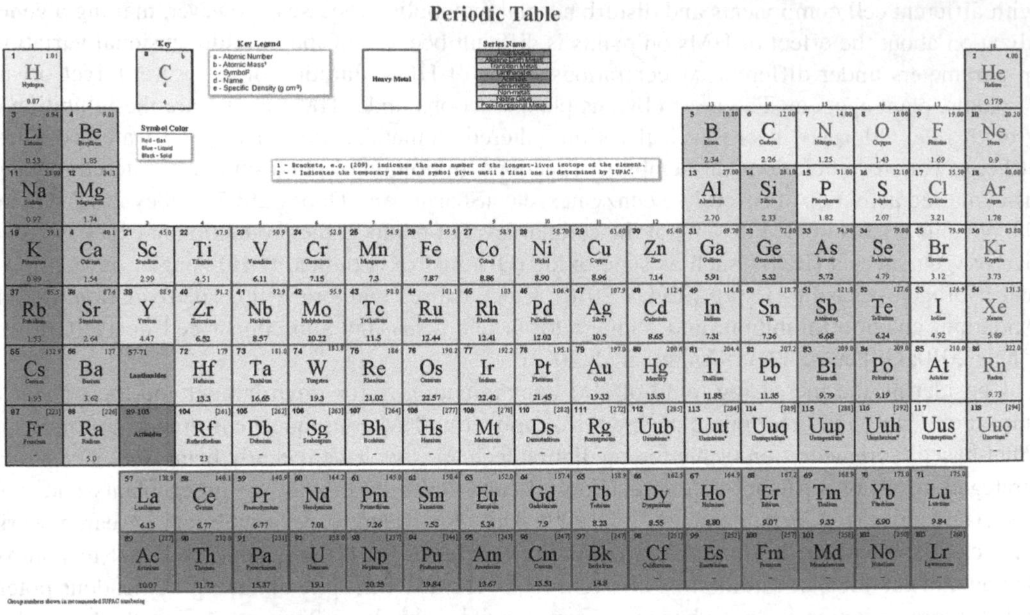

Periodic Table

Period	1	2	3	4	5	6	7	8	9	10	11	12	13	14	15	16	17	18
1	H																	He
2	Li	Be											B	C	N	O	F	Ne
3	Na	Mg											Al	Si	P	S	Cl	Ar
4	K	Ca	Sc	Ti	V	Cr	Mn	Fe	Co	Ni	Cu	Zn	Ga	Ge	As	Se	Br	Kr
5	Rb	Sr	Y	Zr	Nb	Mo	Tc	Ru	Rh	Pd	Ag	Cd	In	Sn	Sb	Te	I	Xe
6	Cs	Ba	Lanthanides	Hf	Ta	W	Re	Os	Ir	Pt	Au	Hg	Tl	Pb	Bi	Po	At	Rn
7	Fr	Ra	Actinides	Rf	Db	Sg	Bh	Hs	Mt	Ds	Rg	Uub	Uut	Uuq	Uup	Uuh	Uus	Uuo

La	Ce	Pr	Nd	Pm	Sm	Eu	Gd	Tb	Dy	Ho	Er	Tm	Yb	Lu
Ac	Th	Pa	U	Np	Pu	Am	Cm	Bk	Cf	Es	Fm	Md	No	Lr

Key Legend: a - Atomic Number; b - Atomic Mass[1]; c - Symbol[2]; d - Name; e - Specific Density (g cm⁻³)

1 - Brackets, e.g. (209), indicates the mass number of the longest-lived isotope of the element.
2 - * Indicates the temporary name and symbol given until a final one is determined by IUPAC.

Group numbers shown in recommended IUPAC numbering

FIGURE 2.1 **(See color insert.)** Periodic table showing the position of different heavy metals (indicated by red borders).

Toxic metals comprise a group of elements that have no biological role in organisms and, in fact, are harmful. Today, mankind is exposed to the highest levels of these metals in recorded history, which is a result of their industrial use; the unrestricted burning of coal, natural gas, and petroleum; and incineration of waste materials worldwide. Toxic metals are now everywhere and affect everyone on planet earth. Often HMs are thought to be synonymous with toxic metals, but lighter metals also have toxicity (e.g., Al), and not all HMs are particularly toxic as some are essential (e.g., Fe, Cu, Zn, Mo, Ni). The definition may also include trace elements when considered in abnormally high, toxic doses. A difference is that there is no beneficial biological role for a toxic metal.

2.3 PHYSICAL AND CHEMICAL NATURES AND ABUNDANCE OF MAJOR HEAVY METALS

2.3.1 Cadmium

Cadmium (Cd) has an atomic number of 48, an atomic weight of 112.4, and a density of 8.65 g cm^{-3}. Cadmium is a soft, silvery white, ductile metal with a faint bluish tinge. It has a melting point of 321°C and a boiling point of 765°C. It belongs to group IIb of elements in the periodic table and, in aqueous solution, has the stable +2 oxidation state. Cadmium is a rare element with a concentration of ~0.1 µg g^{-1} in the lithosphere and is strongly chalcophilic (Callender 2003). Cadmium has assumed importance as an environmental contaminant only within the past 60 years or so. It is commonly released into the arable soil from industrial processes and farming practices and has been ranked no. 7 among the top 20 toxins (Yang et al. 2004). Worldwide production in 1935 totaled 1,000 t yr^{-1} and today is on the order of 21,000 t yr^{-1}. Approximately 7,000 t of Cd are released into the atmosphere annually as a result of anthropogenic activity (mainly nonferrous mining) compared with 840 t from natural sources, such as volcanic eruption (Wright and Welbourn 2002). Bioavailable Cd is best predicted by the free cadmium ion (Cd^{2+}). The tendency for the metal to form chloro complexes in saline environments renders the metal less available from solution and may largely explain the inverse relationship between Cd accumulation and salinity in the estuarine environment (Wright and Welbourn 2002). Generally, the formation of soluble inorganic or organic complexes of Cd reduces Cd uptake by aquatic organisms although there have been some reports of increased Cd uptake in the presence of organic ligands.

2.3.2 Lead

Lead (Pb) is a bluish-white metal of bright luster and is soft, very malleable, ductile, and a poor conductor of electricity. It possesses the atomic number 82, atomic weight 207, and has a specific density of 11.35. Its melting point is 327.5°C, which makes it resistant to corrosion, and thus Pb has been used in the manufacture of metal products for thousands of years (Callender 2003). Lead occurs naturally in trace quantities, and its average concentration in the Earth's crust is about 20 ppm. Weathering and volcanic emissions account for most of the natural processes that mobilize Pb, but human activities are far more significant in the mobilization of Pb than are natural processes (Wright and Welbourn 2002). Early uses of Pb included the construction and application of pipes for the collection, transport, and distribution of water. The term "plumbing" originates from the Latin *plumbum*, for Pb. Lead is now the fifth most commonly used metal in the world. It was used in pipes, drains, and soldering materials for many years. Millions of homes built before 1940 still contain Pb (e.g., in painted surfaces), leading to chronic exposure from weathering, flaking, chalking, and dust. Every year, industry produces about 2.5 million tons of Pb throughout the world; most of this is used for batteries. The remainder is used for cable coverings, plumbing, ammunition, and fuel additives. Other uses are as paint pigments and in PVC plastics, X-ray shielding, crystal glass production, and pesticides (LifeExtension 2011). The usual valence state of Pb is (II), in which state it forms inorganic compounds. Lead can also exist as Pb(IV), forming covalent compounds, the most important of which, from an environmental viewpoint, are the tetraalkyl Pb, especially tetraethyl (Wright and Welbourn 2002). Many of the compounds of Pb are rather insoluble, and most of the metal discharged into water partitions rather rapidly into the suspended and bed sediments. Here, it represents a long-term reservoir that may affect sediment-dwelling organisms and may enter the food chain from this route (Wright and Welbourn 2002). Lead is a cumulative poison. The presence of Pb in drinking water is limited to 0.01 ppm (FAO 1999).

2.3.3 Arsenic

Arsenic (As) is a chemical element with atomic number 33 and atomic mass 74.92, having specific density of 5.73. Arsenic occurs in many minerals, usually in conjunction with sulfur and other metals, and also as a pure elemental crystal. It is the most common cause of acute HM poisoning. Arsenic is released into the environment by the smelting process of Cu, Zn, and Pb, as well as by the manufacturing of chemicals and glasses. Arsine gas is a common by-product produced by the manufacturing of pesticides that contain arsenic. Arsenic may be also be found in water supplies worldwide, leading to exposure in shellfish, cod, and haddock. Other sources are paints, rat poisoning, fungicides, and wood preservatives (LifeExtension 2011). Arsenic is notorious as a toxic element. Its toxicity, however, depends on the chemical (valency) and physical form of the compound, the route by which it enters the body, the dose and duration of exposure, and several other biological parameters (FAO 1999). In soils, As is found in −3, 0, +3, and +5 oxidation states. Its prevalent forms are the inorganic species: arsenate (As[+5]) and arsenite (As [+3]). Arsenic may occur in methylated forms, but these organic species are much less bio-toxic and rare in soils and surface waters (Smith et al. 1998).

2.3.4 Mercury

Mercury (Hg) is an element with atomic number 80, atomic weight 200.59, and a specific density of 13.5; in metallic form, it volatilizes readily at room temperature. The element takes on different chemical states: elemental or metallic mercury (Hg^0), divalent inorganic mercury (Hg^{2+}), and methylmercury (CH_3Hg^+). However, Hg^{2+} forms salts with various anions that are scarcely soluble in water, and in the atmosphere, Hg^{2+} associates readily with particles and water (Wright and Welbourn 2002). Like other metals, Hg occurs naturally, and the absolute amount on the planet does not change. However, its chemical form and location change quite readily, and an appreciation of these changes, in conjunction with the chemical forms, is needed to understand its environmental toxicology. Anthropogenic Hg behaves in exactly the same manner as does the metal that occurs naturally. Although a comparatively rare element, Hg is ubiquitous in the environment, the result of natural geological activity and man-made pollution. The behavior of Hg in the environment depends upon its chemical form and the medium in which it occurs (Wright and Welbourn 2002). Because of its extreme mobility, Hg deposited from a particular pollution source into an ecosystem may be reemitted later into the atmosphere and, in this way, contribute to apparently "natural" sources (Wright and Welbourn 2002). Mercury from natural sources can enter the aquatic environment via weathering, dissolution, and biological processes. Mercury has no known essential biological function. It is highly toxic to the human organism, especially in the form of CH_3Hg^+, because it cannot be excreted and, therefore, acts as a cumulative poison (FAO 1999).

2.3.5 Copper

Copper (Cu) is a rosy-pink transition metal with an atomic number of 29, an atomic weight of 63.546, and it has a density of 8.94 g cm^{-3} (Webelements 2002). This metal is somewhat malleable with a melting point of 1356°C and a boiling point of 2868°C. In aqueous solution, Cu exists primarily in the divalent oxidation state (Cu^{2+}) although some univalent complexes and compounds of Cu do occur in nature (Leckie and Davis 1979). Copper is a moderately abundant HM with a concentration in the lithosphere of about 39 μg g^{-1} (Li 2000; Callender 2003). Copper salts are moderately soluble in water: The pH-dependence of sorption reactions for Cu compounds means that dissolved concentrations of Cu are typically higher at acidic to neutral pH than under alkaline conditions, all other things being equal. Copper ions tend to form strong complexes with organic ligands, displacing more weakly bound cations in mixtures (Wright and Welbourn 2002). Complexation facilitates Cu remaining in solution but usually decreases its biological availability. Copper also forms strong organometal complexes in soils and in sediments. The major processes that result in the mobilization of Cu into the environment are extraction from its ore (mining, milling, and smelting), agriculture, and waste disposal. Soils have become contaminated with Cu by deposition of dust from local sources, such as foundries and smelters, as well as from the application of fungicides and sewage sludge. Aquatic systems similarly receive Cu from the atmosphere and from agricultural runoff, deliberate additions of $CuSO_4$ to control algal blooms, and direct discharge from industrial processes (Wright and Welbourn 2002).

2.3.6 Zinc

Zinc (Zn) is a bluish-white, relatively soft metal with an atomic number of 30 and a density of 7.133 g cm^{-3}. It has an atomic weight of 65.39, a melting point of 419.6°C, and a boiling point of 907°C. Zinc is divalent in all its compounds and is composed of five stable isotopes. It belongs to group IIb of the periodic table, which classifies it as a HM whose geochemical affinity is chalcophilic (Callender 2003). Zinc is the 23rd most abundant element in the Earth's crust. Some soils possess naturally high Zn concentrations. The principal Zn minerals are sulfides, such as sphalerite and wurtzite (ZnS), which usually occur in association with other ores, including Cu, Au, Pb, and Ag. In natural waters, Zn is in the form of the divalent cation Zn^{2+} (hydrated Zn^{2+} at pH between 4 and 7) and in the form of fairly weak complexes (Wright and Welbourn 2002). At low concentration, Zn acts as an essential element for plant life.

2.3.7 Iron

Iron (Fe) has an atomic number of 26, an atomic weight of 55.845, and a density of 7.8 g cm^{-3}. Iron is a lustrous, ductile, malleable, silver-gray metal occupying the group VIII of the periodic table. It is a metal in the first transition series. Iron is chemically active and forms two major series of chemical compounds: the bivalent iron (II), or ferrous (Fe^{2+}), compounds and the trivalent iron (III), or ferric (Fe^{3+}), compounds. Iron is the most common element in the whole planet Earth, forming much of Earth's outer and inner core, and it is the fourth most common element in the Earth's crust, occurring in most rocks and soils. Important ores are oxides and carbonates. From the point of view of environmental chemistry, the single most important aspect of the chemical forms of Fe is the respective properties of Fe^{2+} and Fe^{3+} and their interconversions. Bivalent Fe^{2+} is the more soluble and more toxic form of Fe, but under aerobic conditions, it is readily converted to Fe^{3+}, which is less soluble and thus less toxic (Wright and Welbourn 2002).

2.3.8 Nickel

Nickel (Ni) has an atomic number of 28, an atomic weight of 58.71, and a density of 8.9 g cm^{-3}. It is a silvery-white, malleable metal with a melting point of 1455°C and a boiling point of 2732°C. It has high ductility, good thermal conductivity, moderate strength and hardness, and can be fabricated easily by the procedures that are common to steel (Nriagu 1980). Nickel belongs to group VIIIa and is classified as a transition metal (the end of the first transition series) whose prevalent valence states are 0 and +2. However, the majority of Ni compounds are of the Ni^{2+} species (Callender 2003). Nickel is the most abundant metal in the environment (National Science Foundation 1975). Pristine streams, rivers, and lakes contain 0.2–10 mg L^{-1} total dissolved Ni, and surface water near Ni mines and smelters contain up to 6.4 mg L^{-1} (Wright and Welbourn 2002). Seawater contains approximately 1.5 mg L^{-1} of which approximately 50% is in free ionic form. Nickel is ubiquitous in the environment. Nickel is almost certainly essential for animal and plant nutrition at its lower concentrations. Ni exists in the atmosphere primarily as water-soluble $NiSO_4$, NiO, and complex metal oxides containing Ni. Occupational Safety and Health Administration (OSHA) levels for airborne Ni is 1 mg m^{-3}. Nickel toxicity is highly dependent on the form in which it is introduced into cells. Nickel compounds can be divided into three categories of increasing acute toxicity: water-soluble Ni salts [$NiCl_2$, $NiSO_4$, $Ni(NO_3)_2$, and $Ni(CH_3COO)_2$]; particulate Ni [Ni_3S_2, NiS_2, Ni_7S_6, and $Ni(OH)_2$]; and lipid-soluble Ni carbonyl [$Ni(CO)_4$] (Wright and Welbourn 2002).

2.3.9 Chromium

Chromium (Cr) is a lustrous, brittle, hard metal, which has an atomic number of 24, an atomic weight of 51.996, and a density of 7.14 g cm^{-3}. Crystalline Cr is steel-gray in color, lustrous, and hard with a melting point of 1900°C and a boiling point of 2642°C. It belongs to group VIb of the transition metals, and in aqueous solution, Cr exists primarily in the trivalent (Cr^{3+}) and hexavalent (Cr^{6+}) oxidation states. Chromium has special magnetic properties, and it is the only elemental solid that shows antiferromagnetic ordering at or below room temperature; whereas, above 38°C, it transforms into a paramagnetic state. Chromium is one of the most abundant HMs with a concentration of about 69 μg g^{-1} in the lithosphere (Li 2000). Most rocks and soils contain small amounts of Cr, which remains in a highly insoluble form. However, most of the common soluble forms found in soils are mainly the result of industrial emissions. The major uses of Cr are for chrome alloys; chrome plating; oxidizing agents; corrosion inhibitors; pigments for the textile, glass, and ceramic industries; and in photography. Hexavalent Cr compounds (soluble) are carcinogenic, and the guideline value is 0.05 ppm (FAO 1999).

2.3.10 Vanadium

Vanadium (V) is a soft, silvery-gray, ductile transition metal. The formation of an oxide layer stabilizes the metal against oxidation. The element is found only in chemically combined forms in nature. It has an atomic number of 23, an atomic weight of 50.94, and a density of 6.1 g cm^{-3}. Vanadium is present naturally in the Earth's crust at an average concentration of 150 mg kg^{-1}, which gives it the same abundance as Ni, Cu, Zn, and Pb although the element has a far more even distribution than these. Vanadium in rock is commonly associated with titanium (Ti) and uranium (U) ores. Crude oil and coal contain high levels of V (averaging 50 and 25 mg kg^{-1}, respectively, with a range of 1–1,400 mg kg^{-1}), which can be released to the atmosphere during fuel combustion (Wright and Welbourn 2002).

2.4 SOURCES OF HEAVY METALS

Heavy metal contamination in the environment is a major problem that has been receiving increasing attention for many decades. In nature, HMs are extensively spread and can be found in various background concentrations in all environmental compartments. However, the sources of HMs in the environment and factors influencing their distribution—reactivity, mobility, and toxicity—are numerous (Mishra and Dubey 2005). Meanwhile, there are variations in the metal contents of the soil and water from one location to the other (Adegoke et al. 2009).

The metals found in the environment are derived from a variety of sources (Figure 2.2). The sources of HMs in the environment can be grouped into natural sources and anthropogenic sources.

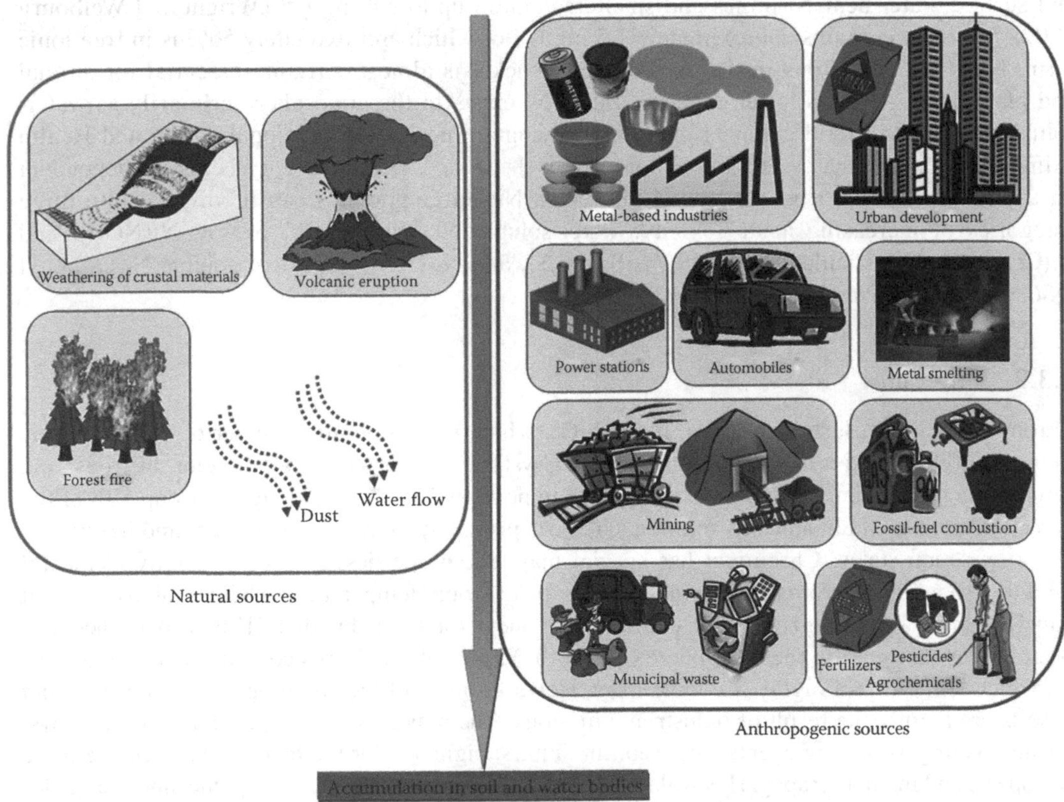

FIGURE 2.2 Major sources of heavy metals in the environment.

2.4.1 Natural Sources

The natural or geogenic sources of HM include the natural weathering of the Earth's crust. Crustal material is either weathered on (dissolved) and eroded from (particulate) the Earth's surface or injected into the Earth's atmosphere by volcanic activity (Callender 2003). These two sources comprise 80% of all the natural sources; forest fires and biogenic sources account for 10% each (Nriagu 1990). Naturally, HM particles come up by soil erosion and are released into the atmosphere as windblown dust. In addition, some particles are released by vegetation. The natural emissions of the five major HMs are 12,000 (Pb); 45,000 (Zn); 1,400 (Cd); 43,000 (Cr); and 29,000 (Ni) metric tons per year, respectively (Nriagu 1990), which indicate an abundant quantity of metals are emitted into the atmosphere from natural sources.

2.4.2 Anthropogenic Sources

There are a multitude of anthropogenic emissions in the environment. Most of the HM occurrences in urban soils tend to originate from anthropogenic sources, such as industrial, urban

TABLE 2.1
Anthropogenic Sources of Some Major Toxic Metals

Metals	Anthropogenic Sources
Cd	Mining, ore dressing, and smelting of nonferrous metals, battery manufacturing, cigarettes, processed and refined foods, large fish, shellfish, tap water, auto exhaust, plated containers, galvanized pipes, air pollution from incineration, occupational exposure
Pb	Residue from the production of Pb electric accumulators, residue and sludge from Pb caster and product industry, tap water, cigarette smoke, hair dyes, paints, inks, glazes, pesticide residues, and occupational exposure in battery manufacture and other industries, waste from the production and application of Pb compounds
As	Mining, ore dressing, and smelting of nonferrous metals, production of As and As compounds, petroleum and chemical industry, pesticides, beer, table salt, tap water, paints, pigments, cosmetics, glass and mirror manufacture, fungicides, insecticides, treated wood and contaminated food, dyestuff and tanning industry
Hg	Production and application of Hg catalyst in chemical industry, Hg battery manufacturing, smelting and restoring of Hg, Hg compound production, pesticide and medicine making, production and application of fluorescent light and Hg lamps, Hg slime from caustic soda production, dental amalgams, large fish, shellfish, medications, manufacture of paper, chlorine, adhesives, fabric softeners, and waxes
Cu	Mining, ore dressing, smelting of nonferrous metals, Cu water pipes, Cu added to tap water, pesticides, intrauterine devices, dental amalgams, nutritional supplements, especially prenatal vitamins, birth control pills, weak adrenal glands, and occupational exposure
Zn	Mining, ore dressing, and smelting of nonferrous metals, metal and plastic electroplate, pigment, beaded paint and rubber working, Zn compound production, Zinoky battery product industry
Ni	Residue from the production of nickeliferous compounds, abandoned nickeliferous catalysts, nickeliferous residue and waste from electroplate technology, nickeliferous waste from analysis, assay, and testing activity, hydrogenated oils (margarine, commercial peanut butter, and shortening), shellfish, air pollution, cigarette smoke, plating, occupational exposure
Cr	Cr compound production, leather-working industry, metal and plastic electroplate, dyestuff and dying by acidic medium, production and application of dyestuff, metal Cr smelting
Al	Cookware, beverages in aluminum cans, tap water, table salt, baking powders, antacids, processed cheese, antiperspirants, bleached flour, antacids, vaccines and other medications, occupational exposure

Source: Wei, S., and Q. Zhou. Trace elements in agro-ecosystem. In M. N. V. Prasad (Ed.), *Trace elements as contaminants and nutrients: Consequences in ecosystems and human health,* Wiley, Hoboken, 2008; Wilson, L. *Toxic metals and human health,* The Center for Development, http://www.drlwilson.com/articles/TOXIC%20METALS.htm, 2011.

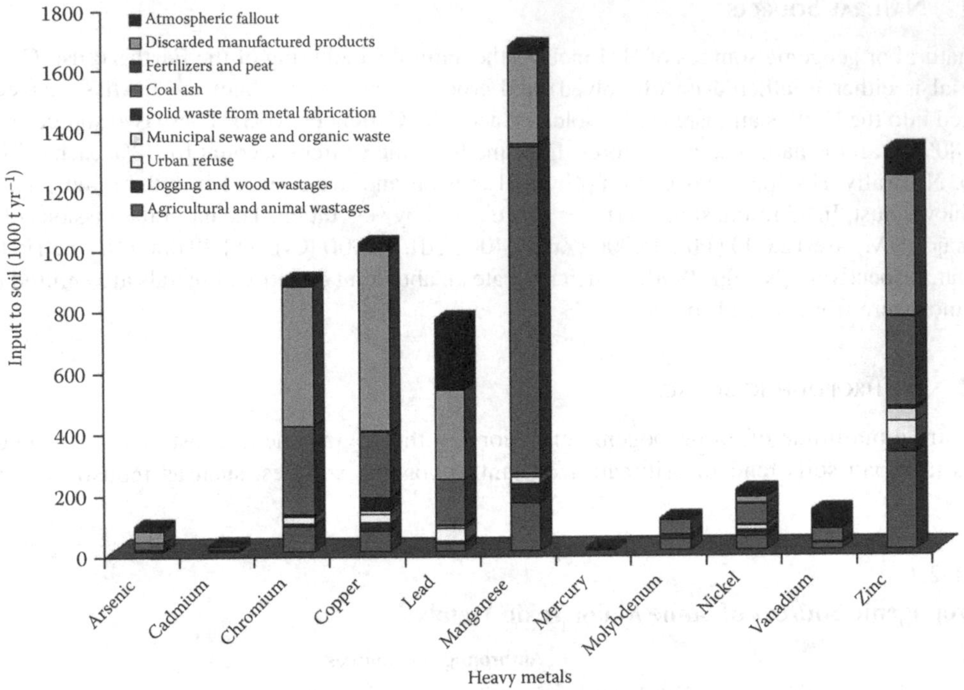

FIGURE 2.3 Worldwide input of heavy metals into soils (1000 tons yr^{-1}). (Nriagu, J. O., and J. M. Pacyna. *Nature, 33*, 1988; Purakayastha, T. J., and P. K. Chhonkar. Phytoremediation of heavy metal contaminated soils. In I. Sherameti, and A. Varma (Eds.), *Soil heavy metals, Soil Biology, 19*. Springer, Berlin, 2010.)

development, and road traffic (Strivastava et al. 2007; Adegoke et al. 2009). The key sources of HMs are mining and smelting. Mining releases HMs into environments such as soil and water as tailings and to the atmosphere as metal-containing dust. Smelting, on the other hand, discharges metals to the environment as a consequence of high-temperature refining processes (Callender 2003). Other important sources of HMs in the terrestrial and aquatic environment include fossil-fuel combustion, municipal-waste disposal, cement production, automobiles, use of commercial fertilizers and pesticides, animal waste, etc. (Nriagu and Pacyna 1988) (Figure 2.2; Table 2.1). Increasing demand for fossil fuels and agrochemicals as well as rapid urbanization is causing more anthropogenic HM inputs (Figure 2.3). However, the two main pathways for HMs to become incorporated into air-soil-sediment-water are transported by air and water (Callender 2003).

2.5 CURRENT STATUS OF HEAVY METALS IN THE WORLD

With the beginning of large-scale metal mining and smelting as well as fossil-fuel combustion in the 20th century, the release of HMs has increased considerably (Callender 2003). The background concentration of HMs in the world and some specific countries are outlined in Table 2.2. Man's impact on the geosphere has been very broad and complex and has often led to irreversible changes. Natural geological and biological alterations of the Earth's surface are generally very slow. On the other hand, man-made or stimulated changes have accumulated extremely quickly in the last decades. These changes disturb the natural balance of the geosphere, which has been formed evolutionarily over a long period of time. These changes most often lead to a degradation of the natural human environment (Papastergios et al. 2004). Although man's impact on the biosphere has been

TABLE 2.2
Heavy Metal Background Concentrations (mg kg^{-1}) in the Soils of the World and Some Specific Countries

Heavy Metals	World	China	Japan	Brazil
Cd	0.06	0.097	–	–
Pb	10–150	13–42	24	26
As	9.36	10.38	–	–
Cu	20	22	350	33
Hg	0.03	0.04	–	–
Ni	40	35	5.4	14
Cr	20.200	<100	2.4	112
Zn	100–300	3–790	23	38

Source: Xie, Z. M., and S. M. Lu, Trace elements and environmental quality. In Q. L. Wu (Ed.), *Micronutrients and biohealth*, Guizhou Science and Technology Press, Guiyan, 2000; Yang, Y. A., and X. E. Yang, Micronutrients in sustainable agriculture. In Q. L. Wu (Ed.), *Micronutrients and biohealth*, Guizhou Science and Technology Press, Guiyan, 2000; Takeda, A., K. Kimura, and S. I. Yamasaki, *Geoderma*, *119*, 2004; Marques, J. J., D. G. Schulze, N. Curi, and S. A. Mertzman, *Geoderma*, *121*, 2004; Wei, S., and Q. Zhou. Trace elements in agro-ecosystem. In M. N. V. Prasad (Ed.), *Trace elements as contaminants and nutrients: Consequences in ecosystems and human health*, Wiley, Hoboken, 2008.

dated from the Neolithic Period, the deterioration of ecosystems because of pollution has become increasingly acute during the latter decades of the 20th century (Kabata-Pendias and Pendias 1992; United Nations 2002).

The enrichment of HMs in the environment can result from both anthropogenic activities and natural processes (Figure 2.2). High concentrations of HMs with geogenic origins in sediments, which are often enriched in refractory minerals, do not imply high-potential toxicity to ecology. Consequently, a clear differentiation of anthropogenic from geogenic HMs is important in evaluating the extent of pollution, preventing further environmental damage, and planning remedial strategies (Xu et al. 2009). A thorough understanding of the source and sink will affect both short- and long-term impact of human activities and natural processes on HM accumulation. The worldwide emission of HMs is presented in Figures 2.4 and 2.5.

Urban soils have some specific properties, such as unpredictable layering, poor structure, and high concentrations of trace elements (Tiller 1992; Manta et al. 2002). Because of the increased population density and rigorous anthropogenic activities, urban soils have been severely disturbed. As a consequence, a large number of environmental problems have emerged, among which the HM pollution remains a foremost issue. HMs can be released in many ways, such as vehicle emission, chemical industry, coal combustion, municipal solid waste, the sedimentation of dust, and suspended substances in the atmosphere (Imperato et al. 2003). The levels of Pb, Cu, and Zn in general are considered to be influenced by traffic sources (Shi et al. 2008; Kadi 2009), and Cd might be associated with industrial activities (Charlesworth et al. 2003). These emissions have continuously added HMs to urban soils, and they will remain present for many years even after the pollution sources have been removed (Xia et al. 2011). However, HM contents also varied with the land-use pattern as studied by Huang et al. (2009) (Table 2.3).

Akan et al. (2010) observed that the sequence of HMs in cultivated soil samples from the Gongulon agricultural site in China was in the order of Zn > Mn > Cd > Pb > Cu > Cr > Fe > Co > As > Ni. The concentrations of HMs showed spatial and temporal variations, which may be ascribed to the variation in HM sources and the quantity of HMs in irrigation water and sewage sludge. This

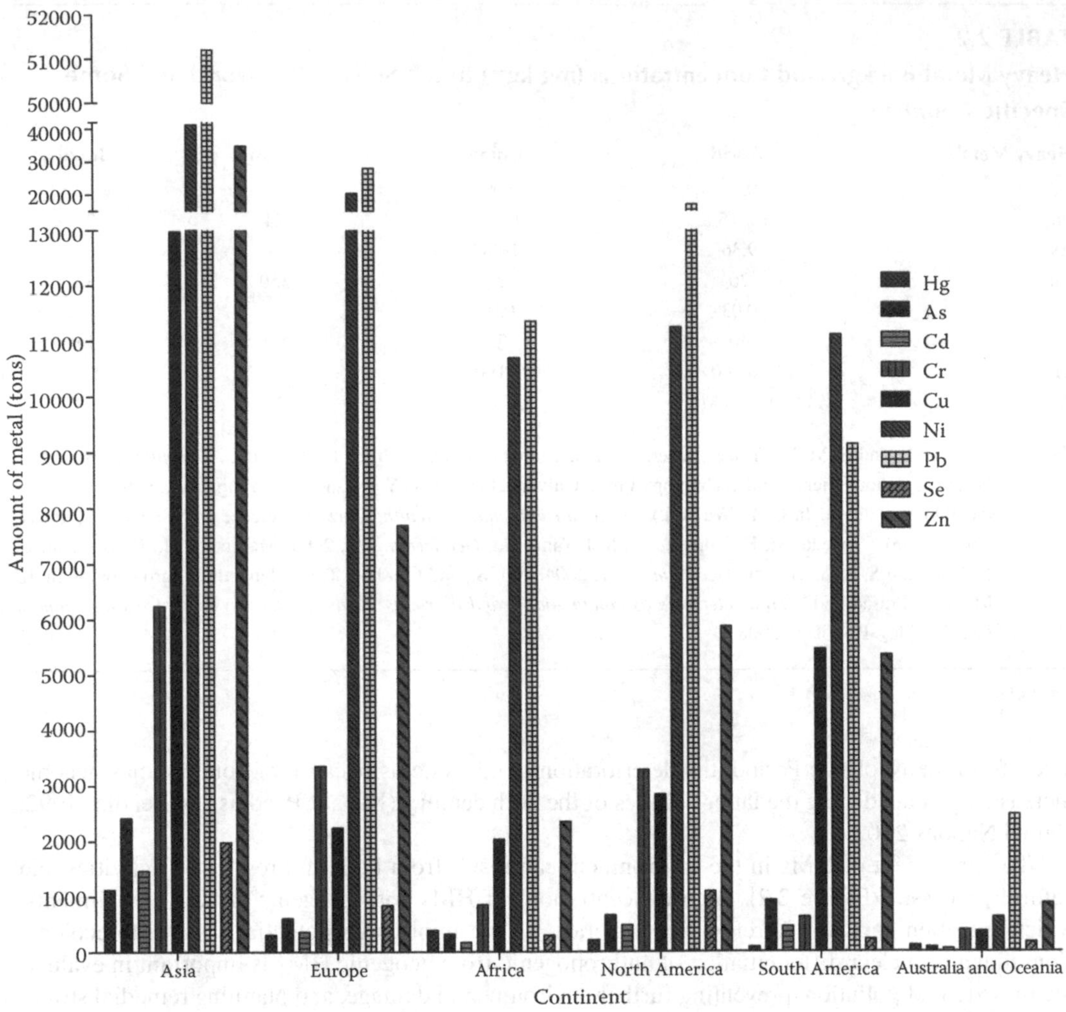

FIGURE 2.4 Worldwide emission of heavy metals from major anthropogenic categories to the atmosphere. (Pacyna, J. M., and E. G. Pacyna, *Environmental Review*, *9*, 2001.)

trend suggests continuous application of sewage sludge and municipal wastewater influenced the soil physicochemical properties (Willett et al. 1984). Evidence that HMs may move in the soil profile was provided by Lund et al. (1976). In their field experiment, the researchers used sludge with a high content of HMs and found that Zn had moved down to 50 cm, Cd to 17 cm, and Ni to 75 cm. Davis et al. (1988) measured the metal distribution in the soil profile in a field experiment where sludge had been applied at a rate of 40 t ha^{-1} and the rainfall rate was around 560 mm per annum over a period of four years. They found a significant movement of Cd, Ni, Pb, and Zn to a depth of 10 cm. Schirado et al. (1986) reported that HMs had a uniform distribution in the soil profile to a depth of 1 m because of their movement. Results such as these tend to have been obtained from the present study where movement of HMs down the soil profile (leaching) to a depth of 15 cm is a result of application of sewage sludge and wastewater.

Over a number of years, atmospheric input of HMs caused by air pollution was very high and until quite recently was on the increase (Nriagu 1979; Bergkvist et al. 1989). During the last decade

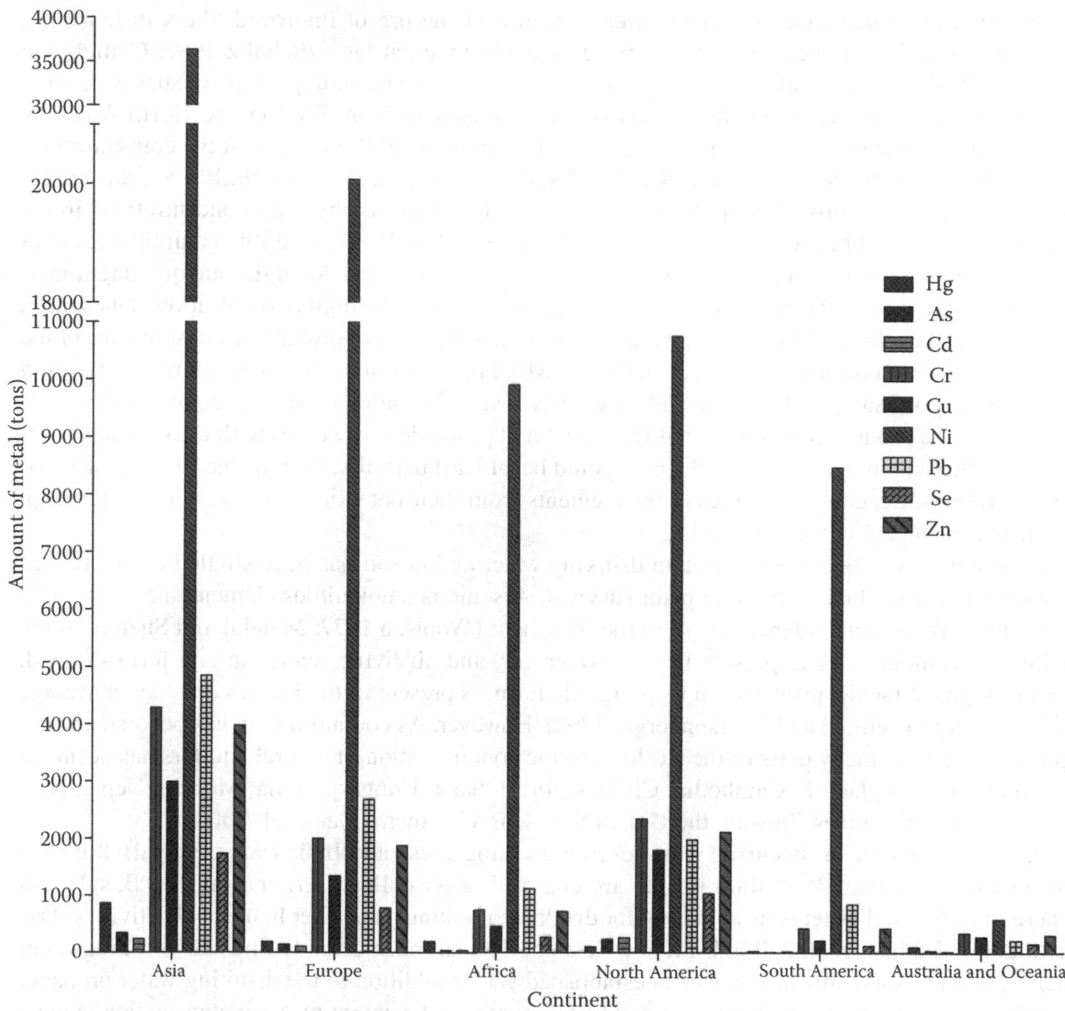

FIGURE 2.5 Worldwide emission of trace metals from combustion of fuels in stationary sources. (Pacyna, J. M., and E. G. Pacyna, *Environmental Review*, 9, 2001.)

TABLE 2.3
Average Heavy Metal content in Soil (mg kg⁻¹) Under Different Land-Use Patterns

	Land-Use Pattern		
Heavy Metals	**Grain Crop Field**	**Vegetable Green House**	**Open Vegetable Fields**
Cd	0.6	0.6	0.6
Pb	30	29.4	28
As	7	7	6
Cu	27.7	29.9	22.3
Zn	72	82	57
Hg	0.1	0.1	0.1
Cr	83	88	79

Source: Huang, S. W., J. Y. Jin, and P. He, *Better Crops*, *93*, 2009.

or two, this input decreased in certain areas because of the use of improved filters in industrial installations and also because of more stringent environmental laws (Schultz 1987; Church and Scudlark 1992; Schulte et al. 1996). Bergkvist et al. (1989) present a comprehensive survey of input and output seepage measurements and ecosystem assessments from Europe and North America. While studying the soil of northern Greece, Papastergios et al. (2004) found that the concentrations of Ca, Mg, K, Fe, Si, S, Al, P, Na, B, Ce, Co, Cs, Ga, Ge, Hg, La, Li, Mo, Ni, Rb, Se, Sn, Sr, Th, U, and W in the topsoils of their study area are mainly influenced by their concentrations in the surrounding rocks. The enrichment of Ag, As, Cd, Cr, Cu, Mn, Pb, Sb, and Zn is mainly a result of the widespread presence of photonic band gap sulfides, Mn, Cd, and As in the surrounding mineralization. Arsenic and Pb show the highest enrichment factors. The high concentration values of Ba and V, as well as those of Mg, K, Fe, Al, and P in some samples, are probably a consequence of the human activities in the area. Arsenic, Cd, Cr, Cu, Mn, Pb, and Zn show high concentration values in almost all topsoil samples from the study area. Because information exists that connects these elements with the production and usage of fertilizers and pesticides, as well as with the combustion of gasoline, these human activities in the area could be, at least partially, responsible for their elevated concentrations. Leaching processes of the elements from their potential sources is the main reason for their enrichment in the local soils.

In recent years, As contamination in drinking water and in soil has been studied extensively as it adversely affects human life and plant survival. Arsenic is a ubiquitous element and is assumed to be the 20th most abundant element in the biosphere (Woolson 1977; Mandal and Suzuki 2002). Being a metalloid, As can present in soil, water, air, and all living matter in any form of solid, liquid, or gas. Arsenic, primarily in its inorganic form, is present in the Earth's crust at an average of 2–5 mg kg^{-1} (Tamaki and Frankenberger 1992). However, As contamination has become a widespread problem in many parts of the world. Arsenic contamination in natural aquifers has occurred in Argentina, Bangladesh, Cambodia, Chile, China, Ghana, Hungary, India, Mexico, Nepal, New Zealand, the Philippines, Taiwan, the United States, and Vietnam (Das et al. 2004).

Arsenic pollution has occurred most severely in Bangladesh and India (West Bengal). It is estimated that more than 35 million people are consuming As-polluted groundwater in Bangladesh where underground water is used mainly for drinking, cooking, and other household activities (Das et al. 2004; Rabbani et al. 2002). Until now, lots of effort has been given to find safe drinking water there, but no suitable measure has been established yet. In addition to the drinking water problem, continued irrigation with As-contaminated water increases the extent of As contamination in agricultural land soil in Bangladesh (Ullah 1998; Alam and Satter 2000; Ali et al. 2003; Islam et al. 2004).

In surveys on As contamination in 60 of 64 districts of Bangladesh, it was observed that many tube wells of shallow depth (less than 100 m) exceed the As concentration level of 0.05 mg L^{-1} (the Bangladesh standard for arsenic in drinking water) in almost all 60 districts (Rahman et al. 2002; BGS/DPHE 2000) (Figure 2.6). In a study conducted by BGS/DPHE (2000), it was observed that approximately 61% of samples exceeded 0.01 mg L^{-1} (WHO guideline for As concentration in drinking water), approximately 45% of samples exceeded 0.05 mg L^{-1}, and 2% exceeded 1 mg L^{-1} of As concentration in a shallow tube well (BGS/DPHE 2000; Hossain 2006). Some studies reported As concentration in uncontaminated land in Bangladesh, which varies from 3–9 mg As kg^{-1} (Ullah 1998; Alam and Satter 2000). On the other hand, elevated As concentrations were observed in many studies in agricultural land soil irrigated with As-contaminated water, which is, in some cases, approximately 10–20 times higher than As concentration in nonirrigated land. Ullah (1998) reported the As concentration in top agricultural land soil (up to 0–30 cm depth) was up to 83 mg As kg^{-1}. However, this finding is not identical with other studies, such as Islam et al. (2005), which found up to 80.9 mg As kg^{-1}, and Alam and Sattar (2000), which found up to 57 mg As kg^{-1} of soil, for samples collected from different districts of Bangladesh.

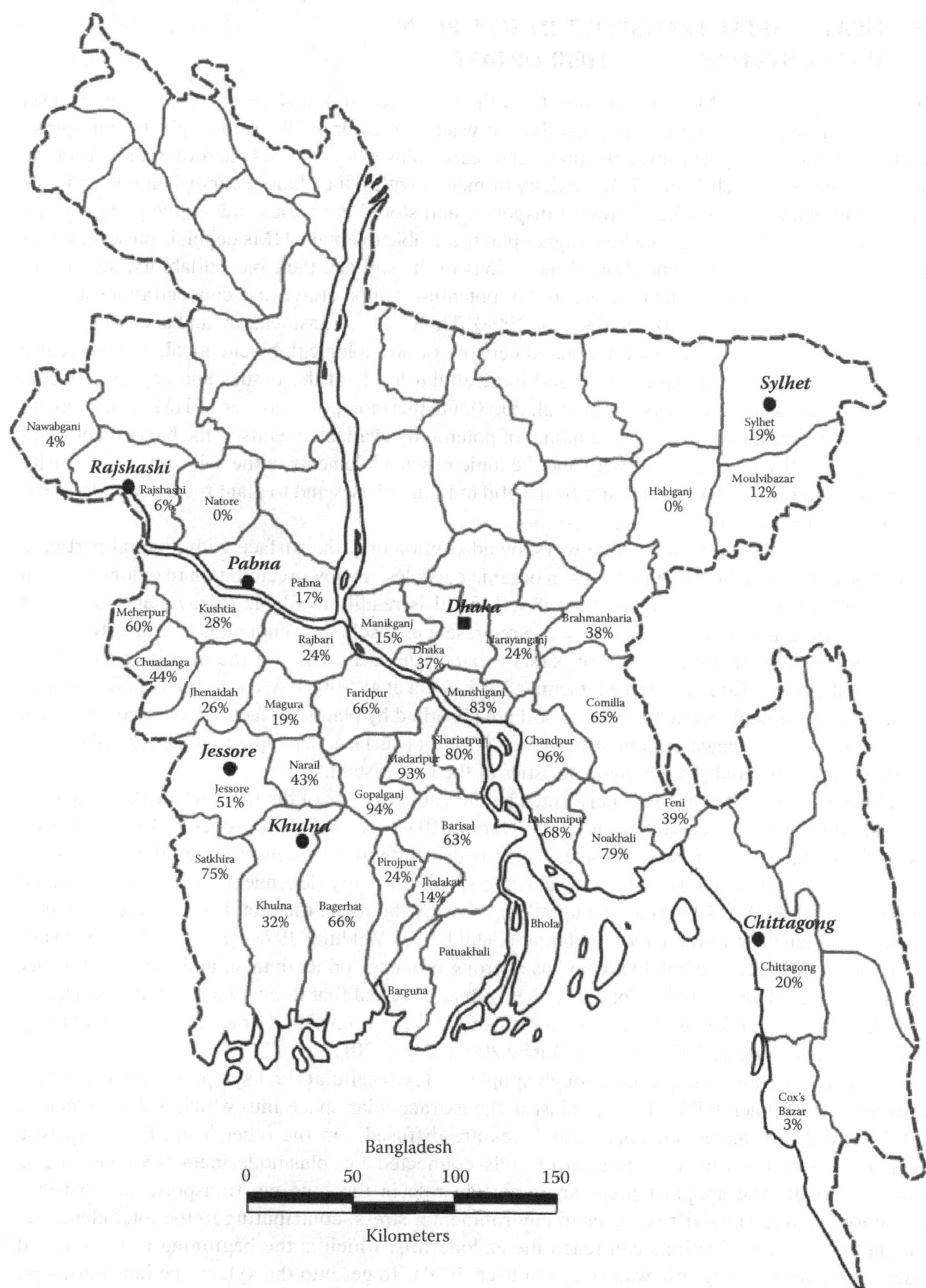

FIGURE 2.6 Percentage of groundwater from the shallow aquifer (less than 150 m deep), exceeding the Bangladesh standard for arsenic of 0.05 mg L^{-1}. (Hossain, M. F., *Agriculture, Ecosystems* and *Environment, 113*, 2006.)

2.6 HEAVY METAL TRANSPORT IN SOIL-PLANT-WATER SYSTEMS AND THEIR UPTAKE

Plants are exposed to HM contamination from the air, water, soil, and sediments. However, HMs can be much more concentrated in soils than in water (Förstner 1979). Higher plants can uptake metals from the atmosphere through shoots and leaves plus entry via roots and rhizomes from the soil (Lyubenova and Schröder 2010). Toxicity of metals within the plant occurs when metals move from soil to plant roots and get further transported and stored in various sites in the plant (Verma and Dubey 2003). The extent to which higher plants are able to uptake HMs depends on several factors. These include the concentration of metal ions in the soil and their bioavailability, modulated by the presence of organic matter, pH, redox potential, temperature, and concentration of other elements (Benavides et al. 2005; Setia et al. 2008). The uptake, translocation, and accumulation of HMs in plants are mediated by an integrated network of physiological, biochemical, and molecular mechanisms and occur at extracellular and intracellular levels of the tissues and organs of plants grown under contaminated sites (Setia et al. 2008). Furthermore, the transfer of HMs from soils to plants depends primarily on the total amount of potentially available metals or the bioavailability of the metal (quantity factor), the activity and the ionic ratios of elements in the soil solution (intensity factor), and the rate of element transfer from solid to liquid phases and to plant roots (reaction kinetics) (Brümmer et al. 1986).

In soils, HMs are retained in three ways: by adsorption onto the surface with mineral particles, by complexation with humic substances in organic particles, and by precipitation reactions (Walton et al. 1994). In general, only a fraction of soil metal is readily available (bioavailable) for plant uptake. Maximum amounts of HMs in soils are present as insoluble compounds (Lasat 2002); however, different rhizopheric activities of plants enhance the availability of the metals and facilitate uptake by the roots (Romheld and Marschner 1986; Setia et al. 2008). After a series of complicated physical and chemical reactions, HMs in soil are absorbed by plants. In fact, metal uptake by roots may take place at the apical region or from the entire root surface, depending on the type of metals, the uptake capacity, and growth characteristics of the root system.

Uptake of most of the metals is performed by the younger parts of the roots where the Casparian strips are not fully developed (Prasad 2004; Dubey 2011). Two different processes have been suggested for metal uptake: passive uptake, which is driven by the concentration gradient across the membrane, and active uptake, which is substrate specific, energy dependent, and carrier mediated (Williams et al. 2000). To initiate the uptake process by the roots, the metal species must occur in soluble form adjacent to the root membrane (Cataldo and Wildung 1978) (Figure 2.7). The availability of metal species in soluble form has a strong influence on its uptake, mobility, and toxicity within the plant (Dubey 2011). Uptake studies of metals reveal that many metal pollutants, such as As, Cd, and Pb, are taken up in the rice roots against the concentration gradient (Shah and Dubey 1995; Verma and Dubey 2003; Jha and Dubey 2004; Dubey 2011).

Solubilized metals enter plants through apoplastic (extracellular) and symplastic (intracellular) pathways (Marschner 1995). The apoplast is the extracellular space into which water molecules and dissolved low-molecular-mass substances are diffused. On the other hand, the symplastic compartment consists of a continuum of cells connected via plasmodesmata (Lyubenova and Schröder 2010). The apoplast plays an important role in the binding, transport, and distribution of ions and in cellular responses to environmental stress, contributing to the total elemental content of the roots. The ions can reach the endodermis, which is the beginning of the internal space, by traveling along this waterway (Nultsch 2001). To get into the xylem, the ions must pass through the endodermis and the Casparian strip (Figure 2.7). The Casparian strip is a waterproof lipophilic surface coating in the radial cylinder of the endodermal cells of the root that plays a role in blocking the passage of soluble minerals and water from the internal symplast through the cell walls (Lyubenova and Schröder 2010). From the root, HMs are transported to the shoot and other parts through the specialized membrane transport processes of the xylem (Salt et al. 1995).

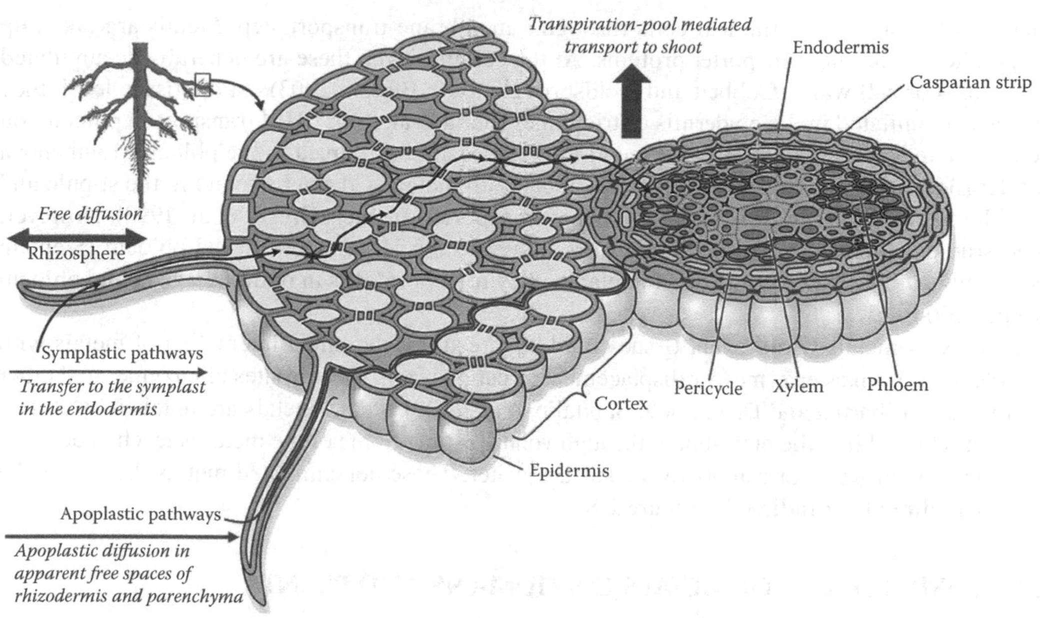

FIGURE 2.7 Uptake pattern of HMs by plant roots. (Lyubenova, L., and P. Schröder. Uptake and effect of heavy metals on the plant detoxification cascade in the presence and absence of organic pollutants. In I. Sherameti, and A. Varma (Eds.), *Soil heavy metals, Soil Biology 9*, Springer, Berlin, 2010.)

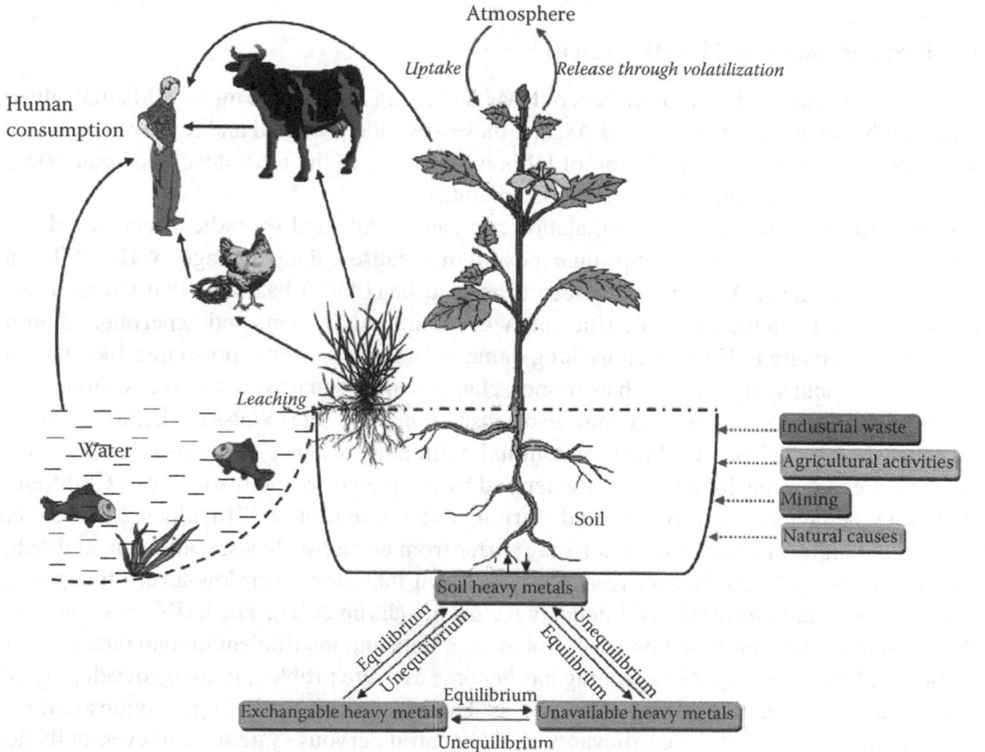

FIGURE 2.8 Interaction among plants, heavy metals, soils, and human activities. (Wei, S., and Q. Zhou. Trace elements in agro-ecosystem. In M. N. V. Prasad (Ed.), *Trace elements as contaminants and nutrients: Consequences in ecosystems and human health*, Wiley, Hoboken, 2008.)

Then it is transported to the leaf cells through a membrane transport step. Metals are taken up by specific membrane transporter proteins. At the cellular level, these are generally accumulated in vacuoles or cell walls (Cobbett and Goldsbrough 2000; Burken 2003). At the tissue level, they may be accumulated in the epidermis or trichomes (Setia et al. 2008). HM transport in phloem, on the other hand, is complicated as the metal ions can easily be coupled to the phloem (Lyubenova and Schröder 2010). For instance, after treatment with metals, Cd can be found in the stipule and in the leaf stalk of *Pisum sativum*, but it is not transported further (Greger et al. 1993). However, some studies elucidated that metal chelators influence the HM content in the phloem (Stephan and Scholz 1993). In the case of aquatic plants, they transport HMs in both the xylem and phloem (Greger 2004).

Toxicity of metals within plant tissue could be a result of the direct interaction of metals with biomolecules, such as enzymes, or displacement of cations from specific sites of enzymes and other biomolecules (Sharma and Dietz 2008; Sandalio et al. 2009). Heavy metals are uptaken by a plant and then released into the atmosphere through volatilization. Some of the metal is leached down to the underground water or run off to the surface water. These contaminated metals then enter the food chain directly or indirectly (Figure 2.8).

2.7 TOXIC EFFECTS OF METALS ON HUMANS AND PLANTS

Toxic HMs cause an inhibitory effect on the growth of animals, plants, fungi, and bacteria and, often, on all four groups. Some metals, such as Cu, Fe, Mn, Ni, and Zn, are essential nutrients for all living organisms but become toxic at higher concentrations. Others, such as Al, Cd, Pb, As, and Hg, do not appear to play any essential role in metabolism (Azevedo 2005).

2.7.1 Toxic Metals and Human Health

There are several acute and chronic effects of HMs on human health. The major HMs that adversely affect human health are Pb, Cd, Hg, and As, and these are widely studied and reviewed by researchers (Järup 2003). Although the emission of HMs is declining in the most developed countries, the adverse effects are increasing in less-developed countries.

Different studies indicate that Cd inhalation can cause acute and sporadic effects (Seidal et al. 1993; Barbee and Prince 1999). Cadmium exposure may cause kidney damage (WHO 1992; Järup 2003). The International Agency for Research on Cancer (IARC) has classified Cd as a human carcinogen (group I) on the basis of sufficient evidence in both humans and experimental animals (IARC 1993). Exposure to Hg may cause lung damage and other chronic poisoning-like neurological and psychological symptoms, such as tremor, changes in personality, restlessness, anxiety, sleep disturbance, and depression. Mercury may also cause kidney damage (Weiss et al. 2002). The toxicity of Pb causes headaches, irritability, abdominal pain, and various symptoms related to the nervous system. Lead encephalopathy is characterized by sleeplessness and restlessness. Children may be affected by behavioral disturbances and learning and concentration difficulties. In severe cases of Pb encephalopathy, the affected person may suffer from acute psychosis, confusion, and reduced consciousness (Järup 2003). Recent research has shown that long-term, low-level Pb exposure in children may also lead to diminished intellectual capacity (Järup 2003). The IARC classified Pb as a possible human carcinogen based on sufficient animal data and insufficient human data (Steenland and Boffetta 2000). Recently, As poisoning has become a severe problem in many developing countries. Inorganic As is acutely toxic, and intake of large quantities leads to gastrointestinal symptoms, severe disturbances of the cardiovascular and central nervous systems, and eventually death. In survivors, bone marrow depression, hemolysis, hepatomegaly, melanosis, polyneuropathy, and encephalopathy may be observed. It may induce peripheral vascular disease, which, in its extreme form, leads to gangrenous changes. Populations exposed to As via drinking water show excess risk of mortality from lung, bladder, and kidney cancer, the risk increasing with increasing exposure. There is also an increased risk of skin cancer and other skin lesions, such as hyperkeratosis and pigmentation changes (Järup 2003).

2.7.2 Metal Toxicity in Plants

Although some HMs (trace elements) are beneficial for plant growth and physiology, after excessive uptake by plants, these elements may participate in some physiological and biochemical reactions that can destroy normal growth of the plant by disturbing absorption, translocation, or synthesis processes. They may combine with some huge molecule, such as nucleic acid, protein, and enzyme, or may substitute the metabolic activities. Therefore, the growth and procreation of the plant is prohibited and leads to death (Wei and Zhou 2008). However, the response of plants to nonessential metals varies across a broad spectrum from tolerance to toxicity with increasing concentration (Baker and Brooks 1989) (Figure 2.9). Too-low doses of trace elements can result in nutrient deficiency with below-optimum growth; as the supply increases, up to a certain point, there is a positive response. This is the range within which the trace element is required as a micronutrient. After the optimum concentration is reached, there is no further positive response, and normally there is a plateau as seen in Figure 2.8. At increasingly higher doses, the trace metal is in excess and begins to have harmful effects, which cause growth rate to decrease (i.e., it has reached concentrations that are toxic). The logical result of further increase in a metal dose beyond that shown on the graph would be death (Wright and Welbourn 2002).

Making a generalization about the effect of HMs on plants is difficult because of the multidimensional variations in parameters under different concentrations, types of HMs, duration of exposure, target organs of plants, plant age, etc. Several physio-biochemical processes in plants cells are affected by HMs (Dubey 2011). The most obvious plant reaction under HM toxicity is the inhibition of growth rate (Sharma and Dubey 2007). Heavy metals also cause chlorosis, necrosis, leaf rolling, inhibition of root growth, stunted plant growth, altered stomatal action, decreased water potential, efflux of cations, alterations in membrane functions, inhibition of photosynthesis, altered metabolism, altered activities of several key enzymes, etc. (Heckathorn et al. 2004;

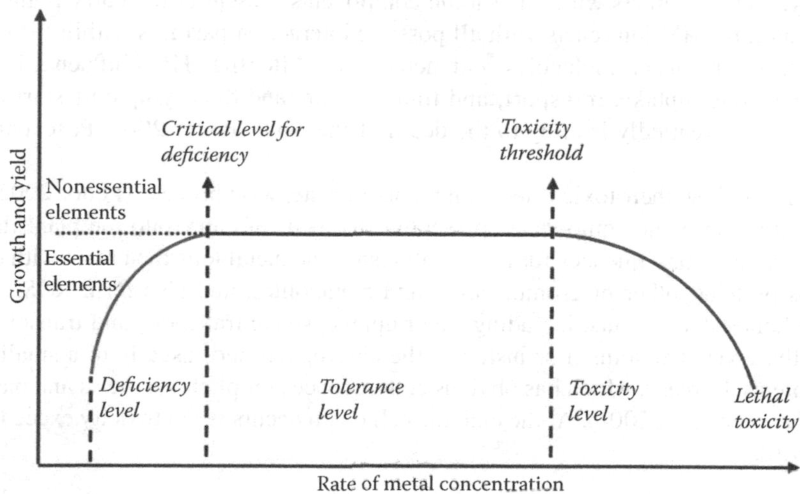

FIGURE 2.9 Relationship between metal concentration and plant growth. (Baker, A. J. M., and R. R. Brooks, *Biorecovery, 1*, 1989.)

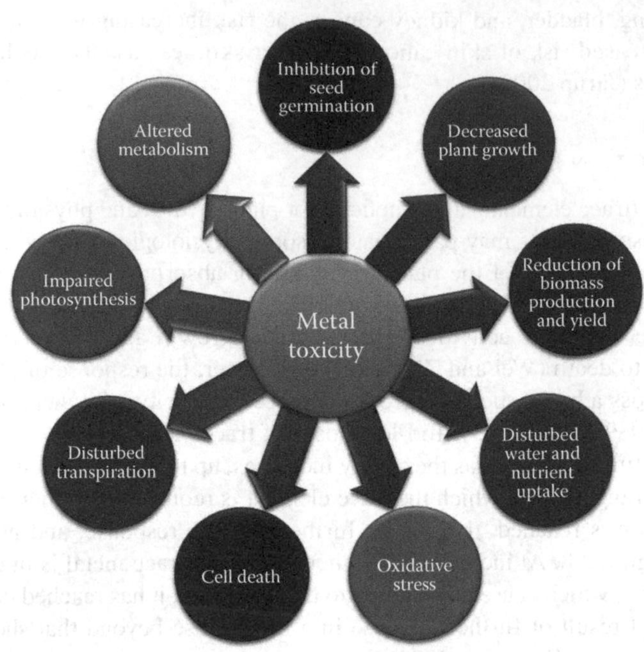

FIGURE 2.10 Common effects of toxic metals in plants.

Sharma and Dubey 2007; Dubey 2011) (Figure 2.10). Seed germination is also severely affected by HMs (Ahsan et al. 2007). HMs inhibit the rate of photosynthesis and respiration and thus inhibit carbohydrate metabolism and their partitioning in growing plants (Llamas et al. 2000; Vinit-Dunand et al. 2002). Direct phytotoxic effects of HMs include their direct interactions with proteins and enzymes; displacement of essential cations from specific binding sites, causing altered metabolism; inhibiting the activities of enzymes, etc. (Sharma and Dubey 2007; Sharma and Dietz 2008). Initially, a HM interacts with other ionic components present at the entry point of a plant root system. Later, the HM ion reacts with all possible interaction partners within the cytoplasm, including proteins, other macromolecules, and metabolites. After that, HMs influence homeostatic events, including water uptake, transport, and transpiration, and thus symptoms start to develop and become visible, eventually leading to the death of the plant (Fodor 2002; Poschenrieder and Barceló 2004).

In general, HMs show their toxic effects on plants in a stepwise process (Fodor 2002). At first, they interact with other ionic components present at the locus of entry into the plant rhizosphere that subsequently have consequences for the metabolism. The metal ions then react with other molecules, such as proteins, other macromolecules, and metabolites, and also form ROS. After that, they influence homeostatic events, including water uptake, solute transport, and transpiration, and start to show the toxic symptoms. For instance, the chlorophyll and, usually to a smaller degree, carotenoid contents decrease, which has obvious consequences for photosynthesis and plant growth (Barceló and Poschenrieder 2004). At the end, the cell death occurs when toxicity exceeds the limit (Appenroth 2010).

2.7.2.1 Germination

Germination rate and seedling growth are largely affected by HM toxicity, and it is often used to assess the abilities of plant tolerance to metal elements (Peralta et al. 2001). Because seed germination is the foremost physiological process affected by toxic elements, the ability of a seed to germinate in a medium containing any metal element would be directly indicative of its level of tolerance to this metal (Peralta et al. 2001). Mahmood et al. (2007) observed that the higher concentrations of Cu, Zn, and Mg (1, 5, and 10 μM) inhibit seed germination and early growth of barley, rice, and wheat seedlings significantly compared to a control. The seed germination of *Echinochloa colona* is reduced to 25% at 200 μM Cr treatment (Rout et al. 2000). In another study, Parr and Taylor (1982) found that high levels (500 ppm) of Cr^{6+} in soil caused a 48% reduction in germination of *Phaseolus vulgaris*. Jain et al. (2000) observed reductions up to 32 and 57% in sugarcane bud germination at 20 and 80 ppm Cr application, respectively. Seed germination of *Medicago sativa* was declined by 23% when exposed to 40 ppm Cr^{6+} treatment (Peralta et al. 2001). In a recent study, Mami et al. (2011) observed a drastic decline in seed germination of tomato (*Lycopersicum esculentum*) seedlings as affected by excess Fe, Pb, and Cu.

2.7.2.2 Plant Growth

Inhibition of plant growth is often used as an indicator for toxic HMs (Lal 2010). HMs either retard the growth of the whole plant or plant parts (Shanker et al. 2005). In general, plant roots show more rapid changes in growth pattern than shoots as they have the direct contact with the contaminated soil (Shah et al. 2010). Different plant studies have demonstrated that HMs cause a significant reduction in plant growth and biomass (Cargnelutti et al. 2006; Israr et al. 2006; Zhou et al. 2007). The reduced growth could be a result of the blocking of cell division or elongation (Patra and Sharma 2000; Zhou et al. 2009). According to Lal (2010), growth inhibition by HMs results from metabolic disorders and direct effects on growth, such as interactions with cell-wall polysaccharides decreasing cell-wall plasticity. Prasad et al. (2001) observed that Cr has a greater effect on the reduction of root length than other metals while studying with *Salix viminalis*. Mokgalaka-Matlala et al. (2008) observed that the root length decreased significantly with increasing concentrations of As in *Prosopis juliflora*. The reduction in plant height is also an obvious effect of HMs that might be mainly a result of reduced root growth and regulation of lesser nutrients and water transport to the aerial parts of the plant (Rout et al. 1997; Shah et al. 2010). Heavy metal toxicity also decreased the biomass of plants. It was reported that HMs, such as Cr and Cd, caused a reduction in biomass production in *Bacopa monnieri* (Tokalioglu and Kartal 2006). Cauliflower (*Brassica oleracea*) plants exposed to 0.5 mM Cr(VI) showed restricted dry biomass production (Chatterjee and Chatterjee 2000). In barley and maize, a significant reduction of biomass was observed when Cr levels in soil reached 100 or 300 mg kg^{-1} (Golovatyj and Bogatyreva 1999). Yildiz (2005) observed a dramatic decrease in dry matter (DM) production in *L. esculentum* and *Zea mays* plants with increasing concentrations of Cd (Yildiz 2005). In a recent study, Chen et al. (2011) observed that both shoot and root weight decreased progressively with increasing Cd concentration for *Brassica campestris* and *Brassica juncea*.

2.7.2.3 Water Relations

The toxic effects of HMs can be shown in plants connected with water relations, especially in the early stage of plant growth. HMs influence membrane transport and inhibit root growth and enzyme activities (Poschenrieder and Barceló 2004; Appenroth 2010). Heavy metals can also decrease the number and size of leaves, stomatal size, and number and diameter of xylem vessels; can increase stomatal resistance, leaf rolling, and leaf abscission; and cause a higher degree of root suberization (Barceló and Poschenrieder 1990; Shah et al. 2010). Many plant studies revealed that the application of HMs increased the stomatal resistance and, in this way, decreased the rate of transpiration. It is assumed that stomatal closure after the application of HMs is a consequence of stress resulting from water deficiency. This leads to increased levels of proline and ABA, which are known indicators of drought stress. Proline is evidently not involved in metal detoxification but in membrane stabilization: K^+ loss and lipid peroxidation were reduced after pretreatment with proline. Poschenrieder and Barceló (2004) stated that HMs have a direct effect on stomata closure, and the roots simply act as an osmometer, producing a hydraulic signal. Instead, roots can influence the water content via chemical signals, especially ABA. In addition, water transport was found to be regulated by an impairment of aquaporins, which is one of the earliest responses to HMs in plants. In cauliflower leaves, Chatterjee and Chatterjee (2000) observed that excess Cr decreased the water potential and transpiration rates and increased diffusive resistance and relative water content.

2.7.2.4 Photosynthesis

One of the most obvious effects of HMs is the reduction of photosynthesis in plants. HMs directly affect the light and dark phases of photosynthesis. They also decrease the photosynthetic pigment content and change stomatal function, which ultimately affects the photosynthesis (Mysliwa-Kurdziel et al. 2004; Appenroth 2010). It seems that nearly all of the components of the photosynthetic apparatus are influenced by almost all HMs, including chlorophyll and carotenoid content, chloroplast membrane structure, light-harvesting and O_2-evolving complexes, photosystems, and constituents of the photosynthetic electron transport chain (Barceló and Poschenrieder 2004). Several enzymes involved in the Calvin cycle are also inhibited, especially RuBisCO and PEP carboxylase, which inhibit photosystem II (PSII) activity and electron transport (Mysliwa-Kurdziel et al. 2004). HMs also change the chloroplast ultrastructure and pigment composition, inhibit the net photosynthetic rate, decrease carboxylation efficiency of RuBisCO, and inhibit photosystem II (PSII) activity and electron transport (Mishra and Dubey 2005; Appenroth 2010).

Chatterjee and Chatterjee (2000) reported a drastic reduction in chlorophyll a and b in cauliflower grown under Co, Cr, and Cu exposures at 0.5 mM following the order of stress, Co > Cu > Cr. Shanker (2003) hypothesized that the decrease in chlorophyll b resulting from Cr could be because of the destabilization and degradation of the proteins of the peripheral part. Šimonova et al. (2007) found a significant decrease in chlorophyll and carotenoid contents under Cd exposure. Chen et al. (2011) found a significant reduction of chlorophyll a and b at 24 mg Cd kg^{-1} soil, which also caused a decline in the net rate of photosynthesis (*Pn*) and stomatal conductance (*Gs*). However, the photosynthetic activity of *Brassica campestris* was more sensitive to Cd stress than that of *Brassica juncea*.

2.7.2.5 Respiration

Some researchers consider that respiration increases under stress; for instance, *Vicia faba* exposed to 1 μM Cd^{2+} showed increased transpiration (Lee et al. 1976). In fact, at toxic concentrations of HMs, respiration is usually inhibited (Lösch 2004). However, it is true that in some plant species the presence of some HMs at lower concentrations increases respiration (Appenroth 2010).

2.7.3 Toxic Effect of Major Heavy Metals in Plants

In fact, the response of plants to HMs depends on several factors, such as uptake and accumulation of metals through binding to extracellular exudates and cell-wall constituents; efflux of HMs from cytoplasm to extranuclear compartments, including vacuoles; complexation of HM ions inside the cell by various substances, for example, organic acids, amino acids, phytochelatins (PCs), and metallothioneins (MTs); accumulation of osmolytes and osmoprotectants and induction of antioxidative enzymes; and activation or modification of plant metabolism to allow adequate functioning of metabolic pathways and rapid repair of damaged cell structures (Cho et al. 2003; John et al. 2009). The plants also showed differential responses to different metals as observed by many plant studies (Table 2.4). It also depends on the concentration of metals to which plants are exposed.

TABLE 2.4
Effect of Toxic Metals on the Growth and Physiology of Plants

Metals	Target Plants	Doses and Duration	Effects	References
Cd	*Raphanus sativus*	100 and 200 mg $CdCl_2$ kg^{-1} soil, 45 d	Decrease in length and fresh and dry weights of shoot and root systems; decrease in leaf number; decrease in chlorophyll, RWC, soluble proteins, and total amino acid content	Farouk et al. (2011)
Cd	*Solanum lycopersicum*	100 μM $CdCl_2$, 90 d	Decrease in carotenoid, chlorophyll, and proline content	Hédiji et al. (2010)
Cd	*Pisum sativum*	50 μM $CdCl_2$, 28 d	Inhibition of growth; reduction in the transpiration, photosynthesis rate, and chlorophyll content of leaves	Sandalio et al. (2001)
Cd	*Carthamus tinctorius*	25–100 μM $CdCl_2$, 4 d	Decreased root and shoot growth	Namdjoyan et al. (2011)
Cd	*Vigna radiata*	25, 50, and 100 mg $CdCl_2$ kg^{-1} soil, 30 d	Decrease in dry weight and leaf area, net photosynthetic rate, and chlorophyll content	Anjum et al. (2011)
Pb	*Brassica juncea*	450–1500 μM Pb $(C_2H_3O_2)_2$, 60 d	Decline in growth, chlorophyll content, carotenoids, and proline content	John et al. (2009)
Pb	*Arabidopsis thaliana*	25–100 μM $Pb(NO_3)_2$, 7 d	Reduction of root growth	Phang et al. (2011)
Pb	*Triticum aestivum*	0.05–1.0 g L^{-1} Pb $(NO_3)_2$, 6 d	Reduction in seed germination and root and shoot growth of seedling	Lamhamdi et al. (2001)
Pb	*Sinapis arvensis*	1500 μM $Pb(NO_3)_2$, 10 d	Inhibition of seed germination and seedling growth	Heidari and Sarani (2011)
Pb	*Oryza sativa*	2000 mg kg^{-1} Pb	Inhibition of growth; decrease in biomass and yield	Liu et al. (2010)
Pb	*Raphanus sativus*	Pb (0.5 mM $Pb(NO_3)_2$, 2 d	Increased proline content	Teklić et al. (2008)
As	*Vigna radiata*	5–100 mg L^{-1} $Na_2HASO_4.7H_2O$, 7 d	Inhibition of germination, root growth, and cell division	Mumthas et al. (2010)
As	*Brassica juncea*	25 μM Na_3AsO_4, 12–96 h	Reduction in root and shoot growth	Khan et al. (2009)
As	*Oryza sativa*	50 and 100 μM Na_2HAsO_4, 10 d; 1000 and 500 μM $NaAsO_2$, 10 d	Decrease in germination percentage, shoot and root elongation, and plant biomass	Shri et al. (2009)
Hg	*Medicago sativa*	1 μM $HgCl_2$, 24 h	Decrease in root length	Zhou et al. (2009)
Cu	*Raphanus sativus*	0.5 mM $Cu(SO_4)_2$, 2 d	Increased proline content	Teklić et al. (2008)
Cu	*Arabidopsis thaliana*	0.5–2.0 mM $CuSO_4.7H_2O$, 3 d	Inhibition of germination and prolonged time required for germination	Gill et al. (2011a)

(continued)

TABLE 2.4 Continued
Effect of Toxic Metals on the Growth and Physiology of Plants

Metals	Target Plants	Doses and Duration	Effects	References
Cu	*Vigna radiata*	$CuSO_4.5H_2O$ 100–250 mg kg^{-1}, 45 d	Decrease in growth and DM production	Manivasagaperumal et al. (2011)
Cu	*Phaseolus vulgaris*	$CuSO_4$ 50 and 75 μM, 3 d	Exhibit chlorosis and necrosis; a dramatic reduction in dry weight production	Bouazizi et al. (2010)
Zn	*Sesbania drummondii*	300 mg L^{-1} $ZnSO_4{\cdot}7H_2O$, 10 d	Inhibition of growth	Israr et al. (2011)
Zn	*Glycine max*	>0.32% $ZnSO_4$	Inhibition of germination; lower seedling growth and fresh weight	Dong et al. (2011)
Zn	*Vigna mungo*	1.50 mM $ZnSO_4.7H_2O$, 7 d	Reduction in fresh and dry weight of plants	Dhankhar and Solanki (2011)
Zn	*Oryza sativa*	2 mM Zn, 7 d	Decrease in root and shoot growth and biomass reduction	Song et al. (2011)
Zn	*Vitis vinifera*	14–35 mM Zn^{2+}	Decreased chlorophyll content and root activity	Yang et al. (2011)
Ni	*Pistia stratiotes*	1 and 10 ppm $NiSO_4$, 6 d	Decrease in leaf chlorophyll content	Singh and Pandey (2011)
Ni	*Zea mays*	100 μM Ni, 5 d	Decrease in plant growth, fresh and DM yield and chlorophyll content	Mishra et al. (2010)
Ni	*Brassica napus*	0.5 mM $NiCl_2{\cdot}6H_2O$, 10 d	Chlorosis and necrosis at leaves; decreased biomass and chlorophyll content	Kazemi et al. (2010)
Ni	*Glycine max*	50–200 μM $NiCl_2$, 10 d	Decrease in fresh and dry mass of root and shoot	El-Shintinawy and El-Ansary (2000)
Ni	*Triticum aestivum*	100 μM Ni, 1–7 d	Decreased fresh weight of root and shoot growth; exhibits chlorosis and necrosis	Gajewska et al. (2009)
Ni	*Triticum aestivum*	100 μM Ni	Decreased chlorophyll content and net photosynthetic rate	Yusuf et al. (2011)
Cr	*Typha angustifolia*	1 mM $K_2Cr_2O_7$, 30 d	Reduction in plant height, shoot and root growth, and dry weight	Bah et al. (2011)
Cr	*Ocimum tenuiflorum*	10–100 μM $K_2Cr_2O_7{\cdot}7H_2O$, 24–72 h	Reduction of biomass and chlorophyll content	Rai et al. (2004)
Cr	*Cucumis melo*	2.5–300 mg L^{-1} $CrCl_3.6H_2O$	Inhibition of germination and seedling growth	Akinci and Akinci (2010)
Al	*Cucumis sativus*	100–2000 μM $Al_2(SO_4)_3$, 10 d	Decrease in electrolyte leakage and chlorophyll content	Pereira et al. (2010)
Al	*Zea mays*	10–60 μM Al^{3+}, 48 h	Decreased relative root growth	Boscolo et al. (2003)
Al	*Pisum sativum*	10 and 50 μM $AlCl_3$, 24 and 48 h	Decreased fresh and dry weight of root and shoot	Panda and Matsumoto (2010)
Al	*Oryza sativa*	160 μM Al^{3+}, 5–20 d	Decreased length of roots and shoots	Sharma and Dubey (2007)

2.7.3.1 Cadmium

Among the effects of different HMs on plants, Cd has widely been studied by researchers. The regulatory limit of Cd in agricultural soil is 100 mg kg^{-1} soil (Salt et al. 1995). But this threshold is continuously exceeded because of several human activities. Cadmium is toxic for most plants at concentrations greater than 5–10 μg Cd per g^{-1} leaf dry weight (White and Brown 2010) except for Cd-hyperaccumulators, which can tolerate Cd concentrations of 100 μg Cd per g^{-1} leaf dry weight (Verbruggen et al. 2009). Despite its high phytotoxicity, Cd is easily taken up by plant roots, is transported to above-ground tissues (DalCorso et al. 2010; Lux et al. 2010; Liu et al. 2010), and enters the food chain (Hall 2002; Gill et al. 2011b). Higher concentrations of Cd also cause a reduction in photosynthesis as well as water and mineral uptake by plants. Plants grown in soil containing high Cd concentrations show visible injuries, such as chlorosis, growth inhibition, and browning of root tips, which ultimately cause death (Wójcik and Tukiendorf 2004; Mohanpuria et al. 2007) (Figure 2.10). It also results in fewer tillers, senescence, and reduced plant growth and biomass (Wu et al. 2003; Cosio et al. 2006). In *Elodea canadensis*, a thinner stem, less-expanded leaves with partial bleaching of green tissues, and 40% internode shortening were observed in response to Cd treatments when compared with control plants (Vecchia et al. 2005). Increasing concentrations of Cd in a liquid culture and pot experiment decreased the germination and root growth of carrot (*Daucus carota*) and radish (*Raphanus sativus*) plants (Chen et al. 2003). Sandalio et al. (2001) reported that roots and leaves of pea (*Pisum sativum*) plants exposed to Cd (50 μM $CdCl_2$, 28 d) displayed a significant inhibition of growth, a reduction in the transpiration and photosynthesis rate and chlorophyll content of leaves, and an alteration in the nutrient status in both roots and leaves. Hédiji et al. (2010) observed a significant decrease in carotenoids and chlorophyll as well as proline content in *L. esculentum* leaves when subjected to 100 μM Cd. Rai et al. (2005) observed a significant decrease in root and shoot length and fresh and dry weight of *Phyllanthus amarus* under Cd stress. Cd inhibited root dry mass and induced changes in biomass allocation patterns without any effect on biomass accumulation at the whole plant level in *Hordeum vulgare* (Vassilev et al. 2004). Cadmium has been shown to affect photosynthetic functions through interacting with the photosynthetic apparatus at various levels of organization and architecture viz. accumulation of metal in the leaf (main photosynthetic organ); partitioning in leaf tissues, such as stomata, mesophyll, and bundle sheath cells; interaction with cytosolic enzymes; and alteration of the functions of chloroplast membranes (Gill et al. 2011b,c). Recently, Anjum, Umar, Iqbal, and Khan (2011b) reported that mung bean (*Vigna radiata*) seedlings treated with Cd (25, 50, and 100 mg $CdCl_2$ kg^{-1} soil) showed significant decrease in the dry weight and leaf area and photosynthetic parameters (net photosynthetic rate and chlorophyll content). The increasing level of Cd in the soil resulted in a gradual decrease in plant dry weight, leaf area, photosynthesis, and chlorophyll content in both *V. radiata* genotypes. At 100 mg Cd kg^{-1} soil, the plant dry weight, leaf area, photosynthesis, and chlorophyll content were reduced by 59.8, 39.8, 30.8, and 40.0%, respectively in Cd-susceptible cv. PS 16 as compared to the control. But in Cd-tolerant cv. Pusa 9531, decreases in these parameters were 26.2, 23.0, 27.5, and 36.3%, respectively. Farouk et al. (2011) observed that Cd at 100 and 150 mg kg^{-1} soil significantly decreased the length, fresh and dry weights of shoot and root systems, and leaf number per plant. Chlorophyll, total sugars, nitrogen, phosphorus, potassium, relative water content, water deficit percentage, soluble proteins, and total amino acid contents were also decreased. In a recent study, researchers observed significant reduction in chlorophyll content and stunted growth in rapeseed seedlings when subjected to Cd stress (Hasanuzzaman and Fujita, unpublished result) (Figure 2.11).

Kachout et al. (2009) showed that exposure of plants to different levels of metal (Cu, Ni, Pb, Zn) reduced DM production and height of shoots in *Atriplex hortensis* and *A. rosea*. The decrease in root growth caused by the toxicity of metals was more severe than the decrease in shoot growth. *Atriplex* plants exhibited a gradual decline in height when exposed to metal: a four-week exposure of *A. hortensis* to 25, 50, 75, and 100% contaminated soil. John et al. (2009) reported that the *Brassica juncea* plant exhibited a decline in growth, chlorophyll content, and carotenoids with Cd

FIGURE 2.11 **(See color insert.)** Cadmium-induced visual symptoms in rapeseed seedlings [control (upper); $CdCl_2$ 0.5 mM, 48 h (lower left); $CdCl_2$ 1 mM, 48 h (lower right)].

and Pb, but Cd was found to be more detrimental than Pb treatment in *B. juncea*. The protein content was decreased by Cd (900 μM) to 95%, 44% by Pb (1500 μM) at the flowering stage. Proline showed an increase at lower concentrations of Cd and Pb, but at higher concentrations, it showed a decrease. More accumulation of Cd and Pb was observed in roots than shoots in *B. juncea*. Cd was found to be more accumulated than Pb, but higher concentrations of Pb hamper Cd absorption.

2.7.3.2 Lead

Liu et al. (2010) showed that a soil Pb concentration of 800 mg kg^{-1} is moderately toxic to rice with 10–30% decreases in plant biomasses. A soil Pb concentration of 2000 mg kg^{-1} is a soil Pb level that severely inhibits rice growth with 20–50% decreases in plant biomasses and significant reductions in grain yields. They also concluded that the genotype indica was generally more sensitive to soil Pb stress than the genotype japonica, especially at earlier growth stages (such as at the tillering stage), and in grain yield. Phang et al. (2011) concluded that Pb exposure (75 μM $Pb(NO_3)_2$, 7 d) to *A. thaliana* seedlings showed a more than 50% inhibition of root growth, which was associated with increased free-radical production by cell wall–accumulated Pb. Lamhamdi et al. (2011) observed a significant reduction in seed germination and seedling growth in wheat plants exposed to Pb. They found 98% germination under control conditions; whereas it was only 71% in the treated seeds at a concentration of 3 mM Pb. Lead exposure also markedly reduced the biomass of seed, root, and shoot growth of the seedlings.

2.7.3.3 Arsenic

Recently As toxicity in plants has come to light as a number of studies indicated its negative impact on plants. Plant height decreases with increasing concentrations of As in irrigation water. The effect of As on leaf number, root length, and biomass production followed a similar pattern as plant height (Ahmed et al. 2006) (Figure 2.12). At higher concentrations, As interferes with plant metabolic processes and can inhibit growth, often leading to death (Jiang and Singh 1994). There are several reports regarding the loss of fresh and dry biomass of roots and shoots, loss of yield and fruit production, and morphological changes when the plants are grown in As-treated soils (Mokgalaka-Matlala et al. 2008; Shaibur et al. 2008; Srivastava et al. 2009). Carbonell-Barrachina et al. (1997) reported that in bean (*Phaseolus vulgaris*) plants leaf dry weight had an average reduction of 50% and fruit production or yield showed even a higher reduction of 84% compared with controls when

FIGURE 2.12 Arsenic-induced visual symptoms in rice seedlings (left) and leaves (right). (Courtesy of Prof. Dr. M. Asaduzzaman.)

As was present in the growing solutions. Shri et al. (2009) demonstrated the effect of As(III) (50 and 100 μM) and As(V) (100 and 500 μM) on growth, oxidative stress, and antioxidant systems in rice seedlings. They found a marked decrease in germination percentage, shoot and root elongation, and plant biomass with As treatments as compared to a control.

2.7.3.4 Mercury

Phytotoxic effects of Hg compounds have been reported in several plants (Schützendübel and Polle 2002; Esteban et al. 2008; Nooraniazad et al. 2010). Increasing reports have shown that Hg can be readily uptaken by higher plants, and it is highly phytotoxic to plant cells (Yadav 2010). Because of its transition properties, mercury is quickly uptaken by plants and thus results in toxicity or even the death of plants (Patra and Sharma 2000; Esteban et al. 2008; Zhou et al. 2009). Many researchers have reported that Hg-induced toxicity in plants results from the binding of its ionic forms (Hg^{2+}) to sulfhydryl (SH) groups of proteins, disruption of structure, and displacement of essential elements (Hall 2002; Schützendübel and Polle 2002; Zhou et al. 2009). However, the extent of the effect depends on the concentration, the mode of application, and the tested plant species. Seeds exposed to Hg may show abnormal germination, which might be a characteristic hypertrophy of the roots and coleoptile of cereal seedlings. Mercury stress may also inhibit growth and DM production of plants (Patra and Sharma 2000). Mercury hampers both light and dark reactions in photosynthesis (Krupa and Baszynski 1995), and it strongly inhibits the electron transport chain, O_2 evolution, and quenching of chlorophyll fluorescence in PS II (Bernier et al. 1992; Lee et al. 1992). Nooraniazad et al. (2010) investigated the effects of Hg toxicity on growth in *Anethum graveolens* under hydroponic conditions with five treatments (0, 5, 10, 15, and 20 μM $HgCl_2$). The results showed that the dry weight of shoot and root, except for treatment of 5 μM, significantly decreased in comparison to the control. The reduction of total chlorophyll in leaves was significant with increasing Hg toxicity.

2.7.3.5 Copper

Copper is considered a micronutrient for plants and plays an important role in CO_2 assimilation and ATP synthesis. It is also an essential component of various proteins, such as plastocyanin of the photosynthetic system and cytochrome oxidase of the respiratory electron transport chain (ETC) (Demirevska-Kepova et al. 2004). At lower concentrations, Cu is required for normal growth and development of plants, but at higher concentrations, it causes toxicity visualized by chlorosis and necrosis, stunting, and inhibition of root and shoot growth (Yruela 2009). It can also disturb protein structure as a result of the unavoidable binding to proteins. Copper toxicity at the cellular level may be a consequence of the binding of Cu to SH groups in proteins, thereby inhibiting enzyme activity or protein function, induction of a deficiency of other essential ions, impaired cell transport processes, and oxidative damage and metabolic disturbances (Tewari et al. 2006; Yruela 2009). While studying orange seedlings, Zhang et al. (2009) observed that chlorophyll content significantly increased at 0.1 μM L^{-1} Cu but decreased at 5, 20, and 40 μM L^{-1}. Carotenoid content decreased with increasing Cu concentrations. The net photosynthetic rate and light saturation point were reduced under Cu stress exceeding 5 μM L^{-1}. The transpiration rate was improved by Cu at a concentration of 0.1 μM L^{-1}, but 5 and 20 μM L^{-1} decreased the transpiration rate and the stomatal conductance of orange plants. At the same time, high Cu concentrations (5, 20, and 40 μmol L^{-1}) restricted photoreduction activity. These findings indicated that lower Cu concentrations improve photosynthesis, and high Cu concentrations inhibit it. In *Phaseolus vulgaris* leaves, Bouazizi et al. (2010) observed that growth and development in expanding leaves was inhibited when they were exposed to 50 or 75 μM $CuSO_4$ for three days. Copper-stressed leaves also exhibited chlorotic symptoms and necrosis at both concentrations but more dramatically at the higher concentration, which was accompanied by a significant reduction in dry weight (Bouazizi et al. 2010). However, because of the exposure, Cu content in the leaves of treated plants was significantly higher than in the control, but this effect was not dose-dependent at the concentrations tested. Recently, Manivasagaperumal et al. (2011) demonstrated that Cu concentrations ($CuSO_4.5H_2O$, 50 mg kg^{-1}) showed a significant increase in the overall growth, DM yield, and nutrient content, and higher concentrations ($CuSO_4.5H_2O$, 100–250 mg kg^{-1}) decreased the growth, DM production, and nutrient content of *Vigna radiata*.

2.7.3.6 Zinc

In general, Zn is beneficial at low concentrations, and for most plant species, the range of this concentration of Zn is 15–100 ppm (Clemens 2006), but it is phytotoxic at high concentrations. Plants exposed to higher levels of Zn may show symptoms similar to other HM toxicities. In most cases, excessive Zn resulted in the inhibition of root growth, induction of chlorosis in young leaves, generation of ROS, and deficiencies of other nutrients (e.g., Fe or Mg) as well as interference with uptake, translocation, and utilization (Marschner 1995; Kramer et al. 2007). Many researchers have studied the toxic effect of Zn on various plant species, such as *Nigella sativa* (El-Ghamery et al. 2003), *Phaseolus vulgaris* (Cuypers et al. 2001), *Triticum aestivum* (El-Ghamery et al. 2003), *Vetiveria zizanioides* (Xu et al. 2009), and *Cajanus cajan* (Madhava Rao and Sresty 2000). Sun, Sha, and Jie (2010a) reported that the content of chlorophyll in melon seedlings decreased with the increase of the Zn toxicity. When the Zn ion concentration was low (50 mmol L^{-1}), there was less impact on the melon (*Cucumis melo*) and no difference with the control, but when the Zn concentration was higher than 100 mmol L^{-1}, the poisoning was more in the melon. Dong et al. (2011) observed the reduction of germination of soybean seeds under excess Zn concentrations. Seedling length, length of major root, and fresh weight of plants were also affected. Dhankhar and Solanki (2011) reported that excess Zn (1.5 mM $ZnSO_4.7H_2O$, 7 d) concentrations resulted in 23.83% and 70.56% decreases in fresh weight and dry weight of *Vigna mungo* seedlings with respect to the control. Recently, Song et al. (2011) examined the toxic effect of high doses of Zn (2 mM) in two contrasting rice (*Oryza sativa* L.) cultivars, TY-167 (Zn-resistant) and cv. FYY-326 (Zn-sensitive), in regards to root and shoot growth. Significant inhibitory effects of high Zn treatment on plant growth were observed. Root length, root surface area, and the amount of root tip of both cultivars were decreased significantly in plants treated with high Zn (2 mM). Root length, root surface area, and amount of root tip in the Zn-resistant cultivar under the treatment of Zn were significantly reduced by 29.7, 24.6, and 27.9%, respectively, as compared to the control. The symptoms of Zn toxicity were typically manifested as a yellow color on the lower leaves starting from the tips and spreading toward the bases of the leaves, which became more severe as the experiment continued. Compared with the control, the high Zn treatment reduced shoot and root biomass by 33.5% and 30.5% in the Zn-resistant cultivar, respectively, and 44.7% and 48.4% in the Zn-sensitive cultivar (Song et al. 2011). Very recently, Israr et al. (2011) reported that growth of *Sesbania drummondii* seedlings was significantly inhibited with Zn treatment, and biomass was reduced by 25.2% compared to the control.

2.7.3.7 Nickel

Nickel is also a beneficial trace element for plants at low concentrations, but it causes toxicity at high concentrations as reported by many researchers. Pandey and Sharma (2002) reported that chlorophyll a content was more reduced than that of chlorophyll b in leaves of Ni-treated cabbage. In addition, Ni also inhibits chlorophyll biosynthesis by creating nutrient imbalances by replacement of Mg^{2+} ions (Molas 2002; Gautam and Pandey 2008). When *Zea mays* plants were exposed to Ni (100 μM, 5 d), different toxicity symptoms were observed by Mishra et al. (2010). The younger expanding leaves showed interveinal chlorosis followed by shrinkage of apical margins of lamina, which turned brown in color. However, in relatively mature leaves, off-white lesions appeared on leaf lamina followed by the development of black dots in these lesions. Hastened senescence in the older leaves of plants receiving an excess supply of Ni was observed. Ni stress also caused a significant decrease in fresh weight (61.08%) and dry weight (50.33%) of plants. Kazemi et al. (2010) found that exposure of *Brassica napus* plants to 0.5 mM $NiCl_2 \cdot 6H_2O$ for 10 days resulted in toxicity symptoms, such as chlorosis and necrosis at the leaves. Treatment with Ni also resulted in a decrease in dry weight of roots (64%) and shoots (52%) and chlorophyll content (61%) of leaves. In a recent study, Singh and Pandey (2011) reported that *Pistia stratiotes* plants exposed to high amounts of Ni (1.0 and 10.0 ppm) showed visible toxicity symptoms, such as wilting, chlorosis in young leaves, browning of root tips, and broken off roots observed six days after treatment. Nickel exposure decreased chlorophyll a and b and total chlorophyll contents. Relative water content decreased at high Ni (1.0 and 10.0 ppm).

2.7.3.8 Chromium

Chromium interferes with several metabolic processes, causing toxicity to plants as exhibited by reduced seed germination or early seedling development (Sharma et al. 1995), root growth and biomass, chlorosis, photosynthetic impairment, and finally, plant death (Scoccianti et al. 2006). Chromium stress is one of the important factors that affect photosynthesis in terms of CO_2 fixation, electron transport, and photophosphorylation (Shanker et al. 2005). Metabolic alteration by Cr stress has also been reported in plants as the direct effect on enzymes and metabolites or by its ability to generate ROS (Shanker et al. 2005). It has been noticed that the 40% inhibition of whole plant photosynthesis in 52-day-old plants at 100 μM Cr(VI) was further enhanced to 65% and 95% after 76 and 89 days of growth, respectively (Bishnoi et al. 1993). High levels (500 ppm) of Cr(VI) in soil reduced germination up to 48% in *Phaseolus vulgaris* (Parr and Taylor 1982). Peralta et al. (2001) found that 40 ppm of Cr(VI) reduced by 23% the ability of seeds of *Medicago sativa* to germinate and grow in the contaminated medium. The reduced germination of seeds under Cr stress could be a depressive effect of Cr on the activity of amylases and on the subsequent transport of sugars to the embryo axes (Zeid 2001). In a glasshouse trial, Sharma and Sharma (1993) reported that after 32 and 96 days, plant height of wheat reduced significantly when sown in sand with 0.5 mM $Na_2Cr_2O_7$. Leaf number per plant was also reduced by 50% with the same dose of Cr (Sharma and Sharma 1993). Tripathi and Tripathi (1999) found that leaf area and biomass of *Albizia lebbek* seedlings was severely affected by a high concentration (200 ppm) of Cr(VI). Zurayk et al. (2001) observed a significant decrease in the dry biomass accumulation of *Portulaca oleracea*. Kocik and Ilavsky (1994) studied the effect of Cr on quality and quantity of biomass in sunflower (*Helianthus annuus*), *Z. mays*, and *Vicia faba* and observed that DM production was not markedly affected by 200 mg kg^{-1} Cr(VI). Rai et al. (2004) reported that excess Cr reduced the biomass accumulation and photosynthesis of *Ocimum tenuiflorum*. Chromium reduced the biomass in a dose- and treatment duration–dependent manner. Chromium also significantly reduced the level of photosynthetic pigments (total chlorophyll, chlorophyll a, chlorophyll b, and carotenoids) in *O. tenuiflorum*. This might be attributed to the toxicity of Cr to the chlorophyll biosynthesis of the test plant. In *Cucumis melo*, excess Cr decreased germination rate, germination index, germination time, and germination uniformity index values in the germination level (Akinci and Akinci 2010). Radicle length, radicle fresh and dry weight, hypocotyll length, hypocotyll fresh and dry weight, growth tolerance index, and seedling relative growth rate were negatively affected by the increased Cr concentrations at the seedling stage. Response of seedlings to Cr was more than that of seed germination. This event is based on the impermeability of seed coats and selectivity of embryos against Cr. Zeng et al. (2010) studied the effects of excess Cr (10, 50, and 100 μM) on the biomass production in two rice varieties viz. Xiushui 113 (low Cr accumulation) and Dan K5 (high Cr accumulation) by rice plants. There was little difference in the dry weight of all plant organs between 10 μM Cr treatment and the control. However, the dry weight of the plants was decreased significantly in the 50 and 100 μM Cr treatments relative to the control. The Cr concentration in each plant organ and the Cr accumulation in the whole plant increased dramatically with the Cr level, irrespective of the genotypes. The biomass accumulation in the two rice genotypes differed significantly in response to Cr stress. Compared to the control, the leaf and stem dry weights of Xiushui 113 decreased by 10.7% and 6.5% more than that of Dan K5 in the 50 μM Cr treatment. Recently, Bah et al. (2011) found that *Typha angustifolia* exposed to 1 mM $K_2Cr_2O_7$ for 30 days significantly decreased plant height and root and shoot growth.

2.7.3.9 Aluminum

Although Al is not a HM, its toxicity to plants has been studied by many researchers. The first evident symptom of Al toxicity is the inhibition of root growth, which occurs after very brief exposure of roots to Al (Vierstra 1987; Dipierro et al. 2005). Pereira et al. (2010) reported that the chlorophyll content of cucumber (*Cucumis sativus*) leaves decreased with increasing concentrations of Al. An inhibition of the chlorophyll content of 60% was observed at 2 mM $Al_2(SO4)_3$. Xu et al. (2011) observed a dose-dependent decrease in root elongation in the seedlings of two wheat varieties—Yangmai-5 (Al-sensitive) and Jian-864 (Al-tolerant)—after exposure to Al, but the former genotype showed greater elongation than the latter. Exposure to 30 μM Al resulted in the most significant difference in the relative root elongations between Yangmai-5 (33.7%) and Jian-864 (71.4%) as compared with the other Al concentrations ranging from 0 to 50 μM. These toxic effects of HMs

TABLE 2.5
Yield Reduction of Crops Resulting from Heavy Metal Toxicity

Metals	Dose	Crop	Yield Reduction	References
Cd	80 mg kg^{-1} $CdCl_2$	*Brassica rapa* cv. Varuna	49.27%	Singh and Brar (2002)
Cd	1 mg L^{-1} $CdCl_2$	*Triticum aestivum* cv. E81513	19.82%	Zhang et al. (2002)
Cd	100 μM $CdCl_2$	*Solanum melongena*	36.56%	Siddhu et al. (2008)
Cd	30 mg L^{-1} $CdCl_2$	*Solanum lycopersicum*	54.16%	Moral et al. (1994)
Pb	400 mg kg^{-1} $[Pb(NO_3)_2]$ with 12 mg kg^{-1} $Cd(NO_3)_2$. $4H_2O$	*Brassica napus*	30%	Chimbira and Moyo (2009)
Pb	1 mM Pb $(NO^3)_2$	*Oryza sativa*	19.51%	Chatterjee et al. (2004)
Pb	800 mg kg^{-1} Pb-COOH	*Oryza sativa* (Japonica) cv. Wu yu jing No. 3	24.53%	Liu et al. (2003)
As	100 mg kg^{-1} Na_2HAsO_4	*Fagopyrum esculentum*	12.5%	Mahmud et al. (2007)
As	50 mg L^{-1} As O	*Amaranthus retroflexus*	89.20%	Choudhury et al. (2008)
As	30 mg kg^{-1} $Na_2HAsO_4.7 H_2O$	*Oryza sativa* cv. BRRI hybrid dhan 1	57.34%	Rahman et al. (2007)
Cu	250 mg kg^{-1} $CuSO_4.5H_2O$	*Vigna radiata*	33.64%	Manivasagaperumal et al. (2011)
Cu	1338 mg kg^{-1}	*Pisum sativum*	15.47%	Wani et al. (2008)
Ni	60 mg kg^{-1} $(NiSO_4{\cdot}7H_2O)$	Lettuce leaves	78.51%	Matraszek et al. (2002)
Ni	60 mg kg^{-1} $(NiSO_4{\cdot}7H_2O)$	Spinach leaves	90.76%	Matraszek et al. (2002)

together attributed to a lower yield of crop plants, which ultimately affects food security. Yield reduction of some crop plants because of metal toxicity is presented in Table 2.5.

2.7.4 Metal Toxicity and Oxidative Stress

There is enough evidence that exposure of plants to excess concentrations of redox-active HMs results in oxidative injury (Anjum et al. 2011a). Metals of biological significance are of two groups, viz. redox-active (Fe, Cu, Cr, Co) and non-redox-active (Cd, Pb, Zn, Ni, Al). Metals with lower redox potentials than those of biological molecules cannot participate in biological redox reactions (Schützendübel and Polle 2002). Redox-active metals, unlike non-redox-active metals, are directly involved in redox reactions in cells (Dietz et al. 1999). HM uptake by transporters and distribution to organelles is followed by ROS generation, stimulated either by HM redox activity or by the effects of a HM on metabolism in a subcellular site-specific manner. HM-dependent activation of plasma-membrane-localized NADPH oxidase also contributes to the release of ROS (Sharma and Dietz 2008). In contrast to physiologically non-redox-active HMs, such as Zn^{2+} and Cd^{2+}, the redox-active HMs Fe, Cu, Cr, V, and Co enable redox reactions in cells. They are involved in the formation of OH· from H_2O_2 via Haber-Weiss and Fenton reactions and initiate non-specific lipid peroxidation (Sharma and Dietz 2008). Lipid peroxidation is also specifically induced by HM-dependent activation of lipoxygenases (LOX) (Montillet et al. 2002). Among the HMs, Cd is the most widely studied in plants. The presence of Cd led to excessive production of ROS causing cell death as a result of oxidative stress, such as membrane lipid peroxidation, protein oxidation, enzyme inhibition, and damage to nucleic acids (Gill and Tuteja 2010; Hasanuzzaman et al. 2012; Gill et al. 2011c). Hasanuzzaman et al. (2012) observed a 60% increase in H_2O_2 and a 134% increase in malondialdehyde (MDA) content in rapeseed seedlings when exposed to Cd stress (1 mM $CdCl_2$, 48 h). Plants exposed to HM stress exhibited an increase in lipid peroxidation as a result of the generation of free radicals (Lozano-Rodríguez, Hernández, Bonay, Carpena-Ruiz 1997; Vanaja et al. 2002). HM also altered the activities of antioxidant enzymes of plants as reported by several researchers (Table 2.6).

When treated with 0, 25, 50, and 100 mg kg^{-1} soil in a greenhouse, pot culture experiment, rapeseed (*Brassica campestris* L.) cv. Pusa Gold plants exhibited different activity of antioxidant enzymes, such as SOD, CAT, APX, and GR (Anjum et al. 2008). A uniform increase in SOD, APX, and GR activity was observed paralleled with a gradual increase in soil-Cd levels. The activity of CAT was noted to initially decrease with 25 mg Cd kg^{-1} soil but increased thereafter. Moreover, maximum increases in the activity of SOD, CAT, and GR were noticed with 100 mg Cd kg^{-1} soil at pre-flowering (98.46%, 21.89%, and 120.34%, respectively). In a recent study, Anjum et al. (2011b) reported that in *Vigna radiata* plants the highest Cd level in soil (100 mg kg^{-1} soil) caused a significantly higher lipid peroxidation and H_2O_2 (69.6 and 113.9% increases in thiobarbituric acid reactive substances, TBARS, and H_2O_2 contents, respectively, over the control). In contrast, the Cd-tolerant cultivar was characterized by a lower degree of lipid peroxidation in terms of lesser increments in TBARS (39.1%) and H_2O_2 (81.0%) contents. Increases in TBARS with increasing Cd concentration has also been previously reported in germinating *Phaseolus vulgaris* seedlings by Somashekaraiah et al. (1992). The accumulation of H_2O_2 after Cd exposure has been detected in the leaves of different plant species, such as *Pisum sativum* (Romero-Puertas et al. 2004), *Arabidopsis thaliana* (Cho and Seo 2005), *Brassica juncea* (Mobin and Khan 2007), *Vigna mungo* (Singh et al. 2008). In *Arabidopsis thaliana*, Phang et al. (2011) reported that Pb exposure induced the generation of ROS and increased the level of lipid hydroperoxide (LOOH). The mean H_2O_2 content of *A. thaliana* seedlings increased progressively with the increasing Pb concentration of the medium. In treatment with 100 μM $Pb(NO_3)_2$, the seedlings contained 2.2-fold more H_2O_2 than the control. LOOH content increased with exposure to increasing Pb concentrations in the medium. In seedlings exposed to 25 μM $Pb(NO_3)_2$, LOOH content was 4.1-fold greater than that of the control. Seedlings treated with 50, 75, and 100 μM $Pb(NO_3)_2$ contained 5.4-, 7.6-, and 9.6-fold, respectively, more LOOH than that of the control. These changes were accompanied by up-regulation of the activity of antioxidative

TABLE 2.6
Effect of Heavy Metals on the Activity of Antioxidant Enzymes in Plants

Metals	Target Plants	Doses and Duration	Enzymes	Increase (+) or Decrease (−)	References
Cd	*Pisum sativum*	50 mM $CdCl_2$, 28 d	CAT, SOD	−	Sandalio et al. (2001)
Cd	*Vigna radiata*	25, 50, and 100 mg $CdCl_2$ kg^{-1} soil, 30 d	SOD, APX, MDHAR, DHAR, GR	+	Anjum et al. (2011)
Cd	*Brassica napus*	1 mM $CdCl_2$, 48 h	CAT, MDHAR, DHAR, APX, GST	− +	Hasanuzzaman et al. (2012)
Pb	*Arabidopsis thaliana*	25–100 μM $Pb(NO_3)_2$, 7 d	SOD, CAT, GR, GPX	+	Phang et al. (2011)
Pb	*Triticum aestivum*	0.05–1 g L^{-1} Pb $(NO_3)_2$, 6 d	APX, POD, SOD, CAT, GST	+	Lamhamdi et al. (2011)
As	*Pteris* spp.	10 mg L^{-1} $Na_2HAsO_4.7H_2O$, 3 d	CAT	+	Kertulis-Tartara et al. (2009)
As	*Brassica juncea*	25 μM Na_3AsO_4, 12–96 h	SOD, APX, GR	+	Khan et al. (2009)
As	*Oryza sativa*	50 and 100 μM Na_2HAsO_4, 10 d 100 and 500 μM $NaAsO_2$, 10 d	SOD, APX, POD, GR	+	Shri et al. (2009)
Hg	*Medicago sativa*	1 μM $HgCl_2$, 24 h	SOD, POD, APX	+ −	Zhou et al. (2009)
Cu	*Raphanus sativus*	0.5 mM $Cu(SO4)_2$, 2 d	CAT	+	Teklić et al. (2008)
Cu	*Phaseolus vulgaris*	50 μM $CuSO_4$, 3 d	GPX, CAT	−	Bouazizi et al. (2010)
Zn	*Oryza sativa*	2 mM Zn, 7 d	CAT, APX	−	Song et al. (2011)
Zn	*Vitis vinifera*	28 mM Zn^{2+}	CAT, POD	−	Yang et al. (2011)
Ni	*Pistia stratiotes*	10 ppm Ni	POD, CAT	+ −	Singh and Pandey (2011)
Ni	*Zea mays*	100 μM Ni, 7 d	CAT, APX, SOD	− +	Mishra et al. (2010)
Ni	*Jatropha curcas* cotyledons	100 and 200 μM $NiCl_2$, 7 d	SOD, POD, CAT	+	Yan et al. (2008)
		400 and 800 μM $NiCl_2$, 7 d	POD, CAT	−	
Ni	*Brassica napus*	0.5 mM $NiCl_2{\cdot}6H_2O$, 10 d	CAT, APX, GPX	−	Kazemi et al. (2010)
Cr	*Typha angustifolia*	1 mM $K_2Cr_2O_7$, 30 d	SOD, POD, APX, GPX	+ −	Bah et al. (2011)
Al	*Cucumis sativus*	1–500 μM $Al_2(SO_4)_3$, 10 d	CAT, APX, SOD	+	Pereira et al. (2010)
		$Al_2(SO_4)_3$ 2 mM, 10 d	APX, SOD	−	
Al	*Avena sativa*	90, 185, 370, and 555 μM $Al_2(SO4)_3$, 5 d	CAT, APX, SOD	+	Pereira et al. (2011)
Al	*Pisum sativum*	10 μM $AlCl_3$, 48 h	APX, CAT, GR, SOD	−	Panda and Matsumoto (2010)

(continued)

TABLE 2.6 Continued
Effect of Heavy Metals on the Activity of Antioxidant Enzymes in Plants

Metals	Target Plants	Doses and Duration	Enzymes	Increase (+) or Decrease (–)	References
Al	*Triticum aestivum*	30 μM $AlCl_3$, 24 h	CAT, APX, MDHAR, DHAR, GR, GPX	+	Xu et al. (2011)
Cr	*Ocimum tenuiflorum*	10–100 μM $K_2Cr_2O_7{\cdot}7H_2O$, 24–72 h	SOD, GPX, CAT, APX	+ –	Rai et al. (2004)
Cr	*Pistia stratiotes*	0.01–10 mM Cr, 24 and 48 h	SOD, GPX, GR, CAT	+ –	Upadhyay and Panda (2010)
Al	Oryza sativa	160 μM Al^{3+}, 5–20 d	SOD, GPX, APX, MDHAR, DHAR, GR	+	Sharma and Dubey (2007)
			CAT, cAPX	–	

enzymes in a dose-dependent manner. In *Triticum aestivum* seedlings, Lamhamdi et al. (2011) found that activity of antioxidant enzymes, such as ascorbate peroxidase (APX), peroxidases (POD), superoxide dismutase (SOD), catalase (CAT), and glutathione *S*-transferase (GST), were generally significantly increased in the presence of $Pb(NO_3)_2$ in a dose-dependent manner (0.05, 0.1, 0.5, 1 g L^{-1}). They also showed that Pb treatment increased lipid peroxidation and enhanced soluble protein concentrations. The level of MDA increased slightly when the concentration was lower than 0.3 mM but increased dramatically at high Pb concentrations (1.5 and 3 mM). When Pb concentration reached 3 mM, MDA content became 7- and 5-fold higher than in control coleoptiles and roots, respectively (Lamhamdi et al. 2011).

Kertulis-Tartara et al. (2009) characterized two antioxidant enzymes, glutathione reductase (GR) and catalase (CAT), in the fronds of *P. vittata*, an As-hyperaccumulating fern, and *P. ensiformis*, an As-sensitive fern. Under As exposure, CAT activity in *P. vittata* was increased by 1.5-fold, but GR activity was unchanged. Similarly, As treatments significantly increased the activity of antioxidative enzymes (SOD, APX, GR), and the contents of antioxidant metabolites, viz. glutathione (GSH) and ascorbate (AsA), which provided an indication regarding the capacity of mustard plants to detoxify the low As level (Khan et al. 2009; Garg and Singla 2011). In rice seedlings, Shri et al. (2009) reported that the MDA was increased significantly with increasing As concentrations. The up-regulation of some antioxidant enzyme activities and the isozymes of SOD, APX, POD, and GR substantiated that As accumulation generated oxidative stress, which was more pronounced in As(III) treatment. Requejo and Tena (2005) reported the effect of As exposure on *Zea mays* L. root proteome and concluded that the induction of oxidative stress is the main process underlying As toxicity in plants. Bouazizi et al. (2010) assayed the toxicity of Cu ($CuSO_4$) in expanding leaves of 14-day-old *Phaseolus vulgaris* seedlings. They observed that MDA content in leaves exposed to Cu revealed the Cu stress did not change this parameter significantly. However, endogenous H_2O_2 was more abundant in stressed leaves, but this effect was only significant at a higher dose (75 μM $CuSO_4$). The activities of GPX and CAT were decreased under 50 μM $CuSO_4$; however, at 75 mM $CuSO_4$, their activities were unchanged compared to the control values. Sun et al. (2010a) reported that with the increase of Zn stress concentration in *C. melo* seedlings, the activities of POD, SOD, and CAT all first increased and then decreased. The content of MDA increased continually upon Zn treatment. Song et al. (2011) observed significantly higher MDA and H_2O_2 contents in rice seedlings subjected to a high dose of Zn (2 mM). Compared with the corresponding control, the concentrations of MDA were significantly increased by treatment with Zn in both rice cultivars. Similarly, the concentration of H_2O_2 increased in both rice cultivars exposed to Zn compared with the control. However, the concentration of H_2O_2 was significantly lower in the resistant cultivar than in the sensitive one. Higher Zn concentration also decreased the activities of SOD, CAT, and APX in both sensitive and tolerant varieties. Nickel, a non-redox reactive metal, cannot generate ROS directly by Fenton-type reaction. However, Ni can cause oxidative stress in plant tissues as indicated by lipid peroxidation (Wang et al. 2001; Rao and Sresty 2004). Exposure to Ni resulted in a severe depletion of GSH (Rao and Sresty 2004), which is believed to be a critical step in Ni-induced ROS accumulation (Schübenzübel and Polle 2002). Hydrogen peroxide is a constituent of oxidative metabolism and is itself a ROS. It has been shown that H_2O_2 content increased significantly with Ni treatment (Boominathan and Doran 2002). In rice seedlings, Lin and Kao (2005) observed that $NiSO_4$ (40 and 60 µM) treatment resulted in increases in H_2O_2, MDA content, and SOD and APX activity. Mishra et al. (2010) observed that both the concentration of H_2O_2 and lipid peroxidation were increased significantly after 14 days of excess Ni (100 µM) treatment. The data obtained after seven days of excess Ni supply were not statistically significant. The activity of POD was increased significantly (25.3%) after seven days of excess Ni supply on a soluble protein basis. On a fresh weight basis, the activity of POD was not altered significantly. Activity of APX also increased at seven days after excess Ni treatment both on a fresh weight basis (324.7%) and a protein basis (415.6%). The activity of APX did not alter after 14 days of excess Ni supply. Moreover, the activity of SOD increased significantly after seven days of excess Ni supply, which, on advancement of duration of excess Ni supply, became nonsignificant both on a fresh weight and soluble protein basis. Kazemi et al. (2010) reported that in Ni-treated plants, the level of LOX activity and MDA, H_2O_2, and proline contents significantly increased while the activity of antioxidant enzymes, such as CAT (31%), GPX (46%), and APX, decreased in leaves and indicated that Ni caused an oxidative stress in *B. napus* plants. According to the findings of Singh and Pandey (2011), oxidative stress followed by antioxidants (CAT, POD activity, and carotenoid content) showed variable responses to different Ni concentrations. The activity of POD increased with an increase in Ni stress. CAT activity also showed an increasing trend with increased nickel concentrations up to 1.0 ppm, which decreased at 10 ppm Ni exposure.

Plants growing in a Cr-stressed environment face a potential risk from ROS, such as O_2, OH·, and H_2O_2. Their presence causes oxidative damage to biomolecules, such as lipids and proteins (Schützendübel and Polle 2002). In an experiment, Rai et al. (2004) demonstrated a significant increase in the MDA content of *O. tenuiflorum* leaves observed initially at 10 µM Cr for 24 h exposure. It was noted that 100 µM Cr (72 h) in a nutrient medium resulted in a fivefold increase in MDA content compared to the control. Upadhyay and Panda (2010) reported that in both roots and shoots of *Pistia stratiotes* Cr produced a marked increase in enzymatic and nonenzymatic antioxidants except in CAT activity, where a strong accumulation of H_2O_2 was indicated, suggesting an imposition of oxidative stress. The results showed an uptake of Cr by *P. stratiotes* and an increase in activity of antioxidants as the concentrations and their duration of treatment increased. An increase in H_2O_2 content was observed with the increase in concentration of Cr^{+6}. The H_2O_2 content increased at a maximum by 32% after 48 h of 10 mM Cr^{+6} treatments in comparison to the control. Very recently, Bah et al. (2011) investigated the modulation of the antioxidant defence system of *Typha angustifolia* after 30 days of exposure of 1 mM Cr and found that the plants showed a high tolerance to HM toxicity with no visual toxic symptoms when exposed to HM stress. A significant increase in SOD and POD activity were recorded in plants subjected to Cr; whereas, APX and GPX activity were decreased.

Although Al itself is not a transition metal and cannot catalyze redox reactions, the involvement of oxidative stress in Al toxicity has been reported (Boscolo et al. 2003). Exposure to Al was found to enhance oxidative stress and was a crucial event in the inhibition of cell growth (Vierstra 1987; Pereira et al. 2011). A number of plant studies have indicated Al-induced oxidative stress, including the increase in enzyme activity related to ROS and lipid peroxidation in various crops, such as *Glycine max* (Cakmak and Horst 1991) and *Pisum sativum* (Yamamoto et al. 2001), and changes in the expression of various genes induced by Al in *Arabidopsis* (Sugimoto and Sakamoto 1997), *Nicotiana tabacum* (Ezaki et al. 2000), and *Triticum aestivum* L. (Hamel et al. 1998). The toxic effects of Al begin in plant roots within minutes of exposure and include root growth inhibition, cytoskeletal damage, alterations of the membrane surface charge, membrane lipid peroxidation, imbalance in Ca^{2+} homeostasis, induction of oxidative stress in plant mitochondria, disruption of the mitochondrial membrane, and several other bioenergetic alterations resulting in cell death (Panda and Matsumoto 2007; Poschenrieder et al. 2008; Panda et al. 2009; Pereira et al. 2011). Sharma and Dubey (2007) suggested that Al^{3+} toxicity is associated with the induction of oxidative stress in rice plants subjected to 80 and 160 μM Al^{3+} for 5–20 days. Al stress significantly decreased the length of roots and shoots. Al^{3+} treatment of 160 μM resulted in increased generation of O_2^- and H_2O_2; an elevated amount of MDA, soluble protein, and GSSG; and decline in concentrations of -SH and AsA. They also observed that Al stress significantly increased the activity of SOD, GPX, and AsA-GSH enzymes (APX, MDHAR, DHAR, and DHAR) while CAT activity decreased. Pereira et al. (2010) showed that when *Cucumis sativus* seedlings were grown at different concentrations of Al ranging from 1 to 2000 μM for 10 days, significant increases in H_2O_2 production were observed in the seedlings, which might be related to the decreased efficiency of the antioxidant system at higher Al concentrations. They also concluded that the antioxidant system was unable to overcome toxicity, resulting in negative effects, such as MDA, protein oxidation, and a decrease in seedling growth. In the study, MDA content in the whole plant increased by 90% at levels of up to 500 μM in comparison with the control while it significantly decreased at 1 and 2 mM to levels near the control. A significant increase (more than 50%) was observed for protein oxidation at all concentrations; whereas, at 2 mM, there was an increase of 84%. The levels of endogenous H_2O_2 increased by about 70% in comparison to control plants at 100 μM $Al_2(SO_4)_3$. At the higher concentrations (1 and 2 mM), there was an increase in H_2O_2 content of about of 34% and 55%, respectively. The presence of Al in the substrate caused an increase in CAT activity of about 18%, 18%, and 20% at concentrations of 10, 100, and 500 μM, respectively (Pereira et al. 2010). On the other hand, CAT activity was reduced to basal levels at 2000 μM when compared with the control. APX activity was increased by about 10%, 68%, and 12% at 10, 100, and 500 μM $Al_2(SO_4)_3$, respectively. Nonetheless, APX activity was inhibited at 1000 and 2000 μM. At the lower concentrations (500 μM $Al_2(SO_4)_3$), a significant increase in SOD activity was observed. However, at the highest Al concentration (2 mM) there was a decrease in SOD levels. Recently, Xu et al. (2011) found that two wheat varieties—Yangmai-5 (Al-sensitive) and Jian-864 (Al-tolerant)—after exposure to Al, increased MDA content and ROS concentrations more significantly in Yangmai-5 than in Jian-864 as compared with Al-free controls. Activities of SOD and POD increased by a greater extent in Yangmai-5; whereas, activities of the other six enzymes (APX, MDHAR, DHAR, GR, CAT, and GPX) increased by a lesser extent in Yangmai-5 than in Jian-864 under Al stress.

2.8 PHYTOREMEDIATION: THE GREEN TECHNOLOGY FOR THE REMOVAL OF HEAVY METALS

HM toxicity is a widespread problem; hence, the remediation of HM–contaminated soils is a task of the utmost importance (Purakayastha and Chhonkar 2010). The idea of employing plants to clean contaminants from the environment is not new. More than 300 years ago, plants were considered for the treatment of wastewater. At the end of the 19th century, *Thlaspi caerulescens* and *Viola calaminaria* were the first plant species reported to accumulate high levels of HMs in their leaves (Baumann 1885). However, a series of interesting scientific discoveries coupled with interdisciplinary research works have developed this idea into a promising environmental technology called phytoremediation (Raskin et al. 1997).

Phytoremediation is defined as "the engineered use of plants *in situ* and *ex situ* for environmental remediation" (Suthersan 2002). This technology has gained recent attention as a strategies to clean up contaminated soil and water. These strategies refer to the use of higher plants and their associated microbiota for the *in situ* treatment of soil, sediment, and groundwater. Fundamental and applied research have clearly demonstrated that selected plant species acquire the genetic potential to remove, degrade, metabolize, or immobilize a wide range of contaminants, including HMs. The phytoremediation is a cost-effective "green" technology based on the use of metal-accumulating plants to remove HMs, including radionuclides, from soil and water. Phytoremediation has recently become the subject of intense public and scientific interest and a topic of many recent reviews.

Phytoremediation has been reported to be an effective, nonintrusive, aesthetically pleasing, socially accepted technology to remediate polluted soils (Alkorta and Garbisu 2001; Weber et al. 2001; Garbisu et al. 2002; Kachout et al. 2010). Biologically based remediation strategies, including phytoremediation, have been estimated to be 4 to 1000 times cheaper, on a per volume basis, than current nonbiological technologies (Sadowsky 1999). The main objective of scientists, agronomists, and engineers dealing with phytoremediation is to exploit, by the most rational way possible, the potential of this natural process. From the technological point of view, phytoremediation is the use of vegetation to decontaminate soils and water from heavy metals and toxic organics. Very often, phytoremediation assumes the joint action of both plants and microorganisms.

2.8.1 Different Kinds of Phytoremediation Mechanisms

Fundamental and applied research has unequivocally demonstrated that selected plant species possess the genetic potential to remove, degrade, metabolize, or immobilize a wide range of

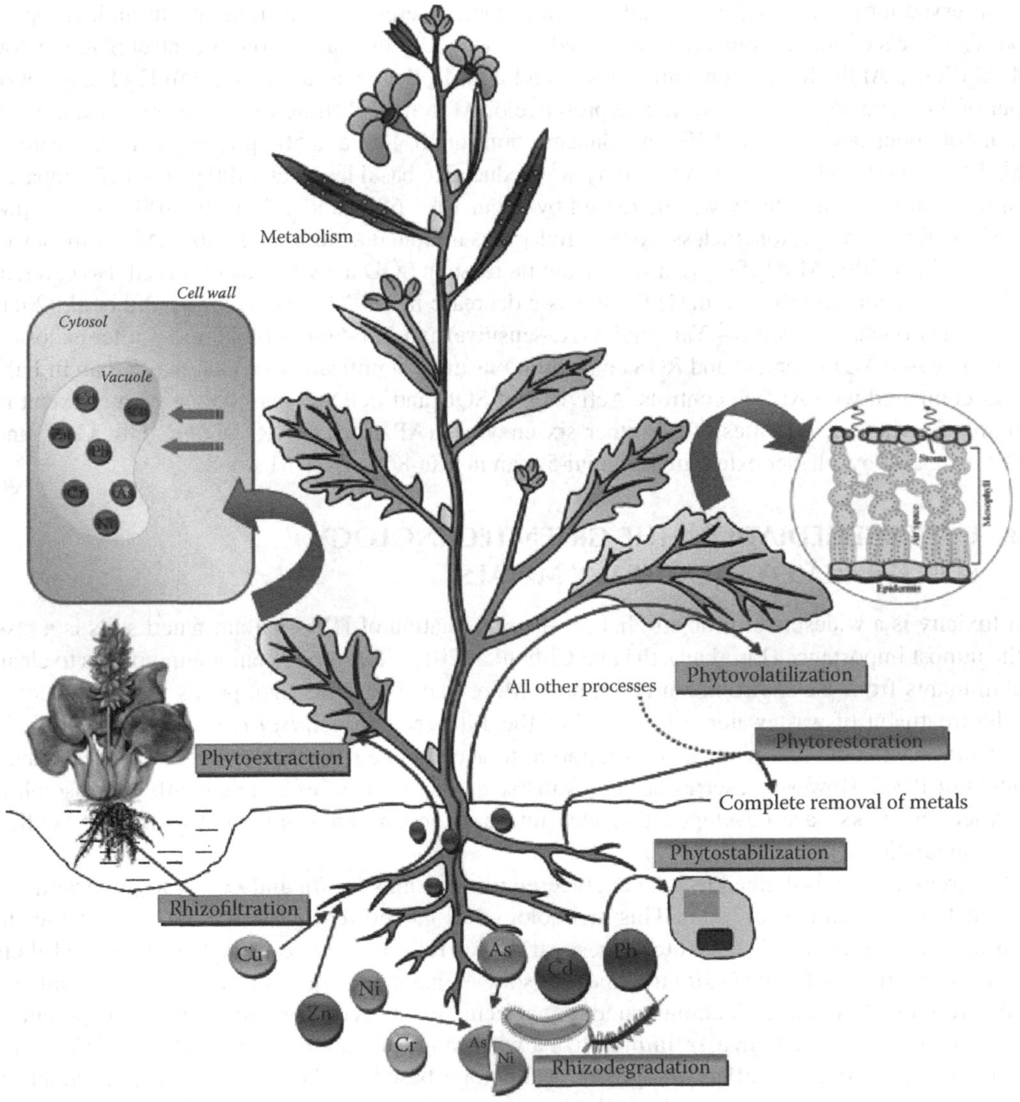

FIGURE 2.13 Process of phytoremediation of toxic metals.

contaminants by different process (Figure 2.13). The phytoremediation process may be divided into the following heads based on the nature of their remediation process:

1. Phytoextraction/Phytoaccumulation
2. Phytostabilization
3. Phytodegradation/Phytotransformation
4. Rhizofiltration
5. Rhizodegradation
6. Phytovolatilization
7. Phytorestoration

2.8.1.1 Phytoextraction/Phytoaccumulation

The use of plants to remove contaminants from the environment and concentrate them in above-ground plant tissue is known as phytoextraction. Phytoextraction is also called phytoaccumulation. Among the phytoremediation mechanisms, phytoextraction is the most important and has been extensively studied. This remediation technique is employed to recover metals from contaminated soils using nonfood crops. In this case, metals are absorbed by roots and subsequently translocated within the above-ground parts of the plants. Some plants possess the genetic and physiological potential to accumulate, translocate, and tolerate high concentrations of metals, which is of concern to the phytoextraction technology. The discovery of some wild plants having an accumulating capacity for high concentrations of metals provided the idea of using such plants to remove HMs from soils. According to Chaney et al. (1997), naturally occurring plants called "metal hyperaccumulators" can accumulate 10–500 times higher levels of toxic metals than cultivated crops. However, some cultivated crop plants also have been found to have an enhanced ability to accumulate metals from hydroponic solutions into their above-ground (harvestable) parts. For instance, *Brassica juncea* (Indian mustard) has an enhanced ability to accumulate metals from hydroponic solutions into its above-ground (harvestable) parts, which concentrates toxic HMs (Pb, Cu, and Ni) to a level up to several percent of its dried shoot biomass (Kumar et al. 1995; Sriprang and Murooka 2007).

The phytoextraction process is of two types: natural or continuous phytoextraction and induced or chelate-assisted phytoextraction. Natural phytoextraction employs the natural ability of the plant to remediate the soil (Salt et al. 1995, 1997). Some plants have been identified that have the potential to take up HMs. At least 45 families have been identified as hyperaccumulator plants; these are Brassicaceae, Fabaceae, Euphorbiaceae, Asteraceae, Lamiaceae, and Scrophulariaceae (Salt et al. 1998; Dushenkov 2003; Saxena and Misra 2010).

In case of induced or chelate-assisted phytoextraction, some oligopeptide ligands, such as phytochelatins (PCs) and metallothioneins (MTs), are induced by toxic metals located in plant cells (Cobbett 2000). These peptides bind with the metal, forming stable complexes, and subsequently neutralize them and minimize the toxicity of the metal ion (Saxena and Misra 2010). Approximately 100 phytochelating ligands have been reported in plant species exposed to HMs (Rauser 1999). MTs are small, gene-encoded, Cys-rich polypeptides. On the other hand, PCs are functionally the same as MTs (Grill et al. 1987). When chelating agents, such as ethylenediamine-tetraacetic acid (EDTA), have been used, they are involved in the uptake of HMs and their detoxification (Saxena and Misra 2010). Both PCs and MTs cause chelation of metals in the cytosol by high-affinity ligands, which is potentially a very important mechanism of HM detoxification and tolerance (Sriprang and Murooka 2007). Shen et al. (2002) used 3.0 mmol kg^{-1} EDTA to treat soil from a mining site in Hong Kong that was heavily contaminated with more than 10,000 mg kg^{-1} of Pb. Application of EDTA in three separate doses was the most effective and enabled the Pb concentration in the shoots of *Brassica rapa* to exceed 5000 mg kg^{-1} of dry plant biomass (Leštan 2006).

2.8.1.1.1 Metal Accumulation by Plants: Potential Tools for the Remediation of Toxic Metals

Among phytoremediation tools, phytoextraction by hyperaccumulator plants is the best way to mitigate toxic metals. The term "hyperaccumulator" was coined for plants that actively take up exceedingly large amounts of one or more HMs from the soil (Brooks et al. 1977; Rascio and Navari-Izzo 2011). Furthermore, the HMs are not retained in the roots but are translocated to the shoot and accumulated in above-ground parts, especially leaves, at concentrations 100- to 1000-fold higher than those found in non-hyperaccumulating species without showing any phytotoxic symptoms (Rascio 1977; Reeves 2006; Rascio and Navari-Izzo 2011). Metal hyperaccumulation capacity is a genetic trait of plants that is present in more than 500 plant species and approximately 0.2% of all angiosperms. Hyperaccumulators are model plants for phytoremediation as they are tolerant to HMs (Sarma 2011).

Considering the pattern of metal accumulation, plants are divided into three main types: accumulator plants that accumulate high quantities of metals primarily in shoots, indicator plants that accumulate metal concentrations in different plant tissues corresponding to the increasing available content of metals in soils, and excluder plants that maintain low metal concentrations in their shoots even if the external metal concentration in the environment is high (i.e., insensitive for uptake and accumulation of metals) (Lal 2010). There are extreme accumulators that can even prosper in contaminated soils and accumulate extremely high levels of trace elements, which are often termed "hyperaccumulators" (Baker 1987; Tlustoš et al. 2006) (Figure 2.14).

Hyperaccumulator plant species posses the capacity to accumulate one or more inorganic elements to levels a hundredfold higher than other species grown under the same conditions and will concentrate more than 10 mg kg^{-1} Hg; 100 mg kg^{-1} Cd; 1000 mg kg^{-1} Co, Cr, Cu, and Pb; and 10,000 mg kg^{-1} Zn and Ni (Baker et al. 2000; Setia et al. 2008). One of the best hyperaccumulators, for example, is *Thlaspi caerulescens* (Brassicaceae), which has been reported to accumulate up to

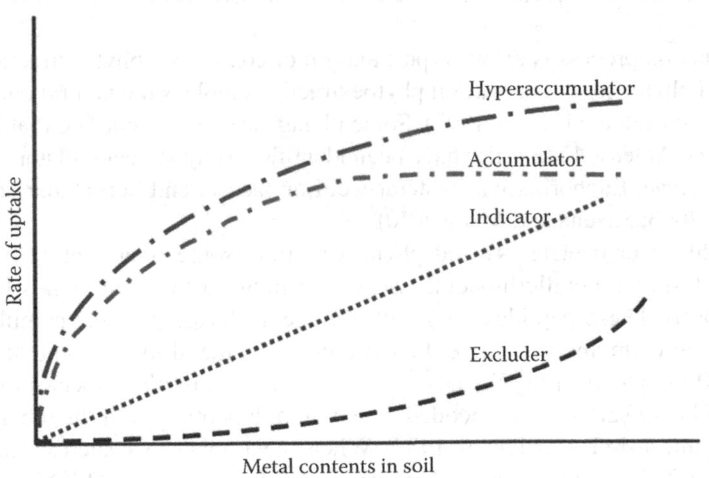

FIGURE 2.14 Metal uptake pattern of different types of metal-accumulating plants. (Adriano, D. C., *Trace elements in terrestrial environments*, Springer, New York, 2001; Ghosh, M., and S. P. Singh, *Applied Ecological and Environmental Research*, *3*, 2005; Tlustoš, P., D. Pavlíková, J. Száková, and J. Balík. Plant accumulation capacity for potentially toxic elements. In J. L. Morel, G. Echevarria, and N. Goncharova (Eds.), *Phytoremediation of metal-contaminated soils*, Springer, Dordrecht, 2006.)

26,000 mg kg^{-1} Zn and up to 22% of soil exchangeable Cd without showing any toxicity (Brown et al. 1995; Gerard et al. 2000). *Brassica juncea* has been found to have a good ability to transport Pb from its roots to its shoots, and it has been found that a Pb concentration of 500 mg L^{-1} is not toxic to *Brassica* species (Henry 2000), which is equivalent to the removal of 4676 kg of Pb ha^{-1} (Henry 2000; Ghosh and Singh 2005).

The efficiency of phytoextraction is determined by the accumulation factor, indicating the ratio of HM concentrations in the plant organs (shoots, roots), in the soil, and in the seasonally harvestable plant biomass (Kvesitadze et al. 2006). There are three processes that regulate the movement of HMs from root tips to xylem (root symplasm). The sequestration of HMs occurs inside cells; they are then subjected to symplastic transport to the stele and subsequent release into the xylem (Figure 2.15). Generally the HM content in various plant organs decreases in the following sequence: root > leaves > stems > inflorescence > seed (Lal 2010). However, metal-accumulating capacity greatly varies with plant species (Table 2.7). In our recent study, we identified several As-hyperaccumulating species, which accumulated different quantities of As in root and shoots (Gani Molla et al. 2010) (Table 2.8).

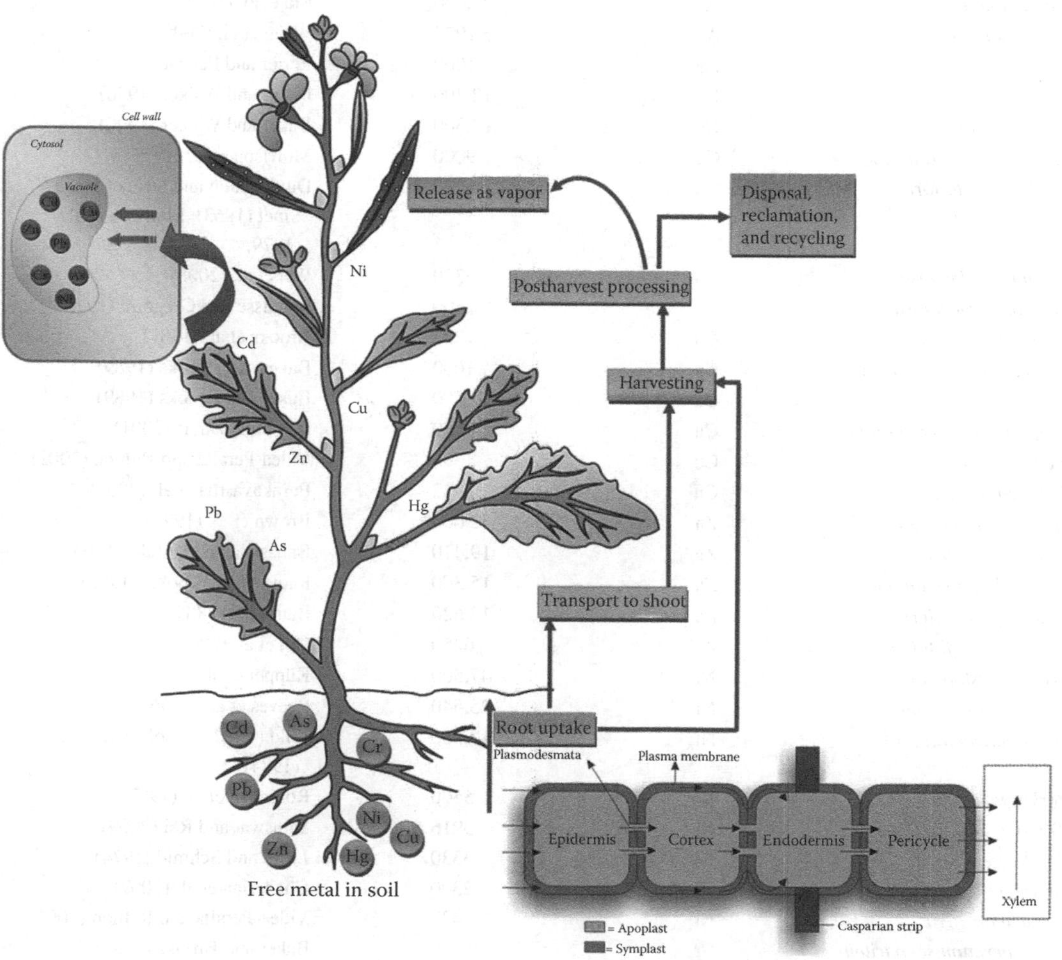

FIGURE 2.15 Phytoextraction of toxic metals.

TABLE 2.7
Some Hyperaccumulator Plant Species and Their Metal-Accumulating Capacities

Plant and Organ	Toxic Metal	Amount (mg kg^{-1})	References
Thlaspi caerulescens	Cd	1800	Baker and Walker (1990)
Medicago sativa	Cd	1079	Videa-Peralta and Ramon (2002)
Sesbania drummondi	Cd	1687	Israr et al. (2006)
Nicotiana tabacum	Cd	40	Evangelou et al. (2004)
Thlaspi caerulescens	Cd	80	Banasova and Horak (2008)
Brassica juncea	Pb	15,000	Blaylock et al. (1997)
Agrostis tenuis	Pb	13,490	Barry and Clark (1978); Williams et al. (1977)
Pisum sativum	Pb	8960	Huang et al. (1997)
Thlaspi rotundifolium	Pb	8200	Baker and Walker (1990)
Arrhenatherum elatius	Pb	1500	Deram and Petit (1997)
Thlaspi goesingense	Pb	2840	Puschenreiter et al. (2001)
Pteris vittata	As	27,000	Wang et al. (2002)
Pteris vittata	As	23,000	Dong (2005)
Pteris vittata	As	23,000	Ma et al. (2001)
Pteris multifida	As	1977	Wang et al. (2006)
Agrostis tenerrima	As	1000	Porter and Peterson (1975)
Ipomoea alpine	Cu	12,300	Baker and Walker (1990)
Ipomea alpine	Cu	12,300	Baker and Walker (1990)
Aeolanthus biformifolius	Cu	9000	Morrison et al. (1979)
Pandiaka metallorum	Cu	6270	Duvigneaud and Denaeyer-De Smet (1963); Malaisse et al. (1979)
Sorghum sudanense	Cu	5330	Wei et al. (2008)
Eragrostis racemosa	Cu	2800	Malaisse and Grégoire (1978)
Triumfetta dekindtiana	Cu	1283	Brooks et al. (1987)
Haumaniastrum robertii	Cu	1000	Baker and Brooks (1989)
Larrea tridentata	Cu	1000	Baker and Brooks (1989)
Streptanthus polygaloides	Cu	120	Boyd and Davis (2001)
Medicago sativa	Cu	85	Videa-Peralta and Ramon (2002)
Brassica juncea	Cu	22	Purakayastha et al. (2008)
Thlaspi caerulescens	Zn	52,000	Brown et al. (1994)
Thlaspi caerulescens	Zn	19,410	Banasova and Horak (2008)
Thlaspi brachypetalum	Zn	15,300	Reeves and Brooks. (1983)
Arabidopsis halleri	Zn	13,620	Baumann (1885)
Potentilla griffithii	Zn	6250	Qiu et al. (2006)
Alyssum lesbiacum	Ni	47,500	Küpper et al. (2001)
Phyllomelia coronata	Ni	25,540	Reeves et al. (1996)
Pearsonia metallifera	Ni	15,350	Wild (1970); Brooks and Yang (1984)
Berkheya coddii	Ni	5500	Robinson et al. (1997)
Brassica juncea	Ni	3916	Saraswat and Rai (2009)
Ruellia geminiflora	Ni	3330	Jaffré and Schmid (1974)
Alyssum bracteatum	Ni	2300	Ghaderian et al. (2007)
Medicago sativa	Ni	437	Videa-Peralta and Ramon (2002)
Leptospermum scoparium	Cr	20,000	Baker and Brooks (1989)
Eichornia crassipes	Cr	6000	Lytle et al. (1998)

(continued)

TABLE 2.7 Continued
Some Hyperaccumulator Plant Species and Their Metal-Accumulating Capacities

Plant and Organ	Toxic Metal	Amount (mg kg^{-1})	References
Zea mays	Cr	2538	Sharma et al. (2003)
Brassica juncea	Cr	1400	Shahandeh and Hossner (2000)
Dicoma niccolifera	Cr	1000	McCutcheon and Schnoor (2003)
Pistia stratiotes	Hg	1000	Baker and Brooks (1989)
Crotalaria cobalticola	Co	3010	Brooks et al. (1987)
Pandiaka metallorum	Co	2131	Morrison (1980); Brooks et al. (1987)
Maytenus pancheriana	Mn	16,370	Reeves and Brooks (1983)
Macademia neurophylla	Mn	200	Baker and Walker (1990)
Thlaspi caerulescens	Mo	1500	Lombi et al. (2001)
Hordeum vulgare	Al	1000	McCutcheon and Schnoor (2003)
Vicia faba	Al	100	McCutcheon and Schnoor (2003)
Brassica juncea	Al	10–1200	Haverkamp et al. (2007)

TABLE 2.8
Identified Naturally Grown Arsenic Hyperaccumulating Plants with Arsenic Concentration in Different Plant Parts

	Quantity of As (mg kg^{-1})		
Name of Sample	**Root**	**Shoot**	
Jussiaea repens	—	46.50	Whole plant
Echinochloa crus-galli	61.25	67.82	Shoot > Root
Xanthium italicum	40.69	23.16	Root > Shoot
Cynodon dactylon	—	67.82	Whole plant
Azolla sp	—	27.65	Whole plant
Oryza sativa	12.10	9.43	Root > Shoot
Pistia stratiotes	12.00	33.70	Shoot > Root
Monochoria hastata	40.78	—	Root > Shoot
Colocasia esculanta	42.08	—	Root
Pteris longifolia	34.47	32.14	Root > Shoot
Cyperus rotundus	11.14	32.49	Shoot > Root
Eichhornia crassipes	67.82	—	—
Alternanthera philoxeroides	—	67.30	—

Source: Gani Molla, M. O., M. A. Islam, and M. Hasanuzzaman, *Journal of Phytology*, 2, 2010.

2.8.1.1.2 Characteristics of a Good Phytoaccumulator Plant

Plants to be used for phytoextraction should have high metal uptake and translocation accumulation capacities, a tolerance to high concentrations of metals without showing toxic symptoms, fast growth patterns, high biomass-producing capacity, and an extensive root system. The success of phytoextraction depends especially on these characteristics (Blaylock et al. 1997; McGrath 1998; Blaylock and Huang 2000). The ability of plants to withstand difficult soil conditions (i.e., soil pH, salinity, soil structure, water content) is also important (Jabeen et al. 2009).

However, there are some limitations with the phytoextraction method. Actual extraction of toxic metals by hyperaccumulators is limited to shallow soil depths (61 cm). If contamination is at substantially greater depths (e.g., 1.8–3.0 m), deep-rooted trees (e.g., poplar) can be used; however, it is not feasible in all cases because of leaf litter and associated toxic residues (Huang et al. 1997). In fact, no plant species has been found that shows a wide spectrum of hyperaccumulation (Leštan 2006). Hyperaccumulators are also mostly slow-growing, low biomass–producing species, lacking in good agronomic characteristics (Cunningham et al. 1995). Therefore, to make this technology feasible, plants having the required criteria should be engineered that have a broad range of phytoextraction.

2.8.1.2 Phytostabilization

Phytostabilization is the removal of soil metals through absorption and accumulation by roots, adsorption onto roots, or precipitation within the root zone. In this case, contaminated metals around the roots become insoluble and/or immobilized. It depends on the roots' ability to limit metal's mobility and bioavailability in the soil (USEPA 2000; Mueller et al. 1999) (Figure 2.16). This process can occur through sorption, precipitation, complexation, or metal valence reduction. It is very effective when rapid immobilization is needed to preserve groundwater and surface water, and disposal of biomass is not required.

Phytostabilization is an indirect way of remediating metal contamination in the environment. The purpose of phytostabilization is not to remove metal contaminants from a site, but rather to stabilize them and prevent migration of metal contaminants into the groundwater or air (Prasad and Freitas 2003). Plants used for phytostabilization are poor translocators of metal from root to shoots, such as grasses, thus minimizing exposure to toxic elements. Phytostabilization is considered a very effective process whenever rapid immobilization is required to conserve groundwater and surface water, and the removal of biomass is not required in this process (Saxena and Misra 2010).

Several researchers studied the possible role of phytostabilization in phytoremediation (Jabeen et al. 2009). Smith and Bradshaw (1992) developed two cultivars of *Agrostis tenuis* and one

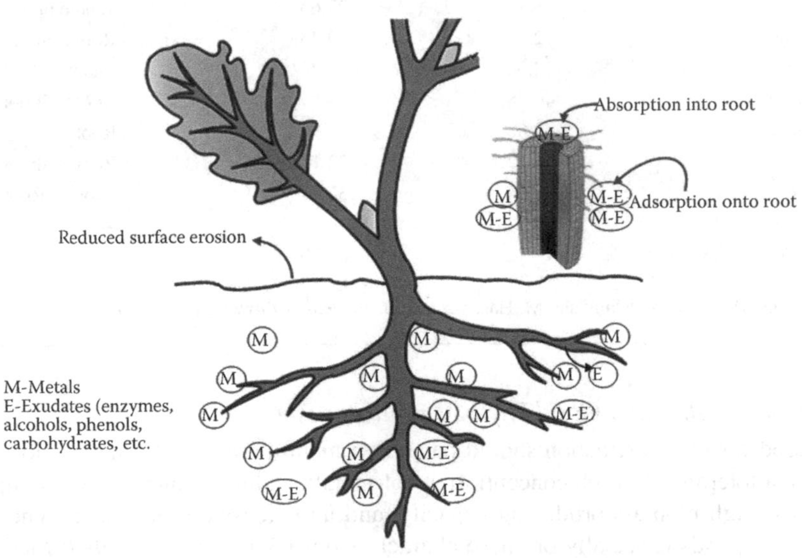

FIGURE 2.16 Phytostabilization of toxic metals.

of *Festuca rubra*, which are now commonly available for phytoremediation of Pb-, Zn-, and Cu-contaminated soils. Phytostabilization, though most effective at sites having fine-textured soils with high organic matter content, can treat a wide range of surface contamination (Cunningham et al. 1995; Berti and Cunningham 2000). Deep-rooting plants could reduce the highly toxic Cr(VI) to Cr(III), which are much less soluble and, therefore, less bioavailable (James 2001). However, its major disadvantage is that the contaminant remains in the soil and so needs regular monitoring (Saxena and Misra 2010).

2.8.1.3 Phytodegradation/Phytotransformation

Phytodegradation is often called phytotransformation. In this process, the contaminants taken up by the plants are broken down into simpler or less-toxic molecules with the help of compounds (such as enzymes) synthesized by plants and then incorporated and/or metabolized into their vascular systems (Saxena and Misra 2010). Plants synthesize a large number of enzymes as a result of primary and secondary metabolism and can quickly uptake and metabolize organic contaminants to less-toxic compounds (Figure 2.17). Plants having a phytodegradation capacity contain enzymes that can catalyze the degradation of contaminants. These enzymes are generally dehalogenases, oxygenases, and reductases (Black 1995; Saxena and Misra 2010).

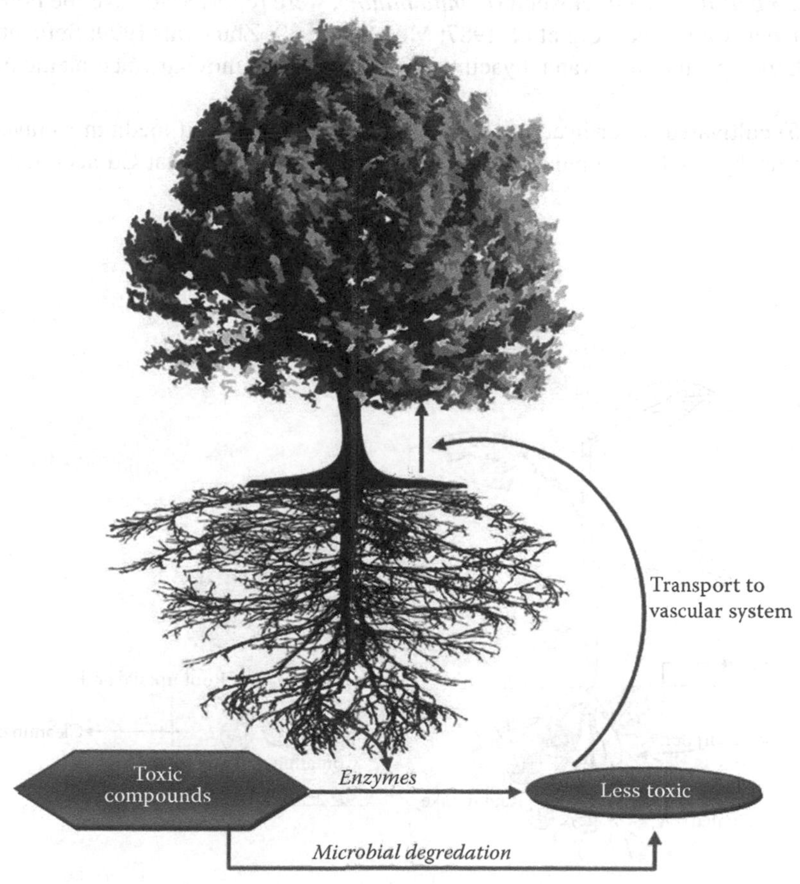

FIGURE 2.17 Phytodegradation of toxic pollutants from soil.

2.8.1.4 Rhizofiltration

Rhizofiltration is a process of phytoremediation using terrestrial and/or aquatic plants to uptake, concentrate, and precipitate metals from contaminated sources through their root systems (Ghosh and Singh 2005) (Figure 2.18). This mechanism highlights the removal of toxic metals derived from aquatic environments, such as damp soil and groundwater, by the rhizosphere. Two strategies are available for its effective exploitation, viz. sorption of metals by the root system and their removal, both of which depend on the physiological and biochemical characteristics of the involved plant. This technology is especially effective because of the ability of some plants to absorb large quantities of metals from soil water, accumulated or passing through the root zone (Kvesitadze et al. 2006). The ideal plants for rhizofiltration should possess extensive root systems, root biomass, and the ability to accumulate and tolerate higher amounts of toxic metals and involve easy handling and a low maintenance cost (Dushenkov and Kapulnik 2000). This technology can be used for HMs, such as Pb, Cd, Cu, Ni, Zn, and Cr, which are primarily held within the roots (Chaudhry et al. 1998; USEPA 2000; Saxena and Misra 2010). A variety of plants, such as *B. juncea*, *N. tabacum*, *H. Annuus*, *Spinacia oleracea*, and *Z. mays*, have been studied for their ability to remove Pb from effluent, with sunflower showing the greatest ability. Mustard has proven to be very effective at removing lead at a wide range of concentrations (4–500 mg L^{-1}) (Raskin and Ensley 2000; Saxena and Misra 2010). Dushenkov et al. (1995) observed that *B. juncea* effectively removed Cd, Cr, Cu, Ni, Pb, and Zn from contaminated sites through their roots. Several aquatic plant species, such as water hyacinth (*Eichhornia crassipes*), pennywort (*Hydrocotyle umbellata*), and duckweed (*Lemna minor*), were reported to have the potentiality to remove HMs from water (Dierberg et al. 1987; Mo et al. 1989; Zhu et al. 1999; Setia et al. 2008). Zhu et al. (1999) reported that water hyacinth is effective in removing trace elements in waste streams.

Fujita (1985) cultivated water hyacinth plants in a Cd-supplemented medium to investigate the reason why water hyacinth accumulates the toxic Cd^{2+}. He observed that Cd accumulated in the

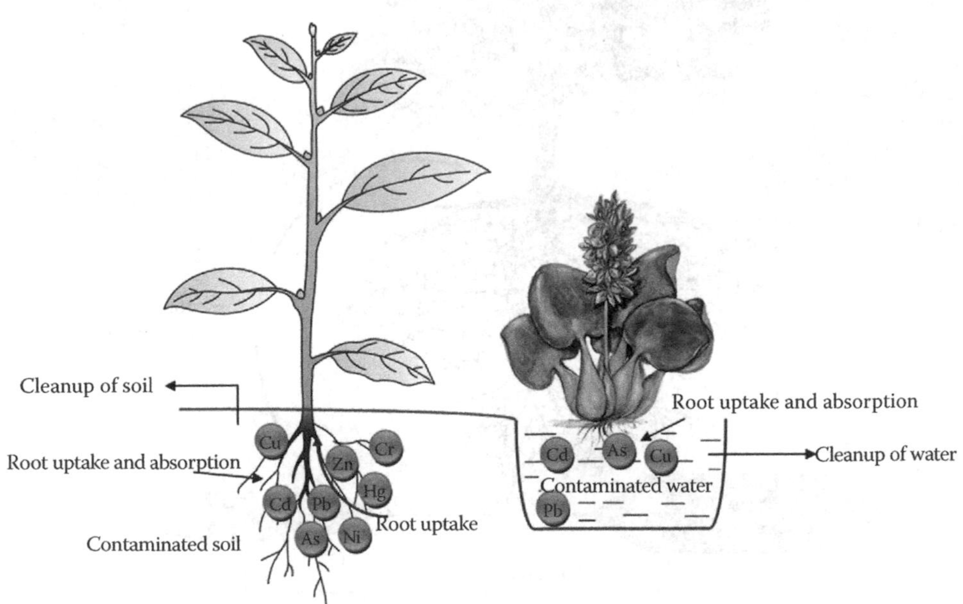

FIGURE 2.18 Rhizofiltration of toxic heavy metals from soil and aquatic environments.

roots against the concentration gradient, mostly as a soluble form in the cytoplasm. Chromatography with Sephadex G-25 and G-50 columns showed that the accumulated Cd was present in two forms with molecular weights of 2300 and 3000. The components carrying Cd showed a high ratio of absorbance at 254 nm to that at 280 nm, which suggests they resemble mammalian Cd-thioneins. These components were not detected in the roots of water hyacinth cultivated in the absence of Cd^{2+}, indicating they are formed in response to the Cd^{2+} supplement. Later, Fujita and Kawanishi (1986) isolated a Cd-binding complex from the root tissue of water hyacinth by chromatography with DEAE-cellulose and Sephadex G-50 columns followed by rechromatography on another DEAE-cellulose column. These results indicate that the water hyacinth root Cd-binding complex is identical to fission yeast Cd-BP1, composed of two each of cadystins A and B and inorganic sulfur. They further confirmed that *Sison amomum*, *Glycine max*, *H. annuus*, *Ipomoea batatas*, *Solanum tuberosum*, and *Coix lachryma-jobi* cultivated in a Cd^{2+}-containing medium also had Cd-binding complexes in the root tissues, which were similar to the complex previously found in water hyacinth roots (Fujita and Kawanishi 1987). The results indicate the widespread existence of complexes similar to fission yeast Cd-BPl in roots of various plants.

However, some researchers reported the limited potentiality of aquatic plants in rhizofiltration processes because of their small, slow-growing roots and the high water content, which complicates their drying, composting, or incineration (Dushenkov et al. 1995).

2.8.1.5 Rhizodegradation

Rhizodegradation is the breakdown of contaminants in the soil through the microbial activity of the rhizosphere (Saxena and Misra 2010) (Figure 2.19). It is also called phytostimulation, rhizosphere biodegradation, enhanced rhizosphere biodegradation, or plant-assisted bioremediation/degradation. This process is slower than phytodegradation. This process is greatly supported by exudates containing a wide range of organic compounds. Rhizodegradation takes place in the rhizosphere by creating beneficial conditions for microbial growth and development. Microorganisms, such as fungi, bacteria, and yeast, consume and assimilate organic matter (fuels and solvents) (Ghosh and Singh 2005). Afterward, these breakdown products are either volatilized or incorporated into the microorganisms and soil matrix of the rhizosphere. However, the microbial population, their activity, and the degradation process depend on the types of plants and the soil condition (Kirk et al. 2005). Grasses with high root density, legumes that fix nitrogen, and alfalfa that has a high evapotranspiration rate are suitable for microbes, and they create a more aerobic environment in the soil

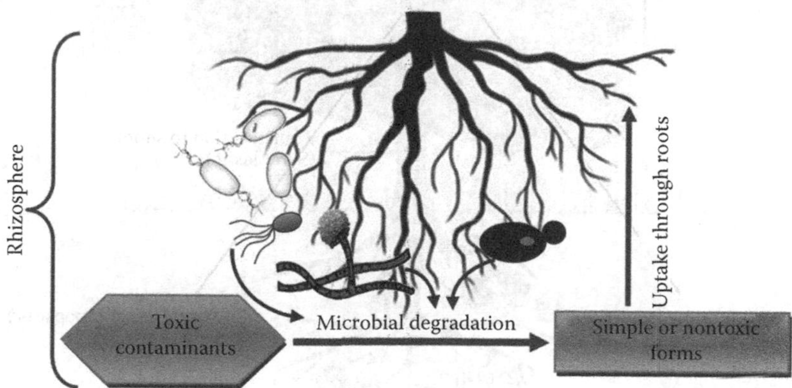

FIGURE 2.19 Rhizofiltration of pollutants. (Suthersan, S. S., *Natural and enhanced remediation system*, Lewis Publishers, Boca Raton, 2002.)

that stimulates microbial activity and, thus, enhances oxidation of organic chemical residues (Peer et al. 2006). In addition, secondary metabolites and other components of root exudates also stimulate microbial activity, a by-product of which may be degradation of organic pollutants (Pieper et al. 2004).

2.8.1.6 Phytovolatilization

Phytovolatilization is a special process of phytoremediation using plants to accumulate toxic compounds from soil followed by transformation into volatile form and release into the atmosphere through transpiration (Figure 2.20). Phytovolatilization occurs as growing trees and other shrubs and herbs take up water and organic and inorganic contaminants (Saxena and Misra 2010). This process is based on the fact that some toxic contaminants can pass through the plants to the leaves and volatilize into the atmosphere at comparatively low concentrations (Mueller et al. 1999). This technology does not eliminate volatile contaminants from the environment but removes them from the soil and groundwater (Kvesitadze et al. 2006). Phytovolatilization is successfully used for the remediation of As-, Hg-, and Se-contaminated soils (Suszcynsky and Shann 1995). Bañuelos (2000) found that some plants grown in high Se media produce volatile Se in the form of dimethylselenide and dimethyldiselenide. This technology has some benefits, such as minimal site disturbance, less erosion, and no need to dispose of metal-contaminated plants. However, there is no control over the migration of metal contaminants that have been removed via volatilization to other areas, which is a major limitation of this technology (Prasad and Freitas 2003; Setia et al. 2008). Some plant species, such as *Brassica juncea*, *Arabidopsis thaliana*, and *Chara canescens*, are reported to possess the capability to absorb HMs and convert them to volatile forms within the plant, which ultimately releases them into the atmosphere (Ghosh and Singh 2005).

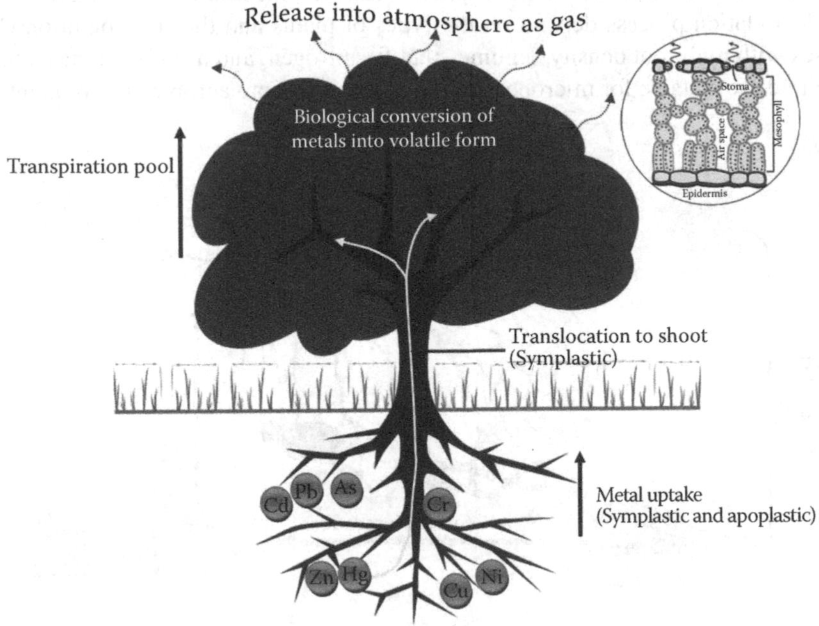

FIGURE 2.20 Phytovolatilization of toxic metals in the environment.

2.8.1.7 Phytorestoration

Phytorestoration is the complete remediation of metal-contaminated soil to make it as suitable for normal activity as non-contaminated soil (Bradshaw 1997). This process of phytoremediation uses plants that are native to the specific area with a view to restoring the land to its natural state. An examination of phytorestoration compared to the other forms of phytoremediation brings to light an important issue: What degree of decontamination do phytoremediation projects aim to achieve? (Peer et al. 2006). There is a vast difference between removing just enough soil pollutants to reach legally defined levels of compliance, remediating soils to a level at which they can be used again, and completely restoring land from its contaminated state to an environmentally uncontaminated state (Peer et al. 2006).

2.8.2 Suitable Plants for Phytoremediation

The phytoremediation potential of a plant mostly depends on how much metal it can uptake without showing any toxic symptoms and its capacity to transform these toxic metals to regular cellular metabolism (Kvesitadze et al. 2006). Before considering a plant for efficient phytoremediation, the following characteristics should be considered:

- Capacity to take up and degrade higher concentrations of toxic metals from the soil and water
- Translocation and sequestration capacity of the uptaken metal to the cellular systems
- Higher capacity to release the exudates to stimulate the growth of microorganisms and secretion of enzymes required for transformation or degradation of toxic contaminants
- Extensive root system and higher growth rate
- Tolerant to adverse environmental conditions
- Capacity to tolerate higher concentrations of toxic metals

However, the best plants for a particular phytoremediation task should be selected based on multiple plant characteristics. Beside these, important environmental factors bearing on the selection of the best remediation technology are soil type and characteristic parameters (pH, average humidity, salt content, metal concentration), the presence of parasites, and the expected amount of precipitation throughout the duration of the remediation process (Kvesitadze et al. 2006).

2.8.3 Limitations of Phytoremediation

Although phytoremediation has several advantages, it also has some limitations. This process is restricted to sites with shallow contamination within the rooting zone of remediative plants. It may take up to several years to remediate a contaminated site. It is restricted to sites with low contaminant concentrations. Harvested plant biomass from phytoextraction may be classified as a hazardous waste; hence, disposal should be done carefully. Climatic conditions are a limiting factor. Introduction of exotic species may affect biodiversity. Consumption of contaminated plant biomass is also a matter of concern (Ghosh and Singh 2005).

2.8.4 Future Perspectives of Phytoremediation

To date, a significant number of research works have been carried out on phytoremediation technologies. Lots of progress has been made. However, the outlook of phytoremediation is still in the research and development phase, and there are many technical barriers that need to be explored. Both agronomic management practices and plant genetic abilities need to be optimized to develop commercially useful practices (Ghosh and Singh 2005). Many efficient metal hyperaccumulator plants are yet to be discovered, and there is a need to know more about their growth physiology

(Raskin et al. 1994). The efficiency of the process, proper understanding of plant HM uptake, and proper disposal of biomass produced is still required (Ghosh and Singh 2005). Therefore, the upcoming challenge for phytoremediation is to further reduce the cost and increase the range of metals amenable to this technology (Setia et al. 2008), which can be achieved by creating superior plant varieties for phytoextraction plants, optimizing agronomic practices for their cultivation, and designing safer and more effective soil amendments (Gleba et al. 1999). In addition, manipulating rhizospheric microorganisms could be used as a strategy for enhancing the establishment of phytodegradation processes (Zhu et al. 2001; Setia et al. 2008). Future work would involve genetic engineering to further improve the spectrum of metal-uptake characteristics by identifying and manipulating the responsible genes for metal accumulation. Very few hyperaccumulator plants have been discovered to date that have the capacity for multiple metal accumulations (Purakayastha and Chhonkar 2010). However, there are some reports indicating metal antagonism may limit uptake from multiplying metal-contaminated soils. Increasing systematic efforts to screen plant materials for these characteristics would surely disclose new hyperaccumulator plants and, thus, new potentials for photoremediation processes (Purakayastha and Chhonkar 2010).

2.9 ROLE OF PHYTOCHELATINS ON HEAVY-METAL TOLERANCE IN PLANTS

Phytochelatins (PCs) include a family of small, enzymatically synthesized peptides having a general structure of $(\gamma\text{-Glu-Cys})_n$-Gly, and these peptides are rapidly synthesized in response to toxic levels of HMs in all tested plants. Recently, researchers are interested in manipulating the expression of PC synthase genes in transgenic lines to enhance metal tolerance and accumulation in plants for potential use in phytoremediation. PCs were termed cadystins when they were identified and characterized in *Schizosaccharomyces pombe* (yeast) (Murasugi et al. 1981; Kondo et al. 1984). After that, it was reported that the major Cd-binding ligands in Cd-intoxicated plant cells are composed of (poly γ-glutamylcysteine)-glycine and were termed phytochelatins (Grill et al. 1985). The structure of PCs are related to GSH, which is $(\gamma\text{-Glu-Cys})_n$-Gly, where n ranges between 2 and 11. Thus, PCs constitute a number of structural species with increasing repetitions of γ-Glu-Cys units.

To avoid the toxic effects of HMs, plants have developed mechanisms to deactivate and scavenge metal ions that penetrate into the cytosol (Lyubenova and Schröder 2010). Numerous studies have reported the presence of PCs in different plant species exposed to HMs (Table 2.9). Cadmium and As effectively induce PC synthesis, and Zn and Ni hardly do so (Grill et al. 1989). PCs contain strongly nucleophilic –SH groups and, thus, can react with many toxic species within cells, such as free radicals, ROS, cytotoxic electrophilic organic xenobiotics, and obviously, HMs (Rabenstein 1989; Grill et al. 2007). Their N-terminal and downstream γ-peptidyl bonds probably serve to protect these thiol peptides from general protease action except specific action of γ-glutamyltranspeptidases. However, the Cd (or other metal) binding peptides formed of both (Glu-Cys)n-Gly or (γ-Glu-Cys) n-Gly have been found indistinguishable (Bae and Mehra 1997; Satofuka et al. 2001). The synthesis of these HM–PC complexes is a vital metabolic process in higher plants. The resulting depletion of GSH in the cytosol is compensated for by the induction of S assimilation and GSH biosynthesis (Rüegsegger et al. 1990; Rüegsegger and Brunold 1992). Glutathione, the substrate for PC synthase (EC 2.3.2.15), is synthesized from its constituent amino acids in two steps; the first step is catalyzed by γ-glutamyl-cys synthase (γ-ECS) and the second one by GSH synthase (GS) (Jabeen et al. 2009). This reaction is strongly induced by HMs. A direct correlation between the properties of the respective metal and the efficacy of inducing the action of the PC enzyme has been reported, according to the following sequence: Cd > Ag > Pb > Cu > Hg > Zn > Sn > Au > As (Lyubenova and Schröder 2010). Besides detoxification, PCs also contribute to the regulation of metal occurrence and homeostasis in plant cells (Lyubenova and Schröder 2010). Metal ions, such as Cu and Zn, play distinct roles in catalytic proteins or structural elements. Hence, PCs play a duel role: On the one hand they complex, detoxify, and store metal ions in the vacuole, and on the other, they guide essential metals to newly synthesized apoenzymes and facilitate contact (Thumann et al. 1991) (Figure 2.21).

TABLE 2.9
Induction of Phytochelatins in Some Higher Plant Species in Response to Different Heavy Metals

Heavy Metals	Tested Plant Species	References
Cd	*Pinus sylvestris, Pinus pinea, Sinapis alba*	Gekeler et al. (1989)
Cd	*Brassica juncea*	Heiss et al. (2003)
Cd	*Brassica oleracea, Lycopersicon esculentum, Zea mays, Eichhornia crassipes*	Grill et al. (1987)
Cd	*Cicer arietinum*	Gupta et al. (2004)
Pb	*Hydrilla verticillata*	Gupta et al. (1995)
Pb	*Vicia faba, Phaseolus vulgaris*	Piechalak et al. (2002)
As	*Cicer arietinum*	Gupta et al. (2002)
As	*Cytisus striatus*	Bleeker et al. (2003)
As	*Pteris vittata*	Zhang et al. (2004)
Cu	*Oryza sativa*	Yan et al. (2000)
Zn	*Rubia tinctorum*	Maitani et al. (1996)
Zn	*Silene vulgaris*	Harmens et al. (1993)
Ni	*Rubia tinctorum*	Maitani et al. (1996)
Hg	*Rubia tinctorum*	Maitani et al. (1996)

Source: Adapted from Grill, E., S. Mishra, S. Srivastava, and R. D. Tripathi, Role of phytochelatins in phytoremediation of heavy metals. In S. N. Singh, and R. D. Tripathi (Eds.), *Environmental bioremediation technology*, Springer, Berlin, 2007.

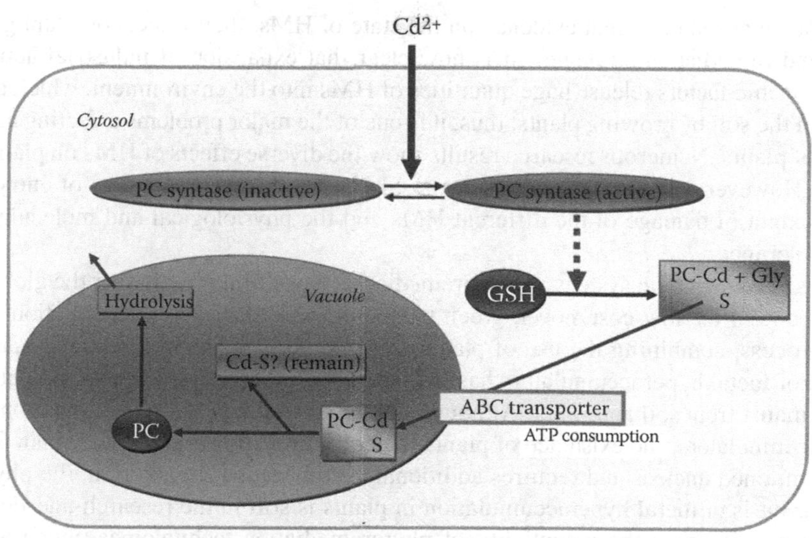

FIGURE 2.21 Control of PC biosynthesis and subcellular transfer of Cd. Cadmium ions penetrate into the cell and activate the synthesis of PCs. Phytochelatins are synthesized from GSH where S may play an important role. The Cd–PC complex is actively transported into the vacuole. The metal is stored there in a different form (i.e., complexed with organic acids) while the PCs are degraded and recycled to the cytosol.

Phytochelatin synthesis in plant cells and PC synthase activity can be induced by a wide variety of metal ions, and PCs are able to bind a number of metal ions *in vitro* through thiolate bonds. However, the only metal-PC complexes that have been isolated from plants contain ions of Cd, Cu, or Ag (Maitani et al. 1996). Cadmium-PC complexes have been extensively studied and are classified as either high or low molecular weight complexes (Goldsbrough 2000). A number of studies demonstrated that glutathione (γ-GluCys-Gly, GSH) is the substrate for synthesis of PCs. An enzyme activity that catalyzes the formation of PCs from GSH has been described in cell-free extracts from a number of plant species (Grill et al. 1989; Klapheck et al. 1995; Chen et al. 1997). This enzyme, PC synthase, transfers γ-GluCys from GSH to an acceptor GSH to produce $(\gamma\text{-GluCys})_2$-Gly (PC2); the same enzyme can add additional γ-GluCys moieties, derived from either GSH or PCs, to PC2 to produce larger PC peptides. PC synthase activity is dependent on the presence of one of a number of free metal ions, e.g., Cd^{2+}, Zn^{2+}, Ag^{+}. Chelation of HM ions, for example, by newly synthesized PCs, inactivates PC synthase, thereby providing an easy method to regulate the synthesis of PCs. The enzyme is expressed in plant roots and stems (Chen et al. 1997) and in plant cells growing in culture (Grill et al. 1989), perhaps providing a constant protective mechanism against HM toxicity.

Different studies revealed that Cd treatments significantly enhanced the synthesis of PCs in plants (Sarma 2011). Sun, Zhou, Xie, and Liu (2010b) reported that the variation in PC productions in root and shoots in two Cd-treated species, viz., *Rorippa globosa* and *Rorippa islandica*, might be used as a biomarker of Cd hyperaccumulation. They also observed that the synthesis of PCs might be related to an increase in the uptake of Cd^{2+} into the cytoplasm. However, they concluded that it was not the primary mechanism for Cd tolerance. Selvam and Wong (2008) reported that *Brassica napus* plants exposed to Cd influenced the synthesis of PCs in the shoot, and it was evident that the detoxification of Cd involves higher molecular-weight thiol complexes in the shoot. *Arabis paniculata* exposed to Cd induced formation of PCs and three unknown thiols in the roots, but none were detected in the shoots (Zeng et al. 2009). However, Zenk (1996) suggested that PCs play a constitutive role in plant metal tolerance, but this role is not conclusive and could just as easily indicate a stress response.

2.10 CONCLUSIONS AND FUTURE PERSPECTIVES

This review has focused on recent evidence on the state of HMs, their effects on plant growth and physiology, and remediation strategies. It is now clear that expansion of industrial activities and other anthropogenic factors release huge quantities of HMs into the environment, which are readily absorbed from the soil by growing plants; thus, it is one of the major problems affecting agricultural productivity of plants. Numerous research results show the diverse effects of HMs on plants directly or indirectly. However, more investigations are needed to reveal the exact sites of entry, mode of actions, and extent of damage of the different HMs and the physiological and molecular bases of metal stress tolerance.

A plant-based remediation system or phytoremediation is well-known all over the globe, and this is considered one of the low-cost, novel, green technologies. In fact, phytoremediation is a multicomponent process, combining the use of plants and microorganisms in landscape management. Identification of metal hyperaccumulators has indicated that plants have a genetic potential to take away contaminants from soil and water. Although numerous findings claim identification of several metal hyperaccumulators, the existence of plants hyperaccumulating metals other than Cd, Ni, Se, and Zn has remained unclear and requires additional corroboration. In addition, the physiological and molecular basis of metal hyperaccumulation in plants is still in the research-and-development phase. In order to improve the potentiality of phytoremediation technologies, much attention is now being paid to the use of transgenic plants with significantly increased remediation potentials compared with parent (unmodified) plants. Increasing systematic efforts to screen plant materials for these traits would obviously develop new hyperaccumulator plants and, thus, new potentials for the phytoremediation process (Purakayastha and Chhonkar 2010). The forthcoming challenge for phytoremediation is to develop a plant with a diverse metal-accumulating capacity in a cost-effective way. Identification of novel genes with high-biomass-yield characteristics and the subsequent development of transgenic plants with superior remediation capacities would be crucial for further research. Future research work would also involve genetic engineering to further improve the spectrum of metal-uptake characteristics by identifying and manipulating the responsible genes for metal accumulation. However, a multigene approach involving a simultaneous transfer of several genes into a suitable candidate plant to remove contaminants of a complex nature is essential. It is also necessary to manipulate metal transporters and their cellular targeting to specific cell organelles, such as vacuoles, to allow for safe sequestration of HMs in locations without disturbing other cellular functions. Furthermore, proteome and DNA array technology may be used for searching for suitable candidate genes/proteins for phytoremediation. Such efforts may lead to a better understanding of metal metabolism in plants and open up important plant applications for environmental cleanup. Overall, the ultimate success will depend upon involvement of a holistic approach to integrate the endeavors of plant physiologists, soil microbiologists, agronomists, biotechnologists, and environmental engineers, which would be an integral remediation option for the current world.

ACKNOWLEDGMENTS

We express our sincere thanks to Prof. Dr. Prasanta C. Bhowmik, University of Massachusetts Amherst, USA, and Prof. Dr. M. N. V. Prasad, University of Hyderabad, India, for his continuous encouragement and constructive suggestion during manuscript preparation. We apologize to all researchers for those parts of their work that were not cited in the manuscripts because of the page limitation.

REFERENCES

Adegoke, J. A., W. B. Agbaje, and O. O. Isaac. (2009). Evaluation of heavy metal status of water and soil at Ikogosi Warm Spring, Ondo State Nigeria. *Ethiopian Journal of Environmental Studies and Management*, *2*, 88–93.

Adriano, D. C. (2001). Trace elements in terrestrial environments. New York: Springer.

Ahmed, S. F. R., K. Killham, and I. Alexander. (2006). Influences of arbuscular fungus *Glomus mosseae* on growth and nutrition of lentil irrigated with arsenic contaminated water. *Plant and Soil*, *258*, 33–41.

Ahsan, N., D. G. Lee, S. H. Lee, K. Y. Kang, J. J. Lee, P. J. Kim, H. S. Yoon, J. S. Kim, and B. H. Lee. (2007). Excess copper induced physiological and proteomic changes in germinating rice seeds. *Chemosphere*, *67*, 1182–1193.

Akan, J. C., F. I. Abdulrahman, O. A. Sodipo, and A. G. Lange. (2010). Physicochemical parameters in soil and vegetable samples from a plant defense against toxic substances in soils. In T. A. Anderson and J. R. Coats (Eds.), *Bioremediation through rhizosphere technology*. Washington, DC: American Chemical Society.

Akinci, I. E., and S. Akinci. (2010). Effect of chromium toxicity on germination and early seedling growth in melon (*Cucumis melo* L.). *African Journal of Biotechnology*, *9*, 4589–4594.

Alam, M. B., and M. A. Satter. (2000). Assessment of arsenic contamination in soils and waters in some areas of Bangladesh. *Water Science and Technology*, *42*, 185–192.

Ali, M. A., A. B. M. Badruzzaman, M. A. Jalil, M. D. Hossain, M. F. Ahmed, A. A. Masud, M. Kamruzzaman, and M. A. Rahman. (2003). Arsenic in plant-soil environment in Bangladesh. In M. F. Ahmed, M. A. Ali, and Z. Adeel (Eds.), *Fate of arsenic in the environment* (85–112). Dhaka: BUET-UNU.

Alkorta, I., and C. Garbisu. (2001). Phytoremediation of organic contaminants. *Bioresource Technology*, *79*, 273–276.

Anjum, N. A., I. Ahmad, I. Mohmood, M. Pacheco, A. C. Duarte, E. Pereira et al. (2012). Modulation of glutathione and its related enzymes in plants' responses to toxic metals and metalloids: A review. *Environmental and Experimental Botany*, *75*, 307–324.

Anjum, N. A., S. Umar, and A. Ahmad. (Eds.). (2011a). *Oxidative stress in plants: Causes, consequences and tolerance*. New Delhi: IK International Publishing House.

Anjum, N. A., S. Umar, A. Ahmad, M. Iqbal, and N. A. Khan. (2008). Ontogenic variation in response of *Brassica campestris* L. to cadmium toxicity. *Journal of Plant Interactions*, *3*, 189–198.

Anjum, N. A., S. Umar, M. Iqbal, and N. A. Khan. (2011b). Cadmium causes oxidative stress in mung bean by affecting the antioxidant enzyme system and ascorbate–glutathione cycle metabolism. *Russian Journal of Plant Physiology*, *58*, 92–99.

Appenroth, K. J. (2010). Definition of "heavy metals" and their role in biological systems. In I. Sherameti, and A. Varma (Eds.), *Soil heavy metals*, *Soil Biol. 9* (19–29). Berlin: Springer.

Azevedo, J. A., and R. A. Azevedo. (2006). Heavy metals and oxidative stress: Where do we go from here? *Communications in Biometry and Crop Science*, *1*, 135–138.

Azevedo, R. A. (2005). Toxic metals in plants. *Brazilian Journal of Plant Physiology*, *17*, 1.

Bae, W., and R. K. Mehra. (1997). Metal-binding characteristics of phytochelatin analog $(Glu\text{-}Cys)_2Gly$. *Journal of Inorganic Biochemistry*, *68*, 201–210.

Bah, A. M., H. Dai, J. Zhao, H. Sun, F. Cao, G. Zhang, and F. Wu. (2011). Effects of cadmium, chromium and lead on growth, metal uptake and antioxidative capacity in *Typha angustifolia*. *Biological Trace Element Research*, *142*, 77–92.

Baker, A. J. M. (1987). Metal tolerance. *New Phytologist*, *106*, 93–111.

Baker, A. J. M., and R. R. Brooks. (1989). Terrestrial higher plants which hyperaccumulate metallic elements: A review of their distribution, ecology and phytochemistry. *Biorecovery*, *1*, 81–126.

Baker, A. J. M., S. P. McGrath, R. D. Reeves, and A. C. Smith. (2000). Metal hyperaccumulator plants: A review of the ecology and physiology of a biological resource for phytoremediation of metal-polluted soils. In N. Terry, and G. Bañuelos (Eds.), *Phytoremediation of contaminated soil and water*. Boca Raton: CRC Press.

Baker, A. J. M., and P. L. Walker. (1990). Ecophysiology of metal uptake by tolerant plants. In A. J. Shaw (Ed.), *Heavy metal tolerance in plants: Evolutionary aspects* (155–177). Boca Raton: CRC Press.

Banasova, V., and O. Horak. (2008). Heavy metal content in *Thlaspi caerulescens* J. et C. Presl growing on metalliferous and non-metalliferous soils in Central Slovakia. *International Journal of Environment and Pollution*, *33*, 133–145.

Bañuelos, G. S. (2000). Phytoextraction of selenium from soils irrigated with selenium-laden effluent. *Plant and Soil*, *224*, 251–258.

Barbee Jr., J. Y., and T. S. Prince. (1999). Acute respiratory distress syndrome in a welder exposed to metal fumes. *Southern Medical Journal*, *92*, 510–512.

Barceló, J., and C. H. Poschenrieder. (1990). Plant water relations as affected by heavy metal stress: A review. *Journal of Plant Nutrition*, *13*, 1–37.

Barceló, J., and C. Poschenrieder. (2004). Structural and ultrastructural changes in heavy metal exposed plants. In M. N. V. Prasad (Ed.), *Heavy metal stress in plants: From biomolecules to ecosystem*, *3rd edn* (223–248), Berlin: Springer.

Barry, S. A. S., and S. C. Clark. (1978). Problems of interpreting the relationship between the amounts of lead and zinc in plants and soil on metalliferous wastes. *New Phytologist*, *81*, 773–783.

Baumann, A. (1885). Das verhalten von zinksalzen gegen pflanzen und im boden: Die Landwirts. *Vers. Stat.*, *31*, 1–53.

Benavides, M. P., S. M. Gallego, and M. L.Tomaro. (2005). Cadmium toxicity in plants. *Brazilian Journal of Plant Physiology*, *17*, 1–18.

Bergkvist, B., L. Folkeson, and D. Berggren. (1989). Fluxes of Cu, Zn, Pb, Cd, Cr, and Ni in temperate forest ecosystems: A literature review. *Water*, *Air* and *Soil Pollution*, *47*, 217–286.

Bernier, M., R. Carpentier, and N. Murata. (1992). Reversal of mercury inhibition in photosystem II by chloride. In Proceedings from IXth International Congress on Photosynthesis. *Photosynthesis Research*, *2*, 97–100.

Berti, W. R., and S. D. Cunningham. (2000). Phytostabilization of metals. In I. Raskin, and B. D. Ensley (Eds.), *Phytoremediation of toxic metals: Using plants to clean up the environment* (71–88). New York: Wiley.

BGS/DPHE. (2000). Groundwater studies for arsenic contamination in Bangladesh, Final Report, Summary, Department of Public Health and Engineering, Government of Bangladesh, DFID, British Geological Survey, Dhaka, Bangladesh.

Bishnoi, N. R., L. K. Chugh, and S. K. Sawhney. (1993). Effect of chromium on photosynthesis, respiration and nitrogen fixation in pea (*Pisum sativum* L) seedlings. *Journal of Plant Physiology*, *142*, 25–30.

Black, H. (1995). Absorbing possibilities: Phytoremediation. *Environmental Health Perspective*, *103*, 1106–1108.

Blaylock, M. J., and J. W. Huang. (2000). Phytoextraction of metals. In I. Raskin, and B. D. Ensley (Eds.), Phytoremediation of toxic metals: Using plants to clean up the environment (53–69). New York: Wiley.

Blaylock, M. J., D. E. Salt, S. Dushenkov, O. Zakharova, C. Gushsman, Y. Kapulnik, B. D. Ensley, and I. Raskin. (1997). Enhanced accumulation of Pb in Indian mustard by soil-applied chelating agents. *Environmental Science and Technology*, *31*, 860–865.

Bleeker, P. M., H. Schat, R. Vooijs, J. A. C. Verkleij, and W. H. O. Ernst. (2003). Mechanisms of arsenate tolerance in *Cytisus striatus*. *New Phytologist*, *157*, 33–38.

Boominathan, R., and P. M. Doran. (2002). Ni-induced oxidative stress in roots of the Ni hyperaccumulator, *Alyssum bertolonii*. *New Phytologist*, *156*, 205–215.

Boscolo, P. R. S., M. Menossi, and R. A. Jorge. (2003). Aluminum-induced oxidative stress in maize. *Phytochemistry*, *62*, 181–189.

Bouazizi, H., H. Jouili, A. Geitmann, and E. El Ferjani. (2010). Copper toxicity in expanding leaves of *Phaseolus vulgaris* L.: Antioxidant enzyme response and nutrient element uptake. *Ecotoxicology and Environmental Safety*, *73*, 1304–1308.

Boyd, R. S., and M. A. Davis. (2001). Metal tolerance and accumulation ability of the Ni hyperaccumulator *Streptanthus polygaloides* Gray (Brassicaceae). *International Journal of Phytoremediation*, *3*, 353–367.

Bradshaw, A. (1997). Restoration of mined lands: Using natural processes. *Ecological Engineering*, *8*, 255–269.

Brooks, R. R., J. Lee, R. D. Reeves, and T. Jaffré. (1977). Detection of nickeliferous rocks by analysis of herbarium specimens of indicator plants. *Journal of Geochemical Explorations*, *7*, 49–57.

Brooks, R. R., S. D. Naidu, F. Malaisse, and J. Lee. (1987). The elemental content of metallophytes from the copper/cobalt deposits of Central Africa. *Bulletin of Society of Royal Botany, Belgique*, *119*, 179–191.

Brooks, R. R., and X. H. Yang. (1984). Elemental levels and relationships in the endemic serpentine flora of the Great Dyke, Zimbabwe, and their significance as controlling factors for this flora. *Taxonomy*, *33*, 392–399.

Brown, S. L., R. L. Chaney, J. S. Angle, and A. J. M. Baker. (1994). Phytoremediation potential of *Thlaspi caerulescens* and bladder campion for zinc and cadmium contaminated soil. *Journal of Environmental Quality*, *23*, 1151–1157.

Brown, S. L., R. L. Chaney, J. S. Angle, and A. J. M. Baker. (1995). Zinc and cadmium uptake by hyperaccumulator *Thlaspi caerulescens* grown in nutrient solution. *Soil Science Society of America Journal*, *59*, 125–133.

Brümmer, G., J. Gerth, and U. Herms. (1986). Heavy metals species, mobility and availability in soils. *Z. Pflanzenernaehr Bodenkd*, *149*, 382–398.

Burken, J. G. (2003). Uptake and metabolism of organic compounds: Green-liver model. In S. C. McCutcheon, and J. L. Schnoor (Eds.), Phytoremediation: Transformation and control of contaminants (59–84). New York: Wiley.

Cakmak, I., and W. J. Horst. (1991). Effect of aluminum on lipid peroxidation, superoxide dismutase, catalase and peroxidase activities in root tips of soybean (*Glycine max*). *Physiologia Plantarum*, *83*, 463–468.

Callender, E. (2003). Heavy metals in the environment: Historical trends. *Treat. Geochem.*, *9*, 67–105.

Carbonell-Barrachina, A. A., F. Burló, A. Burgos-Hernández, E. López, and J. Mataix. (1997). The influence of arsenite concentration on arsenic accumulation in tomato and bean plants. *Scientia Horticulturae*, *71*, 167–176.

Cargnelutti, D., L. A. Tabaldi, R. M. Spanevello, G. O. Jucoski, V. Battisti, M. Redin, C. E. B. Linares, V. L. Dressler, M. M. Flores, F. T. Nicoloso, V. M. Morsch, and M. R. C. Schetinger. (2006). Mercury toxicity induces oxidative stress in growing cucumber seedlings. *Chemosphere*, *65*, 999–1006.

Cataldo, D. A., and R. E. Wildung. (1978). Soil and plant factors influencing the accumulation of heavy metals by plants. *Environmental Health Perspective*, *27*, 149–159.

Chaney, R. L., M. Malik, Y. M. Li, S. L. Brown, E. P. Brewer, J. S. Angle, and A. J. M. Baker. (1997). Phytoremediation of soil metals. *Current Opinion in Biotechnology*, *8*, 279–284.

Charlesworth, S., M. Everett, R. McCarthy, A. Ordonez, and E. Miguel. (2003). Comparative study of heavy metal concentration and distribution in deposited street dust in a large and a small urban area: Birmingham and Coventry, West Midlands, United Kingdom. *Environment International*, *29*, 563–573.

Chatterjee, J., and C. Chatterjee. (2000). Phytotoxicity of cobalt, chromium and copper in cauliflower. *Environmental Pollution*, *109*, 69–74.

Chatterjee, C., B. K. Dube, P. Sinha, and P. Srivastava. (2004). Detrimental effects of lead phytotoxicity on growth, yield, and metabolism of rice. *Communications in Soil Science and Plant Analysis*, *35*, 255–265.

Chaudhry, T. M., W. J. Hayes, A. G. Khan, and C. S. Khoo. (1998). Phytoremediation: Focusing on accumulator plants that remediate metal-contaminated soils. *Australian Journal of Ecotoxicology*, *4*, 37–51.

Chen, X., J. Wang, Y. Shi, M. Q. Zhao, and G. Y. Chi. (2011). Effects of cadmium on growth and photosynthetic activities in pakchoi and mustard. *Botanical Studies*, *52*, 41–46.

Chen, Z., T. E. Young, J. Ling, S. Chang, and D. R. Gallie. (2003). Increasing vitamin C content of plants through enhanced ascorbate recycling. *Proceedings of National Academy of Sciences USA*, *100*, 3525–3530.

Chen, J., J. Zhou, and P. B. Goldsbrough. (1997). Characterization of phytochelatin synthase from tomato. *Physiologia Plantarum*, *101*, 165–172.

Chimbira, C., and D. Z. Moyo. (2009). The effect of single and mixed treatments of lead and cadmium on soil bioavailability and yield of *Brassica napus* irrigated with sewage effluent: A potential human risk. *African Journal of Agricultural Research*, *4*, 359–364.

Cho, M., A. N. Chardonnens, and K. J. Dietz. (2003). Differential heavy metal tolerance of *Arabidopsis halleri* and *Arabidopsis thaliana*: A leaf slice test. *New Phytologist*, *158*, 287–293.

Cho, U., and N. Seo. (2005). Oxidative stress in *Arabidopsis thaliana* exposed to cadmium is due to hydrogen peroxide accumulation. *Plant Science*, *168*, 113–120.

Choudhury, M. R. Q., S. T. Islam, R. Alam, I. Ahmad, W. Zamam, R. Sen, and M. N. Alam. (2008). Effects of arsenic on red amaranth (*Amaranthus retroflexus* L.). *American-Eurasian Journal of Scientific Research*, *3*, 48–53.

Church, T. M., and J. R. Scudlark. (1992). Trace elements in precipitation at the middle Atlantic coast: A successful record since 1982. Proceedings of the Symposium on the deposition and fate of trace metals in our environment.

Clemens, S. (2006). Toxic metal accumulation, responses to exposure and mechanisms of tolerance in plants. *Biochimie*, *88*, 1707–1719.

Cobbett, C. S. (2000). Phytochelatins and their role in heavy metal detoxification. *Plant Physiology*, *123*, 825–832.

Cobbett, C. S., and P. B. Goldsbrough. (2000). Mechanism of metal resistance: Phytochelatins and metallothioneins. In I. Raskin, and B. D. Ensley (Eds.), *Phytoremediation of toxic metals: Using plants to clean up the environment* (247–269). New York: Wiley.

Cosio, C., P. Vollenweider, and C. Keller. (2006). Localization and effects of cadmium in leaves of a cadmium-tolerant willow (*Salix viminalis* L.) I. macrolocalization and phytotoxic effects of cadmium. *Environmental and Experimental Botany*, *58*, 64–74.

Cunningham, S. D., W. R. Berti, and J. W. Huang. (1995). Phytoremediation of contaminated soils. *Trends in Biotechnology*, *13*, 393–397.

Cuypers, A., J. Vangronsveld, and H. Clijsters. (2001). The redox status of plant cells (AsA and GSH) is sensitive to zinc imposed oxidative stress in roots and primary leaves of *Phaseolus vulgaris*. *Plant Physiology and Biochemistry*, *39*, 657–664.

DalCorso, G., S. Farinati, and A. Furini. (2010). Regulatory networks of cadmium stress in plants. *Plant Signaling and Behaviour*, *5*, 663–667.

Das, H. K., A. K. Mitra, P. K. Sengupta, A. Hossain, F. Islam, and G. H. Rabbani. (2004). Arsenic concentrations in rice, vegetables, and fish in Bangladesh: A preliminary study. *Environment International*, *30*, 383–387.

Davis, R. D., C. H. Charlton-Smith, J. H. Stark, and J. A. Campbell. (1988). Distribution of metals in grassland soils following surface applications of sewage sludge. *Environmental Pollution*, *49*, 99–115.

Demirevska-Kepova, K., L. Simova-Stoilova, Z. Stoyanova, R. Holzer, and U. Feller. (2004). Biochemical changes in barley plants after excessive supply of copper and manganese. *Environmental and Experimental Botany*, *52*, 253–266.

Deram, A., and D. Petit. (1997). Ecology of bioaccumulation in *Arrhenatherum elatius* L. (Poaceae) populations: Applications of phytoremediation of zinc, lead and cadmium contaminated soils. *Journal of Experimental Botany*, *48*, 98.

Dhankhar, R., and R. Solanki. (2011). Effect of copper and zinc toxicity on physiological and biochemical parameters in *Vigna mungo* (L.) Hepper. *International Journal of Pharma and Bio Sciences*, *2*, 553–565.

Dierberg, F. E., T. A. Débuts, and J. R. N. A. Goulet. (1987). Removal of copper and lead using a thin-film technique. In K. R. Reddy, and W. H. Smith (Eds.), *Aquatic plants for water treatment and resource recovery* (497–504). Pineville: Magnolia Publishing.

Dietz, K. J., M. Baier, and U. Krämer. (1999). Free radicals and reactive oxygen species as mediators of heavy metal toxicity in plants. In M. N. V. Prasad, and J. Hagemeyer (Eds.), *Heavy metal stress in plants: From molecules to ecosystems* (79–97). Berlin: Springer.

Dipierro, N., D. Mondelli, C. Paciolla, G. Brunetti, and S. Dipierro. (2005). Changes in the ascorbate system in the response of pumpkin (*Cucurbita pepo* L.) roots to aluminum stress. *Journal of Plant Physiology*, *162*, 529–536.

Dong, R. (2005). Molecular cloning and characterization of a phytochelatin synthase gene, *PvPCS1*, from *Pteris vittata* L. *Journal of Industrial Microbiology and Biotechnology*, *32*, 527–533.

Dong, Y., X. Z. Chen, C. S. Xi, and H. Xie. (2011). Effect of $ZnSO_4$ on seed germination of soybean. *Journal of Beijing University Agriculture*, *26*, 61–62.

Dubey, R. S. (2011). Metal toxicity, oxidative stress and antioxidative defense system in plants. In S. D. Gupta (Ed.), *Reactive oxygen species and antioxidants in higher plants* (177–203). Boca Raton: CRC Press.

Dushenkov, D. (2003). Trends in phytoremediation of radionuclides. *Plant and Soil*, *249*, 167–175.

Dushenkov, D., and Y. Kapulnik. (2000). Phytofiltration of metals. In I. Raskin, and B. D. Ensley (Eds.), *Phytoremediation of toxic metals: Using plants to clean up the environment* (89–106). New York: Wiley.

Dushenkov, D., P. B. A. Nanda Kumar, H. Motto, and I. Raskin. (1995). Rhizofiltration: The use of plants to remove heavy metals from aqueous streams. *Environmental Science and Technology*, *29*, 1239–1245.

Duvigneaud, P., and S. Denaeyer-De Smet. (1963). Cuivre et végétation au Katanga. *Bulletin of Society of Royal Botany, Belgique*, *96*, 93–231.

El-Ghamery, A. A., M. A. El-Kholy, and M. A. Abou El-Yousser. (2003). Evaluation of cytological effects of Zn^{2+} in relation to germination and root growth of *Nigella sativa* L. and *Triticum aestivum* L. *Mutation Research*, *537*, 29–41.

El-Shintinawy, F., and A. El-Ansary. (2000). Differential effect of Cd^{2+} and Ni^{2+} on amino acid metabolism in soybean seedlings. *Biologia Plantarum*, *43*, 79–84.

Esteban, E., E. Morenoa, J. Pěnalosa, J. I. Cabrero, M. Millán, and P. Zornoza. (2008). Short and long-term uptake of Hg in white lupin plants: Kinetics and stress indicators. *Environmental and Experimental Botany*, *62*, 316–322.

Evangelou, M. W., H. Daghan, and A. Schaeffer. (2004). The influence of humic acids on the phytoextraction of cadmium from soil. *Environmental Pollution*, *132*, 113–120.

Ezaki, B., R. C. Gardner, Y. Ezaki, and H. Matsumoto. (2000). Expression of aluminum-induced genes in transgenic *Arabidopsis* plants can ameliorate aluminum stress and/or oxidative stress. *Plant Physiology*, *122*, 657–665.

FAO. (1999). Potential pollutants, their sources and their impacts. In J. A. Sciortino, and R. Ravikumar (Eds.), Fishery harbour manual on the prevention of pollution: Bay of Bengal Programme for fisheries management (BOBP/MAG/22). Madras, India.

Farouk, S., A. A. Mosa, A. A. Taha, H. M. Ibrahim, and A. M. El-Gahmery. (2011). Protective effect of humic acid and chitosan on radish (*Raphanus sativus*, L. var. sativus) plants subjected to cadmium stress. *Journal of Stress Physiology and Biochemistry*, *7*, 99–116.

Fodor, F. (2002). Physiological responses of vascular plants to heavy metals. In M. N. V. Prasad, and K. Strzalka (Eds.), *Physiology and biochemistry of metal toxicity and tolerance in plants* (149–177). Dordrecht: Kluwer Academic.

Förstner, U. (1979). Metal transfer between solid and aqueous phases. In U. Förstner, and G. T. W. Wittmann (Eds.), *Metall pollution in the aquatic environment* (197–270). Berlin: Springer.

Fujita, M. (1985). The presence of two Cd-binding components in the roots of water hyacinth cultivated in a Cd^{2+}-containing medium. *Plant and Cell Physiology*, *26*, 295–300.

Fujita, M., and T. Kawanishi. (1986). Purification and characterization of a Cd-binding complex from the root tissue of water hyacinth cultivated in a Cd^{2+}-containing medium. *Plant and Cell Physiology*, *27*, 1317–1325.

Fujita, M., and T. Kawanishi. (1987). Cd-binding complexes from the root tissues of various higher plants cultivated in Cd^{2+}-containing medium. *Plant and Cell Physiology*, *28*, 379–382.

Gajewska, E., M. Wielanek, K. Bergier, and M. Skłodowska. (2009). Nickel induced depression of nitrogen assimilation in wheat roots. *Acta Physiologia Plantarum*, *31*, 1291–1300.

Gani Molla, M. O., M. A. Islam, and M. Hasanuzzaman. (2010). Identification of arsenic hyperaccumulating plants for the development of phytomitigation technology. *Journal of Phytology*, *2*, 41–48.

Garbisu, C., J. Hernandez-Allica, O. Barrutia, I. Alkorta, and J. M. Becerril. (2002). Phytoremediation: A technology using green plants to remove contaminants from polluted areas. *Review in Environmental Health*, *17*, 75–90.

Garg, N., and P. Singla. (2011). Arsenic toxicity in crop plants: physiological effects and tolerance mechanisms. *Environmental Chemistry Letters*, *3*, 303–321.

Gautam, S., and S. N. Pandey. (2008). Growth and biochemical responses of nickel toxicity on leguminous crop (*Lens esculentum*) grown in alluvial soil. *Research in Environment and Life Science*, *1*, 25–28.

Gekeler, W., E. Grill, E. L. Winnacker, and M. H. Zenk. (1989). Survey of the plant kingdom for the ability to bind heavy metals through phytochelatins. *Z. Naturforsch*, *44c*, 361–369.

Gerard, E., G. Echevarria, T. Sterckeman, and J. L. P. Morel. (2000). Availability of Cd to three plant species varying in accumulation pattern. *Journal of Environmental Quality*, *29*, 1117–1123.

Ghaderian, S. M., A. Mohtadi, M. R. Rahiminejad, and A. J. M. Baker. (2007). Nickel and other metal uptake and accumulation by species of *Alyssum* (Brassicaceae) from the ultramafics of Iran. *Environmental Pollution*, *145*, 293–298.

Ghosh, M., and S. P. Singh. (2005). A review on phytoremediation of heavy metals and utilization of its byproducts. *Applied Ecological and Environmental Research*, *3*, 1–18.

Gill, T., V. Dogra, S. Kumar, P. S. Ahuja, and Y. Sreenivasulu. (2011a). Protein dynamics during seed germination under copper stress in *Arabidopsis* over-expressing *Potentilla* superoxide dismutase. *Journal of Plant Research*, *125*, 165–172.

Gill, S. S., N. A. Khan, N. A. Anjum, and N. Tuteja. (2011b). Amelioration of cadmium stress in crop plants by nutrients management: Morphological, physiological and biochemical aspects. *Plant Stress*, *5*, 1–23.

Gill, S. S., N. A. Khan, and N. Tuteja. (2011c). Differential cadmium stress tolerance in five Indian mustard (*Brassica juncea* L.) cultivars: An evaluation of the role of antioxidant machinery. *Plant Signaling and Behaviour*, *6*, 293–300.

Gill, S. S., and N. Tuteja. (2010). Reactive oxygen species and antioxidant machinery in abiotic stress tolerance in crop plants. *Plant Physiology and Biochemistry*, *48*, 909–930.

Gleba, D., N. V. Borisjuk, L. G. Borisjuk, R. Kneer, A. Poulev, M. Skarzhinskaya, S. Dushenkov, S. Logendra, Y. Y. Gleba, and I. Raskin. (1999). Use of plant roots for phytoremediation and molecular farming. *Proceedings of National Academy of Sciences USA*, *96*, 5973–5977.

Goldsbrough, P. (2000). Metal tolerance in plants: The role of phytochelatins and metallothioneins. In N. Terry, and G. Bañuelos (Eds.), *Phytoremediation of contaminated soil and water*. Boca Raton: CRC Press.

Golovatyj, S. E., and E. N. Bogatyreva. (1999). Effect of levels of chromium content in a soil on its distribution in organs of corn plants. *Soil Research and Use of Fertilizers*, *25*, 197–204.

Gratão, P. L., A. Polle, P. J. Lea, and R. A. Azevedo. (2005). Making the life of heavy metal-stressed plants a little easier. *Functional Plant Biology*, *32*, 481–494.

Greger, M. (2004). Metal availability, uptake, transport and accumulation in plants. In M. N. V. Prasad (Ed.), *Heavy metal stress in plants: From biomolecules to ecosystems, 2nd edn.* (1–27). Berlin: Springer.

Greger, M., M. Johansson, A. Stihl, and K. Hamza. (1993). Foliar uptake of Cd by pea (*Pisum sativum*) and suger beet (*Beta vulgaris*). *Physiologia Plantarum*, *88*, 563–570.

Grill, E. L., S. Löeffler, E. L. Winnacker, and M. H. Zenk. (1989). Phytochelatins, the heavy-metal-binding peptides of plants, are synthesized from glutathione by a specific γ-glutamylcysteine dipeptidyl transpeptidase (phytochelatin synthase). *Proceedings of National Academy of Sciences USA*, *86*, 6838–6842.

Grill, E., S. Mishra, S. Srivastava, and R. D. Tripathi. (2007). Role of phytochelatins in phytoremediation of heavy metals. In S. N. Singh, and R. D. Tripathi (Eds.), *Environmental Bioremediation Technology* (101–146). Berlin: Springer.

Grill, E., E. L. Winnacker, and M. H. Zenk. (1985). Phytochelatins: The principal heavy-metal complexing peptides of higher plants. *Science*, *230*, 674–676.

Grill, E., E. L Winnacker, and M. H. Zenk. (1987). Phytochelatins, a class of heavy-metal-binding peptides from plants, are functionally analogous to metallothioneins. *Proceedings of National Academy of Sciences USA*, *84*, 439–443.

Gupta, M., U. N. Rai, R. D. Tripathi, and P. Chandra. (1995). Lead induced changes in glutathione and phytochelatin in *Hydrilla verticillata*. *Chemosphere*, *30*, 2011–2020.

Gupta, D. K., H. Tohoyama, M. Joho, and M. Inouhe. (2002). Possible roles of phytochelatins and glutathione metabolism in cadmium tolerance in chickpea roots. *Journal of Plant Research*, *115*, 429–437.

Gupta, D. K., H. Tohoyama, M. Joho, and M. Inouhe. (2004). Changes in levels of phytochelatins and related metal-binding peptides in chickpea seedlings exposed to arsenic and different heavy metal ions. *Journal of Plant Research*, *117*, 253–256.

Hall, J. L. (2002). Cellular mechanisms for heavy metal detoxification and tolerance. *Journal of Experimental Botany*, *53*, 1–11.

Hamel, F., C. Breton, and M. Houde. (1998). Isolation and characterization of wheat aluminum-regulated genes: Possible involvement of aluminum as a pathogenesis response elicitor. *Planta*, *205*, 531–538.

Harmens, H., P. R. D. Hartog, W. M. Ten Bookum, and J. A. C. Verkleij. (1993). Increased zinc tolerance in *Silene vulgaris* (Moench) Garcke is not due to increased production of phytochelatins. *Plant Physiology*, *103*, 1305–1309.

Hasanuzzaman, M., M. A. Hossain, and M. Fujita. (2012). Exogenous selenium pretreatment protects rapeseed seedlings from cadmium induced oxidative stress by up-regulating the antioxidant defense and methylglyoxal detoxification systems. *Biological Trace Element Research*, DOI: 10.1007/s12011-012-9419-4.

Haverkamp, R. G., A. T. Marshall, and D. van Agterveld. (2007). Pick your carats: Nanoparticles of gold-silver-copper alloy produced *in vivo*. *Journal of Nanoparticle Research*, *9*, 697–700.

Heckathorn, S. A., J. K. Mueller, S. La Guidice, B. Zhu, T. Barrett, B. Blair, and Y. Dong. (2004). Chloroplast small heat-shock proteins protect photosynthesis during heavy metal stress. *American Journal of Botany*, *91*, 1312–1318.

Hédiji, H., W. Djebali, C. Cabasson, M. Maucourt, P. Baldet, A. Bertrand, L. B. Zoghlami, C. Deborde, A. Moing, R. Brouquisse, W. Chaïbi, and P. Gallusci. (2010). Effects of long-term cadmium exposure on growth and metabolomic profile of tomato plants. *Ecotoxicology and Environmental Safety*, *73*, 1965–1974.

Heidari, M., and S. Sarani. (2011). Effects of lead and cadmium on seed germination, seedling growth and antioxidant enzymes activities of mustard (*Sinapis arvensis* L.). *ARPN: Journal of Agricultural and Biological Science*, *6*, 44–47.

Heiss, S., A. Wachter, J. Bogs, C. Cobbett, and T. Rausch. (2003). Phytochelatin synthase (PCS) protein is induced in *Brassica juncea* leaves after prolonged Cd exposure. *Journal of Experimental Botany*, *54*, 1833–1839.

Henry, J. R. (2000). An overview of phytoremediation of lead and mercury. NNEMS Report, 3–9. Washington, DC.

Hossain, M. F. (2006). Arsenic contamination in Bangladesh: An overview. *Agriculture, Ecosystems and Environment*, *113*, 1–16.

Huang, J. W., J. Chen, W. R. Berti, and S. D. Cunningham. (1997). Phytoremediation of lead-contaminated soils: Role of synthetic chelates in lead phytoextraction. *Environmental Science and Technology*, *31*, 800–805.

Huang, S. W., J. Y. Jin, and P. He. (2009). Effects of different patterns of land use on status of heavy metals in agricultural soils. *Better Crops*, *93*, 20–22.

IARC. (1993). Cadmium and cadmium compounds. In *Beryllium, cadmium, mercury and exposure in the glass manufacturing industry*. IARC monographs on the evaluation of carcinogenic risks to humans, 58, 119–237. Lyon: International Agency for Research on Cancer.

Imperato, M., P. Adamo, D. Naimo, M. Arienzo, D. Stanzione, and P. Violante. (2003). Spatial distribution of heavy metals in urban soils of Naples city (Italy). *Environmental Pollution*, *124*, 247–256.

Islam, S. M. A., K. Fukushi, and K. Yamamoto. (2004). Severity of arsenic concentration in soil and arsenic-rich sludge of Bangladesh and potential of their biological removal: A novel approach for a tropical region. Proceedings of the Second International Symposium on Southeast Asian Water Environments, Hanoi, Vietnam.

Islam, S. M. A., K. Fukushi, and K. Yamamoto. (2005). Bioreduction and biomethylation of arsenic in soil of Bangladesh: Novel process for permanent removal of arsenic. Proceedings of the 1st IWA-ASPIRE Conference and Exhibition, Singapore.

Israr, M., A. Jewell, D. Kumar, and S. V. Sahi. (2011). Interactive effects of lead, copper, nickel and zinc on growth, metal uptake and antioxidative metabolism of *Sesbania drummondii*. *Journal of Hazardous Materials*, *186*, 1520–1526.

Israr, M., S. V. Sahi, and J. Jain. (2006). Cadmium accumulation and antioxidative responses in the *Sesbania drummondii* callus. *Archives of Environmental Contamination and Toxicology*, *50*, 121–127.

Jabeen, R., A. Ahmad, and M. Iqbal. (2009). Phytoremediation of heavy metals: Physiological and molecular mechanisms. *Botanical Review*, *75*, 339–364.

Jaffré, T., and M. Schmid. (1974). Accumulation du nickel par une Rubiacée de Nouvelle Calédonie, *Psychotria douarrei* (G. Beauvisage) Däniker. *Compt. Rend. Acad. Sci.*, (Paris), Sér. *278*, 1727–1730.

Jain, R., S. Srivastava, V. K. Madan. (2000). Influence of chromium on growth and cell division of sugarcane. *Indian Journal of Plant Physiology*, *5*, 228–231.

James, B. R. (2001). Remediation-by-reduction strategies for chromate-contaminated soils. *Environmental Geochemistry and Health*, *23*, 175–189.

Järup, L. (2003). Hazards of heavy metal contamination. *British Medical Bulletin*, *68*, 167–182.

Jha, A. B., and R. S. Dubey. (2004). Arsenic exposure alters the activities of key nitrogen assimilatory enzymes in growing rice seedlings. *Plant Growth Regulation*, *43*, 259–268.

Jiang, Q. Q., and B. R. Singh. (1994). Effect of different forms and sources of arsenic on crop yield and arsenic concentration. *Water, Air and Soil Pollution*, *74*, 321–343.

John, R., P. Ahmad, K. Gadgil, and S. Sharma. (2009). Heavy metal toxicity: Effect on plant growth, biochemical parameters and metal accumulation by *Brassica juncea* L. *International Journal of Plant Production*, *3*, 65–76.

Kabata-Pendias, A., and H. Pendias. (1992). *Trace elements in soils and plants, 2nd edn.* Boca Raton: CRC Press.

Kachout, S. S., A. B. Mansoura, J. C. Leclerc, R. Mechergui, M. N. Rejeb, and Z. Ouerghi. (2009). Effects of heavy metals on antioxidant activities of *Atriplex hortensis* and *A. rosea*. *Journal of Food, Agriculture and Environment*, *7*, 938–945.

Kachout, S. S., A. B. Mansoura, J. C. Leclerc, R. Mechergui, M. N. Rejeb, and Z. Ouerghi. (2010). Effects of heavy metals on antioxidant activities of: *Atriplex hortensis* and *A. rosea*. *Journal of Agriculture and Environment*, *9*, 444–457.

Kadi, M. W. (2009). Soil pollution hazardous to environment: A case study on the chemical composition and correlation to automobile traffic of the roadside soil of Jeddah city, Saudi Arabia. *Journal of Hazardous Materials*, *168*, 1280–1283.

Kazemi, N., R. A. Khavari-Nejad, H. Fahimi, S. Saadatmand, and T. Nejad-Sattari. (2010). Effects of exogenous salicylic acid and nitric oxide on lipid peroxidation and antioxidant enzyme activities in leaves of *Brassica napus* L. under nickel stress. *Scientia Horticulturae*, *126*, 402–407.

Kertulis-Tartara, G. M., B. Rathinasapathip, and L. Q. Ma. (2009). Characterization of glutathione reductase and catalase in the fronds of two Pteris ferns upon arsenic exposure. *Plant Physiology and Biochemistry*, *47*, 960–965.

Khan, I., A. Ahmad, and M. Iqbal. (2009). Modulation of antioxidant defence system for arsenic detoxification in Indian mustard. *Ecotoxicology and Environmental Safety*, *72*, 626–634.

Kirk, J., J. Klironomos, H. Lee, and J. T. Trevors. (2005). The effects of perennial ryegrass and alfalfa on microbial abundance and diversity in petroleum contaminated soil. *Environmental Pollution*, *133*, 455–465.

Klapheck, S., S. Schlunz, and L. Bergmann. (1995). Synthesis of phytochelatins and homophytochelatins in *Pisum sativum* L. *Plant Physiology*, *107*, 515–521.

Kocik, K., and J. Ilavsky. (1994). Effect of Sr and Cr on the quantity and quality of the biomass of field crops. Production and utilization of agricultural and forest biomass for energy: Proceedings of a seminar held at Zvolen, Slovakia, 168–178.

Kondo, N., K. Imai, M. Isobe, T. Goto, A. Murasugi, C. Wada-Nakagawa, and Y. Hayashi. (1984). Cadystin A and B, major unit peptides comprising cadmium binding peptides induced in a fission yeast-separation, revision of structures and synthesis. *Tetrahedron Letters*, *25*, 3869–3872.

Kramer, U., I. N. Talke, and M. Hanikenne. (2007). Transition metal transport. *FEBS Letters*, *581*, 2263–2272.

Krupa, Z., and T. Baszynski. (1995). Some aspects of heavy metals toxicity towards photosynthetic apparatus: Direct and indirect effects on light and dark reactions. *Acta Physiolgia Plantarum*, *17*, 177–190.

Kumar, P. B. A. N., V. Dushenkov, H. Motto, and I. Raskin. (1995). Phytoextraction: The use of plants to remove heavy metals from soils. *Environmental Science and Technology*, *29*, 1232–1238.

Küpper, H., E. Lombi, F. J. Zhao, G. Wieshammer, and S. P. McGrath. (2001). Cellular compartmentation of nickel in the hyperaccumulators *Alyssum lesbiacum*, *Alyssum bertolonii* and *Thlaspi goesingense*. *Journal of Experimental Botany*, *52*, 2291–2300.

Kvesitadze, G., G. Khatisashvili, T. Sadunishvili, and J. J. Ramsden. (2006). The ecological importance of plants for contaminated environments. In G. I. Kvesitadze (Ed.), *Biochemical mechanisms of detoxification in higher plants: Basis of phytoremediation* (167–208). Berlin: Springer.

Lal, N. (2010). Molecular mechanisms and genetic basis of heavy metal toxicity and tolerance in plants. In M. Ashraf, M. Ozturk, M. S. A. Ahmad (Eds.), *Plant adaptation and phytoremediation* (35–58). Dordrecht: Springer.

Lamhamdi, M., A. Bakrima, A. Aarab, R. Lafont, and F. Sayah. (2011). Lead phytotoxicity on wheat (*Triticum aestivum* L.) seed germination and seedlings growth. *Comptes Rendus Biologies*, *334*, 118–126.

Lasat, M. M. (2002). Phytoextraction of toxic metals: A review of biological mechanisms. *Journal of Environmental Quality*, *31*, 109–120.

Leckie, J. O., and J. A. Davis. (1979). Aqueous environmental chemistry of copper. In J. O. Nriagu (Ed.), *Copper in the environment* (90–121). New York: Wiley.

Lee, C., H. H. Chang, S. B. Ha, B. Y. Moon, C. B. Lee, and N. Murata. (1992). Mercury induced light dependent alterations of chlorophyll fluorescence kinetics in barley leaf slices. In *Photosynthesis Research*, *4*, 623–626, Proceedings of the IXth International Congress on Photosynthesis.

Lee, K. C., B. A. Cunningham, G. M. Paulson, G. H. Liang, and R. B. Moore. (1976). Effects of cadmium on respiration rates and activities of several enzymes in soybean seedlings. *Plant Physiology*, *36*, 4–6.

Leštan, E. (2006). Ehnanced heavy metal phytoextraction. In M. Mackova, D. Dowling, and T. Macek (Eds.), *Phytoremediation and rhiziremediation: Theoretical background* (115–132). Dordrecht: Springer.

Li, Y. H. (2000). *A compendium of geochemistry*. Princeton: Princeton University Press.

Lide, D. (1992). *CRC handbook of chemistry and physics*, *73rd edn*. Boca Raton: CRC Press.

LifeExtension. (2011). Heavy metal toxicity. Retrieved from www.lef.org.

Lin, Y. C., and C. H. Kao. (2005). Nickel toxicity of rice seedlings: The inductive responses of antioxidant enzymes by $NiSO_4$ in rice roots. *Crop and Environmental Bioinformatics*, *2*, 239–244.

Liu, J., K. Li, J. Xu, Z. Zhang, T. Mac, X. Lu, J. Yang, and Q. Zhu. (2003). Lead toxicity, uptake, and translocation in different rice cultivars. *Plant Science*, *165*, 793–802.

Liu, J., J. Shen, D. Li, and J. Xu. (2010). Toxicity of lead on different rice genotypes. Proceedings of the 4th International Conference on Bioinformatics and Biomedical Engineering (iCBBE). DOI: 10.1109/ICBBE.2010.5516286.

Llamas, A., C. I. Ullrich, and A. Sanz. (2000). Cd^{2+} effects on transmembrane electrical potential difference, respiration and membrane permeability of rice (*Oryza sativa* L.) roots. *Plant and Soil, 219*, 21–28.

Lombi, E., F. J. Zhao, S. J. Dunham, and S. P. McGrath. (2001). Phytoremediation of heavy metal-contaminated soils: Natural hyperaccumulation versus chemically enhanced phytoextraction. *Journal of Environmental Quality, 30*, 1919–1926.

Lösch, R. (2004). Plant mitochondrial respiration under the influence of heavy metals. In M. N. V. Prasad (Ed.), *Heavy metal stress in plants: Biomolecules to ecosystem, 3rd edn.* (182–200). Berlin: Springer.

Lozano-Rodríguez, E., L. E. Hernández, P. Bonay, and R. O. Carpena-Ruiz. (1997). Distribution of cadmium in shoot and root tissues of maize and pea plants: Physiological disturbances. *Journal of Experimental Botany, 306*, 123–128.

Lund, L. J., A. L. Page, and C. O. Nelson. (1976). Movement of heavy metals below sewage disposal pond. *Journal of Environmental Quality, 5*, 330–334.

Lux, A., M. Martinka, P. J. Vaculík. (2010). White root responses to cadmium in the rhizosphere: A review. *Journal of Experimental Botany, 62*, 21–37.

Lytle, C. M., F. W. Lytle, N. Yang, Q. JinHong, D. Hansen, A. Zayed, and N. Terry. (1998). Reduction of Cr(VI) to Cr(III) by wetland plants: Potential for *in situ* heavy metal detoxification. *Environmental Science and Technology, 32*, 3087–3093.

Lyubenova, L., and P. Schröder. (2010). Uptake and effect of heavy metals on the plant detoxification cascade in the presence and absence of organic pollutants. In I. Sherameti, and A. Varma (Eds.), *Soil heavy metals, Soil Biology 9* (65–85). Berlin: Springer.

Ma, L. Q., K. M. Komar, C. Tu, W. Zhang, Y. Cai, E. D. Kenelley. (2001). A fern that hyperaccumulates arsenic. *Nature, 409*, 579.

Madhava Rao, K. V., and T. V. S. Sresty. (2000). Antioxidative parameters in the seedlings of pigeonpea (*Cajanus cajan* (L.) Millspaugh) in response to Zn and Ni stresses. *Plant Science, 157*, 113–128.

Mahmood, T., K. R. Islam, and S. Muhammad. (2007). Toxic effects of heavy metals on early growth and tolerance of cereal crops. *Pakistan Journal of Botany, 39*, 451–462.

Mahmud, R., N. Inouea, S. Y. Kasajima, and R. Shaheen. (2007). Effect of soil arsenic on yield and As and P distribution pattern among plant organs of buckwheat and castor oil plant. Proceedings of the 10th International Symposium on Buckwheat, 297–303.

Maitani, T., H. Kubota, K. Sato, and T. Yamada. (1996). The composition of metals bound to class III metallothionein (phytochelatin and its desglycyl peptide) induced by various metals in root cultures of *Rubia tinctorum. Plant Physiology, 110*, 1145–1150.

Malaisse, F., and J. Grégoire. (1978). Contribution á la phytogéochimie de la Mine de l'Étoile (Shaba, Zaïre). *Bulletin of Society of Royal Botany, Belgique, 111*, 252–260.

Malaisse, F., J. Grégoire. R. R. Brooks, R. S. Morrison, and R. D. Reeves. (1979). Copper and cobalt in vegetation of Fungurume, Shaba Province, Zaïre. *Oikos, 33*, 472–478.

Mami, Y., G. Ahmadi, M. Shahmoradi, and H. R. Ghorbani. (2011). Influence of different concentration of heavy metals on the seed germination and growth of tomato. *African Journal of Environmental Science and Technology, 5*, 420–426.

Mandal, K. M., and K. T. Suzuki. (2002). Arsenic around the world: A review. *Talanta, 58*, 241–235.

Manivasagaperumal, R., P. Vijayarengan, S. Balamurugan, and G. Thiyagarajan. (2011). Effect of copper on growth, dry matter yield and nutrient content of *Vigna radiata* (L.) Wilczek. *Journal of Phytology, 3*, 53–62.

Manta, D. S., M. Angelone, A. Bellanca, R. Neri, and M. Sprovieri. (2002). Heavy metals in urban soils: A case study from the city of Palermo (Sicily), Italy. *Science of the Total Environment, 300*, 229–243.

Marques, J. J., D. G. Schulze, N. Curi, and S. A. Mertzman. (2004). Trace element geochemistry in Brazilian Cerrado soils. *Geoderma, 121*, 31–43.

Marschner, H. (1995). *Mineral nutrition of higher plants*. Cambridge: Academic Press.

Matraszek, R., M. Szymańska, and M. Wróblewska. (2002). Effect of nickel on yielding and mineral composition of the selected vegetables. *Acta Scientiarum Polonorum-Hortorum Cultus, 1*, 13–22.

McCutcheon, S. C., and J. L. Schnoor. (2003). *Phytoremediation: Transformation and control of contaminants*. New Jersey: John Wiley.

McGrath, S. P. (1998). Phytoextraction for soil remediation. In R. R. Brooks (Ed.), *Plants that hyperaccumulate heavy metals* (109–128). New York: CAB International.

Mishra, S., and R. S. Dubey. (2005). Heavy metal toxicity induced alterations in photosynthetic metabolism in plants. In M. Pessarakli (Ed.), *Handbook of photosynthesis, 2nd edn*. (845–863). New York: Taylor and Francis/CRC Press.

Mishra, S., D. Panjwani, B. Mishra, and P. N. Sharma. (2010). Effect of excess nickel on induction of oxidative stress in *Zea mays* L. plants grown in solution culture. *International Journal of Toxicological and Pharmacological Research, 2*, 10–15.

Mo, S. C., D. S. Choi, and J. W. Robinson. (1989). Uptake of mercury from aqueous solution by duckweed: The effect of pH, copper, and humic acid. *Journal of Environmental Health, 24*, 135–146.

Mobin, M., and N. A. Khan. (2007). Photosynthetic activity, pigment composition and antioxidative response of two mustard (*Brassica juncea*) cultivars differing in photosynthetic capacity subjected to cadmium stress. *Journal of Plant Physiology, 164*, 601–610.

Mohanpuria, P., N. K. Rana, and S. K. Yadav. (2007). Cadmium induced oxidative stress influence on glutathione metabolic genes of *Camellia sinensis* (L.) O. Kuntze. *Environmental Toxicology, 22*, 368–374.

Mokgalaka-Matlala, N. S., E. Flores-Tavizön, H. Castillo-Michel, J. R. Peralta-Videa, and J. L. Gardea-Torresdey. (2008). Toxicity of arsenic (III) and (V) on plant growth, element uptake, and total amylolytic activity of mesquite (*Prosopis juliflora × P. velutina*). *International Journal of Phytoremediation, 10*, 47–60.

Molas, J. (2002). Changes of chloroplast ultra structure and total chlorophyll concentration in cabbage leaves caused by excess of organic Ni (II) conmplexes. *Environmental and Experimental Botany, 47*, 115–126.

Montillet, J.-L., J.-P. Agnel, M. Ponchet, F. Vailleau, D. Roby, and C. Trantaphylides. (2002). Lipoxygenase-mediated production of fatty acid hydroperoxides is a specific signature of the hypersensitive reaction in plants. *Plant Physiology and Biochemistry, 40*, 633–639.

Moral, R., I. Gomez, J. Navarro Pedreno, and J. Mataix. (1994). Effects of cadmium on nutrient distribution, yield, and growth of tomato grown in soilless culture. *Journal of Plant Nutrition, 17*, 953–962.

Morrison, R. S. (1980). Aspects of the accumulation of cobalt, copper and nickel by plants. (Ph.D. thesis). Massey University, New Zealand.

Morrison, R. S., R. R. Brooks, R. D. Reeves, and F. Malaisse. (1979). Copper and cobalt uptake by metallophytes from Zaïre. *Plant and Soil, 53*, 535–539.

Mueller, B., S. Rock, D. Gowswami, D. Ensley et al. (1999). Phytoremediation decision tree. Prepared by Interstate Technology and Regulatory Cooperation Work Group, 1–36.

Mumthas, S., A. A. Chidambaram, P. Sundaramoorthy, and K. S. Ganes. (2010). Effect of arsenic and manganese on root growth and cell division in root tip cells of green gram (*Vigna radiata* L.). *Emirats Journal of Food and Agriculture, 22*, 285–297.

Murasugi, A., C. Wada, and Y. Hayashi. (1981). Purification and unique properties in UV and CD spectra of Cd-binding peptides 1 from *Schizosachharomyyces pombe*. *Biochemical and Biophysical Research Communications, 103*, 1021–1028.

Mysliwa-Kurdziel, B., M. N. V. Prasad, and K. Stralka. (2004). Photosynthesis in heavy metal stress plants. In M. N. V. Prasad (Ed.), *Heavy metal stress in plants, 3rd edn*. (146–181). Berlin: Springer.

Namdjoyan, S. H., R. A. Khavari-Nejad, F. Bernard, T. Nejadsattari, and H. Shaker. (2011). Antioxidant defense mechanisms in response to cadmium treatments in two safflower cultivars. *Russian Journal of Plant Physiology, 58*, 467–477.

National Science Foundation. (1975). *Nickel*. Washington, DC: National Academy of Sciences.

Nooraniazad, H., M. R. Hajibagheri, D. Chobineh, and A. K. Ejraee. (2010). The study of $HgCl_2$ toxicity on the growth and some biochemical traits in Dill (*Anethum graveolens* L.) Crop. J. *Plant Science Research, 5*, 19–27.

Nriagu, J. O. (1979). Global inventory of natural and anthropogenic emissions of trace metals to the atmosphere. *Nature, 279*, 409–411.

Nriagu, J. O. (1980). Global cycle and properties of nickel. In J.O. Nriagu (Ed.), *Nickel in the environment* (1–26). New York: Wiley.

Nriagu, J. O. (1990). Global metal pollution: Poisoning the biosphere? *Environment, 32*, 7–33.

Nriagu, J. O., and J. M. Pacyna. (1988). Quantitative assessment of worldwide contamination of air, water, and soils by trace metals. *Nature, 33*, 134–139.

Nultsch, W. (2001). *Allgemeine Botanik*. Stuttgart: Georg Thieme Verlag.

Pacyna, J. M., and E. G. Pacyna. (2001). An assessment of global and regional emissions of trace metals to the atmosphere from anthropogenic sources worldwide. *Environmental Review, 9*, 269–298.

Panda, S. K., F. Baluska, and H. Matsumoto. (2009). Aluminium stress signaling in plants. *Plant Signaling and Behaviour, 4*, 592–597.

Panda, S. K., and H. Matsumoto. (2007). Molecular physiology of aluminum toxicity and tolerance in plants. *Botanical Review, 73*, 326–347.

Panda, S. K, and H. Matsumoto. (2010). Changes in antioxidant gene expression and induction of oxidative stress in pea (*Pisum sativum* L.) under Al stress. *Biometals*, *23*, 753–762.

Pandey, N., and C. P. Sharma. (2002). Effects of heavy metals Co^{2+}, Ni^{2+} and Cd^{2+} on growth and metabolism of cabbage. *Plant Science*, *163*, 753–758.

Papastergios, G., A. Georgakopoulos, J. L. Fernández–Turiel, D. Gimeno, A. Filippidis, A. Kassoli-Fournaraki, and A. Grigoriadou. (2004). Heavy metals and toxic trace elements contents in soils of selected areas of the Kavala prefecture, Northern Greece. *Bulletin of the Geological Society of Greece*, *XXXVI*, 263–272.

Parr, P. D., and F. G. Taylor Jr. (1982). Germination and growth effects of hexavalent chromium in Orocol TL (a corrosion inhibitor) on *Phaseolus vulgaris*. *Environment International*, *7*, 197–202.

Patra, M., and A. Sharma. (2000). Mercury toxicity in plants. *Botanical Review*, *66*, 379–422.

Peer, W. A., I. R. Baxter, E. L. Richards, J. L. Freeman, and A. S. Murphy. (2006). Phytoremediation and hyperaccumulator plants. In M. J. Tamas, and E. Martinoia (Eds.), *Molecular biology of metal homeostasis and detoxification: From microbe to man* (299–340). Berlin: Springer.

Peralta, J. R., J. L. G. Torresdey, K. J. Tiemann, E. Gomez, S. Arteaga, and E. Rascon. (2001). Uptake and effects of five heavy metals on seed germination and plant growth in alfalfa (*Medicago sativa* L.). *Bulletin of Environmental Contamination and Toxicology*, *66*, 727–734.

Pereira, L. B., C. M. de A Mazzanti, D. Cargnelutti, L. V. Rossato, J. F. Gonçalves, N. Calgaroto, V. Dressler, F. T. Nicoloso, L. C. Federizzi, V. M. Morsch, and M. R. C. Schetinger. (2011). Differential responses of oat genotypes: Oxidative stress provoked by aluminum. *Biometals*, *24*, 73–83.

Pereira, L. B., C. M. de A Mazzanti, J. F. Gonçalves, D. Cargnelutti, L. A. Tabaldi, A. G. Becker, N. S. Calgaroto, J. G. Farias, V. Battisti, D. Bohrer, F. T. Nicoloso, V. M. Morsch, and M. R. C. Schetinger. (2010). Aluminum-induced oxidative stress in cucumber. *Plant Physiology and Biochemistry*, *48*, 683–689.

Phang, I. C., D. W. M. Leung, H. H. Taylor, and D. J. Burritt. (2011). Correlation of growth inhibition with accumulation of Pb in cell wall and changes in response to oxidative stress in *Arabidopsis thaliana* seedlings. *Plant Growth Regulation*, *64*, 17–25.

Piechalak, A., B. Tomaszewska, D. Baralkiewicz, and A. Malecka. (2002). Accumulation and detoxification of lead ions in legumes. *Phytochemistry*, *60*, 153–162.

Pieper, D. H., V. A. P. Martins dos Santos, and P. N. Golyshin. (2004). Genomic and mechanistic insights into the biodegredation of organic pollutants. *Current Opinion in Biotechnology*, *15*, 215–224.

Porter, E. K., and P. J. Peterson. (1975). Arsenic accumulation by plants on mine waste (United Kingdom). *Science of the Total Environment*, *4*, 365–371.

Poschenrieder, C., and J. Barceló. (2004). Water relations in heavy metal stressed plants. In M. N. V. Prasad (Ed.), *Heavy metal stress in plants, 3rd edn.* (249–270). Berlin: Springer.

Poschenrieder, C., B. Gunse, I. Corrales, and J. Barceló. (2008). A glance into aluminum toxicity and resistance in plants. *Science of the Total Environment*, *400*, 356–365.

Prasad, M. N. V. (2004). Phytoremediation of metals and radionuclides in the environment: The case for natural hyperaccumulators, metal transporters, soil-amending chelators and transgenic plants. In M. N. V. Prasad (Ed.), *Heavy metal stress in plants: From molecules to ecosystems* (345–391). Berlin: Springer.

Prasad, M. N. V., and H. Freitas. (2003). Metal hyperaccumulation in plants: Biodiversity prospecting for phytoremediation technology. *Electronic Journal of Biotechnology*, *6*, 275–321.

Prasad, M. N. V., M. Greger, and T. Landberg. (2001). *Acacia nilotica* L. bark removes toxic elements from solution: Corroboration from toxicity bioassay using *Salix viminalis* L. in hydroponic system. *International Journal of Phytoremediation*, *3*, 289–300.

Purakayastha, T. J., and P. K. Chhonkar. (2010). Phytoremediation of heavy metal contaminated soils. In I. Sherameti, and A. Varma (Eds.), *Soil heavy metals, Soil Biology*, *19* (389–429). Berlin: Springer.

Purakayastha, T. J., V. Thulasi, S. Bhadraray, P. K. Chhonkar, P. P. Adhikari, and K. Suribabu. (2008). Phytoextraction of zinc, copper, nickel and lead from a contaminated soil by different species of *Brassica*. *International Journal of Phytoremediation*, *10*, 63–74.

Puschenreiter, M., G. Stoger, E. Lombi, O. Horak, and W. W. Wenzel. (2001). Phytoextraction of heavy metal contaminated soils with *Thlaspi goesingense* and *Amaranthus hybridus*: Rhizosphere manipulation using EDTA and ammonium sulfate. *Journal of Plant Nutrition and Soil Science*, *164*, 615–621.

Qiu, R., X. Fang, Y. Tang, S. Du, X. Zeng, and E. Brewer. (2006). Zinc hyperaccumulation and uptake by *Potentilla griffithii* Hook. *International Journal of Phytoremediation*, *8*, 299–310.

Rabbani, G. H., A. K. Chowdhury, S. K. Shaha, and M. Nasir. (2002). Mass arsenic poisoning of ground water in Bangladesh. Proceedings of the Global Health Council Annual Conference, Washington, DC, 2002.

Rabenstein, D. L. (1989). Metal complexes of glutathione and their biological significance. In D. Dolphin, R. Poulson, and O. Avramovic (Eds.), *Glutathione: Chemical, biochemical and medical aspects* (147–186). New York: Wiley.

Rahman, M. A., H. Hasegawa, M. M. Rahman, M. N. Islam, M. A. Majid Miah, and A. Tasmen. (2007). Effect of arsenic on photosynthesis, growth and yield of five widely cultivated rice (*Oryza sativa* L.) varieties in Bangladesh. *Chemosphere, 67*, 1072–1079.

Rahman, M. H., M. M. Rahman, C. Watanabe, and K. Yamamoto. (2002). Arsenic contamination of groundwater in Bangladesh and its remedial measures, Arsenic Contamination in Groundwater: Technical and Policy Dimensions. Proceedings of the UNU-NIES International Workshop. Tokyo: United Nations University (UNU) and Japan National Institute for Environmental Studies (NIES).

Rai, V., S. Khatoon, S. S. Bisht, and S. Mehrotra. (2005). Effect of cadmium on growth, ultramorphology of leaf and secondary metabolites of *Phyllanthus amarus* Schum. and Thonn. *Chemosphere, 61*, 1644–1650.

Rai, V., P. Vajpayee, S. N. Singh, and S. Mehrotra. (2004). Effect of chromium accumulation on photosynthetic pigments, oxidative stress defense system, nitrate reduction, proline level and eugenol content of *Ocimum tenuiflorum* L. *Plant Science, 167*, 1159–1169.

Rao, K. V. M., and T. V. S. Sresty. (2004). Antioxidative parameters in the seedlings of pigeon pea (*Cajanus cajan* (L.) Millspaugh) in response to Zn and Ni stresses. *Plant Science, 157*, 113–118.

Rascio, N. (1977). Metal accumulation by some plants growing on zinc-mine deposits. *Oikos, 29*, 250–253.

Rascio, N., and F. Navari-Izzo. (2011). Heavy metal hyperaccumulating plants: How and why do they do it? And what makes them so interesting? *Plant Science, 180*, 169–181.

Raskin, I., and B. D. Ensley. (2000). *Phytoremediation of toxic metals: Using plants to clean up the environment*. New York: Wiley.

Raskin, I. P., B. A. N. Kumar, S. Dushenkov, M. J. Blaylock, and D. Salt. (1994). Phytoremediation: Using plants to clean up soils and waters contaminated with toxic metals. Emerging Technologies in Hazardous Waste Management VI, ACS Industrial and Engineering Chemistry Division Special Symp. Vol. 1, Atlanta.

Raskin, I., R. D. Smith, and D. E. Salt. (1997). Phytoremediation of metals: Using plants to remove pollutants from the environment. *Current Opinion in Biotechnology, 8*, 221–226.

Rauser, W. E. (1999). Structure and function of metal chelators produced by plants: The case for organic acids, amino acids, phytin, and metallothioneins. *Cell Biochemistry and Biophysics, 31*, 19–48.

Reeves, R. D. (2006). Hyperaccumulation of trace elements by plants. In J. L. Morel, G. Echevarria, and N. Goncharova (Eds.), *Phytoremediation of metal-contaminated soils* (25–52). Dordrecht: Springer.

Reeves, R. D., A. J. M. Baker, A. Borhidi, and R. Berazaín. (1996). Nickel-accumulating plants from the ancient serpentine soils of Cuba. *New Phytologist, 133*, 217–224.

Reeves, R. D., and R. R. Brooks. (1983). European species of *Thlaspi* L. (Cruciferae) as indicators of nickel and zinc. *Journal of Geochemical Explorations, 18*, 275–283.

Requejo, R., and M. Tena. (2005). Proteome analysis of maize roots reveals that oxidative stress is a main contributing factor to plant arsenic toxicity. *Phytochemistry, 66*, 1519–1528.

Robinson, B. H., R. R. Brooks, A. W. Howes, J. H. Kirkman, and P. E. H. Gregg. (1997). The potential of the high biomass nickel hyperaccumulator *Berkheya coddii* for phytoremediation and phytomining. *Journal of Geochemical Explorations, 60*, 115–126.

Romero-Puertas, M. C., M. Rodríguez-Serrano, F. J. Corpas, M. Gomez, L. A. del Río, and L. M. Sandalio. (2004). Cadmium-induced subcellular accumulation of $O_2^{\cdot -}$ and H_2O_2 in pea leaves. *Plant Cell and Environment, 27*, 1122–1134.

Romheld, V., and H. Marschner. (1986). Mobilization of iron in the rhizosphere of different plant species. *Advances in Plant Nutrition, 2*, 155–204.

Rout, G. R., S. Samantaray, and P. Das. (1997). Differential chromium tolerance among eight mung bean cultivars grown in nutrient culture. *Journal of Plant Nutrition, 20*, 473–483.

Rout, G. R., S. Sanghamitra, and P. Das. (2000). Effects of chromium and nickel on germination and growth in tolerant and non-tolerant populations of *Echinochloa colona* (L). *Chemosphere, 40*, 855–859.

Rüegsegger, A., and C. Brunold. (1992). Effect of cadmium on y-glutamylcysteine synthesis in maize seedlings. *Plant Physiology, 99*, 428–433.

Rüegsegger, A., D. Schmutz, and C. Brunold. (1990). Regulation of glutathione synthesis by cadmium in *Pisum satirum* L. *Plant Physiology, 93*, 1579–1584.

Sadowsky, M. J. (1999). Phytoremediation: Past promises and future practices. In C. R. Bell, M. Brylinsky, and P. Johnson-Green (Eds.), *Microbial Biosystems: New Frontiers*. Proceedings of the 8th International Symposium on Microbial Ecology. Halifax, Canada: Atlantic Canada Society for Microbial Ecology.

Salt, D. E., M. Blaylock, P. B. A. N. Kumar, V. Dushenkov, B. D. Ensley, I. Chet, and I. Raskin. (1995). Phytoremediation: A novel strategy for the removal of toxic metals from the environment using plants. *Biotechnology, 13*, 468–475.

Salt, D. E., I. J. Pickering, R. C. Prince, D. Gleba, S. Dushenkov, R. D. Smith, and I. Raskin. (1997). Metal accumulation by aquacultured seedlings of Indian mustard. *Environmental Science and Technology, 31*, 1636–1644.

Salt, D. E., R. D. Smith, and I. Raskin. (1998). Phytoremediation. *Annual Review of Plant Physiology and Plant Molecular Biology, 49*, 643–668.

Sandalio, L. M., H. C. Dalurzo, M. Gómez, M. C. Romero-Puertas, and L. A. del Río. (2001). Cadmium-induced changes in the growth and oxidative metabolism of pea plants. *Journal of Experimental Botany, 52*, 2115–2126.

Sandalio, L. M., M. Rodríguez-Serrano, L. A. del Río, and M. C. Romero-Puertas. (2009). Reactive oxygen species and signaling in cadmium toxicity. In L. A. del Río, and A. Puppo (Eds.), *Signaling and communication in plants* (175–189). Berlin-Heidelberg: Springer.

Saraswat, S., and J. P. N. Rai. (2009). Phytoextraction potential of six plant species grown in multimetal contaminated soil. *Chemistry and Ecology, 25*, 1–11.

Sarma, H. (2011). Metal hyperaccumulation in plants: A review focusing on phytoremediation technology. *Journal of Environmental Science and Technology, 4*, 118–138.

Satofuka, H., T. Fukui, M. Takagi, H. Atomi, T. Imanaka. (2001). Metal-binding properties of phytochelatin-related peptides. *Journal of Inorganic Biochemistry, 86*, 595–602.

Saxena, P., and N. Misra. (2010). Remediation of heavy metal contaminated tropical land. In I. Sherameti, and A. Varma (Eds.), *Soil heavy metals, Soil Biology, 19* (430–477). Dordrecht: Springer.

Schirado, T., I. Vergara, E. B. Schalscha, and P. F. Pratt. (1986). Evidence for movement of heavy metals in a soil irrigated with untreated wastewater. *Journal of Environmental Quality, 15*, 9–12.

Schulte, A., A. Balazs, J. Block, and J. Gehrmann. (1996). Entwicklung der Niederschlags-Deposition von Schwermetallen in West-Deutschland. 1. Blei und Cadmium. *Z Pflanzenern Bodenk, 159*, 377–383.

Schultz, R. (1987). Vergleichende Betrachtung des Schwermetallhaushalts verschiedener Waldökosysteme Norddeutschlands. *Berichte des Forschungszentrums Waldökosysteme/ Waldsterben, Reihe A, Bd, 32*, 217S.

Schützendübel, A., and A. Polle. (2002). Plant responses to abiotic stresses: Heavy metal induced oxidative stress and protection by mycorrhization. *Journal of Experimental Botany, 53*, 1351–1365.

Scoccianti, V., R. Crinelli, B. Tirillini, V. Mancinelli, and A. Speranza. (2006). Uptake and toxicity of Cr(III) in celery seedlings. *Chemosphere, 64*, 1695–1703.

Seidal, K., N. Jorgensen, C. G. Elinder, B. Sjogren, and M. Vahter. (1993). Fatal cadmium-induced pneumonitis. *Scandinavian Journal of Work, Environment and Health, 19*, 429–431.

Selvam, A., and J. W. Wong. (2008). Phytochelatin systhesis and cadmium uptake of *Brassica napus*. *Environmental Technology, 29*, 765–773.

Setia, R. C., K. Navjyot, S. Neelam, and N. Harsh. (2008). Heavy metal toxicity in plants and phytoremediation. In R. C. Setia, H. Nayyar, and N. Setia (Eds.), *Crop improvement: Strategies and applications* (206–218). New Delhi: I.K. International.

Shah, F. U. R., N. Ahmad, K. R. Masood, J. R. Peralta-Videa, and F. U. D. Ahmad. (2010). Heavy metal toxicity in plants. In M. Ashraf, M. Ozturk, and M. S. A. Ahmad (Eds.), *Plant adaptation and phytoremediation* (71–97). Dordrecht: Springer.

Shah, K., and R. S. Dubey. (1995). Effect of cadmium on RNA level as well as activity and molecular forms of ribonuclease in growing rice seedlings. *Plant Physiology and Biochemistry, 33*, 577–584.

Shahandeh, H., and L. R. Hossner. (2000). Plant screening for chromium phytoremediation. *International Journal of Phytoremediation, 2*, 31–51.

Shaibur, M. R., N. Kitajima, R. Sugewara, T. Kondo, S. Alam, S. M. I. Huq, and S. Kawai. (2008). Critical toxicity of arsenic and elemental composition of arsenic-induced chlorosis in hydroponic sorghum. *Water, Air and Soil Pollution, 191*, 279–292.

Shanker, A. K. (2003). Physiological, biochemical and molecular aspects of chromium toxicity and tolerance in selected crops and tree species. (Ph.D. Thesis). Tamil Nadu Agricultural University, Coimbatore, India.

Shanker, A. K., C. Cervantes, H. Loza-Tavera, and S. Avudainayagam. (2005). Chromium toxicity in plants. *Environment International, 31*, 739–753.

Sharma, D. C., C. Chatterjee, and C. P. Sharma. (1995). Chromium accumulation and its effect on wheat (*Triticum aestivum* L. cv. Dh 2204) metabolism. *Plant Science, 111*, 145–151.

Sharma, S. S., and K. J. Dietz. (2008). The relationship between metal toxicity and cellular redox imbalance. *Trends in Plant Science, 14*, 43–50.

Sharma, P., and R. S. Dubey. (2007). Involvement of oxidative stress and role of antioxidative defense system in growing rice seedlings exposed to toxic levels of aluminium. *Plant Cell Reports, 26*, 2027–2038.

Sharma, D. C., and C. P. Sharma. (1993). Chromium uptake and its effects on growth and biological yield of wheat. *Cereal Research Communications*, *21*, 317–321.

Sharma, D. M., C. P. Sharma, and R. D. Tripathi. (2003). Phytotoxic lesions of chromium in maize. *Chemosphere*, *51*, 63–68.

Shen, Z. G., X. D. Li, C. C. Wang, H. M. Chen, and H. Chua. (2002). Lead phytoextraction from contaminated soil with high-biomass plant species. *Journal of Environmental Quality*, *31*, 1893–1900.

Shi, G. T., Z. L. Chen, S. Y. Xu, J. Zhang, L. Wang, C. J. Bi, and J. Y. Teng. (2008). Potentially toxic metal contamination of urban soils and roadside dust in Shanghai, China. *Environmental Pollution*, *156*, 251–260.

Shri, M., S. Kumar, D. Chakrabarty, P. K. Trivedi, S. Mallick, P. Misra, D. Shukla, S. Mishra, S. Srivastava, R. D. Tripathi, and R. Tuli. (2009). Effect of arsenic on growth, oxidative stress, and antioxidant system in rice seedlings. *Ecotoxicology and Environmental Safety*, *72*, 1102–1110.

Siddhu, G., D. S. Sirohi, K. Kashyap, I. A. Khan, and M. A. Khan. (2008). Toxicity of cadmium on the growth and yield of *Solanum melongena* L. *Journal of Environmental Biology*, *29*, 853–857.

Šimonova, E., M. Henselová, E. Masarovičová, and J. Kohanová. (2007). Comparison of tolerance of *Brassica juncia* and *Vigna radiata* to cadmium. *Biologia Plantarum*, *51*, 488–492.

Singh, K., and J. S. Brar. (2002). Genotypic differences in effect of cadmium on yield and nutrient composition in *Brassica* plants. Proceedings of the 7th WCSS Conference, Thailand. Symposium No. 42, Paper 218 (1–7).

Singh, S., N. A. Khan, R. Nazar, and N. A. Anjum. (2008). Photosynthetic traits and activities of antioxidant enzymes in blackgram (*Vigna mungo* L. Hepper) under cadmium stress. *American Journal of Plant Physiology*, *3*, 25–32.

Singh, K., and S. N. Pandey. (2011). Effect of nickel-stresses on uptake, pigments and antioxidative responses of water lettuce, *Pistia stratiotes* L. *Journal of Environmental Biology*, *32*, 391–394.

Smith, R. A. H., and A. D. Bradshaw. (1992). Stabilization of toxic mine wastes by the use of tolerant plant populations. *Transactions of the American Institute of Mining and Metallurgical Engineers*, *81*, A230–A233.

Smith, E., R. Naidu, and A. M. Alston. (1998). Arsenic in the soil environment: A review. *Advances in Agronomy*, *64*, 149–195.

Somashekaraiah, B. V., K. Padmaja, and A. R. K. Prasad. (1992). Phytotoxicity of cadmium ions on germinating seedlings of mung bean (*Phaseolus vulgaris*): Involvement of lipid peroxides in chlorophyll degradation. *Physiologia Plantarum*, *85*, 85–89.

Song, A., P. Li, Z. Li, F. Fan, M. Nikolic, and Y. Liang. (2011). The alleviation of zinc toxicity by silicon is related to zinc transport and antioxidative reactions in rice. *Plant and Soil*, *344*, 319–333.

Sriprang, R., and Y. Murooka. (2007). Accumulation and detoxification of metals by plants and microbes. In S. N. Singh, and R. D. Tripathi (Eds.), *Environmental bioremediation technologies* (77–100). Berlin: Springer.

Srivastava, S., A. K. Srivastava, P. Suprasanna, and S. F. D'Souza. (2009). Comparative biochemical and transcriptional profiling of two contrasting varieties of *Brassica juncea* L. in response to arsenic exposure reveals mechanisms of stress perception and tolerance. *Journal of Experimental Botany*, *181*, 1–13.

Srivastava, R. K., R. P. Tiwari, and P. Bala Ramudu. (2007). Electrokinetic remediation study for cadmium contaminated soil. *Iranian Journal of Environmental Health Science and Engineering*, *4*, 207–214.

Steenland, K., and P. Boffetta. (2000). Lead and cancer in humans: Where are we now? *American Journal of Industrial Medicine*, *38*, 295–299.

Stephan, U. W., and G. Scholz. (1993). Nicotinamin: Mediator of transport of iron and heavy metals in phloem? *Physiologia Plantarum*, *88*, 522–529.

Sugimoto, M., and W. Sakamoto. (1997). Putative phopholipid hydroperoxide glutathione peroxidase gene from *Arabidopsis thaliana* induced by oxidative stress. *Genes, Genetic Systems*, *72*, 311–316.

Sun, T. G., W. Sha, and J. Jie. (2010a). Effects of Zn stress on physiological characteristics of melon seedlings. *Northern Horticulture*, *16*, 51–53.

Sun, Y., Q. Zhou, X. Xie, and R. Liu. (2010b). Spatial, sources and risk assessment of heavy metal contamination of urban soils in typical regions of Shenyang, China. *Journal of Hazardous Materials*, *174*, 455–462.

Suszcynsky, E. M., and J. R. Shann. (1995). Phytotoxicity and accumulation of mercury subjected to different exposure routes. *Environmental and Toxicological Chemistry*, *14*, 61–67.

Suthersan, S. S. (2002). *Natural and enhanced remediation system*. Boca Raton: Lewis Publishers.

Takeda, A., K. Kimura, and S. I. Yamasaki. (2004). Analysis of 57 elements in Japanese soils, with special reference to soil group and agricultural use. *Geoderma*, *119*, 291–307.

Tamaki, S., and W. T. Frankenberger. (1992). Environmental biochemistry of arsenic. *Review of Environmental Contamination and Toxicology*, *124*, 79–110.

Teklić, T., J. T. Hancock, M. Engler, N. Paradiković, V. Cesar, H. Lepeduš, I. Štolfa, and D. Bešlo. (2008). Antioxidative responses in radish (*Raphanus sativus* L.) plants stressed by copper and lead in nutrient solution and soil. *Acta Biologica Cracoviensia Series Botany*, *50*, 79–86.

Tewari, R. K., P. Kumar, and P. N. Sharma. (2006). Antioxidant responses to enhanced generation of superoxide anion radical and hydrogen peroxide in the copper-stressed mulberry plants. *Planta, 223*, 1145–1153.

Thumann, J., E. Grill, E. L. Winnacker, and M. H. Zenk. (1991). Reactivation of metal requiring apoenzymes by phytochelatin–metal complexes. *FEBS Letters, 284*, 66–69.

Tiller, K. G. (1992). Urban soil contamination in Australia. *Australian Journal of Soil Research, 30*, 937–957.

Tlustoš, P., D. Pavlíková, J. Száková, and J. Balík. (2006). Plant accumulation capacity for potentially toxic elements. In J. L. Morel, G. Echevarria, and N. Goncharova (Eds.), *Phytoremediation of metal-contaminated soils* (53–84). Dordrecht: Springer.

Tokalioglu, S., and S. Kartal. (2006). Statistical evaluation of the bioavailability of heavy metals from contaminated soil to vegetables. *Bulletin of Environmental Contamination and Toxicology, 76*, 311–319.

Tripathi, A. K., and S. Tripathi. (1999). Changes in some physiological and biochemical characters in *Albizia lebbek* as bio-indicators of heavy metal toxicity. *Journal of Environmental Biology, 20*, 93–98.

Ullah, S. M. (1998). Arsenic contamination of groundwater and irrigated soils in Bangladesh. Proceedings of the International Conference on Arsenic Pollution on Groundwater in Bangladesh: Causes, Effects and Remedies, Dhaka, Bangladesh.

United Nations. (2002). Johannesburg Summit, Report of the World Summit on Sustainable Development, Johannesburg, South Africa. Retrieved from http://www.johannesburgsummit.org/html/documents/documents.html.

Upadhyay, R., and S. K. Panda. (2010). Influence of chromium salts on increased lipid peroxidation and differential pattern in antioxidant metabolism in *Pistia stratiotes* L. *Brazilian Archives of Biology and Technology, 53*, 1137–1144.

USEPA. (2000). Introduction to phytoremediation. National Risk Management Research Laboratory, Office of Research and Development, EPA/600/R-99/107.

Vanaja, M., N. V. N. Charyulu, and K. V. N. Rao. (2002). Effect of some organic acids and calcium on growth, toxic effect and accumulation of cadmium in *Stigeoclonium tenue* Kutz. *Indian Journal of Plant Physiology, 7*(2), 163–167.

Vassilev, A., J. P. Schwitzguebel, T. Thewys, D. van der Lelie, and J. Vangronsveld. (2004). The use of plants for remediation of metal contaminated soils. *Scientific World Journal, 4*, 9–34.

Vecchia, F. C., N. L. Rocca, I. Moro, S. Faveri, C. Andreoli, and N. Rascio. (2005). Morphogenetic, ultrastructural and physiological damages suffered by submerged leaves of *Elodea canadensis* exposed to cadmium. *Plant Science, 168*, 329–338.

Verbruggen, N., C. Hermans, and H. Schat. (2009). Mechanisms to cope with arsenic or cadmium excess in plants. *Current Opinion in Plant Biology, 12*, 364–372.

Verma, S., and R. S. Dubey. (2003). Lead toxicity induces lipid peroxidation and alters the activities of antioxidant enzymes in growing rice plants. *Plant Science, 164*, 645–655.

Videa-Peralta, J. R., and J. Ramon. (2002). Feasibility of using living alfalfa plants in the phytoextraction of cadmium(II), chromium(VI), copper(II), nickel(II), and zinc(II): Agar and soil studies. (Ph.D. thesis). The University of Texas, El Paso, AAT 3049704.

Vierstra, R. D. (1987). Ubiquitin, a key component in the degradation of plant proteins. *Physiologia Plantarum, 70*, 103–106.

Vinit-Dunand, F., D. Epron, B. Alaoui-Sossé, and P. M. Badot. (2002). Effects of copper on growth and on photosynthesis of mature and expanding leaves in cucumber plants. *Plant Science, 163*, 53–58.

Walton, B. T., E. A. Guthrie, and A. M. Hoylmann. (1994). Toxicant degradation in the rhizosphere. In T. A. Anderson, and J. R. Coats (Eds.), *Bioremediation through rhizosphere technology* (11–26). Washington, DC: American Chemical Society.

Wang, H. H., J. Kang, F. H. Zeng, and M. Y. Jiang. (2001). Effect of nickel at high concentrations on growth activities of enzymes of rice seedlings. *Acta Agronomica Sinica, 27*, 953–957.

Wang. H. B., Z. H. Ye, W. S. Shu, W. C. Li, M. H. Wong, and C. Y. Lan. (2006). Arsenic uptake and accumulation in fern species growing at arsenic-contaminated sites of southern China: Field surveys. *International Journal of Phytoremediation, 8*, 1–11.

Wang, J., F. Zhao, A. A. Meharg, A. Raab, J. Feldmann, and P. S. McGrath. (2002). Mechanisms of arsenic hyperaccumulation in *Pteris vittata*: Uptake kinetics, interactions with phosphate, and arsenic speciation. *Plant Physiology, 130*, 1552–1561.

Wani, P. A., M. S. Khan, and A. Zaidi. (2008). Effects of heavy metal toxicity on growth, symbiosis, seed yield and metal uptake in pea grown in metal amended soil. *Bulletin of Environmental Contamination and Toxicology, 81*, 152–158.

WebElements. (2002). Retrieved from http://www.webelements.com/copper.

Weber, O., R. W. Scholz, R. Bühlmann, and D. Grasmück. (2001). Risk perception of heavy metal soil contamination and attitudes toward decontamination strategies. *Risk Analysis, 21*, 967–977.

Wei, L., C. Luo, X. Li, and Z. Shen. (2008). Copper accumulation and tolerance in *Chrysanthemum coronarium* L. and *Sorghum sudanense* L. *Archives of Environmental Contamination and Toxicology, 55*, 238–246.

Wei, S., and Q. Zhou. (2008). Trace elements in agro-ecosystem. In M. N. V. Prasad (Ed.), *Trace elements as contaminants and nutrients: Consequences in ecosystems and human health* (55–79). Hoboken: Wiley.

Weiss, B., T. W. Clarkson, and W. Simon. (2002). Silent latency periods in methylmercury poisoning and in neurodegenerative disease. *Environmental Health Perspective, 110*, 851–854.

White, P. J., and P. H. Brown. (2010). Plant nutrition for sustainable development and global health. *Annals of Botany, 105*, 1073–1080.

WHO. (1992). Cadmium. *Environmental Health Criteria, Vol. 134*. Geneva: World Health Organization.

Wild, H. (1970). The vegetation of nickel-bearing soils. *Kirkia, 7*, 1–62.

Willett, I. R., P. Jakobsen, K. W. J. Malafant, and W. J. Bond. (1984). Effect of land disposal of lime treated sewage sludge on soil properties and plant growth. Division of water and land resources, CSIRO, Canberra, Div. Rep. 84/3.

Williams, S. T., T. McNeilly, and E. M. H. Wellington. (1977). The decomposition of vegetation growing on metal mine waste. *Soil Biology and Biochemistry, 9*, 271–275.

Williams, L. E., J. K. Pittman, and J. L. Hall. (2000). Emerging mechanisms for heavy metal transport in plants. *Biochimica Biophysica Acta, 1465*, 104–126.

Wilson, L. (2011). Toxic metals and human health. The Center for Development. Retrieved from http://www.drlwilson.com/articles/TOXIC%20METALS.htm.

Wójcik, M., and A. Tukiendorf. (2004). Phytochelatin synthesis and cadmium localization in wild type of *Arabidopsis thaliana. Plant Growth Regulation, 44*, 71–80.

Woolson, E. A. (1977). Generation of alkylarsines from soil. *Weed Science, 25*, 412–416.

Wright, D. A., and P. Welbourn. (2002). *Environmental toxicology*. Cambridge/New York: Cambridge University Press.

Wu, F. B., G. P. Zhang, and P. Dominy. (2003). Four barley genotypes respond differently to cadmium: Lipid peroxidation and activities of antioxidant capacity. *Environmental and Experimental Botany, 50*, 67–77.

Xia, X., X. Chen, R. Liu, and H. Liu. (2011). Heavy metals in urban soils with various types of land use in Beijing, China. *Journal of Hazardous Materials, 186*, 2043–2050.

Xie, Z. M., and S. M. Lu. (2000). Trace elements and environmental quality. In Q. L. Wu (Ed.), *Micronutrients and biohealth* (208–216). Guiyan: Guizhou Science and Technology Press.

Xu, F. J., C. W. Jin, W. J. Liu, Y. S. Zhang, and X. Y. Lin. (2011). Pretreatment with H_2O_2 alleviates aluminum-induced oxidative stress in wheat seedlings. *Journal of Integrative Plant Biology, 53*, 44–53.

Xu, W. H., W. Y. Li, J. He, B. Singh, and Z. Xiong. (2009). Effects of insoluble Zn, Cd, and EDTA on the growth, activities of antioxidant enzymes and uptake of Zn and Cd in *Vetiveria zizanioides. Journal of Environmental Science, 21*, 186–192.

Xu, B., X. Yang, Z. Gu, Y. Zhang, Y. Chen, and Y. Lv. (2009). The trend and extent of heavy metal accumulation over last one hundred years in the Liaodong Bay, China. *Chemosphere, 75*, 442–446.

Yadav, S. K. (2010). Heavy metals toxicity in plants: An overview on the role of glutathione and phytochelatins in heavy metal stress tolerance of plants. *South African Journal of Botany, 76*, 167–179.

Yamamoto, Y., Y. Koayashi, and H. Matsumoto. (2001). Lipid peroxidation is an early symptom triggered by aluminum, but not the primary cause of elongation inhibition in pea roots. *Plant Physiology, 125*, 199–208.

Yan, R., S. Gao, W. Yang, M. Cao, S. Wang, and F. Chen. (2008). Nickel toxicity induced antioxidant enzyme and phenylalanine ammonia-lyase activities in *Jatropha curcas* L. cotyledons. *Plant, Soil and Environment, 54*, 294–300.

Yan, S. L., C. C. Tsay, and Y. R. Chen. (2000). Isolation and characterization of phytochelatin synthase in rice seedlings. *Proceedings of National Science Council, Republic of China (B), 24*, 202–207.

Yang, X. E., X. X. Long, H. B. Ye, Z. L. He, D. V. Calvert, and P. J. Stoffella. (2004). Cadmium tolerance and hyperaccumulation in a new Zn hyperaccumulating plant species (*Sedum alfredii* Hance). *Plant and Soil, 259*, 181–189.

Yang, Y., C. Sun, Y. Yao, Y. Zhang, and V. Achal. (2011). Growth and physiological responses of grape (*Vitis vinifera* "Combier") to excess zinc. *Acta Physiolgiae Plantarum, 33*, 1483–1491.

Yang, Y. A., and X. E. Yang. (2000). Micronutrients in sustainable agriculture. In Q. L. Wu (Ed.), *Micronutrients and biohealth* (120–134). Guiyan: Guizhou Science and Technology Press.

Yildiz, N. (2005). Response of tomato and corn plants to increasing Cd levels in nutrient culture. *Pakistan Journal of Botany, 37*, 593–599.

Yruela, I. (2009). Copper in plants: Acquisition, transport and interactions. *Functional Plant Biology*, *36*, 409–430.

Yusuf, M., Q. Fariduddin, S. Hayat, and A. Ahmad. (2011). Nickel: An overview of uptake, essentiality and toxicity in plants. *Bulletin of Environmental Contamination and Toxicology*, *86*, 1–17.

Zeid, I. M. (2001). Responses of *Phaseolus vulgaris* to chromium and cobalt treatments. *Biologia Plantarum*, *44*, 111–115.

Zeng, F., S. Ali, B. Qiu, F. Wu, and G. Zhang. (2010). Effects of chromium stress on the subcellular distribution and chemical form of Ca, Mg, Fe, and Zn in two rice genotypes. *Journal of Plant Nutrition and Soil Science*, *173*, 135–148.

Zeng, X., L. Q. Ma, R. Qiu, and Y. Tang. (2009). Responses of non-protein thiols to Cd exposure in Cd hyperaccumulator *Arabis paniculata* Franch. *Environmental and Experimental Botany*, *66*, 242–248.

Zenk, M. H. (1996). Heavy metal detoxification in higher plants: A review. *Gene*, *179*, 21–30.

Zhang, W., Y. Cai, K. R. Downum, and L. Q. Ma. (2004). Arsenic complexes in the arsenic hyperaccumlator *Pteris vittata* (Chinese Brake fern). *Journal of Chromatography A*, *1043*, 249–254.

Zhang, G., M. Fukami, and H. Sekimoto. (2002). Influence of cadmium on mineral concentrations and yield components in wheat genotypes differing in Cd tolerance at seedling stage. *Field Crops Research*, *77*, 93–98.

Zhang, G. J., H. Jiang, Q. L. Zheng, J. Chen, D. L. Qiu, and X. H. Liu. (2009). Effect of copper stress on photosynthesis of navel orange seedlings. *Chinese Journal of Eco-Agriculture*, *17*, 130–134.

Zhou, Z. S., K. Guo, A. A. Elbaz, and Z. M. Yang. (2009). Salicylic acid alleviates mercury toxicity by preventing oxidative stress in roots of *Medicago sativa*. *Environmental and Experimental Botany*, *65*, 27–34.

Zhou, Z. S., S. Q. Huang, K. Guo, S. K. Mehta, P. C. Zhang, and Z. M. Yang. (2007). Metabolic adaptations to mercury-induced oxidative stress in roots of *Medicago sativa* L. *Journal of Inorganic Biochemistry*, *101*, 1–9.

Zhu, Y. G., P. Christie, and A. S. Laidlow. (2001). Uptake of Zn by arbuscular mycorrhizal white clover from Zn-contaminated soil. *Chemosphere*, *42*, 193–199.

Zhu, Y. L., A. M. Zayed, J. H. Quian, M. De Souza, and N. Terry. (1999). Phytoaccumulation of trace elements by wetland plants: II. Water hyacinth. *Journal of Environmental Quality*, *28*, 339–344.

Zurayk, R., B. Sukkariyah, and R. Baalbaki. (2001). Common hydrophytes as bioindicators of nickel, chromium and cadmium pollution. *Water Air and Soil Pollution*, *127*, 373–388.

3 Phytotechnology—Remediation of Inorganic Contaminants

Felix A. Aisien, Innocent O. Oboh, and Eki T. Aisien

CONTENTS

3.1 INTRODUCTION

Phytotechnologies are emerging innovative technologies that use plants for effective treatment of a wide variety of contaminants in surface water, groundwater, soil and sediments, sludges, and air through contaminant removal, degradation, or containment. Phytotechnologies have been reported successful in laboratory, bench-scale, and full-scale projects involving a variety of contaminants (Interstate Technical Regulatory Council 2009). The beauty of these technologies is that they have become attractive alternatives to other conventional cleanup technologies, such as excavating soil, pumping and treating contaminated groundwater, electrokinetic systems, chemical treatment, physical barriers, soil vapor extraction, *in situ* oxidation, and *in situ* vitrification, etc., which are generally unsatisfactory. The excellent potential of phytotechnology has been brought into the limelight because of its relatively low capital cost, environmentally friendly, safe nature, and its inherent aesthetic nature. The processes involved in phytotechnology include phytovolatilization, phytodegradation, phytohydraulics, phytoextraction, phytostabilization, rhizodegradation, and phytosequestration (Kramer 2005). These technological strategies can involve remediation, containment, or both. In phytotechnology, plant vegetation is used to isolate or contain, extract, degrade, or immobilize toxic contaminants or pollutants in soil, groundwater, surface water, sediments, sludges, and air. Inorganic contaminants or pollutants in soil and water environments include heavy metals, salts, nutrients, metalloids, and radioactive materials. Also the technology can be used in the treatment of organic contaminants, such as crude oil; petroleum hydrocarbons; volatile organic compounds (VOC); pesticides; polycyclic aromatic hydrocarbons (PAHs); chlorinated compounds; gas condensates; polychlorinated biphenyls (PCBs); benzene, toluene, ethylbenzene, and xylene (BTEX); and explosive compounds (Raskin and Ensley 2000).

Some phytotechnology applications could be primary methods of cleaning up or stabilizing contamination while others will supplement primary remedies. Phytotechnologies may potentially clean up moderate to low levels of selected elemental and organic contaminants over large areas, maintain sites by treating residual contamination after completion of a cleanup, act as a buffer against potential waste releases, aid voluntary cleanup efforts, facilitate nonpoint source pollution control, and offer a more active form of monitored natural attenuation (McCutcheon and Schnoor 2003; Oboh et al. 2011). All phytotechnologies are dependent on the development of healthy and extensive root systems. Phytotechnology promotes a broader understanding of the importance of plants and their beneficial role within both societal and natural systems.

3.2 ADVANTAGES AND DISADVANTAGES OF PHYTOTECHNOLOGY

Phytotechnology is an excellent cleanup technology with various advantages:

- Low-maintenance, passive, *in situ*, self-regulating, solar-driven system
- Potentially applicable in remote locations without utility access
- Decreased air and water emissions as well as secondary waste
- Control of soil erosion, surface water runoff, infiltration, and fugitive dust emissions
- Applicable to simultaneously remediate sites with multiple or mixed contaminants
- It can be used in conjunction with other remediation methods and may be more beneficial than a stand-alone technology
- Habitat creation or restoration provides land reclamation upon completion
- Favorable public perception, increased aesthetics, and reduced noise
- Increasing regulatory approval and standardization
- Carbon dioxide and greenhouse gas sequestration
- Cost-effective and environmentally friendly technology

Also, there are several disadvantages of this technology, which include the following:

- Growth habit of the planted system
- Slow/shallow root penetration of the selected plant(s)
- Limited to available land
- Susceptible to infestation and disease
- Bioaccumulation of contaminants in vegetation
- By-products may be more toxic
- Limited capability for contaminant mass transfer to the treatment zone or root zone
- Unfamiliarity by public or regulatory communities
- May be difficult to establish or maintain vegetation
- Relatively slow in comparison to more active remediation technologies
- Dependence on local climatic conditions and growth season (Schnoor et al. 1995; Prasad 2007).

3.3 APPLICATIONS AND RECENT ADVANCES IN PHYTOTECHNOLOGY

The effectiveness and economic viability of a phytotechnology depend on climate; elevation; precipitation; soil type and quality; type, age, distribution, and concentration of contamination; media;

and viability of the plants and planting system used for each site. Results of research, laboratory studies, and field tests at similar sites can serve as a guide to determine whether phytotechnology is appropriate for a site or not. Successful precedents can help identify appropriate plant species for implementation at a site. After reviewing site characteristics to determine if phytotechnology would be effective, it is then important to select the appropriate phytotechnology mechanism and plant species (Interstate Technical Regulatory Council 2009).

Phytoremediation may be applicable for the remediation of metals, pesticides, solvents, explosives, crude oil, PAHs, and landfill leachates (Aisien et al. 2009; Aisien et al. 2010a). Some plant species have the ability to store metals in their roots. They can be transplanted to sites to filter metals from wastewater. As the roots become saturated with metal contaminants, they can be harvested. Hyperaccumulator plants may be able to remove and store significant amounts of metallic contaminants. Currently, trees are under investigation to determine their ability to remove organic contaminants from groundwater, for translocation and transpiration, and possibly to metabolize them either to CO_2 or plant tissue. Phytoremediation can be used to clean up organic contaminants from surface water, groundwater, leachate, and municipal and industrial wastewater (Aisien et al. 2009; Aisien et al. 2010b). Plants also produce enzymes, such as dehalogenase and oxygenase, which help catalyze degradation. Constructed wetlands have most commonly been used in wastewater treatment for controlling organic matter; nutrients, such as nitrogen and phosphorus; and suspended sediments. The wetlands process is also suitable for controlling trace metals and other toxic materials. Additionally, the treatment has been used to treat acid mine drainage generated by metal or coal mining activities. These wastes typically contain high metal concentrations and are acidic. The process can be adapted to treat neutral and basic tailing solutions.

Although the concept is not entirely new, the area of phytotechnology is rapidly evolving, and novel applications are continuing to emerge. Some examples of phytotechnology applications include the following:

- The use of plants to reduce or solve pollution problems that otherwise would be more harmful to other ecosystems. An example is the use of wetlands and wastewater treatment.
- The replication of ecosystems and plant communities to reduce or solve a pollution problem. Examples are constructed ecosystems, such as ponds and wetlands, for treatment of wastewater or diffuse pollution sources.
- The use of plants to facilitate the recovery of ecosystems after significant disturbances. Examples are coal mine reclamation and the restoration of lakes and rivers.
- The use of plants for societal benefits within the context of a managed ecosystem. Examples include reintegrated agriculture and the management of renewable resources.
- The increased use of plants as sinks for carbon dioxide to mitigate the impact of climate change. Examples of this are reforestation and afforestation.
- The use of plants to augment the natural capacity of urban areas to mitigate pollution impact and moderate energy extremes. An example is the use of rooftop vegetation or "green roofs."

The recent advances in phytotechnology include the following:

- New evidence of phytoremediation effectiveness for PAHs, RDX, ClO_4^-, etc.
- Progress with transgenic plants, for example, use in the field for Se and As removal
- Plume delineation by tree corings
- Emergence of Populus genome database for plant functional genomics
- Assisted natural remediation (amendments are added to the soil in order to accelerate natural processes of remediation) (Van Aken et al. 2005; Adriano et al. 2004; Sohsalam and Sirianuntapiboon 2008; USEPA 2007).

3.4 MECHANISMS OF PHYTOTECHNOLOGY

It should be noted that not all mechanisms are applicable to all contaminants or pollutants or all matrixes. Also, the mechanisms are interrelated and dependent upon plant physiological processes driven by solar energy, rhizospheric processes, and other available precursors. Therefore, in phytechnology applications, multiple mechanisms are involved, depending on the designed application. Six major mechanisms have been implicated as contributing to phytotechnology processes. These are phytoextraction, phytostabilization, phytodegradation, phytovolatilization, rhizodegradation, and phytohydraulic. However, phytoextraction, phytostabilization, phytoimmobilization, and engineered vegetative caps are mechanisms applicable to contaminant cleanup in both soil and water. Phytovolatilization, rhizofiltration, and hydraulic barriers are for soil, and a constructed wetland mechanism is for water (Chaney et al. 2007). The region of activity of phytostabilization, rhizodegradation, and rhizofiltration is the plant root zone, and that of phytoextraction, phytodegradation, and phytovolatilization is the plant tissue.

3.4.1 Phytostabilization

A phytostabilization mechanism is applicable to the cleanup of both organic and inorganic contaminants. It uses certain plant species to immobilize contaminants in the soil, sediments, and groundwater through the absorption and accumulation in the roots, adsorption onto the roots, or precipitation or immobilization within the root zone.

3.4.2 Rhizodegradation

Rhizodegradation, which is also called phytostimulation, rhizosphere biodegradation, or plant-assisted bioremediation or degradation, is the breakdown of contaminants in the soil through the bioactivity that exists in the rhizosphere. This bioactivity is derived from the proteins and enzymes that can be produced and exuded by plants or from soil organisms, such as bacteria, yeast, and fungi. Many of these contaminants can be broken down into harmless products or converted into a source of food and energy for the plants or soil organisms (Donnelly and Fletcher 1994).

3.4.3 Phytodegradation

Phytodegradation, also called phytotransformation, refers to the uptake of organic contaminants from soil, sediments, and water with a subsequent transformation by the plant. In order for a plant to directly degrade, mineralize, or volatilize a compound (see Section 3.4.4, "Phytovolatilization"), it must be able to take that compound up through its roots. Plants transform organic contaminants through various internal, metabolic processes that help catalyze degradation. The contaminants are degraded in the plant with the breakdown products subsequently stored in the vacuole or incorporated into the plant tissues.

3.4.4 Phytovolatilization

Phytovolatilization is the uptake and transpiration of a contaminant by a plant with release of the contaminant or a modified form of the contaminant to the atmosphere from the plant. Phytovolatilization occurs as growing trees and other plants take up water and organic contaminants. Some of these contaminants can pass through the plants to the leaves and volatilize into the atmosphere at comparatively low concentrations.

3.4.5 Phytoextraction

Phytoextraction, also called phytoaccumulation, refers to the use of metal- or salt-accumulating plants that translocate and concentrate these soil contaminants into their roots and above-ground shoots or leaves. Certain plants called hyperaccumulators absorb unusually large amounts of metals in comparison to other plants and the ambient metal concentration.

3.4.6 Phytohydraulics

Phytohydraulics refers to the ability of vegetation to evapotranspire sources of surface water and groundwater. The vertical migration of water from the surface downward can be limited by the water interception capacity of the above-ground canopy and subsequent evapotranspiration through the root system. If water infiltrating from the surface is able to percolate below the root zone, it can recharge groundwater. However, the rate of recharge depends not only on the rooting depth of the species, but on the soil characteristics as well.

3.4.7 Rhizofiltration

Rhizofiltration can be defined as the use of plant roots to absorb, concentrate, and/or precipitate hazardous compounds, particularly heavy metals or radionuclides, from aqueous solutions. Hydroponically cultivated plants rapidly remove heavy metals from water and concentrate them in their roots and shoots. The plants to be used for cleanup are raised in greenhouses with their roots in water rather than in soil.

3.4.8 Rhizodegradation

Rhizodegradation, also called enhanced rhizosphere biodegradation, phytostimulation, or plant-assisted bioremediation or degradation, is the breakdown of contaminants in soil through microbial activity that is enhanced by the presence of the rhizosphere and is a much slower process than phytodegradation. Microorganisms (yeast, fungi, or bacteria) consume and digest organic substances for nutrition and energy (Kamal et al. 2004; Susarla et al. 1999; Kumar et al. 1995).

3.5 ENVIRONMENTAL IMPACT OF PHYTOTECHNOLOGY

Because of the extreme consequences, environmental contamination with heavy metals, particularly lead and mercury, is a significant concern. Now faced with these extensive environmental problems, a cost-effective means of remediation pertinent to the contaminated areas must be found. There are a number of conventional remediation technologies that are employed to remediate environmental contamination with heavy metals, such as solidification, soil washing, and permeable barriers. However, a majority of these technologies are costly to implement and cause further disturbance to the already damaged environment. Phytoremediation is evolving as a cost-effective alternative to high-energy, high-cost conventional methods. It is considered a "green revolution" in the field of innovative cleanup technologies (Henry 2000).

The environmental impact of phytotechnology can generally be divided into five categories:

- Augmenting the adaptive capacity of natural systems to moderate the impact of human activities
- Preventing pollutant releases and environmental degradation
- Controlling pollutant releases and environmental processes to minimize environmental degradation
- Remediation and restoration of degraded ecosystems
- Incorporating indicators of ecosystem health into monitoring and assessment strategies

The integrated ecosystems management component of this focuses on the use of phytotechnology to augment the capacity of natural systems to absorb impact. The prevention component involves the use of phytotechnology to avoid the production and release of environmentally hazardous substances and/or the modification of human activities to minimize damage to the environment; this can include product substitution or the redesign of production processes (Mangkoedihardjo 2011). The control component addresses chronic releases of pollutants and the application of phytotechnology to control and render these substances harmless before they enter the environment. The remediation and restoration component embodies phytotechnology and methods designed to recuperate and improve ecosystems that have declined as a result of naturally induced or anthropogenic effects. The monitoring and assessment component involves the use of phytotechnology to monitor and assess the condition of the environment, including the release of pollutants and other natural or anthropogenic materials of a harmful nature (Prasad et al. 2010; Mangkoedihardjo 2007).

3.6 CONCLUSIONS

- Phytotechnology has been potentially applicable to the cleanup of organic and inorganic contaminants.
- Enormous amounts of research have been done in laboratory, pilot, and field scale.
- Apart from site remediation, phytotechnology can be applied to other situations, such as wastewater treatment or landfill leachate control.
- Excellent and positive results from laboratory studies are difficult to replicate in the field.
- Phytotechnologies are well-suited for large and/or remote sites where traditional methods are not cost-effective or practicable.
- Public acceptance of the technology and the private sector's investment, research, and development will make phytotechnology more efficient.
- It provides a nature-based technology, which can support sustainable development strategies.
- Phytotechnology is innovative but also becoming part of mainstream treatment technology (constructed, wetlands, landfill caps, and ecological restoration).

3.7 FUTURE PERSPECTIVES

Phytotechnology, as a new and promising technology, has gained wide acceptance and is currently an area of active research in environmental decontamination. A good number of plants have already been identified as having potential for phytotechnology applications. Efforts are being made to understand the underlying genetic and biochemical processes involved in contaminant (inorganic and organic) uptake, transport, and storage by hyperaccumulating plants. The knowledge gained from such studies in conjunction with biotechnology has helped to improve, substantially, the phytotechnology capability of plants. For example, new transgenic plants have been developed with improved capacity for inorganic uptake, transport, and accumulation as well as for detoxification of organic pollutants. However, many gaps still persist in our understanding of the processes of plant-microbe interactions, metal accumulation, and ion homeostasis (Salt and Kramer 2000; Eapen et al. 2003; Eapen and D'Souza 2005; Singh et al. 2003).

To obtain further gains, research in the following areas appears to be worth pursuing in the future:

- Manipulation of metal transporters and their cellular targeting of specific cell types, such as vacuoles, to allow for safe compartmentation of heavy metals in locations that do not disturb other cellular functions.
- Genetic manipulation of the chloroplast genome may be, for some plants, an alternative approach to achieve high gene expression while avoiding the risk of transgene escape via pollen.

- Identification of candidate plants with substances that may deter the herbivores from feeding and the subsequent transformation of such plants with altered or improved metal-tolerance capabilities. Such a system will help avoid the transfer of metals to the food chain.
- Genetically modified organisms (GMO) for inventing tolerant plant species to particular environmental pollutants with high removal efficiencies.
- Development of transgenic plants with enhanced plant-microbe interaction or rhizosphere microbial activity. It may be possible either to develop transgenic plants that have the ability to secrete metal-selective ligands capable of solubilizing elements for phytoremediation or to find simple molecules with selective chelation ability that plants can make and secrete into the rhizosphere.
- Bioenergy production feasibilities from the harvested plants.
- Transgenic research in phytoremediation should also address the problem of mixed contamination occurring in many of the polluted sites. A multigene approach involving the simultaneous transfer of several genes into suitable candidate plants may help to remove contaminants of mixed or complex nature.
- Not much data are yet available on the field performance of transgenic plants in phytoremediation. Established field trials are, therefore, urgently needed to make it a commercially viable and acceptable technology.

To further advance our knowledge, phytotechnology research should require organized collaborative studies involving expertise from different fields, such as botany, plant physiology, biochemistry, geochemistry, microbiology, chemical engineering, environmental engineering, agricultural engineering, and genetic engineering among others. In the years to come, as in other areas, plant genetic engineering for improved phytotechnology could also be beneficial. (Moreno et al. 2005; Ruiz et al. 2003; Dominguez-Solis et al. 2004; Kramer 2005; Doty 2008; Ma et al. 2011).

REFERENCES

Adriano, D. C., W. W. Wenzel, J. Vangronsveld, and N. S. Bolan. (2004). Role of assisted natural remediation in environmental cleanup. *Geoderma, 122*, 121–142.

Aisien, F. A., G. E. Oyakhilomen, and E. T. Aisien. (2009a). Biotreatment of brewery waste water using enzymes. *Advanced Materials Research, 64*, 774–778.

Aisien, F. A., I. O. Oboh, and E. T. Aisien. (2009b). Biotreatment of greywater. *Journal of Scientific and Industrial Studies, 7*, 5–8.

Aisien, F. A., O. Faleye, and E. T. Aisien. (2010a). Phytoremediation of heavy metals in aqueous solutions. *Leonardo Journal of Sciences, 17*, 37–46.

Aisien, E. T., C. V. Nwatah, and F. A. Aisien. (2010b). Biological treatment of landfill leachate from Benin City, Nigeria. *Electronic Journal of Environmental, Agricultural and Food Chemistry, 9*, 1701–1705.

Chaney, R. L., J. S. Angle, C. L. Broadhurst, C. A. Peters, R. V. Tappero, and D. L. Sparks. (2007). Improved understanding of hyperaccumulation yields commercial phytoextraction and phytomining technologies. *Journal of Environmental Quality, 36*, 1429–1443.

Dominguez-Solis, J. R., M. C. Lopez-Martin, F. J. Ager, M. D. Ynsa, L. C. Romero, and C. Gotor. (2004). Increased cysteine availability is essential for cadmium tolerance and accumulation in *Arabidopsis thaliana*. *Plant Biotechnology Journal, 2*, 469–476.

Donnelly, P. K., and J. S. Fletcher. (1994). Potential use of mycorrhizal fungi as bioremediation agents. In T. A. Anderson, and J. R. Coats (Eds.), *Bioremediation through Rhizosphere Technology*. Washington, DC: American Chemical Society.

Doty, S. L. (2008). Enhancing phytoremediation through the use of transgenics and endophytes. *New Phytologist, 179*, 318–333.

Eapen, S., K. Suseelan, S. Tivarekar, S. Kotwal, and R. Mitra. (2003). Potential for rhizofiltration of uranium using hairy root cultures of *Brassica juncea* and *Chenopodium amaranticolor*. *Environmental Research, 91*, 127–133.

Eapen, S., and S. F. D'Souza. (2005). Prospects of genetic engineering of plants for phytoremediation of toxic metals. *Biotechnology Advances, 23*, 97–114.

Henry, J. R. (2000). An overview of the phytoremediation of lead and mercury prepared for USEPA. Retrieved from http://clu-in.org.

Interstate Technical Regulatory Council. (2009). *Phytotechnology technical and regulatory guidance and decision trees, revised.*

Kamal, M., A. E. Ghaly, N. Mahmoud, and R. Côté. (2004). Phytoaccumulation of heavy metals by aquatic plants. *Environment International*, *29*, 1029–1039.

Kramer, U. (2005). Phytoremediation: Novel approaches to cleaning up polluted soils. *Current Opinion in Biotechnology*, *16*, 133–141.

Kumar, P. B. A. N., V. Dushenkov, H. Motto, and I. Raskin. (1995). Phytoextraction: The use of plants to remove heavy metals from soils. *Environmental Science and Technology*, *29*, 1232–1238.

Ma, Y., M. N. V. Prasad, M. Rajkumar, and H. Freitas. (2011). Plant growth promoting rhizobacteria and endophytes accelerate phytoremediation of metalliferous soils. *Biotechnology Advances*, *29*, 248–258.

Mangkoedihardjo, S. (2007). Phytotechnology integrity in environmental sanitation for sustainable development. *Journal of Applied Sciences Research*, *3*, 1037–1044.

Mangkoedihardjo, S. (2011). Phytotechnology insight for the flood plains along the river and riparian zone. *International Journal of Academic Research*, *3*(3), I Part.

McCutcheon, S. C., and J. L. Schnoor. (2003). *Phytoremediation: transformation and control of contaminants.* New Jersey: John Wiley & Sons, Inc.

Moreno, F. N., C. W. Anderson, R. B. Stewart, and B. H. Robinson. (2005). Mercury volatilisation and phytoextraction from base-metal mine tailings. *Environmental Pollution*, *136*, 341–352.

Oboh, I. O., E. O. Aluyor, and T. O. K. Audu. (2011). Application of *Luffa cylindrica* in nature form as biosorbents to removal of divalent metals from aqueous solutions: Kinetic and equilibrium study. In S. G. E. Fernando (Ed.), *Waste Water: Treatment and Reutilization*. INTECH Publisher.

Prasad, M. N. V. (2007). Aquatic plants for phytotechnology. In S. N. Singh, and R. D. Tripathi (Eds.), *Environmental bioremediation technologies* (257–274). Netherlands: Springer.

Prasad, M. N. V., H. Freitas, S. Fraenzle, S. Wuenschmann, and B. Markert. (2010). Knowledge explosion in phytotechnologies for environmental solutions. *Environmental Pollution*, *158*, 18–23.

Raskin, I., and B. D. Ensley. (2000). *Phytoremediation of toxic metals: Using plants to clean up the environment*. New York: John Wiley & Sons, Inc.

Ruiz, O. N., Hussein, H. S., Terry, N., and Daniell, H. (2003). Phytoremediation of organomercurial compounds via chloroplast genetic engineering. *Plant Physiology*, *132*, 1344–1352.

Salt, D. E., and U. Kramer. (2000). Mechanisms of metal hyperaccumulation in plants. In I. Raskin, and B. D. Ensley (Eds.), *Phytoremediation of toxic metals: Using plants to clean up the environment* (231–246). New York: Wiley.

Schnoor, J. L., L. A. Light, S. C. McCutcheon, N. L. Wolfe, and L. H. Carriera. (1995). Phytoremediation of organic and nutrient contaminants. *Environment Science and Technology*, *29*, 318–323.

Singh, O. V., S. Labana, G. Pandey, R. Budhiraja, and R. K. Jain. (2003). Phytoremediation: An overview of metallic ion decontamination from soil. *Applied Microbiology and Biotechnology*, *61*, 405–412.

Sohsalam, P., and S. Sirianuntapiboon. (2008). Feasibility of using constructed wetland treatment for molasses wastewater treatment. *Bioresource Technology*, *99*, 5610–5616.

Susarla, S., S. T. Bacchus, N. L. Wolfe, and S. C. McCutcheon. (1999). Phytotransformation of perchlorate using parrot-feather. *Soil & Groundwater Cleanup, Feb/March*, 20–23.

USEPA. (2007). The use of soil amendments for remediation, revitalization and reuse. E.P.A. 542-R-07-013. December.

Van Aken, B., J. M. Yoon, C. L. Just, S. Tanake, L. Brentner, B. Flokstra, and J. L. Schnoor. (2005). Phytoremediation: From the molecular to the field scale. Presented to the International Phytotechnonogy Conference. The University of Iowa, Iowa City, USA.

4 Potential of Constructed Wetland Phytotechnology for Tannery Wastewater Treatment

Cristina S. C. Calheiros, António O. S. S. Rangel, and Paula M. L. Castro

CONTENTS

4.1 INTRODUCTION

In the industrial scenario, a major challenge is to match economical, sustainable wastewater treatment systems that cope with the legal restrictions granting low environmental impacts. The tannery industry is an important sector in the European Union market. This industry covers the treatment of raw materials through the conversion of raw hide or skin, a putrescible material, into leather, a stable material, and finishes, so it can be used in the manufacture of a wide range of consumer products. This industrial sector has been innovative in technology and in higher-quality standards; however, the waste emanating from the production process poses a major concern. In particular, the liquid waste is considered an environmental problem because of its complexity and high organic and inorganic loads. Frequent problems experienced with wastewater treatment systems are that they often work over or under the capacity because of poor design of alterations in production. Efficient wastewater management may be crucial to the optimization of resources and sustainability of the process. The use of a phytotechnology approach, such as constructed wetlands, can be potentially employed to complement the performance of existing systems or as an alternative to a conventional biological treatment.

Constructed wetlands are engineered systems that mimic natural wetland treatment processes in a controlled environment. The wastewater treatment mechanisms encompass a mix of physical, chemical, and biological processes. They have several advantages compared to conventional secondary and advanced wastewater treatment systems, including low cost of construction and maintenance, low energy requirements, and the ability to be established and run by relatively untrained personnel. Also, the systems are usually more flexible and less susceptible to variation in loading rates than conventional treatment systems. Constructed wetlands have proven to be a very effective method for the treatment of municipal wastewater, and several successful cases are described in the literature for industrial wastewater, such as petrochemicals, landfill leachate, food waste, pulp and paper, mines, and agricultural and textile wastewater. However, the application to industrial wastewater has to be carefully analyzed because its composition is frequently highly variable, and the treatment needs are not the same.

A reflection on the use of constructed wetlands to improve water quality with different contaminants with a review on the application to the tannery industry, a subject that has not been intensively studied, is presented here.

4.2 TANNERY INDUSTRY

4.2.1 Industry Characterization

Tanning is the process used to stabilize raw hide or skin into leather, a nonputrescible product. The production process involves a sequence of complex chemical reactions and mechanical procedures. The manufacture of leather is dependent on the animal population and slaughter rate and is related mainly to meat consumption (EC 2011). Tanning hides and skins is one of the oldest crafts known to man. The basic technique of tanning leather dates back to prehistoric times when it was used mainly for crude tents and clothing. Leather production evolved technologically and expanded, being, by medieval times, better organized. Tanneries were situated on the local stream or river, which was used as a source of water for processing and power for the water wheel–driven machines. Plant derivatives, such as oak or sumac bark, were applied in vegetable tanning. Until the end of the 17th century, there were not many changes in leather manufacturing. Nineteenth-century advancements in chemistry were important to the development of the industry, comprising, for example, the use of enzymes and chrome for tanning. This way, salts of chromium were applied to tan the animal skins and hides (COTANCE 2002).

Tanners in Europe have a long tradition of producing all kinds of leather, from bovine and calf leather to sheep and goat leather. Their flexibility and adaptability constitute important assets needed for them to prevail in the market. The enterprises are usually small to medium sized and generally are a family business. On a worldwide scale, it is in locations such as China, the United States of America (USA), Brazil, Argentina, India, Russia, and the European Union (EU) that significant cattle productions are found. Sheepskins originate mostly in China, New Zealand, Australia, the Near East, and the EU. Hides are normally traded in a salted state or as an intermediate product—wet-blue for bovine hides and pickled for ovine skins. The United States remains the EU's top supplier of raw hides and skins, and in terms of major leather production centers in the world apart from the EU, Mexico, Argentina, Brazil, South Korea, China, India, and Pakistan are major players. In terms of distribution in the EU, it is in Italy that this sector is the most important in terms of establishments, employment, production, and turnover, followed by Spain, France, Germany, Portugal, and the United Kingdom accounting for most of the balance of the EU leather industry (EC 2011).

The tanning production process follows a series of operations that aims to create a final product that varies according to the raw materials used and the client requirements. The term "leather" is a general term for hide and skin that retains its original fibrous structure more or less intact and has been treated so it is nonputrescible. The designation of "hide" is attributed to the pelt of a large animal, such as cattle, and "skin" is attributed to the pelt of a small animal, such as a calf, sheep or pig. The generated leather from the production process has specific properties of stability and water and temperature resistance. The production processes in a tannery can be divided into four main categories although this categorization only serves as an indication, and it is possible that a company has other operations: hide and skin storage and beamhouse operations (pre-tanning), tanyard operations (tanning), post-tanning operations, and finishing operations (EC 2011).

4.2.2 Environmental Issues in the Tannery Industry

The tanning industry is considered a sector that has a great potential impact on the environment because it originates high amounts of liquid, solid, and gaseous waste, and it consumes raw materials, such as raw hides, energy, chemicals, and water. The discharge from this type of waste may contain toxic, persistent, and harmful substances. Thanikaivelan et al. (2005) wrote about the recent trends in leather-making processes, problems, and pathways, emphasizing the importance of environmental-protection awareness.

The central focuses for the tannery industry, concerning environmental issues, are water consumption, efficient use of the products in the process, replacement of potentially harmful agents, waste reduction, and recycling. The water issue is very important because most of the operations for leather manufacture are carried out in water. The contamination of water by this sector is not monitored in many places, such as in Albania (Floqi et al. 2007), or effluents are inadvertently discharged into water bodies and open land, such as in Ethiopia, although effluent regulations and guidelines have been developed as part of the Environmental Policy of the Federal Government of Ethiopia (Leta et al. 2004). In tanneries with poor water management, only 50% of the water consumed is actually used in the process (EC 2011). Proper water management is thus a crucial issue in this sector. The development of ecotechnology in order to minimize the environmental impact of this industry in Southern Brazil has been considered (Giannetti et al. 2004).

Efforts are focused on reducing the negative impact on the environment caused by waste throughout its life cycle. Thus, the key to achieving such accomplishments is the awareness of the inputs and outputs of the process regarding the material characteristics, quantities, and environmental impact. The beamhouse, the tanyard, and the post-tanning operations are the stages producing higher amounts of wastewater. Contaminants in the wastewater vary according to the components used during the production process. The particulate matter from dry finishing processes, solvent vapors from the degreasing of sheepskins and coating, and gaseous emissions from wet processing and effluent treatment constitute potential gaseous emissions. Fleshing, splitting, and shaving operations originate the major sources of solid waste. Much of this waste may be classified as by-products as it may be sold as raw materials (EC 2011).

4.3 WASTEWATER IN THE TANNERY SECTOR

4.3.1 Tannery Wastewater Toxicity

Two of the greater problems of modern society are the water shortage and environmental degradation related, in part, to water quality. The wastewater derived from the leather/tannery industry is a powerful source of pollutants. Physical and chemical constituents of leather wastewater depends not only on how the production cycle is carried out but also what it involves in terms of reagents and the type of skin or hide used (Floqi et al. 2007). Wastewater may contain chemical pollutants exhibiting high toxicity with a range of potential toxic effects on the environment (Lofrano et al. 2006). Balasubramanian et al. (1999) reported extensive work on characterization of tannery effluents, indicating five important parameters, viz., BOD (biochemical oxygen demand), COD (chemical oxygen demand), total chromium, sulfide, and TDS (total dissolved solids). Besides that, salinity may also pose a major concern as certain streams from this industry are considered highly saline (EC 2011).

The toxicity of leather-tanning wastewater has been assessed in various ways (Oral et al. 2007), including acute toxicity studies on *Vibrio fischeri* (Cotman et al. 2004), *Daphnia* spp. (Cotman et al. 2004; Lofrano et al. 2006), sea urchin, and marine microalgae (Meriç et al. 2005) and the effects on the development of different plant species (Calheiros et al. 2008b; Calheiros et al. 2012a). Recently, Lofrano et al. (2008) did a multi-approach study to define the characteristics and fluxes of retanning chemicals in terms of BOD_5/COD ratio, UV–VIS absorbance, gas chromatography/mass spectrometry scanning, ecotoxicological endpoints of immobilization (*Daphnia magna* and *Artemia salina*), cell growth inhibition (*Selenastrum capricornutum*), and plant growth index (*Lepidium sativum*) as a combined effect calculated based on inhibition of germination and root length. Karunyal et al. (1994) studied the effects of tannery effluent on the seed germination of *Oryva sativa*, *Acacia holosericea*, and *Leucaena leucocephala* and on leaf area, biomass, and chlorophyll content of *Gossypium hirsutum*, *Vigna mungo*, *Vigna unguiculata*, and *Lycopersicon esculentum*. Germination was inhibited by effluents at the 25% and 50% levels and was prevented at 75% and 100% of tannery effluent. Calheiros et al. (2008b) supplied tannery effluent, with a low treatment level, to wetland plants *Typha latifolia* and *Phragmites australis*. Germination occurred even at effluent concentrations of 100%; whereas germination of *Trifolium pratense* (indicator species) was completely inhibited, almost invariably, at an effluent concentration of 50%. The potential of *P. australis* as a Cr hyperaccumulator was also evaluated. Although phytotoxic signs in the plant were evident, its potential to extract and accumulate this metal on rhizomes is high.

A significant percentage of tanneries, 80–95%, use Cr(III) salts in their tanning processes. Tanning agents can be categorized into three main groups: mineral (chrome) tanning agents, vegetable tanning agents, and alternative tanning agents (syntans, aldehydes, and oil tanning agents). The degree of toxicity of chrome is perhaps one of the most debated issues between the tanning industry and authorities (EC 2011). Cr has several valence states, but the more stable forms are the trivalent Cr(III) and hexavalent Cr(VI) species (Shanker et al. 2005), which are those available for plant uptake (Zurayk et al. 2001). Because Cr is a toxic, nonessential element to plants, they do not possess specific mechanisms for its uptake. This uptake is therefore made through carriers used for the uptake of essential metals for plant metabolism. The toxic effects are primarily dependent on the metal speciation that determines the uptake, translocation, and accumulation (Shanker et al. 2005). The uptake and accumulation of Cr in plants has also been addressed with a view to explore their phytoremediation potential (Zurayk et al. 2001; Sinha et al. 2002; Srisatit and Sengsai 2003; Tiglyene et al. 2005; Calheiros et al. 2008b). According to Sinha et al. (2002), aquatic macrophytes vary greatly in their ability to accumulate Cr in their tissues. High accumulation in the roots of the plants are reported, possibly through Cr immobilization in the vacuoles of the root cells, thus rendering it less toxic, which may be a natural toxicity response of the plant (Shanker et al. 2005). Namasivayam and Höll (2004) studied the Cr(III) removal in tannery wastewater using Chinese Reed (*Miscanthus sinensis*), a fast growing plant, which was shown to be an effective adsorbent. Woody plants have also been targeted for Cr removal. Alves et al. (1993) reported the use of *Pinus sylvestris*, and Shukla et al. (2011) selected five woody plant species, i.e., *Terminalia arjuna*, *Prosopis juliflora*, *Populus alba*, *Eucalyptus tereticornis*, and *Dendrocalamus strictus*, for phytoremediation of tannery sludge dumps at the Common Effluent Treatment Plant in Unnao (Uttar Pradesh), India. The latter reported that the selected woody plants showed a high growth rate and high metal-accumulation capabilities. The concentration of toxic metals was high in the raw tannery sludge, i.e., Fe-1667 > Cr-628 > Zn-592 > Pb-427 > Cu-354 > Mn-210 > Cd-125 > Ni-76 mg/kg^{-1} dw, respectively. After one year of phytoremediation, the level of toxic metals was removed from the tannery sludge up to Cr (70.22)%, Ni (59.21)%, Cd (58.4)%, Fe (49.75)%, Mn (30.95)%, Zn (22.80)%, Cu (20.46)%, and Pb (14.05)%, respectively.

4.3.2 Tannery Wastewater Treatment Processes

Tannery wastewater may be treated in many different ways, and the treatment operations are dictated by the level of contamination in the water, which is directly related to the production process, the legal compliance for water discharge, and the possibility of sending the water to a communal wastewater treatment plant or municipal collector or the possibility of reutilization in the production process. Wastewater treatments comprise physicochemical and biological operations (Durai and Rajasimman 2011). Durai and Rajasimman (2011) presented a review that examines the extent of pollution created by tanneries and the different biological processes available for the treatment and disposal of tannery wastewater. In Europe, tanneries usually discharge their effluent to wastewater treatment plants, which are either municipal treatment plants or plants operated for large tanning complexes, and a few are discharged directly to surface water. Most of them have some kind of treatment before discharging to the sewer (EC 2011).

In general, for wastewater treatment, the segregation of the raw wastewater may be useful to allow preliminary treatment of concentrated effluent in order to facilitate its treatment, reuse, and recycling. A tannery wastewater treatment plant, to be efficient, must be robust enough to be able to tolerate scenarios, such as variations in flow, pollutant loads, and periods of nonfeeding. Many treatment plants are under- or over-dimensioned, or the technology employed for the treatment is expensive to maintain and highly resource consuming. Cost-effective and environmentally friendly options need to be addressed.

4.4 CONSTRUCTED WETLANDS FOR WASTEWATER TREATMENT

4.4.1 General Considerations

Wetlands are land areas that are wet during part or all of the year because of their location in the landscape (Kadlec and Wallace 2009). Under the text of the Ramsar Convention (Article 1.1), wetlands are defined as "areas of marsh, fen, peatland or water, whether natural or artificial, permanent or temporary, with water that is static or flowing, fresh, brackish or salt, including areas of marine water the depth of which at low tide does not exceed six meters" (RCS 2006). Because of wetlands' structure and function, they support a wide range of ecosystems and are among the world's most productive environments, providing a wide array of benefits. Wetland structural components include soil or growth media, macrophytes (macroscopic forms of aquatic vegetation), water detritus, microbes, and a fauna population. Concerning the wetland functions, several physical, chemical, and biological processes occur, such as primary production, water depuration, nutrient and pollution retention, and groundwater recharge (RCS 2006; Kadlec and Wallace 2009).

Constructed wetlands (CWs) are man-made systems that have been conceived to enhance specific characteristics of natural wetland ecosystems for improved treatment capacity. They intend to mimic the optimal treatment conditions of natural wetlands, so they may be applied to different types of wastewater in almost any location for water-quality enhancement (Kadlec and Wallace 2009). In addition, advantages of CWs over natural wetlands include flexibility in sizing, control over the hydraulic pathways, and retention time (Vymazal and Kröpfelová 2008). Their hydrological, physical, and biochemical processes have been reviewed by Scholz and Lee (2005). Also, comprehensive reviews of this technology and its application have been published (Sundaravadivel and Vigneswaran 2001; Vymazal and Kröpfelová 2008).

Unlike conventional treatment systems, no specific design life period is generally prescribed for CWs although it is plausible that treatment wetlands would be ascribed a basic life of 40–50 years with mechanical components being replaced at greater frequency (Kadlec and Wallace 2009). The system lifetime will also depend on its design and building, materials used, pollutant loads, and maintenance and operation (USEPA 2000). There are FWS (free water surface) wetland systems that have retained effectiveness for up to 80 years (Kadlec and Wallace 2009).

In general, CWs are considered to be a cost-effective and technically feasible approach to wastewater treatment (Kadlec et al. 2000; USEPA 2000). The main advantages of CWs are the use of natural processes, the promotion of water reuse and recycling, the ability to tolerate flow fluctuations, that operation and maintenance are required only periodically, low operation and maintenance expenses, that they are generally less expensive to build than other treatments, that it's possible for them to be built in a way that harmoniously fits in the landscape, the aesthetic enhancement of open spaces, and the promotion of wildlife habitat. Also, unlike several biological treatments, wetlands do not generate sludge. The main limitations are the need for larger implementation areas than conventional treatment systems, that they may only be economic where land is available and affordable, that in extreme climates the efficiency may vary seasonally, that the several biological intervenient of the system are sensitive to toxic compounds and loads, and that the complete drought of the systems for long periods may lead to inactivity of the organisms within the system.

There are various CWs design configurations. They may be classified according to the flow pattern in the wetland (FWS flow where the majority of water flows above the substrate surface, subsurface flow (SSF) where the flow is directed through the rooting media with no overland flow, horizontal (HSF) and vertical (VSF) flow) and the life form of the dominant macrophyte (free-floating, emergent, or submerged) (Vymazal and Kröpfelová 2008). Other features may be considered, such as arrangement of the cell (multistage and hybrid systems), type of substrate (gravel, soil, sand, etc.), treatment level of the wastewater (primary, secondary, or tertiary), and the type of loading (continuous or intermittent). The classification of commonly used systems is well described in the literature (Kadlec et al. 2000; Kadlec and Wallace 2009).

CWs have been found to be effective in treating wastewater with removal demands on BOD, SS, nitrogen, and phosphorus, as well as in decreasing the concentration of metals and pathogens from wastewater. Contaminants are removed through a combination of physical, chemical, and biological processes. These processes include sedimentation, filtration, precipitation, adsorption to soil particles, assimilation by plant tissue, and microbial transformations (Vymazal and Kröpfelová 2008). CWs are complex systems that integrate and assemble several components playing different roles in the wetland cosmos. Generally, the basic components of a wetland treatment system are earthwork, media or substrate (in the case of SSF wetlands), liners, outlet and inlet structures, and plants. For instance, in the United States, CWs are dominated by cattails (*Typha* spp.) or bulrush (*Scirpus* spp.) and, in Europe, by common reed (*P. australis*) or grasses (*Phalaris arundinacea* or *Glyceria maxima*) (Kadlec and Wallace 2009). Another important component of wetlands are the microbial communities and aquatic invertebrates that develop naturally. General design considerations are detailed in the literature (USEPA 2000; Brix and Arias 2005; Vymazal and Kröpfelová 2008; Kadlec and Wallace 2009).

4.4.2 Applications in Wastewater Treatment

In Europe, the development of and interest in CWs started in Germany by Käthe Seidel in the early 1950s (Seidel 1953) and by Reinhold Kickuth in 1970s (Kickuth 1977). The historic perspective and development of CWs is well documented in the literature; they adapt to different realities and treatment needs in operation in many places around the globe (Kadlec et al. 2000; Haberl et al. 2003; Vymazal and Kröpfelová 2008; Vymazal 2009; Kadlec and Wallace 2009) with different climate conditions, such as tropical climates (Kaseva and Mbuligwe 2010), hot and semiarid climates (Quanrud et al. 2004), arid climates (Badkoubi et al. 1998), cold climates (Mæhlum and Stålnacke 1999), and temperatures near freezing (Allen et al. 2002). The potential of CWs for wastewater treatment and reuse in developing countries has also been addressed (Kivaisi 2001). In 1999, Nelson et al. (1999) reported the functioning of a wastewater treatment system that was part of the overall strategy for nutrient recycling inside Biosphere 2, a closed ecological system to study Earth's biosphere (Nelson et al. 1993), where the effluent was routed to the irrigation supply for agriculture crops. CWs proved to be useful as an integrated part of the system. CWs were also used for wastewater treatment reestablishment in 2004 Tsunami–affected areas of Thailand (Brix et al. 2007).

CWs have traditionally been used for the treatment of domestic and municipal sewage (Vymazal and Kröpfelová 2008; Kadlec et al. 2000) although, in recent years, there has been an increased spectrum application concerning several types of industrial waters (Kadlec et al. 2000; Vymazal 2009). Vymazal (2009) has undertaken an interesting review on the use of CWs with HSF for various types of wastewater. Also, Haberl et al. (2003) described applications of CWs to olive mill wastewater, groundwater contaminated with hydrocarbon and cyanide, and wastewater with aromatic organic pollutants. CWs have been used to improve urban river water quality, such as the Besòs River in Barcelona (Huertas et al. 2006), and to treat polluted water from the Erh-Ren River in southern Taiwan (Jing and Lin 2004). Besides the above-mentioned applications, CWs may be applied for other purposes, such as the following:

(i) Runoff water: landfill leachate (Nivala et al. 2007), highway runoff (Revitt et al. 2004; Terzakis et al. 2008), aircraft deicing fluid (Castro et al. 2005), greenhouses (Seo et al. 2008), and nursery runoff (Narváez et al. 2011)
(ii) Agricultural wastewater: pig farms (Meers et al. 2008), dairy farms (Mantovi et al. 2003), and fish farms or aquaculture (Lin et al. 2005)
(iii) Municipal and domestic wastewater: domestic wastewater with pharmaceutical compounds (Matamoros et al. 2009), domestic wastewater (Mina et al. 2011), individual houses (Steer et al. 2002; Brix and Arias 2005), tourism units (Calheiros et al. 2011), and schoolhouses (Öövel et al. 2007)
(iv) Industrial wastewater: mining water (Mays and Edwards 2001), food waste (Vrhovšek et al. 1996), petrochemicals (Ji et al. 2007), pulp and paper (Prabu and Udayasoorian 2007), leather tannery (Kucuk et al. 2003; Calheiros et al. 2007, 2008a, 2009a; Dotro et al. 2010b; Dotro et al. 2010c; Kaseva and Mbuligwe, 2010), textiles and tie-dye (Mbuligwe 2005; Bulc and Ojstršek 2008), wineries (Masi et al. 2002), metallurgy (Maine et al. 2006), industrial parks (Haberl et al. 2003), the petrochemical industry (Haberl et al. 2003), the chemical industry (Haberl et al. 2003), and abattoirs (Haberl et al. 2003).

4.5 CONSTRUCTED WETLANDS IN THE TANNERY SECTOR

4.5.1 Applications and Research

It has been claimed that the major potential for CWs technology in the tannery sector is to constitute an alternative to an activated sludge treatment process (Daniels 2001b). CWs must be able to comprise, in the implementation, the removal of less-biodegradable constituents as nutrients.

The use of CWs to treat tannery wastewater is developing, and its application to this industry has not been intensively studied as a potentially cost-effective and environmentally friendly option. The lack of detailed research and information concerning data about CWs' construction, hydraulic operation, and performance efficiency and the role of plants and microorganisms is an important issue that requires more investment. The shortage of information for direct use to the leather manufacturer has also been claimed for a long time (Daniels 1998). Only recently some details related specifically to the construction of reed beds (CWs planted with *P. australis*) for tannery wastewater treatment have been found in literature (Daniels 2007, 2008). Daniels (2008) listed the major technical advantages of using reed beds in tannery wastewater treatment:

- Cost-effective alternative to conventional systems (Renaudin 1996; Tiglyene et al. 2005)
- Estimation of typical capital costs of 10–20% and running costs of about 10% compared with conventional treatment systems, such as activated sludge systems (Daniels 1998, 2007, 2008); cost will vary according to requirements and materials availability (Daniels 2001b)
- Negligible energy costs (Daniels 1998)
- Easily managed and odor-free
- Providing both secondary and polishing treatments (Daniels 1998, 2001b; Kucuk et al. 2003; Calheiros et al. 2007)
- Can be located almost anywhere

It has been referred to previously that a limitation of the use of CWs is the land requirement although, in the case of treating tannery effluent compared to what is needed for conventional secondary treatment and sludge drying, the requirements can be similar (Emmanuel and Anand 2007). Other great advantages that reed beds have are their resilience to shock loadings, stop in the feed, and significant climate variations (Renaudin 1996; Daniels 1998, 2001b; Calheiros et al. 2007, 2009a). CWs have also performed well when subjected to high levels of salinity (Daniels 2001a; Calheiros et al. 2012b). They may be used for tertiary treatment and, although they don't match the performance of membrane treatment, may be equivalent or superior to other options (Daniels 2008).

There are records of studies using CWs for tannery wastewater treatment in countries that include Argentina (Dotro et al. 2010c), Italy (Bragato et al. 2004), India (Daniels 1998; Emmanuel and Anand 2007), Laos PDR (Renaudin 1996; Daniels 2007), México (Aguilar et al. 2008), Morocco (Tiglyene et al. 2005), Portugal (Calheiros et al. 2007, 2008a, 2009a, 2010, 2012b), Tanzania (Kaseva and Mbuligwe 2010), Thailand (Srisatit and Sengsai 2003), Turkey (Kucuk et al. 2003), the UK (Daniels 2007, 2008), and the USA (Dotro et al. 2010b).

In this important perspective, the selection of plants to be part of the CWs has to be critically carried out in order to maximize pollutant removal (Brisson and Chazarenc 2008). Macrophytes are potent tools in the abatement of heavy-metal pollution in aquatic ecosystems receiving industrial effluents and municipal wastewater (Rai 2009). Larue et al. (2010) suggested that macrophytes operate different defense strategies in the presence of xenobiotics. When dealing with industrial wastewater, especially heavily loaded, it is advisable not to use flashy plants. Concerning tannery wastewater treatment in CWs, Calheiros et al. (2007) tested five plant species (*Canna indica*, *T. latifolia*, *P. australis*, *Stenotaphrum secundatum*, and *Iris pseudacorus*), and only *Typha* and *Phragmites* proved to be resilient to the imposed conditions (Figure 4.1). The latter two species had been tested by several authors, such as Calheiros et al. (2009a) (Figure 4.2), Kucuk et al. (2003), and Dotro et al. (2010b,c). Other species, such as *Sarcocornia fruticosa*, *Arundo donax* (Calheiros et al. 2011) (Figure 4.3), and *Scirpus americanus* (Aguilar et al. 2008), have also shown adequacy for this application.

FIGURE 4.1 Constructed wetland pilot units for secondary tannery wastewater treatment in northern Portugal planted with *Canna indica* (U1), *Typha latifolia* (U2), *Phragmites australis* (U3), *Stenotaphrum secundatum* (U4), *Iris pseudacorus* (U5), and an unvegetated control (U6), in Filtralite® MR3-8 substrate.

FIGURE 4.2 Constructed wetland pilot units for secondary tannery wastewater treatment in northern Portugal planted with *Typha latifolia* and *Phragmites australis* in expanded clay substrate.

The understanding of microbial activity in plant rhizospheres and substrates in CWs is important for the understanding of phytoremediation technologies (Franco et al. 2005; Calheiros et al. 2009b,c, 2010). The microbiology inherent to CWs systems treating tannery wastewater has been studied by Calheiros et al. (2009b,c, 2010). Diverse and distinct bacterial communities inhabited each wetland, which was related to the type of plant and substrate present in each CW. Isolates retrieved from the CWs related phylogenetically to *Firmicutes*, *Actinobacteria*, *Bacteroidetes*, *α-*, *β-*, and *γ-Proteobacteria*, *Sphingobacteria*, *Actinobacteria*, and *Bacteroidetes* (Calheiros et al. 2009b,c, 2010). Aguilar et al. (2008), doing a study on identification and characterization of sulfur-oxidizing bacteria in a wetland treating tannery wastewater, reported bacteria isolates belonging to the genera *Actinobacter*, *Alcaligenes*, *Ochrobactrum*, and *Pseudomonas*. Lefebvre et al. (2006) studied the microbial diversity in hypersaline tannery wastewater and found a high diversity in this

FIGURE 4.3 Constructed wetland for polishing tannery wastewater in the center of Portugal planted with *Sarcocornia fruticosa* and *Arundo donax* in an expanded clay substrate.

type of environment, clustering within 193 phylotypes and covering 14 of the 52 divisions of the bacterial domain.

4.5.2 Constructed Wetlands Operation

In the IUE document on recent developments in cleaner production and environmental protection in the world leather sector (IUE 2008), CWs are considered among the current wastewater technologies. A BOD_5 removal efficiency of 85–95% (60–100 mg/L^{-1}) when using CWs after primary treatment is expected. This performance compares with primary or chemical treatment plus aerated facultative lagoons with similar removal efficiencies (85–95% of 60–100 mg/L^{-1}) (IUE 2008). Through the Regional Programme for Pollution Control in the Tanning Industry in Southeast Asia, UNIDO has funded a number of treatment experiments using biotechnological processes, including the reed bed technology. It was pointed out as a viable option to replace the conventional biological treatment step (UNIDO 2001).

For application to tertiary treatment, Aguilar et al. (2008) studied a 450 m^2 artificial wetland with three units serially connected (first and second planted with *Typha* sp. and the third with *Scirpus americanus*). High removal efficiencies were registered for TKN (inlet concentration of 117–267 mg/L^{-1}) with 92–94% and for sulfate (inlet concentration of 1597–2518 mg/L^{-1}) with 88–92%. Kucuk et al. (2003) investigated the performance of a HSF CW planted with *P. australis* for NH_4-N removal, acting as an advanced treatment unit when subject to biologically treated tannery effluents. The COD, phosphorus, and total Cr removal was also evaluated. The experimental results indicated that nearly complete (99%) NH_4-N removal was obtained at seven days hydraulic retention time and at an initial concentration of 20 mg/L^{-1}. However, the COD, PO_4-P, and total Cr removal was low and did not change significantly with the hydraulic retention time. CWs for tannery wastewater operating with different plant species (Calheiros et al. 2007), with different substrates (Calheiros et al. 2008a), in series (Calheiros et al. 2009a), and in parallel (Calheiros et al. 2007) were studied. For instance, the two-stage series of HSF CWs implemented by Calheiros et al. (2009a) with *P. australis* (UP series) and *T. latifolia* (UT series) provided high removal of organics from tannery wastewater, up to 88% of BOD_5 (from an inlet of 420–1000 mg/L^{-1}) and 92% of COD (from an inlet of 808–2449 mg^{-1}), and of other contaminants, such as nitrogen, operating at hydraulic retention times of two, five, and seven days.

Heavy metal pollution in an aquatic ecosystem poses a serious threat to biodiversity, not to mention the health hazards in humans that it may constitute (Rai 2009). Some authors recently tested the potential of CWs for Cr and organic matter removal from tannery wastewater (Table 4.1) (Kucuk et al. 2003; Tiglyene et al. 2005; Aguilar et al. 2008). Dotro et al. (2009, 2010a,b) carried out extensive work related to the evaluation of Cr removal mechanisms in CWs treating tannery wastewater. Mant et al. (2006) investigated the rate of removal of Cr from a synthesized solution and the tolerance of plants to this heavy metal using small experimental units in order to assess the potential of larger CWs to be used in phytoremediation. Results showed that the systems with *Pennisetum purpureum*, *Brachiaria decumbens*, and *P. australis* achieved removal efficiencies of 97–99% for an inlet solution containing 10 and 20 mg Cr per L^{-1} with 97–98% of all chromium taken up on below-ground tissues. However, *P. australis* was found not to be suitable for this purpose in tropical environments because it did not grow well. Bragato et al. (2004) evaluated a FWS CW planted with *P. australis* and *Carex* ssp. in terms of reducing high concentrations of Cr, sulfate, chloride, and sodium. Both macrophytes were able to absorb and store moderate amounts of the pollutant without showing toxic or osmotic stress symptoms. Pollutant wastewater removals of up to 71% for Cr(III) (mean inlet concentration: 0.41 mg/L^{-1}), 62% for Cl^- and SO_4^{2-} (mean inlet concentration: 1805 and 1497 mg/L^{-1}, respectively), and 59% for Na^+ (mean inlet concentration: 1776 mg/L^{-1}) were achieved. Tiglyene et al. (2005) investigated the potential of CWs (capacity: 120 L, diameter: 50 cm) with *P. australis* to remove chromium from concentrated tannery effluent in comparison to unplanted soil under arid climate conditions and concluded that these systems constitute a viable economic alternative

TABLE 4.1
Examples of Use of Constructed Wetlands for Tannery Wastewater

Reference	Country	Type of Treatment	Flow Type	Media Type	Plant Species	Area (m^2)	Flow (m^3 day^{-1})	HRT (day)	Organic Removal (kg BOD ha^{-1} day^{-1})	COD (mg/L)	COD Removal (%)	BOD (mg/L)	BOD Removal (%)	Chromium (mg/L)	Chromium Removal (%)
	–	–	HF	Soil	*G. maxima*	–	–	5	–	1160	87	–	–	–	–
Daniels (1998)	–	–	HF	Soil	*P. australis*	–	–	5	–	1160	70	–	–	–	–
Kucuk et al. (2003)	Turkey	Advanced treatment	HF	Gravel and sand	*P. australis*	378	–	8	–	Max. 210	Max. 30	–	–	0.20	43–55
Kaseva and Mbuligwe (2010)	Tanzania	Secondary	HF	Crushed pumice and limestone	*P. mauritianus*	0.45	0.045 ± 0.005	1.80	–	–	–	–	–	371.70 ± 4448	99.83 ± 0.19
Tiglyene et al. (2005)	Marocco	Chromium removal	VF	Gravel and soil	*P. australis*	0.79	0.05	–	–	530–1216	63–84.6	–	–	2.2–3	97–100
Emmanuel and Anand (2007),	India	Secondary	HF	Sand	*Typha*	–	47.5	–	451	1347	78	649	93	–	–
UNIDO (2001)	India	–	HF	Pebbles	–	–	49.8	–	143	1838	59.1	615	83.4	–	–
Dotro et al. (2010b)	USA	–	HF	Pea gravel	*Typha* spp.	0.31	0.011	4	134	–	–	526	95–99	5	90–99
Bragato et al. (2004)	Italy	–	FWS	Soil	*P. australis*	60	3.8	3 – 3.5	–	–	–	–	–	0.41	Max. 44
	Italy	–	FWS	Soil	*Carex spp.*	60	3.8	3 – 3.5	–	–	–	–	–	0.41	Max. 71
Aguilar et al. (2008)	México	Tertiary	HF	–	*Typha* spp., *S. americanus*	450	28.8	2	–	12,340–17,520	96–98	675–1320	93–95	22–31	99
Calheiros et al. (2007)	Portugal	Secondary	HF	Expanded clay	*T latifolia*	1.2	0.07	3	225–402	1755–2669	57%–73%	740–1300	46–57	0.01	<0.001
Calheiros et al. (2008a)	Portugal	Secondary	HF	Expanded clay	*T latifolia*	1.2	0.1	2	114–306	1751–2100	48–56	620–860	40–48	0.043	<0.001
Calheiros et al. (2009a)	Portugal	Secondary	HF	Expanded clay	*P. australis*	2.4	0.22	2	318–529	1354–2138	57–67	720–1000	48–59	0.370	0.083
Calheiros et al. (2012b)	Portugal	Tertiary	HF	Expanded clay	*A. donax*	72	4	2	6–42	68–285	53–79	16–88	59–90	–	–
	Portugal	Tertiary	HF	Expanded clay	*A. donax*	72	10	1	25–363	190–425	51–69	20–220	60–89	–	–

in comparison to purely chemical approaches. For a COD inlet of 530–1216 mg/L^{-1}, an average removal of 74% was achieved concerning the planted system, and for the unplanted, the average removal was 60.5%. Average removal of 99% was achieved for a total Cr inlet of 534–1000 mg/L^{-1}. Reeds have also been used to dewater tannery sludge where anaerobic conditions and high levels of chrome (21000 mg Cr per L^{-1}) were present (Daniels 2001a,b).

4.6 CONCLUSION

The potential of CWs as a phytotechnology for phytoremediation purposes concerning tannery wastewater treatment was here presented. The need for environmentally friendly and economically feasible wastewater treatment systems is a real demand worldwide. Emphasis was given here to tannery wastewater because the search for the best available technologies to accomplish the legal discharge targets may contribute, in a certain way, to the preservation of this industry, which, in several countries, is considered to be of great importance because of its historical and economic value. From the available literature, detailed documented information on CWs implementation in the tannery industry in several places around the globe and with different treatment purposes, such as organic and inorganic pollutant removal, was compiled here. The data reinforce the selection of CWs as a treatment solution in the sector because, besides all the previously described advantages of these systems, they allow coping with the variation in composition along time. The CWs' performance and the behavior of selected plants have been assessed in pilot scale and field conditions, profiting from the real operational conditions and heterogeneity of the wastewater occurring in the industrial production process. Moreover, the successful application of a phytotechnology, such as CWs, requires an integrated approach for each specific wastewater type, which must consider plant and substrate selection, microflora dynamics, economics, and public acceptance. Also as important is the level of water quality required in each situation. CWs proved to be a viable phytotechnology for reducing the wastewater organic and inorganic (especially chromium) content and, to lower extents, nutrient removal.

ACKNOWLEDGMENTS

Cristina S. C. Calheiros wishes to thank a research grant from Fundação para a Ciência e Tecnologia (FCT), Portugal (SFRH/BPD/63204/2009). Authors thank the support of AdI, PRIME - IDEIA Programme, Project n. 70/00324-Planticurt and FCT Project POCI/AMB/60126/2004.

REFERENCES

Aguilar, J. R. P., J. J. P. Cabriales, and M. M. Vega. (2008). Identification and characterization of sulfur-oxidizing bacteria in an artificial wetland that treats wastewater from a tannery. *International Journal of Phytoremediation*, *10*, 359–370.

Allen, W. C., P. B. Hook, J. A. Biederman, and O. R. Stein. (2002). Temperature and wetland plant species effects on wastewater treatment and root-zone oxidation. *Journal of Environmental Quality*, *31*, 1011–1016.

Alves, M. M., C. G. G. Beça, R. G. Carvalho, J. M. Castanheira, M. C. S. Pereira, and L. A. T. Vasconselos. (1993). Chromium removal in tannery wastewaters "polishing" by *Pinus sylvestris* bark. *Water Research*, *27*, 1333–1338.

Badkoubi, A., H. Ganjidoust, A. Ghaderi, and A. Rajabi. (1998). Performance of a subsurface constructed wetland in Iran. *Water Science and Technology*, *38*, 345–350.

Balasubramanian, S., V. Pugalenthi, K. Anuradha, and S. Chakradhar. (1999). Characterization of tannery effluents and the correlation between TDS, BOD and COD. *Journal of Environmental Science Health*, *2*, 461–478.

Bragato, C., V. Rossignolo, and M. Malagoli. (2004). Evaluation of plant efficiency in a constructed wetland receiving treated tannery wastewater. In Proceedings of the IWA 6th International Conference on Waste Stabilization Ponds and the 9th International Conference on Wetland Systems, Avignon, France.

Brisson, J., and F. Chazarenc. (2008). Maximizing pollutant removal in constructed wetlands: Should we pay more attention to macrophyte species selection? *Science of the Total Environment*, *407*, 3923–3930.

Brix, H., and C. A. Arias. (2005). The use of vertical flow constructed wetlands for on-site treatment of domestic wastewater: New Danish guidelines. *Ecological Engineering*, *25*, 491–500.

Brix, H., H. Koottatep, and C. H. Laugesen. (2007). Wastewater treatment in tsunami affected areas of Thailand by constructed wetlands. *Water Science and Technology*, *56*, 69–74.

Bulc, T. G., and A. Ojstršek. (2008). The use of constructed wetland for dye-rich textile wastewater treatment. *Journal of Hazardous Materials*, *155*, 76–82.

Calheiros, C. S. C., A. O. S. S. Rangel, and P. M. L. Castro. (2007). Constructed wetland systems vegetated with different plants applied to the treatment of tannery wastewater. *Water Research*, *4*, 1790–1798.

Calheiros, C. S. C., A. O. S. S. Rangel, and P. M. L. Castro. (2008a). Evaluation of different substrates to support the growth of *Typha latifolia* in constructed wetlands treating tannery wastewater over long-term operation. *Bioresource Technology*, *99*, 6866–6877.

Calheiros, C. S. C., A. O. S. S. Rangel, and P. M. L. Castro. (2008b). The effects of tannery wastewater on the development of different plant species and chromium accumulation in *Phragmites australis*. *Archives of Environmental Contamination and Toxicology*, *55*, 404–414.

Calheiros, C. S. C., A. O. S. S. Rangel, and P. M. L. Castro. (2009a). Treatment of industrial wastewater with two-stage constructed wetlands planted with *Typha latifolia* and *Phragmites australis*. *Bioresource Technology*, *100*, 3205–3213.

Calheiros, C. S. C., A. F. Duque, A. Moura, I. S. Henriques, A. Correia, A. O. S. S. Rangel, and P. M. L. Castro. (2009b). Changes in the bacterial community structure in two-stage constructed wetlands with different plants for industrial wastewater treatment. *Bioresource Technology*, *100*, 3228–3235.

Calheiros, C. S. C., A. F. Duque, A. Moura, I. S. Henriques, A. Correia, A. O. S. S. Rangel, and P. M. L. Castro. (2009c). Substrate effect on bacterial communities from constructed wetlands planted with *Typha latifolia* treating industrial wastewater. *Ecological Engineering*, *35*, 744–753.

Calheiros, C. S. C., A. Teixeira, C. Pires, A. R. Franco, A. F. Duque, L. F. Crispim, S. C. Moura, and P. M. L. Castro. (2010). Bacterial community dynamics in horizontal flow constructed wetlands with different plants for high salinity industrial wastewater polishing. *Water Research*, *44*, 5032–5038.

Calheiros, C. S. C., V. Bessa, R. B. R. Mesquita, H. Brix, A. O. S. S. Rangel, and P. M. L. Castro. (2011). Implementação de leitos de plantas para o tratamento de águas residuais em unidades de turismo de habitação. XI Congresso Nacional de Engenharia do Ambiente, May, 20–21, 2011, Lisboa, Portugal.

Calheiros, C. S. C., G. Silva, P. V. B. Quitério, L. F. C. Crispim, S. C. Moura, H. Brix, and P. M. L. Castro. (2012a). Toxicity of high salinity tannery wastewater and effects on constructed wetland plants. *International Journal of Phytoremediation*. *14* (7), 669–680.

Calheiros, C. S. C., P. V. B. Quitério, G. Silva, L. F. C. Crispim, S. C. Moura, H. Brix, and P. M. L. Castro. (2012b). Use of constructed wetland systems with *Arundo* and *Sarcocornia* for polishing high salinity tannery wastewater. *Journal of Environmental Management*, *95*, 66–71.

Castro, S., L. C. Davis, and L. E. Erickson. (2005). Natural, cost-effective, and sustainable alternatives for treatment of aircraft deicing fluid waste. *Environmental Progress*, *24*, 26–33.

COTANCE. (2002). Science and Technology in the Tanning Industry: Contributions to sustainable development. Confederation of Tanning Industries of the European Union. European tanners.

Cotman, M., J. Zagorc-Končan, and A. Žgajnar-Gotvajn. (2004). The relation between composition and toxicity of tannery wastewater. *Water Science and Technology*, *49*, 39–46.

Daniels, R. P. (1998). You're now entering the root zone: Investigation: The potential of reed beds for treating waste waters from leather manufacture. *World Leather*, *11*, 48–50.

Daniels, R. P. (2001a). Enter the root-zone: Green technology for the leather manufacturer, Part 1. *World Leather*, *14*, 63–67.

Daniels, R. P. (2001b). Enter the root-zone: Green technology for the leather manufacturer, Part 3. *World Leather*, *14*, 85–88.

Daniels, R. P. (2007). Tannery effluent and reed beds: Working with nature. *Journal of American Leather Chemists Association*, *102*, 248–253.

Daniels, R. P. (2008). Giving nature a chance. *S&V African Leather*, *2*, 17–21.

Dotro, G., P. Palazolo, and D. Larsen. (2009). Chromium fate in constructed wetlands treating tannery wastewaters. *Water Environment Research*, *81*, 617–625.

Dotro, G., D. Larsen, and P. Palazolo. (2010a). Preliminary evaluation of biological and physical-chemical chromium removal mechanisms in gravel media used in constructed wetlands. *Water, Air Soil Pollution*. DOI: 10.1007/s11270-010-0495-9

Dotro, G., D. Larsen, and P. Palazolo. (2010b). Treatment of chromium-bearing wastewaters with constructed wetlands. *Water and Environment Journal.* DOI:10.1111/j.1747-6593.2010.00216.

Dotro, G., O. Tujchneider, M. Paris, A. Faggi, and N. Piovano. (2010c). Constructed wetlands for tannery wastewater treatment in Argentina: First trials. IWA Specialist Group on Use of Macrophytes in Water Pollution Control: Newsletter No. 36, 9–12.

Durai, G., and M. Rajasimman. (2011). Biological treatment of tannery wastewater: A review. *Journal of Environmental Science and Technology*, *4*, 1–17.

EC. (2011). Integrated Pollution Prevention and Control (IPPC), draft reference document on best available techniques for the tanning of hides and skins. BAT Reference Document (BREF). European Commission. Institute for Prospective Technological Studies. European IPPC Bureau, Seville, Spain.

Emmanuel, K. V., and G. Anand. (2007). Reedbeds: Secondary and tertiary performance for tannery and other industrial effluents. *World Leather*, 41–43.

Floqi, T., D. Vezi, and I. Malollari. (2007). Identification and evaluation of water pollution from Albanian tanneries. *Desalination*, *213*, 56–64.

Franco, A. R., C. S. C. Calheiros, C. C. Pacheco, P. De Marco, C. M. Manaia, and P. M. L. Castro. (2005). Isolation and characterization of polymeric galloyl-ester-degrdaing bacteria from a tannery discharge place. *Microbial Ecology*, *50*, 550–556.

Giannetti, B. F., S. H. Bonilla, and C. M. V. B. Almeida. (2004). Developing eco-technologies: A possibility to minimize environmental impact in Southern Brazil. *Journal of Cleaner Production*, *12*, 361–368.

Haberl, R., S. Grego, G. Langergraber, R. H. Kadlec, A.-R. Cicalini, S. Martins Dias, J. M. Novais, S. Aubert, A. Gerth, H. Thomas, and A. Hebner. (2003). Constructed wetlands for the treatment of organic pollutants. *Journal of Soils and Sediments*, *3*, 109–124.

Kaseva, M. E., and S. E. Mbuligwe. (2010). Potential of constructed wetland systems for treating tannery industrial wastewater. *Water Science and Technology*, *61* (4), 1043–1052.

Narváez, L., C. Cunill, R. Cáceres, and O. Marfà. (2011). Design and monitoring of horizontal subsurface-flow constructed wetlands for treating nursery leachates. *Bioresource Technology*, *102*, 6414–6420.

Huertas, E., M. Folch, M. Salgot, I. Gonzalvo, and C. Passarell. (2006). Constructed wetlands effluent for streamflow augmentation in the Besòs River (Spain). *Desalination*, *188*, 141–147.

IUE. (2008). IUE 5: Performance for waste water treatment. IUE document on recent developments in cleaner production and environment protection in world leather sector International Union of Environment (IUE) Commission of International Union of Leather Technologists and Chemists Societies (IULTCS).

Ji, G. D., T. H. Sun, and J. R. Ni. (2007). Surface flow constructed wetland for heavy oil-produced water treatment. *Bioresource Technology*, *98*, 436–441.

Jing, S., and Y. Lin. (2004). Seasonal effect on ammonia nitrogen removal by constructed wetlands treating polluted river water in southern Taiwan. *Environmental Pollution*, *127*, 291–301.

Kadlec, R. H., R. L. Knight, J. Vymazal, H. Brix, P. Cooper, and R. Haberl. (2000). Constructed wetlands for pollution control: Processes, performance, design and operation. IWA Scientific and Technical Report No. 8. IWA Publishing, London, UK.

Kadlec, R. H., and S. Wallace. (2009). *Treatment Wetlands*, 2nd edition. Boca Raton: CRC Press, Taylor & Francis Group.

Karunyal, S., G. Renuga, and K. Paliwal. (1994). Effects of tannery effluent on seed germination, leaf area, biomass and mineral content of some plants. *Bioresource Technology*, *47*, 215–218.

Kickuth, R. (1977). Degradation and incorporation of nutrients from rural wastewaters by plant hydrosphere under limnic conditions. In *Utilization of manure by land spreading*, EUR 5672e, London, 335–343.

Kivaisi, A. K. (2001). The potential for constructed wetlands for wastewater treatment and reuse in developing countries: A review. *Ecological Engineering*, *16*, 545–560.

Kucuk, O. S., F. Sengul, and I. K. Kapdan. (2003). Removal of ammonia from tannery effluents in a reed bed constructed wetland. *Water Science and Technology*, *48*, 179–186.

Larue, C., N. Korboulewsky, R. Wang, and J. Mévy. (2010). Depollution potential of three macrophytes: Exudated, wall-bound and intracellular peroxidase activities plus intracellular phenol concentrations. *Bioresource Technology*, *101*, 7951–7957.

Lefebvre, O., N. Vasudevan, K. Thanasekaran, R. Moletta, and J. J. Godon. (2006). Microbial diversity in hypersaline wastewater: The example of tanneries. *Extremophiles*, *10*, 505–513.

Leta, S., F. Assefa, L. Gumaelius, and G. Dalhammar. (2004). Biological nitrogen and organic matter removal from tannery wastewater in pilot plant operations in Ethiopia. *Applied Microbial Biotechnology*, *66*, 333–339.

Lin, Y., S. Jing, D. Lee, Y. Chang, Y. Chen, and K. Shih. (2005). Performance of a constructed wetland treating intensive shrimp aquaculture wastewater under high hydraulic loading rate. *Environmental Pollution*, *134*, 411–421.

Lofrano, G., V. Belgiorno, M. Gallo, A. Raimo, and S. Meriç. (2006). Toxicity reduction in leather tanning wastewater by improved coagulation flocculation process. *Global NEST Journal*, *8*, 151–158.

Lofrano, G., E. Aydin, F. Russo, M. Guida, V. Belgiorno, and S. Meric. (2008). Characterization, fluxes and toxicity of leather tanning bath chemicals in a large tanning district area (IT). *Water, Air, and Soil Pollution: Focus*, *8*, 529–542.

Mæhlum, T., and P. Stålnacke. (1999). Removal efficiency of three cold-climate constructed wetlands treating domestic wastewater: Effects of temperature, seasons, loading rates and input concentrations. *Water Science and Technology*, *40*, 273–281.

Maine, M. A., N. Suñe, H. Hadad, G. Sánchez, and C. Bonetto. (2006). Nutrient and metal removal in a constructed wetland for wastewater treatment from a metallurgic industry. *Ecological Engineering*, *26*, 341–347.

Mant, C., S. Costa, J. Williams, E. Tambourgi. (2006). Phytoremediation of chromium by model constructed wetland. *Bioresource Technology*, *97*, 1767–1772.

Mantovi, P., M. Marmiroli, E. Maestri, S. Tagliavini, S. Piccinini, and N. Marmiroli. (2003). Application of a horizontal subsurface flow constructed wetland on treatment of dairy parlor wastewater. *Bioresource Technology*, *88*, 85–94.

Masi, F., G. Conte, N. Martinuzzi, and B. Pucci. (2002). Winery high organic content wastewaters treated by constructed wetlands in Mediterranean climate. In Proceedings of the 8th International Conference on Wetland Systems for Water Pollution Control. Arusha, Tanzania. 274–281.

Matamoros, V., C. Arias, H. Brix, and J. M. Bayona. (2009). Preliminary screening of small-scale domestic wastewater treatment systems for removal of pharmaceutical and personal care products. *Water Research*, *43*, 55–62.

Mays, P. A., and G. S. Edwards. (2001). Comparison of heavy metal accumulation in a natural wetland and constructed wetlands receiving acid mine drainage. *Ecological Engineering*, *16*, 487–500.

Mbuligwe, S. E. (2005). Comparative treatment of dye-rich wastewater in engineered wetland systems (EWSs) vegetated with different plants. *Water Research*, *39*, 271–280.

Meers, E., F. M. G. Tack, I. Tolpe, and E. Michels. (2008). Application of a full-scale constructed wetland for tertiary treatment of pig manure: Monitoring results. *Water, Air and Soil Pollution*, *193*, 15–24.

Meriç, S., E. De Nicola, M. Iaccarino, M. Gallo, A. Di Gennaro, G. Morrone, M. Warnau, V. Belgiorno, and G. Pagano. (2005). Toxicity of leather tanning wastewater effluents in sea urchin early development and in marine microalgae. *Chemosphere*, *61*, 208–217.

Mina, I. A.-P., M. Costa, A. Matos, C. S. C. Calheiros, and P. M. L. Castro. (2011). Polishing domestic wastewater on a subsurface flow constructed wetland: Organic matter removal and microbial monitoring. *International Journal of Phytoremediation*, *13*, 947–958.

Namasivayam, C., and W. H. Höll. (2004). Chromium(III) removal in tannery waste waters using Chinese reed (*Miscanthus Sinensis*), a fast growing plant. *European Journal of Wood and Wood Products*, *62*, 74–80.

Nelson, M., T. Burgess, A. Alling, N. Alvarez-Romo, W. Dempster, R. Walford, and J. Allen. (1993). Using a closed ecological system to study earth's biosphere: Initial results from Biosphere 2. *BioScience*, *43*, 225–236.

Nelson, M., M. Finn, C. Wilson, B. Zabel, M. Thillo, P. Hawes, and R. Fernandez. (1999). Bioregenerative recycling of wastewater in Biosphere 2 using a constructed wetland: 2-year results. *Ecological Engineering*, *13*, 189–197.

Nivala, J., M. B. Hoos, C. Cross, S. Wallace, and G. Parkin. (2007). Treatment of landfill leachate using an aerated, horizontal subsurface-flow constructed wetland. *Science of the Total Environment*, *380*, 19–27.

Öövel, M., A. Tooming, T. Mauring, and Ü. Mander. (2007). Schoolhouse wastewater purification in a LWA-filled hybrid constructed wetland in Estonia. *Ecological Engineering*, *29*, 17–26.

Oral, R., S. Meriç, E. D. Nicola, D. Petruzzelli, C. D. Rocca, and G. Pagano. (2007). Multi-species toxicity evaluation of chromium-based leather tannery wastewater. *Desalination*, *211*, 48–57.

Panda, S. K., and S. Choudhury. (2005). Chromium stress in plants. *Brazilian Journal of Plant Physiology*, *17*, 95–102.

Prabu, P. C., and C. U. Udayasoorian. (2007). Treatment of pulp and paper mill effluent using constructed wetland. *Electronic Journal of Environmental, Agricultural and Food Chemistry*, *6*, 1689–1701.

Quanrud, D. M., M. M. Karpiscak, K. E. Lansey, and R. G. Arnold. (2004). Transformation of effluent organic matter during subsurface wetland treatment in the Sonoran desert. *Chemosphere*, *54*, 777–788.

Rai, P. K. (2009). Heavy metal phytoremediation from aquatic ecosystems with special reference to macrophytes. *Critical Reviews in Environmental Science and Technology*, *39*, 697–753.

RCS. (2006). The Ramsar convention manual: A guide to the convention on wetlands (Ramsar, Iran, 1971), 4th Edition. Ramsar Convention Secretariat, RCS, Gland, Switzerland.

Renaudin, J. F. (1996). The application of reed technology to tannery effluent. *World Leather*, 51–52.

Revitt, D. M., R. B. E Shutes, R. H. Jones, M. Forshaw, and B. Winter. (2004). The performances of vegetative treatment systems for highway runoff during dry and wet conditions. *Science of the Total Environment*, 334–335, 261–270.

Scholz, M., and B. Lee. (2005). Constructed wetlands: A review. *The International Journal of Environmental Studies*, *62*, 421–447.

Seidel, K. (1953). Pflanzungen zwischen Gewässern und land. *Journal Max Planck Institute*, 17–20.

Seo, D. C., S. H. Hwang, H. J. Kim, J. S. Cho, H. J. Lee, R. D. DeLaune, A. Jugsujinda, S. T. Lee, J. Y. Seo, and J. S. Heo. (2008). Evaluation of 2- and 3-stage combinations of vertical and horizontal flow constructed wetlands treating greenhouse wastewater. *Ecological Engineering*, *32*, 121–132.

Shanker, A. K., C. Cervantes, H. Loza-Tavera, and S. Avudainayagam. (2005). Chromium toxicity in plants. *Environment International*, *31*, 739–753.

Sharma, D. C., C. P. Sharma, and R. D. Tripathi. (2003). Phytotoxic lesions of chromium in maize. *Chemosphere*, *51*, 63–68.

Shukla, O. P., A. A. Juwarkar, S. K. Singh, S. Khan, and U. N. Rai. (2011). Growth responses and metal accumulation capabilities of woody plants during the phytoremediation of tannery sludge. *Waste Management*, *31*, 115–123.

Sinha, S., R. Saxena, and S. Singh. (2002). Comparative studies on accumulation of Cr from metal solution and tannery effluent under repeated metal exposure by aquatic plants: Its toxic effect. *Environmental Monitoring and Assessment*, *80*, 17–31.

Srisatit, T., and W. Sengsai. (2003). Chromium removal efficiency by *Vetiveria zizanioides* and *Vetiveria nemoralis* in constructed wetlands for tannery post-treatment wastewater. Proceedings of the Third International Conference on Vetiver and Exhibition, Guangzhou, China.

Steer, D., L. Fraser, J. Boddy, and B. Seibert. (2002). Efficiency of small constructed wetlands for subsurface treatment of single-family domestic effluent. *Ecological Engineering*, *18*, 429–440.

Sundaravadivel, M., and S. Vigneswaran. (2001). Constructed wetlands for wastewater treatment. *Critical Reviews in Environmental Science and Technology*, *31*, 351–409.

Terzakis, S., M. S. Fountoulakis, I. Georgaki, D. Albantakis, I. Sabathianakis, A. D. Karathanasis, N. Kalogerakis, and T. Manios. (2008). Constructed wetlands treating highway runoff in the central Mediterranean region. *Chemosphere*, *72*, 141–149.

Thanikaivelan, P., J. R. Rao, B. U. Nair, and T. Ramasami. (2005). Recent trends in leather making: Processes, problems, and pathways. *Critical Reviews in Environmental Science and Technology*, *35*, 37–79.

Tiglyene, S., L. Mandi, and A. Jaouad. (2005). Removal of chromium from tannery wastewater by vertical infiltration beds. *Revue des Sciences de L'Eau*, *18*, 177–198.

UNIDO. (2001). Final report of the mission of Michael Aloy in Chennai, India. United Nations Industrial Development Organization. Regional Programme for Pollution Control in the Tanning Industry in South East Asia. US/RAS/92/120. Vienna, Austria.

USEPA. (2000). A handbook of constructed wetlands, a guide to creating wetlands for agricultural wastewater, domestic wastewater, coal mine drainage, stormwater in the Mid-Atlantic Region, Volume 1: General considerations. United States Environmental Protection Agency, EPA 843B00005. Washington, DC, New York.

Verheijen, L. A. H. M., D. Wiersema, and L. W. Hulshoff Pol. (1996). Management of waste from animal product processing. *Food and Agriculture Organization of the United Nations.*

Vrhovšek, D., V. Kukanja, and T. Bulc. (1996). Constructed wetland (CW) for industrial waste water treatment. *Water Research*, *30*, 2287–2292.

Vymazal, J. (2009). The use constructed wetlands with horizontal sub-surface flow for various types of wastewater. *Ecological Engineering*, *35*, 1–17.

Vymazal, J., and L. Kröpfelová. (2008). Wastewater treatment in constructed wetlands with horizontal subsurface flow. Series: Environmental pollution, volume 14. Springer, Netherlands.

Younger, P. L. (2000). The adoption and adaptation of passive treatment technologies for mine waters in the United Kingdom. *Mine Water and the Environment*, *19*, 84–97.

Zurayk, R., B. Sukkariyah, R. Baalbaki, and D. A Ghanem. (2001). Chromium phytoaccumulation from solution by selected hydrophytes. *International Journal of Phytoremediation*, *3*, 335–350.

5 Phytoremediation of Petroleum Hydrocarbon–Contaminated Soils in Venezuela

Carmen Infante, Ismael Hernández-Valencia, Liliana López, and Marcia Toro

CONTENTS

5.1 INTRODUCTION

Soil contaminated by petroleum hydrocarbons is one of the most studied subjects for the development of remediation technologies because of the wide use of oil crude and, therefore, the high probability of spills on soils, causing severe damage to ecosystems. Among the techniques of rehabilitation of soil impacted by hydrocarbons, there has been an increasing interest in phytoremediation because it is an aesthetically pleasing, passive procedure, which is useful in simultaneously attacking a great variety of contaminants. Additionally, its use is enviromentally friendly, and the cost is lower as compared to other physical or chemical treatments. In tropical lowland areas, such as in Venezuela, phytoremediation is advantageous because of the warm and almost constant temperatures throughout the year, which favor plant growth and microflora activity if water and nutrients are provided in adequate amounts.

Some tropical soils areas are highly weathered with low nutrient contents and poor particle aggregation, so phytoremediation could maintain or improve soil quality by protecting soil from erosion and enhancing organic carbon content, particle aggregation, and water drainage (Cunningham et al. 1996; Pivetz 2001). Venezuela is one of most important petroleum producers, and bioremediation technology has been successfully used in several parts of the country to recover oil-impacted soil and to treat large amounts of waste (Infante et al. 1999; Infante 2006; Infante et al. 2010). Research on phytoremediation is recent, and most of it has been conducted under greenhouse conditions. The main aspects considered are the following: a) field studies for the preselection of potential phytoremediation species; b) effects of petroleum contamination on plant germination, survival, production, and root morphology; and c) the ability of selected plants to reduced O&G (oil and grease) or TPH (total petroleum hydrocarbon) in soil.

In this chapter, we present some of the most relevant studies on phytoremediation in Venezuela; plant species of greatest potential; the influence of type of oil; the role of rhizodegradation, including aspects of microbial activity; mycorrhizae; and the importance of using biomarkers in monitoring biodegradable fractions in the soil.

5.1.1 Selection of Species

Plants growing spontaneously in oil-contaminated soil could be a source of material to be tested for phytoremediation purposes. Merkl et al. (2004) studied the occurrence of plant species in four contaminated sites with heavy and light oil crude located in Venezuela's eastern lowlands, an important petroleum-producing region. They found 57 herbaceous species, seven of them cultivated; legumes (18 species) and grasses (19 species) were the most diverse groups. In order to choose the most promising species, the authors also studied seed propagation, as well as the tiller and root development of selected species. The results show that most of the potentially useful species are indigenous to tropical America. Legumes were the group that propagated more easily; whereas, most of the grasses and herbaceous species could not be propagated successfully. On the other hand, the most favorable root system belonged to some grasses and sedges while, in general, legumes had less-ramified but deeper-reaching roots.

Taking into account these results, Merkl et al. (2005) tested the ability of three legumes (*Calopogonium mucunoides, Centrosema brasilianum*, and *Stylosanthes capitata*) and three grasses (*Brachiaria brizantha, Cyperus aggregatus*, and *Eleusine indica*) to phytoremediate a soil contaminated with 5% w/w of heavy crude oil under greenhouse conditions. Plant biomass production and TPH reduction (total O&G and fractions thereof) were determined after a 90- and a 180-day incubation period. Soil planted with *B. brizantha* and *C. aggregates* showed a significantly lower final O&G concentration than the control. On the other hand, concentration of saturated hydrocarbons was always lower in the planted soil compared to unplanted soil, and *B. brizantha* induced the highest aromatic content reduction while stimulating microbial population and activity. A positive correlation between root biomass production and oil degradation was found, as well as one between oil degradation and root morphology. *B. brizantha* and *C. aggregatus* showed coarser roots, and *B. brizantha* presented a larger root surface area in contaminated soil (Merkl et al. 2005). Additionally, a shift of specific root length and surface area per diameter class toward higher diameters was found. The legumes died before eight weeks under the influence of the contaminant.

It has been suggested that commercial grasses could have great potential for phytoremediation because of their fast growth rates and extended and fasciculated root systems, which improve the rhizosphere's ability to degrade contaminants. Additionally, information regarding their establishment, nutrient requirements, and maintenance is well known, and seeds are readily available. Hernández-Valencia and Mager (2003) assessed the seed germination and biomass production of six commercial grasses in silt-sandy soil contaminated with 3% light oil crude. They found that, after 45 days, the hydrocarbon contamination reduced seed germination and biomass production. *Brachiaria brizantha* and *Panicum maximum* attained the highest seed germination rates in contaminated soil; whereas, *B. brizantha* exhibited the highest above-ground and root production and maximum root length. Both species were chosen to assess their ability to remediate the contaminated soil, and after 240 days, the final O&G content was 1.0% for soil planted with *P. maximum*, 1.2% for soil planted with *B. brizantha*, and 1.6% for the control (Figure 5.1).

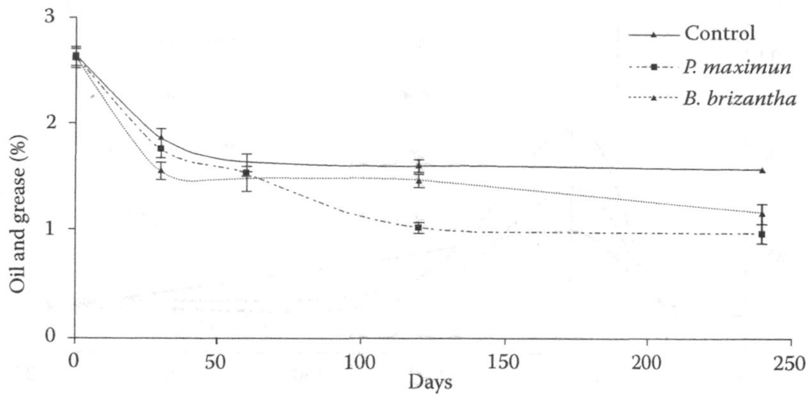

FIGURE 5.1 Changes in O&G content in polluted soils with and without grasses. (Reprinted from Hernández-Valencia, I., and D. Mage, *Bioagro, 15*, 149–155, 2003. With permission.)

5.1.2 Phytoremediation and Rhizodegradation

In case of petroleum-contaminated soil, the main mechanism of phytoremediation is rhizo-biodegradation or phytostimulation (Hutchinson et al. 2003; Merkl et al. 2005). Microbial populations and activity in the rhizosphere can be increased as a result of the presence of root exudates and can result in increased organic contaminant biodegradation in the soil. Additionally, the rhizosphere substantially increases the surface area where active microbial degradation can be stimulated, and the roots provide additional surface area for microbes to grow on and a pathway for oxygen transfer from the environment and, in general, improve the physical proprieties of the soil. Also degradation of the exudates can lead to co-metabolism of contaminants in the rhizosphere (Alkorta and Garbisu 2001).

Although numerous researchers have established that the primary mechanism for the disappearance of both petroleum hydrocarbons and polycyclic aromatics (PAHs) is rhizodegradation (USEPA 2000; Hutchinson et al. 2003), the way the mechanism operates is poorly understood, at least in the tropics. There is some indication that the presence of hydrocarbons may even encourage the proliferation of hydrocarbon-degrading microorganisms at the rhizosphere (Hutchinson et al. 2003; Wang et al. 2008). However, Gerhardt et al. (2009) reported that delimiting the boundary between the rhizosphere and bulk soil is almost impossible. Furthermore, it is impractical to definitively separate fine roots, soil directly in contact with the roots, and rhizosphere soil into different samples for analyses.

In order to test changes in microbial activity in the soil during phytoremediation mediated by *Panicum maximum* and *Brachiaria brizantha*, dehydrogenase activity and microbial biomass C were assessed (Mager 2002). Microbial C biomass showed the same trend for all the treatments considered: a progressive increase until day 60 and then a decrease until day 240 (Figure 5.2). For the same date, statistical differences were not detected between treatments. In the case of dehydrogenase activity, the higher values for all treatments were registered on day 30, and only on day 120 and 240, statistical differences were found between treatments, control, and *P. maximum* being higher than *U. brizantha* (Figure 5.3). Additionally, a weak positive linear correlation between microbial carbon content and dehydrogenase activity ($r = 0.52$, $n = 15$, $p < 0.01$) was found for the control treatment but not for grasses. A higher reduction of O&G content in the presence of grasses is also confirmed. However, the relationship between the stimulation of the microbial activity in the rhizosphere and the increment of TPH degradation is not clear because microbial biomass C

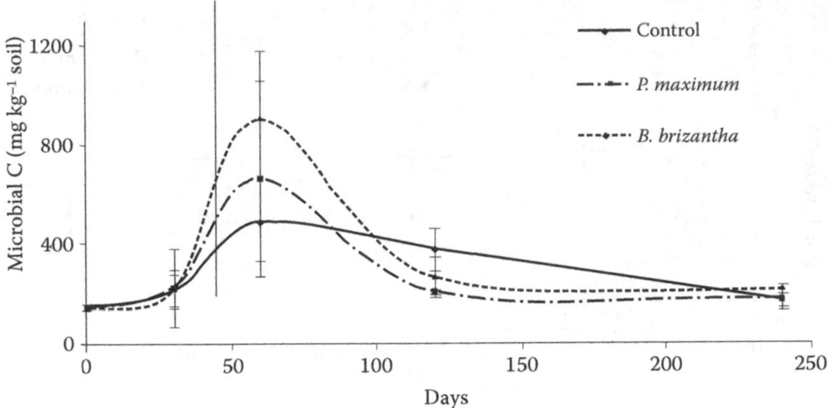

FIGURE 5.2 Changes in microbial C biomass in polluted soils with and without grasses. Bars represent standard deviation. (Reprinted from Mager, D., Evaluación de la capacidad de gramíneas tropicales para fitorremediar suelos contaminados con hidrocarburos de petróleo, Dissertation biology degree. Escuela de Biología, Facultad de Ciencias, Universidad Central de Venezuela, 2000. With permission.)

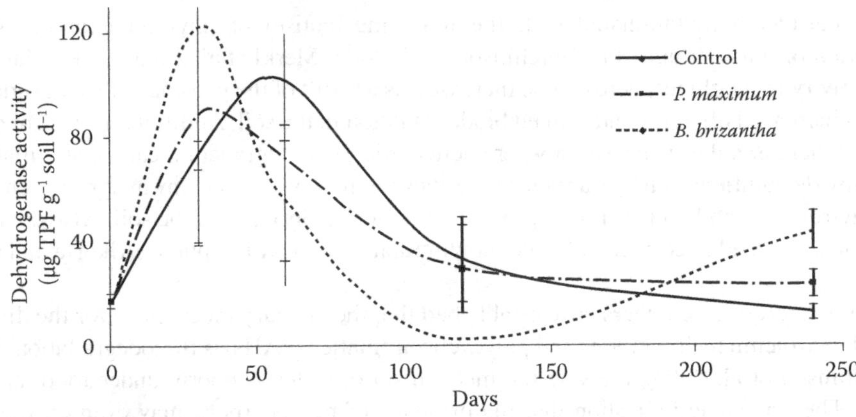

FIGURE 5.3 Changes in dehydrogenase activity in polluted soils with and without grasses. Bars represent standard deviation. (Reprinted from Mager, D., Evaluación de la capacidad de gramíneas tropicales para fitorremediar suelos contaminados con hidrocarburos de petróleo, Dissertation biology degree. Escuela de Biología, Facultad de Ciencias, Universidad Central de Venezuela, 2000. With permission.)

and dehydrogenase activity did not always show differences between soil with and without plants during the experiment. This means there is another factor different from microbial and enzymatic activity in the rhizosphere that exerts a greater influence on soil phytoremediation and, in our case, microbial C biomass and dehydrogenase activity are not good indexes to evaluate the role of microorganisms and enzymes in soil biodecontamination.

While conducting studies on phytoremediation in tropical savanna soils with *Brachiaria brizantha*, Merkl et al. (2006) found that plants caused an inhibition of microbial growth and activity (carbon dioxide evolution rates) with respect to nonrhizosphere soil. The authors concluded that, given the fact that the enhancement of crude oil degradation with *Brachiaria brizantha* could not clearly be correlated to microbial numbers and activity, others factors, such as oxygen availability, plant enzymes, and synergistic degradation by microbial consortia, have to be considered. Although enhanced microbial growth and activity are probably a key component in rhizodegradation, several authors could not find increased microbial numbers or activity in planted soil but, nevertheless, found superior degradation of contaminants (Dzantor et al. 2000; Merkl and Schultze-Kraft 2006).

5.1.3 Types of Hydrocarbon and Phytoremediation

All hydrocarbon constituents will not react in the same manner or degrade at the same rate. In general, smaller, less-complex, or lower-molecular-weight hydrocarbons will be more readily degraded by microorganisms. This means concentrations of straight-chain alkanes might disappear before branched alkanes, which will disappear more quickly in phytoremediation than aromatic hydrocarbons, including PAHs (Elmendorf et al. 1994). Weathered hydrocarbons appear to be more resistant to rhizodegradation, and the vegetation may have a phytostabilization effect instead of breaking down the contaminants.

Phytoremediation greenhouse experiments were carried out to study the effects of heavy crude oil and light contamination. Table 5.1 shows the results of the change of oil crude expressed as a percentage of loss of initial TPH a in sandy-loamy soil planted with *Brachiaria brizantha* (Hochst. ex A. Rich) contaminated with heavy (Merkl et al. 2005) and light crude oil (Hernández-Valencia and Mager 2003) at 50,000 and 30,000 ppm, respectively. In the cases with the heavy crude oil, organic amendments were added to improve the structure and porosity of the soil and, thus, promote the rhizo-biodegradation mechanism. Heavy crude oil–contaminated soils has a tendency toward poor physical conditioning, which is unsuitable for vigorous growth of vegetation and rhizosphere bacteria. It is therefore critical to use amendments to improve the quality of soil before planting.

Data also shows that with light crude the loss of hydrocarbon was significantly higher (50%) than with the heavy crude (22%). These results confirm the important role played by the type of hydrocarbon crude oil in phytoremediation technology. However, although the loss of heavy oil in the phytoremediation test was low (22%), this value is greater than the one recorded in the laboratory microcosm evaluating the bioremediation convention without plants and biostimulation (Infante et al. 2010).

Another important mechanism in phytoremediation is phytostabilization. Perhaps phytostabilization of heavy crude oil could be operating. Some authors suggest hydrocarbons tend to sorb to soil or root surfaces or may be incorporated into the organic material, as in humification (Hutchinson et al. 2003); humification is believed to further reduce the bioavailability of incorporated compounds. The presence of vegetation can cause increased humification, thereby further reducing both the bioavailability and the mobility of these compounds. This process can be thought of as phytostabilization even though concentrations of the compounds might not decline (McCutcheon and Schnoor 2003). Venezuela produces crude oils that rank from less than 10°API in the Orinoco Oil Belt and Boscán Oil Field to more than 40°API in the Campo Rosario Oil Field. However, most of the production and reserves correspond to heavy and extra-heavy crude oils; thus, phytoremediation could be a viable technique, although very slow, for cleaning soils contaminated with heavy crude. More research is needed to understand and improve the process. We aim to study the mechanisms of rhizodegradation and phytostabilization (humification) with this type of oil.

TABLE 5.1
Initial TPH, Residual and Loss of TPH After 180 Days in Soil Planted with *B. brizantha* and Contaminated with Heavy and Light Crude Oil

Crude Oil Type	Initial TPH (ppm)	Residual TPH (ppm)	Loss of TPH (%)	Source
Heavy	50,000	39,000	22	Merkl et al. (2005)
Light	30,000	15,000	50	Hernández-Valencia and Mager (2003)

5.1.4 Mycorrhizae in Oil Contaminated–Soil Remediation

In field experiments, Aprill and Sims (1990) showed that the rhizosphere effect depended on the plant species used. Some of the research on the rhizospheric microbial population has compared only the number of bacteria and fungi, as well as the number of bacteria able to use hydrocarbons, in rhizospheric and nonrhizospheric soil. This consideration was made by other authors, such as Cabello (1997), Leyval and Binet (1998), and Nicolotti and Egli (1998) among others, who focused their interest on the effect of hydrocarbons on mycorrhizae as a significant component of the rhizosphere.

Mycorrhizae are mutualistic symbiotic associations established between certain soil fungi and plant roots. In general, two groups of mycorrhizae can be considered: ectomycorrhizae (EC) and endomycorrhizae (EM). EC can be found in some plant species in temperate zones. Among the families with EC are the Pinaceae, Fagaceae, Betulaceae, etc. in which forest species dominate. Basidiomycetes and ascomycetes fungi, such as *Pisolithus*, *Praxillus*, *Amanita*, and others forms of EC. Several authors have observed that ectomycorrhized plants can stand oil contamination. Nicolotti and Egli (1998) observed that there were EC fungal species that resisted up to a dose of 50 g artificial crude oil per kg^{-1} soil; this fungal species colonized more than 50% of the roots of *Picea abies*. The authors found no effect of oil contamination on EC colonization by *Populus nigra*. Cairney and Meharg (1999) observed that the application of oil to soil in arctic ecosystems caused a significant decrease in colonization by EC in *Salix* sp. The application of crude oil caused a significant reduction in colonization by EC but also produced changes in the dominance of fungal species forming EC.

In the case of EM, arbuscular mycorrhizae (AM) is the most abundant among all types of mycorrhizae. Fungi able to form this symbiosis belong to phylum Glomeromycota, are ubiquitous, and are able to associate with plants of most families within the plant kingdom (Barea and Jeffries 1995). These associations play an important role in the absorption of water, phosphorus, and other elements, such as zinc and copper, that move slowly in the soil solution. AM are of particular importance under stressful environmental conditions, such as water-stressed, saline, or contaminated soils. Several authors report that Glomeromycota fungi help plants to withstand adverse conditions and abiotic stresses (Barea and Jeffries 1995; Leyval et al. 1997; Pozo et al. 2002). AM also contribute to soil aggregation, improving its structure and physical condition (Jastrow and Miller 1991). The potentiality of its application to remediate contaminated soil has been focused by several authors. Cabello (1997) observed a significant decrease of AM propagule density in soil affected by oil spills. However, some plants from contaminated sites maintained AM colonization up to 60%. Leyval and Binet (1998) observed there was no effect on colonization by AM in garlic plants grown in soil amended with 10 g anthracene per kg^{-1} soil; however, the addition of 8 g hydrocarbon per kg^{-1} soil to uncontaminated soil reduced colonization by AM in garlic plants. None of the treatments described had an adverse effect on AM colonization of maize and ryegrass. Cairney and Meharg (1999) suggest that Glomeromycota fungi species show differential sensitivity to oil pollution. For example, the number of spores of the species *Glomus mosseae* increased unlike *Entrophospora infrequens* and *Glomus microcarpum*, whose spores decreased significantly as a result of oil pollution. Volonte et al. (2005) indicated the favorable effect of AM in plants affected by oil. Franco-Ramírez et al. (2007) found that Glomeromycota fungal spores were able to tolerate hydrocarbons in oil-contaminated soil. Liu and Dalpé (2009) found that AM enhanced the nutritional status of leek plants, favoring plant establishment; soil microorganisms cooperated with AM improving the ability of mycorrhized leek to phytoremediate oil-contaminated soil.

5.1.5 Arbuscular Mycorrhizae in Venezuelan Polluted Soils

Two oil industrial areas located in northeastern Venezuela, El Tejero and Orocual, in which soil has been amended with drilling cuttings were studied to observe how the application of these waste products of the oil industry affected soil and plant recovery (Toro and Hernández-Valencia 2000). By the time the study was carried out, oily drilled cuttings were degraded to less than 1% of O&G or TPH, an index considered within acceptable limits, according to Venezuelan environmental regulatory limits for Venezuela (Decreto 2635 República de Venezuela 1998).

It was observed that El Tejero soil had low and patchy plant cover, possibly as a result of soil conditions (lower hydrocarbon concentration, texture, water availability, etc.) that favor its growth locally. In Orocual, the soil had been plowed to improve soil aeration. There was a dense plant cover, and the plants located preferentially in the grooves marked by the plow. In areas with water accumulation, typical aquatic plants belonging to the Hydrophyllaceae and Boraginaceae families were observed. Twenty-eight plant species were identified, most of them typical of disturbed habitats. Table 5.2 shows the list of species and families identified and the presence of symbionts. AM in stained roots (Phillips and Hayman 1970) and/or *Rhizobium* nodules, in the case of legumes, were observed. Seventy-one percent of plants were colonized by AM, and one species showed *Rhizobium*

TABLE 5.2
Taxonomic Classification and Presence of Symbiotic Associations, Arbuscular Mycorhizae, and Rhizobium of Plants Growing in El Tejero and Orocual Soils Contaminated with Oil[a]

Species	Family	Symbiotic Associations
Amaranthus spinosus	Amaranthaceae	AM
Heliotropium indicum	Boraginaceae	AM
Cercidium praecox	Caesalpiniaceae	AM
Senna obtusifolia	Caesalpiniaceae	AM
Senna occidentalis	Caesalpiniaceae	AM
Commelina diffusa	Commelinaceae	Not determined
Bulbostylis capillaries	Cyperaceae	No
Cyperus cuspidatus	Cyperaceae	No
Cyperus odoratus	Cyperaceae	No
Fimbristylis sp.	Cyperaceae	No
Phyllanthus caroliniensis	Euphorbiaceae	Not determined
Aeschynomene brasiliana	Fabaceae	AM
Aeschynomene histrix	Fabaceae	AM
Sesbania exasperata	Fabaceae	AM
Hydrolea spinosa	Hydrophyllaceae	Not determined
Peltaea sessiliflora	Malvaceae	AM
Peltaea trinervis	Malvaceae	AM
Sida angustissima	Malvaceae	AM
Sida rhombifolia	Malvaceae	AM
Acacia sp.	Mimosaceae	AM
Mimosa pigra	Mimosaceae	AM and *Rhizobium* nodules
Brachiaria decumbens	Poaceae	AM
Dactyloctenium aegyptium	Poaceae	AM
Lindernia crustacea	Scrophulariaceae	Not determined
Solanum nigrum	Solanaceae	AM
Solanum torvum	Solanaceae	AM
Mellochia villosa	Sterculiaceae	AM
Lantana cámara	Verbenaceae	AM

Note: AM: arbuscular mycorrhizae; No: absence of symbionts
[a] Toro and Hernandez-Valencia (2000).

nodules. The presence of AM and *Rhizobium* on plants suggests a potential symbiotic association, favoring the establishment of plants that colonize soil contaminated with oil. This highlights the potential of these families as bioremediator plants, especially the mycotrophy of the Malvaceae family, in oil-contaminated soil. The AM affinity of plants could enhance their ability to restore oil-contaminated soil. These results suggest AM help plants to tolerate hydrocarbon contamination in soil. AM have been widely reported in tropical soil, and their use in agricultural and natural ecosystems is well documented (Sieverding 1991); however, very few studies exist on mycorrhizal association for rehabilitation of soil contaminated with oil. Results in Venezuelan soil suggest plants association with AM contribute to their establishment in contaminated soil and may facilitate restoration when it is hydrocarbon-polluted.

5.1.6 Biomarkers

Organic matter from living organisms gives rise to fossil fuels (oil and coal). These fossil fuels required millions of years to form, and this is related to the accumulation and preservation of organic matter. Therefore, the amount and type of organic material preserved and deposited over the course of geological time is also a record of the evolution of species, representing an increase in the diversity and complexity of the organic compounds that make up living organisms. The question arising is the following: How can the idea of living organisms, their diversity, and living conditions be applied to the study of the origin of fossil fuels or of natural or anthropogenic organic matter in recent environments? The answer lies in a group of compounds called biomarkers, which are present in sediments, rocks, and oil extracts and, hence, the precursor-product relationship between the natural product and its analog present in the source rock organic matter (bitumen) and oils (Philp and Oung 1988).

Biomarkers are any organic compounds detected in the geosphere whose basic skeleton suggests an origin related to a natural product (Mackenzie 1984). They are complex organic molecules related to specific natural products, compounds of carbon, hydrogen, and other elements (O, S, N); present in organic extracts obtained from sediments, rocks, and oils; and have a structure similar to their biological precursors (Peters et al. 2005). The biomarkers preservation depends on the stability of the structure of the biological precursor from bacterial attack or temperature effects. Therefore, the biomarker's organic structure is sufficiently stable to remain almost unchanged during organic matter transformation in the geological time needed for fossil fuel formation. For geochemical and environmental studies, the most commonly used biomarkers correspond to compounds with a high degree of taxonomic specificity and high preservation potential, which means they have a limited number of well-defined sources and are recalcitrant to biological changes or chemical activity (Brocks and Summons 2003). Source organisms for these compounds are higher plants, phytoplankton, zooplankton, and bacteria, and their distribution in sediments, rocks, or oils depends on the preservation degree of the organic matter. Their concentration in crude oil and bitumen is low (<1%) and their principal evaluation method is gas chromatography mass spectrometry (GCMS). For identification, the distribution of biomarkers is obtained by mass chromatograms or mass fragmentograms with specific ion mass/charge ratio (m/z) (Peters et al. 2005).

Examples of these compounds are the *n*-alkanes, corresponding to a mixture of products derived from living organisms, together with those generated during the organic matter maturation from fossil fuels. The isoprenoids pristane (2,6,10,14-tetramethylpentadecane) and phytane (2,6,10,14-tetramethylhexadecane) are mainly derived from the phytyl side chain of chlorophyll (Didyk et al. 1978; Tissot and Welte 1984). In addition to chlorophyll, there are other isoprenoids sources; these include unsaturated isoprenoids in zooplankton, higher animals, tocopherols, and archaeal ether lipids (Peters et al. 2005). Terpanes are derivatives of terpenoids and originate from bacterial (prokaryotic) lipid membranes; they include several homologous series represented by acyclic, bicyclic, tricyclic, tetracyclic, and pentacyclic compounds (Peters et al. 2005). Steranes are derived from steroids from the cellular membranes of vertebrates, algae, phytoplankton, and zooplankton, including diatoms (cholesterol C_{27}), microalgae (ergosterol C_{28}), and terrestrial plants (sitosterol and stigmasterol C_{29}). Some microalgae can also synthesize sterols of 29 carbon atoms (Fleck et al. 2002). The steranes most widely used in geochemistry correspond to compounds with 27, 28, and 29 carbon atoms.

5.1.7 Biomarkers and Bioremediation Monitoring

The use of biomarkers in bioremediation studies is based on their susceptibility to alteration by biodegradation or on their resistance to this process (Moldowan et al. 1995). The alteration order may vary according to biodegradation type (aerobic or anaerobic), kinds of bacteria present, and their ability to alter oil components (Wenger and Isaksen 2002). The extent of biodegradation based on the relative abundance of various hydrocarbon classes has been established in biodegradation scales (Peters and Moldowan 1993; Wenger and Isaksen 2002), and that may be adapted to monitoring of the bioremediation of soil contaminated with crude oil (Moldowan et al. 1995), and this can be extended to its application in monitoring laboratory-scale bioremediation.

Bioremediation experiments in soil can be monitored through crude oil biodegradation, using biodegradation indicators, such as variations in SARA composition (saturates, aromatics, resins, and asphaltenes), *n*-alkanes, isoprenoid, terpane, and sterane distribution (Tissot and Welte 1984; Hunt 1995; Peters et al. 2005). Variations in the intensity of the signals obtained by GCMS (*n*-alkanes, isoprenoids, steranes, and terpanes) can establish the alteration order in the saturated hydrocarbon fraction of oil (Moldowan et al. 1995). Variations in signal intensity are used to calculate ratios, such as pristane/n-C_{17}, phytane/n-C_{18}, $C_{23\text{-}3}$/C_{30}-hopane, to monitor the progress of bioremediation (García et al. 2008; Córdova et al. 2011).

Some studies of hydrocarbon biodegradation in soil have been conducted by using such biodegradation schemes in order to monitor the degree of alteration of saturated hydrocarbons, including the sequential hydrocarbon biodegradation in a soil treated with oil under laboratory-controlled conditions (Greenwood et al. 2008); the bioremediation of soil contaminated with petroleum (Greenwood et al. 2009); and the bioremediation of weathered hydrocarbons for soil affected by diverse and very old crude oil spills for *n*-alkanes, monoaromatics, and naphthalene alteration (Gallego et al. 2010). Among other studies carried out, we can mention the study of bioremediation by means of biomarkers for heavy fuel oil spills (Mills et al. 2003; Gallego et al. 2006), and the study of diesel-fuel spills into a sand aquifer using ratios of partially degraded and resistant compounds, including *n*-alkanes (n-C_{17}), pristane, and phytane (Johnston et al. 2007).

In Venezuela, we have used the distribution of biomarkers to monitor bioremediation in soil, conducted with oils of different API gravity, in order to identify variations in the biomarkers concentration with bioremediation progress. Experiments performed over 90 days found that compounds such as *n*-alkanes, isoprenoids, and terpanes were altered (García et al. 2008; Córdova et al. 2011). An example of alteration to a crude oil (27 API) from Barinas Basin, Venezuela, with the progress of bioremediation is presented in Figure 5.4, which was subjected to bioremediation experiments using clay soil. Biomarkers were analyzed in the extracts obtained after 1, 30, and 90 days of bioremediation biodegradation. It was found that *n*-alkanes are significantly depleted, and the unresolved complex mixture (UCM) increases with the bioremediation advance. Additionally, in terpanes fragmentograms, the relative intensity of C_{30}-hopane decreases related to tricyclic terpane $C_{23\text{-}3}$. From CGMS, ratios were calculated for compounds with different biodegradation resistance to determine which are more susceptible to bioremediation. Table 5.3 presents the pristane/n-C_{17}, phytane/n-C_{18}, and $C_{23\text{-}3}$/C_{30}-hopane ratios obtained for original oil and 1-, 30-, and 90-day bioremediation tests. Pristane, phytane, and the tricyclic terpane $C_{23\text{-}3}$ are more resistant to biodegradation than *n*-alkanes C_{17}, C_{18}, and C_{30}-hopane; therefore, the calculated ratios increased with advancing soil remediation.

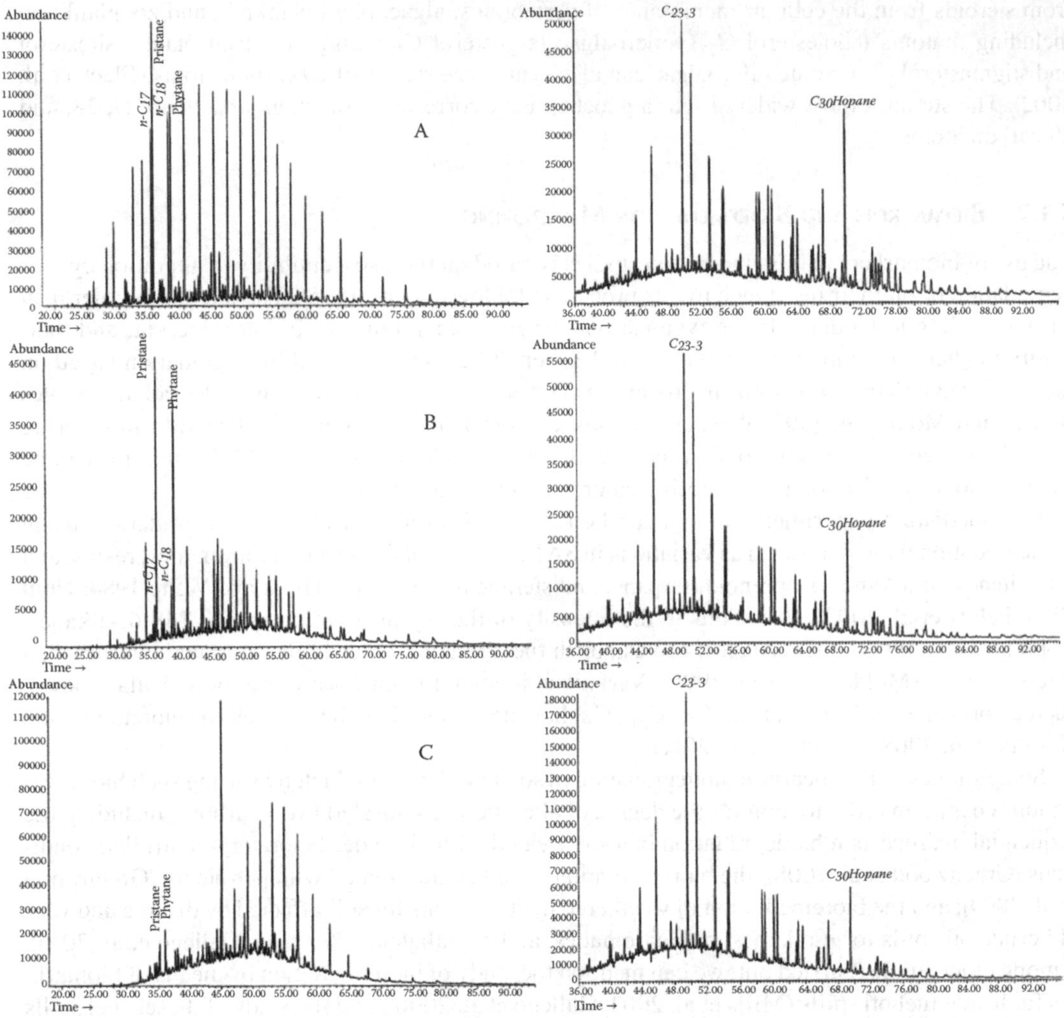

FIGURE 5.4 Mass chromatograms for representative samples for *n*-alkanes, isoprenoids (m/z = 113), and terpanes (m/z = 191) at times 1 day (A), 30 days (B), and 90 days (C). (Reprinted from García, M. G., Estudio de la biodegradación del crudo GF-1X en dos suelos de diferente textura y composición. Universidad Central de Venezuela. Facultad de Ciencias. Escuela de Química. Departamento de Geoquímica. Trabajo Especial de Grado, 2008. With permission.)

TABLE 5.3
Biomarkers Ratio in Oil from a Bioremediation Test

Ratio	Original Oil	1 Day	30 Days	90 Days
Pristane/n-C_{17}	1.8	1.8	8.6	ND[a]
Phytane/n-C_{18}	1.1	1.1	7.9	ND[a]
C_{23-3}/C_{30}-hopane	1.5	0.8	1.4	1.7

Source: García, M. G., Estudio de la biodegradación del crudo GF-1X en dos suelos de diferente textura y composición. Universidad Central de Venezuela. Facultad de Ciencias. Escuela de Química. Departamento de Geoquímica. Trabajo Especial de Grado, 2008.

[a] ND: no determined value resulting from absence of *n*-alkanes C_{17} and C_{18}.

5.1.8 Importance of Assessing Relationships in Soil Pristane/n-C_{17}, Phytane/n-C_{18}, $C_{23\text{-}3}$/C_{30}-Hopane with the Progress of Phytoremediation

The importance of using biomarkers in studies of bioremediation of hydrocarbons in soil is clear because they allow the identification of the effect of microorganisms in oil alteration; however, studies of these compounds have not been implemented in phytoremediation tests. In phytoremediation, the presence of plants and the rhizobiodegradation evolution may be monitored through the biomarkers' alteration with the objective of evaluating the effect of microorganisms associated to plants roots in oil biodegradation. While it is true that for this type of study the amount of organic compounds that may be present is very variable (from plants, microorganisms in plants roots, soils, and oils) and may be complicated, the identification of compounds from different organic fractions, the good discrimination and differentiation based on origin, would allow monitoring the effect of phytoremediation plants on n-alkanes, isoprenoids, steranes, and terpanes from oils. On the other hand, the plants might have an effect of enhancing oil alteration by facilitating the attack by microorganisms of certain compounds (solubilization) or, on the contrary, the plants might inhibit the oil alteration (fixation). To answer these questions, the study of biomarkers in phytoremediation tests is required, and it would allow a better understanding of bioremediation and phytoremediation tests.

5.2 FINAL REMARKS

The results presented above indicate that more research is needed to establish the potential of native and naturalized species, including new species, to remediate petroleum hydrocarbon–polluted soils and use the technique at field scale. Field scale practices require more certainty regarding the plant's tolerance to different types and concentrations of petroleum hydrocarbons, as well as their efficiency to remediate contaminated soils under different ecological conditions (i.e., edaphic, climatic, and biological constraints). Because phytoremediation efficiency is site-specific and varies significantly with environmental factors, issues like nutrient supply and adequate C:N:P ratios, soil textures and humidity, and diverse bulking agents should be considered in future studies. For proper assessment of phytoremediation efficiency, it is important to establish the mass balance in order to quantify how much TPH is volatilized, retained in the soil, or absorbed by plants. Presently, several studies are being carried out with different contaminants, such as crude oil, refined products (e.g., diesel, gasoline), and oil-impregnated drilled cuttings. Additionally, phytoremediation has been assayed as a sole cleanup technique or in combination with land farming. In the latter case, initial decontamination is accomplished through land farming, followed by phytoremediation (phased bioremediation) to treat the more recalcitrant fractions. As stated above, these fractions are little or not toxic; thus, plants can grow more vigorously and develop an extensive rhizosphere needed for effective decontamination and soil stabilization. In conclusion, although biological treatments are effective for the remediation of soil contaminated with medium and light crude oils, much development is still needed to treat heavy and extra-heavy oil contamination, considering that these are the most abundant petroleum types, and their production will increase as long as fossil energy remains the leading energy source.

REFERENCES

Alkorta, I., and C. Garbisu. (2001). Phytoremediation of organic contaminants in soils. *Bioresource Technology*, *79*, 273–276.

Aprill, W., and R. C. Sims. (1990). Evaluation of the use of prairie grasses for stimulating polycyclic aromatic hydrocarbon treatment in soil. *Chemosphere*, *20*, 253–265.

Barea, J. M., and P. Jeffries. (1995). Arbuscular mycorrhizas in sustainable soil-plant systems. In A. Varma, and B. Hock (Eds.), *Mycorrhiza: Structure, function, molecular biology and biotechnology* (521–561). Berlin: Springer-Verlag.

Brocks, J. J., and R. E. Summons. (2003). Sedimentary hydrocarbons, biomarkers for early life. In A. M. Davis, H. D. Holland., and K. K. Turekian (Eds.), *Treatise on geochemistry* (63–115). USA: Elsevier Pergamon.

Cabello, M. N. (1997). Hydrocarbon pollution: Its effect on native arbuscular mycorrhizal fungi (AMF). *FEMS Microbiology Ecology*, *22*, 233–236.

Cairney, J. W. G., and A. A. Meharg. (1999). Influences of anthropogenic pollution on mycorrhizal fungal communities. *Environmental Pollution*, *106*, 169–182.

Córdova, A. C., C. Infante, P. Lugo, and L. López. (2011). Comparación de la eficiencia de extracción con solventes de diferente polaridad, para un crudo mediano a dos concentraciones en un suelo sometido a biorremediación. Paper presented at the biannual meeting of Venezuelan Society of Chemistry, Vargas, Venezuela.

Cunningham, S. D., T. A. Anderson, A. P. Schwab, and F. C. Hsu. (1996). Phytoremediation of soils contaminated with organic pollutants. *Advances in Agronomy*, *56*, 55–114.

Decreto 2635 República de Venezuela. (1998). Normas para el Control de la Recuperación de Materiales Peligrosos y el Manejo de los Desechos Peligrosos. Gaceta Oficial de la República de Venezuela, no. 5212.

Didyk, B. M., B. R. T. Simoneit, S. C. Brassell, and G. Eglinton. (1978). Organic geochemistry indicator of paleoenvironmental conditions of sedimentation. *Nature*, *272*, 216–222.

Dzantor, K., E. T. Chekol, and L. R. Vough. (2000). Feasibility of using forage grasses and legumes for phytoremediation of organic pollutants. *Journal of Environmental Science and Health A*, *35*, 1645–1661.

Elmendorf, D. L., C. E. Haith, G. S. Douglas, and R. C. Prince. (1994). Relative rates of biodegradation of substituted polycyclic aromatic hydrocarbons. In R. E. Hinchee, A. Leeson, L. Semprini, and S. K. Ong (Eds.), *Bioremediation of Chlorinated and Polycyclic Aromatic Hydrocarbons* (188–202). Boca Raton: Lewis Publishers.

Fleck, S., R. Michels, S. Ferry, F. Malartew, P. Elion, and P. Landais. (2002). Organic geochemistry in a sequence stratigraphic framework: The siliciclastic shelf environment of Cretaceous series, SE France. *Organic Geochemistry*, *33*, 1533–1557.

Franco-Ramírez, A., R. Ferrera-Cerrato, L. Varela-Fregoso, J. Pérez-Moreno, and A. Alarcón. (2007). Arbuscular mycorrhizal fungi in chronically petroleum-contaminated soils in Mexico and the effects of petroleum hydrocarbons on spore germination. *Journal of Basic Microbiology*, *47*, 378–383.

Gallego, J. R., E. González-Rojas, A. I. Peláez, J. Sánchez, M. J. García-Martínez, and J. F. Llamas. (2006). Natural attenuation and bioremediation of Prestige fuel oil along the Atlantic coast of Galicia (Spain). *Organic Geochemistry*, *37*, 1869–1884.

Gallego, J. R., C. Sierra, R. Villa, A. I. Peláez, and J. Sánchez. (2010). Weathering processes only partially limit the potential of bioremediation of hydrocarbon-contaminated soils. *Organic Geochemistry*, *41*, 896–900.

García, M. G. (2008). Estudio de la biodegradación del crudo GF-1X en dos suelos de diferente textura y composición. Universidad Central de Venezuela. Facultad de Ciencias. Escuela de Química. Departamento de Geoquímica. Trabajo Especial de Grado.

García, M., C. Infante, and L. López. (2008). Estudio de la biodegradación del crudo Guafita 1X en dos suelos de diferente textura y composición mineralógica. Paper presented at the biannual meeting of the Latino American Association on Organic Geochemistry, Nueva Esparta, Venezuela.

Gerhardt, K. E., X.-D. Huang, B. R. Glick, and B. M. Greenberg. (2009). Phytoremediation and rhizoremediation of organic soil contaminants: Potential and challenges. *Plant Science*, *176*, 20–30.

Greenwood, P. F., S. Wibrow, S. J. George, and M. Tibbett. (2008). Sequential hydrocarbon biodegradation in a soil from arid coastal Australia, treated with oil under laboratory controlled conditions. *Organic Geochemistry*, *39*, 1336–1346.

Greenwood, P. F., S. Wibrow, S. J. George, and M. Tibbett. (2009). Hydrocarbon biodegradation and soil microbial community response to repeated oil exposure. *Organic Geochemistry*, *40*, 293–300.

Hernández-Valencia, I., and D. Mager. (2003). Uso de *Panicum maximum y Brachiaria brizantha* para fitorremediar suelos contaminados con un crudo de petróleo liviano. *Bioagro*, *15*, 149–155.

Hunt, J. M. (1995). *Petroleum Geochemistry and Geology*. New York: W.H. Freeman and Company.

Hutchinson, S. L., A. P. Schwab, and M. K. Banks. (2003). Biodegradation of petroleum hydrocarbons in the rhizosphere. In S. McCutcheon, and J. Schnoor (Eds.), *Phytoremediation: Transformation and Control of Contaminants* (355–386). Hoboken: John Wiley & Sons, Inc.

Infante, C. (2006). Contaminación de Suelos y Recuperación Ecológica en Venezuela. *Acta Biologia Venezuela*, *25*, 43–49.

Infante, C., F. Morales, U. Ehrmann, I. Hernández-Valencia, and N. León. (2010). Hydrocarbon bioremediation and phytoremediation in tropical soils: Venezuelan study case. In G. Płaza (Ed.), *Trends in Bioremediation and Phytoremediation* (429–451). Series: Chemosphere.

Infante, C., M. Romero, A. Arocha, D. Gilbert, and F. Brito. (1999). *In situ* bioremediation of pits from Puerto La Cruz Refinery. In *In situ bioremediation of petroleum hydrocarbon and other organic componunds. Fifth International in situ and on site bioremediation simposium*, B. Allemand, and A. Leeson (Eds.). *5*, 215–219.

Jastrow, J.D., and R. M. Miller. (1991). Methods for assessing the effects of biota in soil structure. *Agriculture, Ecosystems and Environment, 34*, 279–303.

Johnston, C. D., T. P. Bastow, and N. L. Innes. (2007). The use of biodegradation signatures and biomarkers to differentiate spills of petroleum hydrocarbon liquids in the subsurface and estimate natural mass loss. *European Journal of Soil Biology*, *43*, 328–334.

Leyval, C., and P. Binet. (1998). Effect of polyaromatic hydrocarbons in soil on arbuscular mycorrhizal plants. *Journal of Environmental Quality*, *27*, 402–407.

Leyval, C., and K. Haselwandter. (1997). Effect of heavy metal pollution on mycorrhizal colonization and function: physiological, ecological and applied aspects. *Mycorrhiza, 7*, 139–153.

Liu, A., and Y. Dalpé. (2009). Reduction in soil polycyclic aromatic hydrocarbons by arbuscular mycorrhizal leek plants. *International Journal of Phytoremediation*, *11*, 39–52.

Mackenzie, A. S. (1984). Application of biological markers in petroleum geochemistry. In J. Brooks, and D. H. Welte (Eds.), *Advances in Petroleum Geochemistry* (115–214). London: Academic Press.

Mager, D. (2002). Evaluación de la capacidad de gramíneas tropicales para fitorremediar suelos contaminados con hidrocarburos de petróleo. Dissertation biology degree. Escuela de Biología, Facultad de Ciencias. Universidad Central de Venezuela.

McCutcheon, S. C., and J. L. Schnoor. (2003). Overview of phytotransformation and control of wastes. In S. McCutcheon, and J. Schnoor (Eds.), *Phytoremediation: Transformation and Control of Contaminants* (355–386). Hoboken: John Wiley & Sons, Inc.

Merkl, N., and R. Schultze-Kraft. (2006). Influence of the tropical grass *Brachiaria brizantha (Hochst ex A Rich).* Stapf on bacterial community structure in petroleum contaminated soils. *International Journal of Soil Science*, *1*, 108–117.

Merkl, N., R. Schultze-Kraft, and M. Arias. (2006). Effect of the tropical grass *Brachiaria brizantha (Hochst ex A Rich)* Stapf on microbial population and activity in petroleum-contaminated soil. *Microbiological Research*, *161*, 80–91.

Merkl, N., R. Schultze-Kraft, and C. Infante. (2004). Phytoremediation of petroleum-contaminated soils in the tropics: Pre-selection of plant species from eastern Venezuela. *Journal of Applied Botany and Food Quality*, *78*, 185–192.

Merkl, N., R. Schultze-Kraft, and C. Infante. (2005). Assessment of tropical graminoids and legumes for phytoremediation of petroleum contaminated soils. *Water, Air & Soil Pollution*, *15*, 195–209.

Mills, M. A., J. S. Bonner, T. J. McDonald, C. H. Page, and R. L. Autenrieth. (2003). Intrinsic bioremediation of a petroleum-impacted wetland. *Marine Pollution Bulletin*, *46*, 887–899.

Moldowan, J. M., J. Dahl, M. A. McCaffrey, W. J. Smith, and J. Fetzer. (1995). Application of biological markers technology to bioremediation of refinery by-products. *Energy and Fuel*, *9*, 155–162.

Nicolotti, G., and S. Egli. (1998). Soil contamination by crude oil: Impact on the mycorrhizosphere and on revegetation potential of forest trees. *Environmental Pollution*, *99*, 37–43.

Peters, K. E., and M. Moldowan. (1993). The biomarker guide, interpreting molecular fossils in petroleum and ancient sediments. USA: Prentice Hall.

Peters, K. E., C. Walters, and M. Moldowan. (2005). *The Biomarker Guide: Biomarkers and Isotopes in the Environment and Human History*. United Kingdom: Cambridge University Press.

Phillips, J. M., and D. S. Hayman. (1970). Improved procedures for clearing roots and staining parasitic and vesicular-arbuscular mycorrhizal fungi for rapid assessment of infection. *Transactions of the British Mycological Society*, *55*, 158–161.

Philp, R. P., and J. Oung. (1988). Biomarkers. *Analytical Chemistry*, *60*, 887A–896A.

Pivetz, B. E. (2001). Phytoremediation of contaminated soil and ground water at hazardous waste sites. Ground Water Issue. United States Environmental Protection Agency. Office of Solid Waste and Emergency Response. Washington, DC. EPA/540/S-01/500.

Pozo, M. J., S. Slezack-Deschaumes, E. Dumas-Gaudot, S. Ginaninazzi, and C. Azcon-Aguilar. (2002). Plant defense responses induced by arbuscular mycorrhizal fungi. In S. Gianinazzi, H. Schuepp, J. M. Barea, and K. Haselwandter (Eds.), *Mycorrhizal Technology and Agriculture* (103–112). Basel: Birkhäuser-Verlag.

Sieverding, E. (1991). Vesicular-arbuscular mycorrhiza management in tropical agrosystems. Deutsche Gesellschaft für Technische Zusammenarbeiit (GTZ) GmbH, Eschborn, Germany.

Smith, S. E., and D. J. Read. (1997). *Mycorrhizal Symbiosis, 2nd Ed.* Cambridge: Academic Press.

Tissot, B., and D. Welte. (1984). *Petroleum Formation and Occurrence*. New York: Springer-Verlag.

Toro, M., and I. Hernández-Valencia. (2000). Evaluación de plantas capaces de incorporar a los tejidos vegetales o transformar en la rizósfera aceites minerales, con el fin de seleccionar aquellas con capacidad fitorremediadora. *Universidad Central de Venezuela. Informe Técnico presentado a INTEVEP*, 150 pg.

USEPA. (2000). Introduction to Phytoremediation. EPA 600-R-99-107, Office of Research and Development. Retrieved from http://clu-in.org/download/remed/introphyto.pdf.

Volante, A., G. Lingua, P. Cesaro, A. Cresta, and G. Berta. (2005). Influence of three species of arbuscular mycorrhizal fungi on the persistence of aromatic hydrocarbons in contaminated substrates. *Mycorrhiza*, *16*, 43–50.

Wang, J., Z. Zhang, Y. Su, W. He, F. He, and H Song. (2008). Phytoremediation of petroleum polluted soil. *Petroleum Science*, *5*, 167–171.

Wenger, L. M., and G. H. Isaksen. (2002). Control of hydrocarbon seepage intensity on level of biodegradation in sea bottom sediments. *Organic Geochemistry*, *33*, 1277–1292.

6 Fate and Transport Issues Associated with Contaminants and Contaminant By-Products in Phytotechnology

Chris O. Nwoko

CONTENTS

6.1 INTRODUCTION

Plants can remediate various contaminated compartments of the environment (e.g., water, sediments, soil, and air). The technology involved is generally referred to as phytotechnology (Cunningham et al. 1995; Cunningham et al. 1996; Salt et al. 1998). These technologies can be implemented either *in situ* or *ex situ*. Typical organic contaminants that this technology can be used for include petroleum hydrocarbons, gas condensates, crude oil, chlorinated compounds, pesticides, and explosive compounds. Similarly, inorganic contaminants include salts of sodium, potassium (salinity), heavy metals, metalloids, and radioactive materials. In addition, several emerging applications of phytotechnology are being developed, such as the capability of vegetation to utilize atmospheric carbon emissions for greenhouse-gas mitigation.

Increased rates of industrialization and urbanization and changed agricultural practices have enhanced the level of contaminants in the environment with a concomitant effect on human health. Cleaning up of the environment by removal of hazardous contaminants is a crucial problem, which needs multifaceted approaches for reaching suitable solutions. One of the advantages of using plants is that it cleans the environment, especially soils and solutions, coupled with its aesthetic, environmentally friendly, and economic qualities. Plants are autotrophic organisms synthesizing their own food and, therefore, do not require utilizing organic compounds supplied from other sources, such as sources of C, N, and energy.

The specific phytotechnology mechanism used for specific contaminants depends not only on the type of constituent and the media affected, but also on the remediation goals. Typical goals include containment, stabilization, sequestration, assimilation, reduction, detoxification, degradation, metabolization, and/or mineralization. To achieve these goals, the proper phytotechnology system must be designed, developed, and implemented using detailed knowledge of the site layout, soil characteristics, hydrology, climate conditions, analytical needs, operations and maintenance requirements, economics, public perception, and regulatory environment. For effective phytotechnology, there is need to bring to bear knowledge gained in forestry, agriculture, botany, and horticulture to solve environmental problems. Phytotechnology can be used interchangeably with phytoremediation, especially when the goal is to remove contaminants from soil and water compartments.

Several phytoremediation mechanisms are used to address the different environmental conditions that may exist at a site. The specific mechanisms that are exploited depend on several factors, including the specific contaminant, current site conditions, remedial objectives, and regulatory issues. The knowledge of plant physiological processes is the basis for the various phytoremediation mechanisms that can be used to clean up contaminated sites. Specifically, the ability of plant roots to sequester certain inorganic elements in the root zone is known as phytostabilization. Similarly, the exudation of photosynthetic products into the rhizosphere can lead to the phytostabilization of organic compounds. Alternatively, the exuded plant products can also lead to the enhanced biodegradation of organics by the soil organisms—this process is known as rhizodegradation. The ability of plants to take up and transpire large volumes of water from the subsurface also has been exploited in phytotechnology to provide hydraulic control at contaminated sites (Susarla et al. 2002). This hydraulic control can be used to prevent the horizontal migration or vertical leaching of contaminants. During the transpirational uptake of water, dissolved organic and inorganic contaminants in the subsurface can enter into the plant where they are subject to additional phytotechnology mechanisms. Specifically, once inside the plant, organic chemicals are subject to various enzymatic actions leading to the breakdown of the contaminants (Eapen et al. 2007; Watanabe 1997). This mechanism is known as phytodegradation. Similarly, the uptake and accumulation of inorganic elements into the plant tissues is known as phytoaccumulation. The uptake and subsequent transpiration of volatile contaminants through the leaves is known as phytovolatilization. Table 6.1 lists some of the potential phytoremediation mechanisms and the chemical compounds remediated.

Because phytoremediation is a technology that utilizes natural systems to stabilize, sequester, accumulate, degrade, and metabolize contaminants, ecosystems that develop as a result of this process are subject to fate and transport issues. These fate and transport issues are a concern of regulators, stakeholders, and the public and must be addressed before a phytoremediation work can be implemented. This review examined bioavailability and toxicity of contaminants to plants, whether plants grown at contaminated sites pose additional risks for further ecological exposure or food-chain accumulations, and whether the contaminant is transferred into the air or transformed into a more toxic form. At the heart of these issues is whether the contaminants or contaminant by-products are bioavailable or converted into mobile forms that can impact the groundwater system. Influences of transgenic plant species over other biotic communities are also discussed.

TABLE 6.1
Summary of Phytoremediated Chemicals

Mechanism	Process Goal	Chemicals Treated	References
Phytostabilization	Containment	Heavy metals in ponds, phenols and chlorinated solvents	McCutcheon and Schnoor (2003); Newman et al. (1997)
Phytodegradation/ phytotransformation	Remediated by destruction	Nitrobenzene, nitroethane, nitrotoluene, atrazine, chlorinated solvents (chloroform, carbon tetrachloride, hexachloroethane, tetrachloroethene, trichloroethene, dichloroethene, vinyl chloride, trichloroethanol, dichloroethanol, trichloroacetic acid, dichloroacetic acid, monochloroacetic acid, tetrachloromethane, trichloromethane), DDT; dichloroethene; methyl bromide; tetrabromoethene; tetrachloroethane; other chlorine- and phosphorus-based pesticides; polychlorinated biphenols, other phenols, and nitriles	Schnoor et al. (1995); Jacobson et al. (2003)
Phytoaccumulation/extraction	Remediated by extraction and capture	Cd, Cr, Pb, Ni, Zn, radionuclides, BTEX, pentachlorophenol, short chained aliphatic compounds	Horne (2000); Blaylock and Huang (2000)
Phytofiltration	Containment and erosion control	Heavy metals, organics, and radionuclides. Plant nutrients	Horne (2000); Nwoko et al. (2004)
Phytovolatilization	Extraction from media and release to air	Chlorinated solvent, Hg, Se	Terry et al. (1995)
Phytostimulation	Remediation by destruction	Polycyclicaromatic hydrocarbons; BTEX (benzene, ethylbenzene, toluene, and xylenes); other petroleum hydrocarbons; atrazine; alachlor; polychlorinated biphenyl (PCB); tetrachloroethane, trichloroethane, and other organic compounds	Susarla et al. (2002)

Source: Nwoko, C. O., *African Journal of Biotechnology*, 9, 2010.
Note: BTEX, benzene, toluene, ethyl benzene, xylenes; PCB, polychlorinated biphenyl.

6.2 BASIC FATE AND TRANSPORT PROCESSES OF CONTAMINANTS IN THE UNSATURATED ZONE

Physical processes move contaminants from point to point; chemical and biological processes also redistribute contaminants among different phases and chemical forms. Contaminant fate and transport therefore refers to the physical, chemical, and biological processes that impact the movement of the contaminants from point A to point B and how these contaminants may be altered while they are transported. Contaminants rarely move at the rate of groundwater, cytoplasmic fluids, and soil water

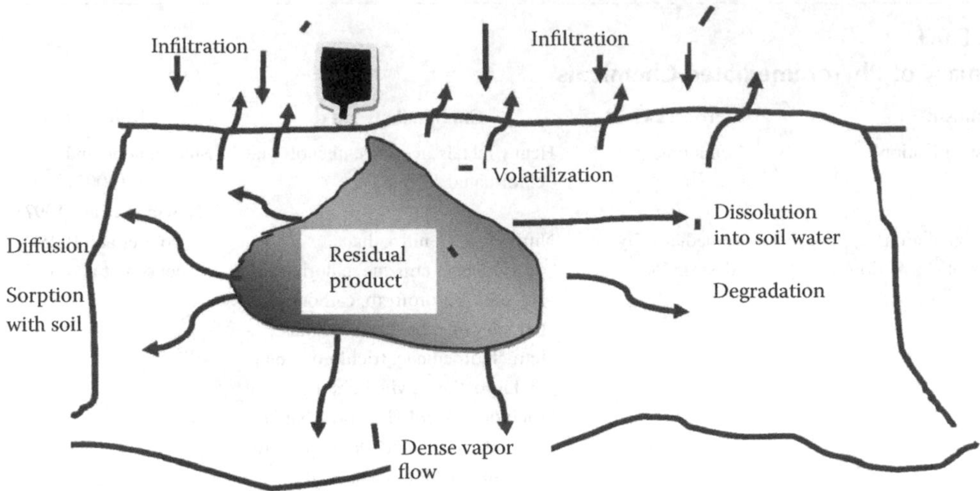

FIGURE 6.1 The basic fate and transport processes in the unsaturated zone.

because of a variety of processes, and they are generally altered when moving through these matrix materials because of various processes. Figure 6.1 shows the basic fate and transport processes in the unsaturated zone.

6.2.1 Transport Processes of Contaminants

Mass transport involves three processes, i.e., advection, dispersion, and retardation:

- Advection (displacement by soil water flow; "go with the flow"): Solutes (e.g., dissolved contaminants) are transported by the bulk portion of the flowing groundwater. Nonreactive (i.e., conservative) solutes are carried at an average rate equal to the average linear velocity of the water.
- Dispersion (spreading of contaminant mass in three dimensions during flow): There are basically two factors responsible, i.e., mechanical and diffusion.
  (a) Mechanical dispersion is a result of water moving and results when
    – flow lines have different lengths, diverging and mixing with each other
    – velocity varies across individual pores because of friction in the pore
    – pore sizes vary
  (b) Diffusion
    – Movement of contaminant as a result of concentration gradient, not flow
- Retardation: Reduction of the average velocity of the contaminant mass relative to the soil water velocity resulting from sorption of the contaminant by the geologic materials. Retardation processes remove contaminants from the groundwater during transport. Thus the contaminant concentration arriving at a certain point at a certain time is less than it would have been for a conservative (nonretarded) contaminant.

6.2.2 Chemical and Biological Processes that Affect Contaminants

This includes the following major processes:

- Volatilization: transfer from liquid phase (or solid) of contaminant to gas phase.
- Dissolution/precipitation: transfer between liquid (e.g., water) and contaminant cosolvency.

- Degradation/transformation (enhanced by biological activity): brings about a change in composition of the contaminant, chemical Processes, e.g., hydrolysis, ion exchange, complexation, oxidation/reduction.
- Abiotic degradation: a chemical transformation mechanism that degrades contaminants without microbial facilitation, can result in partial or complete degradation of contaminants. The rate of abiotic degradation is typically much slower than for biodegradation. It may also result in more toxic by-products than parent compound.
- Biological processes: Biodegradation: reactions involving the degradation of organic compounds and whose rate is controlled by the abundance of microorganisms as well as the water chemistry. An important mechanism for contaminant reduction, but can lead to undesirable daughter products. Geochemical changes may result in mobilization of certain inorganics, such as As, Mn, and Fe. Biodegradation has proved to be more successful for gasoline products (BTEX) than chlorinated solvents.
- Bioaccumulation: the process by which the levels of contamination increase in the tissues of a plant or animal and how the levels of the contamination change (generally increasing) as it passes through the food chain.
- Biological transformation: the change(s) in a chemical via biological process, but not necessarily a breakdown of the chemical.
- Food chain transfers: the exchange of contaminants between plants and animals or between animals and animals during consumption. Can also be from an inorganic substance (soil, sediment, water, etc.) to a plant or animal.

6.3 BIOAVAILABILITY OF CONTAMINANTS

Bioavailability describes the complex processes of mass transfer and uptake of contaminants into soil-living organisms, which are determined by substance properties, soil properties, the biology of the organisms, and climatic influences. The bioavailable contaminant fraction in soil represents the relevant exposure concentration for soil organisms. The bioavailability concept originates from the fact that detrimental effects in exposed organisms and ecosystems are not caused by the entire amount of the chemical compounds released to the environment, but only by a fraction of that bioavailable to them. The soil environment is characterized by a highly complex, three-phase reaction system (soil matrix, soil water, and soil pore space), manifold substance processes (distribution, sorption, transformation, and breakdown), and a high variability in space and time (Gisi et al. 1990; Koehler et al. 1999; Scheffer and Schachtschabel 1989). Figure 6.2 explains the various interactions in bioavailability as

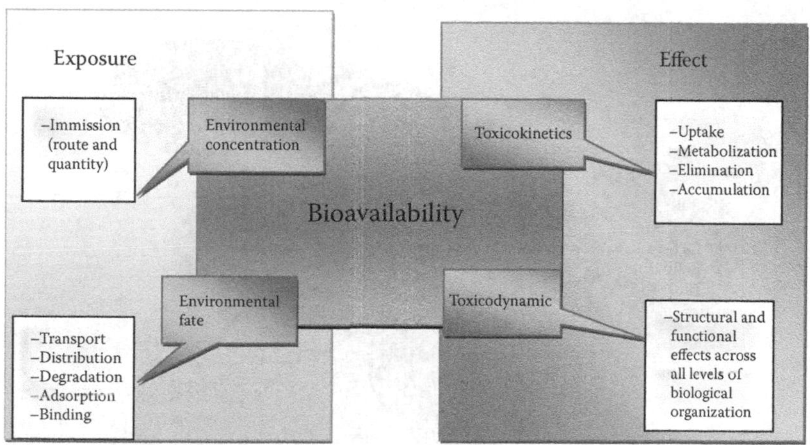

FIGURE 6.2 Interaction and bioavailability concept.

it relates to exposure and the effects of contaminants. There is no uniform strategy for assessment and evaluation of rate of bioavailability of contaminants to soil organisms because of the multiplicity of factors influencing the process. However, it is generally doubtful if it will be possible at all to develop a scientifically sound uniform strategy (Figure 6.3) given the multitude of different soils, their enormous biodiversity, and the large number of environmentally relevant chemicals with different substance properties that result in a complexity of systems that will be very hard to survey, let alone to conceptualize. The insecurity among scientists (resulting from nonexistent knowledge) concerning bioavailability has direct consequences for the regulation of chemicals and for soil protection.

Risk assessments and subsequent regulatory measures (permission of chemicals, threshold values, and remediation goals in soils) are particularly subject to considerable uncertainties as long as nominal and measured total concentrations are the basis of exposure assessment (Peijnenburg et al. 1999).

On a general note, it is difficult to quantify the bioavailability of a contaminant because conditions at the site, such as pH, soil moisture, organic matter content, and the presence (or absence) of other compounds in the soil, can affect bioavailability. The stability of the bioavailable form can also vary, depending on site conditions. Some chemicals can change form readily, and other chemical forms are extremely stable. The recalcitrance of chemicals at a site also influences bioavailability.

Bioavailability is a controversial area in both regulation and remediation. The usual assumption of 100% bioavailability of contaminants, including 100% bioavailability to plants, often overestimates the impact of the contaminant. Solubility of the constituent plays a major role in the bioavailability of contaminants to plants. Research has shown that many organic contaminants do not accumulate in significant amounts in plant tissue because they are minimally water soluble (Susarla et al. 2002). However, some organics can be taken up, particularly those that are phytovolatilized. Furthermore, many inorganics are present in insoluble forms and require the addition of chemical amendments (chelates) in order for them to become more bioavailable. Increasing bioavailability through chemical amendments can also increase the potential for exposure. The USEPA estimated in 1993 that 2–10% of the total mass ingested by animals might be soil (USEPA 1993). This percentage corresponds to 1 to 40 grams per kilogram of body weight per day. Therefore, enhancing the bioavailability of chemicals in the soil can potentially impact wildlife that resides at or near the site even if they do not consume the vegetation. If the animals do consume the vegetation, there is a concern that phytoremediation can increase bioavailability by increasing the accumulation of contaminants in the edible portions of the plant, including the fruits, seeds, and leaves. This potential is greater than if the accumulation occurred only in the stems and roots. Site owners and system designers must address bioavailability on a site-specific basis because it is dependent on the

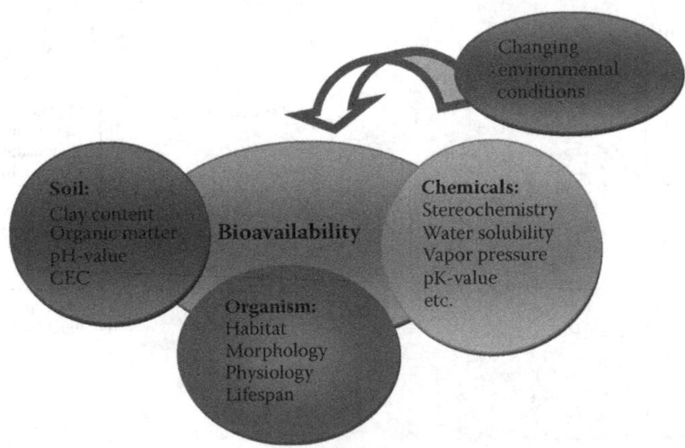

FIGURE 6.3 Complexity of bioavailability processes and parameters.

composition of the contaminant, the type of phytoremediation application, and the conditions at the site. This could include educating concerned parties regarding bioavailability and issues specific to the site.

6.3.1 Factors Influencing Bioavailability

A variety of environmental factors affects or alters the mechanisms of phytoremediation. Soil type and organic matter content can limit the bioavailability of petroleum contaminants. Water content in soil and wetlands affects plant or microbial growth and the availability of oxygen required for aerobic respiration. Temperature affects the rate at which various processes take place. Nutrient availability can influence the rate and extent of degradation in oil-contaminated soil. Finally, sunlight can transform parent compounds into other compounds, which may have different toxicity and bioavailability than the original compounds. These various environmental factors cause *weathering*—the loss of certain fractions of the contaminant mixture—with the end result being that only the more resistant compounds remain in the soil.

6.3.1.1 Soil Structure, Texture, and Organic Matter Content

Soil type is defined according to various characteristics, including structure, texture, and organic matter content. Alexander et al. (1997) identified that phenanthrene may be trapped within the soil and sorbed to the surfaces of *nanopores* (soil pores with diameters <100 nm) that are inaccessible to organisms (i.e., not bioavailable).

Soil texture can also affect phytoremediation efforts by influencing the bioavailability of the contaminant. For example, clay is capable of binding molecules more readily than silt or sand (Brady and Weil 1996). As a result, the bioavailability of contaminants may be lower in soils with high clay contents. In support of this concept, Carmichael and Pfaender (1997) found that soils with larger particles (e.g., sand) typically had greater mineralization of PAHs than soils with smaller particles (e.g., silt and clay), possibly because of the greater bioavailability of the contaminants in the sandy soils. Similarly, Edwards et al. (1982) found that the amount of 14C-anthracene taken up by soybean plants in soil was considerably lower than the amount taken up by plants in nutrient solution. The authors stated that they had anticipated this result because PAHs are known to adsorb to soil constituents and, in doing so, are no longer available for uptake from the soil.

Soil organic matter binds lipophilic compounds, thereby reducing their bioavailability (Cunningham et al. 1996). A high organic carbon content (>5%) in soil usually leads to strong adsorption and, therefore, low availability, and a moderate organic carbon content (1 to 5%) may lead to limited availability (Otten et al. 1997).

Soil type may influence the quality or quantity of root exudates, which may influence phytoremediation efforts. More specifically, research by Taiz and Zeiger (2002) indicates an interrelationship between soil type and levels of amino acids, sugars, and certain enzyme activities in the rhizosphere. On the other hand, Siciliano and Germida (1997) found that the effectiveness of phytoremediation to reduce concentrations of 2-chlorobenzoic acid in three soils from Saskatchewan was not influenced by soil type.

6.3.1.2 Weathering

Weathering processes include volatilization, evapotranspiration, photomodification, hydrolysis, leaching, and biotransformation of the contaminant. These processes selectively reduce the concentration of easily degradable contaminants with the more recalcitrant compounds remaining in the soil. The contaminants left behind are typically nonvolatile or semi-volatile compounds that preferentially partition to soil organic matter or clay particles, which limits their bioavailability and the degree to which they can be degraded (Bossert and Bartha 1984; Cunningham and Ow 1996; Bollag 1992; Cunningham et al. 1996). Carmichael and Pfaender (1997) noted that contaminant bioavailability was a major factor limiting the degradation of weathered (>60 years) PAHs.

6.4 TOXICITY OF CONTAMINANTS TO PLANTS

High concentrations of contaminants may inhibit plant growth and eliminate phytoremediation as a remedial option for site cleanup. A large number of studies, though spread over different crop plants, indicate that excessively absorbed heavy metals interfere with various biochemical, physiological, and structural aspects of plant processes that not only lead to inhibited growth but sometimes result in plant death. The toxic levels of heavy metals affect the structural and permeability properties of inner membranes and organelles; cause inhibition of enzymatic activities, nutrient imbalances, decreases in the rate of photosynthesis and transpiration (Green et al. 2003; Setia et al. 1993; Prasad and Hagemeyer 1999; Azevado et al. 2005); stimulate formation of free radicals and reactive oxygen species, resulting in oxidative stress (Sandalio et al. 2005); suppress seed germination, seedling growth (Beri et al. 1990; Beri and Setia 1996; Setia et al. 1989b), reproductive development (Setia et al. 1988; Setia et al. 1989a), seed yield, and seed quality (Liu and Kottke 2004); and induce deleterious anatomical and ultrastructural changes in crop plants (Setia and Bala 1994; Maruthi Sridhar et al. 2005). Further, consistently increasing levels of different heavy metals in the soil renders the land unsuitable for plant growth and destroys biodiversity. Remediation of soil contaminated with heavy metals is particularly challenging. Conventional engineering-based remediation technology (other than bioremediation) used for *in situ* and *ex situ* remediation of heavy metal–contaminated soil includes solidification and stabilization, soil flushing, electrokinetics, chemical reduction/oxidation, soil washing, low-temperature thermal desorption, incineration, vitrification, pneumatic fracturing, excavation/retrieval, landfill, and disposal (Saxena et al. 1999; Wenzel et al. 1999). But these are prohibitively expensive and often disturb the landscape.

6.4.1 Role of Microorganisms in Reducing Toxicity of Contaminants to Plants

Another role played by microbes involves their ability to reduce the phytotoxicity of contaminants to the point where plants can grow in adverse soil conditions, thereby stimulating the degradation of other, nonphytotoxic contaminants (Siciliano and Germida 1998). In fact, Walton et al. (1994) have hypothesized that the defenses of plants from contaminants may be supplemented through the external degradation of contaminants by microorganisms in the rhizosphere. That is to say, plants and microbes have co-evolved a mutually beneficial strategy for dealing with phytotoxicity, where microorganisms benefit from the plant exudates while the plants benefit from the ability of microorganisms to break down toxic chemicals.

Evidence in support of this hypothesis can be found in several studies. Rasolomanana and Balandreau (1987) found improved growth of rice in soil to which oil residues had been applied. The authors hypothesized that the increased growth resulted from the removal of the oil residues by various bacterial species of the genus *Bacillus*, which used plant exudates to co-metabolize the oil residues in the rhizosphere. Radwan et al. (1995) found the plant *Senecio glaucus* growing along the polluted border of an oil lake in the Kuwaiti desert. The plant roots and adhering sand particles were white and clean while the surface of the transitional zone between the root and shoot was black and polluted. The authors suggested that microbes detoxified contaminants in the rhizosphere, which allowed the plants to survive in the oil-contaminated soil.

6.5 ECOLOGICAL EXPOSURES OF CONTAMINANTS

Chemicals are introduced into the environment intentionally (e.g., fertilizers, pesticides, and herbicides) or unintentionally through accidental spillage or leaks of chemicals used in home and commercial applications (e.g., in wastes from municipal and industrial operations). These chemicals and/or contaminants leak into the soil compartment, thereby opening an exposure pathway to the plants. The uptake of these contaminants by plant roots are influenced by a number of factors, e.g., soil type, soil physicochemical properties, and the contaminant solubility coefficient.

Biomarkers of exposure can include measures of chemical concentrations in plant and animal tissue. Such measures provide insight into the magnitude of chemical exposure that organisms receive from their environment. Measures of biological response, such as biochemical concentrations (e.g., enzymes and ligands), that respond to chemical exposures can also serve as biomarkers of exposure. Examples include histopathological anomalies, such as plant tissue damage from ozone. Chemical stressors can have a detrimental effect on plant and animal communities. Exposure of plants and animals to chemical stressors can lead to increases in tissue concentrations of the chemical stressor in the plants and animals. Once stressor concentrations are above threshold levels, they can affect physiological systems within the plants and animals and can begin to have toxic effects on individuals within the population. These individual effects can lead to changes in the plant and animal community structure when chemical stressor concentrations in the environment reach levels that can affect one or more species or when the population numbers of a key species are detrimentally affected. Biomarkers of exposure, including concentrations of chemical stressors or key biomarkers collected over time within plant and animal tissues, can help to gauge the health of plant and animal communities over time. These biomarkers of chemical exposure, when coupled with other information (e.g., toxicity testing results), can provide a basis for estimating what levels of a chemical stress can and cannot be tolerated in the environment by the plant and animal communities.

Phytoremediation may not provide adequate protection for ecological receptors and could lead to the accumulation of contaminants in the food chain. Specifically, there may be concern that contamination below the ground surface will be transferred into the terrestrial portions of the plants, causing new exposure pathways. This is particularly true for phytoaccumulation, phytodegradation, and phytovolatilization. However, during phytodegradation or phytovolatilization, a small percentage of organic chemicals can remain in the cell structure of the plant (Chappell 1998). Fencing the site to prevent animals from coming into contact with the plants and a maintenance plan to address plant litter can greatly reduce risk of exposure. These issues may also be prevalent for applications utilizing phytostabilization or rhizodegradation when considering soil-borne receptors (i.e., insects, worms, burrowing animals, etc.).

6.5.1 Ecological Risk Assessment: A Requirement for Phytoremediation

Ecological risk assessment is a scientific process that determines the probability of harm to plants and animals that may be exposed to hazardous substances from a contaminated site. It is used

- To study how a plant or animal can become exposed to a contaminant
- To determine if the ecosystem at or near a site will be adversely affected

The process involves five main activities:

(a) *Preliminary problem formulation* identifies each contaminant of concern and its concentration, source, and location at the site. It also identifies how contamination is likely to spread and how plants and animals may be exposed to it, such as breathing vapors, eating, drinking, or skin contact. This information is used to create a diagram or flow chart for the ecosystem that shows the plant and animal species that could be at risk and how the contaminant could get to each species. This model is called a conceptual site model (CSM). The CSM can be modified as the risk assessment progresses. Figure 6.4 shows a Delta-creek aquatic CSM.

(b) *Ecological effects evaluation* involves finding out what amounts of a contaminant cause health problems in the plant and animal species potentially exposed. The evaluation involves reviewing literature, doing field studies, and comparing the effects of the contaminant at other sites to the effect found or anticipated at the site being studied.

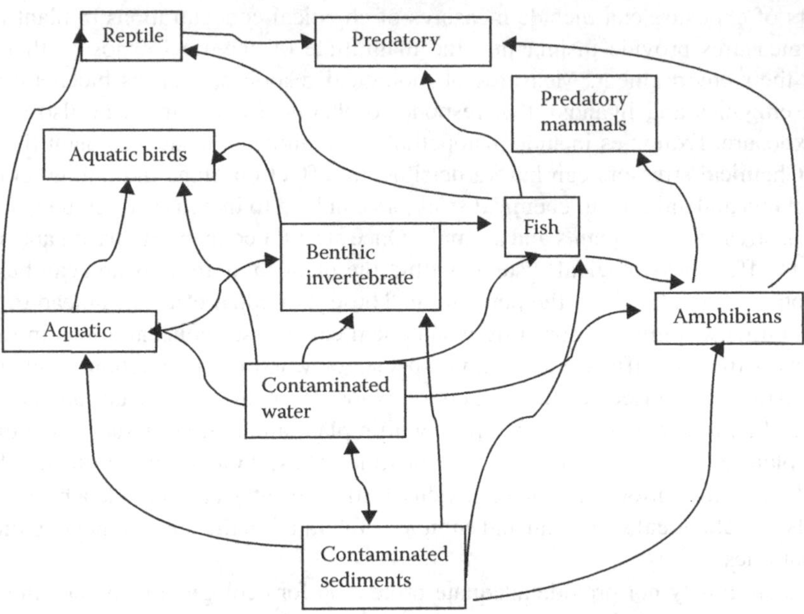

FIGURE 6.4 The delta-creek aquatic conceptual site model (arrows indicate flow of nutrients/energy/contaminants).

(c) *Preliminary risk calculation* uses mathematical formulas to determine the potential risk to each of the species from the expected exposure to a contaminant. These are compared to the highest exposure level at which no adverse effects are known to occur or "risk-based benchmarks."

(d) *Problem formulation* involves refining the preliminary problem formulation using information specific to the site. At this point, the most sensitive or important plant and animal species are selected on which to focus the assessment. The effects on these species are called assessment endpoints.

(e) *Risk characterization* combines information about exposure to the contaminant and the toxicity of the contaminant to determine the level of risk. The characterization also includes assumptions and the levels of uncertainty used. These factors are compared to state regulatory standards in regulation to determine risk-based cleanup levels. The initial screening phase of the ecological risk assessment involves the comparison of contaminant of potential concern (COPC) concentrations in affected media to screening values that identify threshold levels below which adverse effects on receptor organisms are unlikely. These screening values for potential effects on site organisms are referred to as benchmarks. For evaluation of effects on organisms by direct exposure, the value for comparison to benchmark is the concentration of a chemical substance in the medium. For evaluation of exposure through food and/or water ingestion, benchmarks for specific chemical substances are dietary doses associated with various adverse levels documented under experimental conditions. Currently, there are no universally accepted methods for deriving benchmarks used in ecological risk assessments. For different locations and exposure scenarios, multiple approaches and benchmark values have been used in the evaluation of ecological risk. Commonly used benchmarks document toxicological endpoints for individual species and, less frequently, potential effects on several test species or assemblages of organisms.

If a contaminant is shown to be bioavailable, an ecological risk assessment will likely be required. The level of detail required for an ecological risk assessment is site-specific and will vary with the application. For example, a risk assessment for phytostabilization should address the roots and receptors that may ingest or contact them. If a contaminant enters the terrestrial portion of the plant (leaves, stems, branches, etc.) during phytoaccumulation, then pathways through those plant structures will need to be assessed as well.

6.6 UPTAKE OF CONTAMINANTS BY PLANTS

Uptake of pollutants by plant roots is different for organics and inorganics. Organic pollutants are usually man-made and xenobiotic to the plant. As a consequence, there are no transporters for these compounds in plant membranes. Organic pollutants, therefore, tend to move into and within plant tissues driven by simple diffusion, dependent on their chemical properties. An important property of the organic pollutant for plant uptake is its hydrophobicity. Hydrophobicity is expressed as the octanol:water partition coefficient or log K_{ow} (Briggs et al. 1982; Trapp and McFarlane 1995). Organics with a log K_{ow} between 0.5 and 3 are hydrophobic enough to move through the lipid bilayer of the membranes and still water-soluble enough to travel into the cell fluid. If organics are too hydrophilic (log $K_{ow} < 0.5$) they cannot pass through the membranes and never get into the plant; if they are too hydrophobic (log $K_{ow} > 3$) they get stuck in the membranes and cell walls in the periphery of the plant, and organic matter cannot enter the cell fluid. The tendency of organic pollutants to move into plant roots from an external solution is expressed as the root concentration factor (RCF = equilibrium concentration in roots/equilibrium concentration in external solution).

The movement of metals and/or inorganics toward the root surface depends on three factors: a) mass flow because of which the soluble metal ions move from soil solids to root surface (driven by transpiration); b) diffusion of elements along the concentration gradient formed as a result of uptake and thereby depletion of the element in the root vicinity; and c) root interception, where soil volume is displaced by root volume because of root growth (Marschner 1995). The metal uptake by the roots may take place at the apical region or from the entire root surface, depending on the type of element under consideration. Further, the uptake depends on the uptake capacity and growth characteristics of the root system.

There are two pathways for solubilized heavy metals to enter a plant. These are apoplastic (extracellular) and symplastic (intracellular). The apoplast continuum of the root epidermis and cortex is readily permeable to solutes. The metals are first taken into the apoplast of the roots where a significant ion fraction is physically adsorbed at the extracellular negatively charged sites (COO^-) of the root cell walls (Lasat 2000). Then, some of the total amount of metal ions associated with the root cell walls is translocated into the cell. However, the impermeable suberin layers in the cell wall of the root endodermis (Casparian strips) prevent solutes from flowing straight from the root apoplast into the root xylem (Taiz and Zeiger 2002). Therefore, the solutes have to be taken up into the root symplasm before they can enter the xylem apoplast. Metal ions require membrane transporter proteins for their transportation from root endodermis into root xylem (Pilon-Smits 2005).

The plant plasma membrane may be regarded as the first living structure that encounters the heavy-metal toxicity. Because of their charge, metal ions can not move freely across the cellular membranes, which are lipophilic structures. Therefore, ion transport into cells must be mediated by membrane proteins with transport functions, generally known as transporters (Lasat 2000). Several classes of metal transporters are reported in plants that are involved in metal uptake and homeostasis in general and thus could play some role in tolerance (Hall 2002). These include heavy metal CPx-ATPases, the Nramps, and the CDF (cation diffusion facilitator) family (Williams et al. 2000) and the ZIP family (Guerinot 2000). Further, heavy-metal ions, such as Cd, enter the plant cell by transporters for essential cations, such as Fe^{2+} (Thomine et al. 2000). At Nramp, genes in *Arabidopsis* encode the metal transporter, which transports both the metal nutrient iron and the toxic metal cadmium.

Lasat (2000) reported that membrane transporters possess an extracellular binding domain to which the ions attach just before transport and a transmembrane structure, which connects extracellular and intracellular media. The binding domain is receptive only to specific ions and is responsible for transporter specificity. The transmembrane structure facilitates the transfer of bound ions from extracellular space through the hydrophobic environment of the membrane into the cell. The transporters are characterized by certain kinetic parameters, such as transport capacity (V_{max}) and affinity for the ion (K_m).

6.6.1 Translocation of Contaminants to Shoots

Translocation from root to shoot first requires a membrane transport step from root symplast into xylem apoplast. The impermeable suberin layer in the cell wall of the root endodermis (Casparian strip) prevents solutes from flowing straight from the soil solution or root apoplast into the root xylem (Taiz and Zeiger 2002). Organic pollutants pass the membrane between root symplast and xylem apoplast via simple diffusion. The transpiration stream concentration factor (TSCF) is the ratio of the concentration of a compound in the xylem fluid relative to the external solution and is a measure of uptake into the plant shoot. Entry of organic pollutants into the xylem depends on similar passive movement over membranes as they uptake into the plant. Inorganics require membrane transporter proteins to be exported from the root endodermis into the root xylem. Some inorganics are chelated during xylem transport by organic acids (histidine, malate, citrate), nicotianamine, or thiol-rich peptides (Pickering et al. 2000; Von Wiren et al. 1999). For most inorganics, it is still unclear via which transporter proteins they are exported to the root xylem and to which—if any—chelators they are bound during transport. Better knowledge of the transporters and chelators involved in the translocation of inorganics would facilitate the development of transgenics with more efficient phytoextraction capacities.

Bulk flow in the xylem from root to shoot is driven by transpiration from the shoot, which creates a negative pressure in the xylem that pulls up water and solutes (Taiz and Zeiger 2002). Plant transpiration depends on plant properties and environmental conditions. Plant species differ in transpiration rate because of metabolic differences (e.g., C3/C4/CAM photosynthetic pathway) and anatomical differences (e.g., surface-to-volume ratio, stomatal density, rooting depth) (Taiz and Zeiger 2002). Species like poplar are phreatophytes, or water spenders; they have long roots that tap into the groundwater (Dawson and Ehleringer 1991). Mature poplar trees can transpire 200–1000 liters of water per day (Wullschleger et al. 1998). In addition to plant species composition, vegetation height and density affect transpiration as do environmental conditions: Transpiration is generally maximal at high temperature, moderate wind, low relative air humidity, and high light (Taiz and Zeiger 2002). Consequently, phytoremediation mechanisms that rely on translocation and volatilization are most effective in climates with low relative humidity and high evapotranspiration.

6.6.2 Root Level Contaminant Chelation and Sequestration

Plants root exudates (chelators) can affect pollutant solubility and uptake by the plant. Inside plant tissues, such chelator compounds also play a role in tolerance, sequestration, and transport of inorganics and organics (Ross 1994). Phytosiderophores are chelators that facilitate uptake of Fe and perhaps other metals in grasses; they are biosynthesized from nicotianamine, which is composed of three methionines coupled via nonpeptide bonds (Higuchi et al. 1999). Nicotianamine also chelates metals and may facilitate their transport (Stephan et al. 1996; Von Wiren et al. 1999). Organic acids (e.g., citrate, malate, histidine) not only can facilitate uptake of metals into roots but also play a role in transport, sequestration, and tolerance of metals (Salt et al. 1995; Von Wiren et al. 1999). Metals can also be bound by the thiol-rich peptides glutathione (GSH) and phytochelatins (PCs) or by the Cys-rich metallothioneins (MTs) (Cobbett and Goldsbrough 2000). Chelated metals in roots may be stored in the vacuole or exported to the shoot via the xylem. Also, organics may be conjugated and stored or degraded enzymatically. Figure 6.5 explains the transport, sequestration, and tolerance mechanisms of inorganic and organic pollutants in plant cells. Chelation in roots can affect

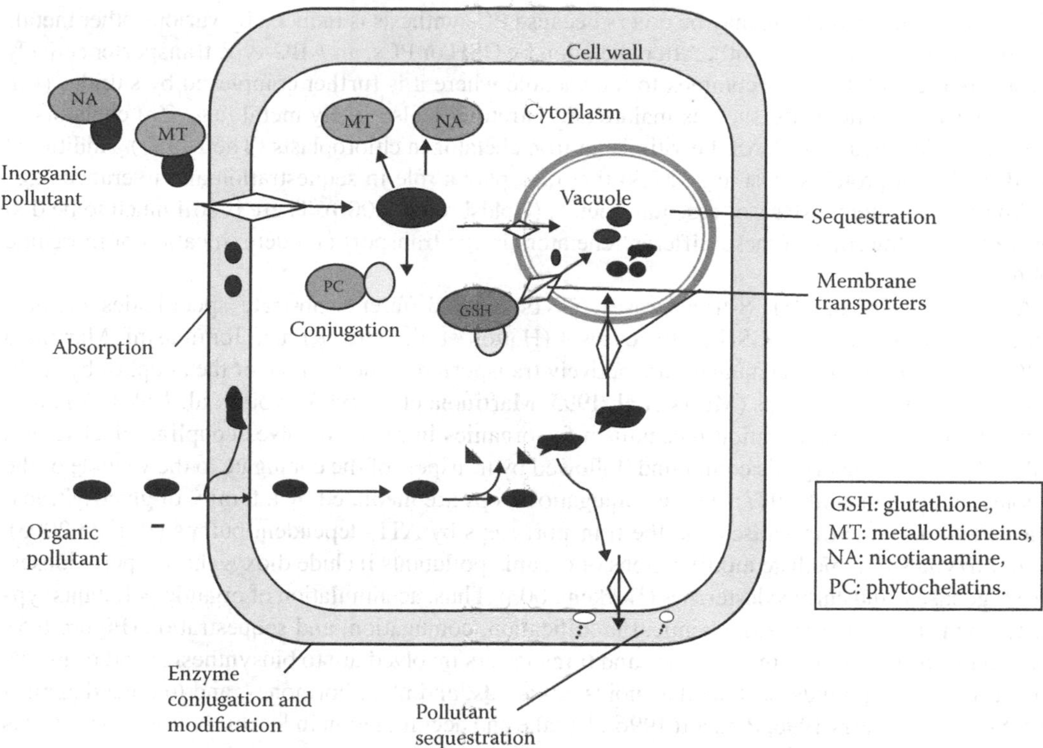

FIGURE 6.5 **(See color insert.)** Transport, sequestration, and tolerance mechanisms of inorganic and organic pollutants in plant cells.

efficiency as it may facilitate root sequestration, translocation, and/or tolerance. Root sequestration may be desirable for phytostabilization (less exposure to wildlife); whereas, export to xylem is desirable for phytoextraction. If chelation is desirable, it may be enhanced by selection or engineering of plants with higher levels of the chelator in question. Root sequestration and export to xylem might be manipulated by overexpression or knockdown of the respective membrane transporters involved. Unfortunately, little is known about these tissue-specific transporters of inorganics. The completion of the sequencing of the *Arabidopsis* and rice genomes should accelerate the analysis of transporter gene families.

6.6.3 Leaf Level Contaminant Chelation and Compartmentation

Translocation of contaminants from roots to leaves involves a number of membrane transport steps. Inorganics are taken up by specific membrane transporter proteins. Organics enter the leaf symplast from the shoot xylem by simple diffusion; the rate depends on the chemical properties of the pollutant. Once inside the leaf symplast, the pollutant may be sequestered in certain plant tissues or cellular locations. In general, toxic pollutants are sequestered in places where they can do the least harm to essential cellular processes. At the cellular level, pollutants are generally accumulated in the vacuole or cell wall (Burken 2003). At the tissue level, they may be accumulated in the epidermis and trichomes (Hale et al. 2001).

When pollutants are sequestered in tissues, they are often bound by chelators or form conjugates (Figure 6.5). Toxic inorganics are usually metals. Chelators that are involved in metal sequestration include the tripeptide GSH (γ-glu-cys-gly) and its oligomers, the PCs. X-ray absorption spectroscopy (XAS) has shown that inorganics that were complexed by PCs *in vivo* include Cd and As

(Pickering et al. 2000); there may be others because PC synthesis is induced by various other metals (Cobbett and Goldsbrough 2000). After chelation by GSH or PCs, an ABC-type transporter actively transports the metal-chelate complex to the vacuole where it is further complexed by sulfides (Lin et al. 2000). Organic acids, such as malate and citrate, are also likely metal (e.g., Zn) chelators in vacuoles, evidenced from XAS. Ferritin is an iron chelator in chloroplasts (Theil 1987). Additional metal-chelating proteins exist (e.g., MTs) that may play a role in sequestration and tolerance (e.g., of Cu) and/or in homeostasis of essential metals (Goldsbrough 2000). There is still much to be discovered about the roles of these different chelators in the transport and detoxification of inorganic pollutants.

A large family of GSH-S-transferases (GSTs) with different substrate specificities mediate conjugation of organics to GSH in the cytosol (Hatton et al. 1996; Kreuz, Tommasini, Martinoia 1996). The glutathione S-conjugates are actively transported to the vacuole or the apoplast by ATP-dependent membrane pumps (Marrs et al. 1995; Martinoia et al. 1993; Wolf et al. 1996). An alternative conjugation-sequestration mechanism for organics in plants involves coupling glucose or a malonyl group to the organic compound, followed by transport of the conjugate to the vacuole or the apoplast (Coleman et al. 1997). These conjugation steps are mediated by a family of glucosyltransferases and malonyltransferases and the transport steps by ATP-dependent pumps (Burken 2003). Other enzymes that mediate modifications of organic pollutants include dioxygenases, peroxidases, peroxygenases, and carboxylesterases (Burken 2003). Thus, accumulation of organic pollutants typically comprises three phases: chemical modification, conjugation, and sequestration (Figure 6.5). Some natural functions of the enzymes and transporters involved are to biosynthesize and transport natural plant compounds, such as flavonoids, alkaloids, and plant hormones, and to defend against biotic stresses (Marrs 1996; Prescott 1996). Uptake and accumulation in leaves without toxic effects are desirable properties for phytoextraction. To maximize these processes, plants may be selected or engineered that have higher levels of transporters involved in the uptake of an inorganic pollutant from the xylem into the leaf symplast.

Similarly, plants with high transporter activities from cytosol to vacuole can be more efficient at storing toxic inorganics (Song et al. 2003; Marrs 1996). Sequestration and tolerance may also be enhanced by selection or engineering of plants with a higher production of leaf chelators or conjugates. This can be mediated by higher levels of enzymes that produce these conjugates, e.g., enzymes synthesizing GSH, PCs, glucose, organic acids, or chelator proteins (Hasegawa et al. 1997; Zhu et al. 1999a; Zhu et al. 1999b). In addition, enzymes that couple the chelator or conjugant to the pollutant (GSH transferases, glucosyltransferases) may be over-expressed (Ezaki et al. 2000) or enzymes that modify organics to make them amenable to conjugation (Didierjean et al. 2002).

From the above, the partitioning of organics to roots and above-ground tissues varies considerably, depending on the chemical in question. Following uptake, organic compounds may have multiple fates; they may be translocated to other plant tissues (Fellows et al. 1996; Schroll et al. 1994) and subsequently volatilized; they may undergo partial or complete degradation (Goel et al. 1997; Newman et al. 1997), or they may be transformed to less toxic compounds and bound in plant tissues to nonavailable forms (Field and Thurman 1996). Schroll et al. (1994) noted that hexachlorobenzene (HBC) and octachlorodibenzo-p-dioxin (OCDD) could be taken up by roots or leaves but that no translocation from roots to shoots or vice versa was observed. On the other hand, roots and foliar uptake of the herbicides chlorobenzene and trichloroacetic (TCA) was followed by translocation in both directions (Wenzel et al. 1999).

Contaminant degradation only applies to organic compounds while inorganics can only be stabilized, moved, and stored. Organic compound degradation can occur in both the rhizosphere zone (root-plant interphase), referred to as rhizodegradation, and within the plant tissue—phytodegradation. Rhizosphere degradation is the breakdown of organic contaminants within the rhizosphere—a zone of increased microbial activity and biomass at the root-soil interphase. Plant roots secrete and slough substances, such as carbohydrates, enzymes, and amino acids, that microbes can utilize as a substrate. Contaminant degradation in the rhizosphere may also result from the additional oxygen transferred from the root system into the soil, causing enhanced aerobic mineralization of organics and stimulation of co-metabolic transformation of chemicals (Anderson et al. 1993).

6.6.4 Degradation and Fate of Trichloroethane

Phytoremediation of trichloroethene (TCE) is a much-studied process, and the remaining uncertainty about its fate illustrates that much still remains to be learned about the metabolic fate of organics in plants. Walton and Anderson (1990) investigated the fate of TCE in a laboratory scale experiment by comparing degradation of TCE in both rhizosphere soil and nonvegetated soil collected from a TCE-contaminated site. The results showed that TCE degraded faster in rhizosphere soils. Higher numbers of methanotrophic bacteria, which have been shown to degrade TCE, were detected in rhizosphere soils and on roots of *Lespedeza cuneata* and *Pinus taeda* than in nonvegetated soils (Brigmon et al. 1999). Orchard et al. (2000) detected TCE metabolites in the roots of hybrid poplar saplings, suggesting rhizosphere degradation, and concluded that the greatest degradation of TCE occurred in the rhizosphere. In phytodegradation, plants produce a large number of enzymes, of which one or more may transform PCE and TCE into daughter products. Although not completely understood, dehalogenase, cytochrome p-450, glutathione-S transferase, methane mono-oxygenase, and monochloroacetic acid are all thought to play a role in chlorinated solvent transformation. Intermediate stable metabolites of these chlorinated compounds include 2,2,2-trichlorethanol, 2,2,2-trichloroacetic acid (TCAA) and 2,2-dichloroacetic acid (DCAA) and have been reportedly found in hybrid poplar (Gordon et al. 1998; Newman et al. 1997; Compton et al. 1998), oak, castor bean, and saw palmetto (Doucette et al. 1998).

Some researchers believe that chlorinated solvents are being metabolized within vegetation; however, the exact mechanism has not been determined yet. Bench-scale laboratory TCE uptake tests with poplar cuttings grown in soil were reported to have measurable amounts of TCE transpired to the air (Newman et al. 1997). A three-year study, commencing with rooted poplar cuttings in a series of constructed, lined, artificial aquifers, evaluated the fate and transport of TCE in the poplar tree. The mature trees were able to remove 99% of the TCE from the groundwater, and less than 9% of the TCE was transpired to the air in the first two years. After two years, TCE was not detected in the air stream. Researchers believe that the mature hybrid poplar tree was dechlorinating the TCE and inferred that degradation in the rhizosphere was not contributing to the loss of TCE (Newman et al. 1999).

An alternate theory about the fate of TCE in poplar trees is that TCE is taken up by suspension cell cultures and is incorporated as a nonvolatile, nonextractable residue (Shang and Gordon 2002). Another investigation of the fate and transport of TCE in carrot, spinach, and tomato plants showed that TCE was taken up, transformed, and bound to plant tissue (Schnabel et al. 1997). This binding, or "sorption," of organic compounds has been linked to plant lipid content and tissue chemistry. Mackay and Gschwend (2000) have studied the sorption of chemicals to wood and developed wood-water partitioning equations. Partitioning onto wood was determined to depend predominantly on the water-lignin partitioning of a compound. Lignin is the chief noncarbohydrate constituent of wood, which binds to cellulose fibers and strengthens the cell walls. Lignin is hydrophobic and shows a strong affinity to hydrophobic organic compounds.

6.6.5 Transformation of Toxic Elements

Another natural mechanism that offers exciting phytoremediation possibilities is the transformation of toxic elements into relatively harmless forms. Many elements (e.g., arsenic, mercury, iron, selenium, chromium) can exist in a variety of states, including different cationic and oxyanionic species and thio- and organo-metallics. These forms vary widely in their transport and accumulation in plants and in their toxicity to humans and other life forms. Mercury offers perhaps the best-understood example of the dangers inherent in one particular species of a heavy metal. Mercury primarily enters the environment either as liquid Hg(0) from industrial and defense-related accidents or as mercury species (Hg[II]) bound to particulate matter from burning coal and trash or from volcanic activity and as complex chemical derivatives released in industrial effluents (Meagher et al. 2000). Although Hg (II) is relatively toxic, it and Hg(0) have seldom been involved in serious incidents of human mercury poisoning without first being transformed into methylmercury (MeHg) (Keating et al. 1997). The world first became aware of the extreme dangers of methylmercury (MeHg) in the 1950s after a large, tragic incident of human mercury poisoning at Minamata Bay, Japan (Harada 1995). In aquatic sediments, various mercury species are efficiently converted to MeHg by anaerobic bacteria (Choi et al. 1994). Unfortunately, MeHg is biomagnified by several orders of magnitude and has a greater toxicity than any other natural mercury compound (Meagher et al. 2000). As a result, the fish-eating predatory animals and humans at the top of the food chain suffer MeHg poisoning (Keating et al. 1997; Boischio and Henshel 1996).

Meagher et al. (2000) and Rugh et al. (1999) made use of two genes from the well-characterized bacterial *mer* operon, *merA* and *merB*, to engineer a mercury transformation and remediation system in plants. The bacterial *merA* gene encodes an NADPH-dependent mercuric ion reductase that converts ionic mercury (Hg [II]) to elemental, metallic mercury (Hg [0]). Metallic mercury is nearly two orders of magnitude less toxic than ionic mercury and is readily eliminated because of its volatility. Diverse plant species expressing *merA* constitutively are resistant to at least 10 times greater concentrations of Hg (II) than those that kill nontransgenic controls (Meagher et al. 2000; Rugh et al. 1996; Meagher and Rugh 1996). These plants volatilize and possibly transpire Hg (0) from their tissues, and they accumulate far less mercury than control plants grown in low concentrations of mercury (Heaton et al. 1998).

The chemical transformation of other toxic elemental pollutants also leads to their remediation. Selenium builds up in irrigation water and contaminates hundreds of square miles of wetlands in the western United States. Selenium and sulfur are nutrients with very similar chemical properties, and their uptake and assimilation proceed through common pathways. Although sulfur is required by all organisms in relatively large concentrations, high levels of selenium are usually toxic. The assimilation of sulfate and selenate is activated by ATP sulfurylase. Selenate is converted to adenosine phosphoselenate (ADP-Se), which is subsequently reduced to selenite. Over-expression of the *Arabidopsis* plastidic ATP sulfurylase (*APS1*) in transgenic Indian mustard results in an increased uptake and assimilation of selenate, increased reduction to selenite, and greater tolerance of selenate (Pilon-Smits et al. 1999).

Inorganic Se can also be volatilized by plants and microorganisms. Volatilization of Se involves assimilation of inorganic Se into the organic selenoaminoacids selenocysteine (SeCys) and selenomethionine (SeMet). The latter can be methylated to form dimethylselenide (DMSe), which is volatile (Terry et al. 2000). Volatilization of the inorganics As and Hg has been demonstrated for microorganisms, but these elements do not appear to be volatilized to significant levels by (nontransgenic) plants (Rugh et al. 1996). Many volatile organic compounds (VOCs) can be volatilized passively by plants. Volatile pollutants with a Henry's law constant Hi > 10–6 that are mobile in both air and water can move readily from the soil via the transpiration stream into the atmosphere (Bromilow and Chamberlain 1995). In this way, plants act like a wick for VOCs to facilitate their diffusion from soil. Examples of organic pollutants that can be volatilized by plants are the chlorinated solvent TCE and the fuel additive methyl tertiary butyl ether MTBE (Newman et al. 1997).

6.7 ECOLOGICAL CONSEQUENCES OF THE USE OF TRANSGENIC PLANTS

To improve the ability of plants to degrade or metabolize contaminants, transgenic plants can be developed by transferring genes from organisms that have the capacity for degradation or mineralization of contaminants to candidate plants, thus overcoming the sexual barrier. Genes involved in the degradation of contaminants can be isolated from bacteria/fungi/animals/plants and introduced into candidate plants using Agrobacterium-mediated or direct DNA methods of gene transfer.

Mammalian cytochrome P450 (CYP) monooxygenases are involved in the metabolism of xenobiotics. Mammalian P450 2E1, a liver enzyme, is known to oxidize a wide range of pollutants, including TCE, ethylene dibromide, carbon tetrachloride chloroform, and vinyl chloride. Hairy root cultures of *Atropa belladonna* with mammalian cytochrome P450 2El gene (Banerjee et al. 2002), when challenged with xenobiotic TCE, showed increased metabolism of TCE. Transgenic tobacco plants having human cytochrome P450 2El also showed increased uptake and debromination of ethylene dibromide and a 640-fold enhanced metabolism of TCE (Doty et al. 2000). Transgenic plants with the potential to degrade explosives, nitroesters, and nitroaromatics were developed by introducing a pentaerythritol tetranitrate reductase gene from bacteria (French et al. 1999). Tobacco plants with a high efficiency for remediating PCP could be developed by introduction of a Mn peroxidase gene from *Coriolus versicolor* (Limura et al. 2002). Transgenic *Arabidopsis thaliana* with an *Escherichia coli* gene encoding a nitroreductase resulted in efficient degradation of TNT (Kurumata et al. 2005). Transfer of a bacterial atrazine chlorohydrolase (atzA) gene to alfalfa, tobacco, and *A. thaliana* resulted in these transgenic plants degrading atrazine—a widely used herbicide (Wang et al. 2005).

It is necessary to determine whether introduction of new traits to plants or crop can make a crop more likely to be persistent (weedy) in an agricultural habitat or more invasive in natural habitats. There are some obvious biological changes expected as a result of the introduction of new genes to another organism, such as changes in tolerance to extremes of temperature, water, soil salinity regimes, and introduction of pest or pathogen resistance, and changes in seed dormancy and propagation characteristics could potentially have significant effects on persistence, invasiveness, and direct or indirect effects on nontarget organisms and ecosystems (Bergelson et al. 1998; Seidler and Levin 1994). Transgenic plants can potentially hybridize with sexually compatible species and have an impact on the environment through the production of hybrids and their progeny. There are four basic elements in determining the likelihood and consequences of gene flow in this way: First is the distance of pollen movement from the transgenic plant. Second is the synchrony of flowering between the crop and the pollen recipient species; third is sexual compatibility between crop and recipient species, and fourth is the ecology of the recipient species. Ramachandran et al. (2000) have investigated the competitive ability of an insect-resistant transgenic oilseed rape variety as compared with nontransgenic oilseed rape in seed mixtures. The transgenic variety was competitively superior when the two varieties were subject to diamondback moth selection pressure in greenhouse experiments and field plots.

The widespread introduction of transgenic hyperaccumulator plants may cause a shift in weed population and thus reduce weed species diversity and ecosystem complexity in the transgenic field and on neighboring farms. The adoption of different transgenic hyperaccumulators in clean-up programs may have different effects on plant and animal diversity in the field and field margins. Radosevich et al. (1992) predicted that shifts in weed species composition and abundance would be exacerbated with the conservative use of the same plant species, providing favorable conditions for growth of particular weeds, insects, and diseases.

6.8 CONCLUSIONS

In general, research is ongoing to clearly identify the degradation pathways of contaminants to carbon dioxide, water, methane, and other basic compounds that become incorporated into organic matter. Therefore, the fate of organics and their degradation products will remain an issue until clear evidence of complete mineralization is adduced. Phytoextraction, phytostabilization, and phytofiltration of inorganics may not completely remove the ecological risk and avoid exposure to the food chain, especially where contaminated sites are not cordoned off from grazers and other herbivores. The use of transgenic plants is very innovative, and the research into more hybrids with superior capabilities for phytoremediation may not be fully exploited because of ethical barriers, ecosystem distortion, and the emergence of super plants. It is recommended that equal consideration be given to degradation products and to the original contaminant when implementing phytoremediation.

REFERENCES

Alexander, M., P. B. Hatzinger, J. W. Kelsey, B. D. Kottler, and K. Nam. (1997). Sequestration and realistic risk from toxic chemicals remaining after bioremediation. *Annals New York Academy of Sciences*, *829*, 1–5.

Anderson, T. A., E. A. Guthrie, and B. T. Walton. (1993). Bioremediation. *Environmental Science and Technology, 27*, 2630–2636.

Azevado, H., C. G. G. Pinto, J. Farnandes, S. Loureiro, and C. Santos. (2005). Cadmium effects on sunflower growth and photosynthesis. *Journal of Plant Nutrition, 28*, 2211–2220.

Banerjee, S., T. Q. Shang, A. M. Wilson et al. (2002). Expression of functional mammalian P450 2EI in hairy root cultures. *Biotechnology and Bioengineering*, *77*, 462–466.

Bergelson, J., C. B. Purrington, and G. Wichmann. (1998). Promiscuity in transgenic plants. *Nature*, *395*, 25.

Beri, A., and R. C. Setia. (1996). Effects of Ni and Cd on seed germination, seedling growth, mobilization of food reserves and activity of some enzymes in *Lens culinaris* Medic (Lentil). In T. A. Sarma, S. S. Saina, M. L. Trivedi, M. Sharma, and B. S. M. P. Singh (Eds.), *Current researches in plant sciences* (85–193). Dehra Dun, India.

Beri, A., R. C. Setia, and N. Setia. (1990). Germination responses of lentil (*Lens culinaris*) seeds to heavy metal ions and plant growth regulators. *Journal of Plant Science Research*, *6*, 24–27.

Blaylock, M. J., and J. W. Huang. (2000). Phytoextraction of metals. In I. Raskin, and B. D. Ensley (Eds.), *Phytoremediation of toxic metals: Using plants to clean up the environment* (53–70). New York: Wiley.

Boischio, A. A., and D. S. Henshel. (1996). Mercury contamination in the Brazilian Amazon: Environmental and occupational aspects. *Water, Air & Soil Pollution*, *80*, 109–107.

Bollag, J. M. (1992). Decontaminating soil with enzymes. *Environmental Science and Technology*, *26*, 1876–1881.

Bossert, I., and R. Bartha. (1984). The fate of petroleum in soil ecosystems. In R. M. Atlas (Ed.), *Petroleum microbiology* (435–473). New York: MacMillan.

Brady, N. C., and R. R. Weil. (1996). *The nature and properties of soils*. New Jersey: Prentice Hall.

Briggs, G. G., R. H. Bromilow, and A. A. Evans. (1982). Relationships between lipophilicity and root uptake and translocation of non-ionized chemicals by barley. *Pesticide Science*, *13*, 405–504.

Brigmon, R. L., T. A. Anderson, and C. B. Filermans. (1999). Methantrophic bacteria in the rhizosphere of trichloroethylene-degrading plants. *International Journal of Phytoremediation*, *1*, 241–253.

Bromilow, R. H., and K. Chamberlain. (1995). Principles governing uptake and transport of chemicals. In S. Trapp, and J. C. McFarlane (Eds.), *Plant contamination: Modeling and simulation of organic chemical processes* (37–68). Boca Raton: Lewis.

Burken, J. G. (2003). Uptake and metabolism of organic compounds: Green-liver model. In S. C. McCutcheon, and J. L. Schnoor (Eds.), *Phytoremediation: Transformation and control of contaminants* (59–84). New York: Wiley.

Carmichael, L. M., and F. K. Pfaender. (1997). Polynuclear aromatic hydrocarbon metabolism in soils: Relationship to soil characteristics and pre-exposure. *Environmental Toxicology and Chemistry*, *16*, 666–675.

Chappell, J. (1998). Phytoremediation of TCE in groundwater using populus. Status report prepared for USEPA, Technology Innovation Office. February. Retrieved from http://cluin.org/products/phytotce.htm.

Choi, S. C., J. T. Chase, and R. Bartha. (1994). Metabolic pathways leading to mercury methylation in *Desulfovibrio desulfuricans* LS. *Applied Environmental Microbiology*, *60*, 4072–4077.

Cobbett, C. S., and P. B. Goldsbrough. (2000). Mechanisms of metal resistance: Phytochelatins and metallothioneins. In I. Raskin, and B. D. Ensley (Eds.), *Phytoremediation of toxic metals: Using plants to clean up the environment* (247–271). New York: Wiley.

Coleman, J. O. D., M. M. A. Blake-Kalff, and T. G. E. Davies. (1997). Detoxification of xenobiotics by plants: Chemical modification and vacuolar compartmentation. *Trends in Plant Science*, *2*, 144–151.

Compton, H., D. Haroski, S. Hirsch, and J. Wrobel. (1998). Pilot-scale use of trees to address VOC contamination. In G. B. Wickramanayake, and R. E. Hinchee (Eds.), *Bioremediation and phytoremediation: Chlorinated and recalcitrant compounds* (245–250). Columbus: Batelle Press.

Cunningham, S. D., T. A. Anderson, P. Schwab, and F. C. Hsu. (1996). Phytoremediation of soils contaminated with organic pollutants. *Advances in Agronomy*, *56*, 55–114.

Cunningham, S. D., W. R. Berti, and J. W. Huang. (1995). Phytoremediation of contaminated soils. *Trends in Biotechnology*, *13*, 393–397.

Cunningham, S. D., and Ow, D. W. (1996). Promises and prospects of phytoremediation. *Plant Physiology*, *110*, 715–719.

Dawson, T. E., and J. R. Ehleringer. (1991). Streamside trees do not use stream water. *Nature*, *350*, 335–337.

Didierjean, L., L. Gondet, R. Perkins et al. (2002). Engineering herbicide metabolism in tobacco and *Arabidopsis* with CYP76B1, a cytochrome P450 enzyme from Jerusalem artichoke. *Plant Physiology, 130*, 179–189.
Doty, S. L., T. Q. Shang, A. M. Wilson et al. (2000). Enhanced metabolism of halogenated hydrocarbons in transgenic plants containing mammalian cytochrome P450 2E1. *Proceedings of National Academy of Sciences USA, 97*, 6287–6291.
Doucette, W., B. Bugbee, S. Hayhurst et al. (1998). Phytoremediation of dissolved-phase trichloroethylene using mature vegetation. In G. B. Wickramanayake, and R. E. Hinchee (Eds.), *Bioremediation and phytoremediation: Chlorinated and recalcitrant compounds* (251–256). Columbia: Batelle Press.
Eapen, S., S. Singh, and S. F. D'Souza. (2007). Advances in development of transgenic plants for remediation of xenobiotic pollutants. *Biotechnological Advances, 25*, 442–451.
Edwards, N. T., B. M. Ross-Todd, and E. G. Garver. (1982). Uptake and metabolism of 14C anthracene by soybean (*Glycine max*). *Environmental and Experimental Botany, 22*, 349–357.
Ezaki, B., R. C. Gardner, Y. Ezaki, and H. Matsumoto. (2000). Expression of aluminum-induced genes in transgenic *Arabidopsis* plants can ameliorate aluminum stress and/or oxidative stress. *Plant Physiology, 122*, 657–665.
Fellows, R. J., S. D. Harvey, and C. C. Ainsworth. (1996). Biotic and abiotic transformation of Amunitions materials (TNT, RDX) by plant and soils: Potentials for attenuation and remediation and remediation of contaminants. IBC International Conference on Phytoremediation, Arlington, VA.
Field, J. A., and E. M. Thurman. (1996). Glutathione conjugation and contaminant transformation. *Environmental Science and Technology, 30*, 1413–1418.
French, C. E., S. J. Rosser, G. J. Davies, S. Nicklin, and N. C. Bruce. (1999). Biodegradation of explosives by transgenic plants expressing pentaerythritol tetranitrate reductase. *Nature Biotechnology, 17*, 491–494.
Gisi, U., R. Schenker, R. Schulin, F. X. Stadelmann, and H. Sticher. (1990). *Bodenökologie*. Thieme: Stuttgart, New York.
Goldsbrough, P. (2000). Metal tolerance in plants: The role of phytochelatins and metallothioneins. In N. Terry, and G. Banuelos (Eds.), *Phytoremediation of contaminated soil and water* (221–234). Boca Raton: Lewis.
Goel, A., G. Kumar, G. F. Payne, and S. K. Dube. (1997). Plant cell biodegradation of a xenobiotic nitrate ester, nitroglycerin. *Nature Biotechnology, 15*, 174–177.
Gordon, M., N. Choe, J. Duffy, G. Ekuan et al. (1998). Phytoremediation of trichloroethylene with hybrid poplars. *Environmental Health Perspective, 106*, 1001–1004.
Green, C., R. Vhaney, and J. Bouwkamp. (2003). Interactions between cadmium and phytotoxic levels of zinc in hard red spring wheat. *Journal of Plant Nutrition, 26*, 417–430.
Guerinot, M. L. (2000). The ZIP family of metal transporters. *Biochemica Biophysica Acta, 1465*, 190–198.
Hale, K. L., S. McGrath, E. Lombi et al. (2001). Molybdenum sequestration in *Brassica*: A role for anthocyanins? *Plant Physiology, 126*, 1391–1402.
Hall, J. L. (2002). Cellular mechanisms for heavy metal detoxification and tolerance. *Journal of Experimental Botany, 53*, 1–11.
Harada, M. (1995). Minamata disease: Methylmercury poisoning in Japan caused by environmental pollution. *Critical Review in Toxicology, 25*, 1–24.
Hasegawa, I., E. Terada, M. Sunairi et al. (1997). Genetic improvement of heavy metal tolerance in plants by transfer of the yeast metallothionein gene (CUP1). *Plant and Soil, 196*, 277–281.
Hatton, P. J., D. Dixon, D. J. Cole, and R. Edwards. (1996). Glutathione transferase activities and herbicide selectivity in maize and associated weed species. *Pesticide Science, 46*, 267–275.
Heaton, A. C. P., C. L. Rugh, N.-J. Wang, and R. B. Meagher. (1998). Phytoremediation of mercury and methylmercury polluted soils using genetically engineered plants. *Journal of Soil Contamination, 7*, 497–509.
Higuchi, K., K. Suzuki, H. Nakanishi, H. Yamaguchi, N. K. Nishizawa, and S. Mori. (1999). Cloning of nicotianamine synthase genes, novel genes involved in the biosynthesis of phytosiderophores. *Plant Physiology, 119*, 471–479.
Horne, A. J. (2000). Phytoremediation by constructed wetlands. In N. Terry, and G. Banuelos (Eds.), *Phytoremediation of contaminated soil and water* (3–40). Boca Raton: Lewis.
Jacobson, M. E., S. Y. Chiang, L. Gueriguian, L. R. Weshtholm, and J. Pierson. (2003). Transformation kinetics of trinitrotoluene conversion in aquatic plants. In S. C. McCutcheon, and J. L. Schnoor (Eds.), *Phytoremediation: Transformation and control of contaminants* (409–427). New York: Wiley.
Keating, M. H., K. R. Mahaffey, R. Schoeny et al. (1997). *Mercury Study Report to Congress*. EPA452/R-9-003. US Environmental Protection Agency, 1997, 1–51.
Koehler, H., K. Mathes, and B. Breckling. (1999). *Bodenökologie interdisziplinär*. Berlin: Springer.

Kreuz, K., R. Tommasini, and E. Martinoia. (1996). Old enzymes for a new job: Herbicide detoxification in plants. *Plant Physiology, 111*, 349–353.

Kurumata, M., M. Takahashi, A. Sakamotoa et al. (2005). Tolerance to and uptake and degradation of 2,4,6-trinitrotoluene (TNT) are enhanced by the expression of a bacterial nitroreductase gene in *Arabidopsis thaliana. Z Naturforsch [C], 60*, 272–278.

Lasat, M. M. (2000). Phytoextraction of metals from contaminated soil: A review of plant/soil/metal interaction and assessment of pertinent agronomic issues. *Journal of Hazardous Substance Research, 2*, 1–14.

Limura, Y., S. Ikeda, T. Sonoki, T. Hayakawa et al. (2002). Expression of a gene for Mn-peroxidase from *Coriolus versicolor* in transgenic tobacco generates potential tools for phytoremediation. *Applied Microbiology Biotechnology, 59*, 246–251.

Lin, Z.-Q., R. S. Schemenauer, V. Cervinka, A. Zayed, A. Lee, and N. Terry. (2000). Selenium volatilization from a soil-plant system for the remediation of contaminated water and soil in the San Joaquin Valley. *Journal of Environmental Quality, 29*, 1048–1056.

Liu, D., and I. Kottke. (2004). Sub-cellular localization of cadmium in root cells of *Allium cepa* by electron energy loss spectroscopy and cytochemistry. *Journal of Bioscience, 29*, 329–335.

Mackay, A., and P. Gschwend. (2000). Sorption of monoaromatic hydrocarbons to wood. *Environmental Science and Technology, 34*, 839–845.

Marrs, K. A. (1996). The functions and regulation of glutathione s-transferases in plants. *Annual Review of Plant Physiology and Plant Molecular Biology, 47*, 127–158.

Marrs, K. A., M. R. Alfenito, A. M. Lloyd, and V. A. Walbot. (1995). Glutathione s-transferase involved in vacuolar transfer encoded by the maize gene Bronze-2. *Nature, 375*, 397–400.

Marschner, H. (1995). *Mineral nutrition of higher plants*. San Diego: Academic.

Martinoia, E., E. Grill, R. Tommasini, K. Kreuz, and N. Amrehin. (1993). ATP-dependent glutathione S-conjugate 'export' pump in the vacuolar membrane of plants. *Nature, 364*, 247–249.

Maruthi Sridhar, B. B., S. V. Diehl, F. X. Han, D. L. Monts, and Y. Su. (2005). Anatomical changes due to uptake and accumulation of Zn and Cd in Indian mustard (*Brassica juncea*). *Environmental and Experimental Botany, 54*, 131–141.

McCutcheon, S. C., and J. L. Schnoor. (2003). Overview of phytotransformation and control of wastes. In S. C. McCutcheon, and J. L. Schnoor (Eds.), *Phytoremediation: Transformation and control of contaminants* (53–58). New York: Wiley.

Meagher, R. B., and C. L. Rugh. (1996). Phytoremediation of heavy metal pollution: Ionic and methyl mercury. In *OECD biotechnology for water use and conservation workshop*. Cocoyoc, Mexico: Organization for Economic Co-Operation and Development.

Meagher, R. B., C. L. Rugh, M. K. Kandasamy, G. Gragson, and N. J. Wang. (2000). Engineered phytoremediation of mercury pollution in soil and water using bacterial genes. In N. Terry, and G. Banuelos (Eds.), *Phytoremediation of contaminated soil and water* (201–221). Boca Raton: Lewis.

Newman, L. A., S. E. Strand, N. Choe, J. Duffy, and G. Ekuan. (1997). Uptake and biotransformation of trichloroethylene by hybrid poplars plant. *Environmental Science and Technology, 31*, 1062–1067.

Newman, L. A., X. Wang, I. A. Muiznieks, G. Ekuan, M. Ruszaj, R. Cortellucci, D. Domroes, G. Karscig, T. Newman, R. S. Crampton, R. A. Hashmonay, M. G. Yost, P. E. Heilman, J. Duffy, M. P. Gordon, and S. E. Strand. (1999). Remediation of trichloroethylene in an artificial aquifer with trees: A controlled field study. *Environmental Science and Technology, 33*, 2257–2265.

Nwoko, C. O. (2010). Trends in phytoremediation of toxic elemental and organic pollutants. *African Journal of Biotechnology, 9*, 6010–6016.

Nwoko, C. O., P. N. Okeke, and N. Ac-Chukwuocha. (2004). Preliminary studies on nutrient removal potential of selected aquatic plants. *Journal of Discovery and Innovation, 16*, 133–136.

Orchard, B., W. Douchette, J. Chard, and B. Bugbee. (2000). Uptake of trichloroethylene by hybrid poplar trees grown hydroponically in flow-through plant growth chambers. *Environmental Toxicology and Chemistry, 19*, 895–903.

Otten, A., A. Alphenaar, C. Pijls, F. Spuij, and H. de Wit. (1997). *In Situ Soil Remediation.* Boston: Kluwer Academic Publishers.

Peijnenburg, W. J. G. M., R. Baerselman, A. C. de Groot, T. Jager, L. Posthuma, and R. P. M. Van Veen. (1999). Relating environmental availability to bioavailability: Soil-type-dependent metal accumulation in the oligochaete *Eisenia andrei*. *Ecotoxicology and Environmental Safety, 44*, 294–310.

Pickering, I. J., R. C. Prince, M. J. George, R. D. Smith, G. N. George, and D. E. Salt. (2000). Reduction and coordination of arsenic in Indian mustard. *Plant Physiology, 122*, 1171–1177.

Pilon-Smits, E. A. H. (2005). Phytoremediation. *Annual Review of Plant Biology, 56*, 15–39.

Pilon-Smits, E. A. H., S. B. Hwang, C. M. Lytle et al. (1999). Overexpression of ATP sulfurylase in *Brassica juncea* leads to increased selenate uptake, reduction and tolerance. *Plant Physiology, 119*, 123–132.

Prasad, M. N. V., and J. Hagemeyer. (1999). *Heavy metal stress in plants: From molecules to ecosystem.* New York: Springer-Verlag, Heidelberg.

Prescott, A. G. (1996). Dioxygenases: Molecular structure and role in plant metabolism. *Annual Review of Plant Physiology and Plant Molecular Biology, 47*, 247–271.

Radosevich, S. R., C. M. Ghersa, and G. Comstock. (1992). Concerns a weeds scientist might have on herbicide-tolerant crops. *Weed Technology, 6*, 635.

Radwan, S., N. Sorkhoh, and I. El-Nemr. (1995). Oil biodegradation around roots. *Nature, 376*, 302.

Ramachandran, S., D. Buntin, J. N. All, and P. L. Raymer. (2000). Intraspecific competition of an insect resistant transgenic canola in seed mixtures. *Agronomy Journal, 92*, 368–378.

Rasolomanana, J. L., and J. Balandreau. (1987). Role de la rhizosphere dans la biodegradation de composes recalcitrants: Cas d'une riziere polluee par des residus petroliers. *Revue D'Ecologie et de Biologie du Sol, 24*, 443–457.

Ross, S. M. (1994). Toxic metals in soil-plant systems. Chichester, England: Wiley.

Rugh, C. L., D. Wilde, N. M. Stack, D. M. Thompson, A. O. Summers, and R. B. Meagher. (1996). Mercuric ion reduction and resistance in transgenic *Arabidopsis thaliana* plants expressing a modified bacterial *merA* gene. *Proceedings of National Academy of Sciences USA, 93*, 3182–3187.

Salt, D. E., M. Blaylock, N. P. B. A. Kumar et al. (1995). Phytoremediation: A novel strategy for the removal of toxic metals from the environment using plants. *Biotechnology, 13*, 468–474.

Salt, D. E., R. Smith, and I. Raskin. (1998). Phytoremediation. *Annual Review of Plant Physiology and Plant Molecular Biology, 49*, 643–668.

Sandalio, L. M., H. C. Dalurzo, M. Gomez, M. C. Romero-Puertas, and L. A. del Rio. (2005). Cadmium induced changes in growth and oxidative metabolism of pea plants. *Journal of Experimental Botany, 52*, 2115–2126.

Saxena, P. K., S. Krishnaraj, T. Dan, M. R. Perras, and N. N. Vettakkoruma-Kankav. (1999). Phytoremediation of metal contaminated and polluted soils. In M. N. V. Prasad, and J. Hagemeyer (Ed.), *Heavy metal stress in plants: From molecules to ecosystems* (305–329). New York: Springer-Verlag, Heidelberg.

Scheffer, F., and P. Schachtschabel. (1989). *Lehrbuch der Bodenkunde*. Enke Verlag, Stuttgart.

Schnabel, W. E., A. C. Dietz, J. G. Burken, J. L. Schnoor, and P. J. Alvarez. (1997). Uptake and transformation of trichloroethylene by edible garden plants. *Water Research, 31*, 816–825.

Schnoor, J. L., L. A. Licht, S. C. McCutcheon, N. L. Wolfe, and L. H. Carreria. (1995). Phytoremediation of organic and nutrient contaminants. *Environmental Science and Technology, 29*, 318–323.

Schroll, R., B. Bierling, G. Cao, U. Dorfler, and M. Lahaniati. (1994). Uptake pathways of organic chemicals from soil by agricultural plants. *Chemosphere, 28*, 297–303.

Seidler, R., and M. Levin. (1994). Potential ecological and non-target effects of transgenic plant gene products on agriculture, silviculture, and natural ecosystems: general introduction. *Molecular Ecology, 3*, 1–3.

Setia, R. C., and R. Bala. (1994). Anatomical changes in root and stem of wheat (*Triticum aestivum* L.) in response to different heavy metals. *Phytomorphology, 44*, 95–104.

Setia, R. C., R. Bala, N. Setia, and C. P. Malik. (1993). Photosynthetic characteristics of heavy metal treated wheat (*Triticum aestivum* L.) plants. *Journal of Plant Science Research, 9*, 47–49.

Setia, R. C., J. Kalia, and C. P. Malik. (1988). Effects of $NiCl_2$ toxicity on stem growth and ear development in *Triticum aestivum* L. *Phytomorphology, 38*, 21–27.

Setia, N., D. Kaur, and R. C. Setia. (1989a). Influence of heavy metals on growth and reproductive behaviour of pea. *Journal of Plant Science Research, 5*, 127–132.

Setia, N., D. Kaur, and R. C. Setia. (1989b). Germination and seedling growth of pea as influenced by Cd and Pb toxicity. *Journal of Plant Science Research, 5*, 137–144.

Shang, T., and M. Gordon. (2002). Transformation of [^{14}C] trichloroethylene by poplar suspension cells. *Chemosphere, 47*, 957–962.

Siciliano, S. D., and J. J. Germida. (1998). Bacterial inoculants of forage grasses enhance degradation of 2-chlorobenzoic acid in soil. *Environmental Toxicology and Chemistry, 16*, 1098–1104.

Song, W., E. J. Sohn, E. Martinoia, Y. J. Lee, Y. Y. Yang et al. (2003). Engineering tolerance and accumulation of lead and cadmium in transgenic plants. *Nature Biotechnology, 21*, 914–919.

Stephan, U. W., I. Schmidke, V. W. Stephan, and G. Scholz. (1996). The nicotianamine molecule is made-to-measure for complexation of metal micronutrients in plants. *Biometals, 9*, 84–90.

Susarla, S., V. F. Medina, C. Steven, and D. McCutcheon. (2002). Phytoremediation: An ecological solution to organic chemical contamination. *Ecological Engineering, 18*, 647–665.

Taiz, L., and E. Zieger. (2002). *Plant Physiology*. Sunderland, MA: Sinauer.

Terry, N., A. M. Zayed, M. P. de Souza, and A. S. Tarun. (2000). Selenium in higher plants. *Annual Review of Plant Physiology and Plant Molecular Biology*, *51*, 401–432.

Terry, N., A. Zayed, E. Pilon-Smits, and D. Hansen. (1995). Can plants solve the selenium problem? In Proceedings of the 14th Annual Symposium, Curr. Top. Plant Biochem. Physiolo. Mol. Biol. Will Plants have a role in Bioremediation? University of Missouri, Columbia. April 19–22, 63–64.

Theil, E. C. (1987). Ferritin: Structure, gene regulation, and cellular function in animals, plants and microorganisms. *Annual Review of Biochemistry*, *56*, 289–315.

Thomine, S., R. Wang, J. M. Ward, N. M. Crawford, and J. I. Schroeder. (2000). Cadmium and iron transport by members of a plant metal transporter family in *Arabidopsis* with homology to Nramp genes. *Proceedings of National Academy of Sciences USA*, *97*, 4991–4996.

Trapp, S., and C. McFarlane. (1995). *Plant contamination: modelling and simulation of organic processes*. Boca Raton, FL: Lewis.

USEPA. (1993). *Wildlife Exposure Factors Handbook. vol. I.* EPA/600/R-93/187a.

Von Wiren, N., S. Klair, S. Bansal et al. (1999). Nicotianamine chelates both FeIII and FeII: Implications for metal transport in plants. *Plant Physiology*, *119*, 1107–1114.

Walton, B., and T. Anderson. (1990). Microbial degradation of trichloroethylene in the rhizosphere: Potential application to biological remediation of waste sites. *Applied and Environmental Microbiology*, *56*, 1012–1016.

Walton, B. T., A. M. Hoylman, M. M. Perez, T. A. Anderson, and T. R. Johnson. (1994). Rhizosphere microbial community as a plant defense against toxic substances in soils. In T. A. Anderson, and J. R. Coats (Eds.), *Bioremediation through rhizosphere technology*. Washington: American Chemical Society.

Wang, L., D. A. Samac, A. Shapir et al. (2005). Biodegradation of atrazine in transgenic plants expressing a modified bacterial atrazine chlorohydrolase (atzA) gene. *Plant Biotechnology Journal*, *3*, 475–486.

Watanabe, M. E. (1997). Phytoremediation on the brink of commercialization. *Environmental Science and Technology*, *31*, 182a–186a.

Wenzel, W. W., E. Lombi, and D. C. Adriano. (1999). Biogeochemical processes in the rhizosphere: Role in phytoremediation of metal-polluted soils. In M. N. V. Prasad, and J. Hagemeyer (Eds.), *Heavy metal stress in plants: From molecules to ecosystems* (271–303). New York: Springer-Verlag, Heidelberg.

Williams, L. E., J. K. Pittman, and J. L. Hall. (2000). Emerging mechanisms for heavy metal transport in plants. *Biochimica et Biophysica Acta*, *77803*, 1–23.

Wolf, A. E., K. J. Dietz, and P. Schroder. (1996). Degradation of glutathione s-conjugates by a carboxypeptidase in the plant vacuole. *FEBS Letters*, *384*, 31–34.

Wullschleger, S., F. Meinzer, and R. A. Vertessy. (1998). A review of whole-plant water use studies in trees. *Tree Physiology*, *18*, 499–512.

Zhu, Y., E. A. H. Pilon-Smits, L. Jouanin, and N. Terry. (1999a). Overexpression of glutathione synthetase in *Brassica juncea* enhances cadmium tolerance and accumulation. *Plant Physiology*, *119*, 73–79.

Zhu, Y., E. A. H. Pilon-Smits, A. Tarun, S. U. Weber, L. Jouanin, and N. Terry. (1999b). Cadmium tolerance and accumulation in Indian mustard is enhanced by overexpressing γ–glutamylcysteine synthetase. *Plant Physiology*, *121*, 1169–1177.

Section II

Genus Brassica *and Contaminants' Remediation*

7 Metals and Metalloids Accumulation Variability in *Brassica* Species: *A Review*

Naser A. Anjum, Sarvajeet S. Gill, Iqbal Ahmad, Armando C. Duarte, Shahid Umar, Nafees A. Khan, and Maria E. Pereira

CONTENTS

7.1 INTRODUCTION

The degradation of soils and sediments with a number of organic and inorganic contaminants on the globe has intensified during the last century as a result of a dramatically rapid increase in several anthropogenic activities, including the use of sludge or municipal compost, pesticides, fertilizers and emissions from municipal waste incinerators, car exhausts, residues from metalliferous mines, and the smelting industry. Major metals and metalloids (called trace metals, TMs hereafter), such as lead (Pb), chromium (Cr), arsenic (As), zinc (Zn), cadmium (Cd), copper (Cu), mercury (Hg), and nickel (Ni), constitute an ill-defined group of inorganic chemical hazards and are those most commonly found at contaminated sites. Unlike organic contaminants, which are oxidized to carbon (IV) oxide by microbial action, most TMs exhibit changes in their chemical forms (speciation) and bioavailability, but they do not undergo microbial or chemical degradation; therefore, once released, these TMs remain persistent for a long time (Adriano 2003). Therefore, worldwide soil and sediment TM contamination has raised severe concerns because of potential movement of these metals up to human food chain (Dahmani-Muller et al. 2001; McGrath et al. 2002).

In the perspective of the possible negative impacts of considerably high levels of TMs on the environmental and human health, a number of concerted efforts have been accomplished for the remediation of contaminated sites. Phytoremediation has been suggested as one of the most promising, environmentally friendly, and potentially cost-effective soil/sediment remediation techniques (Blaylock and Huang 2000). In this innovative remediation technique, plants and their associated microbes are utilized to remove, detoxify, and accumulate varied environmental contaminants. Hence, phytoremediation is widely viewed as ecologically responsible alternative to the environmentally destructive physical remediation methods currently practiced (Meagher 2000). The oleiferous genus *Brassica*, taxonomically placed within the family Brassicaceae (Cruciferae), is the third most important source of vegetable oil in the world after palm and soybean oil. It is pertinent to mention here that the members of the plant family Brassicaceae are distinguished as plants having significant biomass and extraordinary potential to accumulate and tolerate high quantities of a number of TMs (Dushenkov et al. 1995; Kumar et al. 1995; Broadley et al. 2001; Anjum et al. 2012). There are six species of *Brassica* that merit attention for their agronomic, scientific, and economic importance. Among the six species, three are diploid—*B. campestris*, *B. nigra*, and *B. oleracea*—and the other three are amphidiploids: *B. juncea*, *B. napus*, and *B. carinata*. The botanical and genomic relationships between these six species may be represented in the form of a triangle usually known as triangle of U (Nagaharu 1935) (Figure 7.1). According to a recent review by Vamerali et al. (2010), in the last 14 years, the most frequently cited species was *Brassica juncea* (L.) Czern. (148 citations), followed by *Helianthus annuus* (57), and *Brassica napus* and *Zea mays* (both 39 citations). In the ranking, Brassicaceae (*Raphanus sativus*, *Brassica carinata*) and Poaceae (*Festuca* spp., *Lolium* spp., *Hordeum vulgare*) were again represented; whereas, fewer citations were made of Fabaceae, such as soybean, bean, alfalfa and pea. The greater interest in Brassicaceae derives from the fact that research on these species started earlier, together with the interesting concentrations they provide, especially for *Brassica juncea* (L.) Czern. Moreover, among the major species of the genus *Brassica*, the *B. juncea*, *B. nigra*, *B. campestris*, *B. napus*, and *B. oleracea* have been extensively studied for their TM- extraction/remediation and/or tolerance studies (Salt et al. 1995; Diwan et al. 2010a,b). In this context, there exist metals/metalloids- and *Brassica* genotype-dependent differential uptake/accumulation of varied metals/metalloids. Among *B. juncea*, *B. campestris*, *B. carinata*, *B. napus*, and *B. nigra* genotypes tested for different metals/metalloids, *B. juncea* phytoextracted the largest amount of Cu from the soil. Additionally, *B. carinata* showed the highest concentration (mg/kg^{-1}) and uptake ($\mu g\ pot^{-1}$) of Ni and Pb at maturity. Although *B. campestris* showed a higher

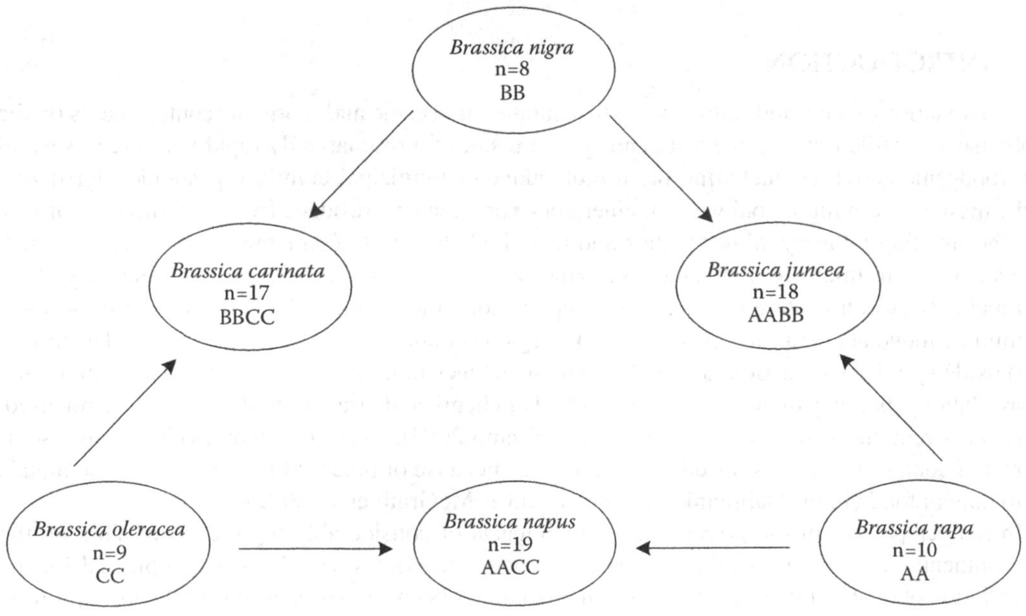

FIGURE 7.1 The "triangle of U diagram" representing the genomic relationships among *Brassica* species (Nagaharu 1935).

concentration of Zn in its shoots (stem plus leaf), *B. carinata* extracted the largest amount of this metal because of its greater biomass production (Purakayastha et al. 2008).

The current paper reports the significance of important *Brassica* species (such as *B. juncea*, *B. napus*, *B. nigra*, *B. campestris*, and *B. oleracea*) in scattered recent toxic metal–remediation studies. The reports presented here may thus be significant for the use of *Brassica* species with extraordinary capacity to accumulate and/or remediate varied TMs so as to clean up varied contaminated environmental compartments.

7.2 *BRASSICA JUNCEA*

Indian mustard (*Brassica juncea*) is one of the most important oilseed crops grown extensively particularly in arid and semiarid regions of the world (Ashraf and McNeilly 2004). Despite the fact that all of the examined crops from the Brassicaceae family do accumulate the metals, *B. juncea* shows the highest ability to accumulate and/or remediate a number of TMs or metalloids. A plethora of studies, including those of Kumar et al. (1995); Salt et al. (1995); Blaylock et al. (1997); Liu et al. (2000); and Clemente et al. (2005) on *B. juncea* reported the significant capacity of this species to extract/accumulate significant quantities of TMs/metalloids, such as As, Cd, Cu, Cr, Ni, Zn, Pb, and Se; thus, this species has a very high potential for phytoremeditation (Salt et al. 1995; Blaylock et al. 1997; Ebbs and Kochian 1998; Henry 2000; Pinto et al. 2009; Diwan et al. 2010). *B. juncea* was reported to possess the ability to eliminate 1.1550 kg Pb from an acre (Henry 2000). Pinto et al. (2009) reported the accumulation of Cd maximally in *B. juncea* roots (versus shoots) for all Cd levels in the soil, which the authors speculated to be a natural protective response of the plants to defend their above-ground parts from Cd toxicity. Furthermore, authors confirmed the performance of *B. juncea* as a phytoextractor for high Cd concentrations by studying the translocation factor (TF), which increased with Cd soil contamination. In another study, Nouairi et al. (2006) observed a TF of 0.3 for *B. juncea* exposed to 11 mg Cd l^{-1} during 15 days in hydroponic solution in between the TF values obtained in this work for *B. juncea* exposed to 5 and 15 mg Cd per kg^{-1} in soil. Ahmad et al. (2010) reported a substantial accumulation of Cd and a relatively higher amount of Cd in the roots than that in the shoots of *B. juncea* seedlings when fed with 100 or 200 mg l^{-1} Cd. Del Rio et al. (2004), found that the content of Pb in the stems of *B. juncea* varied from 6.2 mg to 28.6 mg/kg^{-1} and that of Zn from 94.1 mg/kg^{-1} to 132 mg/kg^{-1}. The obtained values are higher than those found by Ebbs et al. (1997a,b), who examined different species of the *Brassica* family. It has been reported that *B. juncea* accumulates significant amounts of Cr in both shoots and roots at higher soil Cr concentrations despite severe phytotoxic symptoms (Han et al. 2004).

Different cultivars of *B. juncea* have been extensively reported to exhibit differential TM– or metalloid-accumulation/remediation potential. In this context, recently, Bauddh and Singh (2011) reported a Cd dose-dependent increase in Cd accumulation in the seedlings of *B. juncea* cultivars, namely Pusa Jai Kisan, Pusa Bold, Gangotri, PT-1001, and Kranti exposed to 0.0, 0.5, 1.0, 2.0, 3.0, and 5.0 mM $CdCl_2$. At the lower Cd concentration, the accumulation of Cd significantly differed in different cultivars; however, at the higher Cd concentrations, the authors noticed nearly similar accumulation of Cd in all the five cultivars studied. However, the cultivar PJK exhibited the highest Cd accumulation. Gill et al. (2011) reported a differential Cd accumulation potential of five *B. juncea* cultivars, namely Alankar, Varuna, Pusa Bold, Sakha, and RH30 exposed to 0, 25, 50, 100, or 150 mg Cd per kg^{-1} soil. The authors documented significant soil Cd levels dependent on increased accumulation of Cd in the roots and leaves of all the cultivars. For each Cd treatment, its concentration was always higher in the roots in comparison to the leaves in all the cultivars. Maximum Cd accumulation was noted in the roots and leaves of RH30 with less in Alankar under all the Cd treatments. The Cd content in the root and leaves of RH30 was 354.21 and 137.07 μg/g^{-1} dry weight and 273.1 and 70.74 μg/g^{-1} dry weight, respectively, with 150 mg Cd per kg^{-1} soil. In a study by Sharma et al. (2010), *B. juncea* cv. T-59 exhibited a higher accumulation of Cd than TM-4 when treated with 0.5–1.5 mM Cd for 15 days.

Qadir et al. (2004) reported the accumulation of Cd in 10 *B. juncea* cvs.—Vardhan, Pusa Bahar, Pusa Bold, BTO, Pusa Jai Kisan, Agrini, Varuna, Kranti, Vaibhav, and Pusa Basant—exposed to a range of 0.0 to 2.0 mM Cd concentrations for 24, 48, and 72 hours after treatment (HAT). In general, Cd-treated plants showed accumulation of Cd in the leaves, which varied from 2.8 to 6.2, 4 to 6.8, and 4 to 7 μg/g^{-1} dry weight with T4 at 24, 48, and 72 HAT, respectively. Maximum accumulation of Cd was found in *B. juncea* cv. Pusa Jai Kisan (68 μg/g^{-1} dry weight) and the minimum in *B. juncea* cv. Vardhan (53.6 μg/g^{-1} dry weight) with the treatment T4 at 72 HAT. In a study by Diwan et al. (2008), the Cr accumulation potential varied largely among 10 *B. juncea* genotypes—namely, Vardhan, Pusa Bahar, Pusa Bold, BTO, Pusa Jai Kisan, Agrini, Varuna, Kranti, Vaibhav, and Pusa Basant—treated with 10, 25, 50, and 100 μM Cr. At the 100 μM Cr treatment, Pusa Jai Kisan accumulated the maximum amount of Cr (1680 μg Cr/g^{-1} dry weight); whereas, Vardhan accumulated the minimum (107 μg Cr/g^{-1} dry weight).

Bañuelos et al. (2000) studied the extent of Se accumulation when *B. juncea* plants were irrigated with Se-laden effluent or *B. juncea* plants were grown in Se-laden soils. When *B. juncea* was irrigated with Se-laden effluent, plant Se concentrations averaged 21 μg Se/g^{-1} dry mass, and total Se added to soils via effluent decreased by 40%. Plant Se concentrations averaged 75 μg/g^{-1} dry mass for *B. juncea*, and the total Se added to soils prior to planting decreased by 40% when *B. juncea* plants were grown in Se-laden soil. Furthermore, in both studies, plant accumulation of Se accounted for at least 50% of the Se removed from soil planted to *B. juncea*. The authors concluded that although the tested *Brassica* species led to a significant reduction in Se added to soil via use of Se-laden effluent, additional plantings are necessary to further decrease Se content in the soil. Bali et al. (2010) investigated the accumulation and partitioning of Au in *B. juncea* exposed to 100, 1000, and 10,000 ppm Au concentrations for 24, 48, or 72 h. *B. juncea* roots retained 188 mg of Au/g^{-1} dry weight when exposed a 10,000 ppm Au solution for 24 h. The TF, defined as the ratio of metal accumulated in the shoots to metal accumulated in the roots, was between 0.15 and 0.50 in *B. juncea*. Moreover, *B. juncea* roots, in general, showed a greater uptake ratio, defined as the ratio of Au concentration in plant tissues to the concentration in the solution, to a maximum of 906, which suggested the restricted translocation of Au to shoots once absorbed. In another study, Bali et al. (2010) studied the uptake of biogenic Pt and nanoparticle formation in *B. juncea* exposed to 5, 10, 20, 40, and 80 ppm Pt for 24, 48, and 72 h. Across all substrate concentrations, the authors noted a general increase in root and shoot Pt concentration with an increase in substrate concentration, ranging from 0.66 to 24.4 mg Pt/g^{-1} (dry biomass) in the roots to 0.02–0.69 mg Pt/g^{-1} (dry biomass) in the shoots after 24 h. Moreover, the largest quantities of Pt were accumulated at a substrate concentration of 80 ppm and increased with exposure time, ranging from 24.3 at 24 h to 38.5 mg Pt/g^{-1} (dry biomass) at 80 ppm, after 24 and 72 h, respectively. The change in accumulation between experiments at 20 and 40 ppm was also found to be substantial, leading to a nearly sixfold increase in root and shoot Pt concentration after 24 h. The accumulation of 143 and 211 μg As/g^{-1} dry weight of *B. juncea* cv. Pusa Jai Kisan root tissues was reported by Khan et al. (2009) when treated with 5 and 25 μM As respectively for 96 h in a hydroponic culture. The corresponding values of As accumulation in the shoot were 206 and 322 μg As/g^{-1} dry weight, respectively.

Using combined HPLC with an inductively coupled plasma mass spectrometer (ICP-MS), Ogra et al. (2010) documented a dose-dependent increase in Se concentration in *B. juncea* leaves reaching a plateau (11.8 μmol g^{-1}) at the Se exposure concentration of 200 μmol l^{-1}. At the highest concentration of 400 μmol l^{-1}, the incorporation of Se became saturated, and Se toxicity appeared. Te concentration in the leaves was also increased in a dose-dependent manner up to the highest exposure concentration. Te concentration in the leaves exposed to 400 μmol l^{-1} Te was 0.307 μmol g^{-1}, indicating less incorporation of Te than Se in *B. juncea*. Moreover, at Se and Te concentrations of 200 μmol l^{-1}, for which the authors did not observed any adverse effects, Se concentration (11.8 μmol g^{-1}) was found to be 69 times higher than Te concentration (0.171 μmol g^{-1}). In addition, although *B. juncea* effectively accumulated Se, it did not accumulate Te, the same group element as Se. Szollosi et al. (2009) determined the Cd content in germinating *B. juncea* seeds when exposed to 0, 50, 100, and 200 mg l^{-1} Cd concentrations in the dark for 12, 24, 48, and 96 h. The authors documented a Cd level–dependent significant increase in seed Cd content. Seeds of Cd50 accumulated 1.9–3.8, those of Cd100 4.1–6.7, and seeds exposed to 200 mg l^{-1} Cd 4.7–8.6 μM g^{-1} dry weight, depending on exposure time. Moreover, Cd concentration in seeds of Cd200 was 1.5- and fourfold higher after 12 and 96 h compared to those of Cd50. The authors concluded that there exists close relationship between time and concentration used and that Cd content in germinating seeds is affected by both duration and Cd level.

In *B. juncea* cv. Pusa Jai Kisan grown for 15 days in a hydroponic culture supplemented with 0.2, 2.0, and 20.0 μM Cr^{6+}, Pandey et al. (2005) documented a higher restriction of Cr to the roots, and very little of it was transported to the aerial parts. The roots of plants exposed to 0.2, 2.0, and 20.0 μM Cr^{6+} accumulated 76, 410, and 897 μg Cr/g^{-1} dry weight, respectively by day five. Authors observed a maximum accumulation of Cr in the roots of plants grown for 15 days. The amount of Cr detected in the leaves by day five was quite low. Additionally, the maximum amount of Cr (56.5 μg Cr/g^{-1} dry weight) was observed accumulated in leaves of plants grown at 20.0 μM Cr^{6+} for 15 days. In Cr (50, 100, and 200 μM) treated *B. juncea*, Diwan et al. (2010) documented Cr accumulation in the range of 100–300, 224–450, 398–754, and 550–972 μg/g^{-1} dry weight, respectively at 1, 3, 5, and 7 days of 50, 100, and 200 μM Cr treatments; whereas, the authors noted the corresponding figures in the shoots as 80–120, 152–289, 274–419, and 414–504 μg/g^{-1} dry weight. Su et al. (2007) evaluated the capacity of *B. juncea* for uptake and accumulation of radionuclides Cs and Sr natural isotopes. The authors exposed *B. juncea* to different concentrations of Cs (50 and 600 ppm) and Sr (50 and 300 ppm) natural isotopes in solution form for 23 days and noted the bioaccumulation of Cs and Sr in the order of leaves > stems > roots for both Cs- and Sr-treated plants. The highest leaf and root Sr accumulations were observed to be 2708 and 1194 mg/kg^{-1}, respectively; and the highest leaf and root Cs accumulations were 12,251 and 6794 mg/kg^{-1}, respectively. The authors also documented high translocation efficiency for both elements by shoot/root concentration ratios greater than one. Mobin and Khan (2007) documented differential Cd accumulation and partitioning in *B. juncea* cultivars differing in photosynthetic capacity when exposed to 0, 25, 50, and 100 mg Cd per kg^{-1} soil. The authors noticed a greater concentration of Cd in the roots and leaves of *B. juncea* cultivar RH-30 (low photosynthetic capacity) than cv. Varuna (high photosynthetic capacity) at all Cd treatments.

B. juncea cultivars differing in metal tolerances exhibit differential metal-accumulation potentials. When exposed to 0, 25, and 50 μmol l^{-1} Cd, the accumulation of Cd was more in Cd-sensitive cultivar SS2 than in Cd-tolerant Pusa Jai Kisan with the maximum at 50 μmol l^{-1} Cd treatment. Additionally, the cultivar SS2 accumulated about three times more Cd than Pusa Jai Kisan at 50 μmol l^{-1} Cd treatments (Iqbal et al. 2010). Srivastava et al. (2010) reported the accumulation of As in *B. juncea* cvs. TPM-1 (As-sensitive) and TM-4 (As-tolerant) grown in field conditions and exposed to As(V) (50 and 500 μM) and As(III) (25 and 250 μM) stress for a period of seven or 15 days. Arsenic accumulation in the shoots of both varieties was found to be correlative to exposure concentration and duration. The authors observed the maximum As accumulation in response to As(V) in TPM-1 (49 μg/g^{-1} dry weight) followed by TM-4 (30 μg/g^{-1} dry weight) after 15 days. Whereas, in response to As(III), the maximum As accumulation after 15 days was noted as maximum in TPM-1 (37 μg/g^{-1} dry weight) and minimum in TM-4 (33 μg/g^{-1} dry weight). Chaturvedi (2004) observed a variation of Cd concentrations in seeds of *B. juncea* from 3.2–68.5 μg/g^{-1} (with a mean of 31.85 μg/g^{-1}) and in the stem from 3.5 to 128 μg/g^{-1} (with a mean of 51.65 μg/g^{-1}) while that in the root was from 7.2 to 316 μg/g^{-1} (with a mean of 135.29 μg/g^{-1}). Cd was accumulated in the roots in much higher amounts than in the stems, especially in the case of the genotype DHR-9504. Moreover, when treated with 40 mg/kg^{-1} Cd, the concentration of Cd in the roots was found to be about four times higher than in the stems of genotype DHR-9504 and two times higher than in the stems of genotype Varuna. The concentration of Hg was documented in *B. juncea* cvs. Florida Broad Leaf and Long-standing, grown hydroponically in a mercury-spiked solution by Shiyab et al. (2009). The authors observed the highest Hg concentrations in the one-month-old Broad Leaf cultivar after two weeks of Hg treatment (16.7 mg l^{-1}), which were 2570 and 27,900 mg/kg^{-1} in the shoots and roots, respectively. However, Hg concentrations in the shoots and roots of the one-month-old

Long-standing cultivar after two weeks of treatment were 2240 and 18,400 mg/kg^{-1}, respectively, were also observed by the authors.

Diwan et al. (2008) investigated the accumulation of Cr in the *B. juncea* cultivar Pusa Jai Kisan grown on soil contaminated with five levels of Cr, viz., 100, 200, 300, 400, and 800 mg Cr per kg^{-1} soil. Authors observed a Cr dose-dependent progressive increase in Cr accumulation in the roots, stem, and leaves. Cr accumulation of the plant heightened with the onset of the flowering stage of the plant and continued through post-flowering. Cr concentrations in different samples of roots went up to 290–890 mg/kg^{-1} dry weight with 100 mg Cr per kg^{-1} soil. Maximum Cr accumulation was observed with 800 mg Cr per kg^{-1} soil in the stem (4190 mg/kg^{-1} dry weight), leaves (990 mg/kg^{-1} dry weight), and pod wall (12.5 mg/kg^{-1} dry weight). Moreover, the authors noticed only traces of Cr (0.0028–0.086 mg/kg^{-1} dry weight) in seeds in comparison with the Cr concentration of the pod wall, root, stem, and leaf. Uptake and localization of Pb was studied in hydroponically grown *B. juncea* exposed to 3.2, 32, and 217 μM Pb for three days (Meyers et al. 2008). The authors observed Pb-concentration-dependent two to three orders of magnitude greater accumulation of Pb in the roots compared to the shoots. At the highest exposure concentration (500 μM), the mean concentration of Pb in the root tissue was approximately 138 g/kg^{-1}. In spite of these significant tissue concentrations of Pb, the plants continued to grow over the 72 h treatment period although, at exposure concentrations of 50 and 500 μM, a transient loss of turgor in the shoots was noted in most plants within the first 12 h. Moreover, even an exposure concentration of 5 μM did not resulted in macroscopic changes in the shoot at this treatment concentration where concentration of Pb in the roots was approximately 1.3 g/kg^{-1}. Han et al. (2004) investigated the uptake and translocation of Cr in *B. juncea* grown on Cr(III)- and Cr(VI)-contaminated soils. The authors noticed a relative uptake of Cr in *B. juncea* from soil contaminated with either Cr(III) or Cr(VI). Total Cr in the leaves was found to be less than 20–30 mg/kg^{-1} when Cr concentrations in Cr(III)-contaminated soil was less than 600 mg Cr per kg^{-1}, and those in Cr(VI)-contaminated soil was less than 300 mg Cr per kg^{-1}. However, when the Cr in the Cr(III)-contaminated soil increased to 2000 mg/kg^{-1}, Cr concentrations in the leaves and roots also increased up to 2000–3000 and 3000–3300 mg/kg^{-1} (dry weight), respectively. Similarly, the authors observed the Cr concentrations in the leaves and roots to be 600–650 and 1800 mg/kg^{-1}, respectively, when the Cr(VI) addition in the soil was made up to 500 mg/kg^{-1}. Moreover, the authors noted very slow translocation of Cr from the roots to the stems of *B. juncea* grown either in Cr(III)- or Cr(VI)-treated soils. Under simulated field conditions, Sinha et al. (2010) reported accumulation of As, Cr, and Cu in *B. juncea* cv. Pusa Bold grown in contaminated substrates [Cu, Cr(VI), As(III), As(V)]. The authors noted a dose-dependent increase in Cu and Cr accumulation in *B. juncea* organs. In As-treated plants, the accumulation of As in the roots was found to be more than in the upper parts. With an increase in As concentration, the translocation of As(III) from the root to the upper part did not display any trend; in contrast, the authors noted an increase in the translocation of As(V). Maximum accumulation (μg/g^{-1} dry weight) of As(III) and As(V) in the roots was recorded as 438 and 617; in stems, it was 234 and 430; and in leaves, 184 and 334, respectively, in plants grown on As50 mg/kg^{-1} dry weight. Moreover, compared to As(V)-treated plants, accumulation of As was observed more in the plants treated with As(III). The uptake of Cr by the plants was found to be relatively low in comparison with other metals. Chromium accumulation (μg/g^{-1} dry weight) in the leaves was observed to be 26 at Cr 10, which increased to 63 at Cr 50 mg/kg^{-1} dry weight. As was also observed for other metals, the highest accumulation of Cr was more in the roots than in the upper parts. Zaier et al. (2010) investigated the Pb-phytoextraction potential in *B. juncea* seedlings when exposed in a nutrient solution to 0, 200, 400, 800, and 1000 μMPb^{2+} for 21 days. Lead was preferentially accumulated in *B. juncea* roots; whereas, the concentration of sequestered Pb^{2+} in shoots at 1000 μM was 2200 μg/g^{-1}. *Brassica juncea* has been reported to accumulate Zn and especially Pb (Ebbs and Kochian 1998; Epstein et al. 1999) and for its potential for adaptation to Mediterranean climates (Del Río et al. 2000).

Clemente et al. (2005) assessed the multi-metal accumulation potential of two crops of *B. juncea* grown in a field experiment at a site affected by the toxic spillage of acidic, metal-rich waste in

Aznalcó llar (Seville, Spain). The authors documented increased Zn concentrations in the tissues of *B. juncea* in the order roots > stems > leaves. Additionally, Cu, Fe, Mn, and As were also found in higher concentrations in the leaves while the authors could not clearly observe the tendency of Pb accumulation. The concentrations of Zn and Pb in most *B. juncea* tissues exceeded the levels found in plants of *B. juncea* grown in noncontaminated soil from a location in the Guadiamar River area (37.5 μg/g^{-1} Zn; 0.07 μg/g^{-1} Pb) earlier reported by del Río et al. (2000). The accumulation and distribution of Cd and Pb were documented by John et al. (2009) in *B. juncea* treated with six different concentrations of Cd (150, 300, 450, 600, 750, and 900 μM) or Pb (150, 300, 600, 900, 1200, and 1500 μM) at the pre-flowering, flowering, and post-flowering stages. The authors observed the maximum accumulation of Cd (116.32 mg/g^{-1} dry weight of the plant material) in the roots at 750 μM Cd at the pre-flowering stage. While significant decline in Cd accumulation was noted on prolonged exposure (post-flowering stage) of *B. juncea* to higher concentrations of Cd, i.e., 900 μM. Similarly, authors noticed a higher accumulation of Pb in the roots when compared with the shoots. Accumulation of Pb was comparatively lower and reached up to 85.97 mg/g^{-1} dry weight of *B. juncea* at the post-flowering stage. Šimonová et al. (2007) investigated the uptake and accumulation of Cd in plant organs of *B. juncea* L. cv. Vitasso exposed to 6–120 μM Cd. Cd accumulation in roots and shoots and the effects of Cd toxicity on growth were correlated with increasing Cd concentrations in the nutrient solution. *B. juncea* roots accumulated higher amounts of Cd than the shoots. The amount of Cd in the roots of Indian mustard at 120 μM concentration was 5 and 11, respectively. Gupta et al. (2009) documented a *B. juncea* variety and tissue-specific arsenic accumulation pattern when exposed to 50, 150, and 300 μM AsIII concentrations for a period of two to four days. The leaves and roots of both *B. juncea* cv. Varuna and P. Bold exhibited increased metal levels with increasing concentrations of metal in solution. More As concentration was observed in the roots for a shorter duration (two days) as compared to leaves. The leaves showed more accumulation for a higher duration (four days). Concentration of As was higher in the P. Bold variety when compared with the Varuna. In *B. juncea* var. Megarrhiza exposed to lead nitrate ($Pb(NO_3)_2$), uses were in the range of 10^{-5}–10^{-3} M. Liu et al. (2000) reported Pb concentration-dependent differential uptake and accumulation of Pb in the roots, hypocotyls, and shoots. The Pb content in the roots and hypocotyls increased with an increasing solution concentration of Pb^{2+} through 10^{-4} M Pb. Shoots continued to increase for the 10^{-3} M Pb^{2+} treatment. Moreover, the roots in the group treated with 10^{-4} M Pb were reported to accumulate a substantial amount of Pb (15,982.8 μg/g^{-1} dry weight). More than 95% of the Pb accumulated in the treated plants was found in the roots, except for 10^{-3} M Pb, where 24.46 and 14.94% were distributed to the hypocotyls and the shoots, respectively.

7.3 *BRASSICA NAPUS*

Oilseed rape (*Brassica napus* L.) is a bright yellow, flowering member of the family Brassicaceae. It is grown as an important source of edible oil worldwide. It can also be used for biodiesel production to replace part of fossil fuel. Moreover, because of prolific growth, *B. napus* can be grown advantageously for phytoremediation of lands contaminated by industrial waste. *B. napus* and *B. rapa* were reported to accumulate satisfactory quantities of heavy metals (Ebbs et al. 1997; Marchiol et al. 2004).

The available literature reports differential TM-accumulation/remediation potential of *B. napus* under hydroponics, greenhouse, or field conditions. Grispen et al. (2006) reported the shoot and root accumulation of Cd and Zn in *B. napus* grown for four months at two locations, namely Balen and Budel, at metal smelters. At Balen, they found a variation range for shoot Cd of 3.6 to 8.1 mg/kg^{-1} dry weight with an average of 5.6 mg/kg^{-1} dry weight; whereas, at Budel, this variation ranged from 5.2 to 11.3 mg/kg^{-1} dry weight with an average of 7.6 mg/kg^{-1} dry weight. Moreover, the authors documented total extractable Cd at Balen to be approximately more than double the amount at Budel, but the plants grown in Budel showed higher shoot Cd. The plants with a high Cd uptake also appeared to exhibit enhanced Zn uptake. In Balen, shoot Zn ranged from 342 to 1215 mg/g^{-1}

dry weight with an average of 657. In Budel, shoot Zn showed a smaller variation, 743 to 1414 mg/g^{-1} dry weight with an average of 1151. At both locations, *B. napus* accessions were noted to exhibit significant differences, concerning shoot Cd and Zn. The accessions in Balen accumulated an average of 979 mg Cd plant^{-1} within a range of 274 to 1712 mg Cd plant^{-1}; in Budel, an average of 1037 mg Cd plant^{-1} was found with a variation range of 416 to 1621 mg Cd plant^{-1}. A study by Shams et al. (2010) reported a higher accumulation of Cr in leaves when plants were fed with 200 mg l^{-1} solutions compared to 500 mg l^{-1} solutions. In addition, average Cr concentrations in the stems were almost equal to the leaves. Carrier et al. (2003) reported a preferential accumulation of Cd in leaves of *B. napus* grown for 47 days from seeds in reconstituted soil contaminated with different Cd concentrations, ranging from 10 mg/kg^{-1} (subtoxic) to 200 mg/kg^{-1} (very toxic). The authors documented an asymptotic increase in Cd concentration in plant tissues with increasing Cd concentration in the soil. A plateau was observed in leaf Cd with 50 mg/kg^{-1} soil. Interestingly, the plateau level of Cd in the leaves (ca. 260 μg Cd/g^{-1}) was noted to be much higher than that of the stems (ca. 150 μg/g^{-1}). Also, the Cd level in the roots was found to be lower than that measured in the leaves. Moreover, the Cd content of leaf epidermal strips was observed to be substantially lower than the Cd concentration measured in the whole leaves (115 vs. 263 μg/g^{-1}).

Ghnaya et al. (2009) reported Cd and Zn accumulation in *B. napus* cultivars treated with 250 μM Cd and 2000 μM Zn concentrations. In *B. napus* cv. Jumbo, the stems and petioles accumulated the most Cd (64.75 μg/g^{-1} dry weight), followed by the roots (44.5 μg/g^{-1} dry weight), and the limbs displayed the lowest amount of Cd (8.02 μg/g^{-1} dry weight). Here also in *B. napus* cv. Drakkar, the most important content of accumulated Cd was measured in the stems and petioles (51.77 μg/g^{-1} dry weight), and the roots accumulated almost the same amount as did the roots of Jumbo cv. (41.05 μg/g^{-1} dry weight). Also, the Cd level of the limbs was observed to be the weakest (5.45 μg/g^{-1} dry weight). In cv. Cossair also, the most important content of Cd (31.25 μg/g^{-1} dry weight) was registered in the roots; whereas, it was less significant at the level of the stems, petioles, and limbs with respective contents of 17.75 μg/g^{-1} dry weight and 8.27 μg/g^{-1} dry weight. In cv. Pactol, the stems and petioles accumulated 20.05 μg/g^{-1} dry weight; whereas, the roots contained 34.97 μg/g^{-1} dry weight. The authors concluded that the amount of Cd accumulation differed according to the organ and the variety of plant species. In the perspective of Zn accumulation, here also, the stems and petioles of *B. napus* cvs. tested showed the most important content of Zn for Jumbo and Drakkar while roots accumulated the most Zn in Cossair and Pactol.

7.4 *BRASSICA NIGRA*

Angelova and Ivanov (2009) investigated the bio-accumulation and distribution of Pb, Cu, Zn, and Cd in black mustard (*Brassica nigra* Koch) under pot and field conditions using six sets of treatments, namely treatment I—Pb120Cd3Zn540, treatment II—Pb200Cd5Zn900, treatment III—Pb280Cd7Zn1260, treatment IV—Pb400Cd10Zn1800, treatment V—Pb800Cd20Zn3600, and treatment VI—Pb1200Cd30Zn5400. Under field conditions, the authors observed a significant part of accumulated Pb, Cu, Zn, and Cd in *B. nigra* roots; the accumulation was found to be smaller with increasing distance from the source of contamination. The content of Pb in the roots of the *B. nigra* grown at a distance of 0.1 km from the NFMW reached 164.4 mg/kg; the contents of Cu reached 15.7 mg/kg^{-1}; Zn reached up to 426.1 mg/kg^{-1}, and Cd reached up to 7.9 mg/kg^{-1}. Moreover, in the *B. nigra* grown in a region at a distance of 15 km from the NFMW, the same metal accumulation tendencies were preserved, but the reported values were observed to be lower: 3.15 mg/kg^{-1} Pb, 1.6 mg/kg^{-1} Cu, 9.2 mg/kg Zn, and 0.18 mg/kg^{-1} Cd. The authors inferred that large parts of the heavy metals contained in the soil were fixed and accumulated in the roots of the *B. nigra* as a result of the fact that this species forms a well-built root system with a high ability to absorb. Earlier, the accumulation of Cd, Cr, Cu, Ni, Pb, and Zn in significant quantities was also reported in *B. nigra* roots by Dushenkov et al. (1995). Under pot growing conditions, the increase in Pb, Cd, and Zn content in *B. nigra* roots, stems, fruit shells, and seeds was found to be dependent on the concentrations of the studied metals in the soil. In addition, the content of Pb in the roots was noted by the authors to be higher than the content in the epigeal parts of *B. nigra*; however, a contrary tendency was observed for Cd and Zn, thus confirming a differential mobility of Pb, Zn, and Cd. The authors' results showed that the mobility of Pb, Zn, and Cd in the examined plant is not the same. In the case of highly contaminated soil, the authors reported the absorption of major parts of Pb by the *B. nigra* roots and a very small transportation of Pb along the conduction system toward the stems.

7.5 *BRASSICA CAMPESTRIS*

Brassica campestris L. (syn. *B. rapa* L.), commonly known as turnip, turnip rape, or turnip mustard, is a plant widely cultivated as a leaf vegetable, a root vegetable, and an oilseed. *B. campestris* produces significant amounts of biomass, which is advantageous in remediation of contaminated soil. Biomass accumulation can be directed either to the aerial parts or the below-ground parts. The efficiency of *B. campestris* for accumulating contaminants (such as heavy metals) in its below-ground parts makes it a important member for phytoremediation (Gleba et al. 1999). It has been reported that *B. oleracea* is a good candidate for Zn bioaccumulation (Salt et al. 1998). Dheri et al. (2007) conducted an experiment to study the natural potential of *Trigonella foenumgraecum*, *Spinacia oleracea*, and *B. campestris* for cleanup of Cr-contaminated silty loam and sandy soils. It was noted that the concentration of diethylene triamine pentaacetic acid–extractable Cr increased significantly with an increasing rate of Cr application in both soils, but the increase was higher in sandy soil than in silty loam soil. The percent reduction in dry-matter yield (DMY) with increasing levels of added Cr in comparison to the zero-Cr control was highest for *T. foenumgraecum* (49 and 52%), followed by *S. oleracea* (36 and 42%), and lowest for *B. campestris* (29 and 34%) in silty loam soil and sandy soil, respectively. It was interesting to note that the concentration of Cr in both shoots and roots was highest in *B. campestris*, followed by *S. oleracea* and *T. foenumgraecum*. *B. campestris* accumulated four times the amount of Cr in shoots and twice the amount in roots in comparison to *T. foenumgraecum*. It was concluded that the family Cruciferae (*B. campestris*) was most tolerant to Cr toxicity, followed by Chenopodiaceae (*S. oleracea*) and Leguminosae (*T. foenumgraecum*). Because *B. campestris* removed the highest amount of Cr from the soil, it could be used for pytoremediation of mildly Cr-contaminated soils (Dheri et al. 2007). In a greenhouse, pot culture experiment, soil Cd level-dependent significantly increases in Cd content and total Cd content in the root and shoot of *B. campestris* cv. Pusa Gold plants exposed to different Cd levels (25, 50, and 100 mg/kg^{-1} soil) (Anjum et al. 2008). The authors noted that maximum Cd accumulation with the highest Cd level (100 mg/kg^{-1} soil) was in the root (4, 983%) followed by the leaf (2, 634%) and stem (2, 100%).

7.6 *BRASSICA OLERACEA*

The representatives of the Brassicaceae family have potential for phytoremeditation. Like other *Brassica* species, *B. oleracea* was reported to accumulate significant amounts of toxic heavy metals (Kabata-Pendias 2001). In a study, *Brassica* species were tested for their ability to accumulate and transport Pb, and it was noted that the accumulation of Pb in *B. juncea* and *B. oleracea* varied from 1416 to 18,812 μg/g^{-1} dry weight. *B. oleracea* retained a large amount of the absorbed Pb in the roots in comparison to the above-ground parts (Kumar et al. 1994). Three *Brassica* crop species (*B. oleracea*, *B. carinata*, and *B. juncea*) were selected for evaluating their potential for phytoextraction of Pb, Cu, and Zn, and it was noted that *B. oleracea* accumulated 381 mg Zn per kg^{-1} dry weight and 834 mg Cu per kg^{-1} dry weight in the shoots. All the tested *Brassica* species showed a similar trend for Pb accumulation under different soil Pb concentrations. The lesser availability of Pb to these *Brassica* species may be a result of its extreme insolubility and not generally being available for plant uptake in the normal range of soil pH (Gisbert et al. 2006; Gruca-Królikowska and Wacławek 2006). Sun et al. (2007) investigated *Beta vulgaris*, *Brassica juncea*, and *B. oleracea* as potential phytoremediants. Results from this study showed that these plants could extract heavy metals from soil, but the accumulation and translocation of metals differed with the species of plant, categories of heavy metals, and some environmental conditions. Addae et al. (2010) conducted an experiment to examine the potential of *B. oleracea* to remove Cd and Pb from contaminated soils. Different concentrations of Pb (0, 100, 250, 500, and 1000 mg/kg Pb) and Cd (0, 100, 250, and 500 mg/kg Cd) were used. The findings suggested that *B. oleracea* plants tolerated and extracted Cd and Pb from the soil, and the metal uptakes were dose related (Addae et al. 2010).

7.7 CONCLUSIONS AND PERSPECTIVES

Phytoremediation as a natural, solar energy–driven option for the cleaning of contaminated matrices has many potential advantages. Incorporating a range of technologies that use plants to remove, reduce, degrade, or immobilize environmental pollutants from soil and water, phytoremediation as an interdisciplinary technology, thus far, is significantly and efficiently restoring contaminated sites to a relatively clean, nontoxic environment. In this context, different species of the genus *Brassica* such as *Brassica juncea*, *B. nigra*, *B. campestris*, *B. napus*, and *B. oleracea* stand second to none in terms of their importance in the uptake, accumulation, sequestration and/or degradation of varied metals and metalloids. However, in the perspectives of significance as test *Brassica* species, *B. juncea* ranked first, followed by *B. napus* and *B. carinata*.

A lot has been done on the aspects of contaminant uptake, detoxification, and sequestration and consequences using different *Brassica* spp. but mostly in greenhouse, hydroponic systems; assessment of successful experiments under field conditions must be done to get more benefits from the extraordinary metal and metalloid remediation potential of *Brassica* spp. Moreover, deciphering major mechanisms underlying ability of *Brassica* spp. to detoxify and accumulate environmental contaminants, including metals and metalloids, should be the priority of near-future research. In this regard, the identification of novel genes and the subsequent development of transgenic *Brassica* spp. with superior metal- and metalloid-remediation potential and a better understanding of the rhizospheric processes and plant-microbe interactions would significantly benefit *Brassica* spp.–assisted metal- and metalloid-remediation strategies.

ACKNOWLEDGMENTS

NAA (SFRH/BPD/64690/2009), IA, ACD, and EP are grateful to the Portuguese Foundation for Science and Technology (FCT) and the Aveiro University Research Institute/Centre for Environmental and Marine Studies (CESAM) for partial financial support. SSG, SU, and NAK would like to acknowledge the receipt of funds from DBT, DST, and UGC, Govt. of India, New Delhi.

REFERENCES

Addae, C., M. Piva, A. J. Bednar, and M. S. Zaman. (2010). Cadmium and lead bioaccumulation in cabbage plants grown in metal contaminated soils. *Advances in Science and Technology*, *4*, 79–82.

Adriano, D. C. (2003). *Trace elements in terrestrial environments: biogeochemistry, bioavailability and risks of metals*, 2nd ed. Springer: New York.

Ahmad, P., G. Nabi, and M. Ashraf. (2010). Cadmium-induced oxidative damage in mustard [*Brassica juncea* (L.) Czern. and Coss.] plants can be alleviated by salicylic acid. *South African Journal of Botany*, *77*, 36–44.

Angelova, V., and K. Ivanov. (2009). Bio-accumulation and distribution of heavy metals in black mustard (*Brassica nigra* Koch). *Environmental Monitoring and Assessment*, *153*, 449–459.

Anjum, N. A., S. Umar, A. Ahmad, M. Iqbal, and N. A. Khan. (2008). Ontogenic variation in response of *Brassica campestris* L. to cadmium toxicity. *Journal of Plant Interactions*, *3*, 189–198.

Anjum, N. A., I. Ahmad, M. E. Pereira, A. C. Duarte, S. Umar, and N. A. Khan. (2012). *The plant family Brassicaceae: contribution towards phytoremediation*. Environmental Pollution Series No. 21, Springer: Dordrecht, The Netherlands.

Ashraf, M., and T. McNeilly. (2004). Salinity tolerance in Brassica oilseeds. *Critical Reviews in Plant Sciences, 23*, 157–174.

Bali, R., R. Siegele, and A. T. Harris. (2010). Phytoextraction of Au: Uptake, accumulation and cellular distribution in *Medicago sativa* and *Brassica juncea. Chemical Engineering Journal, 156*, 286–297.

Bañuelos, G. S., S. Zambrzuski, and B. Mackey. (2000). Phytoextraction of selenium from soils irrigated with selenium-laden effluent. *Plant and Soil, 224*, 251–258.

Bauddh, K., and R. P. Singh. (2011). Differential toxicity of cadmium to mustard (*Brassica juncea* L.). *Journal of Environmental Biology, 32*, 355–362.

Blaylock, M. J., and J. W. Huang. (2000). Phytoextraction of metals. In I. Raskin, and B. D. Ensley (Eds.), *Phytoremediation of toxic metals: Using plants to clean up the environment* (53–70). New York: John Wiley and Sons.

Blaylock, M. J., D. E. Salt, S. Dushenkov, O. Zakharova, C. Gussman, Y. Kapulnik, et al. (1997). Enhanced accumulation of Pb in Indian mustard by soil-applied chelating agents. *Environmental Science and Technology, 31*, 860–865.

Broadley, M., M. J. Willey, J. C. Wilkins, A. J. M. Baker, A. Mead, and P. J. White. (2001). Phylogenetic variation in heavy metal accumulation in angiosperms. *New Phytologist, 152*, 9–27.

Carrier, P., A. Baryla, and M. Havaux. (2003). Cadmium distribution and microlocalization in oilseed rape (*Brassica napus*) after long-term growth on cadmium-contaminated soil. *Planta, 216*, 939–950.

Chaturvedi, I. (2004). Phytotoxicity of cadmium and its effect on two genotypes of *Brassica juncea* L. *Emirates Journal of Agricultural Sciences, 16*, 1–8.

Clemente, R., D. J. Walker, and M. P. Bernal. (2005). Uptake of heavy metals and As by *Brassica juncea* grown in a contaminated soil in Aznalcollar (Spain): The effect of soil amendments. *Environmental Pollution, 138*, 46–58.

Dahmani-Muller, H., F. van Oort, and M. Balabane. (2001). Metal extraction by *Arabidopsis halleri* grown on an unpolluted soil amended with various metal-bearing solids: A pot experiment. *Environmental Pollution, 114*, 77–84.

Del Rio, M., R. Font, and A. de Haro. (2004). Heavy metal uptake by *Brassica* species growing in the polluted soils of Aznucollar (Southern Spain). *Fresenius Environmental Bulletin, 13*, 1439–1443.

Del Río, M., R. Font, J. M. Fernandez-Martínez, J. Domínguez, and A. de Haro. (2000). Field trials of *Brassica carinata* and *Brassica juncea* in polluted soils of the Guadiamar River area. *Fresenius Environmental Bulletin, 9*, 328–332.

Dheri, G. S., M. S. Brar, and S. S. Malhi. (2007). Comparative phytoremediation of chromium-contaminated soils by Fenugreek, Spinach, and Raya. *Communications in Soil Science and Plant Analysis, 38*, 1655–1672.

Diwan, H., A. Ahmad, and M. Iqbal. (2008). Genotypic variation in the phytoremediation potential of Indian mustard for chromium. *Environmental Management, 41*, 734–741.

Diwan, H., A. Ahmad, and M. Iqbal. (2010a). Chromium-induced modulation in the antioxidant defense system during phenological growth stages of Indian mustard. *International Journal of Phytoremediation, 12*, 142–158.

Diwan, H., I. Khan, A. Ahmad, and M. Iqbal. (2010b). Induction of phytochelatins and antioxidant defence system in *Brassica juncea* and *Vigna radiata* in response to chromium treatments. *Plant Growth Regulation, 61*, 97–107.

Dushenkov, V., P. B. A. N. Kumar, H. Motto, and I. Raskin. (1995). Rhizofiltration: The use of plants to remove heavy metals from aqueous streams. *Environmental Science and Technology, 29*, 1239–1245.

Ebbs, S. D., and L. V. Kochian. (1998). Phytoextraction of zinc by oat (*Avena sativa*), barley (*Hordeum vulgare*), and Indian mustard (*Brassica juncea*). *Environmental Science and Technology, 32*, 802–806.

Ebbs, S. D., M. M. Lasat, D. J. Brady, J. Cornish, R. Gordon, and L. V. Kochian. (1997a). Phytoextraction of cadmium and zinc from contaminated soil. *Journal of Environmental Quality, 26*, 1424–1430.

Ebbs, S. D., M. M. Lasat, D. J. Brady, J. Cornish, R. Gordon, and L. V. Kochian. (1997b). Heavy metals in the environment: Phytoextraction of cadmium and zinc from a contaminated soil. *Journal of Environmental Quality, 26*, 1424–1430.

Epstein, A. L., C. D. Gussman, M. J. Blaylock, U. Yermiyahu, J. W. Huang, Y. Kapulnik, and C. S. Orser. (1999). EDTA and Pb-EDTA accumulation in *Brassica juncea* grown in Pb-amended soil. *Plant and Soil, 208*, 87–94.

Ghnaya, A. B., G. Charles, A. Hourmant, J. B. Hamida, and M. Branchard. (2009). Physiological behaviour of four rapeseed cultivar (*Brassica napus* L.) submitted to metal stress. *Comptes Rendus Biologies*, *332*, 363–370.

Gill, S. S., N. A. Khan, and N. Tuteja. (2011). Differential cadmium stress tolerance in five Indian mustard (*Brassica juncea* L.) cultivars: An evaluation of the role of antioxidant machinery. *Plant Signaling and Behavior*, *6*, 293–300.

Gisbert, C., R. Clemente, J. Navarro-Aviñó, C. Baixauli, A. Ginér, R. Serrano, et al. (2006). Tolerance and accumulation of heavy metals by Brassicaceae species grown in contaminated soils from Mediterranean regions of Spain. *Environmental and Experimental Botany*, *56*, 19–26.

Gleba, D., M. V. Borisjuk, L. G. Borisjuk, R. Kneer, A. Poulev, M. Skarzhinskaya, et al. (1999). Use of plant roots for phytoremediation and molecular farming. *Proceedings of National Academy of Sciences USA*, *96*, 5973–5977.

Grispen, V. M. J., H. J. M. Nelissen, and J. A. C. Verkleij. (2006). Phytoextraction with *Brassica napus* L.: A tool for sustainable management of heavy metal contaminated soils. *Environmental Pollution*, *144*, 77–83.

Gruca-Królikowska, S., and W. Wacławek. (2006). Metale wśrodowisku. Cz. II. Wpływ metali ciężkich na rośliny. *Chemia Dydaktyka Ekologia Metrologia Nr*, *11*, 41–56.

Gupta, M., P. Sharma, N. B. Sarin, and A. K. Sinha. (2009). Differential response of arsenic stress in two varieties of *Brassica juncea* L. *Chemosphere*, *74*, 1201–1208.

Han, F. X., B. B.-M. Sridhar, D. L. Monts, and Y. Su. (2004). Phytoavailability and toxicity of trivalent and hexavalent chromium to *Brassica juncea*. *New Phytologist*, *162*, 189–199.

Henry, J. R. (2000). An overview of phytoremediation of lead and mercury. NNEMS Report. Washington, DC, 3–9.

Iqbal, N., A. Masood, R. Nazar, S. Syeed, and N. A. Khan. (2010). Photosynthesis, growth and antioxidant metabolism in mustard (*Brassica juncea* L.) cultivars differing in cadmium tolerance. *Agricultural Sciences in China*, *9*, 519–527.

John, R., P. Ahmad, K. Gadgil, and S. Sharma. (2009). Heavy metal toxicity: Effect on plant growth, biochemical parameters and metal accumulation by *Brassica juncea* L. *International Journal of Plant Production*, *3*, 65–75.

Kabata-Pendias, A. (2001). *Trace elements in soils and plants*, 3rd ed. Boca Raton: CRC.

Khan, I., A. Ahmad, and M. Iqbal. (2009). Modulation of antioxidant defence system for arsenic detoxification in Indian mustard. *Ecotoxicology and Environmental Safety*, *72*, 626–634.

Kumar, P. B. A. N., V. Dushenkov, H. Motto, and I. Raskin. (1995). Phytoextraction: The use of plants to remove heavy metals from soils. *Environmental Science and Technology*, *29*, 1232–1238.

Kumar, P. B. A. N., S. Dushenkov, D. E. Salt, and I. Raskin. (1994). Crop *Brassicas* and phytoremediation: A novel environmental technology. *Cruciferae Newsletter Eucarpia*, *16*, 18–19.

Liu, D., W. Jiang, C. Liu, C. Xin, and W. Hou. (2000). Uptake and accumulation of lead by roots, hypocotyls and shoots of Indian mustard [*Brassica juncea* (L.)] *Bioresource Technology*, 71, 273–277.

Marchiol, L., S. Assolari, P. Sacco, and G. Zerbi. (2004). Phytoextraction of heavy metals by canola (*Brassica napus*) and radish (*Raphanus sativus*) grown on multicontaminated soil. *Environmental Pollution*, *132*, 21–27.

McGrath, S. P., F. J. Zhao, and E. Lombi. (2002). Phytoremediation of metals, metalloids, and radionuclides. *Advances in Agronomy*, *75*, 1–56.

Meagher, R. B. (2000). Phytoremediation of toxic elements and organic pollutants. *Current Opinion in Plant Biology*, *3*, 153–162.

Meyers, D. E. R., G. J. Auchterlonie, R. I. Webb, and B. Wood. (2008). Uptake and localisation of lead in the root system of *Brassica juncea*. *Environmental Pollution*, *153*, 323–332.

Mobin, M., and N. A. Khan. (2007). Photosynthetic activity, pigment composition and antioxidative response of two mustard (*Brassica juncea*) cultivars differing in photosynthetic capacity subjected to cadmium stress. *Journal of Plant Physiology*, *164*, 601–610.

Nagaharu, U. (1935). Genome analysis in *Brassica* with special reference to the experimental formation of *B. napus* and peculiar mode of fertilization. *Japan Journal of Botany*, *7*, 389–452.

Nouairi, I., W. B. Ammar, N. B. Youssef, D. B. M. Daoud, M. H. Ghorbal, and M. Zarrouk. (2006). Comparative study of cadmium effects on membrane lipid composition of *Brassica juncea* and *Brassica napus* leaves. *Plant Science*, *170*, 511.

Ogra, Y., E. Eita Okubo, and M. Takahira. (2010). Distinct uptake of tellurate from selenate in a selenium accumulator, Indian mustard (*Brassica juncea*). *Metallomics*, *2*, 328–333.

Pandey, V., V. Dixit, and R. Shyam. (2005). Antioxidative responses in relation to growth of mustard (*Brassica juncea* cv. Pusa Jaikisan) plants exposed to hexavalent chromium. *Chemosphere*, *61*, 40–47.

Pinto, A. P., A. S. Alves, A. J. Candeias, A. I. Cardoso, et al. (2009). Cadmium accumulation and antioxidative defences in *Brassica juncea* L. Czern, *Nicotiana tabacum* L. and *Solanum nigrum* L. *International Journal of Environmental Analytical Chemistry*, *89*, 661–676.

Purakayastha, T. J., T. Viswanath, S. Bhadraray, P. K. Chhonkar, P. P. Adhikari, and K. Suribabu. (2008). Phytoextraction of zinc, copper, nickel and lead from a contaminated soil by different species of Brassica. *International Journal of Phytoremediation*, *10*, 61–72.

Qadir, S., M. I. Qureshi, S. Javed, and M. Z. Abdin. (2004). Genotypic variation in phytoremediation potential of *Brassica juncea* cultivars exposed to Cd stress. *Plant Science*, *167*, 1171–1181.

Salt, D. E., M. Blaylock, P. B. A. N. Kumar, V. Dushenkov, B. D. Ensley, et al. (1995). Phytoremediation: A novel strategy for the removal of toxic metals from the environment using plants. *Biotechnology*, *13*, 468–474.

Salt, D. E., R. D. Smith, and I. Raskin. (1998). Phytoremediation. *Annual Review of Plant Physiology and Plant Molecular Biology*, *49*, 643–668.

Shams, K. M., T. Gottfried, F. Axel, M. Sager, T. Peer, A. Bashar, and K. Filip. (2010). Aspects of phytoremediation for chromium contaminated sites using common plants *Urtica dioica*, *Brassica napus* and *Zea mays*. *Plant and Soil*, *328*, 175–189.

Sharma, A., M. Sainger, S. Dwivedi, S. Srivastava, et al. (2010). Genotypic variation in *Brassica juncea* (L.) Czern. cultivars in growth, nitrate assimilation, antioxidant responses and phytoremediation potential during cadmium stress. *Journal of Environmental Biology*, *31*, 773–780.

Shiyab, S., J. Chen, F. X. Han, D. L. Monts, et al. (2009). Phytotoxicity of mercury in Indian mustard (*Brassica juncea* L.). *Ecotoxicology and Environmental Safety*, *72*, 619–625.

Šimonová, E., M. Henselová, E. Masarovičová, and J. Kohanová. (2007). Comparison of tolerance of *Brassica juncea* and *Vigna radiata* to cadmium. *Biologia Plantarum*, *51*, 488–492.

Sinha, S., G. Sinam, R. K. Mishra, and S. Mallick. (2010). Metal accumulation, growth, antioxidants and oil yield of *Brassica juncea* L. exposed to different metals. *Ecotoxicology and Environmental Safety*, *73*, 1352–1361.

Srivastava, S., A. K. Srivastava, P. Suprasanna, and S. F. D'Souza. (2010). Comparative antioxidant profiling of tolerant and sensitive varieties of *Brassica juncea* L. to arsenate and arsenite exposure. *Bulletin of Environmental Contamination and Toxicology*, *84*, 342–346.

Su, Y., B. B. M. Sridhar, F. X. Han, S. V. Diehl, and D. L. Monts. (2007). Effect of bioaccumulation of Cs and Sr natural isotopes on foliar structure and plant spectral reflectance of Indian mustard (*Brassica juncea*). *Water, Air, Soil and Pollution*, *180*, 65–74.

Sun, L., Z. Niu, and T. Sun. (2007). Effects of amendments of N, P, Fe on phytoextraction of Cd, Pb, Cu, and Zn in soil of Zhangshi by mustard, cabbage, and sugar beet. *Environmental Toxicology*, 22, 565–571.

Szollosi, R., I. S. Varga, L. Erdei, and E. Mihalik. (2009). Cadmium-induced oxidative stress and antioxidative mechanisms in germinating Indian mustard (*Brassica juncea* L.) seeds. *Ecotoxicology and Environmental Safety*, *72*, 1337–1342.

Vamerali, T., M. Bandiera, and G. Mosca. (2010). Field crops for phytoremediation of metal-contaminated land: A review. *Environmental Chemistry Letters*, *8*, 1–17.

Zaier, H., T. Ghnaya, A. Lakhdar, R. Baioui, R. Ghabriche, M. Mnasri, S. Sghair, et al. (2010). Comparative study of Pb-phytoextraction potential in *Sesuvium portulacastrum* and *Brassica juncea*: tolerance and accumulation. *Journal of Hazardous Materials*, *183*, 609–615.

8 Oilseed *Brassica napus* and Phytoremediation of Lead

Muhammad Yasin Ashraf, Nazila Azhar, Khalid Mahmood, Rashid Ahmad, and Ejaz Ahmad Waraich

CONTENTS

8.1 INTRODUCTION

Heavy metal pollution is increasing day by day due to worldwide industrialization. This problem is more severe in the third world countries where treatment and disposal of industrial effluents is not properly managed. These untreated chemicals from different industries such as textile, leather, dyeing, steel and medicine (Paivoke 2002) are discharged into the environment. Sewage is added to water reservoirs used to support agricultural production over a long period of time. As a result, both flora and fauna are facing the hazardous effects of soil, air and water pollution. These untreated industrial effluents, in addition to other pollutants, contain large amount of heavy metals such as Pb, Cr, Cd, Ni and Zn (Srikanth and Reddy 1991).

Among heavy metals, lead (Pb) is one of the major antiqualities and has gained considerable importance as a potent environmental pollutant. Lead toxicity has become very significant due to its effect on human health (Juberg et al. 1997, Yang et al. 2000). Lead contamination results from chimneys of factories using Pb, exhaust fumes of automobiles, effluents from storage batteries, industry, mining activities and the smelting of Pb ores, leaded paints, metal plating, fertilizers, pesticides, explosives, gasoline and disposal of industrial effluents and municipal sewage sludge containing Pb and other heavy metals (Chanay and Ryan 1994, Eick et al. 1999). The main source of Pb is automobile exhausts in urban areas which contribute substantially to atmospheric pollution, and plants growing near highways are affected more by Pb pollution than in other localities (Sharma and Dubey 2005) (Figure 8.1). With increasing urbanization sewage sludge contaminated by Pb and other metals is regularly discharged onto field and garden soils (Paivoke 2002). It is reported that

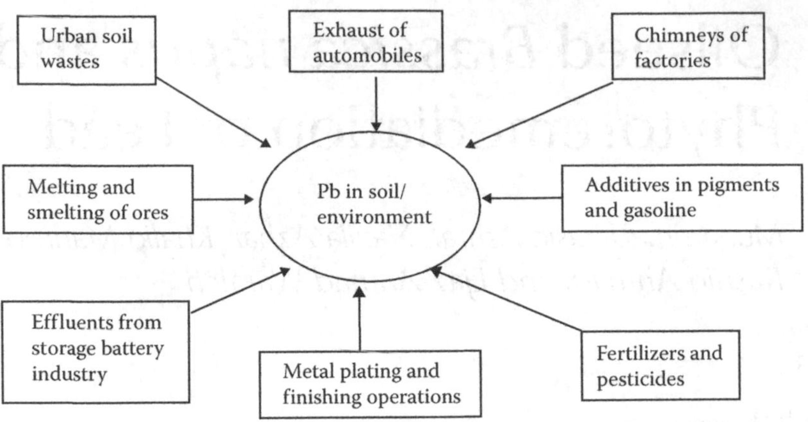

FIGURE 8.1 Sources of lead (Pb) pollution in the environment. (After Sharma, P., and R.S. Dubey, *Brazilian Journal of Plant Physiology* 17: 35–52, 2005.)

Pb-affected soils contain from 400 to 800 mg Pb kg^{-1} soil, whereas in industrialized areas the level may reach up to 1000 mg Pb kg^{-1} soil (Angelone and Bini 1992).

Excessive Pb exposure can cause mental retardation and behavioral disorders in human beings. This exposure in human beings can occur through multiple pathways, through inhalation of air, intake of water, soil or dust, as it is emitted in the environment from automobiles and other vehicles. Lead (Pb) can also enter into the food chain via plants (Wierzbicka and Antosiewiez 1993). Although lead is not an essential element for plant nutrition, it is easily absorbed and accumulated in different parts of the plant, i.e., leaves, stem, roots and seeds, which increases with increase in exogenous Pb levels (Singh et al. 1998). Once in a plant, Pb detrimentally influences plant growth (Iqbal and Shafiq 1998) by hampering enzymatic activities (Javed and Saher 1987), resulting in diverse physiological and biochemical changes, such as a decrease in carotenoids (Fargašová 2001), protein content (Kevresan et al. 2001), nitrate reductase activity (Singh et al. 1997, Kevresan et al. 2001), chlorophyll content (Xiong 1997, Kastori et al. 1998, Fargašová 2001), lamina and mesophyll thickness and diameter of vessels (Kovacevic et al. 1999), together with an increase in concentration of phenols (Lummerzheim et al. 1995, Lavid et al. 2001) and antioxidant activity (Verma and Dubey 2003, Huang et al. 2008).

There is great need to remove Pb from contaminated areas using new techniques. Previously, Pb-contaminated sites have been remediated through engineering-based technologies (Salt et al. 1995). Research has examined the utilization of plants in order to remediate heavy metal-contaminated sites, an approach called phytoremediation (Baker et al. 1994, Raskin et al. 1994). Phytoremediation processes include rhizofiltration, phytostimulation, phytovolatilization, phytostabilization and phytoextraction. Phytostabilization is used where phytoextraction is not possible. In this process, polluted soils are stabilized in order to reduce the flow of contaminants in the environment, whereas through phytoextraction the pollutants are removed from the contaminated area or medium (Susarla et al. 2002).

Phytoextraction seems to be the most promising technique and has attracted attention due to its low cost of implementation and being environment friendly (Salt et al. 1998). Two approaches for the phytoextraction of metals have currently been used. First is the use of hyperaccumulator plants having high metal accumulating capacity (Baker et al. 1994, Brown et al. 1994, Kumar et al. 1995). It has been reported that plants naturally can accumulate more than 100 mg Cd kg^{-1} on dry weight basis, 1000 mg kg^{-1} of Co, Cu, Pb or Ni, or more than 10,000 mg kg^{-1} dry weight of Mn or Zn when grown on contaminated soils (Brooks 1998, Baker et al. 2000, Dahmani-Muller et al. 2000, McGrath et al. 2002, Schmidt 2003). Second approach is the utilization of high biomass plants with a chemically enhanced method of phytoextraction (Salt et al. 1998). The success of phytoextraction is based on biomass production, heavy metals concentration in plant tissues and bioavailability of heavy metals in the rooting medium (McGrath 1998). Many reports are available regarding the uptake and transport of metals in plants that might be enhanced by increasing the heavy metal concentration around roots (Shen et al. 2002, Schmidt 2003).

Naturally hyperaccumulator plants are slow-growing and many of them do not prove suitable for phytoremediation in the field. Generally, the metals taken up by the plants are retained in the roots; however, their limited amount is translocated to the aboveground parts of the plants. The higher biomass producing crops such as sunflower, maize, rapeseed, mustard etc., can be used with appropriate chemical treatments to enhance the translocation from roots to aboveground plant parts (Huang et al. 1997, 2008; Shen et al. 2002).

Brassica napus (canola) can be beneficial for phytoremediation of heavy metals because it produces high aboveground biomass and has fast-growing ability (Figliolia et al. 2002). *Brassica napus* is grown for oil purposes because it has a lower percentage of erucic acid (Sovero 1993) than other *Brassica* species and is the third most important source of vegetable oil in the world after soybean and palm oil. In Pakistan, it was cultivated over an area of 280,600 hectares during 2009–2010 with an average yield of 837 kg ha^{-1} (Government of Pakistan 2010).

Some of the heavy metals are immobile in soils, so the absorption of these metals by plants is very slow, resulting in limited phytoextraction even with the plants with hyperaccumulation ability. Different chemicals are suggested to enhance phytoextraction ability of plants and most of the reports in this regard indicated that chelating agents like ethylenediamine-tetraacetic acid (EDTA), *N*-(2-hydroxyethyl)-ethylenediaminetriacetic acid (HEDTA) and citric acid (CA) are useful in increasing mobility of metals, and thus enhance phytoremediation (Chen and Cutright 2001, Chen et al. 2003, Turan and Esringü 2007). Reports of different workers revealed that EDTA is very effective in improving the phytoremediation of heavy metals from the contaminated soils (Heil et al. 1999, Papassiopi et al. 1999, Azhar et al. 2009). However, phytoextraction efficiency depends on availability of heavy metals in soil, the concentration of EDTA, electrolytes, pH, and soil matrix. Huang et al. (2008) and Blaylock et al. (1997) were able to achieve rapid accumulation of Pb in shoots of Indian mustard (>1% of dry biomass) with EDTA, the most commonly used chelating agent.

Keeping in view these facts, studies were conducted to find out the Pb-tolerant cultivars of canola (*Brassica napus*) and to estimate their Pb accumulation capacity in root and shoot on the basis of which they could be utilized for phytoremediation of contaminated soils. As discussed above, many reports suggest that soil washing and phytoextraction can be accelerated through EDTA utilization. Therefore, this article reports on the enhancement in Pb extraction by *Brassica napus* after applying EDTA in contaminated soils, and presents a review on this idea in the light of available literature.

8.2 EXPERIMENTAL PROCEDURES

The experiment to study the influence of Pb on growth and nutrient uptake was conducted in a net house of NIAB, Faisalabad, Pakistan in plastic pots filled with sandy loam soil having saturation percentage 35, electrical conductivity (EC) 1.25 dS m^{-1} and pH 8.3. Soil was also analyzed for other physico-chemical characteristics (Table 8.1) using the methods described by Jackson (1962) and US-Salinity Staff Handbook-60 (1962). Basal fertilizers (NPK) were applied @ 50 mg P_2O_5 and 50 mg K_2O kg^{-1} soil as DAP and SOP, respectively, while N @ 100 mg N kg^{-1} soil as urea. The experiment was laid out in Completely Randomized Design in factorial arrangement with four Pb treatments: 0, 30, 60 and 90 mg Pb kg^{-1} of soil with three replications and four canola cultivars: Con-II, Con-III, Legend and Shiralee. Lead treatments were applied to 30-day-old seedlings in the form of $PbCl_2$. The plants were watered regularly with tap water having EC 0.973 dS m^{-1} and pH 7.73. Other characteristics of irrigation water are summarized in Table 8.1.

Forty days after the imposition of Pb stress, one plant from each pot was harvested, their roots and shoots separated by cutting at the shoot root junction. Roots were removed from the soil by washing with tap water over 2-mm sieve. Dried ground plant material was analyzed for the estimation of K, Ca, Mg, P, Fe, Mn, Cu, Zn, and Pb. The other two plants were allowed to grow on to

TABLE 8.1
Physical and Chemical Characteristics of the Soil and Irrigation Water Used to Study the Effect of Pb Toxicity on Growth, Nutrient Uptake, and Translocation in Four Canola Cultivars

Soil Characteristics (Units)	Values	Irrigation Water Characteristics	
Soil Texture	Clay-Loam	Characteristics (Units)	Values
EC_e (dSm^{-1})	0.92	EC_e (dSm^{-1})	0.77
pH	7.6	pH	7.8
N (%)	0.5	N (%)	–
P (mg kg^{-1})	55	P (mg L^{-1})	–
Na^+ (mg kg^{-1})	65	Na^+ (mg L^{-1})	7
K^+ (mg kg^{-1})	70	K^+ (mg L^{-1})	0.7
Ca + Mg (mg kg^{-1})	5.76	Ca + Mg (mg L^{-1})	3
CO_3^{2-} (mg kg^{-1})	0.0	CO_3^{2-} (mg L^{-1})	–
HCO_3^- (mg kg^{-1})	4.0	HCO_3^- (mg L^{-1})	2
Organic matter (g kg^{-1})	7.4	Organic matter (g L^{-1})	–
Pb (mg kg^{-1})	0.0	Pb (mg L^{-1})	0.0
Zn (mg kg^{-1})	16.25	Zn (mg L^{-1})	0.0
Fe (mg kg^{-1})	20.45	Fe (mg L^{-1})	0.0
Cu (mg kg^{-1})	3.45	Cu (mg L^{-1})	0.0
Mn (mg kg^{-1})	23.50	Mn (mg L^{1})	0.0

maturity and used for the estimation of yield and fatty acid profile. The data were analyzed statistically using the Fisher's least significant difference test (Steel et al. 1997).

8.3 RESULTS AND DISCUSSION

8.3.1 Lead Toxicity and Plant Growth

The presence of Pb in the growth media significantly reduced the shoot and root dry biomass in all the tested *B. napus* varieties (Figure 8.2), but reduction was more pronounced in the plants of all the varieties grown in soil contaminated with Pb @ 90 mg kg^{-1} soil. Reduction in dry biomass was more severe in root than shoot in all tested varieties. Among the varieties, Con-II produced the highest shoot and root dry biomass and Shiralee the least.

Growth of a plant is the result of increase in metabolic activity and accumulation of dry matter. Decrease in growth due to contamination in soil may be the result of disturbance in metabolic activities. Results of different workers also confirmed that growth media contaminated with heavy metals adversely affected the growth and productivity of plants (Sanchez et al. 1999, Sharma and Sharma 2003, Ashraf et al. 2011) which was due to the higher concentration of heavy metals in soils or irrigation water. Substantial decrease in growth of rye grass was noted due to Pb contamination (Breckle and Kahle 1992, Khan et al. 1999). Other studies also confirmed that Pb toxicity severely affected the growth and biomass in *Azolla pinnata Anabaena* (Kalita and Sharma 1995), bean plants (Sanchez et al. 1999), *Brassica juncea* (Liu et al. 2000) and *Brassica napus* (Ashraf et al. 2011).

The severe decrease in root biomass may be due to disturbance in mineral/nutrient uptake and metabolism as a result of increased Pb concentration in the growth medium (Wensheng et al. 1997, Panda and Choudhary 2005). Results of the present study also confirmed that root contained more Pb than shoot, which may disturb the uptake of essential macro- and micro-nutrients and water absorption, resulting in the greater reduction of root biomass. The findings of Sinha (2000) also showed that accumulation of metals Pb, Cu, Cd, Cr, and Mn was higher in roots than shoots of *Bacopa monnieri*. In contrast to these findings, Fargašová (1999) and Misra and Singh (2000) reported that metals Pb,

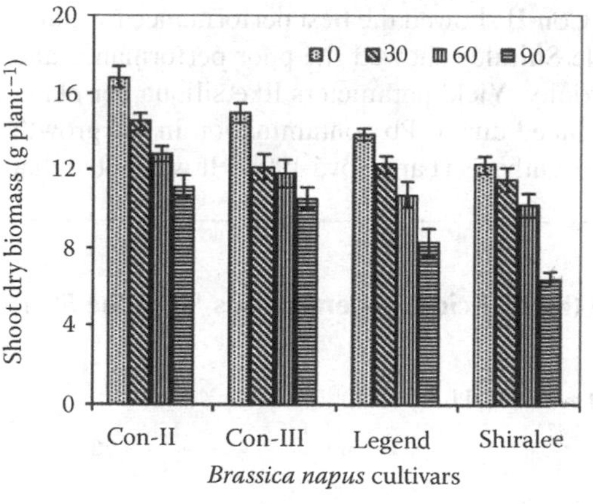

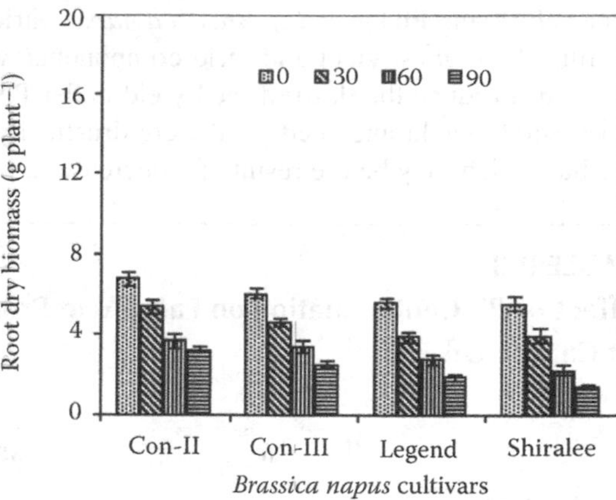

FIGURE 8.2 Growth performance of different *Brassica napus* cultivars grown in the medium contaminated with different levels of Pb (0, 30, 60, 90 mg Pb kg^{-1} soil).

Cd, Cu, and Zn were higher in shoots of *Sinapis alba* and *B. juncea*. From the results of this study and the findings of other scientists, it is very clear that higher concentrations of Pb in the growth media adversely affected the plant growth, resulting in reduction in plant productivity.

8.3.2 Lead Toxicity, Plant Productivity, and Fatty Acid Profile

Numbers of siliqua per plant, 1000-seed weight and seed yield were significantly reduced due to Pb contamination in the growth medium (Table 8.2). The highest reduction was noted at the highest

TABLE 8.2
Effect of Pb Contamination on Yield and Yield Components of *B. napus* Cultivars

Brassica napus	Pb Levels (mg Pb kg^{-1} soil)			
Cultivar	**0**	**30**	**60**	**90**
Con-II	138	133	131	129
Con-III	131	127	124	121
Legend	125	121	119	115
Shiralee	140	113	107	101
Means	133.5	123.5	120.3	116.5
	1000-Seed Weight (g)			
Con-II	3.20	2.93	2.83	2.45
Con-III	2.80	2.63	2.35	1.92
Legend	2.75	2.53	2.15	1.86
Shiralee	3.15	2.20	1.93	1.73
Means	2.97	2.57	2.19	1.99
	Yield $Plant^{-1}$ (g)			
Con-II	3.43	3.15	2.67	2.47
Con-III	2.98	2.46	2.14	1.96
Legend	3.15	2.17	1.70	1.41
Shiralee	3.26	2.10	1.50	1.21
Means	3.21	2.47	2.00	7.07

Pb level (90 mg Pb kg^{-1} soil). *Brassica napus* variety Con-II showed the best performance by maintaining the highest yield and yield components, while Shiralee showed the poor performance and failed to produce the desired seed yield under Pb toxicity. Yield parameters like siliqua per plant, 1000-seed weight and seed yield were drastically reduced due to Pb contamination in the growth media, which may be the result of a decrease in photosynthesis (Fargašová 1998). It was noted that

TABLE 8.3
Effect of Pb Contamination on Fatty Acid Profile (Fatty Acid Concentrations % of the 5%) of Canola Oil

	Pb Levels (mg Pb kg^{-1} soil)			
	0	30	60	90
Palmitic Acid				
Con-II	6.19	6.09	5.73	5.33
Con-III	5.58	4.91	4.83	5.40
Legend	4.72	5.02	6.92	4.84
Shiralee	4.11	5.63	5.39	4.95
Means	5.15	5.41	5.71	5.13
Steric Acid				
Con-II	2.77	1.32	1.19	1.45
Con-III	1.79	1.13	1.94	3.37
Legend	1.14	1.76	1.36	1.31
Shiralee	3.92	1.98	1.49	2.14
Means	2.40	1.54	1.49	2.06
Oleic Acid				
Con-II	55.46	62.86	57.15	60.36
Con-III	43.45	48.99	57.04	50.61
Legend	53.90	44.84	59.15	62.85
Shiralee	56.13	52.13	58.71	40.40
Means	52.23	52.20	58.01	53.55
Linoleic Acid				
Con-II	22.35	21.21	25.21	22.86
Con-III	20.99	21.49	20.35	21.42
Legend	20.17	21.94	25.74	18.34
Shiralee	24.08	22.72	24.04	16.41
Means	21.89	21.84	23.83	19.75
Linolenic Acid				
Con-II	10.18	7.54	8.12	8.25
Con-III	8.27	12.94	12.12	9.02
Legend	14.91	14.61	6.30	6.84
Shiralee	11.73	11.01	8.03	11.72
Means	11.27	11.52	8.64	8.5
Erucic Acid				
Con-II	0.54	0.70	0.94	0.48
Con-III	0.93	0.35	3.83	1.94
Legend	4.21	2.62	Nil	2.02
Shiralee	Nil	4.22	0.84	2.01
Means	1.89	1.97	1.87	1.61

higher concentration of heavy metals in the growth medium suppressed reproductive growth (Arun et al. 2005), thus reduction in seed yield is obvious. Reduction in siliqua per plant and 1000-seed weight is a result of Pb toxicity (Sharma and Dubey 2005). Hussain et al. (2006) also reported severe reduction in yield of mungbean due to Pb and Cr.

As far as fatty acid profile is concerned, it was not significantly changed with Pb toxicity. However, *Brassica napus* variety Con-II contained a slightly higher percentage of palmaetic acid as compared with the other three varieties (Table 8.3). Similarly, different concentrations of Pb in the growth medium did not affect the concentration of steric acid, but the variety Shiralee had slightly higher concentrations of steric acid than others. In contrast to the above, oleic acid concentration increased due to Pb contamination in the growth medium and was the highest at the highest level of Pb (90 mg kg^{-1} soil), the maximum in the case of Con-III. Linoleic acid concentration did not vary with Pb contamination, although Shiralee maintained slightly more linoleic acid than others. In the case of linolenic acid, a higher quantity was noted in plants of Shiralee grown under normal conditions. Erucic acid was higher in Shiralee.

8.3.3 Lead Toxicity and Nutrient Uptake

The pattern for N, P, K contents in the shoots and roots of the tested *B. napus* varieties was similar; they decreased with increasing Pb contamination in the growth media and their minimum concentrations were recorded at the highest level of Pb (Figure 8.3). Phosphorus content in roots was higher than in shoots, while N and K were higher in shoots than in roots of all the *B. napus* varieties. All cultivars responded in a similar way to Pb toxicity, with decrease in N, P, and K content as Pb contamination in the growth media increased. However, varieties can be split into two groups: Con-II and Con-III had higher NPK than Legend and Shiralee.

Slight increase in Ca and Mg concentrations in roots of *B. napus* varieties was noted when grown in Pb-contaminated media, while the reverse was the case for shoots in which both Ca and Mg were reduced as Pb contamination increased from 0 to 90 mg Pb kg^{-1} of soil (Figure 8.4). All the *B. napus* varieties showed similar response to Pb toxicity for Ca and Mg content in roots and shoots. However, on the basis of Ca content in shoot, the varieties can be divided into two groups, i.e. Con-II and Con-III with higher Ca content than Legend and Shiralee.

Lead accumulation both in shoots and roots increased with increasing Pb concentrations in the growth media (Figure 8.5). However, roots contained higher amount of Pb than shoots in all the *B. napus* varieties. The response to Pb accumulation in all tested varieties was similar in shoots and roots.

Due to Pb toxicity, Fe, Zn, Cu and Mn reduced both in shoots and roots of all *B. napus* varieties (Figures 8.6 and 8.7). Iron concentrations in roots were more adversely affected than in shoot in plants grown in Pb-contaminated growth media. All the varieties showed similar trends in response to Pb toxicity for Fe content. Zinc content was higher in root than shoot (Figure 8.6) but a marked reduction was noted in the case of shoots. Varieties Con-II and Con-III had higher Zn and Fe than the varieties Legend and Shiralee.

With increased Pb contamination in the growth media, copper and manganese (Mn) concentrations were reduced in shoots and roots (Figure 8.7). All varieties showed a similar trend of decreasing Cu and Mn concentration under Pb toxicity and variations among them were nonsignificant. However, shoots of all the cultivars maintained higher Cu contents than roots. In contrast, Mn was higher in roots of all *B. napus* varieties.

The nutrients in shoots and roots of all *B. napus* varieties were imbalanced by Pb toxicity. Uptake of N, P and K both in the shoots and roots was disturbed with the increase in Pb contamination from 0 to 90 mg kg^{-1} soil, but reduction in K and N was very severe in roots of all the *B. napus* varieties. Pb contamination in growth medium inhibits the uptake of N and K (Yang et al. 2004). Due to the mobile nature of these elements, all their absorbed quantities are translocated to shoots, because both these nutrients are required in large amounts for metabolic activities (Mamta et al. 1997).

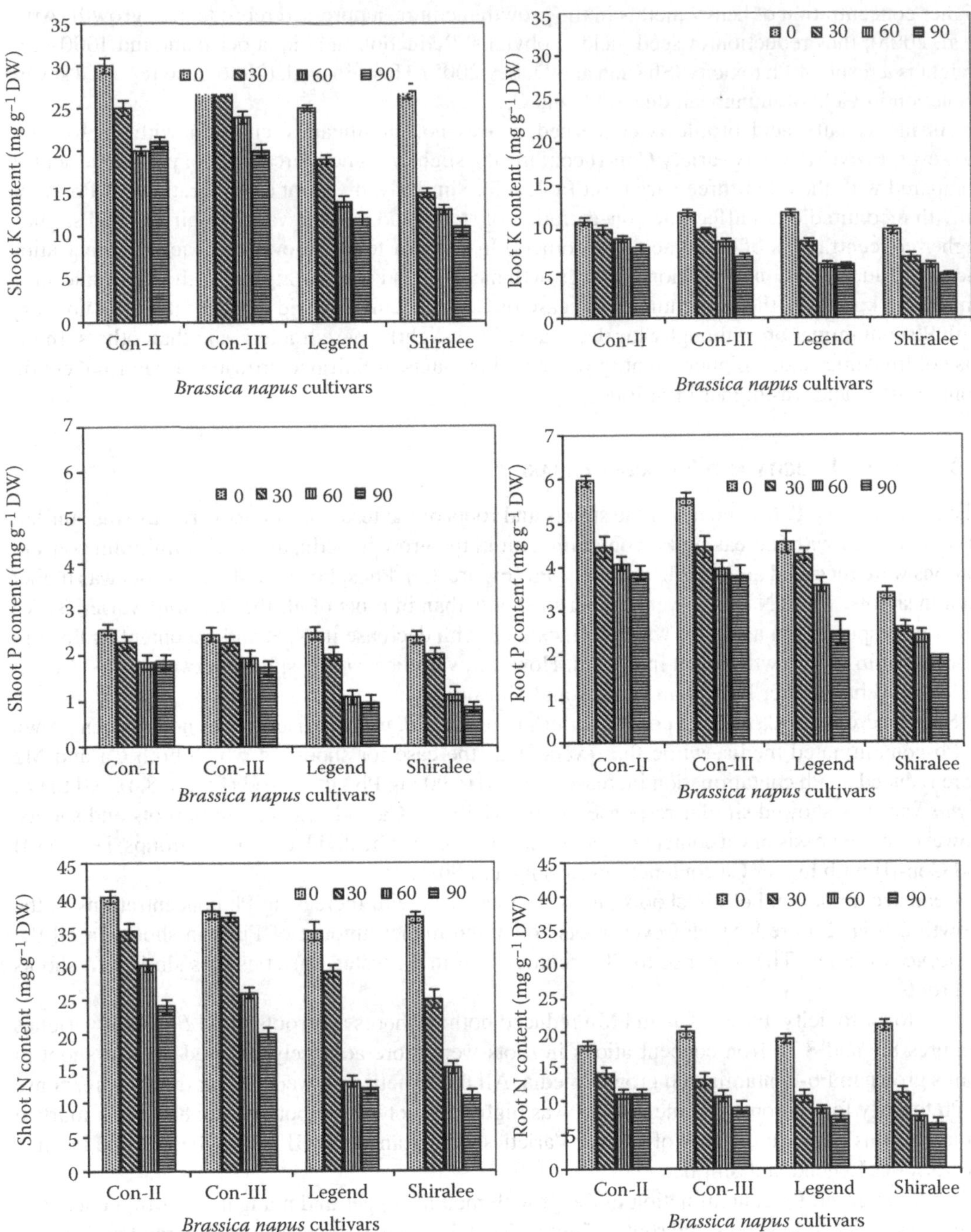

FIGURE 8.3 Influence of Pb toxicity on K, P, and N concentrations in the shoots and roots of four cultivars of *Brassica napus* grown in the medium contaminated with different levels of Pb (0, 30, 60, 90 mg Pb kg^{-1} soil).

Although P content decreased both in shoots and roots, the roots had higher P content than shoots, which indicated that absorption of P was inhibited to some extent. However, its translocation to shoots was severely affected, as a result of which shoots contained a lower amount of P (Kopittke et al. 2007). Nutrient uptake by roots also depends on the selective properties of the plasma membrane (Gussarsson 1994), and higher concentrations of Pb in the growth medium interferes with the

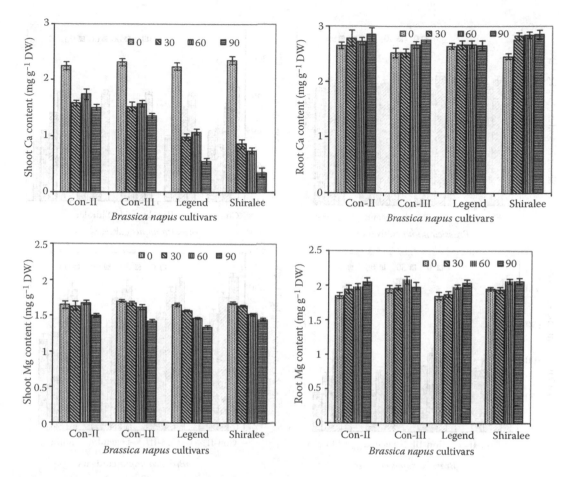

FIGURE 8.4 Effect of Pb contamination in growth medium on Ca and Mg concentrations in the shoots and roots of four cultivars of *Brassica napus* grown in the medium contaminated with different levels of Pb (0, 30, 60, 90 mg Pb kg^{-1} soil).

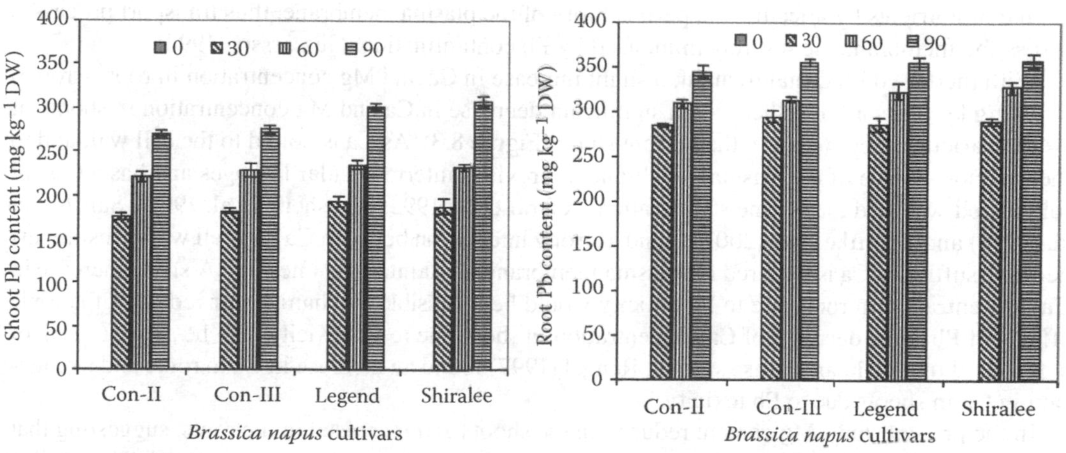

FIGURE 8.5 Effect of Pb contamination in growth medium on Pb accumulation in the shoots and roots of four cultivars of *Brassica napus* grown in the medium contami- nated with different levels of Pb (0, 30, 60, 90 mg Pb kg^{-1} soil).

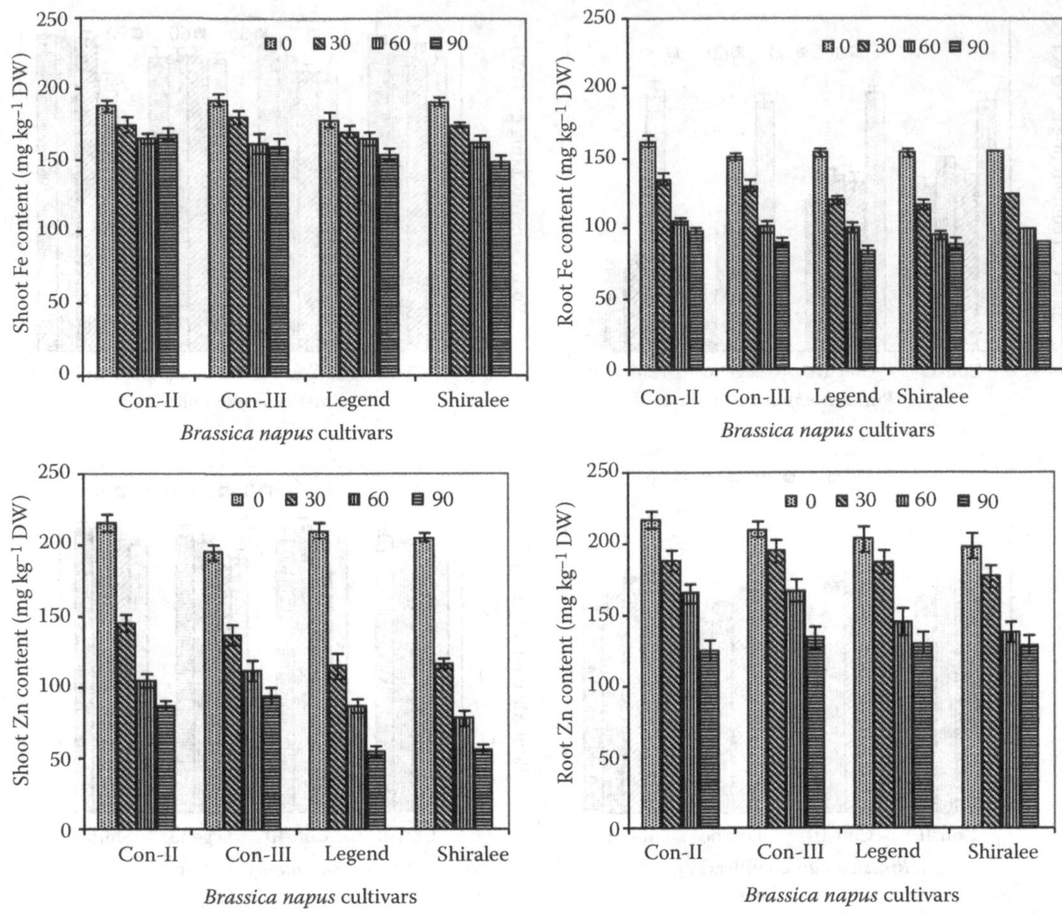

FIGURE 8.6 Effect of Pb toxicity on Fe and Zn concentrations in the shoots and roots of four cultivars of *Brassica napus* grown in the medium contaminated with different levels of Pb (0, 30, 60, 90 mg Pb kg^{-1} soil).

uptake of nutrients by affecting the permeability of the plasma membrane, thus transport processes across the membrane are severely influenced by Pb contamination (Gussarsson 1994).

With increased Pb contamination, a slight increase in Ca and Mg concentration in roots (from 0 to 90 mg kg^{-1} of soil), together with a significant decrease in Ca and Mg concentration in shoots of all the varieties, were found in the present study (Figure 8.3). As Ca is bound to the cell wall and to the exterior surface of the plasma membrane, it provides intermolecular linkages and has a crucial role in cell wall and membrane stabilization (Ashraf et al. 1992, Wensheng et al. 1997). Sanchez et al. (1999) and Kopittke et al. (2007) found a strong interaction between Ca and cell wall constituents because sufficient Ca is required for plasma membrane to maintain its integrity. A slight increase in Ca concentration in roots due to Pb toxicity would be a possible mechanism for reducing the toxic effects of Pb and a decrease of Ca concentration in shoot due to Pb toxicity may be an indication of a damaged intercellular defense system. Rengel (1997) found an increase in Ca in roots and a reduction in Ca in shoots due to Pb toxicity.

In the present study, Mg content reduced in the shoots and increased in the roots, suggesting that uptake of Mg is not affected by Pb toxicity but its translocation to shoots is inhibited. That is why reduction in shoot Mg content was recorded. The reduction in chlorophyll content is also recorded due to Pb toxicity, which itself may be caused by a reduction in translocation of Mg (Huang and Cunningham 1996).

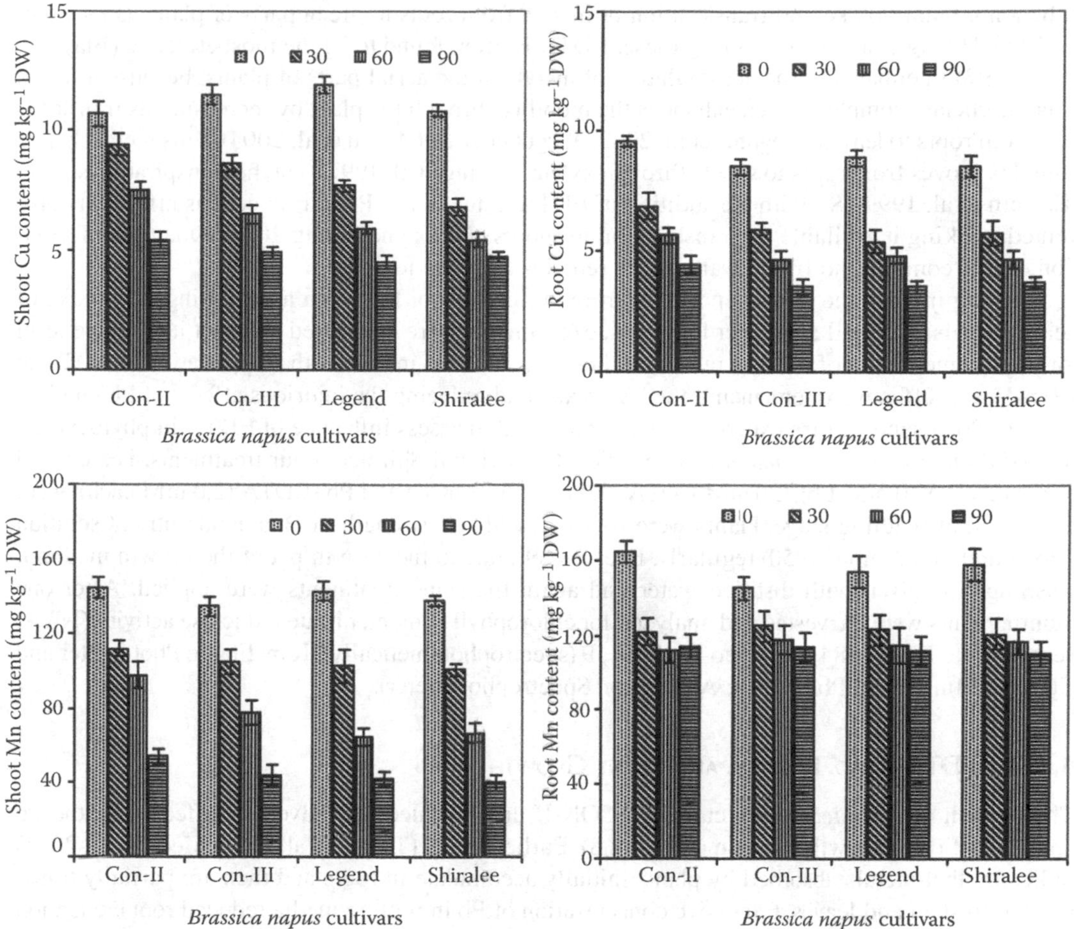

FIGURE 8.7 Effect of Pb toxicity on Cu and Mn concentrations in the shoots and roots of four cultivars of *Brassica napus* grown in the medium contaminated with different levels of Pb (0, 30, 60, 90 mg Pb kg^{-1} soil).

Iron (Fe) was less affected in shoots than roots in all the *B. napus* varieties, which suggested that uptake of Fe was inhibited by the Pb toxicity and whatever Fe was absorbed was translocated to shoots (Ashraf et al. 2011). Although Cu and Mn content was affected by Pb toxicity both in shoots and roots, these were less affected in roots, which means that Pb contamination in the growth media inhibited their uptake more severely than their translocation to shoots. Zinc concentration in roots and shoots of tested *B. napus* varieties decreased significantly due to Pb contamination in growth media (from 0 to 90 mg kg^{-1} soil). However, Zn concentrations in roots seemed less affected by Pb toxicity (Figure 8.6), which indicated that Zn uptake was affected by Pb but its translocation up to shoots was severely inhibited (Lombi et al. 2001, Ashraf et al. 2011).

From the foregoing it can be concluded that presence of Pb in the growth media disturbed the nutrient uptake and translocation which ultimately reduced the plant growth and crop yield. In Pb contaminated growth media, accumulation of Pb was recorded in shoots and roots of all the cultivars so these can be used for the phytoremediation of Pb from soil and water.

8.3.4 Influence of EDTA in Phytoremediation of Lead

Synthetic chelates have long been supplied to plants with micronutrients in both soil and hydroponics, which can enhance phytoextraction by increasing metals' bioavailability, thus dramatically enhancing plant uptake and translocation of metals from roots to green parts of plants (Epstein et al. 1999, Huang et al. 1997). Among these, EDTA is often found to be the most effective (Blaylock et al. 1997) in enhancing the accumulation of metals in the aerial parts of plants, because it forms a metal-chelate complex which enhances the mobility through the plant by increasing its translocation from roots to leaves (Begonia et al. 2002, Turgut et al. 2004, Wu et al. 2004). This metal chelate complex moves from roots to shoots through xylem (Haung et al. 1997), via the transpiration stream (Epstein et al. 1999). So with the addition of EDTA, retention of Pb in root cells is effectively prevented, making it available for translocation to shoots (Jarvis and Leung 2002). Due to the formation of this complex, no freely available Pb remains to create toxicity.

Keeping in view the above reports regarding utilization of EDTA in accelerating the uptake of heavy metals from soil and water by plants, experiments were conducted to select its suitable level for phytoremediation of Pb. The results of these experiments indicated that application of EDTA at 2.0 mM was effective in enhancing the Pb uptake and reducing the injurious effects of Pb on plant growth. So, a sand culture experiment was conducted to assess influence of EDTA in phytoextraction of Pb using two *Brassica napus* varieties, CON-II and Shiralee. Four treatments, i.e., control (0 mM EDTA+0 mM Pb), 2.0 mM EDTA, 2.0 mM Pb ($PbCl_2$) and Pb+EDTA (2.0 mM each), were developed at seedling stage. Plants were irrigated with 1/5 strength of Hoagland nutrient solution (Hoagland and Arnon, 1950) regularly. Every week, due to increase in pH of the growth medium, washing was given with distilled water and again the same treatments were applied. After one month, plants were harvested and analyzed for chlorophyll content, nitrate reductase activity (NRA) according to Sym (1984), N (Mico-Kjaldal) , P (spectrophotometically), K by Flame Photometer and Cu, Zn , Mn, Fe and Pb (Atomic Absorption Spectrophotometer).

8.3.5 EDTA, Lead Toxicity, and Plant Growth

The growth of *Brassica napus* cultivars, CON-II and Shiralee was adversely affected by the Pb contamination in growth medium (Figure 8.8). Earlier works (Turgut et al. 2004, Meers et al. 2005) indicated that metals absorbed by plants initially accumulate in roots and then are partially translocated to stem and leaves. Excessive concentration of Pb in rooting media reduced root elongation which was observed in the present study in *Brassica* cultivars. Reduction in shoot and root growth by the toxicity of lead was also reported in onion (Wierzbicka 1995), Indian mustard (Liu et al. 2000) and *Phaseolus vulgaris* (Geebelen et al. 2002). The inhibition in root elongation is a complex phenomenon influenced by many factors. It may be due to disturbance in plant/water relations. In an environment contaminated with heavy metals the presence of excessive salt concentration in the root medium causes water deficit, a major factor resulting in the reduction in plant growth and development (Wierzbicka 1994).

Application of EDTA alone also produced toxic effects, as reduction in root and shoot dry biomass (Figure 8.8) was noted which can be due to its chelating property as it forms chelate with some micronutrients in the growing medium they are absorbed in amount even higher than their toxic level. By the addition of EDTA to the growth medium contaminated with Pb, the toxic effects of Pb were substantially decreased, resulting in increased root and shoot growth, and their dry biomass. There are many reports on the use of synthetic chelates indicating enhancement in uptake and transport of heavy metals by plants, and proving that EDTA was the most effective chelator for Pb phytoremediation (Huang et al. 1997, Cooper et al. 1999, Lasat 2000). Another concern is the cultivation of commercially utilized crops such as barley, corn, sunflower and *Brassica napus* which, having tolerance for heavy metal ions and high biomass production, are efficient for phytoremediation (Salt et al. 1998, Turan and Esringü 2007). In the present experiments, *Brassica napus* varieties were grown with the addition of Pb + EDTA combinations. EDTA application @ 2 mM Pb and 2 mM EDTA in the growth medium increased plant growth by reducing the toxic effects of Pb on root and root elongation and dry biomass production. The findings of many workers (Turgut et al. 2004, Hajiboland 2005, Tlustoš et al. 2006, Turan and Esringü 2007) confirm the above results.

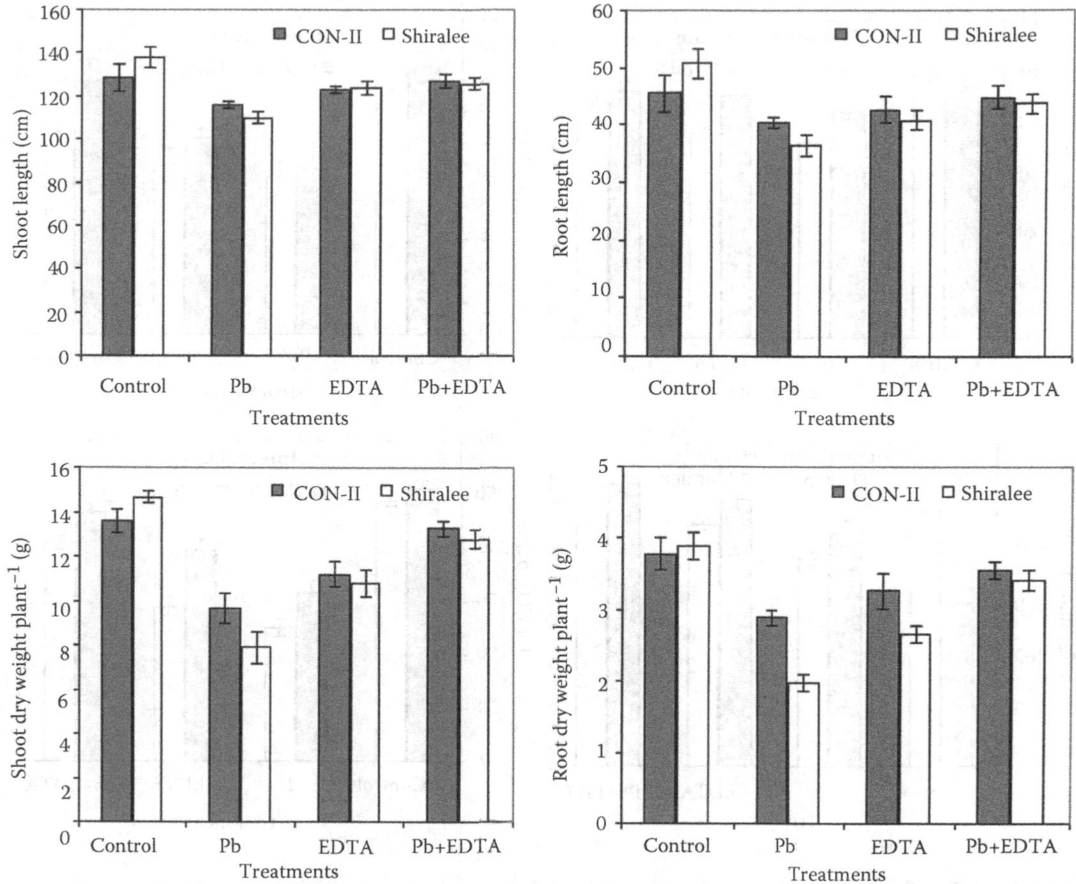

FIGURE 8.8 Growth performance of *Brassica napus* cultivars, CON-II and Shiralee grown under Pb and EDTA contamination.

Brassica napus variety CON-II maintained better growth than Shiralee, which may be due to its genetic potential to tolerate the Pb toxicity. Varietal variation regarding tolerance to heavy metal toxicity in different crops is also reported (Kabata-Pendias and Pendias 1999, Ashraf et al. 2011).

From the present study, it can be concluded that addition of EDTA in the Pb-contaminated medium is effective in minimizing the toxic effects of Pb by converting it into chelated complex resulting in increase in its uptake and Pb accumulation in the cell vacuole. So, EDTA application would be beneficial in accelerating the phytoextraction of Pb through high biomass-producing *Brassica napus* plants. Xu et al. (2007) also found that the addition of EDTA to Pb-contaminated medium is beneficial in alleviating the toxic effects of Pb, and plants showed normal growth (Wu et al. 2004, Yahua et al. 2004).

8.3.6 EDTA, Pb Toxicity, and Photosynthesis

Photosynthesis rate, transpiration, stomatal conductance and enzymatic activities are all adversely affected due to the contamination of excessive Pb in the growth medium (Figure 8.9). Many reports (Kalita and Sharma 1995, Sanchez et al. 1999, Liu et al. 2000) indicated that photosynthesis is reduced due to the presence of Pb in the growing medium, resulting in the reduction in growth and

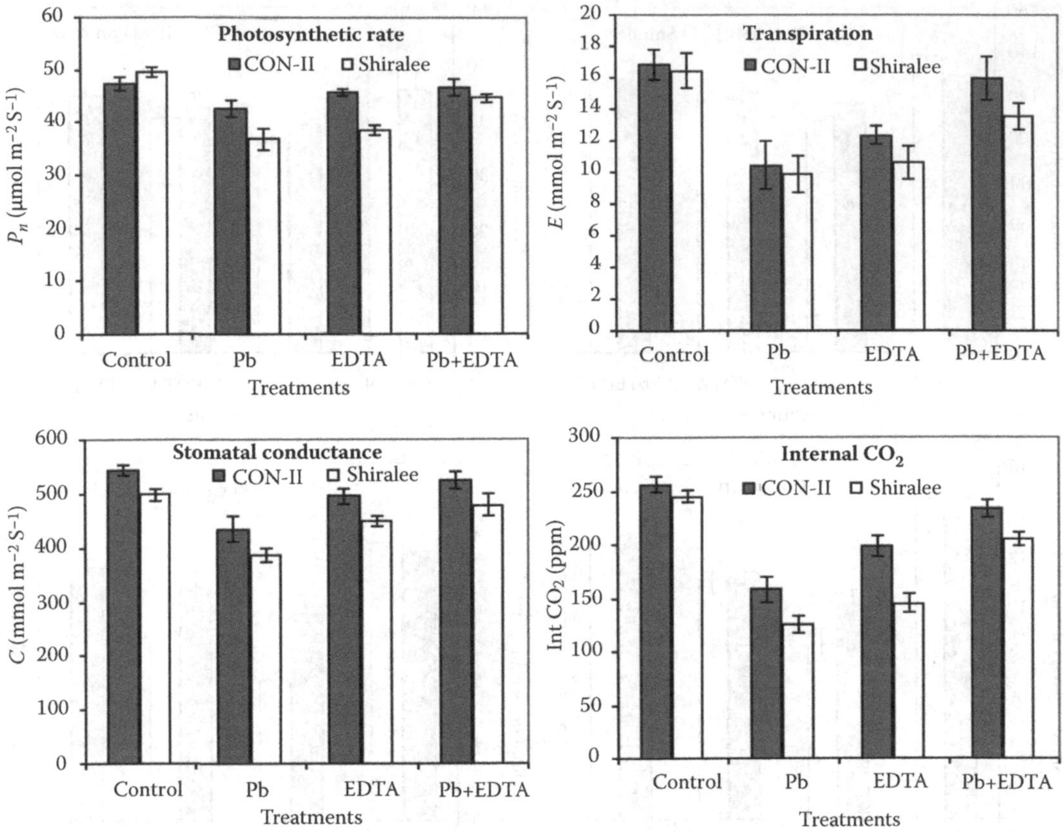

FIGURE 8.9 Photosynthetic rate (P_n), stomatal conductance (C), transpiration (E), and internal CO_2 of *Brassica napus* cultivars grown under Pb and EDTA contamination.

biomass production. Reduction in the rate of photosynthesis was due to the reduction in stomatal conductance and reduced water absorption from the root medium. Water absorption was severely reduced due to the presence of Pb in the growth medium, which also reduced the transpiration and uptake of other necessary solutes essential for stomatal opening, so the required amount of CO_2 was not absorbed by the plants. Thus, reduction in photosynthesis is obvious.

Reduction in transpiration rate due to the presence of Pb, or EDTA alone, was recorded in this study (Figure 8.9), but Pb *plus* EDTA improved it. Reduction in transpiration due to heavy metal contamination in the growth medium was also found by Kastori et al. (1996), but Wu et al. (1999) reported contrary results regarding the addition of EDTA in Pb-contaminated growth medium, in that they found its addition did not influence the transpiration rate. Varietal differences in transpiration, commonly observed in plants (Marchiol et al. 2004), are also very clear in the present findings. *Brassica napus* variety CON-II maintained higher transpiration than Shiralee under all the treatments of Pb, EDTA or their combinations. Our results also confirmed that plants which maintained a higher rate of transpiration had a higher rate of photosynthesis. The decrease in transpiration rate resulted in decrease in stomatal opening/conductance (Kastori et al. 1996, Madhavi and Chyrnulu 1998). Application of EDTA in Pb-contaminated media was effective in enhancing the stomatal conductance (Figure 8.9).

The presence of Pb or EDTA alone in the growth medium decreased the concentration of internal CO_2 which adversely affected the photosynthesis rate in *Brassica* plants. But the addition of EDTA in media having excessive Pb improved internal CO_2 concentration, which subsequently improved the photosynthesis rate. Ashraf et al. (2004) found that internal CO_2 concentration was reduced due to the presence of excessive salt in the growth medium. The reduction in internal CO_2 concentration in plants growing in the medium containing excessive Pb or EDTA is obviously due to higher concentration of salts. *Brassica napus* varieties responded differently for photosynthesis rate under Pb or EDTA stress conditions and cultivar CON-II maintained more internal CO_2 than that of Shiralee. This means CON-II has the genetic ability to consume internal CO_2 if availability of external CO_2 concentration is limited, and thus maintained a higher photosynthesis rate even under adverse conditions. Plants with the ability to utilize the internal CO_2 more efficiently have higher photosynthesis rate and productivity (Ashraf et al. 2005).

Application of EDTA in a growth medium with excessive Pb concentrations improved the rate of photosynthesis. The findings of Fodor et al. (1996) and Madhavi and Chyrnulu (1998) also showed the reduction in the rate of photosynthesis due to heavy metal contamination in the growth medium. Rate of photosynthesis depends on chlorophyll content because it plays a key role in trapping the sunlight and converting it into chemical energy. Chlorophyll content is reduced in the presence of Pb and EDTA alone in the growth medium (Figure 8.10). Reduction in chlorophyll content may result

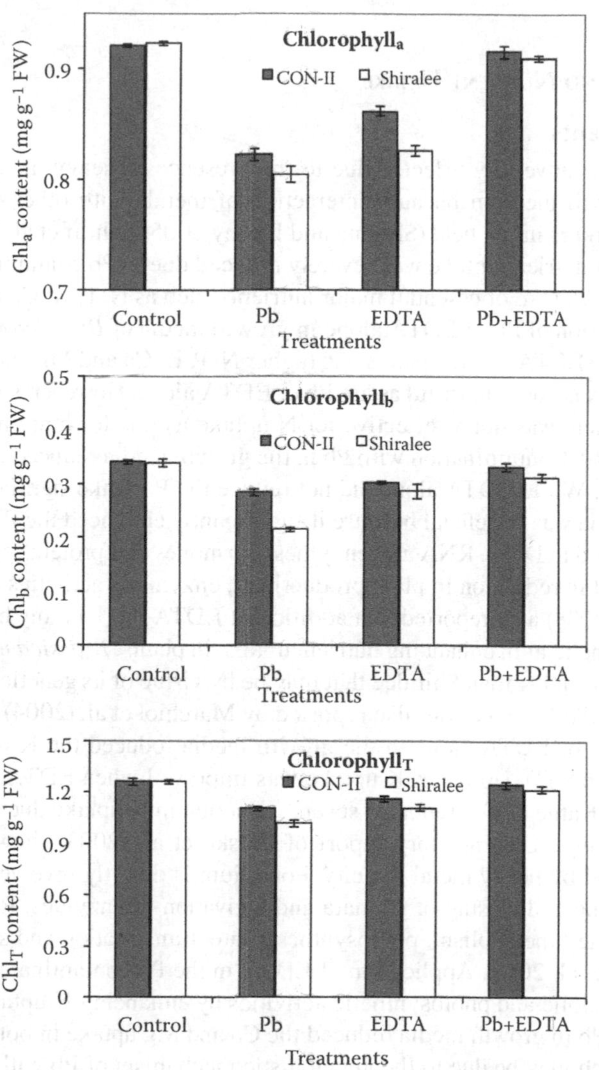

FIGURE 8.10 Changes in chlorophyll a, b, and total of *Brassica napus* cultivars CON-II and Shiralee grown under Pb and EDTA contamination.

in reduction in photosynthesis. The results of Sarvari et al. (1999) substantiate the findings and their results showed a significant reduction in chlorophyll content, resulting in chlorosis in the presence of Pb and EDTA (Fargašová 1999, Ali et al. 2000). Tested *Brassica napus* varieties CON-II and Shiralee contained higher amount of chlorophyll 'a' than chlorophyll 'b'. However, CON-II contained more chlorophyll 'a' than Shiralee; the same was the case with chlorophyll 'b'. Total chlorophyll contents were also affected by the Pb, EDTA and Pb+EDTA application (Figure 8.10). Chlorophyll contents reduced due to Pb toxicity causing impaired Mg and Fe uptake in plants (Sharma and Sharma 2005); reduction in chlorophyll also resulted due to increase in chlorophyllase activity enhanced by Pb contamination (Drazkiewicz 1994). While addition of EDTA in Pb contaminated medium increased the uptake of Mg and Fe by making their chelated complex (Turan and Esringü 2007).

From the results of present study, it can be concluded that chlorophyll contents, rate of photosynthesis, stomatal conductance and transpiration were adversely affected by Pb and EDTA alone, and that all these parameters were improved under Pb+EDTA treatment. These findings clearly indicated that application of EDTA in Pb-contaminated soils/waters is effective in phytoremediation of Pb.

8.3.7 EDTA, Pb, and Nutrient Uptake

8.3.7.1 Macronutrients

Plant nutrient uptake is adversely affected due to the presence of heavy metals in excessive concentration in the growth medium because interaction of metals with other essential elements is observed by many workers in the field (Sharma and Dubey 2005, Ashraf et al. 2011). Results of this study also showed that nutrient uptake was severely reduced due to Pb contamination in the growth medium (Figure 8.11). Uptake of essential major nutrients such as N, P, K, Ca and Mg was severely reduced, with the addition of Pb or EDTA alone in growth media of *Brassica* varieties CON-II and Shiralee. However, Pb+EDTA-treated plants had higher N, P, K, Ca and Mg uptake. This means that the Pb+EDTA complex is not as harmful as are Pb or EDTA alone. However, some reports indicated that Pb+EDTA treatment was not as effective for N uptake as it is for other nutrients (Larbi et al. 2002, Sinha et al. 2006). Contamination with Pb in the growth media reduced the P contents in both the *B. napus* varieties. While EDTA alone did not reduce the P uptake significantly, its addition in Pb-contaminated media was beneficial because it significantly enhanced the P uptake (Figure 8.11). As P is directly involved in DNA, RNA and enzymes, hormones and proteins synthesis, disturbance in P uptake may result in reduction in plant productivity, enzymatic activities and synthesis of hormones. Huang et al. (2008) also reported that addition of EDTA in Pb or any heavy metal contaminated medium was beneficial in enhancing nutrient uptake in plants. *Brassica nupus* variety CON-II maintained higher N, P and K than Shiralee that may be by virtue of its genetic makeup for Pb tolerance. A similar varietal difference was also reported by Marchiol et al. (2004) in canola and radish.

The presence of Pb or EDTA alone in the growth media reduced the K uptake in both the *B. napus* varieties (Figure 8.11). However, K uptake was improved when EDTA and Pb were applied together. Breckle and Kahle (1992) reported severe reduction in K uptake due to heavy metal stress in *Fagus sylvatica* plants. Contradictory report of Ciesko et al. (2003) showed that K uptake in plants was not affected by heavy metal toxicity. Potassium is directly involved in maintaining the water relations, opening and closing of stomata and activation of enzymes. Reduction in its availability adversely affected metabolism, photosynthesis rate, transpiration and stomatal conductance of the plants (Ashraf et al. 2010). Application of EDTA in the Pb contaminated media is beneficial in improving the metabolic and photosynthetic activities by enhancing K uptake.

Contamination of Pb in growth media reduced the Ca and Mg uptake in both the *B. napus* varieties (Figure 8.12) which may be due to the antagonistic mechanism of Pb with Mg and Ca (Walker et al. 1997). Plants growing under Pb+EDTA treatment possessed the highest concentrations of Ca and Mg. This clearly indicated that application of EDTA in Pb-contaminated medium is effective in

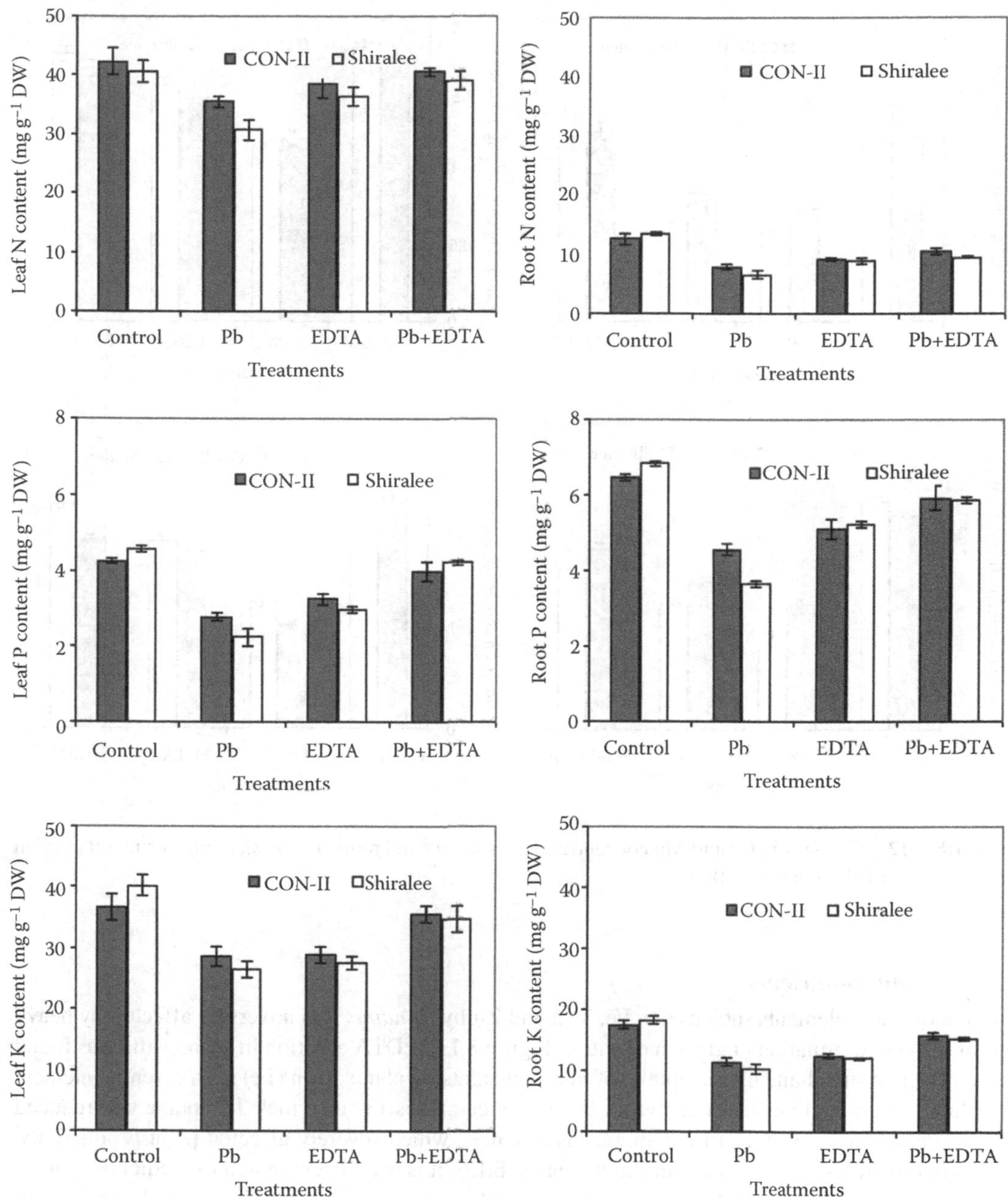

FIGURE 8.11 Effect of Pb and EDTA on N, P, and K concentrations in shoots and roots of *Brassica napus* cultivars.

improving the growth and nutrient uptake (Geebelen et al. 2002). *Brassica napus* cultivar CON-II maintained higher Ca and Mg contents in shoot and roots than Shiralee, which may be due to their genetic makeup as observed by Huang et al. (1996). The addition of EDTA in heavy metal-contaminated medium is effective in improving the nutrient uptake in *Brassica*. So, EDTA application is beneficial in phytoremediation of heavy metals as well as for plant growth and productivity.

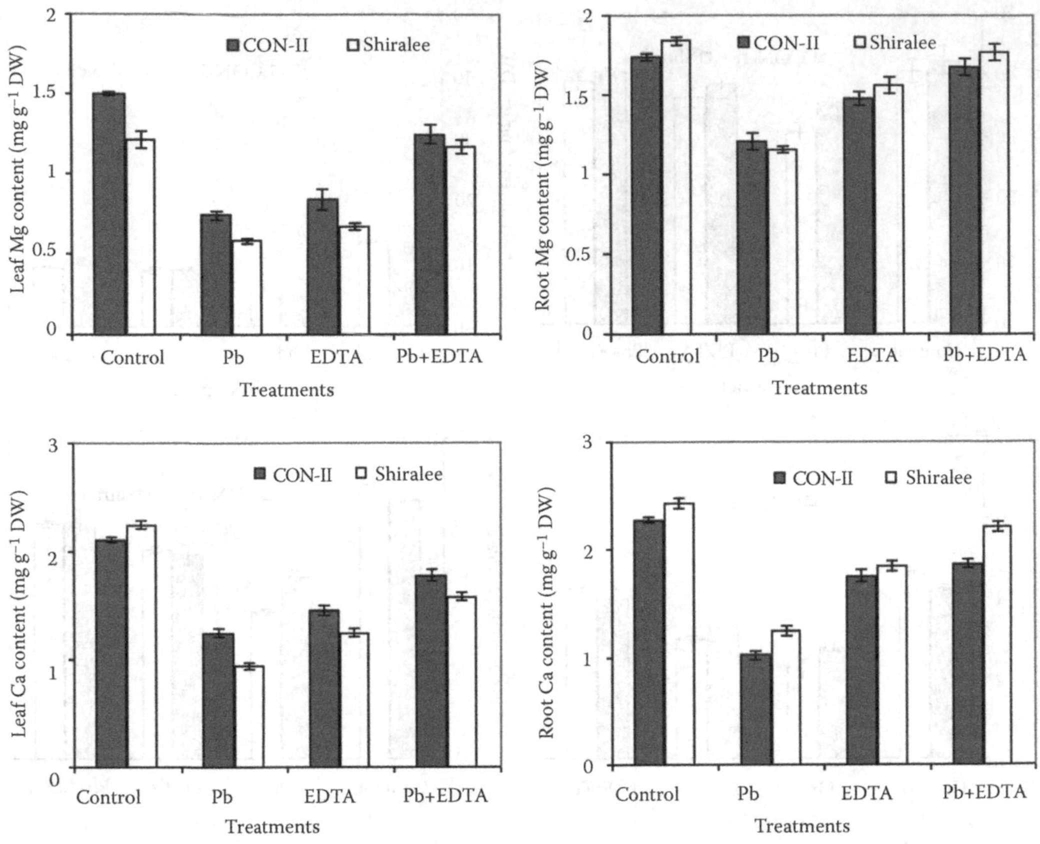

FIGURE 8.12 Changes in Ca and Mg concentrations in shoots and roots of *Brassica napus* cultivars grown under Pb and EDTA contamination.

8.3.7.2 Micronutrients

Uptake of microelements such as Cu, Fe, Mn, and Zn by *B. napus* was adversely affected by heavy metal (Pb) contamination in growth medium (Figure 8.13). EDTA addition in such media was found to be effective in enhancing the uptake of micronutrients by plants. Iron (Fe) is an essential element for plant growth and metabolic activities because it catalyzes 139 enzymes. Its uptake was reduced due to Pb toxicity resulting in metabolic disturbances which severely affected plant productivity. Root contained less Fe content than shoot (Figure 8.13). It is usually estimated to predict the photosynthetic efficiency and catalyzation of enzymes, which were adversely affected by the toxicity of Pb or EDTA when applied alone. However, when Pb and EDTA were applied in combination in growth medium, plants maintained higher photosynthetic efficiency and produced higher biomass (Ashraf et al. 2011). Leaves maintained a higher amount of Fe than roots (Figure 8.13), which is due to the mobile nature of Fe. It is also directly involved in chlorophyll synthesis and enzymatic activities of CAT, POD and glucoxygenase (Holm et al. 1996).

Reduction in Fe uptake decreases the leaf protein, especially membranous proteins (Siedlecka and Krupa 1999, Hall 2002, Barberon et al. 2012). Although results regarding the effect of Fe deficiency on protein contents are not quoted in this article but our findings and literature (Waters et al. 2006, Azhar 2010) support these. Results clearly indicated that addition of EDTA in the Pb-contaminated growth medium enhanced the Fe uptake, which may be the reason for higher Fe content in leaves than roots (Figure 8.13).

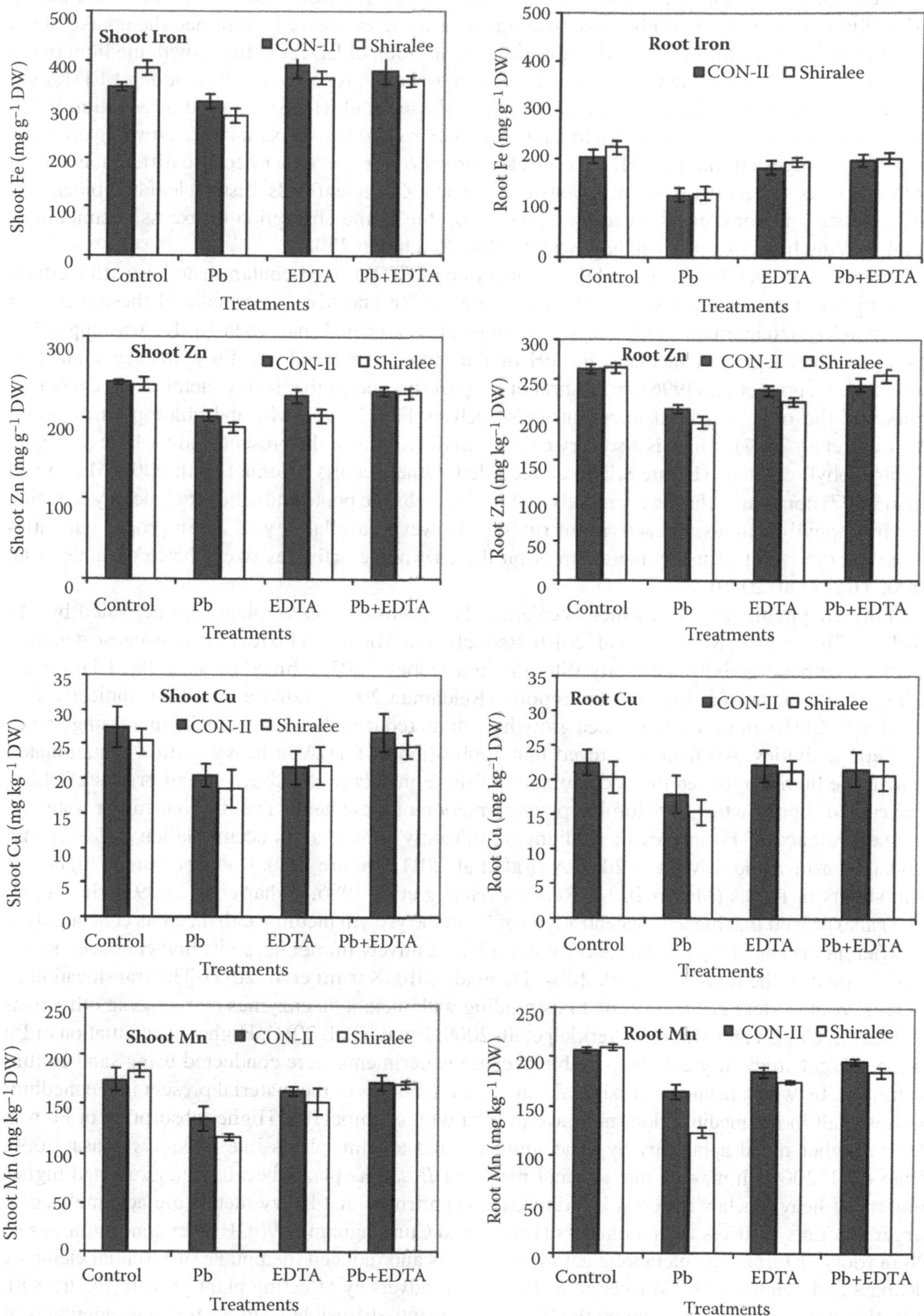

FIGURE 8.13 Changes in Fe, Zn, Cu, and Mn concentrations in shoots and roots of *Brassica napus* cultivars grown under Pb and EDTA contamination.

Results of the present study and literature indicate that Pb decreased the uptake of Cu due to competition between Cu and Pb. Thus, medium containing excessive Pb retarded the uptake of Cu (Sinha et al. 2006, Ashraf et al. 2011). However, the addition of EDTA in the growth medium in the absence of Pb enhanced the uptake of Cu, which may be the result of the formation of EDTA-Cu complex (Székely et al. 2011). On the other hand, Vassel et al. (1998) reported a reduction in Cu uptake when EDTA alone was added to media where *Brassica juncea* plants were grown hydroponically. They suggested that free EDTA was taken up by *B. juncea*, either because different essential micronutrients, especially divalent cations (Fe, Cu and Zn) in leaf cells, disturb leaf cell biochemical reactions and consequently cause cell death, or due to the absorption of excessive amount of metals having higher capacity to bind with EDTA (Reichman 2002).

Gupta and Sinha (2007) reported that application of EDTA in Ni-contaminated growth medium is beneficial in enhancing the absorption by plants of Zn and Mn. The uptake of these metals is pH dependent (Reichman 2002). Present findings also indicated that when EDTA was applied in Pb-contaminated growth medium, the pH of the medium reduced and Pb uptake/accumulation increased. Kupper et al. (1996) and Ashraf et al. (2011) reported that heavy metal stress decreased uptake of the other essential micronutrients such as Fe, Cu, Zn, Mn and chloroplast/pigments (Almeida et al. 2007), which is also very clear from the results of the present study, where reduction in chlorophyll contents (Figure 8.10) was recorded. Other reports (Monnet et al. 2001, Mobin and Khan 2007) confirmed that heavy metals toxicity disturbs the photosynthetic efficiency by reducing the chlorophyll contents and activity of rubisco. However, availability of Zn in proper quantities detoxifies the effect of heavy metal stress on the enzymatic activities like rubisco (Bonnet et al. 2000, Yuan et al. 2009).

Many enzymatic activities which accelerate the photosynthesis in plants are activated by Mn (Gill and Tuteja 2010, Mahmed et al. 2010). Reduction in Mn due to Pb toxicity may cause decrease in photosynthetic activity indirectly (Sharma and Dubey 2005, Ahmad et al. 2008, Islam et al. 2008). However, present findings and reports (Reichman 2002, Azhar et al. 2009) indicated that addition of EDTA in Pb-contaminated growth medium reduced the Pb toxicity by increasing photosynthetic activities. All findings proved that application of ETDA in heavy metals-contaminated growth medium stimulated the accumulation of these metals in the aboveground organs of plants (leaves and shoots) which is helpful in phytoextraction of these heavy metals from soil or water.

The presence of Pb in growth medium significantly enhanced its accumulation in leaves and roots (Srivastava and Srivastava 2010, Ashraf et al. 2011) (Figure 8.14). Roots contained higher Pb than shoots or leaves (Mazen 2004). Reports (Huang et al. 1996, Azhar et al. 2009, Ashraf et al. 2011) also proved that higher concentrations of Pb in the growth medium enhanced its concentration in different organs of the plants. Heavy metal Pb is relatively immobile, so its higher concentration was retained in the roots (Cui et al. 2004, Du et al. 2010, Karimi et al. 2011). The translocation of Pb from root to stem and leaf is due to its binding with metabolic enzymes by replacing other metals like Fe, Cu, Zn and Mn, etc. (Verkleij et al. 2009, Ismail et al. 2011). Higher accumulation of Pb in the present study may also be possible because experiments were conducted using sand culture technique, by which plants can easily take up excess amounts of the material present in the medium because salt in the medium does not have to bind with soil micelles. Higher absorption of Pb was due to higher metal uptake by hyperaccumulator and tolerant plants like *B. napus* (Lasat 2000, Zhao et al. 2000). It may be due to dicot nature of *B. napus* plants because they retained higher amount of heavy metals in roots. Literature also confirmed that heavy metals are accumulated in larger amounts in dicots than monocots (Huang and Cunningham 1996). Higher concentrations of Pb in roots disturbed the metabolic activities of roots and reduced the uptake of essential elements (Kabata and Pendias 1992, Walker et al. 1997), thus adversely affecting plant growth (Figure 8.8). It is, therefore, necessary to change the Pb molecules into diffusible form, so that it accumulates in the cell vacuole and becomes harmless for plants. Reports (Evangelou et al. 2007, Kaushik 2011) indicated that chelated heavy metals can easily be absorbed by plants which are not very toxic to plants, and chelation of heavy metals accelerated their translocation from root to shoot. Present

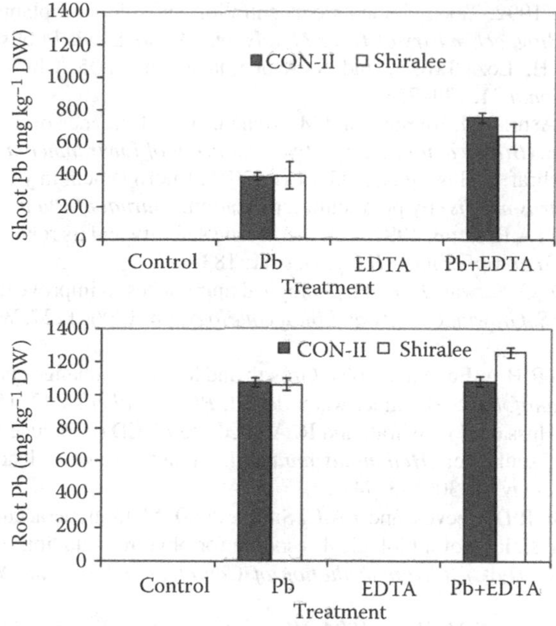

FIGURE 8.14 Accumulation of Pb in shoots and roots of *Brassica napus* cultivars growing under Pb and EDTA contaminated medium

findings also confirmed that addition of EDTA in Pb-contaminated growth medium enhanced Pb accumulation in leaves and roots (Figure 8.14). Accumulation of Pb was higher in leaves than roots of those plants where EDTA was added in Pb-contaminated medium, while the reverse was the case with roots (Jarvis and Leung 2002).

Higher translocation of Pb may be due to reduction in the molecular size of the Pb and EDTA complex formed (Turgut et al. 2004, Wu et al. 2004), which did not disturb the metabolic activities of the plants growing under the influence of Pb-EDTA. Plants growing under this medium produced more biomass, which proved that the addition of EDTA is beneficial in phytoextraction of metals from Pb-contaminated medium.

8.4 CONCLUSIONS

Results presented in this article and the cited literature clearly indicate that Pb contamination in soil and water disturbs the nutrient and water uptake and other metabolic activities, ultimately resulting in plant death. To clean the Pb contamination from soil and water, phytoremediation is the suitable and environmentally friendly approach, which can be accelerated with the application of EDTA to hyperaccumulator or high biomass-producing plants like *Brassica napus*.

REFERENCES

Ahmad, M.S.A., M. Hussain, S. Ijaz, and A.K. Alvi. 2008. Photosynthetic performance of two mung bean (*Vigna radiata*) cultivars under lead and copper stress. *International Journal of Agricultural Biology* 10: 167–172.

Ali, G., P.S. Srivastava, and M. Iqbal. 2000. Influence of cadmium and zinc on growth and photosynthesis of *Bacopa monniera* cultivated *in vitro*. *Biologia Plantarum* 43: 599–601.

Almeida, A.A. F., R.R. Valle, M.S. Mielke, and F.P. Gomes. 2007. Tolerance and prospection of phytoremediator woody species of Cd, Pb, Cu and Cr. *Brazilian Journal of Plant Physiology* 19: 83–93.

Angelone, M., and C. Bini. 1992. Trace elements concentrations in soils and plants of Western Europe. In Adriano, D.C. (Ed.), *Biogeochemistry of Trace Metals*, pp. 19–60. Lewis Publishers, Boca Raton, FL.

Arun, K.S., C. Cervantes, H. Loza-Tavera, and S. Avudainayagam. 2005. Chromium toxicity in plants. *Environment International* 31: 739–753.

Ashraf, M. Y., N. Azhar, M. Ashraf, M. Hussain, and M. Arshad. 2011. Influence of lead on growth and nutrient accumulation in canola (*Brassica napus* L.) cultivars. *Journal of Environmental Biology* 32: 659–666.

Ashraf, M. Y., A. Gul, M. Ashraf, F. Hussain, and G. Ebert. 2010. Improvement in yield and quality of Kinnow (*Citrus deliciosa* x *Citrus nobilis*) by potassium fertilization. *Journal of Plant Nutrition* 33: 1625–1637.

Ashraf, M.Y., A.H. Khan, and A.R. Azmi. 1992. Cell membrane stability and its relation with some physiological process in wheat. *Acta Agronomica Hungarica* 41: 183–191.

Asharf, M.Y., M. Ashraf, and G. Sarwar. 2005. Physiological approaches to improve the plant salt tolerance. In Daris, R. (Ed.), *CROPS. Growth, Quality and Biotechnology*, pp. 1206–1227. WFL Publishers, Helsinki, Finland.

Ashraf, M.Y., A. Khaliq, and R.H.A. Eui-Shik. 2004. Growth and leaf gas exchange characteristics in *Dalbergia sissoo* Roxb. and *D. lantifolia* Roxb. under water deficit. *Photosynthetica* 42: 157–160.

Azhar N., M.Y. Ashraf, M. Hussain, M. Ashraf, and R. Ahmed. 2009. EDTA-induced improvement in growth and water relations of sunflower (*Helianthus annuus* L.) plants grown in lead contaminated medium. *Pakistan Journal of Botany* 41: 3065–3074.

Baker, A.J.M., S.P. McGrath, R.D. Reeves, and J.A.C. Smith. 2000. Metal hyperaccumulator plants: A review of the ecology and physiology of a biological resource for phytoremediation of metal-polluted soils. In Terry, N., Bañuelos, G. (Eds.), *Phytoremediation of Contaminated Soil and Water*, pp. 85–107. CRC Press LLC, USA.

Baker, A.J.M., R.D. Reeves, and A.S.M. Hajar. 1994. Heavy metal accumulation and tolerance in British populations of the metallophyte *Thlaspi caerulescens* J. & C. Presl. (Brassicaceae). *New Phytologist* 127: 61–68.

Barberon, M., E. Zelazny, S. Robert, G. Conéjéro, C. Curie, J. Friml, and G. Vert. 2012. Monoubiquitin-dependent endocytosis of the Iron-Regulated Transporter 1 (IRT1) transporter controls iron uptake in plants. *Proceedings of the National Academy of Sciences USA* 109(21): 8322–8327.

Begonia, M.T.F., A. Begonia, M. Buttler, B. Ighoavodha, and B. Crudup. 2002. Chelate assisted phytoextraction of lead from a contaminated soil using wheat (*Triticum aestivum* L.). *Bulletin of Environmental Contamination and Toxicology* 68: 705–711.

Blaylock, M.J., D.E. Salt, S. Dushenkov, C.D. Gussman, Y. Kapulnik, B.D. Ensley, and I. Raskin. 1997. Enhanced accumulation of Pb in Indian mustard by soil-applied chelating agents. *Environmental Science and Technology* 31: 860–865.

Bonnet, M., O. Camares, and P. Veisseire. 2000. Effects of zinc and influence of *Acremonium lolii* on growth parameters, chlorophyll a fluorescence and antioxidant enzyme activities of ryegrass (*Lolium perenne* L. cv Apollo). *Journal of Experimental Botany* 51: 945–953.

Breckle, S.W., and H. Kahle. 1992. Effects of toxic heavy metals (Pb, Cd) on growth and mineral nutrition of beech (*Fagus sylvatica* L.). *Vegetation* 101: 43–53.

Brooks, R.R. 1998. Phytochemistry of hyperaccumulators. In Brooks, R.R. (Ed.), *Plants that Hyperaccumulate Heavy Metals*, pp. 15–53. CAB International, Wallingford.

Brown, S.L., R.L. Chaney, J.S. Angle, and A.J.M. Baker. 1994. Phytoremediation potential of *Thlaspi caerulescens* and *Bladder campion* for zinc- and cadmium-contaminated soil. *Journal of Environmental Quality* 23: 1151–1157.

Chanay, R.L., and J.A. Ryan. 1994. *Risk based standards for arsenic lead and cadmium in urban soils.* Dechema, Frankfurt, Germany.

Chen, H., and T. Cutright. 2001. EDTA and HEDTA effects on Cd, Cr and Ni uptake by *Helianthus annus. Chemosphere* 45: 21–28.

Chen, Y.X., Q. Lin, Y.M. Luo, Y.F. He, S.J. Zhen, Y.L. Yu, G.M. Tian, and M.H. Wong. 2003. The role of citric acid on the phytoremediation of heavy metal contaminated soil. *Chemosphere* 50: 807–811.

Ciesko, Z., S. Kalembasa, M. Cryszkowski, and E. Rolska. 2003. Effect of soil contamination by cadmium on potassium uptake by plants. *Plac Lodzki* 4: 10–718.

Cooper, E.M., J.T. Sims, S.D. Cunningham, J.W. Huang, and W.R. Berti. 1999. Chelate-assisted phytoextraction of lead from contaminated soils. *Journal of Environmental Quality* 28: 1709–1719.

Cui, Y., Y. Dong, H. Li, and Q. Wang. 2004. Effect of elemental sulphur on solubility of soil heavy metals and their uptake by maize. *Environment International* 30: 323–328.

Dahmani-Muller, H., F. Van-Oort, B. Gelie, and M. Balabane. 2000. Strategies of heavy metal uptake by three plant species growing near a metal smelter. *Environmental Pollution* 109: 231–238.

Drazkiewicz, M. 1994. Chlorophyll-occurrence, functions, mechanism of action, effects of internal and external factors. *Photosynthetica* 30: 321–331.

Du, Y., J. Chang, and X.F. Hu. 2010. Influence of adding Pb to soil on the growth of wheat seedlings. 19th World Congress of Soil Science, *Soil Solutions for a Changing World*, 72:1–6 August 2010, Brisbane, Australia. Published on DVD.

Eick, M.J., J.D. Peak, P.V. Brady, and J.D. Pesek. 1999. Kinetics of lead adsorption and desorption on goethite: Residence time effect. *Soil Science* 164: 28–39.

Epstein, A.L., C.D. Gussman, M.J. Blaylock, U. Yermiyahu, J.W. Huang, Y. Kapulnik, and C.S. Orser. 1999. EDTA and Pb-EDTA accumulation in *Brassica juncea* grown in Pb-amended soil. *Plant and Soil* 208: 87–94.

Evangelou, M.W.H., M. Ebel, and A. Schaeffer. 2007. Chelate-assisted phytoextraction of heavy metals from soil. Effect, mechanism, toxicity, and fate of chelating agents. *Chemosphere* 68: 989–1003.

Fargašová, A. 1999. Phytotoxic effect of Cd, Zn, Pb, Cu and Fe on *Sinapis alba* L. seedlings and their accumulation in roots and shoots. *Biologia Plantarum* 42: 75–80.

Fargašová, A. 2001. Phytotoxic effects of Cd, Zn, Pb, Cu and Fe on *Sinapis alba* L. seedlings and their accumulation in roots and shoots. *Biologia Plantarum* 44: 471–473.

Fargašová, A. 1998. Root growth inhibition, photosynthetic pigments production, and metal accumulation in *Synapis alba* as the parameters for trace metals effect determination. *Bulletin of Environmental Contamination and Toxicology* 61: 762–769.

Figliolia, A., S. Socciarelli, and B. Pennelli. 2002. Capability of *Brassica napus* to accumulate cadmium, zinc and copper from soil. *Acta Biotechnologia* 22: 133–140.

Fodor, F., E. Sarvari, F. Lang, Z. Szigeti, and E. Cseh. 1996. Effect of Pb and Cd on cucumber depending on the Fe complex in the culture solution. *Journal of Plant Physiology* 148: 434–439.

Geebelen, W., J. Vangronsveld, D.C. Adriano, L.C. Van Poucke, and H. Clijsters. 2002. Effects of Pb-EDTA and EDTA on oxidative stress reactions and mineral uptake in *Phaseolus vulgaris*. *Physiologia Plantarum* 115: 377–384.

Gill, S.S., and N. Tuteja. 2010. Reactive oxygen species and antioxidant machinery in abiotic stress tolerance in crop plants. *Plant Physiology and Biochemistry* 48: 909–930.

Government of Pakistan, 2010. *Pakistan Statistical Year Book*. Federal Bureau of Statistics. p. 28.

Gupta, A.K., and S. Sinha. 2007. Phytoextraction capacity of tannery sludge dumping sites. *Bioresource Technology* 98: 1788–1794.

Gussarsson, M. 1994. Cadmium-induced alterations in nutrient composition and growth of *Betula pendula* seedlings: The significance of fine roots as a primary target for cadmium toxicity. *Journal of Plant Nutrition* 17: 2151–2163.

Hajiboland, R. 2005. An evaluation of the efficiency of cultural plants to remove heavy metals from growing medium. *Plant, Soil and Environment* 51: 156–64.

Heil, D.M., Z. Samani, A.T. Hanson, and B. Rudd. 1999. Remediation of lead contaminated soil by EDTA. I. Batch and column studies. *Water, Air and Soil Pollution* 113: 77–95.

Hall, J.L. 2002. Cellular mechanism for heavy metal detoxification and tolerance. *Journal of Experimental Botany* 53: 1–11.

Hoagland, D.R., and D.I. Arnon. 1950. The water-culture method for growing plants without soil. *California Agricultural Experiment Station*: *Circular* 347: 1–32.

Holm, R.H., P. Kenncpohil, and E.I. Solamen. 1996. Structural and functional aspects of metal sites in biology. *Chemical Reviews* 96: 2239–2314.

Huang, H., Li. Tingxuan, T. Shengke, D.K. Gupta, Z. Xizhou, and Y. Xiao. 2008. Role of EDTA in alleviating lead toxicity in accumulator species of *Sedum alfredii* H. *Bioresource Technology* 99: 6088–6096.

Huang, J.W., J. Chen, W.B. Berti, and S.D. Cunningham. 1997. Phytoremediation of lead-contaminated soils; role of synthetic chelates in lead phytoextraction. *Environmental Science and Technology* 31: 800–805.

Hussain, M., M.S.A. Ahmad, and A. Kausar. 2006. Effect of lead and chromium on growth, photosynthetic pigments and yield components in mash bean [*Vigna mungo* (L.) Hepper]. *Pakistan Journal of Botany* 38: 1389–1396.

Iqbal, M.Z., and M. Shafiq. 1998. Effect of lead and cadmium on plants. M.Sc. thesis, Dept. Botany, Univ. Peshawar, Pakistan.

Islam, E., D. Liu, T. Li, X. Yang, X. Jin, Q. Mahmood, S. Tian, and J. Li. 2008. Effect of Pb toxicity on leaf growth, physiology and ultrastructure in the two ecotypes of *Elsholtzia argyi*. *Journal of Hazardous Materials* 154: 914–926.

Ismail, F., M.R. Anjum, A.N. Mamon, and T.G. Kazi. 2011. Trace metal contents of vegetables and fruits of Hyderabad retail market. *Pakistan Journal of Nutrition* 10: 365–372.

Jackson, M.L. 1962. *Soil Chemical Analysis*. Constable, England.

Jarvis, M.D., and D.W.M. Leung. 2002. Chelated lead transport in *Pinus radiata*: An ultrastructural study. *Environmental and Experimental Botany* 48: 21–32.

Javed, I., and M. Saher. 1987. Effect of lead on germination, early seedling growth, soluble protein and acid phosphatase content in *Zea mays* L. *Pakistan Journal of Scientific and Industrial Research* 30: 853–856.

Juberg, D.R., C.F. Kleiman, and S.C. Kwon, 1997. Position paper of the American Council of Science and Health: Lead and human health. *Ecotoxicology and Environmental Safety* 38: 162–180.

Kabata-Pendias, A., and H. Pendias. 1999. *The Biogeochemistry of Trace Metals*. Wydawnictwo Naukowe P.W.N., Warszawa (in Polish).

Kabata-Pendias, A. 2001. *Trace Elements in Soils and Plants*. CRC Press, Boca Raton, Florida, USA.

Karimi, A., H. Khodaverdiloo, M. Sepehri and M.R. Sadaghiani. 2011. Arbuscular mycorrhizal fungi and heavy metal contaminated soils. *African Journal of Microbiology Research* 5: 1571–1576.

Kalita, M.C., and C.M. Sharma. 1995. Effect of lead on growth and nitrogen content of *Azolla pinnata anabena azollae* symbionts. *Neo-Botanica* 3: 123–127.

Kastori, R., M. Petrovic, and N. Petrovic. 1996. Effect of lead on water relations, proline concentration and nitrate reductase activity in sunflower plants. *Acta Agronomica Hungarica* 44: 21–28.

Kastori, R., M. Plesnicar, Z. Sakac, D. Pancovic, and M.I. Arsenijevic. 1998. Effect of excess lead on sunflower growth and photosynthesis. *Journal of Plant Nutrition* 21: 75–85.

Kaushik, S. 2011. Phytoremediation: Use of green plants to remove the pollutants. Online available article: http://www.biotecharticles.com/Environmental-Biotechnology-Article/Phytoremediation-Use-of-green-plants-to-remove-pollutants-704.html.

Kevresan, S., N. Petrovic, M. Popovic, and J. Kandrac. 2001. Nitrogen and protein metabolism in young pea plants as affected by different concentrations of nickel, cadmium, lead, and molybdenum. *Journal of Plant Nutrition* 24: 1633–1644.

Kopittke, P.M., C.J. Asher, R.A. Kopittke and N.W. Menzies. 2007. Toxic effects of Pb^{2+} on growth of cowpea (*Vigna unguiculata*). *Environmental Pollution* 150: 280–287.

Kovacevic, G., R. Kastori, and L.J. Merkulov. 1999. Dry matter and leaf structure in young wheat plants as affected by cadmium, lead and nickel. *Biologia Plantarum* 42: 119–123.

Kumar, P.B.A.N., V. Dushenkov, H. Motto, and I. Raskin. 1995. Phytoextracton: The use of plants to remove heavy metals from soils. *Environmental Science and Technology* 29: 1232–1238.

Larbi, A., F. Morales, A. Abadia, Y. Gogorcena, J.J. Lucena, and J. Abadia. 2002. Effects of Cd and Pb in sugar beet plants grown in nutrient solution: Induced Fe deficiency and growth inhibition. *Functional Plant Biology* 29: 1453–1464.

Lasat, M.M. 2000. Phytoextraction of metals from contaminated soil. A review of plant/soil/metal interaction and assessment of pertinent agronomic issues. *Journal of Hazardous Substances Research* 2: 1–25.

Lavid, N., A. Schwartz, E. Lewinsohn, and E. Tel-Or. 2001. Phenols and phenol oxidases are involved in cadmium accumulation in the water plants *Nymphoides peltata* (*Menyanthaceae*) and *Nymphaeae* (*Nymphaeaceae*). *Planta* 214: 189–195.

Liu, D., W. Jiang, C. Liu, C. Xin, W. Hou, D.H. Liu, W.S. Jiang, C.J. Liu, and W.Q. Hou. 2000. Uptake and accumulation of lead by roots, hypocotyls and shoots of Indian mustard (*Brassica juncea* L.). *Bioresource Technology* 71: 273–277.

Lombi, E., F.J. Zhao, S.J. Dunham, and S.P. McGrath. 2001. Phytoremediation of heavy metal–contaminated soils natural hyperaccumulation versus chemically enhanced phytoextraction. *Journal of Environmental Quality* 30: 1919–1926.

Lummerzheim, M., M. Sandroni, C. Castresana, D. De Oliveira, M.M. Van, D. Roby, and B. Timmerman. 1995. Comparative microscopic and enzymatic characterization of the leaf necrosis induced in *Arabidopsis thaliana* by lead nitrate and by *Xanthomonas campestris* pv. *Campestris* after foliar spray. *Plant, Cell and Environment* 18: 499–509.

Madhavi, R., and N.V.N. Charynlu. 1998. Role of certain chelates like EDTA, gypsum and serpentine soil in reducing the toxic effect of lead, cadmium and mercury on the growth and metabolism of *Trigonella foenum-graecum*. *Plant Physiology and Biochemistry* 25: 95–108.

Mahmed M.F., A.T. Thalooth, and R.Kh.M. Khalifa. 2010. Effect of foliar spraying with uniconazole and micronutrients on yield and nutrients uptake of wheat plants grown under saline condition. *Journal of American Science* 6: 398–404.

Mamta, R., P. Gardre, and M. Jani. 1997. Inhibition of nitrate reductase activity by lead in green bean leaf segments; a mechanistic approach. *Indian Journal of Plant Physiology* 2: 5–9.

Marchiol, L., S. Assdari, P. Sacco, and G. Zerbi. 2004. Phytoextraction of heavy metals by canola (*Brassica napus* L.) and radish (*Raphanus sativus*) grown on multicontaminated soil. *Environmental Pollution* 132: 21–27.

Mazen, A.M.A. 2004. Accumulation of four metals in tissues of *Corchorus olitorius* and possible mechanisms of their tolerance. *Biologia Plantarum* 48: 267–272.

McGrath, S.P. 1998. Phytoextraction for soil remediation. In Brooks, R.R. (Ed.), *Plants that Hyperaccumulate Heavy Metals*, pp. 261–288. CAB International, Wallingford, UK.

McGrath, S.P., F.J. Zhao, and E. Lombi. 2002. Phytoremediation of metals, metalloids and radionuclides. *Advances in Agronomy* 75: 1–56.

Meers, E., A. Ruttens, M. J. Hopgood, D. Samson, and F.M.G. Tack. 2005. Comparison of EDTA and EDDS as potential soil amendments for enhanced phytoextraction of heavy metals. *Chemosphere* 58: 1011–1022.

Misra, S.N., and D.B. Singh. 2000. Accumulation of lead and cadmium in upper parts of mustard (*Brassica juncea*) seedlings. *Indian Journal of Experimental Biology* 38: 814–818.

Mobin, M., and N.A. Khan. 2007. Photosynthetic activity, pigment composition and antioxidative response of two mustard (*Brassica juncea*) cultivars differing in photosynthetic capacity subjected to cadmium stress. *Journal of Plant Physiology* 164: 601–610.

Monnet, F., N. Vaillant, P. Vernay, A. Coudret, H. Sallanon, and A. Hitmi. 2001. Relationship between PSII activity, CO_2 fixation and Zn, Mn and Mg contents of *Lolium perenne* under zinc stress. *Journal of Plant Physiology* 158: 1137–1144.

Panda, S.K., and S. Choudhary. 2005. Chromium stress in plants. *Brazilian Journal of Plant Physiology* 17: 19–102.

Papassiopi, N., S. Tambouris, and A. Kontopoulos. 1999. Removal of heavy metals from calcareous contaminated soils by EDTA leaching. *Water, Air and Soil Pollution* 109: 1–15.

Paivoke, A.E.A. 2002. Soil lead alters phytase activity and mineral nutrient balance of *Pisum sativum*. *Environmental and Experimental Botany* 48: 61–73.

Raskin, I., P.B. Kumar, S. Dushenkov, and D.E. Salt. 1994. Bioconcentration of heavy metals by plants. *Current Opinion in Biotechnology* 5: 285–290.

Reichman, S.M. 2002. The Responses of Plants to Metal Toxicity: A review focusing on Copper, Manganese and Zinc. Published byAustralian Minerals & Energy Environment Foundation as *Occasional Paper No. 14*.

Rengel, Z. 1997. Mechanisms of plant resistance to toxicity of aluminium and heavy metals. In Basra, A.S., Basra, R.K. (Eds.), *Mechanisms of Environmental Stress Resistance in Plants*, pp. 241–276. Harwood Academic, Amsterdam.

Salt, D.E., M. Blaylock, P.B.A.N. Kumar, V. Dushenkov, B.D. Ensley, I. Chet, and I. Raskin. 1995. Phytoremediation: A novel strategy for the removal of toxic metals from the environment using plants. *Biotechnology* 13: 46–475.

Salt, D.E., R.D. Smith, and I. Raskin. 1998. Phytoremediation. *Annual Review of Plant Physiology and Molecular Biology* 49: 643–668.

Sanchez, P.G., L.P. Fernandez, L.T. Trejo, G.G. Aleantra, and J.D. Cruz. 1999. Heavy metal accumulation in beans and its impact on growth and yield in soilless culture. *Acta Horticulturae* 481: 617–623.

Sarvari, E., F. Fodor, E. Cseh, A. Varga, G. Zaray, L. Zolla, G. Horrath, and Z. Szigeti. 1999. Relationship between changes in ion content by leaves and chlorophyll protein composition in cucumber under Cd and Pb stress. *Z Naturforsch* 54c: 746–753.

Sharma, P., and R.S. Dubey. 2005. Lead toxicity in plants. *Brazilian Journal of Plant Physiology* 17: 35–52.

Sharma, D.C., and C.P. Sharma. 2003. Chromium uptake and toxicity effects on growth and metabolic activities in wheat, *Triticum aestivum*. *Indian Journal of Experimental Biology* 34: 689–691.

Shen, Z.G., X.D. Li, H.M. Chen, C.C. Wang, and H. Chau. 2002. Lead phytoremediation from contaminated soil with high biomass plant species. *Journal of Environmental Quality* 31: 1893–1900.

Schmidt, U. 2003. Enhancing phytoextraction: The effect of chemical soil manipulation on mobility, plant accumulation and leaching of heavy metals. *Journal of Environmental Quality* 32: 1939–1954.

Siedlecka, A., and Z. Krupa. 1999. Cd/Fe interaction in higher plants—its consequences for the photosynthetic apparatus. *Photosynthetica* 36: 321–331.

Singh, R.P., S. Dabas, A. Chaudhary, and R. Maheshwari. 1998. Effect of lead on nitrate reductase activity and alleviation of lead toxicity by inorganic salts and 6-benzylaminopurine. *Biologia Plantarum* 40: 399–404.

Singh, R.P., R.D. Tripathi, S.K. Sinha, R. Maheshwari, and H.S. Srivastava. 1997. Response of higher plants to lead-contaminated environment. *Chemosphere* 34: 2467–2493.

Sinha, S. 2000. Accumulation of Cu, Cd, Cr, Mn and Pb from artificially contaminated soil by *Bacopa monnieri*. *Environmental Monitoring and Assessment* 57: 253–264.

Sinha, P., B.K. Dube, P. Srivastava, and C. Chatterjee. 2006. Alteration in uptake and translocation of essential nutrients in cabbage by excess lead. *Chemosphere* 65: 651–656.

Sovero, M. 1993. Rapeseed, a new oilseed crop for the United States. In Janick, J., Simon, J.E. (Eds.), *New Crops*, pp. 302–307. Wiley, New York.

Srikanth, R., and S.R. Reddy. 1991. Lead, chromium and cadmium levels in vegetables grown in urban sewage sludge. *Food Chemistry* 40: 229–234.

Srivastava, N.K., and A.K. Srivastava. 2010. Influence of some heavy metals on growth, alkaloid content and composition in *Catharanthus roseus* L. *Indian Journal of Pharmaceutical Science* 72: 775–778.

Steel, R.G.D., J.H. Torrie, and D.A. Dickey. 1997. *Principles and Procedures of Statistics, A Biometrical Approach*, pp. 178–182. McGraw Hill Co., New York.

Susarla, S., V.F. Medina, and S.C. McCutcheon. 2002. Phytoremediatiom: An ecological solution to organic chemical contamination. *Ecological Engineering* 18: 647–658.

Sym, G.J. 1984. Optimization of the *in-vivo* assay conditions for nitrate reductase in barley (*Hordeum vulgare* L.cv. irgri). *Journal of Science, Food and Agriculture* 35: 725–730.

Székely A., P. Poór, I. Bagi, J. Csiszár, K. Gémes, F. Horváth, and I. Tari. 2011. Effect of EDTA on the growth and copper accumulation of sweet sorghum and sudangrass seedlings. *Acta Biologica Szegediensis* 55: 159–164.

Turan M., and A. Esringü. 2007. Phytoremediation based on canola (*Brassica napus* L.) and Indian mustard (*Brassica juncea* L.) planted on spiked soil by aliquot amount of Cd, Cu, Pb, and Zn. *Plant, Soil and Environment* 53: 7–15.

Tlustoš, P., J. Száková, J. Hrubý, I. Hartman, J. Najmanová, J. Nedělník, D. Pavlíková, and M. Batysta. 2006. Removal of As, Cd, Pb, and Zn from contaminated soil by high biomass-producing plants. *Plant, Soil and Environment* 52: 413–423.

Turgut, C., M.K. Pepe, and T.J. Cutright. 2004. The effect of EDTA and citric acid on phytoremediation of Cd, Cr, and Ni from soil using *Helianthus annuus*. *Environmental Pollution* 131: 147–154.

U.S. Salinity Lab. Staff. 1954. Diagnosis and improvement of saline and alkali soils. USDA *Handbook 60*. Washington DC, USA.

Vassil, A.O., Y. Kapulnik, I. Raskin, and D.E. Salt. 1998. The role of EDTA in lead transport and accumulation by Indian mustard. *Plant Physiology* 117: 447–453.

Verkleij J.C., A. Golan-Goldhirsh, D.M. Antosiewisz, J.P. Schwitzguébel, and P. Schröder. 2009. Dualities in plant tolerance to pollutants and their uptake and translocation to the upper plant parts. *Environmental Experimental and Botany* 67: 10–22.

Verma, S. and R.S. Dubey. 2003. Lead toxicity induces lipid per oxidation and alters the activities of antioxidant enzymes in growing rice plants. *Plant Science* 164: 645–655.

Walker, W.M., J.E. Miller, and J.J. Hassett. 1997. Effect of lead and cadmium upon the calcium, magnesium, potassium and phosphorus concentration in young corn plants. *Soil Science* 124: 145–151.

Waters, B.M., H.H. Chu, D.J. DiDonato, R.J. Roberts, R.B. Eisley, B. Lahner, D.E. Salt, and E.L. Walker. 2006. *Like1* and *Yellow Stripe-Like3* reveal their roles in metal ion homeostasis and loading of metal ions in seeds. *Plant Physiology* 141: 1446–1458.

Wensheng, S., L. Chonytu, and Z. Zhiquan. 1997. Analysis of major constraints on plant colonization at Fankou Pb/Zn mine tailings. *Chinese Journal of Applied Ecology* 8: 314–318.

Wierzbicka, M., and D. Antosiewiez. 1993. How lead can easily enter the food chain. A study of plant roots. *Science of the Total Environment* Suppl. Pt. 1: 423–429.

Wierzbicka, M. 1994. The resumption of metabolic activity in *Allium sepa* L. root tips during treatment with lead salt. *Environmental and Experimental Botany* 34: 173–180.

Wierzbicka, M. 1995. How lead loses its toxicity to plants. *Acta Society Botanica Polonica* 64: 81–90.

Wu, J., F.C. Hsu, and S.D. Cunningham. 1999. Chelate-assisted Pb phytoextraction: Pb availability, uptake and translocation constraints. *Environmental Science and Technology* 33: 1898–1904.

Wu, L.H., Y.M. Luo, Y.R. Xing, and P. Christic. 2004. EDTA-enhanced phytoremediation of heavy metal-contaminated soil with Indian mustard and associated potential leaching risk. *Agriculture, Ecosystem and Environment* 102: 307–318.

Xiong, Z.T. 1997. Bioaccumulation and physiological effects of excess lead roadside pioneer species *Sonchus oleraceus*. *Environmental Pollution* 97: 275–279.

Xu, Y., N. Yamaji, R. Shen, and J.F. Ma. 2007. Sorghum roots are inefficient in uptake of EDTA-chelated lead. *Annals of Botany* 99: 869–875.

Yahua, C., X. Li, and S. Zhenguo. 2004. Leaching and uptake of heavy metals by ten different species of plants during an EDTA-assisted phytoextraction process. *Chemosphere* 57: 187–196.

Yang, X.E., J.X. Liu, W.M. Wang, Z.Q. Ye, and A.C. Luo. 2004. Potassium internal use efficiency relative to growth vigor, potassium distribution, and carbohydrate allocation in rice genotypes. *Journal of Plant Nutrition* 27: 837–852.

Yang, Y.Y., J.Y. Jung., W.Y. Song., H.S. Suh, and Y. Lee. 2000. Identification of rice varieties with high tolerance or sensitivity to lead and characterization of the mechanism of tolerance. *Plant Physiology* 124: 1019–1026.

Yuan, Q.H., G.X. Shi, J. Zhao, H. Zhang, and Q.S. Xu. 2009. Physiological and proteomic analyses of *Alternanthera philoxeroides* under zinc stress. *Russian Journal of Plant Physiology* 56: 495–502.

Zhao, F.J., E. Lombi, T. Breedon, and S.P. McGrath. 2000. Zinc hyper-accumulator and cellular distribution in *Arabidopsis halleri*. *Plant, Cell and Environment* 23: 507–514.

9 Potential for Metal Phytoextraction of *Brassica* Oilseed Species

Guido Fellet, Luca Marchiol, and Giuseppe Zerbi

CONTENTS

9.1 INTRODUCTION

Contamination by heavy metals and organics of environmental matrices, i.e., soil, water, plant and air, is of great concern due to its potential impact on human and animal health. The traditional soil cleanup takes place by means of technologies based on physicochemical approaches. Such technologies include solidification and stabilization, leaching of contaminants by using acid solutions, ion exchange due to electrokinetics, red-ox reactions and excavation and burial of the soil at a hazardous waste site.

All the cleanup technologies based on physicochemical approaches are generally expensive, being power consuming and having detrimental effects. Green remediation (GR) is the practice that considers all the environmental effects of a cleanup process during each phase, and incorporates strategies to maximize the net environmental benefit of the cleanup. The GR reduces the demand placed on the environment during cleanup actions, also known as the "footprint" of remediation, and avoids the potential for collateral environmental damage.

As reported by the USEPA (2008), "GR results in effective cleanups minimizing the environmental and energy footprints of site remediation. Sustainable practices emphasize the need to more closely evaluate core elements of a cleanup project, compare the site-specific value of conservation benefits gained by different strategies of green remediation, and weigh the environmental trade-offs of potential strategies."

9.2 CURRENT PERSPECTIVES FOR PHYTOTECHNOLOGIES

The term "phytotechnologies" is an all-encompassing one that includes a variety of gentle techniques of environmental remediation that are currently being developed and will lead to contaminant degradation, removal, transfer or immobilization. The United Nations Environment Program defined phytotechnologies as "technologies relating to the use of vegetation, to resolve environmental problems in a watershed management, by prevention of landscape degradation, remediation and restoration of degraded ecosystems, control of environmental processes, and monitoring and assessment of environmental quality" (UNEP 2003).

Phytotechnologies are mainly applied *in situ* and can be applied to inorganic contaminants, such as heavy metals, metalloids, radioactive materials, and salts. Organic contaminants, such as hydrocarbons, crude oil, chlorinated compounds, pesticides, and explosive compounds, can be addressed using plant-based methods (ITRC 2009).

Phytotechnologies potentially satisfy several aspects addressed by GR (USEPA 2008), and have become attractive alternatives to conventional cleanup technologies due to relatively low capital costs and the inherently aesthetic nature of planted sites. Moreover, phytotechnologies potentially offer efficient and environmentally friendly solutions for cleanup of contaminated soil and water, improvement of food safety, carbon sequestration, and development of renewable energy sources, all of which contribute to sustainable land use management (Schwitzguébel and Schröder 2009).

Basically, phytotechnologies include some main sub-groups that correspond to mechanisms that enable plants to remove, destroy, transfer, stabilize, or contain contaminants (ITRC 2009). In particular, we consider (i) phytoextraction, which involves the use of plants to remove contaminants from soil; (ii) phytostabilization, which aims at preventing the migration of the soil contaminants by establishing a green cover which requires neither plant harvesting nor disposal; (iii) phytodegradation, applicable to soils contaminated by organics; (iv) rhizofiltration, which takes advantage of the ability of plants to remove contaminants from water and aqueous waste streams.

9.3 WHICH PLANTS SHOULD BE USED FOR METAL PHYTOEXTRACTION?

This is the key question for phytotechnologies. Plant selections must be based on site-specific conditions: concentration of contaminant(s), depth of contamination, climate, altitude, soil salinity, nutrient content, fertility, and plant hardiness are the crucial elements.

A number of plant species have now been found to be capable of accumulating metals in their above-ground tissues at concentrations which are significantly higher than those occurring in the soil. They have been termed hyperaccumulators, when the metal concentrations are 50–100 times higher, depending upon the metal, than in nonaccumulating plants (Baker and Brooks 1989; McGrath and Zhao 2003). For these reasons, they were considered potentially suitable for phytoextraction. However, several drawbacks exist. In fact, hyperaccumulators are highly adapted to adverse soil conditions and generally grow slowly, producing little biomass. This low biomass production precludes their use at a large scale level. Further, hyperaccumulators are highly selective, their metabolism being adapted to an excess of only one or two elements. Neither rotations nor intercroppings can be imagined using hyperaccumulators (Mench et al. 2010).

Therefore, many fast-growing and high biomass plant species have been screened to determine their suitability for phytoextraction. The *Brassicaceae* is a family containing many metal-accumulating species and has received considerable attention since many hyperaccumulators belong to the *Brassicaceae* family (Broadley et al. 2001; Kramer 2010).

One of the most promising nonhyperaccumulator species for phytoextraction is *B. juncea*. Other species of the *Brassica* genus such as *B. campestris*, *B. carinata*, *B. napus*, *B. nigra*, *B. oleracea* and *B. rapa* have also been studied (Kumar et al. 1995; Marchiol et al. 2004; Meers et al. 2005; Gisbert et al. 2006).

Beyond the expected performances of plants against pollutants, it might be interesting to consider plants that can also provide alternative employments other than their disposal as polluted wastes. In this way, phytoextraction could gain more economic value. Some examples regard the (i) cultivation of biomass crops whose metal-rich ash may be suitable for smelter feedstock or (ii) for metal removal (Chaney et al. 2007), composting, production of biofuels, extraction of oils and essential oils (Dickinson et al. 2009; Vangronsveld et al. 2009; Vamerali et al. 2010).

9.4 EXPERIMENTAL ACTIVITIES

9.4.1 Screening *Brassica* Species Cultivars for Seed Germination in Presence of Cd and Cr

The process of phytoextraction is based substantially on the mass transfer of an inorganic pollutant from the bulk soil to the plant biomass. The plant-soil interaction implies that the management of the two elements of the system (plant and soil) should have effects on the efficiency of the process.

The process of metal phytoextraction starts with the seed germination. Seed germination represent an initial and crucial step in the growth cycle of plants. Interferences during this process such as the disturbance due to the high concentrations of heavy metals is likely to impair or retard the seed germination and, as a consequence, the phytoextraction effectiveness.

As the plant establishment is necessary to start a phytoextraction project, it was deemed important to provide data about the germination and early seedlings' growth of different cultivars of plants which have a potential for phytoremediation. Li et al. (2003) discussed some technical options such as direct seeding, seed pelletization and transplanting, for the Ni-hyperaccumulator *Alyssum murale*; the authors conclude that direct seeding is the best solution. No information was found in literature with regard to the high-biomass species, even if some of them can be a viable alternative to the hyperaccumulators for extensive projects. Some accessions of *Brassica juncea,* particularly efficient in the uptake of Pb, were first identified by Kumar et al. (1995). This feature was then compared to other *Brassica* and non-*Brassica* species, demonstrating that *Brassica* species always showed the higher performances. On the other hand a lot of the known hyperaccumulators belong to this family (Guerinot and Salt 2001). Other data around the tolerance of different accessions of *Brassica* species were provided by Ebbs and Kochian (1998) that used lines of *B. juncea* for chelate-assisted phytoextraction.

The aim of this study was to observe the percentage of germination of 41 cultivars of some *Brassica* species in the presence of different concentrations of Cd and Cr.

9.4.1.1 Seed Germination Tests

A factorial experiment of germination was conducted in laboratory conditions. Seeds of 41 cultivars of different *Brassica* species were supplied by the University of Padova (Italy). The cultivars belong to the following species: *Brassica napus* (20 cultivars), *Brassica juncea* (7 cvs.), *Sinapis alba* (6 cvs.), *Raphanus sativus* (5 cvs.), *Brassica rapa sylvestris* (2 cvs.) and *Brassica campestris* (1 cv.). Table 9.1 reports the list of the cultivars.

Germinations were carried out in three 100-seeds replicates of each cultivar placed on Whatman filter paper (Ø 110 mm) in polystyrene Petri dishes (Ø 140 mm) containing 25 mL of Hoagland solution (Brown et al. 1995). The Petri dishes contained a circular plastic support (Ø 110 mm) provided with a central hole. A small piece of filter paper passing through the hole was partially immersed in the nutrient solution; this permitted the raising of the solution which saturated the filter paper with the seeds.

Standard Hoagland solution was used for the controls. Solution pH was adjusted to 6.0 by the addition of KOH as necessary. Cadmium and Cr were provided respectively as cadmium nitrate ($Cd(NO_3)_2$ and chromium dichromate ($K_2Cr_2O_7$), at the concentration of 10, 100, and 1000 μM in the nutrient solution, corresponding respectively to 1.12, 11.2, and 112.1 mg L^{-1} for Cd and 1.03, 10.3, and 104 mg L^{-1} for Cr.

The Petri dishes — placed in an incubator in dark conditions at 25 ± 2°C – were surveyed counting the seedlings which did not showed any symptoms of toxicity; those that died or showed necrotic areas in the hypocotyl were not considered. The percentage of germination 48 h (d2) and 96 h (d4) after the beginning of the experiment was obtained by counting the germinated seeds. A

TABLE 9.1
List of the *Brassica* Cultivars Tested in the Germination Experiment

	Species	Cultivar (cv.)		Species	Cultivar (cv.)
1	*B. napus*	Aladin	21	*B. juncea*	Aurea
2	*B. napus*	Alaska	22	*B. juncea*	Barton
3	*B. napus*	Alice	23	*B. juncea*	Budakalaszi
4	*B. napus*	Comet	24	*B. juncea*	R-3104/59
5	*B. napus*	Emerald 130-040	25	*B. juncea*	Varuna
6	*B. napus*	Forte	26	*B. juncea*	Vittasso
7	*B. napus*	Inca	27	*B. juncea*	WNFP
8	*B. napus*	Kabel	28	*R. sativus*	Arena
9	*B. napus*	Karat	29	*R. sativus*	Colonel
10	*B. napus*	Karibe	30	*R. sativus*	Pegletta
11	*B. napus*	Lisonne 7323311	31	*R. sativus*	Rego
12	*B. napus*	Lizard 3296L	32	*R. sativus*	Rimbo
13	*B. napus*	Orient	33	*S. alba*	Asta
14	*B. napus*	Orkan	34	*S. alba*	Concerta
15	*B. napus*	Oxident	35	*S. alba*	Maxi
16	*B. napus*	Rafaela	36	*S. alba*	Silenda
17	*B. napus*	Rebel	37	*S. alba*	Sirola
18	*B. napus*	Silvia	38	*S. alba*	Vertus × 6256
19	*B. napus*	Sponsor	39	*B. rapa silvestris*	Buko
20	*B. campestris*	Debut	40	*B. rapa silvestris*	Perko

seed was considered to be germinated when its radicle had grown longer than 3 mm. Five days after the start of the experiment, the percentage of healthy seedlings was evaluated.

9.4.1.2 Experimental Design and Data Analysis

The trial was arranged in a factorial design with seven treatments (control, Cd 10 μM, Cd 100 μM, Cd 1000 μM, Cr 10 μM, Cr 100 μM, and Cr 1000 μM) and three replicates per treatment. A two-way ANOVA was carried out to examine the main effects (treatment and cultivar) and the interaction on the experimental parameters; the comparison of means were made by using the Student–Newman–Keuls test ($p < 0.05$). Before running the ANOVA test, percentage data were subjected to the angular transformation. All statistics were computed using CoHort 6.204 (CoHort Software, Monterey, CA).

9.4.1.3 Effects of Cd and Cr on Germination

The effects of Cd and Cr on plant germination varied in response to the increase of the concentration of the metals in the nutrient solution (Table 9.2). The germination of the seeds was more impaired by Cd than Cr at 1000 μM, the highest concentration; the percentage of healthy seedlings was depressed respectively by 25% and 15% when compared to the control for Cd and Cr. At the lower and intermediate concentration of heavy metals, 5 days after the beginning of the experiment

TABLE 9.2
***p* Values from Two-Way ANOVA Table and Average Percentage Values of Germination after 2 and 4 Days and for the Percentage of Healthy Seedlings of Cultivars of the *Brassica* Species Studied**

ANOVA	% 2d	% 4d	% Seedlings
	Main Effects		
Species	.0000 ***	.0000 ***	.0000 ***
Treatment	.9345 ns	.6012 ns	.0000 ***
	Interaction		
Species × Treatment	.4624 ns	.9955 ns	.0000 ***
Post hoc test	(%)	(%)	(%)
	Species		
B. campestris	91.7 abc	92 ab	81 b
B. juncea	88.6 bc	89.4 b	83.5 b
B. napus	82.3 c	86.7 c	76.6 c
B. rapa sylvestris	95.3 ab	96.3 a	91.7 a
R. sativus	93.2 ab	95.5 a	91 a
S. alba	97.6 a	94.2 a	88.1 a
	Treatments		
Control	88.6 a	90.6 a	88.4 a
Cd 10 μM	87.8 a	90.4 a	87.9 a
Cd 100 μM	87.3 a	90.1 a	86.6 a
Cd 1000 μM	86.4 a	89.5 a	63.6 c
Cr 10 μM	87.3 a	90.5 a	87.4 a
Cr 100 μM	86.6 a	89.8 a	86.8 a
Cr 1000 μM	91.5 a	89.2 a	74.9 b

Note: Means with the same letter are not significantly different ($p < 0.05$).
*$P < 0.05$; **$P < 0.01$; ***$P < 0.001$.

TABLE 9.3
Average Percentage of Germination after 2 and 4 Days of the Cultivars Studied

2 Days after Germination			4 Days after Germination		
Species	**cv.**	**%**	**Species**	**cv.**	**%**
S. alba	Concerta	98.5 a	*S. alba*	Concerta	98.9 a
R. sativus	Rego	97.7 ab	*R. sativus*	Rego	98.5 a
S. alba	Silenda	97.6 ab	*S. alba*	Silenda	98.4 a
S. alba	Sirola	97.6 ab	*S. alba*	Sirola	98.3 ab
B. rapa silvestris	Perko	95.8 abc	*R. sativus*	Pegletta	97.4 abc
B. juncea	Aurea	95.5 abcd	*R. sativus*	Rimbo	97.1 abc
R. sativus	Rimbo	94.9 bcde	*R. sativus*	Colonel	97.1 abc
B. juncea	Vitasso	94.8 bcde	*B. rapa silvestris*	Perko	96.7 abc
R. sativus	Colonel	94.8 bcde	*B. juncea*	Aurea	96.1 abcd
B. rapa silvestris	Buko	94.7 bcde	*B. rapa silvestris*	Buko	95.7 abcd
B. napus	Alice	94.5 bcde	*B. juncea*	Vitasso	95.3 abcd
R. sativus	Pegletta	94.4 bcde	*S. alba*	Vertus × 6256	95.3 abcd
B. napus	Lisonne	93.2 cde	*B. napus*	Alice	94.8 bcd
S. alba	Vertus × 6256	92.3 def	*B. napus*	Lisonne	94.3 cde
B. campestris	Debut	91.6 efg	*S. alba*	Maxi	93 def
S. alba	Maxi	89.9 fgh	*B. campestris*	Debut	92 efg
B. juncea	Barton	89.3 ghi	*B. napus*	Karibe	91.3 fgh
B. napus	Orkan	89.2 ghi	*B. napus*	Forte	90.8 fghi
B. napus	Emerald	88.8 ghi	*B. napus*	Emerald	90.7 fghi
B. napus	Karibe	87.9 hij	*B. napus*	Orkan	89.9 fghij
B. juncea	WNFP	87.2 hijk	*B. juncea*	Barton	89.7 fghij
B. napus	Kabel	87.2 hijk	*B. napus*	Rafaela	89.7 fghij
B. juncea	R3104/59	87 hijk	*B. napus*	Kabel	89.6 ghij
B. juncea	Varuna	86.3 ijk	*B. napus*	Lizard	89 ghijk
B. napus	Rafaela	86.3 ijk	*B. juncea*	WNFP	88.6 ghijk
B. napus	Lizard	85.4 jkl	*B. juncea*	R3104/59	88.4 hijk
B. napus	Orient	84.5 klm	*B. napus*	Karat	87.6 ijkl
B. napus	Alaska	84.1 klm	*B. juncea*	Varuna	87.0 jkl
R. sativus	Arena	83.9 klm	*R. sativus*	Arena	87.0 jkl
B. napus	Oxident	83 lm	*B. napus*	Oxident	87.0 jkl
B. napus	Forte	82.8 lm	*B. napus*	Sponsor	85.9 kl
B. napus	Karat	82.8 lm	*B. napus*	Orient	85.7 kl
B. napus	Sponsor	81.5 mn	*B. napus*	Alaska	85.0 l
B. juncea	Budelakaszi	79.8 n	*B. napus*	Rebel	84.4 l
B. napus	Rebel	79.8 n	*S. alba*	Asta	81.2 m
S. alba	Asta	76.8 o	*B. napus*	Aladin	81.0 m
B. napus	Comet	73.8 p	*B. juncea*	Budelakaszi	80.7 m
B. napus	Aladin	70.4 q	*B. napus*	Comet	77.9 n
B. napus	Silvia	64.1 r	*B. napus*	Inca	77.1 n
B. napus	Inca	63.8 r	*B. napus*	Silvia	75.5 n

Note: Means with the same letter are not significantly different ($p < 0.05$).

only light differences were recorded ($LSD_{0.05}$ = 0.959) in comparison to the control. The average percentages of germination after 2 and 4 days, listed in descending order, are reported in Table 9.3.

Regarding the species and their cultivars, it can be noted that *Sinapis alba* cvs. Concerta, Sirola and Silenda had a percentage of germination >80% giving the best performance when compared to other species. Cultivar Asta, having a percentage of germination after 2 days of about 60%, was the lowest among the six cvs. of *Sinapis alba* that we tested (Table 9.3). Except for the cv. Arena (66.5% of germination), *Raphanus sativus* was the most tolerant species in the experimental conditions. In fact, the lowest percentage of germination was 76.7% (Table 9.3). Further, the potential for phytoremediation of *Brassica juncea*, based on the heavy metals tolerance in the germination phase, was confirmed only by cvs. Vitasso and Aurea; on the contrary, the other accessions did not show the same ability (Table 9.3).

The two cultivars of *Brassica rapa sylvestris* we tested gave a good response, unlike *Brassica campestris*. Finally, *Brassica napus* gave the worst response to the exposition to Cd and Cr. None of the twelve cultivars exceeded 71% in germination; moreover, the cultivars Comet, Aladin, Silvia and Inca had a percentage lower than 60% (Table 9.3). The germination process after four days was still in progress; some species showed a delayed germination, without any visual symptoms of metal toxicity (Table 9.3).

The whole data set was then divided by species with the aim to evaluate the behavior of the cvs. within the species. Tables 9.4–9.8 show the *p* values calculated by the ANOVA for the percentage

TABLE 9.4
Leaf Area and Dry Matter Yield in Fractions and in Total Biomass of Control and Treated Plants of *B. juncea* cvs. Barton, Varuna, and WNFP

ANOVA	Leaf Area	Dry Matter: Roots	Dry Matter: Stems/Petioles	Dry Matter: Leaves	Dry Matter: Total Biomass
Main Effects					
Cultivar	.0111 ns	.1147 ns	.1298 ns	.0031 **	.0120 *
Treatment	.0000 ***	.0089 **	.0000 ***	.0000 ***	.0000 ***
Interaction					
Cultivar × Treatment	.1111 ns	.9050 ns	.0206 *	.0195 *	.0144 *
Post hoc test	(cm^2 $plant^{-1}$)	(mg $plant^{-1}$)	(mg $plant^{-1}$)	(mg $plant^{-1}$)	(mg $plant^{-1}$)
Cultivars					
Barton	67.7 b	32.4 a	64.2 a	157 a	254 a
Varuna	74.5 b	33.7 a	65.2 a	130 b	226 ab
WNFP	90.1 a	25.2 a	55.4 a	112 b	192 b
Treatment					
Control	108 a	40.5 a	98.9 a	199 a	334 a
0.1 μM Cd	86.4 a	25.6 ab	65.6 bc	143 ab	234 ab
1 μM Cd	62.8 bc	24.4 b	53.3 cd	104 c	182 c
1.5 μM Cd	64.8 bc	23.3 b	51.1 cd	108 c	182 c
0.1 μM Cr	102 a	37.8 ab	77.8 b	171 ab	287 ab
1 μM Cr	62.8 bc	37.8 ab	48.9 cd	110 c	188 c
1.5 μM Cr	54.9 c	28.9 ab	41.1 d	95.5 c	162 c

Note: Means with the same letter are not significantly different ($p < 0.05$).
$^*P < 0.05$; $^{**}P < 0.01$; $^{***}P < 0.001$.

TABLE 9.5
***p* Values after ANOVA on Cadmium and Chromium Concentration Recorded in the Fractions of *B. juncea* cv. WNFP, Varuna, and Barton**

ANOVA	Roots	Stems	Leaves
	Main Effects		
Cultivar	.359 ns	.007 **	.058 ns
Treatment	.000 ***	.000 ***	.000 ***
	Interaction		
Cultivar × Treatment	.166 ns	0.027 **	.088 ns *

$*P < 0.05$; $**P < 0.01$; $***P < 0.001$.

TABLE 9.6
***p* Values after ANOVA and Means ± Standard Error of Means of Bioconcentration Factor (BF) and Translocation Factor (TF) Calculated for *B. juncea* cv. WNFP, Varuna, and Barton**

ANOVA		BF			TF	
			Main Effects			
Cultivars		.026 **			.375 ns	
Treatments		.0000 ***			.0000 ***	
			Interaction			
Cultivars × Treatment		.059 ns			.429 ns	
Treatments	**Barton**	**Varuna**	**WNFP**	**Barton**	**Varuna**	**WNFP**
0.1 µM Cd	214 ± 14	286 ± 10	256 ± 22	0.06 ± 0.01	0.11 ± 0.01	0.09 ± 0.37
1 µM Cd	142 ± 18	250 ± 26	272 ± 68	0.08 ± 0.03	0.16 ± 0.07	0.16 ± 0.02
1.5 µM Cd	138 ± 15	193 ± 27	120 ± 26	0.07 ± 0.02	0.05 ± 0.01	0.07 ± 0.03
0.1 µM Cr	6.33 ± 0.66	40.8 ± 6.1	38.2 ± 3.5	0.02 ± 0.01	0.01 ± 0.001	0.01 ± 0.002
1 µM Cr	51.5 ± 1.4	44 ± 5.4	55.9 ± 4.0	0.001 ± 0.0002	0.08 ± 0.001	0.007 ± 0.001
1.5 µM Cr	40.3 ± 2.6	50 ± 9.3	48.9 ± 6.4	0.01 ± 0.002	0.001 ± 0.001	0.01 ± 0.0020

$*P < 0.05$; $**P < 0.01$; $***P < 0.001$.

TABLE 9.7
Main Physical and Chemical Parameters of the Control Soil and Carpiano Soil

Parameter	Control Soil	Carpiano Soil
Sand (%)	36.8	77.9
Silt (%)	35.5	17.4
Clay (%)	27.6	4.60
pH (H_2O)	7.3	6.3
EC (dS m^{-1})	0.13	0.48
CEC (cmol(+)/kg)	17.2	14.6
P assimilable (mg kg^{-1})	40	78
Total N (g kg^{-1})	30	3.70
Organic C (g kg^{-1})	17	40

TABLE 9.8
Concentration of Cd, Cu, Pb, and Zn in Control Soil, Carpiano Soil, and the Soil/Sand Mixture

Element	Control Soil (mg kg⁻¹)	Carpiano Soil (mg kg⁻¹)	Soil/Sand (1:1) w/w) (mg kg⁻¹)	Decree 152/06 (mg kg⁻¹)[a]	Decree 152/06 (mg kg⁻¹)[b]
Cd	3.51	70.9	25.6	2	15
Cu	37.9	549	192	120	600
Pb	37.9	1,517	589	100	1,000
Zn	66.1	12,670	4,451	150	1,500

Note: For reference, the metal threshold limits as assigned by Italian legislation (D.Lgs. 152/06) are recorded.
[a] Residential soils.
[b] Industrial soils.

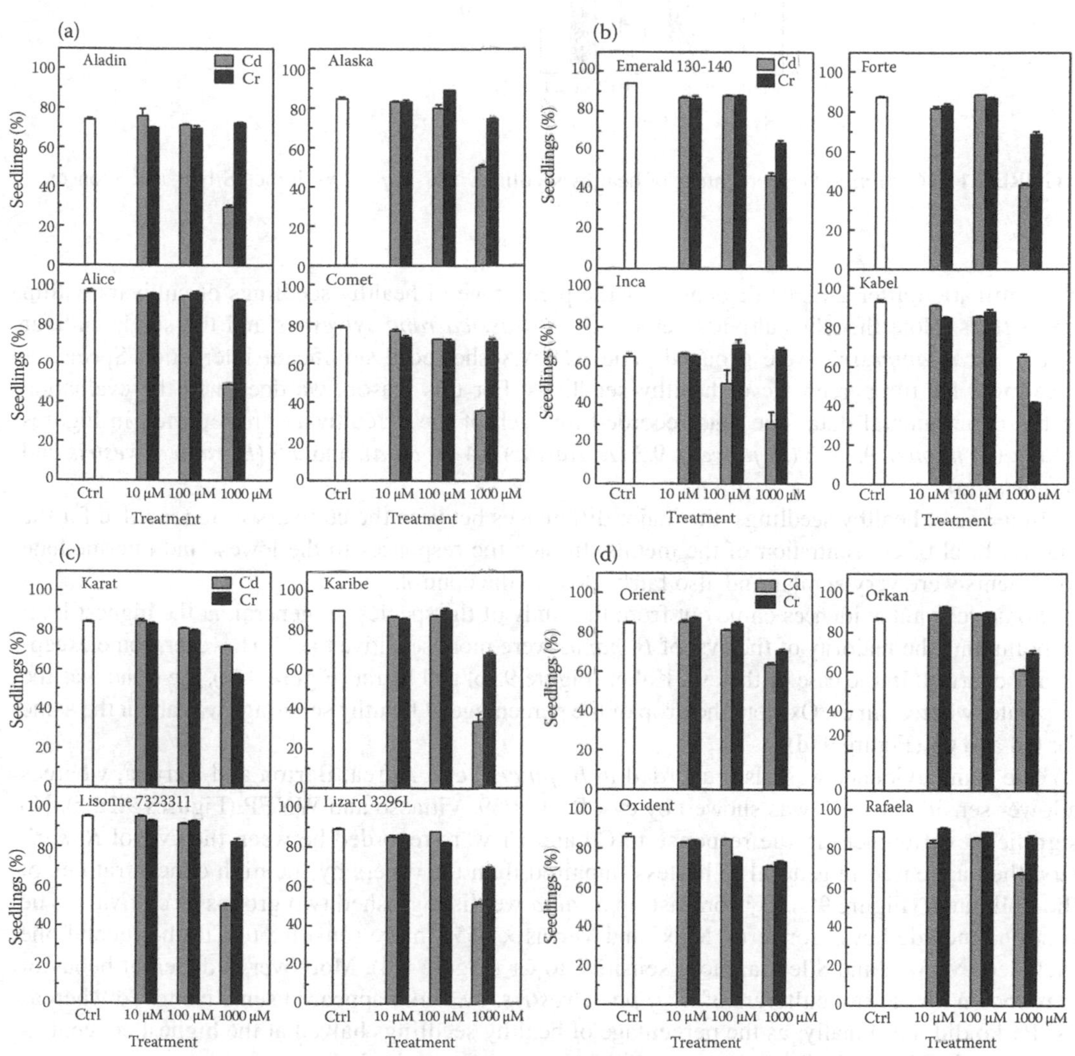

FIGURE 9.1 (a–d) Percentage of healthy seedlings of (a) *B. napus* cvs. Aladin, Alaska, Alice, and Comet; (b) *B. napus* cvs. Emerald 130-040, Forte, Inca, and Kabel; (c) *B. napus* cvs. Karat, Karibe, Lisonne 7323311, and Lizard 3296L; and (d) *B. napus* cvs. Orient, Orkan, Oxident, and Rafaela.

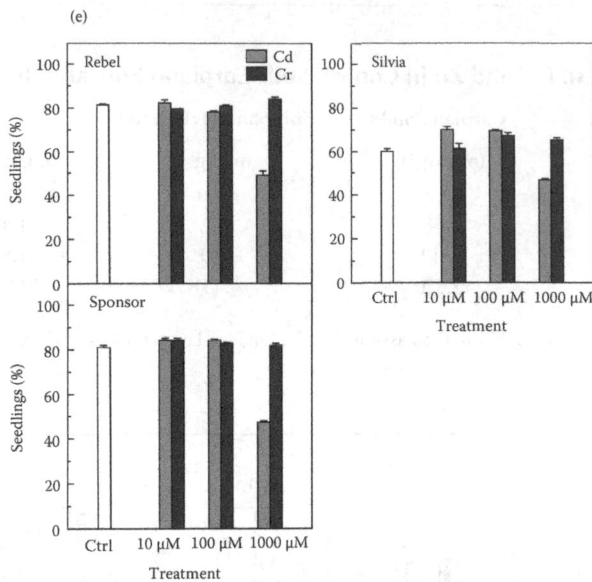

FIGURE 9.1 (*Continued*) (e) Percentage of healthy seedlings of *B. napus* cvs. Rebel, Silvia, and Sponsor.

of germination after 2 and 4 days and for the percentage of healthy seedlings of cultivars within the species. Note that the cultivars belonging to *Brassica rapa sylvestris* and the single cultivar of *Brassica campestris* were grouped. The ANOVA showed a significant interaction Species × Treatment for the percentage of healthy seedlings. For this reason, we deepened the evaluation of the experimental data. The data recorded for each of the 41 cultivars are reported in Figures 9.1a–e (*B. napus*), 9.2a–b (*B. juncea*), 9.3 (*R. sativus*) 9.4 (*S. alba*), and 9.5 (*B. rapa sylvestris* and *B. carinata*).

In terms of healthy seedlings, the major differences between the cultivars were recorded for the highest level of concentration of the metals. In fact, the responses to the lowest and intermediate treatments were very similar and also fairly close to the control.

Some relevant evidences came out from the study of the species. In general, at the highest level of pollutants the majority of the cvs. of *B. napus* were more sensitive to Cd. However, some exceptions occurred. In the case of the cvs. Kabel (Figure 9.1b) and Karat (Figure 9.1c), the trend was the opposite, whereas in cv. Oxident the drop of the percentage of healthy seedlings was about the same for Cd and Cr (Figure 9.1d).

The same evidence was also recorded in *B. juncea*, cvs. Aurea, Barton and Varuna, whereas a lower sensitivity to Cr was showed by cvs. R-3104/59, Vittasso and WNFP (Figures 9.2a,b). No significant differences in the response to Cd and Cr were recorded between the cvs. of *R. sativus*; they appeared in general to be less impaired than the others by the high concentrations of the pollutants (Figure 9.3). By contrast, in *S. alba* we distinguished two groups of cultivars. The first one included cvs. Concerta, Maxi and Vertus × 6556, more sensitive to Cd, the second one included cv. Asta and Silenda, more sensitive to Cr (Figure 9.4). Moreover, a different behavior was shown by the two cultivars of *B. rapa sylvestris*: cv. Buko appeared sensitive to Cd whereas cv. Perko did not. Finally, as the percentage of healthy seedlings halved at the highest concentration of both Cd and Cr, *B. carinata*, confirmed its lower metal tolerance among *Brassica* species tested (Figure 9.5).

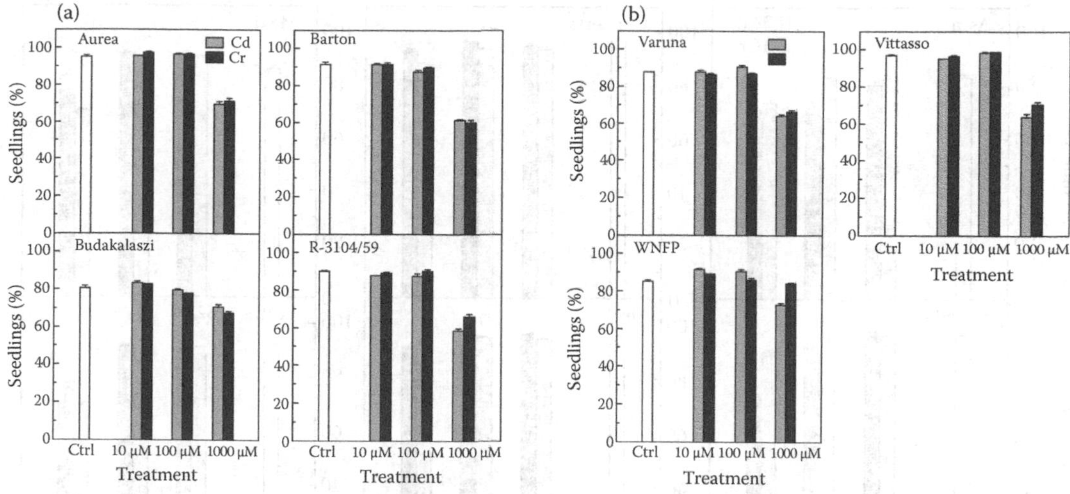

FIGURE 9.2 (a–b) Percentage of healthy seedlings of (a) *B. juncea* cvs. Aurea, Barton, Budakalaszi, and R-3104/59; (b) *B. juncea* cvs. Varuna, Vittasso, and WNFP.

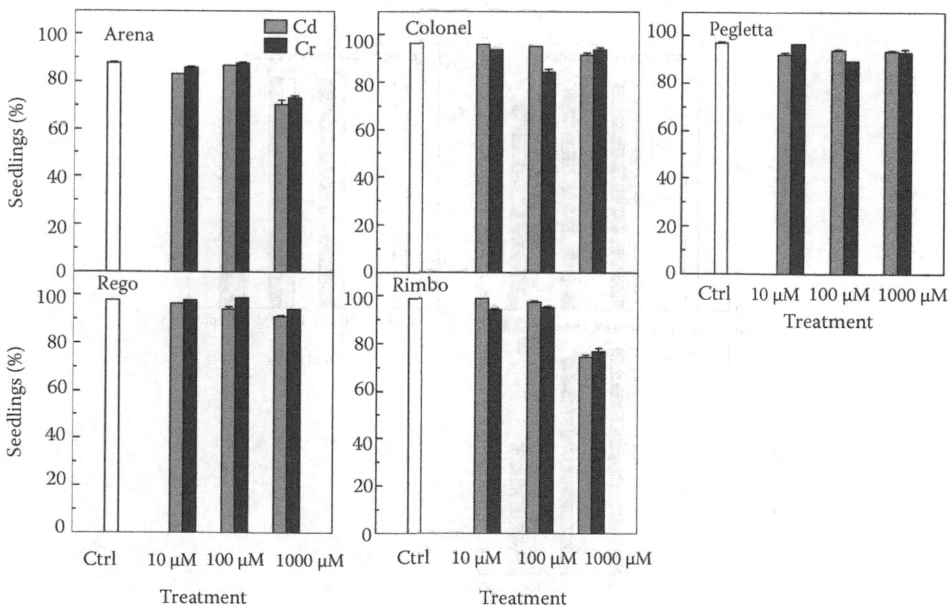

FIGURE 9.3 Percentage of healthy seedlings of *R. sativus* cvs. Arena, Colonel, Pegletta, Rego, and Rimbo.

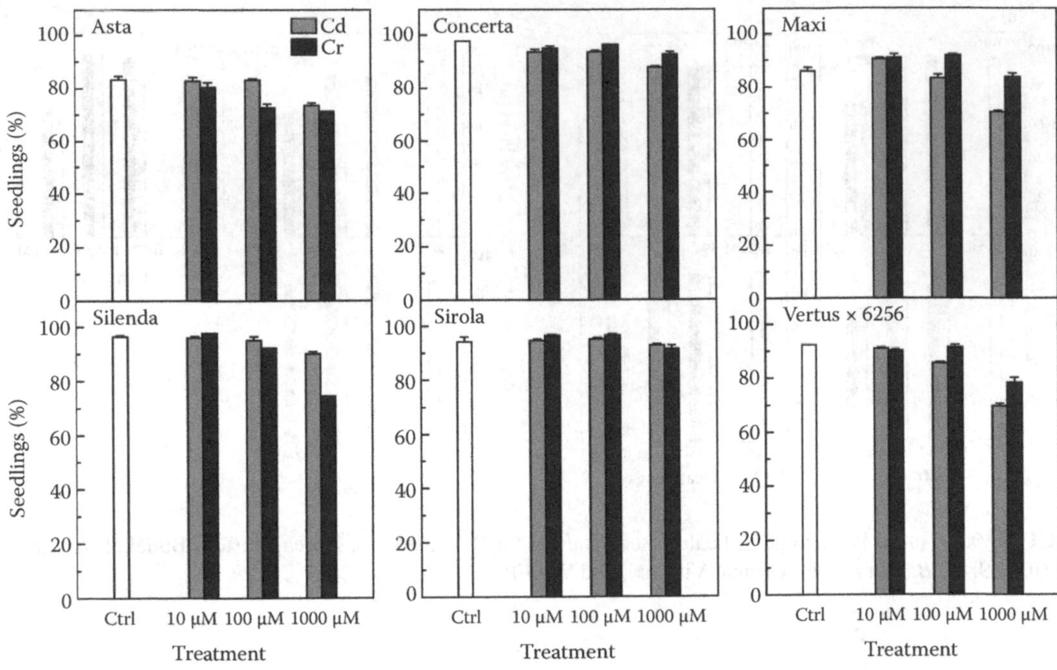

FIGURE 9.4 Percentage of healthy seedlings of *S. alba* cvs. Asta, Concerta, Maxi, Silenda, Sirola, and Vertus × 6256.

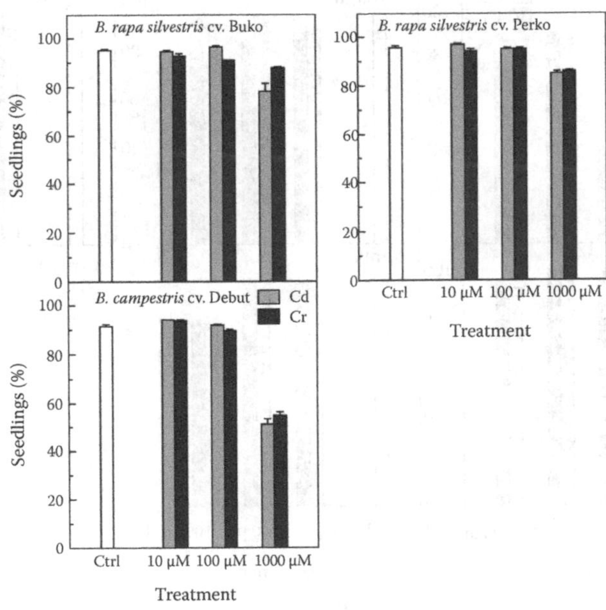

FIGURE 9.5 Percentage of healthy seedlings of *B. rapa sylvestris* cvs. Buko and Perko and *B. campestris* cv. Debut.

9.4.1.4 Practical Implications

In this investigation, we observed the germination and the seedlings' health of 41 cultivars belonging to the *Brassica* family. Even collecting very simple parameters, such as the percentage of germination, useful results can be obtained. We used this experiment as screening in the search of tolerant specie. As regards the issue of direct seeding in phytoremediation projects, it implies that germinating seeds and young seedlings are exposed to the toxic elements from the earliest phases of their growth. A number of experiences have been reported in literature describing hydroponic experiments in which the authors observed seedlings or adult hyperaccumulator or nonhyperaccumulator plants growing in nutrient solution containing heavy metals; but plants were always germinated in absence of metals which were added during the growth cycle of plants. For this reason we made this screening. Some interesting results emerged, confirming the potential for phytotechnologies of several crops belonging to the *Brassica* family.

In conclusion, it seems useful to test the high biomass species, potentially suitable for phytoremediation, for their performance in response to heavy metals since seed germination. This will provide information to the discussion, currently in progress, around the practical aspects of phytoremediation.

9.4.2 Uptake of Cd and Cr by Cultivars of *Brassica juncea* Grown in Hydroponic Conditions

To make phytoremediation a technically viable option for large scale applications, we need plants that are able to guarantee high biomass yield as well as high accumulation of heavy metals in the aerial part of plants. Unfortunately, hyperaccumulator species are not suitable due to their small biomass and slow growth. Alternatively it has been suggested the use of high biomass species.

Brassica juncea was reported as one of the most promising model plants for phytoremediation, as it accumulates more than 400 mg kg^{-1} dry weight Cd in its shoots (Minglin et al. 2005). Moreover, *B. juncea* is fast growing, requires less water, has a relatively high rate of biomass production (at least 10 times higher than other hyperaccumulators), and can also tolerate and accumulate high concentrations of potentially toxic trace elements (Zhu et al. 1999). Even if it is not an hyperaccumulator, *B. juncea* has demonstrated a high tolerance to several heavy metals (Ebbs and Kochian 1998; Jiang et al. 2000; Hammer and Keller 2002). Moreover Dushenkov et al. (1995) reported *B. juncea* as particularly effective in sorbing divalent cations of toxic metals from soil solution. Salt et al. (1997) demonstrated the ability of *B. juncea* to accumulate Cd from hydroponic solution into the above-ground biomass, if compared to other crops. So far, the potential of plants for phytoremediation of Cr has been less explored than Cd. Shahandeh and Hossner (2000a, 2000b), screened a series of crops and observed the effects of organic acids for enhancing Cr uptake from polluted soil; Mei et al. (2002) provided data on Cr^{3+} and Cr^{6+} tolerance and uptake of *B. juncea* grown in nutrient solution.

After the screening phase based on resistance to contaminants during germination, we organized a subsequent experiment to observe the effects of contaminants on growth. Taking into account that among the species studied, *B. juncea* is the one that gave the most interesting results, seedlings of three genotypes were grown hydroponically in the presence of different concentrations of Cd and Cr. The biomass growth response of each genotype and the metal accumulation were investigated.

9.4.2.1 Hydroponic Trial

As the objective of our study was to growth plants in the presence of Cd and Cr since seed germination, preliminary tests were performed in order to find such nontoxic levels of the elements that we have considered.

Cadmium and Cr concentrations in the interval 0.05–100 μM were tested. At high values of metal concentration seed germination took place but then plants showed toxicity symptoms such as purple and necrotic spots on primary leaves and thereafter they died. Such preliminary tests indicated 0.5 μM, 1 μM and 1.5 μM as nontoxic concentration of Cd and Cr for seed germination and the normal growth of the seedlings. Seeds of *Brassica juncea* cvs. Barton, Varuna and WNFP, supplied by *Brassica* collection of the University of Padova (Italy), were germinated and grown for 7 days in small pots with 4 cm deep sand irrigated with a modified 1:10 *Hoagland* solution (Brown et al. 1995). The nutrient solution was the same already used in the previous experiment. Likewise, Cd and Cr were provided respectively as cadmium nitrate ($Cd(NO_3)_2$ and potassium dichromate $K_2Cr_2O_7$, reaching for both elements 0.5, 1 and 1.5 μM in the nutrient solution, corresponding respectively to 0.056, 0.112, and 0.168 mg L^{-1} for Cd, and 0.052, 0.104, and 0.156 mg L^{-1} for Cr.

Seven days after germination the seedlings had fully expanded primary leaves and were transferred to 2.5 L plastic beakers containing aerated nutrient solution. The plants were grown for 21 days on a laboratory bench lit by lamps giving 500 μmols m^{-2} s^{-1} of photosynthetically active radiation (PAR) at the plant top with a 12:12 hours photoperiod; the pots were randomly rotated daily to equalize their light exposure. The ambient temperature was maintained at 22 ± 2°C. In the course of the first week of hydroponic growth, the plants were provided with the same nutrient solution (modified *Hoagland* 1:10). In the following weeks a modified *Hoagland* 1:5 was used. The nutritive solutions were changed every 3 days in order to compensate for water lost through evaporation and plant transpiration.

At the end of the experiment, the total area of the leaves in control and treated plants were measured by means of a leaf area meter (LiCor-3000). Leaves, stems and roots of five plants per treatment were collected having particular care for the root apparatus; fractions were fresh weighed and oven dried (105°C, 24 h) to determine their dry matter. The plant fractions were subjected to a microwave-assisted total digestion (USEPA 1995). Total content of Cd and Cr on the plant fractions was determined by an ICP-OES.

9.4.2.2 Experimental Design and Data Analysis

The trial was arranged in a randomized design with seven treatments (Ctrl, Cd 0.5 μM, Cd 1 μM, Cd 1.5 μM, Cr 0.5 μM, Cr 1 μM, and Cr 1.5 μM) and two pots per treatment; each treatment had 10 plants. Five plants per treatment were sampled. Two–way ANOVA was performed and then Student–Newman–Keuls test ($p < 0.05$) was used for comparison between treatment means. Data are presented as means with standard error. All statistics were computed using CoHort 6.204 (CoHort Software, Monterey, CA).

The potential for phytoextraction of the plants was estimated by calculating the Bioconcentration Factor ($BF = [Me]_{shoots}/[Me]_{medium}$) and the Translocation Factor ($TF = ([Me]_{shoots}/[Me]_{roots})$.

9.4.2.3 How Did the Cultivars of *Brassica juncea* Behave?

As expected, the effects of Cd and Cr on plant biomass growth measured at the end of the experiments varied with the different concentrations of metal in the nutrient solution (Table 9.4). As shown in the table of ANOVA, the cultivars of *B. juncea* performed differently in terms of dry weight of leaves and total biomass (respectively $p = 0.003$ and 0.012). Moreover, they reacted differently to the treatment. In fact, the interaction "Cultivar x Treatment" shown in Table 9.4 was also significant.

As expected, the plants' response to the treatments follows an inverse relationship. On average, the cultivars responded to the increase of the metal concentration reducing the leaf area and the biomass weight of the fractions. The more evident depressive effect on plant growth appeared to be addressed more to Cr than Cd. After these general considerations, to assess how the cultivars behaved, it is necessary to examine more carefully the data. In Figure 9.6, the leaf area per plant and the total biomass are reported (sum of the dry weight of roots, stems and leaves). The biomass yields of control plants and those subjected to the lower and intermediate treatments (0.5 and 1 μM of Cd and Cr, respectively) were more or less the same, and the differences were due to the genetic potential of the cultivars. The cv. Barton showed a certain tolerance to the metals, in fact only in plants subjected to the highest treatment was a significant depression of both leaf area and biomass yield recorded. In the case of cv. Varuna, the plants responded negatively to the presence of Cd in the solution, but then at the increase of the concentration, the decrease of both leaf area and total biomass appeared less marked (Figure 9.6). Finally, cv. WNFP showed a progressive sensitivity to the treatments.

The accumulation of the heavy metals in the plant fractions of the cultivars of *B. juncea* follows the increase of the concentration in the nutrient solution. However, there has been a significant difference in the plants' behavior in relation to the metals in the nutrient solution. As reported in Figure 9.7, Cd appeared less absorbed by plants than Cr. In the roots exposed to 1.5 μM Cd and Cr, the correspondent range of concentration of the elements recorded in the root tissues were 393–627 and 721–770 mg kg^{-1}, respectively. The metal accumulation in the aerial parts of plants was about one order of magnitude lower than in the roots. Moreover, also relevant is the fact that while Cd, after being taken up, is mobilized to the aerial part of the plant, Cr is not. On average, the concentration of Cd is higher in the leaves than in the stems. The average concentrations recorded in the cultivars at 1.5 μM Cd were respectively 66.2 and 53.7 mg kg^{-1}. On the contrary, at the corresponding level of Cr the average concentration values in plants were 5.61 and 22 mg kg^{-1} respectively.

To evaluate the plant's potential for phytoremediation, the Bioconcentration Factor (BF) and the Translocation Factor (TF) were considered. They indicate respectively (i) the ability of a plant to take up and transport the metal/s from the medium (solid or liquid) to the aerial part, and (ii) the rate of mobilization of the metal/s within the plant tissues from the root system to the aerial biomass. The calculated values of BF and TF are reported in Table 9.6. Once again, significant differences between the cultivars ($p = 0.026$) were found, which confirms what was discussed above on the

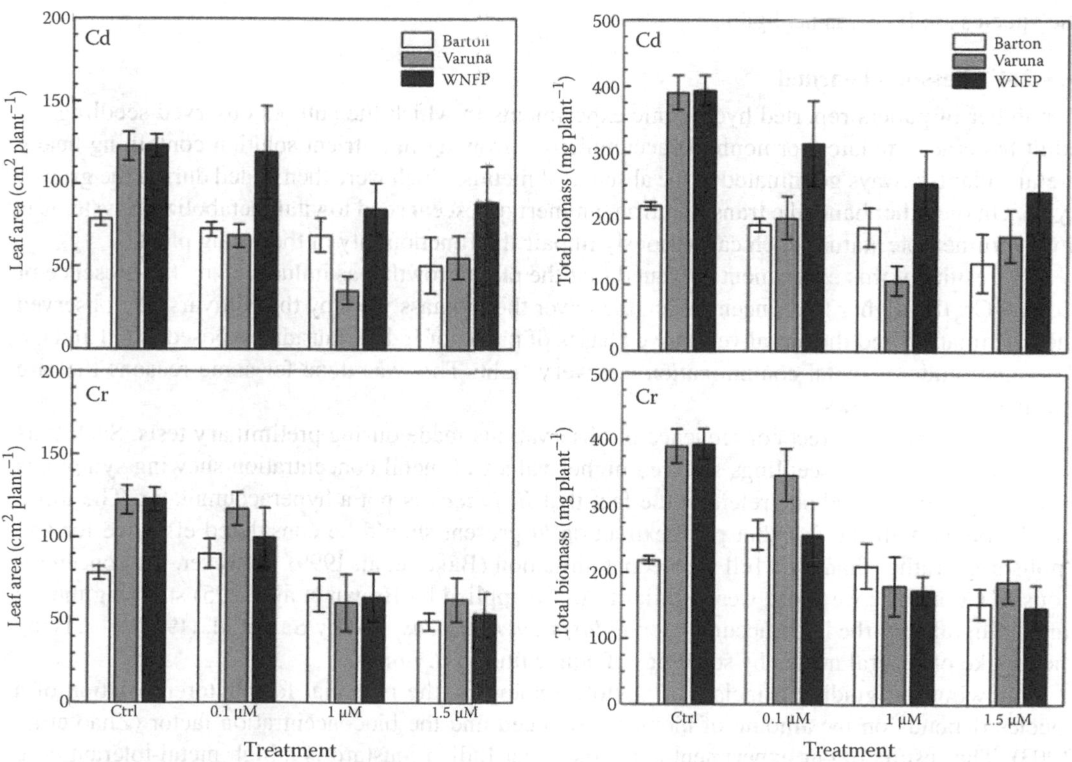

FIGURE 9.6 Leaf area and total biomass yield (mg plant^{-1}) observed in *B. juncea*, cvs. Barton, Varuna, and WNFP grown in hydroponic conditions and exposed to different concentrations of Cd and Cr.

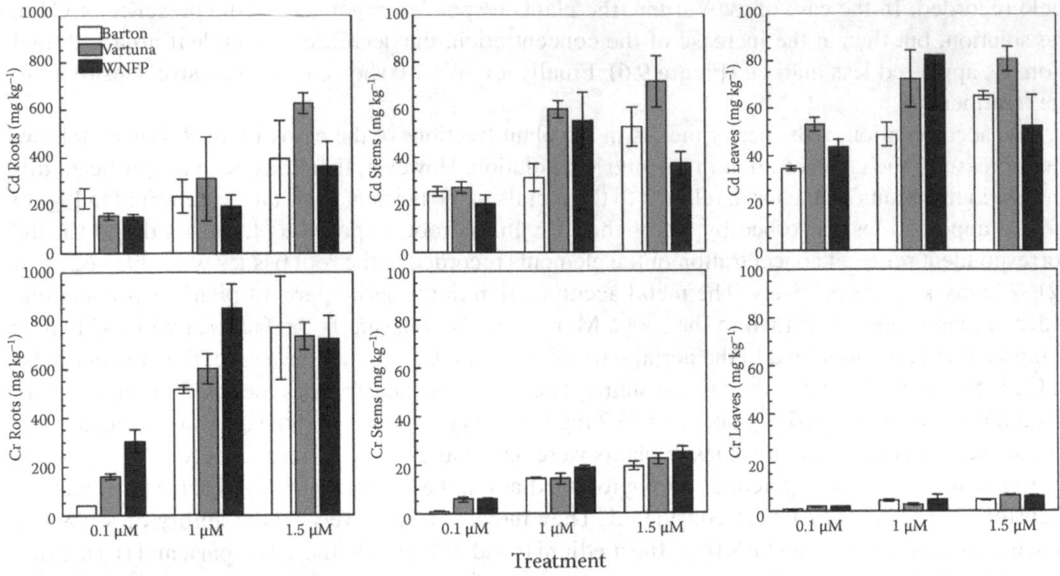

FIGURE 9.7 Cadmium and chromium concentrations (mg kg^{-1}) observed in fractions (roots, stems, and leaves) of *B. juncea*, cvs. Barton, Varuna, and WNFP.

metal concentration in the different plant fractions. The root-to-shoot translocation of Cd and Cr in the species studied was negligible.

9.4.2.4 Lessons Learned

A number of papers reported hydroponic experiments in which the authors observed seedlings or adult hyperaccumulators or nonhyperaccumulators growing in nutrient solution containing heavy metals. Plants always germinated in the absence of metals which were then added during the growth cycle. On the other hand, the transition from an inert quiescent seed to vital metabolizing systems is a very vulnerable status which can seriously impair the functionality of the young plant.

The results of this experiment indicated that the plant growth was influenced by the presence of Cd and Cr: the higher the concentration, the lower the biomass yield by the cultivars. We observed the germination and the initial vegetative phases of plants of Indian mustard exposed to Cd and Cr. The magnitude of metal contamination was very light. This was done for some reasons that are summarized as follows.

The first one is a direct consequence of observations made during preliminary tests. Such tests demonstrated that the seedlings suffered higher values of metal concentration showing symptoms of toxicity. The second one refers to the fact that *B. juncea* is not a hyperaccumulator. The third, finally, deals with the fact that phytoextraction at present should be considered effective for soil "polishing" rather than for a full site decontamination (Baker et al. 1999). However, the concentrations of Cd and Cr we chose were similar to those applied by Brown et al. (1995) studying the Zn and Cd uptake by the hyperaccumulator *Thlaspi caerulescens*, and by Salt et al. (1997) observing the uptake of several metals by seedlings of aquacultured *B. juncea*.

Following the guiding principle of phytotechnologies, the potential for phytoremediation of a species depends on the amount of biomass produced and the bioconcentration factor (Zhao et al. 2003). The results of our experiment confirmed that Indian mustard is a high metal-tolerant crop species, but we also registered poor values of translocation of metals. Hence, the high amount of Cd and Cr concentrated in the root systems did not migrate to the aerial, harvestable, part of the plant.

Therefore, we cannot consider such cultivars for phytoextraction, as they appear to be more suitable for phytostabilization.

9.4.3 Biomass Yield, Photosynthetic Activity, and Metal Uptake by *Raphanus sativus* and Different *Brassica* Species Grown on Multi-Metal Contaminated Soil

Most of the experiments on phytoextraction have been based on hydroponically grown plants; these studies permitted advances on the understanding of metal uptake by hyperaccumulator, accumulator and tolerant plants (Brown et al. 1995a; Pollard and Baker 1996; Salt et al. 1997; Haag-Kerwer et al. 1999; Wenzel et al. 1999; Lasat et al. 2000; Zhao et al. 2000). This type of experiment was also important to determine the uptake efficiency and metal tolerance of potential phytoremediation species.

Another aspect should be considered. Most of the time, anthropogenic soil pollution is not limited to a single pollutant and involves two or more metals (Förstner 1995). Anyway, a limited amount of research has been dedicated to describing the behavior of accumulating plants in the presence of multiple polluting metals, even if this is generally the most frequent case in soil restoration. These studies involve serious difficulties, as plant uptake of heavy metals is subjected to the antagonistic, additive, and synergetic effects that heavy metals exert on each other (Grifferty and Barrington 2000; Baker 2000). Practical aspects of the phytoremediation process should be investigated in existing polluted soils (Robinson 1998), as hydroponic solutions do not approximate to the complexity of field conditions (McGrath et al. 1993; Brown et al. 1994, 1995b). Tests on potted soils could mimic open field conditions, evaluating, in terms of metal phytoextraction, the effects of different agronomic practices, such as soil amendments with substances which modify soil elements' release and plant uptake ability (McGrath 1998).

This trial was undertaken to study the behavior of four *Brassica* species (*B. napus*, *B. juncea*, *R. sativus* and *B. carinata*) grown in pots on a multi-metal contaminated soil. Data on the biomass growth of these species and the basic parameters of carbon dioxide assimilation are reported. To test the potential for phytoextraction of such species the concentration of heavy metals in the plant fractions were also determined, and the bioaccumulation coefficients for each metal were calculated.

9.4.3.1 Characterization of the Polluted Soil

The contaminated soil was collected in an agricultural area at Carpiano (Milano, Italy). The source of pollution in this site had been identified in the use of irrigation water of bad quality over more than a century. The main soil parameters are reported in Table 9.7.

9.4.3.2 Preliminary Tests

Growing plants on the pure contaminated soil during some preliminary tests indicated that they performed poorly compared to control plants. Most of the seedlings were stunted, and purple necrotic spots appeared on the leaves. After a few days, most of them died. Therefore, before starting the experiment, we needed to mitigate the toxicity of the Carpiano's soil. A milder substrate was carefully prepared mixing the polluted soil with sand in 1:1 (w/w) proportion. As control, soil considered to be an agricultural soil (Typic Udifluvent) was collected from the Agricultural Experimental Farm of the University of Udine. The concentration of the metals found in the control and the polluted soil and the soil/sand mixture are reported in Table 9.8. For reference, the threshold values established by Italian legislation are reported as well.

9.4.3.3 Growing Conditions

Seeds of *B. napus* cv. Kabel, *B. juncea* cv. Vittasso, *R. sativus* cv. Rimbo and *B. carinata* cv. BRK13 were sown in 2l plastic pots containing the control soil and the soil-sand mixture. To prevent emergence failures, more seeds were sown in each pot than the number of plants required for the experiment. After the appearance of the first pair of true leaves, the seedlings were thinned out leaving ten plants in each pot. Each pot was irrigated twice a week during the first month and then every two days alternating distilled water and a modified diluted (1:5) Hoagland solution (Brown et al. 1995) to supply mineral nutrients. The plants were grown for 60 days at 22 ± 2°C on a laboratory bench lit by lamps which gave the plant top a photosynthetically active radiation of 500 μmols m^{-2} s^{-1} (12:12-hour photoperiod).

9.4.3.4 Photosynthesis

The measurements of photosynthesis were performed at three different CO_2 air concentrations (100, 400, and 800 ppm respectively). Data on CO_2 assimilation (A) and transpiration (T) were achieved by means of a Li-6400 (LiCor) photosynthesis system.

9.4.3.5 Further Observations

On fully expanded leaves, a measure of relative greenness, proportional to leaf chlorophyll content, was performed by means of SPAD measurements. At the end of the experiment, the total area of the leaves in control and treated plants was measured by means of a leaf area meter (LiCor-3000). Finally, the leaves, stems and roots of five plants per treatment were collected having particular care for the root apparatus; fractions were fresh weighed and oven dried to determine their dry matter. The plant fractions from treated and control pots, and soil samples collected from the same pots were mineralized with concentrated HNO_3. Total content of Cd, Cr, Cu, Ni, Pb, and Zn on the plant and soil samples was determined by an ICP-OES.

9.4.3.6 Experimental Design and Data Analysis

The trial was arranged in a randomized design with two treatments (control and treated) and four pots per treatment, each pot containing 10 plants. Continuous gas exchange measurements were recorded at each CO_2 level once the steady state was reached.

The potential for phytoextraction of the plants was estimated by calculating

(i) The biological accumulation coefficient (BAC = $[Me]_{harvestable\ biomass}/[Me]_{soil}$)
(ii) The bioconcentration factor (BF = $([Me]_{roots}/[Me]_{soil})$
(iii) The translocation factor (TF = $([Me]_{shoots}/[Me]_{roots})$)

9.4.3.7 Plant Growth

During the experiment no remarkable symptoms of metal toxicity were observed. However, the plants' growth on the polluted soil was less than the control ones and the leaves had a paler green color. Table 9.9 reports the data collected by measuring some biometric parameters indicating the plants' performance in terms of biomass growth. The statistics, reported in Table 9.9 as well, demonstrated that the polluted soil had a significant negative effect on plant growth. The plants growing on the polluted substrate showed about 29% less leaf area than that of the controls. In addition to the reduction of the leaf area and the corresponding decrease in leaf biomass, it is likely that negative consequences occurred also at metabolic level. This perhaps explains the statistically significant lower SPAD value of treated plants compared to the control ones. With regard to the species, *B. carinata* were less affected in terms of SPAD; on the opposite side, *B. juncea* and *R. sativus* were the most sensitive (Table 9.9).

The growth of the root apparatus was seriously affected by the presence of pollutants; its growth was about halved in the polluted soil. The Species × Treatment interaction resulted significant at the ANOVA as the species behaved differently. In particular, *B. carinata* was the most sensitive and negatively affected by the soil pollution. Finally, the biomass of the leaves was also negatively affected by the soil pollution. This had strong implications on the quality and the functionality of

TABLE 9.9
p* Values after ANOVA and Mean Values of Leaf Area, Dry Matter Weight of Plant Fraction, and SPAD Index Recorded for *B. napus, B. juncea, R. sativus,* and *B. carinata

ANOVA	Leaf Area	Dry Matter				SPAD Index
		Roots	Stems/Petioles	Leaves	Total	
Main Effects						
Treatment	.0001 ***	.0000 ***	.0000 ***	.0015 **	.0000 ***	0000 ***
Species	.0078 **	.0000 ***	.2404 ns	.0383 *	.0687 ns	0000 ***
Interaction						
Species × Treatment	.0006 **	.0034 ***	.0664 ns	.0179 *	.0996 ns	.7284 ns
Post hoc test	(cm^2 $plant^{-1}$)	(g $plant^{-1}$)	(g $plant^{-1}$)	(g $plant^{-1}$)	(g $plant^{-1}$)	(adimensional)
Treatment						
Control soil	177 a	0.144 a	0.317 a	0.434 a	0.898 a	38.1 a
Polluted soil	126 b	0.074 b	0.203 b	0.333 b	0.610 b	33 b
Species						
B. napus	140 b	0.143 a	0.299 a	0.456 a	0.898 a	39.1 b
B. juncea	138 b	0.114 a	0.233 a	0.357 ab	0.704 a	29.1 c
R. sativus	141 b	0.125 a	0.26 a	0.334 b	0.719 a	30.1 c
B. carinata	188 a	0.057 b	0.248 a	0.393 ab	0.697 a	43.9 a

Note: Letters indicated the results of the *post hoc* means comparison test (Student–Newman–Keuls, $p = 0.05$).
$*p < 0.001$; $**p < 0.01$; $***p < 0.001$.

the photosynthetic apparatus. Among the species, some differences were recorded, although not relevant. Perhaps they were due more to the species' *habitus* than the environmental conditions.

9.4.3.8 Photosynthesis

Differences were evidenced in the CO_2 assimilation by the species that grew on the polluted soil (Figure 9.8). The average values of A (μmol CO_2 m^2 s^{-1}) for treated *B. napus* and *R. sativus* were not significantly different than the control ones. On the contrary, a reduction of respectively 20.2 and 25.3% resulted in *B. juncea* and *B. carinata*. A statistically significant reduction of about the same magnitude in CO_2 assimilation (21.2 and 31.3% respectively) was recorded at 100 ppm, as well.

The WUE, that is the ratio between the carbon dioxide absorbed and the water transpired per leaf surface, contains the variability of two experimental parameters. In this case, *B. carinata* was negatively affected by the polluted soil. A WUE reduction of respectively 42.2, 42.6, and 45.5% was recorded (Figure 9.8). The detrimental effects of multi-contaminated growth media on the plant physiology were studied. It is known that heavy metals affect photosynthesis (Krupa and Baszynsky 1995). This is due to the substitution of Mg in chlorophyll by Hg, Cu, Cd, Ni, Zn, and Pb that leads to a breakdown in photosynthesis (Kupper et al. 1998). Actually, the concomitant presence of metals had a severe influence on the plant growth. The phytoextraction efficiency was reduced as a consequence of the additive effects on the plant metabolism.

9.4.3.9 Metals in Plant Tissues

Figure 9.9 shows the metal concentration in the roots and in the aboveground biomass of *Brassica* species. The species showed a different ability in the uptake, metal accumulation and translocation. Relevant differences between the root and shoots metal concentration were recorded. Hence, a logarithmic scale

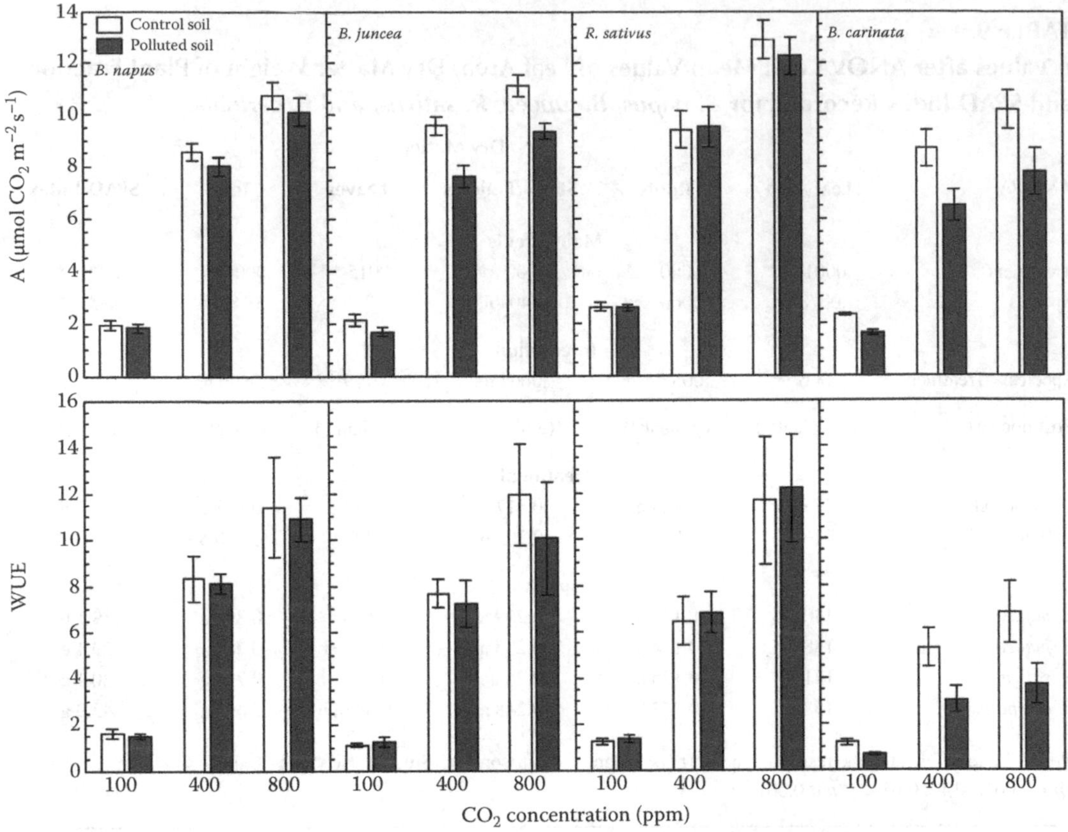

FIGURE 9.8 Photosynthetic activity (A) and water use efficiency (WUE) of *B. napus*, *B. juncea*, *R. sativus*, and *B. carinata* recorded at 100, 400, and 800 ppm of CO_2 concentration. Vertical bars for each box represent the standard error.

was used. The Cu and Pb concentrations in the aerial fractions of the plants were about two orders of magnitude lower than the roots. For these elements the root-to-shoot translocation was negligible (Figure 9.9). In the case of Cd, the difference was about one order of magnitude. Differently to other metals, Zn was accumulated also in the shoots. The range of Zn concentration recorded in roots and shoots were 4,418–1,527 and 969–1,816 mg kg^{-1}, respectively. The most and the least efficient species in accumulating metals in the plant fractions were *R. sativus* and *B. napus*, respectively (Figure 9.9). Among the crops belonging to the *Brassica* family, canola and radish have long been suggested for phytoremediation (Ebbs et al. 1997; Ebbs and Kochian 1997; Bañuelos et al. 1998).

The potential for phytoremediation of the *Brassica* species observed in our experiment was evaluated by calculating the BAC, BF and TF. As a general rule, the plant species with a BAC > 1 are suitable for phytoextraction; those with a BF > 1 and TF < 1 can be considered for phytostabilization (Yoon et al. 2006). The values of such parameters are reported in Table 9.10. In the multi-metal contaminated soil used in our experiment, the *Brassica* species showed a low potential for metal phytoextraction, as suggested by the values of the BAC. A BF > 1 resulted for Cu in *B. juncea*, *R. sativus* and *B. carinata*, whereas BF values of 0.93 and 0.99 were recorded for Zn in *R. sativus* and *B. carinata*, respectively. The TF values were always <1 (Table 9.10). As a general rule, we could consider our species suitable for phytostabilization in soils polluted by Cu and Zn, with the only exception of *B. napus*. These conclusions cannot be generalized. We used a soil contaminated by several metals, therefore we can conclude that in our conditions the *Brassica* species succeeded

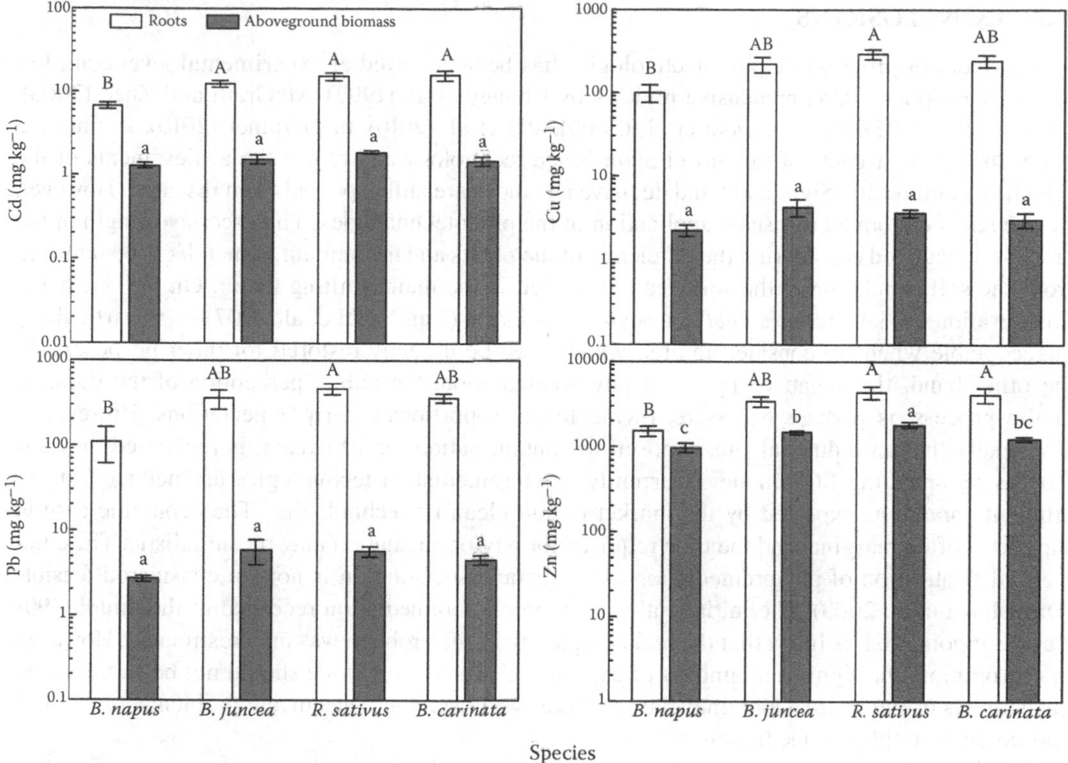

FIGURE 9.9 Concentrations (mg kg^{-1}) of Cd, Cu, Pb, and Zn in the roots and the above-ground biomass of *B. napus*, *B. juncea*, *R. sativus*, and *B. carinata*. Capital and small letters indicated the results of the *post hoc* means comparison test (Student–Newman–Keuls, $p = 0.05$) between the species.

in surviving and extracting metals. They could possibly be employed in slightly polluted soils where their growth would not be impaired and the extraction could be maintained at satisfying levels. Perhaps significant improvements of the efficiency of metal phytoextraction could be obtained to providethe crops with adequate fertilization and amendments in order to modify the bioavailability of metals and their plant uptake (Kaiser et al. 2000; Chaney et al. 2007).

TABLE 9.10
Response of the ANOVA (*p* Value) on the Biological Accumulation Coefficient (BAC), Bioconcentration Factor (BF), and Translocation Factor (TF) and *post hoc* Means Comparison Test (Student–Newman–Keuls, *p* = 0.05) between the Species

ANOVA	BAC Cd	BAC Cu	BAC Pb	BAC Zn	BF Cd	BF Cu	BF Pb	BF Zn	TF Cd	TF Cu	TF Pb	TF Zn
Species	.1083 ns	.2182 ns	.2177 ns	.0022 **	.0089 **	.0478 *	.0331 *	.0359 *	.005 **	.086 ns	.0631 ns	.0866 ns
						post hoc Test						
B. napus	0.05 a	0.011 a	0.005 a	0.217 c	0.26 b	0.52 b	0.19 b	0.34 b	0.194 a	0.025 a	0.033 a	0.66 a
B. juncea	0.061 a	0.022 a	0.010 a	0.333 b	0.49 a	1.16 ab	0.62 ab	0.76 ab	0.124 b	0.020 a	0.017 a	0.448 a
R. sativus	0.073 a	0.019 a	0.010 a	0.408 a	0.60 a	1.53 a	0.80 a	0.99 a	0.124 b	0.012 a	0.013 a	0.442 a
B. carinata	0.057 a	0.016 a	0.008 a	0.283 bc	0.61 a	1.26 ab	0.62 ab	0.93 a	0.093 b	0.012 a	0.012 a	0.321 a

$^{*}P < 0.05$; $^{**}P < 0.01$; $^{***}P < 0.001$.

9.5 CONCLUSIONS

A great deal of progress on phytotechnologies has been achieved at experimental level considering several options. Comprehensive reviews by Chaney et al. (1997), McGrath and Zhao (2003), Pilon-Smits (2005), Vangronsveld et al. (2009), Wu et al. (2010) and Krämer (2010), summarize many important aspects of this novel plant-based technology and report the achievements of the scientific community. Significant and decisive advances are still expected from research. However, we are still far from a large scale application of the phytotechnologies. The excessive length of the process, calculated considering the efficiency of the plants and the amount of metal/s to be removed from the soil in relation to the soil clean up target, is the main limiting factor. On the one hand, this duration — the "*Achilles heel*" of phytoremediation (Van Nevel et al. 2007) — is particularly unacceptable when we consider an area which must be urgently restored for other purposes. On the other hand, the negative "hype" of phytoremediation, the public perception of the duration of the process, is perhaps excessive, owing to disproportionate early expectations. However, it is unlikely that an industrial site, exposed to contamination for 100 years, is reclaimed in a few months by spending little money. Currently, phytoremediation technologies are neither fast nor efficient enough as expected by the market of soil clean up technologies. The economic case to support is often marginal and the time required for phytoremediation may be unrealistic. The commercial application of phytoremediation as a practical site solution is not yet considered feasible (Onwubuya et al. 2009). The initial enthusiasm on phytoremediation recorded in the middle '90s was disappointed. It is likely that the real complexity of the problem was underestimated. However, in our opinion, the significant amount of experience gained worldwide should not be lost. Further investments of intellectual and financial resources will overcome the current problems, restoring a real potential to phytoremediation.

With regard to the expected performances of phytotechnologies, let us take into account two applications: phytoextraction and phytostabilization. From a practical point of view, perhaps the first one is the most fascinating and (theoretically) efficient between the different options offered by the phytotechnologies. The second one has been already successfully tested and it is available for extensive projects.

Despite the intensive research on phytoremediation in the last decade, another widening gap between science and practicality is based on the fact that very few field trials have been realized. So far, unrealistic field scale extrapolations from experimental data from lab and greenhouse trials have raised doubts about the feasibility of metal phytoextraction (Dickinson et al. 2009). For further development, and social and commercial acceptance, there is a clear requirement for up-to-date information on successes and failures of these technologies based on evidence from field scale projects. An inventory of the field trials performed in Europe in the years 2000–2008 indicated that 25 field trials took place in nine European countries. The phytoextraction potentials were evaluated, studying biomass species and, to a lesser extent, hyperaccumulators (SUMATECS 2009). Vamerali et al. (2010) surveyed the literature over the period 1995–2009. They found that the most frequently cited crop-related species were *Brassica juncea* (L.) Czern., followed by *Helianthus annuus* L., *Brassica napus* L., and *Zea mays* L. As regard to the families, *Brassicaceae* (*Raphanus sativus* L., *Brassica carinata* A. Braun), and *Poaceae* (*Festuca* spp., *Lolium* spp., *Hordeum vulgare* L.) were again represented. A few citations considered *Fabaceae* (*Glycine max* L., *Phaseolus vulgaris* L., and *Medicago sativa* L.).

The great interest in *Brassicaceae*, apart from their relatively high ability to bioconcentrate some pollutants, derives from the fact that research on these species started earlier than others, especially for *Brassica juncea* (L.) Czern. On the other hand, plants producing oil or biomass for biofuel, essential oils, fibers and quality hardwood can provide a financial return and are currently considered as the most suited for projects of phytoextraction, particularly when the soil pollution is not extreme (Puschenreiter et al. 2005; Hartley et al. 2009; Schröder et al. 2008; Dickinson et al. 2009; Vangronsveld et al. 2009; Witters et al. 2009; Meers et al. 2010).

ACKNOWLEDGMENTS

The works presented were supported by the Italian Ministry of Research and University (PRIN 1998, PRIN 2003, PRIN 2005). The authors are grateful to Patrizia Zaccheo, University of Milan (Italy) for providing the soil of Carpiano and its chemical characterization. The helpful contribution of Diego Chiabà, Silvia Assolari and Matteo Paladini is also acknowledged.

REFERENCES

Baker, A.J.M., and R.R. Brooks 1989. Terrestrial higher plants which hyperaccumulate metallic elements. A review of their distribution, ecology and phytochemistry. *Biorecovery* 1: 81–126.

Baker, A.J.M. 2000. Phytoremediation: a developing technology for the remediation and decontamination of metal-polluted soils and effluents. In *Atti del XVII Convegno Nazionale della Società Italiana di Chimica Agraria,* ed. A.A.M. Del Re, P. Fusi, R. Izzo, P. Nannipieri, F. Navari-Izzo, R. Pinton, M. Trevisan, Z. Varanini 3–7.

Broadley, M., M.J. Willey, J.C. Wilkins, A.J.M. Baker, A. Mead, and P.J. White. 2001. Phylogenetic variation in heavy metal accumulation in angiosperms. *New Phytologist* 152: 9–27.

Brown, S.L., R.L. Chaney, J.S. Angle, and A.J.M. Baker. 1994. Phytoremediation potential of *Thlaspi caerulescens* and bladder campion for zinc- and cadmium-contaminated soil. *Journal of Environmental Quality* 23: 1151–1157.

Brown, S.L., R.L. Chaney, J.S. Angle, and A.J.M. Baker. 1995a. Zinc and cadmium uptake by hyperaccumulator *Thlaspi caerulescens* grown in nutrient solution. *Soil Science Society of American Journal* 59: 125–133.

Brown, S.L., R.L. Chaney, J.S. Angle, and A.J.M. Baker. 1995b. Zinc and cadmium uptake by hyperaccumulator *Thlaspi caerulescens* and metal tolerant *Silene vulgaris* grown on sludge-amended soil. *Environmental Science and Technology* 29: 1581–1585.

Chaney, R.L., M. Malik, Y.M. Li, S.L. Brown, E.P. Brewer, J.S. Angle, A.J.M. Baker. 1997. Phytoremediation of soil metals. *Current Opinion in Biotechnology* 8: 279–284.

Chaney, R.L., J.S. Angle, C.L. Broadhurst, C.A. Peters, R.V. Tappero, and D.L. Sparks. 2007. Improved understanding of hyperaccumulation yields commercial phytoextraction and phytomining technologies. *Journal of Environmental Quality* 36: 1429–1443.

Dickinson, N.M., A.J.M. Baker, A. Doronila, S. Laidlaw, and R.D. Reeves. 2009. Phytoremediation of inorganics: realism and synergies. *International Journal of Phytoremediation* 11: 97–114.

Diwan, H., A. Ahmad, and M. Iqbal. 2008. Genotypic variation in the phytoremediation potential of Indian mustard for chromium. *Environmental Management* 41: 734–741.

Dushenkov, S., D.D. Vasudev, Y. Kapulnik, D. Gleba, D. Fleisher, K.C. Ting, and B. Ensley. 1997. Removal of uranium from water using terrestrial plants. *Environmental Science and Technology* 31: 3468–3474.

Ebbs, S.D., and L.V. Kochian. 1997. Toxicity of zinc and cadmium to *Brassica* species: implications for phytoremediation. *Journal of Environmental Quality* 26: 776–781.

Ebbs, S.D., and L.V. Kochian. 1998. Phytoextraction of zinc by oat (*Avena sativa*), barley (*Hordeum vulgare*) and Indian mustard (*Brassica juncea*). *Environmental Science and Technology* 32: 802–806.

Förstner, U. 1995. Land contamination by metals: global scope and magnitude of problem. In: *Metal Speciation and Contamination of Soil*, ed. H.E. Allen, C.P. Huang, G.W. Bailey, A.R, Bowers, 1–33. Boca Raton, FL: CRC Press.

Gisbert, C., R. Clemente, J. Navarro-Aviñó, C. Baixauli, A. Giner, R. Serrano, D.J. Walker, and M.P. Bernal. 2006. Tolerance and accumulation of heavy metals by *Brassicaceae* species grown in contaminated soils from Mediterranean regions of Spain. *Environmental and Experimental Botany* 56: 19–27.

Grifferty, A., and S. Barrington. 2000. Zinc uptake by young wheat plants under two transpiration regimes. *Journal of Environmental Quality* 29: 443–446.

Guerinot, M.L., and D.E. Salt. 2001. Fortified foods and phytoremediation. Two sides of the same coin. *Plant Physiology* 125: 164–167.

Haag-Kerwer, A., H.J. Schäfer, S. Heis, C. Walter, and T. Rausch. 1999. Cadmium exposure in *Brassica juncea* causes a decline in transpiration rate and leaf expansion without effect on photosynthesis. *Journal of Experimental Botany* 341: 1827–1835.

Hartley, W., N.M. Dickinson, P. Riby, and N.W. Lepp. 2009. Arsenic mobility in brownfield soils amended with green waste compost or biochar and planted with *Miscanthus*. *Environmental Pollution* 157: 2654–2662.

Hammer, D., and C. Keller. 2002. Changes in the rhizosphere of metal-accumulating plants evidenced by chemical extractants. *Journal of Environmental Quality* 31: 1561–1569.

Interstate Technology & Regulatory Council (ITRC). 2009. *Phytotechnology Technical and Regulatory Guidance and Decision Trees,* Revised. PHYTO-3. Washington, DC: Interstate Technology and Regulatory Council, Phytotechnologies Team, Tech Reg Update.

Jiang, W., D. Liu, and W. Hou. 2000. Hyperaccumulation of lead by roots, hypocotyls, and shoots of *Brassica juncea. Biologia Plantarum* 43: 603–606.

Krämer, U. 2010. Metal hyperaccumulation in plants. *Annual Review of Plant Biology* 61: 517–534.

Krupa, Z., and T. Baszyńsky. 1995. Some aspects of heavy metal toxicity towards photosynthetic apparatus—direct and indirect effects on light and dark reactions. *Acta Physiologia Plantarum* 17: 177–190.

Kumar, N.P.B.A., V. Dushenkov, H. Motto, and I. Raskin. 1995. Phytoextraction: the use of plants to remove heavy metals from soils. *Environmental Science and Technology* 29: 1232–1238.

Lasat, M.M., N.S. Pence, D.F. Gravin, S.D. Ebbs, and L.V. Kochian. 2000. Molecular physiology of zinc transport in the Zn hyperaccumulator *Thlaspi caerulescens. Journal of Experimental Botany* 342: 71–79.

Li, Y.M., R.L. Chaney, E. Brewer, R. Roseberg, J.S. Angle, A.J.M. Baker, R. Reeves, and J. Nelkin. 2003 Development of a technology for commercial phytoextraction of nickel: economic and technical considerations. *Plant and Soil* 249: 107–115.

Marchiol, L., P. Sacco, S. Assolari, and G. Zerbi. 2004. Reclamation of polluted soil: phytoremediation potential of crop-related *Brassica* species. *Water, Air and Soil Pollution* 158: 345–356.

McGrath, S.P., C.N. Sidoli, A.J.M. Baker, and R.D. Reeves. 1993. The potential for the use of metal-accumulating plants for the *in situ* decontamination of metal-polluted soils. In: *Integrated Soil and Sediment Research: A Basis for Proper Protection,* ed. H.J.P. Eijsackers and T. Hamers, 673–676. Dordrecht, The Netherlands: Kluwer Academic Publishers.

McGrath, S.P. 1998. Phytoextraction for soil remediation. In: *Plants that Hyperaccumulate Heavy Metals*, ed. R.R. Brooks, 261–287. Wallington, Oxon, UK: CAB International.

McGrath, S.P., and F.J. Zhao. 2003. Phytoextraction of metals and metalloids from contaminated soils. *Current Opinion in Biotechnology* 14: 277–282.

Meers, E., A. Ruttens, M. Hopgood, E. Lesage, and F.M.G. Tack. 2005. Potential of *Brassica rapa*, *Cannabis sativa*, *Helianthus annuus* and *Zea mays* for phytoextraction of heavy metals from calcareous dredged sediment derived soils. *Chemosphere* 61: 561–572.

Meers, E., S. Van Slycken, K. Adriaensen, A. Ruttens, J. Vangronsveld, G. Du Laing et al. 2010. The use of bio-energy crops (*Zea mays*) for 'phytoattenuation' of heavy metals on moderately contaminated soils: a field experiment. *Chemosphere* 78: 35–41.

Mei, B., J.D. Puryear, and R.J. Newton. 2002. Assessment of Cr tolerance and accumulation in selected plants species. *Plant and Soil* 247: 223–231.

Mench, M., N. Lepp, V. Bert, J.-P. Schwitzguébel et al. 2010. Successes and limitations of phytotechnologies at field scale: outcomes, assessment and outlook from COST Action 859. *Journal of Soils and Sediments* 10: 1039–1070.

Minglin, L., Z. Yuxiu, and C. Tuanyao. 2005. Identification of genes up-regulated in response to Cd exposure in *Brassica juncea* L. *Gene* 363: 151–158.

Onwubuya, K., A. Cundy, M. Puschenreiter, J. Kumpiene, B. Bone, J. Greaves et al. 2009. Developing decision support tools for the selection of "gentle" remediation approaches. *Science of the Total Environment* 407: 6132–6142.

Pilon Smits, E. 2005. Phytoremediation. *Annual Review of Plant Biology* 56: 15–39.

Pollard, J.A., and A.J.M. Baker. 1996. Quantitative genetics of zinc hyperaccumulation in *Thlaspi caerulescens*. *New Phytologist* 132: 113–118.

Puschenreiter, M., O. Horak, W. Friesl, and W. Hartl. 2005. Low-cost agricultural measures to reduce heavy metal transfer into the food chain — a review. *Plant, Soil and Environment* 51: 1–11.

Robinson, B.H., M. Leblanc, D. Petit, R.R. Brooks, J.H. Kirkman, and P.E.H. Gregg. 1998. The potential of *Thlaspi caerulescens* for phytoremediation of contaminated soils. *Plant and Soil* 206: 47–56.

Salt, D.E., M. Blaylock, P.B.A.K. Nanda, V. Dushenkov, B.D. Ensley et al. 1995. Phytoremediation: a novel strategy for the removal of toxic metals from the environment using plants. *Biotechnology* 13: 468–473.

Salt, D.E., I.J. Pickering, R.C. Prince, D. Gleba, V. Dushenkov, R.D. Smith, and I. Raskin. 1997. Metal accumulation by aquacultured seedlings of Indian mustard. *Environmental Science and Technology* 31: 1636–1644.

Santibáñez, C., C. Verdugo, and R. Ginocchio. 2008. Phytostabilization of copper mine tailings with biosolids: Implications for metal uptake and productivity of *Lolium perenne*. *Science of the Total Environment* 395: 1–10.

Schröder, P., R. Herzig, B. Bojnov, A. Ruttens, E. Nehnevajova, S. Stamatiadis et al. 2008. Bioenergy to save the world — producing novel energy plants for growth on abandoned land. *Environmental Science and Pollution Research* 15: 196–204.

Shahandeh, H., and L.R. Hossner. 2000a. Plant screening for chromium phytoremediation. *International Journal of Phytoremediation* 2: 31–51.

Shahandeh, H., and L.R. Hossner. 2000b. Enhancement of Cr(III) phytoremediation. *International Journal of Phytoremediation* 2: 269–286.

SUstainable MAnagement of Trace Element Contaminated Soils (SUMATECS). 2009. Development of a decision tool system and its evaluation for practical application. Final Research Report.

United Nations Environment Programme (UNEP). 2003. Phytotechnologies: A Technical Approach in Environmental Management. *IETC Freshwater Management Series 7*. ISBN No: 92-807-2253-0.

United States Environmental Protection Agency (US EPA). 1995b. EPA Method 3052: Microwave assisted acid digestion of siliceous and organically based matrices. In: *Test Methods for Evaluating Solid Waste*, 3rd ed. Washington, DC.

United States Environmental Protection Agency (US EPA). 2008. Green Remediation: Incorporating Sustainable Environmental Practices into Remediation of Contaminated Sites. EPA 542-R-08-002.

Vamerali, T., M. Bandiera, and G. Mosca. 2010. Field crops for phytoremediation of metal-contaminated land. *Environmental Chemistry Letters* 8: 1–17.

Van Nevel, L., J. Mertens, K. Oorts, and K. Verheyen. 2007. Phytoextraction of metals from soils: how far from practice? *Environmental Pollution* 150: 34–40.

Vangronsveld, J., R. Herzig, N. Weyens, J. Boulet, K. Adriaensen, A. Ruttens, T. Thewys et al. 2009. Phytoremediation of contaminated soils and groundwater: lessons from the field. *Earth and Environmental Science* 16: 765–794.

Wenzel, W., D.C. Adriano, D. Salt, and R. Smith. 1999. Phytoremediation: a plant-microbe-based remediation system. Bioremediation of contaminated soils. *Agronomy Monograph* 37: 457–508.

Witters, N., S. van Slycken, A. Ruttens, K. Adriaensen, E. Meers, L. Meiresonne et al. 2009. Short-rotation coppice of willow for phytoremediation of a metal-contaminated agricultural area: a sustainability assessment. *BioEnergy Research* 2: 144–152.

Wu, G., H. Kang, X. Zhang, H. Shao, L. Chu, and C. Ruan. 2010. A critical review on the bio-removal of hazardous heavy metals from contaminated soils: issues, progress, eco-environmental concerns and opportunities. *Journal of Hazardous Materials* 174: 1–8.

Yoon, J., X. Cao, and O. Zhou. 2006. Accumulation of Pb, Cu, and Zn in native plants growing on a contaminated Florida site. *Science of the Total Environment* 368: 456–464.

Zhu, Y.L., E.A.H. Pilon-Smits, A.S. Tarun, S.U. Weber, L. Jouanin, and N. Terry. 1999. Cadmium tolerance and accumulation in Indian mustard is enhanced by overexpressing g-Glutamylcysteine synthetase. *Plant Physiology* 121: 1169–1177.

10 Phytoremediation Capacity of Brown- and Yellow-Seeded *Brassica carinata*

Xiang Li, Margaret Y. Gruber, Kevin Falk, and Neil Westcott

CONTENTS

10.1 INTRODUCTION: METAL POLLUTION AND BIOSORPTION

The presence of heavy metal ions in emissions, waste waters and watersheds has raised considerable concern in recent years and generated national pollutant release inventories (http://www.ec.gc.ca/inrp-npri/default.asp?lang=En&n=4A577BB9-1). Base metal smelting and refining industries produce cobalt, copper, lead, nickel and zinc. Primary processing generally produces metals from ore concentrates, while secondary processing sources include recycled post-consumer electronic components, metal parts, bars, turnings, sheets, and scrap wire. The manufacturing sector produces metal-containing electrical switches, dental amalgams, fluorescent lights and batteries, and waste incineration and metal goods discarded by consumers also contain these types of materials. Particulate matter in smog and atmospheric emissions (Table 10.1) and metal leachate after acidification of soil and water ecosystems also contribute to heavy metal pollution (reviewed by Suresh and Ravishankar 2004) (http://www.ec.gc.ca/mercure-mercury/default.asp?lang=En&n=CF513593-1).

Metal ion pollution has also generated considerable health concerns and research to find remedial solutions in recent years (Schneegurt et al. 2001; Danil de Namor 2007). Respiration and ingestion of contaminated food sources are two main pathways for many metals to enter humans and animals. Due to atmospheric transport and other forms of dispersal, the Arctic region is a major receptor of heavy metals, such as mercury, cadmium and lead, although water systems close to industrial sites and major population areas are also of concern. Organisms adapted to storing biological energy, such as those in the Arctic, may become accumulators and concentrators of pollutants, which can have broad implication for humans eating local foods (http://www.carc.org/pubs/v18no3/1.htm). For example, chromium (III) is thought to be an essential nutrient required for sugar and fat metabolism in organisms, but long-time contact causes skin allergic reactions (Anderson et al. 1997). Cancer can be caused by chromium oxidization to the more carcinogenic and mutagenic Cr (VI) by MnO_2 in the environment or by soil bacteria (Pickering et al. 2004). Similarly, chromium (Cr), mercury (Hg), cobalt (Co), copper (Cu), zinc (Zn), nickel (Ni), and manganese (Mn) have also been reported as metals toxic to human and animal health

TABLE 10.1
Percentage of Canadian Emissions Released from Base Metal Smelting Sector

		Emissions		
Substance	**Units**	**Base Metal Smelting**[a]	**Canadian Total**[b]	**% Emissions**
[c]Arsenic	tonnes	153	201	76%
Cadmium	tonnes	31	33	94%
Lead	tonnes	196	223	88%
Mercury	kg	1.700	2,949	58%
[c]Nickel	tonnes	258	475	54%
Total Particulate Matter	tonnes	10.757	523,319	2%
Sulphur Dioxide	tonnes	669,967	1,419,520	47%

Source: Data taken from http://www.ec.gc.ca/air/default.asp?lang=En&n=B06262AE-1.
Note: [a]Rubinoff Environmental, Pollution Prevention and Pollution Control Initiatives in the Base Metals Smelting and Refining Sector, February 2004. [b]Environment Canada, 2005. [c]Data from 2002. Remainder of data from 2005.

(Francesconi 2007). Moreover, lithium, a metal ion present in batteries and mood-stabilizing bipolar therapeutics, has also been documented in surface and drinking waters (Zaldivar 1980).

Chemical treatment, chelation, steam segregation, ion-exchange and sludge disposal have been proposed for the remediation of metal-contaminated water (Sims et al. 1985; Juang and Wang 2000). These disposal techniques are classified as either destructive or recovery methods. The chemical treatment of heavy metal effluent by the electroplating industry has been practised for many years, but production of additional pollutants is a drawback when treating the heavy metal with chemicals (U.S. Patent 5967965). Steam segregation and ion-exchange cannot recover metal entirely, and both methods waste a lot of energy (Biffis et al. 2000).

Biosorption is an emerging and attractive technology which involves adsorption of dissolved substances onto a biomaterial (Ahluwalia et al. 2007; Romera et al. 2006; Sekhar et al. 2004). Biosorption materials can then be processed to recover pollutants of value. In recent years, interest has risen in removing heavy metal ions from solution with agricultural byproducts (Tarley et al. 2004; Saeed et al. 2003, 2005; Sharma et al. 2007; Vegliò et al. 2003; Zafar et al. 2007). However, metals recovered from sludge mixed with garbage and some other types of bio-waste adsorbents are likely to go back into solution as organic acids (Lawson et al. 1983). Moreover, parameters such as pH, temperature and concentration affect the biosorption process (Keskinkan et al. 2004; Marin et al. 2002; Martins et al. 2004; King et al. 2007; Ofomaja et al. 2007; Waranusantigul et al. 2003; Yan et al. 2005; Mukhopadhyay et al. 2007).

10.2 YELLOW-SEEDED AND BROWN-SEEDED BRASSICACEAE AND VALUE-ADDED PRODUCT DEMAND

Commercial condiment mustards within the Brassicaceae include *Sinapis alba* L. white (or yellow) mustard, *Brassica juncea* (L.) Czern, Oriental mustard and *Brassica nigra* (L.) Koch black mustard. Some condiment mustards have seed coat mucilage, especially *S. alba*, which is of value in the food processing industry. Others such as *Brassica juncea* do not produce mucilage. Rapeseed/canola processing also produces waste hulls with meal after a crush (~5 × 10^6 metric tonnes of canola meal are produced annually in Canada) (http://www.canolacouncil.org/). Currently these products are only sold as mixed, low value protein supplement in animal feed and food or hulls are discarded. Anti-nutritional, fiber-related phenolics (proanthocyanidins and lignin) within the seed coat (Marles and Gruber 2004), limited protein fractionation, and the lack of suitable markets undermine the potential value of these bio-products.

Brassica carinata (A.) Braun mustard originates in Ethiopia, where it is used both as a leaf vegetable and as an oilseed (Getinet and Rakow 1997). *B. carinata* (Ethiopian mustard) is an allo-tetraploid member of the cruciferous plant family and is composed of two genomes, one originating from *B. nigra* and one from *B. oleracea* (U 1935). The yellow-seeded trait of most *Brassica* species segregates as two or three recessive Mendelian alleles with the exception of the semi-dominant yellow-seeded phenotype of *B. carinata* (Getinet al. 1987; Getinet and Rakow, 1997) and rare *B. napus* lines (Liu et al. 2005; Badani et al. 2006). The basis for gene regulation for this dominant yellow-seeded *B. carinata* phenotype is unknown, except that a range of phenylpropanoid and flavonoid genes are down-regulated in early stages of seed development (Li et al. 2010; Marles et al. 2003). In Arabidopsis, more than 20 recessive yellow-seeded *transparent testa* mutants have been characterized (Lepiniec et al. 2006), but dominant yellow phenotypes have not been described.

B. carinata germplasm is currently being developed in North America as bio-refinery platforms for bio-fuel/lubricants and biopolymers from the oil, as well as protein-based bio-plastics and novel food and feed products from the meal. Taylor et al. (2010) thoroughly reviewed genetic studies, genetic modification work, and breeding efforts to develop *B. carinata* as a bio-refinery and bio-industrial oils platform. Much of this review is focused on the development of transformation and regeneration protocols, target gene isolation, transgene expression and genetic modification for very long-chain fatty acid-enhanced or modified oils. *B. carinata* straws have also been evaluated together with grain straws for energy production (Várhegyi et al. 2009). In particular, new uses for seed coat fiber, lines with reduced hull fiber, high-value protein meal, and valuable co-products are needed to improve the economics of *B. carinata* industrial oilseed processing.

10.3 METAL TOLERANCE AS A VALUE-ADDED TRAIT FOR *BRASSICA CARINATA* (ETHIOPIAN MUSTARD)

Several mustard and wild species of the Brassicaceae family are important sources of edible oil while displaying tolerance to heavy metals. Around the world, Indian mustard (*Brassica juncea*), stinkweed (*Thlaspe spp.*), alyssum (*Alyssum bertolonii*) and *Ipomea alpine* are known to hyper-accumulate a range of heavy metals into roots and shoots (Baker and Brooks 1989; Salt et al. 1995; 1998; Lasat et al. 1996; Hale et al. 2001; Kupper et al. 2001; Ashraf and Mcneilly 2004; reviewed in Suresh and Ravishankar 2004). The cellular compartmentalization of metals within vegetative tissue cell walls and vacuoles has been shown for *Thaspe, Alyssum,* and *B. juncea* (Kramer et al. 2000; Pickering et al. 2000a; Kupper et al. 2001). Mechanisms of uptake in hyper-accumulating *Brassicaceae* include elevated expression of metal transporter genes (Assunção et al. 2001; Pence et al. 2000), excretion of root exudates and organic acid supplementation (Tolra et al. 1996; Huang et al. 1998; Salt et al. 2000), and changes in soil pH (Brown et al. 1994). Metal accumulation has been enhanced in some of these plant species by genetic engineering (Misra and Gedamu 1989; Heaton et al. 1998; Zhu et al. 1999a, 1999b) or by developing metal-tolerant hairy root cultures (Nedelkoska and Doran 2000, 2001). Zinc accumulation has been shown to reduce insect herbivory in Thlaspe (Pollard and Baker 1997). Selenium accumulation in *B. napus* has been tested for its nutritional value to livestock (Banueus and Mayland 2000).

Recently, *Brassica carinata* germplasm has been shown to accumulate arsenic species and copper ions (Irtelli and Narari-izzo 2008; Irtelli et al. 2009), and nicotianamine, histidine and proline were shown to be important copper chelators in this species (Irtelli et al. 2009). A few stress-related genes were also shown to be induced in *B. carinata* after exposure to copper ions (Zheng et al. 2001), while phytochelatin synthase proteins were shown to be induced in *B. juncea* after exposure to cadmium ions (Heiss et al. 2003). Other than this limited number of genes, the transcriptome of mustards exposed to metals is virtually unknown. Seedlings of a brown-seeded *B. carinata* line hyper-accumulated lithium ions and replaced sinapic acid esters and chloroplast lipid with benzoate derivatives, resveratrol and oxylipins (compounds never before found in a *Brassica* species), whereas seedlings of a genetically related *B. carinata* yellow-seeded line 21164 maintained the usual composition before and after exposure to lithium and did not tolerate the high (150 mM) lithium chloride concentrations tolerated by the brown-seeded line (Li et al. 2009). Microarray analysis indicated >89 different genes were differentially expressed more than 20-fold in seedlings of the genetically related *B. carinata* lines grown on lithium chloride, while more than 1083 genes had expression changes greater than two-fold (Li et al. 2009). These transcriptome changes coded for genes affected in defense, primary metabolism, transcription, transportation, secondary metabolism, cytochrome P450, as well as a host of genes with unknown functions. Growth was differentially affected in the two genetically related genotypes (i.e., brown-seeded plants were tolerant and yellow-seeded plants were intolerant) when the lines were grown on cobalt ions (Li and Gruber, unpublished) or Na_2SO_4, the predominant form of salinity found in soils of western North American prairies (Canam et al. submitted). A different profile of phytochemicals and transcripts was induced after treatment of the genetically related pair with lithium compared with Na_2SO_4 (Canam et al. submitted). Nonetheless, these findings suggest plasticity and potential for *Brassica carinata* as a tool in phyto-remediation and salt tolerance.

10.4 *BRASSICA CARINATA* SEEDCOATS AS A BIOSORBENT FOR HEAVY METALS

Since the yellow-seeded and brown-seeded *Brassica carinata* 21164 lines showed differential growth on lithium chloride, cobalt chloride, and Na_2SO_4, their seed coats were tested *in vitro* for potential to adsorb heavy metal ions from aqueous solution. Samples of seed coat were cleaned by air-aspiration, washed with de-ionized water and sized after drying. Seed coats (30, 300 or 600 mg) were transferred into 100 mL Erlenmeyer flasks containing 30 mL aqueous concentrations (up to 1000 ppm) of heavy metal ($CuCl_2$, $MnCl_2$, $HgCl_2$, $NiCl_2$, $ZnCl_2$, $CrCl_3$, or $CoCl_2$). Flasks were stirred for 24 h at room temperature and pH recorded; then seed coats were separated by centrifugation and adsorbed metal concentrations measured using inductively coupled plasma mass spectrometry (ICP-MS) at the University of Saskatchewan.

ICP-MS showed that endogenous metal composition in untreated seed coats of both lines of *B. carinata* is very low (Figure 10.1). Both types of seed coats could bind from 10,000 to 25,000 ppm of heavy metal from a 1,000 ppm solution, depending on the metal (Figure 10.2). Individual differences in sorption order could be summarized as Cu > Zn > Co > Ni > Mn > Cr > Hg for brown seed coats and Zn > Cu > Ni > Co > Mn > Cr > Hg for yellow seed coats. Relatively speaking, yellow seed coats sometimes absorbed more metal than brown seed coat. To determine whether the two seed coat types differed in their ability to bind a range of metal concentrations, a test range of 100, 500, 1,000, 5,000, 10,000, and 15,000 ppm copper chloride was challenged with 300 mg of seed coats in 30 mL water. The results showed that the metal-binding ability of seed coats was dose-dependent (Figure 10.3). In the case of copper, both types of coats could adsorb equivalently to a maximum of 40,000 ppm copper ions.

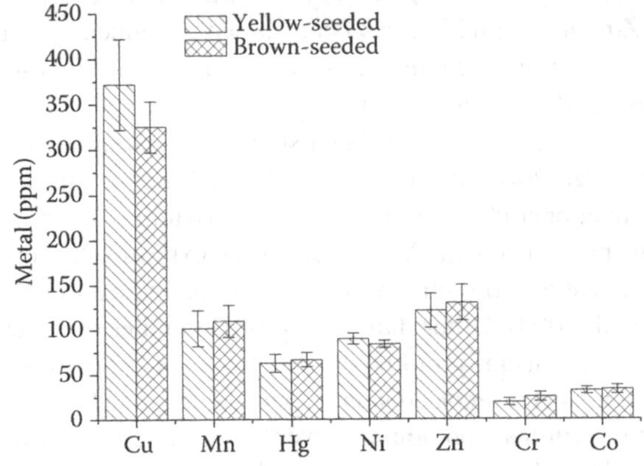

FIGURE 10.1 Metal concentrations detected in 30 mg of seedcoats isolated from brown-seeded and yellow-seeded *Brassica carinata* lines. Values represent the mean ± SD of three independent measurements.

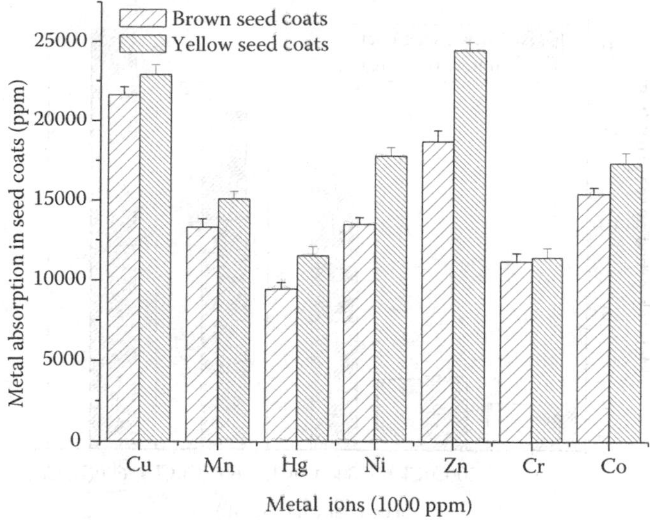

FIGURE 10.2 Absorption of seven different heavy metals to aqueous suspensions of 1000-mg seed coats from genetically related brown-seeded and yellow-seeded *Brassica carinata* lines coats. Values represent the mean ± SD of three independent measurements.

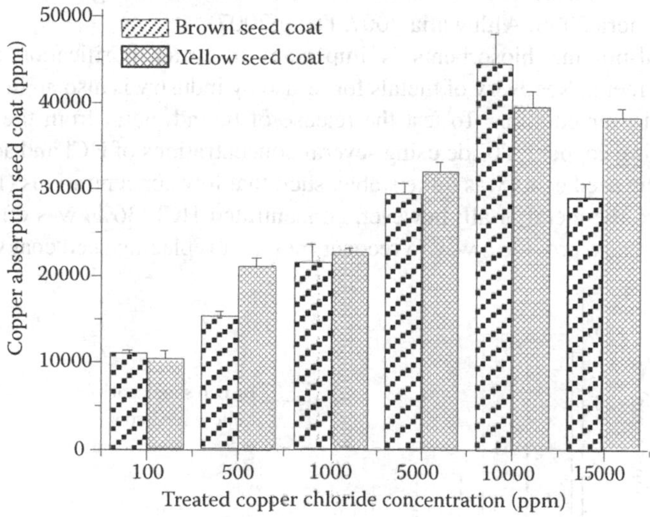

FIGURE 10.3 Copper adsorption in brown-seeded and yellow-seeded *Brassica carinata* seed coats treated with a range of $CuCl_2$ concentrations. Values represent the mean ± SD of three independent measurements.

Copper chloride was also used to explore seed coat:metal ratios. Adding higher amounts of seed coat removes additional copper from aqueous solution (Figure 10.4). The sorption ability was highest at a 20:1 biosorbent:copper ratio (600 mg seed coat), and there were no differences in adsorption between yellow and brown seed coat. However, both types of seed coats tested with 600 mg zinc showed a higher sorption ability than with copper (Figure 10.4). Sorption ability has been shown in other systems to be dependent on temperature, pH, and metal type (Ahluwalia 2007). These adsorption abilities of *B. carinata* seed coats are similar to those of black gram husk (*Cicer arientinum*), rice husks (Tarley and Arruda 2004) and mucilaginous seeds of *Ocimum basilicaum* (Saeed and Iqbal 2003; Melo 2004; Saeed et al. 2005) and 3- to 5-fold higher than those of *Tectona grandis*

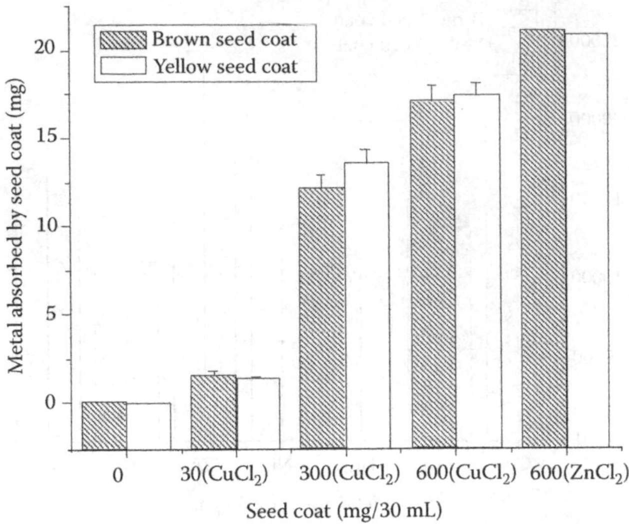

FIGURE 10.4 Concentrations of *B. carinata* seed coats for efficient absorption of 1000 ppm copper chloride and zinc in a 30 mL aqueous solution. Values represent the mean ± SD of three independent measurements.

leaves (Prasanna 2006). Adsorption was also 5-fold lower than some alga and metal-binding fungi (Binupriya 2006; Romera 2006; Ahluwalia 2007; Deng 2007).

Stability of metal-binding biosorbents is important in water purification systems to avoid unplanned release of metal. Recovery of metals for re-use by industry is also an important economic consideration in phyto-remediation. To test the release of bound metal from the *B. carinata* seed coats, we tried releasing copper chloride using several concentrations of HCl and acetic acid. Copper chloride binding to the seed coats was quite stable, such that low concentrations (100 mM) of either acid could not release the metal at all; however, concentrated HCl (36%) was efficient at releasing copper. Ashing would also provide a way to recover metals if replacing seed coat was an option.

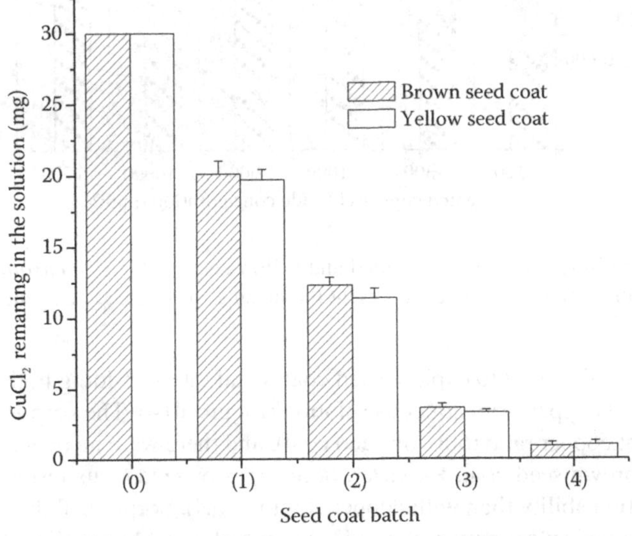

FIGURE 10.5 Step-wise removal of copper ion (1000 ppm) from a 100-mL aqueous solution using four fresh batches of seed coat (300 mg). Values represent the mean ± SD of three independent measurements. (1), (2), (3), and (4) represent step-wise addition of seed coats.

At a 10:1 ratio of biosorbent (seed coat):metal ions, three or four fresh step-wise batches of seed coat removed 99% of the copper from a 1000 ppm solution, such that the remaining aqueous concentration was down to 10 ppm (i.e., mg/L) (Figure 10.5). This is above the recommended acceptable levels for industrial metals in Canadian drinking water system, but well below the levels accepted in soil and brackish industrial water (Table 10.2). In a more limited comparison study with *B. juncea*, ICP-MS results also showed both brown and yellow seed coats of *B. juncea* Oriental mustard could bind similar amounts (from 10,000 to 25,000 ppm) of heavy metals from a 1,000 ppm solution (Figure 10.6), although in this case saturation was not tested. The ranking for metal to binding was very similar to *B. carinata* seed coats, except that zinc and copper binding differed for yellow *B. juncea* seed coats (Figure 10.2). These findings suggest that *Brassica* seed coats could be useful as a pre-filter in municipal water treatment facilities and as primary filter material for remediation of industrial processes. This is particularly applicable to commercial Oriental mustard bran and Ethiopian mustard bran, which currently have no economic value and could be readily and cheaply supplied in Canada.

Changes in seed coat surface features with different metal treatments were observed in *B. carinata* seed coats by scanning electron microscopy (SEM). Seed coats without metal treatment showed a somewhat amorphous reticular rugose (wrinkled) pattern with concave irregular-shaped depressions in periclinal cell walls and raised undulated anticlinal cell walls (Figure 10.7) typical of *B. carinata* (Karcz et al. 2005). Treatment with all seven heavy metals resulted in a much more defined periclinal wall structure with little reticulation (Figure 10.7).

TABLE 10.2
Acceptable Levels of Metals in Canadian Drinking Water and Soils

Metal	Cu	Zn	Cr	Hg	Ni	Mn	Co
*Drinking water (ppm = mg/L)	≤1.0	≤5.0	≤0.05	≤0.001	–	≤0.05	–
**Soil (mg/kg)	30	82	64–87 (total)	–	60	–	–

Source: Data drawn from Guidelines for Canadian Drinking Water Quality Summary Table, Dec. 2010 (http://www.hc-sc.gc.ca/ewh-semt/pubs/water-eau/2010-sum_guide-res_recom/index-eng.php); Canadian Council of Ministers of the Environment, 1997 (www.ccme.ca/assets/pdf/pn_1270_e.pdf); Canadian Council of Ministers of the Environment, 1999a [Chromium(en).pdf], 1999b [Nickel(en).pdf]; Haidary-Monfared, 2011 (http://www.ecologyaction.ca/files/images/file/Community%20Garden%20Heavy%20Metal%20Contamination%20Study.pdf).

Note: mg/L = ppm.

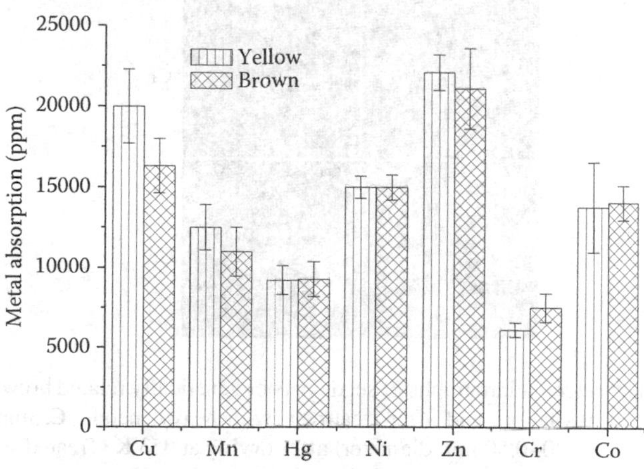

FIGURE 10.6 Metal adsorption in 30 mL volume by seed coats from brown-seeded and yellow-seeded *Brassica juncea* lines. Values represent the mean ± SD of three independent measurements.

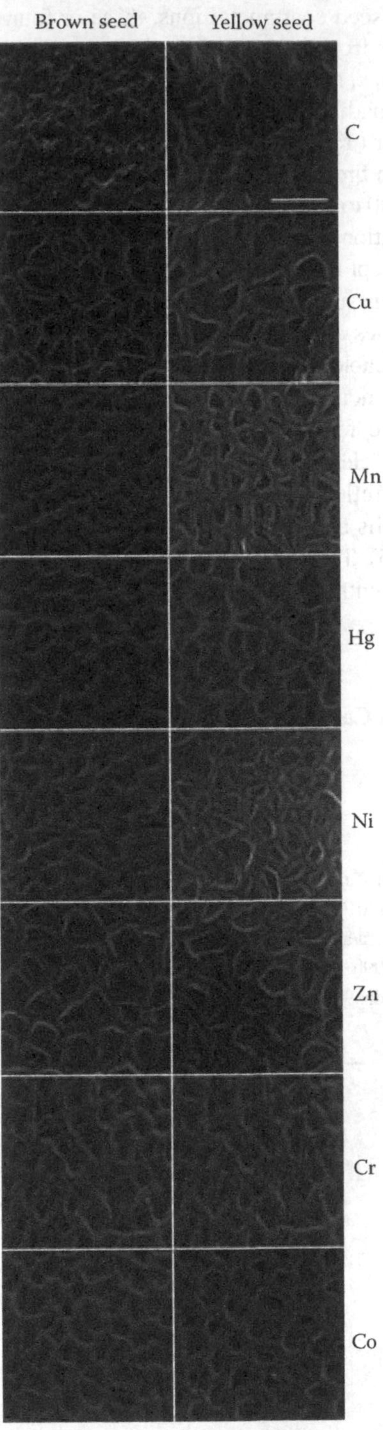

FIGURE 10.7 Scanning electron micrographs of seed coats of genetically related brown-seeded and yellow-seeded *Brassica carinata* lines before and after treatment with heavy metals. C: untreated (control) seed coats. Seed coats were sized (0.30–0.50 mm diameter) after drying at 353 K. Treated seed coats were briefly (moments) immersed in dehydrated alcohol, air-dried, place in aluminum stubs and vacuum coated with gold using IB-2 ion coated chamber. Samples were viewed with a Hitachi S-570 scanning electron microscopy (SEM) at 20 kV. White bar represents 60 micrometers.

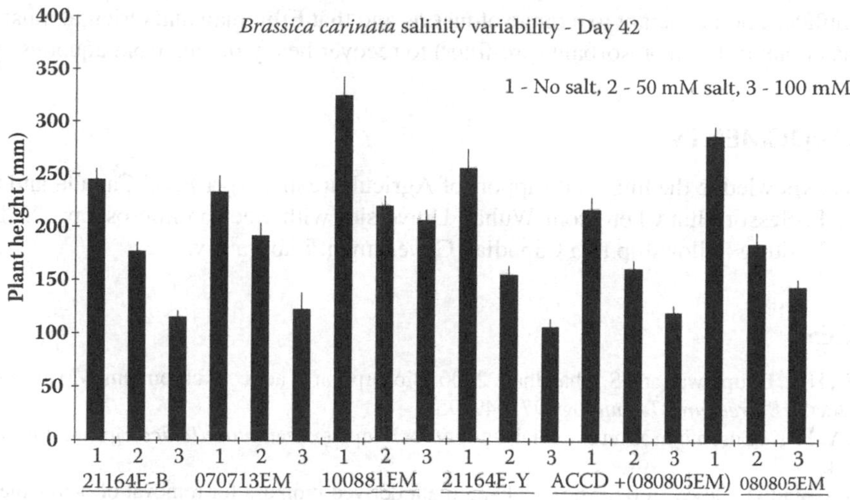

FIGURE 10.8 Variation in seedling height for six *B. carinata* genotypes in the absence and presence of Na_2SO_4. Plants were grown for 21 days without salt in a Saskatoon greenhouse in February 2010 and then treated for an additional 21 days with 0, 50, or 100 mM Na_2SO_4. Values represent the mean ± SD of three independent measurements.

10.5 VARIATION FOR GROWTH AND SALINITY TOLERANCE OF DIVERSE *B. CARINATA* LINES IN Na_2SO_4

Since the brown-seeded *B. carinata* 21164E-B line appeared to tolerate and accumulate metals, unlike the yellow-seeded 21164E-Y line, we tested growth of six *B. carinata* lines from diverse genetic backgrounds for their tolerance to Na_2SO_4. Twenty-one (21) day-old seedlings growing in a soilless mixture supplemented with slow-release fertilizer were watered for an additional 21 days with 200 ml solutions of 0, 50, or 100 mM Na_2SO_4. Biomass (plant height) grown in the absence of salt showed a 33% variation between the six genotypes (Figure 10.8). Addition of 50 or 100 mM Na_2SO_4 showed additional variation for salinity tolerance (Figure 10. 8). Curiously, a bacterial ACC deaminase transgene, which was previously shown to improve tolerance to salt and nickel stress in brown-seeded ACDD+ canola cv. Westar (Stearns et al. 2005; Sergeeva et al. 2006) unexpectedly reduced plant height in the yellow-seeded *B. carinata* genotype 080805EM both in the absence and presence of Na_2SO_4. These data suggest that methods to support plant health and tolerance to metals may differ for yellow-seeded *B. carinata* germplasm compared with brown-seeded germplasm.

10.6 CONCLUSION

B. carinata germplasm is being developed as dry land bio-refinery platforms in North America for obtaining bio-fuel/lubricants and biopolymers from the oil, as well as protein-based bio-plastics and novel food and feed products from the meal. In particular, new uses for seed coat fiber, lines with reduced hull fiber, high-value protein meal, and valuable co-products are sought to improve the economics of industrial oilseed processing. Here, we highlight the potential of this crop in phytoremediation and salinity tolerance. Seed coats of genetically related yellow-seeded and brown-seeded lines bound substantial amounts of heavy metal ions, which re-structured the seed coat. Seedlings of the brown-seeded line tolerated salts of lithium and cobalt compared with the yellow-seeded line, and the brown-seeded line newly accumulated benzoic acid and resveretrol instead of sinapic acid under these conditions. Six diverse *B. carinata* lines showed strong variability for plant height and salinity after 21 days growth on Na_2SO_4. Other *B. carinata* germplasm has shown uptake of arsenic and copper ions. These data and literature review show that Ethiopian mustard hyper-accumulates and is tolerant to a range of metals and that Ethiopian and Oriental mustard seed coat waste has potential as a biosorbant (pre-filter) to recover heavy metals from aqueous systems.

ACKNOWLEDGMENTS

The authors acknowledge the financial support of Agriculture and Agri-Food Canada and the kind assistance of Professor Hun Chen from Wuhan University with electron microscopy. X. Li was a recipient of a Visiting Fellowship to a Canadian Government Laboratory.

REFERENCES

Agarwal, G.S., H.K. Bhuptawat, and S. Chaudhari. 2006. Biosorption of aqueous chromium (VI) by *Tamarindus indica* seeds. *Bioresource Technology* 97: 949–956.

Anderson, R.A. 1997. Chromium as an essential nutrient for humans. *Regulatory Toxicology and Pharmacology* 26: 35–4.

Ahluwalia, S.S., and D. Goyal. 2007. Microbial and plant derived biomass for removal of heavy metals from wastewater. *Bioresource Technology* 98: 2243–2257.

Ashraf, M., and T. McNeilly. 2004. Salinity tolerance in Brassica oilseeds. *Critical Reviews of Plant Science* 23: 157–174.

Assunção, A.G.L., P. da Costa-Martins, S. de Folter, R. Voijs, H. Schat, and M.G.M. Aarts. 2001. Elevated expression of metal transporter genes in three accessions of the metal hyperaccumulator *Thlaspi caerulescens*. *Plant Cell and Environment* 24: 217–226.

Baker, A.M.J., and R.R. Brooks. 1989. Terrestrial higher plants which hyperaccumulate metallic elements: A review of their distribution, ecology and physiochemistry. *Biorecovery* 1: 81–126.

Baneus, G.S., and H.F. Mayland. 2000. Absorption and distribution of selenium in animals consuming canola grown for selenium phytoremediation. *Ecotoxicology and Environmental Safety* 46: 322–328.

Badani, A.G., R.J. Snowdon, B. Wittkop, F.D. Ligsa, R. Baetzel, R. Horn, De Haro, R. Font, F. Lũhs, and W. Friedt. 2006. Colocalization of a partially dominant gene for yellow seed colour with a major QTL influencing acid detergent fiber (ADF) content in different crosses of oilseed rape (*Brassica napus*). *Genome* 149: 1499–1509.

Binupriya, A.R., M. Sathishkumar, K. Swaminathan, E.S. Jeong, S.E. Yun, and S. Pattabi. 2006. Biosorption of metal ions from aqueous solution and electroplating industry wastewater by *Aspergillus japonicus*: phytotoxicity studies. *Bulletin of Environmental Contamination and Toxicology* 77: 219–227.

Biffis A., H. Landes, K. Jerabek, and B. Corain, 2000. Metal palladium dispersed inside macroporous ion-exchange resins: the issue of the accessibility to gaseous reactants. *Journal of Molecular Catalysis A: Chemical* 151: 283–288.

Brown, S.L., R.L. Chaney, J.S. Angle, and A.J.M. Baker. 1994. Zinc and cadmium uptake by *Thlaspi caerulescens* and *Silene vulgaris* in relation to soil metals and soil pH. *Journal of Environmental Quality* 23: 1151–1157.

Canadian Council of Ministers of the Environment. 1997. Canadian soil guidelines for copper. *Environmental and Human Health.* ISBN O-662-25520-8 En 108-4/11-1997E. Winnipeg, MN, Canada.

Canadian Council of Ministers of the Environment. 1999a. Canadian soil quality guidelines for the protection of environmental and human health: Chromium (total 1997) (VI 1999). In: *Canadian environmental quality guidelines. Excerpt. Publication No. 1299.* ISBN 1-896997-34-1.

Canadian Council of Ministers of the Environment. 1999b. Canadian soil quality guidelines for the protection of environmental and human health: Nickel 1999. In: *Canadian environmental quality guidelines. Excerpt. Publication No. 1299.* ISBN 1-896997-34-1.

Deng, L., X. Zhu, X. Wang, Y. Su, and H. Su. 2007. Biosorption of copper (II) from aqueous solutions by green alga *Cladophora fascicularis*. *Biodegradation* 18: 393–402.

Danil de Namor, A.F. 2007. Water purification: from ancient civilization to the XXI Century. *Water Science & Technology: Water Supply* 7: 33–39.

Francesconi K.A. 2007. Toxic metal species and food regulations—making a healthy choice. *Analyst* 132: 17–20.

Getinet, A., G. Rakow, and R. K. Downey. 1987. Seed color inheritance in *Brassica carinata* A. Braun, Cultivar S-67. *Plant Breeding*. 99: 80–82.

Getinet, A., and G. Rakow. 1997. Repression of seed coat pigmentation in Ethiopian mustard. *Canadian Journal of Plant Science* 77: 501–505.

Hale, K.L., S.P. McGrath, E. Lombi, S.M. Stack, N. Terry, I.J. Pickering, G.N. George, and E.A. Pilon-Smits. 2001. Molybdenum sequestration in Brassica species. A role for anthocyanins? *Plant Physiology* 126: 1391–1402.

Heaton, A.C.P., C.L. Rugh, N.J. Wang, and R.B. Meagher. 1998. Phytoremediation of mercury and methyl mercury polluted soils using genetically engineered plants. *Journal of Soil Contamination* 7: 497–509.

Heiss, S., A. Wachter, J. Bogs, C. Cobbett, and T. Rausch. 2003. Phytochelatin synthase (PCS) protein is induced in *Brassica juncea* leaves after prolonged Cd exposure. *Journal of Experimental Botany* 54: 1833–1839.

Haidary-Monfared. 2011. Community garden heavy metal study. (http://www.ecologyaction.ca/files/images/file/Community%20Garden%20Heavy%20Metal%20Contamination%20Study.pdf).

Huang, J.W., M.J. Blayloc, Y. Kapulnik, and B.D. Ensley. 1998. Phytoremediation of uranium-contaminated soils. Role of organic acids in triggering uranium hyperaccumulation in plants. *Environmental Science and Technology* 32: 2004–2008.

Irtelli, B. and F. Navari-izzo. 2008. Uptake kinetics of different arsenic species by *Brassica carinata*. *Plant and Soil* 303: 105–113.

Irtelli, B., W.A. Peprucci, and F. Navari-izzo. 2009. Nicotianamine and histidine/proline are, respectively, the most important copper chelators in xylem sap of *Brassica carinata* under conditions of copper deficiency and excess. *Journal of Experimental Botany* 60: 269–277.

Juang, R.S., and M.S.-W. Wang. 2000. Metal recovery and EDTA recycling from simulated washing effluents of metal-contaminated soils. *Water Research* 34: 3795–3803.

Karcz, J., T. Ksiazczyk, and J. Maluszynska. 2005. Seed coat patterns in rapid cycling *Brassica* forms. *Acta Biologica Cracoviensia Series Botanica* 47: 159–165.

Keskinkan, O., M.Z. Goksu, M. Basibuyuk, and C.F. Forster. 2004. Heavy metal adsorption properties of a submerged aquatic plant (*Ceratophyllum demersum*). *Bioresource Technology* 92: 197–200.

King, P., N. Rakesh, S. Beenalahari, K.Y. Prasanna, and V.S. Prasad. 2007. Removal of lead from aqueous solution using *Syzygium cumini* L.: equilibrium and kinetic studies. *Journal of Hazardous Materials* 142: 340–347.

Kramer, U., I.J. Pickering, R.C. Prince, I. Raskin, and D.E. Salt. 2000. Subcellular localization and speciation of nickel in hyperaccumulator and non-accumulator *Thlaspi* species. *Plant Physiology* 122: 1343–1353.

Kupper, H., E. Lombi, F.J. Zhao, G. Wieshammaer, and S.P. McGrath. 2001. Cellular compartmentalization of nickel in the hyperaccumulators *Alyssum lesbiacum*, *Alyssum bertolonii* and *Thlaspi goesingense*. *Journal Experimental Botany* 52: 2291–2300.

Lawson, P.S., R.M. Sterritt, and J.N. Lester. 1983. Factors affecting the removal of metals during activated sludge wastewater treatment II. The role of mixed liquor biomass. *Archives of Environmental Contaminants and Toxicology* 13: 391–402.

Lasat, M.M., A.J.M. Baker, and V. Kochion. 1996. Physiological characterization of root Zn^{+2} absorption and translocation to shoots in zinc hyperaccumulator and non-hyperaccumulator species of *Thlaspi*. *Plant Physiology* 112: 1715–1722.

Lepiniec, L., I. Debeaujon, J.M. Routaboul, A. Baudry, L. Pourcel, N. Nesi, and M. Caboche. 2006. Genetics and biochemistry of seed flavonoids. *Annual Review of Plant Biology* 57: 405–430.

Li, X., P. Gao, B. Gjetvaj, N.D. Westcott, and M.Y. Gruber. 2009. Analysis of the metabolome and transcriptome of *Brassica carinata* seedlings after lithium chloride exposure. *Plant Science* 177: 68–80.

Li, X., N. Westcott, M. Links, and M. Gruber. 2010. Re-visiting yellow-seeded *Brassica carinata*: seed coat phenolics and the developing silique transcriptome. *Journal of Agriculture and Food Chemistry* 58: 10918–10928.

Liu, X.P., J.X. Tu, B.Y. Chen, and T.D. Fu. 2005. Identification and inheritance of a partially dominant gene for yellow seed colour in *Brassica napus*. *Plant Breeding* 124: 9–12.

Marin, J., and J. Ayele 2002. Removal of some heavy metal cations from aqueous solutions by spruce sawdust. I. Study of the binding mechanism through batch experiments. *Environmental Technology* 23: 1157–1171.

Marles, M.A.S., M.Y. Gruber, G.J. Scoles, and A.D. Muir. 2003. Pigmentation in the developing seed coat and seedling leaves of *Brassica carinata* is controlled at the dihydroflavonol reductase locus. *Phytochemistry* 62: 663–672.

Marles, M.A.S., and M.Y. Gruber. 2004. Histochemical characterization of unextractable seed coat pigments and quantification of extractable lignin in the *Brassicaceae*. *Journal of Science and Food Agriculture* 84: 251–262.

Martins, R.J., R. Pardo, and R.A. Boaventura. 2004. Cadmium (II) and zinc (II) adsorption by the aquatic moss *Fontinalis antipyretica*: effect of temperature, pH and water hardness. *Water Research* 38: 693–699.

Melo, J.S., and S.F. D'Souza. 2004. Removal of chromium by mucilaginous seeds of *Ocimum basilicum*. *Bioresource Technology* 92:151–155.

Misra, S., and L. Gedarmu. 1989. Heavy metal tolerant transgenic *Brassica napus* L. and *Nicotiana tabaccum* L. plants. *Theoretical and Applied Genetics* 78: 161–168.

Mukhopadhyay, M., S.B. Noronha, and G.K. Suraishkumar. 2007. Kinetic modeling for the biosorption of copper by pretreated *Aspergillus niger* biomass. *Bioresource Technology* 98: 1781–1787.

Nedelkoska, T.V., and P.M. Doran. 2000. Hyperaccumulation of cadmium by hairy roots of *Thlaspi caerulescens*. *Biotechnology and Bioengineering* 67: 607–615.

Nedelkoska, T.W., and P.M. Doran. 2001. Hyper-accumulation of nickel by hairy roots of *Alyssum sps*. Comparison with whole regenerated plants. *Biotechnology and Bioengineering* 17: 752–759.

Ofomaja, A.E., and Y.S. Ho. 2007. Effect of pH on cadmium biosorption by *Coconut copra* meal. *Journal of Hazardous Materials* 139: 356–362.

Pence, N.S., P.B. Larsen, S.D. Ebbs, D.L. Letham, M.M. Lasat, D.F. Garvin, D. Eide, and L.V. Kochian. 2000. The molecular physiology of heavy metal transport in the Zn/Cd hyperaccumulator *Thlaspi caerulenscens*. *Proceeding of the National Academy of Science USA* 97: 4956–4960.

Pickering, I.J., R.C. Prince, D.E. Salt, and G.N. George. 2000a. Quantitative, chemically specific imaging of selenium transformation in plants. *Proceedings of the National Academy of Science USA* 97: 10717–10722.

Pickering, I.J., R.C. Prince, M.J. George, R.D. Smith, G.N. George, and D.E. Salt. 2000b. Reduction and coordination of arsenic in Indian mustard. *Plant Physiology* 122: 1171–1177.

Pickering, A., C. Chang, and J.B. Vincent. 2004. Chromium-containing biomimetic cation triaqua-μ3-oxo-μ-hexapropionatotrichromium(III) inhibits colorectal tumor formation in rats. *Journal of Inorganic Biochemistry* 98: 1303–1306.

Pollard, A.J., and A.J.M. Baker. 1997. Determination of herbivory by zinc hyperaccumulation in *Thlaspi caerulenscens* (Brassicaceae). *New Phytologist* 135: 655–658.

Prasanna, K.Y., P. King, and V.S. Prasad. 2006. Equilibrium and kinetic studies for the biosorption system of copper(II) ion from aqueous solution using *Tectona grandis* L.f. leaves powder. *Journal of Hazardous Materials* 137: 1211–1217.

Romera, E., F. González, A. Ballester, M.L. Blázquez, and J.A. Muñoz. 2006. Biosorption with algae: a statistical review. *Critical Reviews in Biotechnology* 26: 223–235.

Saeed, A., M. Iqbal, and M.W. Akhtar. 2005. Removal and recovery of lead (II) from single and multimetal (Cd, Cu, Ni, Zn) solutions by crop milling waste (black gram husk). *Journal of Hazardous Materials* 117: 65–73.

Saeed, A., and M. Iqbal. 2003. Bioremoval of cadmium from aqueous solution by black gram husk (*Cicer arientinum*). *Water Research* 37: 3472–3480.

Salt, D.E., R.C. Prince, I.J. Pickering, and I. Raskin. 1995. Mechanisms of cadmium mobility and accumulation in Indian mustard. *Plant Physiology* 109: 1427–1433.

Salt, D.E., R.D. Smith, and I. Raskin. 1998. Phytoremediation. *Annual Review of Plant Physiology and Plant Molecular Biology* 49: 643–668.

Salt, D.E., N. Kato, Kramer, R.D. Smith, and I. Raskin. 2000. The role of root exudates in nickel hyperaccumulation and tolerance in accumulator and nonaccumulator species of *Thlaspi*. In: Terry, N. and Bañuelos, G.S. (eds.). *Phytoremediation of Contaminated Soil and Water,* 191–202. Berkeley, California: Ann Arbor Press, Inc.

Schneegurt, M.A., J.C. Jain, J.A. Menicucci Jr, S.A. Brown, K.M. Kemner, D.F. Garofalo, M.R. Quallick, C.R. Neal, and C.F. Kulpa. 2001. Biomass by-products for the remediation of wastewaters contaminated with toxic metals. *Environmental Science and Technology* 35: 3786–3791.

Sekhar, K., C.T. Kamala, N.S. Chary, A.R. Sastry, T. Nageswara, and M. Vairamani. 2004. Removal of lead from aqueous solutions using an immobilized biomaterial derived from a plant biomass. *Journal of Hazardous Materials* 108: 111–117.

Sergeeva, E., S. Shah, and B.R. Glick. 2006. Growth of transgenic canola (*Brassica napus* cv. Westar) expressing a bacterial 1-aminocyclopropane-1-carboxylate (ACC) deaminase gene on high concentrations of salt. *World Journal of Microbiology and Biotechnology* 22: 277–282.

Sharma, P., P. Kumari, M.M. Srivastava, and S. Srivastava. 2007. Ternary biosorption studies of Cd (II), Cr (III) and Ni (II) on shelled *Moringa oleifera* seeds. *Bioresource Technolology* 98: 474–477.

Sims, R.C., J.L. Sims, D.L. Sorensen, J. McLean, R. Mashmood, and R.R. Dupont. 1995. Review of in place treatment techniques for contaminated surface soil. *Risk Reduction Research Laboratory Report* EPA-540/2-84-003b 1: 85–124. Cincinnati OH: *US* Environmental Protection Agency.

Stearns, J.C., S. Shah, B.N. Greenberg, D.G. Dixon, and Glick, B.R. 2005. Tolerance of transgenic canola expressing 1-aminocyclopropane-1-carboxylic acid deaminase to growth inhibition by nickel. *Plant Physiology and Biochemistry* 43: 701–708.

Suresh, B., and G.A. Ravishankar. 2004. Phytoremediation—A novel and promising approach for environmental cleanup. *Critical Reviews in Biotechnology* 24: 97–124.

Tarley, C.R., and M.A. Arruda. 2004. Biosorption of heavy metals using rice milling by-products. Characterisation and application for removal of metals from aqueous effluents. *Chemosphere* 54: 987–995.

Tolra, R.P., C. Poschenrieder, and J. Barcelo. 1996. Zinc hyperaccumulation in *Thaspi caerulenscens* II. Influence on organic acids. *Journal of Plant Nutrition* 19: 1541–1550.

Taylor, D., K.C. Falk, C.D. Palmer, J. Hammerlindl, V. Babic, E. Mietkiewska, A. Jadhav, E.-F. Mirrillia, T. Francis, T. Hoffman, E.M. Giblin, V. Katavic, and W.A. Keller. 2010. *Brassica carinata*—a new molecular farming platform for delivering bio-industrial oil feedstocks: case studies of genetic modifications to improve very long-chain fatty acid and oil content in seeds. *Biofuels, Bioproducts and Biorefining* 4: 538–561.

U, N. 1935. Genome analysis in *Brassica* with special reference to the experimental formation of *B. napus* and peculiar mode of fertilization. *Japanese Journal of Botany* 7: 389–452.

Vegliò, F., F. Beolchini, and M. Prisciandaro. 2003. Sorption of copper by olive mill residues. *Water Research* 37: 4895–48903.

Várhegyi, G., H. Chen, and S. Godoy. 2009. Thermal decomposition of wheat, oat, barley and *Brassica carinata* straws. A kinetic study. *Energy and Fuels* 23: 646–652.

Waranusantigul P., P. Pokethitiyook, M. Kruatrachue, and E.S. Upatham. 2003. Kinetics of basic dye (methylene blue) biosorption by giant duckweed (*Spirodela polyrrhiza*). *Environmental Pollution* 125: 385–392.

Yan, C.Z., S.R. Wang, A.Y. Zeng, X.C. Jin, Q.J. Xu, and J.Z. Zhao. 2005. Equilibrium and kinetics of copper(II) biosorption by *Myriophyllum spicatum* L. *Journal of Environmental Science (China).* 17: 1025–1029.

Zafar, M.N., R. Nadeem, and M.A. Hanif. 2007. Biosorption of nickel from protonated rice bran. *Journal of Hazardous Materials* 143: 478–485.

Zaldivar, R. 1980. Geographical gradients of lithium concentration in surface and drinking waters and in plasma of exposed subjects in Northern Chile. *Reviews of Medicine Chile* 108: 822–824.

Zheng, Z.F., T.M. Uchacz, and J.L. Taylor. 2001. Isolation and characterization of novel defence-related genes induced by copper, salicylic acid, methyl jasmonate, abscisic acid and pathogen infection in *B. carinata. Molecular Plant Pathology* 2: 159–169.

Zhu, Y.L., E.A. Pilon-Smits, A.S. Tarun, S.U. Weber, L. Jouanin, and N. Terry. 1999a. Cadmium tolerance and accumulation in Indian mustard is enhanced by overexpression of gamma-glutamylcysteine synthetase. *Plant Physiology* 121: 1169–1178.

Zhu, Y.L., H.E.A. Pilon-Smits, L. Jouanin, and N. Terry. 1999b. Overexpression of glutamine synthetase in Indian mustard enhances cadmium accumulation and tolerance. *Plant Physiology* 119: 73–79.

11 Phytoremediation of Toxic Metals and the Involvement of *Brassica* Species

Aryadeep Roychoudhury, Sreeparna Pradhan, Bodhisatwa Chaudhuri, and Kaushik Das

CONTENTS

11.1 INTRODUCTION

Decades of environmentally detrimental anthropogenic activities, such as urbanization, industrialization, deforestation, combustion of fossil fuels and improper agricultural and mining practices, along with gross neglect of the environment, have resulted in this current perturbing state of the world. Metal pollution may result from point sources such as industrial discharges, or from nonpoint sources like vehicular emissions, soluble salts or the unregulated use of pesticides and fertilizers (Lone et al. 2008). Heavy metals pose a greater threat than most of the other harmful contaminants because of their high persistence in the environment and carcinogenic effects (Lone et al. 2008). Environmental contamination by heavy metals may occur from natural sources such as volcanic eruptions, and from mining, metal industries and phosphate fertilizers (Schutzendubel and Polle 2002). Metals cannot be completely removed by biological means, but can be made innocuous by changing their oxidation states (Lone et al. 2008).

Clean-up of heavy metal-contaminated sites becomes essential to prevent their entry into the food chain and also for land reclamation. Various physical methods such as soil excavation, solidification, soil washing, use of permeable barriers, incineration and pumping, and further treatments are already being used to remediate metal-contaminated soils. These conventional techniques are mostly expensive and in some cases not very efficient. Most importantly, these may cause further secondary damage to the environment in the process of the clean-up (Ghosh and Singh 2005). Hence much research is being done to develop cleaner and greener technological advancements and finding newer, more efficient methods to combat pollution.

11.1.1 Bioremediation

There has been a recent trend towards the use of biological agents to clean contaminated areas, especially in the case of metals which are very difficult to remove by physical means. This use of biological agents to remove environmental contaminants is known as bioremediation. Bioremediation can be carried out by organisms from various taxa like bacteria, fungi and even higher plants, which may be either wild-type or genetically modified (GEMS, genetically engineered microorganisms) in order to restore contaminated sites (Baker et al. 2000; Shah and Nongkynrih 2007). The method is gaining popularity because of its eco-friendly nature and low cost approach along with its aesthetic appeal. A particular organism may be used by itself or in combination with other organisms for efficient bioremediation.

11.1.2 Phytoremediation

Higher plants, either alone or in combination with the associated microbes, are often the most efficient biological systems for bioremediation of polluted soil and water. The use of plants to clean up the environment is known as phytoremediation. The term phytoremediation ("phyto" meaning plant and "remedium" meaning to clean or restore) refers to different kinds of plant-based technologies that use either naturally occurring or genetically engineered plants for cleaning contaminated sites. Although the concept of phytoremediation has been used for over 300 years to treat wastewater (Ghosh and Singh 2005), its application on metal-contaminated sites was reintroduced and developed only in 1983 by Chaney, and the first field trials were conducted on Zn and Cd by Baker in 1991. *Thlaspi caerulescens* and *Viola calaminaria* were the first plant species reported to accumulate high levels of metals in leaves (Hartman 1975).

Phytoremediation takes advantage of the unique, selective and inherent uptake capabilities of plant root systems, together with the translocation, bioaccumulation and pollutant storage/degradation abilities of the entire plant body (Hooda et al. 2007). It is considered to be a "Green Revolution" in the field of innovative clean-up technologies. Besides being aesthetically pleasing, phytoremediation is on average ten times cheaper than other physical, chemical or thermal remediation methods since it is performed *in situ*, is solar driven and can function with minimal maintenance once established (Hooda et al. 2007). Advances in phytoremediation are likely to result from more efficient plant variety selection and soil amendments, as well as by optimizing agronomic practices used for plant cultivation (Sekara et al. 2005). The phytoremediation market is growing rapidly in the U.S. comprising ~\$100–150 million per year (Hooda et al. 2007).

11.2 PHYTOREMEDIATION TECHNOLOGIES

Phytoremediation consists of different plant-based technologies, each having a different mechanism of action for the remediation of metal-polluted soil, sediment or water. These are as follows:

11.2.1 PHYTOEXTRACTION

Phytoextraction, also called phytoaccumulation (United States Protection Agency Reports 2000), refers to the absorption and uptake by plants of large amounts of metal contaminants from the soil and their translocation into the aboveground parts. Plants used for phytoremediation should be hyperaccumulators, i.e., able to accumulate metals in large excess, usually volumes up to 100-fold higher than that present in soil. Such plants can accumulate trace elements, primarily Ni, Zn, Cd, As or Se, in their aboveground tissues without developing any toxicity symptoms (Baker and Brooks 1989). Ideally, they should have fast growth and high biomass. A database called PHYTOREM has been built for hyperaccumulating plants by Environment Canada (McIntyre 2003). A single variant of these hyperaccumulators may be used individually or in combination with many others, depending on the metal needed to be removed and the soil conditions. After a suitable period of growth, the plants are harvested and then incinerated or composted to extract the metals. All waste or incineration ash must be carefully disposed. Although it is the most efficient strategy, phytoextraction is difficult to maintain (Ebbs et al. 1997; Kramer 2005).

Lead is not present in a soluble form in soil and so is not easily absorbed by plant roots. For such cases, chelator-assisted phytoremediation can be used, where chelators such as EDTA (ethylenediamine tetraacetic acid) are used to solubilize poorly available metals in the soil, followed by passive accumulation of metal complexes in plant shoots along with transpiration (Kramer 2005). Phytoextraction efficacy depends both on the concentration of contaminants in the harvested biomass and on the yield (Duquène et al. 2009). The percentage of annual reduction of soil heavy metal concentration can be calculated as: Removal % = (C_{plant} × yield/C_{soil} × m_{soil}) × 100, where C_{plant} is the concentration of metal in shoots (kg ha^{-1}), yield refers to the dry matter biomass yield with one harvest per year (kg ha^{-1}), C_{soil} is the concentration of metal in soil (g ha^{-1}) and m_{soil} is the mass of the contaminated soil layer (kg ha^{-1}) (Duquène et al. 2009). Assuming a rooting depth of 20 cm and a soil density of 1.5 kg dm^{-3}, m_{soil} equals 3,000,000 kg ha^{-1}. Average annual field yields of 12,000 kg dry weight (DW) ha^{-1} year^{-1} were reported for *Brassica juncea* by Duke (1983).

11.2.2 PHYTOSTABILIZATION

Phytostabilization, also known as in-place inactivation, involves the use of certain plant species to immobilize contaminants in the soil and ground water through absorption and accumulation by roots, via adsorption onto roots, or precipitation within the root zone (Sarma 2011). Phytostabilization cannot remove contaminants from a site, but it effectively immobilizes or stabilizes them and makes them unavailable for entry into the food chain. This significantly reduces the risk to human health and the environment. It also decreases the potential migration of contamination through wind erosion, transport of exposed surface soils and leaching of soil contamination to ground water (Prasad 2003). Plants chosen for phytostabilization should not be able to translocate contaminants to the above-ground plant parts suitable for consumption (Prasad 2003). Disposal of hazardous material/biomass is not required in this technique. The presence of plants also reduces soil erosion and decreases the amount of water available in the system (Berti and Cunningham 2000).

11.2.3 PHYTOVOLATILIZATION

Phytovolatilization is the absorption of a contaminant by a plant from the soil through its roots followed by the release of the contaminant or a modified form of the contaminant to the atmosphere via transpiration. Thus, the contaminant that is initially concentrated in the soil is dispersed into the atmosphere, where its concentration remains insignificantly low, posing negligible environmental risk. Phytovolatilization also has the advantage of causing minimal site disturbance, less erosion, and without any need to dispose off of the contaminated plant material (Ghosh and Singh 2005; Prasad 2003). Most of the inorganic contaminants such as heavy metals are nonvolatile. Phytovolatilization however is mainly effective for highly volatile organic contaminants, especially volatile metals like Hg, As and Se. It should be remembered that these metals are toxic even in small quantities and once volatilized, their rate of migration or their translocation to the populated areas cannot be predicted. Thus, this method should be carried out with caution in a properly selected site.

11.2.4 Phytodegradation

Phytodegradation, also called phytotransformation, refers to the breakdown of contaminants, taken up by the plants through metabolic processes within the plant, or the breakdown of contaminants external to the plants through the effect of compounds (such as enzymes) produced by the plants. Pollutants are degraded, incorporated into the plant tissues and used as nutrients (Chaudhry et al. 1998).

11.2.5 Rhizofiltration

Rhizofiltration is primarily used to remediate extracted groundwater, surface water and wastewater with low contaminant concentrations (Henry 2000). It is defined as the use of plants, both terrestrial and aquatic, to absorb, concentrate and precipitate contaminants from polluted aqueous sources in their roots (Ghosh and Singh 2005). Before introducing the plants with large root systems in the actual contaminated area, they are acclimatized in contaminated water collected from a waste site, which is used as their water source. As the roots become saturated with contaminants, they are harvested and either incinerated or composted to recycle the contaminants. Terrestrial plants are generally preferred for rhizofiltration because of their fibrous and much longer root systems, which increase the amount of available root surface area (Henry 2000). However, aquatic plants can also be used for either *in situ* or *ex situ* applications. Moreover, contaminants do not have to be translocated to the shoots. Rhizofiltration can be used for Pb, Cd, Cu, Ni, Zn and Cr, which are primarily retained within the roots (Chaudhry et al. 1998).

11.2.6 Rhizodegradation

Rhizodegradation, also called enhanced rhizosphere biodegradation, phytostimulation or plant-assisted bioremediation/degradation, is the breakdown of contaminants in the soil through microbial activity that is enhanced by the presence of the rhizosphere, and is a much slower process than phytodegradation (UNEP). Various microorganisms like yeast, fungi or bacteria can utilize harmful organic substances such as fuels or solvents as nutrient sources and degrade them into harmless substances. Natural exudates released by the plant roots such as sugars, alcohols and acids containing organic carbon, provide food for soil microorganisms and enhance their activity. The loosening of soil by plant roots and water availability in the rhizosphere also aids rhizodegradation processes (UNEP).

A summarized form of the above technologies is given in Table 11.1. These processes are however not mutually exclusive but occur simultaneously, in an effort to decontaminate the environment maximally. For example, accumulation, stabilization and volatilization of Se have been reported to occur simultaneously in a constructed wetland (Hansen et al. 1998). Therefore, one should not think to implement any of these processes in isolation while aiming for the development of a phytoremediation project. Rather, it is better to focus on an integrated approach, involving the operation of different techniques in a coordinated fashion.

TABLE 11. 1
Phytoremediation Technologies and Mechanisms Involved

Processes	Mechanisms Involved
Phytoextraction	Absorption, translocation and hyperaccumulation
Phytostabilization	Stabilization in soil, unavailable for uptake by plants
Phytovolatilization	Volatilization by leaves
Phytodegradation	Degradation and conversion to utilizable sources
Rhizofiltration	Concentration by roots from aqueous sources
Rhizodegradation	Microbe-enhanced degradation

11.3 HEAVY METALS AND PHYTOTOXICITY: PHYSIOLOGICAL AND BIOCHEMICAL RESPONSES

Although many metals are essential in trace amounts, all metals are harmful at higher concentrations due to the resultant oxidative stress induced by the formation of free radicals. Metals may also be toxic as they can replace other essential metals in pigments or enzymes disrupting their function (Henry 2000). Thus, metals may render contaminated lands unsuitable for plant growth and have an effect on the biodiversity of the region. Heavy metals are defined as metals having atomic weight between 63.54 and 200.59, with density higher than 5g cm^{-3} (Hawkes 1997). About 53 of the 90 naturally occurring elements are heavy metals. Among these metals, Fe, Mo and Mn are important as micronutrients. Zn, Ni, Cu, V, Co, W and Cr are toxic but have high or low importance as trace elements whereas As, Ag, Hg, Sb, Cd, Pb and U have no role as nutrients and seem to elicit a somewhat toxic response in plants (Schutzendubel and Polle 2002). These metals are of particular concern to surface water and soil pollution and have a high affinity for humic acids, organo clays and oxides coated with organic matter (Connell 1984). They exist in colloidal, ionic, particulate and dissolved phase. Unlike many organic contaminants, most metals cannot be eliminated from the environment by chemical or biological transformation (Cunningham and Ow 1996; Prasad 2003). Although the toxicity of certain metals may be decreased by altering their oxidation states, they do not completely degrade and persist in the environment (NRC 1998).

According to their chemical and physical properties, three different molecular mechanisms of metal toxicity can be distinguished (Schutzendubel and Polle 2002):

(a) Most of the heavy metals are transition metals with an incompletely filled ð-orbital present as cations under physiological conditions. Metals with lower redox potentials than those of biological molecules cannot participate in biological redox reactions. Autoxidation of redox active metals such as Fe^{2+} or Cu^{+} results in O_2^- formation and subsequently in the production of H_2O_2 and OH, via Fenton-type reactions. Cellular injury due to this mechanism is identified for iron and copper.

(b) The second mechanism involves the blocking of essential functional groups in biomolecules, i.e., the ability of the heavy metals to bind strongly to oxygen, nitrogen and sulfur atoms. A range of heavy metals (Ag, Hg, Cu, Co, Cd, Zn, Mn, Pb and Ni) tend to form corresponding insoluble sulfides. Thus, heavy metals can inactivate enzymes by binding to Cys residues.

(c) Another mechanism involves displacement of essential metal ions from biomolecules. Divalent cations like Co^{2+}, Ni^{2+} and Zn^{2+} were found to displace Mg^{2+} in RUBISCO and resulted in loss of its activity. Displacement of Ca^{2+} by Cd^{2+} in calmodulin resulted in inhibition of calmodulin-dependent phosphodiesterase activity in radish (Rivetta et al. 1997).

11.3.1 Inhibition of Plant Growth

Both morphogenesis and plant development are especially susceptible to heavy metal stress. The symptoms induced by an elevated level of heavy metals include growth retardation with reduced root/shoot length, plant height, fresh and dry weight of various organs and premature senescence. Symptoms also include adverse effects on seed germination, seedling vigor, plant growth and metabolism by either limiting water transport to growing tissues and leaves or impairing transpiration rate. Inhibition of cell growth or cell expansion might also occur due to direct or indirect action of heavy metals on auxin metabolism or auxin carriers (Breckle 1991).

11.3.2 Damaging Effects on Several Biomolecules

Deleterious effects of heavy metals include altered levels of proteins and amino acids, proteases and peptidases, a decline in total protein content with a substantially induced proteolysis, and the accumulation of amino acids such as proline. Palma et al. (2002) reported that the reduction in total protein content occurs during metal stress. This results from damage to membrane proteins and stimulation of protease activity, leading to proteolysis. The damage is mostly due to the induction of various oxidative stresses, involving peroxidation of polyunsaturated fatty acids of membrane lipids by reactive oxygen species (ROS), which eventually increases membrane fluidity and membrane permeability. This leads to leakage of K^+ ions and other solutes and, finally, cell death. Malondialdehyde (MDA) is one of the several low molecular weight end products formed via decomposition of certain primary and secondary lipid peroxidation products.

Increased ROS production serves as a major source of DNA damage as evidenced through Comet assay, leading to strand breakage, removal of nucleotides accompanied with a variety of modifications in organic bases of nucleotides, and inhibition of DNA repair as a signal of genotoxicity (Bhattacharjee 1998). The lipid composition of thylakoid membranes causes its swelling, the fine structures of chloroplasts degenerate or become distorted, and the affected plants are characterized by the occurrence of large plastoglobuli and disorganized lamellar structures, with reduction in chlorophyll synthesis. The rapid increase in ROS concentration within the plants is termed as the oxidative burst, which depends on the type of the metal, intensity of the metal stress, repeated stress periods and age of the plants (Foy et al. 1978; Vangronsveld and Clijsters 1994).

11.3.3 Effect on Various Enzymes

The activities of key enzymes of various metabolic pathways are influenced by heavy metals. The metals interact with free sulfhydryl (-SH) groups present at the active site of enzymes. The -SH inactivation was suggested to explain the inhibitory effect on the activity of chloroplast enzymes such as RUBISCO and phosphoribulokinase *in vitro*. Chlorophyll biosynthesis may be inhibited due to inhibition of enzymes of porphyrin biosynthesis such as protochlorophyllide reductase, ð-aminolevulinic acid (ALA) synthase and ALA dehydratase (Padmaja et al. 1990).

The activities of enzymes of antioxidative defense systems like superoxide dismutase, catalase, peroxidase and glutathione reductase are inhibited or enhanced during stress, imposed by the majority of heavy metals, due to increased levels of free radicals and peroxides. The increased activity of antioxidative enzymes appears to serve as an important component of the defense mechanism to combat metal-induced oxidative injury. The inhibition of enzyme activity may be due to inhibition of enzyme synthesis or a change in assembly of enzyme subunits at high metal concentrations (Shaw 1995; Shah et al. 2001). For some metals like Cd, increase in acid invertase and sucrose synthase activities, accompanied with decreased level of nonreducing sugars, are noted.

11.3.4 Effect on Various Metabolic Reactions

The processes of nitrogen and phosphorus metabolism in plants are affected by heavy metals. Nitrate assimilation is adversely affected due to the sensitivity of the enzymes nitrate reductase and glutamine synthetase to heavy metals leading to disturbed nitrogen metabolism (Boussama et al. 1999). A decline in the level of total phosphate, along with the inhibition in the activity of phosphorolytic enzymes such as acid phosphatase, alkaline phosphatase and inorganic pyrophosphatase, are responsible for the decreased metabolic activity and inhibited plant growth under the majority of heavy metal stress. A decrease in net photosynthesis is also a marked feature during metal phytotoxicity, primarily due to permanent stomatal closure (due to an increase in abscisic acid content), distorted chloroplast ultrastructure, Photosystem II (PSII)-damages, disturbed electron transport and inhibited activities of Calvin cycle enzymes or CO_2 deficiency in the cells (Clijsters and Van Assche 1985). Even the transport of electrons and protons in mitochondria is affected along with disorganization of the electron transport chain and oxidative phosphorylation. The succinate dehydrogenase complex in the mitochondria, for example, is the primary target susceptible to Cd. High concentration of endogenous metals in plant tissues also lead to the induction in changes in water status, viz., total water content, specific water content, water saturation deficit and transpiration, with undeveloped or defective stomata. The uptake and transport of several mineral nutrients including cations and anions are decreased (Barcelo and Poschenreider 1990). The effect of some of the individual heavy metals on the overall physiology of plants has been represented in Table 11.2.

TABLE 11.2
Effect of Some Heavy Metals on Overall Plant Physiology

Metal	Plant	Physiological Effect	References
Cadmium		Brown and short roots; decrease in root length, plant height and fruiting branch number; fewer tillers, reduced biomass, senescence; decrease in chlorophyll content and biosynthesis, distortion of chlorophyll ultrastructure, chloroplast damage due to increased chlorophyllase activity, accompanied by chlorosis; reduction in water and mineral absorption by roots; impaired uptake of essential elements like Mn and Fe; inhibition of nitrogen fixation and increase in permeability of cell membrane; excess accumulation of MDA and disturbed nutrient metabolism	Bernal and McGrath (1994); Wu and Zhang (2002); Bachir et al. (2004); Benavides et al. (2005); Somashekaraiah et al. (1992)
Arsenate		Interrupted morphological, physiological and biochemical processes including germination, shoot and root growth and biomass production; increased lipid peroxidation; induced generation of ROS through its intraconversion from one ionic form to the other; alterations in superoxide dismutase, catalase and peroxidase activity; DNA damage.	Mylona et al. (1998); Lin et al. (2008); Ahsan et al. (2008); Srivastava et al. (2007); Mishra et al. (2008)
Lead	*Triticum aestivum*	Significant inhibition in seed germination and growth of roots; NADH-dependent generation of extracellular H_2O_2 causing inhibitory effect on seed germination and shoot growth.	Yang et al. (2011)
	Vicia faba	Seedling growth inhibition; Significant induction in lipid peroxidation and HSP70 synthesis; activation in superoxide dismutase, guaiacol peroxidase, ascorbate peroxidase and glutathione (GSH), suggesting the role of glutathione-ascorbate cycle in eliminating toxicity.	Wang et al. (2010)

(*continued*)

TABLE 11.2 (Continued)
Effect of Some Heavy Metals on Overall Plant Physiology

Metal	Plant	Physiological Effect	References
Copper	*Brassica juncea* (var. TM4)	Reduction in photosynthetic pigments and increase in the levels of thiobarbituric acid reactive substances; increase in the activities of superoxide dismutase, ascorbate peroxidase, guaiacol peroxidase and catalase in a concentration and exposure time-dependent manner.	Singh et al. (2010)
Mercury	*Brassica juncea*	A significant phytotoxicity at elevated concentrations (>2 mg L^{-1}); a significant reduction in both biomass and leaf relative water content. Elevated mercury concentrations significantly changed leaf cellular structure: thickly stained areas surrounding the vascular bundles; decreases in the number of palisade and spongy parenchyma cells; and reduced cell size and clotted depositions. The palisade chloroplasts exhibited decreases in their amounts and starch grains as well as a loss of spindle shape. Antioxidant enzymes catalase, peroxidase and superoxide dismutase were the most sensitive indices of mercury-induced oxidative response.	Shiyab et al. (2009)
Aluminum		Severe inhibition of root elongation, roots become stubby and brittle and inefficient in absorption; root apex, i.e., root cap, meristem and elongation zone highly sensitive to Al; strong interaction of Al^{3+} with oxygen donor ligands (proteins, nucleic acids, polysaccharides) results in the inhibition of cell division, cell extension and transport; cellular and ultrastructural changes in leaves, disturbance in chloroplast architecture, increased rates of diffusion resistance, reduction of stomatal aperture, decreased photosynthetic activity leading to chlorosis and necrosis of leaves, total decrease in leaf number and size; induction in lipid peroxidation; disruption of cytoplasmic Ca^{2+} homeostasis; blocking of cell to cell trafficking due to callose accumulation in plasmodesmata.	Mossor-Pietraszewska (2001); Panda et al. (2009)

11.3.5 Metal Ion-Induced Metabolite Accumulation in *Brassica*

Metabolomic changes induced by heavy metals in *B. rapa* have been investigated by subjecting the plants to multiple metal stress, viz., Cu, Fe and Mn stress separately, and utilizing such techniques as ^{1}H NMR and two-dimensional NMR spectra, along with unsupervised and supervised multivariate data analysis including principal component analysis, and partial least square-discriminant analysis (Jahangir et al. 2008). Among the metabolites that showed variation, glucosinolates, phenylpropanoids, and hydroxycinnamic acids conjugated with malates were found to be the discriminating metabolites, as were primary metabolites like carbohydrates and free amino acids (Jahangir et al. 2008). Sugars, glucosinolates and some free amino acids were found in higher amounts in roots. Phenylpropanoids were found to be more concentrated in the leaves. This study showed that the effects of Cu and Fe on plant metabolism were larger than those of Mn, where the plants produced more amino acids, phenolics and glucosinolates. Xiong et al. (2006) showed that the amino acids and phenolics had a metal-chelating effect. This helps in the detoxification response of the plant. The metabolomic changes vary not only according to the type of metal but also according to its concentration; the primary and secondary metabolites increased with increasing metal concentration up to a certain point, beyond which a decrease could be observed (Psotova et al. 2003).

11.4 PLANT SPECIES USED FOR METAL REMEDIATION: HYPERACCUMULATORS OF METALS

The above section dealing with metal phytotoxicity clearly indicates the importance of heavy metal phytoremediation and the necessity of availability of plant species that can tolerate excess metal concentration. Certain plants which grow on metalliferous soil have evolved the capacity to take up and accumulate massive amounts of selected indigenous metals in their shoots and roots at levels that are toxic or inimical to normal plants. The main factors controlling the ability of phytoextraction are plant species, metal availability to plant roots, metal uptake capacity by roots, metal translocation from roots to the harvestable aboveground portions, i.e., shoots, and sequestration of metals, which allows them to tolerate high levels of the element in root and shoot cells by alleviating their toxic effects. Tolerance mechanisms involve internal detoxification and probably also cell compartmentalization and metal complexation (McGrath and Zhao 2003). For several plant species, genetic analysis has demonstrated that tolerance is controlled by some major genes, with additional modifiers determining the level of tolerance (Fusco et al. 2005). Other desirable plant characteristics for phytoextraction include the ability to tolerate adverse soil conditions (i.e., soil pH, salinity, soil structure, water content), the production of a dense root system, ease of establishment, and few disease and insect problems (Prasad 2003).

Chaney (1983) was the first to suggest the use of these "hyperaccumulators" for the phytoremediation of metal-polluted sites. They are species capable of accumulating metals at levels 100-fold greater than those typically measured in common nonaccumulator plants. By definition, a hyperaccumulator must accumulate at least 10 mg g^{-1} Hg; 100 mg g^{-1} (0.01% dry weight) of Cd, As and some other trace metals; 1000 mg g^{-1} (0.1% dry weight) of Co, Cu, Cr and Pb and 10,000 mg g^{-1} (1% dry weight) of Mn, Ni and Zn (Watanabe 1997; Reeves and Baker 2000; Prasad 2003). Hyperaccumulation of heavy metals depends on plant species, soil conditions (pH, organic matter content, cation exchange capacity) and types of heavy metals. Uptake of metals depend on the type and chemical speciation of metal and habitat characteristics of plants, i.e., terrestrial, aquatic, etc. (Sarma 2011). Approximately 400 hyperaccumulator species have been identified from at least 45 families, including the members of Asteraceae, Brassicaceae, Caryophyllaceae, Cyperaceae, Fabaceae, Lamiaceae, Poaceae, Violaceae and Euphorbiaceae (Prasad 2003). The Brassicaceae family contains a large number of hyperaccumulating species with the widest range of metals; these include 87 species from 11 genera. Some hyperaccumulators are highly metal specific, have a small biomass, slow growth habit and require careful management for commercial applications (Gleba et al. 1999). Hence, identification of novel plant species with high biomass yield, along with their ability to tolerate and accumulate multiple metals, has become important to phytoremediation (Sarma 2011). Works on Vetiver grass (*Vetiveria zizanioides*) have shown that this plant is a good candidate for a wide range of phytoremediation purposes (Danh et al. 2009). Metal hyperaccumulation and tolerance are genetically inherited traits. Recent research revealed that *A. thaliana* possesses eight genes encoding members of the type 1B heavy metal-transporting subfamily of the P-type ATPases. Three of these transporters HMA2, HMA3 and HMA4 are closely related to each other, and are most similar in sequence to the other divalent heavy metal cation transporters of prokaryotes (Hussain et al. 2004). Tolerance during hyperaccumulation results from vacuolar compartmentalization and chelation. Metal hyperaccumulating plants also have implications on human health through the food chain and can possibly exhibit elemental allelopathy (metallic compounds leached through plant parts of the hyperaccumulators would suppress the growth of other plants growing in the neighborhood) and resistance against fungal pathogens (Boyd et al. 1994; Prasad 2003).

Hyperaccumulator species have been collected from areas with soil containing greater than usual amount of metals, like polluted areas or areas geographically rich in a particular element. Among the best-known hyperaccumulators is *Thlaspi caerulescens*, commonly known as alpine pennycress. Without showing injury, it accumulated up to 26,000 mg kg^{-1} Zn; and up to 22% of soil exchangeable Cd from contaminated sites (Ghosh and Singh 2005). Various aquatic plants like *Eichhornia crassipes* (water hyacinth), *Lemna minor* (duckweed) and *Azolla pinnata* (water velvet) have been investigated for use in phytoremediation technologies like rhizofiltration, phytodegradation, and phytoextraction (Ghosh and Singh 2005). Farago and Parsons (1994) reported the bioremoval of platinum using *Eichhornia crassipes*. Many of them including *Azolla filliculoides*, *A. pinnata*, *Typha orientalis*, *Salvinia molesta* and *Polygonum hydropiperoides* Michx (smartweed) have been reported as very effective for heavy metal phytoremediation, due to their fast growth and high plant density (Qian et al. 1999). Plants currently used in phytoextraction include *Thlaspi carerulescens*, *Alyssum murale*, *A. lesbiacum* and *A. tenium*, which can accumulate high levels of Zn and Cd in shoots (Ghosh and Singh 2005). Recently, a fern *Pteris vittata* has been shown to accumulate as much as 14,500 mg kg^{-1} As in fronds without showing symptoms of toxicity (Ma et al. 2001). The relevant genes from these hyperaccumulators may be introduced into higher biomass producing nonaccumulators for an improved phytoremediation potential, making the remediation process more commercially viable.

11.4.1 Heavy Metal Remediation: Significance of *Brassica* Species

The Brassicaceae family is distinguished by their ability to accumulate the heavy metals to an extremely high degree (Broadley et al. 2001; Anjum et al. 2012). They have been shown to exhibit differential significant potential to accumulate and transport the heavy metals towards their stems.

11.4.1.1 *Brassica*—A General Account

The genus *Brassica* belongs to the family Brassicaceae (Cruciferae) and is native in the wild in Western Europe, Mediterranean and temperate regions of Asia. *Brassica* species have a long history of breeding and large collections of land races are widely distributed in several continents (Hemingway 1976). As a result of this extensive distribution of *Brassica* throughout the world, a wide range of variability for both morphological and chemical characteristics has been observed. Many of the wild species grow as weeds, especially in North America, South America, and Australia. The genus includes over 30 wild species and hybrids, and numerous additional varieties and hybrids of cultivated origin. Most are annuals or biennials, but some are small shrubs. Almost all parts of some species have been developed for food, including the root (rutabaga, turnips), stems (kohlrabi), leaves (cabbage, brussels sprouts), flowers (cauliflower, broccoli), and seeds (many, including mustard seed and oil-producing rapeseed).

The high level of genetic resemblance between *Brassica* and *Arabidopsis* has allowed *Brassica* to be considered as an alternative model system in the field of plant physiology. Its wide distribution in nature has led to the existence of extremely varied ecotypes (Jahangir 2008). As a result, *Brassica* species are considered to be one of the most important plant models in which to study the interactions between the plant and diverse environmental factors like metals in soil, UV and drought, as well as living organisms such as insects, fungi or bacteria (Lee et al. 2004; Mahuku et al. 1996; Widarto et al. 2006). *Brassica* vegetables are an invaluable source of vegetable oil and proteins and are highly regarded for their nutritional value, containing vitamins, glucosinolates, soluble sugars, fats, carotenoids as well as fibers (Thiyam et al. 2004; Bellostas et al. 2007). They are good sources of health-promoting phytochemicals including phenolics, flavonoids and hydroxycinnamic acids (Vallejo et al. 2004). They provide high amounts of vitamin C and soluble fiber. They contain multiple nutrients with potent anti-cancer properties like 3,3′-Diindolylmethane, sulforaphane and Se.

Brassica vegetables are rich in indole-3-carbinol, which helps in DNA repair and blocks the growth of cancer cells. They are also a good source of the antioxidant, namely carotenoids, broccoli having especially high levels of it. On account of all these nutritional, antioxidative and therapeutic benefits, cultivation of *Brassica* with its adequate productivity constitutes an important issue.

The various species of *Brassica* are *B. carinata* (Abyssinian mustard or Abyssinian cabbage, used to produce biodiesel), *B. elongata* (Elongated mustard), *B. fruticulosa* (Mediterranean cabbage), *B. juncea* (Indian mustard), *B. napus* [Rapeseed, Canola, Rutabaga (Swede Turnip)], *B. narinosa* (Broadbeaked mustard), *B. nigra* (Black mustard), *B. oleracea* (Kale, cabbage, broccoli, cauliflower, Brussels sprouts, Kohlrabi), *B. perviridis* (Mustard spinach), *B. rapa* (= syn *B. campestris*, field mustard) like chinese cabbage, turnip, *B. rupestris* (Brown mustard), *B. septiceps* (Seventop turnip) and *B. tournefortii* (Asian mustard). *B. napus* varieties are mainly used in food applications and also to some extent in the production of biofuel (Grispen et al. 2006). The cultivation of the most popular *Brassica* species (*B. juncea*) extends from India through western Egypt and Central Asia to Europe. Its origin probably occurred in those regions where its parents, *B. nigra* and *B. campestris*, overlap in their distribution (Prasad 2003). The possible centers of origin include Africa, China, Middle East, Southwest Asia and India. Indian mustard is important to researchers, geneticists and plant breeders in particular, because of its unique polyploid genome (Prasad 2003). Several accessions of *B. juncea* have been identified as moderate accumulators of metallic elements and are maintained by the USDA-ARS Plant Introduction Station at Iowa State University, Ames, Iowa. The benefit of using *B. juncea* seed from the plant introduction station is that the genetic integrity of the accessions is preserved through appropriate breeding techniques (Prasad 2003).

So far as phytoremediation is concerned, significant attention is paid to the different species of *Brassica*, which are distinguished as plants characterized by rapid growth, significant biomass and an appreciable capacity to take up high quantities of Cd as well as other toxic heavy metals (Dushenkov et al. 1995; Kumar et al. 1995; Rugh 2004; Ishikawa et al. 2006). Examples are *B. juncea*, *B. nigra*, *B. campestris*, *B. rapa*, *B. napus* and *B. oleracea*. *B. juncea* was shown to absorb, accumulate and transport high levels of heavy metals including Cd, Cr, Cu, Ni, Pb, Zn and Se towards their stems, under certain conditions which particularly enhance the solubility of metals in the soil (Kumar et al. 1995; Salt et al. 1995b; Ebbs and Kochian 1997; Liphadzi and Kirkham 2005). *B. juncea* concentrates Cu, Pb and Zn in its aboveground parts in amounts much higher than those detected in the metal-soluble fractions present in soil contaminated by acid waters and pyritic slurry (del Rio et al. 2000). According to Blaylock et al. (1997) and Ebbs and Kochian (1998), *B. juncea* shows a significant potential for induced phytoremediation and they suggested that it should be used with soils contaminated with heavy metals. Salt et al. (1995b) reported that the rate of metal accumulation in *B. juncea* was mediated by saturable transport system(s) and was inhibited competitively in shoots and noncompetitively in roots by Ca^{2+}, Zn^{2+} and Mn^{2+}. Observations made by Hsiao et al. (2007) reflected the fact that synthetic chelators like EDTA and diethylenetriamine pentaacetate (DTPA) were efficient in increasing the levels of Cr and Ni in the soil solutions over time and also the total uptake (mass removal from soil) of these two metals by *B. juncea* seedlings. However, use of synthetic chelators EDTA and DTPA, which also carries environmental risks, does not assist phytoextraction and considerably reduces plant shoot biomass (Hsiao et al. 2007). In such cases, natural low-molecular-weight organic acids (LMWOAs) like oxalic acid and citric acid could provide an environmentally compatible alternative for phytoremediation of Cr- and Ni-contaminated sites, which may replace the use of synthetic chelators (Hsiao et al. 2007). Do Nascimento et al. (2006) argued that metal accumulation in *B. juncea* could be achieved just as efficiently with degradable LMWOAs than by synthetic chelators. When *B. juncea* plants, grown on multiple metal-contaminated soil (Cd, Cr, Cu, Pb and Zn) for 43 days, followed by treatment with either of the chelators nitrilotriacetate (NTA) or citric acid at 5 mmol kg^{-1} soil were examined, desorption of metals from the soil was found to increase with chelate concentration, NTA being more effective than citric acid in solubilizing the metals. NTA treatment increased shoot metal concentrations by a factor of 2–3. The Cr was detected in the above-ground tissues only after NTA amendment. A significant enhancement of metal uptake was observed in NTA-treated plants for Cu and Zn.

Even though suffering from a moderate biomass reduction, *B. juncea* demonstrated an ability to survive and tolerate more metals at the same time, and in the presence of such severe multiple contamination, a compromise that makes possible the simultaneous uptake of more metal at the expense of biomass reduction cannot be disregarded. A two- to three-fold increase of Cd, Cu, Pb and Zn concentrations in shoots of *B. juncea* was noted after application of 5 mmol kg^{-1} of NTA (Quartacci et al. 2006). Wu et al. (2004) examined whether nutrient enhancement in *B. juncea* could be used to stimulate growth, and thus Cu uptake, from a soil contaminated with Cu. A combination of low N and high P produced a yield increase that was more than adequate to compensate for a slight decrease in Cu concentration, resulting in the highest Cu removal from the contaminated soil (Wu et al. 2004).

Crops like *B. oleracea* are able to accumulate significant quantities of heavy metals (Kabata-Pendias 2001), which may lead to health risks for consumers (Angelova and Ivanov 2009). On application of a structural isomer of EDTA, named S, S-ethylenediamine disuccinic acid (EDDS) at 10 mmol kg^{-1} soil, Grčman et al. (2003) found an uptake increase with a factor of 102 for Pb, 4.7 for Zn and 3.5 for Cd in cabbage leaves, respectively. Shen et al. (2002) reported that 1.5 mmol kg^{-1} of NTA increased Pb concentrations eight-fold in cabbage shoots. NTA also increased Pb translocation from roots to shoots but was shown to be less effective than EDDS for releasing Cu, but better for Zn and Pb (Tandy et al. 2004). Meers et al. (2005) found that of the four agronomic species tested, *B. rapa* exhibited the highest affinity for accumulating Cd and Pb from the soil, either with or without additional use of mobilizing soil amendments. On application of EDDS at 5 mmol kg^{-1} soil, Kos and Lestan (2003) found an 11-fold increase in Pb uptake in *B. rapa* while the soluble Pb fraction increased 12 times. With 10 mmol kg^{-1}, the soluble Pb concentration increased 215-fold, while plant uptake increased 50-fold (Duquène et al. 2009). Marchiol et al. (2004) found that two *Brassica* species (*B. napus* and *Raphanus sativus*) were moderately tolerant when grown on a multi-metal contaminated soil. They concluded that this species could possibly be used with success in marginally polluted soils where their growth would not be impaired, and the extraction of heavy metals could be maintained at satisfying levels.

B. napus is one of the oldest cultivated oil crops in Europe; it is cultivated since the 14th century. It has a higher biomass compared to natural metal hyperaccumulators and this makes *B. napus* very suitable for phytoextraction (Grispen et al. 2006). Phytoextraction with *B. napus* has the potential to become a profitable enterprise when combined with biofuel production. *In situ* experiments were carried out at two contaminated sites in Belgium and the Netherlands. Cd concentrations in 18 different rape seed accessions ranged between 3.6 to 8.1 mg kg^{-1} dry weight (DW) at a total soil Cd concentration of 5.5 mg of Cd kg^{-1} soil DW for the Belgian site, and between 5.2 and 11.3 mg kg^{-1} DW at a total soil Cd concentration of 2.5 mg Cd kg^{-1} soil DW for the Dutch site (Grispen et al. 2006). Ebbs and Kochian (1997) reported Cd concentrations in *B. napus* shoots of 3 mg kg^{-1} DW after a growing period of 3 weeks in a neutral silty soil with total Cd concentrations of 40 mg kg^{-1}. Zn concentrations in the same plants were ~600 mg kg^{-1} DW. No adverse effects on plant growth were observed at these concentrations. Also in case of *B. napus*, a significant genotypic difference was noticed between four varieties regarding Zn and Cd accumulation. For the two varieties, Jumbo and Drakkar, the accumulation was more important in the stems and petioles, whereas this accumulation was at a maximum level in the root system for the two varieties, Cossair and Pactol. Jumbo and Drakkar seemed more efficient in phytoextraction since both showed a significant increase in Zn and Cd accumulation in all parts of the plants, particularly the aerial parts (Ghnaya et al. 2009). In another work by Angelova and Ivanov (2009), the phytoextraction abilities of the *B. nigra*, along with the quantities and the depots of accumulation of Pb, Cu, Zn and Cd in the vegetative and reproductive organs was established. The impact of the combined contamination of the soil with Pb, Zn and Cd on the growth of the *B. nigra* was ascertained. The distribution of the heavy metals without Cu in the organs of the *B. nigra*, when it is grown in conditions of soil contamination, has a selective nature, which decreases in the following order: roots > fruit's shells > stems > seeds. When the *B. nigra* is grown in a non-contaminated area, the order changes to roots > fruit's shells > seeds > stems for the Pb, whereas, regarding Cu, Zn and Cd contaminations, the order changes to fruit's shells > seeds > stems > roots. It was thus proposed that *B. nigra* is tolerant towards the heavy metals and could be successfully grown in regions of low and moderate contamination by Pb, Zn, and Cd, without lowering the quantity and quality of the manufactured production. A relationship is determined between the quantity of the common and the mobile forms of metals on one hand, and their total quantity in the plants in the field, as well as, in the laboratory experiments, on the other.

In vitro breeding and somaclonal variation have been used to improve the potential of *B. juncea* to extract and accumulate toxic metals (Nehnevajova et al. 2007). When *B. juncea* somaclones were regenerated from metal-tolerant callus cells, different phenotypes with improved tolerance to Cd, Zn and Pb were observed under hydroponic conditions. Enhanced metal accumulation occurred in both shoots and roots, whereas metal translocation was limited from roots to shoots, and reduced accumulation resulted in shoots and roots. Several variants showed a significantly higher metal extraction than the control plants. The improvement of metal shoot accumulation of the best regenerant (3× Cd, 1.6× Zn, 1.8× Pb) and metal extraction (6.2× Cd, 3.2× Zn, 3.8× Pb) indicated a successful breeding and selection of *B. juncea*, which could be used for phytoremediation purposes. Purakayastha et al. (2008) compared the heavy metal accumulation capacity of *B. juncea*, *B. campestris*, *B. carinata*, *B. napus* and *B. nigra*. *B. carinata* showed the highest concentration (mg kg^{-1}) as well as uptake (μg pot^{-1}) of Ni and Pb at maturity. Although *B. campestris* showed a higher concentration of Zn in its shoots (stem plus leaf), *B. carinata* extracted the largest amount of this metal due to greater biomass production. *B. carinata* cv. DLSC1 showed the most encouraging prospects, showing greater uptake of Zn, Ni, and Pb, while *B. juncea* cv. Pusa Bold showed the highest uptake of Cu from the soil. *B. napus* also showed promise, as it ranked second with respect to total uptake of Pb, Zn, and Ni, and third for Cu. *B. carinata* was thus reported as a promising phytoextractor for Zn, Ni and Pb, having higher effectiveness than *B. juncea*.

11.4.1.2 Metals-Specific Reports on *Brassica* Species

11.4.1.2.1 Aluminum

The Al tolerance of *B. napus* has been reported to be enhanced by increasing organic acid biosynthesis through overexpression of citrate synthase or malate dehydrogenase genes derived from plants or bacteria (Panda et al. 2009). Another potential mechanism of Al exclusion has been identified through organic acid efflux; this mechanism is the exudation of phenolic compounds. These phenolics form strong complexes with Al^{3+} at neutral pH and were implicated in internal Al detoxification in Al-accumulating species (Matsumoto et al. 1976). The two genes *BnALMT1* and *BnALMT2* from *B. napus* showed Al tolerance (Panda et al. 2009). Basu et al. (1994) reported that transgenic *B. napus* overexpressing *MnSOD* gene acquires an Al resistance phenotype.

11.4.1.2.2 Arsenic

Arsenic (As), a nonessential metalloid has posed serious environmental threats due to its high levels in groundwater in many regions of the world. As contamination may occur from natural sources (through dissolution of As compounds adsorbed onto pyrite ores into water) by geochemical factor(s) as well as from anthropogenic activities (e.g. through the use of insecticides, herbicides and phosphate fertilizers, and from the semi-conductor industry) (Mondal et al. 2006). However, the worst affected regions in Bangladesh and West Bengal from India have been created due to natural processes (Tripathi et al. 2008). Contamination is of serious concern as the metalloid can enter the food chain through irrigation using contaminated groundwater (Meharg 2004). Detoxification of As can occur by chelation with sulfur-containing ligands, like GSH and PCs (Bleeker et al. 2006). The complexes can then be sequestered in vacuoles (Raab et al. 2005). PC synthesis has been observed to increase under As stress in hypertolerant or hyperaccumulator plants (Hartley-Whitaker et al.

2001; Cai et al. 2004; Srivastava et al. 2007; Mishra et al. 2008). PC synthesis under As stress depends on an increase in the synthesis of the precursor GSH. GSH synthesis is further linked to an increase in sulfur assimilation into Cys (Rother et al. 2008). It can thus be hypothesized that exposure to metalloids might lead to sulfur deficiency in plants, and sulfur starvation responses could be induced (Srivastava et al. 2009). *B. juncea* has a significant potential for metalloid accumulation and hence is a suitable candidate for phytoremediation prospects (Gupta et al. 2009). Recent studies by Catarecha et al. (2007) and Abercrombie et al. (2008) showed that As (V)-sensing mechanism acts in opposition to the phosphate-sensing pathways. As (V) exposure represses the genes induced by phosphate starvation by acting as a chemical analog of phosphate. Srivastava et al. (2009) reported that the As-tolerant variety of *B. juncea*, called TPM-1, showed higher accumulation of As upon exposure to both 500 mM As (V) and 250 mM As (III), viz., 49 $\mu g\ g^{-1}$ and 37 $\mu g\ g^{-1}$ DW respectively after 15 days of exposure. This variety also showed better response of thiol metabolism as compared with the responses observed in the As-sensitive variety (TM-4). The stimulated GSH synthesis played a major role in maintaining a high GSH/GSSG ratio in TPM-1. Not only that, significant increases in the levels of nonprotein thiols (NP-SH) and Cys and the activities of serine acetyltransferase and cysteine synthase were observed in TPM-1, which were positively correlated with As accumulation. The increase in the activity of γ-glutamylcysteine synthetase (γ-ECS), a rate-limiting enzyme of GSH biosynthesis, is attributed to an increase in consumption of GSH for the synthesis of PCs. In TPM-1, the sufficient chelation of As with GSH and PCs led to low/no free As. When transcriptional profiling of selected genes that are known to be responsive to sulfur depletion and/or metalloid stress was conducted in As (III)-treated plants, it showed an up regulation of sulfate transporters and auxin and jasmonate biosynthesis pathway genes, whereas there was a down regulation of ethylene biosynthesis and cytokinin-responsive genes only in TPM-1. This suggested that perception of As-induced stress was presumably mediated through an integrated modulation in hormonal functioning that led to both short- and long-term adaptations to combat the stress. The sensitive variety did not show similar coordinated hormonal response. Srivastava et al. (2009) concluded that an early perception of As-induced stress, followed by coordinated responses of various pathways, was responsible for As tolerance in TPM-1. There is evidence that sulfate transporters can also mediate the transport of related oxyanions (Leustek 1996). The higher expression of sulfate transporters could account for greater As accumulation, tolerance and detoxification in TPM-1 than that observed in TM-4.

Differential responses of the two *B. juncea* varieties Varuna and Pusa Bold to arsenite [As (III)] stress has been reported. The As (III) accumulation in both the varieties was dose- and duration-dependent. The increased tolerance in Pusa Bold was suggested to be due to greater induction of superoxide dismutase and guaiacol peroxidase isozymes, activation of MAP kinase (MAPK) and up regulation of *phytochelatin synthase* (*PCS*) gene transcript which is responsible for the production of metal binding peptides (Gupta et al. 2009). Khan et al. (2009) reported that Pusa Jai Kisan variety of *B. juncea* was able to detoxify low As level through induction of antioxidant defense mechanisms utilizing the antioxidative enzymes like superoxide dismutase, ascorbate peroxidase and glutathione reductase, or the antioxidant metabolites (glutathione and ascorbate). The shoots accumulated more As than roots.

The bioaccumulation of As by plants may provide a means of removing this element from contaminated soils and waters. Pickering et al. (2000) have suggested an approach whereby As bioaccumulation in the shoots of *B. juncea* could be promoted. Arsenate (As V) is absorbed by the roots in the form of a phosphate analog, possibly using the phosphate transport mechanism. A small fraction of it is exported to the shoot and aboveground tissues through the xylem, as the oxyanions arsenate (As V) and As (III). In roots, the As (V) is reduced to As (III), and coordinated by three sulfur ligands, which can be modeled as the As (III)-tris-glutathione complex. As (III) having a high affinity for thiols, forms the As (III)-tris-glutathione complex in presence of excess glutathione. Thus the initial reduction of As (V) in *B. juncea* is by glutathione. In the shoot, As is stored as an As (III)-tris-thiolate complex. The majority of the As remains in the roots as an As (III)-tris-thiolate complex. Both glutathione and phytochelatin(s), viz., [PC(s)], probably act as the thiolate donors. Studies showed that the addition of the dithiol As-chelator dimercaptosuccinate to the hydroponic culture medium caused significant increase in As level in leaves. Pickering et al. (2000) suggested that the addition of dimercaptosuccinate to As-contaminated soils offered a way of removing As (V) from contaminated soils, which could be further developed for use as an efficient method of phytoremediation.

Isatis capadocica, a brassica collected from Iranian As-contaminated mine spoils, exhibited As (V) hypertolerance, actively growing at concentrations of >1 mM As (V) in hydroponic solution and accumulating >100 mg kg^{-1} DW of As in its shoots. It also had a shoot: root transfer ratio of >1. The ability to accumulate As was exhibited in both hydroponics and contaminated soils. Tolerance in *Isatis capadocica* was not achieved through suppression of high-affinity phosphate/arsenate root transport, unlike other monocotyledons and dicotyledons. A high percentage (>50%) of As in the tissues was found to be complexed with PCs. However, it is argued that this is a constitutive rather than an adaptive mechanism of tolerance (Karimi et al. 2009).

11.4.1.2.3 Cadmium

Cd is a nonessential element for plant mineral nutrition. However, it is taken up easily by roots and transported through the xylem to the vegetative and reproductive organs. It thus adversely affects nutrient uptake and homeostasis and inhibits root and shoot growth and plant productivity (di Toppi and Gabbrielli 1999; Metwally et al. 2005; Fusco et al. 2005). The roots also start appearing thinner, and send out side roots. The other sensitive responses of higher plants to Cd involve cell wall rigidification and lignification, together with stomatal closure, involving ABA as a signal transduction compound. In nontolerant plants, Cd has been reported to alter RNA synthesis, inhibit ribonuclease activity, chlorophyll synthesis, and absorption and transport of Zn, Fe and Cu, and to interfere with the movement of K^+, Ca^{2+} and ABA in guard cells (Minglin et al. 2005). In some plants, photosystems (PS) I and II are found to be damaged, PSII being more sensitive. The metal also affects PSII reaction centers and electron transport chain by changing enzyme activity and/or protein structure (Haag-Kerwer et al. 1999). Cd may also result in growth inhibition, root damage, chlorosis and affect transpiration (Heiss et al. 2003). The protein denaturation, inhibition of chlorophyll synthesis and damage to photosynthetic apparatus are the other effects of Cd toxicity (Fusco et al. 2005). Studies by Larsson et al. (1998) showed a drop in chlorophyll content and photochemical quantum yield of photosynthesis, after studying seedlings of *B. napus* exposed to Cd under high light intensity for an extended period of time. Chaoui et al. (1997) also reported lipid peroxidation in response to Cd. Cd appeared to accumulate preferentially in roots; part of it is translocated from root to shoot, where it finally accumulates in the leaves (Salt et al. 1995a, b; Clemens et al. 2002). High accumulation was also demonstrated in the trichomes covering the leaf surface (Salt et al. 1995a). It has been reported that *B. juncea* seedlings grown in aquaculture were able to accumulate and remove Cd from contaminated water. Ru et al. (2004) reported a maximum Cd uptake of 577 µg 3 kg^{-1} soil by *B. juncea* shoots grown in a soil, amended with 84 mg kg^{-1} of Cd after 42 days of growth. *B. juncea* was shown to accumulate large amounts of Cd, with bioaccumulation coefficients of up to 1100 in shoots and 6700 in roots at non phytotoxic concentrations of Cd (0.1 µg ml^{-1}) in solution. It can accumulate more than 400 µg g^{-1} DW of Cd in its shoot, which can be easily harvested (Minglin et al. 2005). This was associated with a rapid accumulation of PCs in the root, where the majority of the Cd was coordinated with sulfur ligands, probably as a Cd-S complex. On the other hand, Cd transporting through the xylem sap was coordinated predominantly with oxygen or nitrogen ligands. Cd concentrations in the xylem sap and the rate of Cd accumulation in the leaves showed similar saturation kinetics. Thus it was speculated that the process of Cd transport from solution through the root and into the xylem is mediated by a saturable transport system(s). Cd translocation to the shoot appeared to be driven by transpiration, since abscisic acid (ABA) dramatically reduced Cd accumulation in leaves. Cd was preferentially accumulated in trichomes on the leaf surface and this was assumed to be a possible detoxification mechanism (Salt et al. 1995a, b).

Several Cd-regulated genes have been identified by cDNA-AFLP in *B. juncea*. The expression of an aldehyde dehydrogenase and an RNA-binding protein involved in ABA signalling, induced by Cd in *B. juncea*, corroborates the idea of existing cross-talk between Cd-induced and water stress-induced signaling which also employs ABA as a signal transduction compound (Polle and Schutzendubel 2003). Genes encoding for glutathione S-transferases (GSTs) and cytochrome P450 were reported as responsive to Cd and other stresses (Marrs and Walbot 1997; Suzuki et al. 2001) and are probably functioning in cytotoxic product detoxification. In particular, GSTs catalyze the synthesis of glutathione S-conjugates, allowing them to be recognized for transport into the vacuole. Moreover, in *B. juncea*, changes of expression of a glutathione transporter, in response to Cd exposure, have been reported, also indicating that glutathione plays a prominent role in Cd accumulation and/or detoxification.

Chelation of metals by high-affinity ligands, such as PCs and thiolated peptides, is considered to be a principal mechanism of Cd detoxification in plants. Numerous physiological, biochemical and genetic studies have confirmed that glutathione is the substrate for PC biosynthesis, as has already been discussed (Cobbett 2000b; Cobbett and Goldsbrough 2002). During Cd exposure, glutathione export to the various sink tissues is reduced to meet the high demand of glutathione for PC synthesis during Cd accumulation. A gene encoding for O-acetylserine (thiol) lyase enzyme (OASTL) was up-regulated by Cd exposure. OASTL catalyzes the last step of Cys biosynthesis and *Arabidopsis* plants, overexpressing OASTL, showed high Cd resistance, suggesting that the Cys pool requirement for glutathione biosynthesis is the main factor for tolerance (Dominguez-Solís et al. 2001). The transcription of aquaporins PIP1 and PIP2, seen in *B. juncea* upon exposure to Cd for 24 h, together with the expression of other drought and ABA-responsive genes, strengthens the idea that Cd imposes water stress and that both ABA and Cd act synergistically (Fusco et al. 2005). The CRUSTAL analyses of the PCS clone, BjPCS1 from *B. juncea,* showed 90% identity with *A. thaliana* AtPCS1 and *Thlaspi caerulescens* TcPCS (Heiss et al. 2003). An increase in PCS levels was observed in *B. juncea* leaves after prolonged Cd treatment. It was interesting that this increase was not due to increase in BjPCS1 mRNA, but probably to post-transcriptional regulation (Heiss et al. 2003). Transcript levels of two ARD/ARD family proteins (which are aci-reductone dioxygenase enzymes) were reported to be significantly increased in *B. juncea* 48 h after Cd exposure (~22.4 ppm) in both leaves and roots (Minglin et al. 2005). The Catalase 3 gene (*CAT3*) of *B. juncea* was also found to be up-regulated after 48 h exposure to Cd, which may have a role in regulating the increased intracellular H_2O_2 levels in plants facing Cd stress (Minglin et al. 2005). Diacyl glycerol (DAG) kinase is linked to the stress-signaling molecule, phosphatidic acid (PA), in the lipid-signaling pathway, which plays a role in the rescue of plants from oxidative stress damage. Cd exposure also showed an increase in DAG kinase mRNA expression (3.3-fold) in *B. juncea* roots (Minglin et al. 2005). A transcription factor of the C_3HC_4-type RING zinc-finger protein family was also found to have increased transcript levels in leaves of *B. juncea*, 48 h after exposure to Cd stress (Minglin et al. 2005).

Cd has also been stated to activate the auxin-signaling pathway, which may be involved in producing proteins responsible for Cd tolerance in plants. An auxin-responsive GH3 family mRNA was found to be overexpressed in both the roots and leaves of *B. juncea* under Cd stress (Minglin et al. 2005). The levels of proteins related to pre-RNA processing (of the Nop family), vesicle transport and formation (ARF-like small GTPases/ARF), and stabilization of membrane proteins were also found to be affected by Cd exposure (Minglin et al. 2005). Members of the Nop family may have a role in overcoming the damage of nucleoli or ribonuclease activity induced by Cd. The ARFs (ADP-ribosylation factors) have been suspected to play a regulatory role in coping with multi-stress effects of Cd accumulation in the plant (Minglin et al. 2005). Functional studies of these genes will further help in optimizing their potential for phytoremediation purposes.

Seth et al. (2008) tested the capacity of *B. juncea* to accumulate Cd in the root and shoot, and hence their tolerance to high Cd concentrations, by exposing the plants to low, moderate and high concentrations of Cd. Increase in bioaccumulation was directly proportional to Cd concentration and exposure period. In the case of Cd stress alone, roots showed more susceptibility and metal confinement than shoots. However, Cd stress in conjunction with the chelator EDTA, resulted in desorption of Cd in the cell wall of roots, and translocation from root to the shoot through xylem vessels. Use of EDTA reduced toxicity, as indicated, at the level of biomass production, photosynthetic pigments and protein content. Heavy metal concentration in shoots and roots of both *B. juncea* and *B. napus* increased greatly when EDTA addition levels were increased up to 6 mmol kg^{-1} but decreased beyond it, due to a decrease in total dry matter weight (Turan et al. 2007). Thus, the use of appropriate amount of EDTA in soil, coupled with the use of a suitable plant species, may result in the effective clean-up of Cd contaminated soil. The ability of *B. juncea* to combat Cd-induced oxidative stress was conferred through enhanced level of PCs, reduced glutathione (GSH), nonprotein thiols (NP-SH) and high activity of glutathione reductase (GR), thereby providing sufficient GSH not only for PC synthesis but also for antioxidant function. Enhanced levels of PCs indicated the plant capacity to detoxify the metal via chelation and sequestration in vacuoles. Increase in PC levels in *B. juncea* could be due to activation of phytochelatin synthase (PCS) gene. Further, synthesis of PC was significantly correlated with metal accumulation in *B. juncea*, whereas at higher concentrations, a decline in PC content was observed, due to either degradation or reduction in GSH pool, declined GR activity or inactivation of PCS enzyme. With Cd + EDTA, induction of PC was lower than with Cd alone, due to the binding of free Cd with EDTA, leading to the low availability of free Cd for the induction of PC (Seth et al. 2008).

In another study by Liu et al. (2006) on the uptake of Cd by twenty cultivars of *B. pekinensis* and *B. chinensis*, all the cultivars were found to accumulate more than 150 μg g^{-1} DW of Cd in their shoots, meaning that they served as Cd hyperaccumulators. The maximum Cd concentrations in the shoots of *B. chinensis* and *B. pekinensis* were 278 μg g^{-1} and 255 μg g^{-1} respectively. No visual symptoms of metal toxicity were found on plants grown in Cd-contaminated soil. The amount of Cd phytoextracted in one crop cycle was 99–242 μg kg^{-1} soil by *B. chinensis* cultivars and 155–458 μg kg^{-1} soil by *B. pekinensis* cultivars during the 36-day plant growth period. Between the two species, *B. pekinensis* grew much more rapidly, yielding higher biomass than *B. chinensis*. Shen et al. (2002) earlier showed that 1.9 μg g^{-1} of Cd accumulated in the shoots of *B. chinensis*, grown in soil containing 15 μg g^{-1} of Cd, which was probably due to high concentration of Zn, Cu and Pb in the soil. The *E. coli gshII* gene, encoding the enzyme GSH synthetase, was overexpressed in the cytosol of *B. juncea*, when the transgenic plants were found to accumulate significantly more Cd than the wild-type plants under high Cd concentration. The transgenics also exhibited higher concentration of GSH, PCs, thiol, S and Ca than wild-type plants (Zhu et al. 1999a, b). In another study, a bacterial glutathione reductase was overexpressed in *B. juncea*, targeted to the cytosol or the plastids (Pilon-Smits et al. 2000). The plants exhibited enhanced Cd tolerance at the chloroplast level.

11.4.1.2.4 Chromium

Chromium (Cr) is a heavy metal that causes serious environmental contamination in soil, sediments, and groundwater (Bartlett 1991; Sinha et al. 2002; Shankar et al. 2005). The tanning industry is one of the major contributors of Cr contamination, mostly coming through wastewater containing high amounts of Cr (1.07–7.80 mg l^{-1}) (Diwan et al. 2008). Worldwide anthropogenic discharge of Cr in fresh water bodies has been estimated to be 3550 mt (Nriagu 1990). Cr (VI) is a very toxic, powerful epithelial irritant and an established human carcinogen by International Agency for Research on Cancer (IARC), the Environmental Protection Agency (EPA), and the World Health Organization (WHO). Diwan et al. (2008) focused on phytoremediation potential of *B. juncea* genotypes for Cr (VI) with their responses to Cr accumulation, its phytotoxity and up regulation of antioxidative systems. At 100 μM Cr treatment, Pusa Jai Kisan accumulated the maximum amount of Cr (1680 μg g^{-1} DW), whereas Vardhan accumulated the minimum (107 μg g^{-1} DW). The fully matured plant of Pusa Jai Kisan accumulated considerably high amounts of Cr in its aerial parts (1.3–4.1 mg g^{-1} DW). Although all the genotypes responded to Cr-induced oxidative stress by modulating the non enzymatic antioxidants such as GSH and ascorbate, and enzymatic antioxidants like superoxide dismutase, ascorbate peroxidase and glutathione reductase, the level of induction was the maximum in Pusa Jai Kisan and the minimum in Vardhan. Pusa Jai Kisan, being hypertolerant to Cr-induced stress, was suggested as a viable candidate in the development of phytoremediation technology of Cr-contaminated sites.

11.4.1.2.5 Copper

In a study conducted by Ariyakanon and Winaipanich (2006), the total amount of Cu accumulated in shoots and roots of *B. juncea* was high (3,771 mg kg^{-1}), although the Cu concentration in shoots did not exceed 1,000 mg kg^{-1} DW. Therefore, *B. juncea* could be regarded as a hyperaccumulator for Cu remediation in contaminated soils.

11.4.1.2.6 Lead

Lead (Pb) contamination, mostly caused by anthropogenic activities, has become a major problem in recent years. *B. juncea* has been found to have a good ability to transport Pb from the roots to the shoots. *B. juncea* has phytoextraction coefficient of 1.7 and it has been found that a Pb concentration of 500 mg l^{-1} is not phytotoxic to *Brassica* species (Henry 2000). The phytoextraction coefficient is the ratio of the metal concentration found within the surface biomass of the plant, over the metal concentration found in the soil. Some calculations indicate that *B. juncea* (L.) Czern is capable of removing 1.1550 kg of Pb per acre (Henry 2000). Liu et al. (2000) and Clemente et al. (2005) proposed that despite the fact that *B. juncea* shows a very good ability to eliminate Pb from the soil solution and to accumulate it, the plants convey and absorb small quantities of Pb in their stems. They can also accumulate high levels of other metals, including Zn and Se. The metal-accumulating ability of this plant, coupled with its potential to rapidly produce large quantities of shoot biomass, makes this plant ideal for phytoextraction (Montes-Bayon et al. 2002). In another observation by Salido et al. (2003), it was shown that on application of 10 mmol EDTA kg^{-1} soil, in soil samples containing 338 mg Pb kg^{-1} soil, *B. juncea* plants extracted approximately 32 mg of Pb.

11.4.1.2.7 Mercury

Mercury (Hg), a potent neurotoxin, is released to the environment in significant amounts by both natural processes and anthropogenic activities. The global cumulative Hg production was estimated in 2000 as 0.64 million metric tons, 3.5–6.7% of which is derived from the global coal industry (Han et al. 2002). It has been estimated that the annual anthropogenic input of Hg into the environment is as high as 6×10^6 kg yr^{-1} (Han et al. 2002). Taking into account both anthropogenic and natural sources, a total of about 741×10^6 kg of Hg has been released into the atmosphere, 118×10^6 kg released into water and 806×10^6 kg released into the soil. In the United States, combustion of fossil fuels for power generation is estimated to generate ~30% of the total release of Hg into the atmosphere (Harriss and Hohenemser 1978). Forty US states have issued advisories for methyl mercury on selected water-bodies and 13 states have statewide advisories for some or all sport fish from rivers or lakes (USGS). Hg is a strong phytotoxic and genotoxic metal. It leads to oxidative stress through Fenton-type reactions, generation of free radicals, and induction of expression of genes encoding superoxide dismutase, peroxidase and catalase (Shiyab et al. 2009). Antioxidant enzymes are one of the most sensitive indices for the adaptation and response of *B. juncea* plants to Hg stress. Mercuric ions have serious consequences as they tend to form covalent bonds with DNA. Hg has also been observed to induce sister chromatid exchanges in plant nuclei (Beauford et al. 1977). It has a great affinity for biomolecules containing sulfhydryl (SH) groups (Goyer 2001), and a lower affinity for phosphate, carboxyl, amide and amine groups. This is the reason for its high toxicity.

Shiyab et al. (2008) showed that the two cultivars of *B. juncea*, namely Florida Broadleaf and Longstanding, demonstrated an efficient metabolic defense and adaptation system to Hg-induced oxidative stress. The Hg uptake induced a strong antioxidative response in the two cultivars, with effective generation of catalase to scavenge H_2O_2, resulting in lower H_2O_2 in shoots, especially for higher Hg concentrations. Most of the Hg was accumulated in the roots. Translocations of Hg from roots to shoots were found to be low in the two cultivars. Due to high accumulation of Hg in both shoots and roots, *B. juncea* was proposed to be a potential candidate plant for phytofiltration of contaminated water and phytostabilization of Hg-contaminated soils, in spite of high Hg phytotoxicity (Shiyab et al. 2009). Recently, enhancement of tolerance of *B. juncea* to Hg was achieved through the application of 0.2 mM CO through increased activities of catalase, ascorbate peroxidase and guaiacol peroxidase, as well as the accumulation of proline and nonprotein thiols that can chelate heavy metals (Meng et al. 2011).

11.4.1.2.8 Molybdenum

Molybdenum (Mo) pollution due to mining and the stainless steel industry poses a serious environmental problem at several locations in the U.S., including several Superfund sites (polluted sites in the U.S. designated by the U.S. Environmental Protection Agency for high priority remediation). Plant samples from a mining site near Empire, Colorado were found to contain up to 400 mg kg^{-1} Mo (Trlica and Brown 2000). Phytoremediation seems like a suitable option for remediating Mo in these areas. Studies on *B. juncea* seedlings showed Mo concentration in the vacuoles of the epidermal cells (Hale et al. 2001). When supplied with molybdate, Mo is accumulated in the form of polymolybdate(s) in the plant, which forms complexes with anthocyanins and accumulated as water-soluble blue crystals in the peripheral cell layers. Hale et al. (2001) showed that an anthocyanin-less mutant of *B. rapa,* when supplied with molybdate, did not show accumulation of a blue-colored complex in the peripheral cell layers. Mo concentration or accumulation in shoots by three *B. rapa* varieties was related to the anthocyanin content. Mo is sequestered in the vacuoles of the peripheral cell layers of *Brassica* spp. as a blue Mo-anthocyanin complex. Vacuolar sequestration thus separates Mo from the basic functioning of the cell (Hale et al. 2001). The separation obviously reduces metal toxicity and allows better growth. Faster growth probably enhances Mo accumulation because metal translocation through the xylem is thought to be driven by transpiration (Salt et al. 1995a, b).

11.4.1.2.9 Nickel

Phytoremediation can also be mediated through the use of chelating agents to enhance heavy metal uptake (Brooks and Robinson 1998). Huang et al. (1997) reported that plant uptake of heavy metals could be enhanced by the application of chelators like EDTA, which boost metal absorption by crops through desorption from soils. According to Anderson et al. (1999), EDTA appeared to be effective in inducing accumulation of metals with higher metal solubility. Panwar et al. (2002), while working on *B. juncea* and *B. carinata,* showed that the former was more tolerant to and a higher accumulator of Ni than the latter, irrespective of EDTA application to the soil at the rosette stage. On EDTA application, the Ni concentration doubled in both the species, and the translocation of the metal to the shoot from the root also seemed to be restricted in both the *Brassica* species. In another study, Giordani and Cecchi (2005) showed that cabbage seems effective in Ni uptake and storage. Knowing that the dry matter production of cabbage is 11 tonnes ha^{-1}, the quantity of Ni per ha removed is 341 g.

11.4.1.2.10 Selenium

Anthropogenic Se pollution can result from aqueous discharges from electric power plants, coal ash leachates, refinery effluents and also from industrial wastewater. Se occurs naturally in soils formed from Se-bearing shales (LeDuc et al. 2004). This leads to Se-contaminated irrigation drainage water, one of the most serious agricultural problems in the western United States and other areas with similar environment and geological conditions. Phytoremediation potential for Se removal has been reviewed by Switras (1999). Most plants cannot discriminate between selenate (SeO_4^{2-}) and sulfate (SO_4^{2-}) ions present in soil solution. Consequently, SeO_4^{2-} is easily taken up by the plants and may utilize the sulfur (S) pathway (LeDuc et al. 2006). The incorporation of Se into proteins in plant tissues is believed to be the primary reason for Se toxicity in plants. Proteins containing Se-amino acids may be different in stability and catalytic or regulatory properties from those containing the S-amino acids. Retarded plant growth and impaired protein synthesis results in plants grown under seleniferous soils (Banuelos et al. 1997a, b). The assimilation of SeO_4^{2-} results in Se incorporation into analogs of the S-containing amino acids, Met and Cys. In Se-nonaccumulator plants, selenocystathionine (derived from the seleno amino acid selenocysteine) is metabolized and converted to selenomethionine (Wu et al. 2009). Both selenocysteine and selenomethionine can be incorporated into proteins by substituting for S-containing amino acids Cys and Met. The substitution for Met by selenomethionine may contribute to symptoms of physiological Se toxicity. Therefore, S-accumulating plant species within the family Cruciferae might take up high Se concentrations, if grown on soils containing high concentrations of soluble SeO_4^{2-} (LeDuc et al. 2006). Plants exposed to Se may avoid Se toxicity by preferential incorporation of S-containing amino acids into proteins, rather than their Se isologs (Zayed and Terry 1992; Terry and Zayed 1994). The Se accumulators can accumulate large amounts of Se (>1000 mg Se kg^{-1} DW). Studies have found that *B. juncea* accumulated higher concentrations of Se than other plant species tested (Banuelos et al. 1993). Se phytoremediation has been achieved under field conditions using *B. juncea* (Banuelos et al. 1997a, b), which accumulates Se to hundreds of parts per million (Banuelos and Schrale 1989). Compared to Se accumulator species that limit the incorporation of endogenous selenoamino acid into selenomethionine, the *Brassica* species incorporate some Se into Cys (not detected as free amino acids), which interfered with the formation of disulfide bridges between adjacent polypeptide chains (Brown and Shrift 1982). Because the other free seleno-amino acids were not consistently detectable under these extraction conditions, selenomethionine, relative to the other seleno-amino acids, appeared to be the seleno-amino acid analog not yet stored by the *Brassica* land races under these Se conditions (LeDuc et al. 2004). Se uptake and tolerance also differ widely in the different *Brassica* land races. Se accumulation by different land races of *B. juncea* Czern and Coss and one land race of *B. carinata* was investigated in Se-enriched water and soil cultures containing 2 mg of Se kg^{-1}. In water culture, Se concentrations in the shoot, among the land races, ranged from 501 to 1017 mg of Se kg^{-1} DW, whereas concentrations ranged from 407 to 769 mg of Se kg^{-1} DW in plants grown in Se-laden soil (Banuelos et al. 1997). The assimilation of SeO_4^{2-} by *Brassica* may use the same pathway as SO_4^{2-} to enter the plant. This assumption is consistent with observations that S-accumulating species of Cruciferae often accumulate more Se than other plant species (Mikkelsen et al. 1989). Terry et al. (1992) reported that members of the Brassicaceae are capable of releasing up to 40 g Se ha^{-1} day^{-1} as various gaseous compounds through phytovolatilization.

LeDuc et al. (2004) overexpressed the gene encoding selenocysteine methyltransferase (SMT) from the selenium (Se) hyperaccumulator *Astragalus bisulcatus* in *B. juncea*. SMT detoxifies selenocysteine (SeCys) by methylating it to methylselenocysteine, a nonprotein amino acid. This reduces the intracellular concentrations of SeCys and selenomethionine (Neuhierl and Bock 1996), thereby diminishing the toxic mis-incorporation of Se into proteins. The transgenic plants were found to accumulate more Se in the form of methylselenocysteine than the wild-type. The SMT transgenic seedlings tolerated Se, particularly selenite better than the wild-type, producing three- to seven-fold greater biomass and three-fold longer root lengths. SMT plants also demonstrated increased Se accumulation and volatilization, particularly when the transgenic plants were supplied with selenite. It was thus concluded that the use of SMT transgenic plants to remove Se from seleniferous soils was of little value in soils which predominantly contain selenate. Plants take up selenate actively, accumulating 10- to 20-fold higher Se concentrations than with selenite. The conversion of selenate to selenite becomes slower by rate-limiting amounts of ATP sulfurylase (APS), which catalyzes selenate reduction to organic forms of Se (Pilon-Smits et al. 1999a, b). Thus, plants overexpressing SMT were restricted in their ability to convert selenate to MetSeCys because of their limited capacity for selenate reduction (LeDuc et al. 2006). To enhance the phytoremediation of selenate, double transgenic plants that overexpressed the gene encoding ATP sulfurylase (APS) in addition to SMT, i.e., APS × SMT, were generated (LeDuc et al. 2006). The results showed that there was a substantial improvement in Se accumulation from selenate (4 to 9 times increase) in transgenic plants overexpressing both APS and SMT.

Other superior terrestrial Se volatilizers include cabbage (*B. oleracea* var. *capitata*), broccoli (*B. oleracea* var. *botrytis* cv. Green Valiant), cauliflower (*B. oleracea* var. *cauliflora*), *B. juncea*

Czern. L., and *B. campestris* var. *chinesis*. Once the roots were determined to be the primary site of volatilization, the rate of volatilization from the detopped roots was measured and found to be 1.5 to 5 times faster than that of the intact root.

11.4.1.2.11 Radioactive Elements

Uranium (U), which is sorbed at clay minerals and humus particles, as well as U in connection with the Fe-Mn oxides, is immobile and therefore not available for plants. However, roots can mobilize metals by acidifying with citric acid or other organic acids, through changes of redox conditions or formation of organic complexes. These chelates actually increase the fraction of dissolved metals in the soil solution. *B. juncea* shows a high ability for the accumulation of U in comparison with other plants (Ebbs et al. 1998; Huang et al. 1998; Shahandeh and Hossner 2002a, b; Chang et al. 2005). However, most of the examinations have been accomplished with soil with a high U content (>100 mg U kg^{-1} soil) or nonsoil typical substrates. The capacity of *B. juncea* in the uptake, translocation or elimination of U isotopes (^{238}U, ^{235}U and ^{234}U) and ^{226}Ra, in the presence of chelating agents (EDTA or citrate) and phosphates, has been examined by Tomé et al. (2009). The presence of phosphates enhanced the retention of U by roots, viz., rhizofiltration, but the translocation was poorer. For Ra, the best translocation was in the absence of phosphates. The presence of citrate and EDTA enhanced the translocation and recovery of U by the shoots, and hence led to efficient phytoextraction, but had no clear effect on the transfer of Ra. A slightly better uptake of Ra was noted at neutral pH, although the translocation was lower. Even in the absence of chelating agents, phytoextraction of Ra occurred because of better transfers to the harvested part of the seedlings, especially if phosphate is present. In the presence of phosphate, the use of chelating agents such as citrate or EDTA does not improve the elimination of Ra, so that their addition does not seem advisable. The U concentrations in shoots of *B. juncea* increased 400-fold after application of 5 mmol citric acid kg^{-1} soil (Huang et al. 1998). Shahandeh and Hossner (2002a, b) also found that the addition of 5 mmol kg^{-1} of citric or oxalic acid in soil resulted in a three-fold increase of U content in shoots of *B. juncea*. Duquène et al. (2009) found increase in U uptake several folds by agents like citric acid, NH_4-citrate/citric acid, oxalic acid, EDDS or NTA at a rate of 5 mmol kg^{-1} dry soil.

11.5 PLANT METALLOTHIONEINS, PHYTOCHELATINS, AND PHYTOREMEDIATION USING *BRASSICA* SPECIES

Multiple mechanisms that control plant tolerance and/or adaptation to varied metals and metalloids have been reported. These include cell wall binding, active transport of ions into the vacuole, and formation of complexes with organic acids or peptides. One of the most important mechanisms for metal detoxification in plants appears to be chelation of metals by low molecular-weight proteins such as metallothioneins (MTs) and peptide ligands, the phytochelatins (PCs).

11.5.1 METALLOTHIONEINS

Metallothioneins (MT) are ubiquitous, Cys-rich, low molecular-weight, heavy metal-binding peptides. They are able to bind metal ions through the thiol groups of their Cys residues (Quan et al. 2006). The number of amino acids in plant MTs varies from 45 to 87 (Hassinen et al. 2011). There are about 10 to 17 Cys residues and the number of aromatic amino acids can vary from none to several. Plant MTs are classified into four subfamilies based on amino acid sequences (Guo et al. 2003; Hassinen et al. 2011):

Type 1: They contain six Cys-Xaa-Cys motifs (Xaa is another amino acid) that are distributed equally between two domains. The two domains are separated by a 40-amino acid spacer which is a common feature of plant MTs (Cobbett et al. 2002). This spacer may include aromatic amino acids. It is different from most other MTs in which Cys-rich domains are separated by spacers of less than ten amino acids that do not involve aromatic residues, e.g., MT1a and MT1c of *Arabidopsis thaliana*, and MT1 of *Cicer arietinum* (Hassinen et al. 2011).

Type 2: They have two Cys-rich domains separated by a spacer containing around 40 amino acids. The first Cys-pair is present as a Cys-Cys motif occupying amino acid positions 3 and 4 of these proteins. Generally the sequence of the N-terminal domain remains highly conserved (MSCCGGNCGCS) and a Cys-Gly-Gly-Cys motif is present at the N-terminal end. The C-terminal domain consists of three Cys-Xaa-Cys motifs (Cobbett et al. 2002). The spacer region separating these domains shows more variation between species than Type 1 MTs, e.g., MT2a and MT2b of *A. thaliana* and MT2 of *C. arietinum* (Hassinen et al. 2011).

Type 3: They contain only four Cys residues in the N-terminal domain. The consensus sequence for the first three is Cys-Gly-Asn-Cys-Asp-Cys. The fourth Cys remains within a highly conserved motif, Gln-Cys-Xaa-Lys-Lys-Gly, and does not belong to a pair of cysteines. In the Cys-rich C-terminal domain, six cysteines are arranged in Cys-Xaa-Cys motifs (Cobbett et al. 2002). As with the majority of Type 1 and Type 2 plant MTs, the two domains are separated from each other by approximately 40 amino acids, e.g., MT3 of *A. thaliana* and *Musa acuminate* (Hassinen et al. 2011).

Type 4: They have three Cys-rich domains separated by 10–15 residues, each domain containing five or six conserved Cys residues (Cobbett et al. 2002). Most of the cysteines are present as Cys-Xaa-Cys motifs. Out of the relatively less number of Type 4 MTs that have been identified, it was observed that Type 4 MTs from dicotyledonous plants have 8–10 additional amino acids in the N-terminal domain before the first Cys residue, e.g., MT4a (also called E_c-2) and MT4b (also called E_c-1) of *A. thaliana*, and E_c-1 of *Triticum aestivum* (Hassinen et al. 2011).

MT family peptides show significantly large sequence diversity (Freisinger 2008). Multiple MT isoforms have been found in higher plants (Zhigang et al. 2006). Variable metal-binding properties of different MTs can be attributed to their differential Cys arrangement (Roosens et al. 2005). They are thought to be primarily involved in cellular Cu, Cd (and perhaps Zn) homeostasis (Zhigang et al. 2006). MTs are usually localized in the cytosol and do not seem to be transported into vacuoles (Hassinen et al. 2011). Promoters of various MTs, like $LeMT_B$ of tomato (Whitelaw et al. 1997), PmMT of Douglas-fir (Chatthai et al. 2004), MT3b of poplar (Berta et al. 2009) and Cu- and Zn-inducible ric-MT of rice (Lu et al. 2007) have been found to contain Metal response element (MRE)-like motifs (Hassinen et al. 2010). Zhigang et al. (2006) overexpressed a Type 2 MT protein, named BjMT2, from *B. juncea* in *Arabidopsis thaliana*, under the control of the 35S promoter. The ectopic expression of BjMT2 increased Cu^{2+} and Cd^{2+} tolerance in the seedling stage, based on shoot growth and chlorophyll content. The heavy metal binding status of BjMT2 protein also did not affect its localization within the cytoplasm. However, at normal nutritional essential heavy metal concentrations, the root growth of the BjMT2 transformants was reduced in comparison with the wild-type, suggesting problems in maintaining essential metal homeostasis.

11.5.2 Phytochelatins

Phytochelatins (PCs) represent a group of post-translationally synthesized, short nonprotein, heavy metal binding, sulfur-rich, thiolate peptides $(\gamma\text{-glutamyl cysteinyl})_n$-X, where $n = 2–11$ and X is glycine, serine, β-alanine, glutamate or glutamine (Rauser 1995). PCs are composed of the amino acids Cys, Glu and Gly, with a molecular weight of 1–4 kDa (Zenk 1996). The precursor for PC biosynthesis is glutathione (GSH). The synthesis of PC is considered as a kind of "stress-adaptive reaction" in plants, and plays a major role in detoxification of heavy metals such as As^{5+}, Cd^{2+}, Cu^{2+}, Ag^{+}, Hg^{2+} and Pb^{2+} (Zenk 1996; Rauser 1999; Cobbett 2000a, b; Goldsbrough 2000; Schmoger et al. 2000;

Hall 2002; Raab et al. 2004). Cd binds primarily to the thiol group of Cys residues in the PC peptide and the Cd-PC complex is about 1000 times less toxic to the plant enzymes as compared to free Cd ions (Cobbett 2000a). The use of PCs, as biomarkers for Cd toxicity was proposed by Keltjens and Van Beusichem (1998). Heavy metal ions such as Cd^{2+} enter plant cells via the permeable cell wall and cell membrane, and immediately activate the PC synthase (PCS) enzyme that synthesizes PCs utilizing GSH (Zenk 1996). Thus, PC synthesis is accompanied by a fall in the concentration of GSH. The PC-Cd "low molecular weight" (LMW) complexes are shown to be transported from the cytosol to vacuole across the tonoplast, where they acquire acid labile sulfur (S^{2-}) and form a "high molecular weight" (HMW) complex, with a much higher affinity towards Cd ions (Cobbett 2000a). The complex is inherently stored in vacuoles. Free Cd^{2+} ions can also enter the vacuole by means of $Cd^{2+}/2H^+$ antiport system. Vacuolar compartmentalization of PC-Cd complexes prevents the free circulation of Cd ions in the cytosol preventing damages to the enzymes. Under acidic pH conditions in the vacuole, it is likely that metals are liberated as a result of dissociation from PC-Cd HMW complex. The Cd then becomes complexed with vacuolar organic acids such as citrate, oxalate, malate etc, while the metal free PC molecules are degraded by vacuolar hydrolases into amino acids, and shuttled back into cytosol for new PC synthesis. The Cd-PC content in Cd-tolerant species was 10 to 1000 times higher than the control (Kneer and Zenk 1992). Moreover, the majority of experimental studies have argued that PCs play a role only at high levels of Cd exposure, which are normally not found in natural environments (Cobbett et al. 2002). PCs can also be involved in long-distance transport in both the root-to-shoot and shoot-to-root directions (Chen et al. 2006; Gong et al. 2003). PCs are usually present in large quantities in the phloem sap. This allows them to form stable complexes with Cd. The phloem has been implicated as having an important role in the transport of Cd in the form of thiol–Cd complexes, as Cd and thiols were seen to be concentrated here. The [thiol]/[Cd] stoichiometries suggests that GSH may also have a role in the long-distance transport of Cd (Mendoza-Cózatl et al. 2008).

11.5.3 Role of PCs in *Brassica* Species-Based Metals Remediation

The high Cd accumulation in the shoot of *B. napus* can be regarded as a potential candidate for the phytoextraction of Cd, being accompanied by the accumulation of PC2, PC3 and PC4 (Selvam and Wong 2008). It was seen earlier that rapid *de novo* synthesis of PCs in roots and leaves requires an increased synthesis of GSH, which in turn depends on increased sulfur assimilation. Keeping it in mind, the cDNAs for enzymes involved in sulfur assimilation, i.e., two isoforms each for ATP sulfurylase (ATPS) and APS reductase (APSR), were cloned (Heiss et al. 1999). RNA blot analysis of transcript amounts indicated that upon Cd exposure (25 μM), the expression of ATPS and APSR in roots and leaves of 6-week-old *B. juncea* were strongly increased. Along with the induction of ATPS and APSR mRNAs, Cys concentrations in roots and leaves increased and GSH concentrations dropped (Heiss et al. 1999). Schäfer et al. (1998) observed that in the roots of *B. juncea*, Cd exposure (25 μM) induced a massive formation of PCs, accompanied by only a moderate decrease of the putative PC precursor, GSH. Analysis of expression of the enzymes (involved in GSH synthesis) in the roots, namely OASTL (catalyzing the last step in Cys biosynthesis), γ-glutamylcysteine synthetase (γ-ECS) and glutathione synthetase (GS), showed a moderate increase in OASTL and GS, and a stronger increase in γ-ECS mRNA in roots of Cd-exposed plants Likewise, a strong increase of γ-ECS mRNA in roots and shoots, concomitant with an increase of GSH and PCs, were noted in *B. juncea* with micromolar concentrations of $CuSO_4$. A significant up regulation of γ-ECS mRNA was observed at 25 μM $CuSO_4$, whereas maximum up regulation was obtained at 100 μM $CuSO_4$. $CdSO_4$ at a concentration of 50 μM caused a 72% reduction in shoot growth, without affecting the amounts of γ-ECS mRNAs. At a concentration of 500 μM $ZnSO_4$ did not reduce growth but induced transient increases of γ-ECS mRNAs (Schäfer et al. 1997). Thus it can be predicted that manipulation of the expression of enzymes involved in GSH and PC synthesis may be an effective method to enhance heavy-metal tolerance in plants.

The *E. coli gshII* gene, encoding glutathione synthetase (GS), was overexpressed in the cytosol of *B. juncea*. The transgenic plants accumulated significantly more Cd than the wild type. The shoot Cd concentrations were up to 25% higher and total Cd accumulation per shoot was up to three-fold higher. The GS plants also showed enhanced tolerance to Cd at both the seedling and mature-plant stages. Cd-treated GS plants had higher concentrations of glutathione, PC, thiol, S and Ca than wild type plants (Zhu et al. 1999a, b). Reisinger et al. (2008) showed that the overexpression of either γ-ECS or GS in *B. juncea* transgenics exhibited a significantly higher phytoremediation capacity to tolerate and accumulate a variety of metals or metalloids (particularly As, Cd and Cr) as well as mixed-metal combinations (As, Cd, Zn/As, Pb and Zn). These two enzymes are responsible for GSH formation in plants, which is the first step in the production of PCs. The enhanced metal tolerance was attributable to enhanced production of PCs, sustained by a greater availability of GSH as substrate, as suggested by their higher concentrations of GSH, PC2, PC3 and PC4 compared to wild-type plants. Seth et al. (2008) also showed similar results, where the capacity of *B. juncea* to accumulate and tolerate high concentrations of Cd was through enhanced level of PCs, GSH, NP-SH and glutathione reductase. They predicted that the increase in PC levels in *B. juncea* could be due to activation of *PCS* gene. Further, synthesis of PCs was significantly correlated with metal accumulation in *B. juncea*, whereas at higher concentrations, a decline in PC content was observed due to either degradation or reduction in GSH pool, declined glutathione reductase activity or inactivation of PCS enzyme. With Cd + EDTA, the induction of PCs was lower as compared to Cd alone. Cd chelation with EDTA might have been responsible for the low availability of free Cd, needed for the induction of PCs. Transgenic *B. juncea* expressing an *A. thaliana AtPCS1* gene, encoding PCS, exhibited significantly higher tolerance to Cd and As. Shoots of Cd-treated *PCS* plants had significantly higher concentrations of PCs and thiols than those of wild-type plants (Gasic and Korban 2007a). Moderate expression levels of *AtPCS* improved the ability of *B. juncea* to tolerate certain levels of Cd and Zn, but at the same time did not increase the accumulation potential for these heavy metals (Gasic and Korban 2007b). In *B. napus*, following 24 h of Cd exposure, high levels of PCs were observed in the phloem sap, where the concentration of Cd was four-fold higher, compared to xylem sap. It was suggested that PCs and glutathione (GSH) can function as long-distance carriers of Cd. However, only traces of PCs were detected in xylem sap. Thus they hypothesized that the phloem is a major vascular system for long-distance source-to-sink transport of Cd, as PC–Cd and glutathione–Cd complexes (Mendoza-Cózatl et al. 2008).

Iglesia-Turino et al. (2006) studied the Hg accumulation mechanism in *B. napus* plants grown under a Hg concentration gradient (0 μM–1,000 μM). The roots accumulated two- to 20-fold more Hg than the leaves. The higher accumulation in the roots might be possible because of a greater tolerance to toxic metals in the roots than in the shoots. Only unbound PC_2 appeared in roots, and unbound PC_2 was linearly correlated with Hg accumulation, leading to the conclusion that PC_2 is crucial in extracting Hg from *B. napus* (Gothberg et al. 2004).

11.6 CONCLUSIONS

Heavy metals, which are strongly retained within the soil, without easy breakdown and very little leaching, generally affect the biosphere for long periods of time, posing serious risk to human and animal health. Phytoremediation is a rapidly emerging field and constitutes a sustainable and inexpensive process or a viable alternative to conventional remediation methods. It has no advised effect on soil quality, and can be used in the contaminated soil of large areas. One of the key aspects to the acceptance of phytoextraction and phytostabilization (the two main techniques to remediate heavy metals) pertains to the measurement of their performance, ultimate utilization of by-products and their overall economic viability. Fast growing plants with high biomass and good metal uptake ability, accumulating more than 0.5% of contaminants with a bioconcentration factor greater than 1000, are needed. The review clearly documents that the genus *Brassica* includes several metal hyperaccumulator species which have tremendous potentiality of phytoextraction and heavy metal accumulation, coupled with their added advantage of rapid biomass production. *B. juncea* in this context possesses a remarkable ability to accumulate and translocate Cu, Cr, Cd, Ni, Pb and Zn to the shoots. The capacity of removal of a particular metal, however, varies from species to species. Even within a species, possible genetic variation in phytoextraction potential arises with different accessions, i.e., seeds gathered from different regions may exhibit a different phenotype, since the prevailing environmental factors do influence the natural selection of that species. This has been noted in the case of *B. juncea*. These issues also complicate phytoremediation programs to some extent when using a particular species. The use of phytoremediation requires a very careful approach, and a precise assessment of the type and the level of contamination is essential. A wide range of research is needed prior to the application of this program, particularly for multiple-metal contaminated sites.

REFERENCES

Abercrombie, J.M., M.D. Halfhill, P. Ranjan et al. 2008. Transcriptional responses of *Arabidopsis thaliana* plants to As (V) stress. *BMC Plant Biology* 8: 87.

Anderson, C., A. Deram, D. Petit, R. Brooks, B. Steward, and R. Simcock. 1999. Induced hyperaccumulation: Metal movement and problems, *Proceedings of the 5th International Conference on Biogeochemistry of trace elements*, Vienna, pp. 122–3.

Angelova, V., and K. Ivanov. 2009. Bio-accumulation and distribution of heavy metals in black mustard (*Brassica nigra* Koch). *Environmental Monitoring and Assessment* 153: 449–59.

Anjum, N.A., I. Ahmad, M.E. Pereira, A.C. Duarte, S. Umar, and N.A. Khan. 2012. *The plant family Brassicaceae: Contribution towards phytoremediation*. Environmental Pollution Series No. 21, Springer: Dordrecht, The Netherlands.

Ariyakanon, N., and B. Winaipanich. 2006. Phytoremediation of copper-contaminated soil by *Brassica juncea* (L.) Czern and *Bidens alba* (L.) DC. var. *radiata*. *Journal of Scientific Research Chula University* 31: 49–56.

Ahsan, N., D. Lee, I. Alam et al. 2008. Comparative proteomic study of arsenic-induced differentially expressed proteins in rice roots reveals glutathione plays a central role during As stress. *Proteomics* 8: 3561–76.

Bachir, D.M.L., F.B. Wu, G.P. Zhang, and H.X. Wu. 2004. Genotypic difference in effect of cadmium on the development and mineral concentrations of cotton. *Communications in Soil Science and Plant Analysis* 35: 285–99.

Baker, A.J.M., S.P. McGrath, R.D. Reeves, and J.A.C. Smith. 2000. Metal hyper accumulator plants: a review of the ecology and physiology of a biochemical resource for phytoremediation of metal-polluted soils. In *Phytoremediation of contaminated Soil and Water*, ed. N. Terry and G. Banuelos, 85–107. Boca Raton: FL-Lewis Publishers.

Baker, A.J.M., and R.R. Brooks. 1989. Terrestrial higher plants which hyperaccumulate metallic elements—a review of their distribution, ecology and phytochemistry. *Biorecovery* 1: 81–126.

Banuelos, G.S., H.A. Ajwa, M. Mackey et al. 1997a. Evaluation of different plant species used for phytoremediation of high soil selenium. *Journal of Environmental Quality* 26: 639–46.

Banuelos, G.S., H.A. Ajwa, L. Wu et al. 1997b. Selenium-induced growth reduction in *Brassica* land races considered for phytoremediation. *Ecotoxicology and Environmental Safety* 36: 282–87.

Banuelos, G. S., G. Cardon, B. Mackey et al. 1993. Boron and selenium removal in boron laden soils by four sprinkler-irrigated plant species. *Journal of Environmental Quality* 22: 786–92.

Banuelos, G., and G. Schrale. 1989. Plants that remove selenium from soils. *California Department of Food and Agriculture* 43: 19–20.

Barcelo, J., and C. Poschenreider. 1990. Plant water relation as affected by heavy metal stress. *Journal of Plant Nutrition* 13: 1–37.

Bartlett, R.J. 1991. Chromium cycling in soil and water: links, gaps and methods. *Environmental Health Perspectives* 92: 14–24.

Basu, A., U. Basu, and G.J. Taylor. 1994. Induction of microsomal membrane proteins in roots of an aluminum-resistant cultivar of *Triticum aestivum* L. under conditions of aluminum stress. *Plant Physiology* 104: 1007–13.

Beauford, W., J. Barber, and A.R. Barringer. 1977. Uptake and distribution of mercury within higher plants. *Physiologia Plantarum* 39: 261–5.

Bellostas, N., P. Kachlicki, J.C. Sorensen, and H. Sorensen. 2007. Glucosinolate profiling of seeds and sprouts of *B. oleracea* varieties used for food. *Scientia Horticulturae* 114: 234–42.

Benavides, M.P., S.M. Gallego, and M.L. Tomaro. 2005. Cadmium toxicity in plants. *Brazilian Journal of Plant Physiology* 17: 21–34.

Bernal, M.P., and S.P. McGrath. 1994. Effects of pH and heavy metal concentrations in solution culture on the proton release, growth and elemental composition. *Plant Soil* 166: 83–92.

Berta, M., A. Giovannelli, E. Potenza, M.L. Traversi, and M.L. Racchi. 2009. Type 3 metallothioneins respond to water deficit in leaf and in the cambial zone of white poplar (*Populus alba*). *Journal of Plant Physiology* 166: 521–30.

Berti, W.R., and S.D. Cunningham. 2000. In: *Phytoremediation of Toxic Metals: Using Plants to clean up the Environment*, ed. I. Raskin, 71–88. New York: Wiley-Interscience, John Wiley and Sons, Inc.

Bhattacharjee, S. 1998. Membrane lipid peroxidation, free radical scavengers and ethylene evolution in *Amaranthus* as affected by lead and cadmium. *Biologia Plantarum* 40: 131–5.

Blaylock, M.J., D.E. Salt, S. Dushenkov et al. 1997. Enhanced accumulation of Pb in Indian mustard by soil-applied chelating agents. *Environmental Science and Technology* 31: 860–5.

Bleeker, P.M., H.W.J. Hakvoort, M. Bliek, E. Souer, and H. Schat. 2006. Enhanced arsenate reduction by a CDC25-like tyrosine phosphatase explains increased phytochelatin accumulation in arsenate-tolerant *Holcus lanatus*. *The Plant Journal* 45: 917–29.

Boussama, N., O. Ouariti, and M.H. Ghorbal. 1999. Cd-stress on nitrogen assimilation. *Journal of Plant Physiology* 155: 310–17.

Boyd, R.S., J.J. Shaw, and S.N. Martens. 1994. Nickel hyperaccumulation in *S. polygaloids* (Brassicaceae) as a defense against pathogens. *American Journal of Botany* 81: 294–300.

Breckle, S.W. 1991. Growth under stress: heavy metals. In *Plant roots: The hidden half*, ed. Waisel, Y., A. Eshel, and X. Kafkafi, 351–73. New York: Marcel Dekker.

Broadley, M., M.J. Willey, J.C. Wilkins, A.J.M. Baker, A. Mead, and P.J. White. 2001. Phylogenetic variation in heavy metal accumulation in angiosperms. *New Phytologist* 152: 9–27.

Brooks, R.R., and B.H. Robinson. 1998. The potential use of hyperacuumulators and other plants for phytomining. In *Plants that Hyperaccumulate Heavy Metals*, ed. Brooks, R.R., 327–65. Wallingford, CAB International.

Brown, T.A., and A. Shrift. 1982. Selenium: toxicity and tolerance in higher plants. *Biological Reviews of the Cambridge Philosophical Society* 57: 59–84.

Cai, Y., J. Su, and L.Q. Ma. 2004. Low molecular weight thiols in arsenic hyperaccumulator *Pteris vittata* upon exposure to arsenic and other trace elements. *Environmental Pollution* 129: 69–78.

Catarecha, P., M.D. Segura, J.M. Franco-Zorrilla et al. 2007. A mutant of the *Arabidopsis* phosphate transporter PHT1;1 displays enhanced arsenic accumulation. *The Plant Cell* 19: 1123–33.

Chaney, R.L. 1983. Plant uptake of inorganic waste constitutes. In *Land treatment of hazardous wastes*, ed. Parr, J.F., P.B. Marsh, and J.M. Kla, 50–76. Park Ridge, NJ: Noyes, Data Corp.

Chang, P., K.W. Kim, S. Yoshida, and S.Y. Kim. 2005. Uranium accumulation of crop plants enhanced by citric acid. *Environmental Geochemistry and Health* 27: 529–38.

Chaoui, A., S. Mazoudi, M.H. Ghorbal, and E.E.L. Ferjani. 1997. Cadmium and zinc induction of lipid peroxidation and effects on antioxidant enzyme activities in bean (*Phaseolus vulgaris* L). *Plant Science* 127: 139–47.

Chatthai, M., M. Osusky, L. Osuska, D. Yevtushenko, and S. Misra. 2004. Functional analysis of a Douglas-fir metallothionein-like gene promoter: transient assays in zygotic and somatic embryos and stable transformation in transgenic tobacco. *Planta* 220:118–28.

Chaudhry, T.M., W.J. Hayes, A.G. Khan, and C.S. Khoo. 1998. Phytoremediation—focusing on accumulator plants that remediate metal-contaminated soils. *Australasian Journal of Ecotoxicology* 4: 37–51.

Chen, A., E.A. Komives, and J.I. Schroeder. 2006. An improved grafting technique for mature *Arabidopsis* plants demonstrates long-distance shoot-to-root transport of phytochelatins in *Arabidopsis*. *Plant Physiology* 141: 108–20.

Clemens, S., M.G. Palmgren, and U. Krämer. 2002. A long way ahead: understanding and engineering plant metal accumulation. *Trends in Plant Science* 7: 309–15.

Clemente, R., D.J. Walker, and M.P. Bernal. 2005. Uptake of heavy metals and As by *Brassica juncea* grown in a contaminated soil in Aznalcollar (Spain): the effect of soil amendments. *Environmental Pollution* 136: 46–58.

Clijsters, H., and F. Van Assche. 1985. Inhibition of photosynthesis by heavy metals. *Photosynthesis Research* 7: 31–40.

Cobbett, C.S. 2000a. Phytochelatins and their roles in heavy metal detoxification. *Plant Physiology* 123: 825–32.

Cobbett, C.S. 2000b. Phytochelatin biosynthesis and function in heavy-metal detoxification. *Current Opinion in Plant Biology* 3: 211–6.

Cobbett, C., and P. Goldsbrough. 2002. Phytochelatins and metallothioneins: roles in heavy metal detoxification and homeostasis. *Annual Review of Plant Biology* 53: 159–82.

Connell, D.W., and G.J. Miller. 1984. *Chemistry and Ecotoxicology of Pollution* 444, NY, John Wiley & Sons.

Cunningham, S.D., and D.W. Ow. 1996. Promises and prospects of phytoremediation. *Plant Physiology* 110: 715–9.

Danh, L.T., P. Truong, R. Mammucari, T. Tran, and N. Foster. 2009. Vetiver grass, *Vetiveria zizanioides*: A choice plant for phytoremediation of heavy metals and organic wastes. *International Journal of Phytoremediation* 11: 664–91.

del Rio, M., J. Fernandez-Martinez, J. Dominguez, and A. de Haro. 2000. Field trials of *Brassica carinata* and *Brassica juncea* in polluted soils of the Guadiamar river area. *Fresenius Environmental Bulletin* 9: 328–32.

di Toppi, L.S., and R. Gabbrielli. 1999. Response to cadmium in higher plants. *Environmental and Experimental Botany* 41: 105–30.

Diwan, H., A. Ahmad, and M. Iqbal. 2008. Genotypic variation in the phytoremediation potential of Indian mustard for chromium. *Environmental Management* 41: 734–41.

Do Nascimento, C.W., D. Amarasiriwardena, and B. Xing. 2006. Comparison of natural organic acids and synthetic chelates at enhancing phytoextraction of metals from a multi-metal contaminated soil. *Environmental Pollution* 140: 114–23.

Dominguez-Solís, J.R., G. Gutierrez-Alcalá, J.M. Vega, L.C. Romero, and C. Gotor. 2001. The cytosolic O-acetylserine (thiol) lyase gene is regulated by heavy metals and can function in cadmium tolerance. *Journal of Biological Chemistry* 276: 9297–302.

Duke, J.A. 1983. Handbook of energy crops. In *NewCROP™, the new crop resource online program.* Center for new crops and plant products, Department of horticulture and landscape architecture, Purdue University, Indiana, USA.

Duquène, L., H. Vandenhove, F. Tack, E. Meers, J. Baeten, and J. Wannijn. 2009. Enhanced phytoextraction of uranium and selected heavy metals by Indian mustard and ryegrass using biodegradable soil amendments. *Science of the Total Environment* 407: 1496–505.

Dushenkov, V., P.B.A.N. Kumar, H. Motto, and I. Raskin. 1995. Rhizofiltration: The use of plants to remove heavy metals from aqueous streams. *Environmental Science and Technology* 29: 1239–45.

Ebbs, S.D., and L.V. Kochian. 1997. Toxicity of Zn and Cu to *Brassica* species: implications for phytoremediation. *Journal of Environmental Quality* 26: 776–81.

Ebbs, S.D., M.M. Lasat, D.J. Brady et al. 1997. Phytoextraction of cadmium and zinc from a contaminated soil. *Journal of Environmental Quality* 26: 1424–30.

Ebbs, S.D., and L.V. Kochian. 1998. Phytoextraction of zinc by oat (*Avena sativa*), barley (*Hordeum vulgare*), and Indian mustard (*Brassica juncea*). *Environmental Science and Technology* 32: 802–6.

Ebbs, S.D., D.J. Brady and L.V. Kochian. 1998. Role of uranium speciation in the uptake and translocation of uranium by plants. *Journal of Experimental Botany* 49: 1183–90.

Farago, M.E., and P.J. Parsons. 1994. The effects of various platinum metal species on the water plant *Eichhornia crassipes* (MART.) *Chemical Speciation and Bioavailability* 6: 1–12.

Foy, C.D., R.L. Chaney, and M.C. White. 1978. The physiology of metal toxicity in plants. *Annual Review of Plant Physiology* 29: 511–66.

Freisinger, E. 2008. Plant MTs—long-neglected members of the metallothionein superfamily. *Dalton Transactions* 47: 6663–75.

Fusco, N., L. Micheletto, G. Dal Corso, L. Borgato, and A. Furini. 2005. Identification of cadmium-regulated genes by cDNA-AFLP in the heavy metal accumulator *Brassica juncea* L. *Journal of Experimental Botany* 56: 3017–27.

Gasic, K., and S.S. Korban. 2007a. Transgenic Indian mustard (*Brassica juncea* L.) plants expressing an *Arabodopsis* Phytochelatin synthase (*AtPCS1*) exhibit enhanced As and Cd tolerance. *Plant Molecular Biology* 64: 361–9.

Gasic, K., and S.S. Korban. 2007b. Expression of *Arabidopsis* phytochelatin synthase in Indian mustard (*Brassica juncea*) plants enhances tolerance for Cd and Zn. *Planta* 225:1277–85.

Ghnaya, A.B., G. Charles, A. Hourmant, J.B. Hamida, and M. Branchard. 2009. Physiological behaviour of four rapeseed cultivar (*Brassica napus* L.) submitted to metal stress. *Comptes Rendus Biologies* 332: 363–70.

Ghosh, M., and S.P. Singh. 2005. A Review on Phytoremediation of Heavy Metals and Utilization of its By-products. *Asian Journal on Energy and Environment* 6: 214–31.

Giordani, C., and S. Cecchi. 2005. Phytoremediation of soil polluted by nickel using agricultural crops. *Environmental Management* 36: 675–81.

Gleba, D., N.V. Borisjuk, L. Borisjuk et al. 1999. Use of plant roots for phytoremediation and molecular farming. *Proceedings of the National Academy of Sciences (USA)* 96: 5973–77.

Goldsbrough, P. 2000. Metal tolerance in plants: the role of phytochelatins and metallothioneins. In *Phytoremediation of Contaminated Soil and Water*, ed. Terry, N., and G. Banuelos, 221–33. Boca Raton, FL: CRC Press.

Gong, J.M., D.A. Lee, and J.I. Schroeder. 2003. Long-distance root-to-shoot transport of phytochelatins and cadmium in *Arabidopsis*. *Proceedings of the National Academy of Sciences (USA)* 100: 10,118–23.

Gothberg, A., M. Greger, K. Holm, and B.E. Bengtsson. 2004. Influence of nutrient levels on uptake and effects of mercury, cadmium, and lead in water spinach. *Journal of Environmental Quality* 33: 1247–55.

Goyer, R.A. 2001. Toxic effects of metals. In *Casarett and Doull's Toxicology: The Basic Science of Poisons*, ed. Klaassen, C.D., M.O. Amdur, and J. Doull, 1111. New York: McGraw-Hill.

Grčman, H., D. Vodnik, S. Velikonja-Bolta, and D. Lestan. 2003. Ethylenediaminedisuccinate as a new chelate for environmentally safe enhanced lead phytoextraction. *Journal of Environmental Quality* 32: 500–6.

Grispen, V.M.J., H.J.M. Nelissen, and J.A.C. Verklei. 2006. Phytoextraction with *Brassica napus*: a tool for sustainable management of heavy metal-contaminated soils. *Environmental Pollution* 144: 77–83.

Guo, W.J., W. Bundithya, and P.B. Goldbrough. 2003. Characterization of the *Arabidopsis* metallothionein gene family: tissue-specific expression and induction during senescence and in response to copper. *New Phytologist* 159: 369–81.

Gupta, M., P. Sharma, N.B. Sarin, and A.K. Sinha. 2009. Differential response of arsenic stress in two varieties of *Brassica juncea* L. *Chemosphere* 74: 1201–8.

Haag-Kerwer, A., H.J. Schafer, S. Heiss, C. Walter, and T. Rausch. 1999. Cadmium exposure in *Brassica juncea* causes a decline in transpiration rate and leaf expansion without effect on photosynthesis. *Journal of Experimental Botany* 50: 1827–35.

Hale, K.L., S.P. McGrath, E. Lombi et al. 2001. Molybdenum sequestration in *Brassica* species. A role for anthocyanins? *Plant Physiology* 126: 1391–402.

Hall, J.L. 2002. Cellular mechanisms for heavy metal detoxification and tolerance. *Journal of Experimental Botany* 53: 1–11.

Han, F.X., A. Banin, Y. Su et al. 2002. Industrial age anthropogenic inputs of heavy metals into the pedosphere. *Naturwissenschaften* 89: 497–504.

Hansen, D., P.J. Duda, A. Zayed, and N. Terry. 1998. Selenium removal by constructed wetlands: role of biological volatilization. *Environmental Science and Technology* 32: 591–7.

Harriss, R.C., and C. Hohenemser. 1978. Mercury: measuring and managing the risk. *Environment* 20: 25–36.

Hartley-Whitaker, J., G. Ainsworth, R. Vooijs, W.M. Ten Bookum, H. Schat, and A.A. Meharg. 2001. Phytochelatins are involved in differential arsenate tolerance in *Holcus lanatus* L. *Plant Physiology* 126: 299–306.

Hartman, W.J. Jr. 1975. An evaluation of land treatment of municipal wastewater and physical siting of facility installations, US Department of Army, Washington, DC.

Hassinen, V.H., A.I. Tervahauta, H. Schat, and S.O. Kärenlampi. 2011. Plant metallothioneins-metal chelators with ROS-scavenging activity? *Plant Biology (Stuttgart)* 13: 225–32.

Hawkes, S.J. 1997. What is a "Heavy Metal"? *Journal of Chemical Education* 74: 1374.

Heiss, S., H.J. Schafer, A. Haag-Kerwer, and T. Rausch. 1999. Cloning sulfur assimilation genes of *Brassica juncea* L.: cadmium differentially affects the expression of a putative low-affinity sulfate transporter and isoforms of ATP sulfuryalse and ATP reductase. *Plant Molecular Biology* 39: 847–57.

Heiss, S., A. Wachter, J. Bogs, C. Cobbett, and T. Rausch. 2003. Phytochelatin synthase (PCS) protein is induced in *Brassica juncea* leaves after prolonged Cd exposure. *Journal of Experimental Botany* 54: 1833–9.

Hemingway, J.S. 1976. Mustards: *Brassica* spp. and *Sinapis alba* (*Cruciferae*). In *Evaluation of Crop Plants*, ed. N.W. Simmons, 56–9, London, Longman.

Henry, J.R. 2000. *An Overview of Phytoremediation of Lead and Mercury*, 3–9. Washington, DC, *NNEMS Report*.

Hooda, V. 2007. Phytoremediation of toxic metals from soil and waste water. *Journal of Environmental Biology* 28: 367–76.

Hsiao, K., P. Kao, and Z. Hseu. 2007. Effects of chelators on chromium and nickel uptake by *Brassica juncea* on serpentine-mine tailings for phytoextraction. *Journal of Hazardous Materials* 148: 366–76.

Huang, J.W., M.J. Blaylock, Y.K. Kapulnik, and B.D. Ensley. 1998. Phytoremediation of uranium-contaminated soils: role of organic acids in triggering uranium hyperaccumulation in plants. *Environmental Science and Technology* 32: 2004–8.

Huang, J.W., J. Chen, W.R. Berti, and S.D. Cunninghum. 1997. Phytoremediation of lead-contaminated soils: Role of synthetic chelates in lead phytoextraction. *Environmental Science and Technology* 31: 800–5.

Hussain, D., M.J. Haydon, Y. Wang et al. 2004. P-type ATPase heavy metal transporters with roles in essential zinc homeostasis in *Arabidopsis*. *The Plant Cell* 16: 1327–39.

Iglesia-Turino, S., A. Febrero, O. Jauregui, C. Caldelas, J.L. Araus, and J. Bort. 2006. Detection and quantification of unbound phytochelatin2 in plant extracts of *Brassica napus* grown with different levels of mercury. *Plant Physiology* 142: 742–9.

Ishikawa, S., A.E. Noriharu, M. Murakami, and T. Wagatsuma. 2006. Is *Brassica juncea* a suitable plant for phytoremediation of cadmium in soils with moderately low cadmium concentration? Possibility of using other plant species for Cd-phytoremediation. *Soil Science and Plant Nutrition* 52: 32–42.

Jahangir, M., A. Bayoumi, C. Ibrahim, H. Young, and R. Verpoorte. 2008. Metal ion-inducing metabolite accumulation in *Brassica rapa. Journal of Plant Physiology* 165: 1429–37.

Kabata-Pendias, A. 2001. *Trace elements in soils and plants*, 3rd ed. Boca Raton: CRC.

Karimi, N., S.M. Ghaderian, A. Raab, J. Feldmann, and A.A. Meharg. 2009. An arsenic-accumulating, hypertolerant brassica, *Isatis capadocica. New Phytologist* 184: 41–7.

Keltjens, W.G., and M.L. Van Beusichem. 1998. Phytochelatins as biomarkers for heavy metal toxicity in maize: single metal effects of copper and cadmium. *Journal of Plant Nutrition* 21: 635–48.

Khan, I., A. Ahmad, and M. Iqbal. 2009. Modulation of antioxidant defence system for arsenic detoxification in Indian mustard. *Ecotoxicology and Environmental Safety* 72: 626–34.

Kneer, R., and M.H. Zenk. 1992. Phytochelatins protect plant enzymes from heavy metal poisoning. *Phytochemistry* 31: 2663–7.

Kos, B., and D. Lestan. 2003. Influence of a biodegradable ([S,S]-EDDS) and nondegradable (EDTA) chelate and hydrogel modified soil water sorption capacity on Pb phytoextraction and leaching. *Plant Soil* 253: 403–11.

Kramer, U. 2005. Phytoremediation: novel approaches to cleaning up polluted soils. *Current Opinion in Biotechnology* 16:133–41.

Kumar, P.B.A.N., V. Dushenkov, H. Motto, and I. Raskin. 1995. Phytoextraction: the use of plants to remove heavy metals from soils. *Environmental Science and Technology* 29: 1232–8.

Larsson, E.H., J.F. Bornman, and H. Asp. 1998. Influence of UV-B radiation and Cd^{2+} on chlorophyll fluorescence, growth and nutrient content in *Brassica napus. Journal of Experimental Botany* 49: 1031–9.

LeDuc, D.L., A.S. Tarun, M. Montes-Bayon et al. 2004. Overexpression of selenocysteine methyltransferase in *Arabidopsis* and Indian mustard increases selenium tolerance and accumulation. *Plant Physiology* 135: 377–83.

LeDuc, D.L., M. AbdelSamie, M. Montes-Bayon, C.P. Wu, S.J. Reisinger, and N. Terry. 2006. Overexpressing both ATP sulfurylase and selenocysteine methyltransferase enhances selenium phytoremediation traits in Indian mustard. *Environmental Pollution* 144: 70–6.

Lee, M.K., H.S. Kim, J.S. Kim, S.H. Kim, and Y.D. Park. 2004. *Agrobacterium* mediated transformation system for large-scale production of transgenic Chinese cabbage (*Brassica rapa* L. ssp *pekinensis*) plants for insertional mutagenesis. *Journal of Plant Biology* 47: 300–6.

Leustek, T. 1996. Molecular genetics of sulfate assimilation in plants. *Physiologia Plantarum* 97: 411–9.

Lin, A., X. Zhang, Y.-G. Zhu, and F.-J. Zhao. 2008. Arsenate-induced toxicity: effects on antioxidative enzymes and DNA damage in *Vicia faba. Environmental Toxicology and Chemistry* 27: 413–9.

Liphadzi, M.S., and M.B. Kirkham. 2005. Phytoremediation of soil contaminated with heavy metals: a technology for rehabilitation of the environment. *South African Journal of Botany* 71: 24–37.

Liu, C.P., Z.G. Shen, and X.D. Li. 2006. Uptake of cadmium by different cultivars of *Brassica pekinens* (Lour.) Rupr. and *Brassica chinensis* L. and their potential for phytoremediation. *Bulletin of Environmental Contamination and Toxicology* 76: 732–9.

Liu, D., W. Jiang, C. Liu, C. Xin, and W. How. 2000. Uptake and accumulation of lead by roots, hypocotyls and shoots of Indian mustard (*Brassica juncea* L.). *Bioresource Technology* 71: 273–7.

Lone, M.I., Z. He, P.J. Stoffella, and X. Yang. 2008. Phytoremediation of heavy metal polluted soils and water: progress and perspectives. *Journal of Zhejiang University Science* B 9: 210–20.

Lü, S., H. Gu, X. Yuan et al. 2007. The GUS reporter-aided analysis of the promoter activities of a rice metallothionein gene reveals different regulatory regions responsible for tissue-specific and inducible expression in transgenic *Arabidopsis. Transgenic Research* 16: 177–91.

Ma, L.Q., K.M. Komar, C. Tu, W. Zhang, Y. Cai, and E.D. Kenelley. 2001. Bioremediation: a fern that hyperaccumulates arsenic. *Nature* 409: 579.

Mahuku, G.S., R. Hall, and P.H. Goodwin. 1996. Co-infection and induction of systemic acquired resistance by weakly and highly virulent isolates of *Leptosphaeria maculans* in oilseed rape. *Physiological and Molecular Plant Pathology* 49: 61–72.

Marchiol, L., S. Assolari, P. Sacco, and G. Zerbi. 2004. Phytoextraction of heavy metals by canola (*Brassica napus*) and radish (*Raphanus sativus*) grown on multicontaminated soil. *Environmental Pollution* 132: 21–7.

Marrs, K.A. and V. Walbot. 1997. Expression and RNA splicing of the maize glutathione S-transferase *Bronze2* gene is regulated by cadmium and other stresses. *Plant Physiology* 113: 93–102.

Matsumoto, H., E. Hirasawa, H. Torikai, and E. Takahashi. 1976. Localization of absorbed aluminum in pea root and its binding to nucleic acid. *Plant Cell Physiology* 17:127–37.

McGrath, S.P., and F. Zhao. 2003. Phytoextraction of metals and metalloids from contaminated soils. *Current Opinion in Biotechnology* 14: 277–82.

McIntyre, T. 2003. Phytoremediation of heavy metals from soils. *Advances in Biochemical Engineering/ Biotechnology* 78: 97–123.

Meers, E., A. Ruttens, M. Hopgood, E. Lesage, and F.M.G. Tack. 2005. Potential of *Brassica rapa, Cannabis sativa, Helianthus annuus* and *Zea mays* for phytoextraction of heavy metals from calcareous dredged sediment derived soils. *Chemosphere* 61: 561–72.

Meharg, A.A. 2004. Arsenic in rice—understanding a new disaster for South-East Asia. *Trends in Plant Science* 9: 415–7.

Mendoza-Cózatl D.G., G. David, E. Butko, F. Springer et al. 2008. Identification of high levels of phytochelatins, glutathione and cadmium in the phloem sap of *Brassica napus*. A role for thiolpeptides in the long-distance transport of cadmium and the effect of cadmium on iron translocation. *The Plant Journal* 54: 249–59.

Meng, de K., J. Chen, and Z.M. Yang. 2011. Enhancement of tolerance of Indian mustard (*Brassica juncea*) to mercury by carbon monoxide. *Journal of Hazardous Materials* 186: 1823–9.

Metwally, A., V.I. Safronova, A.A. Belimov, and K. Dietz. 2005. Genotypic variation of the response to cadmium toxicity in *Pisum sativum* L. *Journal of Experimental Botany* 56: 167–78.

Mikkelsen, R.L., A.L. Page, and F.T. Bingham. 1989. Factors affecting selenium accumulation by agricultural crops. In *Selenium in Agriculture and the Environment* ed. Jacobs, L.W., ASA and SSSA, Madison, Wisconsin USA: SSSA Spec. Publ. 23.

Minglin, L., Z. Yuxiu, and C. Tuanyao. 2005. Identification of genes up-regulated in response to Cd exposure in *Brassica juncea* L. *Gene* 363: 151–8.

Mishra, S., S. Srivastava, R.D. Tripathi, and P.K. Trivedi. 2008. Thiol metabolism and antioxidant system complement each other during arsenate detoxification in *Ceratophyllum demersum* L. *Aquatic Toxicology* 86: 205–15.

Mondal, P., C.B. Majumdar, and B. Mohanty. 2006. Laboratory-based approaches for arsenic remediation from contaminated water: recent developments. *Journal of Hazardous Materials* 137: 464–79.

Montes-Bayon, M., E.G. Yanes, C. Ponce de Leon, K. Jayasimhula, A. Stalcup, J. Shann, and J.A. Caruso. 2002. Initial studies of selenium speciation in *Brassica juncea* by LC with ICPMS and ES-MS detection: an approach for phytoremediation studies. *Analytical Chemistry* 74:107–13.

Mossor-Pietraszewska, T. 2001. Effect of Aluminum on plant growth and metabolism. *Acta Biochimica Polonica* 48: 673–86.

Mylona, P.V., A.N. Polidoros, and J.G. Scandalios. 1998. Modulation of antioxidant responses by arsenic in maize. *Free Radical Biology and Medicine* 25: 576–85.

Nehnevajova, E., R. Herzig, K. Erismann, and J. Schwitzguebel. 2007. *In vitro* breeding of *Brassica juncea* L. to enhance metal accumulation and extraction properties. *Plant Cell Reports* 26: 429–37.

Neuhierl, B., and A. Bock. 1996. On the mechanism of selenium tolerance in selenium-accumulating plants. Purification and characterization of a specific selenocysteine methyltransferase from cultured cells of *Astragalus bisulcatus*. *European Journal of Biochemistry* 239: 235–8.

National Research Council (NRC). 1998. Metals and radionuclides: technologies for characterization, remediation, and containment. In *Groundwater and soil clean-up: improving management of persistent contaminants*, 72–128, Washington, DC, National Academy Press.

Nriagu, J.O. 1990. Trace metal pollution of lakes: a global perspective In *Proceedings of 2nd International Conference on Trace Metals in Aquatic Environment.* Sydney, Australia.

Padmaja, K., D.D.K. Prasad, and A.R.K. Prasad. 1990. Inhibition of chlorophyll synthesis in *Phaseolus vulgaris* seedlings by cadmium acetate. *Photosynthetica* 24: 399–405.

Palma, J.M., L.M. Sandalio, F.J. Corpas, M.C. Romero-Puertas, I. MaCarthy, and L.A. Del Rio. 2002. Plant proteases, protein degradation and oxidative stress. Role of peroxisomes. *Plant Physiology and Biochemistry* 40: 521–30.

Panda, S.K., F. Baluska and H. Matsumoto. 2009. Aluminum stress signaling in plants. *Plant Signaling & Behavior* 4: 592–7.

Panwar, B.S., K.S. Ahmed, and S.B. Mittal. 2002. Phytoremediation of nickel-contaminated soils by *Brassica* species. *Environment, Development and Sustainability* 4: 1–6.

Pickering, I.J., R.C. Prince, M.J. George, R.D. Smith, G.N. George, and D.E. Salt. 2000. Reduction and coordination of arsenic in Indian Mustard. *Plant Physiology* 122: 1171–7.

Pilon-Smits, E., M. de Souza, G. Hong, A. Amini, R. Bravo, and N. Terry. 1999a. Selenium volatilization and accumulation by 20 aquatic plant species. *Journal of Environmental Quality* 28: 1011–8.

Pilon-Smits, E., S. Hwang, C.M. Lytle et al. 1999b. Over-expression of ATP sulfurylase in *Brassica juncea* leads to increased selenate uptake, reduction and tolerance. *Plant Physiology* 119: 123–32.

Pilon-Smits, E.A.H., Y.L. Zhu, T. Sears, and N. Terry. 2000. Overexpression of glutathione reductase in *Brassica juncea*: Effects on cadmium accumulation and tolerance. *Physiologia Plantarum* 110: 455–60.

Polle, A., and A. Schützendübel. 2003. Heavy metal signalling in plants: linking cellular and organismic responses. In: Hirt, H., and K. Shinozaki, eds, *Plant Responses to Abiotic Stress*, Vol 4. Springer-Verlag, Berlin, pp. 187–215.

Prasad, M.N.V. 2003. Metal hyperaccumulation in plants—Biodiversity prospecting for phytoremediation technology. *Electronic Journal of Biotechnology* 6.

Psotova, J., J. Lasovsky, and J. Vicar. 2003. Metal chelating properties, electrochemical behaviour, scavenging and cytoprotective activities of six natural phenolics. *Biomedical Papers* 147: 147–53.

Purakayastha, T.J., T. Viswanath, S. Bhadraray, P.K. Chhonkar, P.P. Adhikari, and K. Suribabu. 2008. Phytoextraction of zinc, copper, nickel and lead from a contaminated soil by different species of *Brassica*. *International Journal of Phytoremediation* 10: 61–72.

Qian, J.H., A. Zayed, Y.L. Zhu, Y.U. Mei, and N.P. Terry. 1999. Phytoaccumulation of trace elements by wetland plants. Uptake and accumulation of ten trace elements by twelve plant species. *Journal of Environmental Quality* 28: 1448–55.

Quan, X.Q., H.T. Zhang, L. Shan, and Y.P. Bi. 2006. Advances in plant metallothionein and its heavy metal detoxification mechanisms. *Yi Chuan* 28: 375–82.

Quartacci, M.F., A. Argilla, A.J.M. Baker, and F. Navari-Izzo. 2006. Phytoextraction of metals from a multiply contaminated soil by Indian mustard. *Chemosphere* 63: 918–25.

Raab, A., J. Feldmann, and A.A. Meharg. 2004.The nature of arsenic-phytochelatin complexes in *Holcus lanatus* and *Pteris cretica*. *Plant Physiology* 134: 1113–22.

Raab, A., H. Schat, A.A. Meharg, and J. Feldmann. 2005. Uptake, translocation and transformation of arsenate and arsenite in sunflower (*Helianthus annuus*): formation of arsenic–phytochelatin complexes during exposure to high arsenic concentrations. *New Phytologist* 168: 551–58.

Rauser, W.E. 1995. Phytochelatins and related peptides: structure, biosynthesis, and function. *Plant Physiology* 109: 1141–49.

Shahandeh, H., and L.R. Hossner. 2002a. Enhancement of uranium phytoaccumulation from contaminated soils. *Soil Science* 167: 269–80.

Shahandeh, H., and L.R. Hossner. 2002b. Role of soil properties in phytoaccumulation of uranium. *Water, Air, and Soil Pollution* 141: 165–80.

Shanker, A.K., C. Cervantes, H. Loza-Tavera, and S. Avudainayagam. 2005. Chromium toxicity in plants. *Environmental International* 31: 739–53.

Shaw, B.P. 1995. Effects of mercury and cadmium on the activities of antioxidative enzymes in the seedlings of *Phaseolus aureus*. *Biologia Plantarum* 37: 587–96.

Shen, Z.G., X.D. Li, X.D. Wang, H.M. Chen, and H. Chua. 2002. Lead phytoextraction from contaminated soil with high-biomass plant species. *Journal of Environmental Quality* 31: 1893–900.

Shiyab, S., J. Chen, F.X. Han et al. 2008. Mercury-induced oxidative stress in Indian mustard (*Brassica juncea* L.). *Environmental Toxicology* DOI 10.1002/tox, 462–71.

Shiyab, S., J. Chen, F.X. Han et al. 2009. Phytotoxicity of mercury in Indian mustard (*Brassica juncea* L.). *Ecotoxicology and Environmental Safety* 72: 619–25.

Singh, S., S. Singh, V. Ramachandran, and S. Eapen. 2010. Copper tolerance and response of antioxidative enzymes in axenically grown *Brassica juncea* (L.) plants. *Ecotoxicology and Environmental Safety* 73: 1975–81.

Sinha, S., R. Saxena, and S. Singh. 2002. Comparative study on accumulation of Cr from metal solution and tannery effluent under repeated metal exposure by aquatic plants: its toxic effects. *Chemosphere* 80: 17–31.

Somashekaraiah, B.V., K. Padmaja, and A.R.K. Prasad. 1992. Phytotoxicity of cadmium ions on germinating seedlings of mungbean (*Phaseolus vulgaris*): involvement of lipid peroxides in chlorophyll degradation. *Physiologia Plantarum* 85: 85–9.

Srivastava, S., S. Mishra, R.D. Tripathi, S. Dwivedi, P.K. Trivedi, and P.K. Tandon. 2007. Phytochelatins and antioxidant systems respond differentially during arsenite and arsenate stress in *Hydrilla verticillata* (L.f.) Royle. *Environmental Science and Technology* 41: 2930–6.

Srivastava, S., A.K. Srivastava, P. Suprasanna, and S.F. D'Souza. 2009. Comparative biochemical and transcriptional profiling of two contrasting varieties of *Brassica juncea* L. in response to arsenic exposure reveals mechanisms of stress perception and tolerance. *Journal of Experimental Botany* 60: 3419–31.

Suzuki, N., N. Koizumi, and H. Sano. 2001. Screening of cadmium-responsive genes in *Arabidopsis thaliana*. *Plant, Cell & Environment* 24: 1177–88.

Switras, S. 1999. The potential of phytoremediation techniques for selenium removal. *Restoration and Reclamation Review* 5.

Tandy, S., K. Bossart, R. Mueller et al. 2004. Extraction of heavy metals from soils using biodegradable chelating agents. *Environmental Science and Technology* 38: 937–44.

Terry, N., and A.M. Zayed. 1994. Selenium volatilization by plants. In *Selenium in the Environment*, ed. Frankenberger, W.T. Jr., and S. Benson. New York: Marcel Dekker.

Terry, N., C. Carlson, T.K. Raab, and A. Zayed. 1992. Rates of selenium volatilization among crop species. *Journal of Environmental Quality* 21: 341–44.

Thiyam, U., A. Kuhlmann, H. Stockmann, and K. Schwarz. 2004. Prospects of rapeseed oil by-products with respect to antioxidative potential. *Comptes Rendus Chimie* 7: 611–6.

Tomé, F.V., P.B. Rodríguez, and J.C. Lozano. 2009. The ability of *Helianthus annuus* L. and *Brassica juncea* to uptake and translocate natural uranium and ^{226}Ra under different milieu conditions. *Chemosphere* 74: 293–300.

Tripathi, R.D., S. Srivastava, S. Mishra et al. 2008. Arsenic hazards: strategies for tolerance and remediation by plants. *Trends in Biotechnology* 25: 158–65.

Trlica, M.J., and L.F. Brown. 2000. Reclamation of URAD molybdenum tailing: 20 years of monitoring change. In *Proceedings of the High Altitude Revegetation Workshop, No. 14, Fort Collins, Colorado.*, ed. Keammerer, W., 82–133. Cooperative Extension Resource Center, Colorado State University, Fort Collins, CO.

Turan, M., and A. Esring. 2007. Phytoremediation based on canola (*Brassica napus* L.) and Indian Mustard (*Brassica juncea* L.) planted on spiked soil by aliquot amount of Cd, Cu, Pb, and Zn. *Plant Soil and Environment* 53: 7–15.

UNEP. Phytoremediation: an environmentally sound technology for pollution prevention, control and remediation. *Newsletter and Technical Publications Freshwater Management Series No. 2.*

United States Protection Agency Reports 2000. *Introduction to Phytoremediation.*–EPA 600/R-99/107.

Vallejo, F., A. Gil-Izquierdo, A. Pearez-Vicente, and C. Garciaa-Viguera. 2004. *In vitro* gastrointestinal digestion study of broccoli inflorescence phenolic compounds, glucosinolates, and vitamin C. *Journal of Agriculture and Food Chemistry* 52: 135–8.

Vangronsveld, J., and H. Clijsters. 1994. Toxic effects of metals. In *Plants and the Chemical Elements. Biochemistry, Uptake, Tolerance and Toxicity*, ed. Farago, M.E., 150–77. Weinheim, Germany: VCH Publishers.

Wang, C., Y. Tian, X. Wang, J. Geng, J. Jiang, H. Yu, and C. Wang. 2010. Lead-contaminated soil induced oxidative stress, defense response and its indicative biomarkers in roots of *Vicia faba* seedlings. *Ecotoxicology* 19: 1130–9.

Watanabe, M.E. 1997. Phytoremediation on the brink of commercialization. *Environmental Science and Technology* 31: 182–6.

Whitelaw, C.A., J.A. Le Huquet, D.A. Thurman, and A.B. Tomsett. 1997. The isolation and characterisation of type II metallothionein-like genes from tomato (*Lycopersicon esculentum* L.) *Plant Molecular Biology* 33: 503–11.

Wu, F.B., and G.P. Zhang. 2002. Genotypic differences in effect of Cd on growth and mineral concentrations in barley seedlings. *Bulletin of Environmental Contamination and Toxicology* 69: 219–27.

Wu, L., X. Guo, and G.S. Banuelos. 2009. Accumulation of seleno-amino acids in legume and grass plant species grown in selenium-laden soils. *Environmental Toxicology and Chemistry* 16: 491–7.

Wu, L.H., H. Li, Y.M. Luo, and P. Christie. 2004. Nutrients can enhance phytoremediation of copper-polluted soil by Indian mustard. *Environmental Geochemistry and Health* 26: 331–5.

Xiong, Z.T., L.C. Chao, and B. Geng. 2006. Phytotoxic effects of copper on nitrogen metabolism and plant growth in *Brassica pekinensis. Rupr. Ecotoxicology and Environmental Safety* 64: 273–80.

Yang, Y., Y. Zhang, X. Wei et al. 2011. Comparative antioxidative responses and proline metabolism in two wheat cultivars under short term lead stress. *Ecotoxicology and Environmental Safety* 74: 733–40.

Zayed, A.M., and N. Terry. 1992. Selenium volatilization in broccoli as influenced by sulfate supply. *Journal of Plant Physiology* 140: 646–52.

Zenk, M.H. 1996. Heavy metal detoxification in higher plants: a review. *Gene* 179: 21–30.

Zhigang, A., L. Cuijie, Z. Yuangang et al. 2006. Expression of BjMT2, a metallothionein 2 from *Brassica juncea*, increases copper and cadmium tolerance in *Escherichia coli* and *Arabidopsis thaliana*, but inhibits root elongation in *Arabidopsis thaliana* seedlings. *Journal of Experimental Botany* 57: 3575–82.

Zhu, Y.L., E.A.H. Pilon-Smits, L. Jouanin, and N. Terry. 1999a. Overexpression of glutathione synthetase in Indian mustard enhances cadmium accumulation and tolerance. *Plant Physiology* 119: 73–9.

Zhu, Y.L., E.A.H. Pilon-Smits, A.S. Tarun, S.U. Weber, L. Jouanin, and N. Terry. 1999b. Cadmium tolerance and accumulation in Indian mustard is enhanced by overexpressing γ-glutamylcysteine synthetase. *Plant Physiology* 121:1169–77.

Section III

Other Plant Species and Contaminants' Remediation

12 Phytoremediation of Soils Contaminated by Heavy Metals, Metalloids, and Radioactive Materials Using Vetiver Grass, *Chrysopogon zizanioides*

Luu Thai Danh, Paul Truong, Raffaella Mammucari, Yuan Pu, and Neil R. Foster

CONTENTS

12.1 INTRODUCTION

An increasingly industrialized global economy and rapid rise in world population over the last century have led to dramatically elevated releases of anthropogenic chemicals, particularly heavy metals, into the environment. Heavy metal-contaminated soils have caused serious problems threatening ecological systems and human health, and have recently attracted considerable public attention. Concentrations of heavy metals that have exceeded safety levels in soil should be treated (Baker et al. 1994). There are several methods used for soil remediation, including chemical, physical and biological techniques. Physical treatments involve removal from contaminated sites (soil excavation), deep burial (land filling) and capping, while chemical methods use strong acids and chelators to wash polluted soils. These approaches are expensive, impractical and at times impossible to carry out, as the volume of contaminated materials is very large. Furthermore, they irreversibly affect soil properties, destroy biodiversity and may render the soil useless as a medium for plant growth. Recently, phytoremediation, a diverse collection of plant-based technologies to clean contaminated environments, using either naturally occurring or genetically engineered plants (Cunningham et al. 1997) represents a novel, environmentally friendly and cost-effective technology. Importantly, by-products of phytoremediation can find a range of other uses.

The application of phytoremediation for pollution control, however, has several limitations, including low biomass production, shallow root distribution, difficult cultivation, limited sources, low adaptability and weed potential. Most plants used for phytoremediation are hyperaccumulators, which are nonperennial plants with slow growth and low biomass production, requiring a long-term commitment for remediation. Phytoremediation is mainly limited to the soil levels occupied by the root systems, as the hyperaccumulators generally have a shallow root zone that does not support efficient phytoremediation (Keller et al. 2003). Most hyperaccumulators that are native species and relatively rare taxa, often occur in geographically remote areas with extremely small populations (Baker et al. 1994). The introduction of plants into new environments increases the risk of invasive weeds. Environmental conditions also determine the efficiency of phytoremediation, as the survival and growth of plants are adversely affected by extreme environmental conditions, toxicity and the general conditions of soil in contaminated lands (Northwestern University 2007).

Vetiver grass, *Chrysopogon zizanioides* (L.) Roberty, syn. *Vetiveria zizanioides* (L.) Nash is one of the very few plants that can overcome the afore-mentioned limitations of phytoremediation. Vetiver grass (VG) belongs to the *Poaceae* family, a subfamily of *Panicoideae*, tribe *Andropogonae* and subtribe *Sorghina*, and the genus includes ten species (Bertea and Camusso 2002). It is common to flood plains and stream banks, but can also be found throughout the tropical and subtropical regions of Africa, Asia, America, Australasia and Mediterranean Europe (Maffei 2002). Recently, it has received great interest from scientists and the public for its ability to remove a wide range of inorganic contaminants from soil. In this review, the special characteristics of vetiver grass are evaluated for their phytoremediation potential, and economic incentives to promote application of vetiver for phytoremediation are also discussed.

12.2 EVALUATION OF VETIVER GRASS

Plants used for phytoremediation must satisfy the following requirements: 1) fast growth and high biomass production, 2) widely and deeply distributed root system, 3) adaptability to a wide range of weather conditions, 4) easy cultivation, 5) no potential to become a weed, 6) the ability to accumulate heavy metals. Glasshouse and field studies have demonstrated that vetiver grass can fulfill these conditions.

12.2.1 Biomass Production

A fast growing rate and high biomass production are two of the most important factors determining the efficiency of phytoremediation. Vetiver is a C4 grass that has high growth rates, as indicated by

(a) (b)

FIGURE 12.1 Vigorous growth of vetiver grass in Australia (a) and Asia (b). (From www.vetiver.org.)

high radiation use efficiency of 18 kg ha^{-1} per MJ m^{-1} (Vieritz et al. 2003). It is comparable with other C4 grasses such as maize (*Zea mays* L.) and sugarcane (*Saccharum officinarum*), its growth rate is more than three times higher than that of C3 grasses such as coastal couch grass (*Cynodon dactylon*) (Inman-Bamber 1974; Burton and Hanna 1985; Muchow 1990). High growth rate results in high biomass production of vetiver (Figure 12.1). Its biomass is extremely high under tropical hot and wet conditions, producing more than 100 tons of dry matter ha^{-1} $year^{-1}$ (Truong 2003). In sub-tropical weather, vetiver can produce high dry biomass from 10 to 20 tons after 5–6 months of cultivation (Yang et al. 2003; Shu et al. 2004). This may be due to the fact that vetiver retains high activity of the key enzymes involved in photosynthesis (NADP-MDH and NADP-MET) even when cultivated in temperate climates (Bertea and Camusso 2002).

Biomass production of vetiver can be improved by planting density, the addition of fertilizers, organic matter, and lime under acidic conditions. It has been reported that a planting density of 25 × 25 cm spacing produced the highest biomass production as compared to 50 × 50 and 100 × 100 cm spacing (Pothinam 2006).

12.2.2 Distribution of Root System

The efficiency of phytoremediation depends on the contact between contaminants and the roots of the plants. The latter is determined by the root distribution of the cultivated plants. Generally, the hyperaccumulators have short root systems that confine the heavy metal uptake in the top soil layers, hence limiting the concentrations of contaminants in plants.

Fortunately, vetiver grass possesses a lacework root system that is abundant, complex, and extensive (Figure 12.2). This extensive and thick root system binds the soil and at the same time makes it very difficult to dislodge and extremely tolerant to drought. Due to this massive root system, mature Vetiver grass has a root-to-shoot ratio of approximately 1:1.1, much higher than pasture plants and among the highest of cultivated plants. A field trial carried out in Thailand showed that vetiver roots reached the depth of 1.5 and 2.5 m after one and two and a half months of cultivation, respectively; a higher root density was found at the depth 30–100 cm. The roots were found at soil depth of 4 m and 5 m at 10 and 25 months old, respectively. In another field study in Colombia vetiver grass acquired a total length of 7 meters after 34 months (Tscherning et al. 1995). However, the grass does not penetrate far into the groundwater table. Therefore at locations with high groundwater levels, its root system may not be as long as in drier soil (Van and Truong 2008). The root growth rate of vetiver corresponding to the soil temperature reaches a plateau at 25°C, with approximately 3 cm per day. Although the root extension rate was higher in 35°C, the difference was not statistically significant.

FIGURE 12.2 Root systems of vetiver grass in Thailand and China.

The underground root growth was still detected at low soil temperature (13°C), indicating that vetiver is not dormant at this temperature. It is still possible to establish vetiver under this temperature, although the growing time will be longer (Wang 2000).

The size of vetiver roots is small, with an average of 0.66 mm (range from 0.2 to 1.7 mm) (Cheng et al. 2003). The horizontal spreading of lateral roots was in the range of 0.15–0.29 m, with an average of 0.23 m (Mickovski et al. 2005). Similarly, root growth of vetiver was about 25 cm wide in the study of Nix et al. (2006). After 8 months of cultivation, vetiver produced 0.48 kg of dry roots per plant (Nix et al. 2006). The special characteristics of the vetiver root system allow good contact between vetiver and contaminants, even at deep soil levels.

12.2.3 Adaptability to Soil and Weather Conditions

Heavy metal-contaminated sites exist throughout the world, so the ideal plant used for phytoremediation should be adaptable to a wide range of climatic and soil conditions. Generally, the microenvironments of mine tailings as well as heavy metal-contaminated sites are very stressful to plant growth (Ernst 2005; Audet and Charest 2007), due to extreme temperatures, drought or flooding, low or high pH, low soil fertility, uneven distribution of contaminants, high concentrations of one or several heavy metals, low water-holding capacity, high salinity and lack of normal heterotrophic microbial community. Field and glasshouse studies have demonstrated that vetiver grass is highly adaptable to extreme environmental and soil conditions.

Vetiver grass can survive and grow at extremely cold temperatures, although it is a tropical grass (Xia et al. 1999; Xu and Zhang 1999). The top growth of vetiver grass is terminated under frosty weather conditions but its underground growing points still survive (Truong et al. 2008). Vetiver growth was not affected by severe frost at –11°C in Australia and it survived for a short period at –22°C in northern China. To increase the survival rate during the winter time, vetiver should be planted in early spring so as to allow the full development of vetiver before the following winter (Xu 2005). In addition, vetiver can stand very high temperature up to 55°C (Xia et al. 1999; Xu and Zhang 1999). In particular, it can survive severe fires and fully recover after burning (Figure 12.3).

Vetiver grass is highly adaptable to extreme environmental conditions (Figure 12.4). It can survive the prolonged drought up to 6 months (Truong 1999a). Mine tailings are often characterized by shales, cobbles and pebbles, which have a very low water holding capacity (Ernst 1974). In addition,

(a) (b)

FIGURE 12.3 Strong recovery of vetiver growth after the heavy fires in Vanuatu (a) and Australia (b). (From www.vetiver.org.)

the locations of many mine tailings are in regions where rainfall is limited. The extensive and long root of vetiver grass, as aforementioned, can utilize deep soil moisture supporting the survival of vetiver under extreme drought conditions. Moreover, vetiver was demonstrated to be tolerant to complete submergence under water for more than 120 days, which is much longer than that of other plants (Bahia grass: 60–70 days, carpet grass: 32–40 days, Sour paspalum: 25–32 days, St. Augustine: 18–32 days, Centipede grass: only 7–10 days) (Xia et al. 2003). Similarly, vetiver can survive more than three months under muddy water, as demonstrated in a trial conducted in 2007 to stabilise the Mekong river bank in Cambodia (Toun Van, pers.com). Under partial submergence, it can stand up to eight months in a trial in Venezuela (Figure 12.4).

Most mine tailings and heavy metal-contaminated soils are very low in nutrients, particularly organic matter, nitrogen and phosphorous (Ernst 1974; Ernst and Nelissen 2000). Vetiver grass can survive and grow on such soils, especially infertile soils, in many tropical countries (Siripin 2000), due to the fact that it can establish a strong symbiotic association with a wide range of soil micro-organisms in the rhizosphere (Siripin 2000; Monteiro et al. 2009; Leaungvutiviroj et al. 2010).

The microbes provide nitrogen (nitrogen-fixing bacteria), phosphorous (phosphate-solubilizing bacteria and fungi, mycorrhizal and cellulolytic fungi) and plant growth hormones (plant growth regulator bacteria) for vetiver development. Microbial populations and activities are higher in the

(a) (b)

FIGURE 12.4 Vetiver survival under prolong drought (a) in Australia (note: all native plants were browned off) and submergence of 25 cm for 8 months (b) in Venezuela. (From www.vetiver.org.)

vetiver rhizosphere than outside it. Large numbers of soil microbes have been discovered in the vetiver rhizosphere, ranging from 10^6 to 10^8 cells g^{-1} soil, in which nonsymbiotic nitrogen-fixing bacteria and phosphate-solubilizing microorganisms varied from 10^1 to 10^4 cells g^{-1} soil and the endomycorrhiza were 3–26 spores per 100 g of soil (Siripin 2000). Similar results were found in the recent investigation of Leaungvutiviroj et al. (2010) on microbial population in three problem soils, namely acid sulphate, shallow and saline soils. Cellulolytic bacteria and actinomycetes were the most numerous inhabitants, in the range of 10^6 to 10^8 cells g^{-1} soil, in which cellulolytic fungi, Azotobacter, phosphate-solubilizing bacteria and fungi had similar microbial populations of 10^2 to 10^4 cells g^{-1} soil. Siripin et al. (2000) isolated 35 different nitrogen-fixing bacteria strains from a vetiver root system that was responsible for up to 40% of N content in vetiver. Plant growth hormones produced by microbes promote development of lateral roots and total biomass of vetiver grass. The nitrogen-fixing bacteria, *Azospirillum*, found outside and inside the vetiver roots were demonstrated to produce the plant growth hormone, idole-3-acetic acid (IAA), at concentrations of 30–40 μg/ml in broth media (Patiyuth et al. 2000). Monteirio et al. (2009) found 48, 46 and 49 bacterial strains in the vetiver rhizosphere being responsible for nitrogen fixation, IAA production and phosphate-solubilization, respectively. Interestingly, 25 bacterial strains were found to be involved in three plant growth-promoting characteristics.

Vetiver grass not only survives and grows well on infertile soil, but also improves the soil quality, in terms of nutritional, physical and biological properties, through its symbiotic association with soil micro-organisms (Tables 12.1–12.3). The hostile conditions of heavy metal- contaminated soils impose pressure on the development of soil microorganisms. The cultivation of vetiver grass can increase the microbial populations in such soils.

In arid environments, one common characteristic of soils in mine tailings is high salinity (Mendez and Maier 2008). Generally, plants can tolerate a soil with electrical conductivity (EC) of 4 dS m^{-1} (Marschner 1995). Above this value, soils are considered saline. Soils in arid mine tailings

TABLE 12.1

Changes of Organic Matters (OM) and Macronutrient Contents in Soils after One Year Cultivated with Vetiver Grass

Soil Types	Soil Depth	Treatments	OM (g kg^{-1})	N (%)	P (mg kg^{-1})	K (mg kg^{-1})
Acid sulfate soil	0–30	Non grassed control	23.4	0.14	5.03	163
		Vetiver	24.9	0.21	6.36	214
		Increments (%)	6.4	50	24.44	31.29
	30–60	Non grassed control	11.6	0.12	4.08	133
		Vetiver	12.5	0.15	5.45	147
		Increments (%)	7.76	25.0	33.58	18.05
Shallow soil	0–30	Non grassed control	5.8	0.01	2.07	103
		Vetiver	10.0	0.02	4.82	154
		Increments (%)	72.41	100	132.85	49.51
	30–60	Non grassed control	4.2	0.008	2.39	93
		Vetiver	8.6	0.016	4.52	147
		Increments (%)	104.76	100	118.36	42.7
Saline soil	0–30	Non grassed control	4.9	0.01	1.75	32
		Vetiver	4.4	0.02	3.28	52
		Increments (%)	−10.2	100	87.43	62.5
	30–60	Non grassed control	2.3	0.008	1.49	28
		Vetiver	2.7	0.016	2.41	40
		Increments (%)	17.39	100	61.74	42.86

Source: Adapted from Leaungvutiviroj, C., S. Piriyaprin, P. Limtong, and K. Sasakic, *Applied Soil Ecology* 46: 95–102, 2010.

TABLE 12.2
Physical Properties of the Surface (0–15 cm) Soil after 15 Years under Different Grass Species at Danville in South Africa

Treatment	Bulk Density (tone m^{-3})	Soil Water Retention (%) −33 kPa	Soil Water Retention (%) −1200 kPa	Soil Water Content (%)	Penetrometer Resistance (kPa)	Aggregate Stability (% > 2 mm)	Aggregate Mean Weight Diameter, Dry (mm)
VG	1.23^c	17.32	13.44	15.78^a	386^c	65.4^a	3.2^c
NG	1.44^b	12.64	10.08	11.67^b	517^b	54.2^b	4.7^b
NGC	1.68^a	8.13^a	5.06^c	9.15^c	829^a	21.6^c	7.8^a
LSD (p = 0.05)	0.19	3.77	3.26	2.33	125	9.6	1.44

Source: Adapted from Materechera, S., *19th World Congress of Soil Science, Soil Solutions for a Changing World, Brisbane, Australia, 1–6 August 2010,* 2010.

Note: VG, vetiver grass; NG, natural grass; NGC, non-grassed control.

can be extremely saline with an EC up to 22 dS m^{-1} (Munshower 1994). The high salinity significantly reduces the plant survival and productivity. Vetiver grass can survive in saline soil with EC up to 47.5 dS m^{-1}. Its salinity threshold is at EC = 8 dS m^{-1} and soil EC_{se} values of 20 dS m^{-1} reduce yield by 50%. Vetiver grass is classified in a group of highly salt-tolerant crop and pasture species grown in Australia (Table 12.4).

Vetiver grass has been demonstrated to survive and grow on the soils contaminated with high concentrations of a wide range of heavy metals. The heavy metals in mine tailings and heavy metal-contaminated soils are often concentrated at high levels. Furthermore, they are not evenly distributed, and vary horizontally with soil depth (Geeson et al. 1998; Schwartz et al. 1999; Whiting et al. 2000; Haines 2002; Podar et al. 2004). Such heterogeneity forms hot-spots across a site where concentrations of heavy metals are very high. High concentrations of heavy metals in soils may inhibit plant growth, so may limit application on some sites or some parts of sites. Therefore the survival rate and growth performance of plants are negatively affected at these points, leading to low efficiency of the whole process. A series of single heavy-metal experiments under glasshouse

TABLE 12.3
Influence of Vetiver Grass on Biological Properties of a Surface (0–15 cm) after 15 Years of Cultivation at Danville in South Africa

Treatment	Organic C (%)	Particulate Organic Matter (%)	Microbial Biomass C (mg kg^{-1})	Microbial Quotient	Proportion of Particulate Organic Matter in Whole Soil (%)
VG					
NG	1.64^b	2.67^b	218^b	0.02^b	3.32^b
NGC	0.67^a	1.37^c	96^a	0.01^c	1.16^a
LSD (p = 0.05)	0.75	0.54	9.85	0.007	1.74

Source: Adapted from Materechera, S., *19th World Congress of Soil Science, Soil Solutions for a Changing World, Brisbane, Australia, 1–6 August 2010,* 2010.

TABLE 12.4
Salt Tolerance Level of Vetiver Grass as Compared with Some Crop and Pasture Species Grown in Australia

	Soil EC_{se} (dSm^{-1})	
Species	**Saline Threshold**	**50% Yield Reduction**
Bermuda grass (*Cynodon dactylon*)	6.9	14.7
Rhodes grass (C.V. Pioneer) (*Chloris guyana*)	7.0	22.5
Tall wheat grass (*Thynopyron elongatum*)	7.5	19.4
Cotton (*Gossypium hirsutum*)	7.7	17.3
Barley (*Hordeum vulgare*)	8.0	18.0
Vetiver grass (*Vetiveria zizanioides*)	8.0	20.0

Source: Truong, P., Y.K. Foong, M. Guthrie, and Y.T. Hung, *Handbook of Environmental Engineering* 11: 233–275, 2010.

conditions proved that vetiver has high tolerance to a wide range of heavy metals in soils due to its high threshold levels of these metals in soils (Table 12.5). Most vascular plants are highly sensitive to heavy metal toxicity with very low threshold levels for metals in the soils. In recent studies, vetiver grass has been reported to survive on the soils containing very high concentrations of arsenic, lead, copper, zinc, chromium (Table 12.5).

Vetiver grass tolerates not only high concentrations of individual heavy metals in soils but also combinations of several heavy metals. Many metal-contaminated soils are enriched by more than one element (polymetallic) with different dominance of the various metals (Ernst et al. 2000; Ernst and Nelissen 2000; Walker and Bernal 2004). Natural hyperaccumulators are very often selective for an individual metal. Consequently, the occurrence of multi-heavy metals would strongly affect their productivity and even other metal-tolerant plants. Furthermore, some mine tailings contain very high concentrations of heavy metals that exceed the tolerance ability of even hyperaccumulators,

TABLE 12.5
Threshold Levels of Heavy Metals to Vetiver Growth Based on Single Element Experiment

Heavy Metals	Threshold to Growth of Most Vascular Plants ($mg\ kg^{-1}$)		Threshold to Vetiver Growth ($mg\ kg^{-1}$)	Vetiver Survival under the Highest Levels of Contaminants Reported in the Literature ($mg\ kg^{-1}$ soil)
	Hydroponic Level[a]	**Soil Level[b]**	**Soil Level[c]**	
Arsenic	0.02–7.5	2.0	100–250	959[c]
Boron				180[d]
Cadmium	0.2–9.0	1.5	20–60	60[e]
Copper	0.5–8.0	NA	50–100	2600[f]
Chronium	0.5–10.8	NA	200–600	2290[g]
Lead	NA	NA	>1500	10750[h]
Mercury	NA	NA	>6	17[i]
Nickel	0.5–2.0	7–10	100	100[c]
Selenium	NA	2–14	>74	>74[c]
Zinc	NA	NA	>750	6400[j]

Note: [a]Bowen (1979); [b]Baker and Eldershaw (1993); [c]Truong (1999b); [d]Angin et al. (2008); [e]Minh and Khoa (2009); [f]Castillo et al. (2007); [g]Hoang et al. (2007); [h]Rotkittikhun et al. (2007); [i]Lomonte et al. (2011); [j]Danh et al. (2010).

hence their productivity would be seriously affected. In glasshouse studies, vetiver could survive and grow well on multi-heavy metal-contaminated soils with total Pb, Zn and Cu in the range of 1155–3281.6, 118.3–1583 and 68–1761.8 mg kg^{-1}, respectively (Chiu et al. 2006; Wilde et al. 2005). Under field conditions, vetiver could grow on mine tailing soils containing total Pb, Zn, Cu and Cd of 2078–4164, 2472–4377, 35–174 and 7–32 mg kg^{-1}, respectively (Shu et al. 2002; Yang et al. 2003; Shu et al. 2004; Zhuang et al. 2005).

The survival rates, growth and biomass of vetiver cultivated on heavy metal-contaminated soil can be greatly improved by addition of organic matter (domestic refuse and sewage sludge), inorganic fertilizer and, especially, the combination of organic matter and inorganic fertilizer (Chiu et al. 2006; Wilde et al. 2005; Yang et al. 2003; Shu et al. 2002). The application of organic matter, however, reduced the accumulation of Pb, Zn and Cu in vetiver (Yang et al. 2003; Chiu et al. 2006).

Vetiver grass can tolerate a wide range of pH that makes it adaptable to acidic or alkaline heavy metal-contaminated soils. Old gold mine tailings are often extremely acidic (pH 2.5–3.5) due to high sulfur content (Truong 1999b), and some mine tailings and other heavy metal contaminated soils in temperate environments are often low in pH values (Mench et al. 1994). In contrast, the tailings of fresh gold, bentonite and bauxite have high pH values. Such characteristics would adversely affect the survival and growth performance of many plants, resulting in the failure of the phytoremediation process. Under glasshouse conditions, vetiver grass was reported to grow well on soils with pH ranging from 3.3 to 9.5 (Danh et al. 2009). When adequately supplied with nitrogen and phosphorus fertilisers, vetiver produced excellent growth even under extremely acidic conditions (pH = 3.8) and at a very high level of soil aluminum saturation percentage (68%). Vetiver did not survive an aluminum saturation level (ASL) of 90% with soil pH = 2.0. ASL for vetiver would be between 68 and 90% (Truong and Baker 1997). These results are supported by recent work in which vetiver has been observed to thrive on highly acidic soil, with aluminum saturation percentage as high as 87% (Truong 1999a). Moreover, the growth of vetiver grass was not adversely affected on a bentonite overburden with an exchangeable sodium percentage (ESP) of up to 33% (Bevan et al. 2000). Soils with ESP higher than 15 is considered to be strongly sodic (Northcote and Skene 1972). Under field conditions, vetiver showed excellent growth on old gold tailings (pH = 2.7) and bauxite mine tailings (pH = 12) in Northern Queensland, Australia (Figures 12.5 and 12.6).

Favourable climatic and weather conditions would promote the development of specific diseases and pests. The introduction of hyperaccumulators, which often lack defense mechanisms in new environments, make them more susceptible to attacks of diseases and pests, which then hinder the application of phytoremediation. Fortunately, vetiver grass is nearly free of diseases and pests with a limited cases reported on the world. Vetiver grass is attacked by *Curvularia trifolii* that causes

(a) (b)

FIGURE 12.5 Eight year old gold tailings in Queensland are highly acidic with pH 2.7 (a), and good growth of vetiver 12 months after planting with 20t/ha of lime and 800kg/ha of DAP (b).

(a) (b)

FIGURE 12.6 A large dam wall of bauxite tailings in Queensland is highly caustic with pH 12 (a), and excellent establishment of vetiver after 3 months of cultivation (b).

the leaf blight during the rainy season. The disease can be easily treated by the removal of infected foliage or the application of copper-based fungicides, such as Bordeaux mixture (Islam et al. 2008). Several pests of vetiver have been recorded, namely stem borers (*Chilo* sp.), termites, cicadas and armyworms (Berg et al. 2003). Perhaps the most serious pest threat of vetiver grass comes from stem borer (*Chilo* sp.). The insect can be controlled by pruning or a timely fire to keep the larva from "overwintering" in the vetiver grass stems (Islam et al. 2008). Termites sometimes attack vetiver grass, but mostly happened in arid regions. Normally no treatments are required to control this pest (Islam et al. 2008). The high infestation levels of the cicada, *Amphisalta zelandica*, were reported to affect the growth of vetiver in New Zealand (Miller 2012). In addition, armyworms (*Spodoptera* sp.) are also reported to damage vetiver in Australia. However, this damage is not economically significant as only the tips of vetiver leaves were damaged while the rest of the plants were untouched (Berg et al. 2003).

12.2.4 Cultivation

The availability of plant materials and the ease of cultivation have important roles in the promotion of phytoremediation technology. Vetiver grass has been cultivated at over 100 countries for different purposes, such as environmental protection and essential oil production. Hence the sources of vetiver grass are readily available for phytoremediation purposes in nearly every corner of the world. The suppliers of vetiver grass around the world are listed in the website of The Vetiver Network International (http://www.vetiver.org).

Vetiver grass, *Chrysopogon zizanioides*, produces nonviable seeds, so the propagation is only based on vegetative methods. The establishment of vetiver grass in the field is quite simple. The common way to propagate vetiver is to split mature tillers (shoots) from vetiver clumps or mother plants, which yield bare root slips for immediate planting or propagating in the polybags or plastic containers (Truong et al. 2008). The bare root slips contain two or three shoots (20 cm long) and a part of the crown or roots (5 cm long). After separation, the bare root slips can be dipped into rooting hormone solution, manure slurry (cow or horse tea), clay mud or simply shallow water pools, until new roots appear. Using bare root slips for direct planting is economical, and represents a quick and efficient way to prepare the planting material. However, bare root slips are vulnerable to drying and extreme soil conditions, and require planting on moist soils. To overcome these problems, bare root slips are first propagated in polybags or plastic containers for three to six weeks. When at least three new tillers appear, the plants are ready for transplantation on the field.

FIGURE 12.7 Large-scale cultivation of vetiver by tractors in Australia.

The large scale application of vetiver requires a huge amount of planting materials, that in turn need large-scale nurseries supplying mother plants. In practice, such nurseries are rare. An alternative way to produce the large numbers of vetiver planting materials is by tissue culture (Ruth et al. 2000; Le et al. 2008; Moosikapala and Te-chato 2010). This method involves the culture of small bits of tissue (root tips, young inflorescence, shoot and leaf tissues) in a special medium under aseptic conditions, and the resulting small plantlets are planted in appropriate media until they fully develop into small plants. Furthermore, the large-scale cultivation of vetiver plants can be performed by farm machinery (Figure 12.7).

12.2.5 Weed Potential

One of the most challenging issues of cultivating new plants for phytoremediation in nonnative environments is the risk of introducing hard-to-control weeds. Most hyeraccumulators are native plants, and produce viable seeds that increase the weeding potential in new environments. For vetiver grass, the chance of it becoming a weed is very low, as vetiver grass cultivars from south Indian accessions, deliberately selected for phytoremediation, produce flowers but set no seeds. Furthermore, they produce neither stolons nor rhizomes and have to be established vegetatively by root (crown) subdivisions. The low weeding potential of vetiver grass is clearly demonstrated from several case studies throughout the world. In Fiji, vetiver grass was introduced from India for thatching more than 100 years ago. It has been widely used for soil and water conservation purposes in the sugar industry for over 50 years without showing any signs of invasiveness (Truong and Creighton 1994). An eight-year study conducted in Australia showed that vetiver grass is sterile under various growing conditions (Truong 2002). A new weed risk assessment released by U.S. Department of Agriculture in 2007 showed that nonfertile vetiver cultivars, specifically Sunshine (US) and Monto (Australia) genotypes, are safe for introduction into new environments (http://www.vetiver.org/USA_PIER.htm). Vetiver grass can be destroyed easily either by spraying with glyphosate (Roundup) or by cutting off the plant below the crown (Truong et al. 2008).

12.2.6 Ability to Accumulate High Concentrations of Metals

Recently, vetiver grass has been shown not only to tolerate a soil environment containing high concentrations of a wide range of heavy metals, but also to accumulate high quantities of these metals in its roots and shoots (Table 12.6). The majority of heavy metals are accumulated in the vetiver roots; only small portions are transported into the shoots. Such characteristics make vetiver grass suitable for phytostabilization of heavy metal-contaminated soils.

TABLE 12.6
Highest Concentrations of Heavy Metals Accumulated in the Roots and Shoots of Vetiver Reported in the Literature

	Soil Condition		Hydroponic Condition	
Heavy Metals	**Roots (mg kg^{-1})**	**Shoots (mg kg^{-1})**	**Roots (mg kg^{-1})**	**Shoots (mg kg^{-1})**
Lead	4940[a]	359[a]	≥10,000[de]	≥3350[de]
Zinc	2666[b]	642[b]	>10,000[d]	>10,000[d]
Chromium	1750[c]	18[c]		
Copper	953[f]	65[f]	900[d]	700[d]
Arsenic	268[e]	11.2[c]		
Boron	28[g]	17[g]		
Cadmium	~25[h]	~44[i]	2232[j]	93[j]

Note: [a]Rotkittkhun et al. (2007); [b]Danh et al. (2010); [c]Truong (1999b); [d]Antiochia et al. (2007); [e]Andra et al. (2009a); [f]Castillo et al. (2007); [g]Angin et al. (2008); [h]Lai and Chen (2004); [i]Vo and Le (2009); [j]Aibibu et al. (2010).

12.2.6.1 Lead

Phytoremediation of lead-contaminated soils using vetiver grass has attracted great attention from researchers around the world. This is due to the fact that lead is one of a few heavy metals accumulated in high concentrations in vetiver biomass, and it is very toxic for human health, especially children. Investigations on growth, lead-accumulating potential and characteristics of vetiver grass under glasshouse and field conditions offer valuable knowledge to improve phytoremediation efficiency. Furthermore, recent findings identified the biochemical processes responsible for the observed high Pb tolerance of vetiver.

Vetiver can grow on soils contaminated with high concentrations of Pb, but its survival rate and growth are adversely affected. The growth of vetiver on such contaminated soils can be significantly improved by addition of organic matter, fertilizers and arbuscular mycorrhizal fungi. Under glasshouse conditions, the addition of pig manure was very effective in the improvement of vetiver grass grown on Pb mine soils, while inorganic fertilizers had no effect (Rotkittkhun et al. 2007). Biomass of vetiver grown on Pb/Zn tailing soils treated with sewage sludge and the combination of sewage sludge and inorganic fertilizers was over two and five times, respectively, higher than that of the untreated control plants (Chiu et al. 2006). Similarly, under field conditions Yang et al. (2003) observed that the application of domestic refuse alone and the combination of domestic refuse and inorganic fertilizer (N:P:K = 15:15:15), especially the combination, significantly increased the survival rates and growth of vetiver grass. However, the inorganic fertilizer alone did not improve both the survival rate and growth performance of plants grown on tailings. Recently, a field trial of Pb/Zn mine tailings showed that the addition of refuse compost increased vetiver biomass to more than three times that of the control (Wu et al. 2011). Moreover, the inoculation of vetiver with arbuscular mycorrhizal fungi (AMF) significantly increased growth performance of vetiver grown on Pb-contaminated soils in both glasshouse and field study, and on Pb-contaminated water in a hydroponic study, by a rate of about 30% (Wong et al. 2007; Punamiya et al. 2010; Wu et al. 2011). In short, the application of organic matter together with inorganic fertilizers, is the most effective in improving the growth performance of vetiver grown on Pb-contaminated soils.

The addition of organic matter, however, reduced the amount of Pb accumulated in the biomass of vetiver grown on Pb-contaminated soils. Vetiver grown on Pb-contaminated soils, treated with a wide range of organic matter under field and glasshouse conditions showed a decrease in Pb absorption when compared to the unamended control (Yang et al. 2003; Chiu et al. 2006; Rotkittkhun et al. 2007; Wu et al. 2011). This is due to organic matters in the soils forming complexes with Pb that in turn make it not available for plant uptake. Furthermore, organic matter greatly improves vetiver biomass, leading to the dilution of accumulated Pb concentrations. Similarly, AMF inoculation of vetiver grass decreased Pb uptake (Wong et al. 2007; Wu et al. 2011). In contrast, the application of inorganic fertilizer (NPK) at a high rate significantly increased Pb accumulation by vetiver grass (Rotkittkhun et al. 2007).

The majority of Pb accumulated in vetiver grass is present in the roots, only a small portion is translocated to the shoots of vetiver grown on different growth media (Table 12.7). The solution in the intercellular spaces of vetiver roots has high concentrations of phosphate and carbonate-bicarbonates, and relatively high pH. Such conditions quickly precipitate out the accumulated Pb in the form of phosphates or carbonates, consequently preventing its translocation to the shoots (Danh et al. 2009). High concentrations of Pb in the roots make vetiver grass a good candidate for phytostabilization, a process of using plants for the immobilization of soil contaminants *in situ* (Wild et al. 2005).

Vetiver grass can accumulate very high concentrations of Pb in biomass. It has been demonstrated to be a lead hyperaccumulator that can accumulate Pb at least 1000 mg kg^{-1} DW. Several studies showed that vetiver can uptake over 10,000 and 3000 mg kg^{-1} Pb in roots and shoots, respectively (Antiochio et al. 2007; Andra et al. 2009a). The contents of accumulated lead depend on the bioavailability of Pb in the growth media. as was clearly demonstrated by the study of Rotkittikhun et al. (2007), in which the level of Pb in vetiver roots and shoots increased with extractable concentrations of Pb in the soils. The trend was further supported by comparing the amounts of lead accumulated in the roots of vetiver grown on the contaminated soils, soils irrigated with Pb solution, and hydroponic (Table 12.7), in which the bioavailability of Pb is the highest and lowest in hydroponic and contaminated soils, respectively. Under hydroponic conditions, after 7 days of exposure vetiver could accumulate very high concentrations of Pb in the roots and shoots without phytotoxic symptoms (Table 12.7), indicating the great potential of accumulating Pb in vetiver grass. Therefore, to increase Pb content in vetiver cultivated on contaminated soils, the improvement of Pb bioavailability is the most important practice.

TABLE 12.7
Pb Concentrations in the Roots and Shoots of Vetiver Grass Grown on Different Growth Media

Growing Media	Total Pb Concentrations (mg L^{-1} or mg kg^{-1})	Extractable Pb Concentrations	Pb in Roots (mg kg^{-1} DW)	Pb in Shoots (mg kg^{-1} DW)	References
Hydroponic	0		5.5	5.4	Rotkittikhun et al. (2010)
	10		8140	78.3	
	50		14200	78.6	
	100		19530	111.8	
Hydroponic	400		13200	1700	Andra et al. (2009a)
	1200		19800	3350	
Soils irrigated with lead solution	250 ml of 621 ppm Pb day^{-1} for 30 days		~10000	~4000	Antiochio et al. (2007)
Artificially contaminated soils	113	16.5	18.7	12.5	Rotkittikhun et al. (2007)
	192	43.7	41.1	16.2	
	707	253	474	142	
	10750	1065	4940	359	

The amendments of chelating agents can improve the accumulation of Pb and its translocation from roots to shoots in vetiver grass, as chelating agents can increase the mobility and bioavailability of the metals in the soils. Pb has low solubility in soil environments due to the interaction with various soil components, particularly phosphate. Total soil Pb concentrations at a given site may be high, but the fraction of soluble Pb that plants can extract is low, resulting in low efficiency of phytoremediation. Among chelating agents, ethylenediaminetetraacetic acid (EDTA) has been demonstrated to be the most efficient in the improvement of Pb mobility from various soil environments (Danh et al. 2009). The application of EDTA on Pb-contaminated soils resulted in a significant increase of Pb concentrations in shoots and roots of vetiver grass (Table 12.8). The translocation ratio (TR) of Pb from roots to shoots (the percentage of shoot Pb concentrations versus root Pb concentrations) significantly increased with the applied EDTA, but it generally declined if the rate of EDTA was over 5 mmol kg^{-1}. This may be explained by the fact that the accumulated Pb in the roots exceeds the Pb transporting capacity of vetiver from roots to shoots under treatment of high EDTA levels. The growth of vetiver grass was not adversely affected by the amendments of various EDTA concentrations, although high concentrations of Pb were accumulated in the roots (Table 12.8). EDTA-chelated Pb has low toxicity on vetiver growth. The amount of total thiols (phytochelatins) and catalase activity, two indicators of Pb detoxification in plants, in EDTA-treated vetiver tissues were comparable to cheland unamended controls (Andra et al. 2009b).

TABLE 12.8
Effect of EDTA on Pb Uptake and Translocation Ratio of Vetiver Grass

Pb in Soils (mg kg^{-1} DW)	EDTA (mmol kg^{-1})	Pb in Shoots (mg kg^{-1} DW)	Pb in Roots (mg kg^{-1} DW)	TR (%)	References
500	0	0.82	60.3	1.4	Chen et al. (2004)
	0.5	6.06	83.5	7.3	
	2.5	25.7	200	12.9	
	5.0	42.2	266	15.9	
2500	0	6.52	205.8	3.2	
	0.5	21.3	242	8.8	
	2.5	86.3	464	18.6	
	5.0	160	951	16.8	
5000	0	43	556	7.7	
	0.5	69	871	7.9	
	2.5	127	1440	8.8	
	5.0	243	2280	10.7	
238	0	54	292	18.5	Wilde et al. (2005)
185	5	83	513	16.2	
2078	0	19	136	14	Zhuang et al. (2005)
	6	32	161	19.9	
710	0	~40			Lou et al. (2007)
	2.5	~110			
	0	20	290	6.9	Andra et al. (2009)
	5	~185	~1050	17.6	
	10	~270	~2400	11.3	
	15	480	4460	10.8	

Source: Modified from Danh, L.T., P. Truong, R. Mammucari, T. Tran, and N. Foster, *International Journal of Phytoremediation* 11: 664–691, 2009.

The use of chelate for the improvement of Pb accumulation in vetiver grass, however, can cause secondary environmental pollution in surrounding areas, due to leaching of EDTA and EDTA-Pb complexes into groundwater. To overcome this problem, several aspects should be taken into account in designing a phytoremediation program, particularly the optimum chelate concentration, time and sites of the chelate application, and the root systems of plants (Chen et al. 2004). Vetiver grass possesses a long and massive root system that can penetrate to the deeper layers of the soil (Truong 2000), contributing to the prevention of heavy metal leaching.

The biochemical and physiological processes responsible for the Pb-tolerance mechanism of vetiver grass have been investigated in the recent studies. Heavy metal accumulation often induces oxidative stress in most plants by generating various reactive oxygen species (ROS), such as superoxide anions (O_2^-), hydroxyl radicals (OH^-) and hydroxyl radicals (H_2O_2). In response, plants commonly adopt two different strategies to minimize the toxic effects of metals, namely indirect and direct defense mechanisms. Plants employ antioxidant molecules and antioxidant enzymes for scavenging of various ROS in the indirect method. The direct strategy involves binding heavy metals to plant cell walls (Bringezu et al. 1999) or chelating metals in the cytosol by phytochelatins (Schmöger et al. 2000) or storing metals in vacuoles (Liu and Kottke 2004) in order to avoid cell damage. Vetiver grass has been shown to use both mechanisms for Pb tolerance, related to the activities of antioxidant enzymes, and the production of phytochelatins (PC_n) and Pb-PC_n complex. The enhanced activities of antioxidant enzymes, namely superoxide dismutase (SOD), peroxidase (POD) and catalase (CAT), in vetiver grown on Pb-contaminated soils indicated their indirect role in the defense mechanism by detoxifying the induced ROS (Pang et al. 2003; Andra et al. 2009b). Moreover, upon exposure to lead in hydroponic or soil media, vetiver grass starts to synthesize phytochelatins (PC_n) in the roots and shoots. This was not observed in plants grown on media without Pb (Andra et al. 2009a, 2009b, 2010). Phytochelatin production was positively related to the amount of Pb taken up by plant tissues (Andra et al. 2009b). Concentration of phytochelatins was higher in the roots than shoots, because the amount of Pb translocated to the shoot was minimal when compared with that detected in the root. Possible Pb-PC_n complex production was reported in a series study of Andra et al. (2009a, 2009b, 2010). The data from these experiments indicated that the most probable mechanism for Pb detoxification in vetiver is by synthesizing PC_n and forming Pb-PC_n complexes. From this knowledge, it is possible to improve the growth performance of vetiver cultivated on Pb-contaminated soils without hindering Pb uptake (as observed for the application of organic matter) by increasing the synthesis of phytochelatins. The addition of sulfur, a precursor of phytochelatin, can increase the production of phytochelatins in *Pteris vittata* grown on arsenic-contaminated soils (Wei et al. 2010).

12.2.6.2 Zinc

Vetiver grass has the ability to accumulate high concentrations of Zn in its roots and shoots. Over 10,000 mg kg^{-1} DW (over 1%) was reported in the shoots and roots of vetiver grown on the soils daily irrigated with Zn solution (Antiochia et al. 2007). It is classified as a Zn hyperaccumulator, according to the definition of Barker and Brook (1989).

The content of Zn accumulated in vetiver is mainly determined by the amount of bioavailable Zn in the soils regardless of the total level of Zn contamination (Table 12.9). Under the soil environment, the concentration of plant-available Zn is limited, leading to low levels of Zn accumulated in the vetiver biomass. The highest Zn contents in shoots and roots of vetiver grown on the contaminated soils in the literature were 642 and 2666 mg kg^{-1} DW, respectively (Danh et al. 2010).

To improve Zn bioavailability in the soils, and hence Zn accumulation in vetiver grass, the application of chelating agents is an effective approach. Nitrilotriacetic acid (NTA) was one of the most effective chelating agents to increase the water solubility of Zn in the soils, and therefore the amount of Zn accumulated in vetiver biomass (Chiu et al. 2006). The concentration of the extractable Zn in the soils rapidly increased with the rate of applied NTA, particularly at the rate of 10 and 20 mmol kg^{-1}. The extractable Zn in soils under treatment of 20 mmol kg^{-1} NTA was 2 to 15 times higher than

TABLE 12.9
Level of Zn in Soils and in VZ Roots and Shoots

Total Zn in Soils (mg kg^{-1})	DTPA-Extractable Zn (mg kg^{-1})	Zn in Shoots (mg kg^{-1} DW)	Zn in Roots (mg kg^{-1} DW)	References
1583	834	55	847	Chiu et al. (2006)
2861	219	~170	~750	Wu et al. (2011)
2472	161	144	339	Zhuang et al. (2005)
4377	187	43	316	Yang et al. (2003)
3418	101	30	219	Shu et al. (2004)

with that of N-(2-hydroxyethyl) iminodiacetic acid (HEIDA), hydroxyethylenediaminetriacetic acid (HEDTA), trans-1,2-cyclohexyleneditrilotetraacetic acid (CDTA), diethylene triaminepenta acetic acid, ethylenediaminetriacetic acid, citric acid, ethylenebis (oxyethylenetrinitrilo) tetraacetic acid, and malic acid. Furthermore, this rate increased the concentration of Zn in roots and shoots by about 53% and 136%, respectively, compared to the control treatment. Translocation of Zn from the roots to shoots increased with the presence of NTA, but it was not affected by the different rates of application. EDTA, an effective chelating agent in the improvement of Pb accumulation in vetiver, did not significantly increase Zn content in the shoots and roots of vetiver (Chen 2004; Lai and Chen 2004).

12.2.6.3 Chromium

The study of Shahandeh and Hossner (2000) showed that vetiver grass is the most Cr tolerant plant among 36 plant species, regardless of Cr forms and rates. It can accumulate very high concentration of Cr(VI), up to 10,000 mg kg^{-1} DW in the shoots, but the plants died a few days following the exposure to this ion at a concentration of 500 mg kg^{-1} in the soil. By contrast, it accumulated less Cr(III) in the roots and shoots, but its index of tolerance is very high, over 90%, under treatment of 500 mg kg^{-1} Cr(III) in the soil.

The majority of Cr(III) accumulation in vetiver is retained in the root system, only a small portion of Cr(III) being translocated to the roots. The concentrations of Cr(III) in the roots of vetiver grown on the soils containing 50, 200 and 600 mg kg^{-1} were 404, 1170 and 1750 mg kg^{-1} in roots, respectively, whilst the concentration of Cr in the shoots was less than 20 mg kg^{-1} (Truong 1999b). Similarly, vetiver grown on 100 mg kg^{-1} Cr(III) soils accumulated 3.6 and 89.4 mg kg^{-1} DW in the shoots and roots, respectively.

12.2.6.4 Copper

Vetiver grass has a low potential in accumulating Cu in the roots and shoots. As grown on Cu-contaminated soils under glasshouse conditions, vetiver can accumulate less than 60 and 830 mg kg^{-1} DW of Cu in shoots and roots, respectively (Yang et al. 2003; Wilde et al. 2005; Chiu et al. 2006; Lou et al. 2007; Liu et al. 2009; Danh et al. 2010). The amount of accumulated Cu is positively related to the content of Cu in the soils (total and bioavailable Cu) and the duration of cultivation. Similarly, the study at the Anglo-American El Solado mine in Chile showed that four-month-old vetiver had 69 and 371 mg kg^{-1} in shoots and roots, respectively, whilst vetiver at ten months old absorbed 65 and 953 mg kg^{-1} in the shoots and the roots, respectively (Castillo et al. 2007). These results indicated that older plants retained more Cu in the roots. The higher content of Cu in vetiver is often associated with low biomass.

The majority of Cu is retained in the vetiver roots, only small portion of Cu being transported to the shoots. In general, the translocation ratio (TR) of Cu from roots to shoots (the percentage of Cu concentration in shoots versus in roots) in vetiver is less than 15% (Chiu et al. 2006; Wilde et al. 2005; Yang et al. 2003; Liu et al. 2009; Danh et al. 2010). In contrast, the TR value of up to 83% was recorded in the study of Antiochia et al. (2007), which was due to the fact that vetiver was irrigated daily with Cu solutions containing high levels of bioavailable Cu. Consequently, Cu uptake rate may have been higher than the Cu fixation rate in roots, so a large amount of Cu escaped fixation in roots and was available to be translocated into shoots (Danh et al. 2009).

The amendment of chelating agents can improve Cu uptake in vetiver roots and shoots. HEIDA was demonstrated to be the most effective additive to increase the solubility of Cu in the soils as well as the concentrations of Cu in both roots and shoots of vetiver (Lou et al. 2007; Chiu et al. 2005). EDTA was also effective in increasing the content of Cu in roots (Wilde et al. 2005). On the other hand, the addition of organic matter, such as manure compost and sewage sludge, significantly decreased the concentration of Cu in roots and shoots of vetiver grown on Pb/Zn tailings (Chiu et al. 2006).

12.2.6.5 Arsenic

Vetiver grass has low potential in accumulation of arsenic from the soils. The content of As in the roots and shoots were often less than 300 and 15 mg kg^{-1} DW, respectively (Srisatit et al. 2003; Truong and Baker 1998; Truong et al. 1999b; Chiu et al. 2005). The majority of As were retained in the roots, only small amount being transported to the shoots.

The accumulation of arsenic in vetiver grass can be significantly improved by the addition of chelating agents and organic matters combined with mycorrhizae and azotobacter. Oxalic acid (OA) was the most effective among five tested chelating agents (OA, EDTA, HEDTA, NTA and phytic acid) in increasing As in the soil solution (Lou et al. 2007). It is suggested that OA can increase soil acidity, then mobilize part of As combined with Fe and Mn oxyhydroxides (Huang 1994; Klarup, 1997). Consequently, oxalic acid had the most significant effect on As accumulation in the above-ground parts of VZ, with a three-fold increase compared with the control (Lou et al. 2007). The effects of EDTA, HEDTA, and NTA on arsenate and arsenite ions in the soil solution are minimal because they prefer to form complexes with metal cations (Wongkongkatep et al. 2003). However, in other study NTA was demonstrated to have a significant effect on As solubility, hence As accumulation in vetiver (Chiu et al. 2005). Furthermore, the amendment of dairy waste together with mycorrhizae and azotobacters not only supported the normal growth of vetiver even under treatment of 500 mg As kg^{-1}, but also enhanced As accumulations (Singh et al. 2007). After six months of study, the total arsenic accumulated in VZ reached 185.4 and 100.6 mg kg^{-1} DW in the roots and shoots, respectively, whilst vetiver in the control with no amendments could not survive.

12.2.6.6 Cadmium

Vetiver grass can accumulate low concentrations of Cd. Generally, vetiver grown on Cd-contaminated soils had a low content of Cd accumulated in the shoots and roots, below 25 and 45 mg kg^{-1} DW, respectively. It can be explained by the fact that the levels of Cd, both total and bioavailable Cd, in the soils of those studies were very low (Truong 1999b; Yang et al. 2003; Vo and Le 2009; Xia 2004; Zhuang et al. 2005). On the other hand, vetiver grass cultivated under hydroponic conditions accumulated up to 93 and 2232 mg Cd kg^{-1} DW in the shoots and roots, respectively, at 30 mg L^{-1} treatment (Aibibu et al. 2010). These results indicated that vetiver has a great potential to accumulate Cd, particularly in growth media containing high concentrations of bioavailable Cd. The uptake of Cd in vetiver increased with the level of Cd in the soils or the growth solutions, and the exposure time (Vo 2007; Aibibu et al. 2010). A decrease of biomass production was associated with accumulation of Cd in root and shoot tissues.

The accumulation of Cd by vetiver was much higher in roots than in shoots (Truong 1999b; Yang et al. 2003; Vo and Le 2009; Xia 2004; Zhuang et al. 2005; Aibibu et al. 2010). Retention of high amounts of Cd in the root tissue can be regarded as an important protection mechanism against the diffusion of this heavy metal in plants (Verkleij and Schat 1990). High Cd accumulation in roots is probably owing to the absorption of Cd onto the negatively charged surface of the root cell wall or sequestering within the xylem in the root (Solís-Domínguez et al. 2007). Furthermore, most of the Cd in the medium-root interface tends to be immobilized by root exudates in the root zone, resulting in small amounts of Cd being available for the upward transport (Wang et al. 2008). Restriction of upward movement from roots into shoots can be considered as one of the tolerance mechanisms.

The application of chelating agents, particularly EDTA, did not effectively increase concentration of Cd accumulated in roots and shoots of vetiver grass (Zhuang et al. 2005; Lai and Chen, 2004). Similarly, the treatment of EDTA was not able to enhance Cd accumulation in *Thlaspi caerulescens* (McGrath et al. 2006) and Rumex K-1 (*Rumex patientia* x *R. timschmicus*) (Zhuang et al. 2005). In contrast to the cases of Zn, Pb and Cu, the addition of organic matter did not have a significant effect on Cd uptake (Yang et al. 2003).

12.2.6.7 Boron

The investigation on the ability of vetiver to remove boron from the soils is limited, with, to date, only one study reported in the literature, the work of Angin et al. (2008). In this study, vetiver was grown on the soils artificially contaminated with a range of boron concentrations from 0 to 180 mg B kg^{-1}. The addition of B was shown to have no significant effect on dry matter yield after 90 days of experiments. The content of B was accumulated more in the roots than in the shoots of vetiver. The treatment of 180 mg B kg^{-1} showed that the concentration of B was 28 and 17 mg B kg^{-1} DW in roots and shoots, respectively.

12.2.6.8 Radionuclides

Vetiver grass has been demonstrated to be able to accumulate radioactive elements in the roots and shoots. As grown on artificially uranium-contaminated soils, with concentration of 100 mg kg^{-1}, vetiver accumulated up to 230 and 7 mg U kg^{-1} DW in the roots and shoots, respectively (Shahandeh and Hossner 2002). The content of U in vetiver biomass increased with the exposure time (Roongtanakiat et al. 2010). Furthermore, after one week of treatment vetiver was shown to remove 61% of ^{137}Cs and 94% of ^{90}Sr from solutions containing the individual radionuclide concentration of 5 × 103 k Bq L^{-1} (Singh et al. 2008). As both ^{137}Cs and ^{90}Sr were supplemented together in the solution, 59% of ^{137}Cs and 91% of ^{90}Sr were removed in the same time frame. The locations of ^{137}Cs and ^{90}Sr accumulated in vetiver were different as ^{137}Cs accumulated principally in the roots, ^{90}Sr mostly in shoots. Interestingly, vetiver could efficiently remove radionuclides in low-level nuclear waste solution (7.5 × 104 Bq L^{-1}) to below detection levels within 15 days (Singh et al. 2008).

12.3 ECONOMIC INCENTIVES TO PROMOTE THE USE OF VETIVER GRASS FOR PHYTOREMEDIATION

The application of vetiver grass for phytoremediation of heavy metal-contaminated soils can be further promoted through economic returns from marketable plant produce. Commonly, plants used for phytoremediation often provide no economic return (Baker et al. 2000; Lasat 2002; Robinson et al. 2003; Schmidt 2003). The heavy metal-contaminated lands are a serious problem in developing countries, where governments have limited financial resources and cannot apply expensive physical and chemical remediation technologies. Nor can they provide support for growers to apply phytoextraction. The marketable products obtained from vetiver grass grown on contaminated soils encourage growers to apply this grass for phytoremediation.

12.3.1 Products from Vetiver Roots

The roots of vetiver grass (VG) contain essential oil that finds traditional application in perfumery and medicine. Vetiver oil consists of a complex mixture of over 300 compounds which possess aromatic and biological properties. It is extensively used for blending in oriental types of perfumes, cosmetics and aromatherapy. Recently, vetiver oil has been found to possess antifungal, antibacterial, anticancer, anti-inflammatory and antioxidant activities. It can also be used for the treatment of patients with dementia-related behavior. Such characteristics increase the applicability of VG extracts in the pharmaceutical industry (Danh et al. 2011). Furthermore, vetiver oil can be applied in the food industry as an aromatizing agent for canned asparagus and peas, and in some beverages (Martinez et al. 2004, 2003; Chowdhury et al. 2002; Arctander 1960). Potent topical irritant activity on cockroaches and flies (Jain et al. 1982), and powerful repellent and toxic activities against Formosan subterranean termite (Zhu et al. 2001a, 2001b) have been reported for VG oil extracts.

Economic returns from the commercialization of essential oil extracts obtained from vetiver grass used for phytoremediation purposes can offer a great incentive for recovery of contaminated sites. Vetiver grown on soils contaminated with mild concentrations of Cu (100 mg kg^{-1}) and Zn (1600 mg kg^{-1}) produced essential oils that had yield and chemical compositions comparable to those of control plants grown without the addition of heavy metals. In addition, heavy metals were still retained in plant roots after extraction, leaving essential oil with trace elements (Danh et al. 2010). On the other hand, Pb had an adverse effect on both oil yield and chemical composition of vetiver grown on the contaminated soils, even at the lowest concentration tested (500 mg kg^{-1}). However, the oil yield of vetiver grown under hydroponic conditions increased with the concentration of Pb in the growth media from 0 to 100 mg l^{-1}. Whilst it was not significantly affected by increasing Pb content in the soils up to 1 000 mg kg^{-1} (Rotkittikhun et al. 2010). Moreover, Pb had a slight effect on the chemical constituents of vetiver oil. These results indicated that cultivation of vetiver on Pb-, Zn-, or Cu-contaminated soils can serve the dual purpose of stabilizing the contaminated site and at the same time producing oil with a high commercial value. However, further pot and field trial experiments are required to confirm this hypothesis.

12.3.2 Products from Vetiver Leaves

Vetiver leaves can be used for a wide range of applications, such as handicraft, bio-fuel, green-fuel, mushroom cultivation and animal feed, offering the possibility of an additional income for the growers. However, the use of vetiver grown on heavy metal-contaminated soils is restricted to nonedible products such as handicraft, bio-fuel and green fuel, because heavy metals present in vetiver roots, even at trace amounts, may cause food poisoning for humans through contaminated food chains.

12.3.2.1 Handicrafts

The use of vetiver leaves for handicraft production originated in Thailand. The Department of Industrial Promotion in Thailand developed a way to boil the leaves, and with a needle, remove the sharp-toothed edges before using the flexible leaf to weave a broad array of wonderful products (Chomchalow 2008). Currently, vetiver handicraft production has been successfully practised in Thailand, China, India, Indonesia, Madagascar, Venezuela and Senegal in order to provide a new source of income for vetiver growers in the poor communities. There are a variety of products made with vetiver fiber, including (i) handy accessories such as bags, hats, belts and brooches, (ii) containers such as baskets, pots, boxes, utility bowls, (iii) decorative materials such as clocks, picture frames, lamp shades, dolls, animal figures, flowers, and (iv) home appliances such as chairs, stools, room partitions, and tables (Figure 12.8).

12.3.2.2 Electricity Generation

The phyto-stabilization of heavy metal-contaminated soils by vetiver grass can provide unlimited green fuel for the growers. Vetiver is a perennial grass, and has a high shoot biomass production of more than 50 ton ha^{-1} yr^{-1}. If only the above-ground part of vetiver is harvested and the perennial vetiver plant is left in the ground and allowed to grow back, year after year, a vetiver field is a renewable mine of green fuel with only minimal care requirement. The life-span of vetiver is 60 years which is the typical duration expected for one establishment.

Vetiver green fuel has great potential for environmentally friendly electricity generation in power plant furnace boilers (Figure 12.9). Dry vetiver grass is estimated to generate 7000 BTU lb^{-1}; while

FIGURE 12.8 **(See color insert.)** A variety of products made from vetiver biomass. (From www.vetiver.org.)

petroleum, coal and dry wood typically produce 18,000, 13,000 and 8500 BTU lb^{-1}, respectively. In other words, energy produced by burning two tonnes of dry vetiver is roughly equal to one tonne of coal. If vetiver biomass can be marketed at a third of the price of coal, it will make a good return for the growers. Furthermore, the combustion of fossil fuels for power generation produces CO_2 gas as the main product, which increases the existing inventory of CO_2 in the atmosphere. On the other hand, CO_2 released from the combustion of biomass originates from the current atmosphere in the first place via the photosynthetic pathway. Consequently, biomass combustion does not add or remove CO_2 in the atmosphere.

Furthermore, vetiver biomass can provide an alternative source of cooking fuels for the local communities in developing countries. Firewoods and charcoal, the main fuels used in these regions, has become scarce owing to overexploitation, coupled with deforestation. The cultivation

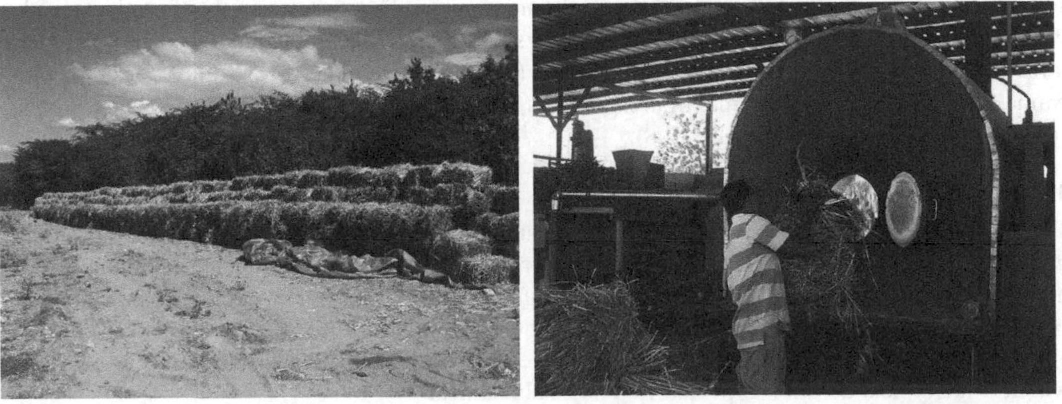

FIGURE 12.9 A simple furnace boiler using vetiver biomass in Dominican Republic. (From www.vetiver.org.)

of vetiver grass, particularly on marginal lands that are not suitable for agricultural production because of heavy metal contamination, could help to alleviate the pressure on the traditional sources of fuel.

12.3.2.3 Bio-Ethanol Production

The by-product of phytoremediation processes, specifically those based on vetiver leaves, can be a good substrate for the production of ethanol. Ethanol can be used as a fuel for vehicles in its pure form or by blending it with gasoline to improve vehicle emissions. The first generation bio-ethanol is made from starch and sugar derived from food crops, such as wheat, corn, sugar cane, and sugar beet. The second generation biofuel is termed cellulosic ethanol, and is based on non-food crops or inedible waste products with a high content of lignocelluloses. The latter generation of bio-ethanol has received great interest because it does not compete with human and domestic animals for food.

The process of producing ethanol from vetiver leaves was investigated in the study of Kuhirum and Punnapayak (2000). Alkali-pre-treated dry leaves were subjected to simultaneous saccharification and fermentation by the addition of cellulose enzyme (extracted from *Trichoderma reesei*) and yeast. The process was performed at 40°C and pH 5.0 and lasted for one week to obtain an ethanol yield of 13%.

12.4 CONCLUSIONS

Vetiver grass possesses nearly all the characteristics of an ideal plant for the phytoremediation of heavy metal-contaminated soils. It has a dense and massive root system, exhibits rapid growth, has a high biomass production, and also possesses high tolerance and adaptability to a wide range of weather, environmental and soil conditions. Furthermore, it can accumulate high concentrations of Pb, Zn and Cr (>1000 mg kg^{-1} DW) and moderate concentrations of Cu and As (100–1000 mg kg^{-1} DW) in the biomass. The majority of heavy metals are retained in the roots, while only small amounts are translocated into the shoots. Such characteristics make vetiver grass suitable for phytostabilization of contaminated sites and prevention of the spreading of contaminants into surrounding areas. The amendment of chelating agents, such as EDTA for Pb and nitrilotriacetic acid for Zn, can significantly improve the accumulation of heavy metals in vetiver as well as their translocation from roots to shoots. Hence, vetiver can be used for phytoextraction, with the aid of chelating agents, by regular removal of vetiver shoots and roots, which will gradually reduce the levels of contaminants in the soils over a long period. The harvested biomass can be used for a range of commercial products, namely essential oils, handicrafts and biofuels, all of which have the potential to provide additional income for the growers. Furthermore, the cultivation of vetiver also improves the soil quality in term of nutritional, physical and chemical properties.

REFERENCES

Aibibu, N., Y. Liu, G. Zeng, X. Wang, B. Chen, H. Song, and L. Xu. 2010. Cadmium accumulation in *Vetiveria zizanioides* and its effects on growth, physiological and biochemical characters. *Bioresource Technology* 101: 6297–6303.

Andra, S.S., R. Datta, D. Sarkar, K.C. Makris, C.P. Mullens, S.V. Sahi, and S.B.H. Bach. 2010. Synthesis of phytochelatins in vetiver grass upon lead exposure in the presence of phosphorus. *Plant Soil* 326: 171–185.

Andra, S.S., R. Datta, D. Sarkar, K.C. Makris, C.P. Mullens, S.V. Sahi, and S.B.H. Bach. 2009a. Induction of lead-binding phytochelatins in vetiver grass [*Vetiveria zizanioides* (L.)]. *Journal of Environmental Quality* 38: 868–877.

Andra, S.S., R. Datta, D. Sarkar, S.K.M. Saminathan, C.P. Mullens, and S.B.H. Bach. 2009b. Analysis of phytochelatin complexes in the lead-tolerant vetiver grass, *Vetiveria zizanioides* (L.), using liquid chromatography and mass spectrometry. *Environmental Pollution* 157: 2173–2183.

Angin, I., M. Turan, Q.M. Ketterings, and A. Cakici. 2008. Humic acid addition enhances B and Pb phytoextraction by Vetiver grass (*Vetiveria zizanioides* (L.). Nash). *Water Air and Soil Pollution* 188: 335–343.

Antiochia, R., L. Campanella, P. Ghezzi, and K. Movassaghi. 2007. The use of vetiver for remediation of heavy metal soil contamination. *Analytical Biochemistry* 388: 947–956.

Audet, P., and C. Charest. 2007. Heavy metal phytoremediation from a meta-analytical perspective. *Environmental Pollution* 147: 231–237.

Baker, A.J.M., S.P. McGrath, C.M.D. Sidoli, and R.D. Reeves. 1994. The possibility of *in situ* heavy metal decontamination of polluted soils using crops of metal-accumulating plants. *Resources, Conservation and Recycling* 11: 41–49.

Baker, A.J.M., and R.R. Brooks. 1989. Terrestrial higher plants which hyperaccumulate metallic elements. A review of their distribution, ecology and phytochemistry. *Biorecovery* 1: 81–126.

Baker, D.E., and V.J. Eldershaw. 1993. Interpreting soil analyses for agricultural land use in Queensland. *Project Report Series Q093014*, QDPI, Brisbane, Australia.

Berg, J.V.D., C. Midega, L.J. Wadhams, and Z.R. Khan. 2003. Can vetiver grass be used to manage insect pests on crops? *The Third International Conference on Vetiver, Guangzhou, China, 6–9 October, 2003.*

Bertea, C.M., and W. Camusso. 2002. Anatomy, biochemistry and physiology. In: *Vetiveria. The Genus Vetiveria*, 19–43. London: Taylor & Francis.

Bevan, O., P. Truong, and M. Wilson. 2000. The Use of Vetiver Grass for Erosion and Sediment Control at the Australian Bentonite Mine in Miles, Queensland. Proc. Fourth Innovative Conf., Australian Minerals and Energy Environment Foundation: *On the Threshold: Research into Practice*. August 2000, Brisbane, Australia, pp. 124–128.

Bowen, H.J.M., ed. 1979. *Plants and the Chemical Elements*. Academic Press, London.

Bringezu, K., O. Lichtenberger, I. Leopold, and D. Neumann. 1999. Heavy metal tolerance of *Silene vulgaris*. *Journal of Plant Physiology* 154: 536–546.

Burton, G.W., and W.W. Hanna. 1985. Bermuda grass. In: *Forages*, 247–254. Iowa State Univ. Press, Ames, Iowa.

Castillo, M., Fonseca, R., and J.R. Candia. 2007. Report on the pilot study on the use of Vetiver grass for Cu mine tailings phytostabilisation at Anglo-American El Solado mine, Chile. *Fundacion Chile* (report in Spanish).

Chen, Y., Z. Shena, and X.D. Lib. 2004. The use of vetiver grass (*Vetiveria zizanioides*) in the phytoremediation of soils contaminated with heavy metals. *Applied Geochemistry* 19: 1553–1565.

Cheng, H., X. Yang, A. Liu, H. Fu, and M. Wan. 2003. A study on the performance and mechanism of soil-reinforcement by herb root system. *The Third International Conference on Vetiver, Guangzhou, China, 6–9 October 2003.*

Chiu, K.K., Z.H. Ye, and M.H. Wong. 2006. Growth of *Vetiveria zizanioides* and *Phragmites australis* on Pb/Zn and Cu mine tailings amended with manure compost and sewage sludge: A greenhouse study. *Bioresource Technology* 97: 158–170.

Chomchalow, N. 2008. Other uses and utilization of vetiver. *First Indian National Vetiver Workshop. Cochin, India, 21–23 February 2008.*

Cunningham, S.D., J.R. Shann, D.E. Crowley, and T.A. Anderson. 1997. Phytoremediation of contaminated water and soil. In: Kruger, E.L., Anderson, T.A. and Coats, J.R. eds. *Phytoremediation of soil and water contaminants*. ACS symposium series 664. Washington, DC, American Chemical Society, p. 2–19.

Danh, L.T., P. Truong, R. Mammucari, T. Tran, and N. Foster. 2009. Vetiver grass, *Vetiveria zizanioides*: a choice plant for phytoremediation of heavy metals and organic wastes. *International Journal of Phytoremediation* 11: 664–691.

Danh, L.T., P. Truong, R. Mammucari, and N. Foster. 2010. Economic Incentive for Applying Vetiver Grass to Remediate Lead, Copper and Zinc Contaminated Soils. *International Journal of Phytoremediation* 13: 47–60.

Ernst, W.H.O., ed. 1974. *Schwermetall vegetation*. der Erde, G. Fischer Verlag, Stuttgart.

Ernst, W.H.O. 2005. Phytoextraction of mine wastes–options and impossibilities. *Chemie der Erde/Geochemistry* 65: 29–42.

Ernst, W.H.O., and H.J.M. Nelissen. 2000. Life-cycle phases of a zinc- and cadmium-resistant ecotype of Silene vulgaris in risk assessment of polymetallic mine soils. *Environmental Pollution* 107: 329–335.

Ernst, W.H.O., H.J.M. Nelissen, and W.M. Ten Bookum. 2000. Combination toxicology of metal-enriched soils: physiological responses of a Zn- and Cd-resistant ecotype of *Silene vulgaris* on polymetallic soils. *Environmental and Experimental Botany* 47: 55–71.

Geeson, N.A., P.W. Abrahams, M.P. Murphy, and I. Thornton. 1998. Fluorine and metal enrichment of soils and pasture herbage in the old mining areas of Derbyshire, UK. *Agriculture, Ecosystems & Environment* 68: 217–231.

Greenfield, J.C., ed. 2002. *Vetiver Grass: An essential grass for conservation of Planet Earth.* Infinity Publishing Co, Haverford, PA, USA.

Haines, B.J. 2002. Zincophilic root foraging in *Thlaspi caerulescens. New Phytologist* 155: 363–372.

Hoang, T.T.T., T.C.L. Tu, and P.Q. Dao. 2007. Progress and results of trials using vetiver for phytoremediation of contaminated canal sludge around Ho Chi Minh City. *Vetiver Workshop. (May 2007)* (in Vietnamese), Hanoi, Vietnam.

Huang, Y.C. 1994. Arsenic distribution in the soils. In: *Arsenic in the environment.* New York: John Wiley.

Inman-Bamber, N.G. 1974. CANEGRO, its history, conceptual basis, present and future uses. *Workshop on Research and Modeling Approaches to Examine Sugarcane Production Opportunities and Constraints, St Lucia, Queensland.*

Islam, M.P., M.K.H. Bhuiyan, and M.Z. Hossain. 2008. Vetiver grass a potential source for rural development in Bangladesh. *Agricultural Engineering International: the CIGR Ejournal. Invited Overview No. 5. Vol. X. December 2008.*

Keller, C., D. Hammer, A. Kayser, W. Richner, M. Brodbeck, and M. Sennhauser. 2003. Root development and heavy metal phytoextraction efficiency: comparison of different plant species in the field. *Plant Soil* 249: 67–81.

Klarup, D.G. 1997. The influence of oxalic acid on release rates of metals from contaminated river sediment. *Science and Total Environment* 204: 223–231.

Lai, H.Y., and Z.S. Chen. 2004. Effect of EDTA on solubility of cadmium, zinc and lead and their uptake by rainbow pink and vetiver grass. *Chemosphere* 55: 421–430.

Le, V.B., T.T. Vo, T.T.U. Nguyen, and V.D. Le. 2008. Low-cost micropropagation of vetiver (*Vetiveria zizanioides* L.). *AU Journal of Technology* 12: 18–24.

Leaungvutiviroj, C., S. Piriyaprin, P. Limtong, and K. Sasakic. 2010. Relationships between soil microorganisms and nutrient contents of *Vetiveria zizanioides* (L.) Nash and *Vetiveria nemoralis* (A.) Camus in some problem soils from Thailand. *Applied Soil Ecology* 46: 95–102.

Liu, D., and I. Kottke. 2004. Subcellular localization of cadmium in the root cells of Allium cepa by electron energy loss spectroscopy and cytochemistry. *Journal of Biosciences* 29: 329–335.

Liu, X., Y. Shen, L. Lou, C. Ding, and Q. Cai. 2009. Copper tolerance of the biomass crops Elephant grass (*Pennisetum purpureum Schumach*), Vetiver grass (*Vetiveria zizanioides*) and the upland reed (*Phragmites australis*) in soil culture. *Biotechnology Advances* 27: 633–640.

Lomonte, C., A. Doronila, D. Gregory, A.J.M. Baker, and S.D. Kolev. 2011. Chelate-assisted phytoextraction of mercury in biosolids. *Science of the Total Environment* 409: 2685–2692.

Lou, L.Q., Z.H. Ye, and M.H. Wong. 2007. Solubility and accumulation of metals in Chinese brake fern, Vetiver and Rostrate Sesbania using chelating agents. *International Journal of Phytoremediation* 9: 325–343.

Maffei, M. 2002. Introduction to the Genus Vetiveria. In: *Vetiveria. The Genus Vetiveria*, 1–18. London: Taylor and Francis.

Mahisarakul, J., P. Snitwongse, and R. Payamanontal. 1996. Characteristics and root distribution studies of some vetiver grass (*Vetiveria zizanioides* L. Nash and *Vetiveria Nemoralis* A. Camus) in Thailand by using P-32 tracer technique. *The First International Conference on Vetiver, Chiang Rai, Thailand, 4–8 February 1996.*

Marschner, H., ed. 1995. *Mineral nutrition of higher plants*. Academic Press Inc., San Diego.

Materechera, S. 2010. Soil physical and biological properties as influenced by growth of vetiver grass (*Vetiveria zizanioides* L.) in a semi-arid environment of South Africa. *19th World Congress of Soil Science, Soil Solutions for a Changing World, Brisbane, Australia, 1–6 August 2010.*

Mench, M.J., V.L. Didier, M. Löfler, A. Gomez, and P. Masson. 1994. A mimicked *in-situ* remediation study of metal-contaminated soils with emphasis on cadmium and lead. *Journal of Environmental Quality* 23: 58–63.

Mendez, M.O., and R.M. Maier. 2008. Phytoremediation of mine tailings in temperate and arid environments. *Reviews in Environmental Science and Biotechnology* 7: 47–59.

Mickovski, S.B., L.P.H. van Beek, and F. Salin. 2005. Uprooting of vetiver uprooting resistance of vetiver grass (*Vetiveria zizanioides*). *Plant and Soil* 278: 33–41.

Miller, D. 2012. Soil temperature and insect damage impacts on vetiver growth. http://www.vetiver.com/NZ_soiltemp.htm. (Accessed 19/06/2012)

Monteiro, J.M., R.E. Vollú, M.R.R. Coelho, C.S. Alviano, A.F. Blank, and L. Seldin. 2009. Comparison of the bacterial community and characterization of plant growth-promoting rhizobacteria from different genotypes of *Chrysopogon zizanioides* (L.) Roberty (Vetiver) rhizospheres. *The Journal of Microbiology* 47: 363–370.

Moosikapala, L., and S. Te-chato. 2010. Application of *in vitro* conservation in *Vetiveria zizanioides* Nash. *Journal of Agricultural Technology* 6: 401–407.

Muchow, R.C., T.R. Sinclair, and J.M. Bennett. 1990. Temperature and solar radiation effects on potential maize yield across locations. *Agronomy Journal* 82: 338–343.

Munshower, F.F., ed. 1994. *Practical handbook of disturbed land revegetation.* Boca Raton: Lewis Publishing.

Nix, K.E., G. Henderson, B.C.R. Zhu, and R.A. Laine. 2006. Evaluation of vetiver grass root growth, oil distribution, and repellency against Formosan subterranean termites. *HortScience* 41: 167–171.

Northcote, K.H., and Skene, J.K.M. 1972. Australian soils with saline and sodic properties. *CSIRO Division of Soil Publication 27.*

Northwestern University, 2007. http://www.civil.northwestern.edu/ehe/html_kag/kimweb/MEOP/Section3.htm.

Pang, J., G.S.Y. Chan, J. Zhang, J. Liang, and M.H. Wong. 2003. Physiological aspects of vetiver grass for rehabilitation in abandoned metalliferous mine wastes. *Chemosphere* 52: 1559–1570.

Patiyuth, S., B. Tangcham, and S. Muanjang. 2000. Studies on N_2-fixing bacteria association with vetiver 1. Biosynthesis of plant growth hormone by Azospirillum. 2. Use of the gusA gene to study Azospirillum. *The Second International Conference on Vetiver, Phetchaburi, Thailand, 18–22 January 2000.*

Podar, D., M.H. Ramsey, and M.J. Hutchings. 2004. Effect of cadmium, zinc and substrate heterogeneity on yield, shoot metal concentration and metal uptake by *Brassica juncea*: implications for human health risk assessment and phytoremediation. *New Phytologist* 163: 313–324.

Pothinam, A. 2006. Vetiver root and soil moisture conservation from vetiver grass establishment on degraded soils. *International Workshop on Sustained Management of the Soil-Rhizosphere System for Efficient Crop Production and Fertilizer Use, Bangkok, Thailand, 16–20 October 2006.*

Punamiya, P., R. Datta, D. Sarkar, S. Barber, M. Patel, and P. Das. 2010. Symbiotic role of *Glomus mosseae* in phytoextraction of lead in vetiver grass, *Chrysopogon zizanioides* (L.). *Journal of Hazardous Materials* 177: 465–474.

Roongtanakiat, N., P. Sudsawad, and N. Ngernvijit. 2010. Uranium absorption ability of sunflower, vetiver and purple guinea grass. *Kasetsart Journal* 44: 182–190.

Rotkittikhun, P., R. Chaiyarat, M. Kruatrachue, P. Pokethitiyook, and A.J.M. Baker. 2007. Growth and lead accumulation by the grasses *Vetiveria zizanioides* and *Thysanolaena maxima* in lead-contaminated soil amended with pig manure and fertilizer: A glasshouse study. *Chemosphere* 66: 45–53.

Rotkittikhun, P., M. Kruatrachue, P. Pokethitiyook, and A.J.M. Baker. 2010. Tolerance and accumulation of lead in *Vetiveria zizanioides* and its effect on oil production. *Journal of Environmental Biology* 31: 329–334.

Ruth, E.L., L. Marianne, E. Charles, H.E. Karl, and W. Bernard. 2000. Compact callus induction and plant regeneration of a non-flowering vetiver from Java. *Plant Cell, Tissue and Organ Culture* 62: 115–123.

Schmöger, M.E.V., M. Oven, and E. Grill. 2000. Detoxification of arsenic by phytochelatins in plants. *Plant Physiology* 122: 793–801.

Schwartz, C., J.L. Morel, S. Saumier, S.N. Whiting, and A.J.M. Baker. 1999. Root development of the zinc-hyperaccumulator *Thlaspi caerulescens* is affected by metal origin, content and localization in the soil. *Plant Soil* 208: 03–115.

Shahandeh, H., and L.R. Hossner. 2000. Plant screening for chromium phytoremediation. *International Journal of Phytoremediation* 2: 31–51.

Shahandeh, H., and L.R. Hossner. 2002. Role of soil properties in phytoaccumulation of uranium. *Water, Air, and Soil Pollution* 141: 165–180.

Shu, W.S., H.P. Xia, Z.Q. Zhang, C.Y. Lan, and M.H. Wong. 2002. Use of vetiver and three other grasses for revegetation of Pb/Zn mine tailings: Field experiment. *International Journal of Phytoremediation* 4: 47–57.

Shu, W.S., Y.L. Zhao, B. Yang, H.P. Xia, and C.Y. Lan. 2004. Accumulation of heavy metals in four grasses grown on lead and zinc mine tailings. *Journal of Environmental Science* 16: 730–434.

Singh, S., S. Eapen, V. Thorat, C.P. Kaushik, K. Raj, and S.F. D'Souzaa. 2008. Phytoremediation of ^{137}cesium and ^{90}strontium from solutions and low-level nuclear waste by *Vetiveria zizanoides*. *Ecotoxicology and Environmental Safety* 69: 306–311.

Singh, S.K., A.A. Juwarkar, S. Kumar, J. Meshram, and M. Fan. 2007. Effect of amendment on phytoextraction of arsenic by *Vetiveria zizanioides* from soil. *International journal of Environmental Science and Technology* 4: 339–344.

Siripin, S., A. Thirathorn, A. Pintarak, and P. Aibcharoen. 2000. Effect of associative nitrogen fixing bacterial inoculation on growth of vetiver grass. *The Second International Conference on Vetiver, Phetchaburi, Thailand, 18–22 January 2000.*

Solís-Domínguez, F.A., M.Z. Gonzalez-Chavez, R. Carrillo-Gonzalez, and R. Rodriguez-Vazquez. 2007. Accumulation and localization of cadmium in *Echinochloa polystachya* grown within a hydroponic system. *Journal of Hazardous Materials* 141: 630–636.

Srisatit, T., T. Kosakul, and D. Dhitivara. 2003. Efficiency of arsenic removal from soil by *Vetiveria zizanioides* and *Vetiveria nemoralis*. *Science Asia* 29: 291–296.

Truong, P. 1999a. Vetiver Grass Technology for flood and stream bank erosion control. *The International Vetiver Workshop, Nanchang, China.*

Truong, P. 1999b. Vetiver Grass Technology for mine rehabilitation. In: *Ground and water bioengineering for erosion control and slope stabilization*, 379–389, New Hampshire USA: Sciences Publishers.

Truong, P. 2000. Vetiver grass for mine site rehabilitation and reclamation. Extended Abstract. *Proc. Remade Lands International Conference, Fremantle, Australia.*

Truong, P. 2002. Vetiver grass technology. In: *Vetiveria. The Genus Vetiveria*, 114–132, London: Taylor and Francis.

Truong, P. 2003. Vetiver system for water quality Improvement. *The Third International Conference on Vetiver, Guangzhou, China, 6–9 October, 2003.*

Truong, P. and C. Creighton. 1994. Report on the potential weed problem of Vetiver grass and its effectiveness in soil erosion control in Fiji. *Division of Land Management, Queensland Department of Primary Industry, Brisbane, Australia.*

Truong, P., and D. Baker. 1997. The role of vetiver grass in the rehabilitation of toxic and contaminated lands in Australia. *The International Vetiver Workshop, Fuzhou, China, 21–26 October 1997.*

Truong, P., and D. Baker. 1998. Vetiver grass system for environmental protection. *Pacific Rim Vetiver Network Tech. Bull. 1.*

Truong, P., T.V. Tran, and E. Pinners. 2008. Vetiver system applications. Proven and green environmental solutions. *The Vetiver Network International, 9–16.*

Truong, P., Y.K. Foong, M. Guthrie, and Y.T. Hung. 2010. Phytoremediation of heavy metal contaminated soils and water using vetiver grass. *Handbook of Environmental Engineering* 11: 233–275.

Tscherning, K., D.E. Leihner, T.H. Hilger, K.M. Mütiller-Sämann, and M.A. El Sharkawy. 1995. Grass barriers in cassava hillside cultivation: Rooting patterns and root growth dynamics. *Field Crops Research* 43: 131–140.

Van, T.T., and P. Truong. 2008. R&D results on unique contributors of vetiver applicable for its use in disaster mitigation purposes in Vietnam. *First Indian National Vetiver Workshop, Cochin, India, 21–23 February 2008.*

Verkleij, J.A.C., and H. Schat. 1990. Mechanisms of metal tolerance in higher plants. In: *Heavy metal tolerance in plants: Evolutionary aspects*, 179–193. Boca Raton: CRC Press.

Vieritz, A., P. Truong, T. Gardner, and C. Smeal. 2003. Modelling Monto Vetiver Growth and Nutrient for Effluent Irrigation Schemes. *The Third International Conference on Vetiver, Guangzhou, China, 6–9 October 2003.*

Vo, V.M. 2007. Uptake potential of cadmium from soil by vetiver grass (*Vetiveria zizanioides* (L.) Nash). *Journal of Science,* Da Nang University, 21. (In Vietnamese).

Vo, V.M., and V.K. Le. 2009. Phytoremediation of cadmium and lead contaminated soil types by vetiver grass. *VNU Journal of Science, Earth Sciences* 25: 98–103.

Walker, B.J., and M.P. Bernal. 2004. The effects of copper and lead on growth and zinc accumulation of *Thlaspi caerulescens* J. and C. Presl: Implication for phytoremediation of contaminated soils. *Water, Air and Soil Pollution* 151: 361–372.

Wang, X., Y.G. Liu, G.M. Zeng, L.Y. Chai, X.C. Song, Z.Y. Min, and X. Xiao. 2008. Subcellular distribution and chemical forms of cadmium in *Bechmeria nivea* (L.) Gaud. *Environmental and Experimental Botany* 62: 389–395.

Wang, Y.W. 2000. The root extension rate of vetiver under different temperature treatments. *The Second International Conference on Vetiver, Phetchaburi, Thailand, 18–22 January 2000.*

Wei, S., L.Q. Ma, U. Saha, S. Mathews, S. Sundaram, B. Rathinasabapathi, and Q. Zhou. 2010. Sulfate and glutathione enhanced arsenic accumulation by arsenic hyperaccumulator *Pteris vittata* L. *Environmental Pollution* 158: 1530–1535.

Whiting, S.N., J.R. Leake, S.P. McGrath, and A.J.M. Baker. 2000. Positive responses to Zn and Cd by roots of the Zn and Cd hyperaccumulator *Thlaspi caerulescens*. *New Phytologist* 145: 199–210.

Wilde, E.W., R.L. Brigmon, and D.L. Dunn. 2005. Phytoextraction of lead from firing range soil by Vetiver grass. *Chemosphere* 61: 1451–1457.

Wong, C.C., S.C. Wu, C.K. Abdul, G. Khan, and M.H. Wong. 2007. The role of mycorrhizae associated with vetiver grown in Pb-/Zn-contaminated soils: Greenhouse study. *Restoration Ecology* 15: 60–67.

Wongkongkatep, J., K. Fukushi, P. Parkpian, R.D. DeLaune, and A. Jugsujinda. 2003. Arsenic uptake by native fern species in Thailand: Effect of chelating agents on hyperaccumulation of arsenic by Pityrogramma calomelanos. *Journal of Environmental Science and Health A* 38: 2773–2784.

Wu, S.C., C.C. Wong, W.S. Shu, A.G. Khan, and M.W. Wong. 2011. Mycorrhizo-remediation of lead/zinc mine tailings using vetiver: A field study. *International Journal of Phytoremediation* 13: 61–74.

Xia, H.P., X. Lu, and H. Ao. 2003. A preliminary report on tolerance of vetiver to submergence. *The Third International Conference on Vetiver. Guangzhou, China, 6–9 October 2003.*

Xia, H.P. 2004. Ecological rehabilitation and phytoremediation with four grasses in oil shale mined land. *Chemosphere* 54: 345–353.

Xia, H.P., H.X. Ao, S.Z. Lui, and D.Q. He. 1999. Application of the vetiver grass bioengineering technology for the prevention of highway slippage in Southern China. *Proceedings of Ground and Water Bioengineering for Erosion Control and Slope Stabilisation, Manila, April 1999.*

Xu, L. 2005. Simple treatments to protect vetiver against cold winters in north subtropics of China. *Vetiverim.* www.vetiver.org/CHN_cold%20planting.pdf.

Xu, L., and J. Zhang. 1999. An overview of the use of vegetation in bioengineering in China. *Proceedings of Ground and Water Bioengineering for Erosion Control and Slope Stabilisation, Manila, Philippines, 19–21 April, 1999.*

Yang, B., W.S. Shu, Z.H. Ye, C.Y. Lan, and M.H. Wong. 2003. Growth and metal accumulation in vetiver and two Sesbania species on lead/zinc mine tailings. *Chemosphere* 52: 1593–1600.

Zhuang, P., Z.H. Ye, C.Y. Lan, Z.W. Xie, and W.S. Shu. 2005. Chemically-assisted phytoextraction of heavy metal contaminated soils using three plant species. *Plant and Soil* 276: 153–162.

Section IV

Enhancing Contaminants' Remediation

13 Effects of Biotic and Abiotic Amendments on Phytoremediation Efficiency Applied to Metal-Polluted Soils

Erik J. Joner

CONTENTS

13.1 INTRODUCTION

Soils heavily polluted with metals are common in and around industrial areas where mining, processing and smelting of ores have taken place. The concentrations of metals like cadmium, (Cd), copper (Cu), lead (Pb) and Zinc (Zn) may be so high that it is difficult to establish plants on these soils to prevent erosion and spreading of pollutants by wind and water. In polluted areas surrounding such industrial sites, crop plants grown on less contaminated soil may grow well, but become too rich in heavy metals to be suitable as food or feed. Alternative soil use, like cultivation of nonfood crops (e.g., fibers or bioenergy), may be a solution, but often the metals contained in such crops will cause problems later in their life cycle. Two solutions then remain: to use biotic or abiotic amendments that reduce metal uptake in plants, or to use amendments that enhance metal uptake, in order to collect aerial plant parts and treat this biomass as hazardous waste. In the latter case, if the ashes are collected and the fumes treated properly, the biomass can still be used for, e.g., bioenergy. This provides a gain from the crop, removes pollutants from the soil and reduces the volume of metal rich waste. The fact that light ash fractions commonly contain more metals than heavy ash fractions (Vervaeke et al. 2006) can be used to treat these ash batches differently, possibly recycle metals in some cases, and further reduce the costs of special treatments.

If a metal-polluted soil is repeatedly cultivated with nonfood crops that are removed, and the plants are chosen for their capacity to extract large amounts of metals, the soil will gradually be depleted of such metals and thereby restored. Such phytoextraction treatments are likely to require decades in order to attain acceptable levels of metals in soil, but the technology is relatively inexpensive, may yield a cash crop during this time (e.g., biofuel) and give the land owners an affordable means of restoring their polluted land. If combined with soil amendments that initially enhance metal uptake in plants, and later amendments that sequester metals irreversibly, the time span required before food crop plants may again be grown on such soil can be reduced substantially.

13.2 INORGANIC AMENDMENTS

Inorganic amendments to metal-polluted soils are well described and most commonly used to immobilize the metals of concern, either by raising pH (lime or ashes) (Ruttens et al. 2010; Trakal et al. 2011), or by sequestering the most labile portion of metals through various adsorption phenomena (clay minerals, phosphates, or different forms of iron) (Lothenbach et al. 1999; Usman et al. 2005; Querol et al. 2006). The amount of lime or ashes added to soil is commonly limited to avoid increasing the soil pH excessively, so as to permit plant growth. Typically, liming targets a soil pH around or slightly above 7, when negative charges in soil become important in binding metal cations (Bolan et al. 2003) without making conditions too alkaline for plants to thrive. Liming thus reduces the concentration of labile or bioavailable metals in the soil solution, as can be measured by leaching of water (Basta and McGowen 2004) or mild soil extract protocols (Janos et al. 2010).

The use of various forms of phosphate, like apatite, rock phosphate or more readily soluble forms like di-ammonium phosphate do reduce the amount of water soluble metals in polluted soil to a high extent, but may release these at a later time following weathering or particle transport to surface waters (Mignardi et al. 2011). They may then also have an undesirable eutrophication effect. In addition to inorganic amendments that immobilize metals in soil, mineral fertilizers are often supplied to ensure optimal plant growth. The secondary effects of metal-immobilizing agents on the availability of plant nutrients may lead to nutrient deficiencies, or in the case of highly soluble phosphates, directly phytotoxic effects (Khan and Jones 2008), and should therefore be adapted to suit specific conditions of the site and the plants used. Complications with matching additions of immobilizing agents and plant nutrients is one of the reasons why organic amendments, continuously releasing plant nutrients through mineralization and symbiotic microorganisms, like mycorrhizal fungi that provide improved access to plant nutrients, have such success in phytroremediation (Donnelly and Fletcher 1994; Alguacil et al. 2011).

13.3 ORGANIC AMENDMENTS

Organic amendments generally have weaker effects on metal immobilization than inorganic amendments (Nwachukwu and Pulford 2009; Zubillaga et al. 2012). The justification for adding organic amendments to metal-polluted soils in the context of phytoremediation is therefore often to enhance plant growth by improving soil structure, water-holding capacity and plant nutrition, or to improve the soil's capacity to retain water to avoid leaching (by enhanced water-holding capacity) or erosion (by improved soil structure) (Park et al. 2011b). Improving the conditions for plant growth is often necessary when soils have been void of vegetation for several years. Soil without vegetation rapidly loses organic matter through mineralization and leaching, leaving a dense and infertile mineral matrix which is a poor substrate for plant growth, and which releases metals to the surrounding environment through leaching. Adding organic matter can counteract several of these negative properties and provide plant nutrients (Alvarenga et al. 2009).

Several types of organic amendments have been tested at different scales, but for field scale applications, inexpensive and easily available amendments like manure, sewage sludge, compost, bark or other biomass-residues have been most common. Recently, the use of biochar (pyrolyzed organic matter) as a soil amendment has also been shown to reduce metal bioavailability and plant uptake of metals like Cd, Cu and Pb (Beesley et al. 2010; Park et al. 2011a). The mechanisms involved include plant growth enhancement and increased pH, but also direct immobilization, as biochar has high specific surface area and strong sorption capacity (Beesley et al. 2011). In a study comparing the capacity of biochar and activated carbon to immobilization of Pb in soil, biochar reduced Pb bioavailability by 80%, whereas activated carbon had no measurable effect (Cao et al. 2011). Both sewage sludge and biochar may however themselves contain a certain amount of heavy metal, and this should be considered prior to their use, to ensure that the amendments are not enhancing the amounts of the metals that are causing concern in the first place.

Chelating agents like EDTA, EDDS, nitriloacetic acid or citrate added directly to soil have been used in order to enhance metal bioavailability and thereby phytoextraction during growth of plants. Bianchi et al. (2008) thus observed enhanced metal mobilization after application of EDTA at approx 0.5 g l^{-1}, but no effects of citric acid. The EDTA-enhanced mobility did result in increased metal concentrations in the seashore finger-grass *Paspalum vaginatum*, but did not affect plant growth. Komarek et al. (2008) working with *Populus* spp. found that EDTA enhanced metal bioavailability, but also that plant growth was reduced. Thus, enhanced metal concentrations in plant tissues were accompanied by a biomass reduction, leading to no net effects on phytoextraction. Similar results were found with sunflower and soil amendments with 10 mM EDTA (ca 3 g kg^{-1}), where metal uptake was increased so much as to induce phytotoxicity and no net phytoextraction (Lesage et al. 2005). These authors also tested citrate as a mobilizing agent, but found that it was metabolized too rapidly, and was unable to surpass the soil pH buffering capacity, even when added in very high amounts (0.4 mol kg^{-1}), with no increase in plant metal uptake as a result. The chelating agent EDDS (SS-ethylenediamine disuccinic acid) seems less efficient than EDTA, as, in sunflower, additions of up to 10 mM (3 g kg^{-1}) only enhanced bioavailability and uptake of Cu and Pb, but not Cd and Zn (Tandy et al. 2006). In a soil rich in clay and organic matter, both EDTA and DTPA had modest effects on metal mobilization and uptake in maize (Meers et al. 2004). A range of organic acids, including citric acid and nitriloacetic acid (NTA), were also tested in this experiment. They were all unable to increase metal uptake in plants, probably due to rapid mineralization. Another experiment with NTA did however demonstrate a ten- to thirty-fold increase in soluble ($NaNO_3$-extractable) Cd and Zn 55 days after application to two different soils (Wenger et al. 2002). Metal uptake in plants, however, only increased 5.5-fold for Zn and 2.5-fold for Cd in one of the soils, with no effects in the other. Although chelating agents have been proved effective in pot experiments, they are less efficient in field applications where root densities are far lower (Neugschwandtner et al. 2008). The latter authors noted that the risk of enhancing metal mobility without simultaneous plant uptake could also enhance leaching and metal contamination of underlying groundwater.

13.4 BIOLOGICAL AMENDMENTS

Plants themselves are of course the primary biological input during phytoremediation, as the entire technology is based on their capacity to bring photosynthetically-fixed carbon and energy into the soil and drive processes that improve soil conditions and extract metals (Wenzel 2009). Apart from plants, earthworms have been suggested as bio-augmenters, but earthworm applications have so far been limited to highly organic substrates and composts to enhance degradation of organic pollutants (Bianchi et al. 2010). Other bio-augmenters are mainly found among microorganisms. These can be divided into free-living and symbiotic organisms. The former are typically rhizosphere bacteria which can enhance plant growth and thereby the outcome of phytoremediation (Belimov et al. 2005; Kuffner et al. 2008; Gurska et al. 2009; De Maria et al. 2011). Some of these studies demonstrate that these rhizosphere bacteria have even had the capacity to enhance phytoextraction yields and transfer metals from roots to shoots (De Maria et al. 2011), and metal mobilization by bacterial metabolites (Kuffner et al. 2008), while others claim that bacteria induce enhanced plant tolerance to metals (Belimov et al. 2005).

The second group of microorganisms reported to be active in phytoremediation are endophytic symbionts living within plants. These may again be sub-divided into organisms living in aerial parts or in roots. Leaf endophytes comprise both fungi and bacteria. The former mainly benefit plants by enhanced growth and reduce herbivory (Clay 1988). Endophytic bacteria, on the other hand, have been shown to alter plant metabolism and thus benefit plants during phytoremediation of chlorinated organics through partial degradation of these compounds inside the plants (Germaine et al. 2006; Weyens et al. 2009). Effects on phytoremediation on metal-contaminated soils have not been reported for either of these groups.

Mycorrhizas are formed by fungal root endophytes, and have a privileged position as a symbiotic interaction in relation to phytoremediation. This is because these fungi influence several aspects that are important for the outcome of phytoremediation: plant establishment, survival and growth, plant nutrient uptake, metal uptake or exclusion, and improvement of plant tolerance towards other types of stress (drought, root pathogens) (Leyval et al. 1997). The reasons why mycorrhizas can provide such a multitude of services is found in the intimacy of the symbiosis where ion transport into the roots is mediated to a large extent by the fungus, which may even filter out or transfer elements selectively (Dehn and Schüepp 1989; Diaz et al. 1996; Cavagnaro et al. 2010; Krznaric et al. 2010). This is most evident in the ectomycorrhizal symbiosis (colonizing trees including, e.g., *Salix* and *Populus* species which are commonly used in phytoremediation), where the roots may be covered completely by fungal tissues (Figure 13.1).

But arbuscular mycorrhizas, colonizing mainly herbaceous plants, as well as certain trees like *Salix* and *Populus* species, also confer a selective transport function that alters elemental composition of their host plants. *Salix* and *Populus* species are indeed interesting because they may be colonized by both of these types of mycorrhiza, and because their use in phytoremediation is so widespread (Pulford and Watson 2003). An experiment comparing the effects of ectomycorrhizal and arbuscular mycorrhizal colonization of *Salix alba* and *Populus nigra* demonstrated that ectomycorrhizal fungi acted as the best barriers against uptake of metals, but that a combination of both types of mycorrhiza led to the highest metal exclusion observed, more than 50% (reduced from 72 to 35 mg Cd kg^{-1} DW in shoots) reduction in uptake of Cd in *S. alba* (Mrnka et al. 2012). An opposite trend was observed for Pb in *P. nigra* colonized by arbuscular mycorrhiza, where mean concentrations in shoots increased from 19 to 29 mg Pb kg^{-1}.

For ectomycorrhiza it has been demonstrated that fungi adapted to elevate metal concentrations also have a higher capacity to reduce metal uptake in their host plants, and to maintain nutrient uptake which was impaired in polluted soil for nonmycorrhizal plants and plants colonized by nonadapted ectomycorrhizal fungi (Krznaric et al. 2010). Additional benefits from ectomycorrhizal fungi include a reduction in indicators of oxidative stress in roots, such stress being one of the major toxicity mechanisms for heavy metals (Schützendübel and Polle 2002).

Arbuscular mycorrhizas are less conspicuous than ectomycorrhizas, as no visible fungal structures are seen at the root surface (Figure 13.2).

Still, these arbuscular mycorrhiza fungi have the capacity to regulate the ion transporters of the roots (Nagy et al. 2009), and even affect the uptake mechanisms for metals/metaloids (Roos and Jakobsen 2008; Christophersen et al. 2009). In an experiment with *Populus alba* and *P. nigra*, Lingua et al. (2008) found that these fungi enhanced growth and reduced Zn transfer to shoots by up to 30% in *P. nigra*, and more so with the fungus *Glomus mosseae* compared with *G. intraradices*. The strong effect was dependent on mycorrhiza being established in plants prior to exposure

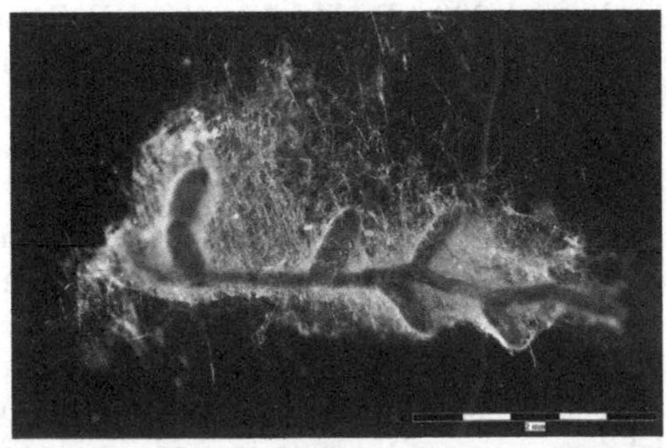

FIGURE 13.1 **(See color insert.)** Ectomycorrhiza on roots of *Populus* spp. showing studded roots covered by a fungal mantle, and extraradical mycelium. (Photo: Martin Vohnik.)

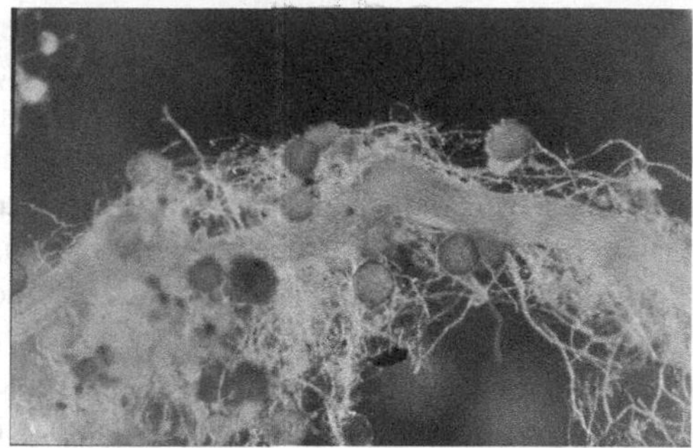

FIGURE 13.2 (See color insert.) Arbuscular mycorrhiza on a clover root, showing extraradical hyphae and spores. (Photo: Iver Jakobsen.)

to contaminants, as in nursery production employing inoculation before transplanting to polluted soil.

Single fungi do however perform less well than fungal consortia, as demonstrated in an experiment with maize and clover, where uptake of Cd, Cu and Zn (but not Pb) was reduced when plants were mycorrhizal, and even more so if they were colonized by a consortium of fungi indigenous to the polluted soil (Joner and Leyval 2001). Similar effects on uptake of Pb have also been observed, both for differently inoculated *Elsholtzia*-plants (Wang et al. 2005) and for maize, in a knock-out experiment where indigenous mycorrhiza formation was inhibited by applying the fungicide Benomyl to a polluted soil (Burke et al. 2000).

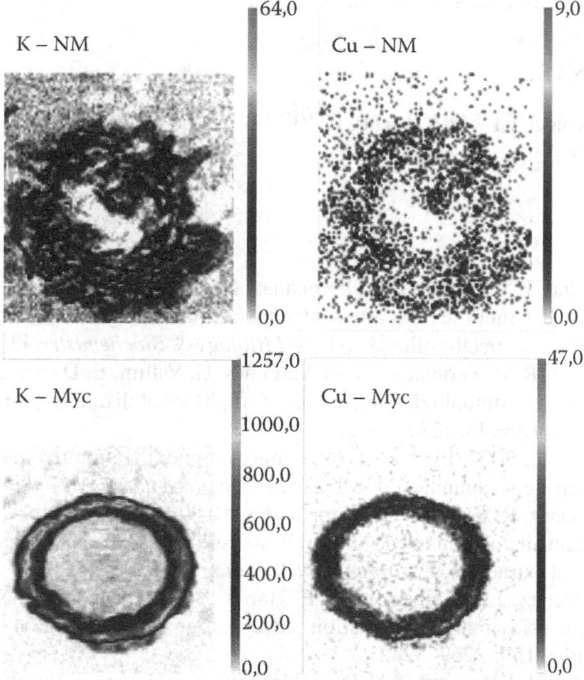

FIGURE 13.3 (See color insert.) Distribution of K and Cu in cross-sections of ectomycorrhizal (Myc) and nonmycorrhizal (NM) pine roots mapped using µ-PIXE imaging. Colors indicate relative concentrations according to the scale bars and show a higher proportion of Cu in the cortex region for mycorrhizal roots than in nonmycorrhizal roots due to fungal accumulation of Cu. (From E. Joner and B. Gouget, unpublished results.)

The mechanisms explaining reduced metal uptake in mycorrhizal plants are multiple. The first encounter between metals and mycorrhizal plants is through the extraradical mycelium. This is abundant for ectomycorrhizas, but may also reach >30 m g^{-1} soil for arbuscular mycorrhizas (Joner and Jakobsen 1994). The extraradical mycelium has been shown to bind significant amounts of metals (Joner et al. 2000), and may even reduce the amount of bioavailable metals in the soil (Janouskova and Pavlikova 2010). Mechanisms that are probably more important may be related to sequestration of metals within fungal mycelium on the root surface and inside the roots (Figure 13.3), precipitation of metal ions with excess phosphate so as to reduce transfer to shoots (mycorrhizas often induce a luxury uptake of P), more efficient export by active transporters due to improved mineral nutrition, or simply dilution through growth enhancement due to improved mineral nutrition of mycorrhizal plants (Blaudez et al. 2000; Frey et al. 2000; Christie et al. 2004; Lee and George 2005).

13.5 CONCLUSIONS

The high variety of factors that can influence plant growth, metal uptake and mobility of metals in soils make phytoremediation a versatile tool that can be adapted to a wide range of situations. The soils to be treated are, however, always different from site to site, and experience from one site does not always easily adapt to a new situation. The amendments available in sufficient amounts and at a reasonable cost may also vary greatly from one area or one country to another, further complicating a pollution management situation. Both the low-cost nature of this technology, the high costs of analyses during post-treatment monitoring, and the delicate issues of site owners' rights to monitoring data, have hampered dissemination of results on phytoremediation efficiency and systematic mapping of successful, and less successful, treatment schemes. This will hopefully emerge as the technology matures and gains recognition.

ACKNOWLEDGMENTS

The support from the project EPTOCOL (no. CZ0092) by the Norwegian Financial Mechanism is gratefully acknowledged.

REFERENCES

Alguacil, M.M., E. Torrecillas, F. Caravaca, D.A. Fernandez, R. Azcon and A. Roldan. 2011. The application of an organic amendment modifies the arbuscular mycorrhizal fungal communities colonizing native seedlings grown in a heavy-metal-polluted soil. *Soil Biology & Biochemistry* 43: 1498–1508.

Alvarenga, P., A.P. Goncalves, R.M. Fernandes, A. de Varennes, G. Vallini, E. Duarte and A.C. Cunha-Queda. 2009. Organic residues as immobilizing agents in aided phytostabilization: (I) Effects on soil chemical characteristics. *Chemosphere* 74: 1292–1300.

Basta, N.T. and S.L. McGowen. 2004. Evaluation of chemical immobilization treatments for reducing heavy metal transport in a smelter-contaminated soil. *Environmental Pollution* 127: 73–82.

Beesley, L., E. Moreno-Jimenez, R. Clemente, N. Lepp and N. Dickinson. 2010. Mobility of arsenic, cadmium and zinc in a multi-element contaminated soil profile assessed by *in situ* soil pore water sampling, column leaching and sequential extraction. *Environmental Pollution* 158: 155–160.

Beesley, L., E. Moreno-Jimenez, J.L. Gomez-Eyles, E. Harris, B. Robinson and T. Sizmur. 2011. A review of biochar's potential role in the remediation, revegetation and restoration of contaminated soils. *Environmental Pollution* 159: 3269–3282.

Belimov, A.A., N. Hontzeas, V.I. Safronova, S.V. Demchinskaya, G. Piluzza, S. Bullitta and B.R. Glick. 2005. Cadmium-tolerant plant growth-promoting bacteria associated with the roots of Indian mustard (*Brassica juncea* L. Czern.). *Soil Biology & Biochemistry* 37: 241–250.

Bianchi, V., G. Masciandaro, B. Ceccanti, S. Doni and R. Iannelli. 2010. Phytoremediation and bio-physical conditioning of dredged marine sediments for their re-use in the environment. *Water, Air & Soil Pollution* 210: 187–195.

Bianchi, V., G. Masciandaro, D. Giraldi, B. Ceccanti and R. Iannelli. 2008. Enhanced heavy metal phytoextraction from marine dredged sediments comparing conventional chelating agents (citric acid and EDTA) with humic substances. *Water, Air, & Soil Pollution* 193: 323–333.

Blaudez, D., B. Botton and M. Chalot. 2000. Cadmium uptake and subcellular compartmentation in the ectomycorrhizal fungus *Paxillus involutus*. *Microbiology* 146: 1109–1117.

Bolan, N.S., D.C. Adriano, P.A. Mani and A. Duraisamy. 2003. Immobilization and phytoavailability of cadmium in variable charge soils. II. Effect of lime addition. *Plant and Soil* 251: 187–198.

Burke, S.C., J.S. Angle, R.L. Chaney and S.D. Cunningham. 2000. Arbuscular mycorrhizae effects on heavy metal uptake by corn. *International Journal of Phytoremediation* 2: 23-29.

Cao, X., L. Ma, Y. Liang, B. Gao and W. Harris. 2011. Simultaneous immobilization of lead and atrazine in contaminated soils using dairy-manure biochar. *Environmental Science & Technology* 45: 4884–4889.

Cavagnaro, T., S. Dickson and F. Smith. 2010. Arbuscular mycorrhizas modify plant responses to soil zinc addition. *Plant and Soil* 329: 307–313.

Christie, P., X.L. Li and B.D. Chen. 2004. Arbuscular mycorrhiza can depress translocation of zinc to shoots of host plants in soils moderately polluted with zinc. *Plant and Soil* 261: 209–217.

Christophersen, H.M., F.A. Smith and S.E. Smith. 2009. Arbuscular mycorrhizal colonization reduces arsenate uptake in barley via downregulation of transporters in the direct epidermal phosphate uptake pathway. *New Phytologist* 184: 962–974.

Clay, K. 1988. Fungal endophytes of grasses: A defensive mutualism between plants and fungi. *Ecology* 69: 10–16.

De Maria, S., A.R. Rivelli, M. Kuffner, A. Sessitsch, W.W. Wenzel, M. Gorfer, J. Strauss and M. Puschenreiter. 2011. Interactions between accumulation of trace elements and macronutrients in *Salix caprea* after inoculation with rhizosphere microorganisms. *Chemosphere* 84: 1256–1261.

Dehn, B. and H. Schüepp. 1989. Influence of VA mycorrhizae on the plant uptake and distribution of heavy metals in plants. *Agriculture, Ecosystems and Environment* 29: 79–83.

Diaz, G., C. Azcon-Aguilar and M. Honrubia. 1996. Influence of arbuscular mycorrhizae on heavy metal (Zn and Pb) uptake and growth of *Lygeum spartum* and *Anthyllis cytisoides*. *Plant and Soil* 180: 241–249.

Donnelly, P.K. and J.S. Fletcher, 1994. Potential use of mycorrhizal fungi as bioremediation agents. In: Anderson, T.A., Coats, J.R. (Eds.). *Bioremediation through rhizosphere technology*, 93–99. Washington: American Chemical Society.

Frey, B., K. Zierold and I. Brunner. 2000. Extracellular complexation of Cd in the Hartig net and cytosolic Zn sequestration in the fungal mantle of *Picea abies Hebeloma crustuliniforme* ectomycorrhizas. *Plant Cell and Environment* 23: 1257–1265.

Germaine, K.J., X. Liu, G.G. Cabellos, J.P. Hogan, D. Ryan and D.N. Dowling. 2006. Bacterial endophyte-enhanced phytoremediation of the organochlorine herbicide 2,4-dichlorophenoxyacetic acid. *FEMS Microbial Ecology* 57: 302–310.

Gurska, J., W. Wang, K.E. Gerhardt, A.M. Khalid, D.M. Isherwood, X.-D. Huang, B.R. Glick and B.M. Greenberg. 2009. Three-year field test of a plant growth promoting rhizobacteria enhanced phytoremediation system at a land farm for treatment of hydrocarbon waste. *Environmental Science & Technology* 43: 4472–4479.

Janos, P., J. Vavrova, L. Herzogova and V. Pilarova. 2010. Effects of inorganic and organic amendments on the mobility (leachability) of heavy metals in contaminated soil: A sequential extraction study. *Geoderma* 159: 335–341.

Janouskova, M. and D. Pavlikova. 2010. Cadmium immobilization in the rhizosphere of arbuscular mycorrhizal plants by the fungal extraradical mycelium. *Plant and Soil* 332: 511–520.

Joner, E.J., R. Briones and C. Leyval. 2000. Metal-binding capacity of arbuscular mycorrhizal mycelium. *Plant and Soil* 226: 227–234.

Joner, E.J. and I. Jakobsen. 1994. Contribution by two arbuscular mycorrhizal fungi to P uptake by cucumber (*Cucumis sativus* L.) from ^{32}P-labelled organic matter during mineralization in soil. *Plant and Soil* 163: 203–209.

Joner, E.J. and C. Leyval. 2001. Time-course of heavy metal uptake in maize and clover as affected by different mycorrhizal inoculation regimes. *Biology & Fertility of Soils* 33: 351–357.

Khan, M.J. and D.L. Jones. 2008. Chemical and organic immobilization treatments for reducing phytoavailability of heavy metals in copper-mine tailings. *Journal of Plant Nutrition and Soil Science* 171: 908–916.

Komarek, M., P. Tlustos, J. Szakova and V. Chrastny. 2008. The use of poplar during a two-year induced phytoextraction of metals from contaminated agricultural soils. *Environmental Pollution* 151: 27–38.

Krznaric, E., J.H.L. Wevers, C. Cloquet, J. Vangronsveld, F. Vanhaecke and J.V. Colpaert. 2010. Zn pollution counteracts Cd toxicity in metal-tolerant ectomycorrhizal fungi and their host plant, *Pinus sylvestris*. *Environmental Microbiology* 12: 2133–2141.

Kuffner, M., M. Puschenreiter, G. Wieshammer, M. Gorfer and A. Sessitsch. 2008. Rhizosphere bacteria affect growth and metal uptake of heavy metal accumulating willows. *Plant and Soil* 304: 35–44.

Lee, Y.J. and E. George. 2005. Contribution of mycorrhizal hyphae to the uptake of metal cations by cucumber plants at two levels of phosphorus supply. *Plant and Soil* 278: 361–370.

Lesage, E., E. Meers, P. Vervaeke, S. Lamsal, M. Hopgood, F.M.G. Tack and M.G. Verloo. 2005. Enhanced phytoextraction: II. Effect of EDTA and citric acid on heavy metal uptake by *Helianthus annuus* from a calcareous soil. *International Journal of Phytoremediation* 7: 143–152.

Leyval, C., K. Turnau and K. Haselwandter. 1997. Effect of heavy metal pollution on mycorrhizal colonization and function: physiological, ecological and applied aspects. *Mycorrhiza* 7: 139–153.

Lingua, G., C. Franchin, V. Todeschini, S. Castiglione, S. Biondi, B. Burlando et al. 2008. Arbuscular mycorrhizal fungi differentially affect the response to high zinc concentrations of two registered poplar clones. *Environmental Pollution* 153: 137–147.

Lothenbach, B., G. Furrer, H. Schurli and R. Schulin. 1999. Immobilization of zinc and cadmium by montmorillonite compounds: Effects of aging and subsequent acidification. *Environmental Science and Technology* 33: 2945–2952.

Meers, E., M. Hopgood, E. Lesage, P. Vervaeke, F.M.G. Tack and M.G. Verloo. 2004. Enhanced phytoextraction: In search of EDTA alternatives. *International Journal of Phytoremediation* 6: 95–109.

Mignardi, S., A. Corami and V. Ferrini. 2011. Evaluation of the effectiveness of phosphate treatment for the remediation of mine waste soils contaminated with Cd, Cu, Pb, and Zn. *Chemosphere* In press: doi: 10.1016/j.chemosphere.2011.1009.1050.

Mrnka, L., M. Kuchár, Z. Cieslarová, P. Matějka, J. Száková, P. Tlustoš and M. Vosátka. 2012. Effects of endo- and ectomycorrhizal fungi on physiological parameters and heavy metals accumulation of two species from the family *Salicaceae*. *Water Air and Soil Pollution*. In press: doi: 10.1007/s11270-11011-10868-11278.

Nagy, R., D. Drissner, N. Amrhein, I. Jakobsen and M. Bucher. 2009. Mycorrhizal phosphate uptake pathway in tomato is phosphorus-repressible and transcriptionally regulated. *New Phytologist* 181: 950–959.

Neugschwandtner, R.W., P. Tlustos, M. Komarek and J. Szakova. 2008. Phytoextraction of Pb and Cd from a contaminated agricultural soil using different EDTA application regimes: laboratory versus field scale measures of efficiency. *Geoderma* 144: 446–454.

Nwachukwu, O.I. and I.D. Pulford. 2009. Soil metal immobilization and ryegrass uptake of lead, copper and zinc as affected by application of organic materials as soil amendments in a short-term greenhouse trial. *Soil Use and Management* 25: 159–167.

Park, J., G. Choppala, N. Bolan, J. Chung and T. Chuasavathi. 2011a. Biochar reduces the bioavailability and phytotoxicity of heavy metals. *Plant and Soil* 348: 439–451.

Park, J.H., D. Lamb, P. Paneerselvam, G. Choppala, N. Bolan and J.W. Chung. 2011b. Role of organic amendments on enhanced bioremediation of heavy metal(loid) contaminated soils. *Journal of Hazardous Materials* 185: 549–574.

Pulford, I.D. and C. Watson. 2003. Phytoremediation of heavy metal-contaminated land by trees—A review. *Environment International* 29: 529–540.

Querol, X., A. Alastuey, N. Moreno, E. Alvarez-Ayuso, A. Garcia-Sanchez, J. Cama, C. Ayora and M. Simon. 2006. Immobilization of heavy metals in polluted soils by the addition of zeolitic material synthesized from coal fly ash. *Chemosphere* 62: 171–180.

Roos, P. and I. Jakobsen. 2008. Arbuscular mycorrhiza reduces phytoextraction of uranium, thorium and other elements from phosphate rock. *Journal of Environmental Radioactivity* 99: 811–819.

Ruttens, A., K. Adriaensen, E. Meers, A. De Vocht, W. Geebelen, R. Carleer, M. Mench and J. Vangronsveld. 2010. Long-term sustainability of metal immobilization by soil amendments: Cyclonic ashes versus lime addition. *Environmental Pollution* 158: 1428–1434.

Schützendübel, A. and A. Polle. 2002. Plant responses to abiotic stresses: Heavy metal-induced oxidative stress and protection by mycorrhization. *Journal of Experimental Botany* 53: 1351–1365.

Tandy, S., R. Schulin and B. Nowack. 2006. Uptake of metals during chelant-assisted phytoextraction with EDDS related to the solubilized metal concentration. *Environmental Science & Technology* 40: 2753–2758.

Trakal, L., M. Neuberg, P. Tlustoš, J. Száková, V. Tejnecký and O. Drábek. 2011. Dolomite limestone application as a chemical immobilization of metal-contaminated soil. *Plant, Soil and Environment* 57: 173–179.

Usman, A., Y. Kuzyakov and K. Stahr. 2005. Effect of clay minerals on immobilization of heavy metals and microbial activity in a sewage sludge-contaminated soil. *Journal of Soils and Sediments* 5: 245–252.

Vervaeke, P., F.M.G. Tack, F. Navez, J. Martin, M.G. Verloo and N. Lust. 2006. Fate of heavy metals during fixed bed downdraft gasification of willow wood harvested from contaminated sites. *Biomass & Bioenergy* 30: 58–65.

Wang, F., X. Lin and R. Yin. 2005. Heavy metal uptake by arbuscular mycorrhizas of *Elsholtzia splendens* and the potential for phytoremediation of contaminated soil. *Plant and Soil* 269: 225–232.

Wenger, K., A. Kayser, S.K. Gupta, G. Furrer and R. Schulin. 2002. Comparison of NTA and elemental sulfur as potential soil amendments in phytoremediation. *Soil and Sediment Contamination* 11: 655–672.

Wenzel, W. 2009. Rhizosphere processes and management in plant-assisted bioremediation (phytoremediation) of soils. *Plant and Soil* 321: 385–408.

Weyens, N., D. van der Lelie, T. Artois, K. Smeets, S. Taghavi, L. Newman, R. Carleer and J. Vangronsveld. 2009. Bioaugmentation with engineered endophytic bacteria improves contaminant fate in phytoremediation. *Environmental Science & Technology* 43: 9413–9418.

Zubillaga, M.S., E. Bressan and R.S. Lavado. 2012. Effects of phytoremediation and application of organic amendment on the mobility of heavy metals in a polluted soil profile. *International Journal of Phytoremediation* 14: 212–220.

14 Phytoremediation: Strategies to Enhance the Potential for Toxic Metal Remediation of *Brassica* Oilseed Species

Mary Varkey, Nand Lal, and Z. H. Khan

CONTENTS

14.1 INTRODUCTION

Pollution of the environment with toxic substances has increased tremendously since the onset of the industrial revolution (Nriagu 1979). Some of these contaminants can interfere with agriculture and also enter the food chain (Dembitsky 2003) by accumulation and biomagnification. The threat of heavy metals to human health is aggravated by their long-term persistence (Shaw 1990). Conventional technical decontamination techniques are too costly to remediate extended areas of contaminated soils, and are also environmentally destructive (Bio-wise 2003). These remediation strategies are an *ex situ* approach and can be damaging to soil structure, ecology and productivity. As an alternative, plant-based bioremediation technologies have received attention as strategies to clean up metal-contaminated soil and water (Raskin et al. 1994; Salt et al. 1995a; Prasad and Freitas 2003; Jadia and Fulekar 2008). All plants have the ability to accumulate from soil and water metals such as Fe, Mn, Zn, Ni, Mo, Mg, and Cu, which are essential for their growth and development. Besides this they can also absorb Cd, Pb, Co, Cr, Ag, Se, Hg which have no biological importance in their physiological activities. The strategies thus adopted have been termed as phytoremediation (Salt et al. 1995a) and refer to the use of green plants and their associated microbiota for *in situ* treatment of water and soil. These remediation strategies have been estimated to be four to a thousand times cheaper per unit volume than nonbiological techniques (Cunningham and Ow 1996). Compounds targeted for phytoremediation strategies include heavy metal salts (Kumar et al. 1995; Salt et al. 1995a), polycyclic aromatic hydrocarbons, polychlorinated biphenyls, herbicides, pesticides (Sadowsky 1998) and radionuclides (Entry et al. 1997).

Phytoremediation is an environmental cleanup technology initially proposed by Utsunamyia (1980) and Chaney (1983). The main types of phytoremediation strategies include stimulation by rhizosphere microorganisms using hyperaccumulators, the use of plants to transform contaminants to less toxic forms, enlarging root mass to increase absorption area, and genetic engineering of plants for improved biodegradation potential and/or improved absorption and accumulation capacity.

Phytoremediation processes can be classified into five categories (Salt et al. 1998): (i) phytoextraction, (ii) rhizofiltration, (iii) phytodegradation, (iv) phytostabilization (phytorestoration), and (v) phytovolatilization.

14.1.1 Phytoextraction

Phytoextraction involves the use of higher plants for removal of contaminants (Kumar et al. 1995; Lasat 2002), transport of metals from the soil into the plant body and their accumulation in harvestable parts and shoots. For this a plant species and its specific genotype should not only accumulate and tolerate high levels of toxic metals, but should also have a rapid growth rate to produce high levels of biomass. Plants that hyperaccumulate metals (hyperaccumulators) have tremendous potential for their application in remediation of metals in the environment (Brooks 1998). The word hyperaccumulator was coined by Brooks and Reeves (Brooks et al. 1977) and it has been defined as metal accumulation exceeding a threshold value of shoot metal concentration of 1% (Zn, Mn) and 0.1% (Ni, Co, Cr, Cu, Pb and Al), 0.01% (Cd and Se) or 0.001% (Hg) of the dry weight shoot biomass (McGrath 1998). The plants that can hyperaccumulate belong to the families of Asteraceae, Brassicaceae, Caryophyllaceae. Cunouniaceae, Flacourtiaceae, Lamiaceae, Poaceae, Violaceae, Euphorbiaceae, Cyperaceae, Fabaceae. Brassicaceae has the largest number of taxa, with 11 genera and 87 species reported to hyperaccumulate Ni, Zn, Cd, Pb and other metals (Baker and Brooks 1989).

For the process of phytoextraction, metal-accumulating plants are cultivated on metal-polluted soil. The roots of established plants absorb metal elements from the soil and translocate them to the shoots (sink), where they accumulate. After sufficient plant growth and metal accumulation, the above-ground plant parts are harvested and suitably disposed (Kumar et al. 1995), and this helps in the removal of contamination. Incineration of harvested material reduces the volume for disposal (Kumar et al. 1995). The valuable metals can be extracted from it for re-use (Cunningham and Ow 1996).

This procedure can take from one to twenty years and is suitable for large areas of contaminated land with a low level of contaminants (Kumar et al. 1995; Blaylock and Huang 2000). This is because plants can only grow in low-contaminated substrate. The success of phytoextraction is dependent on the land condition to allow normal cultivation of crops, and on plant characteristics, such as their ability to accumulate large quantities of biomass and to hyperaccumulate in the plant tissue (Cunningham and Ow 1996; Baylock et al. 1997; Sarma 2011), which helps in phytoremediation.

When roots are in close contact with high metal concentrations they may (a) restrict the uptake, and or (b) immobilize, compartmentalize or detoxify metals in the symplasm by metal-binding compounds. Plants with exclusion strategies can avoid excessive uptake, and one such is the capacity of the cell to bind metal to the cell wall, as seen in *Brassica* spp. by Ramos et al. (2002) and Liu et al. (2007). Uptake and transport of metals can be correlated to special plant membrane proteins which recognize the chemical structure of metals and bind them for transportation (Clarkson and Luttage 1989; Mejare and Bulow 2001). For phytoextraction and accumulation of metals in plant tissues, internal tolerance to resist the cytotoxic burden of metals is important, for which compartmentation in the vacuole and chelation in the cytoplasm are significant mechanisms (Vazquez et al. 1994; Kupper et al. 1999).

Plants possess a range of potential cellular mechanisms that may be involved in detoxification of heavy metals and tolerance to metal stress. These mechanisms include binding to mycorrhiza, cell walls, extracellular exudation, reduced uptake or efflux pumping of metals at plasma membrane or uptake chelation and transport of metal complexes in tissues and sequestering into vacuoles (Guerinot and Salt 2001; Clemens et al. 2002) by tonoplast located transporters. The best known are complexation with phytochelatins (PCs) and metallothioneins (MTs) (Cobbett and Goldsborough 2002). Glutathione (GSH) is the precursor of PCs (Zenk 1996). The metal is carried by PCs and PCs-metal complex into cytoplasm, metal may be bound to an organic acid in the vacuole (Rauser 1995), and PCs is returned to cytoplasm. The vacuole is the site for accumulation of a number of heavy metals like Cd, Zn, Mn and Ni (Ernst et al. 1992; Wagner et al. 1995). Organic acids therein may detoxify the metal (Ebbs et al. 2002) by forming complexes (Ma et al. 2001). Lucarini et al. (1999) reported that *Brassica* vegetables were rich in organic acids. Other low molecular weight chelators like amino acids and nicotiniamine are also used in detoxification and sequestration (Hall 2002). Another approach to detoxifying heavy metals is their transformation to less harmful forms in the plant (Meagher et al. 2000).

Brassica juncea (Indian mustard) has one-third Zn in the tissue, and is more effective for Zn removal than *Thlaspi caerulescens,* a hyperaccumulator of Zn, because it produces ten times more biomass (Ebbs et al. 1997). Soil metals should also be bioavailable for easy absorption. Salt-tolerant plants are more suitable for heavy metal extraction than salt-sensitive ones in salty soils (Zaier et al. 2010). *B. juncea* (L.) Czern. has a high biomass, grows rapidly, and is promising plant for phytoremediation (Terry et al. 1992) It has been used in phytoextraction of a wide range of metals (Salt et al. 1995b; Ebbs and Kochian 1997; Begonia et al. 1998; Ahmad et al. 2001; Panwar et al. 2002; Salido et al. 2003; Lim et al. 2004; Su and Wong 2004; Gupta and Sinha 2006; Ishikawa et al. 2006; Turan and Esringu 2007; Purakayastha et al. 2008; Garg and Kataria 2009). Kumar et al. (1995) determined that the accessions 426308, 211000, 426314 and 182921 of Indian mustard *B. juncea* Czern. are among the best suited for phytoextraction. Other *Brassica* oilseed species like *B. napus, B. carinata, B. campestris, B. nigra* are also used for phytoextraction of metals such as Ni, Zn, Cu, Pb, Cd (Marchiol et al. 2004; Aryakanon and Winaipanich 2006; Quartacci et al. 2009; Zaier et al. 2010).

14.1.2 Rhizofiltration

This refers to the use of plant roots to absorb and accumulate toxic metals from contaminated soil or water. Once the roots are saturated they are harvested, minimizing disturbance. Plants that are efficient in translocating metals to the shoots are generally not used. Rhizofiltration is relatively inexpensive yet potentially more effective than technological methods (Dushenkov et al. 1995; Salt et al. 1995a). Some of the earliest experiments on the use of *Brassica juncea* (Indian mustard) for rhizofiltration of Pb were carried out by Dushenkov et al. (1995). Indian mustard roots are effective in removing Cd, Cr, Cu, Ni, Zn and Pb (Dushenkov et al. 1995). Absorbed amounts ranged from 60 mg Pb/g of dry biomass to 136 for *Brassica juncea.* Rhizofiltration is preferred due to its applicability to many problem metals, its ability to treat high volumes, the reduced need for toxic chemicals, reduced volume of secondary waste, the possibility of recycling, and public acceptance (Dushenkov et al. 1995; Kumar et al. 1995). *B. juncea* roots can accumulate six times more Cd than shoots (Salt et al. 1995b).

A review of literature (Raskin and Ensley 2000; Terry and Banuelos 2000) indicates that the desirable attributes of a plant for rhizofiltration include tolerance to high concentrations of metal, its ability to accumulate high levels of metals and to bioconcentrate the metal, fast growth, high biomass production, and limited root-to-shoot translocation. Aquatic species like *Eichhornia crassipes* (Mart.) (Zhu et al. 1999c) and *Hydrocotyle umbellate* L. (Khilji and Bareen 2008) have the ability to remove heavy metals, but due to their slow-growing roots have limited ability for rhizofilteration. Indian mustard (*B. juncea*) and sunflower (*Helianthus annuus*) have rapidly growing roots and are ideal for rhizofilteration (Brooks and Robinson 1998). Both these species tend to concentrate heavy metals in the root systems and translocate only a small part of the metal burden to the shoots. The uptake of Pb and a few other elements in roots and shoots of hydroponically grown *Thlaspi caerulescens* and *B. juncea* were compared. The root biomass of *Brassica* was found to be far greater than *Thlaspi* and this is the greatest factor in its favor (Brooks and Robinson 1998).

14.1.3 Phytostabilization

This involves absorption and precipitation of contaminants, especially metals, by plants, reducing their mobility and preventing their migration to ground water (leaching) or to wind transport. The aim of phytostabilization is not to remove the metal pollutants from the site but to stabilize them, in order to reduce the risk to living organisms by entry into the food chain (Berti and Cunningham 2000). The three mechanisms within phytostabilization that determine the fate of the contaminants are:

Phytostabilization in the root zone—Immobilization of contaminants in the root zone due to exudations of the roots in the rhizosphere.

Phytostabilization of root membrane—Binding of contaminants to root membrane (surface), not allowing it to enter the plant.

Phytostabilization in the root cells—If contaminants are transported they can be sequestered into the vacuole of root cells preventing translocation (Shilev et al. 2009). Plants chosen for phytostabilization should be poor translocators of metal contaminants to the shoots, and should have a dense shoot and root system. Selected plants should be easy to establish and care for, grow quickly, and be tolerant for metal contaminants. Their dense root systems are able to absorb large volumes of water. In addition, rich canopies of shoots will augment transpiration, so inhibiting precipitation of contaminants into ground water. Fast-transpiring plants maintain an upward flow to prevent downward leaching. Plant root exudates may cause metals to precipitate, converting them to a less bioavailable form.

Indian mustard appeared to have potential for phytostabilization (Salt et al. 1995b). Studies indicated that Indian mustard roots reduced Cr (VI) to Cr (III) (Dushenkov et al. 1995), a process which would promote phytostabilization. *B. juncea* has been shown to reduce leaching of metals from soil by over 98% (Raskin et al. 1994).

14.1.4 Phytodegradation/Phytotransformation

This is a breakdown of organic contaminants existing within or outside plants, with the help of enzymes produced by plants. Pollutants are degraded into simpler substances which are taken up by the plant to help them grow faster. Plants which produce enzymes that metabolize contaminants may be released into the rhizosphere, where they can remain active in contaminant transformation. Enzymes such as peroxidase and nutrilase have been discovered in plant sediments and soils (Schnoor et al. 1995). Uptake and degradation of DDT (dichloro diphenyl trichloro ethane) by *B. juncea* and *Cichorium intybus* (Suresh et al. 2005), and phytotransformation of DDT to its main detoxified metabolites DDD, DDE and DDMU by hairy root cultures, has been reported (Gao et al. 2000; Suresh et al. 2005).

14.1.5 Phytovolatilization

This is the process where plants take up contaminants which are water-soluble and release them into the atmosphere as they transpire the water. The contaminant may become modified along the way as the water travels along the plant's vascular system from the roots to the leaves. The contaminants may then evaporate or volatilize into the air surrounding the plant. Experimental investigations have shown that volatile compounds are released to the atmosphere in significant amounts where plants are grown in soil containing these compounds (Marr et al. 2006; Banuelos and Lin 2007; Nwoko 2010).

Indian mustard has the ability to remove selenium by volatilization (Zayed and Terry 1994; Zayed et al. 1998; De Souza et al. 2000; Montes-Bayon et al. 2002; Pilon Smits et al. 2010). Se removal from polluted water and soils is highly complicated and expensive by conventional methods. Plants remove Se by uptake and accumulation in their tissues, with subsequent volatilization into the atmosphere as harmless gas.

B. juncea (L.) Czern. plant is not an Se hyperaccumulator, but it takes up and volatilizes Se at high rates compared with other plants (Terry et al. 1992). Se, a chemical analog of S, is thought to be assimilated and volatilized by using the same enzymes as S assimilation-pathways (Brown and Shrift 1982; Terry and Zayed 1998). Selenium can thus substitute for S in sulfur metabolic pathways and subsequently be incorporated into amino acids or proteins. This Se incorporation may be the mechanism of metal tolerance in hyperaccumulators (Steven and Schnoor 2003). Among terrestrial plants, several members of the Brassicaceae family which can accumulate S can also accumulate and volatilize Se very efficiently. Indian mustard is the best terrestrial plant for Se phytoremediation (Zayed et al. 1998).

Rhizosphere bacteria seem to be able to enhance selenate volatilization (Zayed and Terry 1994). De Souza et al. (2000) found that it facilitated 35% of Se volatilization and 70% of Se accumulation in plants. A heat-labile proteinaceous compound produced by plant bacteria interaction may be responsible for enhanced Se uptake and volatilization. It is produced by either the plant or the bacteria and may act by converting selenite to Se Met, which is readily volatilized by plants (Zayed et al. 1998). Overexpression of cystathionine-γ-synthase obtained from *Arabidopsis thaliana* enhances selenium volatilization in *B. juncea* (Van Huysen et al. 2003).

14.2 STRATEGIES TO ENHANCE PHYTOREMEDIATION POTENTIAL OF *BRASSICA* SPECIES

Phytoremediation suffers from several limitations: the depth of soil that can be treated by plants, low translocation rate of metals from roots to shoots, etc., and low availability of metals at a given time (Baker et al. 2000). To enhance the amount of toxic metals at the plants' disposal, various approaches are used, either alone or in combinations, which have shown promising results, and some of them have been discussed here.

14.2.1 Use of Chelating Agents

Chemically-induced phytoextraction, mostly with chelating agents added to the soil, significantly enhances metal accumulation by plants, and has been proposed as an alternative for the cleaning up of metal-polluted soils. To minimize phytotoxicity and environmental problems associated with the use of chelating agents, small doses of the chelating agent are gradually applied during the growth period. Synthetic compounds such as DTPA, EDTA (Diethylenetriamine penta acetic acid; ethylene diaminetetra acetic acid) have shown promising results, as reviewed by (Evangelou et al. 2007). Accumulation of several metals — Pb, Cd, Cu, Ni, Zn, As — by Indian mustard plants after applying metal chelates has been reported (Blaylock et al. 1997). *B. napus* can be used for decontamination of affected soils and the addition of chelators increased the efficacy of *B. napus* (Wenzel et al. 2003; Zaier et al. 2010). However, the biodegradability of these compounds is low (Lombi et al. 2001). A much more degradable chelator is EDDS (S,S-ethylenediamine disuccinic acid) which was found to increase the phytoextraction potential of Indian mustard for U, Pb, and Cu by 19-, 34- and 37-fold respectively in the shoots (Lombi et al. 2001; Alkorta et al. 2004). However, metal chelators may be toxic for plant microorganisms (Evangelou et al. 2007). On the basis of the amount applied per hectare, these compounds are expensive (Barona et al. 2001).

14.2.2 Use of Metalicolous Plants

Quartacci et al. (2009) suggested that root exudates played an important role in solubilizing metals in soils and in favoring metal uptake (As, Cd, Cu, Pd and Zn) from multiple metal-contaminated soils. Results showed that the growth of metalicolous populations of *Pinus pinaster, Plantago lanceolata* and *Silene paradoxa* increased the extractable metal levels in the soil, which resulted in a higher accumulation of metals in the above-ground parts of the *B. carinata* plants used for phytoextraction.

14.2.3 Use of Metal-Immobilizing Agents

To overcome phytotoxicity of *B. juncea* plants grown for phytoextraction on multicontaminated soil (Cu, Zn and Ni), some immobilizing agents were evaluated in soil mix by Pedron et al. (2009). The best results for Zn and Ni accumulation was obtained after zeolite addition. It was six times more than in non-treated soil. In the case of Cu the more efficient treatment was $Ca(OH)_2$ where the accumulation was eight times greater (Pedron et al. 2009).

14.2.4 Use of Plant Growth-Promoting Rhizobacteria and Arbuscular Mycorrhizal Fungi

Another promising technique is optimizing the synergistic effect of plants and microorganisms, by coupling phytoextraction with soil augmentation, also called rhizomediation (Kuiper et al. 2004). This technique has been widely developed for the remediation of soil contaminated by organic pollutants (Barac et al. 2004) but not for metals. Uptake of metals can be enhanced by increased mobility of metals with microorganisms producing surfactants (Herman et al. 1995; Mulligan et al. 1999), siderophores (Dubbin and Louise Ander 2003) and organic acids (Di Simine et al. 1998), and enhancement of biomass of plants by associating with Plant Growth Promoting Rhizobacteria (PGPR) (Zhuang et al. 2007) and Arbuscular Mycorrhizal Fungi (AMF) (Khan 2006). Bioaugmentation-assisted phytoextraction is a promising method for phytoremediation and has been reviewed by Lebeau et al. (2000).

Bioaugmentation with bacterial strain *S. mathophila* Sel TEO_2 is suggested to elicit selenite phytoextraction efficacy of *B. juncea* (Ferrari et al. 2009). The association of *B. juncea* with *Pseudomonas flourescens* Pf 27 enhanced soil metal bioavailability of Zn, Cu, Cd and plant growth for successful phytoremediation (Fuloria et al. 2009).

As reviewed by Karami and Shamsuddin (2010), the association of *Achromobacter xylosoxidans, Azotobacter chroococcum* or *Bacillus subtilis* with *B. juncea* improved Ni, Pb, Zn, Cd, and Cu uptake. Cd accumulation was enhanced in *B. napus* when in association with *Pseudomonas napus*. Indole-3-acetic acid (IAA) synthesized by bacteria help in metal uptake. Similarly IAA produced by *Bacillus subtilis* SJ-101 increased Ni concentration in *B. juncea* by a factor of 1.5 (Zaidi et al. 2006). K ascorbate SUD 165, that contains the enzyme ACC deaminase, protected *B. juncea* and *B. campestris* against Ni, Pb and Zn toxicity (Burd et al. 1998; Borgmann 2000). Root elongation of *B. napus* has also been shown to be stimulated by IAA synthesized by PGPR (Sheng and Xia 2006) as well as non-identified rhizobacteria on *B. juncea* roots (Belimov et al. 2005). The accumulation of lead in shoots of *B. juncea* using 0.5 m mol/kg EDTA with electric potential increased by two to four times, compared to EDTA alone (Lim et al. 2004). Somaclonal variations are mutations in plant tissue culture that occur randomly or are induced under *in vitro* conditions. Investigations have been undertaken with *in vitro* breeding and somaclonal variation to develop new variants with increased phytoremediation properties (Delhaize 1996; Jan et al. 1997). *In vitro* culture of *B. juncea* can be easily carried out, and high frequency regeneration and heritable somaclonal variation has been observed (George and Rao, 1980; Jain et al., 1989). Different regeneration capacity between various *B. juncea* genotypes has been reported (Tang et al. 2003). *In vitro* breeding and somaclonal variations have also been used to improve the potential of *B. juncea* to accumulate toxic metals (Nehnevajova et al. 2007).

14.2.5 Genetic Engineering Strategies for Enhancing Phytoremediation

Genetic engineering-based phytoremediation strategies are aimed to improve the plants for effective removal of metals as reviewed by Pilon-Smits and Pilon (2002), Meagher and Heaton (2005) and Fulekar et al. (2009). Engineered plants need to have the following qualities: (a) tolerance to toxic elements, (b) ability to alter their rhizosphere by secreting various enzymes and adjusting pH to enhance absorption, (c) numerous endogenous transporters in roots and roots hairs and xylem, (d) maximum storage capacity in shoots, (e) increased chemical sinks such as organic acids and inbuilt chelators, (f) physical sinks such as sub-cellular vacuoles, epidermal trichome cells which would store large quantities of few toxic pollutants in shoots of various native hyperaccumulators, (g) should be fast-growing with large potential biomass, (h) more strongly metal-accumulating genotypes (Shah and Nongkynrih 2007), and (i) genes to change the oxidation state of heavy metals (Rugh et al. 1996) or that convert metals into less toxic forms. Examples might include genes controlling the synthesis of peptides that sequester metals like phytochelatins e.g. the *Arabidopsis* cad_1 gene. (Howden et al. 1995), genes encoding transport of proteins such as *Arabidopsis* IRTI gene that encodes a protein which can regulate the uptake of metals (Eide et al. 1996), gene-encoding enzymes that change the oxidation of heavy metals like the bacterial mer A gene encoding mercuric oxidase reductase (Rugh et al. 1996), protein $NtCBP_4$ that can modulate plant tolerance of heavy metals (Arazi et al. 1999).

The remedial capacity of plants can be significantly improved by genetic manipulation and plant transformation technologies (Terry et al. 2003; Van Huysen et al. 2004). The identification of unique genes from hyperaccumulators (Danika and Norman 2005) and their transfer to fast-growing species has proved very successful in phytoremediation (De Souza et al. 1998). Yeast metallothionein CUPI metal-binding proteins that confer heavy metal accumulation were introduced into tobacco plants, and Cu and Cd phytoextraction was obtained (Thomas et al. 2003). Improved phytoremediation of Cd by genetically-engineered genes introduced into *B. napus* (Wang and Liu, 2009) and *B. juncea* (Zhu et al. 1999b) have also been reported. Overexpression of two glutathione synthesizing enzymes, γ-glutamyl cysteine synthetase (γ-ECS) or glutathione synthetase (GS) in *B. juncea* showed enhanced Cd tolerance and accumulation (Zhu et al. 1999a, 1999b). The capacity of metal accumulation and tolerance could be enhanced by overexpressing natural or modified genes.

14.2.5.1 Metallothionein and Phytochelatin

The use of metallothionein (MT) and phytochelatin (PC) as natural metal chelators is well documented. MTs have the affinity for binding metal cations (Cobbett and Goldsborough 2002). PCs are also a family of metal-complexing peptides which play a very important role in heavy metal tolerance and chelate these substances (Cobbett and Goldsborough 2002). Plant MT genes have been isolated from several plants including maize, soybean, rice, wheat, tobacco and *B. napus* (Nedkovska and Atanassov 1998). The expression of MTs and PCs in plants to enhance heavy metal accumulation has been reviewed by Mejare and Bulow (2001). Overexpression of chelator gene like MT in oilseed rape resulted in enhanced Cd tolerance (Misra and Gedamu 1989).

14.2.5.2 Metal Transporters

Transporters play an important role in plants by shifting toxic cations across membranes. A number of plant heavy metal transporter families has been identified, such as ZIP (ZRT/IRT like protein), CDF (cation diffusion facilitator), Nramp (natural resistance and macrophage protein) and HMA (heavy metal ATPase) and their hyperexpression may play a role in hyperaccumulation (Guerinot 2000; Pence 2002; Letham et al. 2005). Some of these transporters are specific while others are not. The ZIP family is new gene family of metal ion transport proteins. The members of this family share significant sequence similarity and predicted topology. These proteins are ubiquitous among eukaryotes; closely related genes are found in the genomes of organisms as diverse as fungi, protozoa, plants, nematodes, and humans. The members of this family that have been examined experimentally have been shown to encode metal ion transport proteins. The ZIP family of transporters has been shown to allow Cd, Fe, Mn, and Zn to cross membranes. Other transporters are the ZAT gene of *Arabidopsis* and the ZTP gene of *Thlaspi caerulescens*, which can transport Zn and Ni (Van der Zaal et al. 1999). A comparison of non-accumulator *Arabidopsis thaliana* with zinc and cadmium hyperaccumulator *Arabidopsis halleri* (Weber et al. 2004) indicated that zinc uptake and accumulation appeared to be driven by overexpression of zinc transporter from the ZIP family implicated in zinc influx into cells (Pence et al. 2000). These zinc influx transporters appear to work in combination with overexpressed P type ATP-dependent metal transporters and Nramp ion-transporters (Weber et al. 2004). Improved metal tolerance and accumulation were achieved in several plant species by modifying metal transporters like ZAT, CAX_2, FRE_1/FRE_2 or $NtCBP_4$ (Samuelsen 1998; Arazi et al. 1999; Hirschi et al. 2000; Sunkar et al. 2000).

14.2.5.3 Mercury Tolerance and Volatilization

Mercury (Hg) is one of the most toxic pollutants threatening our health and ecosystem. Genetically-modified plants with bacterial organomercurial lyase (mer B) and mercuric reductase (mer A) genes (Rugh et al. 2000) absorbed elemental Hg II and methyl mercury (MeHg) from the soil, and released volatile Hg(O) from the leaves of *A. thaliana* into the atmosphere (Heaton et al. 1998). If mer A and mer B are expressed in plants like *Brassica* oilseeds these plants could phytoremediate a mercury-contaminated site at relatively low cost compared to conventional methods.

14.2.5.4 Selenium Tolerance and Volatilization

Brassica juncea has been identified as a good candidate for decontamination of Se. Enhanced phytoremediation of Se was achieved by overexpression of enzymes catalyzing rate-limiting steps such as ATP sulfurylase (APS) and systathionine-γ-synthase (CGS) (Pilon-Smits et al. 1999b; Van Huysen et al. 2003, 2004) facilitating the reduction of selenate to selenite. Overexpression of CGS in *Brassica* promoted Se volatilization (Van Huysen et al. 2003). Transgenic Indian mustard *B. juncea* (L.) Czern. lines with overexpressed genes encoding the enzymes adenosine triphosphate sulfurylase (APS), γ-glutamyl-cystein synthetase (ECS) and glutathione synthetase (CS) accumulated 4.3-, 2.8- and 2.3-fold more of Se respectively, compared to wild-type (Banuelos et al. 2005). APS Indian mustard may tolerate metals better because of the thiol-glutathione (GSH) than wild-type (Flocco et al. 2004). Glutathione plays an important role in heavy metal detoxification. GSH can not only form metal complexes to prevent oxidative stress but is also a precursor of phytochelatins (PCs) which bind, detoxify and sequester metal ions to vacuoles.

A gene encoding the enzyme selenocysteine methytransferase (SMT) has been cloned from Se hyperaccumulator *Astragalus bioculatus* (Neuhieral et al. 1999) and when over-expressed in Indian mustard increased Se tolerance accumulation and volatilization (Le Duc 2004). The main advantage of phytovolatilization is that it can completely remove the pollutant without the need for plant disposal. Some of the genes or traits manipulated in transgenic *Brassica* plants for improved tolerance and phytoextraction are outlined in Table 14.1. However, other strategies for engineered phytoremediation with *Brassica* oilseed species have been reported for Se (Montes-Bayon et al. 2002), Hg and As (Meagher and Heaton 2005; Baneulos et al. 2005) and Cd (Wang and Liu 2009).

TABLE 14.1
Some Transgenic *Brassica* (Oilseed) Plants for Improved Phytoremediation

Sl. No.	Plant Genotype	Foreign Gene Introduced	Response Obtained	References
1.	*Brassica juncea* L. cv. 173874	*Arabidopsis* APSI encoding ATP-sulfurylase	Se hyperaccumulation in roots and shoots	Pilon-Smits et al. (1999a) Banuelos et al. (2005)
2.	-do-	*E. coli* gsh II encoding glutathione synthetase (GS)	Cd hyperaccumulation	Zhu et al. (1999a)
3.	-do-	*E. coli* gsh I encoding γ-glutamyl cysteinethione synthetase (GS)	-do-	Zhu et al. (1999b)
4.	-do-	*E. coli* gor gene encoding glutathione reductase (GR)	Cd hyperaccumulation in roots	Pilon-Smits et al. (2000)
5.	*Brassica juncea* L.	Metallothioneine MT_1 gene MT_2 gene	Increased Cd tolerance	Pan et al. (1994) Misra and Gedamu (1989)
6.	-do-	γ-glutamyl cysteine (γ-ECS) and gluthione synthetase (GS)	Increased Cd and Zn accumulation	Bennett et al. (2003)
7.	-do-	Cystathionine-γ synthase CGS gene from *A. thaliana*	Increased Se volatilization	Van Huysen et al. (2003)
8.	-do-	Carboxylase ACC deaminase gene	Protects from deleterious effect of Cd, Co, Cu, Mg, Ni, Pb, Zn	Nie et al. (2002)
9.	-do-	Mer A –CGS gene encoding cystathionine-γ-synthase	Se volatilization rates higher	Van Huysen et al. (2003)
10.	-do-	*A. bisulcatus* SMT gene encoding selenocysteine methyltransferase	-do-	Le Duc et al. (2004)

14.3 CONCLUSIONS

Phytoremediation is an emerging technology and in recent years *Brassica juncea*, *B. carinata* and *B. napus* have been recognized for their suitability for metal phytoremediation. However, work needs to be done on decreasing the length of time needed for phytoremediation, protecting wild life from feeding on plants used for remediation, and commercializing the use of *Brassica* oilseed species in phytoremediation. Developing genetically improved transgenic *Brassica* oilseeds may offer a multiprocess and faster varied environmental contaminants' phytoremedial solution in future.

REFERENCES

Ahmed, K.S., B.S. Panwar, and S.P. Gupta. 2001. Phytoremediation of cadmium contaminated soil by *Brassica* species. *Acta Agronomica Hungarica* 49: 351–60.

Alkorta, I., J. Hernandez-Allica, J.M. Becerril, I. Amezaga, I. Albizu, M. Qnaindia, and C. Garbisu. 2004. Chelate-enhanced phytoremediation of soils polluted with heavy metals. *Reviews in Environmental Science and Biotechnology* 3: 55–70.

Arazi, T., R. Sunker, B. Kaplan, and H.A. Fromm. 1999. Tobacco plasma membrane calmodulin-binding transporter confers Ni^{2+} tolerance and Pb^{2+} hypersensitivity in transgenic plants. *The Plant Journal* 20: 171–82.

Ariyakanon, N., and B. Winaipanich. 2006. Phytoremediation of copper contaminated soil by *Brassica juncea* (L.) Czern. and *Bidens alba* (L.) DC var. *radiata*. *Journal of Scientific Research of Chulalongkorn University* 31: 49–56.

Baker, A.J.M., and R.R. Brooks. 1989. Terrestrial higher plants which hyperaccumulate metallic elements. A review of their distribution, ecology and phytochemistry. *Bio-recovery* 1: 81–126.

Baker, A.J.M., S.P. McGrath, R.D. Reeves, and J.A.C. Smith. 2000. *Metal hyperaccumulator plants*: A review of the ecology and physiology of a biological resource for phytoremediation of metal polluted soils. In: *Phytoremediation of contaminated soil and water*. Terry, N. and G. Banuelos (eds.), 85–107, Lewis Publishers, Boca Raton, FL.

Banuelos, G., N. Terry, D.L. Le Duc, E.A.H. Pilon-Smits, and B. Mackey. 2005. Field trials of transgenic Indian mustard plants shows enhanced phytoremediation of selenium contaminated sediment. *Environmental Science & Technology* 39: 1771–77.

Banuelos, G.S., and Z.Q. Lin. 2007. Acceleration of selenium volatilization in seleniferous agricultural drainage sediments amended with methionine and casein. *Environmental Pollution* 150: 306–12.

Barac, T., S. Taghavi, B. Borremans, A. Provoost, L. Oeyen, J.V. Colpaert, and D. Van der Lelie. 2004. Engineered endophytic bacteria improve phytoremediation of water soluble, volatile organic pollutants. *Nature Biotechnology* 22: 583–88.

Barona, A., I. Aranguiz, and A. Elias. 2001. Metal associations in soils before and after EDTA extractive decontamination: Implications for the effectiveness for further clean-up procedures. *Environmental Pollution* 113: 79–85.

Begonia, G.B., C.D. Davis, M.F.T. Begonia, and C.N. Gray. 1998. Growth responses of Indian mustard [*Brassica juncea* (L.) Czern.] and its phytoextraction of lead from a contaminated soil. *Bulletin of Environmental Contamination and Toxicology* 61: 38–43.

Belimov, A.A., N. Hontzeas, V.I. Safronova, S.V. Demchinskaya, C. Piluzza, S. Bullitta, and B.R. Glick. 2005. Cadmium tolerant plant growth promoting bacteria associated with the roots of Indian mustard *Brassica juncea* (L.) Czern. *Soil Biology and Biochemistry* 37: 241–50.

Berti, W.R., and S.D. Cunningham. 2000. Phytostabilization of metals In: *Phytoremediation of Toxic Metals using Plants to Clean up the Environment*. Raskin, I. and B.D. Ensley (eds.), 71–88, Wiley & Sons, New York.

Bio-Wise. 2003. *Contaminated Land Remediation: A Review of Biological Technology*. London, DTI.

Blaylock, M.J. and J.W. Huang. 2000. Phytoextraction of metals. In: *Phytoremediation of Toxic Metals using Plants to Clean up the Environment.* Raskin, I. and B.D. Ensley (eds.), 53–70, Wiley & Sons, New York.

Blaylock, M.J., D.E. Salt, S. Dushenkov, O. Zakharova, C.D. Gussman, Y. Kapulnik, B.D. Ensley, and I. Raskin. 1997. Enhanced accumulation of Pb Indian mustard by soil applied chelating agents. *Environmental Science & Technology* 31: 860–65.

Bennett, L.E., J.L. Burkhead, K.L. Hale, N. Terry, M, Pilon, and E.A.H. Pilon-Smits. 2003. Bioremediation and biodegradation: Analysis of transgenic Indian mustard plants for phytoremediation of metal contaminated mine tailings. *Journal of Environmental Quality* 32: 432–40.

Borgmann, U. 2000. Methods for assessing the toxicological significance of metals in aquatic ecosystems: bioaccumulation-toxicity relationships, water concentrations and sediment spiking approaches. *Aquatic Ecosystem Health and Management* 3: 277–89.

Brooks, R.R. 1998. *Plants that Hyperaccumulate Heavy Metals. Their Role in Phytoremediation, Microbiology, Archeology, Exploration and Phytomining*. CAB. International, Oxon, UK.

Brooks, R.R., J. Lee, R.D. Reeves, and T. Jaffre. 1977. Detection of nickeliferous rocks by analysis of herbarium specimens of indicator plants. *Journal of Geochemical Exploration* 7: 49–57.

Brooks, R.R., and B.H. Robinson. 1998. Aquatic phytoremediation by accumulator plants. In: *Plants that Hyperaccumulate Heavy Metals: Their Roles in Phytoremediation, Microbiology, Archaecology, Mineral Exploration and Phytomining.* Brooks, R.R. (eds.), 289–312, CAB International, Oxon, UK.

Brown, T.A., and A. Shrift. 1982. Selenium: Toxicity and tolerance in higher plants. *Biological Reviews of the Cambridge Philosophical Society* 57: 59–84.

Burd, G.I., D.C. Dixon, and B.R. Glick. 1998. A plant growth promoting bacterium that decreases nickel toxicity in seedlings. *Applied and Environmental Microbiology* 64: 3663–68.

Chaney, R.L. 1983. Plant uptake of inorganic waste constitutes. In: *Land Treatment Hazardous Wastes.* Parr, J. F., P.B. Marsh, and J.M. Kla, (eds.), 50–76, Park Ridge, NJ: Noyes Data Corporation.

Clarkson, D.T., and U. Luttage. 1989. Mineral nutrition. Divalent cations. Transport, compartmentalization. *Progress in Botany* 51: 93–112.

Clemens, S., M.G. Palmgren, and U. Kraemer. 2002. A long way ahead; understanding and engineering plant metal accumulation. *Trends in Plant Science* 7: 309–15.

Cobbett, C.S., and P.B. Goldsborough. 2002. Phytochelatins and metallothioneins: Roles in heavy metals detoxification and homeostasis. *Annual Review of Plant Biology* 53: 159–82.

Cunningham, S.D., and D.W. Ow. 1996. Promises and prospects of phytoremediation. *Plant Physiology* 110: 715–19.

Danika, L., and Le Duc, T. Norman. 2005. Phytoremediation of toxic trace elements in soil and water. *Journal of Industrial Microbiology and Biotechnology* 32: 514–20.

De Souza, M.P., C.M. Lytle, M.M. Mulholland, M.L. Otte, and N. Terry. 2000. Selenium assimilation and volatilization from dimethyl-selenoniopropionate by Indian mustard. *Plant Physiology* 122: 1281–88.

De Souza, M.P., E.A.H. Pilon-Smits, C.M. Lytle, S. Hwang, J. Tai, T.S.U. Honma, L. Yeh, and N. Terry. 1998. Rate-limiting steps in selenium assimilation and volatilization by Indian mustard. *Plant Physiology* 117: 1487–94.

Delhaize, E. 1996. A metal accumulator mutant of *Arabidopsis thaliana. Plant Physiology* 111: 849–855.

Dembitsky, V. 2003. Natural occurrence of arseno compounds in plants, lichens, fungi, algal species, and microorganisms. *Plant Science* 165: 1177–92.

Di Simine, C.D., J.A. Sayer, and G.M. Gadd. 1998. Solubilization of zinc phosphate by a strain of *Pseudomonas fluorescence* isolated from a forest soil. *Biology and Fertility of Soils* 28: 87–94.

Dubbin, W.E., and E. Louise Ander. 2003. Influence of microbial hydroxamate siderophores on Pb(II) desorption from α-FeOOH. *Applied Geology and Chemistry* 18: 1751–56.

Dushenkov, V., N.P.B.A. Kumar, H. Motto, and I. Raskin. 1995. Rhizofilteration: The use of plants to remove heavy metals from aqueous streams. *Environmental Science & Technology* 29: 1239–45.

Ebbs, S., I. Lau, B. Ahner, and L. Kochian. 2002. Phytochelation synthesis is not responsible for the Cd tolerance in the Zn/Cd hyperaccumulator *Thlaspi caerulescens* (J and C. Preol). *Planta* 214: 635–40.

Ebbs, S.D., and L.V. Kochian. 1997. Toxicity of zinc and copper to *Brassica* species. Implication for phytoremediation. *Journal of Environmental Quality* 26: 776–81.

Ebbs, S.D., M.M. Lasat, D.J. Brady, J. Cornish, R. Gordon, and L.V. Kochian. 1997. Phytoextraction of cadmium and zinc from a contaminated soil. *Journal of Environmental Quality* 26: 1424–30.

Eide, D., M. Broderus, J. Fett, and M.L. Guerinot. 1996. A novel iron regulated metal transporter from plants identified by functional expression in yeast. *Proceedings of National Academy of Sciences, USA* 93: 5624–28.

Entry, J.A., L.S. Watrud, R.S. Manasse, and N.C. Vance. 1997. Phytoremediation and reclamation of soils contaminated with radionuclides. In: *Phyto-Remediation of Soil and Water Contaminants*. Krugler, E.L., T.A. Anderson, and J.R. Coats (eds.), 299–306, *Amer. Chem. Soc.*, Washington, DC.

Ernst, W.H.O., J.A.C. Verkleij, and H. Schat. 1992. Metal tolerance in plants. *Acta Botanica Neerlandica* 41: 229–48.

Evangelou, M.W.H., M. Ebel, and A. Schaeffer. 2007. Chelate-assisted phytoextraction of heavy metals from soil. Effect, mechanism, toxicity and fate of chelating agents. *Chemosphere* 68: 989–1003.

Ferrari, L.S., A.C. Cunha-Queda, P. Alvarenga, S. DiGregoris, and G. Vallini. 2009. Selenite resistant rhizobacteria stimulate $SeO_{(3)}^{(2-)}$ phytoextraction by *Brassica juncea* in bioaugmented water filtering artificial bed. *Environ. Science & Pollution Research* 16: 663–70.

Flocco, C.G., S.D. Lindblom, and E.A.H. Pilon-Smits. 2004. Overexpression of enzymes involved in glutathione synthesis enhances tolerance to organic pollutants in *Brassica juncea. International Journal of Phytochemistry* 6: 289–304.

Fulekar, M.H., Anamika Singh and A.M. Bhawii. 2009. Genetic engineering strategies for enhancing phytoremediation of heavy metals. *African Journal of Biotechnology* 8: 529–35.

Fuloria, A., Shweta Saraswat, and J.P.N. Rai. 2009. Effect of *Pseudomonas fluorescence* on metal phytoextraction from contaminated soil by *Brassica juncea. Chemistry and Ecology* 25: 385–96.

Gao, J., A.W. Garrison, C. Hochamer, C.S. Mazur, and N.L. Wolfe. 2000. New in Research-Phytoremediation. Uptake and Phytotransformation of o1p1-DDT and p1p1-DDT by Axenically cultivated aquatic plants. *Journal of Agricultural & Food Chemistry* 48: 6121–27.

Garg, C., and S.K. Kataria. 2009. Phytoremediation potential of *Raphanus sativus* (L.), *Brassica juncea* (L.) and *Triticum aestivum* (L.) for copper contaminated soil. E-*Proceedings of International Society of System Sciences*, University of Queensland, Brisbane (Australia), July 12–17.

George, L., and P.S. Rao. 1980. *In vitro* regeneration of mustard plants (*Brassica juncea* var. Rai-5) of cotyledon explants from non-irradiated, irradiated and mutagen treated seeds. *Annals of Botany* 46: 107–12.

Grispen, V.M., H.J. Nelissen, and J.A. Verkley. 2006. Phytoextraction with *Brassica napus* L.: a tool for sustainable management of heavy metal contaminated soils. *Environmental Pollution* 144: 77–83.

Guerinot, M.L. 2000. The ZIP family of metal transporters. *Biochimica et Biophysica Acta (BBA)—Biomembranes* 1465: 190–98.

Guerinot, M.L., and D.E. Salt. 2001. Fortified foods and phytoremediation. Two sides of the same coin. *Plant Physiology* 125: 164–67.

Gupta, A.K., and S. Sinha. 2006. Role of *Brassica juncea* (L.) Czern (var. *vaibhav*) in the phytoextraction of Ni from soil amended with fly ash selection of extract ant for metal bioavailability. *Journal of Hazardous Materials* 136: 371–78.

Hall, J.L. 2002. Cellular mechanisms for heavy metal detoxification and tolerance *Journal of Experimental Biology* 53: 1–11.

Heaton, A.C.P., C.L. Rugh, N.J. Wang, and R.B. Meagher. 1998. Phytoremediation of mercury and methylmercury polluted soils using genetically engineered plants. *Journal of Soil Contamination* 7: 497–507.

Herman, D., J. Artiola, and R. Miller. 1995. Removal of cadmium, lead and zinc from soil by a rhamnolipid biosurfactant. *Environmental Science & Technology* 29: 2280–85.

Hirschi, K.D., V.D. Korenkov, N.L. Wilganowski, and G.J. Wagner. 2000. Expression of *Arabidopsis* CAX_2 in tobacco. Altered metal accumulation and increased manganese tolerance. *Plant Physiology* 124: 125–33.

Howden, R., P.B. Goldsborough, C.R. Anderson, and C.S. Cobbett. 1995. Cadmium sensitive, cad_1 mutants of *Arabidopsis thaliana* are phytochelatin deficient. *Plant Physiology* 107: 1059–66.

Ishikawa, S., A.E. Noriharu, M. Murakami, and T. Wagatsuma. 2006. Is *Brassica juncea* a suitable plant for phytoremediation of cadmium in soils with moderately low cadmium contamination?—Possibility of using other plant species of Cd-phytoextraction. *Soil Science and Plant Nutrition* 52: 32–42.

Jadia, C.D., and M.H. Fulekar. 2008. Phytoremediation of heavy metals: Recent techniques. *African Journal of Biotechnology* 8: 921–28.

Jain, R.K., D.R. Sharma, and J.B. Chawdhury. 1989. High frequency regeneration and heritable somaclonal variation in *Brassica juncea. Euphytica* 40: 75–81.

Jan, V.V., C.C. Demacedo, J.M. Kinet, and J. Bouharmont. 1997. Selection of Al-resistant plants from a sensitive rice cultivar using somaclonal variation, *in vitro* and hydroponic cultures. *Euphytica* 97: 303–10.

Karami, A., and Z.H. Shamsuddin. 2010. Phytoremediation of heavy metals with several efficiency enhancer methods. *African Journal of Biotechnology* 9: 3689–98.

Khan, A.G. 2006. Mycorrhizoremediation—an enhanced from of phytoremediation. *Journal of Zhejiang University Science-B*. 7: 503–14.

Khilji, S., and Firdaus-e-Bareen. 2008. Rhizofilteration of heavy metals from the tannery sludge by the anchored hydrophyte, *Hydrocotyle umbellata* L. *African Journal of Biotechnology* 7: 3711–17.

Kuiper, I., L.V. Kravchenko, G.V. Bloemberg, and B.J. Lugtenberg. 2004. Rhizoremediation: A beneficial plant microbe interaction. *Molecular Plant- Microbe Interactions* 17: 6–15.

Kumar, N.P.B.A., V. Dushenkov, H. Motto, and I. Raskin. 1995. Phytoextraction: The use of plants to remove heavy metals from soils. *Environmental Science & Technology* 92: 1232–38.

Kupper, H., F. Zhao, and S.P. McGrath. 1999. Cellular compartmentation of zinc in leaves of the hyperaccumulator *Thlaspi caerulescens. Plant Physio*logy 119: 305–11.

Lasat, M.M. 2002. Phytoextraction of Toxic metals: A review of biological mechanisms. *Journal of Environmental Quality* 31: 109–20.

Le Duc, D.L., A.S. Tarun, M. Montes-Bayon, J. Meija, M.F. Malit, C.P. Wu, M. Abdel Samie, C.Y. Chiang, A. Tagmount, M. De Souza, B. Neuhierl, A. Bock, J. Caruso, and N. Terry. 2004. Overexpression of selenocysteine methyl transfers in *Arabidopsis* and Indian mustard increases selenium tolerance and accumulation. *Plant Physiology* 135: 377–83.

Lebeau, T., A. Braud, and K. Jezeguel. 2000. Performance of augmentation assisted phytoextraction applied to metal contaminated soils: A review. www.aseanenenvironment.info/abstract/41016970.pdf.

Letham, D.L.D., N.S. Pence, M.M. Lasat, and L.V. Kochian. 2005. Molecular and physiological investigations of *Thlaspi caerulescens*, a Zn/Cd hyperaccumulator. In: *Roots and Soil Management: Interactions between Roots and the Soil*. Zobel, R.F. and S.F. Wright (eds.), 95–106, Agronomy Society of American Monographs.

Lim, J.M., A.L. Salido, and D.J. Butcher. 2004. Phytoremediation of lead using Indian mustard (*Brassica juncea*) with EDTA and electrodics. *Microchemical Journal* 76: 3–9.

Liu, C.P., Z.G. Shen, and X.D. Li. 2007. Accumulation and detoxification of cadmium in *Brassica pekinensis* and *B. chinensis. Biologia Plantarum* 51: 116–20.

Lombi, E., F.J. Zhao, S.J. Dunham, and S.P. McGrath. 2001. Phytoremediation of heavy metal contaminated soils: Natural hyperaccumulation versus chemically enhanced phytoextraction. *Journal of Environmental Quality* 30: 1919–26.

Lucarini, M., R. Canali, M. Cappelloni, G. Lullo, Di, and G. Lombardi-Boccia. 1999. *In vitro* calcium availability from *Brassica Vegetables* (*Brassica oleracea* L.) and as consumed in composite dishes. *Food Chemistry* 64: 519–23.

Ma, L.Q., K.M. Komar, and C. Tu. 2001. A fern that accumulates arsenic. *Nature* 409: 579.

Marchiol, L., S. Asssolari, P. Sacco, and G. Zerbi. 2004. Phytoremediation of heavy metals by canola (*Brassica napus*) and radish (*Raphanus sativus*) grown on multi contaminated soil. *Environmental Pollution* 132: 21–27.

Marr, L.C., C.B. Elizabeth, R.C. Anderson, M.A. Widdowson, and J.T. Novak. 2006. Direct volatilization of naphthalene to the atmosphere at a phytoremediation site. *Environmental Science & Technology* 40: 5560–66.

McGrath, S.P. 1998. Phytoextraction for soil remediation In: *Plants that Hyperaccumulate Heavy Metals: Their Role in Phytoremediation, Microbiology, Archeology, Exploration and Phytomining*. Brooks R.R. (ed.), 261–88, CAB International, New York.

Meagher, R.B., and A.C. Heaton. 2005. Strategies for engineered phytoremediation of toxic element pollution, mercury and arsenic. *Journal of Industrial Microbiology and Biotechnology* 32: 502–13.

Meagher, R.B., C.L. Rugh, M.K. Kandasamy, G. Gragson, and N.J. Wang. 2000. Engineered phytoremediation of mercury pollution in soil and water using bacterial genes. In: *Phytoremediation of Contaminated Soil and Water*. Terry, N. and G. Banuelos (eds.), 201–21, Lewis Publishers, Boca Raton, FL.

Mejare, M., and L. Bulow. 2001. Metal binding proteins and peptides in bioremediation and phytoremediation of heavy metals. *Trends in Biotechnology* 19: 67–73.

Misra, S., and L. Gedamu. 1989. Heavy metal tolerant transgenic *Brassica napus* L. and *Nicotiana tabaccum* L. plants. *Theoretical and Applied Genetics* 78: 161–68.

Montes-Bayon, M., E.G. Yanes, C. Ponce de Leon, K. Jayasimhulu, A. Stalcup, J. Shann, and J.A. Caruso. 2002. Initial studies of selenium speciation in *Brassica juncea* by LC with ICPMS and ES-MS detection: an approach for phytoremediation studies. *Analytical Chemistry* 74: 107–13.

Mulligan, C.N., R.N. Yong, B.F. Gibbs, S. James, and H.P.J. Benett. 1999. Metal removal from contaminated soils and sediments by biosurfactant surfaction. *Environmental Science & Technology* 33: 3812–20.

Nedkovska, M., and A. Atanassov. 1998. Metallothionein genes and expression for heavy metal resistance. *Biotechnology* 11: 11–16.

Nehnevajova, E., R. Herzig, K.H. Erismann, and J.P. Schwitzguebel. 2007. *In vitro* breeding of *Brassica juncea* L. to enhance accumulation and extraction properties. *Plant Cell Reports* 26: 429–37.

Neuhieral, B., M. Thanbichler, F. Lottspeich, and A. Bock. 1999. A family of S-methyl methionine dependent thiol/selenol methyl transferase. *Journal of Biological Chemistry* 274: 5407–14.

Nie, L., S. Shah, A. Rashid, G.I. Burd, G. Dixon, and B.R. Glick. 2002. Phytoremediation of arsenate contaminated soil by transgenic *Canola* and plant growth promoting bacterium *Enterobacter cloacae* CAL_2. *Plant Physiology & Biochemistry* 40: 355–61.

Nriagu, J.O. 1979. Global inventory of natural anthropogenic emissions of trace metals to the atmosphere. *Nature* 279: 409–11.

Nwoko, C.O. 2010. Trends in phytoremediation of toxic elemental and organic pollutants. *African Journal of Biotechnology* 9: 6010–16.

Pan, A.H., M.Z. Yang, F. Tie, L.G. Li, Z.L. Chen, and B. Ru. 1994. Expression of mouse metallothionein-1 gene confers cadmium resistance in transgenic tobacco plants. *Plant Molecular Biology* 24: 341–51.

Panwar, B.S., K.S. Ahmed, and S.B. Mittal. 2002. Phytoremedration of nickel contaminated soils by *Brassica* sps. *Environment, Development and Sustainability* 4: 1–16.

Pedron, F., C. Petruzzelli, M. Barbafieri, and E. Tassi. 2009. Strategies to use phytoextraction in very acidic soil contaminated by heavy metals. *Chemosphere* 75: 808–14.

Pence, N.S. 2002. Molecular physiology of Zn/Cd hyperaccumulation in *Thlaspi caerulescens*. Ph.D. Thesis, Cornell University, U.S.A.

Pence, N.S., P.B. Larsen, S.D. Ebbs, D.L. Letham, M.M. Lasat, D.F. Garvin, D. Eide, and L.V. Kochian. 2000. The molecular physiology of heavy metal transport in the Zn/Cd hyperaccumulator *Thlaspi caerulescens*. *Proceedings of National Academy of Sciences, USA* 97: 4956–60.

Pilon-Smits, E.A.H., C. Banuelos, and D.R. Parker. 2010. *Uptake, Metabolism and Volatilization of Selenium by Terrestrial Plants*. ARS Public.

Pilon-Smits, E.A.H., M.P. De Souza, G. Hong, A. Amini, T. Leustek, N. Terry, and R.O. Bravo. 1999a. Selenium volatilization and accumulation by twenty aquatic plant species. *Journal of Environmental Quality*. 28: 1011–17.

Pilon-Smits E.A.H., S. Hwang, C.M. Lutle, Y. Zhu, J.C. Tai, R.C. Bravo, Y. Chen, T. Leustek, and N. Terry. 1999b. Overexpression of ATP-sulfurylase in Indian mustard leads to increased selenate uptake, reduction and tolerance. *Plant Physiology* 119: 123–32.

Pilon-Smits E.A.H., and M. Pilon. 2002. Phytoremediation of metals using transgenic plants. *Critical Reviews in Plant Science* 21: 439–56.

Pilon-Smits E.A.H., Y.L. Zhu, T. Sears, and N. Terry. 2000. Overexpression of glutathione reductase in *Brassica juncea*. Effects of cadmium accumulation and tolerance. *Physiologia Plantarum* 110: 445–60.

Prasad, M.N.V., and H.M. Freitas. 2003. Metal hyperaccumulation in plants. Biodiversity prospecting for phytoremediation technology. *Electronic Journal of Biotechnology* 6: 275–321.

Purakayastha, T.J., T. Viswanath, S. Bhadraray, P.K. Chhonkar, P.P. Adhikari, and K. Suribabu. 2008. Phytoextraction of zinc, copper, nickel and lead from a contaminated soil by different species of *Brassica*. *International Journal of Phytoremediation* 10: 61–72.

Quartacci, M.F., B. Irtelli, C. Gonnelli, R. Gabbrielli, and F. Navari-Izzo. 2009. Naturally assisted metal phytoextraction by *Brassica carinata*: Role of root exudate. *Environmental Pollution* 157: 2697–703.

Ramos, I., E. Esteban, J.J. Lucena, and A. Garate. 2002 Cadmium uptake and subcellular distribution in plants of *Lactuca* sp. Cd-Mn interaction. *Plant Science* 162: 761–67.

Raskin, I., and B.D. Ensley. 2000. *Phytoremediation of Toxic Metals: Using Plants to Clean up the Environment.* John Wiley & Sons, New York.

Raskin, I., P.B.A.N. Kumar, S. Dushenko, and D.E. Salt. 1994. Bioconcentration of heavy metals by plants. *Current Opinion in Biotechnology* 5: 285–90.

Rauser, W.E. 1995. Phytochelatins and related peptides. *Plant Physiology* 109: 1141–49.

Rugh, C.L., S.P. Bizily, and R.B. Meagher. 2000. Phytoremediation of environmental mercury pollution. In: *Phytoremediation of toxic metals: Using plants to clean-up the environmental*. Ensley, B. and Raskin, I. (eds.), 151–196, John Wiley & Sons, New York.

Rugh, C.L., H.D. Wilde, N.M. Stack, D.M. Thompson, A.O. Summers, and R.B. Meagher. 1996. Mercuric ion reduction and resistance in transgenic *Arabidopsis thaliana* plants expressing a modified bacterial mer A gene. *Proceedings of National Academy of Sciences, USA* 9: 3182–87.

Sadowshky, M.J. 1998. Phytoremediation, past promises and future practices. In: *Abstracts of Eighth International Symposium on Microbial Ecology* (ISME-8), 9–14 Aug. 1998, 288–89, Halifax, Canada.

Salido, A.L., K.L. Hasty, J. Lim, and D.J. Butcher. 2003. Phytoremediation of arsenic and lead in contaminated soil using Chinese brake ferns (*Pteris vittata*) and Indian mustard (*Brassica juncea*). *International Journal of Phytoremediation* 5: 89–103.

Salt, D.E., M. Blaylock, N.P.B.A. Kumar, V. Dushenkov, B.D. Ensley, and I. Raskin. 1995a. Phytoremediation: A novel strategy for the removal of toxic metals from the environment using plants. *Biotechnology* 13: 468–74.

Salt, D.E., R.C. Prince, I.J. Pickering, and I. Raskin. 1995b. Mechanisms of cadmium mobility and accumulation in Indian mustard. *Plant Physiology* 109: 1427–33.

Salt, D.E., R.D. Smith, and I. Raskin. 1998. Phytoremediation. *Annual Review Plant Physiology and Plant Molecular Biology* 49: 643–68.

Samuelsen, A.I., R.C. Martin, D.W.S. Mok, and C.M. Mok. 1998. Expression of the yeast FRE genes in transgenic tobacco. *Plant Physiology* 118: 51–58.

Sarma, H. 2011. Metal hyperaccumulation in plants: A review focusing on phytoremediation technology *Journal of Environmental Science & Technology* 4: 118–38.

Schnoor, J.L., L.A. Licht, S.C. Mc Cutcheon, N.L. Wolfe, and L.H. Carreira. 1995. Phytoremediation of organic and nutrient contaminants. *Environmental Science & Technology* 29: 318–23.

Shah, K., and J.M. Nongkynrih. 2007. Metal hyperaccumulation and bioremediation. *Biologia Plantarum* 51: 618–34.

Shaw, A.J. 1990. Heavy metal tolerance in plants: Evolutionary aspects. CRC Press, Boca Raton, FL.

Sheng, X.F., and J.J. Xia. 2006. Improvement of rape *Brassica napus* plant growth and cadmium uptake by cadmium-resistant bacteria. *Chemosphere* 64: 1036–1042.

Shilev, S., M. Benllock, Palomares, and E.D. Sancho. 2009. Phytoremediation of metal contaminated soil for improving food safety. In: *Predictive Modeling and Risk Assessment*. Costa, R. and K. Kristbergsson (ed.), Springer Sci.+Business Media, New York.

Steven, C.M., and J.L. Schnoor. 2003. *Phytoremediation Transformation and control of Contaminants*. John Wiley & Sons, New York.

Su, D.C., and J.W.C. Wong. 2004. Selection of mustard oil seed rape (*Brassica juncea* L.) for phytoremediation of cadmium contaminated soil. *Bulletin of Environmental Contamination and Toxicology* 72: 991–98.

Sunkar, R., B. Kaplan, N. Bouche, T. Arazi, D. Dolev, I.N. Talke, J.M. Frans, F.J.M. Maathuis, D. Sanders, D. Bouchez, and H. Fromm. 2000. Expression of a truncated tobacco $NtCBP_4$ channel in transgenic plants and disruption of the homologous *Arabidopsis* $CNGC_1$ gene confer Pb^{2+} tolerance. *The Plant Journal* 24: 533–42.

Suresh, B., P.D. Sherkhane, S. Kale, S. Eapen, and G.A. Ravi Shanker. 2005. Uptake and degradation of DDT by hairy root cultures of *Cichorium intybus* and *Brassica juncea*. *Chemosphere* 61: 1288–92.

Tang, G.X., W.J. Zhou, H.Z. Li, B.Z. Mao, Z.H. He, and K. Yoneyama,. 2003. Medium explants and genotype factors influencing shoot regeneration in oil seed *Brassica* sp. *Journal of Agronomy and Crop Science* 189: 351–58.

Terry, N., and G. Banuelos. 2000. *Phytoremediation of Contaminated Soil and Water*. Lewis Publishers, New York.

Terry, N., C. Carlson, T.K. Raab, and A.M. Zayed. 1992. Rates of selenium volatilization among crop species. *Journal of Environmental Quality* 21: 341–344.

Terry, N., S.V. Sambukumar, and D.L. Le Duc. 2003. Biotechnological approaches for enhancing phytoremediation of heavy metals and metalloids. *Acta Biotechnologica* 23: 281–88.

Terry, N., and A. Zayed. 1998. Phytoremediation of selenium. In: *Environment Chemistry of Selenium*. Frankenberger, W.T. Jr. and R.A. Enberg (eds.), Marcel Dekker Inc., New York.

Thomas, J.C., E.C. Davies, F.K. Malick, C. Endreszi, C.R. William, M. Abbas, S. Petella, K. Swisher, M. Perron, R. Edwards, P. Ostenkowski, N. Urban czyk, W.N. Wiesend, and K.S. Murray. 2003. Yeast metallothione in transgenic tobacco promotes copper uptake from contaminated soils. *Biotechnology Progress* 19: 273–80.

Turan, M., and A. Esringu. 2007. Phytoremediation based on canola (*Brassica napus* L.) and Indian mustard (*Brassica juncea* L.) planted on spiked soil by aliquot amount of Cd, Cu, Pb and Zn. *Plant, Soil and Environment* 53: 7–15.

Utsunamyia, T. 1980. Japanese patent application no: 55-72959.

Van der Zaal, B.J., L.W. Neuteboom, J.E. Pinas, A.N. Chardonnes, H. Schat, J.A.C. Verkleij, and P.J.J. Hooykaas. 1999. Overexpression of a novel *Arabidopsis* gene related to putative zinc–transporter genes from animals can lead to enhanced zinc resistance and accumulation. *Plant Physiology* 119: 1047–55.

Van Huysen, T., S. Abdel-Ghany, K.L. Hale, D. Le Duc, N. Terry, and E.A.H. Pilon-Smits. 2003. Overexpression of cystathionine-γ-synthase enhances selenium volatilization in *Brassica juncea*. *Planta* 218: 71–78.

Van Huysen, T., N. Terry, and E.A.H. Pilon-Smits. 2004. Exploring the selenium phytoremediation potential of transgenic Indian mustard overexpressing ATP sulfurylase or cystathionine-γ-synthase. *International Journal of Phytoremediation* 6: 1–8.

Vazquez, M.D., C. Poschenrieder, J. Barcelo, A.J.M. Baker, P. Halton, and G.H. Cope. 1994. Compartmentation of zinc in roots and leaves of zinc hyperaccumulator *Thlaspi caerulescens* (J & C Presl.). *Botanica Acta* 107: 243–50.

Wagner, G.J., D. Salt, G. Gries, K. Donachie, R. Wang, and X. Yan. 1995. Biochemical studies of heavy metal transport in plants In: *Proceedings/Abstracts of the 14th Annual symposium, Current Topics in Plant Biochemistry, Physiology and Molecular Biology*, 21–22, Univ. of Missouri, Columbia.

Wang, B., and L. Liu. 2009. Improved phytoremediation of oilseed rape (*Brassica napus*) by *Trichoderma* mutant constructed by restriction enzyme-mediated integration (REMI) in cadmium polluted soil. *Chemosphere* 74: 1400–3.

Weber, M., E. Harada, C.V. Vess, E. Van Roepenack-Lahaye, and S. Clemens. 2004. Comparative microarray analysis of *Arabidopsis thaliana* and *Arabidopsis halleri* roots identifies niotinamine synthase, a ZIP transporter and other genes as metal hyperaccumulation factors. *The Plant Journal* 37: 269–81.

Wenzel, W.W., R. Unterbrunner, P. Sommer, and P. Sacco. 2003. Chelate-assisted phytoextraction using canola (*Brassica napus* L.) in outdoors pot and field-lysimeter experiments. *Plant and Soil* 249: 83–96.

Zaidi, S., S. Usmani, B.R. Singh, and J. Musarrat. 2006. Significance of *Bacillus subtilis* strain SJ-101 as a bioinoculant for concurrent plant growth promotion and nickel accumulation in *Brassica juncea*. *Chemosphere* 64: 991–97.

Zaier, H., T. Ghnaya, K. Ben-Rejeb, A. Lakhdar, S. Rejeb, and F. Jamal. 2010. Effects of EDTA on phytoextraction of heavy metals (Zn, Mn and Pb) from sludge amended soil with *Brassica napus*. *Bioresource Technology* 101: 3978–83.

Zayed, A.M., C.M. Lytte, and N. Terry. 1998. Accumulation and volatilization of different chemical sps. of selenium by plants. *Planta* 206: 284–92.

Zayed, A.M., and N. Terry. 1994. Selenium volatilization in root and shoot effects of shoot removal and sulfate level. *Journal of Plant Physiology* 143: 8–14.

Zenk, M.H. 1996. Heavy metal detoxification in higher plants—a review. *Gene* 179: 21–30.

Zhu, Y.L., E.A.H. Pilon-Smits, L. Jouanin, and N. Terry. 1999a. Overexpression of glutathione synthetase in Indian mustard enhances cadmium accumulation and tolerance. *Plant Physiology* 119: 73–79.

Zhu, Y.L., E.A.H. Pilon-Smits, A.S. Tarun, S.U. Weber, L. Jouanin, and N. Terry. 1999b. Cadmium tolerance and accumulation in Indian mustard is enhanced by overexpressing-γ-glutamylcysteine synthetase. *Plant Physiology* 121: 1169–77.

Zhu, Y.L., A.M. Zayed, J.H. Qian, M. De Souza, and N. Terry. 1999c. Phytoaccumulation of trace elements by wetland plants: II. Water Hyacinth (*Eichhornia crassipens*). *Journal of Environmental Quality* 28: 339–44.

Zhuang, X., J. Chen, H. Shim, and Z. Bai. 2007. New advances in plant growth promoting rhizobacteria for bioremediation. *Environment International* 33: 406–13.

15 Enhanced Phytoextraction Using *Brassica* Oilseeds
Role of Chelates

Zhao Zhongqiu and Liu Xiaona

CONTENTS

15.1 INTRODUCTION

With the rapid development of industrialization, urbanization, and human activities, more and more soils were polluted by heavy metals. Heavy metal-contaminated soil has been a widespread global problem. The restoration of heavy metal–contaminated soils is a global hot issue. Currently, there are physical, chemical, and biological remediation technologies, but the traditional physical or chemical methods can be very costly and also destructive to the soil. Phytoextraction, one way of bioremediation, is recognized the green eco-friendly *in situ* remediation method (Zaier et al. 2010b). However, the efficiency of phytoextraction is generally considered as too slow duo to the inherent characteristics of hyperaccumulator plants: the hyperaccumulator plants grows slowly, the long growth cycle and the small biomass, leading to the small quantity of heavy metals in the shoot, affecting the actual practice value. So the researchers begin to study the more effective method: using the general plant with fast growing and large biomass and other assistive technologies to strengthen the extraction, such as the chemistry-plant and the microorganism-plant joint reparation (Kim et al. 2010). In the chemistry-plant joint remediation, the chelating agents are widely used. The chelate can increase the bioavailability of heavy metals and the plant absorption, improving the effectiveness of the phytoextraction.

15.2 CHELATE AND ITS CHEMICAL AFFINITY FOR METALS

Chelate is a macromolecular compound, with the multidentate ligands, which has the ability to chelate with a variety of metal ions, and could combine with the metal ions in the soil particles to form metal chelates (Zheng et al. 2009; Zaier et al. 2010a). Through the combination of chelating agents and heavy metal ions, chelate can change the existing forms of the heavy metals in soil and improve the bioavailability of heavy metals, to enhance the effect of plant extraction.

Chelate can be divided into two categories: The first category is the natural small molecule organic acids, such as citric acid, oxalic acid, tartaric acid, malic acid, MDA. The other category is multicarboxyl amino acids, such as EDTA, EDDS, NTA. Different chelating agents have a different affinity for different metals, which is related to the stability of metal chelates. The higher the stability factor, the greater ability of chelate to activate the corresponding metals (Ding et al. 2009).

After the chelating agent is added into the soil, it can chelate with the heavy metals, forming the water-soluble metal-chelate complex, changing the occurrence form of heavy metals in the soil, increasing the bioavailability of heavy metals, and thus can enhance the plants absorption of target heavy metals (Jalali et al. 2007; Meers et al. 2008; Sarkar et al. 2008). The main mechanisms of heavy metal phytoextraction enhanced by the addition of chelating agents include the interactions between plant and soil. The process in the soil is with the chelating agent being added into the soil, heavy metals will be resolved from the soil particles to soil solution, thus greatly increasing the possibility of the plant absorption of heavy metals. The process in the plant is the root absorbs the heavy metals in the soil solution, then heavy metals are transferred and stored (Hu et al. 2010).

15.2.1 EDTA and Other Synthetic Aminopolycarboxylic Acids

A lot of research has shown that the chelating agent aminopolycarboxylic acid (APCAs), ethylene diamine tetraacetate (EDTA) can greatly increase the absorption and plant accumulation of heavy metals, improving the effectiveness of the phytoextraction. But EDTA is found not suitable for long-term use, owing to its high environmental persistence and nonbiodegradabe property, which may lead to secondary contamination. In the late 1980s and early 1990s EDTA was suggested as a chelating agent for inducing the phytoextraction process. In many literatures, pot experiments described the influence of EDTA that has ranged from no significance to over 100-fold enhanced accumulation (Evangelou et al. 2007). The wide variation maybe due to the metal species, the metal content in the soil, the soil itself, the amount of EDTA applied, as well as the time and frequency of EDTA applied. Although the uptake is increased by the application of EDTA, while the heavy metals amount mobilized in the soil is higher than the uptake. Although EDTA has been proved to be effective in enhancing phytoextraction, EDTA and EDTA-heavy metal complexes are toxic to soil microorganisms (Evangelou et al. 2007) and plants by severely decreasing shoot biomass (Epstein et al. 1999; Chen and Cutright 2001). Meanwhile, its prolonged presence in the soil increases the leaching risk of heavy metals. In a soil experiment of Meers et al. (2005), after EDTA application, the mobilized metals (Zn, Cu, Cd, Ni) did not or decrease very slightly in 40 d. This shows the high environmental risk of EDTA in soil.

A number of different synthetic chelates such as hydroxylethylene diamine tetraacetic acid (HEDTA), diethylene triamino pentaacetic acid (DTPA), *trans*-1,2-cyclohexylene dinitrilo tetraacetic acid (CDTA), ethylenebis[oxyethylenetrinitrilo] tetraacetic acid (EGTA), ethylenediamine-*N,N'*bis(*o*-hydroxyphenyl)acetic acid (EDDHA), *N*-(2-hydroxyethyl) iminodiacetic acid (HEIDA), and *N,N'*-di(2-hydroybenzyl) ethylene diamine *N,N'*-diacetic acid (HBED) have been tested. As in the case of these chelating agents, including EDTA, the uptake efficiency depends on the studied plants and the specific heavy metals in the soil (Evangelou et al. 2007). Huang et al. (1997) reported the order in increasing Pb accumulation in both pea (*P. sativum L.*) and corn (*Z. mays*): EDTA > HEDTA > DTPA > EGTA > EDDHA. The effectiveness depends not only on the chelating agent and heavy metal but also on the plant. This finding is in agreement with Huang et al. (1997), Sekhar et al. (2005) and Shen et al. (2002). These chelating agents have shown the partially good effect in inducing the uptake by the plants, but have also shown the toxicity, which is comparable to that of EDTA. Shen et al. (2002), Chen and Cutright (2001) and Huang et al. (1997) noticed, after applying EDTA, HEDTA, or DTPA, a reduction in the biomass of the studied plants.

15.2.2 Ethylene Diamine Disuccinates and Other Natural Aminopolycarboxylic Acids

In the last about 10 years, several experiments have been performed with the addition of Ethylene diamine disuccinate (EDDS), which has proven to be effective in enhancing the uptake of several metals. Similar to EDTA, EDDS increased the uptake of heavy metals, however, relating to the phytoavailable amounts of heavy metals in the soil, only a fraction of the mobilized metals are absorbed by the plants and translocated to the shoot. In a study by Luo et al. (2005), by adding 5 mmol/kg of EDTA and EDDS to the soil, the uptake of Cu, Cd, Zn and Pb increased significantly, while EDDS was more effective at increasing the concentration of Cu and Zn in shoots, while EDTA was more effective with respect to the metals Pb and Cd. These findings are concurrent with the results of Meers et al. (2005), Hauser et al. (2005) and Kos and Leštan (2004). The toxicity of EDDS depends on the plants studied. Meers et al. (2005) found no significant decrease in *Heliunthus annuus* dry weight by the application of EDDS (applying one week before harvesting) in concentrations between 1.77 and 8.88 mmol/kg. In such a short time period, the toxicity effects may not be visible. However, in the study of Luo et al. (2005), the dry weight of corn (*Z. mays*) and bean (*P. vulgaris* L.) decreased significantly through the application of 5 mmol/kg EDDS. At the end of the experiment, 14 d after the application of chelates, plants showed chlorosis and necrosis. EDDS is a biodegradable agent, relating to EDTA, which is friendly to the surrounding environment.

Besides EDDS, NTA (nitrilotriacetic acid) is also a biodegradable chelating agent. In spite of its expected positive properties, few studies have been performed with NTA as the ligand to assist phytoextraction. In the study of Meers et al. (2005), NTA (1.8 mmol/kg) did not significantly increase the uptake of Zn, Cu, or Cd in comparison to the control in sunflower (*H. annuus*), but the uptake of Ni was increased 2.5-fold. In phytoextraction experiments, comparing to the control and the NTA (2.6 mmol/kg) treatment, Robinson et al. (2000) observed no significant difference in the Cd concentration both for leaves or stems. As NTA, the amount of mobilized heavy metals in soil is higher than the amount extracted by the plants. Although Quartacci et al. (2007) proved that EDDS is more effective than NTA in the phytoextraction of a soil contaminated by As, Cd, Cu, Pb and Zn. The treatment with 5 mmol/kg EDDS accumulated 157, 122 and 129 mg/kg of Cu, Pb and Zn, respectively, which was about 2-fold that of the corresponding NTA treatments. Recently, the other new biodegradable chelates, iminodisuccinic acid (IDSA) and aspartic acid diethoxvsuccinate (AES) has been found effective in phytoextracting Zn and Cd of contaminated soils, so it may be another alternative for the unbiodegradable chelates (Zhao et al. 2010a,b). The performed studies have shown the potential of chelates to increase the bioavailability of heavy metals in soil and to increase the translocation from the roots to shoots, but have also revealed many drawbacks and have left many questions unanswered. The plants are only able to take up a small proportion of the phytoavailable heavy metals in the soil, leaving a high degree of mobile metals in the soil, increasing the risk of leaching of metals. Additionally, studies so far have shown that not all plants react in the same way to the chelating agents applied. They vary in toxicity-symptoms and uptake effectiveness depending on the added chelating agent, the amount applied and the soil conditions. Thus, the detailed application needs the specific analysis (Evangelou et al. 2007).

15.3 CHELATE-ASSISTED PHYTOEXTRACTION USING *BRASSICA* OILSEEDS

It is greatly important to select suitable plants for chelate-assisted phytoextraction. Compared to other plants, *Brassica* plants, dry-land species, has a short life, large biomass on the aboveground, strong tolerance to heavy metals and can accumulate a variety of heavy metals (Pb, Cu, and Zn) at the same time from contaminated soils together with reasonable biomass yields (Sarkar et al. 2008), which has a great potential in phytoremediation. There is an important practical significance in using *Brassica* oilseeds to be the repair plant.

In the compound pollution soil with Cd and Pb, Guo et al. (2009) found that, in the polluted soil with high concentrations of Pb and Cd made the Indian mustard (*B. juncea*) and rape showing the symptoms of poisoning-small plant, slow growth and yellowing leaves, even the individual rape died, while all of Indian mustard can complete the growth period, showing a strong patience. Meanwhile, for the effect of Cd, the aboveground biomass of Indian mustard and rape significantly reduced, however, the Indian mustard biomass is larger, about 1.1- to 2-fold, than that which is very beneficial to phytoremediation. With the increasing of concentration of Pb and Cd, the Indian mustard showed the properties of hyperaccumulator. The purification of Cd and Pb of *Brassica* oilseeds was 0.35%–9.22% and 0.015%–0.356%, respectively, which was 2.1–3.5 times and 1.4–5.5 times higher than the same treatment of the rape, respectively.

Quartacci et al. (2007) tested the potential of nine different species of accumulate metals (As, Cd, Cu, Pb and Zn) in the shoots, showing that of the nine species, *Brassica carinata* accumulated the highest amounts of metals in shoots without suffering a significant biomass reduction. Studies by Moreno et al. (2005) on the volatilization of mercury using Indian mustard (*B. juncea*) demonstrated that phytovolatilization of mercury is an effective and economical technique, with maximum extraction yield of 25 g Hg h^{-1}. Table 15.1 gives the outlook the effects of the addition of different chelators on the mobilization and uptake of heavy metals by *Brassica* plants in various studies.

TABLE 15.1
Effects of the Addition of Different Chelators on the Mobilization and Uptake of Heavy Metals by *Brassica* Plants in Various Studies

Chelators	Amount Added (mmol/kg)	Heavy Metal	Plant Available Heavy Metal in Soil	Heavy Metal Uptake	Plants	Side Effects	References
EDTA	3	Pb	~23-fold (H_2O-extractable)	26-fold	Indian mustard (*B. juncea*)	No side effect (≧10mmol/kg biomass reduction)	Epstein et al. (1999)
	0.13	Cd	400-fold (H_2O-extractable)	Up to 2-fold	Indian mustard (*B. juncea*)	Biomass reduction	Jiang et al. (2003)
	4 times 10	Cd	5400-fold	3-fold	Chinese cabbage (*B. rapa*)	Biomass reduction, necrosis and chlorosis	Vodnik et al. (2003)
		Zn	50-fold (H_2O-leaching)	3-fold			
		Pb	3500-fold	59.7-fold			
	3	Cu	30-fold	2.6-fold	Indian mustard (*B. juncea*)	No toxicity symptoms	Wu et al. (2004)
		Zn	1.3-fold	Not sig.			
		Pb	5.7-fold	2.8-fold			
		Cd	Not sig. (NH_3NO_4-extractable)	Not sig.			
EGTA	5 10	Pb	Several 100-fold (H_2O-extractable)	Several 1000-fold	Indian mustard (*B. juncea*)	Biomass reduction	Blaylock et al. (1997)
CDTA	5 10	Pb	Several 100-fold (H_2O-extractable)	Several 1000-Fold	Indian mustard (*B. juncea*)	Biomass reduction	Blaylock et al. (1997)

(continued)

TABLE 15.1 (Continued)
Effects of the Addition of Different Chelators on the Mobilization and Uptake of Heavy Metals by *Brassica* Plants in Various Studies

Chelators	Amount Added (mmol/kg)	Heavy Metal	Plant Available Heavy Metal in Soil	Heavy Metal Uptake	Plants	Side Effects	References
EDDS	4 times 10	Pb Cd Zn	250-fold Cd 5000-fold 4-fold (H_2O-leaching)	10.3-fold 3-fold 3-fold	Chinese cabbage (*B. rapa*)	Biomass reduction, chlorosis necrosis	Grčman et al. (2003)
NTA	Several treatments of 4.2 and 8.4	Cd Zn Cu	78-fold 37-fold 9-fold	2- to 3-fold	Indian mustard (*B. juncea*)	No toxicity symptoms	Kayser et al. (2000)
	10 20	Cd	8-fold 14-fold	2-fold 3.3-fold	Indian mustard (*B. juncea*)	No toxicity symptoms	Quartacci et al. (2005)
Citric acid	10 20	Cd	No sig. change 1.5-fold (H_2O-extractable)	1.5-fold 3.5-fold	Indian mustard (*B. juncea*)	No toxicity symptoms	Quartacci et al. (2005)
	20	U	200-fold (H_2O-extractable)	1000-fold 900-fold 700-fold	Indian mustard (*B. juncea*) Chinese cabbage (*B. chinesis*) Chinese mustard (*B. narinosa*)	No toxicity symptoms	Huang et al. (1998)

Source: Evangelou, M.W.H., M. Ebel, and A. Schaeffer, *Chemosphere* 68: 989–1003, 2007.

15.4 POTENTIAL MECHANISMS FOR CHELATE-ASSISTED PHYTOEXTRACTION BY *BRASSICA* OILSEEDS

Brassica oilseeds have BJPCS1 genes that can improve its tolerance to heavy metals, in the plant phytochelatins (PC) with heavy metals chelation, and further transported to the vacuole storage, reducing the concentration of heavy metals in the cytoplasm, achieving the detoxifying effect. In a study of Sun (2010), the phytochelatin synthase gene come from the heavy metal hyperaccumulator *Brassica* oilseeds was added into the tobacco, and the result of PCR and Northern proved that the gene has been integrated into the tobacco genome and expression at the transcriptional level. In the stress condition with three kinds of metals (Cd, Zn, and Ni), transgenic tobacco plants proline, soluble sugcr content, MDA content, relative conductivity and chlorophyll content and other indicators are better than untransformed control plants, indicating that BJPCS1 increase the resistance to heavy metals of tobacco.

Accordingly, *Brassica* oilseeds show high potential to tolerate and accumulate metal in their tissues by triggering mechanisms for toxic metals detoxification such adequate compartmentation in the vacuole or in cell wall and peptide detoxification with phytochelatines.

15.5 IMPLICATIONS FOR PHYTOREMEDIATION OF CHELATE-ASSISTED PHYTOEXTRACTION BY *BRASSICA* OILSEEDS

Phytoremediation is an important technology to remediate the contaminated soils, which we should promote more mature, making the value of practical applications greater. On the one hand, we have to seek more environmentally friendly chelating agents, which do not cause too much damage to the plant and the microorganisms, and do not bring the secondary pollution to the surrounding environment. On the other hand, we should aim for looking for more plants with fast-growing, large biomass and strong tolerance to heavy metals, and then analyze the mechanism for these characteristics, and give the explanation on the molecular, determining the specific genes, and further through the genetic engineering, making more general plants have this characteristic, at that condition, the quantity of plants that can be used to phytoextraction maybe expand, thus promoting the application of chelator in phytoextraction. Besides, we should actively explore the technologies that can be used to strengthen the effectiveness of phytoextraction, and study the possibility of combining multiple technologies that can promote the viability and efficiency of phytoextraction, achieving the aim of better remediation of contaminated soils.

15.6 CONCLUSIONS

Numbers of studies have shown the potential of chelators to increase the bioavailability of heavy metals in soil and thereby to increase heavy metals accumulations in plants. *Brassica* plants are very popularly used plants in the studies of chelators assisted phytoextraction due to its large biomass in underground and strong tolerance to heavy metals and accordingly high effectiveness. Chelator assisted phytoextraction has described as a promising technique to remediate heavy metal–contaminated soils. However, the main drawback, heavy metal leaching and second contamination of underground water, has still been the barrier for its practice for site remediation. Continue discovering more environmentally friendly chelators and exploring combining multiple technologies that can promote the efficiency of phytoextraction and solve the problem heavy metal leaching may be the important issue for ongoing of this field.

REFERENCES

Blaylock, M.J., D.E. Salt, S. Dushenkov, O. Zakharova, C. Gussman, Y. Kapulnik, B.D. Ensley, and I. Raskin. 1997. Enhanced accumulation of Pb in Indian mustard by soil-applied chelating agents. *Environmental Science & Technology* 31: 860–865.

Chen, H., and T. Cutright. 2001. EDTA and HEDTA effects on Cd, Cr, and Ni uptake by *Helianthus annuus*. *Chemosphere* 45: 21–28.

Ding, Z.H., X. Hu, and D.Q. Yin. 2009. Application of chelants in remediation of heavy metals-contaminated soil (in Chinese). *Ecology and Environmental Sciences* 18(2): 777–782.

Epstein, A.L., C.D. Gussman, M.J. Blaylock, U. Yermiyahu, J.W. Huang, Y. Kapulnik, and C.S. Orser. 1999. EDTA and Pb-EDTA accumulation in *Brassica juncea* grown in Pb-amended soil. *Plant and Soil* 208: 87–94.

Evangelou, M.W.H., M. Ebel, and A. Schaeffer. 2007. Chelate assisted phytoextraction of heavy metals from soil: Effect, mechanism, toxicity, and fate of chelating agents. *Chemosphere* 68: 989–1003.

Grčman, H., D. Vodnik, S. Velikonja-Bolta, and D. Leštan. 2003. Ethylenediaminedissuccinate as a new chelate for environmentally safe enhanced lead phytoextraction. *Journal of Environmental Quality* 32: 500–506.

Guo, Y.J., B.W. Li, and H. Yang. 2009. Study on the effects of Cadmium and Lead absorption and accumulation by *Brassica juncea* and its phytoremediation efficiency. *Journal of soil and water conservation* 23(4): 130–135.

Hauser, L., S. Tandy, R. Schulin, and B. Nowack. 2005. Column extraction of heavy metals from soils using the biodegradable chelating agent EDDS. *Environmental Science & Technology* 39: 6819–6824.

Hsiao, K.H., P.H. Kao, and Z.Y. Hseu. 2007. Effects of chelators on chromium and nickel uptake by *Brassica juncea* on serpentine-mine tailings for phytoextraction. *Journal of Hazardous Materials* 148: 366–376.

Hu, Y.H., S.H. Wei, Q.X. Zhou, J. Zhan, L.H. Ma, R.C. Niu, Y.M. Li, and S.S. Wang. 2010. Application of chelator in phytoremediation of heavy metals contaminated soils: a review. *Journal of Agro-Environment Science* 29(11): 2055–2063.

Huang, J.W., M.J. Blaylock, Y. Kapulnik, and B.D. Ensley. 1998. Phytoremediation of uranium-contaminated soils: role of organic acids in triggering uranium hyperaccumulation in plants. *Environmental Science & Technology* 32: 2004–2008.

Huang, J.W., J. Chen, W.B. Berti, and S.D. Cunningham. 1997. Phytoremediation of lead-contaminated soils: role of synthetic chelates in lead phytoextraction. *Environmental Science & Technology* 31: 800–805.

Jalali, M., and Z.V. Khanlari. 2007. Redistribution of fractions of zinc, cadmium, nickel, copper, and lead in contaminated calcareous soils treated with EDTA. *Archives of Environmental Contamination and Toxicology* 53: 519–532.

Jiang, X.J., Y.M. Luo, Q.G. Zhao, A.J.M. Baker, P. Christie, and M.H. Wong. 2003. Soil Cd availability to indian mustard and environmental risk following EDTA addition to Cd-contaminated soil. *Chemosphere* 50: 813–818.

Kayser, A,, K. Wenger, A. Keller, W. Attinger, H.R. Felix, S.K. Gupta, and R. Schulin 2000. Enhancement of phytoextraction of Zn, Cd and Cu from calcareous soil: the use of NTA and sulfur amendments. *Environmental Science & Technology* 34: 1778–1783.

Kim, K.R., G. Owens, and S.I. Kwon. 2010. Influence of Indian mustard (*Brassica juncea*) on rhizosphere soil solution chemistry in long-term contaminated soils: A rhizobox study. *Journal of Environmental Sciences* 22(1): 98–105.

Kos, B., and D. Leštan. 2003. Phytoextraction of lead, zinc and cadmium from soil by selected plants. *Plant, Soil and Environment* 49: 548–553.

Kos, B., and D. Leštan. 2004. Chelator induced phytoextraction and *in situ* soil washing of Cu. *Environmental Pollution* 134: 333–339.

Lai, H.Y., and Z.S. Chen. 2005. The EDTA effect on phytoextraction of single and combined metals-contaminated soils using rainbow pink (*Dianthus chinensis*). *Chemosphere* 80: 1062–1071.

Luo, C., Z. Shen, and X. Li. 2005. Enhanced phytoextraction of Cu, Pb, Zn and Cd with EDTA and EDDS. *Chemosphere* 59: 1–11.

Meers, E., A. Ruttens, M.J. Hopgood, D. Samson, and F.M.G. Tack. 2005. Comparison of EDTA and EDDS as potential soil amendments for enhanced phytoextraction of heavy metals. *Chemosphere* 58: 1011–1022.

Meers, E, F.M.G. Tack, and M.G. Verloo. 2008. Degradability of ethylenediaminedisuccinic acid (EDDS) in metal contaminated soils: Implications for its use soil remediation. *Chemosphere* 70: 358–363.

Quartacci, M.F., A.J.M. Baker, and F. Navari-Izzo. 2005. Nitriloacetate- and citric acid-assisted phytoextraction of cadmium by Indian mustard (*Brassica juncea (L.) Czernj, Brassicaceae*). *Chemosphere* 59: 1249–1255.

Quartacci, M.F., B. Irtelli, A.J.M. Baker, and F. Navari-Izzo. 2007. The use of NTA and EDDS for enhanced phytoextraction of metals from a multiply contaminated soil by *Brassica arinata*. *Chemosphere* 68: 1920–1928.

Robinson, B.H., T.M. Millis, D. Petit, L.E. Fung, S.R. Green and B.E. Clothier. 2000. Natural and induced cadmium-accumulation in poplar and willow: implications for phytoremediation. *Plant and Soil* 227: 301–306.

Sarkar, D., S.S. Andra, S.K.M. Saminathan, and R. Datta. 2008. Chelant-aided enhancement of lead mobilization in residential soils. *Environmental Pollution* 156: 1139–1148.

Sekhar, K.C., C.T. Kamala, N.S. Chary, V. Balaram, and G.Garcia. 2005. Potential of Hemidesmus indicus for the phytoextraction of lead from industrially contaminated soils. *Chemosphere* 58: 507–514.

Shen, Z.G., X.D. Li, C.C. Wang, H.M. Chen, and H. Chua. 2002. Lead phytoextraction from contaminated soils with high biomass plant species. *Journal of Environmental Quality* 31: 1893–1900.

Sun, T. 2010. *Heavy metal stress adaptation of Auxin regulating gene BjGH3.1 and BjEXPA1 in Indian mustard* (in Chinese). Beijing: PhD, Graduate University of Chinese Academy of Sciences.

Vodnik, D., S. Velikonja-Bolta, and D. Leštan. 2003. Ethylenediaminedissuccinate as a new chelate for environmentally safe enhanced lead phytoextraction. *Journal of Environmental Quality* 32: 500–506.

Wu, L.H., Y.M. Luo, X.R. Xing, and P. Christie. 2004. EDTA-enhanced phytoremediation of heavy metal contaminated soil with Indian mustard and associated potential leaching risk. *Agriculture, Ecosystems & Environment* 102: 307–318.

Zaier, H., T. Ghnaya, K. Ben Rejeb, A. Lakhdar, S. Rejeb, and F. Jemal. 2010b. Effects of EDTA on phytoextraction of heavy metals (Zn, Mn and Pb) from sludge-amended soil with *Brassica napus*. *Bioresource Technology* 101: 3978–3983.

Zaier, H., T. Ghnaya, A. Lakhdar, R. Baioui, R. Ghabriche, M. Mnasri, S. Sghair, S. Lutts, and C. Abdelly. 2010a. Comparative study of Pb-phytoextraction potential in *Sesuvium portulacastrum* and *Brassica juncea*: Tolerance and accumulation. *Journal of Hazardous Materials* 183: 609–615.

Zhao, Z.Q., M.Z. Xi, Y.Z. Huang, Z.K. Bai, and G.Y. Jiang. 2010b. The potential of new biodegradable chelator AES for phytoextraction of heavy metals in contaminated soils (in Chinese). *Environmental Chemistry* 29: 407–411.

Zhao, Z.Q., M.Z. Xi, G.Y Jiang, X.N. Liu, Z.K. Bai, and Y.Z. Huang. 2010a. Effects of IDSA, EDDS and EDTA on heavy metals accumulation in hydroponically grown maize (*Zea mays, L.*). *Journal of Hazardous Materials* 181: 455–459.

Zheng, X.L., and K. Zhu. 2009. The application of chelating agents in the phytoremediation of heavy metal contaminated soils. *Environmental Science and Management* 34(8): 106–109.

16 Organic Acid–Assisted Phytoremediation in Salt Marshes

From Hydroponics to Field Mesocosm Trials

M. Caçador, B. Duarte, and J. Freitas

CONTENTS

16.1 INTRODUCTION

In highly industrialized estuaries, there is also a large input of heavy metals, which are accumulated in salt marsh sediments. These high inputs make salt marshes key zones for the biogeochemistry of the estuary, but also for metal cycling. When accumulated in salt marsh sediments, metals can become adsorbed to the sediment constituents and taken up by plant roots and translocated to above-ground plant organs. This plant metal uptake is a very important factor in estuarine remediation. *Halimione portulacoides* and *Spartina maritima* are two of the more abundant species in the Mediterranean salt marshes. Plants are known to exudate by their roots low-molecular-weight organic acids (LMWOA) to scavenge metallic elements important for their metabolism and/or in other cases to maintain these metallic elements outside their tissues to avoid toxicity cases. This work explores LMWOA's natural interaction with metallic ions and how they can influence the phytoremediation potential of these two halophytes. Two trials are addressed in this chapter, starting in an hydroponic greenhouse trial to test the potential enhancement of the pytoremediation process using *H. portulacoides*, assisted by citric acid addition. Another assisted phytoremediation trial intended to scale up the hydroponic trial to more field realistic conditions. In this trial intact cores of *S. maritima* individuals and sediments were added with several LMWOA. With this experiment became evident that with the addition of citric or acetic acid the phytoremediation potential of this halophyte can be greatly increased. Both these trial showed that assisted phytoremediation is a promising technique for enhancing the remediative process in contaminated salt marshes.

16.1.1 Salt Marshes as Sinks of Heavy Metals

Salt marshes located in estuaries frequently receive large inputs of nutrients, particulate and dissolved organic matter (Tobias et al. 2001). This high-nutrient input makes salt marsh one of the most productive ecosystems of the planet. In highly industrialized estuaries, there is also a large input of heavy metals that are accumulated in salt marsh sediments (Doyle and Otte 1997). These high inputs make salt marshes key zones for the biogeochemistry of the estuary, but also for metal cycling (Weis and Weis 2004; Caçador et al. 2009; Duarte et al. 2010). When accumulated in salt marsh sediments, metals can become adsorbed to the sediment constituents and taken up by plant roots and translocated to aboveground plant organs. This plant metal uptake is a very important factor in estuarine remediation. It reduces the input of metals back into the water column and may provide a long-term sink (Caçador et al. 2009; Duarte et al. 2008). This uptake depends on various factors namely sediment physical and chemical characteristics and metal speciation (Reboreda and Caçador 2007a; Du Laing et al. 2009). The chemical speciation is very variable along the salt marsh and affected by the plant coverage (Reboreda and Caçador 2007b). Plants can modify metal speciation throughout several processes, like radial oxygen loss (ROL) and exudation of organic acids (Sundby et al. 1998; Jacob and Otte 2004; Duarte et al. 2007).

16.1.2 Halophyte Biogeochemical Effect on Metal Mobility

Two of the most important and abundant species present in Tagus salt marshes are *H. portulacoides* (Caryophylaller: Chenopodiaceae) and *S. maritima* (Poales: Poaceae). Previous studies (Caçador et al. 2000; Padinha et al. 2000; Caçador et al. 2009; Duarte et al. 2010) showed that the major pool of metals in the salt marsh is the sediment and the major living pool is the root tissue of the halophytes. Although this general compartmentation, there are important differences among the halophytes. Given the effect of biomass, pools of Cu, Cd and Pb in the stems and leaves of *H. portulacoides* often are significantly higher when compared to the same parts of *S. maritima* (Figure 16.1). Other studies also found lower values in *Spartinaalterniflora* when compared with *Phragmitesaustralis*

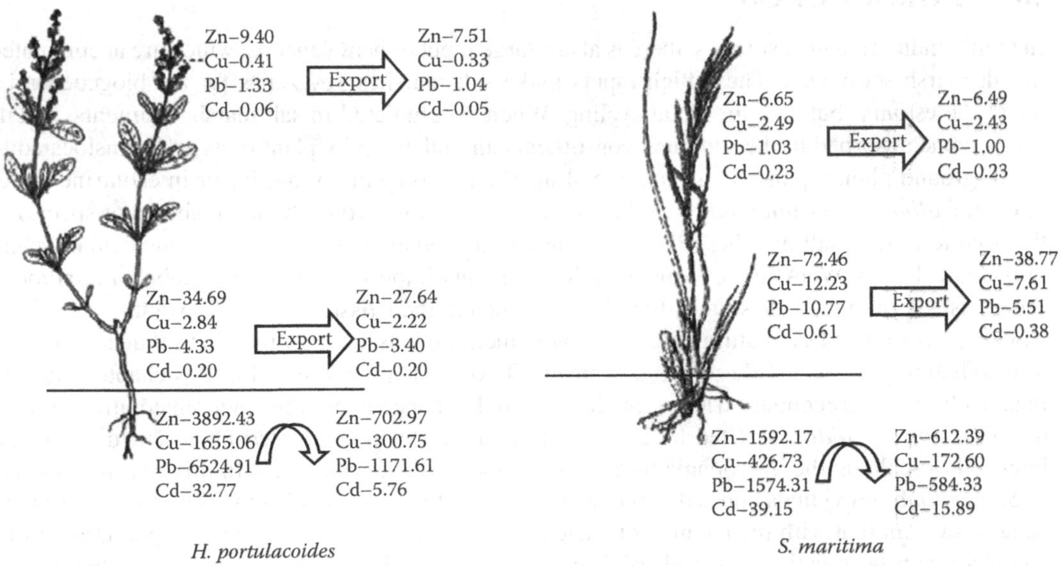

FIGURE 16.1 Metal primary accumulation (MPA, mg) and losses due to litter generation for the studied species and plant organs.

(Windham et al. 2003). As a result of this distribution, the overall amount of metals in above-ground parts of plants from those areas colonized by *H. portulacoides* was significantly higher than in areas colonized by *S. maritima* (Reboreda and Caçador 2007a). Considering similar amounts of heavy metals along a specific salt marsh these differential accumulation suggest a specific influence of the species root system in the sediment biogeochemistry and metal mobility.

16.2 ASSISTED PHYTOREMEDIATION: ENHANCING A NATURAL PROCESS

Many of the so-called heavy metals are elements with important roles in plant metabolism, such as Fe, Mn, Cu, Ni and Zn. Plants have developed strategies to acquire these elements form the rhizosphere medium, developing ionic channels, specific protein carriers and other organic transporters (Marschner 1995). Included in this last category are LMWOAs, which are also described as exudates. Through the exudation of a wide variety of compounds, roots may regulate the soil microbial community in their immediate vicinity, cope with herbivores, encourage beneficial symbioses, change the chemical and physical properties of the soil, and inhibit the growth of competing plant species (Nardi et al. 2000). The ability to secrete a vast array of compounds into the rhizosphere is one of the most remarkable metabolic features of plant roots, with nearly 5% to 21% of all photosynthetically fixed carbon being transferred to the rhizosphere through root exudates (Marschner 1995). Although this large amount of carbon based molecules deflected from photosynthesis, LMWOA correspond only to a very small part of the variety of molecules exudated by the root system (Mucha et al. 2010). Having in mind that one of the major functions of these molecules is mineral acquisition, several studies began to emerge were these small amounts are enhanced either by chemical additions or genetic engineering (Duarte et al. 2007, 2011; Rugh 2004; Evangelou et al. 2007; Chen et al. 2006 and several others). These supplementations intent, not only, to improve the harvest and uptake of metals from the rhizosphere to the root tissues, but also enhance its translocation to the aerial organs, so it can be harvested and extracted. This remediative process was denominated assisted phytoremediation (Figure 16.2).

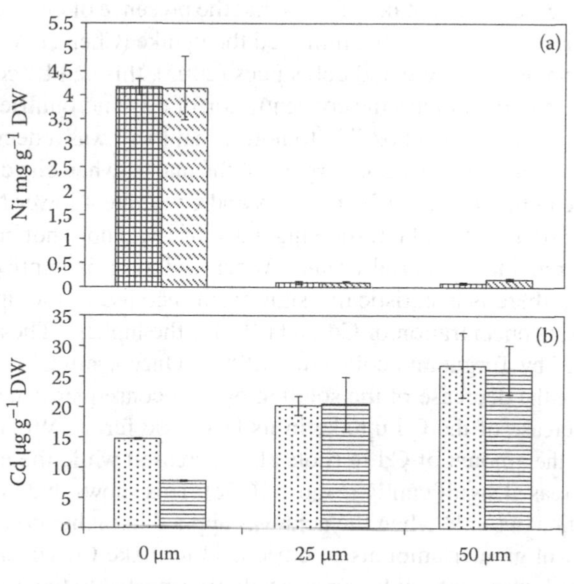

FIGURE 16.2 Aerial organs metal concentrations upon the application of different organic acids.

16.3 HYDROPONICAL TRIALS: A STARTING POINT

Basic research on the development of phytoremediation technologies has utilized the centuries old practice of growing plants in nutrient solutions (water containing fertilizers), with or without the use of a solid medium to provide mechanical support, called hydroponics (hydro = water, ponos = labor, i.e., working water). The hydroponic technique offers a number of advantages in plant screening tests: (1) bioavailability of the contaminant is greatly increased; (2) there is better control of the environment (i.e., root zone, nutrient feeding, light, temperature, humidity, and air composition); and (3) a simpler root-zone environment is created to study plant-mediated processes without the complexity introduced by soil minerals. In the salt marsh specific case, one of the species that is widely abundant and with a high rate of success in hydroponical propagation is *H. portulacoides* (L.) Allen (Chenopodeaceae). Its fast-growing metabolism and high production of root biomass allied to its high metal uptake and accumulation (Caçador et al. 2009; Duarte et al. 2010) make it suitable for hydroponical-assisted phytoremediation trials (Duarte et al. 2007). It can be propagated using prop technique, using its nods to induce root development by simply emerging it on a Hoagland nutrient solution. This technique was already employed in previous studies, for example for studying the effect of the application of citric acid on the uptake and translocation of cadmium and nickel, to heavy metals with appreciated concentrations in the sediments colonized by *H. portulacoides* at Tagus Estuary. These studies showed that, for example, the application of citric acid with any of the concentrations of nickel leads to a statistically significant decrease in the uptake of nickel as shown in Figure 16.1a. This uptake decreased by approximately 97% with the application of citric acid in both concentrations. This decrease in the uptake is contradictory to past studies, in which the nickel uptake is enhanced via Ni complexation with citric acid (Boominathan and Doran 2003). This might be due to one of two reasons. One is that citric acid complexation increased the mobility of nickel leading to an excess of nickel mobilization, turning this metal from a nutrient state to a phytotoxic form (Pushenreiter et al. 2001). This increase of mobility is also directly related to an increase of the solution acidification, caused by the application of citric acid. The pH values determined in previous monitoring studies of soil samples between *H. portulacoides* roots are near 6.5 and 6 in nonvegetated sediment (Prasad 2001). In our experiment the pH values are between 4 and 5, showing that the acidification of the solutions is due to the citric acid added. The other possible reason for this decrease in the uptake of Ni is that the presence of citric acid resulted in the formation of citric acid–nickel complexes that inhibited the uptake (Chen et al. 2003). Comparing this data with the results obtained by Ingle and colleagues (2005), this can be considered as avoidance behavior, allowing the plant to maintain the concentration of this micronutrient below excessive levels. Observing Figure 16.3a, it is also possible to notice that there was a decrease of approximately 90% of the nickel concentration in the aerial organs of the plants, when citric acid was applied. This is a result of a decrease in the uptake of Ni by roots and consequent lower Ni concentrations to be transported to the upper parts of the plant, showing that citric acid does not have any effect on nickel translocation from the roots to the aerial organs. When cadmium is applied with 25 ?M of citric acid, it can be seen that there is a statistically significant increase in the uptake (Figure 16.4b) of about 27% for the lowest concentration of Cd and 61% for the highest. These results are consistent with the results obtained by Turgut and colleagues (2004). Once again Chen et al. (2003) attributes this to a consequence of the decrease of the solution pH and consequent increase of mobility, leading to a sustainable increase of the Cd uptake in its less toxic forms. When these citric acid concentrations are higher, the amount of Cd in roots also increased while the content Cd of the aerial organs of the plant decreased significantly (Figure 16.3b). This shows that citric acid treatment also inhibited Cd translocation but only when the acid was applied in higher concentrations. This might be due to the necessity of greater amounts of citric acid to make Cd widely available and phytotoxic, contrarily to nickel, therefore not being so easily transported to the upper organs of the plant. According to previous studies (Belimov et al. 2003; Rosens et al. 2003), this is clearly a mechanism of tolerance to cadmium. In these studies, it has been stated that for cadmium tolerant species, there

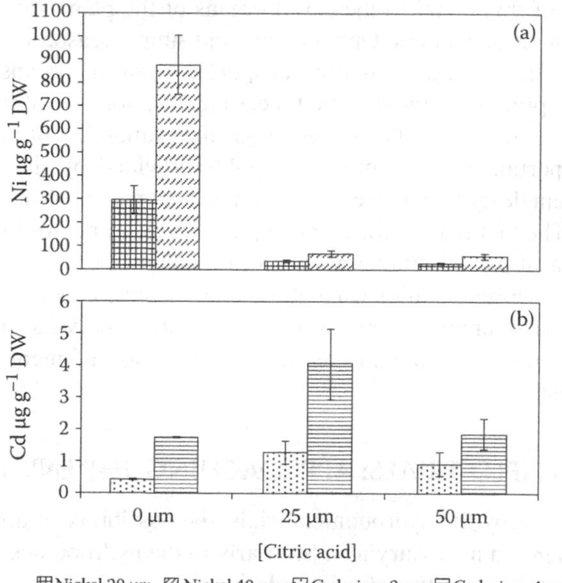

FIGURE 16.3 Nickel and cadmium aerial organs accumulation with the application of different concentrations of citric acid.

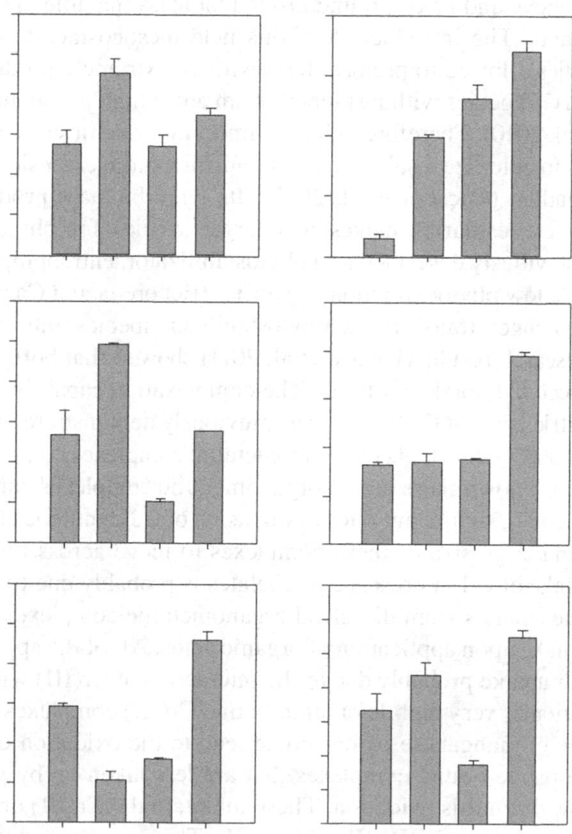

FIGURE 16.4 Nickel and cadmium root accumulation with the application of different concentrations of citric acid.

is a high translocation of this metal to the aerial organs of the plant, this fact being considered a tolerance mechanism by these authors. One possible and simple reason for this is that nickel being an essential nutrient for the plant is normally transported to all the organs, while Cd not being an essential nutrient is not preferentially absorbed, needing chelators to enter in greater amounts, but also becoming phytotoxic when the chelator increases the availability of the metal.

Although these important but preliminary insights provided by the hydroponical trials, the advantages of this system described above can somehow become disadvantages when transposed to the real environment. The fact that all the metal supplied is available for the trial since there are no interactions with sediment particles/minerals and/or organic matter can be useful to study toxicity effects and maximum phytoremediation capacities using extreme concentrations, but in field this is not so linear. The salt marsh environment is very variable principally due to the tidal flooding, with repercussions in high amounts of organic matter, metal inputs sediment interactions and uncontrolled root environment.

16.4 MESOCOSM FIELD TRIALS: APPROACHING THE ENVIRONMENT

With the insights provided by the hydroponical trials, the possibility of application of LMWOA in field mesocosmos becomes a necessary test. Similarly to the hydroponics, this method also has its advantages such as the (1) possibility of using a large number of species, (2) introducing environmental factors such as the sediment and all its components, (3) utilization of naturally contaminated sediments with realistic heavy metal forms. One of the basic principles of phytoremediation is the choice of the species. Typically a good phytoremediator should be a species that produces high amounts of both above and belowground ground biomass that tolerates and accumulates high amounts of contaminants. The introduction of this field mesocosmos trial scheme in halophyte phytoremediation studies allowed to preform tests with an extremely productive species, *S. maritima* (Loisel). This is a C4 species with fast metabolism and a high production of biomass (Caçador et al. 2009; Duarte et al. 2010). Therefore, it has an important role in salt marsh establishment. It is one of the first species to colonize a salt marsh, establishing large extensions of *S. maritima* banks scattered along the mudflat (Caçador et al. 2007). Its large biomass production and widespread occurrence in Portuguese estuaries makes it a target species for phytoremediation purposes. However, it has been previously described as a phytostabilizator, entrapping metals in its rhizosediments which results in a low phytoextraction capability (Reboreda and Caçador 2007a,b). This fact poses a different challenge: transform a phytostabilizer species into a phytoextractor using LMWOA. Previous research results (Duarte et al. 2011) showed that both the application of citric and acetic acid enhanced Cd uptake by roots. The complexation capability of acetic acid for Cd^{2+} is lower than that of citric acid for Cd^{2+}. This was previously demonstrated by Han et al. 2005, who concluded that lower stability constants of organometallic complexes coincided with the greater the metal uptake. To what is known there are no organometallic complexes carriers in the root plasma membrane (Bell et al. 2003), significant rates of diffusion of Cd-synthetic chelates are unlikely, due to the high polarity and large size of these complexes to move across the lipid layer (Han et al. 2006). The higher uptake of Cd in presence of chelates is probably due to the fact that plant roots are able to liberate trace metals from dissolved organometallic complexes (Nor and Cheng 1986). Cr exhibited higher uptake upon application of organic acids. All of the applications showed important increases in the Cr uptake probably due to the interaction of Cr(III) with these organic ligands resulting in the formation of very mobile organic-bound Cr(III) complexes (Srivastava et al. 1999). Although the presence of manganese oxides could lead to the oxidation of Cr(III) to Cr(VI) and formation of Cr(VI) organic-bound complexes that are less taken up by plants, there are several side reactions that slow down this oxidation. These maintain the Cr(III) organic-bound complexes as the more abundant species of Cr(III) (Bartlett 1991). This is in agreement with the present data, which shows a high increase of Cr uptake upon organic acids application, due to the formation of this highly mobile organic-bound Cr(III) complexes. As for Cd, due to the size and polarity of

Pb(II)-organic ligand complexes absorption is also very unlikely to occur. However, it is possible that the diffusion of Pb-acetate complex was more rapid than the diffusion of Pb^{2+} alone, because of the high charge density of this ion. This will allow these complexes to move more rapidly towards the roots. Once these complexes arrive at the root surface, this surface will promote their dissociation into free Pb^{2+} which is more easily absorbed or taken up by roots. The mechanism is very similar to the one verified for Cd (Wan et al. 2007). Only acetic acid application promoted an effective enhancement of Pb uptake, probably due to the low stability constant of complex and consequent higher facilitation for the plant uptake (Han et al. 2005). As Cu, Ni and Zn are nutrients for the plants, there are very few references of enhanced phytoextraction of these elements using LMWOA. Nascimento (2006) also found an increase in the uptake of these metals upon citric and acetic acid application, although comparatively in smaller amounts. As seen for Cd also citric acid enhanced Zn uptake by roots, which was also observed in a leaching trial preformed by Schwab et al. 2008. These authors observed that effluents from sediments treated with citric acid had higher concentrations of Zn. Similarly, Ahumada et al. (2001) found the same effect for Cu phytoextraction with the application of citric acid. In this work the higher plant uptake of Cu from soils treated with citric acid was attributed to the solution of the Cu bound to carbonates and to manganese oxides. Previous works in salt marshes showed that Ni speciation has a strong seasonal pattern (Duarte et al. 2008). In the autumn, this metal is mostly found in the less adsorbed chemical fractions of the sediments (Duarte et al. 2008) and easily removed by acetic acid, as described in the Tessier fractioning scheme (Tessier 1979). As shown by the results, the higher amounts of phytoextraction were verified with the application of acetic acid, while with the application of the remaining tested LMWOA there was no significant effect. This can be attributed to the higher solubility of the more abundant chemical fractions of Ni, and consequent higher phytoavailability and root uptake. Also the translocation of Cd to the aerial organs was only higher in plants treated with citric and acetic acid. The translocation of this heavy metal to the photosynthetic organs is very unlikely to occur at high concentrations. Chen et al. (2003) suggested that LMWOA, like citric and acetic acid, could alleviate the Cd toxicity and stimulate it transportation from root to shoot. Previous work showed that most of the Cr found in the roots was present in the form of Cr acetate (Duarte et al. 2010). This form is mostly stored in the cortex of the roots and lately translocated to the aboveground parts by conversion into Cr oxalate. In the plants treated with acetic acid, there was a very high increase in Cr content not only in the roots but also in the shoots. This should probably be attributed to this mechanism of acetate-oxalate conversion, in contrast to what was found in the other LMWOA treatments where only the root Cr content increased. Although there was no significant effect on Ni concentrations in the roots when citric acid is applied, this metal concentration in the above-ground organs was very high when compared to the control. This may be due to the fact that Ni translocation is mainly mediated by citric acid (Duarte et al. 2007). Although there wasn't an evident increase of Ni concentration in roots, it is possible to observe an increase of its total concentration in the whole plants upon application of citric acid. This is probably due to the rapid translocation of citrate-Ni complexes to the aboveground organs, maintaining the root Ni concentration low. In the sediments amended with acetic acid, the high concentrations verified in the aerial organs can be attributed to a higher uptake of this metal and consequent dilution in order to low its toxicity. This could also be the mechanism involving Pb translocation, once the Pb-acetate complexes are dissociated in the root surface. The high Pb uptake by roots could Pb to a phytotoxic situation if it was not readily allocated and redistributed to the aerial and belowground organs. Although the effective increase in Zn and Cu root uptake, upon citric acid and citric or acetic application respectively, there was no significant translocation increase in neither of the cases. This indicates that the complexes between metals and organic acids are dissociated or degraded in the root system, restoring their normal mobility into the aboveground organs. Because both these elements are essential nutrients for plants their circulation inside the plants is mediated by specific transporters (Marschener 1995) without need of chelating agents or coordination with LMWOA (Figure 16.5).

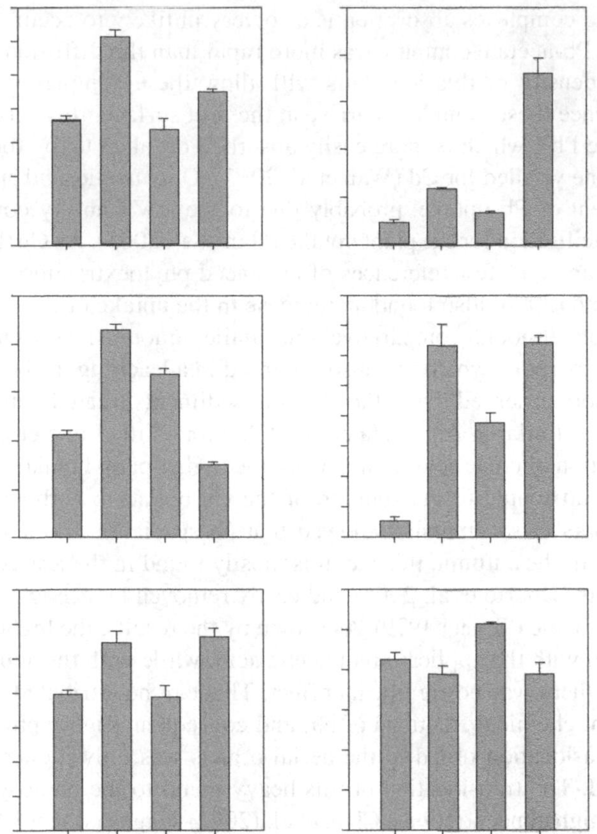

FIGURE 16.5 Root metal concentrations upon the application of different organic acids.

Although *S. maritima* is known to stabilize metal contaminants in its rhizosediments, it is possible to enhance its metal uptake with the application of LMWOA. Particularly, citric acid and, more markedly, acetic acid were found to be efficient enhancers of the phytoextraction process, although this process does not have the same efficiency for all tested metals. For example, the 10-fold increase in Cr uptake is observed when acetic acid is applied against a very low uptake of Pb in all tested conditions.

16.5 CONCLUSIONS

The potential environmental risk must be recognized at the early stage of LMWOA application. With this application the labile complexes associated metals could be absorbed and taken up directly by plants. The introduction of new techniques that could facilitate the decomposition of these organometal complexes in short term, then the proportion of free ions would increase and the uptake by plants would also be enhanced, thus minimizing the environmental risk. Metals can therefore be removed of the system by harvesting of the above-ground biomass. The high above-ground productivity is by this fact another strong point in favour of this application, allowing a large pool of metals available for harvesting. This points to out to a research need to make the use of these environmentally-friendly phytoextraction enhancers feasible for commercial phytoextraction. In addition, the use of natural compounds in contraposition to synthetic chelates sounds better for the public acceptance of phytoextraction as a technology to clean up metal-polluted soils.

REFERENCES

Ahumada, I., J. Mendoza, P. Escudero, and A. Loreto. 2001. Effect of acetate, citrate, and lactate incorporation on distribution of Cd and Cu chemical forms in soil. *Communications in Soil Science and Plant Analysis* 32: 771–785.

Belimov, A., V. Safronova, E. Tsyganov, Y. Borisnov, A. Kozhemyakov, V. Stepanok, M. Martenson, V. Gianinazzi-Pearson, and A. Tikhonovich. 2003. Genetic variability in tolerance to cadmium and accumulation of heavy metals in pea (*Pisum sativum* L). *Euphytica* 131: 25–35.

Bell, P., M. McLaughlin, G. Cozens, D. Stevens, G. Owens, and H. South. 2003. Plant uptake of 14C-EDTA, 14CCitrate, and 14C-Histidine from chelator-buffered and conventional hydroponic solutions. *Plant and Soil* 253: 311–319.

Boominathan, R., and P. M. Doran. 2003. Organic acid complexation, heavy metal distribution and the effect of ATPase inhibition in hairy roots of hyperaccumulator plant species. *Journal of Biotechnology* 101: 131–146.

Caçador, I., M. Caetano, B. Duarte, and C. Vale. 2009. Stock and losses of trace metals from salt marsh plants. *Marine Environmental Research* 67: 75–82.

Caçador, I., C. Vale, and F. Catarino. 2000. Seasonal variation of Zn, Pb, Cu and Cd concentrations in the root–sediment system of Spartina maritima and Halimione portulacoides from Tagus estuary salt marshes. *Marine Environmental Research* 49: 279–290.

Chen, Y., Q. Lin, Y. Luo, Y. He, S. Zhen, Y. Yu, G. Tian, and M. Wong. 2003. The role of citric acid on the phytoremediation of heavy metal contaminated soil. *Chemosphere* 50: 507–811.

Chen, Y., Y. Wang, W. Wu, Q. Lin, and S. Xue. 2006. Impacts of chelate-assisted phytoremediation on microbial community composition in the rhizosphere of a copper accumulator and non-accumulator. *Science of The Total Environment* 356: 247–255.

Doyle, M., and M. Otte. 1997. Organism-induced accumulation of Fe, Zn and AS in wetland soils. *Environmental Pollution* 96: 1–11.

Du Laing, G., J. Rinklebe, B. Vandecasteele, E. Meers, and F. Tack. 2009. Heavy metal mobility and availability in estuarine and riverine floodplain soils and sediments: a review. *Science of Total Environment* 407: 3972–3985.

Duarte, B., M. Caetano, P. Almeida, C. Vale, and I. Caçador. 2010. Accumulation and biological cycling of heavy metal in the root-sediment system of four salt marsh species, from Tagus estuary (Portugal). *Environmental Pollution* 158: 1661–1668.

Duarte, B., M. Delgado, and I. Caçador. 2007. The role of citric acid in cadmium and nickel uptake and translocation, in Halimione portulacoides. *Chemosphere*, 69: 836–840.

Duarte, B., J. Freitas, and I. Caçador. 2011. The role of organic acids in assisted phytoremediation processes of salt marsh sediments. *Hydrobiologia* 674: 169–177.

Duarte, B., R. Reboreda, and I. Caçador. 2008. Seasonal variation of Extracellular Enzymatic Activity (EEA) and its influence on metal speciation in a polluted salt marsh. *Chemosphere*, 73: 1056–1063.

Evangelou, M., M. Ebel, and A. Schaeffer. 2007. Chelate assisted phytoextraction of heavy metals from soil. Effect, mechanism, toxicity, and fate of chelating agents. *Chemosphere* 68: 989–1003.

Han, F., X. Shan, S. Zang, B. Wen, and G. Owens. 2006. Enhanced Cd accumulation in maize roots—the impact of organic acids. *Plant and Soil* 289: 355–368.

Han, F., X. Shan, J. Zhang, Y. Xie, Z. Pei, S. Zhang, Y. Zhu, and B. Wen. 2005 Organic acids promote the uptake of lanthanum by barley roots. *New Phytologist* 165: 481–492.

Ingle, R., S. Mugford, J. Rees, M. Campbell, and A. Smith. 2005. Constitutively high expression of the histidine biosynthetic pathway contributes to nickel tolerance in hyperaccumulator plants. *Plant Cell* 17: 2089–2106.

Jacob, D., and M. Otte. 2004. Influence of Typha latifolia and fertilization on metal mobility in two different Pb Zn mine tailing types. *Science of Total Environment* 333: 9–24.

Marschner, H. 1995. *Mineral Nutrition in Higher Plants*, 2nd Ed. Academic Press Limited, London.

Mucha, A., M. Almeida, A. Bordalo, and M. Vasconcelos. 2010. LMWOA (low molecular weigh organic acid) exudation by salt marsh plants: natural variation and response to Cu contamination. *Estuarine, Coastal and Shelf Science* 88: 63–70.

Nardi, S., G. Concheri, D. Pizzeghello, A. Sturaro, R. Rella, and G. Parvoli. 2000. Soil organic matter mobilization by root exudates. *Chemosphere* 5: 653–658.

Nascimento, C. 2006. Organic acids effects on desorption of heavy metals from a contaminated soil. *Scientia Agricola* 63: 276–280.

Nor, M., and H. Cheng. 1986. Chemical speciation and bioavailability of Cu: uptake and accumulation by Eichornia. *Environmental Toxicology and Chemistry* 5: 941–947.

Nzengung, V. 2007. *Using Hydroponic Bioreactors to Assess Phytoremediation Potential of Perchlorate.* Methods in Biotechnology, Vol. 23 (III), 221–232. Humana Press Inc., Totowa, EUA.

Padinha, C., R. Santos, and M. Brown. 2000. Evaluating environmental contamination in Ria Formosa (Portugal) using stress indexes of *Spartina maritima*. *Marine Environmental Research* 49: 67–78.

Prasad, M. 2003. Phytoremediation of metal-polluted ecosystems: hype for commercialization. *Russian Journal of Plant Physiology* 50: 764–780.

Pushenreiter, M., G. Stöger, E. Lombi, O. Horak, and W. Wenzel. 2001. Phytoextraction of heavy metal contaminated soils with *Thlaspi goesingense* and *Amaranthus hybridus*: rhizosphere manipulation using EDTA and ammonium sulphate. *Journal of Plant Nutrition and Soil Science* 164: 615–621.

Reboreda, R., and I. Caçador. 2007a. Halophyte vegetation influences in salt marsh retention capacity for heavy metals. *Environmental Pollution* 146: 147–154.

Reboreda, R., and I. Caçador. 2007b. Cu, Zn and Pb speciation in salt marsh sediments colonized by Halimione portulacoides and Spartina maritima. *Chemosphere* 69: 1655–1661.

Rosens, N., N. Verbruggen, P. Meerts, P. Ximenez-Embun, and J. Smith. 2003. Natural variation in cadmium tolerance and its relationship to metal hyperaccumulation for seven populations of *Thlaspi caerulescens* from Western Europe. *Plant Cell Environment* 26: 1657–1672.

Rugh, C. 2004. Genetically engineered phytoremediation: one man's trash is another man's transgene. *Trends in Biotechnology* 22: 496–498.

Schwab, A., D. Zhu, and M. Banks. 2008. Influence of organic acids on the transport of heavy metals in soil. *Chemosphere* 72: 986–994.

Srivastava, S., S. Prakash, and M. Srivastava. 1999. Chromim mobilization and plant availabity—the impact of organic complexing ligands. *Plant and Soil* 212: 203–208.

Sundby, B., C. Vale, I. Caçador, F. Catarino, M.J. Madureira, and M. Caetano. 1998. Metal-rich concretions on the roots of salt marsh plants: mechanism and rate of formation. *Limnology and Oceanography* 43: 245–252.

Tessier, A. 1979. Sequential extraction procedure for the speciation of particulate trace metals. *Analytical Chemistry* 51: 844–851.

Tobias, C., S. Macko, I. Anderson, and E. Canuel. 2001. Tracking the fate of a high concentration groundwater nitrate plume through a fringing marsh: a combined groundwater tracer and *in situ* isotope enrichment study. *Limnology and Oceanography* 46: 1977–1989.

Turgut, C., M.K. Pepe, and T.J. Cutright. 2004. The effect of EDTA and citric acid on phytoremediation of Cd, Cr and Ni from soil using *Helianthus annuus*. *Environmental Pollution* 131: 147–154.

Wan, H., X. Shan, T. Liu, Y. Xie, B. Wen, S. Zhang, F. Han, and M. van Genuchten. 2007. Organic acides enhance the uptake of Pb by wheat roots. *Planta* 225: 1483–1494.

Weis, J., and P. Weis. 2004. Metal uptake, transport and release by wetland plants: implications for phytoremediation and restoration. *Environmental International* 30: 685–700.

Windham, L., J. Weis, and P. Weis. 2003. Uptake and distribution of metals in two dominant salt marsh macrophytes, *Spartina alterniflora* (cordgrass) and *Phragmites australis* (common reed). *Estuarine, Coastal and Shelf Science* 56: 63–72.

17 Plant–Microbe Enabled Contaminant Removal in the Rhizosphere

L. M. Stout and K. Nüsslein

CONTENTS

17.1 INTRODUCTION: PLANT–MICROBE–CONTAMINANT INTERACTIONS IN THE RHIZOSPHERE

While many plants and bacteria have their own mechanisms for dealing with heavy metal or organic contaminants, the interaction of plants and microorganisms may be the primary method of organic contaminant removal, and these interactions can increase or decrease heavy metal accumulation in plants, depending on the nature of the plant–microbe interaction. Because phytoremediation is a relatively new technology, understanding mechanisms of plant–microbe interactions in removing contaminants from the environment is still not completely characterized. There have, however, been advances in our understanding of how plant-associated microorganisms play a role in contaminant removal from the environment as the technology matures. A set of terms has emerged to reflect our understanding that the process often does not involve plants alone, but a reliance on microorganisms or the symbiosis of plants and microorganisms. Phytoremediation was previously considered to be a separate technology from bioremediation, which refers to contaminant-degrading or removing microorganisms that can be found in many soils, sediments, and water. The application of nonnative microbes to contaminated sites has not been entirely successful for several reasons, including the inability of introduced microorganisms to compete with indigenous soil microbes, inability of introduced microbes to grow at depths or in conditions where the contaminant is located, low nutrient availability for microbial growth, poorly bioavailable contaminants, use of carbon sources other than the contaminant by the microorganisms, and high concentrations of toxic compounds that will inhibit microbial growth (for a detailed review, see Gerhardt et al. 2009). Phytoremediation, the use of plants to remove contaminants from the environment, has been viewed as a technology with great promise, and the use of plants to support microorganisms in contaminant removal may be more effective than microorganisms alone. Among the advantages of phytoremediation techniques are a low capital and maintenance cost, aesthetically pleasing with minimal soil disturbance, and wide public acceptance. This is countered by possible disadvantages such as the potentially time consuming nature of phytoremediation, the limited application to sites with low to moderate contamination and only in shallow areas, and the potential risk to wildlife when contaminants are taken up and stored in plant tissues (Doty 2008).

Phytoremediation includes any of several technologies for detoxifying the environment with wild type or genetically modified plants (Kraemer 2005). These include phytoextraction, where plants accumulate contaminants in shoots, which can be harvested; phytovolatilization, where plants accumulate contaminants, transform them, and release volatile byproducts into the environment; phytostabilization, where plants are used to immobilize contaminants in the soil or sediment; and rhizofiltration, where plant roots absorb or precipitate, and thus filter, aquatic or hydroponic systems (Salt et al. 1995; Flathman and Lanza 1998; Kraemer 2005).

For removal of organic contaminants, phytoremediation can also include phytodegradation, where plant enzymes alone degrade organic compounds (Gerhardt et al. 2009; Glick 2010). Other terms make reference to activities in the rhizosphere, and to root-associated microorganisms. These include rhizoremediation, which is defined as the degradation of contaminants in the rhizosphere, and includes interactions of roots, root exudates, rhizosphere microorganisms, and soil (Gerhardt et al. 2009). Plant-assisted bioremediation (Salt et al. 1995), or microbe-assisted phytoremediation (Salt et al. 1995), both refer to the use of plant roots and their associated rhizosphere microbes, whether naturally occurring or seeded, to remove contaminants. Rhizodegradation or phytostimulation refers to rhizosphere microorganisms using their own enzymes to degrade contaminants, but they are stimulated by or in need of the relationship with the plant (Kuiper et al. 2004; Glick 2010). The U.S. Environmental Protection Agency (USEPA) groups these together as phytotechnologies and maintains a website dedicated to summarizing field- and large greenhouse-scale applications of these plant-based technologies at http://www.clu-in.org/products/phyto/.

Whereas for organic contaminants, the role of microorganisms may be obvious, in that in some cases they are responsible for the degradation of the contaminant, in the case of inorganic contaminants, which cannot be degraded, microbial involvement has been more difficult to specify. In their paper describing bacterial enhancement of Se and Hg uptake by wetland plants, De Souza et al. (1999a) proposed several possible mechanisms for microbial involvement in contaminant accumulation, including bacterial stimulation of plant metal uptake compounds such as siderophores; bacterial root growth promotion increasing the root surface area; bacterial transformation of elements into more soluble forms; or bacterial stimulation of plant transporters that may transport essential elements as well as heavy metals (in the case of selenate, the sulfate transporter). van der Lelie et al. (2000) related the basis of this plant–microbe interaction to bacterial metal resistance, since the bioavailability of metals could be altered by bacterial expression of resistance systems.

The current chapter discusses degradation of organic contaminants by rhizosphere bacteria; microbial roles in rhizofiltration of metals from aquatic and hydroponic systems; and the role of bacterial extracellular polymeric substance production and chelators, stimulation of plant root elongation, changes in rhizosphere pH, and microbial metal resistance as mechanisms for plant–microbe removal or degradation of contaminants.

17.2 PLANT–MICROBIAL ORGANIC CONTAMINANT DEGRADATION

Many microorganisms are capable of degrading organic contaminants, including bacteria, especially members of the genus *Pseudomonas*, and mycorrhizal fungi (Vosatka et al. 2006; Andreoni and Zaccheo 2010). The most successful strategies for remediation of organic contaminants have relied on microbe-assisted phytoremediation, using the relationship between plants and rhizosphere microbes to enhance the contaminant removal process (Gerhardt et al. 2009). Here we discuss some of the known plant-associated microorganisms involved in recent studies of contaminant degradation and their application.

Microbe-assisted phytoremediation, or plant-assisted rhizoremediation, relies in large part on degradative enzymes found in bacteria or fungi (Table 17.1). Organic contaminants, due to their complex structures, may persist in the environment and can be resistant to natural breakdown processes (McGuinness and Dowling 2009).

Organic contaminants come in many varieties, and include petroleum products, pesticides and herbicides, and polychlorinated biphenyls (PCBs). In one study of a petroleum hydrocarbon, diesel fuel, the contaminant was found to affect the growth of Italian ryegrass plants, although inoculation of plants with a *Pseudomonas* strain or a *Pantoea* strain resulted in greater growth of the plants. Plants inoculated with the *Pantoea* strain, which showed ACC deaminase activity and was thus able to lower the production of the plant stress hormone ethylene, showed greater growth than those inoculated with the *Pseudomonas* strain that did not possess this gene. As far as degradation of the contaminant, plants inoculated with the *Pantoea* strain showed the greatest hydrocarbon degradation (20% in loamy soil for the control, compared to 56% degradation in soils inoculated with *Pseudomonas* and 62% degradation for the soil inoculated with the *Pantoea* strain. This showed that inoculation with a bacterial strain providing with it a particular set of genes to support plant growth also allowed for more removal of contaminant (Afzal et al. 2011). Another study using Italian ryegrass, along with birdsfoot trefoil, and rhizosphere bacteria detected alkane-degrading genes *alkB* or cytochrome P450 alkane hydroxylases. In this study, the location of these genes within the genomes of isolates was also tested, and they were distributed on chromosomes or plasmids (Yousaf et al. 2010). Distribution on plasmids suggests that these genes can be easily moved between microbial isolates. When total petroleum hydrocarbons in salt marshes were studied for removal by the plants *Juncus maritimus*, *Phragmites australis*, *Triglochin striata*, and *Spartina patens*, a high number of potential hydrocarbon degrading bacteria were found in the rhizospheres of these plants, together with a high concentration of hydrocarbons, possibly due to the release of organic root exudates, which may complex the organic contaminants, and keep them close to the plant (Ribeiro et al. 2010). This could make these compounds more available for microbial degradation in the rhizosphere. Furthermore, *Phragmites australis* exhibited high degradation capabilities, suggesting some utility for this invasive salt marsh plant. Another study of diesel fuel remediation focused on the microbial populations in the rhizosphere. Four different groups of microorganisms, including total bacteria, phenanthrene degraders, diesel degraders, and pristine degraders, were enumerated from different treatments that included different plants and soil that was contaminated with diesel fuel or uncontaminated soil. Increased populations of bacteria were seen in contaminated soil, and it was shown that the total bacterial counts were responding more to the different root exudates produced by the plant. The phenanthrene and pristane degraders responded more to the presence of the contaminant, and diesel fuel degraders responded equally to contaminant or plant root exudates. The authors proposed a "phytoremediation benefit model" where in cases where a contaminant plume may sit underneath the zone of initial plant growth, the plant will support populations of bacteria that are potential hydrocarbon degraders, due to the organic root exudates. As the plant grows, the roots will reach into the contaminated area and bring contaminant degrading bacteria with them (Jones et al. 2004).

In another study focusing on removal of petroleum hydrocarbons, natural soil/rhizosphere communities were studied in setups with three different plants (ryegrass, summer vetch, or white mustard), and contaminated or uncontaminated soil containing a mixture of petroleum hydrocarbons, including the recalcitrant polycyclic aromatic hydrocarbons (PAHs). Roots were analyzed after plant growth to examine microbial types and numbers found in each treatment. In the mustard and vetch treatments, rhizosphere bacteria were most abundant, although numbers were still lower than in treatments with no contaminant. At the end of experiments, a number of potential hydrocarbon degraders such as *Flavobacterium johnsoniae* and *Pseudomonas* spp. were found in the

TABLE 17.1
Selection of Recent Experiments on Plant–Microbe Degradation of Organic Contaminants

Contaminant	Microbe	Mechanism	Plant	% Contaminant Reduction	References
Diesel	*Pseudomonas* strain ITRI53, *Pantoea* strain BTRH79	Alkane degradation genes-Alkane monooxygenase, cytP450 alkane hydroxylase	Italian Ryegrass	62	Afzal et al. (2011)
PAHs, total petroleum hydrocarbons (TPH)	Rhizosphere community		Summer vetch, white mustard	TPH: mustard 84.3, vetch 80.7	Liste and Felgentreu (2006)
PAHs	Rhizosphere community		Perennial ryegrass	Tot PAH: 52	Olson et al. (2007)
Hexadecane Diesel PAHs	Rhizosphere, endophyte communities	Alkane monooxygenase, catechol 2,3-dioxygenase, naphthalene inducible dioxygenase, phenanthrene dioxygenase, naphthalene dioxygenase	Alfalfa, Altai wild rye, tall wheat grass, perennial rye grass, Nuttal's salt meadow grass	Hexadecane: AWR, TWG 30%; phenanthrene: AWR 46, TWG 3	Phillips et al. (2008)
2,4-D	*Pseudomonas putida* VM1450		Pea	93	Germaine et al. (2006)
HCH	Rhizosphere community		Portuguese broom, velvet grass	α,γ: 98 β, δ: 44, 51	Kidd et al. (2008)

rhizospheres of particularly the mustard and vetch plants in contaminated treatments. In these samples, the remaining total hydrocarbons were significantly lower than in unplanted controls, although PAH degradation did not appear different in the planted vs. unplanted samples. The authors point out that these PAHs are persistent organic pollutants that over time may become less available as they are sequestered in the soil, and other more easily attacked petroleum hydrocarbons may be degraded first (Liste and Felgentreu 2006). In another study of PAHs that had been aged in soil, a suite of plants, including 18 species within 8 families, were grown in soil contaminated with PAHs. Samples were taken at several time points throughout the experiment to examine PAH degradation, including analysis of 16 different PAHs, and microbial enumeration. Grasses were most effective at reducing total PAH levels in the soils. Also, the types of PAHs were examined, and the PAHs with more rings (such as benzo[*b*]fluoranthene, a 5-ring PAH) decreased less over time than those with fewer rings, as they are more recalcitrant to microbial degradation. This study highlighted some of the differences in PAH removal between different plant families, which is an important consideration in phytoremediation. These differences in PAH removal may be due to compounds released by plant roots, which may attract different microbes or have different effects on PAH availability (Olson et al. 2007). In a study of the ability of two different clones of birch trees to degrade a variety of PAHs in different soils, the protein expression of the trees and changes in the associated microbial communities were studied. One birch clone was less susceptible to stress and its proteome revealed expression of fewer stress proteins. The less stress-tolerant member of the birch clones eliminated more PAHs in sandy soils, and this sample revealed the most diverse microbial community, perhaps contributing to the greater degradation (Tervahauta et al. 2009).

Endophytic bacteria, those bacteria that live inside the tissues of plants, rather than on the surface, may also be important contributors to organic contaminant degradation. In a study of soils contaminated with BTEX compounds and the PAHs naphthalene, phenanthrene, and pyrene, along with the plants perennial ryegrass (*Lolium perenne* L.), alfalfa (*Medicago sativa* L.), and a mix of local plant species including tall wheat grass (*Agropyron elongatum*), Altai wild rye (*Elymus angustus* Trin.), and Nuttal's salt meadow grass (*Puccinellia nuttalliana*), hydrocarbon degradation and microbial community potential for degradation were studied. Differences between endophytes and rhizosphere bacteria were recorded. The type of plant influenced the nature of the endophytic community, more than that of the rhizosphere community. All endophytic communities had high hydrocarbon degradation potential, and the degradation capabilities of the endophyte and rhizosphere communities may be directly linked (Phillips et al. 2008). In other studies involving organic contaminants and endophytes, a site contaminated with trichloroethylene (TCE) was used as a study site, and was planted with hybrid poplar trees inoculated with a TCE-degrading strain of *Pseudomonas putida*, a poplar root endophyte. Because TCE is a compound that often undergoes evapotranspiration, where the compound is only partially degraded into another toxic volatile compound, the goal was to improve degradation of the compound. In the study, evapotranspiration was lowered and TCE was degraded in plants inoculated with the endophytes. However, as the TCE degradation genes were plasmid-borne, as soon as TCE concentrations became too low to maintain selective pressure, the function was lost (Weyens et al. 2009). The location of genes involved in contaminant degradation must be a consideration when engineered or naturally occurring strains of microorganisms with plasmid-borne capabilities are used in phytoremediation.

Like PAHs, other organic contaminants are of great concern because of their persistence in the environment due to relative difficulty for degradation. One of these compounds is the organochlorine molecule 1,2,3,4,5,6-hexachlorcyclohexane (HCH), which is any combination of eight different isomers of this compound (De Souza et al. 1999b). This has been a widely used pesticide since the 1940s, with some mixtures now banned in many parts of the world. Like other organochlorines including DDT and PCBs, these molecules may have estrogenic effects and can be linked to tumor formation (De Souza et al. 1999b). In a study using HCH-contaminated soil and the plants common velvet grass (*Holcus lanatus* L.) and Portuguese broom (*Cytisus striatus*), plants were grown in soils with different levels of contamination, and plants, microorganisms, bulk and rhizosphere soil were analyzed after plant growth. Soil concentrations of the contaminant class $\sum$-HCH (the sum of α, β, δ, and γ isomers) were reduced after plant growth, but individually, alpha and gamma isomers were especially reduced, and alpha isomers were notably more reduced in rhizosphere soil compared to bulk soil. It is thought that there may be selective enrichment for microbes in the rhizosphere that can degrade this particular isomer (Kidd et al. 2008).

Some of the most notoriously persistent organic pollutants may be the polychlorinated biphenyls (PCBs). PCBs are a class of chemicals that include different "congeners," or molecules that differ in the number and position of chlorines on biphenyl rings. These compounds are no longer produced, but are still found in the environment from sources including paint pigments, plastics, fluids for transformers and capacitors, and lubricants, and still pose serious environmental threats (Safe 1994; Pieper and Seeger 2008). While PCB degrading microorganisms have been found, the difficulty for microorganisms to degrade these compounds on their own is well known, and strategies for stimulating microbial degradation of PCBs using plants has shown promise (Leigh et al. 2006). The theory of PCB degrader enrichment is that the compound biphenyl, which has no chlorine substitutions, can be added to a medium to stimulate PCB degradation. Yet, in practice biphenyl is not easy to work with, and so other compounds that may stimulate PCB degradation have been sought. Some of these compounds, such as flavonoids and terpenes, are released by plants as root exudates, and use of these plant metabolites has been termed the "analogue enrichment hypothesis" (Leigh et al. 2006). In a study using the plants *Nicotianum tabacum*, *Solanum nigrum*, *Armoracia rusticana*, and *Salix caprea* grown in soil from a PCB-contaminated site, rhizosphere and bulk soil bacteria were examined after a period of plant growth, and differences in microbial degradative capabilities were observed. *S. caprea* and *A. rusticana* best promoted the growth of potential PCB degraders, as determined by the numbers of bacteria that were able to grow on biphenyl, and by the presence of the biphenyl degradation gene *bphA1*. This may be due to plant-specific differences in root exudate compounds (Ionescu et al. 2009). In a review of degradation of toxic organic compounds, McGuinness and Dowling (McGuinness and Dowling 2009) highlighted some of the metabolic capabilities of microorganisms with regard to PCBs. Anaerobic bacteria can remove some chlorines from PCBs by reductive dechlorination, after which aerobic bacteria, such as those found in the rhizosphere, including *Burkholderia xenovorans* LB400 and *Rhodococcus* sp. RHA1, can begin to cleave rings to produce benzoates and pentanoic acids that may be degraded by yet other bacteria.

Like HCH mentioned above, other pesticides and herbicides have become increasingly common in the environment. For their removal from the environment, like that of other organic compounds, phytoremediation and its associated technologies may be a viable option. For a review of pesticide remediation, and a discussion of pesticide-degrading microorganisms and plants, see Hussain et al. (2009). Many pesticides can be degraded by rhizosphere microorganisms, and some can be taken up by plants, where they may be translocated to shoot biomass, or sequestered in the roots. These chemicals may then be degraded by enzymes from the plant (Hussain et al. 2009).

One herbicide that has received increasing attention is glyphosate, which replaced atrazine after its use was banned (Villenueve et al. 2011), and which has been used heavily, especially since the introduction of Roundup™ Ready crops in 1996 (Johal and Huber 2009). While microbial degradation of this compound has been known since the 1980s (Jacob et al. 1988), there have not been many studies showing plant–microbe remediation of this contaminant. In their reviews, Scott et al. (2008) focus on enzymes capable of degrading pesticides, and point out that glyphosate can be degraded to aminomethylphosphonate (AMPA) by glyphosate oxidase. This is an enzyme from *Pseudomonas* sp. LBr and also *Agrobacterium* strain T10, which can be transgenically expressed in plants for degradation *in planta* (Scott et al. 2008). Other microorganisms, including *Pseudomonas* strain PG2982 and *Arthrobacter* sp. GLP-1, degrade glyphosate to glycine by a separate pathway (Jacob et al. 1988). Whether the enzyme is produced in plants or in the bacteria, these genes are typically found in soil bacteria that may also be associated with plants and so plant-assisted rhizoremediation may also occur.

17.3 A SPECIAL CONCERN: EMERGING ORGANIC CONTAMINANTS AND THEIR REMOVAL

There has been concern over organic contaminants such as those discussed above for some time now, and many of these contaminants are considered priority pollutants (http://www.atsdr.cdc.gov/cercla/07list.html). Within the last decade a new group of compounds, classified as emerging contaminants, have quickly gained importance. These are chemical pollutants, biotic and abiotic, that are not new to us or to the environment, but, by definition, are previously unknown as contaminants and were not included in routine monitoring programs. Recent studies suggest that emerging contaminants may be harming aquatic life, and may have an effect on human health. These include compounds such as pharmaceuticals (estrogens, antibiotics, other prescription and nonprescription drugs), illicit drugs, personal care products (triclosan, surfactants, nanomaterials, e.g., sunscreens), detergents (fluorescent brighteners), and newly discovered degradation byproducts of pesticides (Mastroianni et al. 2010). Some of these chemicals have been shown to affect fish and amphibian populations, entering waterways after passing through water treatment intact (McConnell and Sparling 2010). The first major sampling effort for these chemicals in waterways occurred in 1999–2000, where they were detected in 80% of streams sampled across the United States (Kolpin et al. 2002). While many of these chemicals are entering waterways through personal use, a major source of emerging contaminants to waterways is hospital waste, where pharmaceuticals, radionuclides, solvents, and disinfectants are commonly found in wastewater streams. These chemicals are not necessarily persistent organic pollutants, and may be found in low concentrations, but their constant introduction into the environment may allow them to cause harm even if they do not accumulate in tissues (Verlicchi et al. 2010).

For the remediation of these compounds, biological treatment options have been explored. Most of these have focused on activated sludge processes and membrane bioreactors, and depending on the compound, these have been more or less effective compared to other treatments. Nitrifying bacteria have been reported to play a major role in the degradation of pharmaceuticals (Verlicchi et al. 2010). There are few studies of treatment options of these as yet unregulated pollutants that involve plant–microbe systems, but several recent studies suggest this may be an option. Hijosa-Valsero et al. (37) focused on antibiotics and the use of constructed wetlands to remove them. Different wetland designs were set up with different plant types, and these wetland treatments were compared to a wastewater treatment plant. The wastewater treatment plant was able to effectively eliminate doxycycline and sulfamethoxazole, whereas constructed wetlands were able to effectively remove sulfamethoxazole as well as trimethoprim. Interestingly, some antibiotic removal only occurred in certain constructed wetlands: amoxicillin in a free-water subsurface flow wetland with *Typha augustifolia*; doxycycline with free-water *T. augustifolia* and also in a *Phragmites australis*-floating macrophyte system; erythromycin was removed in a *P. australis*-subsurface flow system; and ampicillin could be removed in a *T. augustifolia*-floating macrophyte system. While no parallel microbial investigations were done, it was suggested that in some cases rhizosphere communities could play a role in antibiotic degradation (Hijosa-Valsero et al. 2011). Another study combined examination of antibiotic removal with excess plant nutrient removal from swine waste in a constructed wetland, testing three different varieties of Italian ryegrass (varieties Dryan, Tachimasari, and Waseyutaka) for removal ability (Xian et al. 2010). Nitrogen and Phosphorus were removed by 80% or greater and 70% or greater, respectively, with the Dryan variety performing best in both cases. Sulfonamide antibiotics were removed up to 99%, indicating that this was an effective method. N and P removal were attributed to plants and their associated root microorganisms, and loss of antibiotic was attributed to biodegradation, although it is not clear if plant or microbial degradation, or some combination, were most effective.

Besides antibiotics, another major group of emerging contaminants are endocrine disruptors. This group of compounds behaves similarly to hormones and interferes with biological functions such as reproduction and development. Endocrine disruptors include synthetic estrogens (Bradley et al. 2009), and even PCBs and pesticides (Seeger et al. 2010). One endocrine disruptor is bisphenol A, or BPA. This xeno-estrogen is used in preparation of polycarbonates and resins, and also as a stabilizer for plastics (Loffredo et al. 2010). Because it can leach out of these products, there is concern over its use in food grade plastic bottles and toys. In a study testing the ability of several grasses and horticultural species to degrade BPA, plants were grown axenically by sterilizing seed surfaces, or plants were grown with associated microorganisms. Some plant species, including radish and tall fescue, showed significantly greater degradation in the experiments with microbes, whereas perennial ryegrass and marrow plant did not show significant differences between the axenic and septic experiments (Loffredo et al. 2010).

In a related study by Reinhold et al. the removal of eight different emerging contaminants by duckweed, *Lemna minor*, in a constructed wetland was investigated. The contaminants included the pharmaceutical compound ibuprofen, the personal care products triclosan and DEET, and pesticides including 2,4-D and atrazine (Reinhold et al. 2010). Duckweed was able to directly uptake the antidepressant pharmaceutical fluoxetine, degrade triclosan and 2,4-D, and the presence of duckweed increased the microbial transformation of ibuprofen. Another study of personal care products in the environment showed that several wetland plants, *Sesbania herbacea*, *Bidens frondosa*, and *Eclipta prostrata*, were affected differently by triclosan, an antimicrobial additive to numerous consumer products. Germination was affected in some of the plants, and root length was affected to some degree for all plants. *B. frondosa* seemed to accumulate triclosan at the roots, where methyl triclosan was formed (Stevens et al. 2009). This suggests triclosan as a problem for constructed wetlands, but the appropriate plant–microbe combination could lessen toxicity to the plants.

In addition to the currently active research discussion on organic emerging contaminants, there are also numerous examples of inorganic emerging contaminants. For instance, the USEPA prepared a fact sheet on the element Tungsten, a metal which was previously thought to be not easily soluble in water and was therefore considered stable in soil. Recent studies suggest that this may not be the case (USEPA 2008). Nanomaterials, a broad term that refers to particles of 100 nm or less, can include metals or carbon-based materials. Elemental nanoparticles, such as rare earth element compounds and metal oxides, are used in microelectronics and have even been shown to improve crop growth, but their toxicity is still largely unknown (Lopez-Moreno et al. 2010). Cerium oxide nanoparticles were studied for their effects on plant germination and root and shoot growth in several crop plants, including alfalfa, corn, cucumber, and tomato. Germination in the presence of nanoparticles was significantly reduced for tomato, cucumber, and corn, while root growth was promoted in cucumber and corn, and reduced in alfalfa and tomato, and shoots were elongated in all plants (Lopez-Moreno et al. 2010). Nanoparticles were able to accumulate in plants.

17.4 PLANTS AND MICROBES IN HEAVY METALS REMOVAL: PLANT GROWTH PROMOTION AND ROOT ELONGATION

Phytoremediation of metal contaminants was once thought to be a primarily plant-based activity, and it is only in more recent years that the role of microorganisms in the phytoremediation of heavy metals has also been addressed. Still, many unanswered questions remain surrounding the role of microbes in the phytoremediation of metals. One reason that microorganisms have been overlooked in this process is the finding that certain plants have the ability to accumulate exceptionally high levels of heavy metals and sequester them in their tissues. Plants that can accumulate exceptionally high concentrations of metals are termed *hyperaccumulators*, and around 450 hyperaccumulator species have been identified so far (Prasad and Freitas 2003; Fones et al. 2010).

One plant that has been studied for its ability to remove Cd from surface waters is the aquatic plant *Lemna minor* (Debusk, Jr., and Schwartz 1996; Zayed et al. 1998; Hasar and Obek 2001; Wang et al. 2002). This plant may be a good choice for remediation projects, because it is a known Cd accumulator, and due to its rapid growth rate and ease of removal, it can be harvested at regular intervals (weekly or biweekly), keeping the metal from continuous reintroduction into the ecosystem (Debusk Jr. and Schwartz 1996).

While studies have shown that *L. minor* can accumulate high levels of Cd, it has also been shown that the plant cannot tolerate high concentrations of Cd in the surrounding water. *L. minor* has been used to assess toxicity in the environment due to its sensitivity to metals. Signs of toxicity in *L. minor* may include chlorosis, stunting of growth, colony separation, and root detachment (American Public Health Association, American Water Works Association, and Water Environment Federation 1992). Agencies continue to develop methods to use the plant as a toxicity indicator. Wang (1990) explained seemingly contradictory observations of tolerance and sensitivity as a function of the adaptability of this plant, and this may also be the case for other known hyperaccumulators or accumulators (Figure 17.1).

Hyperaccumulators may be defined based on the bioconcentration factor (BCF), or the ability to accumulate metals in plant tissues. For instance, the ability to accumulate greater than 1000 times the concentration of Cd (based on concentration of metal in dry weight of plant) than that in the surrounding medium would be considered hyperaccumulation (Zayed et al. 1998). Recently, the ability to hyperaccumulate has been attributed to the plant's self-defense strategy (Fones et al. 2010) and the molecular alterations of the associated hypertolerance became a focus of comparative genomics studies (Hanikenne and Nouet 2011). The 'elemental defence hypothesis' suggests that metal hyperaccumulation provides a defence against pathogens. When Fones et al. (Fones et al. 2010) grew *Thlaspi caerulescens* plants in the presence of increasing concentrations of zinc, nickel, and cadmium, and subsequently inoculated them with the bacterial pathogen *Pseudomonas syringae*, growth of the bacterium was increasingly inhibited as the concentration of each metal rose. In addition, endophytic bacteria collected from the leaves of plants growing at the site of a former zinc mine exhibited a higher level of zinc tolerance than the most tolerant *P. syringae* mutant tested.

One idea of how microorganisms participate in metal removal is that bacteria promote plant growth, thus increasing surface area of the plant and enabling increased metal uptake. Certain compounds produced by bacteria have been shown to promote plant growth, including siderophores. Siderophores, iron chelating compounds produced by microorganisms, may promote plant growth, even in the presence of heavy metals (Burd et al. 2000; Tripathi et al. 2005). Bacterial production of siderophores may protect plants from heavy metal toxicity, increasing plant growth but decreasing accumulation of metals by plants. Stout and Nüsslein observed that the addition of rhizosphere isolates to cultures of *Lemna minor* increased root length, and actually decreased accumulation of Cd in plants. Root elongation did not seem to be directly related to which bacteria could produce siderophores, although some of the isolates that produced siderophores allowed higher accumulation

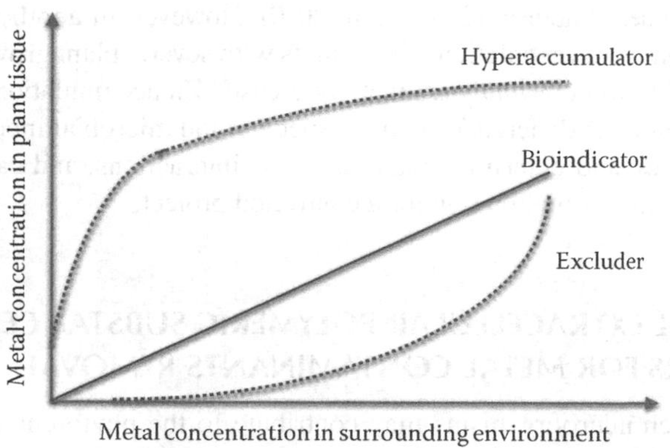

FIGURE 17.1 Hyperaccumulating plants take up high amounts of heavy metals from the environment compared to the opposite case, the excluders. Bioindicator plants are those with a linear response to environmental concentrations of bioavailable metals.

of Cd in plants (Stout et al. 2010). In a study of auxin- and siderophore-producing strains of the metal-resistant soil bacterial genus *Streptomyces* the bacteria were exposed to Cd and Ni, and production of boh compounds was observed. It was found that not only were siderophores able to chelate iron, but certain siderophores were produced that could chelate nickel and cadmium. Auxins and siderophores could be produced simultaneously, but the presence of metals decreased auxin production, although this was less pronounced so long as siderophores were also being produced (in comparison to a siderophore-deficient mutant). The authors concluded that siderophores, by chelating the metals, promoted auxin synthesis and could therefore increase plant growth-promoting effects of auxins, which could improve phytoremediation (Dimkpa et al. 2008). For a review of the diversity of siderophores and siderophore-producing microorganisms, and the functions of microbial siderophores in plant growth (see Rajkumar et al. 2010).

Another plant growth-promoting compound that has been studied in relation to heavy metals is 1-aminocyclopropane-1-carboxylic acid deaminase (ACC deaminase). ACC is an intermediate of ethylene produced by plants under stress, and bacteria that produce ACC deaminase can lower the levels of ethylene in plants, promoting plant growth (Dell'Amico et al. 2005). Belimov et al. (2001) found that bacteria containing ACC deaminase improved plant growth under conditions of heavy metal elevation. Another study, however, found that ACC deaminase expression of a plant growth-promoting *P. putida* strain (UW4) was not directly related to nickel stress, but expression of other proteins involved in amino acid synthesis, protein folding, DNA replication, cell division, and cell communication were all down-regulated as a response to nickel stress (Cheng et al. 2009).

Bacteria as well as plants can produce the auxin indole-3-acetic acid (IAA). Rajkumar and Freitas (Rajkumar and Freitas 2008) suggested that IAA indirectly promotes metal accumulation in plants by increasing plant biomass. Grandlic et al. found that 76% of plant growth-promoting isolates from plants grown in mine tailings were able to produce IAA when supplemented with Tryptophan, and these bacteria could promote growth of plants growing in mine tailings for phytostabilization (Grandlic et al. 2008). While this study did not examine the metal accumulation in plants, and focused more on the ability of the rhizobacteria to promote biomass, some studies with plant growth-promoting rhizosphere bacteria have shown that rather than allowing the plant to accumulate higher amounts of metal, they function to enhance plant growth and may stabilize contaminants in the soil, but they may have a more protective effect on the plant allowing it to accumulate less contaminant. One example of this is a study using the plant *Lupinus luteus* in soil contaminated with multiple metals (Cu, Cd, Pb, As, Zn). This plant was grown in combination with plant growth-promoting rhizobacteria including *Bradyrhizobium* sp. 750 and others isolated from the contaminated site. While As was excluded from the plant tissues Zn and the other metals were readily accumulated, yet accumulation was significantly less in cases where plants were inoculated with the rhizosphere bacteria (Dary et al. 2010). However, in another study, using the plant *Orychophragmus violaceus*, inoculation of the plants with several plant growth-promoting bacterial strains, especially a *Flavobacterium* sp., greatly increased Zn accumulation in the plants (He et al. 2010). This again shows that differences in plant species (and microbial inoculum species) can vary greatly in their abilities, and determine plant–microbe interactions, and care must be taken when choosing the plant/microbe combination for a controlled project.

17.5 MICROBIAL EXTRACELLULAR POLYMERIC SUBSTANCES AND CHELATORS FOR METAL CONTAMINANTS REMOVAL

Another way in which microorganisms may contribute to the phytoremediation of metals is the production of microbial extracellular polymeric substances (EPS). Kunito et al. (2001) examined rhizosphere bacteria from *Phragmites* grown in copper. They found the EPS production was greater for rhizosphere bacteria compared to nonrhizosphere bacteria. Due to copper binding to bacterial EPS, the rhizosphere soil may become less toxic to bacteria and also to plants.

Siderophores, which may contribute to plant growth promotion, may also chelate or sequester metals other than iron (Dimkpa et al. 2008; Rajkumar et al. 2010), and so may not only promote plant growth by production of compounds such as auxins, but could directly contribute to metal removal from the environment. Cd was reported to induce siderophore production in a plant-associated bacterium isolated from wastewater in India, *Pseudomonas* sp. KUCd1. This strain was resistant to 8 mM Cd and was shown to accumulate Cd intracellularly. When mustard or pumpkin plants were inoculated with this strain in Cd contaminated soil, plant growth increased, and Cd accumulation in the plant was reduced (Sinha and Mukherjee 2008). Another study showed that when siderophore-producing *P. aeruginosa* was inoculated into soils, the bioavailability of Cr and Pb increased, and maize shoots accumulated higher amounts of the metals (Braud et al. 2009).

17.6 MICROBIAL METAL RESISTANCE AND METAL CONTAMINANTS REMOVAL

Metal resistance has been described as a necessity for plant-associated bacteria in contaminated environments (Salt et al. 1999). van der Lelie et al. (2000) related plant metal uptake to bacterial metal resistance, since the bioavailability of metals could be altered by expression of bacterial metal resistance systems. Faisal and Hasnain (2005) inoculated sunflower plants with chromium resistant bacteria, and found that plant growth in chromium was improved by inoculation, although inoculated plants accumulated less chromium than uninoculated plants. In another study, oat plants (*Avena sativa*) were inoculated with copper-resistant isolates from the rhizosphere, and plants were grown in copper contaminated soil. Plants grown with the isolates showed greater plant growth and copper uptake, especially with two of the isolates, *Stenotrophomonas maltophilia* and *Acinetobacter calcoaceticus* (Andreazza et al. 2010). Jiang et al. (2008) isolated 21 strains of potentially metal resistant bacteria, the most promising being a heavy metal-resistant strain of *Burkholderia* (J62), and showed that it was resistant to multiple heavy metals and antibiotics. The isolate produced siderophores, IAA, and ACC deaminase and was able to solubilize phosphate, a trait also linked to plant growth promotion. It was able to mobilize Cd and Pb, and when mustard, tomato, or maize was grown in soil inoculated with the bacterium, maize and tomato showed increased biomass. Concentrations of Cd and Pb increased in roots of maize when grown with the strain, whereas tomato accumulated higher concentrations of Cd and Pb in its shoots, compared to uninoculated plants (Jiang et al. 2008). Similarly, in a study by Kuffner et al. (2008), 10 rhizosphere bacteria were isolated from heavy metal accumulating willow trees, and were assessed for IAA, ACC deaminase, and siderophore production, as well as how effectively they solubilized Zn, Pb, and Cd. None of the strains produced ACC deaminase, but four produced siderophores, two produced IAA, and one produced both. When grown with plants in contaminated soil, the IAA producers did not promote plant growth and the siderophore producers did not mobilize more metals. One strain, *Agromyces terreus*, mobilized twice the amount of metals than the control, yet it did not produce siderophores or IAA, indicating that other factors may be involved, as the authors suggest perhaps other organic compounds. A study by Li et al. also examined metal resistant bacteria, strains of *Pseudomonas* that had been isolated from contaminated sediments. These bacteria were tested for resistance to multiple metals and antibiotics, as well as phosphate solubilization, and IAA and siderophore production. One of these strains was then inoculated into copper-contaminated soil with sunflower or maize, and in both cases, the plants took up significantly more copper when the bacterium was added to the soil. Addition of the bacterium caused biomass increase in maize but not sunflower. The ability of this bacterium to solubilize phosphate and produce IAA meant that it had growth-promoting capabilities, but the authors point out that expression of these characteristics by bacteria may be plant-dependent (Li and Ramakrishna 2011).

A copper-resistant bacterium, CCNWRS33-2, closely related to *Agrobacterium tumefaciens* LMG 196, was isolated from the root nodules of the plant *Lespedeza cuneata*, in mine tailings of a gold mine in China. The copper resistance gene *copA* was amplified from this organism (Wei et al. 2009). Its metal resistance, and its relationship as a plant symbiont, suggests that it could play a role in metal phytoremediation. Ma et al. (2009) isolated a copper-resistant plant growth-promoting bacterium, named *Achromobacter xylosoxidans* Ax10, from copper mine soil. This bacterium was able to produce IAA and solubilize phosphate, and when inoculated with the plant *Brassica juncea* increased plant biomass and allowed greater accumulation of copper in both roots and shoots of the plant (Ma et al. 2009). Interestingly, one of the better-studied bacteria with regard to metal resistance systems is *A. xylosoxidans* 31A, which contains plasmid-borne nickel, cadmium, and cobalt resistance (*nccA*) (Schmidt and Schlegel 1994) and another nickel resistance locus, *nreB* (Grass et al. 2001), and so it may be possible that this new copper-resistant *A. xylosoxidans* strain has related similar mechanisms. As some of the above studies show, bacterial metal resistance may be necessary for microbial survival in contaminated environments, but other mechanisms, such as production of plant growth-promoting compounds, are often employed by metal-resistant bacteria in phytoremediation.

17.7 MICROBIAL ALTERATION OF RHIZOSPHERE pH AND METAL CONTAMINANTS REMOVAL

Another way in which microbes could be involved in phytoremediation of metals could be the lowering of pH in the rhizosphere by bacteria, which would make metals more soluble. The rhizosphere pH could be lowered by processes listed above. Bravin et al. (2009) pointed out that rhizosphere pH can also be raised by biological activity; in extremely low pH environments, roots alkalized the rhizosphere, making copper less bioavailable. Abou-Shanab et al. hypothesized that rhizosphere bacteria of *Alyssum murale* lowered rhizosphere pH, solubilizing metal for hyperaccumulation of Ni by the plant (Abou-Shanab et al. 2003). Li et al. isolated a bacterium from the rhizosphere of *Sedum alfredii*, a hyperaccumulator plant. The bacterium, a strain of *Burkholderia cepacia*, was capable of mobilizing Cd and Zn, by causing a dramatic change in rhizosphere pH, from about pH 7 to pH 3, most likely by releasing organic acids (Li et al. 2010). In another example, a chemical was added to the rhizosphere, which caused a change in microbial composition and subsequent changes in pH values. In this study, Shi et al. (2011) examined what happened to rice rhizospheres in copper contaminated soils following sulfur addition. Subsequent to sulfur addition, the pH was lowered, and the microbial community changed. The more available Cu proved toxic to the rice plants, especially at higher concentrations (Shi et al. 2011), which points out that phytoremediation is most effective when facing lower concentrations of contaminants. In another study that shows quite different results, chromium accumulation by rice plants and rhizosphere pH were examined. Here rhizosphere pH decreased with increasing Cr levels in solutions. Organic acids were again suggested as the reason for this, and the production of six different organic acids was examined. Two of these organic acids, oxalic and malic acid, had a strong positive correlation with pH, and with the accumulation of Cr in plants (Zeng et al. 2008).

While bacteria have been well studied in the rhizosphere, fungi are also an important contribution to the total rhizosphere microbial community. Furthermore, there are examples of fungal involvement in phytoremediation of metals. For example, in a study of mesquite plants grown in mine tailings, plants inoculated with arbuscular mycorrhizal fungi (AMF) showed increased biomass and root length, but shoot accumulation of metals was low, indicating that these plants and fungi are better adapted for phytostabilization. Also, the presence of the inoculated fungi influenced the bacterial community structure in the rhizosphere, and it is not known if this is a direct effect of the fungi or some indirect influence of the fungi on the plant (Solis-Dominguez et al. 2011).

17.8 GENETICALLY MODIFIED PLANT–MICROBE PHYTOTECHNOLOGIES

The biotechnological potential of plant–microbe relationships for use in phytoremediation has included research on transgenic plants and microorganisms. Many engineered plant–microbe systems for phytoremediation have been used to degrade organic contaminants, with genes from microorganisms enhancing degradation by the plant. A review by Scott et al. points out that microorganisms are the sources of many of the genes used to degrade pesticides and herbicides, such as glyphosate, atrazine, endosulfan, and a host of others, and remediation strategies are underway to express these genes *in planta*, or to augment treatments with the microorganisms containing the genes (Scott et al. 2008). In a study using endophytic bacterial strains, it was found that the toluene-degrading endophyte *Burkholderia cepacia* VM1330 was able to increase degradation of toluene and provide protection against phytotoxicity. While this particular strain was genetically engineered to contain a marker to trace the bacterium in yellow lupine plants, the authors suggested that a naturally occurring endophyte could be used, in which case it would not be genetically engineered (Barac et al. 2004). In the 2009 study by Weyens et al., the researchers used a natural endophyte of poplar that was able to degrade TCE, and the endophyte's metabolic capabilities were spread naturally through horizontal gene transfer, thus removing concern for the potential harm of a genetically engineered microorganism (Weyens et al. 2009). In a study of PCB degradation by tobacco, the *bphC* gene from *Pseudomonas testosteroni* B-356, which encodes the enzyme in the biphenyl degradation pathway that cleaves biphenyl, was inserted successfully into the plant genome (Novakova et al. 2009). In an example of fungal genes being inserted into plants, a glutathione transferase, involved in the metabolism of organic pollutant transformation, was transferred from the fungus *Trichoderma virens* to tobacco to increase plant tolerance to and degradation of anthracene, a capability which was observed to be greater than in the wild type plants (Dixit et al. 2011).

While most transgenic plants used in phytoremediation rely on genes from bacteria or fungi, some studies have focused on the mammalian liver enzyme cytochrome P450. Because this enzyme is known to metabolize low molecular weight compounds such as TCE and vinyl chloride, it can potentially degrade these compounds if expressed in plant tissue. Transgenic tobacco was created and exposed to several of these volatile organic compounds, and the transgenic tobacco was able to more efficiently remove several of these compounds, compared to wild type plants (James et al. 2008). Numerous studies have also focused on the combination of transgenic plants and metal exposure. Parkash Dhanker et al. (2003) found that tobacco plants expressing the bacterial arsenate reductase gene, *arsC*, were more tolerant to and accumulated more Cd (Dhankher et al. 2003). Che et al. (2003) engineered transgenic cottonwood trees (*Populus deltoides*) to express bacterial *merB* genes, and found that plants were more resistant to organic Hg compounds than wild-type plants. Hussein et al. engineered transgenic tobacco plants through the chloroplast genome with both *merA* and *merB* genes, and saw fewer toxic effects of Hg and more Hg accumulation than in wild type plants (Hussein et al. 2007).

Plant–microbe symbioses also have been exploited in transgenics. Wu et al. (2006) manufactured a synthetic phytochelatin analog that was expressed in *Pseudomonas putida* to increase Cd binding, and this engineered bacterium was then added to sunflower roots to increase Cd accumulation and lessen Cd toxicity in plants. A study by Moontongchoon et al. focused on genetically altered water spinach (*Ipomoea aquatica*). Here, expression of genes from sulfate assimilation pathways in these transgenes provided elevated Cd tolerance, which enabled this system to remediate metals in high sulfate environments (Moontongchoon et al. 2008).

17.9 CONCLUSIONS AND PERSPECTIVES

The examples presented in this chapter have focused on single contaminants or groups of contaminants (mixtures of different organic contaminants, or mixtures of different metals). However, contaminated sites in the environment are rarely so easily separated into only organics or only metals, and several studies have addressed mixtures of inorganic and organic contaminants. Beesley et al. (2010) pointed out the difficulties of working with multicontaminant sites, in that any amendments made to try to improve phytoremediation might have contradictory effects on the mobility and availability of different contaminants. In their study, they used biochar, a carbon-rich charcoal, to try to remediate soil contaminated with several metals as well as PAH's. While biochar treatment did prove successful, plant–microbial solutions are now also gaining ground. In a study by Chaturvedi et al. (2008), rhizosphere bacteria were isolated from *Phragmites australis*, and were assessed for their potential to treat distillery effluent in a constructed wetland. Distillery effluent waste consists of phenols, sulfide, heavy metals, and other organic material. Antibiotic and metal resistances were studied in the isolates, and it was suggested that antibiotic resistance could help in detoxification of the organic contaminants, whereas metal resistance is a prerequisite for metal removal, and so microbes with multiple antibiotic and metal resistances, such as carried by plasmids, would be ideal candidates for this type of mixed contamination. In another co-contaminant study, Li et al. (2008) studied the combined effects of pyrene and copper contamination. At pyrene concentrations of 50–500 mg/kg with 0 or 200 mg/kg Cu, pyrene was decreased in the soil, and Cu became bioavailable to be accumulated by maize. At higher Cu concentrations, however, pyrene was not decreased, most likely due to changes in the microbial community as a result of high copper levels, or even in response to changes in plant root exudates due to plant stress.

Overall, it is now well acknowledged that plants and microbes are not separate systems, but work together in soils and aquatic environments to remove contaminants, even though the mechanisms of how they work together to do this efficiently or even synergistically are still not completely clear. It is likely that certain types of plants and microorganisms have closer relationships with each other than others, and strategic consideration of the type of plant, the types of microoganisms, the particular site characteristics, and the nature of the contamination to be treated all need to be taken into account when planning phytoremediation projects.

REFERENCES

Abou-Shanab, R.A., T.A. Delorme, J.S. Angle, R.L. Chaney, K. Ghanem, H. Moawad, and H.A. Ghozlan. 2003. Phenotypic characterization of microbes in the rhizosphere of *Alyssum murale*. *International Journal of Phytoremediation* 5: 367–79.

Afzal, M., S. Yousaf, T.G. Reichenauer, M. Kuffner, and A. Sessitsch. 2011. Soil type affects plant colonization, activity and catabolic gene expression of inoculated bacterial strains during phytoremediation of diesel. *Journal of Hazard Materials* 186: 1568–75.

American Public Health Association, American Water Works Association, and Water Environment Federation, eds. 1992. *Standard Methods for the Examination of Water and Wastewater*, 18th ed. Washington, DC: APHA, AWWA, WEF.

Andreazza, R., B.C. Okeke, M.R. Lambais, L. Bortolon, G.W. de Melo, and F.A. Camargo. 2010. Bacterial stimulation of copper phytoaccumulation by bioaugmentation with rhizosphere bacteria. *Chemosphere* 81: 1149–54.

Andreoni, V., and P. Zaccheo. 2010. Potential for the use of rhizobacteria in the sustainable management of contaminated soils. In *Plant Adaptation and Phytoremediation*, 313–34. New York: Springer.

Barac, T., S. Taghavi, B. Borremans, A. Provoost, L. Oeyen, J.V. Colpaert, J. Vangronsveld, and D. van der Lelie. 2004. Engineered endophytic bacteria improve phytoremediation of water-soluble, volatile, organic pollutants. *Nature Biotechnology* 22: 583–8.

Beesley, L., E. Moreno-Jiménez, and J.L. Gomez-Eyles. 2010. Effects of biochar and greenwaste compost amendments on mobility, bioavailability and toxicity of inorganic and organic contaminants in a multi-element polluted soil. *Environmental Pollution* 158: 2282–7.

Belimov, A.A., V.I. Safronova, T.A. Sergeyeva, T.N. Egorova, V.A. Matveyeva, V.E. Tsyganov, A.Y. Borisov, I.A. Tikhonovich, C. Kluge, A. Preisfeld, K.J. Dietz, and V.V. Stepanok. 2001. Characterization of plant growth promoting rhizobacteria isolated from polluted soils and containing 1-aminocyclopropane-1-carboxylate deaminase. *Canadian Journal of Microbiology* 47: 642–52.

Bradley, P.M., L.B. Barber, F.H. Chapelle, J.L. Gray, D.W. Kolpin, and P.B. McMahon. 2009. Biodegradation of 17β-estradiol, estrone and testosterone in stream sediments. *Environmental Science and Technology* 43: 1902–10.

Braud, A., K. Jezequel, S. Bazot, and T. Lebeau. 2009. Enhanced phytoextraction of an agricultural Cr- and Pb-contaminated soil by bioaugmentation with siderophore-producing bacteria. *Chemosphere* 74: 280–6.

Bravin, M.N., P. Tentscher, J. Rose, and P. Hinsinger. 2009. Rhizosphere pH gradient controls copper availability in a strongly acidic soil. *Environmental Science and Technology* 43: 5686–91.

Burd, G.I., D.G. Dixon, and B.R. Glick. 2000. Plant growth-promoting bacteria that decrease heavy metal toxicity in plants. *Canadian Journal of Microbiology* 46: 237–45.

Chaturvedi, S., R. Chandra, and V. Rai. 2008. Multiple antibiotic resistance patterns of rhizospheric bacteria isolated from *Phragmites australis* growing in constructed wetland for distillery effluent treatment. *Journal of Environmental Biology* 29: 117–24.

Che, D., R.B. Meagher, A.C.P. Heaton, A. Lima, C.L. Rugh, and S.A. Merkle. 2003. Expression of mercuric ion reductase in Eastern cottonwood (*Populus deltoides*) confers mercuric ion resistance. *Plant Biotechnology Journal* 1: 311–19.

Cheng, Z., Y.Y. Wei, W.W. Sung, B.R. Glick, and B.J. McConkey. 2009. Proteomic analysis of the response of the plant growth-promoting bacterium *Pseudomonas putida* UW4 to nickel stress. *Proteome Science* 7: 18.

Dary, M., M.A. Chamber-Pérez, A.J. Palomares, and E. Pajuelo. 2010. "*In situ*" phytostabilisation of heavy metal polluted soils using *Lupinus luteus* inoculated with metal resistant plant-growth promoting rhizobacteria. *Journal of Hazardous Materials* 177: 323–30.

De Souza, M.P., C.P.A. Huang, N. Chee, and N. Terry. 1999a. Rhizosphere Bacteria enhance the accumulation of selenium and mercury in wetland plants. *Planta* 209: 259–63.

De Souza, M.P., D. Chu, M. Zhao, A.M. Zayed, S.E. Ruzin, D. Schichnes, and N. Terry. 1999b. Rhizosphere bacteria enhance selenium accumulation and volatilization by indian mustard. *Plant Physiology* 119: 565–74.

Debusk, T.A., Laughlin R.B. Jr., and L.N. Schwartz. 1996. Retention and compartmentalization of lead and cadmium in wetland microcosms. *Water Research* 30: 2707–16.

Dell'Amico, E., L. Cavalca, and V. Andreoni. 2005. Analysis of rhizobacterial communities in perennial Graminaceae from polluted water meadow soil, and screening of metal-resistant, potentially plant growth-promoting bacteria. *FEMS Microbiology Ecology* 52: 153–62.

Dhankher, O.P., N.A. Shasti, B.P. Rosen, M. Fuhrmann, and R.B. Meagher. 2003. Increased cadmium tolerance and accumulation by plants expressing bacterial arsenate reductase. *New Phytologist* 159: 431–41.

Dimkpa, C.O., A. Svatos, P. Dabrowska, A. Schmidt, W. Boland, and E. Kothe. 2008. Involvement of siderophores in the reduction of metal-induced inhibition of auxin synthesis in *Streptomyces* spp. *Chemosphere* 74: 19–25.

Dixit, P., P.K. Mukherjee, P.D. Sherkhane, S.P. Kale, and S. Eapen. 2011. Enhanced tolerance and remediation of anthracene by transgenic tobacco plants expressing a fungal glutathione transferase gene. *Journal of Hazardous Materials* 192: 270–6.

Doty, S.L. 2008. Enhancing phytoremediation through the use of transgenics and endophytes. *New Phytologist* 179: 318–333.

Faisal, M., and S. Hasnain. 2005. Chromate resistant *Bacillus cereus* augments sunflower growth by reducing toxicity of Cr(VI). *Journal of Plant Biology* 48: 187–94.

Flathman, P.E., and G.R. Lanza. 1998. Phytoremediation: Current Views on an Emerging Green Technology. *Journal of Soil Contamination* 7: 415–32.

Fones, H., C.A. Davis, A. Rico, F. Fang, J.A. Smith, and G.M. Preston. 2010. Metal hyperaccumulation armors plants against disease. *PLoS Pathogens* 6: e1001093.

Gerhardt, K.E., X.D. Huang, B.R. Glick, and B.M. Greenberg. 2009. Phytoremediation and rhizoremediation of organic soil contaminants: potential and challenges. *Plant Science* 176: 20–30.

Germaine, K.J., X. Liu, G.G. Cabellos, J.P. Hogan, D. Ryan, and D.N. Dowling. 2006. Bacterial endophyte-enhanced phytoremediation of the organochlorine herbicide 2,4-dichlorophenoxyacetic acid. *FEMS Microbiology Ecology* 57: 302–10.

Glick, B.R. 2010. Using soil bacteria to facilitate phytoremediation. *Biotechnology Advances* 28: 367–74.

Grandlic, C.J., M.O. Mendez, J. Chorover, B. Machado, and R.M. Maier. 2008. Plant growth-promoting bacteria for phytostabilization of mine tailings. *Environmental Science and Technology* 42: 2079–84.

Grass, G., B. Fan, B.P. Rosen, K. Lemke, H.G. Schlegel, and C. Rensing. 2001. NreB from *Achromobacter xylosoxidans* 31A Is a nickel-induced transporter conferring nickel resistance. *Journal of Bacteriology* 183: 2803–7.

Hanikenne, M., and C. Nouet. 2011. Metal hyperaccumulation and hypertolerance: a model for plant evolutionary genomics. *Current Opinion in Plant Biology*14: 252–9.

Hasar, H., and E. Obek. 2001. Removal of toxic metals from aqueous solution by duckweed (*Lemna minor* L): Role of harvesting and adsorption isotherms. *Arabian Journal of Science and Engineering* 26: 47–54.

He, C.Q., G.E. Tan, X. Liang, W. Du, Y.L. Chen, G.Y. Zhi, and Y. Zhu. 2010. Effect of Zn-tolerant bacterial strains on growth and Zn accumulation in *Orychophragmus violaceus*. *Applied Soil Ecology* 44: 1–5.

Hijosa-Valsero, M., G. Fink, M.P. Schlusener, R. Sidrach-Cardona, J. Martin-Villacorta, T. Ternes, and E. Becares. 2011. Removal of antibiotics from urban wastewater by constructed wetland optimization. *Chemosphere* 83: 713–9.

Hussain, S., T. Siddique, M. Arshad, and M. Saleem. 2009. Bioremediation and phytoremediation of pesticides: recent advances. *Critical Reviews in Environmental Science and Technology* 39: 843–907.

Hussein, H.S., O.N. Ruiz, N. Terry, and H. Daniell. 2007. Phytoremediation of mercury and organomercurials in chloroplast transgenic plants: enhanced root uptake, translocation to shoots, and volatilization. *Environmental Science and Technology* 41: 8439–46.

Ionescu, M., K. Beranova, V. Dudkova, L. Kochankova, K. Demnerova, T. Macek, and M. Mackova. 2009. Isolation and characterization of different plant associated bacteria and their potential to degrade polychlorinated biphenyls. *International Biodeterioration and Biodegradation* 63: 667–72.

Jacob, G.S., J.R. Garbow, L.E. Hallas, N.M. Kimack, G.M. Kishore, and J. Schaefer. 1988. Metabolism of glyphosate in *Pseudomonas* sp. strain LBr. *Applied and Environmental Microbiology* 54: 2953–8.

James, C.A., G. Xin, S.L. Doty, and S.E. Strand. 2008. Degradation of low molecular weight volatile organic compounds by plants genetically modified with mammalian cytochrome P450 2E1. *Environmental Science and Technology* 42: 289–93.

Jiang, C.Y., X.F. Sheng, M. Qian, and Q.Y. Wang. 2008. Isolation and characterization of a heavy metal-resistant *Burkholderia* sp. from heavy metal-contaminated paddy field soil and its potential in promoting plant growth and heavy metal accumulation in metal-polluted soil. *Chemosphere* 72: 157–64.

Johal, G.S., and D.M. Huber. 2009. Glyphosate effects on diseases of plants. *European Journal of Agronomy* 31: 144–52.

Jones, R.K., W.H. Sun, C.S. Tang, and F.M. Robert. 2004. Phytoremediation of petroleum hydrocarbons in tropical coastal soils. II. Microbial response to plant roots and contaminant. *Environmental Science and Pollution Research International* 11: 340–6.

Kidd, P.S., A. Prieto-Fernández, C. Monterroso, and M.J. Acea. 2008. Rhizosphere microbial community and hexachlorocyclohexane degradative potential in contrasting plant species. *Plant and Soil* 302: 233–247.

Kolpin, D.W., E.T. Furlong, M.T. Meyer, E.M. Thurman, S.D. Zaugg, L.B. Barber, and H.T. Buxton. 2002. Pharmaceuticals, hormones, and other organic wastewater contaminants in U.S. streams, 1999–2000: a national reconnaissance. *Environmental Science and Technology* 36: 1202–11.

Kraemer, U. 2005. Phytoremediation: novel approaches to cleaning up polluted soils. *Current Opinion in Biotechnology* 16: 1–9.

Kuffner, M., M. Puschenreiter, G. Wieshammer, M. Gorfer, and A. Sessitsch. 2008. Rhizosphere bacteria affect growth and metal uptake of heavy metal accumulating willows. *Plant and Soil* 304: 35–44.

Kuiper, I., E.L. Lagendijk, G.V. Bloemberg, and B.J. Lugtenberg. 2004. Rhizoremediation: a beneficial plant–microbe interaction. *Molecular Plant Microbe Interactions* 17: 6–15.

Kunito, T., K. Saeki, K. Nagaoka, H. Oyaizu, and S. Matsumoto. 2001. Characterization of copper-resistant bacterial community in rhizosphere of highly copper-contaminated soil. *European Journal of Soil Biology* 37: 95–102.

Leigh, M.B., P. Prouzova, M. Mackova, T. Macek, D.P. Nagle, and J.S. Fletcher. 2006. Polychlorinated biphenyl (PCB)-degrading bacteria associated with trees in a PCB-contaminated site. *Applied and Environmental Microbiology* 72: 2331–42.

Li, K., and W. Ramakrishna. 2011. Effect of multiple metal resistant bacteria from contaminated lake sediments on metal accumulation and plant growth. *Journal of Hazardous Materials* 189: 531–9.

Li, W.C., Z.H. Ye, and M.H. Wong. 2010. Metal mobilization and production of short-chain organic acids by rhizosphere bacteria associated with a Cd/Zn hyperaccumulating plant, *Sedum alfredii*. *Plant and Soil* 326: 453–67.

Lin, Q., K.L. Shen, H.M. Zhao, and W.H. Li. 2008. Growth response of *Zea mays* L. in pyrene-copper co-contaminated soil and the fate of pollutants. *Journal of Hazardous Materials* 150: 515–21.

Liste, H.H., and D. Felgentreu. 2006. Crop growth, culturable bacteria, and degradation of petrol hydrocarbons (PHCs) in a long-term contaminated field soil. *Applied Soil Ecology* 31: 43–52.

Loffredo, E., C. Eliana Gattullo, A. Traversa, and N. Senesi. 2010. Potential of various herbaceous species to remove the endocrine disruptor bisphenol A from aqueous media. *Chemosphere* 80: 1274–80.

Lopez-Moreno, M.L., G. de la Rosa, J.A. Hernandez-Viezcas, J.R. Peralta-Videa, and J.L. Gardea-Torresdey. 2010. X-ray absorption spectroscopy (XAS) corroboration of the uptake and storage of CeO(2) nanoparticles and assessment of their differential toxicity in four edible plant species. *Journal of Agricultural and Food Chemistry* 58: 3689–93.

Ma, Y., M. Rajkumar, and H. Freitas. 2009. Inoculation of plant growth promoting bacterium *Achromobacter xylosoxidans* strain Ax10 for the improvement of copper phytoextraction by *Brassica juncea*. *Journal of Environmental Management* 90: 831–7.

Mastroianni, N., M. López de Alda, and D. Barceló. 2010. Emerging organic contaminants in aquatic environments: state-of-the-art and recent scientific contributions. *Contributions to Science* 6: 193–97.

McConnell, Laura L., and Donald W. Sparling. 2010. Emerging contaminants and their potential effects on amphibians and reptiles. In *Ecotoxicology of Amphibians and Reptiles*, 487–509. Boca Raton, FL: CRC Press.

McGuinness, M., and D. Dowling. 2009. Plant-associated bacterial degradation of toxic organic compounds in soil. *International Journal of Environmental Research and Public Health* 6: 2226–47.

Moontongchoon, P., S. Chadchawan, N. Leepipatpiboon, A. Akaracharanya, A. Shinmyo, and H. Sano. 2008. Cadmium-tolerance of transgenic *Ipomoea aquatica* expressing serine acetyltransferase and cysteine synthase. *Plant Biotechnology* 25: 201–03.

Novakova, M., M. Mackova, Z. Chrastilova, J. Viktorova, M. Szekeres, K. Demnerova, and T. Macek. 2009. Cloning the bacterial *bphC* gene into *Nicotiana tabacum* to improve the efficiency of PCB phytoremediation. *Biotechnology and Bioengineering* 102: 29–37.

Olson, P.E., A. Castro, M. Joern, N.M. DuTeau, E.A. Pilon-Smits, and K.F. Reardon. 2007. Comparison of plant families in a greenhouse phytoremediation study on an aged polycyclic aromatic hydrocarbon-contaminated soil. *Journal of Environmental Quality* 36: 1461–9.

Phillips, L.A., J.J. Germida, R.E. Farrell, and C.W. Greer. 2008. Hyrocarbon degradation potential and activity of endophytic bacteria associated with prairie plants. *Soil Biology and Biochemistry* 40: 3054–64.

Pieper, D.H., and M. Seeger. 2008. Bacterial metabolism of polychlorinated biphenyls. *Journal of Molecular Microbiology and Biotechnology* 15: 121–38.

Prasad, M.N.V., and H.M.D. Freitas. 2003. Metal hyperaccumulation in plants—biodiversity prospecting for phytoremediation technology. *Electronic Journal of Biotechnology* 6: 285–321.

Rajkumar, M., N. Ae, M.N. Prasad, and H. Freitas. 2010. Potential of siderophore-producing bacteria for improving heavy metal phytoextraction. *Trends in Biotechnology* 28: 142–9.

Rajkumar, M., and H. Freitas. 2008. Effects of inoculation of plant-growth promoting bacteria on Ni uptake by Indian mustard. *Bioresource Technology* 99: 3491–8.

Reinhold, D., S. Vishwanathan, J.J. Park, D. Oh, and F.M. Saunders. 2010. Assessment of plant-driven removal of emerging organic pollutants by duckweed. *Chemosphere* 80: 687–92.

Ribeiro, H., A.P. Mucha, C.M. Almeida, and A.A. Bordalo. 2010. Hydrocarbon degradation potential of salt marsh plant–microorganisms associations. *Biodegradation* 22: 729–39.

Safe, S.H. 1994. Polychlorinated biphenyls (PCBs): environmental impact, biochemical and toxic responses, and implications for risk assessment. *Critical Reviews in Toxicology* 24: 87–149.

Salt, D.E., N. Benhamou, M. Leszczyniecka, and I. Raskin. 1999. A possible role for rhizobacteria in water treatment by plant roots. *International Journal of Phytoremediation* 1: 67–79.

Salt, D.E., M. Blaylock, N.P. Kumar, V. Dushenkov, B.D. Ensley, I. Chet, and I. Raskin. 1995. Phytoremediation: A Novel Strategy for the Removal of Toxic Metals from the Environment using Plants-Review. *BioTechnology* 13: 468–74.

Schmidt, T., and H.G. Schlegel. 1994. Combined nickel-cobalt-cadmium resistance encoded by the *ncc* locus of *Alcaligenes xylosoxidans* 31A. *Journal of Bacteriology* 176: 7045–54.

Scott, C., G. Pandey, C.L. Hartley, C.J. Jackson, M.J. Cheesman, M.C. Taylor, R. Pandey, J.L. Khurana, M. Teese, C.W. Coppin, K.M. Weir, R.K. Jain, R. Lal, R.J. Russell, and J.G. Oakeshott. 2008. The enzymatic basis for pesticide remediation. *Indian Journal of Microbiology* 48: 65–79.

Seeger, M., M. Hernández, V. Méndez, B. Ponce, M. Córdova, and M. González. 2010. Bacterial degradation and bioremediation of chlorinated herbicides and biphenyls. *Journal of Soil Science and Plant Nutrition* 10: 320–32.

Shi, J.Y., H.R. Lin, X.F. Yuan, X.C. Chen, C.F. Shen, and Y.X. Chen. 2011. Enhancement of copper availability and microbial community changes in rice rhizospheres affected by sulfur. *Molecules* 16: 1409–17.

Sinha, S., and S.K. Mukherjee. 2008. Cadmium-induced siderophore production by a high Cd-resistant bacterial strain relieved Cd toxicity in plants through root colonization. *Current Microbiology* 56: 55–60.

Solis-Dominguez, F.A., A. Valentin-Vargas, J. Chorover, and R.M. Maier. 2011. Effect of arbuscular mycorrhizal fungi on plant biomass and the rhizosphere microbial community structure of mesquite grown in acidic lead/zinc mine tailings. *Science of the Total Environment* 409: 1009–16.

Stevens, K.J., S.Y. Kim, S. Adhikari, V. Vadapalli, and B.J. Venables. 2009. Effects of triclosan on seed germination and seedling development of three wetland plants: *Sesbania herbacea*, *Eclipta prostrata*, and *Bidens frondosa*. *Environmental Toxicology and Chemistry* 28: 2598–609.

Tervahauta, A.I., C. Fortelius, M. Tuomainen, M.L. Akerman, K. Rantalainen, T. Sipila, S.J. Lehesranta, K.M. Koistinen, S. Karenlampi, and K. Yrjala. 2009. Effect of birch (*Betula* spp.) and associated rhizoidal bacteria on the degradation of soil polyaromatic hydrocarbons, PAH-induced changes in birch proteome and bacterial community. *Environmental Pollution* 157: 341–6.

Tripathi, M., H.P. Munot, Y. Shouche, J.M. Meyer, and R. Goel. 2005. Isolation and functional characterization of siderophore-producing lead- and cadmium-resistant *Pseudomonas putida* KNP9. *Current Microbiology* 50: 233–7.

United States Environmental Protection Agency. 2008. Emerging contaminant—Tungsten fact sheet. United States Environmental Protection Agency.

van der Lelie, D., P. Corbisier, L. Diels, A. Gills, C. Lodewyckx, M. Mergeay, S. Taghavi, N. Spelmans, and J. Vangronsveld. 2000. The role of bacteria in the phytoremediation of heavy metals. In *Phytoremediation of Contaminated Soil and Water*, 265–281. Boca Raton, FL: Lewis Publishers.

Verlicchi, P., A. Galletti, M. Petrovic, and D. Barceló. 2010. Hospital effluents as a source of emerging pollutants: an overview of micropollutants and sustainable treatment options. *Journal of Hydrology* 389: 416–28.

Villenueve, A., S. Larroudé, and J.F. Humbert. 2011. Herbicide contamination of freshwater ecosystems: impact on microbial communities. In *Pesticides: formulations, effects, fate*, 285–312. InTech.

Vosátka, M., J. Rydlová, R. Sudová, and M. Vohnik. 2006. Mycorrhizal fungi as helping agents in phytoremediation of degraded and contaminated soils. In *Phytoremediation and Rhizoremediation (Focus on Biotechnology)*, 237–58. Dordrecht, the Netherlands: Springer.

Wang, Q., Y. Cui, and Y. Dong. 2002. Phytoremediation of polluted waters: Potentials and prospects of wetland plants. *Acta Biotechnologica* 22: 199–208.

Wang, W. 1990. Literature Review on Duckweed Toxicity Testing. *Environmental Research* 52: 7–22.

Wei, G., L. Fan, W. Zhu, Y. Fu, J. Yu, and M. Tang. 2009. Isolation and characterization of the heavy metal resistant bacteria CCNWRS33-2 isolated from root nodule of *Lespedeza cuneata* in gold mine tailings in China. *Journal of Hazardous Materials* 162: 50–6.

Weyens, N., D. van der Lelie, T. Artois, K. Smeets, S. Taghavi, L. Newman, R. Carleer, and J. Vangronsveld. 2009. Bioaugmentation with engineered endophytic bacteria improves contaminant fate in phytoremediation. *Environmental Science and Technology* 43: 9413–8.

Wu, C.H., T.K. Wood, A. Mulchandani, and W. Chen. 2006. Engineering plant–microbe symbiosis for rhizoremediation of heavy metals. *Applied and Environmental Microbiology* 72: 1129–34.

Xian, Q., L. Hu, H. Chen, Z. Chang, and H. Zou. 2010. Removal of nutrients and veterinary antibiotics from swine wastewater by a constructed macrophyte floating bed system. *Journal of Environmental Management* 91: 2657–61.

Yousaf, S., V. Andria, T.G. Reichenauer, K. Smalla, and A. Sessitsch. 2010. Phylogenetic and functional diversity of alkane degrading bacteria associated with Italian ryegrass (*Lolium multiflorum*) and Birdsfoot trefoil (*Lotus corniculatus*) in a petroleum oil-contaminated environment. *Journal of Hazardous Materials* 184: 523–32.

Zayed, A., S. Gowthaman, and N. Terry. 1998. Phytoaccumulation of trace elements by wetland plants: I. Duckweed. *Journal of Environmental Quality* 27: 715–21.

Zeng, F., S. Chen, Y. Miao, F. Wu, and G. Zhang. 2008. Changes of organic acid exudation and rhizosphere pH in rice plants under chromium stress. *Environmental Pollution* 155: 284–9.

18 *Brassica* Oilseeds–Microbe Interactions and Toxic Metals Remediation

Lixiang Cao

CONTENTS

18.1 INTRODUCTION

Trace metals contamination due to natural and anthropogenic sources is a global environmental concern. A number of concerted efforts have been made to decontaminate the polluted biosphere. Keeping in view the nonbiodegradability of majority of inorganic environmental contaminants, microbial metals bioremediation is an efficient strategy due to its low cost, high efficiency, and eco-friendly nature. Plant-associated microorganisms include endophytic, phyllospheric, and rhizosphere microorganisms. Endophytic microorganisms can be defined as microorganisms that colonize the internal tissue of the plant without causing visible external sign of infection or a negative effect of the host (Weyens et al. 2009). Because endophytic microorganisms (including bacteria and fungi) can proliferate within the plant tissue, they are likely to interact closely with their host and therefore face less competition for nutrients and are more protected from adverse changes in the environment than bacteria in the rhizosphere and phyllosphere. Endophytic microorganisms exhibit tremendous diversity in both plant hosts and microbial taxa such as bacteria and fungi. The phyllosphere is the external regions of plant parts that are above ground, including leaves, stems, blossoms and fruits. Because the majority of the surface area available for colonization is located on the leaves (the dominant tissue of the phyllosphere), the external region of plant regions was called phyllosphere parts. Microorganisms residing in the phyllosphere are exposed to large and rapid fluctuations in temperature, solar radiation and water availability. In the review, the phyllosphere microorganisms were not included. The unique ability of hyperaccumulator plants to accumulate excessive amounts of metals or metalloids are related to transport systems of the root tissues (Wenzel et al. 2009). Root exudates are believed to have a major influence on the diversity of microorganisms (Weyens et al. 2009). More recently, attention has concentrated on the plant-growth promoting capacity of endophytes and a close relationship exists between microorganisms living in the rhizosphere and those inside the plant roots (endophytes). Plant roots are the main site of endophytic colonization. Root colonization by microorganisms was described to involve several stage (Taghavi et al. 2010): in the initial step, bacteria move toward the plant roots either passively via soil water fluxes, or actively via specific induction of flagellar activity by plant-released compounds; second, a nonspecific adsorption of bacteria to roots occurs, followed by anchoring that results in firm attachment of bacteria to the root surface. Specific or complex interactions between the bacterium and the host plant, such as the secretion of root exudates may arise resulting in the induction of bacterial gene expression. Finally, endophytes can enter the plant at sites of tissue damage, which naturally arise as the result of plant growth through root hairs and at epidermal conjunction. In addition, plant exudates given off through these wounds provide a nutrient source for the colonization microorganisms and thus create favorable conditions. After entering the plant, endophytes must establish themselves once established, they can induce plant resistance to pathogen and insects or promote plant growth by producing plant growth regulating substances such as indole acetic acid (IAA), cytokinins or metabolize the stress ethylene precursor 1-aminocyclopropane-1-carboxylic acid (ACC). The plant-microbe partnerships are beneficial for increasing biomass production of host plants. The rapeseed meal is high in nitrogen (6% wt/wt), stimulates 100-fold increases in populations of resident *Streptomyces* species, and suppresses fungal infection of roots subsequently cultivated in the amended soil (Cohen et al. 2004).

It is well known that members of the cruciferae, which include major commercial crops such as Chinese cabbage, broccoli and rapeseed, are known as nonmycorrhizal plants and no other types of fungal symbionts were found in these plant species (Usuki and Narisawa 2007). So the endophytes showed important influence on the growth of rapeseed in metal-contaminated soils. The review focused on the interactions between rhizosphere microorganisms (including root endophytes) of rapeseed and their hosts, and then their potentials for phytoremediation of metal contaminated soils were also discussed.

18.2 BACTERIAL AND FUNGAL COMMUNITIES OF *BRASSICA* OILSEEDS

Over 300 rhizoplane and 220 endophytic bacteria were randomly selected from *B. napus* L. in Saskatchewan, Canada, by agar-solidified trypticase soy broth. Based on fatty acid methylester (FAME) profiles, 18 bacterial genera were identified with a similarity index > 0.3, but 73% of the identified isolates belonged to four genera: *Bacillus* (29%), *Flavobacterium* (12%), *Micrococcus* (20%) and *Rathayibacter* (12%) (Germida et al. 1998). The endophytic community had a lower Shannon–Weaver diversity index compared to the rhizoplane, and a higher proportion of *Bacillus*, *Flavobacterium*, *Micrococcus*, and *Rathayibacter* genera compared to rhizoplane populations. Genera identified in the endophytic isolates were also found in the rhizoplane isolates. So endophytes are a subset of the rhizoplane community (Germida et al. 1998). The composition and diversity of the bacterial community associated with plant roots is influenced by a variety of plant factors such as root density and exudation. Rhizoplane communities of wheat and canola grown at the same field site differ in their taxonomic composition (Germida et al. 1998). Further work suggests that the selective effect of plants on the rhizosphere community can occur even at the cultivar level.

Approximately 2300 bacteria were selected from root-endophytic and rhizosphere bacterial communities associated with three canola cultivars (Parkland, *B. rapa*; Excel, *B. napus*; and Quest, *B. napus*) grown at two field sites and identified based on FAME profiles (Siciliano and Germida 1999). Fewer *Bacillus*, *Micrococcus*, and *Varioverax* isolates, and more *Flavobacterium* and *Pseudomonas* isolates were found in the root interior of Quest compared to Excel or Parkland. Furthermore, fewer *Arthrobacter* and *Bacillus* isolates were recovered from the rhizosphere of Quest compared to Excel or Parkland. The root endophytic bacterial community of the transgenic cultivar, Quest, was separated by principal component analysis from the other cultivars, and exhibited a lower diversity compared to Excel or Parkland. The rhizosphere of all cultivars yield more *Arthrobacter*, *Aureobacterium*, and *Bacillus* isolates, but fewer *Micrococcus*, *Variovorax*, and *Xanthomonas* isolates compared to the root interior (Siciliano and Germida 1999).

Phylogenetic analysis of the 16S rRNA gene clone library from the rhizoplane of oilseed rape (*B. napus* cv. Westar) revealed considerable differences from the corresponding nucleotide sequences of cultured bacteria. The 16S rRNA gene clone library was dominated by α-*Proteobacteria* and bacteria of the *Cytophaga–Flavobacterium–Bacteroides* phylum (51% and 30%, respectively), less than 17% of the cultured bacteria belonged to these two groups. More than 64% of the cultivated isolates were allocated to the β- and γ-subclasses of the *Proteobacteria*, which were present in the clone library at about 14%. Most of the clones of the α-*Proteobacteria* of the library showed highest similarity to *Bradyrhizobium* sp. No such bacteria were found in the culture collection. Similarly, the second dominant group of the clone library comprising members of the *Cytophaga–Flavobacterium–Bacteroides* phylum was represented in the culture collection by a single isolate. The phylogenetic analysis of isolates of the culture collection clearly emphasized the need to use different growth media for recovery of rhizoplane bacteria (Kaiser et al. 2001). So the bacterial communities should be analyzed by different methods including cultivation-dependent and cultivation-independent approaches. It has been demonstrated that microorganisms present in the rhizosphere and within oilseed rape tissues affect the susceptibility to its wilt pathogen *Verticillium longisporum* (Granér et al. 2003).

Cultivar differences in terms of resistance to pathogen may partly be due to qualitative and quantitative differences in their endophytic microbial populations and its endophytic microbial populations. The result is helpful to study the resistance of oilseed rape to metals. A successful outcome of biological control in the field demands a better understanding of the complex microbial interactions in plants. Therefore, the potential of endophytic strains naturally colonizing plants should be given greater attention in plant breeding and calls for further studies of microbial ecology and pathology. One important factor required for an optimal performance of an introduced endophytic microorganism is, however, considered to be the relationship between plant genotype and effective colonization of the host.

Fungal endophytes residing in the internal tissues of living plants occur in almost every plant on earth from the arctic to the tropics (Aly et al. 2011). A major limitation of cultivation-dependent studies for unraveling diversity of endophytes is due to the bias toward fast-growing ubiquitous species, wherease rare species with minor competitive strength and more specialized requirements may remain undiscovered. For some technique used for isolation of bacteria did not favor growth of fungi, few fungal colonizers were studied in the oilseed rape materials. The presence of both fungi and bacteria from *B. napus* was found by using rape root extract-based media (Granér et al. 2003). It was shown that endophytes possessing suitable metal sequestration or chelation systems were able to increase host plant tolerance to presence of heavy metal, thereby assisting their hosts to survive in contaminated soil (Aly et al. 2011). Dark septate endophytic (DSE) fungi are defined as conidial or sterile ascomycetous fungi, which colonize living plant root tissues intracellularly and intercelularly without causing any apparent negative effect, such as tissue disorganization or forming any typical mycorrhizal structures (Usuki and Narisawa 2007). Root-associated dark septate endophytic fungi, including *Mycelium radicis atrovirens* (MRA) have been reported from approximately 600 plant species of 320 genera in 114 families and may have a potentially important functional relationship with host plants (Ohki et al. 2002). Chinese cabbage (*B. campestris* L.) could be colonized by endophytic fungus *Heteroconium chaetospira*. Three weeks post inoculation, some hyphae became irregular lobed and formed microsclerotia within host epidermal cells of healthy plants. In stunted plants, hyphae formed closely packed masses of fungal cells within host epidermal cells, but conidiophores rarely broke through the cell walls to produce conidia (Ohki et al. 2002). Chinese cabbage could not use some amino acids while *Heteroconium chaetospira*-treated plants were able to use all of the nitrogen forms provided and the fungus obtained carbon, mainly as sucrose, from the host plant, so the existence of a fungus establishing a mutualistic association with the Chinese cabbage. The endophytic fungi *Phialocephala fortinii*, *Heteroconium chaetospira* could most effectively inhibit the development of *Verticillium* yellows with reductions in the percentages of external and internal disease symptoms of 84 and 88%, respectively (Narisawa et al. 2004). The fungal population of rape needs to be further studied.

18.3 PROPERTIES OF RHIZOSPHERE AND ENDOPHYTIC MICROORGANISMS

The individual bacterial isolates or mixtures of bacteria that originated from symptomless oilseed rape significantly improved seed germination, seedling length and plant growth of oilseed rape (*B. napus* L. cv. Casino), and when used for seed treatment significantly reduced disease symptoms caused by vascular wilt pathogen *Verticillium dahliae* (Nejad and Johnson 2000). A screening of fungal isolates for biologically active secondary metabolites (antibacterial, antifungal, herbicidal) showed that the proportion of endophytic isolates that produced herbicidally active substances was three times that of the soil isolates and twice that of the phytopathogenic fungi, and in the cases in which the concentrations of phenolic metabolites were higher in the roots infected with an endophyte than those infected with a pathogen. So the endophyte-host interactions involve constant mutual antagonisms at least in part based on the secondary metabolites the partners produce, whereas, the pathogen-host interaction is imbalanced and results in disease and the interaction between the endophyte and its host is a balanced antagonism (Schulz et al. 1999). The metabolites (such as IAA, cytokins and herbicidal metabolites) produced by endophytes are involved in the interactions between endophyte and host.

Brassica juncea, grown as leafy vegetables, are known to accumulate large amounts of heavy metals in their shoots and roots because of their high biomass and root proliferation. The reduction in root and shoot growth corresponded with the amounts of extractable Cd in the soils. The total content of Cd in the crops increased gradually as the rate of applied Cd rose and the roots accumulated much higher amounts than the shoot. The relationship of Cd with Zn and Fe was synergistic in both roots and shoots at the lower rates, but antagonistic at higher Cd application rates (Sidhu and Khurana 2010). It had been showed that *B. juncea* was less able to accumulate Cd in shoots compared with hydroponically cultured rice and sugar beet, and was even less effective when grown in soil culture. No significant decrease in soil Cd concentration by *B. juncea*. So it was suggested that *B. juncea* does not offer much promise for phytoextraction of Cd from soils with relatively low contamination (Ishikawa et al. 2006). The solubility of lead in soil is limited due to complexation with organic matter, sorption on clays and oxides, and precipitation as carbonates, hydroxides, and phosphates and dramatic increase in the accumulation of Cd, Cu, Ni, Pb, and Zn from soil in the presence of added synthetic chelates and EDTA was the most effective in increasing Pb accumulation in shoots of *B. juncea* (Epstein et al. 1999). However, if the excessive soluble Pb were not completely taken up by plants, further groundwater contamination by EDTA-Pb will occur.

One hundred rhizobacteria isolated from the rhizosphere of *Brassica* species were screened for their growth promoting activity in *B. napus* L. revealed that 58% of the rhizobacteria increased root length (up to 139%), 39% enhanced shoot length (up to 78%), and shoot weight (up to 72%) of *B. napus*; the pot trials revealed that inoculation with selected PGPR increased plant height, root length, number of branches per plant, stem diameter, number of pods per plants, 1000-grain weight, grain yield, and oil content over a range of 7%–57% above the uninoculated control (Asghar et al. 2004). It had been showed that the rhizosphere isolates were more efficient producers of auxins than soil isolates not associated with plant roots, and seed inoculation with auxin-producing rhizobacteria significantly increased plant height (up to 56.5%), stem diameter (up to 11.0%), number of branches (up to 35.7%), number of pods per plant (up to 26.7%), 1000-grain weight (up to 33.9%), grain yield (up to 45.4%) and oil content (up to 5.6%) over the uninoculated *B. juncea* control (Asghar et al. 2002). Many endophytic bacteria can establish themselves in plants other than their original hosts, indicating a lack of host specificity. This absence of host specificity in endophytes has been reported for other members of the same plant family from which they are isolated and for members of other families. *B. oleracea* inoculated with *Enterobacter* sp. strain 35 isolated from sugarcane had a significantly greater fresh weight than uninoculated plants, and the fresh weight of plants inoculated with *Herbaspirillum* sp. strain B501 tended to be higher than that of uninoculated plants. *Enterobacter* sp. strain 35 had larger bacterial populations and greater acetylene reduction activity than those inoculated with *Herbaspirillum* sp. strain B501 (Zakria et al. 2008).

Phytoremediation of metals and other inorganic compounds may take one of several forms: phytoextraction, the absorption and concentration of metals from the soil into the roots and shoots of the plant; rhizofiltration, the use of plant roots to remove metals from effluents; phytostabilization, the use of plants to reduce the spread of metals in the environment; or phytovolatilization, the uptake and release into the atmosphere of volatile materials such as mercury or arsentic-containing compounds (Glick 2003). Even oilseed rape that are relatively tolerant of various environmental contaminants often remain small in the presence of the metal contaminants. Anthropogenic soil pollution is usually not limited to a single pollutant, several metals may be present at high concentration in soils, however, the *B. napus* (canola) are moderately tolerant to heavy metals and the species showed relatively low phytoremediation potential of multicontaminated soils (Marchiol et al. 2004).

Plant growth-promoting bacteria that facilitate the proliferation of various plants especially under environmentally stressful conditions may be added to the roots of plants. The bacteria could lower the level of growth-inhibiting stress ethylene within the plant and also to provide the plant with iron from the soil. The inoculation of plants with bacteria could significantly increase in both the number of seeds that germinate and the amount of biomass that the plants are able to attain, and a much faster and more efficient phytoremediation process could be achieved by inoculation of plant growth-promoting bacteria (Glick 2003). Densely packed mycorrhizal shealth and phenolic interhyphal material can protect plant roots from direct contact with pollutant (Wenzel 2009). While the large surface and cation exchange capacity of extramatrical mycelia may reduce bioavailable metal concentrations through their substantial adsorption capacity. The structure of the fungal shealth and the density and the density and surface area of the mycelium are likely to be important characteristics determining the efficiency of an ectomycorrhizal association to withstand metal toxicity and to protect the host plant from metal contact (Wenzel et al. 2009).

Selection, traditional breeding and genetic engineering of plants focus on increasing pollutant tolerant, root and shoot biomass, root architecture and morphology, pollutant uptake properties were used to enhance the phytoremediation process, other approaches of management of microbial consortia: the selection and engineering of microorganisms with capabilities for metal tolerance, beneficial effects on the phytoremediation crops or modifying effect on the metal availability were also to improve the phytoremediation efficiency (Kidd et al. 2009). Metal bioavailability can be defined as the fraction of the total metal content of the soil that can interact with a biological target. In the soil solution, elements are present as free uncomplexed ions, ion pairs, ions complexed with organic anions, and ions complexed with organic macromolecules and inorganic colloids. The most important metal pools in the solid phase include the exchange complex metals complexed by organic matter, sorbed onto or occluded within oxides and clay minerals co-precipitated with secondary pedogenic minerals (e.g. Al, Fe, Mn oxides, carbonates and phosphates, sulfides) or as part of the crystal lattices of primary minerals (Kidd et al. 2009). Availability to plants is governed by the pseudo-equilibrium between aqueous and solid soil phases rather than by the total metal content. It have been showed that metal uptake by plants is dependent on metal availability in the soil (McGrath et al. 2009). The availability of metals for plant uptake is greatly restricted by their adsorption to solid soil fractions. Chelant-assisted phytoextraction has been used to improve the effectiveness of conventional phytoextraction of metals contaminated soils by dissolving target metals from soil and making them more available for plant uptake and translocation to harvestable above-ground parts of high biomass crops (Quartacci et al. 2009). Aminopolycarboxylic acids such as nitrilotriacetic acid (NTA) and (S, S)-ethylenediaminedisuccinic acid (EDDS) have been proposed as an alternative to EDTA and other persistent synthetic chelants. Even through the application of such biodegradable chelants could minimize the risks of potential off-site migration of metals, either in surface runoff or by leaching into groundwater, the accumulation of metals such as Cd, Cu, and Zn increased only by a factor of 2–3, although their solubility in soil increased by a much higher factor. Therefore efforts have been focused on screening plant species that are more sensitive to chelant treatments and on developing new phytotechnologies that will help reduce and/or eliminate the amount of chelant applied to the field decreasing the environmental risk of mobilized metals. Root exudates may influence the availability of nutrients by a direct impact on the uptake of metals by acidification, chelation, precipitation and redox reactions as well as by an indirect impact through their effects on microbial activity, the physical and chemical properties of the rhizosphere and root growth patterns. The growth of the metallicolous populations (*Pinus pinaster*, *Plantago lanceolata*, *Silene paradoxa*) increased the extractable metal levels in the soil, which resulted in a higher accumulation of metals in the above-ground parts of *B. carinata* (Quartacci et al. 2009). Interactions between root exudates and soil components prevented the increase of water-soluble organometallic chelates in the rhizosphere, suggesting that the organic compounds exuded by roots can rapidly become sorbed to the soil solid phase. The reductions observed in water-soluble or labile pools of metals generally accounted for a low percentage (<10%) of the total metal uptake by the plants, indicating that the metals are mainly acquired from less available pools.

18.4 MICROBIAL EFFECTS ON METALS-/METALLOIDS-BIOAVAILABILITY IN THE RHIZOSPHERE

Microbial activity strongly influences metal speciation and transport in the environment. Different organisms exhibit diverse responses to toxic ions, which confer upon them a certain range of metal tolerance. Eukaryotes are more sensitive to metal toxicity than bacteria and their typical mechanism for regulating intracellular metal ion concentrations is the expression of metallothioneins (MTs), a family of metal-chelating protein. Bacterial MTs are functionally homologous to the short (approximately 60 amino acids) cysteine-rich eukaryotic metal-binding proteins (Valls and Lorenzo 2002). The production of MTs as the main mechanism of tolerance to metals is exceptional in the bacterial world. A number of specific resistance mechanisms, including active efflux and sequestration or transformation to other chemical species, become functional at concentrations above the homeostatic or nontoxic levels. These tolerance mechanisms are often plasmid-borne, which facilitates dispersion from cell to cell (Valls and Lorenzo 2002). Some metal-binding peptides have been introduced into bacteria to increase brightness in metal bioaccumulation (Valls and Lorenzo 2002).

In addition to their use as biosorbents, bacteria can be used to efficiently immobilize certain heavy metals through their capacity to reduce these elements to a lower redox state, producing less bioactive metal species, sometimes precipitation and biosorption are overlapping phenomena and it can be difficult to assign the contribution of each to metal immobilization (Newman and Reynolds 2002). Microbiological metal precipitation is a wide spread activity that is either the result of a dissimilatory reduction or the secondary consequence of metabolic processes unrelated to the transformed metals, indirect reduction by formation of metal sulfide and phosphates or dissimilatory metal reduction can be used for decontamination (Lloyd and Lovley 2001).

Bacteria exhibit a number of enzymatic activities that transform certain metal species through oxidation, reduction, methylation and alkylation. The enzymatic transformation also lead to metal precipitation and immobilization, other enzymatic reactions that generate less poisonous metal species such as mercury and arsenic also could be applied to bioremediation (Lloyd and Lovley 2001). Some bacteria and fungi could reduce metals as mechanisms to improve resistance to metals and could accumulate silver, platinum, or gold nanoparticles in their intracellular spaces. The cytoplasmic and periplasmic hydrogenases were suggested to play an important role in the transformation (Durán et al. 2011). As a rapid, cost effective and eco-friendly methods to produce metal nanoparticles, biological production of metal nanoparticles has been studied by other researchers. However, *B. juncea* had a limited capacity to reduce metal ions to form metal nanoparticles. It was suggested that the capacity was related to the electrochemical potential and depends on the reduction potential of the metals. At a potential greater than 0 V, the metal accumulation is limited by reducing capacity of the plant *B. juncea* (Durán et al. 2011).

Microorganisms can increase solubility and change speciation of metals/metalloids by producing organic ligands via microbial decomposition of soil organic matter and exudation of metabolites and microbial siderophores that can complex cationic metals or desorb anionic species by ligand exchange (Wenzel 2009). Depending on the surface charge of soil minerals and below metal-specific pH values, siderophores produced by microbes or plants may immobilize cadmium, copper or zinc cationic. The complexity of interactions between organic ligands and metals was demonstrated by modeling of copper solubility and transport in the root zone in the absence and presence of organic ligands, showing that ligands do not necessarily increase the solubility and bioavailability of metals, and pH was a major control of mobilization and immobilization. Enhanced dissolution of metal (metalloid) compounds or desorption triggered by ion competition between the metal and proton may arise from heterotrophic proton efflux via plasma membrane H^+-ATPases or from dissociation of carboxylic acid accumulated from carbon dioxide respiration (Wenzel 2009). Microbially mediated reduction and oxidation processes can also modify the solubility of metals and metalloids. Dissimilatory in metal reducing bacteria can use Cr^{6+}, Fe^{3+}, Mn^{4+}, Hg^{2+}, Se^{4+}, U^{6+} as terminal electron acceptor. The reduction results in mobilization of iron and manganese, however, can immobilize uranium, chromium, and selenium. The metal solubility can also be subject to co-dissolution if adsorbed to or occluded in sesquioxides if it is not directly affected by changes of the redox potential (Wenzel 2009). Microorganisms can immobilize metals or metalloids in various other ways. They can take up the elements and accumulate them in their biomass via intracellular sequestration or precipitation, or adsorb them onto cell walls or exopolymers released into their surroundings, and mineralization of dissolved metal-organic complexes, the microbial processes are also the causes of microbially mediated immobilization (Wenzel 2009).

The fungi appear to be generally more tolerant to metals than bacteria (Kidd et al. 2009). The fungi can efficiently explore the soil microsites that are not accessible for plant roots due to their small diameter. The fungi could compete with roots and other microorganisms for water and metal uptake, protect the roots from direct interaction with the metals and impeded metal transport through increased soil hydrophobicity (Wenzel 2009). The phytoremediation practices may benefit from microbial processes but in turn may influence the composition and function of the microbial consortia in the rhizosphere of oilseed rape.

18.5 RHIZOSPHERE AND ENDOPHYTIC MICROORGANISMS IMPACT ON OILSEED RAPE GROWTH AND PHYTOREMEDIATION POTENTIAL

The success of phytoremediation is strongly determined by the amount of plant biomass present and the concentration of heavy metals in plant tissues. The high uptake and efficient root-to-shoot transport system endowed with enhanced metal tolerance provide hyperaccumulators with a high potential detoxification capacity, however, most hyperaccumulators are not suitable for phytoremediation applications in the field owing to their small biomass and slow growth (Rajkumar et al. 2009). As elevated levels of metals are toxic to most plants leading to impaired metabolism and reduced plant growth, the potential for metal phytoextraction is highly restricted and necessitates the development of other phytoremediation strategies for heavy metal contaminated soils. The interface between microorganisms and plant roots is considered to greatly influence the growth and survival of plants. So exploiting rhizosphere microorganisms to reduce metal toxicity to plants have been investigated to improve phytoremediation efficiencies, furthermore, the discovery of rhizosphere microorganisms resistant to heavy metals or promote plant growth have raised hopes for ecologically friendly and cost-effective strategies toward reclamation of heavy metal polluted soils. The exploitation of metal-resistant rhizosphere or endophytic microorganisms is important for host plants as they can provide nutrients such as iron to plants, which could reduce the deleterious effects of metal contamination, further more, metabolites produced by the microorganisms could enhance metal bioavailability in the rhizosphere of plants (Rajkumar et al. 2010). The resulting increase in trace metal uptake by the plants might enhance the effectiveness of phytoextraction processes of contaminated soil. Developing new methods for either enhancing or reducing the bioavailability of metal contaminants in the rhizosphere as well as improving plant establishment, growth, and health could significantly speed up the process of bioremediation techniques (Ma et al. 2010).

The plants of *B. juncea* L. cv. Pusa Bold grown on contaminated substrates (Cu^{2+}, Cr^{6+}, As^{3+}, As^{5+}) have shown translocation of metals to the upper part and its sequestration in the leaves without significantly affecting on oil yield, except for Cr^{6+} and higher concentration of As^{5+}, compared to control. Decrease in the oil content in As^{5+} treated plants was observed in a dose dependent manner; however, maximum decrease was recorded in Cr^{6+} treated plants (Sinha et al. 2010). Among all the metal treatments, Cr^{6+} was the most toxic as evident from the decrease in oil content, growth parameters and antioxidants (Sinha et al. 2009).

The plant hormone ethylene, most popularly associated with ripening, plays a major role regulating seed germination, seedling growth, leaf and petal abscission, organ senescence, stress, and pathogen responses throughout the entire life of the plant. Low levels of ethylene appear to enhance root extension, higher levels of ethylene produced by fast growing roots can lead to inhibition of root elongation and cause proliferation of small lateral roots (Madhaiyan et al. 2006). In ethylene synthesis, methionine is converted to S-adenosylmethionin (SAM), 1-aminocyclopropane-1-carboxylate (ACC) and ethylene in three consecutive reactions catalyzed by the enzymes SAM-synthetase, ACC-synthase and ACC-oxidase, respectively. It was postulated that much of the ACC produced by the latter reaction may be exuded from seed or root exudates. The ACC in the exudates may be taken up by bacteria and was hydrolyzed by ACC deaminase to ammonia and α-ketobutyrate. The mechanism effectively reduces the amount of ethylene evolved by the plant due to more ACC is exuded by the plant (Arshad et al. 2007). The ability to promote root elongation by microorganisms is a direct consequence of their ACC deaminase. Lower amounts of ACC were present in the tissues of canola seeds treated with ACC deaminase containing *Methylobacterium fujisawaense* strains than in control seeds. The activities of ACC oxidase, the enzyme catalyzing conversion of ACC to ethylene remained lower in *M. fujisawaense* treated seedlings. The amount of ACC was reduced due to bacterial ACC deaminase activity. However, the increased activities of ACC synthase in the tissue extracts of the treated seedlings might be due to bacterial indole-3-acetic acid (IAA) (Madhaiyan et al. 2006). The presence of plant growth-promoting rhizobacteria, including nitrogen-fixing bacteria and phosphate and potassium solubilizers, stimulated *B. juncea* growth and protected the plant from Pb, Cu, Cd, and Zn toxicity. Inoculation with rhizobacteria had little influence on the metal concentrations in plant tissues, however, produced a much higher above-ground biomass and altered metal bioavailability in the soil (Wu et al. 2006a). Inoculation of Ni-resistant *Pseudomonas* sp. Ps29c and *Bacillus megaterium* Bm4c strains promoted *B. juncea* growth and protected the plant from Ni toxicity. Inoculation with the strains had little influence on the accumulation of Ni in root and shoot system, but produced a much larger aboveground biomass. The strains protect the plants against the inhibitory effects of nickel, probably due to the production of IAA, siderophore and solubilization of phosphate (Rajkumar and Freitas 2008). Both Cr^{2+}- and Ni^{2+}-resistant *Enterobacter aerogenes* and *Rahnella aquatilis* strains were siderophore producing and found capable of stimulating plant biomass and enhance phytoextraction of Ni and Cr from fly ash by metal accumulating *B. juncea*. The concurrent production of siderophores, ACC deaminase, IAA and phosphate solubilization revealed their plant growth promotion potential (Kumar et al. 2009). The inoculation of *Achromobacter xylosoxidans* significantly increased the root length, shoot length, fresh weight and dry weight of *B. juncea* plants compared to the control. This effect can be attributed to the utilization of ACC, production of IAA and solubilization of phosphate. The *A. xylosoxidans* Ax10 inoculation significantly improved Cu uptake by *B. juncea* (Ma et al. 2009a). The strains *Pseudomonas* sp. SRI2, *Psychrobacter* sp. SRS8, and *Bacillus* sp. SN9 isolated from the nonrhizosphere and rhizosphere soils of *Alyssum serpyllifolium* and *Astragalus incanus* at a serpentine site were inoculated with *B. juncea* and *B. oxyrrhina* based on their ability to solubilize Ni in soil. Further assessment on plant growth-promoting parameters revealed the intrinsic ability to produce IAA, siderophores, utilize ACC, and solubilize insoluble phosphate. Inoculation of plants with the Ni mobilizing strains increased the biomass of both *B. juncea* and *B. oxyrrhina*. The

Pseudomonas sp. SRI2, *Psychrobacter* sp. SRS8, and *Bacillus* sp. SN9 showed maximum increase in the biomass of the test plants. The strain SN9 significantly increased the Ni concentration in the root and shoot tissues of *B. juncea* and *B. oxyrrhina* and a significantly positive correlation was observed between the bacterial Ni mobilization in soil and the total Ni uptake in both plant species (Ma et al. 2009b). Compared with control treatment, inoculation of plant growth-promoting bacteria (PGPB) strains significantly increased the concentrations of bioavailable Ni. In soil contaminated with 450 mg kg^{-1} Ni, *Psychrobacter* sp. SRA2 significantly increased the fresh (351%) and dry biomass (285%) of the *B. juncea* plants. Whereas *Psychrobacter* sp. SRA1 and *Bacillus cereus* SRA10 significantly increased the accumulation of Ni in the root and shoot tissues of *B. juncea* compared with noninoculated controls. A significant increase was also noted for growth parameters of the *B. oxyrrhina* plants when the seeds were treated with strain SRA2. This effect might be attributed to the utilization of ACC, solubilization of phosphate and production of IAA. The inoculation of Ni mobilizing strains *Psychrobacter* sp. SRA1 and *B. cereus* SRA10 increases the efficiency of phytoextraction directly by enhancing the metal accumulation in plant tissues and the efficient *Psychrobacter* sp. SRA2 increases indirectly by promoting the growth of *B. juncea* and *B. oxyrrhina* (Ma et al. 2009c). The Ni-tolerant *B. subtilis* strains SJ-101 also exhibited the capability of producing IAA and solubilizing inorganic phosphate. The plants exposed to $NiCl_2$ (1750 mg kg^{-1}) in soil bioaugmented with strain SJ-101 could accumulate 0.147% Ni in dry biomass of the plants; the control plants accumulated 0.094% Ni under the same conditions. Furthermore, the beneficial effects of strain SJ-101 with significant increase in the plant growth attributes in untreated control soil and the protective effect of the strain SJ-101 against Ni phytotoxicity was evident in plants grown in soil treated with $NiCl_2$ in concentration range of 250–1750 mg kg^{-1} (Zaidi et al. 2006).

The cadmium-tolerant *Variovorax paradoxus*, *Rhodococcus* sp. and *Flavobacterium* sp. were capable of stimulating root elongation of *B. juncea* seedlings either in the presence or absence of toxic Cd concentrations, a positive correlation between the *in vitro* ACC deaminase activity of the bacteria and their stimulating effect on root elongation and utilization of ACC is an important bacterial trait determining root growth promotion (Belimov et al. 2005). The *Enterobacter* sp. NBRIK28 and its siderophore overproducing mutant were capable of stimulating plant biomass and enhance phytoextraction of Ni, Zn, and Cr from fly ash by *B. juncea* plants than wild type (Kumar et al. 2008). The multimetal and antibiotic-resistant *Burkholderia* sp. J62 was found to significantly increase the biomass of test plants and increase in tissue Pb and Cd contents varied from 38% to 192% and from 5% to 191% in inoculated plants growing in heavy metal-contaminated soils compared to the uninoculated control, respectively. The isolate could produce IAA, siderophore, ACC deaminase, and increased bacterial solubilization of lead and cadmium in solutions culture and in soils (Jiang et al. 2008a).

The cadmium-resistant bacterial strains showed variable amounts of water soluble Cd released from cadmium carbonate. Based on cadmium-resistance, bio-activation of $CdCO_3$ and growth-promoting activity, inoculation of strains increase root dry weight and shoot dry weight of rape (*B. napus*) and an increase in cadmium content varying from 16 to 74%, compared to the noninoculated control, was observed in rape plants cultivated in soil treated with 100 mg kg^{-1} Cd and inoculated with the isolates (Sheng and Xia 2006). The cadmium-resistant *Pseudomonas tolaasii* ACC23, *Pseudomonas fluorescens* ACC9, *Alcaligenes* sp. ZN4, and *Mycobacterium* sp. ACC14 synthesized ACC deaminase *in vitro* when 0.4 mM Cd^{2+} was added to the growth medium. About 34% up to 97% increases in root elongation of inoculated canola (*B. napus*) seedlings compared to the control plants. The strains did not influence the specific accumulation of cadmium in the root and shoot systems, but all increased the plant biomass and consequently the total cadmium accumulation (Dell'Amico et al. 2008).

Two lead (Pb)-resistant endophytic bacteria *Pseudomonas fluorescens* G10 and *Microbacterium* sp. G16 exhibited different multiple heavy metal and antibiotic resistance and increased water-soluble Pb in solution and in Pb-added soil: increases in root elongation, biomass production and total Pb uptake in the bacteria-inoculated rape plants were obtained compared to the control (Sheng et al. 2008a). The rhamnolipid biosurfactant produced by *Pseudomonas aeruginosa* BS2 removes not only the leachable or available fraction of Cd and Pb but also the bound metals as compared to tap water which removed the mobile fraction only. Washing of contaminated soil with tap water revealed about 2.7% of Cd and 9.8% of Pb in contaminated soil was in freely available or weakly bound forms. However, washing with rhamnolipid removed 92% of Cd and 88% of Pb and the pH of leachates treated with dirhamnolipid solution was lower than that with tap water (Juwarkar et al. 2007). However, lipopeptide biosurfactant-producing *Bacillus* sp. J119 exhibit different multiple heavy metal (Pb, Cd, Cu, Ni, and Zn) and antibiotic resistance characteristics. Cd treatment did not significantly decreased growth of rape, tomato, and maize plant. However, in the Cd-added soil, above ground biomass and root dry weights of tomatoes were increased by 24 and 59%, respectively, in live bacterial inoculation compared to dead bacterial inoculation control. In the soil treated with 50 mg kg^{-1} Cd, increase in above-ground tissue Cd content varied from 39% to 70% in live bacteria-inoculated plants compared to dead bacterium inoculated control (Sheng et al. 2008b). the plant type influenced root colonization activity of the introduced strain and bacterial effects on plant growth and Cd accumulation. The transgenic canola (*B. napus*) seeds that constitutively express the *Enterobacter cloacae* Uw4 ACC deaminase gene could increase 40% more seed germination compared to the nontransformed seeds in the presence of arsenate, however the extent or rate of seed germination was not significantly affected by treating the seeds with an ACC deaminase-containing bacterium. Although bacterial ACC deaminase can act as a sink to lower seed or root ACC levels, the activity of the enzyme is induced rather slowly because of its complex mode of regulation (Nie et al. 2002). However, when grown in the presence of arsenate, the fresh and dry weights of roots and shoots of transgenic canola, especially when they were treated with *E. cloacae*, were much higher than nontransformed canola. Other properties of the strain such as producing IAA, siderophores, and antibiotics may contribute to the result (Nie et al. 2002). Besides transition of gene between bacteria, transgenic plants have been constructed for higher remediation efficiency. The expression of ACC deaminase in the plant exhibit several advantages against in the bacteria (Zhuang et al. 2007): (i) during the initial stages of seed germination, the bacterial ACC deaminase activity is likely to be much lower than the activity in transgenic plants; (ii) it can constantly stimulate plant growth, which lead to a higher metal accumulation; (iii) in some cases, an increase in the shoot/root ratio; (iv) promoting metal uptake of certain fast-growing plants for the substitution of slow-growing hyperaccumulators.

The intracellular sequestration (bioaccumulation) of the metal by chelating proteins or peptides (metallothioneins, phytochelatins, glutathione) or other molecules (polyphosphates) to prevent interactions with metal-sensitive cellular targets (Gutiérrez et al. 2009). The introduction and /or overexpression of metal binding proteins have been widely exploited to increase the metal binding capacity, tolerance or accumulation of bacteria and plants. Modification in the biosynthesis of phytochelatins in plants has recently been accomplished to enhance the metal accumulation, whereas in bacteria, various peptides consisting of metal-binding amino acids (mainly histidine and cysteine residues) have been studied for enhanced heavy metal accumulation by bacteria (Mejáre and Bülow 2001). In the aim of enhancing the metal tolerance, sequestration or accumulation of bacteria, the high metal-binding capacity of metallothioneins (MTs) has been widely exploited and MTs from various sources have been expressed intracellularly in *Escherichia coli* including monkey MT, yeast MT, human MT-2, mouse MT-1, rainbow trout MT, plant MT (Mejáre and Bülow 2001). However, intracellular expression of MTs have problems with the stability of the expressed heterologous proteins. Expression the metal binders on the cell surface where they will not interfere with intracellular activities. Expression of MTs on the surface of *E. coli* showed an up to 1.8-fold increase in Cd^{2+} binding capacity, however, display MTs did not contribute to an increase in the accumulation of Cu^{2+} and Zn^{2+} (Kotrba et al. 1999). Bacteria such as *E. coli* and *Moraxella* sp. expressing metal-binding peptide (EC20, with 20 cysteins) on the cell surface or intracellular have been shown to accumulate up to 25-fold- more cadmium or mercury than the wild-type strain (Wu et al. 2006b). Expression of EC20 in a rhizobacterium *Pseudomonas putida* 06909, not only improved cadmium binding but also alleviated the cellular toxicity of cadmium and inoculation of sunflower roots with the engineered rhizobacterium resulted in a marked decrease in cadmium phytotoxicity and a 40% increase in the cadmium accumulation in the plant root (Wu et al. 2006b). The introduced transgenic bacteria had no effect on the function and structure of the bacterial community in bulk soil, although they enhanced biodegradation of polychlorinated biphenyl (PCB) as determined by chemical analysis. However, the transgenic bacteria affected the development of functionally and genetically distinct bacterial communities in the rhizosphere (de Cárcer et al. 2007). The effect of transgenic bacteria on the function and structure of the bacterial community were not studied. Numerous attempts to engineer the production of MTs in plants to increase metal tolerance and/or accumulation have been reported. However, it is mainly the model plant species that have been genetically engineered. The engineering of crop plants or high biomass plant such as oilseed rape would be a further step toward the development of metal accumulating plants for phytoremediation (Mejáre and Bülow 2001).

The Cr^{6+}-resistant bacteria *Pseudomonas* sp. PsA4 and *Bacillus* sp. Ba32 both produce siderophore, IAA and solubilize soil phosphate. Inoculation of the strains promoted the growth of *B. juncea* at 95.3 and 198.3 mg kg^{-1} Cr^{6+}. However, the strain did not influence the quantity of accumulation of chromium in root and shoot system. The strains protect the plants against the inhibitory effects of chromium, probably, due to production of IAA, siderophores and solubilization of phosphate (Rajkumar et al. 2006).

Selenium was directly absorbed by the spider plant (*Chlorophytum comosum*) roots when supplemented with Se^{6+}, but a combination of passive and direct absorption occurred when supplemented with Se^{4+} due to the partial oxidation of Se^{4+} to Se^{6+} in the rhizosphere. Higher molecular weight selenium species were more prevalent in the roots of plants supplemented with Se^{4+}, but in the leaves of plants supplemented with Se^{6+} due to an increased translocation rate. When supplemented as As^{3+}, arsenic is proposed to be passively absorbed as As^{3+} and partially oxidized to As^{5+} in the plant root (Afton et al. 2009). Oxyanions of arsenic and selenium can be used in microbial anaerobic respiration as terminal electron acceptors and the arsenate and selenate respiring bacteria are widespread and metabolically active in nature (Stolz and Oremland 1999). *B. juncea* could accumulate high tissue Se concentrations and volatilizes Se in relatively nontoxic forms, such as dimethyselenide. The presence of bacteria in the rhizosphere of *B. juncea* was necessary to achieve the best rates of plant Se accumulation and volatilization of selenate (de Souza et al. 1999). Compared with axenic controls, plants inoculated with rhizosphere bacteria had 5-fold higher Se concentration in roots (the site of volatilizaton) and 4-fold higher rates of Se volatilization (de Souza et al. 1999).

From the biosafety viewpoint, not all naturally occurring soil bacteria are ideal as bioremediation agents. *Burkholderia cepacia* has potential as an agent for bioremediation and for biological control of phytopathogens. However, it is a human pathogen known to be involved in cystic fibrosis and it is resistant to multiple antibiotics, and this has lead to rejection by the Environmental Protection Agency (EPA) of its use as an environmental agent (Davison 2005). The special conditions of bioremediation might select for bacteria with undesirable properties. The *Pseudomonas* efflux pumps are responsible for the pumping out of metals and antibiotics and a number of transposons carrying resistance to antibiotics and metals. The possibility that *in situ* remediation of metal contaminated site could simultaneously select for antibiotic-resistant bacteria must be considered (Davision 2005).

The fungi showed more tolerance and higher biomass than bacteria, however the effect of fungi except for arbuscular mycorrhizal fungi on the phytoextraction of metals were seldom reported (Pawlowska et al. 2000). Fungal endophytes are fungi that reside in internal tissues of living plants without causing any immediate overt negative effects, but may turn pathogenic during host senescence. The majority of endophytes are horizontally transmitted to their host plants through airbore spores. In contrast, some endophytes may also be vertically transmitted to the next plant generations via seeds once inside host tissue. They assume a quiescent state either for the whole lifetime of the host plant or for an extended period of time, i.e., until environmental conditions are favorable for the fungus or the ontogenetic state of the host changes to the advantage of the fungus (Aly et al. 2011).

Endophytic fungal communities generally demonstrate single host specificity at the plant species level, which can be further influenced by microhabitat and microclimatic conditions, the endophytic fungi may also exhibit organ and tissue specificity as a result of their adaptation to different physiological conditions in plants and comparisons of foliar and root endophyte communities showed little overlap (Aly et al. 2011). Endophytic fungi may also increase host fitness and competitive abilities by increasing germination success and growth rate through evolving biochemical pathways to produce plant growth hormones such as indole-3-acetic acid, indole-3-acetonitrile, and cytokinins, or enhancing the absorption of nutritional elements by the host (Aly et al. 2011). Furthermore, it was shown that endophytes possessing suitable degradation pathways or metal sequestration or chelation systems were able to increase host plant tolerance to presence of heavy metal, thereby assisting their hosts to survive in contaminated soil (Aly et al. 2011). The fungus *Trichoderma atroviride* F6 resistant to Cd^{2+} and Ni^{2+} significantly alleviated the cellular toxicity of cadmium and nickel to *B. juncea* (L.) Coss. var. *foliosa*. Inoculation of plants with the fungal strain F6 resulted a 110%, 40%, and 170% increase in fresh weight in Cd-, Ni-, and Cd–Ni-contaminated soils, respectively, and the fungal treated plants grown in Cd-Ni combination contaminated soils showed higher phytoextraction efficiency than those in Cd or Ni contaminated soils (Cao et al. 2008). Other *Trichoderma* sp. H8 and *Aspergillus* sp. G16 strains also promoted the growth of *B. juncea* (L.) Coss. var. *foliosa* in the Cd, Ni and Cd-Ni contaminated soil (Jiang et al. 2008b). Some endophytic yeasts (the fungi reproducing asexually by budding from single cells, with absent or reduced hyphal states) have been isolated from plants except for oilseed rape. The yeasts *Rhodotorula mucilaginosa* are able to promote the growth of maize by producing plant auxins such as indole-3-acetic acid and indole-3-pyruvic acid (Xin et al. 2009). However, the effects of yeasts on plant growth in metal contaminated soils were not reported.

18.6 CONCLUSIONS AND FUTURE PROSPECTS

The recent researches of microorganisms on the remediation of contaminated soils showed a brilliant prospect for the successive studies; however, breakthroughs in the field are still very difficult to achieve.

1. Almost all the previous works on bioremediation with microorganisms were carried out in laboratory or greenhouse, how the remediation efficiency will change in the field is till unknown. In order to implement these microbe-assisted phytoremediation in the field level, intensive future research is needed on understanding the diversity and ecology of plant-associated microorganisms in multiple metal-contaminated soils, further understanding of the role of indigenous microorganisms that have been cultured and enriched in the laboratory on phytoremediation potential of various plants in metal contaminated soils would provide helps for improving the technology.
2. The complexity and heterogeneity of soils contaminated with multiple metals, metalloids, and organic compounds requires the design of integrated phytoremediation systems that combine different processes and approaches. Co-cropping different species may enhance the overall capacities of a phytoremediation system to explore the contaminated soil volume, address different pollutants, and support differential microbial consortia in their rhizosphere, however, we still know little about the process, mechanism and how the microorganisms really interact with plant roots and other bacteria, fungi (including yeasts). It is obvious that the complexity of interactions in the plant-microbe-soil-pollutant systems requires substantial further research efforts to improve our understanding of the rhizosphere processes involved.
3. Some effort is currently being devoted to obtaining microorganisms better adapted to the actual conditions in which biotreatment are to be performed. If microbial lines are not living as plant endophytes, the strains are at a competitive disadvantage. It can be assumed

that it is relatively straightforward to equip endophytes via horizontal gene transfer with appropriate degradation pathways because many endophytes are closely related to environmental strains that carry similar degradation pathways or heavy metal resistance on mobile DNA element. In this way, endophytes can be specifically tailored for a number of different mixed contaminants (Weyens et al. 2009). The endophytes might be better adapted to the actual contaminated soils.

ACKNOWLEDGMENTS

The authors gratefully acknowledge support from the grants of the Chinese National Natural Science Foundation (Nos. 51039007 and 50779080), and the Fundamental Research Funds for the Central Universities.

REFERENCES

Afton, S.E., B. Catron, J.A. Ceruso et al. 2009. Elucidating the selenium and arsenic metabolic pathways following exposure to the non-hyperaccumulating *Chlorophytum comosum*, spider plant. *Journal of Experimental Botany* 60: 1289–97.

Aly, A.H., A. Debbab, and P. Proksch. 2011. Fungal endophytes: unique plant inhabitants with great promises. *Applied Microbiology and Biotechnology* 90: 1829–1845.

Arshad, M., M. Saleem, and S. Hussain. 2007. Perspectives of bacterial ACC deaminase in phytoremediation. *Trends in Biotechnology* 25: 356–61.

Asghar, H.N., Z.A. Zahir, M. Arshad et al. 2002. Relationship between *in vitro* production of auxins by rhizobacteria and their growth-promoting activities in *Brassica juncea* L. *Biology and Fertility of Soils* 35: 231–237.

Asghar, H.N., Z.A. Zahir, and M. Arshad. 2004. Screening rhizobacteria for improving the growth, yield, and oil content of canola (*Brassica napus* L). *Australian Journal of Agricultural Research* 55: 187–94.

Belimov, A.A., N. Hontzeas, V.I. Safronova et al. 2005. Cadmium-tolerant plant growth-promoting bacteria associated with the roots of Indian mustard (*Brassica juncea* L. Czern). *Soil Biology and Biochemistry* 37: 241–50.

Cao, L., M. Jiang, Z. Zeng et al. 2008. *Trichoderma atroviride* F6 improves phytoextraction efficiency of mustard (*Brassia juncea* (L) Coss. var. *foliosa* Bailey) in Cd, Ni contaminated soils. *Chemosphere* 71: 1769–1773.

de Cárcer, D.A., M. Martín, M. Mackova et al. 2007. The introduction of genetically modified microorganisms designed for rhizoremediation induces changes on native bacteria in the rhizosphere but not in the surrounding soil. *ISME Journal* 1: 215–23.

Cohen, M.F., H. Yamasaki, and M. Mazzola. 2004. Bioremediation of soils by plant-microbe systems. *International Journal of Green Energy* 1: 301–12.

Davison, J. 2005. Risk mitigation of genetically modified bacteria and plants designed for bioremediation. *Journal of Industrial Microbiology and Biotechnology* 32: 639–50.

Dell'Amico, E., L. Cavalca, V. Andreoni et al. 2008. Improvement of *Brassica napus* growth under cadmium stress by cadmium-resistant rhizobacteria. *Soil Biology and Biochemistry* 40: 74–84.

Durán, N., P.D. Marcato, M. Durán et al. 2011. Mechanistic aspects in the biogenic synthesis of extracellular metal nanoparticles by peptides, bacteria, fungi, and plants. *Applied Microbiology and Biotechnology* 90: 1609–24.

Epstein, A.L., C.D. Gussman, M.J. Blaylock et al. 1999. EDTA and Pb-EDTA accumulation in *Brassica juncea* grown in Pb-amended soil. *Plant and Soil* 208: 87–94.

Germida, J.J., S.D. Siciliano, J.R. de Freitas et al. 1998. Diversity of root-associated bacteria associated with field-grown canola (*Brassica napus* L.) and wheat (*Triticum aestivum* L.). *FEMS Microbiology and Ecology* 26: 43–50.

Glick, B.R. 2003. Phytoremediation: synergistic use of plants and bacteria to clean up the environment. *Biotechnology Advances* 21: 383–93.

Granér, G., P. Persson, J. Meijer et al. 2003. A study on microbial diversity in the different cultivars of *Brassica napus* in relation to its wilt pathogen, *Verticillium longisporum*. *FEMS Microbiology Letters* 224: 269–76.

Gutiérrez, J.C., F. Amaro, and A. Martín-González. 2009. From heavy metal-binders to biosensors: ciliate metallothioneins discussed. *Bioessays* 31: 805–16.

Ishikawa, S., N. Ae, M. Murakami et al. 2006. Is *Brassica juncea* a suitable plant for phytoremediation of cadmium in soils with moderately low cadmium contamination?- Possibility of using other plant species for Cd-phytoextraction. *Soil Science and Plant Nutrition* 52: 32–42.

Jiang, C., X. Sheng, M. Qian et al. 2008a. Isolation and characterization of a heavy metal-resistant *Burkholderia* sp. from heavy metal-contaminated paddy field soil and its potential in promoting plant growth and heavy metal accumulation in metal-polluted soil. *Chemosphere* 72: 157–64.

Jiang, M., L. Cao, and R. Zhang. 2008b. Effects of Acacia (*Acacia auriculaeformis* A. Cunn)-associated fungi on mustard (*Brassica juncea* (L.) Coss. var. *foliosa* Bailey) growth in Cd- and Ni-contaminated soils. *Letters in Applied Microbiology* 47: 561–65.

Juwarkar, A.A., A. Nair, K.V. Dubey et al. 2007. Biosurfactant technology for remediation of cadmium and lead contaminated soils. *Chemosphere* 68: 1996–2002.

Kaiser, O., A. Pühler, and W. Selbitschka. 2001. Phylogenetic analysis of microbial diversity in the rhizoplane of oilseed rape (*Brassica napus* cv. Westar) employing cultivation-dependent and cultivation-independent approaches. *Microbiological Ecology* 42: 136–49.

Kidd, P., J. Barceló, M.P. Bernal et al. 2009. Trace element behaviour at the root-soil interface: implications in phytoremediation. *Environmental and Experimental Botany* 67: 243–59.

Kotrba, P., L. Doleckova, V. de Lorenzo et al. 1999. Enhanced bioaccumulation of heavy metal ions by bacterial cells due to surface display of short metal binding peptides. *Applied Environmental Microbiology* 65: 1092–8.

Kumar, K.V., N. Singh, H.M. Behl et al. 2008. Influence of plant growth promoting bacteria and its mutant on heavy metal toxicity in *Brassica juncea* grown in fly ash amended soil. *Chemosphere* 72: 678–83.

Kumar, K.V., S. Srivastava, N. Singh et al. 2009. Role of metal resistant plant growth promoting bacteria in ameliorating fly ash to the growth of *Brassica juncea*. *Journal of Hazardous Materials* 170: 51–57.

Lioyd, J., and D.R. Lovley. 2001. Microbial detoxification of metals and radionuclides. *Current Opinion in Biotechnology* 12: 248–53.

Ma, Y., M. Rajkumar, and H. Freitas. 2009a. Inoculation of plant growth promoting bacterium *Achromobacter xylosoxidans* strain Ax10 for the improvement of copper phytoextraction by *Brassica juncea*. *Journal of Environmental Management* 83: 1–7.

Ma, Y., M. Rajkumar, and H. Freitas. 2009b. Isolation and characterization of Ni mobilizing PGPB from serpentine soils and their potential in promoting plant growth and Ni accumulation by *Brassica* spp. *Chemosphere* 75:719–25.

Ma, Y., M. Rajkumar, and H. Freitas. 2009c. Improvement of plant growth and nickel uptake by nickel resistant-plant-growth promoting bacteria. *Journal of Hazardous Materials* 166: 1154–61.

Ma, Y., M.N.V. Prasad, M. Rajkumar et al. 2010. Plant growth promoting rhizobacteria and endophytes accelerate phytoremediation of metalliferous soils. *Biotechnology Advances* doi:10.1016/j.biotechadv .2010.12.001.

Madhaiyan, M., S. Poonguzhali, J. Ryu et al. 2006. Regulation of ethylene levels in canola (*Brassica campestris*) by 1-aminocyclopropane-1-carboxylate deaminase-containing *Methylobacterium fujisawaense*. *Planta* 224: 268–78.

Marchiol, L., S. Assolari, P. Sacco et al. 2004. Phytoextraction of heavy metals by canola (*Brassica napus*) and radish (*Raphanus sativus*) grown on multi contamianted soil. *Environmental Pollution* 132: 21–27.

McGrath, S.P., F.J. Zhao, and E. Lombi. 2001. Plant and rhizosphere processes involved in phytoremediation of metal-contaminated soils. *Plant and Soil* 232: 207–14.

Mejáre, M., and L. Bülow. 2001. Metal-binding proteins and peptides in bioremediation and phytoremediation of heavy metals. *Trends in Biotechnology* 19: 67–73.

Narisawa, W., F. Usuki, and T. Hashiba. 2004. Control of *Verticillium* yellows in Chinese cabbage by the dark septate endophytic fungus LtVB3. *Phytopathology* 94: 412–18.

Nejad, P., and P.A. Johnson. 2000. Endophytic bacteria induce growth promotion and wilt disease suppression in oilseed rape and tomato. *Biological Control* 18: 208–15.

Newman, L.A., and C.M. Reynolds. 2005. Bcateria and phytoremediation: new uses for endophytic bacteria in plants. *Trends in Biotechnology* 23: 6–8.

Nie, L., S. Shah, A. Rashid et al. 2002. Phytoremediation of arsenate contaminated soil by transgenic canola and the plant growth-promoting bacterium *Enterobacter cloacae* CAL2. *Plant Physiology and Biochemistry* 40: 355–61.

Ohki, T., H. Masuya, M. Yonezawa et al. 2002. Colonization process of the root endophytic fungus *Heteroconium chaetospira* in roots of chinese cabbage. *Mycoscience* 43: 191–4.

Pawlowska, T.E., R.L. Chaney, M. Chin et al. 2000. Effects of metal phytoextraction practices on the ingigenous community of arbuscular mycorrhizal fungi at a metal-contaminated landfill. *Applied Environmental Microbiology* 66: 2526–30.

Quartacci, M.F., B. Irtelli, C. Gonnelli et al. 2009. Naturally-assisted metal phytoextraction by *Brassica carinata*: role of root exudates. *Environmental Pollution* 157: 2697–703.

Rajkumar, M., and H. Freitas. 2008. Effects of inoculation of plant-growth promoting bacteria on Ni uptake by Indian mustard. *Bioresource Technology* 99: 3491–98.

Rajkumar, M., R. Nagendran, K.J. Lee et al. 2006. Influence of plant growth promoting bacteria and Cr^{6+} on the growth of Indian mustard. *Chemosphere* 62: 741–8.

Rajkumar, M., N. Ae, and H. Freitas. 2009. Endophytic bacteria and their potential to enhance heavy metal phytoextraction. *Chemosphere* 77: 153–60.

Rajkumar, M., N. Ae, M.N.V. Prasad et al. 2010. Potential of siderophore-producing bacteria for improving heavy metal phytoextraction. *Trends in Biotechnology* 28: 142–8.

Schulz, B., A.-K. Römmert, U. Dammann et al. 1999. The endophyte-host interaction: a balanced antagonism? *Mycological Research* 103: 1275–83.

Sheng, X-F., and J.-J. Xia. 2006. Improvement of rape (*Brassica napus*) plant growth and cadmium uptake by cadmium-resistant bacteria. *Chemosphere* 64: 1036–42.

Sheng, X.-F., J.-J. Xia, C.-Y. Jiang et al. 2008a. Characterization of heavy metal-resistant endophytic bacteria from rape (*Brassica napus*) roots and their potential in promoting the growth and lead accumulation of rape. *Environmental Pollution* 156: 1164–70.

Sheng, X., L. He, Q. Wang et al. 2008b. Effect of inoculation of biosurfactant-producing *Bacillus* sp. J119 on plant growth and cadmium uptake in a cadmium-amended soil. *Journal of Hazardous Materials* 155: 17–22.

Siciliano, S.D., and J.J. Germida. 1999. Taxonomic diversity of bacteria associated with the roots of field-grown transgenic *Brassica napus* cv. Quest, compared to the non-transgenic *B. napus* cv. Excel and *B. rapa* cv. Parkland. *FEMS Microbiology and Ecology* 29: 263–72.

Sidhu, V.P.S., and M.P.S. Khurana. 2010. Effect of cadmium-contaminated soils on dry matter yield and mineral composition of raya (*Brassica juncea*) and spinach (*Spinacia oleracea*). *Acta Agronomica Hungarica* 58: 407–17.

Sinha, S., S. Singh, S. Mallick et al. 2009. Role of antioxidants in Cr tolerance of three crop plants: metal accumulation in seeds. *Ecotoxicology and Environmental Safety* 72: 1111–21.

Sinha, S., G. Sinam, R.K. Mishra et al. 2010. Metal accumulation, growth, antioxidants and oil yield of *Brassica juncea* L. exposed to different metals. *Ecotoxicology and Environmental Safety* 73: 1352–61.

de Souza, M.P., D. Chu, M. Zhao et al. 1999. Rhizosphere bacteria enhance selenium accumulation and volatilization by Indian mustard. *Plant Physiology* 119: 565–73.

Stolz, J.F., and R.S. Oremland. 1999. Bacterial respiration of arsenic and selenium. *FEMS Microbiology Review* 23: 615–27.

Taghavi, S., D. van der Lelie, A. Hoffman et al. 2010. Genome sequence of the plant growth promoting endophytic bacterium *Enterobacter* sp. 638. *PLoS Genetics* doi:10.1371/journal.pgen.1000943.

Usuki, F., and K. Narisawa. 2007. A mutualistic symbiosis between a dark septate endophytic fungus, *Heteroconium chaetospira*, and a nonmycorrhizal plant, Chinese cabbage. *Mycologia* 99: 175–84.

Valls, M., and V. de Lorenzo. 2002. Exploiting the genetic and biochemical capacities of bacteria for the remediation of heavy metal pollution. *FEMS Microbiology Review* 26: 327–38.

Wenzel, W.W. 2009. Rhizosphere processes and management in plant-assisted bioremediation (phytoremediation) of soils. *Plant and Soil* 321: 385–408.

Weyens, N., D. van der Lelie, S. Taghavi et al. 2009. Exploiting plant-microbe partnerships to improve biomass production and remediation. *Trends in Biotechnology* 27: 591–8.

Wu, S.C., K.C. Cheung, Y.M. Luo et al. 2006a. Effects of inoculation of plant growth-promoting rhizobacteria on metal uptake by *Brassica juncea*. *Environmental Pollution* 140: 124–35.

Wu, C.H., T.K. Wood, A. Mulchandani et al. 2006b. Engineering plant-microbe symbiosis for rhizoremediation of heavy metals. *Applied Environmental Microbiology* 72: 1129–34.

Xin, G., D. Glawe, and S.L. Doty. 2009. Characterization of three endophytic, indole-3-acetic acid-producing yeasts occurring in *Populus* trees. *Mycological Research* 113: 973–80.

Zaidi, S., S. Usmani, B.R. Singh et al. 2006. Significance of *Bacillus subtilis* strain SJ-101 as a bioinoculant for concurrent plant growth promotion and nickel accumulation in *Brassica juncea*. *Chemosphere* 64: 991–7.

Zakria, M., A. Ohsako, Y. Saeki et al. 2008. Colonization and growth promotion characteristics of *Enterobacter* sp. and *Herbaspirillum* sp. on *Brassica oleracea*. *Soil Science and Plant Nutrition* 54: 507–16.

Zhuang, X., J. Chen, and H. Shim. 2007. New advances in plant growth-promoting rhizobacteria for bioremediation. *Environment International* 33: 406–13.

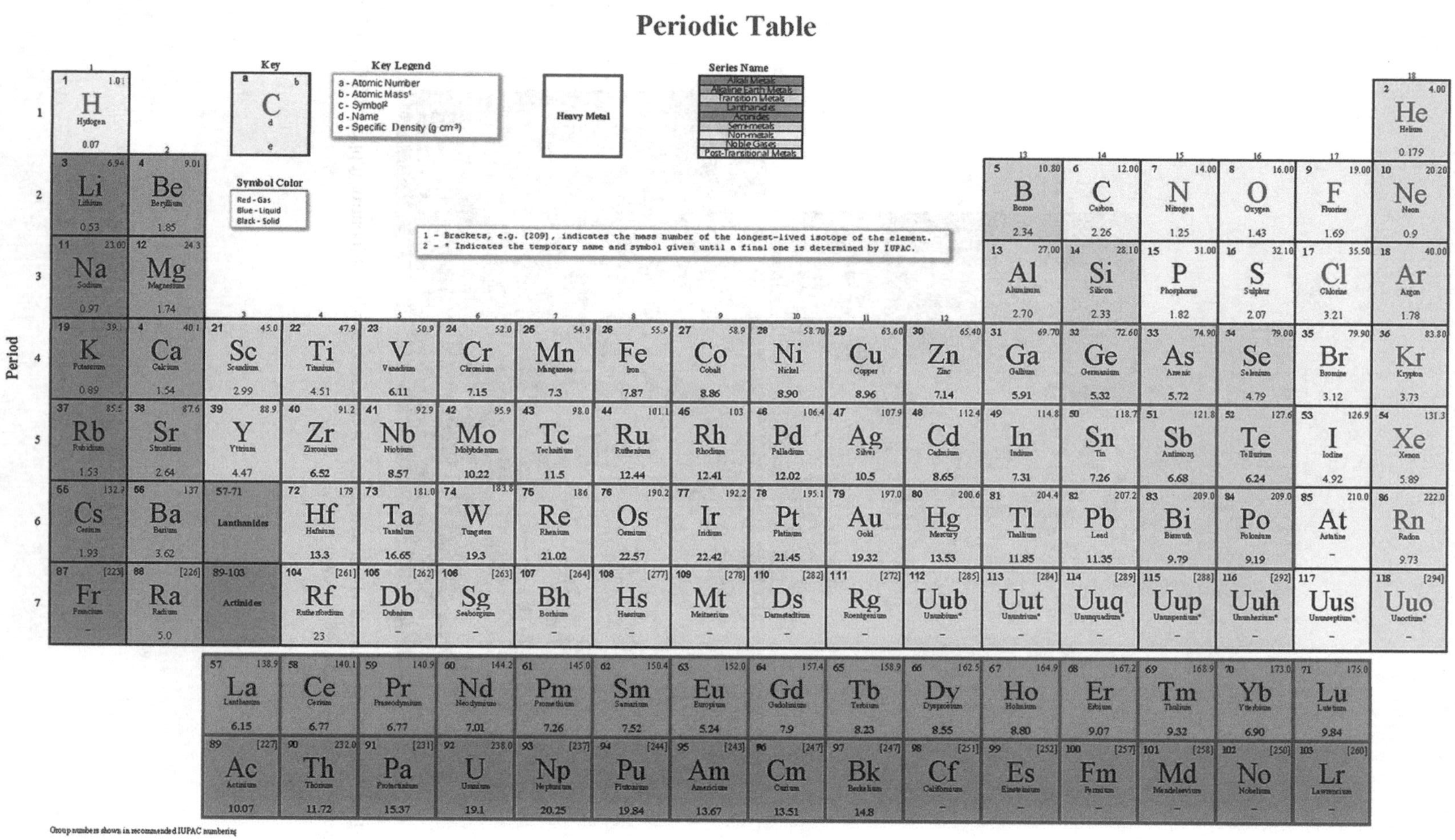

Periodic Table

Key: a (Atomic Number), b (Atomic Mass), C (Symbol), d (Name), e (Specific Density)

Key Legend
a - Atomic Number
b - Atomic Mass[1]
c - Symbol[2]
d - Name
e - Specific Density (g cm⁻³)

Heavy Metal

Series Name: Alkali Metals; Alkaline Earth Metals; Transition Metals; Lanthanides; Actinides; Semi-metals; Non-metals; Noble Gases; Post-Transitional Metals

Symbol Color: Red - Gas; Blue - Liquid; Black - Solid

1 - Brackets, e.g. [209], indicates the mass number of the longest-lived isotope of the element.
2 - * Indicates the temporary name and symbol given until a final one is determined by IUPAC.

Period	1	2	3	4	5	6	7	8	9	10	11	12	13	14	15	16	17	18
1	1 1.01 H Hydrogen 0.07																	2 4.00 He Helium 0.179
2	3 6.94 Li Lithium 0.53	4 9.01 Be Beryllium 1.85											5 10.80 B Boron 2.34	6 12.00 C Carbon 2.26	7 14.00 N Nitrogen 1.25	8 16.00 O Oxygen 1.43	9 19.00 F Fluorine 1.69	10 20.20 Ne Neon 0.9
3	11 23.00 Na Sodium 0.97	12 24.3 Mg Magnesium 1.74											13 27.00 Al Aluminum 2.70	14 28.10 Si Silicon 2.33	15 31.00 P Phosphorus 1.82	16 32.10 S Sulphur 2.07	17 35.50 Cl Chlorine 3.21	18 40.00 Ar Argon 1.78
4	19 39.1 K Potassium 0.89	4 40.1 Ca Calcium 1.54	21 45.0 Sc Scandium 2.99	22 47.9 Ti Titanium 4.51	23 50.9 V Vanadium 6.11	24 52.0 Cr Chromium 7.15	25 54.9 Mn Manganese 7.3	26 55.9 Fe Iron 7.87	27 58.9 Co Cobalt 8.86	28 58.70 Ni Nickel 8.90	29 63.60 Cu Copper 8.96	30 65.40 Zn Zinc 7.14	31 69.70 Ga Gallium 5.91	32 72.60 Ge Germanium 5.32	33 74.90 As Arsenic 5.72	34 79.00 Se Selenium 4.79	35 79.90 Br Bromine 3.12	36 83.80 Kr Krypton 3.73
5	37 85.5 Rb Rubidium 1.53	38 87.6 Sr Strontium 2.64	39 88.9 Y Yttrium 4.47	40 91.2 Zr Zirconium 6.52	41 92.9 Nb Niobium 8.57	42 95.9 Mo Molybdenum 10.22	43 98.0 Tc Technetium 11.5	44 101.1 Ru Ruthenium 12.44	45 103 Rh Rhodium 12.41	46 106.4 Pd Palladium 12.02	47 107.9 Ag Silver 10.5	48 112.4 Cd Cadmium 8.65	49 114.8 In Indium 7.31	50 118.7 Sn Tin 7.26	51 121.8 Sb Antimony 6.68	52 127.6 Te Tellurium 6.24	53 126.9 I Iodine 4.92	54 131.3 Xe Xenon 5.89
6	55 132.9 Cs Cesium 1.93	56 137 Ba Barium 3.62	57-71 Lanthanides	72 179 Hf Hafnium 13.3	73 181.0 Ta Tantalum 16.65	74 183.8 W Tungsten 19.3	75 186 Re Rhenium 21.02	76 190.2 Os Osmium 22.57	77 192.2 Ir Iridium 22.42	78 195.1 Pt Platinum 21.45	79 197.0 Au Gold 19.32	80 200.6 Hg Mercury 13.53	81 204.4 Tl Thallium 11.85	82 207.2 Pb Lead 11.35	83 209.0 Bi Bismuth 9.79	84 209.0 Po Polonium 9.19	85 210.0 At Astatine -	86 222.0 Rn Radon 9.73
7	87 [223] Fr Francium -	88 [226] Ra Radium 5.0	89-103 Actinides	104 [261] Rf Rutherfordium 23	105 [262] Db Dubnium -	106 [263] Sg Seaborgium -	107 [264] Bh Bohrium -	108 [277] Hs Hassium -	109 [278] Mt Meitnerium -	110 [282] Ds Darmstadtium -	111 [272] Rg Roentgenium -	112 [285] Uub Ununbium* -	113 [284] Uut Ununtrium* -	114 [289] Uuq Ununquadium* -	115 [288] Uup Ununpentium* -	116 [292] Uuh Ununhexium* -	117 Uus Ununseptium* -	118 [294] Uuo Ununoctium* -

57 138.9 La Lanthanum 6.15	58 140.1 Ce Cerium 6.77	59 140.9 Pr Praseodymium 6.77	60 144.2 Nd Neodymium 7.01	61 145.0 Pm Promethium 7.26	62 150.4 Sm Samarium 7.52	63 152.0 Eu Europium 5.24	64 157.4 Gd Gadolinium 7.9	65 158.9 Tb Terbium 8.23	66 162.5 Dy Dysprosium 8.55	67 164.9 Ho Holmium 8.80	68 167.2 Er Erbium 9.07	69 168.9 Tm Thulium 9.32	70 173.0 Yb Ytterbium 6.90	71 175.0 Lu Lutetium 9.84
89 [227] Ac Actinium 10.07	90 232.0 Th Thorium 11.72	91 [231] Pa Protactinium 15.37	92 238.0 U Uranium 19.1	93 [237] Np Neptunium 20.25	94 [244] Pu Plutonium 19.84	95 [243] Am Americium 13.67	96 [247] Cm Curium 13.51	97 [247] Bk Berkelium 14.8	98 [251] Cf Californium -	99 [252] Es Einsteinium -	100 [257] Fm Fermium -	101 [258] Md Mendelevium -	102 [250] No Nobelium -	103 [260] Lr Lawrencium -

Group numbers shown in recommended IUPAC numbering

FIGURE 2.1 Periodic table showing the position of different heavy metals (indicated by red borders).

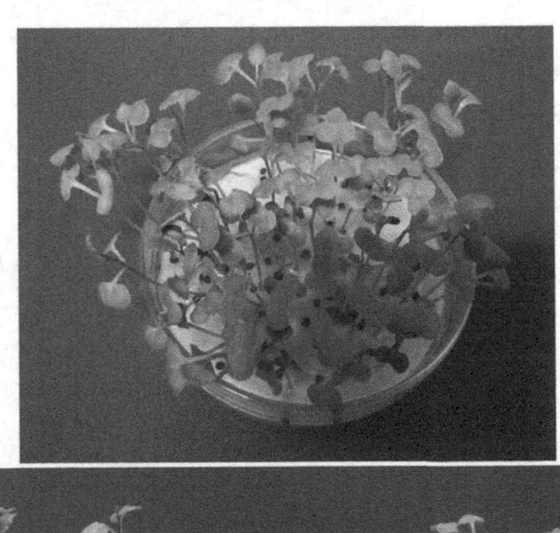

FIGURE 2.11 Cadmium-induced visual symptoms in rapeseed seedlings [control (upper); $CdCl_2$ 0.5 mM, 48 h (lower left); $CdCl_2$ 1 mM, 48 h (lower right)].

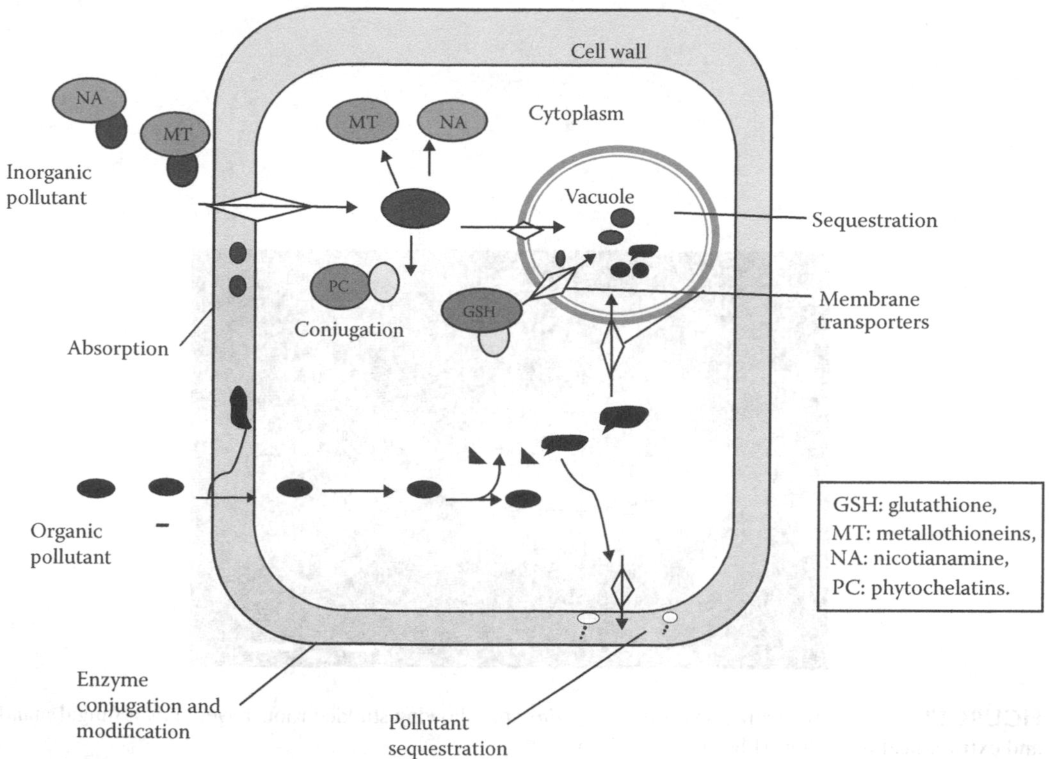

FIGURE 6.5 Transport, sequestration, and tolerance mechanisms of inorganic and organic pollutants in plant cells.

FIGURE 12.8 A variety of products made from vetiver biomass. (From www.vetiver.org.)

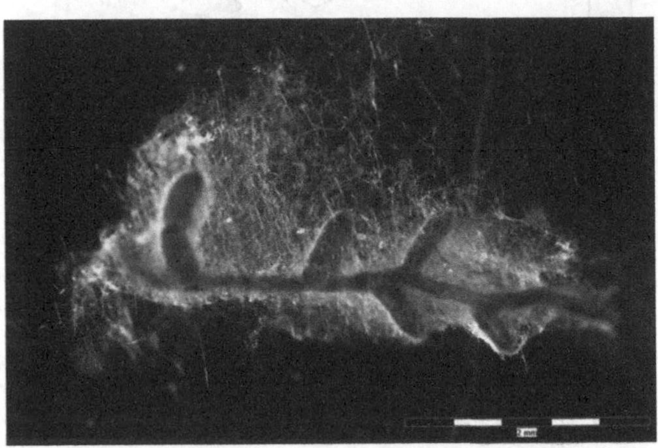

FIGURE 13.1 Ectomycorrhiza on roots of *Populus* spp. showing studded roots covered by a fungal mantle, and extraradical mycelium. (Photo: Martin Vohnik.)

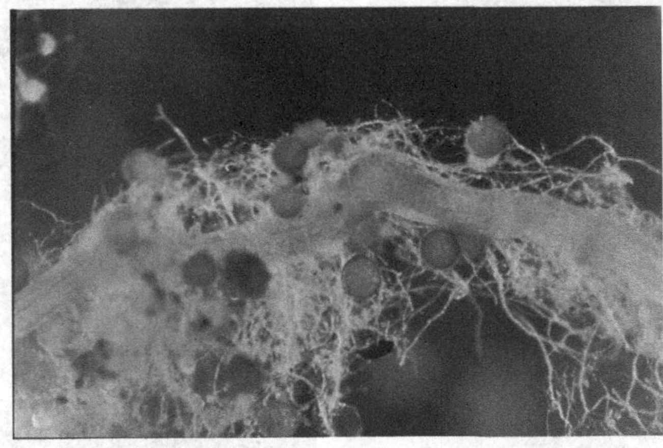

FIGURE 13.2 Arbuscular mycorrhiza on a clover root, showing extraradical hyphae and spores. (Photo: Iver Jakobsen.)

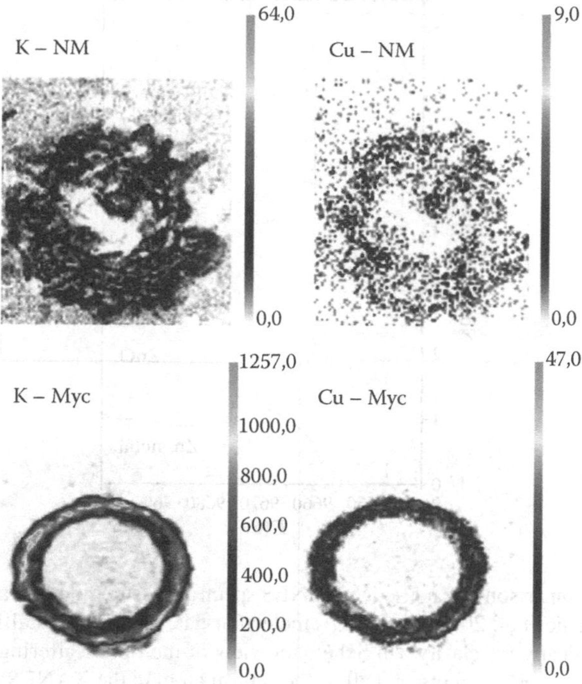

FIGURE 13.3 Distribution of K and Cu in cross-sections of ectomycorrhizal (Myc) and nonmycorrhizal (NM) pine roots mapped using µ-PIXE imaging. Colors indicate relative concentrations according to the scale bars and show a higher proportion of Cu in the cortex region for mycorrhizal roots than in nonmycorrhizal roots due to fungal accumulation of Cu. (From E. Joner and B. Gouget, unpublished results.)

FIGURE 25.2 *Minuartia verna*, a Pb and Zn ore indicator from the river bed Gailitz, Arnoldstein, Austria.

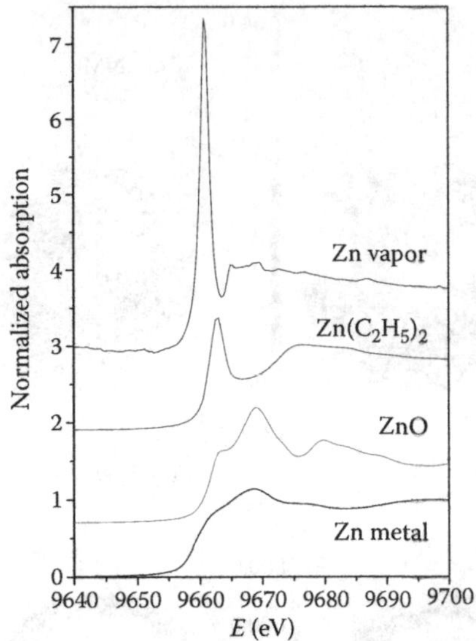

FIGURE 25.22 (a) Comparison of Zn K-edge XANES spectra measured on free and bound Zn atoms: Zn in monatomic vapor (Mihelič et al. 2002), $Zn(C_2H_5)_2$ molecular gas, in solid (crystalline) ZnO and in Zn metal. Spectra are shifted vertically for clarity. (b) Schematic view of multiple scattering of slow photoelectron on neighbor atoms after photo-effect in inner shell of the central atom in the XANES energy range.

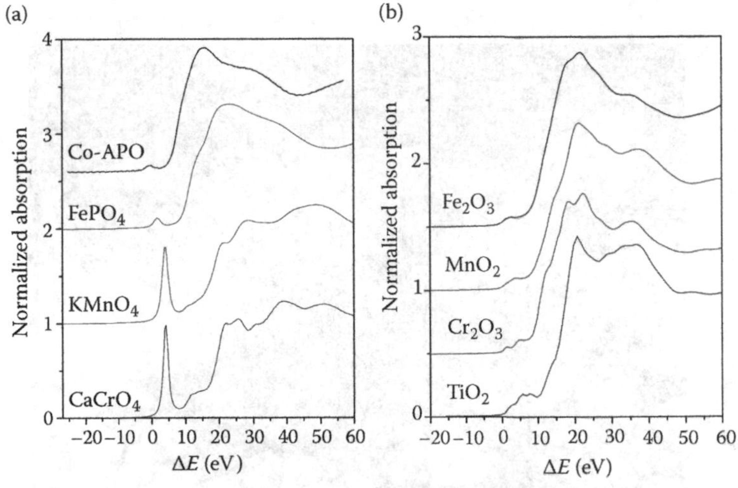

FIGURE 25.24 (a) K-edge XANES spectra of tetrahedrally coordinated cations Ca, Mn, Fe and Co in $CaCrO_4$, $KMnO_4$, $FePO_4$ and Co-APO compounds. (b) K-edge XANES spectra of octahedrally coordinated cations Ti, Cr, Mn and Fe in TiO_2-anatase, Cr_2O_3, MnO_2 and Fe_2O_3 compounds. Energy scale is relative to the energy position of the K-edge of corresponding metal.

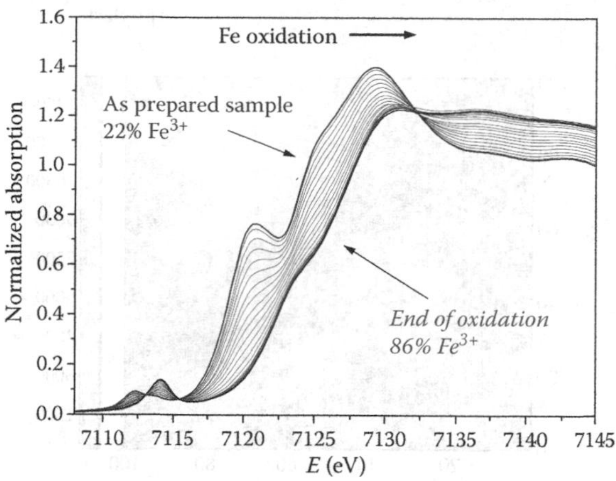

FIGURE 25.25 Fe K-edge XANES spectra measured *in-situ* on the $Li_{2-x}Fe_{0.8}Mn_{0.2}SiO_4$ cathode material for Li-ion batteries during oxidation of Fe^{2+} to Fe^{3+}, as a consequence of Li extraction (Dominko et al. 2010). As prepared sample contained 22% of Fe^{3+}, after oxidation 86% of Fe atoms were in trivalent state.

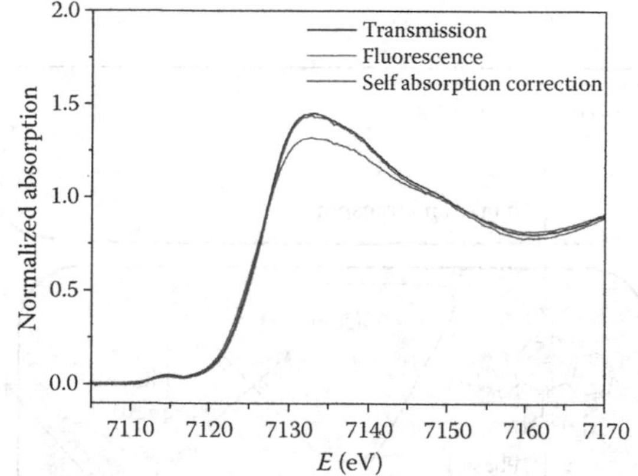

FIGURE 25.27 Fe XANES spectra measured on iron gal ink on the cellulose simultaneously in transmission (blue line) and fluorescence detection mode (red line—without SA correction, magenta line—after SA correction) (a), and their absorption derivatives (b). (From Arčon, I. et al., *X-ray Spectrometry*, 36, 199–205, 2007. With permission.)

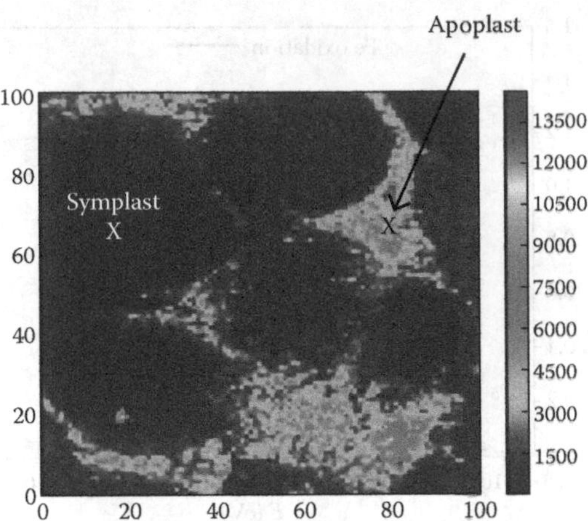

FIGURE 25.29 (a) μ-XRF Cd elemental map (100 × 100 μm^2) recorded on mesophyll tissue of Cd hyperaccumulator plant *Thlaspi praecox*, treated with 300 μM $CdSO_4$. The map was measured at ID21 beamline of ESRF with a monochromatic micro-beam (3550 eV) with 1 μm^2 resolution. (b) Cd L3-edge XANES spectrum (black line) measured in Cd rich region in cell wall of mesophyll tissue (marked with X). The spectrum can be completely described with a linear combination fit (magenta dashed line) of reference XANES spectra measured on the complex Cd-pectin (red line) and Cd-glutathione (blue line).

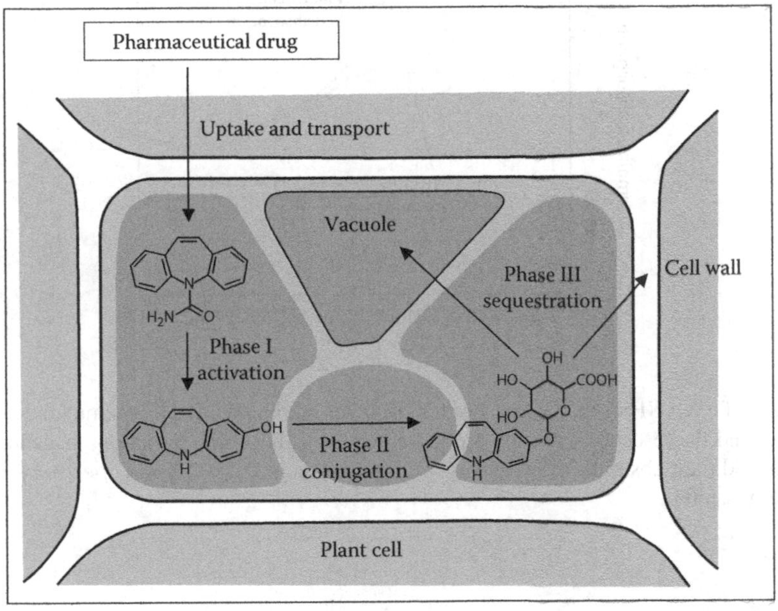

FIGURE 28.2 The three phases of the *green liver* model: Hypothetical pathway representing the metabolism of carbamazepine in plant tissues: phase I: Activation of carbamazepine by hydroxylation and hydrolysis to 2-hydroxyiminostilbene; phase II: Conjugation with a glucose to form an *O*-glucuronide; phase III: Sequestration of the conjugate into the cell wall or in the vacuole. (Adapted from Van Aken, B., *Trends in Biotechnology*, 26: 5225–228, 2008; Celiz, M.D., T.S.O. Jerry, and S. Daina, *Environmental Toxicology and Chemistry*, 28: 122473–2484, 2009.)

19 Plant Growth-Promoting Bacteria and Metals Phytoremediation

Elisa Gamalero and Bernard R. Glick

CONTENTS

19.1 INTRODUCTION

There are hundreds of sources of heavy metal pollution, including both natural and human ones. In particular, anthropic activities such as mining, metal working industries, combustion of fossil fuels, disposal of ash residues from coal combustion, vehicular traffic and the use of pesticides and fertilizers in agriculture (Clemens 2006) significantly contribute to heavy metal pollution, which in turn, impacts negatively on both human and environmental health. While some heavy metals are important as micronutrients, the majority of them are toxic for plants and microorganisms (Schützendübel and Polle 2002).

The use of plants to clean up heavy metal polluted soils, i.e., so-called phytoremediation (Salt et al. 1995), represents an environmental-friendly and cost-effective alternative to traditional remediation approaches such as soil removal or chemical and physical extraction. Although phytoremediation is a clean technology that could easily be accepted by a concerned public, only a small number of plant species can naturally tolerate/accumulate heavy metals. In addition, many of the plants that are most effective at removing metals from the soil are characterized by their small size and slow growth, thus reducing their remediation potential and restricting their use in this technology (Khan et al. 2000). Therefore, to be considered effective for soil remediation, plants must be tolerant to one or more pollutants, highly competitive, fast growing and produce a high biomass (Gamalero et al. 2009a).

Plants interact with heavy metals from the environment in various ways including (i) stabilization of the heavy metal in soil by reducing its bioavailability (phytostabilization); (ii) extraction, transport and accumulation of heavy metals in plant tissues (phytoextraction); and (iii) transformation of heavy metal into volatile forms (phytovolatilization) (Pilon-Smits 2005).

19.1.1 Plant Growth-Promoting Bacteria

The literature is replete with examples of the use of soil bacteria to promote the growth of a wide variety of plants under both controlled laboratory and field conditions (reviewed by Reed and Glick 2004). These bacteria, known as plant growth-promoting bacteria (PGPB) typically improve plant growth by either direct stimulation of plant growth and development or through biocontrol of plant pathogenic organisms (i.e., suppressive activity against soil-borne diseases) (Glick 1995). The beneficial effect of PGPB on plants is mediated by a range of mechanisms including improvement of mineral nutrition (generally based on nitrogen fixation, phosphate solubilization, or iron sequestration by the PGPB), enhancement of plant tolerance to biotic and/or abiotic stress (via the modulation of plant ethylene levels), modification of root architecture (mediated by the synthesis of phytohormones), as well as suppression of soil-borne diseases (through the synthesis of antibiotics, lytic enzymes and/or siderophores) (Glick 1995; Glick et al. 1999; Kloepper et al. 1989; Gamalero et al. 2002). These bacterial capabilities may sustain and improve plant growth and development, under a variety of environmental conditions included stressed ones.

19.1.2 Historical and Current Uses of Plant Growth-Promoting Bacteria

In addition to their use as crop inoculants in agriculture (Reed and Glick 2004), PGPB have recently been considered for their potential in environmental applications such as bioremediation of wastewater (De Bashan et al. 2004; Hernandez et al. 2006), biofertilizer of plants in degraded desertic (Bashan et al. 1999; Jackson et al. 1991) or extremely dry (Puente et al. 2004) areas and helpers in the restoration of soil polluted by organic (Gerhardt et al. 2009; Alarcón et al. 2008; Al-Awadhi et al. 2009; Sheng et al. 2008; Germaine et al. 2009; Gurska et al. 2009) or inorganic (Belimov et al. 2001, 2005; Glick 2003; Dell'Amico et al. 2008) xenobiotics in assisted phytoremediation schemes.

19.2 PLANT GROWTH-PROMOTING BACTERIAL COLONIZATION OF PLANT TISSUES

Soil bacteria can "sense" and move toward the root following the chemo-attractant effect given by root exudation. Therefore, bacterial cells establish and proliferate where root exudates are more favourable for bacterial growth or are released at a higher rate. In these specific sites, single bacterial cells bind to the root and start to replicate forming little clusters, elongated or thick microcolonies and finally, a true biofilm embedded in an extracellular matrix (Fujishige et al. 2006) in which cellular activity is regulated by quorum-sensing (Watnick and Kolter 2000). Interestingly, not only the cell organization (Gamalero et al. 2004; Pivato et al. 2008), but also the cell culturability and viability (Gamalero et al. 2005) differ according to the localization within the root zone.

Besides chemotaxis to exudates, bacterial traits such as the presence of one or more flagella, the release of signal molecules, as well as the production of specific compounds/enzymes are involved in the colonization processes (Compant et al. 2010). Plant-associated bacteria establish a more or less intimate relationship with the root system according to their preferential localization either inside (endophytic bacteria) or outside the plant tissue, both on the root surface (rhizoplane) and as part of the volume of soil immediately surrounding the root system (rhizosphere) (Lingua et al. 2007). Moreover, some PGPB are able to penetrate the root both passively, mainly through the cracks formed in lateral root junctions (Sørensen and Sessitsch 2006), and actively by releasing lytic enzymes (Compant et al. 2005; Reinhold-Hurek et al. 2006). Once they enter the root, bacteria quickly spread and proliferate often reaching a high cell density (Barraquio et al. 1997). In addition, it has been demonstrated that bacterial mutants affected in the *O*-antigenic side chain of the lipopolysaccharide lose the ability to colonize the root interior (Duijff et al. 1997).

19.3 PLANT GROWTH-PROMOTING BACTERIAL TRAITS FACILITATING PLANT NUTRIENTS ACQUISITION

Once bacterial cells are established at the appropriate density, plant growth promotion by bacteria through the provision of nutrients that are not sufficiently available in the soil can occur. In particular, phosphorous, nitrogen and iron are three elements that are essential for plant nutrition but have limited bioavailability in the soil. Due to the high reactivity of phosphate (P) ions with a number of soil constituents this macronutrient is the least mobile and available macronutrient to plants in most soil conditions, and it is often reported to be a major or even the prime limiting factor for plant growth (Feng et al. 2004). Solubilization of P by PGPB occurs through the synthesis of low-molecular-weight organic acids (Bnayahu 1991; Rodriguez et al. 2004) or chelating substances (Sperber 1958), and the release of H^+ (Illmer and Schinner 1992), while the mineralization of organic P take place through the synthesis of phosphatases (phosphomonoesterase, phosphodiesterase, and phosphotriesterase), all of which catalyze the hydrolysis of phosphoric esters (Rodriguez and Fraga 1999).

Despite nitrogen's abundance in the atmosphere, it must first be reduced to ammonia, before it can be metabolized by plants and utilized to become an integral component of proteins, nucleic acids and other biomolecules (Bøckman 1997). *Rhizobia* and other free-living nitrogen fixing bacteria are the most important microorganisms that are currently used as bacterial inoculants in agriculture in order to improve the nitrogen content of plants (Gamalero and Glick, in press).

Although iron is the fourth most abundant element on earth (Ma, 2005), in the soil, it is generally unavailable for direct assimilation by plants and microorganisms. This is because ferric ion, Fe^{3+}, the predominant form of iron in nature, is only sparingly soluble (~10^{-18} M at pH 7.4) so that the amount of iron present in soil that is biologically available is much too low to support either plant or microbial growth. Consequently, these organisms produce low-molecular-weight compounds that bind to and sequester the available iron. In bacteria, cellular iron deficiency induces the synthesis of low-molecular-weight siderophores, molecules with an extraordinarily high affinity for Fe^{+3} (i.e., K_a ranging from 10^{23} to 10^{52}) as well as membrane receptors that can bind the Fe-siderophore complex, thereby allowing iron uptake by microorganisms (Neilands 1981; O'Sullivan and O'Gara 1992). In addition, a portion of the iron–bacterial siderophore complex may be taken up directly by plants (e.g., Wang et al. 1993). In this regard, environmental conditions, especially iron concentration and pH, the composition of the microbial population and hence the spectrum of siderophores available in the rhizosphere, and the iron acquisition mechanisms of the host plant, all influence iron uptake from microbial siderophores by plants.

19.4 PLANT GROWTH-PROMOTING BACTERIAL TRAITS MODIFYING PLANT HORMONE LEVELS

When plants are exposed to environmental stress, such as heavy metal pollution, they quickly respond by synthesizing a small amount of ethylene that triggers a protective response by the plant, based on the transcription of pathogenesis-related genes and induction of acquired resistance (Ciardi et al. 2000; Van Loon and Glick 2004). The concentration of ethylene that is produced by the plant becomes elevated to a greater extent if the stress becomes chronic or intense. In this condition, senescence, chlorosis, and abscission leading to a significant inhibition of plant growth and survival occur (Abeles et al. 1992).

In 1978, Honma and Shimomura isolated an enzyme catalyzing the degradation of the ethylene precursor, 1-aminocyclopropane-1-carboxylic acid (ACC), to ammonia and α-ketobutyrate, from *Pseudomonas* sp. strain ACP. This enzyme, known as ACC deaminase is relatively common in soil bacteria (Blaha et al. 2006; Duan et al. 2009). When soil bacteria expressing ACC deaminase colonize the roots of plants exposed to stress, the amount of ethylene produced by the plant is reduced. As a consequence, the ethylene concentration does not reach levels that might otherwise be inhibitory for plant growth (Glick 1995) and bacteria expressing ACC deaminase improve plant growth in the presence of a variety of environmental stresses including flooding (Grichko and Glick 2001; Farwell et al. 2007), pollution by organic toxicants (Saleh et al. 2004; Huang et al. 2004a,b; Reed and Glick 2005) and by heavy metals and metalloids (Burd et al. 1998, 2000; Belimov et al. 2001, 2005; Nie et al. 2002; Glick 2003; Reed and Glick 2005; Farwell et al. 2006; Rodriguez et al. 2008), salinity (Mayak et al. 2004b; Saravanakumar and Samiyappan 2006; Cheng et al. 2007; Gamalero et al. 2010), drought (Mayak et al. 2004a), and phytopathogen attack (Wang et al. 2000; Hao et al. 2007).

19.5 MECHANISMS USED BY PLANT GROWTH-PROMOTING BACTERIA TO FACILITATE METAL PHYTOREMEDIATION

In addition to facilitating the growth of plants as an adjunct to agricultural practice, plant growth-promoting bacteria can promote plant growth in the presence of inhibitory levels of metal contaminants. As a consequence, plant growth-promoting bacteria may become partners with plants in various phytoremediation strategies (most commonly phytoextraction or the removal of metals from the soil and their uptake by the plant). In this regard, a number of bacterial traits may contribute to metal phytoremediation including siderophore production, IAA biosynthesis, ACC deaminase activity, and the synthesis and secretion of certain organic acids.

19.5.1 SIDEROPHORES

Using one genera of bacteria as an example, *Pseudomonas* spp. produces yellow-green, water soluble, fluorescent pigments collectively called pyoverdines, composed of a quinoleinic chromophore bound together with a peptide and an acyl chain, conferring a characteristic fluorescence to the bacterial colonies (Meyer and Abdallah 1978). About 100 different pyoverdines have been identified (Budzikiewicz 2004; Meyer et al. 2008) and represent about 20% of the known microbial siderophores (Boukhalfa and Crumbliss 2002). Pyoverdine-mediated iron uptake confers a competitive advantage to fluorescent pseudomonads over other microorganisms (Mirleau et al. 2000, 2001).

Plant iron nutrition improvement by soil bacteria is may be more important when the plant is exposed to an environmental stress such as heavy metal pollution. By supplying iron to plants exposed to high levels of various metal contaminants, siderophores help to alleviate some of the stress imposed on the plants by the metal contaminants (Diels et al. 2002; Belimov et al. 2005; Braud et al. 2006). For example, the bacterium *Kluyvera ascorbata* SUD165, a plant growth-promoting bacterium able to synthesize siderophores can protect canola, Indian mustard, and tomato from metal (nickel, lead and zinc) toxicity (Burd et al. 1998, 2000). In addition, a siderophore-overproducing mutant, *K. ascorbata* SUD165/26, of this bacterium provided even greater protection, as indicated by the enhanced biomass and chlorophyll content in plants cultivated in nickel contaminated soil (Burd et al. 2000).

In another experiment, two siderophore-producing bacterial strains reduced Zn uptake by willow (*Salix caprea*), a result that was interpreted as suggesting that bacterial siderophores may bind to metals from soil and inhibit their uptake by plants. On the other hand, enhancement of Zn and Cd uptake was observed in willow inoculated with a *Streptomyces* strain that did not produce siderophores, suggesting the importance of physiological traits other than siderophore production for metal accumulation by willow (Kuffner et al. 2008). These seemingly contradictory results regarding the effect of siderophores in the presence of high concentrations of metals may be understood within the context of the ability of some (but not all) siderophores to bind tightly to metals other than iron. Moreover, this complexity likely also reflects differences in soil composition, metal type and concentration, and the particular siderophore(s) and plant(s) utilized.

19.5.2 Indole-3-Acetic Acid

Although several naturally occurring auxins have been reported in the literature, indole-3-acetic acid (IAA) is the most common and the most studied auxin, and frequently the terms auxin and IAA are used interchangeably. IAA influences a large number of diverse cellular functions and as a result it is considered to be a critically important regulator of plant growth and development. IAA has been implicated in the orientation of root and shoot growth in response to light and gravity (Kaufman et al. 1995), in differentiation of plant vascular tissue (Aloni 2006), in apical dominance (Tamas 1995), in the initiation of plant lateral and adventitious roots (Gaspar et al. 1996; Malamy and Benfey 1997), in stimulation of plant cell division (Kende and Zeevaart 1997), and in elongation growth in plant stems and roots (Yang et al. 1993; Kende and Zeevaart 1997). Besides influencing division, extension, and differentiation of plant cells and tissues, auxins stimulate seed and tuber germination; increase the rate of xylem and root development; control processes of vegetative growth; initiate lateral and adventitious roots; mediate responses to light and gravity, florescence, and fructification of plants; and also affect photosynthesis, pigment formation, biosynthesis of various metabolites, and resistance to stressful conditions (Tsakelova et al. 2006).

Production of auxin is widespread among plant-associated bacteria. Some of these bacteria are phytopathogens while others promote plant growth. Moreover, several different pathways for auxin biosynthesis have been identified in rhizobacteria (Patten and Glick 1996). The impact of bacterial IAA on plants ranges from positive to negative effects according to the amount of auxin available to the plant and to the sensitivity of the host plant to the hormone. In addition, in plant roots, endogenous IAA may be suboptimal or optimal for growth (Pilet and Saugy 1987) and additional IAA from bacteria could alter the plant's auxin level to either optimal or supraoptimal, resulting in either plant growth promotion or inhibition, respectively.

One of the main effects of bacterial IAA on plants is the enhancement of lateral and adventitious rooting, leading to improved mineral and nutrient uptake and increased root exudation that in turn stimulates bacterial proliferation on the roots (Dobbelaere et al. 1999; Lambrecht et al. 2000; Steenhoudt and Vanderleyden 2000). The role of IAA synthesized by the PGPB *Pseudomonas putida* GR12-2, which produces relatively low levels of this phytohormone, in the development of the canola roots has been studied following the construction of an IAA-deficient mutant of this bacterial strain (Patten and Glick 2002). In these experiments, seed inoculation with wild-type *P. putida* GR12-2 induced the formation of tap roots that were 35-50% longer than the roots from seeds treated with the IAA-deficient mutant and the roots from uninoculated seeds. Conversely, inoculation of mung bean cuttings with the mutant *aux1* of *P. putida* GR12-2 (Xie et al. 1996), which overproduces IAA approximately fourfold, yielded a greater number of much shorter roots compared to plants treated with the wild-type strain of this bacterium (Mayak et al. 1999). In this case, the data indicate that the bacterial IAA incorporated by the plant stimulated the transcription of the gene encoding the enzyme ACC synthase, resulting in increased synthesis of ACC (Jackson 1991), and a subsequent rise in ethylene, which inhibited root elongation. Therefore, the production of IAA alone does not account for growth promotion capacity of *P. putida* GR12-2 (Xie et al. 1996). Rather, the body of evidence suggests that the complex interaction of bacterial IAA and bacterial ACC deaminase is largely responsible for bacterial promotion of plant growth in both the absence and presence of abiotic environmental stresses such as the presence of metals (Glick et al. 2007).

Several reports are consistent with the proposed role of IAA in alleviating certain abiotic stresses. For example, bacterial IAA was reported to stimulate increased root and shoot length of wheat seedlings exposed to high levels of saline (Egamberdieva 2009). An increased tolerance of *Medicago truncatula* against salt stress was also observed in plants nodulated by the IAA-overproducing strain *Sinorhizobium meliloti* DR-64 (Bianco and Defez 2009); plants inoculated with this mutant accumulated a higher than normal amount of proline, and showed enhanced levels of the antioxidant enzymes compared with plants inoculated with the parental strain. In addition, in several studies where plant growth-promoting bacteria were used as adjuncts to metal phytoremediation protocols, researchers concluded that one of the key factors contributing to the ability of the bacterium to facilitate phytoremediation (or at least plant growth in the presence of the metal) was the availability of IAA (Glick 2010).

19.5.3 ACC Deaminase

The plant hormone ethylene controls a wide range of behaviors and responses of vascular plants including fruit ripening, seed germination, tissue differentiation, the formation of root and shoot primordia, lateral bud development, leaf abscission, flower wilting, and the response of plants to both biotic and abiotic stresses, including responses that both turn on a plant's defenses and those that result in increases in plant senescence (Abeles et al. 1992; Arshad and Frankenberger 2002; Matoo and Suttle 1991). On the other hand, when plant growth-promoting bacteria that express the enzyme ACC deaminase are bound to either the seed coat or root of a developing plant, this may act as a mechanism for ensuring that the ethylene level within the plant does not become elevated to the point where plant (both root and shoot) growth is impaired. By facilitating the formation of longer roots and protecting stressed plants from some of the deleterious effects of the phytohormone ethylene, these bacteria can enhance the survival and protect the yield of a variety of plants, especially during the first few days after the seeds are planted and plants are most vulnerable to environmental damage (Glick 2004).

For metal phytoremediation schemes to be successful, it is necessary for plants to take up metal from the soil, effectively translocate the metal from the roots to the leaves and shoots, and attain a reasonably high level of biomass. Unfortunately, even plants that satisfy the first two criteria mentioned above and accumulate relatively large amounts of metal per unit of biomass (these plants are called hyperaccumulators) generally tend to grow slowly and attain only a small size. That is, even for metal-hyperaccumulating plants, high concentrations of metal contaminants are inhibitory to plant growth. This growth inhibition is likely the combined effect of the decrease in the amount of iron available to the plant in the presence of high concentrations of metal (discussed above) and the synthesis of stress ethylene by the plant (discussed below).

Based on a model that was initially developed to describe how plant growth-promoting bacteria that express the enzyme ACC deaminase facilitate the rooting of young seedlings (Glick et al. 1998), it was predicted that the plant-inhibitory effects of many ethylene-causing environmental stresses (both biotic and abiotic) would be significantly decreased if plants were first treated with ACC deaminase-expressing bacteria (Glick 2004). To test this idea, ACC deaminase-expressing bacteria were isolated from nickel-contaminated soil and then used to treat plants grown in the presence of high concentrations of various metals (Burd et al. 1998, 2000). In this case, the addition of the selected bacteria significantly increased the growth of several different plants in the presence of otherwise inhibitory levels of metals in the soil, and the bacteria apparently did this by lowering the level of ethylene that was produced by the plant.

19.5.4 Other Mechanisms

In addition to the three above-mentioned mechanisms of plant growth promotion (siderophores, IAA and ACC deaminase), several other bacterial traits might facilitate metal phytoremediation.

For example, one of the problems often encountered when attempting to use plants to remove contaminating metals from soil is that metals may be bound quite tightly to soil particles so that they cannot are not readily taken up into plants (Gamalero et al. 2009a). These results are in stark contrast to laboratory studies where metal-spiked soils are often employed and the bioavailability of the metal is not an issue. In this regard, some researchers have suggested that it may be useful to employ bacteria that have an active phosphate solubilization system and that this system may help to make some metals that are bound to soil particles more bioavailable thereby assisting metal uptake by plants (Ma et al. 2011a and b; Rodríguez and Fraga 1999). For example, some soil bacteria can solubilize mineral phosphates via the production and secretion of organic acids such as gluconic acid (Rodríguez and Fraga 1999).

As well, it has been suggested that the use of bacteria that produce certain biosurfactants may be an important adjunct to some metal phytoremediation protocols. It is envisioned that the biosurfactants secreted by the bacteria may decrease the tight binding between some metals and soil particles and thereby help to make the metals more bioavailable (Sheng et al. 2008a). In addition to the mechanisms utilized by plant growth-promoting bacteria, it is important to bear in mind that in nature the roots of approximately 90% of terrestrial plants are colonized by mycorrhizal fungi. Thus, in order to fully understand the optimal conditions used by PGPB to facilitate the phytoremediation of metals in the field, it is ultimately necessary to take into account the interactions of both the plant and the added bacteria with (endogenous or added) mycorrhizae (Gamalero et al. 2009a).

19.6 EXAMPLES OF THE USE OF PLANT GROWTH-PROMOTING BACTERIA IN METAL PHYTOREMEDIATION PROTOCOLS

The initial observations that plant growth-promoting bacteria can facilitate the growth of plants in the presence of high levels of contaminating metals (Burd et al. 1998 and 2000), and that plant metal accumulation from contaminated soil can sometimes be increased by the addition of plant growth-promoting bacteria (Höflich and Metz 1997; de Souza et al. 1999a, 1999b) were all made approximately 12–15 years ago. Since then, the use of plant growth-promoting bacteria as an adjunct to various metal phytoremediation protocols has become quite common and has been reviewed extensively (Glick 2003, 2010; Pilon-Smits and Freeman 2006; Gamalero et al. 2009a; Khan et al. 2009; Rajkumar et al. 2009, 2010; Ma et al. 2011a; Pacwa-Plociniczak et al. 2011). Notwithstanding the many practical difficulties that scientists have encountered when attempting to remove contaminating metals from various soils, interest in this approach continues to grow.

Table 19.1 summarizes a large number of studies where the addition of plant growth-promoting bacteria has been reported to facilitate metal phytoremediation. A detailed examination of the data summarized in this table (and in the papers cited therein) allows a number of listed below drawn generalizations:

(a) Nearly all of the studies of metal phytoremediation that include the use of plant growth-promoting bacteria have been done in a laboratory or greenhouse setting. Given the large number of moderately successful laboratory studies, it is time to take this technology to the field, notwithstanding the regulatory hurdles that exist in many jurisdictions regarding the deliberate release of bacteria to the environment;
(b) Only a relatively small number of metals have been tested in these studies. Most studies have been limited to Cd, Cu, Cr, Ni, Pb, Se or Zn, with a very small number of studies examining As or Hg;
(c) Despite the many reports about the effectiveness of several so-called "metal hyper-accumulating plants" (Cobbett and Goldsbrough 2002; Freeman et al. 2004, 2005; Krämer et al. 1996) in being able to accommodate high metal concentrations in their tissues, most of the studies reported in Table 19.1 have used a limited number of "moderately-accumulating" plants, most notably *Brassican juncea* and *Brassica napus*. This likely

TABLE 19.1
Facilitation of Metal Phytoremediation by Plant Growth-Promoting Bacteria

Bacterium	Plant	Metal	Laboratory or Field	References
Burkholderia cepacia, *B. vietnamiensis*	Yellow lupine	Ni + toluene	L	Weyens et al. (2011)
Mixture including *Rhizobia* spp.	*Leucaena leucophala*	Zn, Cd	F	Saraswat et al. (2011)
Pseudomonas putida, *Chryseobacterium joostei*, C. sp., *Tsukamurella strandjordae*, *E. coli*, *Sphingomonas* sp., *Curtobacterium pusillum*	*Arabidopsis halleri*	Zn, Cd	L	Farinati et al. (2011)
Psychrobacter sp.	*Ricinus communis* (Castor bean) and *Helianthus annuus* (sunflower)	Ni	L	Ma et al. (2011)
Micrococcus roseus, *Bacillus endophyticus*, *Paenibacillus macerans*, *Bacillus pumilus*	*Brassica juncea* (Indian mustard)	Fly ash containing Al, Si, Fe, B, Ni, and Zn	L	Kumari and Singh (2011)
Pseudomonas sp., *Pantoea ananatis*, *Enterobacter ludwigii*, *Ralstonia* sp., *Pseudomonas thivervalensis*	*Brassica napus* (canola)	Cu	L	Zhang et al. (2011)
Bacillus mycoides	*Amaranthus retroflexus*, *Helianthus annus*, *Medicago sativa*	Cd	L	Motesharezadeh et al. (2010)
Rhizobium leguminosarum, *Azotobacter chroococum*	*Zea mays* (corn)	Pb	L	Hadi and Bano (2010)
Bacillus pumilus, *Azospirillum brasilense*	*Atriplex lentiformis* (quailbush)	Mine tailings	L	de-Bashan et al. (2010)
Flavobacterium sp.	Orychophragmus violaceus	Zn	L	He et al. (2010)
Enterobacter sp., *Pseudomonas* sp., *Xanthamonadaceae* sp., *Sanguibacter* sp., *Clostridium aminovalericum*, *Stenotrophomonas* sp.	*Nicotiana tabacum* (tobacco)	Cd	L	Mastretta et al. (2009)
Variovorax papadoxus	*Brassica juncea*	Cd	L	Belimov and Wenzel (2009)
Pseudomonas sp., *Bacillus* sp.	*Lycopersicon esculentum* (tomato)	Cd, Pb	L	He et al. (2009)
Streptomyces tendae	*Helianthus annuus*	Cd	L	Dimkpa et al. (2009)
Achromobacter xylosoxidans	*Brassica juncea*	Cu	L	Ma et al. (2009)
Enterobacter aerogenes, *Rahnella aquatilis*	*Brassica juncea*	Cr, Ni	L	Kumar et al. (2009)

(*continued*)

TABLE 19.1 (Continued)
Facilitation of Metal Phytoremediation by Plant Growth-Promoting Bacteria

Bacterium	Plant	Metal	Laboratory or Field	References
Pseudomonas sp., *Janthinobacterium lividum*, *Serratia marcesans*, *Flavobacterium frigidimaris*, *Streptomyces* sp., *Agromyces terreus*	*Salix caprea* (willow)	Cd, Zn	L	Kuffner et al. (2008)
Pseudomonas tolaasii, *Pseudomonas fluorescens*, *Alcaligenes* sp., *Mycobacterium* sp.	*Brassica napus*	Cd	L	Dell'Amico et al. (2008)
Bacillus thuringiensis, *Bacillus licheniformis*, *Bacillus biosubtyl*	*Brassica juncea*	Cd, Cr, Se	L	Hussein (2008)
Pseudomonas putida	*Brassica napus*	Ni	L	Rodriguez et al. (2008)
Pseudomonas fluorescens, *Microbacterium* sp.	*Brassica napus*	Pb	L	Sheng et al. (2008a)
Bacillus sp.	*Brassica napus*, *Zea mays*, *Sorghum vulgare* var. *sudanense* (sudan grass), *Lycopersicon esculentum*	Cd	L	Sheng et al. (2008b)
Bacillus edaphicus	*Brassica juncea*	Pb	L	Sheng et al. (2008c)
Bacillus subtilis, *Bacillus pumilus*, *Pseudomonas pseudoalcaligenes*, *Brevibacterium halotolerans*	*Zea mays*, *Sorghum bicolor*	Cr, Cu, Pb, Zn	L	Abou-Shanab et al. (2008)
Burkholderia sp.	*Brassica juncea*, *Zea mays*, *Lycopersicon esculentum*	Cd, Pb	L	Jiang et al. (2008)
Enterobacter sp.	*Brassica juncea*	Cr, Ni, Zn	L	Kumar et al. (2008)
Pseudomonas aeruginosa	*Vigna mungo*	Cd	L	Ganesan (2008)
Pseudomonas sp., *Bacillus* sp.	*Brassica juncea*	Ni		Rajkumar and Freitas (2008a)
Pseudomonas sp., *Pseudomonas jessenii*	*Ricinus communis*	Cu, Ni, Zn		Rajkumar and Freitas (2008b)
Proteus vulgaris	*Cajanus cajan* (pigeon pea)	Cu		Rani et al. (2008)
Pseudomonas sp.	*Cicer arietinum* (chick pea)	Ni		Tank and Saraf (2008)
Rhizobium sp.	*Pisum sativum* (pea)	Ni, Zn		Wani et al. (2008)
Bradyrhizobium sp.	*Vigna radiata* (mung bean)	Ni, Zn		Wani et al. (2007)
Burkholderia cepacia	*Sedum alfredii*	Cd, Zn		Li et al. (2007)
Mesorhizobium huakuii subsp. *rengei*	*Astragalus sinicus* (Chinese milk vetch)	Cd		Ike et al. (2007)

(*continued*)

TABLE 19.1 (Continued)
Facilitation of Metal Phytoremediation by Plant Growth-Promoting Bacteria

Bacterium	Plant	Metal	Laboratory or Field	References
Pseudomonas diminuta, *Brevundimonas diminuta*, *Nitrobacteria irancium*, *Ochrobacterum anthropi*, *Bacillus cereus*	*Eichhornia crassipes* (water hyacinth)	Cr		Abou-Shanab et al. (2007)
Nine different nickel-resistant bacterial strains	Alyssum murale	Ni		Abou-Shanab et al. (2006)
Sinorhizobium sp.	*Brassica juncea*	Pb		Di Gregorio et al. (2006)
Pseudomonas brassicacearum, *Pseudomonas marginalis*	*Pisum sativum*	Cd	L	Safranova et al. (2006)
Pseudomonas putida	*Helianthus annuus*	Cd		Wu et al. (2006a)
Azotobacter chroococcum, *Bacillus megaterium*, *Bacillus mucilaginosus*	*Brassica juncea*	Cu, Pb, Zn		Wu et al. (2006b)
Pseudomonas fluorescens, *Pseudomonas putida*	*Brassica napus*	Ni		Ashour et al. (2006)
Pseudomonas fluorescens	*Helianthus annuus*	As		Shilev et al. (2006)
Pseudomonas sp., *Bacillus* sp.	*Brassica juncea*	Cr		Rajkumar et al. (2006)
Pseudomonas putida	*Brassica napus*	Ni	F	Farwell et al. (2006)
Pseudomonas sp., *Bacillus* sp., *Xanthomonas* sp.	*Brassica napus*	Cd	L	Sheng and Xia (2006)
Bacillus subtilis	*Brassica juncea*	Ni	L	Zaidi et al. (2006)
Rhizosphere bacteria	*Graminaceae* grasses	Cd, Ni, Zn	L	Dell'Amico et al. (2005)
Pseudomonas putida	*Vigna radiata* (mung bean)	Cd, Pb		Tripathi et al. (2005)
Variovorax paradoxus, *Rhodoccus* sp., *Flavobacterium* sp.	*Brassica juncea*	Cd	L	Belimov et al. (2005)
Pseudomonas aspleni	*Brassica napus*	Cu	L	Reed and Glick (2005)
Microbacterium arabinogalactanolyticum	*Alyssum murale*	Ni		Abou-Shanab et al. (2003)
Brevibacillus sp.	*Trifolium pratense* (red clover)	Pb		Vivas et al. (2003)
Enterobacter cloacae	*Brassica napus*	As	L	Nie et al. (2002)
Rhizosphere bacteria	*Thlaspi caerulescens* (Alpine pennygrass)	Zn	L	Whiting et al. (2001)
Kluyvera ascorbata	*Brassica napus, Brassica juncea, Lycopersicon esculentum*	Ni, Pb, Zn	L	Burd et al. (2000)
Rhizosphere bacteria	*Brassica juncea*	Se	L	de Souza et al. (1999a)
Rhizosphere bacteria	*Scirpus robustus* (saltmarsh bulrush), *Polypogon monspeliensis* (rabbitfoot grass)	Hg, Se	L	de Souza et al. (1999b)
Kluyvera ascorbata	*Brassica napus*	Ni	L	Burd et al. (1998)

reflects the ability of *Brassican juncea* and *Brassica napus* to attain a much greater biomass than most of the hyperaccumulators and thereby to remove greater amounts of metal from the soil;

(d) The majority of the plant growth-promoting bacteria that were used in these experiments are either pseudomonads or bacilli. This may be a reflection of the relative abundance of these organisms in many soils. Alternatively, this choice may reflect the fact that plant growth-promotion traits are more common in these than in other soil bacteria. These traits include the bacterial synthesis of siderophores, IAA, ACC deaminase, organic acids and biosurfactants;

(e) While both rhizospheric and endophytic PGPB strains have been tested in various phytoremediation protocols, in the final analysis, endophytes are likely to be better adjuncts to phytoremediation than rhizosphereic bacteria. This is a consequence of the decreased environmental sensitivity and vulnerability of the endophytic strains.

19.7 CONCLUSIONS AND PERSPECTIVES

Perhaps the major scientific issue that remains before this technology can be commercialized on a wide scale is the (usually) limited bioavailability of many metal contaminants. However, by judicious selection of phosphate solubilizing and biosurfactant-producing PGPB strains it should be possible to overcome this limitation. However, given the reluctance of many jurisdictions to approve the deliberate release of bacteria to the environment, there is also a significant political impediment to the commercialization of metal phytoremediation. This having been said, successful field trials where metals are removed from contaminated sites should go a long way to convincing both the public and government regulators to actively pursue the development of this technology.

ACKNOWLEDGMENTS

The work from our laboratories that is reviewed here was supported by funds from the Italian Ministry of University and Research (EG) and the Natural Sciences and Engineering Research Council of Canada (BRG).

REFERENCES

Abeles, F.B., P.W. Morgan, and M.E. Saltveit, Jr. 1992. *Ethylene in Plant Biology*, 2nd ed., Academic Press, New York.

Alarcón, A., F.T. Davies Jr, R.L. Autenrieth, and D.A. Zuberer. 2008. Arbuscular mycorrhiza and petroleum-degrading microorganisms enhance phytoremediation of petroleum-contaminated soil. *International Journal of Phytoremediation* 10: 251–263.

Al-Awadhi, H., I. El-Nemr, H. Mahmoud, N.A. Sorkhoh, and S. Radwan. 2009. Plant-associated bacteria as tools for the phytoremediation of oily nitrogen-poor soils. *International Journal of Phytoremediation* 11: 11–27.

Aloni, R. E. Aloni, M. Langhans, and C. I. Ullrich. 2006. Role of cytokinin and auxin in shaping root architecture: regulating vascular differentiation, lateral root initiation, root apical dominance and root gravitropism. *Annals of Botany* 97: 883–893.

Arshad, M., and W.T. Frankenberger. 2002. *Ethylene: Agricultural Sources and Applications*, Kluwer Academic/Plenum Publishers, New York.

Barraquio, W.L., L. Revilla, and J.K. Ladha. 1997. Isolation of endophytic diazotrophic bacteria from wetland rice. *Plant and Soil* 194: 15–24.

Bashan, Y., A. Rojas, and M.E. Puente. 1999. Improved establishment and development of three cacti species inoculated with *Azospirillum brasilense* transplanted into disturbed urban desert soil. *Canadian Journal of Microbiology* 45: 441–451.

Belimov, A.A., V.I. Safronova, T.A. Sergeyeva, T.N. Egorova, V.A. Matveyeva et al. 2001. Characterization of plant growth promoting rhizobacteria isolated from polluted soils and containing 1-aminocyclopropane-1-carboxylate deaminase. *Canadian Journal of Microbiology* 47: 642–652.

Belimov, A.A., N. Hontzeas, V.I. Safronova, S.V. Demchinskaya, G. Piluzza et al. 2005. Cadmium-tolerant plant growth-promoting bacteria associated with the roots of Indian mustard (*Brassica juncea* L. Czern.). *Soil Biology and Biochemistry* 37: 241–250.

Bianco, C., and R. Defez. 2009. *Medicago truncatula* improves salt tolerance when nodulated by an indole-3-acetic acid-overproducing *Sinorhizobium meliloti* strain. *Journal of Experimental Botany* 60: 3097v3107.

Blaha, D., C. Prigent-Combaret, M.S. Mirza, and Y. Moënne-Loccoz. 2006. Phylogeny of the 1-aminocyclopropane-1-carboxylic acid deaminase-encoding gene *acdS* in phytobeneficial and pathogenic *Proteobacteria* and relation with strain biogeography. *FEMS Microbiology Ecology* 56: 455–470.

Bnayahu, B.Y. 1991. Root excretions and their environmental effects: influence on availability of phosphorus. In: *Plant Roots: The Hidden Half*, Y. Waisel, A. Eshel, and U. Kafkafi, eds. Marcel Decker, New York, pp. 529–557.

Bøckman, O.C. 1997. Fertilizers and biological nitrogen fixation as sources of plant nutrients: perspectives for future agriculture. *Plant and Soil* 194: 11–14.

Boukhalfa, H., and A.L. Crumbliss. 2002. Chemical aspects of siderophore mediated iron transport. *BioMetals* 15:325–339.

Braud, A., K. Jézéquel, E. Vieille, A. Tritter, and T. Lebeau. 2006. Changes in extractability of Cr and Pb in a polycontaminated soil after bioaugmentation with microbial producers of biosurfactants, organic acids and siderophores. *Water, Air and Soil Pollution* 6: 261–279.

Budzikiewicz, H. 2004. Siderophores of the Pseudomonadaceae *sensu stricto* (fluorescent and non fluorescent *Pseudomonas* spp.). In: *Progress in the Chemistry of Organic Natural Products*, W. Herz, H. Falk, and G.W. Kirby, eds. Springer, Vienna, pp. 81–237.

Burd, G.I., D.G. Dixon, and B.R. Glick. 1998. A plant growth-promoting bacterium that decreases nickel toxicity in plant seedlings. *Applied and Environmental Microbiology* 64: 3663–3668.

Burd, G.I., D.G. Dixon, and B.R. Glick. 2000. Plant growth-promoting bacteria that decrease heavy metal toxicity in plants. *Canadian Journal of Microbiology* 46: 237–245.

Cheng, Z., E. Park, and B.R. Glick. 2007. 1-Aminocyclopropane-1-carboxylate deaminase from *Pseudomonas putida* UW4 facilitates the growth of canola in the presence of salt. *Canadian Journal of Microbiology* 53: 912–918.

Ciardi, J.A., D.M. Tieman, S.T. Lund, J.B. Jones et al. 2000. Response to *Xanthomonas campestris* pv. *vesicatoria* in tomato involves regulation of ethylene receptor gene expression. *Plant Physiology* 123: 81–92.

Clemens, S. 2006. Toxic metal accumulation, responses to exposure and mechanisms of tolerance in plants. *Biochimie* 88: 1707–1719.

Cobbett, C., and P. Goldsbrough. 2002. Phytochelatins and metallothioneins: Roles in heavy metal detoxification and homeostasis. *Annual Review Plant Biology* 53: 159–182.

Compant, S., B. Reiter, A. Sessitsch, J. Nowak, C. Clément, and E.A. Barka. 2005. Endophytic colonization of *Vitis vinifera* L. by plant growth promoting bacterium *Burkholderia* sp strain PsJN. *Applied and Environmental Microbiology* 71: 1685–1693.

Compant, S., C. Clément, and A. Sessitsch. 2010. Colonization of plant growth-promoting bacteria in the rhizo- and endosphere of plants: importance, mechanisms involved and future prospects. *Soil Biology and Biochemistry* 42: 669–678.

de Souza, M.P., D. Chu, M. Zhou, A.M. Zayed, S.E. Ruzin et al. 1999a. Rhizosphere bacteria enhance selenium accumulation and volatilization by Indian mustard. *Plant Physiology* 119: 565–573.

de Souza, M.P., C.P.A. Huang, N. Chee, and N. Terry. 1999b. Rhizosphere bacteria enhance the accumulation of selenium and mercury in wetland plants. *Planta* 209: 259–263.

de-Bashan L.E., J.-P. Hernandez, T. Morey, and Y. Bashan. 2004. Microalgae growth-promoting bacteria as "helpers" for microalgae: a novel approach for removing ammonium and phosphorus from municipal wastewater. *Water Research* 38: 466–474.

Dell'Amico, E., L. Cavalca, and V. Andreoni. 2008. Improvement of *Brassica napus* growth under cadmium stress by cadmium-resistant rhizobacteria. *Soil Biology and Biochemistry* 40: 74–84.

Diels, L., N. van der Lelie, and L. Bastiaens. 2002. New developments in treatment of heavy metal contaminated soils. *Reviews in Environmental Science and Biotechnology* 1: 75–82.

Dobbelaere, S., A. Croonenborghs, A. Thys, A. Vande Broek, and J. Vanderleyden. 1999. Phytostimulatory effect of *Azospirillum brasilense* wild type and mutant strains altered in IAA production on wheat. *Plant and Soil* 212: 155–164.

Duan, J., K.M. Müller, T.C. Charles, S. Vesely, and B.R. Glick. 2009. 1-Aminocyclopropane-1-carboxylate (ACC) deaminase genes in *Rhizobia* from southern Saskatchewan. *Microbial Ecology* 57: 423–436.

Duijff, B.J., V. Gianinazzi-Pearson, and P. Lemanceau. 1997. Involvement of the outer membrane lipolysaccharides in the endophytic colonization of tomato roots by biocontrol *Pseudomonas fluorescens* strain WCS417r. *New Phytologist* 135: 325–334.

Egamberdieva, D. 2009. Alleviation of salt stress by plant growth regulators and IAA producing bacteria in wheat. *Acta Physiologiae Plantarum* 31: 861–864.

Farwell, A. J., S. Vesely, V. Nero, H. Rodriguez et al. 2006. The use of transgenic canola (*Brassica napus*) and plant growth-promoting bacteria to enhance plant biomass at a nickel-contaminated field site. *Plant and Soil* 288: 309–318.

Farwell, A.J., S. Vesely, V. Nero, H. Rodriguez, K. McCormack et al. 2007. Tolerance of transgenic canola plants (*Brassica napus*) amended with plant growth-promoting bacteria to flooding stress at a metal-contaminated field site. *Environmental Pollution* 147: 540–545.

Feng, K., H.M. Lu, H.J. Sheng, X.L. Wang, and J. Ma. 2004. Effect of organic ligands on biological availability of inorganic phosphorus in soils. *Pedosphere* 14: 85–92.

Freeman, J.L., M.W. Persans, K. Nieman, C. Albrecht, W. Peer et al. 2004. Increased glutathione biosynthesis plays a role in nickel tolerance in *Thlaspi* nickel hyperaccumulators. *Plant Cell* 16: 2176–2191.

Freeman, J.L., D. Garcia, D. Kim, A. Hopf, and D.E. Salt. 2005. Constitutively elevated salicylic acid signals glutathione-mediated nickel tolerance in *Thlaspi* nickel hyperaccumulators. *Plant Physiology* 137: 1082–1091.

Fujishige, N.A., N.N. Kapadia, and A.M. Hirsh. 2006. A feeling for the micro-organism: structure on a small scale. Biofilms on plant roots. *Botanical Journal of the Linnean Society* 150: 79–884.

Gamalero, E., M.G. Martinotti, A. Trotta, P. Lemanceau, and G. Berta. 2002. Morphogenetic modifications induced by *Pseudomonas fluorescens* A6RI and *Glomus mosseae* BEG12 in the root system of tomato differ according to plant growth conditions. *New Phytologist* 155: 293–300.

Gamalero, E., G. Lingua, F.G. Caprì, A. Fusconi, G. Berta, and P. Lemanceau. 2004. Colonization pattern of primary tomato roots by *Pseudomonas fluorescens* A6RI characterized by dilution plating, flow cytometry, fluorescence, confocal and scanning electron microscopy. *FEMS Microbiology Ecology* 48: 79–87.

Gamalero, E., G. Lingua, R. Tombolini, L. Avidano, B. Pivato, and G. Berta. 2005. Colonization of tomato root seedling by *Pseudomonas fluorescens* 92rkG5: spatio-temporal dynamics, localization, organization, viability and culturability. *Microbial Ecology* 50: 289–297.

Gamalero, E., G. Berta, and B.R. Glick. 2009a. Effects of plant growth promoting bacteria and AM fungi on the response of plants to heavy metal stress. *Canadian Journal of Microbiology* 55: 501–514.

Gamalero, E., G. Berta, N. Massa, B.R. Glick, and G. Lingua. 2010. Interactions between *Pseudomonas putida* UW4 and *Gigaspora rosea* BEG9 and their consequences for the growth of cucumber under salt stress conditions. *Journal of Applied Microbiology* 108(1): 236–245.

Gamalero, E., and B.R. Glick. *in press*. Mechanisms used by plant growth-promoting bacteria. In: *Bacteria in Agrobiology*, D.K. Maheshwari, ed. Springer, Germany.

Gaspar, T., C. Kevers, C. Penel, H. Greppin, D.M. Reid et al. 1996. Plant hormones and plant growth regulators in plant tissue culture. *In Vitro Cellular and Developmental Biology—Plant* 32: 272–289.

Gerhardt, K.E., X.D. Huang, B.R. Glick, and B.M. Greenberg. 2009. Phytoremediation and rhizoremediation of organic soil contaminants: potential and challenges. *Plant Science* 176: 20–30.

Germaine, K.J., K. Keogh, D. Ryan, and D.N. Dowling. 2009. Bacterial endophyte-mediated naphthalene phytoprotection and phytoremediation. *FEMS Microbiology Letters* 296: 226–234.

Glick, B.R. 1995. The enhancement of plant-growth by free-living bacteria. *Canadian Journal of Microbiology* 41: 109–117.

Glick, B.R., D.M. Penrose, and J. Li. 1998. A model for the lowering of plant ethylene concentrations by plant growth-promoting bacteria. *Journal of Theoretical Biology* 190: 63–68.

Glick, B.R., C.L. Patten, G. Holguin, and D.M. Penrose. 1999. Biochemical and genetic mechanisms used by plant growth-promoting bacteria. Imperial College Press, London.

Glick, B.R. 2003. Phytoremediation: Synergistic use of plants and bacteria to clean up the environment. *Biotechnology Advances* 21: 383–393.

Glick, B.R. 2004. Bacterial ACC deaminase and the alleviation of plant stress. *Advances in Applied Microbiology* 56:291–312.

Glick, B.R., Z. Cheng, J. Czarny, and J. Duan. 2007. Promotion of plant growth by ACC deaminase-containing soil bacteria. *European Journal of Plant Pathology* 119: 329–339.

Glick, B.R. 2010. Using soil bacteria to facilitate phytoremediation. *Biotechnology Advances* 28: 367–374.

Grichko, V.P., and B.R. Glick. 2001. Amelioration of flooding stress by ACC deaminase-containing plant growth-promoting bacteria. *Plant Physiology and Biochemistry* 39: 11–17.

Gurska, J., W. Wang, K.E. Gerhardt, A.M. Khalid, D.M. Isherwood, X.D. Huang et al. 2009. Three year field test of a plant growth-promoting rhizobacteria enhanced phytoremediation system at a land farm for treatment of hydrocarbon waste. *Environmental Science Technology* 43: 4472–4479.

Hao, Y., T.C. Charles, and B.R. Glick. 2007. ACC deaminase from plant growth promoting bacteria affects crown gall development. *Canadian Journal of Microbiology* 53: 1291–1299.

Hernandez, J.P., L.E. de-Bashan, and Y. Bashan. 2006. Starvation enhances phosphorus removal from wastewater by the microalga *Chlorella* spp. co-immobilized with *Azospirillum brasilense*. *Enzyme and Microbial Technology* 38: 190–198.

Höflich, G., and R. Metz. 1997. Interaction of plant-microorganism-association in heavy metal containing soils from sewage farms. *Bodenkultur* 48: 238–247.

Honma, M., and T. Shimomura. 1978. Metabolism of 1-aminocyclopropane-1-carboxylic acid. *Agricultural Biology and Chemistry* 42: 1825–1831.

Huang, X.D., Y. El-Alawi, D.M. Penrose, B.R. Glick, and B.M. Greenberg. 2004a. Responses of three grass species to creosote during phytoremediation. *Environmental Pollution* 130: 453–463.

Huang, X.D., Y. El-Alawi, D.M. Penrose, B.R. Glick, and B.M. Greenberg. 2004b. A multi-process phytoremediation system for removal of polycyclic aromatic hydrocarbons from contaminated soils. *Environmental Pollution* 130: 465–476.

Illmer, P., and F. Schinner. 1992. Solubilization of inorganic phosphates by microorganisms isolated from forest soil. *Soil Biology and Biochemistry* 24: 389–395

Jackson, M.B. 1991. Ethylene in root growth and development. In: *The Plant Hormone Ethylene*, A.K. Mattoo and J.C. Suttle, eds. CRC Press, Boca Raton, FL, pp. 169–181.

Jackson, L.L., J.R. McAuliffe, and B.A. Roundy. 1991. Desert restoration. *Restoration and Management Notes* 9: 71–80.

Kaufman, P.B., L.-L. Wu, T.G. Brock, and D. Kim. 1995. Hormones and their orientation of growth. In: *Plant Hormones*, P.J. Davies, ed. Kluwer Academic Publishers, Dordrecht, The Netherlands, pp. 547–571.

Kende, H., and J.A.D. Zeevaart. 1997. The five "classical" hormones. *Plant Cell* 9: 1197–1210.

Khan, A.G., C. Kuek, T.M. Chaudhry, C.S. Khoo, and W.J. Hayes. 2000. Role of plants, mycorrhizae and phytochelators in heavy metal contaminated land remediation. *Chemosphere* 41: 197–207.

Khan, M.S., A. Zaidi, P.A. Wani, and M. Oves. 2009. Role of plant growth-promoting rhizobacteria in the remediation of metal contaminated soils. *Environmental Chemistry Letters* 7: 1–19.

Kloepper, J.W., R. Lifshitz, and R.M. Zablotowitz. 1989. Free living bacteria inocula for enhancing crop productivity. *Trends in Biotechnology* 7: 39–43.

Krämer, U., J.D. Cotter-Howells, J.M. Charnock, A.J.M. Baker, and J.A.C. Smith. 1996. Free histidine as a metal chelator in plants that accumulate nickel. *Nature* 379: 635–638.

Kuffner, M., M. Puschenreiter, G. Wieshammer, M. Gorfer, and A. Sessitsch. 2008. Rhizosphere bacteria affect growth and metal uptake of heavy metal accumulating willows. *Plant and Soil* 304: 35–44.

Lambrecht, M., Y. Okon, A. Vande Broek, and J. Vanderleyden. 2000. Indole-3-acetic acid: a reciprocal signaling molecule in bacteria-plant interactions. *Trends in Microbiology* 8: 298–300.

Lingua, G., E. Gamalero, P. Lemanceau, and G. Berta. 2007. Colonization of plant roots by pseudomonads and AM fungi: a dynamic phenomenon, affecting plant growth and health. In: *Mycorrhiza 2nd Edition*, Varma A., ed. Springer-Verlag, pp. 601–626.

Ma, J.F. 2005. Plant root responses to three abundant soil minerals: Silicon, aluminum and iron. *Critical Reviews in Plant Sciences* 24: 267–281.

Ma, Y., M.N.V. Prasad, M. Rajkumar, and H. Freitas. 2011a. Plant growth-promoting rhizobacteria and endophytes accelerate phytoremediation of metalliferous soils. *Biotechnology Advances* 29: 248–258.

Ma, Y., M. Rajkumar, J.A.F. Vicente, and H. Freitas. 2011b. Inoculation of Ni-resistant plant growth promoting bacterium *Psychrobacter* sp. strain SRS8 for the improvement of nickel phytoextraction by energy crops. *International Journal of Phytoremediation* 13: 126–139.

Malamy, J.E., and P.N. Benfey. 1997. Down and out in *Arabidopsis*: the formation of lateral roots. *Trends Plant Science* 2: 390–396.

Matoo, A.K., and C.S. Suttle (eds.). 1991. *The Plant Hormone Ethylene*. CRC Press, Boca Raton, FL.

Mayak, S., T. Tirosh, and B.R. Glick. 1999. Effect of wild type and mutant plant growth-promoting rhizobacteria on the rooting of mung been cuttings. *Journal of Plant Growth Regulator* 18: 49–53.

Mayak, S., T. Tirosh, and B.R. Glick. 2004a. Plant growth-promoting bacteria that confer resistance to water stress in tomato and pepper. *Plant Science* 166: 525–530.

Mayak, S., T. Tirosh, and B.R. Glick. 2004b. Plant growth-promoting bacteria confer resistance in tomato plants to salt stress. *Plant Physiology and Biochemistry* 42: 565–572.

Meyer, J.M., and M.A. Abdallah. 1978. The fluorescent pigment of *Pseudomonas fluorescens*: biosynthesis, purification and physico-chemical properties. *Journal of General Microbiology* 107: 319–328.

Meyer, J.M., C. Gruffaz, V. Raharinosy, I. Bezverbnaya, M.H. Schäfer, and H. Budzikiewicz. 2008. Siderotyping of fluorescent *Pseudomonas*: molecular mass determination by mass spectrometry as a powerful pyoverdine siderotyping method. *BioMetals* 21: 259–271.

Mirleau, P., S. Delorme, L. Philippot, J.M. Meyer, S. Mazurier, and P. Lemanceau. 2000. Fitness in soil and rhizosphere of *Pseudomonas fluorescens* C7R12 compared with a C7R12 mutant affected in pyoverdine synthesis and uptake. *FEMS Microbiology Ecology* 34: 35–44.

Mirleau, P., L. Philippot, T. Corberand, and P. Lemanceau. 2001. Involvement of nitrate reductase and pyoverdine in competitiveness of *Pseudomonas fluorescens* strain C7R12 in soil. *Applied and Environmental Microbiology* 67: 2627–2635.

Neilands, J.B. 1981. Iron adsorption and transport in microorganisms. *Annual Review Nutrition* 1: 27–46.

Nie, L., S. Shah, G.I. Burd, D.G. Dixon, and B.R. Glick. 2002. Phytoremediation of arsenate contaminated soil by transgenic canola and the plant growth-promoting bacterium *Enterobacter cloacae* CAL2. *Plant Physiology and Biochemistry* 40: 355–361.

O'Sullivan, D.J. and F. O'Gara. 1992. Traits of fluorescent *Pseudomonas* spp. involved in suppression of plant root pathogens. *Microbiological Reviews* 56: 662–676.

Pacwa-Plociniczak, M., G.A. Plaza, Z. Piotrowska-Seget, and S.S. Cameotra. 2011. Environmental applications of biosurfactants: recent advances. *International Journal of Molecular Sciences* 12: 633–654.

Patten, C., and B.R. Glick. 1996. Bacterial biosynthesis of indole-3-acetic acid. *Canadian Journal Microbiology* 42: 207–220.

Patten, C.L., and B.R. Glick. 2002. The role of bacterial indoleacetic acid in the development of the host plant root system. *Applied and Environmental Microbiology* 68: 3795–3801.

Pilet, P.E., and M. Saugy. 1987. Effect on root growth of endogenous and applied IAA and ABA. *Plant Physiology* 83: 33–38.

Pilon-Smits, E. 2005. Phytoremediation. *Annual Review Plant Biology* 56: 15–39.

Pilon-Smits, E.A.H., and J.L. Freeman. 2006. Environmental cleanup using plants: biotechnological advances and ecological considerations. *Frontiers in Ecology and the Environment* 4: 203–210.

Pivato, B., E. Gamalero, P.Lemanceau, and G. Berta. 2008. Cell organization of *Pseudomonas fluorescens* C7R12 on adventitious roots of *Medicago truncatula* as affected by arbuscular mycorrhiza. *FEMS Microbiology Letters* 289: 173–180.

Puente, M. E., C.Y. Li, and Y. Bashan. 2004. Microbial populations and activities in the rhizoplane of rock-weathering desert plants. II. Growth promotion of cactus seedlings. *Plant Biology* 6: 643–650.

Rajkumar, M., N. Ae, and H. Freitas. 2009. Endophytic bacteria and their potential to enhance heavy metal phytoremediation. *Chemosphere* 77: 153–160.

Rajkumar, M., N. Ae, M.N.V. Prasad, and H. Freitas. 2010. Potential of siderophore-producing bacteria for improving heavy metal phytoextraction. *Trends in Biotechnology* 28: 142–149.

Reed, M.L.E., and B.R. Glick. 2004. Applications of free living plant growth-promoting rhizobacteria. *Anton van Leeuwenhoek* 86: 1–25.

Reed, M.L.E., and B.R. Glick. 2005. Growth of canola (*Brassica napus*) in the presence of plant growth-promoting bacteria and either copper or polycyclic aromatic hydrocarbons. *Canadian Journal of Microbiology* 51: 1061–1069.

Reinhold-Hurek, B., T. Maes, S. Gemmer, M. Van Montagu, and T. Hurek. 2006. An endoglucanase is involved in infection of rice roots by the not-cellulose-metabolizing endophyte *Azoarcus* Sp. strain BH72. *Molecular Plant-Microbe Interactions* 19: 181–188.

Rodríguez, H., and R. Fraga. 1999. Phosphate solubilizing bacteria and their role in plant growth promotion. *Biotechnology Advances* 17: 319–339.

Rodriguez, H., T. Gonzalez, I. Goire, and Y. Bashan. 2004. Gluconic acid production and phosphate solubilization by the plant growth-promoting bacterium *Azospirillum* spp. *Naturwissenschaften* 91: 552–555.

Rodríguez, H., S. Vessely, S. Shah, and B.R. Glick. 2008. Effect of a nickel-tolerant ACC deaminase-producing *Pseudomonas* strain on growth of nontransformed and transgenic canola lines. *Current Microbiology* 57: 170–174.

Saleh, S., X.-D. Huang, B.M. Greenberg, and B.R. Glick. 2004. Phytoremediation of persistent organic contaminants in the environment. In: *Applied Bioremediation and Phytoremediation*, (Series: Soil Biology, Vol. 1), A. Singh and O. Ward, eds. Springer-Verlag, Berlin, pp. 115–134.

Salt, D.E., M. Blaylock, P.B.A.N. Kumar, V. Dushenkov, B.D. Ensley et al. 1995. Phytoremediation: a novel strategy for the removal of toxic metals from the environment using plants. *BioTechnology* 13: 468–474.

Saravanakumar, D., and R. Samiyappan. 2006. ACC deaminase from *Pseudomonas fluorescens* mediated saline resistance in groundnut (*Arachis hypogea*) plants. *Journal of Applied Microbiology* 102: 1283–1292.

Schützendübel, A., and A. Polle. 2002. Plant responses to abiotic stresses: heavy metal induced oxidative stress and protection by mycorrhization. *Journal of Experimental Botany* 53: 1351–1365.

Sheng, X.-F., J.-J. Xia, C.-Y. Jiang, L.-H. He, and M. Qian. 2008. Characterization of heavy metal-resistant endophytic bacteria from rape (*Brassica napus*) roots and their potential in promoting the growth and lead accumulation of rape. *Environmental Pollution* 156: 1164–1170.

Sørensen, J., and A. Sessitsch. 2006. Plant-associated bacteria lifestyle and molecular interactions. In: *Modern Soil Microbiology (2nd edn.)*, van Elsas, J. D. et al., eds. CRC Press, pp. 211–236.

Sperber, J.I. 1958. The incidence of apatite-solubilizing organisms in the rhizosphere and soil. *Australian Journal of Agricultural Research* 9: 778–781.

Steenhoudt, O., and J. Vanderleyden. 2000. *Azospirillum*, a free-living nitrogen-fixing bacterium closely associated with grasses: genetic, biochemical and ecological aspects. *FEMS Microbiology Review* 24: 487–506.

Tamas, I.A. 1995. Hormonal regulation of apical dominance. In: *Plant Hormones*, P.J. Davies, ed. Kluwer Academic Publishers, Dordrecht, The Netherlands, pp. 572–797.

Tsavkelova, E.A., S.Y. Klimova, T.A. Cherdyntseva, and A.I. Netrusov. 2006. Microbial producers of plant growth stimulators and their practical use: a review. *Applied Biochemistry Microbiology* 42: 117–126.

Van Loon, L.C., and B.R. Glick. 2004. Increased plant fitness by rhizobacteria. In: *Molecular Ecotoxicology of Plants, Ecological Studies*, Vol. 170, H. Sandermann, ed. Springer-Verlag, Berlin/Heidelberg, pp. 177–205.

Wang, Y., H.N. Brown, D.E. Crowley, and P.J. Szaniszlo. 1993. Evidence for direct utilization of a siderophore, ferrioxamine B, in axenically grown cucumber. *Plant, Cell and Environment* 16: 579–585.

Wang, C., E. Knill, B.R. Glick, and G. Defago. 2000. Effect of transferring 1-aminocyclopropane–1-carboxylic acid (ACC) deaminase genes into *Pseudomonas fluorescens* strain CHA0 and its *gacA* derivative CHA96 on their growth-promoting and disease-suppressive capacities. *Canadian Journal of Microbiology* 46(10): 898–907.

Watnick, P., and R. Kolter. 2000. Biofilm, city of microbes. *Journal of Bacteriology* 182: 2675–2679.

Yang, T., D.M. Law, and P.J. Davies. 1993. Magnitude and kinetics of stem elongation induced by exogenous indole-3-acetic acid in intact light-grown pea seedlings. *Plant Physiology* 102: 717–724.

Xie, H., J.J. Pasternak, and B.R. Glick. 1996. Isolation and characterization of mutants of the plant growth-promoting rhizobacterium *Pseudomonas putida* GR12-2 that overproduce indoleacetic acid. *Current Microbiology* 32: 67–71.

Zaidi, S., S. Usmani, B.R. Singh, and J. Musarrat. 2006. Significance of *Bacillus subtilis* strain SJ-101 as a bioinoculant for concurrent plant growth promotion and nickel accumulation in *Brassica juncea*. *Chemosphere* 64: 991–997.

20 Plant Growth Regulators and Improvements in Phytoremediation Process Efficiency

Studies on Metal Contaminated Soils

Meri Barbafieri, Jose R. Peralta-Videa, Francesca Pedron, and Jorge L. Gardea-Torresdey

CONTENTS

20.1 INTRODUCTION

20.1.1 Importance of Plant Growth Regulators in Different Agronomic Fields

Plant growth regulators (PGRs) are organic compounds that in small amounts alter the growth of a plant or plant part. The term PGR includes naturally occurring plant growth substances, phytohormones, as well as synthetic compounds or chemical analogies. PGRs affect flowering, aging, root growth, distortion and killing of leaves, stems, and other parts, prevention or promotion of stem elongation, color enhancement of fruit, prevention of leafing and/or leaf fall, among others. They act as plant hormones would do; interacting with the plant's physiological processes (Arteca 1996).

Phytohormones are substances naturally produced by plants, which control normal plant functions such as root growth, fruit set and drop, growth and other developmental processes. Phytohormones are classically defined as small, mobile compounds that, in trace quantities, influence growth and development in tissues distant from the site of synthesis. The first five hormones discovered (auxin, cytokinin, gibberellins, abscisic acid [ABA], and ethylene) have been joined more recently by others that more or less fit into the classic definition, such as polyamines, brassinosteroids, jasmonic acid, salicylic acid, and peptide hormones.

While metabolism provides the power and building blocks for plant life, plant hormones play a crucial role in controlling the way plants grow and develop. Hormones regulate the speed of growth of the individual parts and integrate these parts to produce the form that we recognize as a plant (Davies 2010). Since the 1940s, both natural and synthetic growth regulators have been increasingly used in agriculture and horticulture to modify crop plants by controlling plant developmental processes from germination through vegetative growth. PGRs also control reproductive development, maturity, senescence (or aging), and postharvest preservation. In the category of biosynthetic bioregulators, the primary focus has been on herbicides. Estimates of the complexity and the costs associated with the discovery of new PGRs are many times greater than for herbicides. Hence, the weed-killing aspects of plant growth regulation have overshadowed the other uses of PGRs in crop production. The use of PGRs in agricultural production within the United States began in the 1930s. The first discovery and use of PGRs was with acetylene and ethylene, which enhanced flower production in pineapples. Subsequently, the use of PGRs has grown exponentially to become a major component of agricultural commodity production. In the area of ornamental crops, PGRs can act as growth retardant (particularly for hedges, trees, shrubs and groundcovers) with temporary stops of shoot elongation and promoting lateral branching. Table 20.1 shows some selected plant regulator classes, their associated functions, and practical uses.

20.1.2 Why PGRs Can Be Important in Phytoremediation/Phytoextraction?

Phytoremediation is a "phytotechnology" that uses plants as "depollution agents" to counter the damaging effects of human activity on ecology and human health. This technique has been the subject of many extensive research programs worldwide since the 1990s. On one hand, it helps to clean

TABLE 20.1
PGR Class, Associated Function(s), and Practical Uses

Class	Function(s)	Practical Uses
Auxins	Shoot elongation	Thin tree fruit, increase rooting and flower formation
Gibberellins	Stimulate cell division and elongation	Increase stalk length, increase flower and fruit size
Cytokinins	Stimulate cell division	Prolong storage life of flowers and vegetables and stimulate bud initiation and root growth
Ethylene generators	Ripening	Induce uniform ripening in fruit and vegetables
Growth inhibitors	Stops growth	Promote flower production by shortening internodes
Growth retardants	Slows growth	Retard tobacco sucker growth

and preserve essential resources, such as water, soil, and air in a particular zone. On the other hand, it reproduces greenery, which is beneficial for the landscape and for biodiversity, making it pleasant for the inhabitants. It also has economic advantages with respect to other technologies.

PGRs could benefit phytoremediation in various ways, increasing the efficiency of a selected "depolluting plant." As an example, it is deliverable for a "depolluting plant" (1) to increase survival and development in stressful conditions; contaminated sites are often quite poor in fertility regarding soil structure and nutrient content. (2) To increase tolerance to toxic effects of different contaminants (organics and/or inorganics) at the level of concentration present in contaminated media. (3) To increase the rate of contaminant treatment (increasing degradation of organics, or increasing biomass production and metal uptake, and accumulation in the upper plant in case of inorganics). The way in which PGRs could be used to manipulate plants is schematically shown in Figure 20.1.

PGRs could provide phytoremediation management professionals with tools that give them greater flexibility in their management programs and allow more efficient use of labor and equipment. This chapter includes a description of these natural chemicals, how they work, and how they can be valuable aids in the technology of phytoremediation. Emphasis has also been placed on recent research studying their application in phytoremediation, as a very new area of research involving both plant physiologists and environmental problems. To our knowledge, few studies have been conducted to verify the possible use of PGRs for phytoremediation purposes (see Section 20.3).

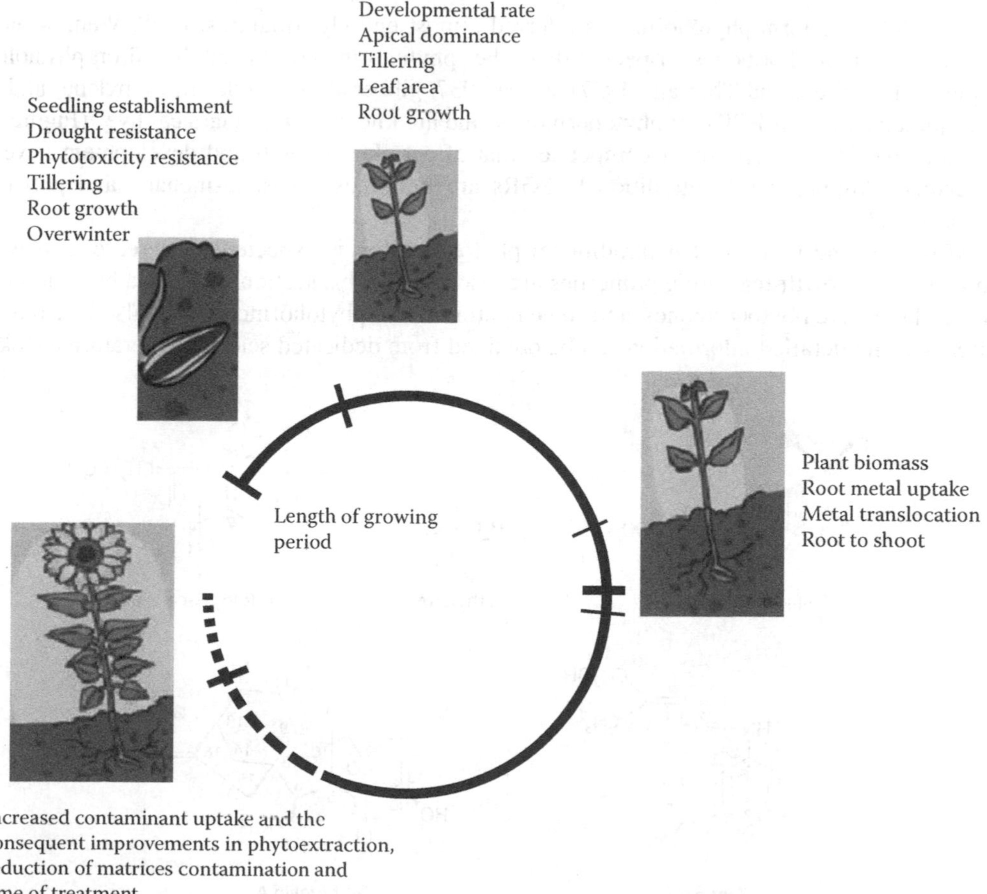

FIGURE 20.1 Possible targets for manipulation of plants for phytoextraction purposes by PGRs. (Adapted from Basara A.S., *Plant Growth Regulators in Agriculture and Horticulture*, The Haworth Press, 2000, 59.)

Trying to figure out the right PGR can be difficult when dealing with phytoremediating plants; however, that challenge is twice as difficult when dealing with phytoremediating plants for phytoextraction. In this approach, plants capable of accumulating high levels of metals are grown on contaminated soil. At maturity, metal-enriched aboveground biomass is harvested and a fraction of soil metal contamination removed. The efficiency of phytoextraction process is a challenge as depends on factors such as metal availability for uptake into roots (phytoavailability) and plant ability to intercept, absorb, transfer and accumulate metals in shoot (Ernst 1996; Barbafieri 2000). Ultimately, the potential for phytoextraction depends on the interaction between three main factors: soil, metal, and plant. This interaction makes the phytoextraction process a real challenge as each factor is governed by intrinsic processes not fully understood, yet. This underlines the importance of understanding the mechanism and process that governs metal uptake and accumulation in plants, bringing the success of phytoextraction as an environmental cleanup technology (Lasat 2002). Research on the role of PGRs for phytoremediation is still in its infancy. Here we report the little that is known and comment on future directions. The chapter is intended to serve as a source of information for students and professionals in plant science and environmental technology (phytotechnology); offering information on plant hormones and their role in plants with a focus on their potential application in phytoremediation.

20.2 WHAT ARE PLANT GROWTH FACTORS?

Before 1937, the term phytohormones referred almost entirely to auxins. F. W. Wents's pioneering work on plant hormones suspected that other phytohormones existed, based on physiological experiments (Went and Thimann 1937). Since 1937, gibberellins, cytokinins, ethylene, and ABA have joined auxins as PGRs, or phytohormones, and are known as the "classical five" (Figure 20.2).

More recently, several other compounds that affect plant growth and development have been described. Among the "nontraditional" PGRs are brassinosteroids, jasmonate, and polyamines (Figure 20.3).

Moreover, the number of nontraditional phytohormones is expected to increase as new compounds with growth-regulating properties are discovered. This section includes a brief description of the classic five phytohormones, and some nontraditional phytohormones will also be cited. More extensive and detailed information can be obtained from dedicated scientific literature (books and

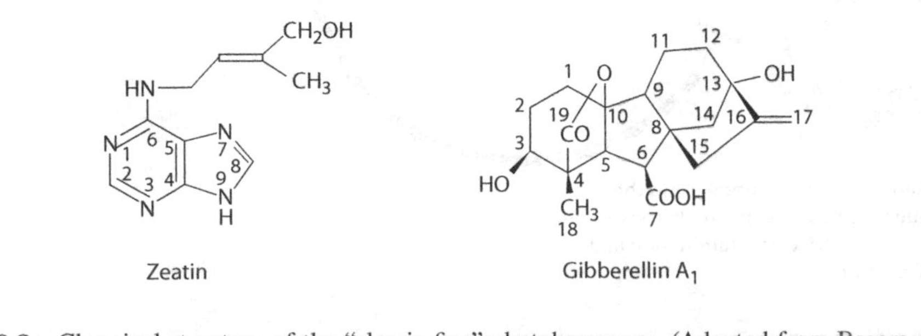

S-(+)-Abscisic acid

Ethylene

Indole-3-acetic acid

Zeatin

Gibberellin A_1

FIGURE 20.2 Chemical structure of the "classic five" phytohormones. (Adapted from Basara A.S., *Plant Growth Regulators in Agriculture and Horticulture*, The Haworth Press, 2000, 3.)

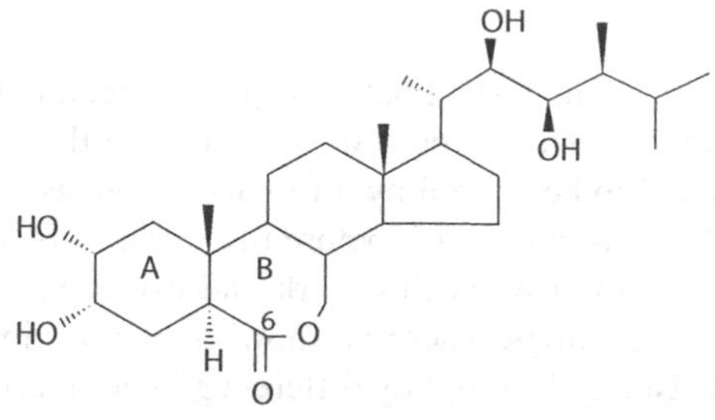

Brassinolide

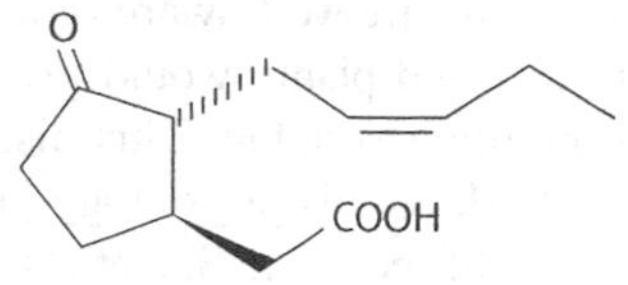

Jasmonic acid

$H_2N\text{-}(CH_2)_4\text{-}NH_2$

Putrescine

$H_2N\text{-}(CH_2)_4\text{-}NH\text{-}(CH_2)_3\text{-}NH_2$

Spermidine

$H_2N\text{-}(CH_2)_3\text{-}NH\text{-}(CH_2)_4\text{-}NH\text{-}(CH_2)_3\text{-}NH_2$

Spermine

FIGURE 20.3 Chemical structure of the "nontraditional" phytohormones. (Adapted from Basara A.S., *Plant Growth Regulators in Agriculture and Horticulture*, The Haworth Press, 2000, 4.)

articles). The literature cited is not intended to be absolutely comprehensive, but is just a starting point for those interested in the topic.

20.2.1 Abscisic Acid

ABA is a sequiterpene composed of three isopropene units (Figure 20.2). ABA is the PGR that controls plants and seed dormancy in winter and helps plants to shed its old leaves. ABA is also produced in the roots in response to decreased soil water potential and other situations in which the plant may be under stress. It is then translocated to the leaves, where it rapidly alters the osmotic potential of stomatal guard cells, causing them to shrink and stomata to close. The ABA-induced stomatal closure reduces transpiration; thus, preventing further water loss from leaves in times of low water availability. Seed germination is inhibited by ABA in antagonism with gibberellins. ABA also prevents loss of seed dormancy (Davies 1995; Mauseth 1991; Raven 1992; Salisbury and Ross 1992).

20.2.2 Cytokinins

Cytokinin (CK) is the generic name used to designate a plant-growth substance that plays a major role in cell division and cell differentiation. Such compounds induce cell division and organogenesis in plant cell cultures and affect many other physiological and developmental processes in plants. The cytokinins comprise a group of substances that are a collection of naturally occurring, N6-substituted adenine derivatives that were first identified more than 40 years ago as essential factors for plant cell division (Barciszewski et al. 2000; Roef and Onckelen 2010). Cytokinins are usually produced in roots, young fruits, and seeds. They enter the shoot organs via the xylem. Cytokinins enhance the resistance of plants to various forms of stress such as salinity and high temperature. They regulate plant growth under drought conditions and delay senescence of intact plants. Furthermore, they also might be involved in coordinating certain metabolic activities, sink capacities and the mobilization of storage products in different parts of leguminous and nonleguminous plants during development (Barciszewski et al. 2000). Cytokinins are also destined to control the stomatal opening. The most common form of naturally occurring cytokinin in plants today is called zeatin that was isolated from corn, *Zea mays* (Figure 20.2). We must keep in mind that the response will vary depending on the type of cytokinin and plant species.

20.2.3 Gibberellins

Farmers in Asia are well aware of a disease of rice plants called bakanae (foolish seedling) disease in Japan. Infected plants would grow excessively taller than the rest of the rice in the paddy, fall over and be unharvestable. This disease was found to be caused by a fungus known as *Gibberella fujikuroi*. Work on this physiology problem in Japan occurred just before the world wars in the first half of the 20th century. Scientists in the 1950s rediscovered this work and extracted a range of chemicals that elicit the growth response in rice seedlings. These are now known as gibberellins. Like auxin, gibberellin promotes cell elongation. Nearly 100 slightly different gibberellins have been chemically identified. Gibberellins are known to be regulators of many phases of higher plant development, including seed germination, stem growth, induction of flowering, pollen development and fruit growth. Chemically, all of the gibberellins are based on a kaurene carbon skeleton (Figure 20.2).

20.2.4 Auxins

Auxin, or indole acetic acid (IAA), is an essential, multifunctional plant hormone that influences virtually every aspect of plant growth and development. When this group of PGRs was first discovered, it included the weed killer 2,4D. Although auxin-dependent growth is evident in all plant tissues, it is synthesized primarily in the apical regions of the shoot and is then transported in a polar fashion to other sites. Auxin produced in apical buds tends to inhibit the activation of buds lower on the stems. This is known as apical dominance. This effect lessens with distance from the shoot tip. The only natural auxin is IAA. Cytokinins counter the apical dominance of auxins. Auxins help in cell elongation (by softening the cell wall) and growth of roots and shoots, especially roots (Cleland 2010). They are widely used commercially to help root formation in cuttings and transplants; however, at high concentrations they inhibit plant growth.

20.2.5 Ethylene

Ethylene is a gaseous plant growth substance that has been shown to be involved in numerous aspects of plant growth and development. Since it is a gas, it can readily be diffused out of the tissues; thus, no particular metabolism is essential for its removal.

According to literature, ethylene is able to stimulate leaf and flower senescence, induce leaf abscission, seed germination, growth of adventitious roots during flooding, and root hair growth (increasing the efficiency of water and mineral absorption). In addition, it affects gravitropism, inhibit stem growth and stimulate stem and cell broadening as well as lateral branch growth after the seedling stage, and inhibit shoot growth and stomatal closing. This not occurs in some water plants or habitually flooded ones such as some rice varieties, where the opposite occurs (conserving CO_2 and O_2). It is reported that plant growth-promoting rhizobacteria (PGPR) influence plant biomass production (Hall et al. 1996) by lowering plant ethylene synthesis.

20.2.6 Brassinosteroids

Brassinosteroids (BRs) are a group of naturally occurring polyhydroxy steroids (Figure 20.3). More than 60 BRs have been identified in many plants including dicots, monocots, gymnosperms, green algae and ferns (Sakurai and Fujioka 1994). These were first explored nearly 40 years ago when Mitchell et al. (1970) reported promotion of stem elongation and cell division by the treatment of organic extracts of rapeseed (*Brassica napus*) pollen. Brassinolide was the first brassinosteroid isolated in 1979 when it was shown that pollen from *B. napus* could promote stem elongation and cell divisions, and the biologically active molecule was isolated (Bajguz 2010).

Brassinosteroids are involved in numerous plant processes such as promotion of cell expansion and cell elongation (where they work with auxin to do so). They have an unclear role in cell division and cell wall regeneration and promote vascular differentiation. They are necessary for pollen elongation and pollen tube formation, accelerate senescence in dying tissue cultured cells, and can provide some protection to plants during chilling and drought stress. Brassinosteroids have been reported to counteract both abiotic and biotic stress in plants as well as metal uptake and toxicity (Sharma and Bardawwaj 2007; Sharma et al. 2008). Application of brassinosteroids to cucumbers was shown to increase metabolism and removal of pesticides, which could be beneficial for reducing human ingestion of residual pesticides from nonorganically grown vegetables (Xiao et al. 2009). Exogenous brassinosteroid application has shown many and different uses in agriculture as well as in plant propagation (Gomes 2011). A review on the potential impact of brassinosteroids on phytoremediation has been recently published by Barbafieri and Tassi (2011).

20.2.7 Jasmonates

Jasmonic acid (JA) is derived from linoleic acid (Figure 20.3). Early studies showed that exogenous JA can promote senescence and regulate growth (Arteca 1996). The major function of JA, and its various metabolites, is to regulate plant responses to abiotic and biotic stress as well as plant growth and development. Regulated plant growth and development processes include growth inhibition, senescence, tendril coiling, flower development and leaf abscission. It is also responsible for tuber formation in potatoes, yams, and onions. It also has an important role in response to wounding of plants and systemic acquired resistance.

20.2.8 Polyamines

The polyamines putrescine, spermidine, and spermine (Figure 20.3) are involved in numerous processes associated with plant growth by affecting cell division (Bagni et al. 1982) and cell development (Slocum et al. 1984). Their biological activity is attributed to their cationic nature.

20.3 EVIDENCE OF THE USE OF PLANT GROWTH REGULATORS IN PHYTOREMEDIATION/PHYTOEXTRACTION

Although most of the plants absorb metals, only a few species are considered apt for phytoremediation/phytoextraction purposes. The feasibility of using this technique rests on the accumulation and distribution of contaminants in plant tissues for subsequent reclamation. Metal bioavailability and plant absorption capacity must converge for absorption through roots and subsequent translocation to leaves. Bioavailability of metals in soil can be achieved by lowering the pH or through the utilization of chelating agents such as ethylenediaminetetraacetic acid (EDTA) and nitrilotriacetic acid (NTA), among others (Epstein et al. 1999; Nowack and VanBriesen 2005). Fuentes (2001) stated that mobilization of metals with chelating agents could pose a problem for groundwater. This researcher suggested the use of PGR to increase rate of growth and plant biomass production, hence increasing metal accumulation. Reports on the use of auxins, cytokinins, mixture of auxins–cytokinins, and brassinosteroids have shown the potential use of PGR on phytoremediation of heavy metals. No reports were found on the use of PGR for nanoparticles uptake.

20.3.1 Auxins

Fuentes (2001) was, to our knowledge, one of the first to investigate the PGR-assisted improvement of metal accumulation by plants. He used NAA and IBA in corn plants (*Zea mays*) grown in a mine tailing soil impacted with Pb, Cu, Mn, Zn, and Fe. Several treatments including foliar spray combined with soil applications of the same PGR, or PGR combined with EDTA plus soil pH adjustments and fertilization were assayed (Table 20.2). No results were reported for EDTA treatments.

TABLE 20.2
PGRs Used for Phytoremediation Purposes in Metal Contaminated Soil or Spiked Solutions

PGR	Concentration	Application Form	Amendment	Plant Species	Substrate	Main Results	References
IBA NAA	10 100 1000 μg/mL	Dip and spray	EDTA at 10 or 20 g/L	Corn	Mine site soil high in Pb and Fe	NAA increased 21% accumulation of Pb and Fe compared to IBA	Fuentes (2001)
NAA	700 μg/mL	Spray	50 or 100 mL/kg soil of NPK at a ratio of 5.6:1:3.1	Corn	1 kg of mine site soil in pots	100 mL nutrients increased Cu, Pb, and Fe accumulation by 108, 150, and 174%	Fuentes (2001)
IAA	3 and 6 mg/L	Applied to soil and spray to leaves	EDTA 1 g/kg soil	Sunflower	Soil sludge or composted soil	With/without EDTA, in composted soil IAA, increased Cd and Pb in leaves. Without EDTA in soil sludge IAA increased Mn and Ni in leaves	Liphadzi et al. (2006)
IAA, GA, Kinetin	1, 10, and 100 μm	Added to nutrient solution	EDTA 0.2 mM	Alfalfa	Hydroponics	Kinetin plus EDTA increased Pb in leaves by 17-, 43-, and 67-fold, respectively compared to leaves of plants treated with Pb alone; and 2-, 5-, and 8-fold compared to plants treated with Pb/EDTA	Lopez et al. (2007a)
IAA-kinetin	100 μM each	Added to soil	EDTA 0.8 mM	Alfalfa	Soil spiked with Pb at 80 mg/kg	Plants treated with EDTA plus IAA-kinetin had significantly more Pb in stems and leaves compared to Pb alone and Pb/EDTA	Lopez et al. (2009)
Kinetin	100 μM	Added to soil	CDTA 2.5 mM cystein 0.5 mM NTA 5 mM	*Prosopis* sp.	Soil spiked with As(III) 30 mg/kg and As(V) at 50 mg/kg	Kinetin plus cysteine increased total As in roots by 36% from As(III) and 65% from As(V). While kinetin plus CDTA increased total As in roots by 20% from As(III) and by 100% from As(V)	Lopez et al. (2008)
IAA	250 and 500 μM	Added to hydroponic solution		Corn	Hydroponic solution containing Pb at 20 μM	IAA at 500 μM increased Pb in roots by 144% five days after treatment application; however, Pb in shoot decreased 47%	Wang et al. (2007)

Citokin (several cytokinins) Phytagro (mixture of phytohormones) Kinetin	20 mg/L	Sprayed at 22nd, 28th, and 33rd days after germination	EDTA 2 mM/kg soil in four diluted doses (one per starting the 23rd day after germination)	Sunflower	50 g of contaminated soil plus 150 g of inert silica pellets	EDTA plus PGRs increased biomass production. Cytokin plus EDTA increased Pb accumulation >600% in leaves and >300% in stems compared to control. Accumulation of Zn in leaves increased >170%	Tassi et al. (2008)
Kinetin	250 μM	Solution applied to soil 15 days after germination		Mexican Palo Verde	1800 g of soil contaminated with Cr(III) at 60 mg/kg and Cr(VI) at 10 mg/kg	Compared to controls, kinetin increased total Cr accumulation by 45, 103, and 72% in root, stem, and leaf of Cr(III) treated plants and 53, 129, and 168% in Cr(VI) treated plants, respectively. Kinetin increased ascorbate peroxidase enzyme activity in Cr(VI) treated plants	Zhao et al. (2011)
IAA and NAA	1, 10, and 100 μM	Solution applied to 15-day-old plants in nutrient solution	EDTA 100 mg/L	Rattlebush (*S. drummondii*)	Half strength Hoagland solution spiked with Pb at 500 mg/L	Compared to control, NAA increased Pb in shoots by 654% and IAA by 415%. NAA plus EDTA increased Pb in shots 1349% and IAA plus EDTA by 1252%	Israr and Sahi (2008)
IAA	10^{-10} M	Applied to six-day-old plants in nutrient solution	EDDS 500 μM	Sunflower	10% Hoagland solution spiked with Zn (15 μM) or Pb (2.5 μM)	IAA plus EDDS increased Zn in shoots. Pb was not detected	Fässler et al. (2010)
28-Homobrassinolide	10^{-10} to 10^{-6} M	Seeds were soaked in the HBL solutions for 8 h		Indian mustard	Acid washed sand spiked with Cu at 50, 100, 150 mg/kg	HBL treated plants had roots and shoots ~62% and 93~ longer; 47% more leaf area and 93% higher dry mass compared to not HBL	Fariduddin et al. (2009)
IAA, GA, zeatin, and ET	1, 0.1, 0.01, 0.001 μM	Salt-stressed seeds soaked in NaCl solutions		Wheat	Petri dishes containing seeds stressed with NaCl at 500 mM	The highest germination percentage was obtained with IAA at 0.1 μM (97%), ET at and GA 0.01 μM (89% and 88%, respect.), and zeatin at 0.001 μM (68%)	Egamberdieva (2009)

Conversely, foliar spray of NAA at 1000 μg/mL increased Cu, Pb, and Fe concentrations by 80, 171, and 167%, respectively; however, 45% of the plants died and biomass production decreased by 70%. In further experiments he found that NAA sprayed at 800 μg/mL, pH adjusted to 4.0 and fertilization with N:P:K to have a ratio of 5.6:1:3.1, produced an increase in Cu, Pb, and Fe accumulation of 108, 150, and 174%, respectively, with a biomass increase of 34%.

In a study with sunflower (*Helianthus annuus* L.) cultivated in soil with metals from sewage sludge or composted soil, IAA was applied as spray on the leaves and added to the soil in combination with EDTA at 1 g per kg of soil (Liphadzi et al. 2006). These researchers demonstrated that with/without EDTA, leaves of plants grown in the composted soil and treated with IAA had more Cd and Pb than leaves without IAA treatment. On the other hand, leaves of plants grown without EDTA in the soil with sludge had more Mn and Ni compared to plants not treated with IAA. An interesting result was that IAA did not increase root growth when EDTA was part of the treatment. IAA and NAA have demonstrated to significantly increase the translocation of Pb from root to shoot in *Sesbania drummondii*, a shrub proven to be a Pb hyperaccumulator. Twenty day old seedlings grown in hydroponics were treated for 10 days with different treatments including the PGR IAA and NAA, and EDTA. Compared to plants treated with Pb alone, both IAA and NAA at 100 μM increased Pb accumulation in *S. drummondii* shoots. However, the addition of EDTA at 100 mg/L further increased the Pb accumulation, compared to Pb translocation from Pb alone. Growth data showed that at 100 mM of IAA or NAA, growth was comparable to plants grown with Pb alone (Israr and Sahi 2008). IAA has also been combined with ethylenediaminedisuccinic acid (EDDS) to improve the phytoremediation ability of sunflower. Combinations of IAA at 10^{-10} M and EDDS at 500 μM were applied to Pb or Zn stressed plants grown in hydroponics. The uptake of Zn significantly increased in plants treated with IAA plus EDDS in comparison to EDDS only treated plants. The authors concluded that IAA can alleviate toxic effects of Pb and Zn and improves the phytoextraction potential of sunflower (Fässler et al. 2010).

20.3.2 Cytokinins

Cytokinins in combination with EDTA were used to increase metal uptake by sunflower plants grown in soil from a former processing gas-plant site. Plants treated with cytokine (a mixture of different and naturally occurring cytokinins) plus EDTA accumulated 890% and 330% more Pb and Zn, respectively. The authors concluded that cytokinins can increase the phytoextraction potential of sunflower by increasing the biomass production and plant transpiration (Tassi et al. 2008).

In a more recent study, Zhao et al. (2010) applied kinetin to 15-day old Mexican Palo Verde (*Parkinsonia aculeata*, MPV) plants grown in soil contaminated with Cr(III) and Cr(VI). Thirty days after treatment application, plants were harvested and processed for ICP-OES analysis. The data showed that total Cr concentration in MPV tissues significantly increased compared to plants not treated with kinetin. For Cr(III) treated plants the increase in Cr accumulation in roots, stems, and leaves was of 45%, 103%, and 72%, respectively, compared to plants not treated with kinetin. For Cr(VI) treated plants the increase in Cr concentrations was 53%, 129%, and 168% in roots, stems, and leaves, respectively, compared to Cr(VI) alone. Cytokinins have also showed potential use for the increasing of Ni phytoextraction capability in *Alyssum murale*, a well known Ni hyperaccumulator (Cassina et al. 2011). Seventy day-old *A. murale* plants were transplanted to plastic pots containing four kg of serpentine soil with 1521 mg/kg Ni (approximately 25 mg/kg diethylenetriaminepentaacetic acid (DPTA)-extractable). Twenty days after transplants, plants were treated every six days and harvested after three foliar, in soil, and combined soil and foliar applications of a mixture of cytokinins. Although the phytoextraction data indicated a significant tendency to increase Ni phytoextracted in the combined treatment (foliar + soil), no significant differences were found due to a high variability in Ni concentration in shoots.

20.3.3 Mixture of Auxins–Cytokinins

In a series of experiments Lopez et al. (2007a,b, 2008 2009) studied the use of PGR in order to increase metal(loid)s uptake by alfalfa (*Medicago sativa*) and *Prosopis* sp. plants. These researchers treated alfalfa plants in hydroponics with Pb at 40 mg/L plus EDTA at 0.2 mM (equimolar to Pb) and IAA, GA, and kinetin. They found that plants treated with EDTA plus kinetin significantly increased the amount of Pb in leaves compared to plants treated with Pb alone or plants treated with Pb/EDTA (Table 20.2). In addition, the leaves of plants treated with EDTA plus the IAA–kinetin mixture at 100 μM each, had approximately 9500 mg of Pb/kg of dry weight biomass (Lopez et al. 2007a). These researchers evaluated the stress of the treatments through the catalase (CAT), ascorbate peroxidase (APOX), and total amylase activities (TAA). Results showed that only the treatment including Pb plus EDTA and kinetin caused toxicity to the APOX enzyme. However, all treatments caused toxicity to the amylase enzyme in alfalfa seedlings (Lopez et al. 2007b). Based on the results from hydroponics, an experiment was performed in soil treated with Pb at 80 mg/kg (from $Pb(NO_3)_2$ EDTA at 0.8 mM and a mixture of IAA plus kinetin at 100 μM each). Fifteen days after treatment application, plants were harvested and analyzed by inductively coupled plasma-optical emission spectrometry. Results showed that in roots of plants treated with Pb alone, Pb–EDTA, and Pb–EDTA-IAA-kinetin had 160, 140, and 150 mg Pb kg^{-1} DW, respectively. However, only plants treated with EDTA and EDTA–IAA-kinetin showed Pb in stems [(78 and 142 mg kg^{-1} DW) and leaves (92 and 127 mg kg^{-1} DW), respectively]. Moreover, EDTA and EDTA–IAA-kinetin treated plants had significantly more zinc and manganese in leaves. The x-ray absorption spectroscopic studies showed that Pb in leaves had the same oxidation state as the one of the Pb applied to the soil (Pb(II)) (Lopez et al. 2009). In another study, *Prosopis* sp. plants (shrub/tree native to desert areas of the southwestern United States and northern Mexico) were grown in soil treated with arsenic(III) at 30 mg/kg and arsenic(V) at 50 mg/kg. One month old plants were treated with cysteine, cyclohexylenedinitrotetraacetic acid (CDTA), nitrilotriacetic acid, and kinetin. Fifteen days after treatment application, plants were harvested and processed for As analysis. The data showed that, compared to plants treated with As only, roots of plants treated kinetin plus CDTA or cysteine significantly increased total As concentration in tissues, but the increase was higher with As(V) treatments. Additionally, plants treated with As(V) and cysteine plus kinetin had roots of similar size to control plants, which suggest that these compounds increased the mesquite tolerance to As(V) (Lopez et al. 2008). Wang et al. (2007) reported that in hydroponically grown corn seedlings cultivated for five days with IAA and Pb, accumulation of Pb increased in roots, but decreased in shoots.

20.3.4 Brassinosteroids

Brassinosteroids are a new type of plant hormones that are universally distributed in plants; however, the biological function in plants of these compounds is not well understood (Bajguz and Tretyn 2003). Reports indicate that brassinosteroids increase the tolerance of plants to several types of stress (Fariduddin et al. 2009). 28-homobrassinolide (HBL) was applied to Indian mustard (*Brasscia juncea*) seeds that were sand grown under copper stress and harvested 30 days after sowing. Although the authors did not determine the uptake of copper, they reported that HBL improved the growth, photosynthetic parameters and antioxidant enzyme activity that might be responsible to overcome the toxic effects of copper in *B. juncea*.

PGRs have also shown to have a role in alleviating seed dormancy enforced by salt stress. Wheat seeds immersed in a 100 mM NaCl solution (concentration that produces a great inhibitory effect on wheat germination) were treated with IAA, GA, zeatin, and ET. The germination percentage of control was 97%, with NaCl 46%, and with 0.01 μM concentration of the PGR the germination percentages were 96, 89, 76, and 60% for IAA, ET, GA, and zeatin, respectively. The data clearly showed that the PGR substantially alleviated the germination inhibition imposed by salinity. In addition, at all concentrations the PGR promoted the root and shoot elongation (Egamberdieva 2009).

20.4 CONCLUSIONS AND FUTURE PERSPECTIVES

Major conclusions drawn from the current chapter as well as future perspectives may be listed as follows: (1) PGRs known to increase the efficiency of the phytoremediation process, and (2) the available literature that up to now has been produced by the research performed on this scenario. It is clearly shown that PGRs, alone or combined with chelating agents, plus pH adjustment, dramatically increase the uptake and translocation of metal(loid)s by plants. Reports from experiments in hydroponics, sand or soil have shown that PGRs increase the accumulation of potentially toxic elements at high percent, even several folds, compared to not PGR treatments. However, to the authors' knowledge, there are no reports on the use of PGRs in phytoextraction/phytoremediation projects. Perhaps one reason could be the lack of cost/benefits analysis in the available literature. Although used at low concentrations, PGRs are expensive. Cost/benefits analyses are in urgent need to boost the use of PGRs in large scale phytoextraction/phytoremediation projects. Moreover, the potential use of agrochemicals for phytoremediation still remains unexplored. It requires specific tests focused on verifying effectiveness, and represents an open door that could lead to interesting positive results for increasing the phytoremediation technologies efficiency and applicability.

Important thing to underline is that plant characteristics often altered by PGRs can vary greatly from one experiment to another, and conflicting results are easily obtained. This is because the effects of PGRs are greatly dependent on: (1) application time in relation to plant growth stage; (2) its interaction and effect on partitioning of current photosynthates; (3) their dependence on weather conditions, especially occurrence of stress, and last but not the least, (4) genotypic differences in response to PGRs. Furthermore, this chapter raises the possibility that the effects of PGRs could provide positive interactions with plants, making them more efficient depolluting agents for a more effective application of phytoremediation technologies.

ACKNOWLEDGMENTS

This material is based upon work supported by the National Science Foundation and the Environmental Protection Agency under Cooperative Agreement Number DBI-0830117. Any opinions, findings, and conclusions or recommendations expressed in this material are those of the author(s) and do not necessarily reflect the views of the National Science Foundation or the Environmental Protection Agency. This work has not been subjected to EPA review and no official endorsement should be inferred. The authors also acknowledge the USDA grant number 2008-38422-19138 and 2011-38422-3085, and the NSF Grant # CHE-0840525. J. L. Gardea-Torresdey acknowledges the Dudley family for the Endowed Research Professorship in Chemistry. M. Barbafieri acknowledges the Italian National Research Council (CNR) for supporting studies on phytoremediation in the frame of ISE-CNR "Remediation of Contaminated Sites" line of research. The authors also acknowledge Ms. Margarita Medina for her assistance in the editing of the manuscript.

REFERENCES

Arteca, N.R. 1996. *Plant Growth Substances: Principles and Applications*. New York: Chapman & Hall.

Bagni N., D. Serafini-Fracassini, and P. Torrigiani 1982. Polyamines and cellular growth processes in higher plants. In *Plant Growth Substances*, 473–482. London: Academic Press.

Bajguz, A., and A. Tretyn. 2003. The chemical characteristic and distribution of brassinosteroids in plants. *Phytochemistry* 62: 1027–1046.

Bajguz, A. 2010. Brassinosteroids: occurrence and chemical structures in plants. In: *Brassinosteroids: A Class of Plant Hormone*, 1–28. The Netherlands: Springer.

Barbafieri, M. 2001. The importance of nickel phytoavailable chemical species characterization in soil for phytoremediation applicability. *International Journal of Phytoremediation* 2: 105–115.

Barbafieri, M., and E. Tassi. 2011. Brassinosteroids for phytoremediation application. In *Brassinosteroids: A Class of Plant Hormone*, 403–437. The Netherlands: Springer.

Barciszewski, J., G. Siboska, S.I.S. Rattan and B.F.C.Clark. 2000. Occurrence, biosynthesis and properties of kinetin (N-6-furfuryladenine). *Plant Growth Regulation* 32: 257–265.

Basara, A.S. 2000. *Plant Growth Regulators in Agriculture and Horticulture*, The Haworth Press.

Cassina, L., E. Tassi, E. Morelli, L. Giorgetti, D. Remorini, R.L. Chaney, and M. Barbafieri. 2011. Exogenous cytokinin treatments of a Ni hyperaccumulator, *Alyssum murale*, grown in a serpentine soil: implications for phytoextraction. *International Journal of Phytoremediation* 13(S1): 90–101.

Cleland, R.E. 2010. Auxin and cell elongation. In *Plant Hormones: Biosynthesis, Signal Transduction, Action*, 204–220. The Netherlands: Springer.

Davies, P.J. 2010. The plant hormones: their nature, occurrence, and functions. In *Plant Hormones: Biosynthesis, Signal Transduction, Action*, 1–15. The Netherlands: Springer.

Egamberdieva, D. 2009. Alleviation of salt stress by plant growth regulators and IAA producing bacteria in wheat. *Acta Physiologia Plantarum* 31: 861–864.

Epstein, A.L., C.D. Gussman, and M. J. Blaylock. 1999. EDTA and Pb–EDTA accumulation in *Brassica juncea* grown in a Pb–amended soil. *Plant and Soil* 208: 87–94.

Ernst, W.H.O. 1996. Bioavailability of heavy metals and decontamination of soil by plants. *Applied Geochemistry* 11: 163–167.

Fariduddin, Q., M. Yusuf, S. Hayat, and A. Ahmad. 2009. Effect of 28-homobrassinolide on antioxidant capacity and photosynthesis in *Brassica juncea* plants exposed to different levels of copper. *Environment Experimental Botany* 66: 418–424.

Fässler, E., M.W. Evangelou, B.H. Robinson, and R. Schulin. 2010. Effects of indole-3-acetic acid (IAA) on sunflower growth and heavy metal uptake in combination with ethylene diamine disuccinic acid (EDDS). *Chemosphere* 80: 901–907.

Fuentes, H.D. 2001. Studies in the use of plant growth regulators on phytoremediation. PhD Thesis, University of Western Sydney, Hawkesbury; College of Science, Technology and Environment; School of Science, Food and Horticulture, Aust., 181 p.

Gomes, M.M.A. 2011. Physiological effects related to brassinosteroid application in plants. In *Brassinosteroids: A Class of Plant Hormone*, 193–242. The Netherlands: Springer.

Hall, J.A., D. Peirson, S. Ghosh, and B.R. Glick. 1996. Root elongation in various agronomic species by the growth promoting rhizobacterium *Pseudomonas putida* GR12-2. *Israel Journal of Plant Sciences* 44: 37–42.

Israr, M., and S.V. Sahi. 2008. Promising role of plant hormones in translocation of lead in *Sesbania drummondii* shoots. *Environmental Pollution* 153: 29–36.

Lasat, M.M. 2002. Phytoextraction of toxic metals: a review of biological mechanism. *Journal of Environmental Quality* 31:109–120.

Liphadzi, M.S., M.B. Kirkham, and G.M. Paulsen. 2006. Auxin-enhanced root growth for phytoremediation of sewage-sludge amended soil. *Environmental Technology* 27: 695–704.

Lopez, M.L., J.R. Peralta-Videa, J.G. Parsons, T. Benitez, and J.L. Gardea-Torresdey. 2007a. Gibberellic acid, kinetin, and the mixture indole-3-acetic acid-kinetin assisted with EDTA induced lead hyperaccumulation in alfalfa plants. *Environmental Science & Technology* 41(23): 8165–8170.

Lopez M.L., J.R. Peralta-Videa, H. Castillo-Michel, A. Martinez-Martinez, and J.L. Gardea-Torresdey. 2007b. Lead toxicity in alfalfa plants exposed to phytohormones and ethylenediaminetetracetic acid monitored by peroxidase, catalase and amylase activities. *Environmental Toxicology & Chemistry* 26: 2717–2723.

Lopez, M.L., J.R. Peralta-Video, J.G. Parsons, and J.L. Gardea-Torresdey. 2008. Concentration and biotransformation of arsenic by *Prosopis* sp. grown in soil treated with chelating agents and phytohormones. *Environmental Chemistry* 5: 320–331.

Lopez, M.L., J.R. Peralta-Videa, J.G. Parsons, and J.L. Gardea-Torresdey. 2009. Plant growth, uptake and translocation of lead, micro, and macronutrients in alfalfa treated with indole-3-acetic acid, kinetin, and ethylenediaminetetraacetic acid. *International Journal of Phytoremediation* 11(2): 131–149.

Mauseth, J.D. 1991. *Botany: An Introduction to Plant Biology*, 348–415. Philadelphia: Saunders.

Mitchell, J.W., N.B. Mandava, J.F. Worley, J.R. Plimmer, and M.V. Smith. 1970. Brassins: a new family of plant hormones from rape pollen. *Nature* 225: 1065–1066.

Nowack, V.-B. 2005. Chelating agents in the environment. In *Biogeochemistry of Chelating Agents*, 1–18. Washington, DC: American Chemical Society.

Raven, P.H., R.F., Evert, and S.E. Eichhorn. 1992. *Biology of Plants*, 545–572. New York: Worth.

Roef, L., and V.H. Onckelen. 2010. Cytokinin regulation of the cell division cycle. In *Plant Hormones: Biosynthesis, Signal Transduction, Action*, 241–261. The Netherlands: Springer.

Sakurai, A., and S. Fujioka. 1994. The current status of physiology and biochemistry of brassinosteroids: a review. *Journal of Plant Growth Regulation* 13: 147–159.

Salisbury, F.B., and C.W. Ross. 1992. *Plant Physiology*, 357–407, 531–548. Belmont, CA: Wadsworth.

Sharma, P., and R. Bhardwaj. (2007). Effects of 24-Epibrassinolide on growth and metal uptake in "*Brassica juncea*" L. under copper metal stress. *Acta Physiologiae Plantarum* 29(3): 259–263.

Sharma, P., R. Bhardwaj, H.K. Arora, H.K. Arora, and A. Kumar. 2008. Effects of 28-homobrassinolide on nickel uptake, protein content and antioxidative defence system in "*Brassica juncea*." *Biologia Plantarum* 52 (4): 767–770.

Slocum, R.D., R. Kaur-Sawehney, and A.W. Galston. 1984. The physiology and biochemistry of polyamines in plants. *Archives of Biochemistry and Biophysics* 235(2):283–303.

Tassi, E., J. Pouget, G. Petruzzelli, and M. Barbafieri. 2008. The effects of exogenous plant growth regulators in the phytoextraction of heavy metals. *Chemosphere* 71: 66–73.

Xia X.J., Y. Zhang, J.X. J.X. Wu, J.T. Wang, Y.H. Zhou, K. Shi, Y.L. Yu, and J.Q. Yu. 2009. Brassinosteroids promote metabolism of pesticides in cucumber. *Journal of Agricultural and Food Chemistry* 57(18): 8406–8413.

Wang, H.H., X.Q. Shan, B. Wen, G. Owens, J. Fang, and S.Z. Zhang. 2007. Effect of indole-3-acetic acid on lead accumulation in maize (*Zea mays* L.) seedlings and the relevant antioxidant response. *Environmental and Experimental Botany* 61: 246–253.

Went, F.W., and K.V. Thimann. 1937. *Phytohormones*. New York: Macmillan.

Zhao, Y., J.R. Peralta-Videa, M.L. Lopez-Moreno, M. Ren, and J.L. Gardea-Torresdey. 2011. Kinetin increases chromium absorption, modulates its distribution, and changes the activity of catalase and ascorbate peroxidase in Mexican Palo Verde. *Environmental Science & Technology* 45(3): 1082–1087.

21 Remediation of Sites Contaminated with Persistent Organic Pollutants

Role of Bacteria

Ondrej Uhlik, Lucie Musilova, Michal Strejcek, Petra Lovecká, Tomas Macek, and Martina Mackova

CONTENTS

21.1 INTRODUCTION

Microorganisms are involved in global cycling of all elements which makes them essential for maintaining equilibrium on our planet. In addition, some microorganisms dispose of unique abilities to survive in harsh or extreme environments and to assimilate or degrade unusual or toxic substrates, including environmental contaminants. From this point of view, bacteria are of outstanding importance due to their diversity, versatility, and abilities to adapt very fast to changing conditions and selective pressure (Mackova et al. 2006). All these traits make them ideal candidates to funnel pollutants into natural biogeochemical cycles. Therefore, the use of microorganisms for the clean-up of toxic substances from the environment (microbial bioremediation) has been driving more and more attention. Bioremediation indeed is a cheaper and much more ecologically sound way of removing toxic compounds compared to conventional technologies. These are typically very expensive and destructive, including solidification and stabilization, soil flushing, soil washing, electrokinetics, chemical reduction/oxidation, low-temperature thermal desorption, incineration, vitrification, pneumatic fracturing, excavation/retrieval, landfill, and disposal (Demnerova et al. 2005).

21.2 BIODEGRADATION OF ORGANIC POLLUTANTS

Environmental contamination poses a serious risk to both ecosystem functioning and human health. Most of the problems have been caused by large-scale production and use as well as improper handling and disposal of certain synthetic organic chemicals. These include polychlorinated biphenyls (PCBs), petroleum chemicals including polycyclic aromatic hydrocarbons (PAHs) and BTEX (benzene, toluene, ethylbenzene, and xylenes), pesticides, explosives, industrial solvents, dyes and pigments, dioxins, furans, and flame retardants (Chaudhry et al. 2005; McGuinness and Dowling 2009). In 2001, PCBs, nine chlorinated organic pesticides, dioxins, and furans were listed as persistent organic compounds (POPs) by the Stockholm Convention on Persistent Organic Pollutants, which aims to eliminate or restrict the production and use of these pollutants. More recently, in 2010, nine additional POPs were added to the Convention, including other pesticides or brominated flame retardants.

The properties these organic chemicals have in common are especially their low water solubility and low biodegradability rates. These are the characteristics predetermining their persistence in the environment. Many POPs are taken up by water organisms and thus enter the food chain where they accumulate and magnify (bioaccumulation and biomagnification). In addition, some POPs not only pose a threat to human health in the neighborhood of the contaminated area but volatilize and migrate to places far from where they were originally released.

Although biodegradation pathways are incredibly diverse, which makes it impossible to describe them all within this text, similar structures can be found among many pollutants. Most POPs are aromatic compounds bearing one or more halogen atoms, usually chlorine. Therefore, in general, bacteria use similar strategies to degrade most of the compounds. Some of the most common strategies will be described here.

Halogen atoms are usually removed by dehalogenases. Aromatic structures are biotransformed by activation of the ring and its subsequent cleavage. Under aerobic conditions, such an activation occurs mostly via hydroxylation performed either by monooxygenases or dioxygenases (Gerhardt et al. 2009). Under anaerobic conditions, the ring is activated by reduction. Also the cleavage mechanisms differ under aerobic and anaerobic conditions being oxygenolytic or hydrolytic with catechol and benzoyl-CoA being the most frequent intermediates, respectively (Madsen 2008). Following sections will talk about key microbial, especially bacterial, enzymes for the biotransformation of most pollutants.

21.2.1 Aromatic Ring Hydroxylating Dioxygenases

Aromatic ring hydroxylating dioxygenases (ARHDs) catalyze the insertion of molecular oxygen into aromatic ring while forming arene diols. Structurally, ARHDs consist of a terminal oxygenase component and an electron transport system. Terminal oxygenases are either homooligomeric (α_n) or heterooligomeric ($\alpha_n\beta_n$) structures with α subunit (large subunit) being the catalytic component. This subunit contains a substrate-binding site and two conserved regions—a Rieske-type [2Fe-2S] cluster and a mononuclear nonheme iron oxygen activation center. The electron transport system consisting of a ferredoxin and a flavoprotein reductase or a combined ferredoxin-NAD(P)H-reductase transfers reducing equivalents from NAD(P)H to the oxygenase components (Kweon et al. 2008; Nam et al. 2001; Peng et al. 2010; Pieper and Seeger, 2008). ARHDs were originally classified into three groups based on the number of constituent components and the nature of the redox centers (Batie et al. 1992). Later, in 2001, a new clustering system was proposed based on the homology of amino acid sequences of terminal oxygenase components (Nam et al. 2001). Recently, with a vast amount of new sequence information on ARHDs being accumulated, they have been reclassified into five distinct types (Kweon et al. 2008). This classification system reflects sequence information as well as interactions between the enzyme components.

ARHDs are key enzymes in the biodegradation of many aromatic pollutants, including BTEX, PAHs, or lower chlorinated biphenyls. In general, ARHDs have broad substrate specificity and therefore can transform more aromatic compounds. Even their designation is rather historical than based on specificity (Pieper and Seeger 2008). Benzene, toluene, and ethylbenzene are biotransformed by hydroxylation and subsequent dehydrogenation of the diol to catechol or its derivatives (Figure 21.1). The ARHDs involved are benzene 1,2-dioxygenase, toluene dioxygenase, and ethylbenzene dioxygenase, respectively. ARHD are involved also in *o*- and *p*-xylene degradation—*o*-xylene 3,4-dioxygenase and *o*-xylene 4,5-dioxygenase produce 3,4- and 4,5-dimethylcatechol,

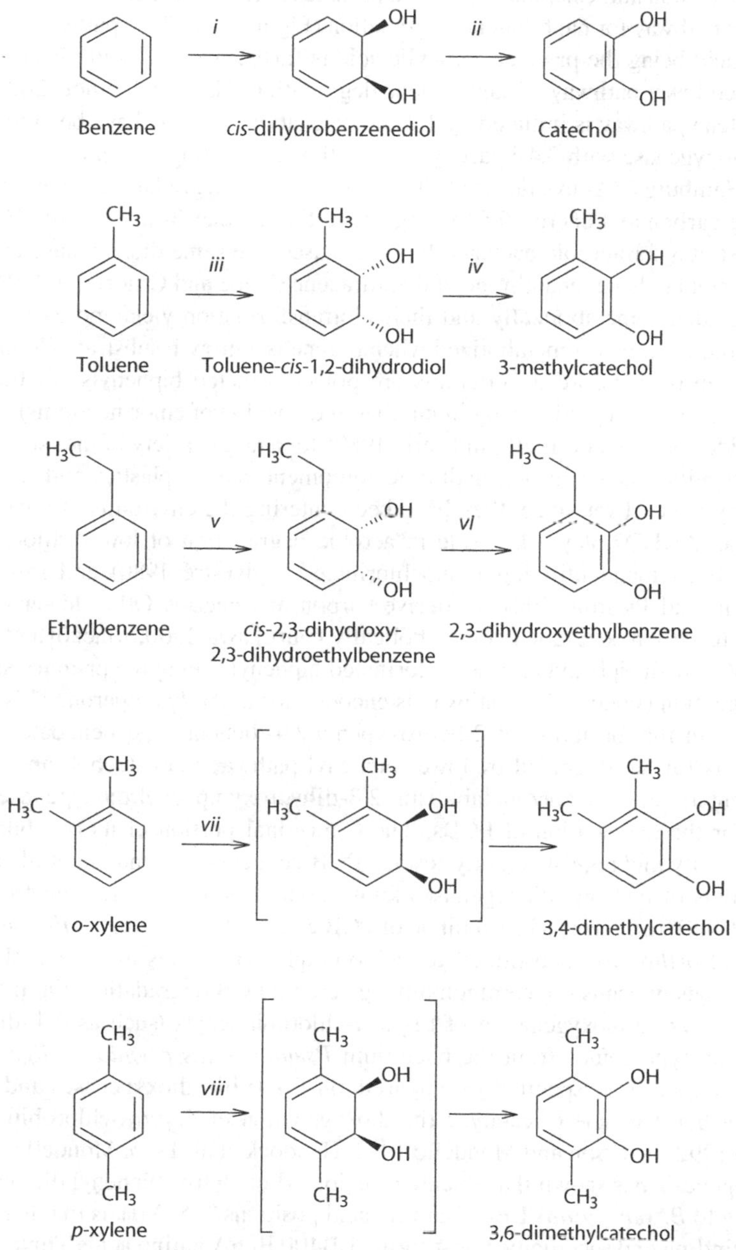

FIGURE 21.1 Biotransformation of BTEX compounds. The enzymes involved in each reaction step are benzene 1,2-dioxygenase (i), *cis*-1,2-dihydrobenzene-1,2-diol dehydrogenase (ii), toluene dioxygenase (iii), toluene dihydrodiol dehydrogenase (iv), ethylbenzene dioxygenase (v), ethylbenzene glycol dehydrogenase (vi), *o*-xylene 3,4-dioxygenase (vii), *p*-xylene dioxygenase (viii).

respectively, and 3,6-dimethylcatechol is produced from *p*-xylene by *p*-xylene dioxygenase (Figure 21.1) (Jang et al. 2005).

PAHs, found in the environment after the disposal of coal processing wastes, petroleum sludge, creosote, asphalt, and other wood preservative wastes (Habe and Omori 2003), are also transformed by ARHDs. The simplest PAH, naphthalene, which is often used a model compound to study the metabolism of PAHs, is hydroxylated by naphthalene dioxygenase. In most cases, naphthalene dioxygenases hydroxylate the aromatic ring in 1,2-positions while forming *cis*-1,2-dihydroxy-1,2-dihydronaphthalene (Eaton and Chapman 1992). Naphthalene 1,2-dioxygenase is the first enzyme of the upper catabolic pathway for naphthalene degradation (Figure 21.2). This pathway consists of 6 steps with salicylic acid being the product. Salicylic acid is further metabolized via catechol or gentisic acid by so called lower pathway of naphthalene degradation (Habe and Omori 2003). Although the upper naphthalene pathway is initiated by 1,2-dioxygenation, bacteria have been found that possess naphthalene dioxygenase with 3,4-hydroxylating activity, including model organism *Bacillus thermoleovorans* Hamburg 2 (Annweiler et al. 2000). Anyway, biodegradation of naphthalene is assimilative, yielding carbon and energy for the organism. Some other 3- and 4-ring PAHs can also act as growth substrates for aerobic bacteria. In many cases, the same dioxygenase can transform not only naphthalene but also phenanthrene and anthracene (Habe and Omori 2003). Higher PAHs are usually biodegraded cometabolically and their biotransformation yields no carbon or energy. For instance, benzo(*a*)pyrene is cometabolized when pyrene is a growth substrate (Johnsen et al. 2005).

Another group of widespread pollutants are polychlorinated biphenyls (PCBs). These chlorinated derivatives of biphenyl (differing in number and position of chlorine atoms) were largely produced worldwide between late 1920s and early 1980s for a large variety of industrial uses, including dielectric applications, heat transfer, hydraulic equipment, paints, plastics and rubber products, or carbonless copy paper. Ever since, they have been entering the environment contaminating many sites worldwide. ARHDs play a key role in aerobic degradation of lower chlorinated biphenyls. Bacteria have been found utilizing monochlorinated (Sylvestre 1980) and rarely dichlorinated congeners (Kim and Picardal 2001) to derive carbon and energy. Other lower chlorinated PCB congeners are usually biodegraded cometabolically (Furukawa 2000; Mackova et al. 2010; Pieper and Seeger 2008) with biphenyl or monochlorinated biphenyls being the primary substrates. Upper biphenyl degradation (Figure 21.3) pathway is encoded for in the *bph* operon. This 4-step catabolic pathway results in the formation of 2-hydroxypenta-2,4-dienoate and benzoate (or chlorobenzoate). Benzoate is further degraded by lower biphenyl pathway via catechol, and 2-hydroxypenta-2,4-dienoate enters intermediary metabolism. 2,3-dihydroxybiphenyl dioxygenase is considered a key enzyme for the degradation of PCBs. The C-terminal portion of its α subunit is crucial for substrate specificity and regiospecificity toward PCB congeners (Mondello et al. 1997). There are two basic models of biphenyl dioxygenases known. The first one originating from *Burkholderia xenovorans* LB400 transforms a broad range of PCB congeners, including *ortho*-substituted dichlorobiphenyls and *ortho-meta*-substituted tetrachlorobiphenyls. It has also been shown to degrade hexachlorbiphenyls which is not common among aerobic PCB degradation. On the other hand this type cannot catalyze a dioxygenation of di-*para*-chlorobiphenyls (such as 4,4′-dichlorobiphenyl). The other model type comes from the bacterium *Pseudomonas pseudoalcaligenes* KF707. This type has narrower substrate specificity compared to LB400-like dioxygenases and lacks 3,4-dioxygenase activity but it is able to catalyze the dioxygenation of di-*para*-chlorobiphenyls (Erickson and Mondello 1992; Erickson and Mondello 1993; Haddock et al. 1995; Mondello et al. 1997). Site-directed mutagenesis has shown that so called region III of BphA, biphenyl dioxygenase α subunit (corresponding to *B. xenovorans* LB400 amino acid positions 335–341), is in particular responsible for substrate binding and specificity. Substituting LB400 BphA amino acids Thr_{335}, Phe_{336} with Ala, Met resulted in mutant BphA with increased substrate specificity and altered regiospecificity toward chlorobiphenyls. Conclusions of these studies suggest that interactions between the chlorine substitutes on the phenyl ring and specific amino acid residues of the protein influence the orientation of the phenyl ring inside the catalytic pocket (Barriault et al. 2004; Barriault and Sylvestre 2004).

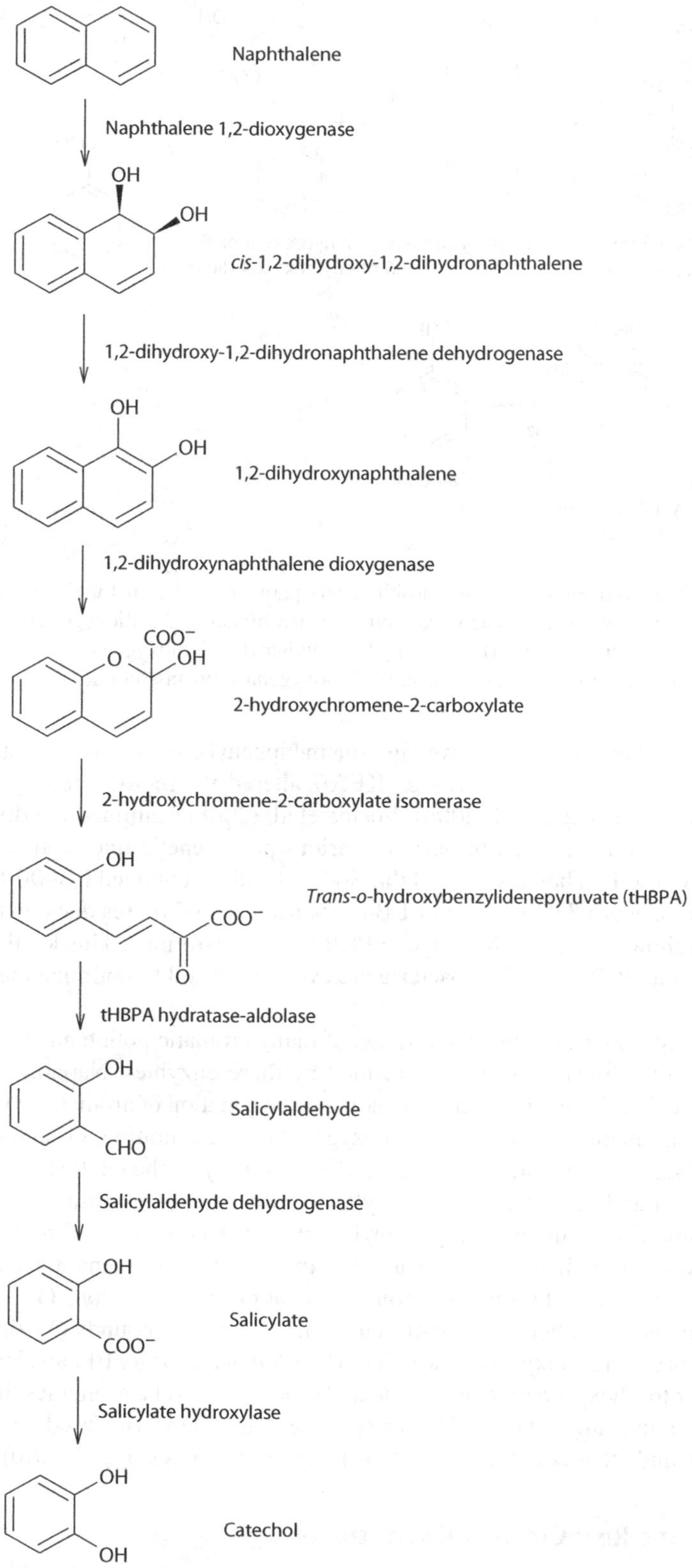

Naphthalene

Naphthalene 1,2-dioxygenase

cis-1,2-dihydroxy-1,2-dihydronaphthalene

1,2-dihydroxy-1,2-dihydronaphthalene dehydrogenase

1,2-dihydroxynaphthalene

1,2-dihydroxynaphthalene dioxygenase

2-hydroxychromene-2-carboxylate

2-hydroxychromene-2-carboxylate isomerase

Trans-o-hydroxybenzylidenepyruvate (tHBPA)

tHBPA hydratase-aldolase

Salicylaldehyde

Salicylaldehyde dehydrogenase

Salicylate

Salicylate hydroxylase

Catechol

FIGURE 21.2 Upper catabolic pathway of naphthalene and transformation of salicylate into catechol.

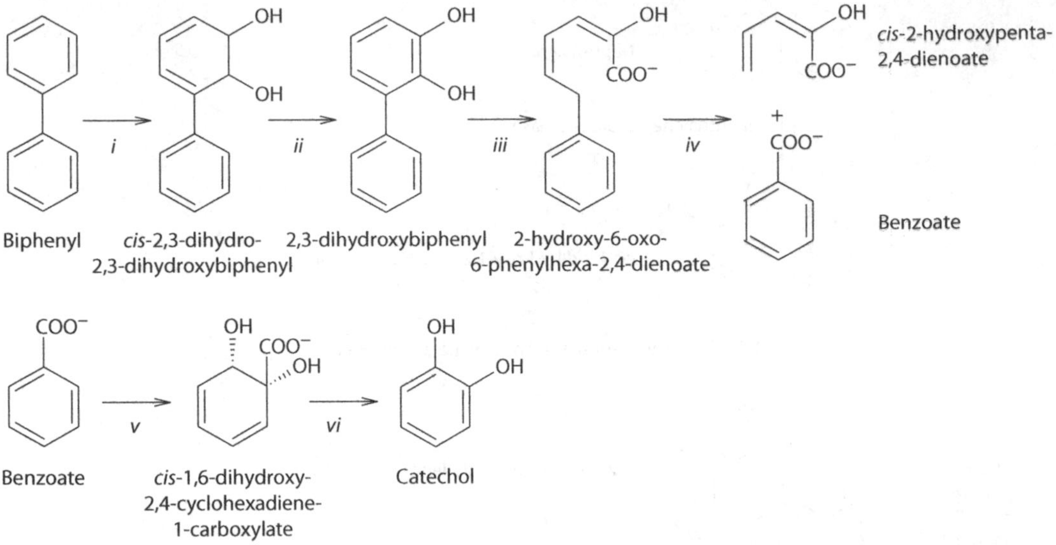

Biphenyl → (i) → *cis*-2,3-dihydro-2,3-dihydroxybiphenyl → (ii) → 2,3-dihydroxybiphenyl → (iii) → 2-hydroxy-6-oxo-6-phenylhexa-2,4-dienoate → (iv) → *cis*-2-hydroxypenta-2,4-dienoate + Benzoate

Benzoate → (v) → *cis*-1,6-dihydroxy-2,4-cyclohexadiene-1-carboxylate → (vi) → Catechol

FIGURE 21.3 Upper biphenyl (and 4-chlorobiphenyl) pathway and transformation of benzoic acid into catechol. The enzymes involved in each reaction step are biphenyl 2,3-dioxygenase (i), 2,3-dihydro-2,3-dihydroxybiphenyl dehydrogenase (ii), 2,3-dihydroxybiphenyl 1,2-dioxygenase (iii), 2-hydroxy-6-oxo-6-phenylhexa-2,4-dienoate hydrolase (iv), benzoate 1,2-dioxygenase (v), cis-diol dehydrogenase (vi).

Similarly, changing Thr_{376} near the active-site iron in biphenyl dioxygenase α subunit of other model organism *Pseudomonas pseudoalcaligenes* KF707 altered the regiospecificity and expanded the substrate specificity (Suenaga et al. 2006). Vezina et al. (2008) examined the diversity of biphenyl dioxygenase α subunit in cultured bacteria of various phylogenetic lines and in soil microflora of PCB-contaminated soils. They discovered that soil DNA also contained LB400-like sequences and also those where residues 335 and 336 of LB400 were replaced by residues that previous enzyme engineering had shown to extend the range of PCB to be transformed. This led the authors to a possible conclusion that PCB-degrading bacteria are evolving in soil to optimize their PCB-degrading capacity.

ARHDs are key enzymes in the degradation of many aromatic pollutants. Especially important is the vast range of substrates being transformed by these enzymes. Naphthalene dioxygenase is known to catalyze 76 different reactions, including dioxygenation of aromatic, substituted aromatic, and heterocyclic aromatic compounds, monooxygenation, desaturation, O- and N-dealkylation, or sulfoxidation of various substrates (Gao et al. 2010). Similarly, biphenyl dioxygenase can oxygenate except for biphenyl and certain chlorobiphenyls also other benzene or diphenyl skeletons, including those substituted with methyl, ethyl, vinyl, carboxyl, halogens, or nitro groups, and *cis*-diol bicyclic- or tricyclic-fused heterocyclic aromatics. In addition to oxygenation activity, this enzyme catalyzes oxygenolytic dehalogenation, monooxygenation, sulfoxidation, O- and N-dealkylation and desaturation for a number of aromatic and benzocyclic compounds (Vezina et al. 2008). Of special concern is toluene dioxygenase which in *Pseudomonas putida* F1 can catalyze 109 different reactions. Except for dioxygenation of toluene and other monocyclic aromatics, it can perform their monooxygenation and sulfoxidation. This enzyme can also transform fused and linked aromatics, aliphatic olefins, and other compounds, such as indan or indole (Gao et al. 2010).

21.2.2 Aromatic Ring Cleaving Dioxygenases

Another crucial step in aerobic degradation of aromatic xenobiotics is the ring cleavage. This process is performed by extradiol or intradiol aromatic ring cleaving dioxygenases (ARCDs). The

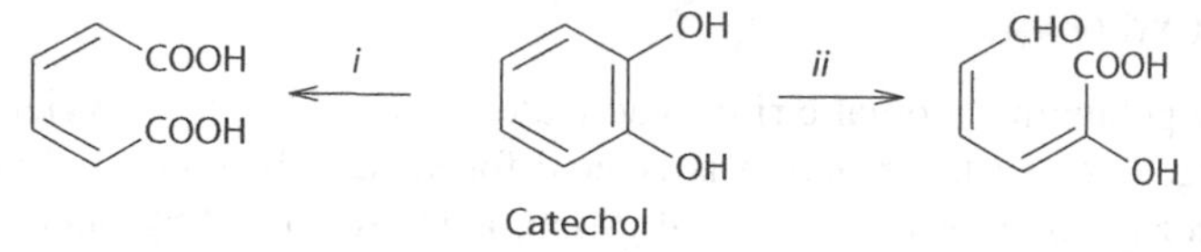

FIGURE 21.4 Intradiol (i) and extradiol (ii) cleavage of catechol.

former cleave the aromatic ring adjacent (*meta*) to the hydroxyl substituents using nonheme Fe(II) whereas the latter between (*ortho*) the substituents using nonheme Fe(III) (Figure 21.4). ARCDs, similarly to ARHDs, often determine the specificity of a degradation pathway (Vaillancourt et al. 2006). In general, extradiol ARCDs are more versatile cleaving much wider range of substrates than intradiol ones. Evolutionary, intradiol and extradiol ARCDs belong to distinct classes of proteins with no significant sequence or structural similarities. Based on structural folds, intradiol ARCDs represent one group only, whereas extradiol ARCDs cluster in three different groups (Eltis and Bolin 1996; Vaillancourt et al. 2006).

A typical substrate of ARCDs is a catecholic metabolite possessing hydroxyl substituents on two adjacent carbon atoms (Eltis and Bolin 1996). ARCDs represent the critical step of ring fission after ARHDs had activated the thermodynamically stable benzene ring (Vaillancourt et al. 2006). They are, therefore, involved in the degradation pathways of many POPs. Examples can be (i) BTEX or PAH catabolic pathways where ARCDs cleave catechol or its derivatives and initiate thereby the β-ketoadipate pathway (Harwood and Parales, 1996) (Figure 21.5), or (ii) biphenyl catabolic pathway where they cleave 2,3-dihydroxybiphenyl (Figure 21.3). Genes encoding for ARCDs are usually organized in clusters with other catabolic genes, which enable the stepwise construction of novel catabolic pathways by the reorganization of genetic segments and hence acquire the ability to use novel compounds as carbon and/or nitrogen sources (Reams and Neidle, 2004). For instance, catechol dioxygenases often cluster in supraoperonic arrangements with benzoate degradative genes, such as in pollutant-degrading bacterium *Cupriavidus necator* JMP134 (Perez-Pantoja et al. 2008), *Acinetobacter* sp. ADP1 (Reams and Neidle 2004), or other bacteria. 2,3-dihydroxybiphenyl 1,2-dioxygenase responsible for the ring cleavage in biphenyl/chlorobiphenyl degradation pathway also clusters with other genes needed to transform biphenyl into benzoate and *cis*-2-hydroxypenta-2,4-dienoate. This cluster is called *bph* operon and can be found in most well-investigated Gram-negative PCB-degrading strains (Furukawa et al. 1989; Furukawa 2000; Pieper 2005; Pieper and Seeger 2008). In Gram positive bacteria, however, *bph* genes are organized independently in three clusters (Taguchi et al. 2004). In many PCB-degrading rhodococci more *bphC* genes were detected encoding for 2,3-dihydroxybiphenyl 1,2-dioxygenase (Asturias and Timmis 1993; Maeda et al. 1995; Hauschild et al. 1996; Kosono et al. 1997; Arai et al. 1998; Sakai et al. 2002; McKay et al. 2003). Multiplicity of these genes shows the complexity of biodegradation pathways in rhodococci as well as the crucial role of ARCDs in the biodegradation of a wide variety of environmental pollutants.

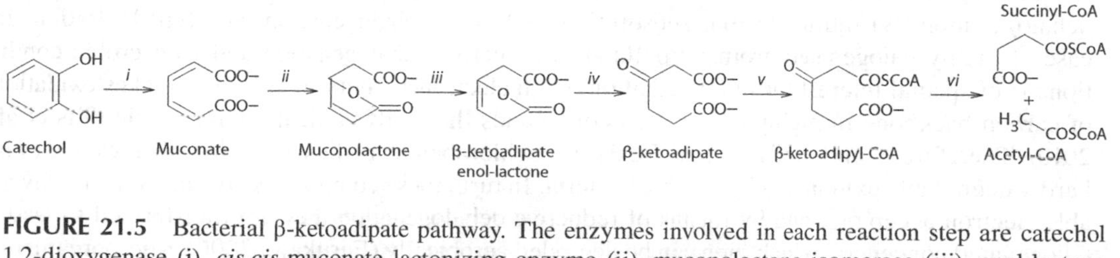

FIGURE 21.5 Bacterial β-ketoadipate pathway. The enzymes involved in each reaction step are catechol 1,2-dioxygenase (i), *cis,cis*-muconate lactonizing enzyme (ii), muconolactone isomerase (iii), enol-lactone hydrolase (iv), β-ketoadipate:succinyl-CoA transferase (v), β-ketoadipyl-CoA thiolase (vi).

21.2.3 Monooxygenases

In some cases, the pollutant's aromatic ring is activated by means of hydroxylating monooxygenase rather than dioxygenase. Such a reaction is typical for some substituted pollutants. For instance, phenol, which already contains one hydroxyl group, is hydroxylated by phenol 2-monooxygenase into catechol (Izzo et al. 2011). Toluene, when not biotransformed by means of dioxygenase, can be biotransformed by toluene 2-monooxygenase (Shields et al. 1995) or toluene 3-monooxygenase (Olsen et al. 1994) as the first step during 3-methylcatechol formation, or by toluene 4-monooxygenase in a pathway resulting in 4-hydroxybenzoate formation (Yen et al. 1991). Benzoate can also be biotransformed to 4-hydroxybenozate by benzoate 4-monooxygenase. Such a biotransformation of benzoate is typical for yeasts (*Rhodotorula graminis*) and fungi (*Aspergillus niger*) although it was reported in some pseudomonads as well (Gao et al. 2010).

In general, bacterial monooxygenases are nonheme diiron enzymes transcribed from single operons that code for four to six polypeptides. The hydroxylating activity is bound to assembled complex of α and β subunits and, in some cases, an additional γ subunit. Other components include NADH oxidoreductase with an N-terminal chloroplast-type ferredoxin domain and a C-terminal reductase domain with FAD- and NAD(P)-ribose binding regions, a small effector or coupling protein with no prosthetic groups, and sometimes, a Rieske-type ferredoxin protein. Bacterial monooxygenases can be divided into four groups—the soluble methane monooxygenases, the Amo alkene monooxygenase of *Rhodococcus corallinus* B-276, the phenol hydroxylases, and the four-component alkene/aromatic monooxygenases (Leahy et al. 2003; Notomista et al. 2003).

A very important class of monooxygenases involved in biodegradation of POPs are cytochrome P450 monooxygenases. They are not only present in most aerobic bacteria but also in all fungi and plants (Gerhardt et al. 2009). P450 systems have been shown to oxidize many pollutants, including PAHs (England et al. 1998), chlorinated phenols (Jones et al. 2001), PCBs (Wolkers et al. 1998), or explosives (Doty 2008). Unlike other monooxygenases, P450s use heme where heme iron serves as a source or sink of electrons during electron transfer.

21.2.4 Dehalogenation Reactions

Persistent Organic Pollutants (POPs) are mostly halogenated compounds. The presence of halogen atoms usually increases both recalcitrance to biodegradation and the risk of forming toxic intermediates during biotransformation of the compound. Dehalogenation reactions, which are pivotal during microbial degradation of halogenated pollutants, are therefore of special interest for bioremediation (Janssen et al. 2001). A wide variety of enzymatic systems that exist in nature are believed to have been developed in order to enable microorganisms to degrade haloorganic compounds of natural origin. These biogenic organohalides include both chlorinated and brominated structures that are produced in both terrestrial and marine systems, respectively (Smidt and de Vos 2004). Biotransformation of organohalides is based on using these compounds as terminal electron acceptors, carbon sources, or their cometabolic degradation (Janssen et al. 2001).

There are seven mechanisms of halogen removal known (Fetzner and Lingens 1994)—(i) reductive dehalogenation, (ii) oxygenolytic dehalogenation, (iii) hydrolytic dehalogenation, (iv) thiolytic dehalogenation, (v) intramolecular substitution, (vi) dehydrohalogenation, and (vii) hydration. In case of highly halogenated aromatic pollutants, reductive dehalogenation under anaerobic conditions is of special interest for bioremediation. High electronegativity of chlorines makes oxidation of carbon backbone in highly chlorinated compounds thermodynamically unfavorable (Taş et al. 2010). Therefore, compounds such as PCBs, hexachlorobenzene, dioxins, chlorophenols, etc., are hardly degradable oxidatively by aerobic bacteria. In turn, these compounds are energetically favorable electron acceptors, and by means of reductive dehalogenation they are transformed to lesser halogenated congeners, which then can be degraded aerobically (Furukawa 2006). Microorganisms performing reductive dehalogenation can either metabolize organohalides cometabolically or, in case of halorespiring bacteria, couple the dehalogenation to energy conservation. For these purposes, halorespiring bacteria employ reductive dehalogenases—cobalamin cofactor and iron–sulfur clusters containing enzymes associated with the cytoplasmic membrane (Furukawa 2006; Futagami et al. 2008; Janssen et al. 2001). The first halorespiring organism to be discovered was *Desulfomonile tiedjei* DCB-1 capable of respiring chlorobenzoates (Mohn and Tiedje 1992). Ever since, many halorespiring bacteria have been identified (reviewed by Furukawa 2006), including *Dehalospirilum*, *Desulfitobacterium*, *Dehalobacter*, or *Dehalococcoides*. Among these organisms, *Dehalococcoides* spp. are of an outstanding importance. They represent highly specialized bacterial genus because they are restricted to conserving energy for growth solely by halorespiration, and growth on other electron acceptors has not been reported yet. *Dehalococcoides* spp. have been found to transform toxic perchloroethylene, PCBs, chlorophenols, chlorobenzenes, polychlorinated dibenzo-*p*-dioxins, and chloroethanes. This variety of transformed compounds is likely to be contributed by very high numbers of reductive dehalogenase (*rdh*) genes in genomes of *Dehalococcoides* spp. For instance, genomes of sequenced strains *Dehalococcoides ethenogenes* 195, *Dehalococcoides* sp. CBDB1, and *Dehalococcoides* sp. VS contain 17, 32, and 36 *rdh* genes, respectively. Considering the fact *Dehalococcoides* spp. have some of the smallest genomes among free-living bacteria (less than 1.5 Mbp), these numbers are pretty stunning (reviewed by Taş et al. 2010).

21.3 CONTRIBUTION OF PLANTS TO MICROBIAL BIOREMEDIATION

In the environment, microbial bioremediation is often stimulated by the presence of plants. Dynamic synergy exists between plant roots and rhizosphere microorganisms (Chaudhry et al. 2005)—the area in close proximity of roots (rhizosphere) is considered to be a "hotspot" of plant-microbe interactions. The numbers of microorganisms in the rhizosphere compared to bulk soil are several orders of magnitude higher. Processes such as root exudation or root turnover provide microbial growth substrates in the form of sugars, organic acids, etc. A vastly wide range of plant secondary metabolites, which are also exuded by roots, are argued to be responsible for pollutant degrading capabilities of microorganisms. The facts that almost all pollutants are more or less structurally similar to some plant secondary metabolites and pollutant-degrading bacteria are found in virtually every gram of soil support the explanation that pollutant-degrading abilities evolved from the continued supply of plant secondary metabolites (Singer et al. 2003; Singer et al. 2004; Singer 2006). Preceding research has shown that plants can promote the mobilization of pollutants into the rhizosphere where they occur at increased concentrations and are more susceptible to biodegradation (Yi and Crowley 2007). This happens not only because of elevated numbers of microorganisms in the rhizosphere, but also because some plant secondary metabolites released into soil promote microbial cometabolism of many POPs (Leigh et al. 2006) or act as inducers of degradative enzymes. For example, many terpenes have relatively high antimicrobial activities, and their presence is thought to induce cytochrome P450 monooxygenase-mediated detoxification which results in fortuitous degradation of pollutants (Singer 2006). In addition, roots exude surfactants and phytochemicals that increase the bioavailability of pollutants with low solubility (Yi and Crowley 2007). The rhizosphere is an environment usually richer in oxygen essential for the activity of dioxygenases and monooxygenases that are often involved in biodegradation processes (Leigh et al. 2002). Plants contribute to remediation also by taking up and transforming contaminants or storing them in above-ground parts in a process called phytoremediation (Macek et al. 2000). Plants also host many bacteria in their endosphere. These endophytic bacteria have been shown to contribute to biodegradation of toxic organic compounds in contaminated soil and could have potential for improving phytoremediation (McGuinness and Dowling 2009).

Studies have been published showing that plants themselves, however, have limited abilities to mineralize pollutants. In case of PCBs, monohydroxylated and/or dihydroxylated derivatives are formed without the aromatic ring being cleaved (Rezek et al. 2007; Rezek et al. 2008). These transformed compounds have higher solubility and can be even more toxic than the original ones. The key step in mineralization of these compounds, aromatic structure break-up, can be mediated by microorganisms. Thus, the inability of plants to cleave aromatic rings can be overcome by the preparation of transgenic plants bearing bacterial genes encoding for enzymes known to cleave such structures (Novakova et al. 2009). Similarly transgenic plants translating enzymes with broad substrate specificity, such as toluene dioxygenase, have been prepared in order to foster remediation potential of plants (Novakova et al. 2007).

21.4 STUDYING MICROORGANISMS INVOLVED IN BIOREMEDIATION

Key issues associated with understanding microbial bioremediation include the identification and metabolic characterization of microbes present in the contaminated environment, and an understanding of global microbial community shifts in response to changing biotic and abiotic factors (Lovley 2003). However, until recently much of this information had been very difficult to elucidate and rarely achieved. A key step in the understanding of how to foster intrinsic microbial bioremediation is thorough investigation of the diversity of "naturally working" microorganisms. Cultivation-based techniques are limited because only about 1% of microbes are able to be routinely cultivated under laboratory conditions (Lozupone and Knight 2008; Zhang and Xu 2008). This makes it very difficult to link contaminant transformation to phylogenetic identity of active microbes and has therefore become a main challenge in experiments. The recent development and use of cutting edge molecular techniques has made it possible to link the function and structure of active communities. Stable isotope probing (SIP) is an elegant tool which enables researchers to trace metabolically active microbial populations. It involves tracking the incorporation of heavy stable isotopes, usually ^{13}C or ^{15}N, from specific substrates into phylogenetically informative biomarkers (DNA, RNA, fatty acids, proteins) associated with microbes that assimilate the substrate. SIP is thus applicable as a tool for the identification of microbes and genes that play key roles in given habitats' metabolic processes. After stable isotope labeled compounds have been pulsed into a microbial community and active cells utilizing the substrate have become heavily labeled, biomarkers are recovered and analyzed (Dumont and Murrell 2005; Uhlik et al. 2009). Currently, DNA-SIP is of interest because of recent possibilities of metagenomic analyses. Metagenomics is presently undergoing a rapid development due to advances in high-throughput sequencing (e.g., 454 pyrosequencing) and constantly improving bioinformatic tools. The main aims of metagenomics are to identify the role of populations within the ecosystem, especially those that have not yet been cultured, and to reconstruct their important metabolic pathways (Vieites et al. 2009). The first experiments in which shotgun sequencing was employed showed that metagenomics may be capable of revealing the entire metabolic potential of the microbial system but will hardly succeed in linking individual functional genes (hence functions) to certain populations. This may be achieved only in case of very dominant species or in case of communities with a very low diversity (Chen and Murrell 2010). However, even populations with low abundance can play significant roles in the community functioning. Targeting metagenomics to specific subpopulations which are likely to contain the genes of interest may overcome this obstacle, as can be achieved with stable isotope probing. This was demonstrated by Schwarz and colleagues (Schwarz et al. 2006) who isolated genes encoding for coenzyme B12-dependent glycerol dehydratases. The source for this key enzyme for the anaerobic dehydration of glycerol were the enrichment cultures of glycerol-fermenting microorganisms from a sediment sample. When metagenomic library construction was preceded by SIP, the frequency of target genes was increased almost four fold. Additionally, metagenomic exploration of active populations can clarify the metabolic capabilities as well as regulatory mechanisms within microbes. These properties can be subjected to genetic manipulations with the aim of improving the efficacy of bioremediation. The potential of SIP combined with metagenomics to contribute to bioremediation and other biotechnological applications is therefore vast.

Alternatively, next-generation sequencing can be applied to 16S rRNA genes (or functional genes) amplicon analyses. Unlike analyzing clone libraries, amplicon pyrosequencing enables one to sample the community much more thoroughly and provides greater sensitivity (Engelbrektson et al. 2010). Therefore, amplicon pyrosequencing (referred to as gene-targeted metagenomics) is suitable for analyzing how different stimuli influence microbial communities.

Metagenomic analyses can be performed, in addition to high-throughput sequencing, by high-throughput microarrays. Recently, they have been used to analyze microbial communities and monitor environmental biogeochemical processes. GeoChip microarrays (He et al. 2010) contain about 28,000 50-meric sequences covering approximately 57,000 genes encoding for enzymes responsible for element (C, N, P, S) cycling, energy gaining processes, heavy metals resistance, antibiotics resistance, degradation of pollutants, and *gyrB* genes. Marker *gyrB* encoding for gyrase β-subunit is used instead of more common 16S rRNA genes as these usually do not provide resolution below genus level. *gyrB* can be used to differentiate even closely related species. GeoChip microarrays can be therefore used to study structure, dynamics, and potential metabolic activity of microbial communities and their variations depending on different stimuli.

Since the first development of SIP, this technique has progressed rapidly and has greatly broadened the field of microbial ecology by detecting novel clades with no cultured representatives and identifying novel functional gene sequences (Uhlik et al. 2009). When combined with metagenomics, SIP becomes a powerful tool for understanding the functional community dynamics of entire microbial systems (Chen and Murrell 2010). On the other hand as with any method, SIP has a few potential limitations such as cross-feeding and enrichment. In the context of biodegradation of organic substances, SIP is one of the first techniques to connect microbial functional and phylogenetic identities. By employing SIP with metagenomics, researchers can gain a deeper insight to the microbial community and use these findings to work toward more efficient bioremediation of contaminated sites.

While metagenomics has brought new potential to DNA-SIP, use of biomarkers other than DNA for SIP has progressed as well. The use of mRNA-SIP has recently been successfully demonstrated (Huang et al. 2009). In their study, authors recovered the key phylogenetic and functional RNAs and proved that RNA-SIP can be used for studying the transcriptomes in addition to taxonomic identities of targeted guilds. Although the use of mRNA for determining metabolic activity within the community has already been demonstrated, it can often be problematic due to low yields of mRNA retrieved from environmental samples, difficulties to separate it from other RNA types, and instability of mRNA (Neufeld et al. 2007; Simon and Daniel 2009). Once these problems are overcome, mRNA-SIP could be combined with metatranscriptomic approaches in a similar manner in which DNA-SIP is combined with metagenomics. Constructing a metagenomic library with cDNA obtained from the isotopically labeled transcriptome would produce a library containing expressed genes only from the active microbial populations of interest.

Although modern microbial ecology techniques have brought a much deeper insight into microbial diversity, cultivation still has not been overcome as a means to associate functions to newly discovered microbes and to test physiologically potential functions eventually deduced from gene amplification or metagenomic data. As with all molecular techniques in microbial ecology, including 16S rRNA ribotyping and metagenomics, there are limitations which need to be addressed: PCR biases cannot be ruled out when 16S rRNA gene analyses are performed, and metagenomic analyses are likely to be misinterpreted by inappropriate predictions caused by annotation errors in databases. To overcome these potential issues, the genomes of many more cultivated bacteria should be investigated to provide valuable references for metagenomic annotation and prediction; this also highlights the importance of continued cultivation efforts (Lopez-Garcia and Moreira 2008). Therefore, as SIP and metagenomic studies are conducted, they should be combined with the parallel development of cultivation strategies in order to provide a clear, multidimensional view into the diversity and function of the microbial world.

21.5 CONCLUSIONS

The ever-increasing body of knowledge on bacteria important for bioremediation is caused by constant development of microbiology tools. Until the end of the last century, bioremediation research focused on isolates or enrichment cultures. As a result, overall knowledge of structural and functional diversity of dehalorespiring populations, which are difficult to culture, and their enzymes is rather fragmental compared to our understanding of aerobic degradation. Recently, molecular biology tools have broadened the scope in which bioremediation could be studied. However, the persistence of organic pollutants in the environment suggests that under most environmental conditions the bacteria are not effective enough in transforming the xenobiotic compounds and that our understanding of complexity of bioremediation processes is still not sufficient to stimulate bioremediation. Therefore, in order to promote bioremediation, further and much deeper research in this area is required, such as investigating the diversity of metabolic pathways of environmental microbes, understanding the mutual relationships among microbes as well as between plants (and plant compounds) and microbes. The very recently developed tools and techniques employed in environmental microbiology, such as stable isotope probing, next generation sequencing, and bioinformatic software tools should help scientists reach this goal.

ACKNOWLEDGMENTS

This work was supported by The Ministry of Education, Youth and Sports of the Czech Republic (grants NPVII 2B06156 and ME 10041, and research projects MSM 6046137305 and Z 40550506) and Grant Agency of the Czech Republic (grant 525/09/1058).

REFERENCES

Annweiler, E., H.H. Richnow, G. Antranikian, S. Hebenbrock, C. Garms, S. Franke, W. Francke, and W. Michaelis. 2000. Naphthalene degradation and incorporation of naphthalene-derived carbon into biomass by the thermophile Bacillus thermoleovorans. *Applied and Environmental Microbiology* 66: 518–523.

Arai, H., S. Kosono, K. Taguchi, M. Maeda, E. Song, F. Fuji, S.Y. Chung, and T. Kudo. 1998. Two sets of biphenyl and PCB degradation genes on a linear plasmid in Rhodococcus erythropolis TA421. *Journal of Fermentation and Bioengineering* 86: 595–599.

Asturias, J.A., and K.N. Timmis. 1993. Three different 2,3-dihydroxybiphenyl-1,2-dioxygenase genes in the gram-positive polychlorobiphenyl-degrading bacterium *Rhodococcus globerulus* P6. *Journal of Bacteriology* 175: 4631–4640.

Barriault, D., and M. Sylvestre. 2004. Evolution of the biphenyl dioxygenase BphA from *Burkholderia xenovorans* LB400 by random mutagenesis of multiple sites in region III. *Journal of Biological Chemistry* 279: 47480–47488.

Barriault, D., F. Lepine, M. Mohammadi, S. Milot, N. Leberre, and M. Sylvestre. 2004. Revisiting the regiospecificity of *Burkholderia xenovorans* LB400 biphenyl dioxygenase toward 2,2′-dichlorobiphenyl and 2,3,2′,3′-tetrachlorobiphenyl. *Journal of Biological Chemistry* 279: 47489–47496.

Batie, C.J., D.P. Ballou, and C.C. Correll. 1992. Phthalate Dioxygenase Reductase and Related Flavin-Iron-Sulfur-Containing Electron Transferases. In *Chemistry and Biochemistry of Flavoenzymes*, 543–556. Boca Raton, FL: CRC Press.

Chaudhry, Q., M. Blom-Zandstra, S. Gupta, and E.J. Joner. 2005. Utilising the synergy between plants and rhizosphere microorganisms to enhance breakdown of organic pollutants in the environment. *Environmental Science and Pollution Research International* 12: 34–48.

Chen, Y., and J.C. Murrell. 2010. When metagenomics meets stable-isotope probing: progress and perspectives. *Trends in Microbiology* 18: 157–163.

Demnerová, K., M. Macková, V. Spěváková, K. Beranová, L. Kochánková, P. Lovecká, E. Ryšlavá, and T. Macek. 2005. Two approaches to biological decontamination of groundwater and soil polluted by aromatics—characterization of microbial populations. *International Microbiology* 8: 205–211.

Doty, S.L. 2008. Enhancing phytoremediation through the use of transgenics and endophytes. *New Phytologist* 179: 318–333.

Dumont, M.G., and J.C. Murrell. 2005. Stable isotope probing—linking microbial identity to function. *Nature Reviews Microbiology* 3: 499–504.

Eaton, R.W., and P.J. Chapman. 1992. Bacterial metabolism of naphthalene: construction and use of recombinant bacteria to study ring cleavage of 1,2-dihydroxynaphthalene and subsequent reactions. *Journal of Bacteriology* 174: 7542–7554.

Eltis, L.D., and J.T. Bolin. 1996. Evolutionary relationships among extradiol dioxygenases. *Journal of Bacteriology* 178: 5930–5937.

Engelbrektson, A., V. Kunin, K.C. Wrighton, N. Zvenigorodsky, F. Chen, H. Ochman, and P. Hugenholtz. 2010. Experimental factors affecting PCR-based estimates of microbial species richness and evenness. *ISME Journal* 4: 642–647.

England, P.A., C.F. Harford-Cross, J.A. Stevenson, D.A. Rouch, and L.L. Wong. 1998. The oxidation of naphthalene and pyrene by cytochrome P450(cam). *FEBS Letters* 424: 271–274.

Erickson, B.D., and F.J. Mondello. 1992. Nucleotide sequencing and transcriptional mapping of the genes encoding biphenyl dioxygenase, a multicomponent polychlorinated-biphenyl-degrading enzyme in *Pseudomonas* strain LB400. *Journal of Bacteriology* 174: 2903–2912.

Erickson, B.D., and F.J. Mondello. 1993. Enhanced biodegradation of polychlorinated biphenyls after site-directed mutagenesis of a biphenyl dioxygenase gene. *Applied and Environmental Microbiology* 59: 3858–3862.

Fetzner, S., and F. Lingens. 1994. Bacterial dehalogenases: biochemistry, genetics, and biotechnological applications. *Microbiological Reviews* 58: 641–685.

Furukawa, K. 2000. Biochemical and genetic bases of microbial degradation of polychlorinated biphenyls (PCBs). *Journal of General and Applied Microbiology* 46: 283–296.

Furukawa, K. 2006. Oxygenases and dehalogenases: molecular approaches to efficient degradation of chlorinated environmental pollutants. *Bioscience Biotechnology and Biochemistry* 70: 2335–2348.

Furukawa, K., N. Hayase, K. Taira, and N. Tomizuka. 1989. Molecular relationship of chromosomal genes encoding biphenyl/polychlorinated biphenyl catabolism: some soil bacteria possess a highly conserved *bph* operon. *Journal of Bacteriology* 171: 5467–5472.

Futagami, T., M. Goto, and K. Furukawa. 2008. Biochemical and genetic bases of dehalorespiration. *The Chemical Record* 8: 1–12.

Gao, J., L.B. Ellis, and L.P. Wackett. 2010. The University of Minnesota Biocatalysis/Biodegradation Database: improving public access. *Nucleic Acids Research* 38: D488–491.

Gerhardt, K.E., X.D. Huang, B.R. Glick, and B.M. Greenberg. 2009. Phytoremediation and rhizoremediation of organic soil contaminants: potential and challenges. *Plant Science* 176: 20–30.

Habe, H., and T. Omori. 2003. Genetics of polycyclic aromatic hydrocarbon metabolism in diverse aerobic bacteria. *Bioscience, Biotechnology, and Biochemistry* 67: 225–243.

Haddock, J.D., J.R. Horton, and D.T. Gibson. 1995. Dihydroxylation and dechlorination of chlorinated biphenyls by purified biphenyl 2,3-dioxygenase from *Pseudomonas* sp. strain LB400. *Journal of Bacteriology* 177: 20–26.

Harwood, C.S., and R.E. Parales. 1996. The *b*-ketoadipate pathway and the biology of self-identity. *Annual Review of Microbiology* 50: 553–590.

Hauschild, J.E., E. Masai, K. Sugiyama, T. Hatta, K. Kimbara, M. Fukuda, and K. Yano. 1996. Identification of an alternative 2,3-dihydroxybiphenyl 1,2-dioxygenase in *Rhodococcus* sp. strain RHA1 and cloning of the gene. *Applied and Environmental Microbiology* 62: 2940–2946.

He, Z., Y. Deng, J.D. Van Nostrand, Q. Tu, M. Xu, C.L. Hemme, X. Li, L. Wu, T.J. Gentry, Y. Yin, J. Liebich, T.C. Hazen, and J. Zhou. 2010. GeoChip 3.0 as a high-throughput tool for analyzing microbial community composition, structure and functional activity. *ISME Journal* 4: 1167–1179.

Huang, W.E., A. Ferguson, A.C. Singer, K. Lawson, I.P. Thompson, R.M. Kalin, M.J. Larkin, M.J. Bailey, and A.S. Whiteley. 2009. Resolving genetic functions within microbial populations: *in situ* analyses using rRNA and mRNA stable isotope probing coupled with single-cell raman-fluorescence *in situ* hybridization. *Applied and Environmental Microbiology* 75: 234–241.

Izzo, V., G. Leo, R. Scognamiglio, L. Troncone, L. Birolo, and A. Di Donato. 2011. PHK from phenol hydroxylase of *Pseudomonas* sp. OX1. Insight into the role of an accessory protein in bacterial multicomponent monooxygenases. *Archives of Biochemistry and Biophysics* 505: 48–59.

Jang, J.Y., D. Kim, H.W. Bae, K.Y. Choi, J.C. Chae, G.J. Zylstra, Y.M. Kim, and E. Kim. 2005. Isolation and characterization of a *Rhodococcus* species strain able to grow on *ortho-* and *para-*xylene. *Journal of Microbiology* 43: 325–330.

Janssen, D.B., J.E. Oppentocht, and G.J. Poelarends. 2001. Microbial dehalogenation. *Current Opinion in Biotechnology* 12: 254–258.

Johnsen, A.R., L.Y. Wick, and H. Harms. 2005. Principles of microbial PAH-degradation in soil. *Environmental Pollution* 133: 71–84.

Jones, J.P., E.J. O'Hare, and L.L. Wong. 2001. Oxidation of polychlorinated benzenes by genetically engineered CYP101 (cytochrome P450(cam)). *European Journal of Biochemistry* 268: 1460–1467.

Kim, S., and F. Picardal. 2001. Microbial growth on dichlorobiphenyls chlorinated on both rings as a sole carbon and energy source. *Applied and Environmental Microbiology* 67: 1953–1955.

Kosono, S., M. Maeda, F. Fuji, H. Arai, and T. Kudo. 1997. Three of the seven *bphC* genes of *Rhodococcus erythropolis* TA421, isolated from a termite ecosystem, are located on an indigenous plasmid associated with biphenyl degradation. *Applied and Environmental Microbiology* 63: 3282–3285.

Kweon, O., S.J. Kim, S. Baek, J.C. Chae, M.D. Adjei, D.H. Baek, Y.C. Kim, and C.E. Cerniglia. 2008. A new classification system for bacterial Rieske non-heme iron aromatic ring-hydroxylating oxygenases. *BMC Biochemistry* 9: 11.

Leahy, J.G., P.J. Batchelor, and S.M. Morcomb. 2003. Evolution of the soluble diiron monooxygenases. *FEMS Microbiology Reviews* 27: 449–479.

Leigh, M.B., J.S. Fletcher, X. Fu, and F.J. Schmitz. 2002. Root turnover: an important source of microbial substrates in rhizosphere remediation of recalcitrant contaminants. *Environmental Science & Technology* 36: 1579–1583.

Leigh, M.B., P. Prouzová, M. Macková, T. Macek, D.P. Nagle, and J.S. Fletcher. 2006. Polychlorinated biphenyl (PCB)-degrading bacteria associated with trees in a PCB-contaminated site. *Applied and Environmental Microbiology* 72: 2331–2342.

Lopez-Garcia, P., and D. Moreira. 2008. Tracking microbial biodiversity through molecular and genomic ecology. *Research in Microbiology* 159: 67–73.

Lovley, D.R. 2003. Cleaning up with genomics: applying molecular biology to bioremediation. *Nature Reviews Microbiology* 1: 35–44.

Lozupone, C.A., and R. Knight. 2008. Species divergence and the measurement of microbial diversity. *FEMS Microbiology Reviews* 32: 557–578.

Macek, T., M. Macková, and J. Káš. 2000. Exploitation of plants for the removal of organics in environmental remediation. *Biotechnology Advances* 18: 23–34.

Macková, M., D. Dowling, and T. Macek, eds. 2006. *Phytoremediation and Rhizoremediation. Theoretical Background.* Dordrecht, Netherlands: Springer.

Macková, M., O. Uhlík, P. Lovecká, J. Viktorová, M. Nováková, K. Demnerová, M. Sylvestre, and T. Macek. 2010. Bacterial Degradation of Polychlorinated Biphenyls. In *Geomicrobiology: Molecular and Environmental Perspective*, 347–366. Dordrecht, Netherlands: Springer.

Madsen, E.L. 2008. *Environmental microbiology. From genomes to biogeochemistry.* Malden, MA, USA: Blackwell Publishing.

Maeda, M., S.Y. Chung, E. Song, and T. Kudo. 1995. Multiple genes encoding 2,3-dihydroxybiphenyl 1,2-dioxygenase in the gram-positive polychlorinated biphenyl-degrading bacterium *Rhodococcus erythropolis* TA421, isolated from a termite ecosystem. *Applied and Environmental Microbiology* 61: 549–555.

McGuinness, M., and D. Dowling. 2009. Plant-associated bacterial degradation of toxic organic compounds in soil. *International Journal of Environmental Research and Public Health* 6: 2226–2247.

McKay, D.B., M. Prucha, W. Reineke, K.N. Timmis, and D.H. Pieper. 2003. Substrate specificity and expression of three 2,3-dihydroxybiphenyl 1,2-dioxygenases from *Rhodococcus globerulus* strain P6. *Journal of Bacteriology* 185: 2944–2951.

Mohn, W.W., and J.M. Tiedje. 1992. Microbial reductive dehalogenation. *Microbiological Reviews* 56: 482–507.

Mondello, F.J., M.P. Turcich, J.H. Lobos, and B.D. Erickson. 1997. Identification and modification of biphenyl dioxygenase sequences that determine the specificity of polychlorinated biphenyl degradation. *Applied and Environmental Microbiologyl* 63: 3096–3103.

Nam, J.W., H. Nojiri, T. Yoshida, H. Habe, H. Yamane, and T. Omori. 2001. New classification system for oxygenase components involved in ring-hydroxylating oxygenations. *Bioscience, Biotechnology, and Biochemistry* 65: 254–263.

Neufeld, J.D., M.G. Dumont, J. Vohra, and J.C. Murrell. 2007. Methodological considerations for the use of stable isotope probing in microbial ecology. *Microbial Ecology* 53: 435–442.

Notomista, E., A. Lahm, A. Di Donato, and A. Tramontano. 2003. Evolution of bacterial and archaeal multicomponent monooxygenases. *Journal of Molecular Evolution* 56: 435–445.

Nováková, M., M. Macková, M. Sylvestre, J. Prokešová, and T. Macek. 2007. Preparation of transgenic plants with *todC1C2* genes—cloning of bacterial *todC1C2* genes into plants to improve their phytoremediation abilities. *Listy Cukrovarnicke a Reparske* 123: 323–324.

Nováková, M., M. Macková, Z. Chrastilová, J. Viktorová, M. Szekeres, K. Demnerová, and T. Macek. 2009. Cloning the bacterial *bphC* gene into *Nicotiana tabacum* to improve the efficiency of PCB phytoremediation. *Biotechnology and Bioengineering* 102: 29–37.

Olsen, R.H., J.J. Kukor, and B. Kaphammer. 1994. A novel toluene-3-monooxygenase pathway cloned from *Pseudomonas pickettii* PKO1. *Journal of Bacteriology* 176: 3749–3756.

Peng, R.-H., A.-S. Xiong, Y. Xue, X.-Y. Fu, F. Gao, W. Zhao, Y.-S. Tian, and Q.-H. Yao. 2010. A profile of ring-hydroxylating oxygenases that degrade aromatic pollutants. In *Reviews of Environmental Contamination and Toxicology Volume 206*, 65–94: Springer, New York.

Perez-Pantoja, D., R. De la Iglesia, D.H. Pieper, and B. Gonzalez. 2008. Metabolic reconstruction of aromatic compounds degradation from the genome of the amazing pollutant-degrading bacterium *Cupriavidus necator* JMP134. *FEMS Microbiology Reviews* 32: 736–794.

Pieper, D.H. 2005. Aerobic degradation of polychlorinated biphenyls. *Applied Microbiology and Biotechnology* 67: 170–191.

Pieper, D.H., and M. Seeger. 2008. Bacterial metabolism of polychlorinated biphenyls. *Journal of Molecular Microbiology and Biotechnology* 15: 121–138.

Reams, A.B., and E.L. Neidle. 2004. Selection for gene clustering by tandem duplication. *Annual Review of Microbiology* 58: 119–142.

Rezek, J., T. Macek, M. Macková, and J. Tříska. 2007. Plant metabolites of polychlorinated biphenyls in hairy root culture of black nightshade *Solanum nigrum* SNC-9O. *Chemosphere* 69: 1221–1227.

Rezek, J., T. Macek, M. Macková, J. Tříska, and K. Růžičková. 2008. Hydroxy-PCBs, methoxy-PCBs and hydroxy-methoxy-PCBs: metabolites of polychlorinated biphenyls formed *in vitro* by tobacco cells. *Environmental Science & Technology* 42: 5746–5751.

Sakai, M., E. Masai, H. Asami, K. Sugiyama, K. Kimbara, and M. Fukuda. 2002. Diversity of 2,3-dihydroxybiphenyl dioxygenase genes in a strong PCB degrader, *Rhodococcus* sp. strain RHA1. *Journal of Bioscience and Bioengineering* 93: 421–427.

Shields, M.S., M.J. Reagin, R.R. Gerger, R. Campbell, and C. Somerville. 1995. TOM, a new aromatic degradative plasmid from *Burkholderia* (*Pseudomonas*) *cepacia* G4. *Applied and Environmental Microbiology* 61: 1352–1356.

Schwarz, S., T. Waschkowitz, and R. Daniel. 2006. Enhancement of gene detection frequencies by combining DNA-based stable-isotope probing with the construction of metagenomic DNA libraries. *World Journal of Microbiology & Biotechnology* 22: 363–367.

Simon, C., and R. Daniel. 2009. Achievements and new knowledge unraveled by metagenomic approaches. *Applied Microbiology and Biotechnology* 85: 265–276.

Singer, A.C. 2006. Bioremediation and phytoremediation from mechanistic and ecological perspectives. In *Focus on Biotechnology*, 5–21. Dordrecht, Netherlands: Springer.

Singer, A.C., D.E. Crowley, and I.P. Thompson. 2003. Secondary plant metabolites in phytoremediation and biotransformation. *Trends in Biotechnology* 21: 123–130.

Singer, A.C., I.P. Thompson, and M.J. Bailey. 2004. The tritrophic trinity: a source of pollutant-degrading enzymes and its implications for phytoremediation. *Current Opinion in Microbiology* 7: 239–244.

Smidt, H., and W.M. de Vos. 2004. Anaerobic microbial dehalogenation. *Annual Review of Microbiology* 58: 43–73.

Suenaga, H., M. Goto, and K. Furukawa. 2006. Active-site engineering of biphenyl dioxygenase: effect of substituted amino acids on substrate specificity and regiospecificity. *Applied Microbiology and Biotechnology* 71: 168–176.

Sylvestre, M. 1980. Isolation method for bacterial isolates capable of growth on *p*-chlorobiphenyl. *Applied and Environmental Microbiology* 39: 1223–1224.

Taguchi, K., M. Motoyama, and T. Kudo. 2004. Multiplicity of 2,3-dihydroxybiphenyl dioxygenase genes in the Gram-positive polychlorinated biphenyl degrading bacterium *Rhodococcus rhodochrous* K37. *Bioscience Biotechnology and Biochemistry* 68: 787–795.

Taş, N., M.H.A. Van Eekert, W.M. De Vos, and H. Smidt. 2010. The little bacteria that can—diversity, genomics and ecophysiology of '*Dehalococcoides*' spp. in contaminated environments. *Microbial Biotechnology* 3: 389–402.

Uhlík, O., K. Ječná, M.B. Leigh, M. Macková, and T. Macek. 2009. DNA-based stable isotope probing: a link between community structure and function. *Science of the Total Environment* 407: 3611–3619.

Vaillancourt, F.H., J.T. Bolin, and L.D. Eltis. 2006. The ins and outs of ring-cleaving dioxygenases. *Critical Reviews in Biochemistry and Molecular Biology* 41: 241–267.

Vézina, J., D. Barriault, and M. Sylvestre. 2008. Diversity of the C-terminal portion of the biphenyl dioxygenase large subunit. *Journal of Molecular Microbiology and Biotechnology* 15: 139–151.

Vieites, J.M., M.E. Guazzaroni, A. Beloqui, P.N. Golyshin, and M. Ferrer. 2009. Metagenomics approaches in systems microbiology. *FEMS Microbiology Review* 33: 236–255.

Wolkers, J., I.C. Burkow, C. Lydersen, S. Dahle, M. Monshouwer, and R.F. Witkamp. 1998. Congener specific PCB and polychlorinated camphene (toxaphene) levels in Svalbard ringed seals (*Phoca hispida*) in relation to sex, age, condition and cytochrome P450 enzyme activity. *Science of the Total Environment* 216: 1–11.

Yen, K.M., M.R. Karl, L.M. Blatt, M.J. Simon, R.B. Winter, P.R. Fausset, H.S. Lu, A.A. Harcourt, and K.K. Chen. 1991. Cloning and characterization of a *Pseudomonas mendocina* KR1 gene cluster encoding toluene-4-monooxygenase. *Journal of Bacteriology* 173: 5315–5327.

Yi, H., and D.E. Crowley. 2007. Biostimulation of PAH degradation with plants containing high concentrations of linoleic acid. *Environmental Science & Technology* 41: 4382–4388.

Zhang, L., and Z. Xu. 2008. Assessing bacterial diversity in soil. *Journal of Soils and Sediments* 8: 379–388.

22 Using Endophytes to Enhance Phytoremediation

Zareen Khan and Sharon Doty

CONTENTS

22.1 INTRODUCTION

Industrialization and other human activities have led to release of toxins into the environment, and our planet is becoming polluted with a wide variety of pollutants, most of which are carcinogenic. Most polluted sites often contain a mixture of both organic and inorganic pollutants (Ensley 2000). It is estimated that around \$6–8 billion is spent each year on environmental cleanup in the United States (Glass 1999) and around \$25–50 billion globally (Tsao 2003). The most commonly used methods for clean up are the engineering-based methods such as pump and treat, excavation, soil washing, capping and burning. These are environmentally invasive and extremely expensive. Biological methods (bioremediation), such as bioaugmentation using special microbes, are also being used to treat contaminated sites. Bioremediation can occur on its own or can be spurred on via addition of superior strains of pollutant-degrading microbes. However, there are multiple challenges to bioremediation. For example, use of additional substrates or co-metabolites makes the whole process expensive. Although microbes can utilize a wide variety of organic compounds thereby converting them to nontoxic forms, survival at a particular site, bioavailability of the pollutant, and the presence of inducers to activate expression of necessary enzymes are some factors that need to be considered before bioremediation can be used.

A more recent method that is gaining popularity is phytoremediation. Phytoremediation is the use of plants for the removal of contamination either indirectly by facilitating microbial activity or directly through uptake, sequestration or degradation of pollutants (reviewed extensively by Meagher 2000; Pilon-Smith 2002; Newman and Reynolds 2005; Audet and Charest 2007). Government agencies and environmental consultancies are increasingly using phytoremediation, spending about \$100–150 million/yr (Pilon-Smith and Freeman 2006). The use of plants to remediate recalcitrant compounds in soil represents a low cost alternative when contrasted with expensive, environmentally invasive mechanical methods (Sung et al. 2003), sometimes even yielding useful products such as wood or biofuel. The plant root system offers several benefits including production of exudates, promoting movement of water and gases through the soil and promoting microbial activity, which are important factors influencing contaminant availability for uptake and its degradation. Because of its excellent benefits over traditional treatment methods, phytoremediation could become a major technology for remediation projects.

Phytoremediation makes use of different processes for environmental clean up. *Phytoextraction* is removal of pollutants by the roots followed by translocation to aboveground plant tissues, which are subsequently harvested. This is mainly used for metals, metalloids, and radionuclides. Certain plant species grown at heavy metal contaminated sites have been found to accumulate unusually high concentrations of heavy metals without impacting their growth and development. These plants are called hyper accumulators and the idea of using them for soil metal remediation has increasingly being examined as a potential practical and a more cost effective technology than other traditional methods. Recent reviews on many aspects of metal phytoremediation are available by Doty (2008), Dhankher et al. (2010), and Rajkumar et al. (2010). *Phytodegradation* is a process by which the pollutants are partially or completely degraded due to internal or secreted plant enzymes. This can take place inside the plant or within the rhizosphere of the plant. *Phytostimulation* also involves enzymatic breakdown but through microbial activity. The plant root zone has rich microbial populations that help breaking down contaminants for uptake and degradation. *Phytovolatilization* is the transpiration of the contaminant into the atmosphere through leaves. For volatile contaminants, the compounds are dispersed unaltered into the atmosphere, whereas certain inorganics can be transformed enzymatically into volatile forms during uptake and release. *Rhizofiltration* is a process using roots to filter contaminants from wastewaters. In *phytostabilization*, the root exudates immobilize and reduce the availability of soil pollutants thereby preventing leaching into the ground or surface water sources and also reducing erosion and production of contaminated dust.

Phytoremediation has been used to treat a variety of pollutants including chlorinated solvents, hydrocarbons, explosives, polycyclic aromatic hydrocarbons, polycyclic biphenyls, toxic metals and other organics. Although most of the studies have been done to demonstrate the success of phytoremediation, its wide application is still in its nascent stages. Much effort still has to be directed toward an understanding of the basic mechanisms and toward improving knowledge of the applications (Marmiroli et al. 2006). A primary disadvantage when compared to traditional clean-up methods is that it is considered too slow and seasonal. Another limitation is that for phytoremediation to be effective the pollutant should be bioavailable to be taken up and degraded by the plants. Soil properties, toxicity level, and climate are other factors to be considered in applying phytoremediation at a contaminated site. Sometimes even plants that are tolerant to the presence of these contaminants remain small due to the toxicity of the pollutants they are accumulating or the toxic end products of their degradation. For example, in the case of hyper accumulators, most of them identified so far have either slow growth or small biomass which restricts the employment of this technology. Phytoremediation is also limited by root depth because the plants have to be able to reach the pollutant. For this reason, deep rooted trees like poplars and willows are the "trees of choice" for phytoremediation projects.

To overcome these limitations, two different approaches have been studied. One is by using transgenic plants (reviewed in Doty 2008; Dhanker 2011) and a more recent approach is by using endophytes.

22.1.1 Endophyte-Assisted Phytoremediation

Endophytes include microbes from bacterial and eukaryotic domains of life that reside inside the plant and live asymptomatically. Burkholderiaceae, Pseudomonaceae and Enterobacteriaceae are among the most common families of cultivable endophytic species found. Beneficial microbes provide several benefits to the plants including plant growth promotion, increased nutrient uptake, protection against pathogens, increased tolerance to environmental stresses (drought, heat, salt, contaminant) with some even fixing nitrogen for the plant (Rodriguez et al. 2009; Doty 2010; Redman et al. 2011). Recently, researchers have focused on exploiting the benefits of endophytes to overcome constraints of phytoremediation. Several studies demonstrated that bacterial endophytes aid phytoremediation by direct degradation of contaminants and/or by reducing toxicity to the plants. Many authors have recently highlighted the benefits of bacterial endophytes in phytoremediation (Newman and Reynolds 2005; Bacon and Hinton 2007; Doty 2008; Ryan et al. 2008; Weyens et al. 2009a; Capuana 2011).

In this chapter, we summarize the current knowledge about the contribution of endophytes to degradation of different classes of pollutants, and their potential for phytoremediation applications.

22.2 MAJOR EXAMPLES OF ENDOPHYTE-ASSISTED PHYTOREMEDIATION

22.2.1 Explosives

Explosives like TNT and RDX have been listed as priority pollutants by USEPA and are the most widespread conventional explosives. In a study by van Aken and colleagues (2004) the authors isolated a pink pigmented bacterium from hybrid poplar *P. deltoides X P. nigra* DN34, that was capable of transforming TNT and RDX and HMX to carbon dioxide. It was interesting to note that this strain was isolated from poplars not previously exposed to TNT, RDX or HMX suggesting a possibility of finding endophytes with abilities to degrade this class of pollutants. Since the structures of TNT and RDX are similar to lignin precursors one would expect that the endophytes might have the natural tendency to degrade explosives. Indeed, several endophytes isolated from poplar were found to be resistant to high levels of TNT (Doty, unpublished).

22.2.2 Petroleum and Associated Compounds

Endophytes have been shown to assist in phytoremediation of petroleum compounds. Siciliano et al. (2001) isolated and studied endophytes from a petroleum contaminated site. Some observations were that the genes for hydrocarbon degradation were more common in endophytic bacteria. Similarly, in other field sites containing nitroaromatics, genes encoding nitrotoluene degradation were more prevalent in endophytes when compared to the rhizospheric bacteria. This suggests that there may be recruitment of these microbes by the plants to protect it from contaminant stress. Other interesting results from their study were that the numbers of endophytes were dependent on the concentration of chemicals in the soil and it was also plant species specific. This study shows that pollutant-degrading microorganisms may enhance a plants' adaptation to contaminants such as petroleum hydrocarbons. Studies need to be conducted to directly compare plants with and without the endophytes, however, in order to determine if the endophytes are actually benefitting these plants. In another study, degradation of BTEX by endophytes from poplar trees growing on a BTEX contaminated was demonstrated by Moore et al. (2006). This is another example that suggests that there is a selection of pollutant-degrading endophytes by plants to alleviate contaminant stress. From simply looking at the diversity of endophytes, research has moved to testing the effects of a specific endophyte for a specific pollutant. This was demonstrated by Van der Lelie's group on toluene, an important component of BTEX, which is a major environmental soil and groundwater and are usually found near petroleum and natural gas production sites. In their work (Barac et al. 2004), the pTOM plasmid from *Burkholderia cepacia* (G4) was naturally inserted into the lupine endophyte *B. cepacia* BU0072 via bacterial conjugation. The inoculated yellow lupine plants, with the altered endophyte, were tolerant to toluene, resulting in reduced toxicity to the plants and growth, at lethal concentrations of toluene (1000mg/l). The inoculated plants did not show any toxic effects whereas the uninoculated plants showed growth reduction with as little as 100mg/l of toluene. Evapotranspiration of toluene was reduced by 50%–70% by the inoculated plants. This was a groundbreaking study on how engineered endophytes could be applied to improve phytoremediation. Questions about instability of the modified endophyte was answered by Taghavi et al. (2005) who demonstrated that horizontal gene transfer could be used to adapt the natural endophytic communities. In their study, in plants, horizontal transfer of the toluene-degrading gene from the endophytic strain *Burkholderia cepacia* VM1468 occurred into the different members of the endogenous community, and the inoculated poplar had increased tolerance to toluene and evapotranspired less toluene than the inoculated controls. These are excellent studies demonstrating that endophytes equipped with the appropriate degradation pathway can suppress phytotoxicity and evapotranspiration of volatile organic chemicals.

22.2.3 Trichloroethylene

Trichloroethylene (TCE) is one of the most prevalent volatile groundwater contaminants (Shang et al. 2001; Halsey et al. 2005; Behrens et al. 2008; Lee et al. 2008). Hence its removal is a major priority for many contaminated sites all over the industrialized world. Several studies demonstrated effective degradation of TCE by using endophytes. Changes in the microbial community in poplar trees in response to TCE exposure was observed by Weyens et al. (2009b) which correlated with the reports by Siciliano et al. (2001) who also noticed similar responses in that there were more petroleum degrading endophytes in plants growing on petroleum contaminated site. An example of endophyte assisted phytoremediation of TCE was shown by Weyens et al. (2010a). When poplar trees were inoculated with the TCE-degrading poplar endophyte *Pseudomonas putida* W619-TCE, it resulted in 90% reduction in evapotranspiration of TCE. The endophyte established successfully, dominated as a root endophyte and subsequently transferred the TCE degradation ability to the natural endophytes of poplar. The TCE-degrading endophytes remained as the dominant strains and were lost under nonselective conditions. Experiments in the green house showed that the endophyte *Pseudomonas putida* W619-TCE not only protected the host from phytotoxicity but also reduced the amounts of TCE accumulation in plant tissue with decreased evapotranspiration. These studies are encouraging and can be used to apply in more field tests. However use of genetically modified strains could face public criticism thereby hindering its use for phytoremediation applications. A more benign approach could be to look for naturally occurring microbes or endophytes living inside plants inhabiting contaminated areas. In such recent studies, microbes such as *Rhodococcus* sp. L4 (that was induced by plant oils) (Suttinun et al. 2009) and a *Bacillus* sp. 2479 (that was isolated from a hazardous waste site) (Dey and Roy 2009) were found to degrade TCE aerobically. Use of such microbes to inoculate poplars and willows could be a cost effective and environmental friendly alternative to treat sites contaminated with TCE. In an effort to isolate TCE degraders, our lab discovered a novel endophyte, PDN3, isolated from poplar, capable of aerobically metabolizing TCE to chloride ion (Kang et al. 2012). Experiments are currently underway to test for increased uptake and degradation of TCE in inoculated poplar and willow.

22.2.4 Polyaromatic Hydrocarbons

Polyaromatic hydrocarbons (PAHs) are another group of widespread environmental contaminants. Sixteen PAHs are listed as priority pollutants by the USEPA because of their carcinogenic properties and prevalence. These are hydrophobic and phytotoxic. Recently Germaine et al. (2009) used a genetically enhanced endophytic strain of the poplar endophyte *Pseudomonas putida* VM1441 to protect pea plants against phytotoxicity of naphthalene (a low molecular weight PAH). The inoculated plants had higher naphthalene (40%) degradation compared with the uninoculated plants in artificially contaminated soil. This was a good proof of concept study showing degradation of this important class of chemicals by endophytes. In an experiment done in our lab (Khan, unpublished), we have shown that certain endophytes of poplar have the ability to grow in PAHs including naphthalene, phenanthrene and pyrene. Experiments are underway to test if these endophytes can effectively degrade these compounds when inoculated into poplars and willows.

22.2.5 Herbicides

There is only one study done so far on endophyte assisted phytoremediation of herbicides (Germaine et al. 2006). The authors isolated bacterial endophytes with the ability to degrade 2,4 Dichlorophenoxyacetic acid (2,4 D) from hybrid cottonwood and inoculated pea plants. The inoculated pea plants with a 2,4-D degrading endophyte showed high removal of 2,4-D from the soil compared to the uninoculated controls. Also there was no 2,4-D detected in the aerial parts of the plants indicating that it was degraded by the endophyte within the plant. Also, the inoculated plants had higher biomass than the uninoculated controls thereby the endophyte provided a growth promoting and a protective effect to the pea plants.

22.2.6 Metals

Heavy metal contamination in soil is one of the world's major environmental problems. Phytoremediation offers more benefits than the conventional methods to accumulate toxic metals from soil. But slow growth and low biomass of plants may limit the efficiency of phytoremediation (Burd et al. 1998). Use of plant growth promoting bacteria in metal phytoextraction has been shown in many studies (Glick 2010). It has been observed that heavy metal resistant endophytes are present in various hyper accumulator plants growing on heavy metal contaminated soil (Rajkumar et al. 2010). The possibilities of using endophytes for phytoremediation of metals has been demonstrated by Lodewyckx et al. (2001) where the authors showed that endophytes of yellow lupin, genetically constructed for nickel resistance, increased nickel accumulation and tolerance of inoculated plants. In a recent study by Mastreta et al. (2009), the authors showed that endophytes isolated from *Nicotiana tobacum* reduced cadmium toxicity by increasing uptake of metals like zinc and iron by plants. Sheng et al. (2008) showed that lead resistant endophytes isolated from *Brassica napus* enhanced phytoremediation potential by promoting plant growth and lead uptake by rape. Inoculation of copper (Cu) tolerant bacteria in the plant rhizosphere increased Cu accumulation by a copper accumulator plant- *Elsholtzia splendens*, generally growing on copper mines (Chen et al. 2005). In a study by Trotta et al. (2006), the authors showed higher uptake of arsenic (As) by the As hyperaccumulating fern inoculated with arbuscular fungi (intimately associated with plant roots). In a recent study by Ren (2011), an endophytic fungus when inoculated to *Lolium arundinaceum* Darbyshire ex. Schreb plants improved the phytoextraction of cadmium. Endophyte infected plants not only had higher biomass, but also enhanced Cd accumulation and improved Cd transport from root to shoot. In a nickel and TCE co-contaminated scenario, an engineered endophyte (*Burkholderia cepacia* VM1468) sequestered Ni five times more than the uninoculated controls (Weyens et al. 2010b). Thus it is clear that the benefits of endophytes and plants can be successfully tried for removal of toxic metals from contaminated soils.

22.3 CONCLUSIONS

With increasing studies on using endophytes for phytoremediation of a variety of compounds, this technology has great potential for field applications. Diverse microorganisms live inside the plant and their associations can be exploited to improve the plants' abilities to deal with contaminants. Use of endophytic inoculations for plant assisted remediation also offers great potential. Studies showing inoculations of engineered endophytes for increased pollutant tolerance and uptake are promising. However, their field application will need more knowledge including a better understanding of the interactions. Instead, using natural endophytic isolates possessing pollutant degrading abilities would be the best case scenario. Strategies should be designed for effective re-colonization of endophytic bacteria and more pilot scale studies should be undertaken demonstrating the use of this technology to application on real contaminated sites. For optimum phytoremediation, it is necessary for plants to grow as large as possible in the presence of various contaminants. Therefore, making use of endophytes with the dual traits of pollutant degradation as well as phytohormone production would be highly valuable.

REFERENCES

Audet, P., and C. Charest. 2007. Heavy metal phytoremediation from a meta-analytical perspective. *Environmental Pollution* 147(1): 231–237.

Bacon, C.W., and D.M. Hinton. 2007. Bacterial endophytes: The endophytic niche, its occupants, and its utility. In: *Plant Associated Bacteria*, ed. S.S. Gnanamanickam, 155–194. Springer, New York.

Barac, T., S. Taghavi, B. Borremans et al. 2004. Engineered endophytic bacteria improve phytoremediation of water-soluble, volatile, organic pollutants. *Nature Biotechnology* 22: 583–588.

Behrens, S.M.F., P. Azizian, J. McMurdie et al. 2008. Monitoring abundance and expression of "Dehalococcoides" species chloroethene-reductive dehalogenases in a tetrachloroethene-dechlorinating flow column. *Applied and Environmental Microbiology* 74(18): 5695–5703.

Burd, G.I., D.G. Dixon, and B.R. Glick. 1998. A plant growth promoting bacterium that decreases nickel toxicity in plant seedlings. *Applied and Environmental Microbiology* 64: 3663–3668.

Capuana, M. 2011. Heavy metals and woody plants-biotechnologies for phytoremediation. *Forest Biogeosciences and Forestry* 4(1): 7–15.

Chen, Y.X., Y. P. Wang, Q. Lin et al. 2005. Effect of copper-tolerant rhizosphere bacteria on mobility of copper in soil and copper accumulation by *Elsholtzia splendens. Environment International* 131: 861–866.

Dey, K., and P. Roy. 2009. Degradation of trichloroethylene by *Bacillus* sp.: Isolation strategy, strain characteristics, and cell immobilization. *Current Microbiology* 59: 256–260.

Dhankher, O.M., E.A.H. Pilon-Smits, R.B. Meagher et al. 2011. Biotechnological approaches for phytoremediation. In: *Plant Biotechnology and Agriculture: Prospects for the 21st century*, A. Altman and P.M. Haegawa eds., Elsevier Press (in press).

Doty, S.L. 2008. Enhancing phytoremediation through the use of transgenics and endophytes. *New Phytologist* 179: 318–333.

Doty, S.L. 2010. Nitrogen-fixing endophytic bacteria for improved plant growth. In: *Bacteria in Agrobiology,* ed. D.K. Maheshwari, Springer.

Ensley, B.D. 2000. Rationale for use of phytoremediation. In: *Phytoremediation of Toxic Metals: Using Plants to Clean Up the Environment,* ed. I. Raskin and B.D. Ensley. New York: Wiley.

Germaine, K.J., X. Liu, G.G. Cabellos et al. 2006. Bacterial endophyte-enhanced phytoremediation of the organochlorine herbicide 2,4-dichlorophenoxyacetic acid. *FEMS Microbiology Ecology* 57: 302–310.

Germaine, K.J., Keogh, E., Ryan, D. et al. 2009. Bacterial endophyte-mediated naphthalene phytoprotection and phytoremediation. *FEMS Microbiology Letters* 296: 226–234.

Glass, D. 2000. *U.S and International Markets for Phytoremediation*, D. Glass Associates, Inc.: Needham, MA.

Glick, B.R. 2010. Using soil bacteria to facilitate phytoremediation. *Biotechnology advances* 28(3): 367–374.

Halsey, K.H., L. A. Sayavedra-Soto, P.J. Bottomley et al. 2005. Trichloroethylene degradation by butane-oxidizing bacteria causes a spectrum of toxic effects. *Applied Microbiology Biotechnology* 68(6):794–801.

Kang, J.W., Z. Khan, and S.L. Doty. 2012. Biodegradation of trichloroethylene by an endophyte of hybrid poplar. *Applied and Environmental Microbiology* 78(9): 3504.

Lee, P.K.H., T. W. Macbeth, K. S. Sorenson et al. 2008. Quantifying genes and transcripts to assess the *in situ* physiology of "Dehalococcoides" spp. in a trichloroethene-contaminated groundwater site. *Applied and Environmental Microbiology* 74(9): 2728–273.

Lodewyckx, C., S. Taghavi, M. Mergeay et al. 2001. The effect of recombinant heavy metal resistant endophytic bacteria in heavy metal uptake by their host plant. *International Journal of Phytoremediation* 3: 173–187.

Marmiroli, N., M. Marmiroli, E. Maestri. 2006. Phytoremediation and phytotechnologies: a review for the present and the future. In: *Soil and Water Pollution Monitoring, Protection and Remediation*, ed. I. Twardowska et al., 3–23. Springer: The Netherlands.

Mastretta, C., T. Barac, J. Vangronsveld et al. 2006. Endophytic bacteria and their potential application to improve the phytoremediation of contaminated environments. *Biotechnology and Genetic Engineering Reviews* 23: 175–206.

Meagher, R.B. 2000. Phytoremediation of toxic elemental and organic pollutants. *Current Opinion in Plant Biology* 3: 153–162.

Moore, P.F., T. Barac, B. Borremanns et al. 2006. Endophytic bacterial diversity in poplar trees growing on BTEX-contaminated site: the characterization of isolates with potential to enhance phytoremediation. *Systematic and Applied Microbiology* 29: 539–556.

Newman, L.A., and C.M. Reynolds. 2005. Bacteria and phytoremediation: new uses for endophytic bacteria in plants. *Trends in Biotechnology* 23: 6–8.

Pilon-Smith, E.A.H., and M. Pilon. 2002. Phytoremediation of metals using transgenic plants. *Critical Reviews in Plant Sciences* 21: 439–456.

Pilon-Smith, E.A.H., and J.L. Freeman. 2006. Environmental cleanup using plants: biotechnological advances and ecological considerations. *Frontiers in Ecology and the Environment* 4: 203–210.

Rajkumar, M., Ae N., Freitas, H. et al. 2010. Endophytic bacteria and their potential to enhance heavy metal phytoextraction. *Chemosphere* 77: 153–156.

Redman R.S., Y.O. Kim, C.J.D.A. Woodward et al. 2011. Increased fitness of rice plants to abiotic stress via habitat adapted symbiosis: a strategy for mitigating impacts of climate change. *PLoS ONE* 6(7): e14823. Doi: 10.1371/journal.pone.0014823.

Ren, A., C. Li, and Y. Gao. 2011. Endophytic fungus improves growth and metal uptake of *Lolium arundinaceum* Darbyshire ex Schreb. *International Journal of Phytoremediation* 13(3): 233–243.

Rodriguez, R.J., J.F.J. White, A.E. Arnold et al. 2009. Fungal endophytes: diversity and functional roles. *New Phytologist* 182(2): 314–330.

Ryan, R.P., K. Germaine, A. Franks et al. 2008. Bacterial endophytes: recent developments and applications. *FEMS Microbiology Letters* 278: 1–9.

Shang, T.Q., D.L. Doty, A.M. Wilson et al. 2001. Trichloroethylene oxidative metabolism in plants: The trichloroethanol pathway. *Phytochemistry* 58(7): 1055–1065.

Sheng, X.F., J.J. Xia, C.Y. Jiang et al. 2008. Characterization of heavy metal-resistant endophytic bacteria from rape (*Brassica napus)* roots and their potential in promoting the growth and lead accumulation of rape. *Environmental Pollution* 156: 1164–1170.

Siciliano, S.D., N. Fortin, A. Mihoc et al. 2001. Selection of specific endophytic bacterial genotypes by plants in response to soil contamination. *Appllied and Environmental Microbiology* 66: 4673–4678.

Sung, K., C.L. Munster, R. Rhykerd et al. 2003. The use of vegetation to remediate soil freshly contaminated by recalcitrant contaminants. *Water Research* 37: 2408–2418.

Suttinun, O., R. Müller, and E. Luepromchai. 2009. Trichloroethylene cometabolic degradation by *Rhodococcus* sp. L4 induced with plant essential oils. *Biodegradation* 20: 281–291.

Taghavi, S., T. Barac, B. Greenberg et al. 2005. Horizontal gene transfer to endogenous endophytic bacteria from Poplar improves phytoremediation of toluene. *Appllied and Environmental Microbiology* 71: 8500–8505.

Trotta, A., P. Falaschi, L. Cornara et al. 2006. Arbuscular mycorrhizae increase the arsenic translocation factor in the As hyperaccumulating fern *Pteris vittata* L. *Chemosphere* 65: 74–81.

Tsao, D.T. 2003. Overview of phytotechnologies. *Advances in Biochemical Engineering Biotechnology* 78: 1–50.

vanAken, B., C.M. Peres, S.L. Doty et al. 2004. Methylobacterium populi sp. nov., a novel aerobic, pink-pigmented, facultatively methylotrophic, methane-utilizing bacterium isolated from poplar tree. *International Journal of Systematic and Evolutionary Microbiology* 54: 1191–1196.

Weyens, N., D. van der Lelie, S. Taghavi et al. 2009a. Phytoremediation: plant-endophyte partnerships take the challenge. *Current Opinion in Biotechnology* 20(2): 1848–1854.

Weyens, N., S. Taghavi, T. Barac et al. 2009b. Bacteria associated with oak and ash on a TCE-contaminated site: Characterization of isolates with potential to avoid evapotranspiration of TCE. *Environmental Science and Pollution Research International* 16(7): 830–843.

Weyens, N., S. Truyens, J. Dupae et al. 2010a. Potential of the TCE-degrading endophyte *Pseudomonas putida* W619-TCE to improve plant growth and reduce TCE phytotoxicity and evapotranspiration in poplar cuttings. *Environmental Pollution* 158(9): 2915–2919.

Weyens, N., S. Croes, J. Dupae et al. 2010b. Endophytic bacteria improve phytoremediation of Ni and TCE co-contamination. *Environmental Pollution* 158: 2422–2427.

23 Genetically Modified Plants Designed for Phytoremediation of Toxic Organic and Inorganic Contaminants

Tomas Macek, Martina Novakova, Pavel Kotrba, Jitka Viktorova, Petra Lovecká, Jan Fiser, Miroslava Vrbová, Eva Tejklová, Jitka Najmanova, Katerina Demnerova, and Martina Mackova

CONTENTS

23.1 INTRODUCTION

The use of plants or plant products to restore or stabilize contaminated sites, collectively known as phytoremediation, takes advantage of the natural abilities of plants to take up, accumulate, store, or degrade organic and inorganic substances (McIntyre 2003). Phytoremediation is of public acceptance and is an aesthetically pleasant, solar-energy driven, passive technique that can be used to clean up sites with shallow, low to moderate levels of contamination (Schnoor et al. 1995; Macek et al. 2000; Kotrba et al. 2009). Phytoremediation is not only a growing science; it's also a growing industry. This technique can be used along with or, in some cases, in place of mechanical cleanup methods (Kayser 1998). Phytoremediation possesses some particularly important advantages over bioremediation using microorganisms: the capability of autotrophic plants to produce high biomass with low nutriet requirements; the capacity to reduce the spread of pollutants through water and wind erosion; and a better public acceptance. Phytoremediation uses different plant processes and mechanisms normally involved in the accumulation, complexation, volatilization, and degradation of organic and inorganic pollutants (Macek et al. 2000). Plants also produce various beneficial root exudates which support the proliferation of soil microflora, participating in remediation, especially at the rhizosphere, as well as specific chelating agents mobilizing elements in bioavailable forms (Kotrba et al. 2009, 2011).

An ideal phytoremediator would have: high tolerance to the pollutant; the ability to either degrade or concentrate the contaminant at high levels in the biomass; extensive and branched root systems; the capacity to absorb large amounts of water from the soil; and fast growth rates, high levels of biomass, a wide geographic distribution, is easy to cultivate; and is relatively easy to harvest. Although several species can tolerate and grow in some contaminated sites, these species typically grow very slowly, often produce very low levels of biomass, and are adapted to very specific environmental conditions. Trees that have extensive root systems, high biomass, and low agricultural inputs requirements tolerate some pollutants poorly and do not accumulate them. Most conventional plants therefore fail to meet the requirements for successful phytoremediators. Phytoremediation efficiency of plants can be substantially improved using genetic engineering technologies. The introduction of novel traits for the uptake and accumulation of pollutants into high biomass plants is proving to be a successful strategy for the development of improved phytoremediators. Recent research is directed to overexpression of genes whose protein products are involved in metal uptake, transport, and sequestration, or act as enzymes involved in the degradation of hazardous organics or design systems containing special endophytes (Kramer 2005; Macek et al. 2008; Doty 2008; Kotrba et al. 2009).

23.2 DESIGN OF TRANSGENIC PLANTS FOR PHYTOREMEDIATION OF ORGANIC POLLUTANTS

The design of transgenic plants for phytoremediation of organic pollutants is based on the facts, how plants are able to metabolize organic pollutants themselves. Wild type plants can take up the organic compounds and metabolize them with similar strategy and mechanisms as human liver does. This process includes activation of a compound (oxidation, reduction, hydrolysis), further conjugation with sugars, glutathione or amino acids and the last step represents storage in the vacuole, cell wall or lignin. Organic pollutants are thus transformed into nonphytotoxic compounds, not every time nontoxic for mammals or man. The design of transgenic plants with increased abilities to degrade organic pollutants is therefore focused on insertion of foreign genes (bacterial, human, yeasts) for enzymes further metabolizing desired pollutant. Because of the complexity of organic pollutants degradation, there have been proposed several approaches with the aim to degrade the organic pollutant as far as it can be degraded; and the most important task, to degrade it to the nontoxic compound.

23.2.1 Phytoremediation of TCE and Aliphatic Pollutants

The major part in metabolism of toxic pollutants play monooxygenases like cytochrome P450 and peroxidases (Chroma et al. 2002). These enzymes can be found also in plant cells. The effort was therefore focused to overexpression of these enzymes in plants or use the genes from different organisms to support plant enzyme systems. There have been prepared transgenic plants of *Nicotiana tabacum* containing human cytochrome P450 2E1 that dramatically increased the metabolism of TCE 640 times (Doty et al. 2000). These transgenic plants also debrominated dibromoethylene. Doty et al. (2007) also prepared hybrid poplar plants (*Populus tremula* × *Populus alba*) with the rabbit gene for cytochrome P450 CYP2E1. Transgenic poplar trees exhibited increased metabolism even 100× of TCE, vinylchloride, chloroform, benzene and tetrachloromethane in comparison with wild type plants. The rabbit gene for cytochrome P450 2E1 was also introduced into *Atropa belladonna* with the help of *Agrobacterium rhizogenes*. Transgenic "hairy root" culture showed also increased metabolism of TCE (Banerjee et al. 2002).

To obtain metabolization of organic pollutants, mostly bacterial genes from well known bacterial degradation pathways are used to be cloned into plant genomes. This was shown in the case of insertion of bacterial *dhaA* gene into *Nicotiana tabacum* genome for phytoremediation of 1-chlorobutane (Uchida et al. 2005). Expressed DhaA enzyme in transgenic plants dechlorinated 1-chlorobutane into 1-butanol. Mena-Benitez et al. (2008) prepared also transgenic *N. tabacum* dehalogenating aliphatic hydrocarbon, 1,2-diochloroethane, into 2-chloroethanol with the help of expressed enzyme dehalogenalkane dehalogenase DhlA (bacterial *dhlA* gene). 2-chloroethanol was further metabolized by plant endogenous enzymes into 2-chloroacetaldehyde which was the substrate for another recombinant enzyme produced in transgenic plant, dehalogenase DhlB of halogenated carboxyacids. Chloroacetic acid was dehalogenated into glycolate entering the glyoxalate cycle.

23.2.2 Phytoremediation of TNT and RDX Explosives

Trinitrotoluene (TNT) is an explosive difficult to degrade, it has very toxic properties and together with hexahydro-1,3,5-trinitro-1,3,5-triazin (RDX) belongs to "priority pollutants" of "U.S. Environmental Protection Agency" EPA (Van Aken 2009). Otherwise, some microorganisms use TNT as their carbon source (Ramos et al. 2005), bioremediation of TNT is very slow and only a few microorganisms can fully mineralize TNT (Esteve-Nunez et al. 2001; Robertson and Jjemba 2005; Rylott and Bruce 2009). Some prepared transgenic *N. tabacum* used *onr* gene coding for bacterial pentaerythritol tetranitrate reductase to enhance the tolerance to TNT (French et al. 1999), some used *nfsI* gene encoding bacterial nitroreductase to metabolize TNT into 2-hydroxyl-4,6-dinitrotoluene, which is further reduced into 2-amino-4,6-dinitrotoluene (Hannink et al. 2001, 2007; Rylott and Bruce 2009). Bacterial gene for nitroreductase (*pnrA* gene) was also used (Van Dillewijn et al. 2008) for preparation of transgenic aspen accumulating higher amount of TNT in comparison to wild type trees. Concerning TNT, Gandia-Herrero (2008) overexpressed *743B4* or *73C1* genes coding for glycosyltransferase in *A. thaliana*. The overexpression of glycosyltransferase caused increased formation of TNT intermediates conjugates. Even here, in the case of phytoremediation of RDX explosive, the gene for cytochrome P450 degrading RDX (XplA) was used for introduction into *A. thaliana* (Rylott et al. 2006). Transgenic plants were more tolerant to RDX and degraded RDX. Cytochrome P450 XplA was also used (Jackson et al. 2007) for preparation of transgenic *A. thaliana* containing *xplA* and *xplB* (encoding reductase) genes. Transgenic plants removed RDX faster from liquid media than wild type plants and also faster than transgenic plants containing only *xplA* gene.

23.2.3 Phytoremediation of Herbicides and Pesticides

As almost in every case, genes for cytochrome P450 (like *CYP1A1, CYP2B6* or *CYP2C19*) were used to enhance also phytoremediation of herbicides (Inui and Ohkawa 2005; Kawahigashi et al. 2005, 2006, 2007, 2008). Didierjean (2002) cloned *Helianthus tuberosus CYP76B1* gene encoding cytochrome P450 into *N. tabacum* and *A. thaliana*. Transgenic plants were 20× more tolerant to linuron and 10 times more tolerant to isoproturon and chlorotoluron. There have been prepared also transgenic plants with specific feature. Wang et al. (2005) cloned modified bacterial *atzA* gene for atrazinchlorohydrolase AtzA catalyzing hydrolytic dechloration of atrazine into hydroxyatrazin into the genome of *A. thaliana, N. tabacum* and *Medicago sativa*. Other specific transformation was performed using bacterial *ophc2* gene coding hydrolase of organophosphates (Wang et al. 2008). Unexpected transformation which is mostly performed for enhanced phytoremediation of inorganic pollutants is transformation of plants with the gene for glutathione-S-transferase (maize gene into tobacco plants) (Karavangeli et al. 2005) or *gshI* and *gsh2* genes into *B. juncea* (Flocco et al. 2004). Transgenic plants expressing maize glutathione S-transferase I exhibit significant catabolic activity for the chloroacetanilide herbicide alachlor and appear to be involved in its detoxifying process (Karavangeli et al. 2005). Recent experiments have shown the importance of glutathione S-transferase in organic pollutant detoxification (Schroder et al. 2008).

23.2.4 Phytoremediation of Chlorocatechol

Chlorocatechol is a key intermediate in the degradation of aromatic hydrocarbons in contaminated soil. To degrade such compound (Shimizu et al. 2002) transgenic maize with bacterial *cbnA* gene encoding chlorocatechol-1,2-dioxygenase was designed Transgenic plant could therefore help in phytoremediation and rhizoremediation processes because of 3-chlorocatechol transformation into less toxic 2-chloromuconate. Similar approach was investigated in the case of transgenic *A. thaliana* with *Plesiomonas tfdC* gene. Expressed chlorocatechol-1,2-dioxygenase caused conversion of catechol into *cis,cis*-muconate in transgenic plants.

23.2.5 Phytoremediation of Phenol and Pentachlorophenol

At the beginning researchers prepared transgenic plants using fungal potential to degrade phenolic compounds. Iimura et al. (2002) cloned gene of white rot fungi *Coriolus versicolor* for manganese peroxidase into *N. tabacum*. Transgenic plants degraded pentachlorophenol (PCP) 2× more than wild type plants. Sonoki et al. (2005) cloned *cvL3* gene from *C. versicolor* into *N. tabacum*. Excretion of laccase from transgenic plants into the media decreased the PCP concentration and also bisphenol A concentration in medium. Recently another approach was used to enhance the phytoremediation of phenol. Tomato *tpx1* and *tpx2* genes for peroxidases isoenzymes were cloned (Oller et al. 2005). Transgenic plants removed more phenol from media and showed increased activities of peroxidases.

23.2.6 Phytoremediation of Polychlorinated Biphenyls

Polychlorinated biphenyls (PCBs) are persistent pollutants; otherwise, some bacteria are able to metabolize them, e.g., by aerobic pathway, or the so-called upper degradation pathway. There are several designs for preparation of transgenic plants for phytoremediation of PCBs (Macek et al. 2008; Sylvestre et al. 2009; Van Aken 2009). Mohammadi et al. 2007 prepared transgenic plants of *N. tabacum* expressing *Burkholderia xenovorans* LB400 *bphAE* (BphAE enzyme) or *bphF* (BphF ferredoxin) or *bphG* (BphG ferredoxin reductase) genes encoding multienzyme complex of dioxygenase ISP_{BPH} (BphAEFG) (Sylvestre et al. 2009). The activity of individual components was verified after reconstitution with other recombinant components expressed in bacteria. We have also implicated preparation of transgenic plants for PCB phytoremediation, therefore the *Pandoraea pnomenusa* B-356 *bphC* gene was cloned in *N. tabacum* genome. *BphC* gene encodes the 2,3-dihydroxybiphenyl-1,2-dioxygenase that can cleave the aromatic structure of 2,3-dihydroxybiphenyl. This step cannot be performed by endogenous plant enzyme because of lack of proper dioxygenases in eukaryotic organism. Prepared transgenic plants showed higher tolerance to 2,3-dihydroxybiphenyl and to Delor103 (mixture of PCB congeners). The experiment with transgenic plants expressing BphC enzyme showed also higher decrease of 2,3-dihydroxybiphenyl from media compared to wild type plants (Novakova et al. 2009). Similarly, Uchida et al. (2005) transferred bacterial *dbfB* gene encoding 2,2′,3-trihydroxybiphenyl-1,2-dioxygenase into *A. thaliana*. Expressed enzyme cleaved also aromatic structure of 2,3-dihydroxybiphenyl and was expressed either in cytoplasm or in apoplast.

23.2.7 Phytoremediation of Toluene

Recently, James et al. (2008) prepared transgenic *N. tabacum* cv. Xanthii with the human gene for cytochrome P450 2E1. Transgenic plants showed higher elimination of trichloroethylene, vinylchloride, tetrachlormethane, benzene, toluene, chloroform and bromodichloromethane in hydroponic environment than wild type plants. Transgenic plants had therefore lower transpiration of toluene into the air. Other approach how to phytoremediate toluene was designed in our laboratory. We cloned bacterial *todC1C2* genes encoding terminal dioxygenase of ISP_{TOL} (toluene-2,3-dioxygenase) into the genome of *N. tabacum* (Novakova et al. 2007). Prepared transgenic plants should be able to oxidize not only toluene but also biphenyl.

Another alternative way how to decrease the amount of toluene from the environment was studied by different authors (Barac et al. 2004; Newman and Reynolds 2005; Taghavi et al. 2005). They prepared genetically engineered endophytic bacteria with pTOM plasmid, therefore the degradation of toluene is performed *in planta* and the transpiration rate of toluene by the plant is lowered. Plasmid pTOM was transferred from *Burkholderia cepacia* G4 (Barac et al. 2004) or *Burkholderia cepacia* BU61 (Taghavi et al. 2005) and introduced into endophytic bacteria of *Lupinus luteus* L. *Burkholderia cepacia* BU0072 (Barac et al. 2004; Taghavi et al. 2005). These recombinant endophytic bacteria were inoculated to poplar trees (Taghavi et al. 2005) or *Lupinus luteus* L. (Barac et al. 2004). Inoculated plants were more tolerant to toluene and the toluene transpiration was 50%–70% lower in the case of yellow lupin.

23.3 DESIGN OF TRANSGENIC PLANTS FOR PHYTOREMEDIATION OF INORGANIC POLLUTANTS

Requirement for plants removing heavy metals and metalloids is to grow fast in the contaminated environment, to be resistant, able to accumulate toxic metals and transfer cations or oxyanions into the harvestable (above ground) parts, or transform them into less-toxic forms. Prerequisite to the efficient accumulation of metal is its mobilization from soil, efficient metal, uptake mechanism, cellular capability to maintain homeostasis of essential metals and competence to detoxify (over) accumulated metal species (Clemens et al. 2002). As many metallic species exert their toxic effect also by induction of reactive oxygen species and other free radicals, their elimination is another challenge faced by the cell (Boominathan and Doran 2003; Foyer and Noctor 2005). The most important parameter for selection of suitable plants is not the tolerance of the plant to heavy metals, but the effectiveness in their accumulation. In addition to accumulation capacity, biomass production must be considered in order to determine the total metal uptake.

The plants are considered as hyperaccumulating if capable of depositing in their organs metal(loid) species at concentrations 50 to 500 times higher than those in the soil substrate (Clemens 2006; Clemens et al. 2002). Another definition uses the term hyperaccumulator for species containing at least 100 times higher concentrations of a particular element than other species growing over an underlying substrate with the same characteristics (Brooks 1998). Specifically, the currently accepted concentration limits in shoot tissues of hyperaccumulators on a dry-weight basis are 0.1 wt.% for most metals, except, for example, for zinc (1 wt.%), cadmium (0.01 wt.%) or gold (0.0001 wt.%) (Baker et al. 2000). Enhancement of the metabolic abilities of plants can be achieved by traditional breeding, protoplast fusion, and the direct insertion of novel genes. Genetic engineering methods are widely used for the improvement of different crop plants.

At present the main goal is metal-hyperaccumulation traits that might be introduced into fast growing, high-biomass plants or vice-versa (Dorlhac de Borne et al. 1998). Three different engineering approaches are being employed to develop plants suitable for phytoremediation. These include: (i) increasing the number of metal transporters along with modulation of the specificity of the metal(loid) uptake system; (ii) enhancing intracellular ligand production and the efficiency of metal targeting into vacuoles, both to keep accumulated metal(loid) in a safe form and not disturbing cellular processes; and (iii) biochemical transformation of metal(loid)s to less toxic, volatile forms.

Plants have developed their own systems for binding heavy metals based largely on the synthesis of phytochelatins, described by Grill et al. 1989. Heavy metal binding in plants is normally achieved, as reviewed by Kotrba et al. (1999, 2009), Kraemer et al. (2001) or described by Bailey et al. (2003), by phytochelatins and phytosiderophores. Many studies have concentrated on hyperaccumulator plants (most studied being cadmium hyperaccumulator, *Thlaspi caerulescens*) in order to clarify the mechanisms of hyperaccumulation, transforming them, e.g., by *Agrobacterium rhizogenes* to obtain hairy roots (Nedelkoska and Doran 2000). The problem is very complex and seems full understanding and engineering plant metal accumulation is a long way ahead (Clemens et al. 2002). Heavy-metal binding peptides and proteins in plants have been studied already for decades; there is a large choice of such compounds (Kotrba et al. 1999, 2009). Many proteins need metals for their proper function, while three groups are involved in heavy metal homeostasis in plants—metallothioneins, phytochelatins and glutathione. Different strategies are used to obtain GM-plants with improved properties and have been summarized in reviews (Pilon-Smits 2000; Kraemer and Chardonnens 2001; Pilon-Smits and Pilon 2002; Macek et al. 2008; Kotrba et al. 2009).

Modifications of levels of enzymes normally present, e.g., overexpression of glutathione synthetase in *Brassica juncea* enhances cadmium tolerance and accumulation (Pilon-Smits et al. 1999). Overexpression of ATP sulfurylase in Indian mustard lead to increased selenium uptake, reduction and tolerance Banuelos et al. (2005). Field trial of transgenic Indian mustard plants showed enhanced phytoremediation of selenium-contaminated sediments (Pilon-Smits et al. 1999). Engineering tolerance and hyperaccumulation of arsenic in plants by combining arsenate reductase and gamma-glutamylcysteine synthetase expression was described by Dhankher et al. (2002). Over expression of phytochelatin synthase in *Arabidopsis* lead to enhanced arsenic tolerance and cadmium hypersensitivity (Li et al. 2004).

Introduction of modified bacterial genes represents a successful approach tested in *Arabidopsis* and in trees, lead to mercuric ion reduction and resistance in transgenic plants (Meagher 2000). A very different way to enhance metal uptake uses cloning of nicotianamine synthase genes (Higuchi et al. 1999), involved in the synthesis of phytosiderophores. The occurrence of the essential metal binding amino acid nicotianamine in plants and microorganisms was described already in early eighties by Rudolph et al. (1985). Secondary metabolites or organic acids seem to have a still underestimated role in metal uptake, for example aluminium tolerance in transgenic plants could be achieved by alteration of citrate synthesis (de la Fuente et al. 1997). In the last decade, many attempts were made to prepare transgenic plants bearing genes coding for metal-binding proteins, the metallothioneins (MTs), of different origin—human MT, animal MT, and yeast MT genes (Dorlhac de Borne 1998; Macek et al. 1999, 2002; Bailey et al. 2003). Cadmium partitioning in transgenic tobacco plants expressing a mammalian metallothionein gene is described in detail by Dorlhac de Borne et al. (1998). The authors found that the Cd content is increased in roots but decreased in leaves. Further enhancement of the performance of transgene products can be expected by implanting an additional heavy metal binding site, like a polyhistidine tail, known for its high affinity for heavy metals. Such a gene coding for six histidines was obtained from a commercial plasmid pTrc and cloned into an *A. tumefaciens* vector for delivery into plants (Macek et al. 1999). The transgenic plants were tested, with preliminary results showing two to three times higher Cd accumulation in tobacco plants bearing a construct with polyhistidine bound to the yeast metallothionein gene in comparison with control tobacco plants (Macek et al. 2002). Already preceding the transformation of plants, the different constructs were tested in *E. coli* in a series of growth curves, and together with improved resistance, a substantial increase of Cd accumulation was found. The obtained GM plants were tested in contaminated soil and showed enhanced transfer factors of cadmium to above ground biomass than nontransgenic plants (Pavlikova et al. 2004).

The transport of essential metals or alkali cations across plasma and organellar membranes by means of primary and secondary active transporters is of central importance in the metal homeostasis network in plants. There is a growing number of studies on the plant metal transporters of different families (Krämer et al. 2007) and the relatively broad substrate (metal) specificity of particular transporters make them a promising tool to improve toxic metal uptake for phytoremediation. Recent advances in the identification and functional evaluation of transporters in model plants *A. thaliana* and *N. tabacum*, and an understanding of the mechanisms and regulation of transport events in the accumulators *Arabidopsis harlei*, *B. juncea* and *T. caerulescens*, thus offer great promise for the manipulation of candidate plants (Krämer et al. 2007; Muthukumar et al. 2007; Milner and Kochian 2008).

Concerning the treatment of heavy metal contaminated fields, an interesting approach is atmosphere. Rugh et al. (1998) described the transformation of poplar proembryogenic masses by microprojectile bombardment with three modified *merA* constructs. They obtained plantlets releasing elemental mercury at a high rate. The results indicated that plants expressing modified merA constructs may provide a means for the phytoremediation of mercury pollution. The acquisition of metal(loid)s also depends on the ability of a plant to solubilize and mobilize elements in the soil. Increased attention should thus be devoted to modifications that enhance the capacity to secrete metal-complexing exudates such as siderophores and organic acids into the rhizosphere and implementation of the cognate metal-complex transport mechanism (Macek et al. 2008; Eapen and D'Souza 2005). An understanding of the complex plant–microbe interactions in the rhizosphere would further allow for the genetic modification of plants and their microbial symbionts to further promote the mobilization of metal(loid) species of interest (Kotrba et al. 2009).

23.4 CONCLUSIONS AND FUTURE PROSPECTS

The transgenic technology has progressed from single gene engineering to multigene engineering and successfully demonstrated for the enhanced degradation and remediation of many persistent hazardous pollutants. Although the use of biotechnology to develop transgenic plants with improved potential for efficient, clean, cheap, and sustainable bioremediation technologies is very promising, several challenges remain.

- Phytoremediation technologies are currently available for only a small subset of pollutants, and many sites are contaminated with several chemicals. Therefore, phytoremediators need to be engineered with multiple stacked genes in order to meet the requirements of specific sites.
- Phytoremediation technology is still at an early development stage, and field testing of transgenic plants for phytoremediation is very limited. Biosafety concerns need to be properly addressed, and strategies to prevent gene flow into wild species need to be developed.
- The true costs of benefits of phytoremediation with biotech plants must be determined.
- Political will and funding are required, both to pursue basic research into phytoremediation, and to implement novel strategies.

Unlike with transgenic crops intended for use as human or animal foods, with phytoremediation plants the issues of food safety, allergenicity and labelling are not relevant and the use of herbicide, insect or virus resistance genes also do not appear likely at present (Davison 2005). Thus the most serious risk concerns that of gene flow from cultivated plants to wild relatives, and this situation would need to be monitored. Plants modified for an improved tolerance to toxic metal(loid)s have indeed a selective advantage at contaminated sites. The possibilities of some transformation of the natural flora by cross-pollination, effective over long distances, the risk of invasion of privileged plants, and the potential loss of diversity, should be taken into account. With respect to this, the potential negative effects on related soil microorganisms, herbivores and other organisms along the food chain should not be forgotten. One hypothesis was described by Linacre et al. (2003) who noted that the risks of entry of metals with engineered accumulator would often be low, because such plants would be in isolated industrial districts, rather than in agricultural areas. The threat of crossing with the relatives and uncontrolled spreading of pollen or seeds could be substantially reduced, when plants used for phytoextraction could be harvested before flowering (Kotrba et al. 2009).

Different plant species may survive higher concentrations of hazardous wastes than many microorganisms used for bioremediation. This phenomenon was described not only for hyperaccumulators like alpine pennycress, indian mustard or Leguminosae milkvetch, Asian stonecrop, but also for industrial crop plants like flax (Baker et al. 2000; Smykalova et al. 2010; Bjelková et al. 2011). Phytoremediation also yields other benefits including carbon sequestration, soil stabilization, and the possibility of biofuel or fiber production. It increases the amount of organic carbon in the soil, which can stimulate microbial activity and augment the rhizospheric degradation of the pollutants. The development of phytoremediation technologies for the plant-based clean-up of contaminated soils is therefore of significant interest. Use of transgenics could enhance and improve bioremediation efforts but increased understanding of the molecular basis of the pathways involved in the degradation and accumulation of pollutants, enzymatic processes involved in plant tolerance and detoxification of xenobiotics can provide new directions for manipulating plants with superior remediation potential. A better understanding is needed. Further analysis and discovery of genes suitable for phytoremediation is essential.

REFERENCES

Bailey, N.J.C., M. Oven, E. Holmes, J.K. Nicholson, and M.H. Zenk. 2003. Metabolomic analysis of the consequences of cadmium exposure in *Silene cucubalus* cell cultures via H-1 NMR spectroscopy and chemometrics. *Phytochemistry* 62: 851–858.

Baker, A., S. McGrath, R. Reeves, and J. Smith. 2000. Metal hyperaccumulator plants: a review of the ecology and physiology of a biological resource for phytoremediation of metal polluted soils. In Terry, N. and G.S. Bañuelos, eds. *Phytoremediation of Contaminated Soil and Water*, 85–107. CRC Press, Boca Raton.

Banerjee, S., T.Q. Shang et al. 2002. Expression of functional mammalian P450 2E1 in hairy root cultures. *Biotechnology and Bioengineering* 77: 462–466.

Banuelos, G. 2005. Field trial of transgenic indian mustard plants shows enhanced phytoremediation of selenium-contaminated sediments. *Environmental Science and Technology* 39: 1171–1177.

Barac, T., S. Taghavi et al. 2004. Engineered endophytic bacteria improve phytoremediation of water-soluble, volatile, organic pollutants. *Nature Biotechnology* 22: 583–588.

Boominathan, R., and P.M. Doran. 2003. Cadmium tolerance and antioxidative defenses in hairy roots of the cadmium hyperaccumulator, *Thlaspi caerulescens. Biotechnology and Bioengineering* 83: 158–167.

Bjelkova, M., V. Gencurova, and M. Griga. 2011. Accumulation of cadmium by flax and linseed cultivars in field-simulated conditions: A potential for phytoremediation of Cd-contaminated soils *Industrial Crops and Products* 33: 761–774.

Brooks, R.R. 1998. General introduction. In R.R. Brooks, ed. *Plants that Hyperaccumulate Heavy Metals*, 1–14. CABI, Wallingford.

Cherian, S., and M.M. Oliveira. 2005. Transgenic Plants in Phytoremediation: Recent Advances and New Possibilities. *Environmental Science and Technology* 39: 9377–9390.

Chromá, L., T. Macek et al. 2002. Decolorization of RBBR by plant cells and correlation with the transformation of PCBs. *Chemosphere* 49: 739–748.

Chromá, L., M. Macková, P. Kucerova, C. in der Wiesche, J. Burkhard, and T. Macek. 2002. Enzymes in plant metabolism of PCBs and PAHs. *Acta Biotechnologica* 22: 35–41.

Clemens, S., M. Palmgren, and U. Krämer. 2002. A long way ahead: understanding and engineering plant metal accumulation. *Trends in Plant Science* 7: 309–315.

Clemens, S. 2006. Toxic metal accumulation, responses to exposure and mechanisms of tolerance in plants. *Biochimie* 88: 1707–1719.

Davison, J. 2002. Monitoring horizontal gene transfer. *Nature Biotechnology* 22: 1349.

de la Fuente, J.M., V. Ramírez-Rodríguez, J.L. Cabrera-Ponce, and L. Herrera-Estrella. 1997. Aluminum tolerance in transgenic plants by alteration of citrate synthesis. *Science* 276: 1566–1568.

Dhankher, O.P. 2002. Engineering tolerance and hyperaccumulation of arsenic in plants by combining arsenate reductase and gamma-glutamylcysteine synthetase expression. *Nature Biotechnology* 20: 1140–1145.

Doty, S.L., and C.A. James. 2007. Enhanced phytoremediation of volatile environmental pollutants with transgenic trees. *Procedings of National Academy of Sciences USA* 104: 16,816–16,821.

Doty, S.L., and T.Q. Shang. 2000. Enhanced metabolism of halogenated hydrocarbons in transgenic plants containing mammalian cytochrome P450 2E1. *Procedings of National Academy of Sciences USA* 97: 6287–6291.

Doty, S.L. 2008. Enhancing phytoremediation through the use of transgenics and endophytes. *New Phytologist* 179: 318–333.

Didierjean, L. 2002. Engineering herbicide metabolism in tobacco and Arabidopsis with CYP76B1, a cytochrome P450 enzyme from Jerusalem Artichoke. *Plant Physiology* 130: 179–189.

Dorlhac de Borne, F.D., T. Elmayan, Ch. de Roton, L. de Hys, and M. Tepfer. 1998. Cadmium partitioning in trans-genic tobacco plants expressing a mammalian metallothionein gene. *Molecular Breeding* 4: 83–90.

Eapen, S., and S. D'Souza. 2005. Prospects of genetic engineering of plants for phytoremediation of toxic metals. *Biotechnology Advances* 23: 97–114.

Esteve-Nunez, A., A. Caballero et al. 2001. Biological degradation of 2,4,6-trinitrotoluene. *Microbiology and Molecular Biololgy Review* 65: 335–352.

Flocco, C.G., S.D. Lindblom et al. 2004. Overexpression of enzymes involved in glutathione synthesis enhances tolerance to organic pollutants in *Brassica juncea. International Journal of Phytoremediation* 6: 289–304.

Francova, K., M. Mackova, T. Macek, and M. Sylvestre. 2004. Ability of bacterial biphenyl dioxygenases from *Burkholderia* sp. LB400 and *Comamonas testosteroni* B-356 to catalyse oxygenation of ortho-hydroxy-biphenyls formed from PCBs by plants. *Environmental Pollution* 127: 41–48.

French, C.E., S.J. Rosser et al. 1999. Biodegradation of explosives by transgenic plants expressing pentaerythritol tetranitrate reductase. *Nature Biotechnology* 17: 491–494.

Gandia-Herrero, F., A. Lorenz et al. 2008. Detoxification of the explosive 2,4,6-trinitrotoluene in Arabidopsis: discovery of bifunctional O- and C-glucosyltransferases. *Plant Journal* 56: 963–974.

Hannink, N., S.J. Rosser et al. 2001. Phytodetoxification of TNT by transgenic plants expressing a bacterial nitroreductase. *Nature Biotechnology* 19: 1168–1172.

Hannink, N.K., M. Subramanian et al. 2007. Enhanced transformation of tnt by tobacco plants expressing a bacterial nitroreductase. *International Journal of Phytoremediation* 9: 385–401.

Higuchi, K., K. Suzuki, H. Nakanishi, H. Yamaguchi, N.K. Nishizawa, and S. Mori. 1999. Cloning of nicotianamine synthase genes, novel genes involved in the synthesis of phytosiderophores. *Plant Physiology* 119: 471–479.

Iimura, Y., S. Ikeda et al. 2002. Expression of a gene for Mn-peroxidase from Coriolus versicolor in transgenic tobacco generates potential tools for phytoremediation. *Applied Microbiology and Biotechnology* 59: 246–251.

Inui, H., and H. Ohkawa. 2005. Herbicide resistance in transgenic plants with mammalian P450 monooxygenase genes. *Pest Management Science* 61: 286–291.

Jackson, R.G., E.L. Rylott et al. 2007. Exploring the biochemical properties and remediation applications of the unusual explosive-degrading P450 system XplA/B. *Proceedings of the National Academy of Sciences USA* 104: 16822–16827.

James, C.A., G. Xin et al. 2008. Degradation of low molecular weight volatile organic compounds by plants genetically modified with mammalian cytochrome P450 2E1. *Environmental Science and Technology* 42: 289–293.

Karavangeli, M., N.E. Labrou et al. 2005. Development of transgenic tobacco plants overexpressing maize glutathione S-transferase I for chloroacetanilide herbicides phytoremediation. *Biomolecular Engineering* 22: 121–128.

Kawahigashi, H., S. Hirose et al. 2005. Transgenic rice plants expressing human CYP1A1 remediate the triazine herbicides atrazine and simazine. *Journal of Agriculture and Food Chemistry* 53: 8557–8564.

Kawahigashi, H., S. Hirose et al. 2006. Phytoremediation of the herbicides atrazine and metolachlor by transgenic rice plants expressing human CYP1A1, CYP2B6, and CYP2C19. *Journal of Agriculture and Food Chemistry* 54(8): 2985–2991.

Kawahigashi, H., S. Hirose et al. 2007. Herbicide resistance of transgenic rice plants expressing human CYP1A1. *Biotechnology Advances* 25: 75–84.

Kawahigashi, H., S. Hirose, et al. 2008. Transgenic rice plants expressing human p450 genes involved in xenobiotic metabolism for phytoremediation. *Journal of Molecular Microbiology and Biotechnology* 15: 212–219.

Kayser, A., and H.-R. Felix. 1998. Five years of phytoremediation in the field. In H. Timmis, ed. *Book of Ext Abstr Int Workshop–Innovative Potential of Advanced Biological Systems for Remediation*, 81–86. Technical University Hamburg-Harburg, Germany.

Kotrba, P., T. Macek, and T. Ruml. 1999. Heavy metal-binding peptides and proteins in plants, a review. *Collection of Czech Chemical Communications* 64: 1057–1086.

Kotrba, P., J. Najmanova, T. Macek, T. Ruml, and M. Mackova. 2009. Genetically modified plants in phytoremediation of heavy metal and metalloid soil and sediment pollution. *Biotechnology Advances* 27: 799–810.

Kotrba P., M. Mackova, and T. Macek. 2011. Transgenic approaches to improve phytoremediation of heavy metal polluted soils, str. 409–438. In Khan, M.S., Zaidi, A., Goel, R., Musarrat, J., eds. *Biomanagement of Metal-Contaminated Soils*, ISBN 978-94-007-1913-2, Springer, Dordrecht.

Krämer, U. 2005. Phytoremediation: novel approaches to cleaning up polluted soils. *Current Opinion in Biotechnology* 16: 133–141.

Kraemer, U., and A. Chardonnens. 2001. The use of transgenic plants in the bioremediation of soils contaminated with trace elements. *Applied Microbiology and Biotechnology* 55: 661–672.

Krämer, U., I. Talke, and M. Hanikenne. 2007. Transition metal transport. *FEBS Letters* 581: 2263–2272.

Linacre, N.A., S.N. Whiting, A.J.M. Baker, S. Angle, and P.K. Ades. 2003. Transgenics and phytoremediation: the need for an integrated risk assessment, management, and communication strategy. *International Journal of Phytoremediation* 3: 181–185.

Li, Y., O.P. Dhankher, L. Carreira, D. Lee, A. Chen, J.I. Schroeder, R.S. Balish, and R.B. Meagher. 2004. Overexpression of phytochelatin synthase in Arabidopsis leads to enhanced arsenic tolerance and cadmium hypersensitivity. *Plant Cell Physiology* 45: 1787–1797.

Macek, T., and M. Mackova. 1999. Phytoremediation—the use of plants to remove xenobiotics and pollutants from the environment, including transgenic plants tailored for this purpose. *Biologia* 54: 70–73.

Macek, T., M. Mackova, and J. Kas. 2000. Exploitation of plants for the removal of organics in environmental remediation. *Biotechnology Advances* 18: 23–35.

Macek, T., M. Mackova, D. Pavlikova, J. Szakova J., M. Truksa, A. Singh-Cundy, P. Kotrba, N. Yancey, and W.H. Scouten. 2002. Accumulation of cadmium by transgenic tobacco. *Acta Biotechnologica* 22: 101–106.

Macek, T., P. Kotrba et al. 2008. Novel roles for genetically modified plants in environmental protection. *Trends in Biotechnology* 26: 146–152.

Meagher, R.B. 2000. Phytoremediation of toxic elemental and organic pollutants. *Current. Opinion in Plant Biology* 3: 153–162.

Mena-Benitez, G.L., F. Gandia-Herrero et al. 2008. Engineering a catabolic pathway in plants for the degradation of 1,2-dichloroethane. *Plant Physiology* 147: 1192–1198.

McIntyre, T. 2003. Phytoremediation of heavy metals from soils. *Advances in Biocheical Engineering and Biotechnology* 78: 97–123.

Milner, M.J., and L.V. Kochian. 2008. Investigating heavy-metal hyperaccumulation using *Thlaspi caerulescens* as a model system. *Annals of Botany* 102: 3–13.

Mohammadi, M., V. Chalavi et al. 2007. Expression of bacterial biphenyl-chlorobiphenyl dioxygenase genes in tobacco plants. *Biotechnology and Bioengineering* 97: 496–505.

Muthukumar, B., B. Yakubov, and D.E. Salt. 2007. Transcriptional activation and localization of expression of Brassica juncea putative metal transport protein BjMTP1. *BMC Plant Biology* 7: 32.

Nedelkoska, T.V., and P.M. Doran. 2000. Hyperaccumulation of cadmium by hairy roots of *Thlaspi caerulescens*. *Biotechnology and Bioengineering* 67: 607–615.

Newman, L.A., and C.M. Reynolds. 2005. Bacteria and phytoremediation: new uses for endophytic bacteria in plants. *Trends in Biotechnology* 23: 6–8.

Novakova, M., M. Mackova et al. 2009. Cloning the bacterial bphC gene into Nicotiana tabacum to improve the efficiency of PCB phytoremediation. *Biotechnology and Bioengineering* 102: 29–37.

Novakova, M., M. Mackova et al. 2007. Preparation of transgenic plants with todC1C2 genes—Cloning of bacterial todC1C2 genes into plants to improve their phytoremediation abilities. *Listy Cukrovarnicke a Reparske* 123: 323–324.

Oller, A.L.W., E. Agostini, M.A. Talano, C. Capozucca, S.R. Milrad, H.A. Tigier et al. 2005. Overexpression of a basic peroxidase in transgenic tomato (*Lycopersicon esculentum* Mill. cv. Pera) hairy roots increases phytoremediation of phenol. *Plant Science* 169: 1102–1111.

Pavlikova, D., T. Macek, M. Mackova, M. Sura, J. Szakova, and P. Tlustos. 2004. The evaluation of cadmium, zinc, and nickel accumulation ability of transgenic tobacco bearing different transgenes. *Plant, Soil and Environment* 50: 513–517.

Pilon-Smits, E., Y.L. Zhu, T. Sears, and N. Terry. 2000. Overexpression of glutathione reductase in *Brassica juncea*: effects on cadmium accumulation and tolerance. *Physiologia Plantarum* 110: 455–460.

Pilon-Smith, E., and M. Pilon. 2002. Phytoremediation of metals using transgenic plants. *Critical Review in Plant Science* 21: 439–456.

Ramos, J.L., M.M. González-Pérez et al. 2005. Bioremediation of polynitrated aromatic compounds: plants and microbes put up a fight. *Current Opinion in Biotechnology* 16: 275–281.

Robertson, B.K., and P.K. Jjemba. 2005. Enhanced bioavailability of sorbed 2,4,6-trinitrotoluene (TNT) by a bacterial consortium. *Chemosphere* 58: 263–270.

Rugh, C.L., J.F. Senecoff, R.B. Meagher, and S.A. Merkle. 1990. Development of transgenic yellow poplar for mercury phytoremediation. *Nature Biotechnology* 16: 925–928.

Rylott, E.L., and N.C. Bruce. 2009. Plants disarm soil: engineering plants for the phytoremediation of explosives. *Trends in Biotechnology* 27: 73–81.

Rylott, E.L., R.G. Jackson et al. 2006. An explosive-degrading cytochrome P450 activity and its targeted application for the phytoremediation of RDX. *Nature Biotechnology* 24: 216–219.

Shimizu, M., T. Kimura et al. 2002. Molecular breeding of transgenic rice plants expressing a bacterial chlorocatechol dioxygenase gene. *Applied Environmental Microbiology* 68: 4061–4066.

Schroder, P., D. Daubner et al. 2008. Phytoremediation of organic xenobiotics—glutathione dependent detoxification in *Phragmites* plants from European treatment sites. *Bioresource Technology* 99: 7183–7191.

Smykalova, I., M. Vrbova, E. Tejklova, M. Vetrovcova, and M. Griga. 2010. Large scale screening of heavy metal tolerance in flax/linseed (*Linum usitatissimum* L.) tested *in vitro*. *Industrial Crops Production* 32: 527–533.

Schnoor, J.L., L.A. Licht, S.C. McCutcheon, N.L. Wolfe, and L.H. Carreira. 1995. Phytoremediation of organic contaminants. *Environmental Science and Technology* 29: 318–323.

Sonoki, T., S. Kajita et al. 2005. Transgenic tobacco expressing fungal laccase promotes the detoxification of environmental pollutants. *Applied Microbiology and Biotechnology* 67: 138–142.

Sylvestre, M., T. Macek et al. 2009. Transgenic plants to improve rhizoremediation of polychlorinated biphenyls (PCBs). *Current Opinion in Biotechnology* 20: 242–247.

Taghavi, S., T. Barac et al. 2005. Horizontal gene transfer to endogenous endophytic bacteria from poplar improves phytoremediation of toluene. *Applied and Environmental Microbiology* 71: 8500–8505.

Uchida, E., T. Ouchi et al. 2005. Secretion of bacterial xenobiotic-degrading enzymes from transgenic plants by an apoplastic expressional system: an applicability for phytoremediation. *Environmental Science and Technology* 39: 7671–7677.

Van Aken, B. 2009. Transgenic plants for enhanced phytoremediation of toxic explosives. *Current Opinion in Biotechnology* 20: 231–236.

Van Dillewijn, P., J.L. Couselo et al. 2008. Bioremediation of 2,4,6-trinitrotoluene by bacterial nitroreductase expressing transgenic aspen. *Environmental Science and Technology* 42: 7405–7410.

Wang, L., D.A. Samac et al. 2005. Biodegradation of atrazine in transgenic plants expressing a modified bacterial atrazine chlorohydrolase (atzA) gene. *Plant Biotechnology Journal* 3: 475–486.

Wang, X., N. Wu et al. 2008. Phytodegradation of organophosphorus compounds by transgenic plants expressing a bacterial organophosphorus hydrolase. *Biochemical and Biophysical Research Communications* 365: 453–458.

Section V

Plants' Contaminants Tolerance

24 Utilization of Different Aspects Associated with Cadmium Tolerance in Plants to Compare Sensitive and Bioindicator Species

Marisol Castrillo, Beatriz Pernia, Andrea De Sousa, and Rosa Reyes

CONTENTS

24.1 INTRODUCTION

Heavy metal contamination is an environmental problem and is currently evaluated in terms of its health risks for human beings (Dal Corso et al. 2010). Cadmium (Cd) is a toxic heavy metal found in soil, sediments, air, and water. Volcanic eruptions, forest burn-off, and wind transportation of particles from the soil are some of the natural sources of cadmium. However, in the last few years, Cd contamination has increased as a consequence of industrial activities such as mining, metal smelting, electroplating, use and purification of Cd, fossil fuel burn-off, use of phosphatised fertilizers (Cupit et al. 2002), and manufacture of batteries, pigments and plastics (Popova et al. 2009). Cadmium's high toxicity and solubility with water has been widely recognized. It can form chemical compounds in combination with other elements such as chlorides, oxides, and sulfides, which bond with soil particles, and remain in the soil for many years. The relevance of Cd contamination and its consequences increases in the face of its ecological, nutritional, and environmental implications. Stress caused by heavy metals is one of the major problems affecting plant agricultural productivity. On the other hand, natural flora shows some relative differences in its tolerance capacity to heavy metals. Some plants can grow in a soil enriched with toxic levels of heavy metals, while others cannot (Yadav 2010). Several effects of cadmium exposure in plants, such as chlorosis, growth reduction, brown coloration of the root apex (Burzynski and Klobus 2004; Wojcik et al. 2005), chlorophyll content and photosynthetic rate reduction (Haag-Kewer et al. 1999; Burzynski and Klobus 2004; Mobin and Khan 2007; Tukaj et al. 2007; Monteiro et al. 2009) have been reported. Decrease in ribulose 1,5–bis-P regeneration, photosystem II quantic efficiency (Pankovic et al. 2000), and Rubisco activity (DiCagno et al. 2001) can be found among the photosynthetic parameters affected by cadmium exposure, as well as transpiration reduction (Haag-Kewer et al. 1999), and polymorphism induction in DNA (Aina et al. 2004; Liu et al. 2005; Aina et al. 2007). A relevant aspect of cadmium's toxicity is its chemical similarity to essential elements such as Zn, Fe y Ca, producing irregularities in their homeostasis or causing their displacement from proteins (Verbruggen et al. 2009).

Cadmium toxicity raises oxidative stress by increasing the levels of reactive oxygen species (ROS) (Smirnoff 1995; Romero-Puertas et al. 2004). These species can damage the cells through lipid peroxidation, and DNA and protein oxidation, and also initiate responses, such as new gene expression and nucleic acid damage (Di Cagno et al. 2001; Gill and Tuteja 2010). ROS have been proposed as oxidative signalling, associated with an environmental sensorial function, for the induction of the appropriate adjustments in genetic expression, metabolism and physiology (Foyer and Noctor 2005). The signalling routes, responsible for the sensors and the transduction of the metal signal inside the cell, control the activation of the transcription factors and the gene expression, preparing plants to counteract against the stress caused by heavy metals (Dal Corso et al. 2010). Lipid peroxidation, considered the most harmful process that living organisms can experience, decreases the membrane's fluidity, raising cell leakage. Another effect of Cd treatment is the significant increase of lipid peroxide accumulation in different plant species.

24.1.1 Defining Biomarkers, Tolerance, and Sensitivity

24.1.1.1 Biomarkers

Biomarkers are a measurement for biochemical, cellular, physiological, genetic, morphological, and structural parameters that indicate evidence of pollution exposure (Markert et al. 2003). They offer biologically relevant information on the potential impact that toxic pollutants can have on the health of living organisms (van der Oost et al. 1996). In addition, biomarkers can be used as early signs of stress (Vangronsveld et al. 1998). Increase in lipid peroxidation, variations in the chlorophyll/carotenoid ratio, increase in jasmonic acid, nicotianamine, thiols, and reduced glutathione (GSH) concentration, appearance of chelating peptides and phytochelatins, and increase of antioxidant enzyme activities or inhibition are among the signs considered as biomarkers (Keltjens and van Beusichem 1998; Pernia et al. 2008). In the case of exposure to heavy metals, some plant alterations at the physiological level can be considered biomarkers too.

24.1.1.2 Tolerance

Tolerance was defined as the "desired resistance of an organism or community to unfavourable abiotic (climate, radiation, pollutants) or biotic factors (parasites, pathogens), when adaptive physiological changes (e.g., enzyme induction, immune response) can be observed" (Markert et al. 2003). Nonetheless, we suggest a variation to this approach: Tolerance is defined by the different responses of adaptability (genetic, biochemical, biophysical, physiological, structural, ultra-structural and morphological), that organisms put into action to cope with biotic and abiotic stress factors.

24.1.1.3 Sensitivity

Sensitivity was defined as follows: "Sensitivity of an organism or a community means its susceptibility to biotic or abiotic change" (Markert et al. 2003). We, on our part, suggest the following approach: sensitivity in an organism induces responses to the damage caused by an environmental stress factor. This damage could be totally or partially reversible, or irreversible and unsustainable in time.

24.2 ASPECTS ASSOCIATED WITH CADMIUM TOLERANCE IN PLANTS

24.2.1 Physiological and Biochemical Characteristics of Biomarkers Significant for Plant Cadmium Tolerance

24.2.1.1 Metal Homeostasis in Plants

Metal homeostasis has been defined as the "property of an organism that regulates its internal metal environment (cellular and organellar metal concentration) so as to maintain a stable and constant condition" (Hassan and Aarts 2011). In extreme conditions of deficiency or high concentrations, plant metal homeostasis helps their survival. It includes processes of mobility, chelation and sequestration. The mobility and absorption of the metal from the soil, compartmentalization and sequestration inside the root, the load and transportation in the xylem, the transfer between the source of metal and the aerial parts of the plant and its sequestration in the cells are processes that have an influence in the homeostasis of a metallic element (Clemens et al. 2002). Evidence of a higher absorption of cadmium, associated with transportation between the roots and the shoot, have been reported (Li et al. 2011).

24.2.1.2 Cadmium Absorption through Roots

Metal accumulation depends on several processes: soil metal mobility, penetration through the roots, metal compartmentalization and sequestration inside the root cells, transport efficiency through the xylem, metal distribution in the aerial parts, and finally, sequestration and storage in the leaf cells (Hassan and Aarts 2011). Several families have been reported in regard to the transporters: the CPx-ATPasas (Solioz and Vulpe 1996; Williams et al. 2000); the protein antiport cation/H^+ family (Maser et al. 2001); the NRAMP (Natural Resistance Associated Macrophage Protein) (Curie et al. 2000; Rogers et al. 2000; Thomine et al. 2000; William et al. 2000; Nevo et al. 2006); and the ZIP (Zinc-regulated transporter, Iron-regulated transporter Proteins) (Grotz et al. 1998; Pence et al. 2000; Leitenmaier et al. 2011). Generally, cadmium is predominantly kept in the roots (Conn and Gilliham 2010; Ramos et al. 2002; Souza and Rauser 2003).

24.2.1.3 Cadmium Mobility from the Root to the Shoot

Metals penetrating via symplast are capable of reaching the xylem; however, metal sequestration inside the root cells, or symplastic transportation to the stele and liberation inside the xylem, can occur. Once the metals have entered the root cells, they become less toxic by joining the phytochelatins and forming the PC-metal-S^{+2} complex. This complex is subsequently transported to the vacuoles through a transporter of the ABC family, causing the sequestration and transportation of Cd in the plants (Bovet et al. 2005). The increase of the metal load in the xylem and its transportation to the shoot is a physiological key for metal detoxification and storage. The gene responsible for the increase of the metal load in the xylem is HMA4, which encodes P_{1B}-type ATPase located in the exterior of the plasmatic membrane, expressing itself at the level of stele and root and conducting the metal to the exterior of the cell (Hassan and Aarts 2011). Cd movement from the root to the shoot occurs through the xylem, conducted by leaf transpiration (Hart et al. 1998; Wojcik et al. 2006). Mobility via xylem and phloem has been reported in wheat (Herren and Feller 1997). On the other hand, high rates of translocation from the root to the aerial organs have been reported in

Sedum alfredii (hyperaccumulator specie), possibly due to increase of xylem load (Lu et al. 2008). Once in the xylem, metals form chelates with organic acids (Senden et al. 1995). Citrate content in the shoot is correlated with the Cd translocation from the roots to the shoot, suggesting its role in Cd transportation through the xylem vessels (Zorrig et al. 2010).

Complexes are carried to the leaves where they penetrate the cells via symplast through transporters, and are distributed through the leaves via symplast or apoplast (Clemens et al. 2002). In whole rice plants, it has been reported that nodes are the central places where Cd transference occurs, from the xylem to the phloem, playing a paramount function in the route from the soil to the grains at filling stage (Fujimaki et al. 2010). Also, there has been reported a significant Cd-induced increase of nicotianamine (NA) in the phloem fluid of *Brassica napus* plants (Mendoza-Cozalt et al. 2008). High levels of Cd and Phytochelatins (PCs) in the phloem fluid are comparable with those occurring in the xylem fluid, suggesting a transport of Cd also through the phloem. Likewise, significant mobility of Cd in the phloem has been reported in potato cultivars (Dunbar et al. 2003). On the other hand, high [PCs]/[Cd] and [glutathione]/[Cd] quotients in the phloem fluid have been reported to be indicative of PCs and glutathione (GHS) capability of transporting Cd to long distances. Cd is transported as complexes PC-Cd y GSH-Cd. In xylem fluid, only traces of PC have been detected (Mendoza-Cozalt et al. 2008).

24.2.1.4 Defense and Detoxification Mechanisms through Chelation and Sequestration

Antioxidant enzymes—Antioxidant metabolism is an important aspect of the defense system. Levels of ROS are eliminated by enzymatic antioxidants (Apel and Hirt 2004; Gill and Tuteja 2010; Pernia et al. 2008; Smirnoff 1995), such as superoxide dismutase enzymes (SOD); ascorbate peroxidase (APX); catalase (CAT); glutathione reductase (GR); glutathione transferase (GST); monodehydroascorbate reductase (MDHAR); dehydroascorbate reductase (DHAR); guaiacol peroxidase (GPX). In plants of different species, under cadmium-induced stress, SOD activity has been detected after exposure to the element. Moreover, decrease and increase of CAT has been reported in plants of different species exposed to Cd, and increases in APX and GR in plants of some species.

Ascorbic Acid—Ascorbic acid (ASH) is the most abundant, powerful, water-soluble antioxidant capable of lowering the damage caused by ROS. The ASH cycle is integrated through its oxidized forms monodehydroascorbate (MDHA) and dehydroascorbate (DHA). DHA is reduced to ASH by GSH (Apel and Hirt 2004; Di Cagno et al. 2001; Gill and Tuteja 2010; Smirnof 1995).

Glutathione—Tripeptide glutathione (GSH, γ-Glu-Cys-Gly) GSH is one of the crucial metabolites in plants, and is considered to be the most important intercellular defense against oxidative ROS-induced damages. It performs a key function in physiological processes such as sulfate transportation, detoxification, and expression of genes responding to stress (Xiang et al. 2001). It is the main ROS eliminator through the ascorbate-glutathione cycle (ASH-GSH) (Foyer and Halliwell 1976; Smirnof 1995; Apel and Hirt 2004). As stress increases and lingers, concentrations of GSH can decrease and the redox state moves toward oxidation, causing a state of deterioration in the system (Tausz et al. 2004; Verbruggen et al. 2009). GHS is precursor for the synthesis phytochelatins (PCs) (Yadav 2010) which have a function in the control of heavy metals. High concentration of GSH is correlated to plants ability to endure with oxidative stress induced by metals (Gill and Tuteja 2010). Due to its diverse crucial functions in oxidative metabolism it is considered one of the main biomarkers (Pernia et al. 2008).

Phytochelatins—One of the main heavy metal-binding ligands in plants is the phytochelatin group (PCs), a family of peptides, rich in cysteine. They synthesize from GSH (Steffens 1990; Kneer and Zenk 1992; Zenk 1996; Cobbett 2000) in a transpeptide reaction, catalyzed by the phytochelatin synthase enzyme (Cobbett and Goldsbrough 2002; Verbruggen et al. 2009; Yadav 2010). Its vital function in heavy metal detoxification has been reported for ionic homeostasis maintenance (Cobbett 2000). Phytochelatins form complexes with metallic ions and sequestrate them at the vacuole (Cobbett and Goldsbrough 2002). They are also considered one of the main biomarkers (Pernia et al. 2008).

Proline—It is a powerful antioxidant and potential inhibitor of cellular programmed death. Besides its functions in osmoprotection, it also stabilizes proteins, chelates metals and inhibits lipid peroxidation (Gill and Tuteja 2010; Szabados and Savoure 2010).

Oxalate crystals—They can participate in heavy metal detoxification (Franceschi and Nakata 2005), particularly from cadmium. Cd can be incorporated inside the oxalate crystals (Mazen and El Maghraby 1998). Sequestration's main location is the epidermis (Ma et al. 2005) and the trichomes (Choi et al. 2001; Küpers et al. 2000). Some detoxification–sequestration mechanisms in the aerial organs of hyperaccumulator plants consist in the formation of complexes with heavy metal ligands, and/or their removal from the cytosol toward the vacuoles and cellular walls (Rascio and Navari-Izzo 2010). In *Lactuca sativa* plants exposed to Cd, it has been reported that the greater percentage present in leaves was of Cd bonded to the cellular wall (Ramos et al. 2002).

24.3 THE CONCEPT OF BIOINDICATOR: DIFFERENT REPORTED INTERPRETATIONS

The bioindicator has been outlined as "A bioindicator is an organism (or part of an organism or a community of organisms) that contains information on the quality of the environment" (Markert et al. 2003). A bioindicator should have taxonomic identity, wide distribution, known ecology, abundance, experimentation simplicity, responses to environmental stress factors (Markert et al. 2003). Bioindicator species are used as signalling agents for environmental quality, and to detect the presence, concentration or effect of a polluting agent. Presence or absence of these agents can be determined by measurable and analyzable visual symptoms. Several vegetable species have been used as bioindicators, and tests have been developed to evaluate the toxicity of environmental pollutants in vegetable organisms. Bioindication must consider ecological differentiation and the selection of resistant ecotypes as natural, long-term processes (Ernst 2003). Bioindication is the analysis of living systems' informative structure, from simple organisms to complex ecosystems. To define environmental quality or environmental risk, bioindicators' main function is to determine general physiological effects, and not the measures of environmental concentrations of the stress-causing agents (Fränzle 2003, 41). When using a bioindicator, it is important to establish clearly the environmental stress factor it is sensitive to. For example, plant sensitivity to heavy metals depends on the combination of physiological, biochemical, genetic, morphological, anatomical, and phenological characteristics (Mulder and Breuret 2003).

Bioindicators are defined in several different ways, some definitions are even mutually exclusive. This term can be defined with relative freedom, which is why it is recommended that the terms and characteristics of the indicator used in each particular study be established with clarity (Heink and Kowarik 2010). It has been pointed out that plants with growth sensitivity to metal exposure, and plants with the capacity to accumulate metal, are potential bioindicators (Mhatre and Pankhurst 1997). In this regard, we have encountered the use of the term "bioindicator" (it was also pointed out by Markert et al. 2003) with different meanings: (a) There is one meaning referring to species of plants that show symptoms of cadmium exposure, for example, bean seedlings (*Phaseolus aureus* Roxb.). They have been proposed as bioindicators of soil and water cadmium pollution. As their most sensitive parameters, root growth reduction, leaf pigment reduction, and particularly, putrescine accumulation in the roots before growth reduction, have been reported (Geuns et al. 1997). In barley seedlings, used as bioindicators of cadmium pollution, root growth inhibition, soluble protein content reduction, and DNA polymorphism at the root apex have been observed (Liu et al. 2005). Responses offered by cadmium sensitive specie *Arabis alpine*, after exposure to 10 μM $CdCl_2$, also lead to the conclusion of its bioindicator character (Bovet et al. 2006). (b) Another meaning involves bioaccumulator species with great tolerance to pollution. They are defined in the following terms: "bioindicator plants, as they resist certain complexes, being capable of absorbing and accumulating them in measurable quantities" (Capó 2002). Plant species located in a polluted area and tolerant to the pollution agent, as metal accumulation bioindicator moss, fall into this category (Pesch and Schroeder 2006; Siebert et al. 1996). Also in this category are metal hyperaccumulator plants. *Thlaspi. caerulescens*, *Thlaspi praecox*, *Arabidopsis halleri*, *Sedum alfredii*, and *Solanum nigrum* have been described as Cd hyperaccumulator (Rascio and Navari-Izzo 2010). (c) Finally, a third meaning attributes bioindication to certain parameters that are capable of indicating pollution, such as pollen quality (in terms of absorptivity and viability) (Calzoni et al. 2007). *Robinia pseudoacacia* bark appears to be a better bioindicator for long-term accumulator pollution, while its leaves are better bioindicators of short-term accumulator tendency (Samecka-Cymerman 2009). Pollutants such as Fe, Cu, Zn, Pb, Cr, Cd and Al, have been reported in *Pinus massoniana* needles (Kuang et al. 2007). Lichen (*Hypogymnia physodes* L.) was used as bioindicator for environmental pollution in several national parks in Poland (Sawicka-Kapusta and Rakowska 1993).

Departing from the definitions suggested in the present work, and considering the different meanings of the term 'Bioindicator,' we present the following rating of species, based on results and conclusions reported. [Species rating, according to the results and conclusions obtained, organised in terms of their reported response capacity, based on absorption, mobility, defence and detoxification metabolic mechanisms, and plant cadmium chelation and sequestration as forms of cadmium homeostasis management.] (Tables 24.1–24.5).

TABLE 24.1
Plant Species Reported as Hyperaccumulator and Tolerant Bioindicator

Plant Species	References
Thlaspi caerulescens	Brown et al. (1995), Lombi et al. (2000), Wojcik et al. (2005)
Arabidopsis halleri	Kupper et al. (2000)
Sedum alfredii	Yang et al. (2004)
T. praecox	Vogel et al. (2005)
Solanum nigrum	Sun et al. (2006)

TABLE 24.2
Plant Species Reported as Hyperaccumulators

Plant Species	References
Lonicera japonica	Liu et al. (2009) *potential*
Bidens pilosa	Sun et al. (2009) *potential*
Picris divaricata	Ying et al. (2010)
Solanum photeinocarpum	Zhang et al. (2011)
Phytolacca Americana	Zhao et al. (2011)

TABLE 24.3
Some Plant Species Reported as Cadmium Accumulator-Tolerant, Considered as Bioindicators

Plant Species	References
Helianthus annuus L.	Gallegos et al. (1996)
Phragmites australis and *Typha latifolia*	Fediuc and Erdei (2002)
Phragmites australis	Ali et al. (2004)
Allium schoenoprasum L.	Barazani et al. (2004)
Salsola kali	De la Rosa et al. (2004)
Sesuvium portulacastrum (more tolerant; *Mesembryanthemum crystallinum*, more sensitive)	Ghnaya et al. (2005)
Prosopis juliflora	Senthilkumar et al. (2005)
Brassica juncea (L.) Czern (more accumulator)	Gisber et al. (2006)
B. carinata A. Br.	
B. oleracea L.	
Oryza sativa	Aina et al. (2007)
Arabis paniculata	Zeng et al. (2009)
Atriplex halimus subsp. *Schweinfurthii*	Nedjimi and Daoud (2009)
Solanum lycopersicum L. cv (acclimation to 20 μM Cd for a long period)	Hediji et al. (2010)
Zea mays and *Hordeum vulgare* (accumulated Cd in roots)	Puerta-Mejias et al. (2010)
Arthrocnemum macrostachyum	Redondo-Gomez et al. (2010)
Saccharum spp. (sugarcane)	Sereno et al. (2010)
Groenlandia densa	Yılmaz and Parlak (2010)
Vetiveria zizanioides	Aibibu et al. (2010)
Calotropis procera	D'Souza et al. (2010)
Amaranthus hybridus L.	Zhang et al. (2010)
Lactuca sativa	Zorrig et al. (2010)

TABLE 24.4
Plant Species Reported with Different Cadmium Accumulation and Tolerance Capacity, Some Suggested as Bioindicators

Plant Species	References
Phaseolus vulgaris L.	Fuhrer (1982), Hardiman et al. (1984), Chaoui et al. (1997), Smeets et al. (2005)
Phaseolus aureus Roxb (recommended as bioindicator)	Geuns et al. (1997)
Typha domingensis	Mendelssohn et al. (2001)
Spartina alterniflora	
Arabis alpine (did not show hyper-accumulation at 10 μM Cd; qualified as a bioindicator species)	Bovet et al. (2006)
Hordeum vulgare (recommended as bioindicator. Cd indices changes in DNA)	Liu et al. (2005)
Triticum aestivum L. cv. Albimonte	Ranieri et al. (2005)
Helianthus annuus L.	Niu et al. (2007)
Brassica juncea L.	
Medicago sativa L.	
Ricinus communis (sunflower showed higher ability for accumulation and biomass)	
Pisum sativum L.	Agrawal and Mishra (2008)
Helianthus annuus L. (considered highly tolerant, sensitive to Cd 0.5–1 mM in initial states)	Groppa et al. (2008)

TABLE 24.5
Plant Species Reported as Sensitive to Cadmium Exposure

Plant species	References
Pisum sativum L cv. Frisson (Cd short-term exposure causes chromosomic aberration to de 0.25 a 1 μM Cd. Higher concentrations produce mitosis blockade)	Fusconi et al. (2005)
Arabidopsis thaliana (Cd sensitive, non accumulator)	Bovet et al. (2005), Van de Mortel (2008)
Arachys hypogaea L. (Protein content, proline, chlorophyll, nitrate reductasa and nitrite reductase activity reduction; SOD activity increase in leaves and roots showing high sensitivity to low cadmium concentrations)	Dinakar et al. (2008)
Phaseolus vulgaris L. (Germination percentage reduction; embryo and biomass distribution growth. Inhibition of α-amilasa and invertasas activities: acid soluble, neutral soluble and bond to the cellular wall. Loss of solutes related to high MDA content and increase in lipoxygenase activity.)	Sfaxi-Bousbih et al. (2010)
Solanum torvum Sw. cv. Torubamubiga (Low Cd accumulation specie. Low Cd concentrations of Cd 0.1 μM generate oxidative stress.)	Yamaguchi et al. (2010)

24.4 CONCLUSION AND PERSPECTIVES

A wide range of plasticity in Cd tolerance can be observed, oscillating from hyperaccumulator phenotypes of tolerant plant species, to species with a high sensitivity degree, unable to tolerate low cadmium concentrations. Inside this range, there are different tolerance levels, ranging from sensitive plants (nontolerant, nonaccumulator) to plants capable of absorbing and accumulating different amounts of Cd in the roots. There, they activate absorption and transport mechanisms, and mechanisms of loading in the root's xylem. After transport to the aerial part the protection processes in the different organs occur, associated with the antioxidant mechanisms of chelation and sequestration allowing full tolerance capacity development and thus, hyperaccumulation and higher activity and effectiveness in all protection mechanisms. This leads to plant survival, with limited growth, however, due to low biomass. In relation to sensitive species it must be pointed out that some reports make reference to the relevance of the plant's age. For some species reported as sensitive to cadmium exposure, experimentation was carried out on seedlings, for example, *Phaseolus vulgaris* L., (Sfaxi-Bousbih et al. 2010); *Pisum sativum* L. cv. Frisson (Fusconi et al. 2005); *Arachys hypogaea* L. (Dinakar et al. 2008). Plant sensitivity can change with growth and development. Other species treated with cadmium were older. In 60-day-old plants of the specie *Solanum torvum* Sw. cv. Torubamubiga (Yamaguchi et al. 2010), exposed to 0.1 μM, low accumulation of cd and oxidative stress was observed. In regards to the both nonaccumulator and sensitive specie *Arabidopsis thaliana*, some ABC transporters were identified, being differentially regulated after treatment with Cd in 28 day-year-old plants (Bovet et al. 2005). In another report on 21-day-old *A. thaliana* plants, transcriptional analysis indicates specific responses to cadmium exposure (Van de Mortel 2008). In plants affected by stress due to cadmium exposure, different degrees of responses associated with metal tolerance have been observed.

Continuous study of species (invasive and agricultural) growing in natural environments, in environments with high concentrations of heavy metals, and in halophyte environments will open new ways of knowledge and comprehension of the deciding mechanisms involved in tolerance increase and Cd accumulation in plant species. This will lead to their use for phytoextraction, and soil and environment cleansing from this harmful heavy metal. The degree of tolerance in plants will be determined by the absorption of cadmium through the roots, connected to diverse sequestration and chelation mechanisms (either in the cellular walls or bonded to PCs), to the transport to the vacuoles or to aerial organs and their sequestration and chelation there, and to the increase of the antioxidant systems in cells, root tissue and leaves. Therefore, the main difference between sensitive and bioindicator plants is their higher or lower efficiency in putting these tolerance mechanisms into action.

REFERENCES

Aibibu, N., Y. Liu, G. Zeng, X. Wang, B. Chen, H. Song, and L. Xu. 2010. Cadmium accumulation in *Vetiveria zizanioides* and its effects on growth. physiological and biochemical characters. *Bioresource Technology* 101:6297–6303.

Agrawal, S.B., S. Mishra. 2008. Effects of supplemental ultraviolet-B and cadmium on growth, antioxidants and yield of Pisum sativum L. *Ecotoxicology and Environmental Safety* 72:610–618.

Bovet, L., U. Feller, and E. Martinoia.2005. Possible involvement of plant ABC transporters in cadmium detoxification: a cDNA sub-microarray approach. *Environment International.* 31:263–267.

Bovet, L., P.M. Kammer, M. Meylan-Bettex, R. Guadagnuolo, and V. Matera. 2006. Cadmium accumulation capacities of *Arabis alpine* under environmental conditions. *Environmental and Experimental Botany* 57:80–88.

Briviba, K., L.O. Klotz, and H. Sies. 1997. Toxic and signaling effects of photochemically or chemically generated singlet oxygen in biological systems. *Journal Biology Chemistry* 378:1259–1265.

Brown, S.L., R.L. Chaney, J.S. Angle, and A.J.M. Baker. 1995. Zinc and cadmium uptake by hyperaccumulator *Thlaspi caerulescens* grown in nutrient solution. *Soil Science Society American Journal* 59, 125–133.

Burzynski, M., and G. Klobus. 2004. Changes of photosynthetic parameters in cucumber leaves under Cu, Cd and Pb stress. *Photosynthetica* 42:505–510.

Calzoni, G.L., F. Antognoni, E. Pari, P. Fonti, A. Gnes, and A. Speranza. 2007. Active biomonitoring of heavy metal pollution using *Rosa rugosa* plants." *Environmental Pollution.* 149:239–245.

Capó, M. 2002. *Principios de Ecotoxicología.* Madrid: McGraw-Hill.

Chaoui, A., S. Mazhoudi, M.H. Ghorbal, and E. El Ferjani. 1997. Cadmium and zinc induction of lipid peroxidation and effects on antioxidants enzyme activities in bean (*Phaseolus vulgaris* L.). *Plant Science* 127:139–147.

Choi, Y.-E., E. Harada, M. Wada, H. Tsuboi, and Y. Morita. 2001. Detoxification of cadmium in tobacco plants: formation and active excretion of crystals containing cadmium and calcium through trichomes. *Planta* 213:45–50.

Clemens, S., M.G. Palmgren, and U. Kramer. 2002. A long way ahead: understanding and engineering plant metal accumulation. *Trends in Plant Science* 7:309–315.

Cobbett, C.S. 2000. Phytochelatins and their Role in Heavy Metal. *Plant Physiology* 123:825–832.

Cobbett, C., and P. Goldsbrough. 2002. Phytochelatins and metallothioneins: roles in heavy metal detoxification and homeostasis. *Annual Review of Plant Biology* 53:159–182.

Conn, S., and M. Gilliham. 2010. Comparative physiology of elemental distributions in plants *Annals of Botany.* 105:1081–1102.

Cupit, M., O. Larsson, C. de meeûs, G.H. Eduljee, and M. Hutton. 2002. Assessment and management of risks arising from exposure to cadmium in fertilizers II. *Science of the Total Environment* 291:189–206.

Curie, C., J.M. Alonso, M.L.E. Jean, J.R. Ecker, and J.F. Briat. 2000. Involvement of NRAMP1 from *Arabidopsis thaliana* in iron transport. *Biochemical Journal* 347:749–755.

Cuypers, A., K. Smeets, J. Ruytinx, K. Opdenakker, E. Keunen, T. Remans, N. Horemans, N. Vanhoudt, S. Van Sanden, F. Van Belleghem, Y. Guisez, J. Colpaert, and J. Vangronsveld. 2010. The cellular redox state as a modulator in cadmium and copper responses in *Arabidopsis thaliana* seedlings. *Journal of Plant Physiology* 168(4):309–316.

Dal Corso, G., S. Farinati, and A. Furini. 2010. Regulatory networks of cadmium stress in plants. *Plant Signaling and Behaviour* 5:663–667.

De la Rosa, G., J.R. Peralta-Videa, M. Montes, J.G. Parsons, I. Cano-Aguilera, and J.L. Gardea-Torresdey. 2004. Cadmium uptake and translocation in tumbleweed (*Salsola kali*), a potential Cd-hyperaccumulator desert plant species: ICP/OES and XAS studies. *Chemosphere* 55:1159–1168.

Di Cagno, R., L. Guidi, L. De Gara. and G.F. Soldatini. 2001. Combined cadmium and ozone treatments affect photosynthesis and ascorbate-dependent defences in sunflower. *New Phytologist* 151:627–636.

Dinakar, N., P.C. Nagajyothi, S. Suresh, Y. Udaykiran, and T. Damodharam. 2008. Phytotoxycity of cadmium on protein, proline and antioxidant enzyme activities in growing. *Arachis hypogaea* L. seedlings. *Journal of Environmental Sciences* 20:199–206.

D'Souza, R.,V. Mayank, M. Jamson, and S.P. Manoj. 2010. Identification of *Calotropis procera* L. as a potential phytoaccumulator of heavy metals from contaminated soils in Urban North Central India. *Journal of Hazardous Materials* 184:457–464.

Dunbar, K.R., M.J. McLaughlin, and R.J. Reid. 2003. The uptake and partitioning of cadmium in two cultivars of potato (*Solanum tuberosum* L.). *Journal of Experimental Botany* 54:349–354.

Ernst, W.H.O. 2003. The use of higher plants as bioindicators *Trace Metals and other Contaminants in the Environment* 6. In *Bioindicators and Biomonitors: Principles, Concepts, and Applications* edited by B.A. Markert, A.M. Breure and H.G. Zechmeister, 423–463. New York: Elsevier Science.

Fediuc, E., L. Erdei. 2002. Physiological and biochemical aspects of cadmium toxicity and protective mechanisms induced in *Phragmites australis* and *Typha latifolia*. *Jourrnal of Plant Physiology* 159:265–271.

Foyer, C., and B. Halliwell. 1976. The presence of glutathione and glutathione reductase in chloroplasts: a proposed role in ascorbic acid metabolism. *Planta* 133:21–25.

Foyer, C., and G. Noctor. 2005. Oxidant and antioxidant signaling in plants: a revaluation of the concept of oxidative stress in a physiological context. *Plant, Cell and Environment* 28:1056–1071.

Franceschi, V.R., and P.A. Nakata. 2005. Calcium oxalate in plants: formation and function. *Annual Review of Plant Biology* 56:41–71.

Fränzle, O. 2003. Bioindicators and environmental stress assessment. *Trace Metals and other Contaminants in the Environment,* 6. In *Bioindicators and Biomonitors: Principles, Concepts, Applications*. Edited by B.A. Markert, A.M. Breure, H.G. Zechmeister. 41–84. New York: Elsevier Science.

Fuhrer, J. 1982. Early effects of excess cadmium uptake in *Phaseolus vulgaris. Plant, Cell and Environment* 5:263–270.

Fujimaki, S., N. Suzui, N.S. Ishioka, N. Kawachi, M. Ito, S. Chino, and S. Nakamura. 2010. Tracing cadmium from culture to spikelet: noninvasive imaging and quantitative characterization of absorption, transport, and accumulation of cadmium in an intact rice plant. *Plant Physiology* 152:1796–1806.

Fusconi, A., O. Ombretta Repetto, E. Bona, N. Massa, C. Gallo, E. Dumas-Gaudot, and G. Berta. 2006. Effects of cadmium on meristem activity and nucleus ploidy in roots of *Pisum sativum* L. cv. Frisson seedlings. *Environmental and Experimental Botany* 58:253–260.

Gallego, S.M., M.P. Benavides, and M.L. Tomaro. 1996. Effect of heavy metal ion excess on sunflower leaves evidence for involvement of oxidative stress. *Plant Science* 121:151–155.

Geuns, J.M.C., A.J.F. Cuypers, T. Michiels, J.V. Colpaert, A. Van Laere, K.A.O. Van den Broeck, and C.H.A. Vandecasteel. 1997. Mung bean seedlings as bio-indicators for soil and water contamination by cadmium. *Science of Total Environment* 203:183–197.

Ghnaya, T., I. Nouairi, I. Slama, D. Messedia, C. Grignon, C. Abdelly, and M.H. Ghorbel. 2005. Cadmium effects on growth and mineral nutrition of two halophytes: *Sesuvium portulacastrum* and *Mesembryanthemum crystallinum. Journal of Plant Physiology* 162:1133–1140.

Gill, S.S., and N. Tuteja. 2010. Reactive oxygen species and antioxidant machinery in abiotic stress tolerance in crop plants. *Plant Physiology and Biochemistry* 48:909–930.

Gisbert, C., R. Clemente, J. Navarro-Aviño, C. Baixauli, A. Giner, R. Serrano, D.J. Walker, and M.P. Bernal. 2006. Tolerance and accumulation of heavy metals by Brassicaceae species grown in contaminated soils from Mediterranean regions of Spain. *Environmental and Experimental Botany* 56:19–27.

Groppa, M.D., M.S. Zawoznik, M.L. Tomaro, and M.P. Benavides. 2008. Inhibition of root growth and polyamine metabolism in sunflower (*Helianthus annuus*) seedlings under cadmium and copper stress. *Biology Trace Element Research* 126:246–256.

Grotz, N., T. Fox, E. Connolly, W. Park, M.L. Guerinot, and D. Eide. 1998. Identification of a family of zinc transporter genes from *Arabidopsis* that respond to zinc deficiency. *Proceedings of the National Academy of Sciences USA* 95:7220–7224.

Haag-Kerwer, A., H.J. Schäfer, S. Heiss, C. Walter, and T. Rausch. 1999. Cadmium exposure in *Brassica juncea* causes a decline in transpiration rate and leaf expansion without effect on photosynthesis. *Journal of Experimental Botany* 50:1827–1835.

Hardiman, R.T., B. Jacoby, and A. Banin. 1984. Factors affecting the distribution of cadmium, copper and lead and their effect upon yield and zinc content in bush beans (*Phaseolus vulgaris* L.). *Plant and Soil* 81:17–27.

Hart, J.J., R.M. Welch, W.A. Norvell, L.A. Sullivan, and L.V. Kochian. 1998. Characterization of cadmium binding, uptake, and translocation in intact seedlings of bread and durum wheat cultivars. *Plant Physiology* 116:1413–1420.

Hassan, Z., and M.G.M. Aarts. 2011. Opportunities and feasibilities for biotechnological improvements of Zn, Cd or Ni tolerance and accumulation in plants. *Environmental and Experimental Botany* 72:53–63.

Heink, U., and I. Kowarik, 2010. What are indicators? On the definition of indicators and environmental planning. *Ecological Indicators* 10:584–593.

Hediji, H., W. Djebali, C. Cabasson, M. Maucourt, P. Baldet, A. Bertrand, L.B. Zoghlami, C. Deborde, A. Moing, R. Brouquisse, W. Chaıbi, and P. Gallusci. 2010. Effects of long-term cadmium exposure on growth and metabolomic profile of tomato plants. *Ecotoxicology and Environmental Safety* 73(8): 1965–1974.

Henson, M.C., and P.J. Chedrese. 2004. Endocrine disruption by cadmium, a common environmental toxicant with paradoxical effects on reproduction. *Experimental Biology and Medicine* 229:383–392.

Herren, T., and U.R.S. Feller. 1997. Transport of cadmium via xylem and phloem in maturing wheat shoot: comparison with the translocation of zinc, strontium and rubidium. *Annals of Botany* 80:623–628.

Keltjens, W.J., and M.L. van Beusichem. 1998. Phytochelatins as biomarkers for heavy metal stress in maize (*Zea mays*L.) and wheat (*Triticum aestivum* L.): combined effects of copper and cadmium. *Plant and Soil* 203:119–126.

Kneer, R., and M. Zenk. 1992. Phytochelatins protect plant enzymes from heavy metal poisoning. *Phytochemistry* 31:2663–2667.

Kuang, Y.W., A.Z. Wen, G.Y. Zhou, and S.Z. Liu. 2007. Distribution of elements in needles of *Pinus massoniana* (Lamb) was uneven and affected by needle age. *Environmental Pollution* 145:146–153.

Kupper, H., E. Lombi, F.-J. Zhao, and S.P. McGrath. 2000. Cellular compartmentation of cadmium and zinc in relation to other elements in the hyperaccumulator *Arabidopsis halleri*. *Planta* 212:75–84.

Li, J. T., B. Liao, R. Zhu, Z.Y. Dai, C.Y. Lan, and W.S. Shu. 2011. Characteristic of Cd uptake, translocation and accumulation in a novel Cd-accumulating tree, star fruit (*Averrhoa carambola* L., Oxidalidaceae). *Environmental and Experimental Botany* 71:352–358.

Liu, Z., X. He, W. Chen, F. Yuan, K. Yan, and D. Tao. 2009. Accumulation and tolerance characteristics of cadmium in a potential hyperaccumulator—*Lonicera japonica* Thunb. *Journal of Hazardous Materials* 169:170–175.

Liu, W., P.J. Li, X.M. Qi, Q.X. Zhou, L. Zheng, T.H. Sun, and Y.S. Yang. 2005. DNA changes in barley (*Hordeum vulgare*) seedlings induced by cadmium pollution using RAPD analysis. *Chemosphere* 61:158–167.

Lombi, E., F.J. Zhao, S.J. Dunham, and S. P. McGrath. 2000. Cadmium accumulation in populations of *Thlaspi caerulescens* and *Thlaspi goesingense. New Phytologist* 145:11–20.

Lu, L., S. Tian, X. Yang, X. Wang, P. Brown, T. Li, and Z. He. 2008. Enhanced root-to-shoot translocation of cadmium in the hyperaccumulating ecotype of *Sedum alfredii*. *Journal of Experimental Botany* 59: 3203–3213.

Ma, J.F., D. Ueno, F.J. Zhao, and S.P. McGrath. 2005. Subcellular localisation of Cd and Zn in the leaves of a Cd-hyperaccumulating ecotype of *Thlaspi caerulescens*. *Planta* 220:731–736.

Markert, B.A., A.M. Breure, and H.G. Zechmeister. 2003. Definitions, strategies and principles for bioindication/biomonitoring of the environment. *Trace Metals and other Contaminants in the Environment.* In *Bioindicators and Biomonitors: Principles, Concepts, and Applications* edited by B.A. Markert, A.M. Breure, H.G. Zechmeister. 3–39. New York: Elsevier Science.

Mazen, A.M.A., and O.M.O. El-Maghraby. 1998. Accumulation of cadmium, lead and strontium, and a role of calcium oxalate in water hyacinth tolerance. *Biologia Plantarum* 40:411–17.

Maser, P., S. Thomine, J. Schroeder, J.M. Ward, K. Hirschi, H. Sze, I.N. Talke, A. Amtmann, F.J.M. Maathuis, D. Sanders, J.F. Harper, J. Tchieu, M. Gribskov, M.W. Persans, D.E. Salt, S.A. Kim, and M.L. Guerinot. 2001. Phylogenetic relationships within cation transporter families of *Arabidopsis*. *Plant Physiology* 126:1646–1667.

Mendelssohn, I.A., K.L. McKee, and T. Kong. 2001. A comparison of physiological indicators of sublethal cadmium stress in wetland plants. *Environmental and Experimental Botany* 46:263–275.

Mendoza-Cozalt, D.G., E. Butko, F. Springer, J.W. Torpey, E.A. Kornives, J. Kehr, and J.I. Schroeder. 2008. Identification of high levels of phytocheltins, glutathione, and cadmium in the phloem sap of *Brassica napus*. A role for thiol-peptides in the long distance transport of cadmium and the effect of cadmium on iron transport. *Plant Journal* 54:249–259.

Mhatre, G.N., and C.E. Pankhurst. 1997. Bioindicators to Detect Contamination of Soils with Special Reference to Heavy Metals. In *Biological Indicators of Soil Health* edited by C. Pankhurst, B.M. Doube, and U.V.S.R. Gupta, 349–369. Wallingford: CAB International.

Mobin, M., and N.A. Khan. 2007. Photosynthetic activity, pigment composition and antioxidative response of two mustard (*Brassica juncea*) cultivars differing in photosynthetic capacity subjected to cadmium stress. *Journal of Plant Physiology* 164:601–610.

Monteiro, M.S., C. Santos, A.M.V.M. Soares, and R.M. Mann. 2009. Assessment of biomarkers of cadmium stress in lettuce. *Ecotoxicology and Environmental Safety* 72:811–818.

Mulder, Ch., and A.M. Breure. 2003. Plant biodiversity and environmental stress. Trace metals and other contaminants in the environment. 6. In *Bioindicators and Biomonitors: Principles, Concepts, and Applications* edited by Markert, B.A. A.M. Breure, H.G. Zechmeister 501–525. New York: Elsevier Science.

Nedjimi, B., and Y. Dao. 2009. Cadmium accumulationin *Atriplex halimus* subsp. *schweinfurthii* and its influence on growth, proline, root hydraulic conductivity and nutrient uptake. *Flora* 204:316–324.

Niu, Z., L. Sun, T. Sun, Y. Li, and H. Wang. 2007. Evaluation of phytoextracting cadmium and lead by sunflower, ricinus, alfalfa and mustard in hydroponic culture. *Journal of Environmental Sciences* 19:961–967.

Pankovic, D., M. Plesnicar, I. Arsenijevic-Maksimovic, N. Patrovic, Z. Sakac, and R. Kastori. 2000. Effects of nitrogen nutrition on photosynthesis in Cd-treated sunflower plants. *Annuals of Botany* 6:841–846.

Pence, N.S., P.B. Larsen, S.D. Ebbs, D.L. Lethman, M.M. Lasat, D.F. Garvin, D. Eide, and L.V. Kochian. 2000. The molecular physiology of heavy metal transport in the Zn/Cd hyperaccumulator *Thlaspi caerulescens*. *Proceedings of the National Academy of Sciences USA* 97:4956–4960.

Pernia, B., A. De Sousa, R. Reyes, and M. Castrillo. 2008. Biomarcadores de contaminaci6n por cadmio en las plantas. *Interciencia*. 33:112–119.

Pesch, R., and W. Schroeder. 2006. Mosses as bioindicators for metal accumulation: Statistical aggregation of measurement data to exposure indices. *Ecological Indicators* 6:137–152.

Popova, L.P., L.T. Maslenkova, Y. Rusina, R.Y. Yordanova, A. Ivanova, A.P. Krantev, G. Szalai, and T. Janda. 2009. Exogenous treatment with salicylic acid attenuates cadmium toxicity in pea seedlings. *Plant Physiology and Biochemistry* 47:224–231.

Puertas-Mejias, M.A., B. Ruiz-Dıez, and M. Fernandez-Pascual. 2010. Effect of cadmium ion excess over cell structure and functioning of *Zea mays* and *Hordeum vulgare*. *Biochemical Systematics and Ecology* 38: 285–291.

Ramos, I., E. Esteban, J.J. Lucena, and A. Garate. 2002. Cadmium uptake and subcellular distribution in plants of *Lactuca* sp. Cd-Mn interaction. *Plant Science* 162:761–767.

Ranieri, A., A. Castagna, F. Scebba, M. Careri, I. Zagnoni, G. Predieri, M. Pagliari, and L. Sanita di Toppi. 2005. Oxidative stress and phytochelatin characterisation in bread wheat exposed to cadmium excess. *Plant Physiology and Biochemistry* 43:45–54.

Rascio, N., and F. Navari-Izzo. 2010. Heavy metal hyperaccumulating plants: How and why do they do it? And what makes them so interesting? *Plant Science* 180:169–181.

Redondo-Gómez, S., E. Mateos-Naranjo, and L. Andrades-Moreno. 2010. Accumulation and tolerante characteristics of cadmium in a halophytic Cd-hyperaccumulator, *Arthrocnemum macrostachyum*. *Journal of Hazardous Materials* 184: 299–307.

Rogers, E.E., D.J. Eide, and M.L. Guerinot. 2000. Altered selectivity in an *Arabidopsis* metal transporter. *Proceedings of the National Academy of Sciences USA* 97:12356–12360.

Romero-Puertas, M.C., M. Rodriguez-Serrano, F.J. Corpas, M. Gomez, L.A. Del Rio, and L.M. Sandalio. 2004. Cadmium-induced subcellular accumulator of O_2^- and H_2O_2 in pea leaves. *Plant, Cell and Environment* 27:1122–1134.

Samecka-Cymerman, A., A. Stankiewicz, K. Kolon, and A.J. Kempers. 2009. Self-organizing feature map (rural networks) as a tool to select the best indicator of road traffic pollution (soil, leaves or bark of *Robinia pseudoacacia* L.). *Environmental Pollution* 157:2061–2065.

Sawicka-Kapusta, K., and A. Rakowska. 1993. Heavy metal contamination in Polish national parks. *Science of The Total Environment* 134:161–166.

Senden, M.H.M.N., A.J.G.M. Van Der Meer, T.G. Verburger, and H.T. Wolterbeek. 1995. Citric acid in tomato plant roots and its effect on cadmium uptake and distribution. *Plant and Soil* 171:333–339.

Senthilkumar, P., W.S.P.M. Prince, S. Sivakumar, and C.V. Subbhuraam. 2005. *Prosopis juliflora*—A green solution to decontaminate heavy metal (Cu and Cd) contaminated soils. *Chemosphere* 60:1493–1496.

Sereno, M.L., R.S. Almeida, D.S. Nishimura, and A. Figueira. 2007. Response of sugarcane to increasing concentrations of copper and cadmium and expression of metallothionein genes. *Journal of Plant Physiology* 164:1499–1515.

Sfaxi-Bousbih, A., A. Chaoui, and E. El Ferjani. 2010. Cadmium impairs mineral and carbohydrate mobilization during the germination of bean seeds. *Ecotoxicology and Environmental Safety* 73:1123–1129.

Siebert, A., I. Bruns, G.J. Krauss, J. Miersch, and B. Markert. 1996. The use of the aquatic moss *Fontinalis antipyretica* L. ex Hedw. As a bioindicator for heavy metals:1. Fundamental investigations into heavy metal accumulation in *Fontinalis antipyretica* L. ex Hedw. *Science of The Total Environment* 177: 137–144.

Smeets, K., A. Cuypers, A. Lambrechts, B. Semane, P. Hoet, A. Van Laere, and J. Vangronsveld. 2005. Induction of oxidative stress and antioxidative mechanisms in *Phaseolus vulgaris* after Cd application. *Plant Physiology and Biochemistry* 43:437–444.

Smirnoff, N. 1995. Antioxidant systems and plant response to the environment In *Environment and Plant Metabolism flexibility and acclimation* edited by N. Smirnoff, 217–243. Oxford: BIOS Scientific Publication Limited.

Solioz, M., and C. Vulpe. 1996. CPx-type ATPases: a class of P-type ATPases that pump heavy metals. *Trends in Biochemical Sciences* 21:237–241.

Souza, J.F., and W. Rauser. 2003. Maize and radish sequester excess cadmium and zinc in different ways. *Plant Science* 165:1009–1022.

Steffens, J.C. 1990. The heavy metals-binding peptides of plants. *Annual Review of Plant Physiology and Plant Molecular Biology* 41:553–575.

Sun, Y., Q. Zhou, L. Wang, and W. Liu. 2009. Cadmium tolerance and accumulation characteristics of *Biden pilosa* L. as a potential hyperaccumulator. *Journal of Hazardous Materials* 161:808–814.

Sun, R., Q. Zhou, C. Jin. 2006. Cadmium accumulation in relation to organic acids in leaves of *Solanum nigrum* L. as a newly found cadmium hyperaccumulator. *Plant Soil* 285:125–134.

Szabados, L., and A. Savoure. 2010. Proline: a multifunctional amino acid. *Trends in Plant Science* 15: 89–97.

Tausz, M., H. Sircelj, and D. Grill. 2004. The glutathione system as a stress marker in plant ecophysiology: is a stress-response concept valid? *Journal of Experimental Botany* 404:1955–1962.

Thomine, S., R. Wang, J.M. Ward, N.M. Crawford, and J.I. Schroeder. 2000. Cadmium and iron transport by members of a plant metal transporter family in Arabidopsis with homology to Nramp genes. *Proceedings of the National Academy of Sciences USA* 97:4991–4996.

Tukaj, Z., A. Bascik-Remisiewicz, T. Skowronski, and C. Tukaj. 2007. Cadmium effect on the growth, photosynthesis, ultrastructure and phytochelatin content of green microalga *Scenedesmus armatus*: A study at low and elevated CO_2 concentration. *Environmental and Experimental Botany* 60:291–299.

Van de Mortel, J.E., H. Schat, P.D. Moerland, E.V.L.V. Themaat, S.V. Entm, H. Blankestijn, A. Ghandilyan, S. Tsiatsiani, and M.G.M. Aarts. 2008. Expression differences for genes involved in lignin, glutathione and sulphate metabolism in response to cadmium in *Arabidopsis thaliana* and the related Zn/Cd-hyperaccumulator *Thlaspi caerulescens*. *Plant, Cell and Environment* 31:301–324.

Verbruggen, N., C. Hermans, and H. Schat. 2009. Mechanisms to cope with arsenic or cadmium excess in plants. *Current Opinion in Plant Biology* 12:364–372.

Vogel-Mikus, K., M. Regvar, J. Mesjasz-Przybyłowicz, W.J. Przybyłowicz, J. Simcic, P. Pelicon, and M. Budnar. 2005. Spatial distribution of cadmium in leaves of metal hyperaccumulating *Thlaspi praecox* using micro-PIXE. *New Phytologist* 179:712–721.

Williams, L.E., J.K. Pittman, and J.L. Hall. 2000. Emerging mechanisms for heavy metal transport in plants. *Biochimica et Biophysica Acta* 1465:104–126.

Wojcik, M., E. Skorzynska-Polit, and A. Tukiendorf. 2006. Organic acids accumulation and antioxidant enzyme activities in *Thlaspi caerulescens* under Zn and Cd stress. *Plant Growth Regulation* 48:145–155.

Wojcik, M., J. Vangronsveld, J. D'Haen, A. Tukiendorf. 2005. Cadmium tolerance in *Thlaspi caerulescens*. II. Localization of cadmium in *Thlaspi caerulescens*. *Environmental Experimental Botany* 53:163–171.

Xiang, C., B.L. Werner, E.M. Christensen, and D.J. Oliver. 2001. The biological functions of glutathione revisited in Arabidopsis transgenic plants with altered glutathione levels. *Plant Physiology* 126:564–574.

Yadav, S.K. 2010. Heavy metals toxicity in plants: An overview on the role of glutathione and phytoxhelatins in heavy metlas stress tolerance of plants. *South African Journal of Botany* 76:167–179.

Yamaguchi. H., H. Fukuoka, T. Arao, A. Ohyama, T. Nunome, K. Miyatake, and S. Negoro. 2010. Gene expression analysis in cadmium-stressed roots of a low cadmium-accumulating solanaceous plant, *Solanum torvum*. *Journal Experimental Botany* 61:423–437.

Yang, X.E., X.X. Long, H.B. Ye, Z.L. He, D.V. Calvert, and P.J. Stoffella. 2004. Cadmium tolerance and hyperaccumulation in a new Zn hyperaccumulating plant species (*Sedum alfredii* Hance). *Plant and Soil* 259: 181–189.

Yılmaz, D.D., and K.U. Parlak. 2010. Changes in proline accumulation and antioxidative enzyme activities in *Groenlandia densa* under cadmium stress. *Ecological Indicators* 11:417–423.

Ying, R.R., R.L. Qiu, Y.T. Tang, P.J., Hu, H. Qiu, H.R. Chen, T.H. Shi, and J.L. Morel. 2010. Cadmium tolerance of carbon assimilation enzymes and chloroplast in Zn/Cd hyperaccumulator *Picris divaricata*. *Journal of Plant Physiology* 167:81–87.

Zeng, X., L. Ma, R. Qiu, and Y. Tang. 2009. Responses of non-protein thiols to Cd exposure in Cd hyperaccumulator *Arabis paniculata* Franch. *Environmental and Experimental Botany* 66:242–248.

Zenk, M.H. 1996. Heavy metal detoxification in higher plants- a review. *Gene* 179:21–30.

Zhang, X., S. Zhang, X. Xua, T. Li, G. Gong, G. Jia, Y. Li, and L. Denga. 2010. Tolerance and accumulation characteristics of cadmium in *Amaranthus hybridus* L. *Journal of Hazardous Materials* 180:303–308.

Zhang, X., H. Xia, Z. Li, P. Zhuang, and B. Gao. 2011. Identification of a new potential Cd hyperaccumulator *Solanum photeinocarpum* by soil seed bank-metal concentration gradient method. *Journal of Hazardous Materials* 189:414–419.

Zhao, L., Y.L. Sun, S.X. Cui, M. Chen, H.M. Yang, H.M. Liu, T.Y. Chai, and F. Huang. 2011. Cd-induced changes in leaf proteome of the hyperaccumulator plant *Phytolacca americana*. *Chemosphere* 85: 56–66.

Zorrig, W., A. Rouached, Z. Shahzad, C. Abdelly, J.C. Davidian, and P. Berthomieu. 2010. Identification of three relationships linking cadmium accumulation to cadmium tolerance and zinc and citrate accumulation in lettuce. *Journal of Plant Physiology* 167:1239–1247.

25 Analytical Tools for Exploring Metal Accumulation and Tolerance in Plants

Katarina Vogel-Mikuš, Iztok Arčon, Peter Kump, Primož Pelicon, Marijan Nečemer, Primož Vavpetič, Špela Koren, and Marjana Regvar

CONTENTS

25.1 INTRODUCTION

The origin of naturally occurring metals is in rocks, soils and sediments, where they are primarily trapped in stable, insoluble forms. Yet, through natural processes small amounts of metals can be mobilized and allowed to circulate in water, soil and air, through biogeochemical cycles that keep distribution of any given metal within an ecosystem at relatively constant concentrations over time. Their constant cycling is especially important for the biosphere, since certain amounts of metals are essential for organisms, but can be highly toxic when present in excess. With mining of ores, in addition to smelting and other purification practices, metals have been rapidly released from their more stable insoluble forms to less stable, soluble forms and released into the environment. So now days metal contents released into the environment by anthropogenic activities highly overwhelm natural biogeochemical cycles (Singh 2005).

Humans have been introducing trace metals into the environment since they first gained knowledge on their numerous useful properties. Recent archaeological findings have revealed that humans have been using metals since the Neolithic period. Mining and processing of metals began in around 8,000 BCE, 10,000 years ago. The first metals that were used by humans were copper and gold. This is understandable, as both are easy to see and easy to process using simple tools. Since then, humans have used various means to extract these and also other metal compounds from the earth's crust. Although the use of metals has brought numerous benefits to human society, we have had to embrace the harsh consequences of metal pollution (Singh 2005). Early on, metal pollution only affected human populations in the nearest vicinity of the source of metal mining and smelting activities; at the turn of 19th century, with the start of Industrial Revolution, however, metal pollutants were distributed over wide areas by means of air and water causing visible detrimental effects to numerous ecosystems (Singh 2005).

Increased metal concentrations present in the environment pose threat to the living organisms from microorganisms and plants to animals and humans, because they interfere with vital biological processes, including photosynthesis and respiration. By replacement of essential minerals in structural and functional organic molecules, especially enzymes, binding to free thiol and amino groups of proteins (especially Cd, Hg, Pb, Zn have high affinity for binding to free thiol or amino groups), or induction of free radical formation (Cu, Fe) (Figure 25.1), they cause malfunctioning of important biological molecules and oxidative stress.

Due to pollution and degradation of ecosystems worldwide, there was a growing need to develop powerful analytical tools for monitoring trace element concentrations in the biosphere and its abiotic environment with the common goal to be able to assess metal bioavailability, toxicity and risk relationships, and to revert to the environmentally friendly technologies, namely phytoremediation, which use suitable plant species for cleaning and remediation of metal polluted and degraded ecosystems (Thompson-Eagle and Frankenberger 1990; Wenzel et al. 1999; Lombi et al. 2000; Mulligan et al. 2001; Barceló and Poschenrieder 2003; Kidd et al. 2009).

Plants have been related to human exploitation connected with metal mining already from the 5th century BC, when a connection between vegetation and the minerals located underground was first noticed in China. There were particular plants that thrived on and indicated areas rich in copper, nickel, zinc, and possibly gold, though the latter has never been confirmed. Also in the

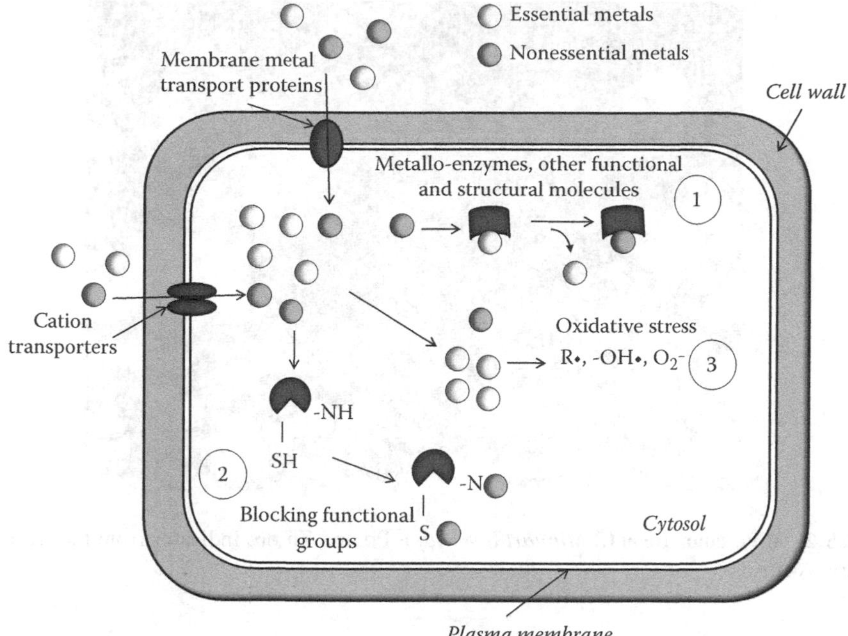

FIGURE 25.1 Schematic presentation of the main mechanisms of metal toxicity in plant cells. Non-essential metals as well as essential metals when present in excess can: 1) Replace essential metals in metallo-enzymes and other structural and functional molecules; 2) Interact with functional groups (-SH, -NH, -OH, -COOH...) of proteins and enzymes and block them; 3) Cause oxidative stress in the cells by the formation of free radicals.

West the early starts of geobotany, which depends solely on visual observation of the vegetation cover, dates back at least to Roman times (Brooks 1998). However the famous mediaeval metallurgist and miner Georgius Agricola who published his celebrated *De Re Metallica* (Of Metallic Matters) in 1556, describing vegetation cover that can be found over concealed ore bodies could be acknowledged as the world's first practical geobotanist. In the same century Thalius (1588) determined *Minuartia verna* (Figure 25.2) as an ore indicator in Harz Mountains in Germany (adapted after Brooks 1998).

In contrast to geobotany, biogeochemistry that depends on advances in analytical chemistry before it could be developed, dates back only to 1938 when the Soviet scientist S.M. Tkalich observed that iron content of tundra plants reflected the concentration of this element in the soil (adapted after Brooks 1998). After the Second World War, when rapid analytical and data handling techniques became available, geochemical and biogeochemical investigations were undertaken in many parts of the world mainly for mineral exploration (Cole 1965; Cole and Smith 1984). These have identified an increasing number of mineral indicator and accumulator plant species, and find out that some plants, so called metallophytes, are endemic to particular naturally metal enriched regions. The examples of communities of indicator plants occupying naturally toxic ground over copper, lead-zinc and nickel deposits in Africa, Australia, Europe and the USA are in detail described by Cole and Smith (1984), and Brooks (1998). Besides revealing mineralization in sub-surface bedrock, including economic ore-bodies, indicator plants and anomalous vegetation communities can also be found in areas contaminated by man's activities.

Soon it has become clear that plants occurring on sites with elevated metal concentrations present in soils have evolved peculiar physiological mechanisms which enable them to

FIGURE 25.2 **(See color insert.)** *Minuartia verna*, a Pb and Zn ore indicator from the river bed Gailitz, Arnoldstein, Austria.

survive in such hostile environments. The majority of plant species growing at the sites with elevated metal concentrations exclude metals from their tissues or retain metals in their roots (so called excluders) while a small proportion of plants known as (hyper)accumulators are known to (hyper)accumulate extremely high concentrations of different metal in their shoots (Baker 1981). Brooks and his colleagues first coined the term "hyperaccumulation" in 1977, however, the pioneer scientist dealing with hyperaccumulating plants was A. Bauman, who already in 1885 reported over 1% of zinc in dry shoots of *Viola calaminaria* and *Thlaspi calaminare*. Later, in the 1930s, O. A. Beath and his coworkers discovered hyperaccumulation of selenium in *Astragalus* plants from western USA, although horse hoofs loosing disease connected with feeding on poisonous plants (selenium accumulators) was already described in the 13th century by Marco Pollo. Finally, credit must be given to Italian scientists O. Vergano Gambi from University of Florence, and C. Minguzzi who in 1948 discovered the unusual hyperaccumulation of nickel by the Tuscan serpentine plant *Alyssum bertolonii*. Nevertheless few decades has to pass before the applicability of metal hyperaccumulators was introduced to the "green" movements in late 1980s and 1990s of the past century (adapted after Brooks 1998). Since then, the outstanding ability of particular plant species to (hyper)accumulate and tolerate extremely high concentration of metals in their shoots without being affected by the metal toxicity has drawn attention of many plant physiologists, and in the past 15 years the scientists have been challenged to reveal underlying physiological mechanisms of metal (hyper)accumulation and tolerance from organismal down to tissue, cellular and molecular levels. In depth, understanding of the uptake, transport, localization and speciation of metals in plants is critical for understanding of metal metabolic pathways within plants. This can be provided by the use of complementary analytical techniques spanning from "bulk" analyses of metal concentrations in plant organs to those enabling imaging of metal distribution in plant tissues and cells, and techniques enabling determination of speciation and local ligand environment. The techniques addressed in this chapter therefore include X-ray fluorescence and X-ray absorption techniques: energy dispersive X-ray fluorescence spectrometry (EDXRF), micro-proton induced X-ray emission (PIXE), synchrotron micro-X-ray fluorescence spectroscopy (SR-micro-XRF) and X-ray absorption spectroscopy (extended X-ray absorption fine structure [EXAFS], and X-ray absorption near-edge structure [XANES]) using synchrotron light.

25.2 TOOLS FOR DETERMINATION OF METAL UPTAKE IN PLANT TISSUES

25.2.1 X-RAY FLUORESCENCE

Determination of metal concentrations in plant organs (roots, stems, leaves, flowers, seeds) followed by assigning metal uptake strategies (exclusion, accumulation) is a prerequisite to further study metal uptake and tolerance mechanisms at tissue, cellular and molecular levels. In view of the growing needs of global environmental protection and also to minimize the relevant research costs, it is important that the analytical procedures for determination of "bulk" elemental concentrations in soil, water and biological materials are accurate, reliable and reproducible. In addition, analytical techniques have to be accessible (cheap), with simple sample preparation procedures, enabling analysis of many replicates in reasonable time, therefore providing more accurate estimation of the processes going on in soil-plant systems. The first part of this chapter focuses on the main characteristics and sample preparation protocols of X-ray fluorescence-based analytical techniques for "bulk" sample analyses, namely energy dispersive X-ray fluorescence spectrometry (EDXRF) and total reflection X-ray fluorescence spectrometry (TXRF). At present, EDXRF and TXRF may be far less frequently applied for analyses of element concentrations in soil, water, air and biological materials than, for example, atomic absorption spectroscopy (AAS) and/or inductively coupled plasma atomic-emission spectroscopy (ICP-AES). However, with the rapid development of detecting and signal processing systems they can offer low cost, fast, sensitive and accurate element analysis which is particularly advantageous from the economic and environmental protection points of view (Nečemer et al. 2008, 2011).

Besides standard energy dispersive X-ray fluorescence analysis (EDXRF) enabling "bulk" elemental analyses in different materials, the process of X-ray fluorescence also represents fundaments of highly sophisticated particle induced (micro-PIXE) and synchrotron-based micro-X-ray fluorescence (micro-XRF) spectroscopy techniques that enable visualization of spatial localization and distribution of particular elements in biological samples (Vogel-Mikuš et al. 2007, 2008a, b, 2010; Kaulich et al. 2009). Therefore an overview of basic principles of X-ray fluorescence process and spectroscopy is provided in order to better understand the working principles of analytical tools described in the present chapter.

25.2.1.1 Basic Principles

X-ray Radiation—X-rays are electromagnetic waves with a spectrum wavelengths spanning from about 80 nm (about 15 eV) down to about 0.001 nm (about 1.2 MeV), overlapping to some extent the region of γ-rays. Electromagnetic radiation (usually above 1 MeV) generated by nuclear processes is called γ-radiation while the radiation below 80 nm wavelength generated by electrons slowed down in the outer field of an atomic nucleus or by transitions between bound states of electrons in the electronic shells of an atom is called X-radiation.

The history of X-ray fluorescence (XRF) dates back to the accidental discovery of X-rays in 1895 by the German physicist Wilhelm Conrad Roentgen. While studying cathode rays in a high-voltage, gaseous-discharge tube, Roentgen observed that even though the experimental tube was encased in a black cardboard box, a barium-platino-cyanide screen, which was lying adjacent to the experiment, emitted fluorescent light whenever the tube was in operation. This was possible because the energies of X-ray photons are of the same order of magnitude as the binding energies of inner-shell electrons (K, L, M, N, …), and therefore they can be used to excite and/ or probe these atomic levels (Janssens 2004).

Interaction of X-rays with Matter -When an X-ray beam passes through matter, some photons are absorbed inside the material or scattered away from the original direction (Figure 25.3).

The intensity (I_0) of an X-ray beam passing through a layer of thickness (d) and density (ρ) is reduced to intensity (I) according to the Lambert-Beer law (Equation 25.1)

$$I = I_0 e^{-\mu \rho d} \tag{25.1}$$

The number of photons per second (the intensity) is reduced, but their energy remains generally unchanged. The term (μ) is called the *mass attenuation coefficient* and has the dimensions of

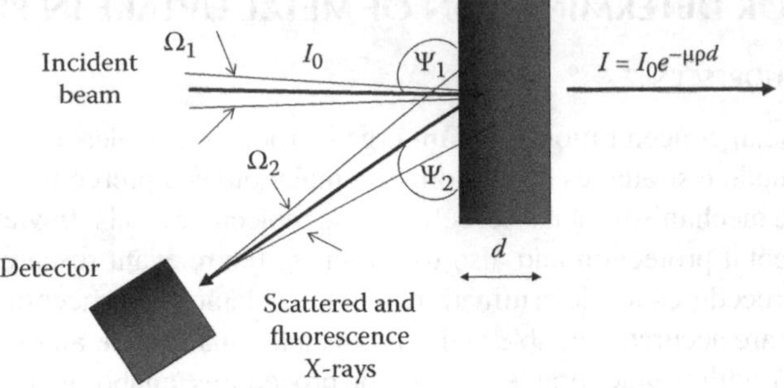

FIGURE 25.3 Interactions of X-rays with matter. (Adapted from Nečemer, M. et al., Use X-ray fluorescence-based analytical techniques in phytoremediation. In *Handbook of Phytoremediation, (Environmental Science, Engineering and Technology)*, 331–358, 2011, New York: Nova Science Publishers.)

cm^2g^{-1}. The product ($\mu_L = \mu\rho$) is called *linear absorption coefficient* and is expressed in cm^{-1}. $\mu(E)$ is sometimes also called the *total cross-section for X-ray absorption* at energy (E).

25.2.1.2 X-ray Fluorescence Process

X-ray fluorescence occurs after the excitation (ionization) of atoms in the tightly bound inner K and L atomic shells with energies that must exceed the binding energies of the K and L electrons through

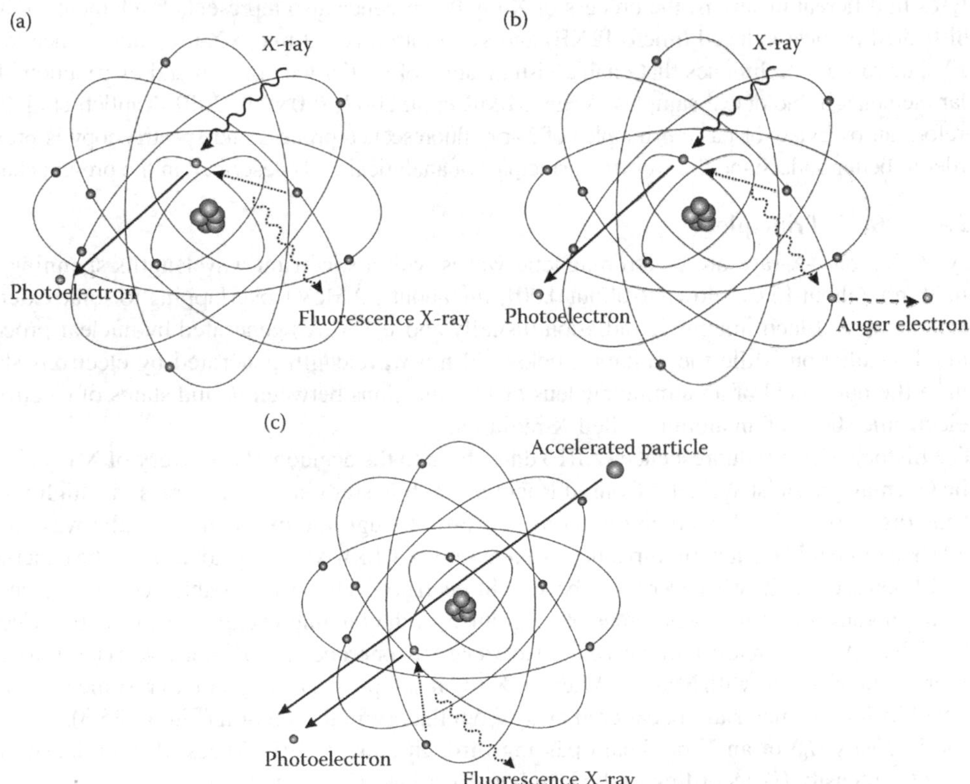

FIGURE 25.4 Interaction of an atom with X-rays (a) *photoelectric effect*; (b) *Auger effect*; or charged particles c) electromagnetic Coulomb interaction. (Adapted from Nečemer, M. et al., Use X-ray fluorescence-based analytical techniques in phytoremediation. In *Handbook of Phytoremediation, (Environmental Science, Engineering and Technology)*, 331–358, 2011, New York: Nova Science Publishers.)

the process called the *photoelectric effect* (Markowicz 1993) (Figure 25.4a). Beside the excitation of atoms by photoelectric process with X-ray radiation emitted from different X-ray sources (e.g., X-ray tube, radioisotope source, synchrotron radiation source) the atoms can be also excited by accelerated charged particles via the *Coulomb interaction* (electrons, protons, α particles) (Figure 25.4c).

In the photoelectric absorption process, an X-ray photon from the X-ray source is completely absorbed by an atom and an electron (called a *photoelectron*, Figure 25.4a) is ejected from one of the inner (K or L) shells (Figure 25.5). In the excitation process a part of the photon energy is used to overcome the binding energy (Φ) of the electron, and the rest is left to the electron in the form of the kinetic energy. After the interaction with an X-ray photon, the atom (in fact the ion) is left in an excited state with a vacancy created in one of its inner (K or L) shells. This atom almost immediately returns to a more stable electronic configuration in the process called *relaxation or de-excitation*, in which the electrons from outer shells (L, M or N) fill the vacancy created in the K or L shells (Figure 25.5). Since during relaxation the electrons pass from a higher to a lower energy state, the difference in energies of respective energy states can be emitted in the form of *characteristic X-ray photons* (K, L, or M) (Figure 25.4a; Figure 25.5), or this energy can be absorbed by an weakly bound electron in one of the outer shells (L, M, or N) (Figure 25.4b; Figure 25.5). This electron is then ejected from the atom as an Auger electron (Figure 25.4b). Characteristic K, L and M X-ray photons are, according to their energy, typical for a particular element, and this is the basis for elemental characterization by X-ray spectrometry (Markowicz 1993). The ratio between the number of emitted characteristic X-rays and the total number of inner shell vacancies in a particular atomic shell that gave rise to them, is called the *fluorescence yield* of that shell (e.g., ω_K). For light elements ($Z < 20$) predominately Auger electrons are produced during relaxation after K-shell ionization ($\omega_K < 0.2$) (Figure 25.4b), while medium and high Z elements preferentially relax in a radiative manner ($0.2 < \omega_K < 1.0$) (Figure 25.4a).

When atoms are excited by accelerated charged particles (Figure 25.4c), these charged particles tear off the electrons from the inner atom shells through the electromagnetic Coulomb interaction, resulting, similarly to the photoelectric effect, in the electron vacancies in particular inner electron shells. The vacancies are then filled with electrons from outer shells and the difference in kinetic energy is emitted in a form of characteristic K, L or M lines.

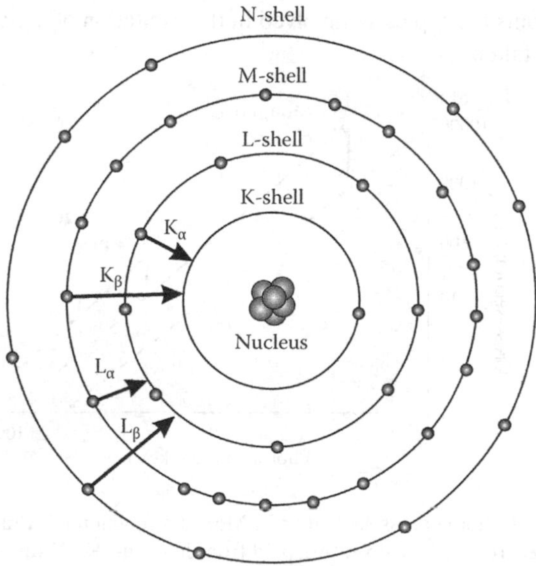

FIGURE 25.5 Electronic transitions in an excited rubidium atom, where one of the electrons in the K shell is missing.

Photoelectric absorption can only occur if the energy of the photon (E) is equal or higher than the binding energy (Φ) of an electron. For example, the X-ray photon with the energy of 22.16 keV (Ag-Kα from Cd-109 radioisotopic source) can eject a K-electron (Φ_K = 12.66 keV) or an L_3-electron (Φ_{L3} = 1.433 keV) from the selenium (Se) atom, but the 5.9 keV photon (Mn-Kα from the Fe-55 radioisotope source) can only eject an electron from the L-shell of the Se atom. Since photoelectric absorption can occur at each of the (excitable) energy levels of the atom, the *total photoelectric cross-section* σ_i is the sum of the (sub)shell-specific contributions (Equation 25.2).

$$\begin{aligned}\sigma_i &= \sigma_{i,K} + \sigma_{i,L} + \sigma_{i,M} + \cdots \\ &= \sigma_{i,K} + (\sigma_{i,L1} + \sigma_{i,L2} + \sigma_{i,L3}) + (\sigma_{i,M1} + \sigma_{i,M2} + \cdots + \sigma_{i,M5}) + \cdots\end{aligned} \quad (25.2)$$

In the case for example of Mo (Figure 25.6) at high energy, e.g., >50 keV, the probability of ejecting a K-electron is rather low and that of ejecting an L_3-electron is even lower. As the energy of the X-ray photon decreases, the cross-section increases, i.e., more vacancies are created. At a binding energy Φ_{KMo} = 19.99 keV, there is an abrupt decrease in the cross-section, because X-rays with lower energy can no longer eject electrons from the K-shell. However, these photons continue to interact with the (more weakly bound) electrons in the L- and M-shells. The discontinuities of the photoelectric cross-section are called *absorption edges*. The ratio of the cross-section just above and just below the absorption edge is called the jump ratio (r). As XRF is the result of selective absorption of radiation, followed by spontaneous emission, an efficient absorption process is required. An element can therefore be determined with high sensitivity by means of XRF when the exciting radiation has its maximum intensity at energy just above the K or L-edge (for heavier elements) of that element (Janssens 2004; Nečemer et al. 2011).

Soon after the discovery of X-rays, in 1913, Henry Moseley established the relation between the atomic number (Z) and the specific X-ray wavelength (λ) of an element (Equation 25.3),

$$1/\lambda = K(Z - s)^2 \quad (25.3)$$

where, K and s are constants; s is the shielding constant and takes value close to one, and K has a different value for each of the line series considered (e.g., the K_α-lines, the L_α-lines,...) (Figure 25.5). Each unique atom has a number of available electrons that can take part in the energy transfer and since millions of atoms are typically involved in the excitation of a given specimen, all possible de-excitation routes are taken.

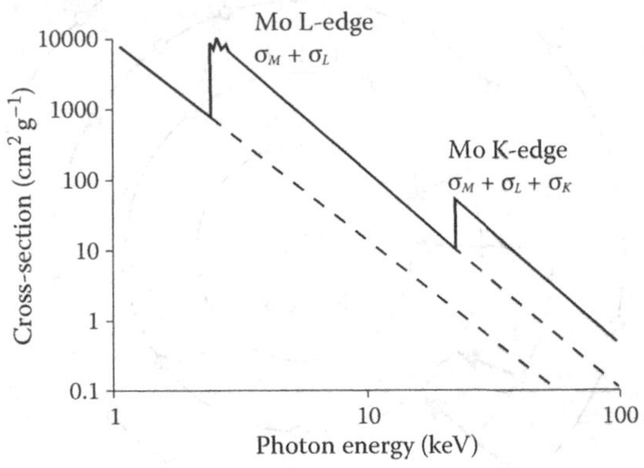

FIGURE 25.6 Variation of photon cross-section (σ in Mo) as a function of X-ray photon energy. The K, L1, L2 and L3 absorption edges are clearly visible. (Adapted from Janssens, K., X-ray based methods of analysis. In *Non-destructive microanalysis of cultural heritage materials*, 129–226, 2004, Amsterdam: Elsevier. Nečemer, M. et al., Use X-ray fluorescence-based analytical techniques in phytoremediation. In *Handbook of Phytoremediation, (Environmental Science, Engineering and Technology)*, 331–358, 2011, New York: Nova Science Publishers.)

25.2.2 Standard Energy-Dispersive X-ray Fluorescence Analysis

After discoveries of Charles G. Barkla who in 1909 found a connection between X-rays radiating from a sample and the atomic weight of the sample and already mentioned observations of Moseley in 1913, which helped to count the elements with the use of X-rays, the potential of the X-ray fluorescence technique was quickly realized, with half of the Nobel Prizes in Physics given to the development in X-rays physics from 1914 to 1924. Originally X-ray spectroscopy used accelerated electrons as an excitation source, but the requirements such as a high vacuum, electrically conducting specimens, and volatility of the sample posed major roadblocks. To overcome these problems an X-ray source was used to promote the fluorescent emission in the sample. Excitation of the sample by this method introduced roadblocks of its own, by lowering the efficiency of photon excitation and requiring instrumentation with complex detection components. Despite these disadvantages, the fluorescent emission of X-rays would provide the most powerful tool for the analyst using commercial instruments. From the 1950s to 1960s nearly all the X-ray spectrometers were wavelength dispersive spectrometers. In a wavelength dispersive spectrometer, a crystal separates the wavelengths of the fluorescence from the sample, similar to grating spectrometers for visible light. Although the earliest commercial XRF devices used simple air path conditions, machines were soon developed utilizing helium or vacuum paths, permitting the detection of lighter elements. In the 1960's, XRF devices began to use lithium fluoride crystals for diffraction and chromium or rhodium target X-ray tubes to excite longer wavelengths. This development was quickly followed by that of multichannel spectrometers for the simultaneous energy measurement of many elements (Beckhoff et al. 2006).

The other X-ray fluorescence spectrometer available at that time was the electron microprobe, which uses a focused electron beam to excite X-rays in a solid sample as small as 10^{-12} cm^3 with an energy dispersive proportional detector. The first microprobe was built by R. Castaing in 1951 and became commercially available in 1958 (Beckhoff et al. 2006).

In the early 1970s, energy dispersive spectrometers became available and used Li-drifted silicon or germanium detectors. The advantage of these instruments brought the ability to measure the entire X-ray fluorescence spectrum simultaneously. With the help of computers, deconvolution methods were developed to extract the net intensities of overlapping individual X-rays lines (Beckhoff et al. 2006). An XRF device was even included on the Apollo 15 and 16 missions (LearnXRF 2011) and Mars Pathfinder mission in 1996–1997 (Pantazis et al. 2010). While in 1970s the XRF spectrometry applications in environmental sciences demonstrated some amateurism, by the end of the 20th century the mentioned applications were the most published X-ray analysis results in the scientific literature. X-ray fluorescence spectrometry proved to be very efficient tool also in life sciences for analysis of major and minor mineral constituents of organisms with its nondestructiveness and simple sample preparation that allows fast analysis of large number of samples (Nečemer et al. 2008, 2009, 2011).

25.2.2.1 XRF Instrumentation

Conventional energy-dispersive X-ray fluorescence (EDXRF) spectrometers consist of only two basic units—the excitation source and the spectrometer or detection system (Figure 25.7).

In this case, the resolution of the EDXRF system depends directly on the resolution of the detector. Typically, a semiconductor detector of high intrinsic resolution is employed [Si(Li)]. The use of this type of detector allows one to record an electronic signal (voltage pulse) processed by the preamplifier and amplifier, which is proportional to the energy of the detected photon dissipated within the sensitive volume of the detector. An analogue to digital converter (ADC) and a multi-channel analyzer can be then used to sort, integrate, store, and display the detected pulses into an X-ray spectrum (Figure 25.8). In recently built systems, digital pulse acquiring, shaping and processing is used instead of analogue, which is much faster with greatly reduced dead times from transient overload signals (Warburton et al. 2000). Using this configuration, all of the X-rays emitted by the sample are collected at a very high rate irrespective of their size. In addition, this configuration also

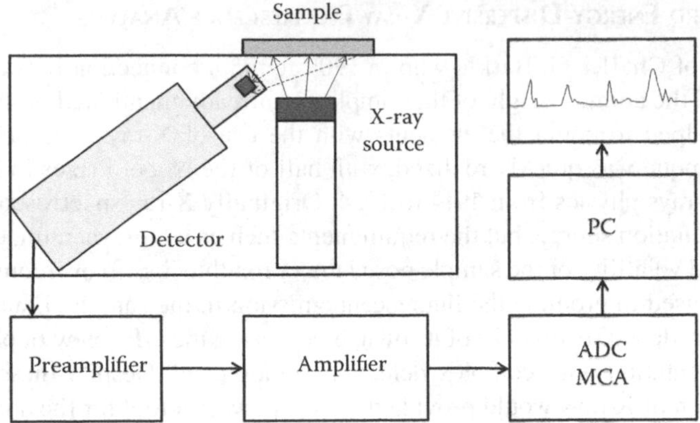

FIGURE 25.7 Diagram of an XRF system composed of X-ray source (X-ray tube or radioisotopic source) energy dispersive detector, preamplifier, amplifier, multi-channel analyzer and PC.

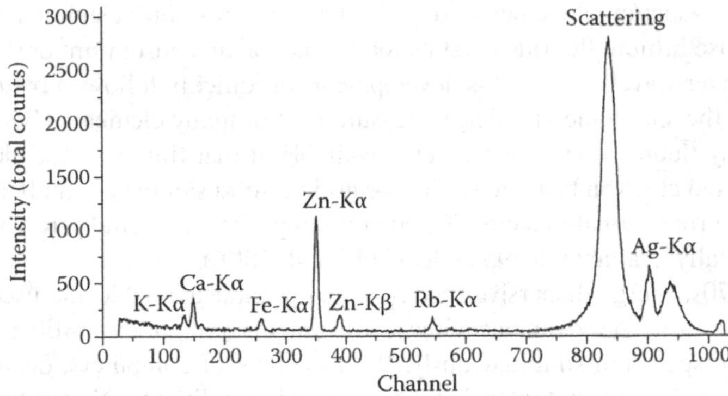

FIGURE 25.8 X-ray fluorescence spectrum of dried leaves of Zn hyperaccumulating pennycress *Thlaspi caerulescens* recorded after excitation with Cd-109 radioisotope source.

enables high speed acquisition and display of spectral data. EDXRF systems are therefore classified according to the type of excitation source, the geometry of excitation and the type of energy dispersive detector installed (Margui et al. 2009).

25.2.2.2 Excitation Sources

Excitation of the elements in the sample can be performed using almost monochromatic radioisotope sources, or partially monochromatic and polychromatic radiation from an X-ray tube in a secondary target or direct mode of operation. Table 25.1 lists radioisotope sources typically used in EDXRF analysis.

The most commonly used sources include Fe-55, Co-57, Cd-109, and Am-241. Each of these emits radiation at specific energy levels and therefore efficiently excites elements within a specific atomic number range. As a result, no single radioisotope source is sufficient for exciting the entire range of elements of interest in environmental analysis, and many instruments use two or three sources to maximize element range. The half-life of a source is important, especially for Fe-55, Co-57, and Cd-109 sources. With half-lives as short as 270 days some sources may have to be replaced after a few years when their intensity decreases to a level too low to provide adequate excitation of the elements of interest (Kalnicky and Singhvi 2001). Alternatively, an X-ray tube can be used to irradiate the sample with characteristic and continuum X-rays. X-ray tubes can be air cooled

TABLE 25.1
Commonly Used Radioisotope Sources Used in EDXRF Analysis

Isotope	Half-Life	Useful Radiation	Energy (keV)	X-rays Excited Efficiency
Fe-55	2.7 years	Mn-K X-rays	5.9	Al-Cr
Co-57	270 days	Fe-K X-rays	6.4	<Cr
		γ-rays	14.4;122;136	
Cd-109	1.3 years	Ag-K X-rays	21.1	K-Tc
Am-241	470 years	Np-L X-rays	14, 21	Sn-Tm
		γ-rays	59	

Source: Adapted from Nečemer, M., P. Kump, and K. Vogel-Mikuš, *Handbook of Phytoremediation,* (*Environmental Science, Engineering and Technology*), 331–358, New York: Nova Science Publishers, 2011.

(low power, 3–50 W) or water cooled (high power, 2 kW), with different anodes, such as Ag (22.1 keV), Rh (20.2 keV), Mo (17.4 keV), and Cr (5.4 keV) (Nečemer et al. 2011). XRF measurements may be performed in different geometries. When thick specimens are excited by the continuous X-ray spectrum from an X-ray tube, the sensitivity of the method is lowered because of the relatively high background from the scattered continuous radiation from the sample, or its substrate, in the spectral region of the fluorescent radiation. A secondary target irradiation geometry using different metal targets as Mo, Rh, Cr, etc. can be used to partly monochromatize the X-ray tube radiation, and thus to decrease this background scattering (Jaklevic and Giauque 1993; Nečemer et al. 2011).

25.2.2.3 Detectors

The X-ray detector converts the energies of the X-ray fluorescence photons into voltage pulses that can be counted to provide a measurement of the total X-ray flux. X-ray detectors are typically "proportional" devices where the absorbed energy of the incident X-ray photon in a sensitive volume of the detector determines the size of the output voltage. A polychromatic fluorescence beam of radiation incident upon the detector produces a spectrum, with a pulse height distribution proportional to the energy distribution of the incident fluorescence radiation beam. A multichannel analyser collects the spectrum and enables discrimination between characteristic fluorescence lines of different elements, depending on the resolution of the detector (Kalnicky and Singhvi 2001). The three most common types of detectors are: the gas flow proportional detector, the scintillation detector, and semiconductor detectors (Si(Li), SiPIN, SiDrift and Hyperpure Ge detectors). These detectors differ in resolution and intrinsic efficiency. Resolution is the ability of the detector to separate X-rays of different energies, and is important for minimizing spectral interferences and overlapping, while the efficiency depends on the absorption of X-rays in the sensitive volume of the detector. Semiconductor detectors have the best resolution and are preferred for EDXRF instruments. These detectors may require liquid nitrogen as a coolant or employ electric cooling via built-in Peltier elements (Nečemer et al. 2011).

The selection of the detector is very important. In cheaper spectrometers, a radioactive source and a proportional detector (gas) can be used. However, one shortcoming of such a device is the poor energy resolution of the detector (800–1.000 eV at 5.9 keV), making the quantification of the insufficiently resolved characteristic X-ray lines in the measured spectrum quite difficult. This problem can be overcome by using semiconductor detectors such as Si(Li), Si PIN or Si drift (SDD) detectors, which have energy resolutions of around 120 eV to 140 eV at 5.9 keV (Nečemer et al. 2011).

25.2.2.4 Sampling and Sample Preparation for EDXRF

Sampling and sample preparation represent two the most critical steps in the analysis of environmental samples, regardless of the analytical method applied. To accurately characterize site conditions, the samples collected must be representative of the site or area under investigation (Margui et al. 2009). Representative soil or vegetal sampling ensures that a sample or group of samples accurately reflects the concentration of the contaminant(s) of concern at a given time and location. Analytical results from representative samples reflect the variation in contaminant presence and concentration range throughout a site. Parameters affecting the variability of the results of representative samples include: (1) geologic and plant material variability, (2) contaminant concentration variability, (3) collection and preparation variability, and (4) analytical variability (Nečemer et al. 2011).

Use of analytical techniques such as AAS or ICP-AES usually requires sample-preparation procedures involving total destruction of the matrix by chemical treatment. Sample dissolution is usually a demanding, time-consuming step that sometimes limits application of the analytical procedures in environmental studies and quality-control processes. Commonly, dry ashing (involving combustion of the sample) and wet digestion (involving digestion with strong acids) have been used to destroy the organic matter and dissolve the analytes in such matrices (Margui et al. 2009). Compared to dry-ashing methods, wet-mineralization procedures using acid digestion present a wide range of options, depending on the choice of reagents and their mixtures as well as the devices used for the procedure (Margui et al. 2009). Especially demanding is wet digestion of soil and plant samples containing silicon, as in such cases, the use of hydrofluoric acid is essential to achieve total destruction of the matrix and thus determination of the total concentrations of analytes. In the last 15 years, classical open systems (digestions at atmospheric pressure) using conventional sources of heating (e.g., sand baths and hot plates) were gradually replaced by digestion procedures using microwave ovens and closed vessels, since in open systems there were problems with losses of volatile elements, such as Hg, Cd and others. In addition, microwave assisted procedures can also shorten the digestion procedure and reduce the amount of reagents employed, as well as avoid analyte losses and contamination from other samples or from the surroundings (Margui et al. 2009). To sum up, the choice of the best decomposition procedure for soil and vegetal samples should be preceded by verification of the procedure for each specific matrix and analyte under study. This becomes quite difficult in environmental studies where several or many soil and plant samples are used as pollution indicators for different metals, and in such cases, the application of techniques that obviate matrix destruction become even more attractive. For this reason, study of the suitability of other methods for direct and multi-elemental analysis of soil and vegetal samples has increased in recent years, from instrumental neutron activation analysis (INAA) to simple X-ray fluorescence-based techniques. INAA is based on measuring the radioactivity produced by neutron reactions on naturally-occurring nuclides (Nečemer et al. 2008), but the most serious shortcomings of INAA are its high costs, the availability of a nuclear reactor for irradiation and the rather long time of analysis imposed by the cooling periods for the decay of interfering activated short-lived radionuclides, although the method is very sensitive and accurate. In addition, INAA does not allow determination of some environmentally-important elements (e.g., Pb) (Margui et al. 2009). When summing up the advantages and disadvantages of INAA, the method is not very suitable for routine application in environmental studies. Most X-ray fluorescence (XRF) techniques, on the other hand, comply with the desirable features for analysis of soil and vegetal specimens, including: i) the possibility of performing analysis directly on solid samples; ii) simultaneous multi-element capability; iii) the possibility of performing qualitative, semi-quantitative and quantitative determinations; iv) a wide dynamic range; v) high throughput, and vi) low cost per determination (Nečemer et al. 2011).

The main drawbacks of XRF instrumentation restricting its more frequent use for environmental purposes have been its limited sensitivity for some important pollutant elements (e.g., Cd, Pb and Hg) and a somewhat poorer precision and accuracy compared to atomic spectroscopic techniques (Nečemer et al. 2008, 2011). The main source of uncertainty in quantitative XRF analysis originates from the inhomogeneity of the sample in a rather small surface part of the sample (due to the absorption of excitation and fluorescence X-rays; it is usually just few mg cm^{-2}), which contributes to the measured fluorescence intensities. Nevertheless, there have recently been improvements in XRF instrumentation (e.g., development of spectrometers using digital signal processing and enhancement of X-ray production with better designs for excitation-detection geometry), which have added the advantage of increased instrumental sensitivity, thus allowing improvements in both precision and productivity. These improvements have therefore increased the possibility of the use of XRF spectroscopy as a technique in the environmental and life science field (Margui et al. 2009; Nečemer et al. 2011).

Both solid and liquid samples can be analysed by EDXRF. In the case of solid samples, no special chemical treatment of the sample is necessary. Determination of the composition of solid samples, without any sample preparation, is possible for samples that are homogeneous in all three dimensions and with a flat surface. This is the case for direct analysis of metals and alloys and that has been one of the main applications of XRF. However, most environmental solid materials (e.g., soil and biological materials) require sample pre-treatment to make them homogeneous and to ensure the quality and the reproducibility of measurements (Margui et al. 2009; Nečemer et al. 2011). Commonly, this procedure is based on crushing or grinding the materials into fine powder followed by pelletization at high pressure, which is the most frequent method of preparing soil and vegetal samples for analysis by XRF techniques (Margui et al. 2009). Prior to grinding, soil or vegetal samples are usually oven dried at 60–100°C, or in the case of vegetal samples freeze dried to remove their water content. In reducing the soil or vegetal material to a fine powder, grinding or milling is usually employed with concomitant problems and contamination arising from the grinding matrix. Particularly for trace elements, precautions should be taken by choosing suitable materials (e.g., agate, silicon carbide, boron carbide, and tungsten carbide). Agate grinding may introduce significant contamination into biological material by Ti, V, Cr, Mn, Fe and Pb (Margui et al. 2009), but this is the case in all analytical procedures, because for AAS and ICP-AES analysis the samples should also be well ground and homogenized prior to acid digestion. For EDXRF analysis approximately 100–200 mg of solid, well ground and homogenized (soil or vegetal) material is sufficient; however any inhomogeneity of the pulverized solid sample can influence the accuracy of the measurement, especially in the case of lighter elements (Nečemer et al. 2008). In addition care should be taken during pellet preparation to assure a uniform thickness in order to avoid bias due to inhomogeneity.

For liquid samples, 100–1000 ml of solution is required, and the elements are concentrated from the sample by precipitation. Several precipitation agents are available for this purpose. For instance, the reagent ammonium pyrrolidine dithiocarbamate (APDC) can be used for the precipitation of Cu, Fe, Ni and Pb. Note that any particular precipitating agent can selectively precipitate only certain specific elements. Therefore, only these specific elements can be determined in liquid samples using this method of sample preparation. The precipitated elements are separated from the liquid phase by filtration, and the precipitate that is deposited on the filter is measured directly by the EDXRF system. Due to the pre-concentration of the precipitated elements, the limits of detection for these elements decrease to a few 10 $\mu g\ l^{-1}$, which cannot be achieved in the analysis of solid samples. This approach is especially suitable for monitoring contaminating elements in water (Nečemer et al. 2011).

25.2.3 Quantitative Elemental Analysis by EDX-ray Fluorescence Spectroscopy

25.2.3.1 Basic Principles

As previously described, X-ray fluorescence spectroscopy is based on X-ray excitation of atoms in the sample material by the photo-effect process, followed by radiative decay of the excited atoms, i.e., by emission of characteristic X-rays of the particular atoms. In the relaxation process the radiative and Auger transitions compete and the fluorescent yield determines the probability of radiative transition. The fluorescent yield favours the radiative decay of heavier atoms. The intensities of radiative transitions of atoms in the sample, measured by the X-ray spectrometer, are then used in qualitative and quantitative analyses of the elemental composition of the sample. The measured intensities of characteristic X-rays depend on the mode of excitation (radioisotope, X-ray tube or synchrotron excitation), on the fundamental physical constants which determine the probabilities of photo-effect excitation of atoms, the probability of radiative decay (fluorescent yield), the absorption of excitation radiation and that emitted within the sample while penetrating toward the detector, and finally on the detector efficiency and the geometry of the excitation-detection experiment. The process of X-ray fluorescence is well understood and presented in many books and papers (Tertian and Claisse 1982; Rousseau 1984; He and Van Espen 1991; Van Dyck et al. 1986; Nečemer et al. 2011). In order to better understand quantitative X-ray fluorescence analysis basic principles, adapted after Nečemer et al. (2011), are summarized below.

The relation between the measured characteristic intensities and concentrations of respective atoms in the sample is established using the above mentioned fundamental parameters and experimental conditions. In the case of monochromatic excitation by energy E_1 in the K shell of atoms, the respective relation is as follows (Equation 25.4).

$$I_i = S_i c_i T_i(c_1, c_2, \ldots c_n) Hi(c_1, c_2, \ldots c_n), \tag{25.4}$$

Where I_i: measured fluorescent intensity; S_i: elemental sensitivity (slope of calibration curve in the case of a thin or diluted sample); T_i: absorption correction factor (depends on sample composition); H_i: enhancement correction factor (depends on sample composition) and c_i: is the concentration of element i;

$$S_i = GK_i \tag{25.5}$$

$$G = \overline{A_0 \Omega_1 \Omega_2 \operatorname{cosec} \psi_1} \tag{25.6}$$

$$K_i = \sigma_i^{ph}(E_1)\left(1 - \frac{1}{J_k}\right)_i \omega_i^k f_i^{K\alpha} \varepsilon_{rel}(E_i) \tag{25.7}$$

In the above equations (5–7), A_0: activity of the excitation source; Ω_1,Ω_2: solid angles at the sample from the source and detector, respectively; $\sigma_i^{ph}(E_1)$: photo effect cross-section at energy E_1 in element i; $\left(1 - \frac{1}{J_k}\right)_i$: relative probability for excitation of K-shell of element i; ω_i^k: fluorescent yield for K-shell of element i; $f_i^{K\alpha}$: relative transition probability for K_a X-ray of element i; $\varepsilon_{rel}(E_i)$: relative detector efficiency for characteristic X-rays of energy E_i of element i;

The combined absorption of primary and fluorescence X-rays in the sample is determined as follows (Equation 25.8):

$$\overline{a_{i,s}} = \mu_s(E_1)\overline{\operatorname{cosec}\psi_1} + \mu_s(E_i)\overline{\operatorname{cosec}\psi_2} \tag{25.8}$$

$\mu_s(E_1)$: absorption cross section in the total sample at excitation energy E_1; $\mu_s(E_i)$: absorption cross section in the total sample at characteristic energy E_i of element i;

The expressions for the absorption and enhancement correction factors T_i and H_i are as follows:

$$T_i(c_1, \ldots c_n) = \frac{1 - \exp(-\overline{a_{i,s}}\rho d)}{\overline{a_{i,s}}} \tag{25.9}$$

$$H_i(c_1, c_2, \ldots c_n) = 1 + \Sigma \rho_{i,k}(c_1, \ldots c_i, \ldots c_k, \ldots c_n) c_k \tag{25.10}$$

In the case of a polychromatic excitation defined by the distribution $w(E_j)$ of primary X-rays, which is usually calculated (Pella et al. 1985), the excitation and absorption or enhancement of fluorescent radiation must be treated together. Factorization of the basic equation is impossible, and the evaluation of particular concentrations becomes more complicated. The basic equation becomes:

$$I_i = GK_i \sum_j \left[\sigma_i^{ph}(E_j) w(E_j) T_i(E_1, E_j) \left(1 + \sum_k \rho_{i,k}(E_j) c_k\right)\right] \tag{25.11}$$

Where $\rho_{i,k}$ are contributions to the enhanced intensity of element "*i*" by excitation of the fluorescent radiation of elements "*k*". The summation is performed over all the elements in the sample which could enhance element "*i*". This factor depends on the composition of the sample through absorption in all elements of the sample. In this case the constant K_i is no longer expressed as above but is defined as:

$$K_i = \left(1 - \frac{1}{J_k}\right)_i \omega_i^k f_i^{K\alpha} \varepsilon_{rel}(E_i) \tag{25.12}$$

Exact expressions for the above mentioned correction factors, as well as for the expressions of K_i for L-series X-rays can be found in Tertian and Claisse (1982), Rousseau (1984), He and Van Espen (1991) and Van Dyck et al. (1986).

25.2.3.2 Starting the Quantitative XRF-Analysis: Principal Problems and Necessary Assumptions

The theoretical background of the X-ray fluorescence process is well known and supported by the above equations. The measured fluorescent intensities represent starting data for the quantification procedure. But there is a basic problem, namely, when using the above equations the concentration of the element can be determined from the measured intensity only, if the composition of the sample is known. Namely the fluorescent intensity depends not only on the concentration of the respective element but also on concentrations of all other elements in the sample, which attenuate the excitation and fluorescent radiation in the sample before it excites the atoms within the sample and when the emitted fluorescent radiation penetrates toward the surface of the sample in the direction of the detector. In this case, but only if all the elements in the sample respond by a fluorescent signal in the spectrum, the concentrations can be obtained from the respective set of equations (4–12) by iteration (the number of unknowns c_i in this case equals the number of equations). But in almost all other cases a problem exists due to the unknown part of the sample, namely that part of the sample which does not give a response in the fluorescence spectrum (light elements like H, C, O, F, and sometimes also Na, Mg, Al, Si, etc., depending on the excitation, low fluorescence yield, and detector efficiency). These elements, which comprise the so-called residual or dark or low-Z matrix, additionally attenuate the excitation and the measured fluorescent radiation (Nečemer et al. 2011). Different approaches to quantification are therefore applied to solve this problem:

i) In the case of known composition of the residual or dark or low Z part of the sample matrix (i.e., oxides, cellulose, alumo-silicates, etc.), the concentrations can be calculated by iteration of a set of equations considering additional absorption in the preselected residual matrix (Nečemer et al. 2011);
ii) Use of measured intensities of the scattered excitation radiation in the spectrum enables assessment of the composition of the sample matrix (Van Dyck and Van Grieken 1980; Nečemer et al. 2011);
iii) Additional measurement of absorption performed on the sample by the transmission-emission method (Markowicz and Van Grieken 1993; Nečemer et al. 2011).

The first approach leads to semi-quantitative analysis. Namely in most cases the selected composition of the residual matrix is only a more or less good guess or approximation and therefore it is in principle not possible to say that the result is quantitative (determined only from measured quantities and fundamental constants) (Nečemer et al. 2011).

The second approach uses the scattered primary radiation from the sample, which is usually measured together with the fluorescence. Scattering is in principle a rather complicated physical process, dependent on the geometry of the experiment, on the thickness and on the average atomic number of the atoms in the sample. This correction is usually applied to rather thin samples, for which the absorption corrections are rather small and therefore uncertainty of these corrections only little affect the uncertainty of the results (Nečemer et al. 2011).

On the other hand, the absorption process is a straightforward process and yields good experimental results. In two of the models of our approach to quantitative analysis we utilized absorption measurements in the sample by the transmission-emission method at a single energy, usually at 8.04 keV or 17.44 keV, corresponding to application of Cu or Mo radiators (Markowicz and Van Grieken 1993; Nečemer et al. 2011). By the iteration of the set of equations (4–12) for the measured elements and including the absorption in a selected residual matrix, the concentrations of measured elements could be obtained. The absorption in this particular sample at energy of 8.04 or 17.44 keV is calculated and compared with the measured value. If the values do not coincide, further iteration, selecting gradually larger absorption in the selected residual matrix then leads toward the final values of concentrations of the measured elements and also to a correct residual matrix, so that the measured absorption at a particular energy in the sample coincides with the calculated one (Nečemer et al. 2011).

25.2.3.3 Calibrations

In any model or approach to quantitative analysis it is first necessary to calibrate the XRF system, in order to evaluate the geometrical constant G as defined in equation (6). For this purpose a set of thin samples (Van Espen and Adams 1981) or thick pure metals or samples of stable chemical compounds (Yap et al. 1987) are measured and the sensitivities S_i are calculated using equations (5) to (9), employing the known compositions of selected samples. It should be mentioned that the uncertainty of a such calibration can greatly influence the quantification of the results. To evaluate the uncertainty of the calibration, we preceded in a somehow different way from that of other authors Markowicz et al. (1992), and evaluated the geometrical constant G by equations (5–7) from the calculated sensitivities (Nečemer et al. 2011). The evaluated constant G by definition should be the same for all calibrated elements. Therefore the calculated standard deviation of the average value of G in principle determines the uncertainty of the calibration procedure, which includes the uncertainty of the measured intensities, uncertainties of fundamental parameters in constant K_i, as well as uncertainties in the absorption coefficients used in the calculations (McMaster et al. 1986; Nečemer et al. 2011). This uncertainty represents a part of the total uncertainty of the complete quantification procedure. In most cases this method of calibration yields an average geometrical constant G with an uncertainty of 2% to 5%. In the case of the polychromatic excitation, the uncertainty can also reach 5% to 10%, due to the uncertainty of the calculated distribution of the excitation X-rays $w(E_j)$ (Pella et al. 1985). In the quantification procedure the experimental geometrical constants for particular calibrated elements are usually used rather than the average geometrical constant (Nečemer et al. 2011).

25.2.3.4 Detection Limits

In quantitative XRF analysis, it is important to determine the limit of detection (LOD). It is usually accepted that the minimal detectable intensity of a spectral line must exceed by a factor 3 the standard deviation of the integrated background under the spectral line. According to this definition the minimal detectable mass or LOD in grams for a definite element is expressed as:

$$\mathrm{LOD}[g] = \frac{3\left(\frac{\sqrt{B}}{t}\right)}{S} \tag{25.13}$$

Sensitivity S is expressed as the signal count rate per gram of sample, B is the integrated background in counts under the spectral line, and t is the time of measurement. It is important to stress that the sensitivity and therefore the LOD depend very much on the absorption in the sample matrix. But the explicit dependence of the LOD on the time of measurement is not quite appropriate (Kump 1997; Nečemer et al. 2011). Therefore the above relation can be written in somewhat different form, which shows that the accepted LOD as such is quite arbitrary and also requires some additional data to really correctly determine or assess the limit of detection:

$$\mathrm{LOD}[g] = 3\left(\frac{I_B}{S}\right)\left(\frac{\sqrt{B}}{B}\right) \tag{25.14}$$

From this expression it is evident that the LOD depends on the relative standard deviation of the background under the spectral line and not only on the measuring time. But since the sensitivity S is also defined as the signal count rate of the spectral line corresponding to the sample mass m_0 (I_0/m_0), the LOD at 33% relative standard deviation of the background can be expressed as:

$$\mathrm{LOD}[g] = \left(\frac{I_B}{I_0}\right)m_0 \tag{25.15}$$

This confirms that the LOD depends only on the ratio of background to signal count rates (Kump 1997; Nečemer et al. 2011).

25.2.4 Total Reflection X-ray Fluorescence

25.2.4.1 TXRF-Excitation Module

The basic fundamentals of total reflection XRF spectrometry (TXRF) are similar to those of EDXRF, although they have quite different excitation modes. In TXRF systems, extremely small amounts of the sample (just few µl) are first deposited as a liquid solution on the optically smooth substrate, which is usually quartz and then dried. The dried residues are then excited by a well-collimated X-ray beam at an angle smaller than the angle of total reflection for the substrate (<1.8 mrad for quartz) (Klockenkämper 1997; Kump et al. 1997; Nečemer et al. 2011). In this case, the majority of the incident X-ray radiation is totally reflected from the quartz surface, and only a minor part of it is absorbed by the deposited sample to excite fluorescence. The penetration of the incident X-ray beam into the reflecting material is drastically reduced under these conditions, and the scattered and fluorescence radiation contributed by the carrier in this geometry is therefore negligible. Consequently, the background radiation due to scattering on a small amount of sample is very low, significantly increasing the sensitivity of TXRF when compared to standard XRF spectrometry (Schwenke and Knoth 1993; Kump et al. 1997).

The sensitivity level, however, still strongly depends on the atomic number of the element, although it does extend down to a few 10 ppb (10 µg kg^{-1}) dry weight. To achieve the described excitation conditions, a special total reflection module is required to shape the excitation beam from the X-ray tube into a suitable form that will excite a small amount of the dried sample residue placed on the quartz substrate. Although there are many expensive commercially available TXRF spectrometers, there are also several cheaper laboratory-built systems that exist worldwide. EDXRF systems, which usually have a fine-focus Mo anode X-ray tube, an X-ray generator, a semiconductor X-ray detector and spectroscopy electronics, may be equipped with a total reflection module provided by the Atom Institute (Vienna). In this way, a cheap alternative to the commercial TXRF system can be built and used for multi-element analyses of different environmental samples (Kump 1997; Nečemer et al. 2011).

25.2.4.2 Sample Preparation and Quantification

For the TXRF analyses the sample material in form of solution must be prepared. A solid ground sample thus requires destructive treatment using wet or dry digestion procedures. This process usually utilizes a decomposition procedure with a small amount of ground material (0.1–0.2 g), and involves the application of a mixture of mineral acids followed by microwave digestion. The resulting solution can be analysed by the TXRF after the addition of an internal standard, which is usually a Ga, V or Y in the form of a standard AAS solution. A small amount of this decomposed sample solution (10 µl) is then pipetted onto a quartz substrate, dried in a desiccator, and measured (Nečemer et al. 2011). As in the case of EDXRF, the TXRF method enables multi-element analysis. Usually, with a Mo anode excitation tube, elements from $Z = 16$ (S) to $Z = 92$ (U) can be determined in the concentration range of a few percent to a few mg kg^{-1}. The determination of lighter elements like Na, Al and Mg is possible in a vacuum and with the application of a Si drift detector with thin beryllium or polymer window. For analysis of silicon the substrate should be optically flat Plexiglas instead of quartz.

Since only a very small amount of dry sample is analysed by TXRF, the relative sensitivity is about one to two orders of magnitude better than for EDXRF, although the absolute sensitivity of TXRF is very good and reaches few pictograms, in comparison to a few micrograms for EDXRF. The main advantage of TXRF over EDXRF is the possibility of rapidly analysing a larger number of liquid samples (i.e., waters, different soil extracts for determining the soluble fractions of metals extracted by $CaCl_2$ or NH_4-acetate) and samples that may be prepared by a simple procedure (i.e., by the dilution of soluble samples, like bee honey in water or juices, milk and vines) (Nečemer et al. 2009, 2011). TXRF is also very suitable for the analysis of very small amounts of biological samples, like plant xylem sap, where only a small amount of material is available. In the latter case the TXRF is actually the only method which can provide multi-element analysis of a very small amount of the sample (Nečemer et al. 2011).

25.3 EXPLORING METAL-LOCALIZATION AND DISTRIBUTION IN PLANT TISSUES AND CELLS

Metal pollution is frequently resulting in changes of uptake and distribution of several mineral nutrients and metals in plant tissues and cells, hence affecting their primary functions. To lower their effects, the mechanisms of ion homeostasis in plants seem to exert a tendency for compartmentalization of excess metals in tissues and cell parts with the lowest metabolic activity. Bulk elemental analyses do not allow for specific analysis of mechanisms contributing to metal tolerance and detoxification on the cellular and sub-cellular levels, therefore multi-elemental imaging techniques have to be applied when their accumulation and distribution patterns within plant tissues are questioned and the mechanisms underlying metal detoxification and compartmentalization on the cellular, tissue and organ levels are being described. Due to the complex morphology and highly heterogeneous chemical composition of biological systems, discriminating qualitatively and quantitatively metal composition and their local distribution, and correlating them to the cellular and sub-cellular biological structures, continues to be a challenge (Kaulich et al. 2009).

Determination of the distribution and chemical state of elemental constituents within biological systems at sub-cellular level down to trace level concentrations is of growing importance for gaining new insights about the highly complex functions of elements within the particular biological tissues or cells. Different analytical techniques have been developed in the past years, which are complementary in terms of lateral resolution, chemical sensitivity, quantitative analysis, depth profiling or bulk sensitivity and detection of elemental isotopes (Kaulich et al. 2009). Among those techniques, energy dispersive X-ray analysis (EDX) in a transmission electron microscope provides the highest lateral resolution (<10 nm), but low to moderate chemical sensitivity (0.01–0.1 wt %) and requires the specimen to be analysed in vacuum and to be sectioned to (ultra)thin slices. Secondary ion milling spectroscopy has very high chemical sensitivity (ppb–ppm range) and high spatial resolution (<100 nm) and allows detection of elemental isotopes, but quantification of data is very difficult. Electron energy-loss spectroscopy has very high lateral resolution (<10 nm) and chemical sensitivity in the ppm range, but requires sectioning of the specimen, and like secondary ion milling spectroscopy and EDX is only surface sensitive and does not provide bulk information (Mills et al. 2005; Kaulich et al. 2009). A distinct limitation of the electron-based techniques is also that they require conducting surfaces and high vacuum environment (Kaulich et al. 2009).

X-ray spectrometric techniques by means of excitation of an inner shell electrons and emission of quanta of characteristic fluorescence X-rays was first proposed in 1928 and was well established during the 20th century (Jenkins 1999; Kaulich et al. 2009). Nowadays micro-proton-induced X-ray emission (PIXE) based on excitation of the atoms in the sample using focused proton beam and micro-X-ray fluorescent spectroscopy based on excitation of the atoms in the sample using focused X-ray light, proved to be the most suitable techniques for element localization studies in biological tissues. Micro-PIXE achieves today's lateral resolution in the sub-micron range and it is fully quantitative. With development of cryostats at the new generation microprobes (Tylko et al. 2007) it allows scanning of frozen-hydrated tissues, therefore avoiding artefacts which may appear during freeze-drying of the samples (Tylko et al. 2007). On the other hand micro-X-ray fluorescence (XRF) spectroscopy emerged with the development of high brilliant and energy-tunable third generation synchrotron radiation sources and the possibility of focusing such X-ray light down to sub-micron probes (Kaulich et al. 2009). The main advantage of synchrotron X-ray microprobes is, in addition to high lateral resolution mapping element of distribution in X-ray fluorescence mode, probing chemical environment of elements of interest at sub-micron scale by recording X-ray absorption near edge structure (XANES) spectra, or across absorption-edge imaging (AAEI) (Kaulich et al. 2011). X-ray microprobes offers also other modes; for example full field transmission X-ray microscopy, scanning X-ray microscopy, absorption and phase contrast imaging which aim to resolve sample morphological characteristics that could be then related to element distribution maps (reviewed in detail in Kaulich et al. 2011).

25.3.1 Micro-Proton-Induced X-ray Emission Spectroscopy

The proton-induced X-ray emission spectroscopy (PIXE) was first proposed in 1970 by Sven Johansson of Lund University, Sweden, and developed over the next few years with his colleagues Roland Akselsson and Thomas B. Johansson (Johansson et al. 1970).

PIXE is analytical X-ray fluorescence-based detection method that has a high elemental sensitivity and is mostly used for measurements of trace elements in various types of materials. The limits of detection range from 0.1 to 1 $\mu g\ g^{-1}$ (ppm) for mid-Z elements ($20 < Z < 40$), and generally well below 10 ppm elsewhere in the periodic system of elements from Na to U (Johansson and Campbell 1988). High elemental sensitivity of PIXE is a result of a low physical continuous background in the X-ray spectra induced by protons. The physical background in PIXE consists dominantly of the bremsstrahlung of the protons and the bremsstrahlung of secondary electrons (Ishii et al. 2005). The standard PIXE method includes a proton beam with a diameter of several millimetres that is used to induce X-ray fluorescence in the sample, and its applications extend from geology to archeometry and air aerosol particulate studies, and to homogenized biological materials.

The capabilities of the PIXE technique are greatly extended for its application with a focused proton beam. High energy focused proton beam set-ups are frequently referred to as nuclear microprobes (Breese et al. 1996). The charged particle lenses, among them most frequently magnetic quadrupole lenses, are used to focus the proton beam down to a sub-micrometer diameter. By scanning of the focused proton beam in a raster mode, and by the detection of the induced X-rays, lateral element distribution maps can be measured within samples. The acronym of this method is micro-PIXE, which shows many methodological similarities with the energy-dispersive X-ray analysis (EDX) that is available with scanning electron microscopes (SEMs). In contrast, micro-PIXE has approximately two orders of magnitude higher element sensitivity (Legge and Cholewa 1994). As the method of micro-PIXE is in most cases run in parallel with the methods based on the detection of elastically scattered protons from the sample (Rutherford Backscattering Spectroscopy, RBS) and with the measurement of the energy loss of protons in the sample (scanning transmission ion microscopy, STIM), the thickness and the light-element matrix composition of the tissue can be determined, allowing full quantification of the element contents in the sample.

Elemental mapping of biological tissues has been identified in its early days of micro-PIXE developments as one of the most promising applications of high-energy focused ion beams. Engineering efforts in the improvements of the accelerators, ion lenses, detectors, data acquisition systems and data processing software, enabled micro-PIXE to became a technique of choice in mapping element distribution in biological tissues in the cases, where high elemental sensitivity, high lateral resolution and full quantification are required simultaneously. The availability of the method is improving significantly in the last past years, as several laboratories are offering the method to the external users based on a peer-review proposal system, similarly as synchrotron radiation sources. Micro-PIXE is currently sharing the community of users from biomedical field who require high resolution tissue element mapping, together with micro-X-ray Fluorescence (micro-XRF) method available at synchrotron radiation facilities and with table-top method of Laser Ablation ICP-MS. As an example of micro-PIXE set up, Figure 25.9 depicts nuclear microprobe setup at Jožef Stefan Institute, Ljubljana, Slovenia, recently upgraded with On-Off Axis STIM detection system.

Photon yield dY_i of the element i from a thin target slice with thickness of dz depends on the mass fraction of the element i, which we here define as the density ratio ρ_i/ρ, could be derived as (Equation 25.16) (Johansson and Campbell 1988).

$$dY_i = \frac{\Delta\Omega}{4\pi} N_p N_A \frac{\varepsilon_i \eta_i}{M_i} \sigma_i^x(z) e^{-\mu_i \zeta(z)} \rho_i dz \tag{25.16}$$

In above equation, $\Delta\Omega$ stands for the detector solid angle, ε_i for the detector efficiency, η_i for the transmission of the absorbers in front of the detector, N_p for the number of protons that hit the sample during the measurement, M_i for the molar mass of element i, and N_A for the Avogadro number. Cross section for production of fluorescence X-rays is noted by $\sigma_i^x(z)$ and the part $e^{-\mu_i \zeta(z)}$ describes the probability that an X-ray photon is not absorbed in the sample material with the X-ray attenuation coefficient μ_i on the exit path $\zeta(z)$ toward the X-ray detector.

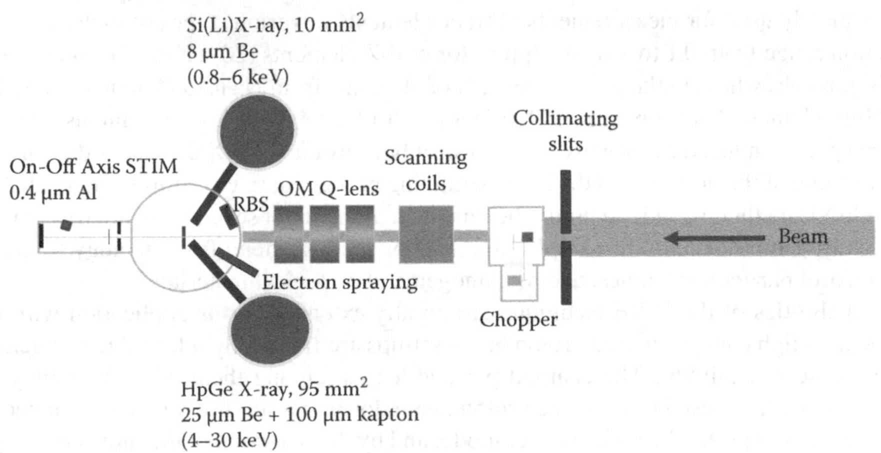

FIGURE 25.9 Micro-PIXE setup at Jožef Stefan Institute, Ljubljana, recently upgraded with On-Off Axis STIM detection system.

The thickness of tissue slices used in the micro-PIXE analysis is normally selected to cover from one up to five cell diameters. This corresponds to a few micrometers of dried tissue after the freeze-drying or up to 60 micrometers of frozen hydrated tissue. Proton beam of 3 MeV penetrates such a sample and losses a fraction of its initial energy. In a PIXE methodology, such samples are considered as samples of intermediate thickness. During the penetration of the sample, the proton beam is continuously losing its energy, as is described by the value of stopping power $S(E)$ (Ziegler 2011).

The yield from a sample of finite thickness over the entire depth could be presented as the integral over the proton energy:

$$Y_i = \frac{\Delta\Omega}{4\pi} N_p N_A \frac{\varepsilon_i \eta_i}{M_i} x_i \int_{E_f}^{E_0} \frac{\sigma_i^x(E)}{S(E)} e^{-\mu_i \zeta(E)} \rho_i dE \tag{25.17}$$

Here, the x_i denotes the mass fraction of the element i, E_0 the proton initial energy (3 MeV in our case) and E_f the proton exit energy after passing the sample. Determination of the mass fraction x_i by micro-PIXE requires information on the total number of protons that hit the sample during the measurement, information on the proton exit energy, sample matrix composition that plays a role in both, X-ray absorption, as well as in the proton beam stopping, and exact characterization of the detection efficiency of the X-ray detector. When the light element matrix of the sample is unknown, Rutherford backscattering spectroscopy (RBS) can reveal information about the element composition of the tissue matrix (including C, O and N) as well as its thickness. RBS spectra could be acquired in parallel with the PIXE spectra using multi-detector approach. For the quantitative element analysis of thin sections of biological materials, the proton exit energy is determined in parallel with micro-PIXE by Scanning transmission ion microscopy (STIM). STIM in its basic form should be measured in the low current mode and with ion detector positioned in the beam to gain optimal energy resolution. This introduces very demanding sequence of analytical work, consisting of prior STIM measurement in a low current mode and consequent switching to a high-current mode for micro-PIXE analysis of the same tissue region. In addition, the tissue integrity at the end of the micro-PIXE analysis is again tested in the low current mode with STIM (Vogel-Mikuš et al. 2008a). Very effective idea that allows high-resolution STIM to be measured in parallel with micro-PIXE originated from the Lund group (Pallon et al. 2004). In the so-called On-Off Axis STIM method, ion detector measures the energy of the protons scattered from a thin mono-elemental foil positioned behind the sample (Figure 25.10).

In addition to STIM and RBS for full quantification of the measured spectra, proton dose which hits the sample during raster scanning has to be determined correctly. For precise proton dose determination at JSI microprobe, an in-beam chopping device (gold-plated graphite; beam intersection frequency,

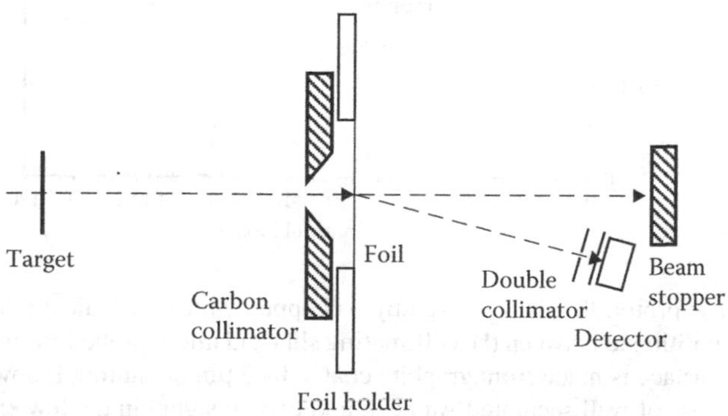

FIGURE 25.10 Schematic drawing of an On-Off Axis STIM detection system, proposed by Lund group (Pallon et al. 2004), which is used at Jožef Stefan Institute, Ljubljana, to obtain sample thickness information in parallel with PIXE and RBS in a high-current mode.

approximately 10 Hz) (Figure 25.11) was implemented in the beam line after the last collimation of the beam, before it hits the sample, thus making the system insensitive to beam intensity fluctuations. The spectrum of backscattered protons from the chopper is recorded in parallel with the PIXE spectra in a list-mode. This high-energy part of the spectrum consists of protons scattered from the Au layer and appears as a separate peak. During off-line data processing, the spectrum accumulated at the in-beam chopper over an arbitrary scanning area could be extracted from the list-mode results simultaneously with PIXE spectra, and used for dose information (Vogel-Mikuš et al. 2009a).

Trace element allocation by micro-PIXE in environmental studies can provide information on the uptake and localization of heavy metals in biological tissues (Schneider et al. 1999; Orlic et al. 2003; Bhatia et al. 2004; Vogel-Mikuš et al. 2007, 2008a, 2008b, 2009a, 2009b). Micro-PIXE element maps for the elements that are dominant in tissue physiology (K, Ca, S and P) usually provide high-contrast morphology images, which are easily correlated with images obtained by optical and electron microscopy. Where the trace element concentrations exceed 50 ppm, elemental images can still provide excellent allocation information. However, the trace elements can be also

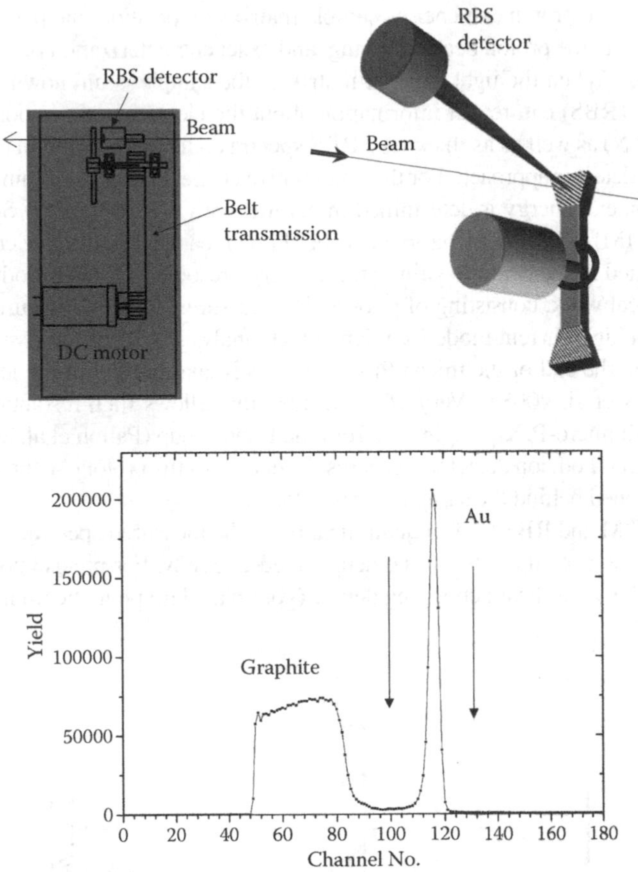

FIGURE 25.11 Above: proton flux is measured by a chopper, which periodically intersects the collimated beam. The chopper is positioned between the collimating slits and micro probe forming triplet quadruple lens. Beam-facing chopper surface is made from graphite coated by 3 μm of Au foil. Below: spectrum of the back-scattering protons consists of well separated Au peak and carbon signal in the low energy region. Integrated yield in the Au peak is proportional to the proton flux. (Reprinted from *Nucl. Instrum. Methods Phys. Res. B. Beam Interact. Mater. Atoms*, 267, Vogel-Mikuš, K., Pelicon, P., Vavpetič, P., Kreft, I., and Regvar, M., Elemental analysis of edible grains by micro-PIXE: common buckwheat case study, 2884–2889, Copyright 2009, with permission from Elsevier.)

present in concentrations that are of the order of 1 ppm or below. In such cases, the corresponding trace element distribution maps might not provide useful information on the allocation due to poor spectral statistics. Using an off-line analysis of the micro-PIXE spectra extracted from distinctive morphological structures, the resulting data can still provide reliable information on the trace element concentrations in the selected part of the tissue when the spectra are accumulated with sufficient statistics. Using the software package GEOPIXE II (Ryan 2001), the list-mode data could be processed and the results could be, among several other options provided, extracted in a form of linear profiles, which in general require lower spectral statistics in comparison to the presentation in the form of a two-dimensional map.

25.3.1.1 Sample Preparation

The key starting point in micro-PIXE analysis on biological materials is sample preparation. A procedure developed over the last two decades is most frequently used for biological tissues containing large quantities of water, which include tissue cryo-fixation, cryo-cutting and freeze-drying (Schneider et al. 2002; Vogel-Mikuš et al. 2008b, 2009b). This method provides excellent results for element distributions at the tissue and single cell levels, and it has been described in detail by Vogel-Mikuš et al. (2009b).

As far as the elemental imaging of sub-cellular structures is concerned, even though freeze-drying allows the ice formed during the rapid freezing to sublime from the samples (under vacuum, at low temperatures), the withdrawal of water can still result in intracellular morphology deformation. This can include shrinking of the cellular membranes, loosening of intracellular structures, and the sticking of these to the cellular membranes and cell walls. To avoid freeze-drying, several attempts have been made to provide controlled water substitution using other agents. The resulting element distributions, however, indicate significant element redistribution from the original state (Mesjasz-Przybyłowicz and Przybyłowicz 2002; Kachenko et al. 2008). Thus, these substitution procedures are effective for morphological studies with optical or electron microscopy, but must be avoided in any case for the element distribution studies in biological tissues.

An advanced methodology for sample preparation without freeze-drying is to keep the tissue frozen after cryo-fixation, and then to transfer it to the vacuum chamber for analysis using a cryostage, where it is analysed in the frozen hydrated state (Tylko et al. 2007). This approach is already in use with several X-ray micro-beams in synchrotron laboratories (Fahrni 2007). Indeed, this method of tissue preparation is the most promising direction to be taken for tissue treatment in the new generation of nuclear nano-probes, where PIXE is expected to be available with sub 500-nm beams. At the same time, it is the sample preparation technique of choice for the development of three-dimensional (3D) X-ray imaging methods with nuclear microprobes that are emerging from STIM-PIXE (Habchi et al. 2006) and from 3D PIXE with a confocal capillary set-up (Karydas et al. 2007).

25.3.1.2 Data Processing and Evaluation

Several open problems still remain with the micro-PIXE methodology, and these need to be resolved before commercially available packages can integrate all of the required steps in the data processing. When the element mapping of a tissue is prepared in the form of a list mode file, the quantification needs to take into account sample thickness variations, sample matrix composition, and precise energy-dependent efficiency calibration of the X-ray detectors, among other aspects. Quantitative elemental maps should be produced by deconvolution of the background and interference peaks, and several packages do satisfactory cover some of the aspects that are required. The GEOPIXE II package provides deconvolution mapping for thick samples (Ryan 2001) and is commercially available. The PIXEKLM (Uzonyi and Szabo 2005) and BIOPIXE (Scheloske and Schneider 2002) programmes provide deconvolution mapping and handle thin samples of nonuniform thickness, but these are not available for the community of users. In biomedical research with the nuclear microprobe of Jožef Stefan Institute, the PIXE analysis software GUPIXWIN (Campbell et al. 2000) is used for the quantitative element analysis of selected morphological structures within a biological tissue, and it incorporates the thickness information measured by STIM (Vogel-Mikuš et al. 2008a).

At the end of the procedure the resulting elemental maps should be interpreted correctly. The interpretation based on visual appearance may be demanding, because human eyes are not sensitive enough for tiny differences in (re)distribution of elements within plant tissues caused for example by metal treatments. Computerized colocalization analysis that was developed first for confocal microscopy can be of great help in interpretation of the results. Programs like GEOPIXE II have already incorporated plug-ins to perform basic analyses of correlation and colocalization. In addition, quantitative 8 or 16-bit maps can be analyzed also by colocalization analysis plug-ins incorporated in freely accessible ImageJ program (WCIF-ImageJ 2011), which offers calculation of basic Pearson's, Menders and Intensity correlation coefficients together with PCA analysis (Tolra et al. 2010). In addition it has to be emphasized that in the case of colocalization of particular metals and elements in anionic form (e.g., Cd and Cl) this does not necessary mean that metals form complexes with these particular anions in tissues. Therefore additional methods like XANES and EXAFS has to be used for the purpose of resolving complexation and chemical environment of metals in biological tissues.

25.3.1.3 Instrumentation and Examples

The required instrumentation for micro-PIXE includes an ion accelerator, ion lenses, a scanning system, detectors and acquisition software. The method is becoming readily available to external

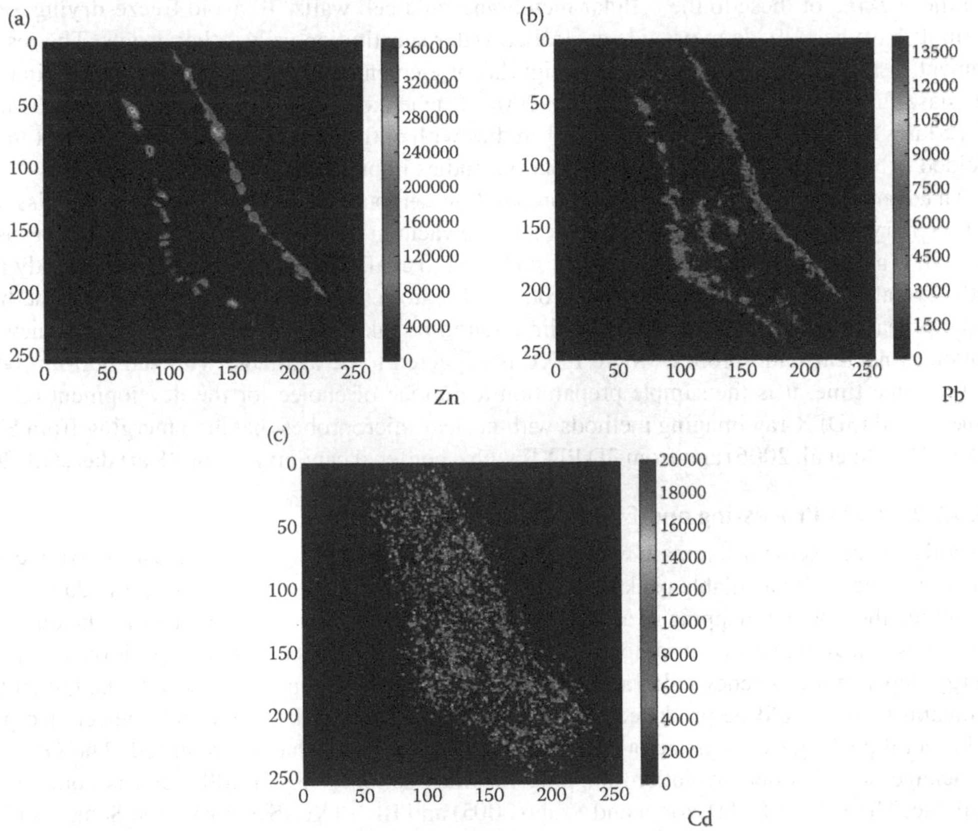

FIGURE 25.12 Qantitative Zn, Pb, and Cd distribution maps of a leaf cross-section of metal hyperaccumulator *Thlaspi praecox*. Measurements were taken from Vogel-Mikuš et al. (2008a) and processed with GEOPIXE II software for quantitative analysis and true elemental imaging. Concentrations showed by vertical bars are in μg g^{-1}.

users in several laboratories as a standard technique for quantitative element mapping in biological tissue. Although micro-PIXE is a multi-element technique, where all of the elements are detected simultaneously, the duration of any single measurement depends strongly on the targeted trace element(s). When the interest is primarily on the distribution of light or mid-Z elements, the time required for a single measurement ranges from 10 minutes to several hours. In extreme cases, such as with cadmium distribution in plant tissue (Vogel-Mikuš et al. 2008a), a single measurement may take up to two days. Figure 25.12 illustrates an example of the use of micro-PIXE for element localization in plant tissues. This includes quantitative elemental maps of a leaf cross-section from the Cd/Zn hyperaccumulator *Thlaspi praecox* (Wulf.) growing in highly Pb-, Cd- and Zn-polluted soil. The Zn map illustrates the epidermal cells that are enriched with Zn, the Pb map shows enriched epidermis and vascular colenchyma tissues. Finally, the Cd map demonstrates even distribution of Cd within the leaf tissues (Vogel-Mikuš et al. 2008a).

The typical running costs of the micro-PIXE method range from €200 per hour to €1,500 to €4,000 per running day. At several micro-PIXE facilities, access is granted after evaluation and selection of scientific proposals.

25.3.2 Synchrotron Micro-X-ray Fluorescence Spectroscopy

Synchrotron radiation (SR) related techniques, utilizing high and also low energy X-rays (Kaulich et al. 2011), have become widely used research tools for environmental research to explore plant–water–soil relationships (Lombi and Susini 2009). There are approximately 50 synchrotron facilities located throughout the world and access to such research institutions is accomplished by competitive general user proposals. In most cases, beam-time at these tax-payer funded facilities is free, resulting in travel costs as the primary expense. Among large variety of synchrotron techniques employed in this chapter the major focus is given to micro-X-ray spectroscopy techniques which are more in details reviewed also by Kaulich et al. (2011).

Synchrotron micro-X-ray spectroscopy (SR-micro-XRF) has become available only recently with development of the third generation high brilliance synchrotron facilities together with the development of X-ray focusing optics and high-through output detecting systems. SR-micro-XRF is a rapid, nondestructive technique for determination of elements in a wide variety of samples in the ppm or ppb concentration range. The high intensity, linear polarization and natural collimation of synchrotron radiation contribute to the high sensitivity and achievable spatial resolution (below 1 μm) of SR-micro-XRF (Lombi and Susini 2009; Kaulich et al. 2011). The physical processes behind SR-micro-XRF are identical as previously described for EDXRF, i.e., there is a high probability that an X-ray photon interacting with an atom will eject a core level electron when the energy of the impinging X-rays is approximately equivalent to or slightly greater that the binding energy of the core level (Figure 25.4). The de-excitation of the atom via fluorescence X-ray production and the measurement of integrated intensity of the X-ray fluorescence spectrum are then directly related to the element concentration.

XRF techniques therefore permit mapping and quantification, of the elements present in a sample in a similar way as it is possible with nonsynchrotron techniques such as EDXRF as already described above, with all accompanying problems and challenges. The advantage of SR-XRF lays in its sensitivity, due to the high photon flux available, the possibility of beam tunability and weak scattering. Furthermore, as it will be discussed below, the absorption edge of an element is related to the chemical environment of the absorbing atom and its oxidation state. Therefore, by selecting appropriate energies of the incoming X-ray photons, it is possible to generate chemical maps of an element in relation to its oxidation state and chemical bonding (Lombi and Sussini 2009). In addition to chemical information based on X-ray Fluorescence (XRF) or micro-X-ray absorption (micro-XAS) spectroscopy, X-ray microscopes operated at the synchrotron facilities can also provide simultaneously morphological information, through absorption and phase contrast imaging as in detail described by Kaulich et al. (2011).

The advantage of the higher penetration depth of X-rays compared to electrons has opened unique opportunities for investigation of biological samples with sub-micrometer lateral resolution Kaulich et al. (2011). In comparison to micro-proton induced X-ray emission in X-ray fluorescence microscopy the atoms in the samples are usually excited by monochromatic X-rays of different energies. X-ray microprobes can be roughly divided to soft and hard X-ray microprobes, first operating in the range of 0.12–12 keV and second operating at range over 12 keV. Laboratory and synchrotron-based XRF instrumentation typically uses hard X-rays and only a few soft X-ray XRF instruments have been reported (Kaulich et al. 2011). The use of soft X-rays for XRF analysis has mainly been limited by low fluorescence yield of low-Z elements and the unavailability of suited detectors and electronics. Such a low-energy XRF system operated in the soft X-ray regime is especially suited for bio-related research as it gives access to the elemental distribution of low-Z elements carbon, nitrogen, oxygen, fluorine, iron, zinc, magnesium and other elements with fundamental importance for metabolism in biological systems on the cellular or sub-cellular level (Kaulich et al. 2009, 2011).

25.3.2.1 Basic Principles

Synchrotron light is the electromagnetic radiation emitted when electrons, moving at velocities close to the speed of light, are forced to change direction under the action of a magnetic field (Lombi and Susini 2009). Although natural synchrotron radiation from charged particles spiralling around magnetic-field lines in space is as old as the stars—for example the light we see from the Crab Nebula, short-wavelength synchrotron radiation generated by relativistic electrons in circular accelerators is only a half-century old. The first observation, literally since it was visible light that was seen, occurred at the General Electric Research Laboratory in Schenectady, New York in 1947. The first generation synchrotron facilities were originally constructed as particle accelerators for the high energy physics community to conduct experiments on the fundamental properties of matter. Soon after their commissioning it was realized that large quantities of electromagnetic radiation, including X-rays, were produced. In the 60 years since, SR has become a premier research tool for the study of matter in all its variety, and dedicated facilities around the world were built and constantly evolved to provide this light in ever more useful forms (Robinson 2009; Lombi and Susini 2009).

A key component of a synchrotron is the storage ring where SR is emitted in a narrow cone in the forward direction, at a tangent to the electron's orbit. The higher the kinetic energy (i.e., the speed) of the electrons, the narrower the emission cone becomes. The spectrum of the emitted radiation also shifts toward shorter wavelengths as the electron energy increases. In so-called third generation light sources, the storage ring is designed to include special magnetic structures known as insertion devices (undulators and wigglers). Insertion devices generate specially shaped magnetic fields that drive electrons into an oscillating trajectory for linearly polarized light or sometimes a spiral trajectory for circularly polarized light. Each bend acts like a source radiating along the axis of the insertion device, hence the light is very intense and in some cases takes on near-laser-like brightness (Lombi and Susini 2009).

To summarize, synchrotron light has a number of unique properties as: i) high brightness: synchrotron light is hundreds of thousands of times more intense than that from conventional X-ray tubes and is naturally highly collimated; ii) wide energy spectrum: synchrotron light is emitted with energies ranging from infrared light to hard X-rays. Furthermore, the emitted light is tunable; iii) highly polarized: the synchrotron emits highly polarized radiation, which can be linear, circular or elliptical; and iv) time-structured emission: nano-second long light pulses enable time-resolved studies (Lombi and Susini 2009).

25.3.2.2 General X-ray Microprobe Characteristics and Set-up

As already described above, conventional XRF is typically performed on homogenized, centimeter-sized samples with a laboratory X-ray tube source. The principal advantage of using synchrotron radiation for XRF analysis is that it allows the spatial resolution of the method to be reduced down to the (sub)micrometer level. There are several reasons why this is possible. As already mentioned, the synchrotron radiation is several orders more intense than X-rays from tube sources. Second, the synchrotron beam is well-collimated, so that the intensity remains high at considerable distances from the source. This means that simple apertures and focusing mirrors can be used to produce small, intense beams. Third, synchrotron radiation is highly linearly polarized which allows background from scattered radiation to be minimized by the geometry of the experiment (Lanzirotti and Sutton 2002). Synchrotron micro-XRF (SR-micro-XRF) is complementary to other microanalysis techniques, such as EDX analysis, micro-PIXE, LA-ICP-MS and secondary ion mass spectrometry (SIMS). Each of these techniques is optimized for particular applications, elements, or sample types. The attractiveness of SR-micro-XRF lies in its capability for nondestructive, trace level analyses of a wide range of elements with high spatial resolution. Another advantage is the low power deposition, a particularly important consideration when analyzing volatile-rich specimens or biological materials. For a given fluorescent signal, X-rays deposit between 10^{-3} and 10^{-5} times less energy than charged particles. The SR-micro-XRF probes are therefore particularly well suited for: i) trace element analysis of nanogram samples (e.g., various types of particles, aerosols, and inclusions), and ii) characterization of trace element distributions with high spatial resolution (e.g., diffusion profiles, chemical zonation, impurity distribution, and compositional mapping).

Pioneered by Horowitz and Howell (1972), SR-micro-XRF probes are nowadays incorporated in most of the second and third generation synchrotron facilities. Their performances benefits from the latest developments in X-ray optics, detectors and samples environments (Lombi and Sussini 2009). SR-micro-XRF instruments rely upon a very specific 45°/45° geometry where the sample is placed at 45° to the incident beam and X-ray detectors are typically placed in the plane of the storage ring and at 90° to the incident beam. The incoming beam is naturally polarized in the horizontal plane, therefore this geometry minimizes the contribution of elastically scattered primary X-rays. Furthermore, the full control of both tunability and spectral bandwidth of the incoming monochromatic radiation minimize the radiation damage without compromising the signal-to-noise ratio and allow accurate quantification (within 3–5%; Vincze et al. 2004; Lombi and Sussini 2009). Hard X-rays probe deeper into samples and relax a number of issues associated with attenuation of fluorescence X-rays in the air path and space around the sample. As depicted in Figure 25.13, the most advanced hard X-ray microprobes offer a multi-modal microanalysis platform where SR-micro-XRF, SR-micro-XAS and SR-micro-XRD can be combined on the same instrument (Lombi and Sussini 2009). All these techniques share

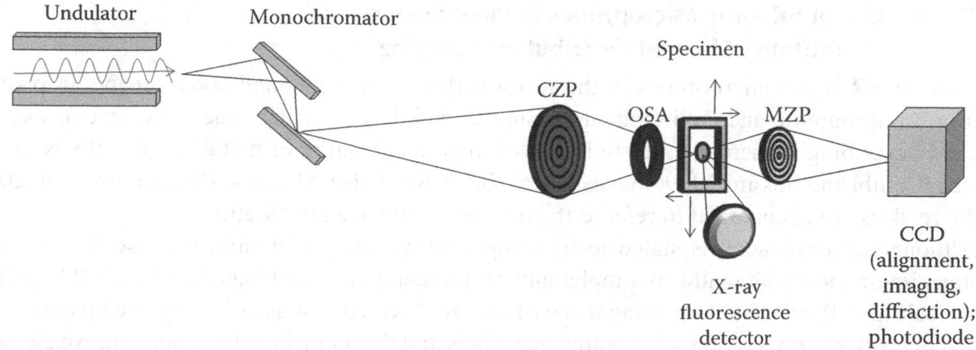

FIGURE 25.13 A layout of X-ray microprobe beamline, providing simultaneous access to various signals. Synchrotron light produced by undulator is monochromatized by means of silicon crystals and pass further to Fresnel zone plate and order selecting aperture (OSA), focusing the beam in the level of sample focal plane. Emitted fluorescence X-rays are then collected by energy dispersive detector, while transmitted X-ray light can be detected either by photodiode or different charge coupled devices (CCD). Hard X-ray microprobe can operate in air, while soft X-ray microprobes operate in vacuum (Adapted from Lombi, E. et al., *PLos One*, 6, e20626, 2011b; Kaulich, B. et al., *Journal of the Royal Society Interface*, 6, 641–647, 2009; Kaulich, B. et al., *Journal of Physics: Condensed Matter*, 23, 083002, 2011.)

the same experimental strategies where a sample is typically mounted on a motorized vertical holder and then translated in the x and y directions through the beam path. Some X-ray microprobe beamlines also have visible, electron or Raman microscope imaging capabilities so that interesting features on a sample can be easily identified and interrogated by X-ray techniques. Elemental concentrations or chemically specific information is then collected at each pixel to generate corresponding maps. Multi-keV X-ray microscopy often suffers from a low contrast due to a weak absorption of the analyzed samples. Furthermore, the use of absorption contrast can subject the specimen to high radiation doses leading to possible structural changes. Even for radiation of hard materials, many of the samples imaged using fluorescence yield are insufficiently absorbing to provide high contrast images in transmission mode. Therefore accurate morphological localization of trace elements is difficult or even impossible. The development of phase contrast methods fully compatible with detection in fluorescence yield is therefore essential. Several strategies have been successfully developed and are now routinely used, and include differential phase contrast (DPC) using configured detectors (Hornberger et al. 2007, 2008), differential interferential contrast (DIC) with a configured zone-plate (Wilhein et al. 2001) or with aperture alignment (Kaulich et al. 2002; Lombi and Sussini 2009).

Low-energy XRF setup with multiple silicon drift detectors (SDDs) for low-Z element detection coupled to a fast read-out electron multiplied charge-coupled device (CCD) camera that allows simultaneous collection of the XRF emission signal and the analysis of the specimen's morphology in bright-field, differential phase and dark-field contrast can be found for example at TwinMic beamline at Elettra Trieste (Kaulich et al. 2009). The TwinMic X-ray spectromicroscope at Elettra (Kaulich et al. 2006) was constructed by the concerted effort of eight European facilities and institutions and includes various imaging modes including full-field imaging (Niemann et al. 1976) and raster-scanning (Rarback et al. 1987), and different contrast techniques such as bright-field, dark-field, differential phase and interference contrast Kaulich et al. (2011). Principles and applications of X-ray microscopy have recently been reviewed by Howells et al. (2007). The analytical potential of the instrument is based on X-ray near-edge absorption spectroscopy, across absorption edge imaging and recently implemented LEXRF. X-ray microprobes are formed by means of diffractive focusing optics and the lateral resolution achievable with this instrument is dependent on the imaging mode 0.05–0.5 mm. Operated in the 280–2200 eV photon energy range, TwinMic accesses the Wolter water window and major low-Z elements (K-edge: Z Ľ 5–9, L-edge Ľ 10–37) of interest for bio-related and other applications (Kaulich et al. 2009).

25.3.2.3 Use of SR X-ray Microprobes in Plant Science and Quantitative Element Distribution Mapping

The use of SR X-ray microprobes in the investigation of nutrients and contaminants in plants in relation to agronomic and environmental issues is rapidly increasing. There are several excellent reviews describing numerous case studies of element localization of metals in metalloids in plant tissues (Lombi and Sussini 2009; Punshon et al. 2009; Kaulich et al. 2009, 2011; Lombi et al. 2011a), so the readers are encouraged to refer to them for more precise information.

Although quantification is stated to be comparatively straightforward, because the physics of photons interaction with matter is simpler and well understood Lombi and Sussini (2009), publications dealing with elemental mapping using SR-micro-XRF consist usually only of elemental distribution maps based on K- and L- X-ray line intensities and do not include fully quantitative elemental maps. In order to be able to quantify the data obtained from mapping element distribution in thin and middle thick biological samples at the at SR-micro-XRF beamlines, numerous conditions have to be fulfilled, starting from stability of the beam flux, precise calibration with suitable standard reference materials, precise measuring of the photon beam current hitting and leaving the sample in order to be able to determine sample thickness, dead time correction of spectrum measurements etc., making quantification of the data quite demanding.

An example of quantitative element mapping with SR-micro-XRF is presented in Figure 25.14. Plants of Cd hyperaccumulator *Thlaspi praecox* were grown in hydroponics in modified Hoagland

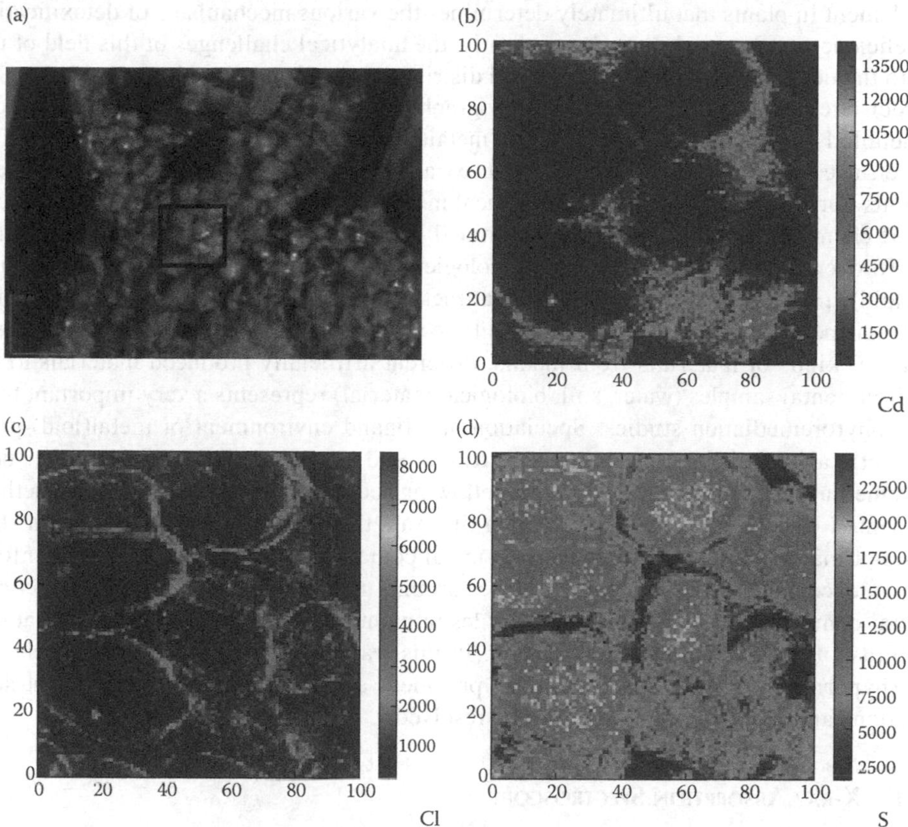

FIGURE 25.14 Quantitative element distribution SR-micro-XRF elemental maps (100 × 100 μm^2) recorded on mesophyll tissue of Cd hyperaccumulating plant *Thlaspi praecox,* treated with 300 μM $CdSO_4$, The maps was measured at ID21 beamline of ESRF with a monochromatic micro-beam (3550 eV) with 1 μm^2 resolution in the region represented by black square in upper left video-microscope photography. Cd (cadmium), S (sulfur), Cl (chlorine). Vertical bar represents concentrations in μg g^{-1} calculated using software developed by P. Kump, IJS.

solution with added 300 μM Cd. Leaves were cryo-fixed, cryo-cut to 20 μm cuttings and freeze-dried (Vogel-Mikuš et al. 2008a). Mesophyll cells were then raster scanned using excitation energy of 3.55 keV in order to well excite Cd-L_{III} line (Cd-L_{III} electron binding energy = 3.538; Cd-L_{III} emission line energy = 3.13 keV) and avoid excitation of K-Kα line (K-Kα electron binding energy = 3.538 keV, K-Kα = 3.31 keV) which strongly overlaps with Cd-L_{III} line. Before quantitative analysis the system was calibrated using thin pure metal standards. Sample thickness was estimated from the scattering peaks and cellulose was taken as an average dark matrix on the basis of the measured absorption on the sample, as described above in a part which deals with basic principles of quantitative analysis by EDXRF. Quantitative analysis was performed by the software developed by P. Kump at Jožef Stefan Institute, Ljubljana, Slovenia.

25.4 EXPLORING LIGAND ENVIRONMENT OF METALS IN PLANT TISSUES

It is increasingly recognized that the availability and toxicity of a given metal(loid) contaminant or micronutrient varies substantially in relation to its speciation and chemical environment (Scheckel et al. 2009). Also, in the case of plants, it is the speciation as related to the distribution of an element in plants that ultimately determines the various mechanisms of detoxification, toxicity/deficiency, and availability. Consequently, the analytical challenges in this field of research relate to the determination of speciation and distribution *in planta* (Lombi et al. 2011a). Sensitive and precise techniques that couple chromatography with mass spectrometric detection, can provide detailed information with regards to metal(loid) speciation (e.g., Feldmann et al. 2009). However, use of chemicals during different extraction procedures and destruction of cell compartments during homogenization of biological materials can sometimes drastically alter speciation and chemical environment of the examined metal(loids) when compared to that in *in vivo* stage. To overcome this limitation, methodologies that are able to investigate the speciation of metal(loid)s *in situ* (and if possible *in vivo*) attract increasing interest (Lombi et al. 2011a). X-ray absorption spectroscopy (XAS) with its ability to sense elemental speciation and ligand environment in all kinds of materials from metals, different artificially produced materials to all kind of environmental samples (water, soil, biological material) represents a very important technique also in phytoremediation studies. Speciation and ligand environment of metal(loid) pollutants can directly alter their bioavailability and toxicity and consequentially influence the success of phytoremediation practices applied. In the following sections the principles of XAS methods will be presented with practical examples which illustrate the possibilities for their use in structural analysis of materials, with the emphasis on metal contaminants in ecosystems. The information on the valence states and the local structures around metal cations can provide direct evidence for metal complexation with organic molecules in plant and animal tissues or information about availability of toxic metals in soil or water. In this way, insights into tolerance mechanisms in metal (hyper)accumulating species can be provided, and important questions about soil-plant interactions and metal accumulation can be resolved.

25.4.1 X-Ray Absorption Spectroscopy

High resolution X-ray absorption spectroscopy (XAS) with synchrotron radiation is a powerful experimental method for the investigation of atomic and molecular structures of materials which enables identification of the local structure around atoms of a selected type in the sample (Wong et al. 1984; Koningsberger and Prins 1988; Rehr and Albers 2000; Bunker 2010). In XANES (X-ray absorption near-edge structure), the valence state of the selected type of the atom in the sample and the local symmetry of its unoccupied orbitals can be deduced from the information hidden in the shape and energy shift of the X-ray absorption edge. In EXAFS (extended X-ray absorption fine structure), number and species of neighbor atoms, their distance from the selected atom and the thermal or structural disorder of their positions can be determined from the oscillatory part of the absorption coefficient above K or L absorption edge. The analysis can be applied to crystalline, nanostructural, or amorphous materials, liquids and molecular gases. EXAFS is often the only practical way to study the arrangement of atoms in materials without long range order, where traditional diffraction techniques cannot be used.

25.4.1.1 Basic Principles

With bright X-ray synchrotron radiation sources, where high-flux monochromatic X-ray beams with the energy resolution $\Delta E/E$ of the order of 10^{-4} are easily obtainable, measurements of high quality absorption spectra can be obtained in a short time. The basic experiment is very simple (Figure 25.15): a thin homogeneous sample of the investigated material with the optimal absorption thickness (μd) of about 2 is prepared, and the intensities of the incident and the transmitted X-ray beam are recorded in the stepwise progression of the incident photon energy.

In a typical synchrotron radiation experimental set-up (Figure 25.16), ionization cells monitor the intensity of incident (I_0) and transmitted (I_1) monochromatic photon beam through the sample. With the well-known exponential attenuation of X-rays in a homogeneous medium current (I_1) is proportional to the current (I_0) as (Equation 25.18).

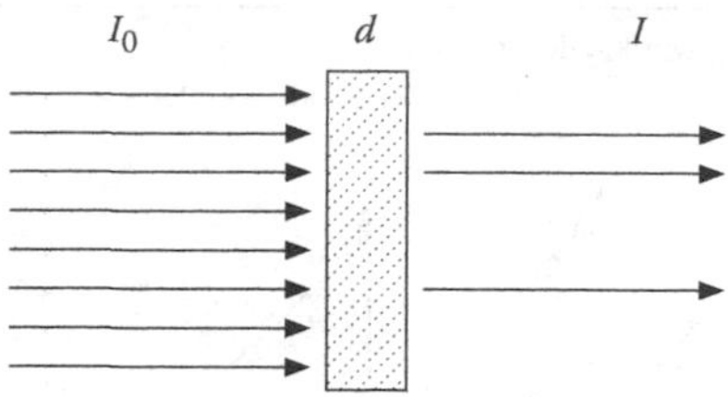

FIGURE 25.15 Measurement of the total absorption coefficient on the sample of thickness (d) by a monochromatic X-ray beam in transmission detection mode.

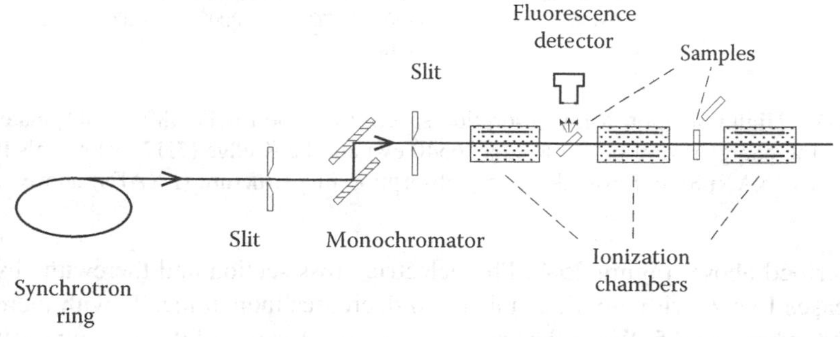

FIGURE 25.16 A schematic representation of X-ray absorption spectroscopy experiment with synchrotron radiation.

$$I_1 = I_0 e^{-\mu(E)d} \tag{25.18}$$

The absorption coefficient $\mu(E)$ at a given photon energy E can be obtained from the relation

$$\mu(E) = \frac{\ln\left(\frac{I_0}{I_1}\right)}{d}, \tag{25.19}$$

where, d is the sample thickness.

The energy dependence of the absorption coefficient is collected by a stepwise scan of the photon energy in the monochromatic beam with the Bragg monochromator. Exact energy calibration is established with simultaneous absorption measurements on a reference metal foil placed between the second and the third ionization chamber.

In case of diluted samples with small amount of investigated element or in case of thin films deposited on thick substrates, standard transmission detection modern cannot be used. Instead, we can exploit fluorescence or total electron yield (TEY) detection mode, where fluorescence photons or the emitted electrons from the sample are monitored, instead of the transmitted beam. The intensity of the detected signal is linearly proportional to the absorption coefficient of the investigated element. The investigated thickness of the sample is proportional to the penetration depth of X-ray beam in case of fluorescence detection, while for TEY only a thin layer at the surface of the sample is monitored, typically about 100 nm or less, limited by average electron path length in the sample.

The dominant process in the X-ray absorption at photon energies below 100 keV is photo effect, whereby the photon is completely absorbed, transferring its energy to the ejected photoelectron as

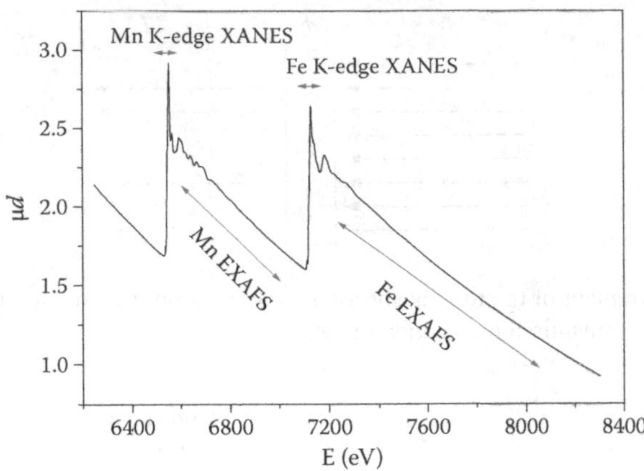

FIGURE 25.17 High resolution X-ray absorption spectrum of the $Li_2(Fe_{0.5}Mn_{0.5})SiO_4$ material (Dominko et al. 2010) in the energy region of Mn K-edge (6540 eV) and Fe K-edge (7112 eV) reveals the structure of absorption edges (XANES) and extended X-ray absorption fine structure (EXAFS) above each absorption edge.

already described above (Figure 25.4). Photoelectric cross section and therewith absorption coefficient increases like Z^5 with atomic number and decreases monotonically with increasing photon energy as $E^{-\alpha}$, where $\alpha \sim 3.5$. When the photon energy reaches one of the deep inner-shell ionization energies of the atom, a sharp jump (absorption edge) marks the opening of an additional photoabsorption channel (Figures 25.6 and 25.17).

The detailed shape of the edge and of the X-ray absorption spectrum above it contains useful structural information. Immediately above the absorption edge, in a range from about 50 eV of up to about 1000 eV, a rich fine structure is superposed onto the smooth energy dependence (Figure 25.17). The structure is called Extended X-ray Absorption Fine Structure (EXAFS) and can be used to determine the number and species of neighbor atoms, their distance from the absorbing atom and the thermal or structural disorder of their positions (Koningsberger and Prins 1988; Bunker 2010). The energy region of the absorption edge within about 50 eV of the threshold, the so called X-ray Absorption Near-Edge Structure (XANES) contains information on chemical bonds and the site symmetry of the absorbing atom (Koningsberger and Prins 1988; Stoehr 1992).

25.4.1.2 Extended X-ray Absorption Fine Structure

EXAFS appears above the absorption edges whenever the absorbing atom is closely surrounded by other atoms, i.e., in solid state, in liquids or in molecular gasses. Only in case of free atoms, as for example in noble gasses or monatomic vapors (Figure 25.18), there is no EXAFS component in the absorption spectrum (Kodre et al. 2002, 2006, 2010).

EXAFS arises from the wavelike nature of the final photoelectron state. When an X-ray photon is absorbed an inner shell electron is preferentially ejected as a photoelectron with kinetic energy equal to the difference between the photon energy E and the inner-shell binding energy E_0. According to quantum theory this photoelectron can be visualized as an outgoing spherical wave centred at the excited atom. The photoelectron wavevector ($k = 2\pi/\lambda$) is given by (Equation 25.20):

$$k = \sqrt{\frac{2m(E - E_0)}{\nabla^2}} . \tag{25.20}$$

where m is the electron mass and ∇ is the Planck constant divided by 2π. This electron wave is scattered by neighbor atoms, and the scattered waves are superposed to the initial outgoing wave. The interference of the initial and scattered waves at the absorbing atom affects the probability for

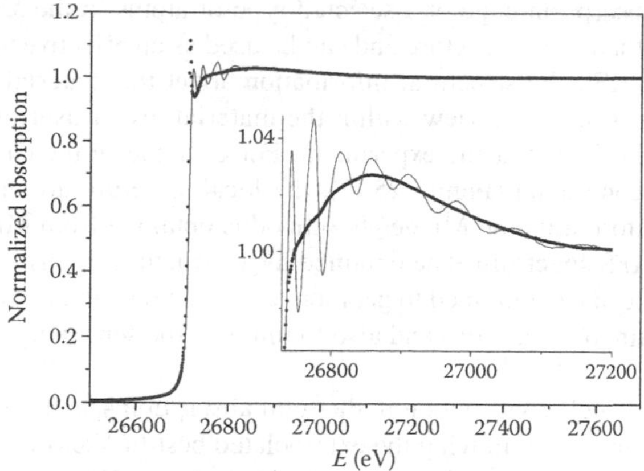

FIGURE 25.18 Normalized K-shell absorption spectra of Cd metal (solid line) and Cd vapor measured at 800°C (dots) in the energy range of Cd K-edge (26,711 eV). Superposition of the EXAFS signal onto the atomic absorption background measured on the monoatomic gas of the same element is illustrated in the inset. (From Kodre, A. et al., *Radiation Physics and Chemistry*, 75, 188–194, 2006. With permission.)

photo-effect. With the increasing photon energy the wavevector of the photoelectron wave increases, leading to alternating constructive and destructive interference (Figure 25.19).

The emitted photoelectron in the process of photo-effect acts as a detection wave, sensing the immediate vicinity of the absorbing atom (nearest neighbors typically up to about 5 Å from absorbing atom) and the information is stored in the resulting EXAFS oscillations. EXAFS spectrum

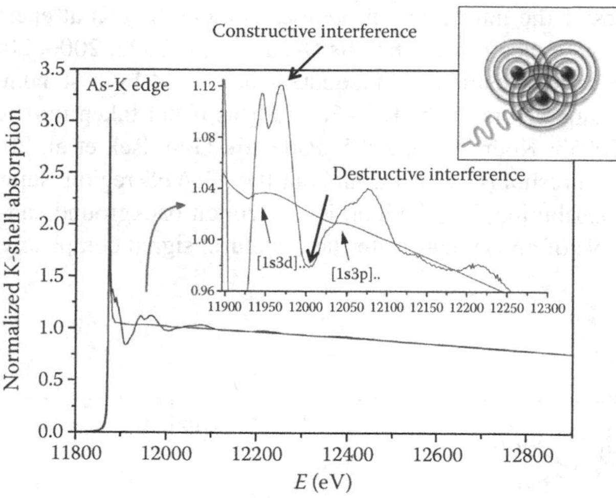

FIGURE 25.19 The normalized As K-shell absorption spectrum of scorodite (solid line) (Arčon et al. 2005a) and the normalized As K-edge atomic absorption background spectrum (middle line) determined from the absorption measurements on arsine gas AsH_3 (Prešeren et al. 2001). Schematic view of the EXAFS process illustrates the origin of EXAFS oscillations due to the interference of outgoing and backscattered photoelectron wave. The atomic absorption background incorporates all the collective intra-atomic effects including the sharp features of the multielectron photoexcitations (As[1s3d] and As[1s3p] indicated in the inset) at energy thresholds for such simultaneous two or more electron excitations (Kodre et al. 2002, 2006, 2010), which in some cases cannot be adequately described by a standard spline function and may introduce systematic errors in the EXAFS analysis if not taken into account. (From Prešeren, R. et al., *Journal of Synchrotron Radiation*, 8, 279–281, 2001; Arčon, I. et al., *Physica Scripta*, T115, 235–236, 2005; Padežnik Gomilšek, J. et al., *Journal of Synchrotron Radiation*, 18, 557–563, 2011. With permission.)

measured above the absorption edge of a selected type of atoms in the sample therefore contains scalar information on their local structure and can be used as an effective tool in structural analysis of materials. With EXAFS the structural information about the material can be obtained simultaneously from different points of view within the material, by measuring EXAFS spectra of all constituent elements, amenable at the experiment. For example in the case of the $Li_2(Fe_{0.5}Mn_{0.5})SiO_4$ nanocrystalline compound (Figure 25.17), the local structure around Fe and Mn cations is detected separately: information on Mn neighborhood is obtained from Mn EXAFS and Fe neighborhood from Fe EXAFS spectrum. The obtained local structural information of the two elements is complementary and can be combined to get more complete insight into the structure of the investigated material (Dominko et al. 2010) and also to improve the sensitivity of the method (Makovec et al. 2009).

The relative K- or L-shell contribution in the total absorption spectra measured in transmission detection mode is obtained by removing the extrapolated best-fit Victoreen (or just linear) function $\mu_p(E)$ determined in the pre-edge region, typically from about –250 eV to about –50 eV relative to the edge energy position. The oscillatory part of the absorption coefficient μ above the edge is then separated from the atomic absorption background μ_o, approximated, per standard, by a smooth best fit spline function (Figure 25.20), and normalized to a unit edge jump μ_K by conventional normalization (extrapolating the post-edge spline background, determined in the range from about 100 to 1000 eV) (Wong et al. 1984; Koningsberger and Prins 1988; Ravel and Newville 2005), defining the EXAFS signal (Equation 25.21):

$$\chi = \frac{(\mu - \mu_o)}{\mu_K} \tag{25.21}$$

The atomic absorption background μ_o incorporates all the collective intra-atomic effects, including the sharp features of the multielectron photoexcitations (MPE) at energy thresholds for such simultaneous two or more electron excitations (Kodre et al. 2002, 2006, 2010) (Figures 25.18 and 25.19), which in some cases cannot be adequately described by a standard spline function and may introduce systematic errors in the EXAFS analysis if not taken into account (Prešeren et al. 2001; Arčon et al. 2005b; Kodre et al. 2005, Padežnik Gomilšek et al. 2011). Depending on the atomic number, these thresholds occur throughout the EXAFS region, superposed onto the structural signal. Their contribution, i.e., the atomic absorption background can ideally be determined on a monatomic sample of an element, with the structural signal completely absent. However very

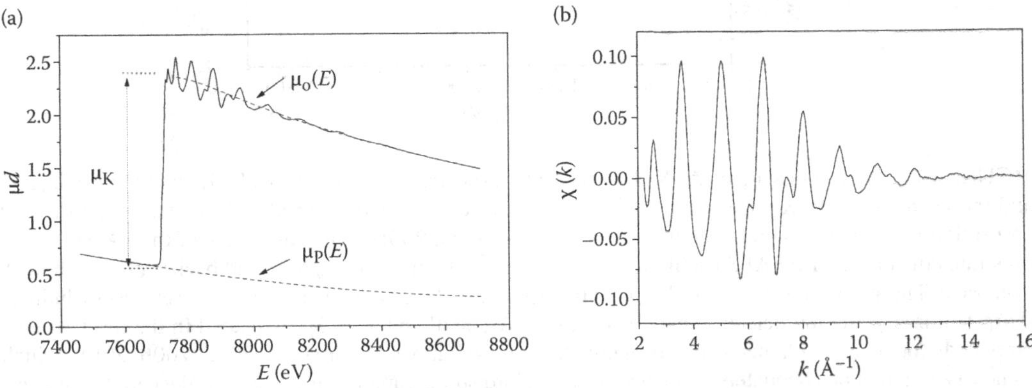

FIGURE 25.20 (a) Extraction of the Co K-edge EXAFS signal from absorption spectrum $\mu(E)$, illustrated on the example of XAS spectrum measured on metallic Co foil. (b) The EXAFS spectrum $\chi(k)$ of Co metal.

few elements are amenable to such measurement. For materials with a strong EXAFS signal, the spline approximation to atomic absorption background is mostly satisfactory (Figure 25.20). For disordered materials with weak EXAFS signal the sharp MPE features can prevail. In such a case, the adoption of the exact atomic absorption background is mandatory for precise extraction of oscillatory EXAFS signal (Padežnik Gomilšek et al. 1999; Kodre et al. 2005).

The EXAFS signal $\chi(k)$ can be described by a sum of sine terms as the function of the wavevector k. If we consider only the contribution of single scattering of the photoelectron from the surrounding atoms, then each term represents a contribution of a spherical shell of equivalent atoms at a distance R_i from the absorbing atom (Koningsberger and Prins 1988):

$$\chi(k) = \sum_i A_i(k)\sin(2kR_i + \delta_i), \tag{25.22}$$

with the atom-specific phase shift δ_i and the amplitude factor:

$$A_i(k) = \frac{N_i}{kR_i^2} S_0^2 F_i(k) e^{(-2k^2\sigma_i^2)} e^{-R_i\lambda_i}, \tag{25.23}$$

where, N_i is the number of atoms in the shell, $F_i(k)$ the corresponding magnitude of the photoelectron backscattering amplitude, σ_I^2 Debye-Waller factor, measuring the thermal and structural disorder in the i-th shell, and λ_i is the mean free path of the photoelectron. Additional amplitude reduction factor S_o^2 is introduced to approximately describe the average effect of multielectron photoexcitations on the EXAFS amplitude. Its value is typically between 0.75 and 0.95 for different absorbing elements and it is transferable between EXAFS spectra of the same absorbing atom measured in different compounds (Roy and Dorssen 1997; Padežnik Gomilšek et al. 1999; Kodre et al. 2001).

By Fourier transformation (FT) of the measured EXAFS signal the contributions of individual shells of atoms are separated visually (Figure 25.21). Each peak in FT magnitude spectra represents one sine term in the EXAFS signal. FT EXAFS can be regarded as an (approximate) radial

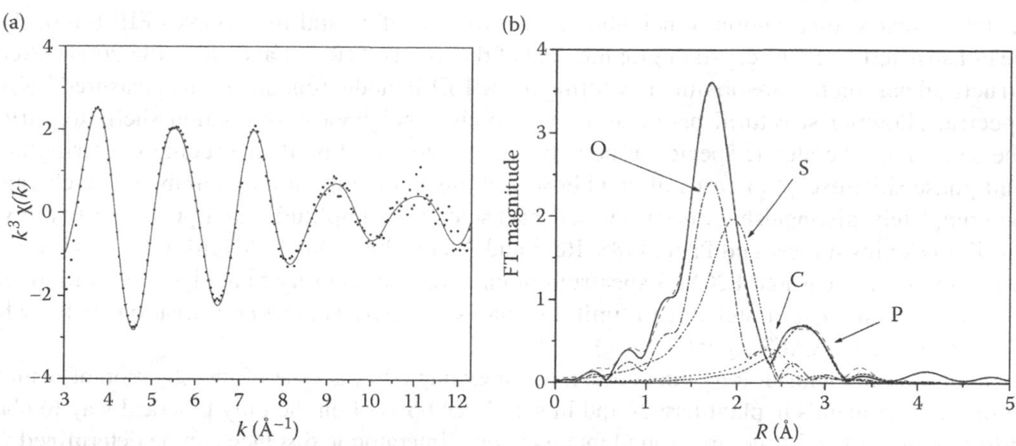

FIGURE 25.21 (a) The k³ weighted Cd K-edge EXAFS spectrum measured on embryos of Cd hyperaccumulator plant *Thlaspi praecox* (dots), together with best-fit EXAFS model (solid line). (b) Fourier transform magnitude of the k³ weighted Cd EXAFS spectrum (experiment—heavy line; best-fit EXAFS model—gray-dashed line), and the contributions of individual Cd neighbor shells composed of two oxygen atoms at 2.19 Å (dash-dotted line); two sulfur atoms at 2.49 Å (dash-dot-dot line); three carbon atoms at 2.95 Å (dotted line), and two phosphorus atoms at 3.32 Å (short-dashed line). (From Vogel-Mikuš, K. et al., *Plant and Soil*, 331, 439–451, 2010. With permission.)

distribution of the neighbors. Peaks appear at the corresponding atom positions R_i, or more precisely, shifted by a few tenth of an Å to lower value due to a k dependence of the phase shift $\delta_i(k)$. Despite the useful visualization of the local neighborhood of the absorbing atom offered by FT magnitude spectrum, the quantitative structural information on the local environment (number and species of neighboring atoms in a given shell, their distance from the absorbing atom and their thermal or structural disorder) can be obtained only by a quantitative EXAFS analysis in which the model EXAFS function is fitted to the measured EXAFS spectra in real (k) or in Fourier transform (R) space (Figure 25.21) (Koningsberger and Prins 1988; Rehr and Albers 2000; Ravel and Newville 2005).

The fitting interval in k space should be as large as possible to improve the sensitivity of the method and reduce uncertainty of the best fit structural parameters. The amplitude of EXAFS signal diminishes roughly as k^{-3}, so in practice the upper limit for the useful k interval for the analysis depends on the signal-to-noise ratio in the measured EXAFS spectrum. Typical k interval amenable for the analysis ranges from about 3 Å^{-1} up to about 14 Å^{-1}. Only rarely (for crystalline samples measured at low temperatures) the quality of measured EXAFS signal allows the analysis up to k = 18 Å^{-1} or beyond.

The theoretical basis of the EXAFS method is firmly established and the necessary electron scattering data ($F_i(k)$, $\delta_i(k)$ and λ_i) can be calculated *ab-initio* with sufficient accuracy (Rehr et al. 1992; Rehr and Albers 2000; Di Cicco 1995), so that modelling of the structure is possible. Several computer programs have been developed for the quantitative analysis, which take into account single scattering as well as multiple scattering contributions to the EXAFS signal (Di Cicco 1995; Benfatto et al. 2001; Ravel and Newville 2005).

Multiple-scattering (MS) paths are those in which the photo-electron scatters from more than one atom before returning to the central atom. The total amplitude of MS depends on the angles in the photoelectron path. Triangular MS paths with scattering angles between 45° and 135° are relatively week and may often be neglected in the EXAFS analysis. Linear MS paths with scattering angles around 180° are, on the other hand, very strong, because the photoelectron is focussed through one atom to the next. Contributions of such strong linear MS scattering, if they appear within in the analysed R region, must always be included in quantitative EXAFS analysis. Model EXAFS function can be constructed *ab-initio*, for example with frequently used FEFF6 program code (Rehr and Albers 2000) from a set of scattering paths of the photoelectron obtained in a tentative spatial distribution of neighbor atoms. In case of crystalline samples FEFF model can be constructed from the crystallographic data of the unit cell (Ravel and Newville 2005). Precise structural parameters are obtained by fitting of the FEFF model function to the measured EXAFS spectra, allowing structural parameters of individual neighbor coordination shells to vary. In the same way, the atomic species of neighbors are recognized by their specific scattering factor and phase shift except in cases of neighboring atoms with close atomic numbers, which cannot be completely distinguished, because their backscattering amplitudes and phase shifts are very similar (Koningsberger and Prins 1988; Rehr and Albers 2000; Vogel-Mikuš et al. 2010). Fourier filtering of the measured EXAFS spectrum often gives a possibility to analyse structure of each coordination shell separately, or to limit the analysis to only nearest coordination shells, which produce largest EXAFS signal.

For materials without long range order, as for example in the case of investigation of complexation of toxic metals in plant tissues and in soil, EXAFS is often the only practical way to obtain information on local structure around metal cations. Interatomic distance can be determined with very high accuracy (typical uncertainties below 1%), while for the number of neighbors and the corresponding Debye-Waller factor lower precision (~10%) is only attainable, due to correlations between the two parameters. Although the obtained structural information is scalar (providing only data about the radial distribution of neighbors, without any direct information on their angular distribution), it may still be used very efficiently in the process of identification of molecular structures, especially if combined with other complementary spectroscopic methods like FTIR or UV-vis (Arčon et al. 2006; Leita et al. 2009; Bele et al. 1998).

EXAFS provides average structural information on the selected type of atom in the sample and can be best used to analyse individual (single phase) materials. For environmental or biological samples, where mixtures of several unknown compounds of the same element are typically present, it is not easy to interpret the EXAFS signal and identify structural information of each compound. Generally it is not possible to unambiguously decompose the measured EXAFS signal into contributions of two or more individual binding sites. However, in many practical cases, we can still draw some useful conclusions from such mixed structural signal (Arčon et al. 2005a; van Elteren et al. 2006; Vogel-Mikuš et al. 2010).

For illustration, we may consider the example shown in Figure 25.21, where Cd K-edge EXAFS analysis was used to obtain information on the local structure around Cd cations bound in plant embryos of Cd hyperaccumulating species *Thlaspi praecox*, in order to identify the Cd complexes responsible for Cd immobilization (Vogel-Mikuš et al. 2010). EXAFS provided structural information (number of Cd neighbors, interatomic distances and Debye-Waller factors) only of nearest Cd coordination shells up to about 4 Å from Cd atom. More distant neighbors could not be detected because of large disorder in more distant coordination shells (or in a hydration shell), which smears out their contribution to EXAFS signal. This is typically the case also in other studies of metal complexes with organic molecules. Two oxygen and three sulfur atoms were identified in the first coordination shell of Cd cations bound in embryos, with Cd-O distances of 2.19 Å, and Cd-S distances of 2.49 Å. The Cd-O and Cd-S distances are in good agreement with those in the reference compounds $3(CdSO_4) \times 8H_2O$ and in Cd-thiol complexes (Frenkel et al. 2001). A relatively weak signal of three carbon neighbors was indicated in the second coordination shell at 2.95 Å and two phosphorous atoms at a distance of 3.32 Å. It can be expected that Cd in the embryo's tissue is coordinated with several different molecules, so the structural parameters represent an average Cd environment. Even though, an unambiguous decomposition of the measured Cd EXAFS signal into contributions of individual binding sites is not possible, the results indicate that Cd is partly bound to thiol groups (Cd-S-C—coordination), partly to molecules rich in phosphorous such as phytate (Cd-O-P- coordination), and partly to free carboxyl and hydroxyl groups (Cd-O-C- coordination). Further information about possible mechanisms of Cd immobilization in the plant were obtained by comparison of Cd EXAFS spectra measured in the different plant organs and tissues (roots, shoots, epidermis, mesophyll, veins), especially by focusing on similarities and differences in the respective data. For example, the Cd-O-P- coordination was found only in embryos and whole seeds. Another significant difference was a much higher percentage of sulfur ligands in embryonic tissues compared to other plant tissues. In such a comparative study additional strength of EXAFS analysis can be exploited: a simultaneous fit of all the measured EXAFS spectra. In the simultaneous relaxation, some of the parameters within the groups of similar spectra can be constrained to common values. In this way, the uncertainties of parameters in the fit are minimized (Ravel and Newville 2005; Makovec et al. 2009; Vogel-Mikuš et al. 2010).

25.4.1.3 X-ray Absorption Near Edge Structure

In XANES (X-ray Absorption Near Edge Structure) analysis, the shape of the absorption edge itself is examined in high resolution scan (Figure 25.22). In the energy region near absorption edge the slow photoelectron probes the empty electronic levels of the material. The resulting pre-edge and edge structure within about 50 eV of the threshold is rich in chemical and structural information (Wong et al. 1984; Koningsberger and Prins 1988; Stoehr 1992; Rehr and Albers 2000). K and L edge XANES spectra of monoatomic vapors can be completely described by pre-edge resonances due to transitions from inner shell (1s, 2s or 2p) initial state to Rydberg final states below ionization threshold and absorption edge (arc tg) due to transitions into continuum states above ionization threshold (Mihelič et al. 2002; Kodre et al. 2002; Kodre et al. 2006; Padežnik Gomilšek et al. 2008) (Figure 25.23a). In bound atoms in molecules or in solid state, this simple picture is blurred by much more complex configurations of possible photoelectron final states (the structure of the conducting bands in the metal, unoccupied valence bands in crystalline solids or valence and quasi-bound orbitals in

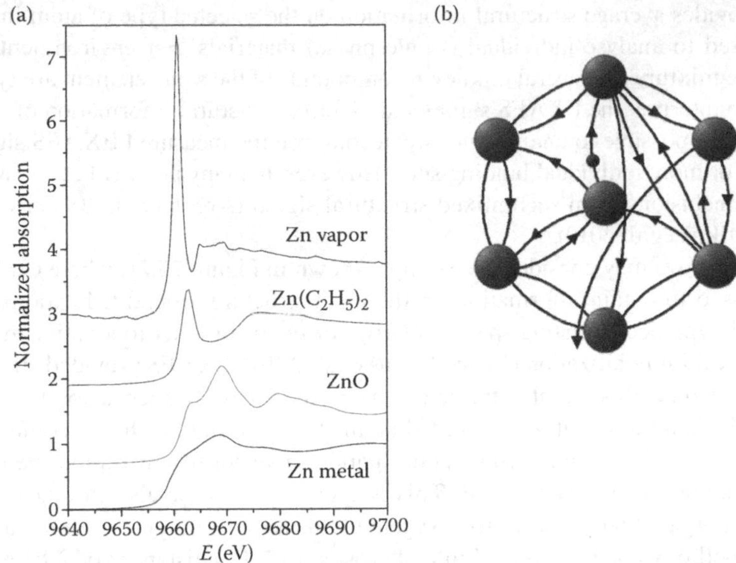

FIGURE 25.22 **(See color insert.)** (a) Comparison of Zn K-edge XANES spectra measured on free and bound Zn atoms: Zn in monatomic vapor (Mihelič et al. 2002), $Zn(C_2H_5)_2$ molecular gas, in solid (crystalline) ZnO and in Zn metal. Spectra are shifted vertically for clarity. (b) Schematic view of multiple scattering of slow photoelectron on neighbor atoms after photo-effect in inner shell of the central atom in the XANES energy range.

molecules) where multiple scattering of photoelectron on neighbor atoms must be taken into account (Della Longa et al. 2001; Rehr and Ankudinov 2005) (Figure 25.22b).

Some program codes were developed to calculate *ab-initio* XANES spectra for a given configuration of atoms in a small cluster. FEFF8 (Rehr and Albers 2000; Rehr and Ankudinov 2005) for example is known to give good results especially for crystalline materials (Figure 25.23b), MXAN

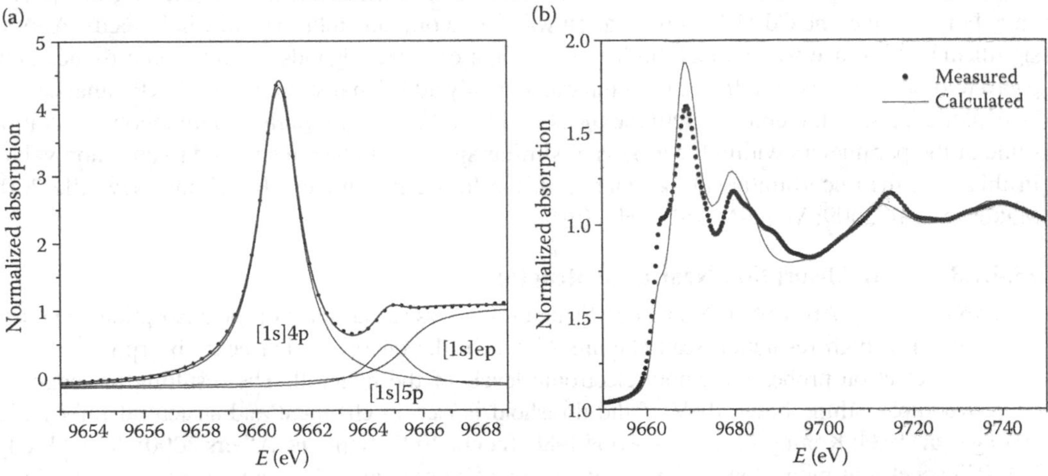

FIGURE 25.23 (a) Decomposition of atomic Zn K-edge XANES spectrum into pre-edge resonances due to transitions from Zn 1s initial state to Rydberg final states 4p and 5p below 1s ionization threshold, and absorption edge (arc tg) due to transitions into continuum states εp above 1s ionization threshold (Mihelič et al. 2002). (b) The comparison of the measured XANES spectrum of crystalline ZnO with a calculated spectrum, obtained by FEFF8 program (Rehr and Albers 2000). The calculation based on a muffin-tin approximation included a cluster of 180 atoms.

on the other hand, which takes into account molecular dynamics of liquids, is particularly powerful in the XANES analysis of metal cations in solutions (Benfatto et al. 2001) and in biological samples (Sarangi et al. 2005; D'Angelo et al. 2010). However, in many cases the theoretical picture is too complex to allow a routine *ab-initio* XANES analysis, therefore simplified approaches, the so called "fingerprint methods" are often used. Namely, some pre-edge or edge features in the XANES spectra can reliably be used as fingerprints in identification of the symmetry of the absorbing atom (Wong et al. 1984; Farges et al. 1997; Rueff et al. 2004; Farges 2005).

The effect of site symmetry in case of K-edge XANES spectra is illustrated in Figure 25.24 by some standard Ti, Co, Mn, Cr and Fe compounds. Tetrahedrally coordinated cations, lacking an inversion centre (Ca, in $CaCrO_4$, Mn in $KMnO_4$, Co Co-APO and Fe in $FePO_4$), exhibit a characteristic isolated pre-edge peak, which can be assigned to the electron transition from 1s to an unoccupied tetrahedral state $3t_2$, and a shoulder on the edge slope assigned to the electron transition $1s \rightarrow 4t_2$, which form as a consequence of strong mixing of metal 3d and oxygen 2p orbitals in such sites with low symmetry. On the other hand, octahedrally coordinated cations that possess an inversion centre (Cr in Cr_2O_3, Fe in Fe_2O_3, Ti in TiO_2-anatase, Cr in Cr_2O_3 and Mn in MnO_2), exhibit a different edge shape with two weak resonances in the pre-edge region assigned to transitions of 1s electron into antibonding orbitals with octahedral symmetry (Wong et al. 1984; Bianconi et al. 1991; Fernandez-Garcia 2002; Pantelouris et al. 2004). The change of K-edge shape for a progression from a distorted octahedral to a square planar coordination is also significant, as shown for the case of Cu compounds (Lytle et al. 1988).

The binding energies of the valence orbitals and therefore the energy position of the edge and the pre-edge features are correlated with the valence state of the absorbing atom in the sample. This effect can be used to deduce the valence state of atoms. With increasing oxidation state each absorption feature in the XANES spectrum is shifted to higher energies (Wong et al. 1984, Farges et al. 1997; Rueff et al. 2004; Farges 2005). The effect is illustrated in Figure 25.25 by an example of *in-situ* XANES analysis of nanostructured $Li_{2-x}Fe_{0.8}Mn_{0.2}SiO_4$ cathode material for Li-ion batteries (Dominko et al. 2010). Fe XANES were measured repeatedly during battery charging to obtain information on Fe valence changes in the process of lithium extraction during battery charging. The Fe K-edge shifts to higher energies as Fe^{2+} is gradually oxidized to Fe^{3+}.

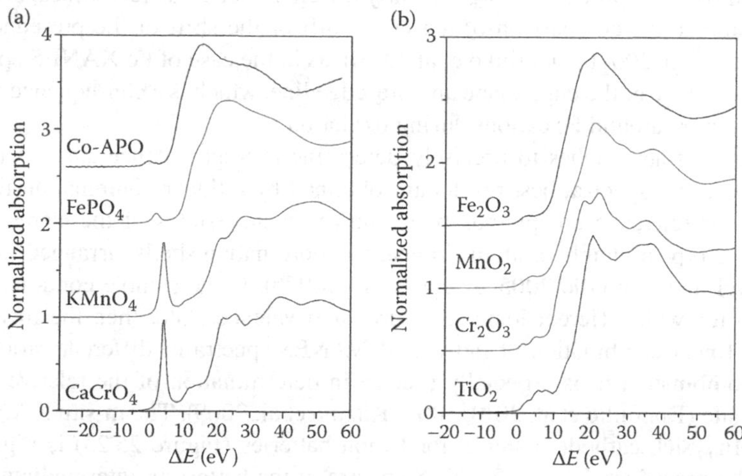

FIGURE 25.24 **(See color insert.)** (a) K-edge XANES spectra of tetrahedrally coordinated cations Ca, Mn, Fe and Co in $CaCrO_4$, $KMnO_4$, $FePO_4$ and Co-APO compounds. (b) K-edge XANES spectra of octahedrally coordinated cations Ti, Cr, Mn and Fe in TiO_2-anatase, Cr_2O_3, MnO_2 and Fe_2O_3 compounds. Energy scale is relative to the energy position of the K-edge of corresponding metal.

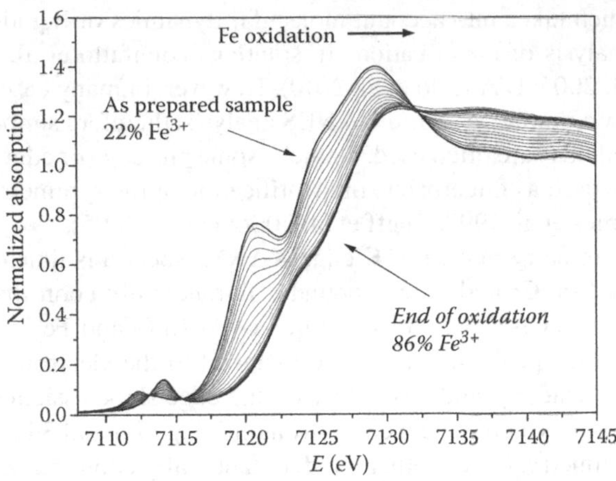

FIGURE 25.25 **(See color insert.)** Fe K-edge XANES spectra measured *in-situ* on the $Li_{2-x}Fe_{0.8}Mn_{0.2}SiO_4$ cathode material for Li-ion batteries during oxidation of Fe^{2+} to Fe^{3+}, as a consequence of Li extraction (Dominko et al. 2010). As prepared sample contained 22% of Fe^{3+}, after oxidation 86% of Fe atoms were in trivalent state.

The largest shifts, up to a few eV per oxidation state, are observed at the edge position. Shifts of the pre-edge peaks are considerably smaller, of the order of a few tenths of eV (Wong et al. 1984; Farges et al. 1997). For atoms with the nearest neighbors of the same chemical species a linear relation between the edge shift and the valence state was established (Wong et al. 1984; Lytle et al. 1988; Arčon et al. 1998; Arčon et al. 2007a). The energy shift depends strongly on the electronegativity of the neighboring atoms, so the linear law can only be used to determine the valence state of the atom for systems having the same ligand type. Smaller deviations from the empirical linear law have also been reported for compounds with different site symmetries of the cation under investigation (Pantelouris et al. 2004; Arčon et al. 2007b).

If the similarity of the edge profiles allows, a comparison of the energy shift of the K-edge or pre-edge resonances in different samples can be used as an effective and precise tool to determine changes of the valence state of the atom (Wong et al. 1984; Farges et al. 1997; Arčon et al. 2007a). However, when the edge profiles are significantly different, due to different local structures around investigated cations, the comparison of the edge shift or the shift of the pre-edge resonances is hindered (Arčon et al. 2007b; Dominko et al. 2010), as in the case of Fe XANES spectra on Figure 25.25. Note the change of the edge shape and pre-edge line, which is a consequence of deformations of the local structure around Fe cations during oxidation.

Among different approaches to precisely determine average valence state of the atom in the sample from XANES spectra, best results are obtained by a linear combination fit with XANES spectra of proper reference compounds with known valence states of the element, with similar symmetry, same type of neighbor atoms in nearest coordination shells, arranged in a similar local structure (Ravel and Newville 2005; Arčon et al. 2007b). If the sample contains same cation in two or more sites with different local structures and valence state, then the measured XANES spectrum is a linear combination of individual XANES spectra of different cation sites. In such cases linear combination fit is especially precise in determination of the relative amounts of the cation at each site (Dominko et al. 2010, 2009; Küzma et al. 2009). The in-situ XANES experiment on $Li_{2-x}Fe_{0.8}Mn_{0.2}SiO_4$ cathode material for Li-ion batteries (Figure 25.25) is a good example to demonstrate the procedure. The Fe XANES spectra of the battery in intermediate states could be completely described by a linear combination of the XANES spectrum of the starting (as prepared) material before charging and the spectrum at the highest oxidation state after charging. The quality of fit is illustrated in Figure 25.26. In this way it is possible to determine the relative amount of each component with a precision better than 1%.

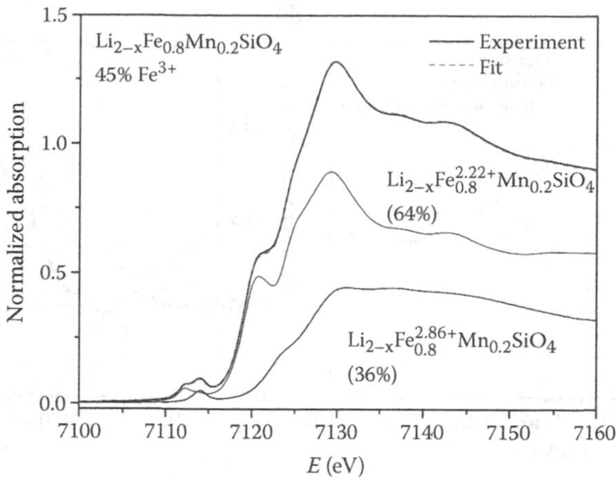

FIGURE 25.26 Fe K-edge XANES spectrum measured on the $Li_{2-x}Fe_{0.8}Mn_{0.2}SiO_4$ cathode material in an intermediate state during oxidation of Fe^{2+} to Fe^{3+} (from Fig. 25), containing 45% of Fe^{3+} (solid black line) (Dominko et al. 2010). The spectrum can be completely described with a linear combination fit (dashed line) of XANES spectrum measured on the same material in as prepared state before oxidation, which contained 22% of Fe^{3+} (light gray line) and XANES spectrum measured after oxidation, containing 86% of Fe^{3+} (dark gray line).

25.4.1.4 Systematic Errors in XANES Analysis

As demonstrated, K-edge XANES spectroscopy can be a reliable and sensitive tool in determination of average valence state of the atom in the sample if a linear combination of proper reference compounds with similar composition and similar local symmetry of the investigated atom neighborhood is used. However, special care should be given to avoid or compensate systematic experimental errors. Absolute energy calibration of measured XANES spectra with accuracy of ±0.1 eV or better must be provided. Apparent shift of the absorption edge to lower energies and a consequent underestimation of average valence in the sample may appear due to inhomogeneity of the sample when measurements are performed in transmission detection mode (Stern and Kim 1981; Lu and Stern 1983; Arčon et al. 2007b) or due to self-absorption effects which appear when absorption spectra are detected in fluorescence detection mode (Koningsberger and Prins 1988; Bunker 2010).

The absorption spectra measured in fluorescence detection mode can be corrected for the self-absorption (SA) effect (Bunker 2010), for example by a procedure implemented in ATHENA (Ravel and Newville 2005). Comparison of absorption spectra measured simultaneously in transmission and fluorescence detection mode (Figure 25.27a) show that if SA effect in fluorescence spectra is not corrected, an apparent shift of the absorption edge of the order of a few tenths of eV to lower energies appears due to the edge profile deformation (best observed in the absorption derivatives of the edge profiles (Figure 25.27b). The effect is more prominent in the case of concentrated samples. After the SA correction, the fluorescence spectra are identical to those measured in transmission mode.

To avoid or diminish similar reduction of the XANES amplitude and a consequent apparent shift of the absorption edge to lower energies due to inhomogeneity of the sample, when measurements are performed in transmission detection mode, special care should be given to sample preparation (Bunker 2010). If sample is not perfectly homogeneous, the incident beam size should be reduced and positioned on the most homogeneous spot on the sample. Numerical corrections to compensate inhomogeneity effects in XANES spectra would require a precise knowledge of holes and cracks in the sample and its surface roughness on the spot where SR beam is transmitted through the sample, which is practically impossible.

At bright synchrotron radiation sources XANES spectra can be measured with micro-focusing beam (μ-XANES) to provide distribution of the investigated element and a 2-D map of changes in

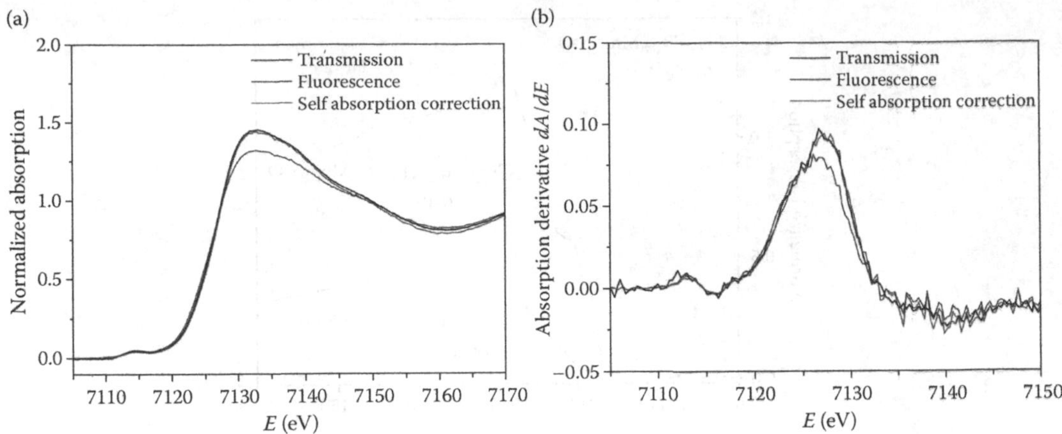

FIGURE 25.27 **(See color insert.)** Fe XANES spectra measured on iron gal ink on the cellulose simultaneously in transmission (blue line) and fluorescence detection mode (red line—without SA correction, magenta line—after SA correction) (a), and their absorption derivatives (b). (From Arčon, I. et al., *X-ray Spectrometry*, 36, 199–205, 2007. With permission.)

valence state on different parts of the sample (Isaure et al. 2006; Hettiarachchi et al. 2006). A lateral resolution below 1 micrometer can be achieved. However, the method is limited by the fact that too high flux density of the micro-focused X-ray beam can cause radiation damage to some materials due to the high absorbed dose of ionizing radiation (Kanngießer et al. 2004). The effect is illustrated on Figure 25.28 for the case of Fe micro-XANES measurements on iron gal ink on the cellulose. High absorbed dose of X-rays during measurements caused reduction of Fe^{3+} to Fe^{2+} and the Fe K-edge shifted to lower energies in consecutive scans. This can easily happen also when determining chemical speciation and ligand environment of metal(loids) in biological tissues. Therefore measurement of XANES spectra on such samples should be performed at low temperatures and with short dwell times. More repetitions of the same scan are recommended to check for the effects of radiation damage.

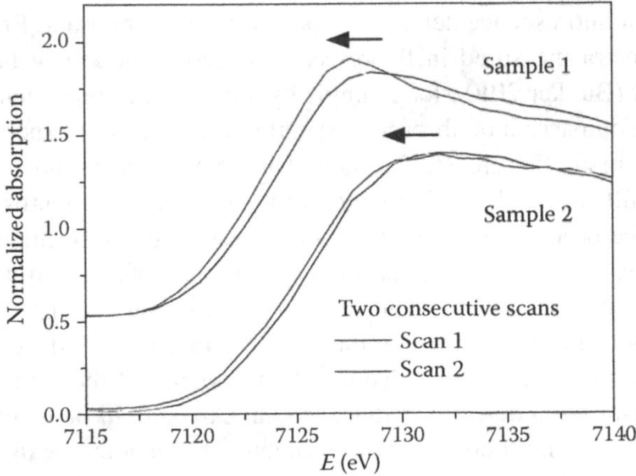

FIGURE 25.28 Effect of extremely high X-ray beam flux density (>10^{12} photons/s/mm^2) with microfocused beam on Fe XANES spectrum measured on iron gal ink on the cellulose (Arčon et al. 2007b). The Fe K absorption edge shifted to lower energies in consecutive scans, due to reduction of Fe^{3+} to Fe^{2+}, because of extremely high absorbed dose of X-rays.

25.4.1.5 X-ray Absorption Spectroscopy Experiments

Several beamlines at different synchrotron radiation facilities are dedicated to EXAFS and XANES experiments. The range of the elements amenable to EXAFS analysis depends on the monochromator and on the other X-ray optical components of the beamline (e.g., mirrors, filters). The low-energy limit of the most widely used Si(111) monochromators is around 2 keV, translating to $Z = 16$ (K edge of sulfur at 2472 eV). With special monochromator crystals, this technique can be extended to about 1 keV (K edge of aluminium at 1560 eV). The upper energy limit depends on the characteristics of the synchrotron radiation source and on the monochromator crystals available. The Si(111) crystals are typically used to about 20 keV, while Si(311) or Si(511) are used to obtain a monochromatic beam beyond 20 keV. Monochromator crystals with higher Miller indices can provide better energy resolution of the monochromatic beam.

Within the available energy span offered by the beamline, the K edge, or in the case of heavier elements ($Z > 50$), also the L_3 absorption edge, of the element can be used for EXAFS analysis. The switch to L_3 is not recommended for elements below antimony (L_3 edge, 4.132 keV; K edge, 30.491 keV), since the range of the L_3 EXAFS signal is too short due to the cut-off by the subsequent L_2 edge (4.381 keV), such that the spatial resolution of this method is seriously impaired. In such cases only XANES analysis can be performed.

In a basic EXAFS experiment, the sensitivity to the content of the element under investigation in the sample is not very high: the element measured must contribute at least a few percent to the total absorption to produce a significant edge jump, and thereupon a meaningful structural EXAFS signal above the absorption edge. Weaker signals tend to be drowned in the statistical noise of the beam. To optimize the signal-to-noise ratio, the samples should be prepared with a total absorption thickness of about 2 above the absorption edge of the element under investigation (Lee et al. 1981). In the fluorescence detection mode, the sensitivity is improved by one or two orders of magnitude (Koningsberger and Prins 1988). Fluorescence EXAFS spectra, with a good enough signal-to noise-ratio that allows a reliable quantitative analysis, can be measured on samples containing at least few hundred ppm of the investigated element in the sample, as usual for the majority of metal(loids) accumulated in biological tissues.

The scan of a standard EXAFS spectrum measured in transmission detection mode requires from about 30 minutes to 1 hour with a standard XAS synchrotron beamline. In fluorescence detection mode, significantly longer detection times are needed to obtain the necessary signal-to-noise ratio in the spectrum. For biological tissues with low metal(loid) concentrations measurements can take up to 8 hours or more to assure sufficient statistics for reliably analysis.

On modern synchrotron X-ray sources with high brilliance, much faster detection modes have been developed for studying chemical reactions in real time: quick EXAFS (Frahm 1989) and dispersive EXAFS (Pascarelli et al. 1999) detection. Currently, 100 ms for a scan is possible, with a promise of a 10-fold improvement with the next generation of coherent X-ray sources.

Bright synchrotron radiation sources make possible also a powerful combination of microscopy and X-ray absorption spectroscopy. There are several beamlines at different SR laboratories (ESRF, HASYLAB, ELETTRA ...) equipped with high energy resolution Bragg monchromator and microfocusing optics (capillary lenses, zone plates, or Kirkpatric–Baez mirrors), with lateral resolution below 1 micron. A 2D distribution of the elements under investigation can be performed with micro-XRF as already described above and complemented with XAS spectra measured at selected spots on the sample. The micro-spectroscopy combination is very useful for biological samples, since it allows mapping of element distribution and chemical state (valence state and local atomic structure) of selected elements on the sub-cellular level (Isaure et al. 2006; Hettiarachchi et al. 2006; Proost et al. 2004). An example is illustrated on Figure 25.29, where the distribution of Cd within the cells of different tissuc of Cd hyperaccumulator plant *Thlaspi praecox* was measured with micro X-ray fluorescence (μ-XRF) (see Figure 25.14 above) with 1 μm^2 spatial resolution on ID21 beamline of ESRF. The chemical form of the metal was then established by micro X-ray absorption near edge structure (μ-XANES) at the Cd L_{III} edge on the different spots within the cells.

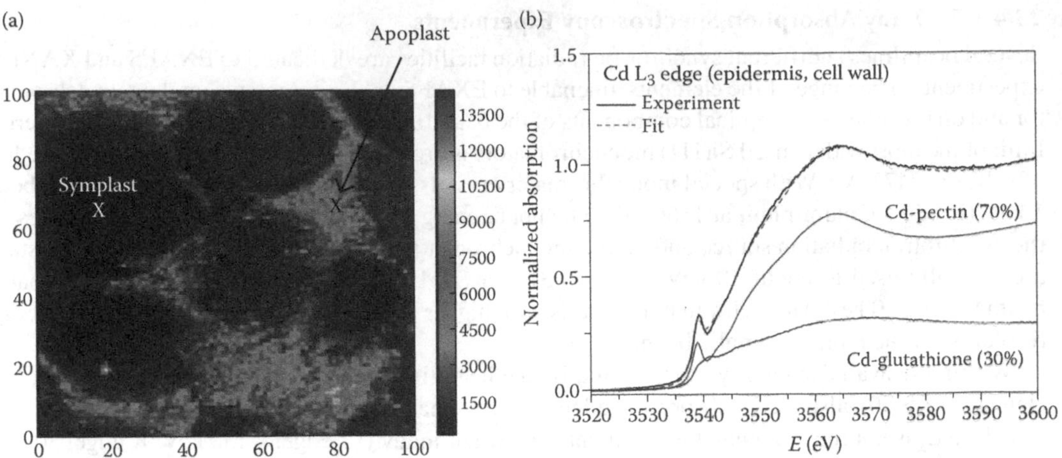

FIGURE 25.29 (See color insert.) (a) μ-XRF Cd elemental map (100 × 100 μm^2) recorded on mesophyll tissue of Cd hyperaccumulator plant *Thlaspi praecox*, treated with 300 μM $CdSO_4$. The map was measured at ID21 beamline of ESRF with a monochromatic micro-beam (3550 eV) with 1 μm^2 resolution. (b) Cd L3-edge XANES spectrum (black line) measured in Cd rich region in cell wall of mesophyll tissue (marked with X). The spectrum can be completely described with a linear combination fit (magenta dashed line) of reference XANES spectra measured on the complex Cd-pectin (red line) and Cd-glutathione (blue line).

25.4.1.6 EXAFS and XANES in Practice

In this chapter some additional examples from the field of environmental pollution with toxic metals are selected to illustrate the use of EXAFS and XANES analysis. In the first example we used X-ray absorption spectroscopy in combination with UV-Vis spectroscopy and DPS voltammetry, to examine the fate of eco-toxic hexavalent chromium in the environment (Leita et al. 2009). After entering the soil, Cr^{6+} remains thermodynamically metastable in the pore solution and is generally much more mobile in soil than Cr^{3+}. Hexavalent chromium is a strong oxidant and can be readily reduced to Cr^{3+} in the presence of electron donors, including soil organic matter. However, it is known that in favourable physical conditions Cr^{6+} may persist, especially in organic soils, for prolonged periods of time, even years. Our work focused on the elucidation of the possible interaction of Cr^{6+} with humic acids (HA). HAs are carbon-rich polydispersed polyanionic biopolymers. Chemically, HAs behave as supramolecules, which are able to polymerize and aggregate, form micelles, and also form supramolecular ensembles with other compounds. HAs represent a strongly pH-dependent reservoir of electron donors/acceptors, which can hypothetically contribute to reduction of several inorganic and organic contaminants. The UV-Vis analysis of the interaction of Cr^{6+} with HAs in water solution at pH 6.5 indicated that HAs promoted the probability of intramolecular O–Cr^{6+} electronic charge transfer. Also, the results of DPS voltammetry, obtained during a stepwise addition of HAs to K_2CrO_4 solution showed a strong interaction of CrO_4^{2-} complex with HAs, which indicated the formation of Cr^{6+}–HAs micelles. So we decided to use X-ray absorption spectroscopy to directly probe eventual structural distortion of tetrahedral CrO_4^{2-} complex as a consequence of interaction with HAs. We measured Cr XANES and EXAFS spectra of K_2CrO_4 water solution at pH 6.5 with or without HAs (Figure 25.30). The characteristic pre-edge resonance clearly indicated that Cr in solution is tetrahedrally coordinated. From the energy position of the pre-edge resonance and the Cr K-edge we could deduce that the valence state of chromium in both samples was Cr^{6+}. Even after detailed inspection we did not find any significant difference between the spectra of the chromate with and without HAs, which clearly demonstrated that local symmetry and valence state of hexavalent Cr in the aqueous solution did not change due to interaction of CrO_4^{2-} with HAs and that its valence state remained 6+.

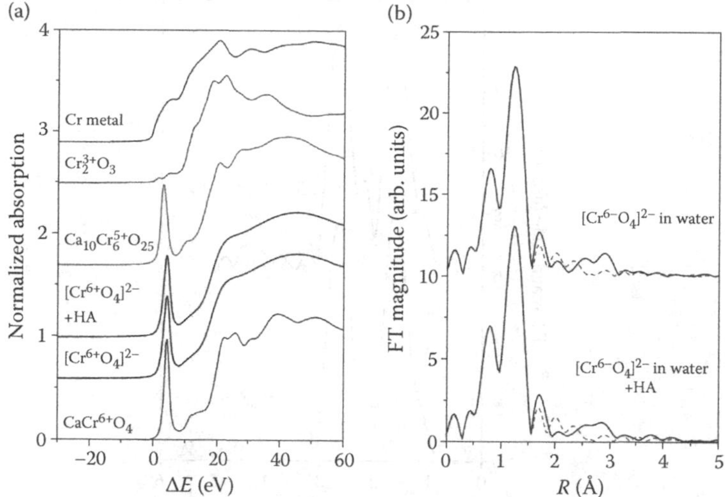

FIGURE 25.30 (a) Normalized Cr K-edge profiles of K_2CrO_4 water solution before and after addition of HAs and reference samples with Cr valence state 0, 3+, 5+ and 6+. The energy scale is set relative to the Cr K-edge in metal (5989.0 eV). (b) Fourier transforms of Cr EXAFS spectra of K_2CrO_4 solution before and after addition of HAs (experiment: solid line; best fit EXAFS model: dashed line) (Leita et al. 2009). The spectra in both graphs are shifted vertically for clarity.

In order to verify whether the interaction between HAs and CrO_4^{2-} has involved any structural modifications, we analysed also Cr K-edge EXAFS spectra. The comparison of the EXAFS spectra of K_2CrO_4 water solution before and after addition of HAs showed no structural changes in the local Cr neighborhood: both spectra were identical within the noise level. Fourier transform of EXAFS spectra (Figure 25.30b) exhibited a single peak that originated from scattering of the photoelectron on the first coordination shell of Cr neighbor atoms. As expected, no signal from more distant Cr neighbors could be detected. Contributions to EXAFS signal from water molecules in the hydration shell around CrO_4^{2-} complex, or from eventual HA atoms in the vicinity of CrO_4^{2-} complex, are too weak to detect because of large disorder, which smears out their EXAFS signal. The quantitative EXAFS analysis showed that there are four oxygen atoms around Cr at the distance of 1.61 Å, thus confirming that interaction with HAs had not induced any structural modifications of the tetrahedral CrO_4^{2-} complex in the solution. From these results we were able to conclude that the presence of HAs did not favor reduction of Cr^{6+} to Cr^{3+}. The interaction between Cr^{6+} and HAs rather leads to the formation of Cr^{6+}–HAs micelles via supramolecular chemical processes, which can explain the relatively high persistence of eco-toxic hexavalent chromium in soils.

In a similar study, we used the Fe K-edge EXAFS analysis in combination with UV-Vis spectrometry to examine the fate of iron cyanide, another hazardous and toxic compound, in the environment (Arčon et al. 2006). More specifically, we examined the capacity of HAs to interact with the $[Fe(CN)_6]^{3-}$ complex, which could prevent their leaching from soils and the consequent risk of contamination of water table. UV-vis spectra recorded during a stepwise additions of HAs to a 1 mM water solution of $[Fe(CN)_6]^{3-}$ evidenced the interaction between ferricynide complex and HAs. In order to verify whether the interaction between HAs and ferricyanide involved structural modifications on the ferricyanide complex we carried out Fe K-edge EXAFS analyses. We measured EXAFS spectra on 1 mM $K_3Fe(CN)_6$ water solution before and after addition of humic acids. Already by the qualitative comparison of the measured spectra (Figure 25.31) it was possible to deduce that no significant structural modifications of the ferricyanide complex were induced by the interaction with HAs. Namely, the two spectra were practically identical within noise level, i.e., no structural changes in the local Fe neighborhood were indicated.

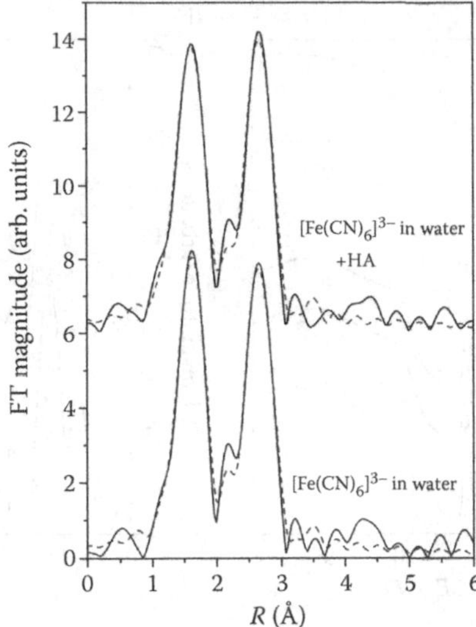

FIGURE 25.31 Fourier transforms of Fe EXAFS spectra of $[Fe(CN)_6]^{3-}$ complex in water solution before and after addition of HAs (experiment: solid line; best fit EXAFS model: dashed line) (Arčon et al. 2006). The spectra are shifted vertically for clarity.

The structure of $[Fe(CN)_6]^{3-}$ complex is octahedral with Fe atom in the central position and Fe-C-N bond angles of 180°. Quantitative EXAFS analysis for such distribution of neighbor atoms is more complex than in previous case of tetrahedral CrO_4^{2-} complex, since it needs to take into account not only single scattering of the photoelectron on nearest neighbor shells, but also strong multiple scattering within the complex due to focusing effects of the photoelectron on collinear arrangement of Fe neighbors (Fe-C-N) (Hayakawa et al. 2004).

Fourier transform of Fe EXAFS spectrum of $[Fe(CN)_6]^{3-}$ complex in water solution exhibits two peaks (Figure 25.31): first originates from single scattering of the photoelectron on six C neighbors in first coordination sphere, while the second peak is composed of two contributions—single scattering on N neighbors in the second coordination shell and strong MS contributions within collinear Fe-C-N coordination.

The results of EXAFS analysis showed that the structural parameters of the ferricyanide complex before and after the addition of HAs are unchanged in both samples: Fe atoms are bound to six carbon atoms in the first coordination shell at a distance of 1.94 ± 0.01 Å and six nitrogen atoms at a distance of 3.12 ± 0.04 Å. Even though complementary studies with UV-Vis spectrometry indicated the interaction between humic acids and ferricyanide complex which led to a formation of ferricyanide-humo micelles, EXAFS results showed that the interaction did not imply changes in the original structure of ferricyanide complex.

25.5 CONCLUSIONS

- Standard energy-dispersive X-ray fluorescence (EDXRF) spectroscopy and total-reflection X-ray spectroscopy are techniques suitable for fast, cheap and reliable determination of bulk metal(loid) concentrations in soil, biologic materials and waters. However, the mentioned techniques are not appropriate for determination of lower concentrations of the light elements (e.g., Na, Mg Al, Si).
- Micro-proton induced X-ray emission (PIXE) or synchrotron micro-X-ray fluorescence spectroscopy (micro-XRF) are highly sophisticated methods available only at the accelerators or synchrotron sources. They enable visualization of metal(loid) spatial

distribution in biological tissues and therefore studies of the basic toxicity and tolerance mechanisms in all kind of organisms at tissue and cellular levels. However, the key point in these analyses is the sample preparation, in which element distribution needs to be preserved. For now, rapid freezing followed by controlled freeze-drying is the most appropriate method of sample preparation, although the development of techniques allowing the measurement of samples in a frozen hydrated state is necessary to reach a sub-cellular level.

- Synchrotron radiation X-ray absorption methods EXAFS (extended X-ray absorption fine structure) and XANES (X-ray absorption near edge structure) are well established and frequently used tools for characterization of atomic and molecular structure of materials. XANES spectra provide information on the valence state and local symmetry of the investigated atom, while EXAFS is used to determine number and species of neighbor atoms, their distance from the selected atom and the thermal or structural disorder of their positions. The analysis can be applied to crystalline, nanostructural or amorphous materials, organic tissues, liquids and molecular gases. The methods are very successfully applied in the field of biological samples and environmental research (pollution with heavy metals). For materials without long range order, as for example in the case of investigation of complexation of toxic metals in plant tissues and in soil, EXAFS and XANES are often the only practical way to obtain information on local structure around metal cations. Although the obtained structural information provides only data about the radial distribution of neighbors, without any direct information on their angular distribution, it can be used very efficiently in the process of identification of molecular structures, especially if combined with other complementary spectroscopic methods like FTIR, UV-vis or EDXRF. Bright synchrotron radiation sources make possible also a powerful combination of microscopy and X-ray absorption spectroscopy, the so called micro-spectroscopy methods (micro-XANES, micro-EXAFS, micro-XRF), which are efficiently used for biological sample, since they allow mapping of chemical state (valence state and local atomic structure) of selected elements in plant tissues on the sub-cellular level.

ACKNOWLEDGMENTS

This work was supported by the Ministry of Education, Science and Sport of Slovenia, the Slovenian Research Agency, and by DESY and the European Community's Seventh Framework Programme (FP7/2007-2013) ELISA (European Light Sources Activities) under grant agreement no. 226716. Access to synchrotron radiation facilities of HASYLAB (projects II-20080058 EC and I-20110082 EC), ESRF in Grenoble (projects EC-398 and EC-719), ELETTRA in Trieste (projects 20095090, 20085196, 20105073, 20110127, 20110086). IAEA CRP projects "Applications of Synchrotron Radiation for Environmental Sciences and Materials Research for Development of Environmentally Friendly Resources," research contract No. 16796 and "Microanalytical Techniques based on Nuclear Spectrometry for Environmental Monitoring and Material Studies" research agreement No. 15955 are acknowledged.

REFERENCES

Arčon, I., B. Mirtič, and A. Kodre. 1998. Determination of valence states of chromium in calcium chromates by using X-ray absorption near-edge structure (XANES) spectroscopy. *Journal of the American Ceramic Society* 81: 222–224.

Arčon, I., J.T. van Elteren, H.J. Glass, A. Kodre, and Z. Šlejkovec. 2005a. EXAFS and XANES study of Arsenic in contaminated soil. *X-ray Spectrometry* 34: 435–438.

Arčon, I., A. Kodre, J. Padežnik Gomilšek, M. Hribar, and A. Mihelič. 2005b. Cs L-edge EXAFS atomic absorption background. *Physica Scripta* T115: 235–236.

Arčon, I., A. Pastrello, L. Catalano, M. De Nobili, P. Cantone, and L. Leita. 2006. Interaction between Fe-Cyanide Complex and Humic Acids. *Environmental Chemistry Letters* 4: 191–194.

Arčon, I., A. Benčan, A. Kodre, and M. Kosec. 2007a. X-ray absorption spectroscopy analysis of Ru in La_2RuO_5. *X-ray Spectrometry* 30: 301–304.

Arčon, I., J. Kolar, A. Kodre, A. Hanžel, and M. Strlič. 2007b. XANES analysis of Fe valence in iron gall inks. *X-ray Spectrometry* 36: 199–205.

Baker, A.J.M. 1981. Accumulators and excluders—strategies in the response of plants to heavy metals. *Journal of Plant Nutrition* 3: 643–654.

Barceló, J., and C. Poschenrieder. 2003. Phytoremediation: principles and perspectives. *Contributions to Science* 2: 333–444.

Beckhoff, B., B. Kangisser, N. Langhoff, R. Wedell, and H. Wolf. 2006. *Handbook of practical X-ray fluorescence analysis*. Berlin Heidelberg: Springer-Verlag.

Bele, M., A. Kodre, I. Arčon, J. Grdadolnik, S. Pejovnik, and J.O. Besenhard. 1998. Adsorption of cetyltrimethylammonium bromide on carbon black from aqueous solution. *Carbon* 36: 1207–1212.

Benfatto, M., A. Congiu-Castellano, A. Daniele, and S. Della Longa. 2001. MXAN: a new software procedure to perform geometrical fitting of experimental XANES spectra. *Journal of Synchrotron Radiation* 8: 267–269.

Bhatia, N.P., K.B.Walsh, I. Orlic, R. Siegele, N. Ashwath, and A.J.M. Baker. 2004. Studies on spatial distribution of nickel in leaves and stems of the metal hyperaccumulator *Stackhousia tryonii* using nuclear microprobe (micro-PIXE) and EDXS techniques. *Functional Plant Biology* 31: 1061–1074.

Bianconi, A., J. Garcia, M. Benfatto, A. Marcelli, C.R. Natoli, and M.F. Ruiz-Lopez. 1991. Multielectron excitations in the K-edge X-ray absorption edge spectra of V, Cr and Mn 3d0 compounds with tetrahedral coordination. *Physical Review B* 43: 6885–6892.

Breese, M.B.H., D.N. Jamieson, and P.J.C. King. 1996. *Materials Analysis Using a Nuclear Microprobe*. New York: John Wiley & Sons.

Brooks, R.R. 1998. Geobotany and hyperaccumulators. In *Plants that Hyperaccumulate Heavy Metals. Their Role in Phytoremediation, Microbiology, Archaeology, Mineral Exploration and Phytomining*, 55–95. New York: CAB International.

Bunker, G. 2010. *A practical guide to X-ray absorption fine structure spectroscopy*. Cambridge: Cambridge University Press.

Campbell, J.L., T.L. Hopman, J.A. Maxwell, and Z. Nejedly. 2000. The Guelph PIXE software package III: Alternative proton database. *Nuclear Instruments & Methods in Physics Research Section B-beam Interactions with Materials and Atoms* 170: 193–204.

Cole, M.M., and R.F. Smith. 1984. Vegetation as indicator of environmental pollution. *Transactions/Institute of British Geographers* (N.S.) 9: 477–493.

Cole, M.M. 1965. *Biogeography in the service of man: with particular reference to the underdeveloped lands: an inaugural lecture*. London: Bedford College.

D'Angelo, P., S. Della Longa, A. Arcovito, M. Anselmi, A. Di Nola, and G. Chillemi. 2010. Dynamic investigation of protein metal active sites: Interplay of XANES and molecular dynamics simulations. *Journal of the American Chemical Society* 132: 14901–14910.

Della Longa, S., A. Arcovito, M. Girasole, J.L. Hazemann, and M. Benfato. 2001. Quantitative analysis of X-ray absorption near edge structure data by a full multiple scattering procedure: the Fe-CO geometry in photolyzed carbonmonoxy-myoglobin single crystal. *Physical Review Letters* 87: 15550–15551.

Di Cicco, A. 1995. EXAFS multiple-scattering data-analysis: GNXAS methodology and applications. *Physica B* 117: 125–128.

Dominko, R., I. Arčon, A. Kodre, D. Hanzel, and M. Gaberšček. 2009. *In-situ* XAS study on Li_2MnSiO_4 and Li_2FeSiO_4 cathode materials. *Journal of Power Sources* 189: 51–58.

Dominko, R., C. Sirisopanaporn, C. Masquelier, D. Hanzel, I. Arčon, and M. Gaberšček. 2010. On the origin of the electrochemical capacity of $Li_2Fe_{0.8}Mn_{0.2}SiO_4$. *The Journal of The Electrochemical Society* 157: A1309–A1316.

Fahrni, C.J. 2007. Biological applications of X-ray fluorescence microscopy: exploring the subcellular topography and speciation of transition metals. *Current Opinion in Chemical Biology* 11: 121–127.

Farges, F., G.E. Brown, and J.J. Rehr. 1997. Ti K-edge XANES studies of Ti coordination and disorder in oxide Compounds: Comparison between theory and experiment. *Physical Review B* 56: 1809–1819.

Farges, F. 2005. Ab-initio and experimental pre-edge investigations of the Mn K-edge XANES in oxide-type materials. *Physical Review B* 71: 1–14.

Feldmann, J., P. Salaun, and E. Lombi. 2009. Critical review perspective: elemental speciation analysis methods in environmental chemistry—moving towards methodological integration. *Environmental Chemistry* 6: 275–289.

Fernandez-Garcia, M. 2002. XANES analysis of catality systems under reaction conditions. *Catalysis Reviews-Science and Engineering* 44: 59–121.

Frahm, R. 1989. New method for time-dependent X-ray absorption studies. *Review of Scientific Instruments* 60: 2515–2518.

Frenkel, A., A. Vairavamurthy, and M. Newville. 2001. A study of the coordination environment in aqueous cadmium-thiol complexes by EXAFS spectroscopy: experimental vs theoretical standards. *Journal of Synchrotron Radiation* 8: 668–671.

Habchi, C., D.T. Nguyen, G. Devès, S. Incerti, L. Lemelle, P. Le Van Vang, Ph. Moretto, R. Ortega, H. Seznec, A. Sakellariou, C. Sergeant, A. Simionovici, M.D. Ynsa, E. Gontier, M. Heiss, T. Pouthier, A. Boudou, and F. Rebillat. 2006. Three-dimensional densitometry imaging of diatom cells using STIM tomography. *Nuclear Instruments and Methods in Physics Research Section B: Beam Interactions with Materials and Atoms* 249: 653–659.

Hayakawa, K., K. Hatada, P. D'Angelo, S. Della Longa, C.R. Natoli, and M. Benfatto. 2004. Full quantitative multiple-scattering analysis of X-ray absorption spectra: Application to potassium hexacyanoferrat(II) and -(III) complexes. *Journal of the American Chemical Society* 126: 15618–15623.

He, F., and P. Van Espen. 1991. General approach for quantitative energy dispersive X-ray fluorescence analysis based on fundamental parameters. *Analytical Chemistry* 63: 2237–2224.

Hettiarachchi, G.M., K.G. Scheckel, J.A. Ryan, S.R. Sutton, and M. Newville. 2006. μ-XANES and μ-XRF investigations of metal-binding mechanisms in biosolids. *Journal of Environmental Quality* 35: 342–351.

Hornberger, B., M. Feser, and C. Jacobsen. 2007. Quantitative amplitude and phase contrast imaging in a scanning transmission X-ray microscope. *Ultramicroscopy* 107: 644–655.

Hornberger, B., M.D. de Jonge, M. Feser, P. Holl, C. Holzner, C. Jacobsen, D. Legnini, D. Paterson, P. Rehak, L. Strüder, and S. Vogt. 2008. Differential phase contrast with a segmented detector in a scanning X-ray microprobe. *Journal of Synchrotron Ra*diation 15: 355–362.

Horowitz, P., and J.A. Howell. 1972. A scanning X-ray microscope using synchrotron radiation. *Science* 10: 608–611.

Howells, M., C. Jacobsen, T. Warwick, and A. Van den Bos. 2007. Principles and applications of zone plate X-ray microscopes. In *Science of microscopy*, 835–926. New York: Springer.

Isaure, M.P., B. Fayard, G. Sarret, S. Pairis, and J. Bourguignon. 2006. Localization and chemical forms of cadmium in plant samples by combining analytical electron microscopy and X-ray spectromicroscopy. *Spectrochimica Acta Part B—Atomic Spectroscopy* 61: 1242–1252.

Ishii, K., H. Yamazaki, S. Matsuyama, W. Galster, T. Satoh, and M. Budnar. 2005. Contribution of atomic bremsstrahlung in PIXE spectra and screening effect in atomic bremsstrahlung. *X-ray Spectrometry* 34: 363–365.

Jaklevic, J.M., and R.D. Giauque. 1993. Energy-dispersive X-ray fluorescence analysis using X-ray tube excitation. In *Handbook of X-ray Spectronomy*, 151–180. New York: Marcel Dekker Inc.

Janssens, K. 2004. X-ray based methods of analysis. In *Non-destructive microanalysis of cultural heritage materials*, 129–226. Amsterdam: Elsevier.

Jenkins, R. 1999. *X-ray Fluorescence Spectrometry*. New York: John Wiley & Sons Inc.

Johansson, T.B., R. Akselsson, and S.A.E. Johansson. 1970. X-ray analysis: Elemental trace analysis at the 10^{-12} g level. *Nuclear Instruments & Methods in Physics Research Section B—Beam Interactions with Materials and Atoms* 84: 141–143.

Johansson, S.A.E., and J.L. Campbell. 1988. *PIXE: A novel technique for elemental analysis*. New York: John Wiley & Sons.

Kachenko, A.G., R. Siegele, N.P. Bhatia, B. Singh, and M. Ionescu. 2008. Evaluation of specimen preparation techniques for micro-PIXE localisation of elements in hyperaccumulating plants. *Nuclear Instruments & Methods in Physics Research Section B—Beam Interactions with Materials and Atoms* 266: 1598–1604.

Kalnicky, D.J., and R. Singhvi. 2001. Field portable XRF analysis of environmental samples. *Journal of Hazardous Materials* 93: 93–122.

Kanngießer, B., O. Hahn, M. Wilke, B. Nekat, W. Malzer, and A. Erko. 2004. Investigation of oxidation and migration processes of inorganic compounds in ink-corroded manuscripts. *Spectrochimica Acta Part B—Atomic Spectroscopy* 59: 1511–1516.

Karydas, A.G., D. Sokaras, C. Zarkadas, N. Grlj, P. Pelicon, M. Žitnik, R. Schütz, W. Malzer, and B. Kanngießer. 2007. 3D Micro PIXE—a new technique for depth-resolved elemental analysis. *Journal of Analytical Atomic Spectrometry* 22: 1260–1265.

Kaulich, B., T. Wilhein, E.D. Fabrizio, F. Romanato, M. Altissimo, S. Cabrini, B. Fayard, and J. Susini. 2002. Differential interference contrast X-ray microscopy with twin zone plates. *Journal of the Optical Society of America A-Optics Image Science and Vision* 19: 797–806.

Kaulich, B., D. Bacescu, J. Susini, C. David, E. Di Fabrizio, G.R. Morrison, P. Charalambous, J. Thieme, T. Wilhein, J. Kovac, D. Cocco, M. Salome, O. Dhez, T. Weitkamp, S. Cabrini, D. Cojo, A. Gianoncelli, U. Vogt, M. Podnar, M. Zangrando, M. Zacchigna, and M. Kiskinova. 2006. A European twin X-ray microscopy station commissioned at ELETTRA. In *Proc. 8th Int. Conf. X-ray microscopy. Conf. Proc. Series IPAP, no. 7*, 22–25. Tokyo: Institute of Pure and Applied Physics.

Kaulich, B., A. Gianoncelli, A. Beran, D. Eichert, I. Kreft, P. Pongrac, M. Regvar, K. Vogel-Mikuš, and M. Kiskinova. 2009. Low-energy X-ray fluorescence microscopy opening new opportunities for bio-related research. *Journal of the Royal Society Interface* 6: 641–647.

Kaulich, B., P. Thibault, A. Gianoncelli, and M. Kiskinova. 2011. Transmission and emission X-ray microscopy: Operation modes, contrast mechanisms and applications. *Journal of Physics: Condensed Matter* 23: 083002.

Kidd, P., J. Barceló, M.P. Bernal, F. Navarri-Izzo, C. Poschenrieder, S. Shilev, R. Clemente, and C. Monterosso. 2009. Trace element behaviour at the root-soil interface: Implications in phytoremediation. *Environmental and Experimental Botany* 67: 243–259.

Klockenkämper, R. 1997. *Total-reflection X-ray fluorescence Analysis* (*Chemical analysis: A series of monographs on analytical chemistry and its applications*). New York: John Wiley & Sons Inc.

Kodre, A., R. Prešeren, I. Arčon, J. Padeznik Gomilšek, and M. Borowski. 2001. A study of transferability of atomic background on EXAFS spectra of simple gaseous compounds of As. *Journal of Synchrotron Radiation* 8: 282–284.

Kodre, A., I. Arčon, J. Padežnik Gomilšek, R. Prešeren, and R. Frahm. 2002. Multielectron excitations in X-ray absorption spectra of Rb and Kr. *Journal of Physics B—Atomic Molecular and Optical Physics* 35: 3497–3518.

Kodre, A., I. Arčon, J. Padežnik Gomilšek, and A. Mihelič. 2005. Atomic absorption background in EXAFS spectra of Rb in inter-alkaline alloys. *Physica Scripta* T115: 218–220.

Kodre, A., J. Padežnik Gomilšek, A. Mihelič, and I. Arčon. 2006. X-ray absorption in atomic Cd in the K edge region. *Radiation Physics and Chemistry* 75: 188–194.

Kodre, A., J. Padeznik Gomilsek, I. Arčon, and G. Aquilanti. 2010. X-ray atomic absorption of cesium and xenon in the L-edge region, *Physical Review A* 82: 022513.

Koningsberger, D.C., and R. Prins. 1988. *X-ray Absorption, Principles, Techniques of EXAFS, SEXAFS and XANES*. New York: John Wiley & Sons.

Kump, P. 1997. Some considerations on the definition of the limit of detection in X-ray fluorescence spectrometry. *Spectrochimica Acta Part B—Atomic Spectroscopy* 52: 405–408.

Kump, P., M. Nečemer, and M. Veber. 1997. Determination of trace elements in mineral water using total-reflection X-ray fluorescence spectrometry after pre-concentration with ammonium pyrrolidinedithiocarbamate. *X-ray Spectrometry* 26: 232–236.

Küzma, M., R. Dominko, D. Hanžel, A. Kodre, I. Arčon, A. Meden, and M. GaberščEk. 2009. Detailed *in situ* investigation of the electrochemical processes in Li_2FeTiO_4 cathodes. *Journal of the Electrochemical Society* 156: 809–816.

Lanzirotti, A., and S. Sutton. 2002. *Hard X-Ray Microprobe. Guidebook to Imaging and Microspectroscopy at the NSLS: Imaging and Microspectroscopy at the National Synchrotron Light Source: Techniques.* http://www.nsls.bnl.gov/newsroom/publications/otherpubs/imaging/hardxraymicroprobe.pdf (accessed November 16, 2011).

LearnXRF. 2009. LearnXRF.com—Your Source for Learning XRF. http://www.learnxrf.com (accessed November 16, 2011).

Lee, P.A., P.H. Citrin, P. Eisenberger, and B.M. Kincaid. 1981. Extended X-ray absorption fine structure—Its strengths and limitations as a structural tool. *Reviews of Modern Physics* 53: 769–806.

Legge, G.J.F., and M. Cholewa. 1994. The principles of proton probe microanalysis in biology. *Scanning Microscopy Supplement* 8: 295–315.

Leita, L., A. Margon, A. Pastrello, I. Arčon, M. Contin, and D. Mosetti. 2009. Soil humic acids may favour the persistence of hexavalent chromium in soil. *Environmental Pollution* 157: 1862–1866.

Lombi, E., F.J. Zhao, S.J. Dunham, and S.P. McGrath. 2000. Cadmium accumulation in populations of *Thlaspi caerulescens* and *Thlaspi goesingense*. *New Phytologist* 145: 11–20.

Lombi, E., and J. Susini. 2009. Synchrotron-based techniques for plant and soil science: opportunities, challenges and future perspectives. *Plant and Soil* 320: 1–35.

Lombi, E., K.G. Scheckel, and I.M. Kempsond. 2011a. *In situ* analysis of metal(loid)s in plants: State of the art and artefacts. *Environmental and Experimental Botany* 72: 3–17.

Lombi, E., M.D. de Jonge, E. Donner, P.M. Kopittke, D.L. Howard, R. Kirkham, C.G. Ryan, and D. Paterson. 2011b. Fast X-ray fluorescence microtomography of hydrated biological samples. *PLos One* 6: e20626.

Lu, K., and E.A. Stern. 1983. Size effect of powdered sample on EXAFS amplitude. *Nuclear Instruments and Methods* 212: 475–478.

Lytle, F.W., R.B. Greegor, and A.J. Panson. 1988. Discussion of X-ray-absorption near-edge structure: Application to Cu in the high-Tc superconductors $La_{1.8}Sr_{0.2}CuO_4$ and $YBa_2Cu_3O_7$, *Physical Review B* 37: 1550–1562.

Makovec, D., A. Kodre, I. Arčon, and M. Drofenik. 2009. Structure of manganese zinc ferrite spinel nanoparticles prepared with co-precipitation in reversed microemulsions. *Journal of Nanoparticle Research* 11: 1145–1158.

Marguí, E., M. Queralt, and M. Hidalgo. 2009. Application of X-ray fluorescence spectrometry to determination and quantification of metals in vegetal material. *Trends in Analytical Chemistry* 28: 362–372.

Markowicz, A., N. Haselberger, and P. Mulenga. 1992. Accuracy of calibration procedure for energy-dispersive X-ray fluorescence spectrometry. *X-ray Spectrometry* 21: 271–276.

Markowicz, A.A. 1993. X-ray physics. *In Handbook of X-ray Spectronomy*, 1–74. New York: Marcell Dekker Inc.

Markowicz, A.A., and R.E. Van Grieken. 1993. Quantification in XRF Analysis of Intermediate—Thickness Samples. In *Handbook of X-ray Spectroscopy: Methods and Techniques*, 339–358. New York: Marcell Dekker Inc.

McMaster, W.H., N.K. Delgrand, J.H. Mallet, and J.H. Hubbell. 1968. *Compilation of X-ray cross sections, UCRL report 50174*. USA: Lawrence Livermore Lab., University of California.

Mesjasz-Przybyłowicz, J., and W.J. Przybyłowicz. 2002. Micro-PIXE in plant sciences: Present status and perspectives. *Nuclear Instruments & Methods in Physics Research Section B—Beam Interactions with Materials and Atoms* 189: 470–481.

Mihelič, A., A. Kodre, I. Arčon, J. Padežnik Gomilšek, and M. Borowski. 2002. A double cell for X-ray absorption spectrometry of atomic Zn. *Nuclear Instruments & Methods in Physics Research Section B—Beam Interactions with Materials and Atoms* 196: 194–197.

Mills, E.N.C., M.L. Parker, N. Wellner, G. Toole, K. Feeney, and P.R. Shewry. 2005. Chemical imaging: The distribution of ions and molecules in developing and mature wheat grain. *Journal of Cereal Science* 41: 193–201.

Mulligan, C.N., R.N. Yong, and B.F. Gibbs. 2001. Remediation technologies for metal contaminated soils and groundwater: an evaluation. *Engineering Geology* 60: 193–207.

Nečemer, M., P. Kump, J. Ščančar, R. Jaćimović, J. Simčič, P. Pelicon, M. Budnar, Z. Jeran, P. Pongrac, M. Regvar, and K. Vogel-Mikuš. 2008. Application of X-ray fluorescence analytical techniques in phytoremediation and plant biology studies. *Spectrochimica Acta Part B-Atomic Spectroscopy* 63: 1240–1247.

Nečemer, M., I.J. Košir, P. Kump, U. Kropf, M. Korošec, J. Bertoncelj, N. Ogrinc, and T. Golob. 2009. Application of total reflection X-ray spectrometry in combination with chemometric methods for determination of the botanical origin of Slovenian honey. *Journal of Agricultural and Food Chemistry* 57: 4409–4414.

Nečemer, M., P. Kump, and K. Vogel-Mikuš. 2011. Use X-ray fluorescence-based analytical techniques in phytoremediation. In *Handbook of phytoremediation, (Environmental science, engineering and technology)*, 331–358. New York: Nova Science Publishers.

Niemann, B, D. Rudolph, and G. Schmahl. 1976. X-ray microscopy with synchrotron radiation. *Applied Optics* 15: 1883–1884.

Orlic, I., R. Siegele, K. Hammerton, R.A. Jeffree, and D.D. Cohen. 2003. Nuclear microprobe analysis of lead profile in crocodile bones. *Nuclear Instruments & Methods in Physics Research Section B—Beam Interactions with Materials and Atoms* 210: 330–335.

Padežnik Gomilšek, J., A. Kodre, I. Arčon, A.M. Loireau-Lozac'h, and S. Bénazeth. 1999. Multielectron photoexcitations in X-ray absorption spectra of 4p elements. *Physical Review A* 59: 3078–3081.

Padežnik Gomilšek, J., A. Kodre, I. Arčon, and V. Nemanič. 2008. X-ray absorption in atomic potassium. *Nuclear Instruments & Methods in Physics Research Section B—Beam Interactions with Materials and Atoms* 266: 677–680.

Padežnik Gomilšek, J., A. Kodre, I. Arčon, S. De Panfilis, and D. Makovec. 2011. Atomic absorption background of Ba in EXAFS analysis of $BaFe_{12}O_{19}$ nanoparticles. *Journal of Synchrotron Radiation* 18: 557–563.

Pallon, J.,V. Auzelyte, M. Elfman, M. Gramer, P. Kristiansson, K. Malmqvist, C. Nilsson, A. Shariff, and M. Wegdén. 2004. An off-axis STIM procedure for precise mass determination and imaging. *Nuclear Instruments & Methods in Physics Research Section B—Beam Interactions with Materials and Atoms* 219–220: 988–993.

Pantazis, T., J. Pantazis, A. Huber, and R. Redus. 2010. The historical development of the thermoelectrically cooled X-ray detector and its impact on the portable and hand-held XRF industries (February 2009). *X-ray Spectrometry* 39: 90–97.

Pantelouris, A., H. Modrow, M. Pantelouris, J. Hormes, and D. Reinen. 2004. The influence of coordination geometry and valency on the K-edge absorption near-edge spectra of selected chromium compounds. *Chemical Physics* 300: 13–22.

Pascarelli, S., T. Neisius, and S. De Panfillis. 1999. Turbo-XAS: dispersive XAS using sequential acquisition. *Journal of Synchrotron Radiation* 6: 1044–1050.

Pella, P.A., L. Feng, and J.A. Small. 1985. Analytical algorithm for calculation of spectral distributions of X-ray tubes for quantitative X-ray fluorescence analysis. *X-ray Spectrometry* 14: 125–135.

Prešeren, R., A. Kodre, I. Arčon, and M. Borowski. 2001. Atomic background and EXAFS of gaseous hydrides of Ge, As, Se and Br. *Journal of Synchrotron Radiation* 8: 279–281.

Proost, K., K. Janssens, B. Wagner, E. Bulska, and M. Schreiner. 2004. Determination of localized Fe^{2+}/Fe^{3+} ratios in inks of historic documents by means of μ-XANES. *Nuclear Instruments & Methods in Physics Research Section B—Beam Interactions with Materials and Atoms* 213: 723–728.

Punshon, T., M.L. Guerinot, and A. Lanzirotti. 2009. Using synchrotron X-ray fluorescence microprobes in the study of metal homeostasis in plants. *Annals of Botany* 103: 665–672.

Rarback, H., J.M. Kenney, J. Kirz, M. Howells, P. Chang, P.J. Coane, R. Feder, and P.J. Houzego. 1987. Recent results from the Stony Brook scanning microscope. In *X-ray microscopy*, Springer Series in Optical Sciences 43, 203–215. Berlin: Springer.

Ravel, B., and M. Newville. 2005. ATHENA, ARTEMIS, HEPHAESTUS: data analysis for X-ray absorption spectroscopy using IFEFFIT. *Journal of Synchrotron Radiation* 12: 537–541.

Rehr, J.J., R.C. Albers, and S.I. Zabinsky. 1992. High-order multiple-scattering calculations of X-ray-absorption fine structure. *Physical Review Letters* 69: 3397–3400.

Rehr, J.J., and R.C. Albers. 2000. Theoretical approaches to X-ray absorption fine structure. *Reviews of Modern Physics* 72: 621–654.

Rehr, J.J., and A.L. Ankudinov. 2005. Progress in the Theory and Interpretation of XANES Coordination. *Chemical Reviews* 249: 131–140.

Robinson, A. 2009. History of synchrotron radiation. In *X-ray data booklet.* Berkeley California: Lawrence Berkeley National Laboratory, University of California.

Rousseau, R.M. 1984. Fundamental algorithm between concentration and intensity in XRF analysis, 1—Theory. *X-ray Spectrometry* 13: 115–120.

Roy, M., S.J. Gurman, and G. van Dorssen. 1997. The amplitude reduction factor in EXAFS. Proceedings of the 9th International Conference on X-ray Absorption Fine Structure. *Journal de Physique IV France* 7, C2-151–C2-152.

Rueff, J.P., L. Journel, P.E. Petit, and F. Farges. 2004. Fe K-pre-edges as revealed by resonant X-ray emission. *Physical Review B* 69: 235107.

Ryan, C.G. 2001. Developments in dynamic analysis for quantitative PIXE true elemental imaging. *Nuclear Instruments & Methods in Physics Research Section B—Beam Interactions with Materials and Atoms* 181: 170–179.

Sarangi, R., M. Benfatto, K. Hayakawa, L. Bubacco, E.I. Solomon, K.O. Hodgson, and B. Hedman. 2005. MXAN analysis of the XANES energy region of a mononuclear copper complex: Applications to bioinorganic systems. *Inorganic Chemistry* 44: 9652–9659.

Scheckel, K.G., R.L Chaney, N.T. Basta, and J.A. Ryan. 2009. Advances in assessing bioavailability of metal(loids) in contaminated soils. *Advances in Agronomy* 104: 1–52.

Scheloske, S., and T. Schneider. 2002. BIOPIXE: A new PIXE-data software package to analyse quantitative elemental distributions of inhomogeneous samples. *Nuclear Instruments & Methods in Physics Research Section B—Beam Interactions with Materials and Atoms* 189: 148–152.

Schneider, T., A. Haag-Kerwer, M. Maetz, M. Niecke, B. Povh, T. Rausch, and A. Schuszler. 1999. Micro-PIXE studies of elemental distribution in Cd-accumulating *Brassica juncea* L. *Nuclear Instruments & Methods in Physics Research Section B—Beam Interactions with Materials and Atoms* 158: 329–334.

Schneider, T., S. Sheloske, and B. Povh. 2002. A method for cryosectioning of plant roots for proton microprobe analysis. *International Journal of PIXE* 12: 101–107.

Schwenke, H., and J. Knoth. 1993. Total reflection XRF. In *Handbook of X-ray Spectronomy*, 453–490. New York, Marcel Dekker Inc.

Singh, V.P. 2005. Metal toxicity in animals and plants. New Delhi: Sarup & Sons.

Stern, E.A., and K. Kim. 1981. Thickness effect on the extended X-ray-absorption-fine-structure amplitude. *Physical Review B* 23: 3781–3787.

Stoehr, J. 1992. NEXAFS Spectroscopy. Springer series in surface sciences 25. Heidelberg: Springer-Verlag.

Tertian, R., and F. Claisse. 1982. *Principles of Quantitative X-ray Fluorescence Analysis*. London-Philadelphia-Rheine: Heyden & Son Ltd.

Thompson-Eagle, E.T., and J.W.T. Frankenberger. 1990. Protein-mediated selenium biomethylation in evaporation pond water. *Environmental Toxicology and Chemistry* 9: 1453–1560.

Tolrà, R.P., K. Vogel-Mikuš, R. Hajiboland, P. Kump, P. Pongrac, B. Kaulich, A. Gianoncelli, V. Babin, J. Barceló, M. Regvar, and C. Poschenrieder. 2011. Localization of aluminium in tea (*Camelia sinensis*) leaves using low energy X-ray fluorescence spectro-microscopy. *Journal of Plant Research* 124: 165–172.

Tylko, G., J. Mesjasz-Przybyłowicz, and W.J. Przybyłowicz. 2007. X-ray microanalysis of biological material in the frozen-hydrated state by PIXE. *Microscopy Research and Technique* 70: 55–68.

Uzonyi, I., and Gy. Szabó, 2005. PIXEKLM-TPI—a software package for quantitative elemental imaging with nuclear microprobe. *Nuclear Microprobe Technology and Applications* 231: 156–161.

Van Dyck, P.M., and R.E. Van Grieken. 1980. Absorption correction via scattered radiation in energy-dispersive X-ray fluorescence analysis for samples of variable composition and thickness. *Analytical Chemistry* 52: 1859–1864.

Van Dyck, P.M., S.B. Torok, and R.E. Van Grieken. 1986. Enhancement effect in X-ray fluorescence analysis of environmental samples of medium thickness. *Analytical Chemistry* 58: 1761–1766.

van Elteren, J.T., Z. Šlejkovec, I. Arčon, and H.J. Glass. 2006. An interdisciplinary physical-chemical approach for characterisation of arsenic in a calciner residue dump in Cornwall (UK). *Environmental Pollution* 139: 477–488.

Van Espen, P., and F. Adams. 1981. Calibration of tube excited energy-dispersive X-ray spectrometers with thin film standards and with fundamental constants. *X-ray Spectrometry* 10: 64–68.

Vincze, L., B. Vekemans, F.E. Brenker, G. Falkenberg, K. Rickers, A. Somogyi, M. Kersten, and F. Adams. 2004. Three-dimensional trace element analysis by confocal X-ray microfluorescence imaging. *Analitical Chemistry* 76: 6786–679.

Vogel-Mikuš, K., P. Pongrac, P. Kump, M. Nečemer, J. Simčič, J. Pelicon, M. Budnar, B. Povh, and M. Regvar. 2007. Localisation and quantification of elements within seeds of Cd/Zn hyperaccumulator *Thlaspi praecox* by micro-PIXE. *Environmental Pollution* 147: 50–59.

Vogel-Mikuš, K., J. Simčič, J. Pelicon, M. Budnar, P. Kump, M. Nečemer, J. Mesjasz-Przybyłowicz, W. Przybyłowicz, and M. Regvar. 2008a. Comparison of essential and non-essential element distribution in leaves of the Cd/Zn hyperaccumulator *Thlaspi praecox* as revealed by micro-PIXE. *Plant, Cell and Environment* 31: 1484–1496.

Vogel-Mikuš, K., M. Regvar, J. Mesjasz-Przybyłowicz, W. Przybyłowicz, J. Simčič, P. Pelicon, and M. Budnar. 2008b. Spatial distribution of Cd in leaves of metal hyperaccumulating *Thlaspi praecox* using micro-PIXE. *New Phytologist* 179: 712–721.

Vogel-Mikuš, K., P. Pelicon, P. Vavpetič, I. Kreft, and M. Regvar. 2009a. Elemental analysis of edible grains by micro-PIXE: common buckwheat case study. *Nuclear Instruments & Methods in Physics Research Section B—Beam Interactions with Materials and Atoms* 267: 2884–2889.

Vogel-Mikuš, K., P. Pongrac, P. Pelicon, P. Vavpetič, B. Povh, H. Bothe, and M. Regvar. 2009b. Micro-PIXE Analysis for Localisation and Quantification of Elements in Roots of Mycorrhizal Metal-Tolerant Plants. In *Soil Biology* 18, 227–242. Berlin, Heidelberg: Springer.

Vogel-Mikuš, K., I. Arčon, and A. Kodre. 2010. Complexation of cadmium in seeds and vegetative tissues of the cadmium hyperaccumulator *Thlaspi praecox* as studied by X-ray absorption spectroscopy. *Plant and Soil* 331: 439–451.

WCIF-ImageJ. 2006. Online Manual for the WCIF-ImageJ collection, Colocalisation analysis: http://www.uhnresearch.ca/facilities/wcif/imagej/colour_analysis.htm (accessed November 16, 2011).

Warburton, W.K., M. Momayezi, B. Hubbard-Nelson, and W. Skulski. 2000. Digital pulse processing: new possibilities in nuclear spectroscopy. *Applied Radiation and Isotopes* 53: 913–920.

Wenzel, W.W., D.C. Adriano, D. Salt, and R. Smith. 1999. Phytoremediation: a plant-microbe-based remediation system. In *Bioremediation of contaminated soils*. Madison WI: ASA Monograph 37: 273–303.

Wilhein, T., B. Kaulich, E. Di Fabrizio, F. Romanato, S. Cabrini, and J. Susini. 2001. Differential interference contrast X-ray microscopy with submicron resolution. *Applied Physics Letters* 78: 2082–2084.

Wong, J., F.W. Lytle, R.P. Messmer, and D.H. Maylotte. 1984. K-edge spectra of selected vanadium compounds. *Physical Review B* 30: 5596–5610.

Yap, C.T., P. Kump, S.M Tang, and L. Wijesinghe. 1987. Calibration of the radioisotope-excited X-ray spectrometer with thick standards. *Applied Spectroscopy* 41: 80–85.

Ziegler, J.F., SRIM & TRIM. 2011. The Stopping and Range of Ions in Matter. Downloadable software which describes the transport properties of ions in matter. http://www.srim.org/ (accessed November 15, 2011).

26 Metals and Metalloids Detoxification Mechanisms in Plants *Physiological and Biochemical Aspects*

Palaniswamy Thangavel, Ganapathi Sridevi, Naser A. Anjum, Iqbal Ahmad, and Maria E. Pereira

CONTENTS

26.1 INTRODUCTION

Plants are continuously confronted with the harsh environmental conditions including toxic metals and metalloids contamination. Anthropogenic and especially industrial activities have impacted severely in our environment during the last century through releasing major toxic metals and metalloids to varied environmental compartments. Increasing emissions of metals/metalloids are dangerous because they may get ultimately into the food chain with risks for human health (Angelone and Bini 1992).

Therefore, exploring metals/metalloids stress-tolerant plant species is urgent to reclaim metals/metalloids-contaminated sites so as to make them suitable for recultivation/revegetation. In this context, however, biotechnological efforts are underway to improve plant stress tolerance and the ability to extract pollutants from the soil with the aim of using plants for soil clean-up. To devise new strategies for phytoremediation and improved tolerance, it is important to understand the basic principles as to how the metals/metalloids are detoxified at whole plant and cellular levels. In fact, plants have been reported to adopt a number of vital strategies such as immobilization, exclusion, chelation and compartmentalization of the metal ions in addition to inducing major enzymatic and nonenzymatic components of antioxidant defense system to detoxify metals/metalloids stress (Cobbett 2000; Schützendübel and Polle 2002). The current chapter aimed to discuss the significant physiological and biochemical metals-detoxification mechanisms in plants. Furthermore, as the members of plant family Brassicaceae stand second to none in terms of metals-accumulation and/or remediation potential and they exhibit love for plant nutrient sulfur, this chapter also critically discusses the role of sulfur-containing metabolites in metal detoxification mechanism in oilseed rapes in the light of available recent reports.

26.2 METAL COMPLEXATION BY LIGANDS

Organic acids are natural chelators synthesized by plants and they play a key role in heavy metal cellular homeostasis (Rauser 1999). Since only small amounts of metal exist as free ions, they are believed to be bound by LMW ligands such as organic acids (e.g., malic, citric, oxalic acids). Organic acids chelate free metal ions in the cytoplasm, thereby participating in exclusion mechanism and increase the rate of diffusion of chelates to vacuole (Boominathan and Doran 2003). In *Brassica*, the LMW organic acids that act as chelating agents are increased after metal stress (Seth et al. 2008). A significant linear relationship was observed between shoot organic acid concentration and Cd content in *B. juncea* grown under Cd treatments. This provided the evidence that Cd induces the organic acid formation and also their role in immobilization and metal detoxification (Irtelli and Navari-Izzo 2006). Plant vacuoles are major repositories for organic acids. Thus, the association between the metals and organic acids clearly vivid that metal detoxification occurs by vacuole sequestration.

Two-thirds of the Cd ligands belong to thiol groups (Cd-S-C-) in intact seeds and isolated embryos of *T. praecox*. Cd is mainly coordinated with O-ligands in mature and senescent leaves and with S-ligands in younger leaves of *T. caerulescens*. A higher proportion of S-ligands bound by the metal are stored in the mesophyll while the sequestration into the epidermis is less feasible (Küpper et al. 2004). The microproton-induced X-ray emission (micro-PIXE) localization studies in the Cd treated *T. praecox* revealed that Cd accumulates in the photosynthetically active mesophyll symplast, where it is bound to S-ligands (Vogel-Mikuš et al. 2008). On the contrary, in the vacuole, the weaker O-ligands (e.g., organic acids) play a dominant role in metal detoxification. In seeds, a high proportion of S-ligands are present indicating that the Cd detoxification is similar to those in young vegetative tissues (Küpper et al. 2004). The phloem is the main source of transport during seed development, as phloem sap is more an alkaline pH environment than xylem sap. Hence, the Cd may be loaded into seeds as Cd-thiolate complex, a nontoxic compound with strong Cd-S coordination. The phloem sap of *B. napus* contained mainly Cd complexed with thiols (Mendoza-Cózatl et al. 2008). Almost 60% ligands in the veins of *T. praecox* were thiol groups while the remaining were O-ligands. The question whether particular Cd binding ligands (O or S) dominate in epidermal or mesophyll leaf tissues is unanswered. At least two routes of loading and storage of Cd exist in the *T. praecox* seeds. The S (thiolate) ligand involves transport and loading Cd-thiolate complexes and the O (phytate) ligand for Zn transport and loading (Tauris et al. 2009). The Cd can be transported to seeds as Cd-nicotinamine complex and stored as Cd-phytate in embryonic tissues. In roots, the Cd is present in lowest concentration as compared to the other vegetative organs and tissues, thus binding 79% of Cd to O-ligands (Cd-O-C-coordination) of cell wall components or by sequestration into vacuole via organic acids. The remaining 21% of Cd is bound to S-ligands (Cd-S-C-coordination). The epidermis of mature leaf tissues possessed 80% of O-ligands and the remaining 20% were S-ligands. A mixture of O and S-ligands existed in mesophyll but there were slight differences in the ratio between the S and O-ligands (Vogel-Mikuš et al. 2010).

The same metal has different oxidation mechanisms at different oxidation states. The uptake of As(V) in roots of Indian mustard via the phosphate transport channel of which small fraction is transported as oxyanions of As(V) and As(III) to the shoot through the xylem (Pickering et al. 2000). Hence, the As accumulated more as As(V) rather than As(III) in *B. juncea* possibly due to difference in the transport mechanism. The As(V) is transported via phosphate transporters whereas As(III) is via aquaporin nodulin 26-like intrinsic proteins (NIPs) (Zhao et al. 2009). The majority of the metalloid exists as AsIII-tris-thiolate complex in the roots and AsIII-tris-glutathione in shoots. The Cr(VI) transport is an active mechanism that involves essential anions such as SO_4^{2-}. There is a high accumulation of Cr(VI) in the roots which may be due to less transport of Cr from the root to aerial parts in *B. juncea* plants (Han et al. 2004) and Cr immobilization in the vacuoles of root cells (Shanker et al. 2005).

26.3 ENZYMATIC ANTIOXIDANTS AND PLANT METAL STRESS TOLERANCE

Environmental stresses including metal stress, results in the formation of reactive oxygen species (ROS), such as superoxide radicals (O_2^-), hydrogen peroxide (H_2O_2) and hydroxyl radicals and followed by the electron transport chain impairment (Noctor and Foyer 1998; Anjum et al. 2012). Several metal ions such as Zn, Cd, Cu, Pb and Se results in the peroxidation of plasma and chloroplast membrane lipids via ROS generation and thereby altering the membrane properties thus resulting in the negative effects on plant growth (Halliwell and Gutteridge 1999). These free radicals also oxidize photosynthetic pigments, proteins and nucleic acids thus causing a great damage to the plant cells (Reddy et al. 2004). In plants, antioxidant defense system falls into two classes: (i) lipid soluble low molecular weight antioxidants like β-carotene etc. and water soluble reductants like GSH, ascorbic acid (AsA) and (ii) enzymatic antioxidants like superoxide dismutase (SOD), catalase (CAT), ascorbate peroxidase (APX) and glutathione reductase (GR). Other enzymes like dehydroascorbate reductase (DHAR) and monodehydroascorbate reductase (MDHAR) are involved in detoxification of H_2O_2 and peroxides (Gill et al. 2011; Anjum et al. 2011, 2012).

In plants, the vital steps in the AsA-GSH cycle are catalyzed by GR i.e., the conversion of GSH from the oxidized state to reduced state, GSSG (Noctor and Foyer 1998). GR plays an important role in maintaining the GSH:GSSG and AsA:DHA ratios and thus protects the plants from oxidative stress. Upon Cd treatment, increased GR activity was evident in *B. juncea* (L.) Czern. & Coss. However, it was suppressed by salicylic acid (SA) treatment (Ahmad et al. 2011). The different electrophilic substrates of GSH were catalyzed by glutathione S-transferase (GST). GSTs play a key role in the detoxification of xenobiotics (Edwards and Dixon 2004) and water soluble and less toxic substances are further transported to vacuole for further degradation (Marrs 1996). A variety of environmental stresses including heavy metals induces the GST activity (Dixit et al. 2001; Edwards and Dixon 2004; Gajewska and Skłodowska 2010).

Glutathione peroxidase (GPx) utilizes the GSH pool and catalyzes the reduction of hydroperoxides, thereby protecting the cell against oxidative damage (Halliwell and Gutteridge 1999). The response of GPx to heavy metal exposure is diverse and depends upon the type of metal, species/plant tissue, duration of metal exposure, co-occurring antioxidant and metal detoxification mechanisms (Dixit et al. 2001). The adaptation of GPx response persists over longer period of exposure of metals. If the exposure is prolonged, the biological response for a particular concentration of metal exposed may decline, i.e., due to adaptation. This may be due to the induced alternative detoxification/depuration mechanisms to restore homeostasis (Wu et al. 2005).

SOD constitutes the 'first line of defense' catalyzing the dismutation of superoxide radical; one O^{2-} is reduced to hydrogen peroxide (H_2O_2) while the other is oxidized to O_2. The SOD also decreases the risk of hydroxyl radical formation through the Habere–Weiss type reaction. H_2O_2 content is increased markedly in mustard upon Cd treatment (Ahmad et al. 2011). The Cd treated rape seedlings showed increasing tissue H_2O_2 levels in both roots and leaves as a result of oxidative stress (Filek et al. 2008). GSH as an antioxidant regulates H_2O_2 concentration by controlling the redox status of the plant cells. The H_2O_2 should be removed by CAT by directly dismutating H_2O_2 into H_2O and O_2. The increased H_2O_2 levels in *Lepidum sativum* are correlated with SOD activity in Cd stress. The APX levels are induced via AsA-GSH cycle in which AsA acts as hydrogen donor (Gill et al. 2011). AsA also plays a very promising role in scavenging free radicals together with GSH. Enhanced levels of AsA in roots were reported in *B. juncea* grown on tannery sludge amended soil (Singh and Sinha 2005) and *B. campestris* irrigated with effluents (Chandra et al. 2009). In plants, the H_2O_2 is reduced by APX to form two molecules of monodehydroascorbate (MDHA) which is further reduced to AsA by MDHAR with the help of NADH/NADPH, electron donors. As MDHA possess short lifetime, it should be rapidly reduced to AsA by DHAR with the help of GSH. Hence, DHAR and MDHAR plays a key role in maintaining the AsA pool and important for oxidative stress tolerance (Eltayeb et al. 2007). Also, GR catalyzes the oxidized GSSG to reduced GSH thus maintain a high ratio of GSH:GSSG (Noctor et al. 1998), crucial for the protection against oxidative damage (Srivastava et al. 2005).

The *L. sativum* plants (Brassicaceae) showed increased levels of SOD, CAT, APX and GR in leaves upon Cd treatment reflecting the efficiency of antioxidant mechanism (Gill et al. 2012). The Cd tolerant *B. juncea* cultivar, Pusa Jai Kisan had increased activity of CAT, APX, GR and higher content of GSH and AsA than sensitive cultivar SS2 exhibited efficient antioxidant metabolism for removal of Cd induced ROS (Iqbal et al. 2010). The decreased activities of antioxidant enzymes like CAT, GPX and APX was evident in canola plants subjected to excess Ni (Kazemi et al. 2010). These enzymes possess Fe in their structure. Thus, high Ni concentrations have shown to decrease Fe content (Pandey and Sharma 2002) leading to deficiency in Fe for biosynthesis of these enzymes. The mustard plants exposed to 5 mmol dm^{-3} Zn showed increased activities of APX and DHAR (Prasad et al. 1999). However, *B. juncea* cv. Pusa Jai Kisan and Pusa bold subjected to Cr stress under hydroponic conditions showed as decreased SOD, APX and GR activities (Pandey et al. 2005; Sinha et al. 2010). Overall, the Cr induced oxidative damage is more pronounced in *B. juncea* leading to the increased production of ROS but the antioxidant enzymes fail to scavenge them efficiently thus leading to toxicity (Sinha et al. 2010). Unlike other metals, As strongly induces antioxidant enzymes even at low metal concentrations. The As(III) treated *B. juncea* cv. Pusa bold showed an insignificant increase in GPX activity (Gupta et al. 2009). The tolerance and mobility of As(III) may be due to the sequestration of the metals and binding to the sulfhydryl group in mustard plants.

Lipid peroxidation is a prospective biomarker of metal stress. One of the potential indicators of oxidative stress is malondialdehyde (MDA), a major cytotoxic product of lipid peroxidation (Skórzyńska-Polit and Krupa 2006). The level of free radical production is indicated by the level of MDA formation. Excessive accumulation of metals in the roots and leaves of *Brassica* sp. grown on tannery sludge amended soil lead to the increase in MDA content (Singh and Sinha 2005; Chandra et al. 2009). There is a high level of MDA content in As treated mustard plants due to the presence of ROS generated via the conversion of As(V) to As(III), that may damage DNA, proteins and lipids (Mascher et al. 2002; Sinha et al. 2010). Similarly, an increase in MDA content was reported in *Brassica* and *Cajanus* that possess high Zn content (Alia et al. 1995).

26.4 ROLE OF OTHER IMPORTANT FACTORS FOR METAL TOLERANCE

In general, the primary and the secondary metabolites have three major functions, i.e., metal binding, antioxidant defense and signaling (Sharma and Dietz 2006). Amino acids and phenolics are reported to have a metal-chelating effect (Xiong et al. 2006), indicating that the observed increase in amino acids and phenylpropanoids might be a detoxification response of the plant. Metal toxicity leads to the accumulation of proline thus act as an osmoregulator and protects enzymes against denaturation and stabilization of protein synthesis. Matysik et al. (2002) reported the role of proline against heavy metal toxicity in *B. napus* L. cv. PF. There is an increase in the total amino acid content in the leaves of Cu-exposed *B. pekinensis* suggesting the possible role of amino acids in metal detoxification (Xiong et al. 2006).

26.4.1 Salicylic Acid

Salicylic acid (SA) acts as a signaling molecule involved in the specific responses to various stresses. There are many earlier evidences that SA can ameliorate the heavy metal toxicity in plants (Zhou et al. 2009). In mustard plants, Cd treatment caused an increase in the intrinsic SA levels in the leaves. However, the accumulation of intrinsic SA levels is decreased in Cd treated mustard plants due to the exogenous application of SA. Exogenous application of SA results in consequent changes in a number of biochemical compounds thus beneficial in action. The level of H_2O_2 is decreased upon SA treatment and also they reduce the Cd-induced injuries in mustard plants (Ahmad et al. 2011). SA treatment leads to a small decrease in H_2O_2 content in pea seedlings subjected to Cd stress (Popova et al. 2009). This may be due to the counteracting effect of SA as an antioxidant, thereby leading to the reduction in the generation of H_2O_2 under Cd stress. SA application along with Ni ions substantially increased in the activities of peroxidases like CAT, GPX and APX whereas decreased in lipoxygenase (LOX) activity, MDA and H_2O_2 contents in canola plants (Kazemi et al. 2010).

26.4.2 Nitric Oxide

Nitric oxide (NO) also acts as signaling molecule in scavenging ROS thus protecting the cells from oxidative damage (Arasimowicz and Floryszak-Wieczorek 2007). NO reduces the oxidative stress and induces the degradation of proline in canola plants thus mitigating the damage caused by Ni stress. Further, the application of sodium nitroprusside (SNP), a NO donor significantly improved the growth and chlorophyll content in Ni treated canola plants (Kazemi et al. 2010).

26.4.3 Phenolics

The phenolic composition also plays a role in Cd tolerance and the chemical properties of phenolic compounds render them to participate in scavenging ROS. The higher total phenolic acids in Cd-treated *B. juncea* plants suggested the amelioration of Cd induced oxidative stress (Navari-Izzo and Quartacci 2001). The increase of phenolic compounds is dependent on both the type of metal and its concentration, which correlates with their chelating activity, hydroxyl (OH^-) scavenging capacity, reduction potential and cytoprotectivity (Psotova et al. 2003).

26.4.4 Metallothioneins and Selenoproteins

Metallothionines (MTs) are a group of LMW (6–7 kDa) Cys-rich proteins. They chelate the metals through their Cys thiol groups and protect the cells from the toxic effects of metals. Similar to PCs, MTs are also direct gene products. MTs are involved in Cu and Zn homeostasis, but some studies provided evidence for the tolerance mediated by MTs to Cd. The MT gene expression can be induced by certain metal ions too (Lee et al. 2004; An et al. 2006). The action of Cd on rape cell could be coupled either with the ion transporters or binding with MTs (Filek et al. 2008).

Selenium (Se) is similar to S and assimilated by S metabolic pathways. Many organisms require Se for maintaining hormonal balance, antioxidation and also a component of seleno-tRNAs (Stadtman 1990). Se enhances the tolerance of plants against oxidative stress by decreasing lipid peroxidation and also spontaneous disproportion of superoxide radicals (Hartikainen et al. 2000). Se can be substituted instead of S in biological systems. As rape plants are known for S uptake, Se and S are interchangeable. Among the different chemical forms of Se, selenate (SeO_4) and selenomethionine (SeMeth), the most abundant bioavailable form are readily taken up by rape plants via sulfate transporters. The SeO_4 may be transported to leaves and chloroplasts (Leustek 2002) where

SeO_4 is reduced to selenite (SeO_3) and selenide, finally incorporated into selenocysteine (SeCys) and SeMeth (Terry et al. 2000; Ellis and Salt 2003; Sors et al. 2005). These two seleno-amino acids replace Cys and Met and get misincorporated into proteins, thereby leading to Se toxicity (Statdman 1996). The key enzyme, selenocysteine methyltransferase (SMT) methylates SeCys and prevents them incorporated into proteins thus preventing toxicity (Neuhierl and Bock 1996). The *B. juncea* plants overexpressing SMT gene from *Astragalus bisulcatus* (Neuhierl and Bock 1996) showed enhanced Se tolerance and accumulation when SeO_3 was supplied to plants rather than SeO_4 (Ellis et al. 2004; LeDuc et al. 2004). This SMT enzyme is localized in the chloroplast, the principal site of Se and S assimilation in plants (Sors et al. 2009). The Brassicaceae hyperaccumulator, *Stanleya pinnata* mainly accumulates methylselenocysteine (MeSeCys) similar to *A. bisulcatus* (Freeman et al. 2006). It is not yet clear that which mechanism may contribute Se hyperaccumulation or tolerance in *A. bisulcatus* or any other Se hyperaccumulators (Sors et al. 2005, 2009) in addition to the methylation of SeCys contributed by SMT. There exists variation in Se accumulation within the species of *S. pinnata* and *S. albescens* (Parker et al. 2003) with respect to toxicity concentration, ROS scavenging of metabolites and up regulation of S assimilation pathway genes. In *S. pinnata*, the presence of Se prevents the S starvation but *S. albescens* suffers S deficiency in presence of Se (Freeman et al. 2009).

The selenoproteins, 1-SeMeth and 1-SeCys are involved in heavy metal detoxification and protect the cellular systems (Filek et al. 2008). Increase in Se concentration is accompanied by an increase in APX and GR activities to some extent in oilseed rape (Kąkleswski et al. 2008). Se is involved in the protective role by removing Cd^{2+} from the metabolically active sites and reduction of oxygen radicals. Se addition reversed the effect of Cd-induced changes such as decrease in fresh mass and lipid unsaturation and peroxidation. Plasmalemma play a crucial role in metal tolerance and the membrane unsaturation increased upon Cd addition. Se counteracts the Cd induced changes in fatty acid saturation. As Se and Cd compete for specific binding sites, this partly explains the protective role of Se against Cd toxicity (Filek et al. 2009). The decreased fatty acid unsaturation in Cd stressed rape also resulted in increased fluorescence depolarization, i.e., decreased membrane fluidity. As a result, there is hindrance in the activity of potential protein membrane-Cd transporters. Also, the strong affinity of Cd toward N and S molecules and results in the formation of bridges, thereby leading to the inhibition of active sites of membrane enzymes and/or conformational change in the macromolecules (Mishra et al. 2006). This further restrains the metabolite transportation into chloroplast. The presence of Se can restore the chloroplast membrane fluidity. In addition, the Se did prevent the DNA methylation in rape seedlings due to high Cd concentrations. Nevertheless, the protective mechanisms of Se metabolites/conjugates on vacuolar channels/transporters are yet to be explored in detail.

26.4.5 Glucosinolates

Glucosinolates (GSLs) are low molecular mass S and N-containing secondary metabolites that occur in at least 15 dicotyledonous taxa. Within this group, the Brassicaceae accounts for most species of agricultural relevance (Schnug 1993). GSLs are supposed to play a role as an intermediary metabolic S storage *via* an enzymatic recycling (Schnug 1993; Schnug and Haneklaus 1993). After enzymatic cleavage, GSLs yield thiocyanates and SO_4^{2-}, which can be further on utilized for the synthesis of primary products (e.g., Cys, Met and GSH) under conditions of S starvation (Schnug 1993). Their basic structure is synthesized from α-amino acids, e.g., Met in the case of alkenyl, thio, sulfinyl and sulfonyl GSLs; tryptophan in the case of indole GSLs. The first stable products in the GSL biosynthetic pathway are hydroxylated amino acids, which are the precursors of aldoximes (Underhill 1980). In the next step, the thiol group of Cys is transferred to aldoxime synthesizing a thiohydroximic acid. The thiohydroximic acid is then glucosylated by the action of thiohydroximate-glucosyltransferase to lead to desulfoglucosinolates. After the transfer of SO_4^{2-} from 3′-phosphoadenosine-5′-phosphosulfate (PAPS) by a sulfotransferase, GSLs derive. From this basic structure, different GSLs are derived by the action of specific enzymes through elongation and hydroxylation of the side chain. GSLs play a role in plant defense. The enzyme myrosinase degrades them into various products that are toxic to pathogens and insect pests (Bones and Rossiter 2006). A significant linear relationship between Cd hyperaccumulation and the production of aromatic GSL sinalbin in *T. praecox* indicates a Cd-induced enhancement of the shikimate pathway (Tolrá et al. 2006) and favors of this pathway in SA mediated defense mechanism deserves further investigation.

The potential crops from Brassicacea family (*B. oleracea* and *B. napus*) have been used for Se phytoextraction purposes in Central California (Bañuelos 2002). The utilization of harvested Se-enriched phyto-products enables us to overcome the challenge faced for the disposal of plant materials enriched with Se after phytoextraction process (Bañuelos 2006). In his field experiments in Red Rock Ranch, the low GSL concentration (21 μmol g^{-1}) is an important quality of the canola seed meal, because the GSLs break down into toxic aglucones and their bitter taste results in reduced feed intake (Bañuelos 2006). The low GSL concentration of canola compared to other cultivars of rapeseed constitutes the major improvement in the meal quality. In contrast to canola, mustard seed meal contains high concentrations of GSLs, which are natural toxicants, being associated with goiter and liver damage when consumed in large quantities. Lower levels of defensive GSLs found in Ni hyperaccumulator *Streptanthus polygaloides* (Davis and Boyd 2000) and in Zn hyperaccumulator *T. caerulescens* (Tolrá et al. 2001), when compared with congener nonhyperaccumulator species, support the view of a trade off between metal hyperaccumulation and secondary metabolite synthesis.

26.4.6 Brassinosteroids

The *Brassica* family contains naturally occurring plant steroidal compound, Brassinosteroids (BRs) that is present in all parts of the plant including roots (Bajguz and Tretyn 2003). They provide a wide spectrum of tolerance against various biotic and abiotic stresses including heavy metal exposure (Janeczko et al. 2005). The toxic effect generated by Cd in *B. juncea* plants was ameliorated by the application of 28-homobrassinolide (Hayat et al. 2007) in terms of reduced Cd uptake (Fariduddin et al. 2009). This strategy can be used for revegetation and/or phytostabilization of contaminated soil where the process involved in avoidance of metal transfer to plants and animals for higher trophic levels.

26.4.7 Polyamines

Polyamines, such as putrescine (Put), spermidine (Spd), and spermine (Spm) are polybasic aliphatic amines (N containing metabolites) that play major role in plants for various physiological and developmental processes. Due to their potential binding ability to negatively charged macromolecules and membranes, they are implicated in a wide range of basic cellular regulatory processes, including DNA replication, transcription, translation, cell division, modulation of enzyme activities, cellular cation-anion balance and membrane stability (Bouchereau et al. 1999). Polyamine contents are altered in response to the exposure to heavy metals (Franchin et al. 2007; Thangavel et al. 2007). Up-regulation of polyamine metabolism has been reported for poplars exposed to high Zn or Cu concentrations under *in vitro* (Franchin et al. 2007) or greenhouse (Lingua et al. 2008) conditions, and has been shown to correlate with the extent of metal tolerance. A specific role of polyamines in plants under metal stress is not yet known. However, there is a strong possibility that they can effectively stabilize and protect the membrane systems against the toxic effects of metal ions particularly the redox active metals. Toxicity of Cd is largely determined by its affinity for sulfydryl groups in proteins, peptides and amino acids. On the contrary, the behaviour of other trace elements like Zn and Ni is largely based on a high affinity to O and N-containing ligands (Sharma and Dietz 2006; Verkleij et al. 2009). Polyamines themselves have also been suggested to function as metal chelators (Løvaas 1996). de Agazio et al. (2000) have shown that exogenous addition of polyamines could increase the hyperaccumulation ability of plants because of the formation of metal-polyamine complex, as in the case with maize seedlings, which accumulated 2.5-fold more lead accumulation when fed with Spm (2 mM) compared with that of control. Shevyakova et al. (2008) reported that the polyamine chelate with Ni in rapeseed plant and improves the Ni accumulation ability in plant.

26.5 SULFUR NUTRITION, SULFUR-CONTAINING METABOLITES AND METAL TOLERANCE: EXAMPLE OF GENUS *BRASSICA*

Rape seed crops belong to the *Brassica* genus, one of the oldest cultivated plants commonly known as Crucifers. A long history with ten thousand year old traces of cultivation renders this group being highly variable in phytochemistry and use. The well known species in the oil seed rape family include *Brassica napus* (rape seed, swede/rutabaga), *Brassica rapa* (turnip, Chinese cabbage), *Brassica juncea* (Indian/brown mustard) and *Brassica carinata*. Oilseed rape crops generally grow in cool weather with a temperature between 14 and 21°C. Depending upon the variety, the maximum and minimum temperatures are 30 and 4°C, respectively (Wurr et al. 1996). These crops require deep, well-drained, fertile, friable, sandy or silty loam soils with a neutral pH of 6.5 for mineral soils and 5.8 for organic soils (Dixon 2007). *B. napus* and *B. rapa* are the most predominant species in Canada, northern Europe and China. The most dominant oilseed crop in India and some parts of China is *B. juncea* while *B. carinata* is cultivated mostly in Ethiopia. The importance of oilseed rape has been increased in recent years as they are used for human nutrition, regenerative source of energy, animal nutrition with a source of high energy and protein content in the form of rape cake and meal, green manure and as a forage crop. They serve as a good source of antioxidants as they have high phenolics and glucosinolate content (Jahangir et al. 2008).

Oilseed rape has become a plant of major agro-economic importance and the production depends mainly on climate, soil fertility, fertilizer input, species and variety of rape crop. However, the growth and productivity of the crop is decreased in most areas due to the susceptibility to a number of biotic and abiotic stresses including heavy metals. Heavy metals enters into the environment through various anthropogenic and nonanthropogenic sources such as power stations, heating systems, waste incinerators, urban transportation, cement factories, phosphate fertilizers, pesticides, animal manures, sewage sludge and industrial wastes (Sanità di Toppi and Gabbrielli 1999). Excess uptake of metal ions is toxic to plants upon exceeding the threshold concentration that varies strongly among plant species and ecotypes and also with the properties of metals. Heavy metal accumulation in plants results in a variety of toxicity symptoms such as chlorosis, wilting, growth reduction and to cell death. In addition, they also interfere with cellular metabolism including carbohydrate metabolism, nitrate absorption and reduction, water balance and photosynthesis (Nagajyoti et al. 2010). The accumulation of heavy metals in plants may involve risks for human health through food chain transfer. Plants respond to metal toxicity by triggering a wide range of cellular defense mechanisms. They include restricted uptake and/or translocation, exclusion, compartmentalization and synthesis of heavy metal-binding factors like peptides, stress proteins and formation of metal complexes to detoxify metals (Sanità di Toppi and Gabrielli 1999). As the removal of such metals from soil by physical and chemical methods is laborious and too expensive, an environmental friendly green approach, phytoremediation is being currently deployed throughout the world to manage the contaminated soils. Nutrient management is another possible way to overcome heavy metal toxicity. Excessive toxic metals affect the regulation of sulfur (S) uptake and assimilation by reacting with S metabolites and possibly induced the synthesis of S-rich phytochelatins (PCs) and metallothioneins (MTs) (Ernst et al. 2008). Hence, S assimilation is crucial for survival of plant under heavy metal stress. As S metabolism of plant offers several advantages, special attention has been given to the regulatory mechanism of S uptake and assimilation, synthesis of S-rich peptides like nonprotein thiols for metal tolerance mechanisms in *Brassica* species.

26.5.1 Sulfur and Metals Stress

Sulfur is mostly present as inorganic sulfate (SO_4^{2-}) that is uptaken from the soil, reduced to sulfide and incorporated into various biomolecules in the S assimilation pathway (Leustek et al. 2000; Kopriva 2006). The S assimilation pathway is present in plants, algae and many microorganisms except in Metazoans (Patron et al. 2008). The S assimilation in plants involve (i) the uptake of SO_4^{2-} from the soil which is facilitated with the help of sulfate transporters; (ii) activation of SO_4^{2-} by ATP and catalyzed by ATP-sulfurylase (ATP-S) or APS-kinase to yield sulfate-activated forms adenosine-5′-phosphosulfate (APS) in the "bound" sulfite pathway and 3′-phosphoadenosine-5′-phosphosulfate (PAPS) in the "free" sulfite pathway; (iii) the activated SO_4^{2-} is directly reduced to sulfide (S^{2-}) and (iv) finally the S^{2-} is incorporated into the synthesized L-cysteine (Leustek et al. 2000; Na and Salt 2011). The genes involved in S uptake and assimilation as well as their corresponding sulfate transporters have been reported and cloned (Kopriva et al. 2009). The family of sulfate transporters comprising of 14 genes from *Arabidopsis* have been divided into five groups based on their sequence similarity (Hawkesford 2003; Buchner et al. 2004). In addition, the sulfate transporters have been identified from other plants such as *Oryza sativa, Sorghum bicolor*, etc. (Kopriva et al. 2009). All the plant genomes sequenced so far, possess all the five groups of sulfate transporters. Different groups possess specific functions, while the relation among the isoforms is much less understood in plant species other than *Arabidopsis*. The S uptake and assimilation is demand-driven process regulated by the actual S requirement in plants (Hawkesford and De Kok 2006). Under normal conditions, the uptake, distribution and assimilation of S are strictly regulated, where the rate of S uptake will be equal to S domain to maintain growth (Hawkesford and De Kok 2006; Koralewska et al. 2007, 2008; Rouached et al. 2009). However, this may not be true in Brassica species as they require high S for its growth. The greater proportion of uptaken S is stored as SO_4^{2-} in shoot rather than completely metabolized (Koralewska et al. 2007, 2008, 2009a,b). Hence, the increased S demand for the plants subjected to heavy metal toxicity is met out presumably by the synthesis of S rich metal-chelating compounds (Ernst et al. 2008). ATP-S is considered as the rate limiting step in initiating S metabolism (Hofgen et al. 2001). A higher expression of *ATP-S* is necessary for maintaining optimal level of GSH and proper functioning of cell. Overexpression of *ATPS* in *B. juncea* increases the basal tolerance to As, Cd, Cu and Zn, and results in higher levels of Cd, Cu and S in shoots of transgenic plants (Wangeline et al. 2004).

Exposure to heavy metals might affect the uptake and assimilation of SO_4^{2-} in plants as a result of reaction with S metabolites and also due to the inductions of S-rich metabolites like PCs and MTs (Ernst et al. 2008). The uptake of S and its distribution may be regulated at transcriptional, translational and/or post-translational level. The metal exposure were found to up-regulate the *Sultr1;1* and *Sultr2;1* (encode sulfate transporters) which are responsible for the primary uptake of SO_4^{2-} by the root. The *Sultr2;1* activity and uptake may be related to metabolic demand (Koralewska et al. 2008; Rouached et al. 2008, 2009). Upon SO_4^{2-} deprivation, the *Sultr1;1* expression is highly up-regulated in root and the *in situ* SO_4^{2-} concentration trigger them (Buchner et al. 2004; Koralewska et al. 2007, 2008, 2009a,b; Parmar et al. 2007; Rouached et al. 2008, 2009; Stuiver et al. 2009). During S deficiency, the plants try to remobilize and redistribute the available S from the vacuoles by up-regulating the constitutive sulfate transporter, *Sultr4;1* and inducible *Sultr4;2* (Buchner et al. 2004; Parmar et al. 2007; Koralewska et al. 2009a,b; Stuiver et al. 2009). The content and distribution of S compounds were affected in the root and shoots of Chinese cabbage subjected to Cu stress. The high concentrations of Cu up-regulated the expression of sulfate transporters (*Sultr1;2* and *Sultr4;1*) followed by an enhanced S content in the shoots (Shahbaz et al. 2010). The impact of S nutrition on the S assimilation and the expression of sulfate transporters in *Brassica* species have been reported in detail (Buchner et al. 2004; Koralewska et al. 2007, 2008, 2009a,b; Parmar et al. 2007; Stuiver et al. 2009). In *B. napus* and in *B. juncea*, transcript levels of *Sultr2;2* (*LAST*) decrease in roots, but increase in shoots with elevating Cd exposure levels (Heiss et al. 1999; Sun et al. 2007a). The authors suggest that such a decrease of this transporter activity may keep sufficient S in the roots for Cd detoxification as evidenced by increasing levels of S^{2-}, GSH and nonprotein thiols (Sun et al. 2007a). The increased GSH content and S application was found to protect the dry mass and photosynthesis in Cd-treated *B. campestris* (Anjum et al. 2008b).

The S assimilation pathway has also been reported to coordinate along with N and carbon (C) assimilation in crop plants. The nitrate reductase (NR) and N content was found to be decreased during higher Cd treatments disrupting the coordination between C, N and S metabolism and vice versa at low Cd concentration. Therefore, engineering N assimilation pathway is crucial for the growth and development of Brassicacea plants on highly contaminated soils. Further, the better understanding of the molecular mechanism between C, N and S need to be explored in future.

26.5.2 Sulfur-Containing Metabolites

Plant nutrition plays a major role for crop productivity and yield high quality feed. Oilseed rape cultivation requires some of the major nutrients like S, nitrogen (N), phosphorus (P) and potassium (K). The N:S molar ratio of 25:1 should be strictly coordinated and balanced since both are required for the synthesis of amino acids and involved in protein synthesis (Hesse et al. 2004). Besides N, the next essential macronutrient, S plays a vital role in plant growth and development, as they are present in most of the major metabolic compounds. Among the total S in plants, up to 90% is bound in the S containing amino acids, cysteine (Cys), cystine and methionine (Met) (Giovanelli et al. 1980). In addition, S is also a major component in glutathionine (GSH), co-enzymes (biotin, thiamine pyrophosphate), iron-S proteins like thioredoxins, sulfolipids, glucosinolates (GSL) and polysaccharides (Thompson et al. 1986). The adequate supply of S enhances the photosynthetic machinery while the deficiency regulates the chlorophyll content of leaves and photosynthetic enzymes (Lunde et al. 2008). Similar to other Brassicaceae, oilseed rape has a greater requirement for S than other crops like wheat or maize. For example, nearly ~16 kg of S is required for the production of 1 tonne rape seeds (McGrath and Zhao 1996). In the early 1980s, the first report of S deficiency in oilseed rape was reported (Schnug and Pissarek 1982) and severe S deficiency adversely affects the yield and quality of arable crops (Schnung and Haneklaus 1998). The S metabolism in plants offers efficient mechanisms to tackle various stresses including S compounds like Cys, Met, GSH, GSLs and phytoalexins (Anjum et al. 2008a; Ernst et al. 2008; Hardulak et al. 2011).

26.5.3 Cysteine

Individual amino acids act as potential ligands for metal detoxification and tolerance. Thus, determination of individual amino acids could serve as a potential biochemical indicator for metal tolerance. The first stable organic product containing S in the reduced form is Cys. Two enzymatic reactions are involved in the incorporation of S^{2-} into Cys. The enzyme serine acetyltransferase (SAT), acetylates L-serine by acetyl-CoA to form O-acetylserine (OAS). OAS provides the link between S and N assimilation pathways and also regulates Cys biosynthesis (Giovanelli 1990; Leustek et al. 2000; Anjum et al. 2008a; Na and Salt 2011). In the second step, O-acetylserine (thiol)lyase (OAS-TL) catalyzes OAS in reaction with S^{2-} to form L-cysteine. In bacteria (Kredich and Tomkins 1966) and plants (Wirtz et al. 2001), SAT and OAS-TL form a multi-enzyme complex known as cysteine synthase. Cys biosynthesis occurs in plants while animals take them in the diet as they are not able to synthesize it. Cys represents the starting point for the synthesis of Met, GSH and various other S-containing metabolites (Na and Salt 2011). Several studies reported the emission of hydrogen sulfide (H_2S) may occur from higher plants before or after Cys synthesis in a side reaction (Rennenberg 1989; Schroeder 1993). In bacteria and many higher plants, the Cys is degraded to H_2S

and alanine or to H_2S, ammonium and pyruvate by L-cysteine desulfhydrase (Giovanelli et al. 1980; Rennenberg et al. 1987; Burandt et al. 2001).

Cys is a key metabolite in antioxidant defense and metal sequestration. The pool of free Cys was also identified as one of the component involved in plant resistance. The root and leaves of wheat and mustard plants irrigated with effluents was found to have increased Cys content. The increased Cys content in *B. campestris* corroborated with the level of tolerance in metal-treated plants (Chandra et al. 2009). After prolonged period of exposure, the magnitude of Cys content decreased due to the reduction in the activities of SO_4^{2-} reduction enzymes ATP-S and APS under metal stress (Nussbaum et al. 1988). The difference between the tolerant and sensitive plants depend upon the ability to produce/utilize Cys rather than PCs. Cys and other low molecular weight thiols have been implicated in the ability of *Thlaspi caerulescens* to hyperaccumulate Cd (Hernández-Allica et al. 2006). Cd can easily form Cd-thiol complexes as they have high affinity to thiol groups (Vairavamurthy et al. 2000). The relative Cd tolerance in *B. juncea* could also be related to increase in synthesis of Cys along with γ-glutamylcysteine (γ-EC) and PCs than in *B. napus*. The metal-induced Cys synthesis under acute Cd stress supported the survival of *Arabidopsis* (Dominguez-Solis et al. 2004). A significant positive relationship between Ni concentration and amino acids histidine (His) and Cys in canola leaves and xylem sap indicates the ability of Ni tolerance in terms of enhanced root to shoot translocation (Ali et al. 2010). Freeman et al. (2004) observed that GSH, Cys and OAS are strongly correlated with the ability to hyperaccumulate Ni^{2+} in various *Thlaspi* hyperaccumulators.

26.5.4 Glutathione

Glutathione accounts for only approximately 2% of the total organic S in plants (Ernst et al. 2008), with up to 50% of the organic S being present in the form of GSL (Tolrà et al. 2006). GSH (γ-glutamyl-cysteinyl-glycine) is a major nonprotein thiol abundant in plants and involved in plant defense by the production of PCs and resistance to metals as direct chelators (Mendoza-Cózatl and Moreno-Sánchez 2006). GSH acts as a signal in regulating S nutrition. The synthesis of GSH involves two enzyme-catalyzed ATP-dependent steps (i) γ-glutamylcysteine synthetase (γ-ECS) catalyze the formation of peptide bond between L-glutamate and Cys to form γ-EC and (ii) glutathione synthetase (GS) forms the peptide bond between α-amino group of glycine and cysteinyl carboxyl group of γ-EC (Mendoza-Cózatl and Moreno-Sánchez 2006). In *A. thaliana* and *B. juncea*, γ-ECS is exclusively confined to the plastids, whereas GS is found in both plastids and cytosol (Wachter et al. 2005). Moreover, like in animals, the activity of regulatory enzyme, γ-ECS is regulated by GSH in feedback mechanism, a central point in GSH synthesis. The rate limiting factor for GSH biosynthesis is Cys, a final product in the S assimilation pathway. GSH has been detected in all cell compartments, including the cytosol, chloroplasts, endoplasmic reticulum, vacuoles and mitochondria.

Earlier studies have demonstrated that an increase in metal concentration in metal-stressed plants, the levels of GSH may decrease (Wójcik et al. 2005; Lima et al. 2006) while few others have reported no change in GSH content (Wójcik and Tukiendorf 2004; Mishra et al. 2006; Thangavel et al. 2007). Yang et al. (2009) reported the GSH-Cd and PC_2-Cd was mainly existed after 4 days of Cd exposure in mustard leaves and roots, respectively. The pentavalent As dimethylarsonous acid (DMA) bind to GSH upon activation by S^{2-}, and result in the formation of dimethylarsinothioyl glutathione (DMAS-GS) complex in *B. oleracea* (Raab et al. 2007). Apart from metal chelation, the other physiological functions of GSH include regulation of intracellular redox status, ROS scavenger, and transport of GSH-conjugated amino acids (Noctor and Foyer 1998). The ability of plants to tolerate metal-induced oxidative stress is correlated with the GSH content. GSH upon oxidation is converted to GSSG and it is converted again to GSH to perform normal physiological functions in plant cells. The ability of plants to tolerate metal-induced oxidative stress is correlated with the increased GSH content and the low GSH levels may be due to limited PC synthesis that leads to sensitivity for heavy metals. Hence, the manipulation of GSH biosynthesis in plants under stress resulted in enhanced tolerance to ROS (Xiang et al. 2001).

26.5.5 Phytochelatins

The toxicity of metals mainly depends upon their intracellular localization and binding to other ligands and organelles. To avoid the harmful effects of metal toxicity, they should be kept in the cytoplasm at low concentration. This can be achieved by increased binding of metals to cell wall or reduced uptake via cation/proton exchangers or by effective efflux of metal out of the cell (Tong et al. 2004). The alternative strategy in metal detoxification mechanism is to deactivate the metal ion entering the cytoplasm either through chelation or compartmentalization and/or converting it into less toxic form. As vacuole serves as the main storage site for metals (Verbruggen et al. 2009), the compartmentalization of metals to vacuoles is also a part of metal tolerance mechanisms in some hyperaccumulators. The metal detoxification enables the normal functioning of metabolic process in metal-hyperaccumulating plants (Küpper et al. 2004) thereby resulting in successful seed production, a final proof in the metal tolerance mechanism (Ernst 2006). The sequestration and storage of nonlabile metal-organic complexes in vacuoles are the main detoxification mechanism at cellular and sub-cellular levels (Vogel-Mikuš et al. 2010; Qiu et al. 2011).

The PCs comprises of a family of Cys-rich small peptides consisting of three amino acids. The general structure of PCs is (γ-Glu-Cys)*n*-Gly, where *n* ranges from 2 to 11 (Grill et al. 1985). Instead of the terminal Gly residue, some plants also possess variants with β-Ala (homo-PCs), Ser (hydroxymethyl-PCs) or Glu (iso-PCs). While in some plant species and yeasts, PC variants without the C-terminal Gly (desGly-PCs) residues have been reported (Zenk 1996; Oven et al. 2002). The PCs are synthesized within minutes in response to heavy metals (Cobbett, 2000) and the metal-PC complexes are nontoxic. Among these metals, Cd is the strongest inducer of PCs while Cu, Zn, Pb and Ni are less effective (Maitani et al. 1996; Thangavel et al. 2007). The γ-glutamylcysteine dipeptidyl transpeptidase enzyme commonly known as phytochelatin synthase (PCS) is involved in the synthesis of PCs from GSH (Grill et al. 1989). The PCS is a constitutive enzyme regulated by metals and metalloids both at transcriptional and translational levels. The precursors of PC synthesis along with heavy metals activate the PC synthesis by forming low molecular weight (LMW) complex. The Cd-PC LMW complexes in the cytoplasm acquire S^{2-} at the tonoplast to form high molecular weight (HMW) complexes that can be transported into the vacuole (Salt et al. 1995). Inside the vacuole, the Cd-PC complex dissociates due to the acidic condition and the Cd^{2+} chelates with organic acids like citrate, oxalate and malate.

PCs play a prominent role in Cd detoxification and compartmentalization. The species from *Brassica* family are well known metal accumulators, especially *B. juncea.* The increased synthesis of PCs and GSH enhanced the accumulation of Cd and tolerance in Indian mustard (Zhu et al. 1999). The PCs and GSH function as long-distance carriers of Cd in *B. juncea*. The two species, *B. napus* and *B. juncea* was compared to determine the level of Cd-induced stress in relation to the PC accumulation. Both responded to PC production from the lowest Cd treatment. The PC production was about four times more in *B. juncea* than *B. napus* but the leaves of *B. juncea* had only half that of *B. napus*. In spite that both species had similar Cd levels, *B. juncea* was more tolerant to Cd over the life time than *B. napus*. The difference in tolerance cannot be explained only in the terms of Cd concentrations or PCs production, yet some other mechanism is also involved in the increased protection of *B. juncea* from Cd toxicity (Gadapathi and Macfie 2006). The PC accumulation is more of a function as a result of Cd induced stress rather than Cd tolerance and their relationship is not straightforward. It is not necessary to expect that PC concentration should be always higher in metal-tolerant plants than metal-sensitive plants. This also depends upon the uptake of Cd by the plants. The Cd tolerance also depends upon the shuttling of PC-Cd complexes to or to reuse apo-PCs from the vacuole. The higher S^{2-} levels (4–5 times) in roots of basal metal-tolerant *B. napus* after 3 days of Cd exposure (Sun et al. 2007a) can help to stabilize Cd-PC complexes in Cd tolerant plants (Verkleij et al. 2003). In addition, the GSH and other LMW thiols play a role in long distance transport in *T. praecox* rather than PCs. PCs were detected in plants exposed to acute Cd stress but not in field collected or greenhouse grown plants where they are exposed chronically to lower Cd levels (Wójcik et al. 2005). The induction of considerable amount of PCs was detectable in control red spruce cell suspension cultures grown in medium containing basal level of Zn as required for growth (Thangavel et al. 2007). It has been hypothesized that besides their functions in detoxification, PCs also play an important role in intracellular metal homeostasis. However, Tennstedt et al. (2009) revealed that Zn^{2+} elicited PC_2 accumulation in *Arabidopsis thaliana* roots and concluded that PC formation is essential for Zn tolerance by providing the driving force for the Zn accumulation. The role of PCs in heavy metal tolerance has been confirmed in many plants by deploying transgenic approaches. The overexpression of *γ-ECS* gene from *Escherichia coli* (Zhu et al. 1999) and *glutathione reductase* gene (Pilon-Smits et al. 2000) in *B. juncea* resulted in the increased biosynthesis of GSH and PCs, thereby increasing the Cd tolerance. In addition to GSH biosynthetic genes, overexpression of APS have also been reported to contain higher levels of GSH and total thiols in Indian mustard (Van Huysen et al. 2004).

The detoxification mechanism of arsenic (As) was tested in five different varieties of *B. juncea* (*viz*., Varuna, Pusa bold, Pusa Jaikisan, Pusa Agrini and Pusa Jaganath) and the two varieties Varuna and Pusa bold were least affected or tolerant varieties. The detailed characterization of these varieties revealed that at low As levels the antioxidant enzymes act as scavengers in removing As and PCs play a major role at high As concentrations (Gupta et al. 2009). A number of the hyperaccumulating plants belong to the Brassicaceae family including *Alyssum*, *Thlaspi*, *Arabidopsis* and *Brassica* species. *B. napus* serves as a suitable candidate for phytoextraction due to the following reasons: (i) the combination of phytoextraction with other applications like biofuel production will be a profitable enterprise, (ii) the genetic relatedness to *A. thaliana* (85% sequence conservation in DNA coding regions) and *T. caerulescens* helps to gain knowledge on metal uptake mechanisms and (iii) and higher biomass compared to other natural hyperaccumulators. Nearly, 77 accessions of *B. napus* were tested for their Cd extraction potential under controlled greenhouse conditions, of which 18 were tested in field conditions for determining Cd and Zn extraction potential. The phloem sap of *B. napus* contained high levels of PCs, GSH and Cd suggesting that these thiols may be carriers of Cd in the long-distance phloem transport (Mendoza-Cózatl et al. 2008). Membrane transport systems are likely to play a central role in the translocation process. Many gene families that are involved in metal transport especially for Cd and Zn have been reviewed in Krämer et al. (2007) and Verkleij et al. (2009). As the ionic radii of Cd (0.97 A°) and Ca (0.99 A°) are almost identical, the Cd^{2+} ions are suggested to be transported through Ca^{2+} channels (Bondgaard and Bjerregaard 2005). The intense accumulation of Ca in the mesophyll symplast of field-collected *T. praecox* is connected with the Cd sequestration in mesophyll vacuoles (Vogel-Mikuš et al. 2008). Cd utilizes the Ca transport systems (putative Ca^{2+}/H^+ antiporters, encoded by CAX1 and CAX2) for the transport into vacuole (Fox and Guerinot 1998; Verbruggen et al. 2009). The plants that possess shoot/root (S/R) ratio >1 can be used for phytoextraction of metals from soil/growing media. *B. napus* cultivars have better ability to retain Ni^{2+} in the roots as they have low S/R Ni ratio and possibly contributing to Ni^{2+} tolerance in terms of binding and sequestering it in the vacuoles (Ali et al. 2010).

26.5.6 N:S Ratio and Metal Tolerance

Canola has the ability to accumulate Se under increasing sulfate-salinity conditions and to tolerate increasing soil B concentrations in the soil (Rhoades 1984). Plants belonging to *Brassica* species could reduce total Se up to 40% under green house conditions and up to 20% under field conditions after one growing season (Bañuelos 2000). As the phytoextraction of contaminants depends on the shoot biomass production, agronomic practices will also need to be improved to optimize the metal uptake. Previous studies have demonstrated that fertilization can enhance the efficiency of phytoremediation of Cd and Zn through the increased biomass production of hyperaccumulators (Schwartz et al. 2003; Sirguey et al. 2006). The balance of N to S is crucial for improved canola production (Jackson 2000; Rathke et al. 2006). Canola will respond to S only in the presence of N, whereas using N alone in S deficient soils may not improve crop growth (Soil Factsheet 1985).

N and S are both involved in protein biosynthesis; thus, a shortage in the S supply also limits the utilization of applied N (Cecotti 1996). Broccoli and rapeseed, both the *Brassica* species showed an average N:S ratio of 9:1 (Zhao et al. 1997; Schonhof et al. 2007). Malhi et al. (2007) suggested that *Brassica* oil seed species/cultivars showed a positive response to S fertilizer for seed yield and oil content. As S and Se had a similar biogeochemical analogs in soil, the high SO_4^{2-} concentration in the drainage water likely inhibited plant uptake of selenite and kept plant Se concentrations under 7 mg kg^{-1} (Bañuelos 2006).

There is a great demand for N in canola because the presence of this nutrient is greater in seeds and plant tissues compared to most grain crops. Moreover, canola plants have strong tap roots that penetrate compacted soil layers so that it can efficiently extract the soil mineral N rather than linola (Hocking et al. 2002). The biomass production is also increased in canola and Indian mustard by N fertilization except linola (Hocking et al. 1997). Although both shoot and root growth was increased with N application in *B. napus*, shoot growth was favored over root growth under hydroponic conditions (Bruns et al. 1990) and pot experiments (Svečnjak and Rengel 2006). Nitrogen deficiency caused a decrease in canola leaf number (Pinkerton 1998). Further, the uptake of N and seed yield of winter oilseed rape enormously depends on soil properties rather than a little changes in soil N dynamics (MacDonald et al. 1989).

Variation in the form of mineral N fertilizer had no effect on the oil content, fatty acid composition and GSL concentration was observed in oilseed rape (Behrens 2002). More importantly, the Se concentration in the meal was less than 2 mg kg^{-1} DM. Further, the higher total N content (>4% by weight) of canola seed meal was used as a source of green manure. Similarly, earlier research by Bañuelos and his colleagues demonstrated that incorporating Se-enriched vegetative *Brassica* material to soils as an organic source of Se, increased plant Se concentration in alfalfa and other forages under green house and field conditions, respectively (Bañuelos et al. 1991, 1992). In addition to Se, Se-enriched plant materials also supplied significant amounts of S, P and micronutrients to the growing fodder crops (Dhillon et al. 2007).

The uptake of both S and N is mutually regulated and synergistic under optimal level conditions while it becomes antagonistic under the extreme conditions of excessive level of N or S. The maximum yield responses in canola were observed under balanced N and S conditions (Jackson 2000). On the other hand, the yields of oilseed rape seeds responded not only to the total amount of fertilizer N, but also the distribution pattern within the applied N rate. Fismes et al. (2000) indicated that S fertilization is required to improve N-use efficiency and thereby maintaining a sufficient oil level and fatty acid quality in oilseed rape. As more natural occurring S is absorbed by the plant with optimal N fertilizer (N:S ratio), S content will be lowered as compared to the initial S levels in soil. This reduction in S content in the soil will lead to an increased S uptake of Se (Se and S are analogs of one another). Schonhof et al. (2007) reported that insufficient S supply in associated with optimal N supply could thus lead to accumulation of OAS and reduced Cys synthesis, resulting in a lack of precursors for GSL synthesis.

26.6 FUTURE PERSPECTIVES

PC accumulation was higher in nonmetallicolocus plants than in hypertolerant plants of *T. caerulescens*, suggesting that PC-based sequestration is not essential for constitutive tolerance or hypertolerance of metals. Furthermore, adaptive Cd hypertolerance is not dependent on PC-based sequestration (Ebbs et al. 2002; Schat et al. 2002). Upon exposure to Cd for 7 days, no PC accumulation was found in any tissues of Cd/Zn hyperaccumulator, *Sedum alfredii* from mine population, while PCs were rapidly synthesized in leaves, stems, and roots of nonmine plants (Sun et al. 2007b). On the contrary, Zhang et al. (2008) reported that PC formation could be induced in the roots, stems and leaves of *S. alfredii* exposed to 400 μM Cd for 5 days and no PCs were found in any part of the plant when exposed to 1600 μM Zn. Although the plants have been collected from the same Pb/Zn mining area in Quzhou City, Zhejiang Province and same pre-column derivatization method was used, controversial results have been reported (Sun et al. 2007b). Hence, it is necessary to explore the entire detoxification mechanisms not only in *S. alfredii*, but also in other hyperaccumulators. McNear et al. (2010) determined the mechanisms underlying the Ni complexation with O and N donor ligands for transport and accumulation of Ni in hyperaccumulator *Alyssum murale*. In addition to defense against oxidative stress in plants, polyamines may also involve the metal tolerance in terms of metal binding in plants (de Agazio et al. 2000; Wen et al. 2010). However, there is a lack of information about the characterization of polyamine-metal complex in plants even in nonhyperaccumulators. The *Brassica* species serves as an alternative model system in the field of plant biology as they possess genetic resemblance to *Arabidopsis* (Abdel-Farid et al. 2007). This will pave a new dimension for many more unanswered questions in elucidating the underlying mechanisms involved in metal hyperaccumulation in plants.

REFERENCES

Abdel-Farid, I.B., H.K. Kim, Y.H. Choi, and R. Verpoorte. 2007. Metabolic characterization of *Brassica rapa* leaves by NMR spectroscopy. *Journal of Agricultural and Food Chemistry* 55: 7936–43.

Ahmad, P., G. Nabi, and M. Ashraf. 2011. Cadmium-induced oxidative damage in mustard [*Brassica juncea* (L.) Czern. & Coss.] plants can be alleviated by salicylic acid. *South African Journal of Botany* 77: 36–44.

Ali, M.A., M. Ashraf, and H.R. Athar. 2010. Influence of nickel stress on growth and some important physiological/biochemical attributes in some diverse canola (*Brassica napus* L.) cultivars. *Journal of Hazardous Materials* 172: 964–69.

Alia, K.V., S.K. Prasad, and P.P. Saradhi. 1995. Effect of Zn on free radicals and proline in *Brassica* and *Cajanus*. *Phytochemistry* 39: 45–7.

An, Z.G., C.J. Li, Y.Y. Zu, Y.J. Du, A. Wachter, R. Gromes, and T. Rausch. 2006. Expression of BjMT2, a metallothionein 2 from *Brassica juncea*, increases copper and cadmium tolerance in *Escherichia coli* and *Arabidopsis thaliana*, but inhibits root elongation in *Arabidopsis thaliana* seedlings. *Journal of Experimental Botany* 57: 3575–82.

Angelone M, and Bini C. 1992. Trace elements concentrations in soils and plants of Western Europe. In A.C. Adriano, ed. Biogcochemistry of trace metals. Boca Raton, FL: Lewis Publishers, 19–60.

Anjum, N.A., S. Umar, A. Ahmad, M. Iqbal, and N.A. Khan. 2008b. Sulphur protects mustard (*Brassica campestris* L.) from cadmium toxicity by improving leaf ascorbate and glutathione. *Plant Growth Regulation* 54: 271–79.

Anjum, N.A., S. Umar, S. Singh, R. Nazar, and N.A. Khan. 2008a. Sulfur assimilation and cadmium tolerance in plants. In N.A. Khan et al., eds., *Sulfur Assimilation and Abiotic Stress in Plants*, 271–302. Springer-Verlag, Berlin, Heidelberg.

Anjum, N.A., S. Umar, and A. Ahmad. Editors. 2011. *Oxidative stress in plants: Causes, consequences and tolerance*. New Delhi: IK International Publishing House.

Anjum, N.A., I. Ahamd, I. Mohmood, M. Pacheco, A.C. Duarte, E. Pereira et al. 2012. Modulation of glutathione and its related enzymes in plants' responses to toxic metals and metalloids—A review. *Environmental and Experimental Botany* 75: 307–324.

Arasimowicz, M., and J. Floryszak-Wieczorek. 2007. Nitric oxide as a bioactive signaling molecule in plant stress responses. *Plant Science* 172: 876–87.

Bajguz, A., and A. Tretyn. 2003. The chemical structures and occurrence of brassinosteroids in plants. In S. Hayat, and A. Ahmad, eds., *Brassinosteroids: Bioactivity and Crop Productivity*, 1–44. Kluwer Academic Publishers, Dordrecht.

Bañuelos, G.S. 2000. Factors influencing field phytoremediation of selenium-laden soils. In N. Terry, and G.S. Bañuelos, eds., *Phytoremediation of Contaminated Soil and Water*, 41–59. CRC Press LLC, Boca Raton, FL.

Bañuelos, G.S. 2002. Irrigation of broccoli and canola with boron and selenium-laden effluent. *Journal of Environmental Quality* 31: 1802–8.

Bañuelos, G.S. 2006. Phyto-products may be essential for sustainability and implementation of phytoremediation. *Environmental Pollution* 144: 19–23.

Bañuelos, G.S., R. Mead, and S. Akohoue. 1991. Adding selenium-enriched plant tissue to soil causes the accumulation of selenium in alfalfa. *Journal of Plant Nutrition* 14: 701–13.

Bañuelos, G.S., R. Mead, L. Wu, P. Beuselinck, and S. Akohoue. 1992. Differential selenium accumulation among forage plant species grown in soils amended with selenium-enriched plant tissue. *Journal of Soil and Water Conservation* 47: 338–42.

Behrens, T. 2002. Stickstoffeffizienz von Winterraps (*Brassica napus* L.) in Abhängigkeit von der Sorte sowie einer in Menge. In *Zeit und Form variierten Stickstoffdüngung*, Cuvillier Verlag, Gottingen.

Bondgaard, M., and P. Bjerregaard. 2005. Association between cadmium and calcium uptake and distribution during the moult cycle of female shore crabs, *Carcinus maenas*: an *in vivo* study. *Aquatic Toxicology* 72: 17–28.

Bones, A.M., and J.T. Rossiter. 1996. The myrosinase–glucosinolate system, its organisation and biochemistry. *Physiologia Plantarum* 97: 194–208.

Boominathan, R., and P.M. Doran. 2003. Organic acid complexation, heavy metal distribution and the effect of ATPase inhibition in hairy roots of hyperaccumulator plant species. *Journal of Biotechnology* 101: 131–46.

Bouchereau, A., A. Aziz, F. Larher, and J. Martin-Tanguy. 1999. Polyamines and environmental challenges: recent development. *Plant Science* 140: 103–25.

Bruce, T.J., and J.A. Pickett. 2007. Plant defence signalling induced by biotic attacks. *Current Opinion in Plant Biology* 10: 387–92.

Bruns, G., R. Kuchenbuch, and J. Jung. 1990. Influence of triazole plant growth regulator on root and shoot development and nitrogen utilization of oilseed rape (*Brassica napus* L.). *Journal of Agronomy and Crop Science* 165: 257–62.

Buchner, P., H. Takahashi, and M.J. Hawkesford. 2004. Plant sulphate transporters: co-ordination of uptake, intracellular and long-distance transport. *Journal of Experimental Botany* 55: 1765–73.

Burandt, P., J. Papenbrock, A. Schmidt, E. Bloem, S. Haneklaus, and E. Schnug. 2001. Genotypical differences in total sulphur contents and cysteine desulphhydrase activities in *Brassica napus* L. *Phyton* 41: 75–86.

Cecotti, S.P. 1996. Plant nutrient sulphur—a review of nutrient balance, environmental impact and fertilizers. *Fertilizer Research* 43: 117–25.

Chandra, R., R.N. Bharagava, S. Yadav, and D. Mohan. 2009. Accumulation and distribution of toxic metals in wheat (*Triticum aestivum* L.) and Indian mustard (*Brassica campestris* L.) irrigated with distillery and tannery effluents. *Journal of Hazardous Materials* 162: 1514–21.

Cobbett, C.S. 2000. Phytochelatins and their roles in heavy metal detoxification. *Plant Physiology* 123: 825–33.

Davis, M.A., and R.S. Boyd. 2000. Dynamics of Ni-based defence and organic defences in the Ni hyperaccumulator *Streptanthus polygaloides* Gray (Brassicaceae). *New Phytologist* 146: 211–17.

de Agazio, M., E. Rea, A. Fruggiero, and M. Zacchini. 2000. Spermine treatment improves lead accumulation and translocation in maize. *12th FESSP Congress on Plant Physiology and Biochemistry* 38: 188.

Dhillon, S.K., B.K. Hundal, and K.S. Dhillon. 2007. Bioavailability of selenium to forage crops in a sandy loam soil amended with Se-rich plant materials. *Chemosphere* 66: 1734–43.

Dixit, V., V. Pandey, and R. Shyam. 2001. Differential antioxidative responses to cadmium in roots and leaves of pea (*Pisum sativum* L. cv. Azad.). *Journal of Experimental Botany* 52: 1101–09.

Dixon, G.R. 2007. *Vegetable Brassicas and Related Crucifers*. Oxfordshire, UK: CABI Publishing.

Dominguez-Solis, J.R., M.C. Lopez-Martin, F.J. Ager, M.D. Ynsa, L.C. Romero, and C. Gotor. 2004. Increased cysteine availability is essential for cadmium tolerance and accumulation in *Arabidopsis thaliana*. *Plant Biotechnology Journal* 2: 469–76.

Ebbs, S., I. Lau, B. Ahner, and L. Kochian. 2002. Phytochelatin synthesis is not responsible for Cd tolerance in the Zn/Cd hyperaccumulator *Thlaspi caerulescens* (J&C Presl). *Planta* 214: 635–40.

Edwards, R., and D.D. Dixon. 2004. Metabolism of natural and xenobiotic substrates by the plant glutathione S-transferase super family. In H. Sandermann, ed., *Molecular Ecotoxicology of Plants*, 17–50. Ecological Studies, Vol. 170. Springer-Verlag, Berlin, Heidelberg.

Ellis, D.R., and D.E. Salt. 2003. Plants, selenium and human health. *Current Opinion in Plant Biology* 6: 273–79.

Ellis, D.R., T.G. Sors, D.G. Brunk, C. Albrecht, C. Orser, B. Lahner, K.V. Wood, H.H. Harris, I.J. Pickering, and D.E. Salt. 2004. Production of Se-methylselenocysteine in transgenic plants expressing selenocysteine methyltransferase. *BMC Plant Biology* 4: 1.

Eltayeb, A.E., N. Kawano, G.H. Badawi, H. Kaminaka, T. Sanekata, T. Shibahara, S. Inanaga, and K. Tanaka. 2007. Overexpression of monodehydroascorbate reductase in transgenic tobacco confers enhanced tolerance to ozone, salt and polyethylene glycol stresses. *Planta* 225: 1255–64.

Ernst, W.H.O. 2006. Evolution of metal tolerance in higher plants. *Forest Snow and Landscape Research* 80: 251–74.

Ernst, W.H.O., G.J. Krauss, J.A.C. Verklej, and D. Wesenberg. 2008. Interaction of heavy metals with the sulphur metabolism in angiosperms from an ecological point of view. *Plant Cell and Environment* 31: 123–43.

Fariduddin, Q., M. Yusuf, S. Hayat, and A. Ahmad. 2009. Effect of 28-homobrassinolide on antioxidant capacity and photosynthesis in *Brassica juncea* plants exposed to different levels of copper. *Environmental and Experimental Botany* 66: 418–24.

Filek, M., M. Zembala, H. Hartikainen, Z. Miszalski, A. Kornaś, R. Wietecka-Posłuszny, and P. Walas. 2009. Changes in wheat plastid membrane properties induced by cadmium and selenium in presence/absence of 2,4-dichlorophenoxyacetic acid. *Plant Cell Tissue and Organ Culture* 96: 19–28.

Filek, M., R. Keskinen, H. Hartikainen, I. Szarejko, A. Janiak, Z. Miszalski, and A. Golda. 2008. The protective role of selenium in rape seedlings subjected to cadmium stress. *Journal of Plant Physiology* 165: 833–44.

Fismes, J., P.C. Vong, A. Guckert, and E. Frossard. 2000. Influence of sulfur on apparent N-use efficiency, yield and quality of oilseed rape (*Brassica napus* L.) grown on a calcareous soils. *European Journal of Agronomy* 12: 127–41.

Fox, T.C., and M.L. Guerinot. 1998. Molecular biology of cation transport in plants. *Annual Review of Plant Physiology and Plant Molecular Biology* 49: 669–96.

Franchin, C., T. Fossati, E. Pasquini, G. Lingua, S. Castiglione, P. Torrigiani, and S. Biondi. 2007. High concentrations of zinc and copper induce differential polyamine responses in micropropagated white poplar (*Populus alba*). *Physiologia Plantarum* 130: 77–90.

Freeman, J.L., C.F. Quinn, S.D. Lindblom, E.M. Klamper, and E.A.H. Pilon-Smits. 2009. Selenium protects the hyperaccumulator *Stanleya pinnata* against black tailed prairie dog herbivory in native habitats. *American Journal of Botany* 96: 1075–85.

Freeman, J.L., L.H. Zhang, M.A. Marcus, S. Fakra, and E.A.H. Pilon-Smits. 2006. Spatial imaging, speciation and quantification of selenium in the hyperaccumulator plants *Astragalus bisulcatus* and *Stanleya pinnata*. *Plant Physiology* 142: 124–34.

Freeman, J.L., M.W. Persans, and K. Nieman. 2004. Increased glutathione biosynthesis plays a role in nickel tolerance in *Thlaspi* nickel hyperaccumulators. *The Plant Cell* 16: 2176–91.

Gadapathi, W.R., and S.M. Macfie. 2006. Phytochelatins are only partially correlated with Cd stress in two species of Brassica. *Plant Science* 170: 471–80.

Gajewska, E., and M. Skłodowska. 2010. Differential effect of equal copper, cadmium and nickel concentration on biochemical reactions in wheat seedlings. *Ecotoxicology and Environmental Safety* 73: 996–1003.

Gill, S.S., N.A. Khan, N.A. Anjum, and N. Tuteja. 2011. Amelioration of cadmium stress in crop plants by nutrients management: morphological, physiological and biochemical aspects. In N.A. Anjum, F. Lopez-Lauri, eds., *Plant Nutrition and Abiotic Stress Tolerance III, Plant Stress* 5 (Special Issue 1): 1–23.

Gill, S.S., N.A. Khan, N.A. Anjum, and N. Tuteja. 2012. Cadmium at high dose perturbs growth, photosynthesis and nitrogen metabolism while at low dose it up regulates sulfur assimilation and antioxidant machinery in garden cress (*Lepidum sativum* L.). *Plant Science* 182: 112–20.

Giovanelli, J. 1990. Regulatory aspects of cysteine and methionine biosynthesis. In L.J. De Kok et al., eds., *Sulphur Nutrition and Sulphur Assimilation in Higher Plans: Fundamental, Environmental and Agricultural Aspects*, 33–48. SPB Academic Publishers, The Netherlands.

Giovanelli, J.S., S.H. Mudd, and A.H. Datko. 1980. Sulphur amino acids in plants. In B.J. Miflin, ed, *The Biochemistry of Plants* Vol. 5, 453–505. Academic Press, New York.

Grill, E., and E.-L. Winnacker, and M.H. Zenk. 1985. Phytochelatins: the principle heavy metal complexing peptides of higher plants. *Science* 230: 674–76.

Grill, E., S. Loffler, E.-L. Winnaker, and M.H. Zenk. 1989. Phytochelatins, the heavy metal-binding peptides of plants, are synthesized from glutathione by a specific gamma-glutamylcysteine dipeptidyl transpeptidase (phytochelatin synthase). *Proceedings of the National Academy of Sciences USA* 86: 6838–42.

Gupta, M., P. Sharma, N.B. Sarin, and A.K. Sinha. 2009. Differential response of arsenic stress in two varieties of *Brassica juncea* L. *Chemosphere* 74: 1201–08.

Halliwell, B., and J.M.C. Gutteridge. 1999. *Free Radicals in Biology and Medicine*. London: Oxford University Press.

Han, F.X., B.B. Maruthi Sridhar, D.L. Monts, and Y. Su. 2004. Phytoavailability and toxicity of trivalent and hexavalent chromium to *Brassica juncea*. *New Phytologist* 162: 489–99.

Hardulak, L.A., M.L. Preuss, and J.M. Jez. 2011. Sulfur metabolism as a support system for plant heavy metal tolerance. In I. Sherameti, and A. Varma, eds., *Detoxification of Heavy Metals*, *Soil Biology* 30, 289–301. Springer-Verlag, Berlin, Heidelberg.

Hartikainen, H., T. Xue, and V. Piironen. 2000. Selenium as an antioxidant and pro-oxidant in ryegrass. *Plant and Soil* 225: 193–200.

Hawkesford, M.J. 2003. Transporter gene families in plants: the sulphate transporter gene family—redundancy or specialization? *Physiologia Plantarum* 117: 155–63.

Hawkesford, M.J., and L.J. De Kok. 2006. Managing sulfur metabolism in plants. *Plant Cell and Environment* 29: 382–95.

Hayat, S., B. Ali, S.A. Hasan, and A. Ahmad. 2007. Brassinosteroid enhanced the level of antioxidants under cadmium stress in *Brassica juncea*. *Environmental and Experimental Botany* 60: 33–41.

Heiss, S., H.J. Schafer, A. Haag-Kerwer, and T. Rausch. 1999. Cloning sulfur assimilation genes of *Brassica juncea* L.: cadmium differentially affects the expression of a putative low-affinity sulfate transporter and isoforms of ATP sulfurylase and APS reductase. *Plant Molecular Biology* 39: 847–57.

Hernández-Allica, J., C. Garbisu, J.M. Becerril, O. Barrutia, J.I. Garcıá-Plazaola, F.J. Zhao, and S.P. McGrath. 2006. Synthesis of low molecular weight thiols in response to Cd exposure in *Thlaspi caerulescens*. *Plant Cell and Environment* 29: 1422–29.

Hesse, H., V. Nikiforova, B. Gakiere, and R. Hoefgen. 2004. Molecular analysis and control of cysteine biosynthesis: integration of nitrogen and sulphur metabolism. *Journal of Experimental Botany* 55: 1283–92.

Hocking, P.J., J.A. Kirkegaard, J.F. Angus, A. Bernardi, and L.M. Mason. 2002. Comparison of canola, Indian mustard and Linola in two contrasting environments. III. Effects of nitrogen fertilizer on nitrogen uptake by plants and on soil nitrogen extraction. *Field Crops Research* 79: 153–72.

Hocking, P.J., J.A. Kirkegaard, J.F. Angus, A.H. Gibson, and E.A. Koetz. 1997. Comparison of canola, Indian mustard and Linola in two contrasting environments. I. Effects of nitrogen fertilizer on dry-matter production, seed yield and seed quality. *Field Crops Research* 49: 107–25.

Hofgen, R., O. Kreft, L. Willmitzer, and H. Hesse. 2001. Manipulation of thiol contents in plants. *Amino Acids* 20: 291–99.

Iqbal, N., A. Masood, R. Nazar, S. Syeed, and N.A. Khan. 2010. Photosynthesis, growth and antioxidant metabolism in mustard (*Brassica juncea* L.) cultivars differing in cadmium tolerance. *Agricultural Sciences in China* 9: 519–27.

Irtella, B., and F. Navari-Izzo. 2006. Influence of sodium nitrilotriacetate (NTA) and citric acid on phenolic and organic acids in *Brassica juncea* grown in excess of cadmium. *Chemosphere* 65: 1348–54.

Jackson, G.D. 2000. Effects of nitrogen and sulfur on canola yield and nutrient uptake. *Agronomy Journal* 92: 644–49.

Jahangir, M., I.B. Abdel-Farid, Y.H. Choi, and R. Verpoorte. 2008. Metal-ion inducing metabolite accumulation in *Brassica rapa*. *Journal of Plant Physiology* 165: 1429–37.

Janeczko, A., J. Koscielniak, M. Pilipowicz, G. Szarek-Lukaszewska, and A. Skoczowski. 2005. Protection of winter rape photosystem 2 by 24-epibrassinolide under cadmium stress. *Photosynthetica* 43: 293–98.

Kąkleswski, K., J. Nowak, and M. Ligocki. 2008. Effects of selenium content in green parts of plants on the amount of ATP and ascorbate-glutathione cycle enzyme activity at various growth stages of wheat and oilseed rape. *Journal of Plant Physiology* 165: 1011–22.

Kazemi, N., R.A. Khavari-Nejad, H. Fahimi, S. Saadatmand, and T. Nejad-Sattari. 2010. Effects of exogenous salicylic acid and nitric oxide on lipid peroxidation and antioxidant enzyme activities in leaves of *Brassica napus* L. under nickel stress. *Scientia Horticulturae* 126: 402–07.

Kopriva, S. 2006. Regulation of sulfate assimilation in *Arabidopsis* and beyond. *Annals of Botany* 97: 479–95.

Kopriva, S., S.G. Mugford, C. Matthewman, and A. Koprivova. 2009. Plant sulfate assimilation genes: redundancy versus specialization. *Plant Cell Reports* 28: 1769–80.

Koralewska, A., C.E.E. Stuiver, F.S. Posthumus, M.J. Hawkesford, and L.J. De Kok. 2009b. The upregulated sulfate uptake capacity, but not the expression of sulfate transporters is strictly controlled by the shoot sink capacity for sulfur in curly kale. In A. Sirko, L.J. De Kok, S. Haneklaus, M.J. Hawkesford, H. Rennenberg, K. Saito, E. Schnug, and I. Stulen, eds., *Sulfur metabolism in plants: regulatory aspects, significance of sulfur in the food chain, agriculture and the environment*, 61–68. Weikersheim: Margraf Publishers.

Koralewska, A., C.E.E. Stuiver, F.S. Posthumus, S. Kopriva, M.J. Hawkesford, and L.J. De Kok. 2008. Regulation of sulfate uptake, expression of the sulfate transporters Sultr1;1 and Sultr1;2, and APS reductase in Chinese cabbage (*Brassica pekinensis*) as affected by atmospheric H_2S nutrition and sulfate deprivation. *Functional Plant Biology* 35: 318–27.

Koralewska, A., F.S. Posthumus, C.E.E. Stuiver, P. Buchner, M.J. Hawkesford, and L.J. De Kok. 2007. The characteristic high sulfate content in *Brassica oleracea* is controlled by the expression and activity of sulfate transporters. *Plant Biology* 9: 654–61.

Koralewska, A., P. Buchner, C.E.E. Stuiver, F.S. Posthumus, S. Kopriva, M.J. Hawkesford, and L.J. De Kok. 2009a. Expression and activity of sulfate transporters and APS reductase in curly kale in response to sulfate deprivation and re-supply. *Journal of Plant Physiology* 166: 168–79.

Krämer, U., I.N. Talke, and M. Hanikenne. 2007. Transition metal transport. *FEBS Letters* 581: 2263–72.

Kredich, N.M., and G.M. Tomkins. 1966. The enzymatic synthesis of L-cysteine in *Escherichia coli* and *Salmonella thyphimurium*. *The Journal of Biological Chemistry* 244: 4955–65.

Küpper, H., A. Mijovilovich, W. Meyer-Klaucke, and P.H.M. Kroneck. 2004. Tissue- and age-dependent differences in the complexation of cadmium and zinc in the cadmium/zinc hyperaccumulator *Thlaspi caerulescens* (Ganges ecotype) revealed by X-ray absorption spectroscopy. *Plant Physiology* 134: 748–57.

LeDuc, D.L., A.S. Tarun, M. Montes-Bayon, J. Meija, M.F. Malit, C.P. Wu, M. AbdelSamie, C.Y. Chiang, A. Tagmount, M. de Souza, B. Neuhierl, A. Böck, J. Caruso, and N. Terry. 2004. Overexpression of selenocysteine methyltransferase in *Arabidopsis* and Indian mustard increases selenium tolerance and accumulation. *Plant Physiology* 135: 377–83.

Lee, J., D. Shim, W.Y. Song, I. Hwang, and Y. Lee. 2004. *Arabidopsis* metallothioneins 2a and 3 enhance resistance to cadmium when expressed in *Vicia faba* guard cells. *Plant Molecular Biology* 54: 805–15.

Leustek, T. 2002. Sulfate metabolism. In C.R. Somerville, and E.M. Meyerowitz, eds., *The Arabidopsis Book*. American Society of Plant Biologists, Rockville, MD, USA.

Leustek, T., M.N. Martin, J.-A. Bick, and J.P. Davies. 2000. Pathways and regulation of sulphur metabolism revealed through molecular and genetic studies. *Annual Review of Plant Physiology and Plant Molecular Biology* 51: 141–65.

Lima, A.I.S., S.I.A. Pereira, E.M.A.P. Figueira, G.C.N. Caldeira, and H.D.Q.M. Caldeira. 2006. Cadmium detoxification in roots of *Pisum sativum* seedlings: relationship between toxicity levels, thiol pool alterations and growth. *Environmental and Experimental Botany* 55: 149–62.

Lingua, G., C. Franchin, V. Todeschini, S. Castiglione, S. Biondi, B. Burlando, V. Parravicini, P. Torrigiani, and G. Berta. 2008. Arbuscular mycorrhizal fungi differentially affect the response to high zinc concentrations of two registered poplar clones. *Environmental Pollution* 153: 137–47.

Løvaas, E. 1996. Antioxidant and metal-chelating effects of polyamines. In H. Sies, ed, *Advances in pharmacology. Antioxidants in disease mechanisms and therapy*, Volume 38. 119–49. New York: Academic Press.

Lunde, C., A. Zygadlo, H.T. Simonsen, P.L. Nielsen, A. Blennow, and A. Haldrup. 2008. Sulfur starvation in rice: the effect on photosynthesis, carbohydrate metabolism and oxidative stress protective pathways. *Physiologia Plantarum* 134: 508–21.

MacDonald, A.J., D.S. Powlson, P.R. Poulton, and D.S. Jenkinson. 1989. Unused fertilizer nitrogen in arable soils—Its contribution to nitrate leaching. *Journal of the Science of Food and Agriculture* 46: 407–19.

Maitani, T., H. Kubota, K. Sato, and T. Yamada. 1996. The composition of metals bound to class III metallothionein (phytochelatins and its desglycyl peptide) induced by various metals in root cultures of *Rubia tinctorum*. *Plant Physiology* 110: 1145–50.

Malhi, S.S., Y. Gan, and J.P. Raney. 2007. Yield, seed quality, and sulfur uptake of *Brassica* oilseed crops in response to sulfur fertilization. *Agronomy Journal* 99: 570–77.

Marrs, K.A. 1996. The functions and regulation of glutathione S-transferases in plants. *Annual Review of Plant Physiology and Plant Molecular Biology* 47: 127–58.

Mascher, R., B. Lippmann, S. Holzinger, and H. Bergmann. 2002. Arsenate toxicity: effects on oxidative stress response molecules and enzymes in red clover plants. *Plant Science* 163: 961–69.

Matysik, J., Bhalu Alia, and B.P. Mohanty. 2002. Molecular mechanisms of quenching of reactive oxygen species by proline under stress in plants. *Current Science* 82: 525–32.

McGrath, S., and J. Zhao. 1996. Sulphur uptake, yield responses and interaction between nitrogen and sulphur in winter oilseed rape (*Brassica napus*). *Journal of Agricultural Science* 126: 53–62.

McNear, Jr. D.H., R.L. Chaney, and D.L. Sparks. 2010. The hyperaccumulator *Alyssum murale* uses complexation with nitrogen and oxygen donor ligands for Ni transport and storage. *Phytochemistry* 71: 188–200.

Mendoza-Cózatl, D.G., and R. Moreno-Sánchez. 2006. Control of glutathione and phytochelatin synthesis under cadmium stress. Pathway modeling for plants. *Journal of Theoretical Biology* 238: 919–36.

Mendoza-Cózatl, D.G., F. Butko, F. Springer, J.W. Torpey, E.A. Komives, J. Kehr, and J.I. Schroeder. 2008. Identification of high levels of phytochelatins, glutathione and cadmium in the phloem sap of *Brassica napus*. A role for thiol-peptides in the long distance transport of cadmium and the effect of cadmium on iron translocation. *The Plant Journal* 54: 249–59.

Mishra, S., S. Srivastava, R.D. Tripathi, R. Govindarajan, S.V. Kuriakose, and M.N.V. Prasad. 2006. Phytochelatin synthesis and response of antioxidants during cadmium stress in *Bacopa monnieri* L. *Plant Physiology and Biochemistry* 44: 25–37.

Na, G.N., and D.E. Salt. 2011. The role of sulfur assimilation and sulfur-containing compounds in trace element homeostasis in plants. *Environmental and Experimental Botany* 72: 18–25.

Nagajyoti, P.C., K.D. Lee, and T.V.M. Sreekanth. 2010. Heavy metals, occurrence and toxicity for plants: a review. *Environmental Chemistry Letters* 8: 199–216.

Navari-Izzo, F., and M.F. Quartacci. 2001. Phytoremediation of metals. Tolerance mechanisms against oxidative stress. *Minerva Biotechnology* 13: 73–83.

Neuhierl, B., and A. Bock. 1996. On the mechanism of selenium tolerance in selenium-accumulating plants: purification and characterization of a specific selenocysteine methyltransferase from cultured cells of *Astragalus bisulcatus*. *European Journal of Biochemistry* 239: 235–38.

Noctor, G., and C.H. Foyer. 1998. Ascorbate and glutathione: keeping active oxygen under control. *Annual Review of Plant Physiology and Plant Molecular Biology* 49: 249–79.

Nussbaum, S., D. Schmutz, and C. Brunold. 1988. Regulation of assimilatory sulfate reduction by cadmium in *Zea mays* L. *Plant Physiology* 88: 1407–10.

Oven, M., J.E. Page, H. Zenk, and T.M. Kutchan. 2002. Molecular characterization of the homo-phytochelatin synthase of soybean *Glycine max*. *The Journal of Biological Chemistry* 277: 4747–54.

Parmar, S., P. Buchner, and M.J. Hawkesford. 2007. Leaf developmental stages affects sulfate depletion and specific sulfate transporter expression during sulfur deprivation in *Brassica napus* L. *Plant Biology* 9: 647–53.

Pandey, N., and C.P. Sharma. 2002. Effect of heavy metals Co^{2+}, Ni^{2+} and Cd^{2+} on growth and metabolism of cabbage. *Plant Science* 163: 753–58.

Pandey, V., V. Dixit, and R. Shyam. 2005. Antioxidative responses in relation to growth of mustard (*Brassica juncea* cv. Pusa Jaikisan) plants exposed to hexavalent chromium. *Chemosphere* 61: 40–7.

Parker, D.R., J.F. Laura, T.W. Varvel, D.N. Thomason, and Y. Zhang. 2003. Selenium phytoremediation potential of *Stanleya pinnata*. *Plant and Soil* 249: 157–65.

Patron, N.J., D.G. Durnford, and S. Kopriva. 2008. Sulfate assimilation in eukaryotes: fusions, relocations and lateral transfers. *BMC Evolutionary Biology* 8: 39.

Pickering, I.J., R.C. Prince, M.J. George, R.D. Smith, G.N. George, and D.E. Salt. 2000. Reduction and coordination of arsenic in Indian mustard. *Plant Physiology* 122: 1171–77.

Pietrini, F., M.A. Iannelli, S. Pasqualini, and A. Massacci. 2003. Interaction of cadmium with glutathione and photosynthesis in developing leaves and chloroplasts of *Phragmites australis* (Cav.) Trin. ex Steudel. *Plant Physiology* 133: 829–37.

Pilon Smits, E.A.H., Y.L. Zhu, T. Sears, and N. Terry. 2000. Overexpression of glutathione reductase in *Brassica juncea*: effects on cadmium accumulation and tolerance. *Physiologia Plantarum* 110: 455–60.

Pinkerton, A. 1998. Critical sulfur concentrations in oilseed rape (*Brassica napus*) in relation to nitrogen supply and to plant age. *Australian Journal of Experimental Agriculture* 38: 511–22.

Popova, L.P., L.T. Maslenkova, R.Y. Yordanova, A.P. Ivanova, A.P. Krantev, G. Szalai, and T. Janda. 2009. Exogenous treatment with salicylic acid attenuates cadmium toxicity in pea seedlings. *Plant Physiology and Biochemistry* 47: 224–31.

Prasad, K.V.S.K., P.P. Saradhi, and P. Sharmila. 1999. Concerted action of antioxidant enzymes and curtailed growth under zinc toxicity in *Brassica juncea*. *Environmental and Experimental Botany* 42: 1–10.

Psotova, J., J. Lasovsky, and J. Vicar. 2003. Metal chelating properties, electrochemical behaviour, scavenging and cytoprotective activities of six natural phenolics. *Biomedical Papers* 147: 147–53.

Qiu, R.-L., P. Thangavel, P.-J. Hu, P. Senthilkumar, R.-R. Ying, and Y.-T. Tang. 2011. Interaction of cadmium and zinc on accumulation and sub-cellular distribution in leaves of hyperaccumulator *Potentilla griffithii*. *Journal of Hazardous Materials* 186: 1425–30.

Raab, A., S.H. Wright, M. Jaspars, A.A. Meharg, and J. Feldmann. 2007. Pentavalent arsenic can bind to biomolecules. *Angewandte Chemie International Edition* 46: 2594–97.

Rathke, G.-W., T. Behrens, and W. Diepenbrock. 2006. Integrated nitrogen management strategies to improve seed yield, oil content and nitrogen efficiency of winter oilseed rape (*Brassica napus* L.): A review. *Agriculture Ecosystems and Environment* 117: 80–108.

Rauser, W.E. 1999. Structure and function of metal chelators produced by plants. The case for organic acids, amino acids, phytin and metallothioneins. *Cell Biochemistry and Biophysics* 31: 19–48.

Reddy, A.R., K.V. Chaitanya, and M. Vivekanandan. 2004. Drought-induced responses of photosynthesis and antioxidant metabolism in higher plants. *Journal of Plant Physiology* 161: 1189–1202.

Rennenberg, H. 1989. Synthesis and emission of hydrogen sulphide by higher plants. In E.S. Saltzman, and W.J. Cooper, eds., *Biogenic Sulphur in the Environment*, 44–57. ACS Symposium Series Vol. 393.

Rennenberg, H., N. Arabatzis, and I. Grundel. 1987. Cysteine desulphhydrase activity in higher plants: evidence for the action of L- and D-cysteine specific enzymes. *Phytochemistry* 26: 1583–89.

Rhoades, J.D. 1984. New strategy for using saline waters for irrigation. In J.A. Replogle, and K.G. Renard, eds., *Water Today and Tomorrow*, 231–236. Proceedings of Special Conference of the Irrigation and Drainage Division, Flagstaff, AZ, 24–26 July 1984. ASCE, New York.

Rouached, H., D. Secco, and A.B. Arpat. 2009. Getting the most sulfate from soil: regulation of sulfate uptake transporters in *Arabidopsis*. *Journal of Plant Physiology* 166: 893–902.

Rouached, H., M. Wirtz, R. Alary, R. Hell, A. Bulak Arpat, J.-C. Davidian, F. Pierre, and B. Pierre. 2008. Differential regulation of the expression of two high-affinity sulfate transporters, SULTR1.1 and SULTR1.2 in *Arabidopsis*. *Plant Physiology* 147: 897–911.

Salt, D.E., R.C. Prince, J.I. Pickering, and I. Raskin. 1995. Mechanism of cadmium mobility and accumulation in Indian mustard. *Plant Physiology* 109: 1427–33.

Sanità di Toppi, L., and R. Gabbrielli. 1999. Response to cadmium in higher plants. *Environmental and Experimental Botany* 41: 105–30.

Schat, H., M. Llugany, R. Vooijs, J. Hartley-Whitaker, and P.M. Bleeker. 2002. The role of phytochelatins in constitutive and adaptive heavy metal tolerances in hyperaccumulator and non-hyperaccumulator metallophytes. *Journal of Experimental Botany* 53: 2381–92.

Schnug, E. 1993. Physiological functions and environmental relevance of sulphur-containing secondary metabolites. In De Kok et al., eds., *Sulphur Nutrition and Sulphur Assimilation in Higher Plants: Regulatory, Agricultural and Environmental Aspects*, 179–90. SPB Academic Publishers, The Netherlands.

Schnug, E., and H.P. Pissarek. 1982. Kalium und Schwefel, Minimumfaktoren des schleswig-holsteinischen Rapsanbaus. *Kali-Briefe (Büntehof)* 16: 77–84.

Schnug, E., and S. Haneklaus. 1993. Physiological background of different sulphur utilisation in *Brassica napus* varieties. *Aspects of Applied Biology* 34: 211–18.

Schnug, E., and S. Haneklaus. 1998. Diagnosis of sulphur nutrition. In E. Schnug, and H. Beringer, eds., *Sulphur in Agroecosystems*, 1–38. Kluwer Academic Publishers. The Netherlands.

Schonhof, I., D. Blankenburg, S. Müller, and A. Krumbein. 2007. Sulfur and nitrogen supply influence growth, product appearance, and glucosinolate concentration in broccoli. *Journal of Plant Nutrition and Soil Science* 170: 65–72.

Schroeder, P. 1993. Plants as source of atmospheric sulphur. In L.J. De Kok et al., eds., *Sulphur Nutrition and Sulphur Assimilation in Higher Plants: Regulatory, Agricultural and Environmental Aspects*, 253–270. SPB Academy Publishers, The Netherlands.

Schützendübel, A., and A. Polle. 2002. Plant responses to abiotic stresses: heavy metal-induced oxidative stress and protection by mycorrhization. *Journal of Experimental Botany* 53: 1351–1365.

Schwartz, C., G. Echevarria, and J.L. Morel. 2003. Phytoextraction of cadmium with *Thlaspi caerulescens*. *Plant and Soil* 249: 27–35.

Seth, C.S., P.K. Chaturvedi, and V. Misra. 2008. The role of phytochelatins and antioxidants in tolerance to Cd accumulation in *Brassica juncea* L. *Ecotoxicology and Environmental Safety* 71: 76–85.

Shahbaz, M., M.H. Tseng, C.E.E. Stuiver, A. Koralewska, F.S. Posthumus, J.H. Venema, S. Parmar, H. Schat, M.J. Hawkesford, and L.J. De Kok. 2010. Copper exposure interferes with the regulation of the uptake, distribution and metabolism of sulfate in Chinese cabbage. *Journal of Plant Physiology* 167: 438–46.

Shanker, A.K., C. Cervantes, H. Loza-Tavera, and S. Avudainayagam. 2005. Chromium toxicity in plants. *Environment International* 31: 739–53.

Sharma, S.S., and K.J. Dietz. 2006. The significance of amino acids and amino acid-derived molecules in plant responses and adaptation to heavy metal stress. *Journal of Experimental Botany* 57: 711–26.

Shevyakova, N.I., E.N. Il'ina, and Vl.V. Kuznetsov. 2008. Polyamines increase plant potential for phytoremediation of soils polluted with heavy metals. *Doklady Biological Sciences* 423: 457–60.

Singh, S., and S. Sinha. 2005. Accumulation of metals and its effects in *Brassica juncea* (L.) Czern.(cv. Rohini) grown on various amendments of tannery waste. *Ecotoxicology and Environmental Safety* 62: 118–27.

Sinha, S., G. Sinam, R.K. Mishra, S. Mallick. 2010. Metal accumulation, growth, antioxidants and oil yield of *Brassica juncea* L. exposed to different metals. *Ecotoxicology and Environmental Safety* 73: 1352–61.

Sirguey, C., C. Schwartz, and J.L. Morel. 2006. Response of *Thlaspi caerulescens* to nitrogen, phosphorus and sulfur fertilisation. *International Journal of Phytoremediation* 8: 149–61.

Skórzyńska-Polit, E., and Z. Krupa. 2006. Lipid peroxidation in cadmium-treated *Phaseolus coccineus* plants. *Archives of Environmental Contamination and Toxicology* 50: 482–87.

Soil Factsheet. 1985. *Fertilizer Management for Canola in Central B.C.* Ministry of Agriculture and Food, British Columbia, Canada.

Sors, T.G., C.P. Martin, D.E. Salt. 2009. Characterization of selenocysteine methyltransferases from *Astragalus* species with contrasting selenium accumulation capacity. *The Plant Journal* 59: 110–22.

Sors, T.G., D.R. Ellis, and D.E. Salt. 2005. Selenium uptake, translocation, assimilation and metabolic fate in plants. *Photosynthesis Research* 86: 373–89.

Srivastava, M., L.Q. Ma, N. Singh, and S. Singh. 2005. Antioxidant responses of hyperaccumulator and sensitive fern species to arsenic. *Journal of Experimental Botany* 56: 1335–42.

Stadtman, T.C. 1990. Selenium biochemistry. *Annual Review of Biochemistry* 59: 111–27.

Stuiver, C.E.E., A. Koralewska, F.S. Posthumus, and L.J. De Kok. 2009. Impact of sulfur deprivation on root formation, and activity and expression of sulfate transporters in Chinese cabbage. In A. Sirko, L.J. De Kok, S. Haneklaus, M.J. Hawkesford, H. Rennenberg, K. Saito, E. Schnug, and I. Stulen, eds., *Sulfur metabolism in plants: regulatory aspects, significance of sulfur in the food chain, agriculture and the environment*, 89–92. Weikersheim: Margraf Publishers.

Sun, Q., Z.H. Ye, X.R. Wang, and M.H. Wong. 2007b. Cadmium hyperaccumulation leads to an increase of glutathione rather than phytochelatins in the cadmium hyperaccumulator *Sedum alfredii*. *Journal of Plant Physiology* 164: 1489–98.

Sun, X.M., B. Lu, S.Q. Huang, S.K. Mehta, L.L. Xu, and Z.M. Yang. 2007a. Coordinated expression of sulfate transporters and its relation with sulfur metabolites in *Brassica napus* exposed to cadmium. *Botanical Studies* 48: 43–54.

Svečnjak, Z., and Z. Rengel. 2006. Canola cultivars differ in nitrogen utilization efficiency at vegetative stage. *Field Crops Research* 97: 221–26.

Tauris, B., S. Borg, P.L. Gregersen, and P.B. Holm. 2009. A roadmap for zinc trafficking in the developing barley grain based on laser capture microdissection and gene expression profiling. *Journal of Experimental Botany* 60: 1333–47.

Tennstedt, P., D. Peisker, C. Böttcher, A. Trampczynska, and S. Clemens. 2009. Phytochelatin synthesis is essential for the detoxification of excess zinc and contributes significantly to the accumulation of zinc. *Plant Physiology* 149: 938–48.

Terry, N., A.M. Zayed, M.P. de Souza, and A.S. Tarun. 2000. Selenium in higher plants. *Annual Review of Plant Biology* 51: 401–32.

Thangavel, P., S. Long, and R. Minocha. 2007. Changes in phytochelatins and their biosynthetic intermediates in red spruce (*Picea rubens* Sarg.) cell suspension cultures under cadmium and zinc stress. *Plant Cell Tissue and Organ Culture* 88: 201–16.

Thompson, J.F., I.K. Smith, and J.T. Madison. 1986. Sulphur metabolism in plants. In M.A. Tabatabai et al., eds., *Sulphur in Agriculture*, 57–123. American Society of Agronomy Inc. Publisher, Madison, Wisconsin USA.

Tolrà, R., C. Poschenrieder, R. Alonso, D. Barceló, and J. Barceló. 2001. Influence of zinc hyperaccumulation on glucosinolates in *Thlaspi caerulescens*. *New Phytologist* 151: 621–26.

Tolrá, R., P. Pongrac, C. Poschenrieder, K. Vogel-Mikuš, M. Regvar, and J. Barceló. 2006. Distinctive effects of cadmium on glucosinolate profiles in Cd hyperaccumulator *Thlaspi praecox* and non-hyperaccumulator *Thlaspi arvense*. *Plant and Soil* 288: 333–41.

Tong, Y.-P., R. Kneer, and Y.-G. Zhu. 2004. Vacuolar compartmentalization: a second-generation approach to engineering plants for phytoremediation. *Trends in Plant Science* 9: 7–9.

Underhill, E.W. 1980. Glucosinolates. In E.A. Bell, and B.V. Charlwood, eds., *Encyclopaedia of Plant Physiology Vol. 8 Secondary Plant Products*, 493–511. Springer-Verlag, Berlin, Germany.

Vairavamurthy, M.A., W.S. Goldenberg, S. Ouyana, and S. Khalid. 2000. The interaction of hydrophilic thiols with cadmium: investigation with a simple model, 3-mercaptopropionic acid. *Marine Chemistry* 70: 181–89.

Van Huysen, T., N. Terry, and E.A.H. Pilon-Smits. 2004. Exploring the selenium phytoremediation potential of transgenic Indian mustard overexpressing ATP sulfurylase or cystathionine-gamma-synthase. *International Journal of Phytoremediation* 6: 111–18.

Verbruggen, N., C. Hermans, and H. Schat. 2009. Molecular mechanisms of metal hyperaccumulation in plants. *New Phytologist* 181: 759–76.

Verkleij, J.A.C., A. Golan-Goldhirsh, D.M. Antosiewisz, J-P. Schwitzguébel, and P. Schröder. 2009. Dualities in plant tolerance to pollutants and their uptake and translocation to the upper plant parts. *Environmental and Experimental Botany* 67: 10–22.

Verkleij, J.A.C., F.E.C. Sneller, and H. Schat. 2003. Metallothioneins and phytochelatins: ecophysiological aspects. In Y.P. Abrol, and A. Ahmad, eds., *Sulphur in Plants*, 163–76. Kluwer Academic Publishers, Dordrecht, The Netherlands.

Vogel-Mikuš, K., I. Arčon, and A. Kodre. 2010. Complexation of cadmium in seeds and vegetative tissues of the cadmium hyperaccumulator *Thlaspi praecox* as studied by X-ray absorption spectroscopy. *Plant and Soil* 331: 439–51.

Vogel-Mikuš, K., J. Simčič, P. Pelicon, M. Budnar, P. Kump, M. Nečemer, J. Mesjasz-Przybyłowicz, W.J. Przybyłowicz, and M. Regvar. 2008. Comparison of essential and non-essential element distribution in leaves of the Cd/Zn hyperaccumulator *Thlaspi praecox* as revealed by micro-PIXE. *Plant Cell and Environment* 31: 1484–96.

Wachter, A., S. Wolf, H. Steininger, J. Bogs, and T. Rausch. 2005. Differential targeting of GSH1 and GSH2 is achieved by multiple transcription initiation: implications for the compartmentation of glutathione biosynthesis in the Brassicaceaee. *The Plant Journal* 41: 15–30.

Wangeline, A.L., J.L. Burkhead, K.L. Hale, S.D. Lindblom, N. Terry, M. Pilon, and E.A.H. Pilon-Smits. 2004. Overexpression of ATP sulfurylase in Indian mustard: effects on tolerance and accumulation of twelve metals. *Journal of Environmental Quality* 33: 54–60.

Wen, X-P., Y. Ban, H. Inoue, N. Matsuda, and T. Moriguchi. 2010. Spermidine levels are implicated in heavy metal tolerance in a *spermidine synthase* overexpressing transgenic European pear by exerting antioxidant activities. *Transgenic Research* 19: 91–103.

Wirtz, M., O. Berkowitz, M. Droux, and R. Hell. 2001. The cysteine synthase complex from plants. Mitochondrial serine acetyltransferase from *Arabidopsis thaliana* carries a bifunctional domain for catalysis and protein-protein interaction. *European Journal of Biochemistry* 268: 686–93.

Wójcik, M., and A. Tukiendorf. 2004. Phytochelatin synthesis and cadmium localization in wild type of *Arabidopsis thaliana. Plant Growth Regulation* 44: 71–80.

Wójcik, M., J. Vangronsveld, and A. Tukiendorf. 2005. Cadmium tolerance in *Thlaspi caerulescens* I. Growth parameters, metal accumulation and phytochelatin synthesis in response to cadmium. *Environmental and Experimental Botany* 53: 151–61.

Wu, R.S.S., W.H.L. Siu, and P.K.S. Shin. 2005. Induction, adaptation and recovery of biological responses: Implications for environmental monitoring. *Marine Pollution Bulletin* 51: 623–34.

Wurr, D.C.E., J.R. Fellows, and K. Phelps. 1996. Investigating trends in vegetable crop response to increasing temperature associated with climate change. *Scientia Horticulturae* 66: 255–63.

Xiang, C., B.L. Werner, E.M. Christensen, and D.J. Oliver. 2001. The biological functions of glutathione revisited in *Arabidopsis* transgenic plants with altered glutathione levels. *Plant Physiology* 126: 564–74.

Xiong, Z.T., C. Chao Liu, and B. Geng. 2006. Phytotoxic effects of copper on nitrogen metabolism and plant growth in *Brassica pekinensis* Rupr. *Ecotoxicology and Environmental Safety* 64: 273–80.

Yang, H.-X., W. Liu, B. Li, W. Wei, H.-J. Zhang, and D.-Y. Chen. 2009. Speciation analysis of cadmium in Indian mustard (*Brassica juncea*) by size exclusion chromatography-high performance liquid chromatography-inductively coupled mass spectrometry. *Chinese Journal of Analytical Chemistry* 37: 1511–14.

Zenk, M.H. 1996. Heavy metal detoxification in higher plants—a review. *Gene* 179: 21–30.

Zhang, Z., X. Gao, and B. Qiu. 2008. Detection of phytochelatins in the hyperaccumulator *Sedum alfredii* exposed to cadmium and lead. *Phytochemistry* 69: 911–18.

Zhao, F.J., J.F. Ma, A.A. Meharg, and S.P. McGrath. 2009. Arsenic uptake and metabolism in plants. *New Phytologist* 181: 777–94.

Zhao, F.J., P.J.A. Whiters, P.J. Evans, J. Monaghan, S.E. Salomon, P.R. Shewry, and S.P. McGrath. 1997. Sulphur nutrition: An important factor for the quality of wheat and rapeseed. *Soil Science and Plant Nutrition* 43: 1137–42.

Zhou, Z.S., K. Guo, A.A. Elbaz, and Z.M. Yang. 2009. Salicylic acid alleviates mercury toxicity by preventing oxidative stress in roots of *Medicago sativa. Environmental and Experimental Botany* 65: 27–34.

Zhu, Y.L., E.A.H. Pilon-Smits, A.S. Tarun, S.U. Weber, L. Jouanin, and N. Terry. 1999. Cadmium tolerance and accumulation in Indian mustard is enhanced by overexpressing γ-glutamylcysteine synthetase. *Plant Physiology* 121: 1169–77.

27 Studies on Phytoextraction Processes and Some Plants' Reactions to Uptake and Hyperaccumulation of Substances

Andrew Agbontalor Erakhrumen

CONTENTS

27.1 INTRODUCTION

Phytoremediation can be broadly defined as the utilization of vascular plants, algae, and fungi to control, breakdown, or remove wastes, or to encourage degradation of contaminants in the rhizosphere, or root region of the plant (McCutcheon and Schnoor 2003). In other words, it refers to a diverse collection of plant-based technologies that use either naturally occurring or genetically engineered plants for cleaning contaminated environments (Cunningham et al. 1997; Flathman and Lanza 1998).

While the term "phytoremediation" is a recent coinage, its application as a method of cleaning up contaminated medium, through the application of living and growth processes of plants and associated microbiota is not a totally new concept. Earlier and recent reports that corroborate this assertion are available in literature (e.g., Baumann 1885; Byers 1935; Timofeev-Resovsky et al. 1962; Brooks et al. 1977; Chaney 1983; Baker et al. 1991; Salt et al. 1995a; Cunningham et al. 1997; Brooks 1998; Henry 2000; Erakhrumen 2007). Presently, this concept is getting increasingly popular globally.

This increasing global popularity and public acceptance of phytoremediation technologies is not unlikely to be connected to the fact that remediation of polluted medium through these technologies have been reported to be environmentally friendly, potentially very effective, and less expensive method than the physical and chemical remediation techniques for the clean-up of a broad spectrum of pollutants, done *in-situ*, preserving the top soil (if the contaminated medium is soil), reduces the amount of hazardous materials generated during clean-up, coupled with the possible recycling of valuable elements, in the case of phytomining. Studies have shown that phytoremediation consists of different plant-based techniques/mechanisms of action for the remediation of polluted/

contaminated medium among which are phytoextraction also referred to as phytoaccumulation or phytomining (Gosh and Singh 2005; Claus et al. 2007), rhizofiltration also referred to as phytofiltration (Tomé et al. 2008; Yang et al. 2008), phytostabilization or immobilization (Vazquez et al. 2006; Ehsan et al. 2009), rhizodegradation/phytostimulation (Dietz and Schnoor 2001; Chaudhry et al. 2005), phytodegradation/phytotransformation (Newman and Reynolds 2004; Subramanian et al. 2006), phytovolatilization (Rugh 2004; Ayotamuno and Kogbara 2007), hydraulic control/plume control (Robinson et al. 2003a; Widdowson et al. 2005), among others not yet described or well established.

Phytoextraction appears to be the most commonly recognized of all phytoremediation technologies, consequently, the terms "phytoremediation" and "phytoextraction" are sometimes incorrectly used as synonyms (Prasad and Freitas 2003), hence, it is worthy of note at this stage that phytoremediation, as earlier stated, is a concept that encompasses many types of technologies that use plant to clean-up contaminated media while phytoextraction is a specific clean-up technology. Phytoextraction is the ability of plants to remove pollutants particularly metals and other compounds in the range of 100- to 1000-fold the levels normally found in most plant species and translocate them into the above-ground harvestable biomass (leaves, stems and other plant tissues) most especially through a process termed hyperaccumulation.

Presently, most of the reports on the progress made in phytoremediation have been on the removal of metals from contaminated medium (e.g., Salt et al. 1995a; Blaylock and Huang 2000) with phytoextraction processes being the mostly reported to be very efficient in this regard. Phytoextraction has been shown to be effective in the removal of different types of contaminants/pollutants from contaminated media (soils, groundwater, sediments and sludges, water), including the removal of radionuclides from these media (Erakhrumen 2011). In addition, most of the plant species capable of phytoextraction are reported to be highly tolerant to specific contaminants; nevertheless, some other studies have shown that many of these plants react in different ways to absorption and hyperaccumulation of contaminants in their tissues. Therefore, there is the need to be acquainted with current information concerning the available scientific understanding in this regard. In line with the foregoing, this article that is focussed on phytoextraction studies is intended to highlight the current trends in achieving phytoremediation through phytoextraction processes including mechanisms used by some plants to tolerate excess amount of noxious substances in their tissues with a conclusion that suggest some tips for improving phytoextraction and associated processes.

27.2 PHYTOEXTRACTION: A VITAL PHYTOREMEDIATION TECHNOLOGY

Pollutants/contaminants of interest to be removed from a contaminated medium can either be organic or elemental. Examples of organic pollutants that are potentially important target are polychlorinated biphenyls (PCBs) such as dioxin; polycyclic aromatic hydrocarbons (PAHs) such as benzoapyrene; nitroaromatics such as trinitrotoluene (TNT); and linear halogenated hydrocarbons such as trichloroethylene (TCE). Many of these compounds are not only toxic and teratogenic but are also carcinogenic (Meagher 2000). According to Cunningham et al. (1996) the goal of phytoremediation is to completely mineralize organic contaminants into relatively nontoxic constituents, such as carbon dioxide, nitrate, chlorine, and ammonia.

However, elemental pollutants are essentially immutable by any biological or physical process short of nuclear fission and fusion, so their remediation presents special scientific and technical problems. Elemental pollutants include toxic heavy metals and radionuclides, such as arsenic, cadmium, cesium, chromium, lead, mercury, strontium, technetium, tritium, and uranium (Meagher 2000) among others. With a few notable exceptions, the best scenarios for the phytoremediation of elemental pollutants involve plants extracting and translocating a toxic cation or oxyanion to above-ground tissues for later harvest; converting the element to a less toxic chemical species (i.e., transformation); or at the very least sequestering the element in roots to prevent leaching from the site (Meagher 2000).

Certain plant species are able to survive and reproduce on metalliferous soils. Such species are divided into two main groups, viz. pseudo-metallophytes that can grow on both contaminated and noncontaminated soils and absolute metallophytes that grow only on metal- contaminated and naturally metal rich soil (Baker 1987). Some among these species have developed the ability to accumulate massive amounts of the indigenous metals in their tissues without exhibiting symptoms of toxicity (Reeves and Brooks 1983a; Baker and Brooks 1989; Baker et al. 1991; Entry et al. 1999). These vascular plants that have the capacity for extreme metal absorption and accumulation are termed hyperaccumulators because they accumulate appreciable quantities of metal in their tissues regardless of the concentration of metal in the soil (or other contaminated media), as long as the metal in question is present (Prasad and Freitas 2003). Hyperaccumulation of noxious substances such as heavy metals is a complex phenomenon which involves several steps such as (1) transport of metals across plasma membrane of root cells, (2) xylem loading and translocation, and (3) detoxification and sequestration of metals at the whole plants and cellular levels (Lombi et al. 2002).

The term hyperaccumulator was originally used to describe plants that acquired an inordinately high concentration of nickel on a dry weight basis (Brooks et al. 1977) but later extended to other elements such as heavy metals like cadmium, cobalt, copper, lead, selenium, zinc and others. Natural phytoextraction process depends on hyperaccumulating plants that are naturally able to grow on soils rich in metals based on genetic and physiological capacity of specialized plants to accumulate metals unlike induced metal uptake (Salt et al. 1998). It is noteworthy that in line with strong hyperaccumulation definition, this ability has to be proven in plants' natural habitat, not in artificial conditions (Reeves and Baker 2000). The definition of a hyperaccumulator has to take into consideration not only the metal concentration in the above-ground biomass, but also the metal concentration in the soil (Gonzaga 2006). Therefore, both bioaccumulation factor (BF) and translocation factor (TF) have to be considered while evaluating whether a particular plant is a hyperaccumulator (Ma et al. 2001).

The term BF, defined as the ratio of metal concentration in plant biomass to that in the soil, has been used to determine the effectiveness of plants in removing metals from soil while the term TF, defined as the ratio of metal concentrations in the shoots to those in the roots of a plant, has also been used to determine the effectiveness of plants in translocating metals from the roots to the shoots (Tu and Ma 2002). Thus, plant species with high BF and TF values were considered suitable for phytoextraction (Yoon et al. 2006; Li et al. 2007). Hyperaccumulator plants are usually found on metalliferous soils, where the natural exposure to a surplus of various metals has driven the evolution of metal hyperaccumulation as well as plant resistance to heavy metals (Ernst 1998). Discovery of these species that hyperaccumulate specific contaminant has actually boosted this type of phytoremediation technology. Brooks (1998) described hyperaccumulators using the following criteria: exceeding the hyperaccumulation threshold: concentration 100 times higher than in normal plants for each metal of interest, i.e., (mg kg^{-1}): 100 Cd; 1000 As, Ni; 10,000 Zn, Mn, and so on; bioconcentration factor >1 (concentration of the element in the plant > concentration in the soil); and translocation factor >1 (element concentration in the above ground part of the plant > than in roots). In other words, hyperaccumulation is usually defined as the concentration of a metal ion to >0.1–1% of the dry weight of the plant (Baker 1999). Metal pollution is particularly a challenge because apart from the fact that they are different from other toxic substances in that they are neither created nor destroyed by humans, all metals are toxic at higher concentrations as they cause oxidative stress by formation of free radicals, irrespective of the fact that many of them may be essential for plant growth at low concentration. They may also be toxic as they can replace essential metals in pigments or enzymes disrupting their function (Henry 2000). Therefore, if polluted by metals, land may be made unsuitable for plant growth and thereby destroy the biodiversity. Some of these metals, e.g., lead, depending upon the reactant surface, pH, redox potential, and other factors can, for instance, bind tightly to the soil (Kumar et al. 1995; Cunningham and Ow 1996; Blaylock et al. 1997) with a retention time of 150 to 5,000 years (Friedland 1990; Kumar et al. 1995).

Owing to the fact that most elemental pollutants are immutable and cannot be made completely nontoxic, the final goal of most phytoremediation strategies is efficient hyperaccumulation in harvestable above-ground plant tissues (Meagher 2000). Phytoextraction is suitable for the removal of contaminants from diffusely polluted areas contaminated at shallow depths with low to moderate concentration of contaminants (Kumar et al. 1995; Rulkens 1998; Blaylock and Huang 2000) thus; the selection of contaminated medium that is conducive to phytoextraction should be of primary importance as plant growth may not be sustained in heavily polluted medium also noting that the knowledge needed for proper cultivation of many of the reported hyperaccumulators is still lacking (Alkorta et al. 2004).

Different types of phytoextraction methods have been investigated but two basic strategies are mostly reported in literature. These are: (i) continuous phytoextraction or natural hyperaccumulation in which the removal of contaminants depends on the natural ability of the plant to remediate or take up the contaminants from the contaminated medium and (ii) induced or chelate assisted phytoextraction or induced or assisted hyperaccumulation in which a conditioning fluid with artificial chelates or other acidifying agents are added to increase the mobility and easy uptake of metal contaminants by the plant to be used for phytoremediation. The main characteristics of these two phytoextraction systems are summarized in Table 27.1 according to Nascimento and Xing (2006).

Many of the naturally hyperaccumulating plants tend to grow slowly and usually produce low biomass coupled with the challenges posed by low mobility and bioavailability of some heavy metals to natural phytoextraction process (Komárek et al. 2007). This might have prompted researches aimed at the use of high biomass producing plants that can be chemically treated to enhance translocation of an element from roots to shoots. Apart from this, in order to improve solubility of heavy metals in the soil solution and its subsequent uptake by plants, synthetic chelates can be used to enhance phytoextraction. Among these are chelates such as ethylenediaminedio-hyroxyphenylacetic acid (EDDHA), ethylenediaminedisuccinate (EDDS), ethylenediaminetetraacetic acid (EDTA), ethyleneglycol-O,O′-bis-[2-amino-ethyl]-N,N,N′,N′,-tetra acetic acid (EGTA), nitrilotriacetic acid (NTA), pyridine-2-6-dicarboxylic acid (PDA), citric acid, nitric acid, hydrochloric acid and fluorosilicic acid (Cooper et al. 1999; Romkens et al. 2002; Tandy et al. 2006), and so on. It has also been noted that the solubility and transport of many heavy metals into roots is increased in acidic soils, i.e., metals uptake may be increased due to decreasing pH (Brown et al. 1994) thus, plants and soils can be manipulated to increase or decrease the uptake of chemical substances under otherwise toxic acidic conditions. However, soil acidification and addition of chelating agents should be done with utmost care to prevent a situation whereby increased solubility of some toxic metals may lead to their being leached into the groundwater (Sun et al. 2001; Romkens et al. 2002; Wenzel et al. 2003;

TABLE 27.1
Main Characteristics of the Two Strategies of Phytoextraction of Metals from Soils

Natural Phytoextraction	Chemically Assisted Phytoextraction
Plants naturally hyperaccumulate metals	Plants are normally metal excluders
Slow growing, low biomass production	Fast growing, high biomass plants
Natural ability to extract high amount of metals from soils	Synthetic chelators and organic acids are used to enhance metal uptake
Efficient translocation of metals from roots to shoots	Chemical amendments increase the metal transfer from roots to shoots
High tolerance; survival with high concentrations of metals in tissues	Low tolerance to metals; the increase in absorption leads to plant death
No environmental drawback regarding leaching of metals	Risk of leaching of metal chelates to groundwater

Source: Nascimento, C.W.A., and B. Xing, *Sci. Agric. (Piracicaba, Braz.)* 63 (3): 299–311, 2006.

Madrid et al. 2003; Chen et al. 2004) particularly beyond the reach of the roots of the phytoextracting plants.

Irrespective of the fact that recent research verified the possibility of using biodegradable chelates for soil washing processes (Tandy et al. 2004) and to enhance phytoextraction of heavy metal contaminated soils (Kos and Leštan 2003; Tandy et al. 2006; Cao et al. 2007) there are reports of persistence of metallic complexes in the soil water for several weeks in experiments using chelates such as EDTA to induce phytoextraction (Lombi et al. 2001; Wenzel et al. 2003) even as it has been found that EDTA was often the most effective among many of the synthetic chelates used in enhancing phytoextraction of heavy metals by plants (Blaylock et al. 1997). Apart from this, synthetic chelators such as EDTA are barely degradable by micro-organisms and have been observed by Nascimento et al. (2006) to be toxic and drastically reduce the growth of Indian mustard grown on a metal multi-contaminated soil.

In general, soil pH seems to have the greatest effect of any single factor on the solubility or retention of metals in soils (Baker and Walker 1990; McNeil and Waring 1992; Henry 2000; Ghosh and Singh 2005) as it has been noted that it can be manipulated to increase metal phytoavailability in the soil (Schmidt 2003) although, other soil factors such as amount of metal (Garcia-Miragaya 1984), cation exchange capacity (Martinez and Motto 2000), organic carbon content (Elliot et al. 1986), the oxidation state of the mineral components, the redox potential of the system (Connell and Miller 1984), presence of other metals (Greger 1999) and so on, play their roles. Plant uptake capacity toward heavy metals and, as a consequence, metal removal efficiency from soil, is limited by the bioavailable metal fraction presence and abundance (Peijnenburg and Jager 2003). Therefore, for the success of phytoextraction, contaminants should be bioavailable, or subject to absorption by plant roots and translocated to the harvestable above-ground portions. It is noteworthy at this stage that phytoextraction is not an easy technology just consisting of picking up some hyperaccumulating plants and placing them in the metal polluted area. On the contrary, it is highly technical, requiring expert project designers with plenty of field experience that carefully choose the proper species and cultivars for particular metals (and combinations of them) and regions, and manage the entire system to maximize pollutant removal efficiency (Alkorta et al. 2004).

Indeed, increasing root uptake is a major step to successful removal of noxious substances from a contaminated medium through phytoextraction processes. Thus, efforts aimed at improving factors that will enhance the ability of phytoextraction at the root region, such as increased and sustained production of specific root exudates that may be identified to be related to increase uptake of noxious substance should be intensified. These kinds of efforts targeting root exudates as natural chelators is expected to aid efficient phytoextraction and also overcome environmental constraints associated with chemically assisted phytoextraction.

Once a hyperaccumulator has been identified to be suitable for the removal of particular pollutants from a medium, the plant is either raised hydroponically and transplanted into the polluted medium or may be seeded or transplanted into polluted soil and cultivated using established agricultural practices and allowed to grow and absorb/concentrate the pollutants in their roots and above-ground biomass. After sufficient plant growth and hyperaccumulation, the whole or above-ground portions of the plant are harvested and removed, resulting in the permanent removal of pollutants from the site. Planting and harvesting may be repeated, as required, to reduce contaminant levels to allowable limits/acceptable level limits (Kumar et al. 1995; Padmavathiamma and Li 2007). Phytoextraction is considered to be a long-term process that can take several years/decades (Robinson et al. 2003b) with the time required for remediation depending on the type and extent of contamination, the length of the growing season, and the efficiency of metal removal by plants, but normally ranges from 1 to 20 years (Kumar et al. 1995; Blaylock and Huang 2000) although, the acceptable duration of phytoextraction process is considered to be not more than 5 to 10 years (Robinson et al. 1998; Blaylock and Huang 2000).

Several authors have considered hyperaccumulation as more important trait than biomass yield (Chaney et al. 1997), while others do not accept this opinion (Ebbs et al. 1997; Ernst 1998; Kayser et al. 2000). Nonetheless, it is important to note that hyperaccumulating plants that can combine high biomass production with high accumulation of pollutants of interest in the harvestable parts of the plant (Macek et al. 2004) including ability to thrive in toxic environments with little maintenance are likely to be very useful for phytoextraction. It is necessary that plants for phytoextraction be tolerant to the targeted pollutants and be efficient at translocating them from roots to the harvestable above-ground portions of the plant (Blaylock and Huang 2000). Furthermore, other desirable plant characteristics include the ability to tolerate difficult soil conditions (e.g., soil pH, salinity, soil structure, water content), rapid growth rate, the production of a dense and deep root system, element selectivity, ease of care and establishment, and few disease and insect problems, method of harvesting (Baker et al. 1994; Cunningham and Ow 1996), among others. Nevertheless, no single plant species have been described to presently possess all of these and other desirable traits. It may be possible that high biomass nonaccumulating plants that are fast-growing could be modified using transgenic approach to achieve these properties (Clemens et al. 2002), thus, finding such and other plants with desirable traits continue to be the focus of many plant-breeding and genetic-engineering research efforts (Kärenlampi et al. 2000; Prasad and Freitas 2003; Alkorta et al. 2004) also noting that researches on possible genetic manipulations of plant rhizosphere to enhance phytoextraction will be necessary. It is however, important to be mindful of the recent ever-increasing public concern over the release of genetically modified organisms.

27.3 PHYTOEXTRACTION MECHANISMS ADAPTED BY HYPERACCUMULATING PLANTS

In order to grow and complete their life cycle, plants need both essential macro-nutrients and micronutrients in various quantities. Thus, they have evolved highly specific mechanisms to take up, translocate, store and utilize these nutrients as required. However, the mechanisms and extent of uptake is selective, plants preferentially acquiring some ions over others with ion uptake selectivity sometimes depending upon the structure and properties of membrane transporters (Lasat 2000), although, a detailed knowledge of these mechanisms (uptake, translocation, storage, utilization, etc.) in most plants are presently required as they are incomplete.

Membrane transporters are certain types of proteins with transport functions that mediate ion transport into cells, since ions such as metals, cannot move freely across cellular membranes, which are lipophilic structures, owing to their charge, also noting that uptake of some metals may not actually be mediated by membrane transporters (Shrift and Ulrish 1976; Abrams et al. 1990; Arvy 1993). It is also known that functions of many plant transporters that are central to phytorextraction still remain largely uncharacterized. Different types of plant species are known to react differently to the presence of certain elements/compounds in a medium. For instance, in the case of high concentration of metals in a medium, some plants, (known as excluders) prevent metal uptake into root cells and therefore have little potential for metal extraction. Another group of plants (known as accumulators) allow uptake and bioaccumulation of extremely high concentration of metals in their tissues while some other plants (also known as indicators) show poor control over metal uptake and transport processes, consequently, the extent of metal accumulation reflects metal concentration in the medium.

There are several possible ways by which noxious substances can get into the plant tissues. However, the roots of plants are the main pathways through which these substances enter into the plant body. Uptake of noxious substances into root cells, the point of entry into living plant tissues, is a step of major importance for the process of phytoextraction as it has been noted that it is possible for a plant exhibiting significant metal uptake through the roots to express a limited capacity for phytoextraction (Lasat 2000) through the inability of the accumulated substances to be transported to the shoots. This implies that majority of the metals taken up by many of such plants are accumulated in the roots (Carbonell-Barrachina et al. 1994; Pickering et al. 2000). Therefore, current knowledge regarding this process, particularly in line with some recent advances in the knowledge of root biology (Waisel et al. 1996; Walker et al. 2003) and how root exudates as well as plant-produced root organic acids and introduced synthetic chelators amendments influence phytoextraction is important noting that adsorption processes are orders of magnitude less efficient in soils than in liquid medium (Dushenkov et al. 1995; Raskin et al. 1997) because root surfaces must compete for nutrients with diverse particulate soil materials, for example, clays and humic acids (Meagher 2000). Solute transport from the external parts of the root to the central root xylem where the material is carried to the shoots takes place through two major pathways: the apoplastic (cell wall space between cell membranes) and the symplastic (crossing many cell membranes along the path). In the apoplastic pathway, the presence of the lipophilic Casparian strip at the root endodermis disrupts the apoplastic water flow and directs it to cross cell plasma membranes at least twice, where selective transport as well as passive permeation of solutes occur (Tanton and Crowdy 1972). Under normal plant physiological conditions, the Casparian strip guarded pathway accounts for over 99% of water flow through the roots (Hanson et al. 1985). However, the pathway not guarded by the endodermal Casparian strip which constitutes less than 1% of the total water movement may however, significantly increase in percentage under stress conditions, such as anaerobic conditions (Hanson et al. 1985), salinity stress (Yeo et al. 1987), root disturbance (Moon et al. 1986), and high phosphorus nutrition (Skinner and Radin 1994), among others. This process is uncertain under synthetic chelate-induced phytoextraction. However, hydrophilic compounds favour the apoplastic pathway, whereas lipophilic compounds favour the symplastic pathway (Wu et al. 1999).

The movement of most metals from external solution to the cell walls is mostly a nonmetabolic, passive process, driven by diffusion or mass flow (Marschner 1995), although, some metals such as selenium in the form of selenate is accumulated in plant cells against its likely electrochemical potential gradient through a process of active transport (Brown and Shrift 1982). Some metals can also be taken up metabolically in low proportion with some external factors such as light intensity, temperature, among others, influencing it. Metals stop at first on root surface and then a portion of ions that penetrate into the roots are bound in cell walls while the rest is accumulated in the intercellular spaces (Wierzbicka 1987). Root surfaces, which have evolved specifically to adsorb elemental nutrients from soil and pore water, have extraordinarily large surface areas (Dittmer 1937) and high-affinity chemical receptors (Dushenkov et al. 1995; Salt et al. 1995b; Dushenkof et al. 1997; Raskin et al. 1997; Heaton et al. 1998; Salt and Krämer 1999). In the process of adsorption, root surfaces bind many elemental pollutants as well as nutrients. Transmembrane transporters possess an extracellular binding domain to which the ions attach just before the transport, and a transmembrane structure which connects extracellular and intracellular media. The binding domain is receptive only to specific ions and is responsible for transporter specificity. The transmembrane structure facilitates the transfer of bound ions from extracellular space through the hydrophobic environment of the membrane into the cell (Lasat 2000). In general, solutes have to be taken up into the root symplasm before they can enter the xylem (Tester and Leigh 2001). In some instances, for metals to reach xylem vessels, metal uptake takes place at the younger parts of the root, where Casparian strips are not well developed (Hardiman et al. 1984). The root uptake rate and xylem steady-state concentration of a compound depend on the compound's lipophilicity (Wu et al. 1999). Subsequent to metal uptake into the root symplasm, three processes govern the movement of metals from the root into the xylem: sequestration of metals inside root cells, symplastic transport into the stele and release into the xylem.

The absorbed metals are unloaded from the xylem parenchyma into mature xylem vessels, but this process is still not well understood (Salt and Krämer 1999). After metal's loading to the xylem, movement of sap containing it from the root to shoot, termed translocation, is believed to be primarily driven by transpiration and root pressure. Once inside the plant, most metals are too insoluble to move freely in the vascular system, so they usually form carbonate, sulfate or phosphate precipitates immobilizing them in apoplastic (extracellular) and symplastic (intracellular) compartments (Raskin et al. 1997). Indeed, the phytochemistry involved in metal transport and storage, for instance, seems to vary considerably depending on the type of the metal treatment imposed on plants and plant species (Brooks et al. 1998).

Water that evaporates from the above-ground part of plants serves as a pump that assists in absorbing nutrients and other substances in the soil into plant roots through a process termed evapotranspiration. Plants transpire water to move nutrients from the soil solution to leaves and stems, where photosynthesis occurs. Translocation efficiency of many compounds from root periphery to xylem has been measured, and this information is most likely applicable to phytoextraction (Briggs et al. 1982; Hsu et al. 1990). However, for a successful phytoextraction to occur, noxious substances of interest must be transported from the root to the harvestable above-ground parts of the plants (Macek et al. 2004). Presently, knowledge on rhizospheric processes mediated by root exudates particularly as it concerns phytoextraction is still limited, as the fate of exudates in the rhizosphere and the nature of reactions involved in phytoextraction and transport of noxious substances such as metals by plants are not yet fully understood (Nascimento and Xing 2006). However, it is recognized that some of these exudates contribute significantly to the accumulation of metals in some plants. Certain chemical compounds likely to occur in the rhizosphere are clearly associated with increase of metals uptake from soil and their translocation to shoots (Mench and Martin 1991; Salt et al. 1995b; Krishnamurti et al. 1997; Lin et al. 2003; Wenzel et al. 2003). Among these compounds are the low molecular-weight organic acids thought to be probably the most important exudates in natural phytoextraction systems. They influence the acquisition of metals by either forming complexes with metal ions or decreasing the pH around the roots and altering soil characteristics (Nascimento and Xing 2006). Capacity to complex rather than decreasing the pH has been shown to be the main factor related to mobilization of metals in soil and their accumulation in plants (Bernal et al. 1994; McGrath et al. 1997; Gupta et al. 2000; Quartacci et al. 2005). It is also thought that the efficiency of organic acids released by roots to mobilize metals, for instance, from soil seems to depend upon the rate of biodegradation (Krishnamurti et al. 1997; Renella et al. 2004). The biodegradation process is under the control of the soil's microbial community, which is also not yet fully understood (Ryan et al. 2001), but the process of consumption of organic acids by micro-organisms is probably an important process in reducing their effectiveness in complexing metals around the plant roots. Indirect effects of root exudates on microbial activity, rhizosphere physical properties and root growth dynamics may also influence ion solubility and uptake in various ways (Marschner 1995; Walker et al. 2003). It is also thought that metal accumulators may enhance metal solubility by releasing organic chelators from the roots. However, only few reports on the involvement of specific exudates in the uptake and accumulation of potentially toxic metals by plants are presently available with the exudation rates and chemical composition of exudates of hyperaccumulator species being presently virtually unknown with no conclusive evidence so far that hyperaccumulators exude specific chelators in the rhizosphere to enhance phytoextraction (Nascimento and Xing 2006).

The ability of hyperaccumulating plants to mobilize and subsequently uptake noxious substances, particularly metals, as a result of depending wholly on root exudates as well as plant-produced root organic acids, would have sounded as a better phytoextraction technique, most especially to the public when compared to techniques that use synthetic chelators, but presently, a large body of literature has demonstrated the lower effectiveness of natural organic acids on metals mobilization and subsequent plant uptake as compared to synthetic chelators, for example, in the case of lead phytoextraction (Salt et al. 1995b; Gupta et al. 2000; Lombi et al. 2001; Wu et al. 2003; Kos and Leštan 2004). Metal solubility and availability are both dependent on soil characteristics and are strongly influenced by pH and the degree of complexation with soluble ligands (Kaschl et al. 2002). Metals exist in soil in various pools: in solution as ionic or organically complexed species; on exchange sites of reactive soil components; complexed with organic matter; occluded in Fe, Al, and Mn oxides and hydroxides; entrapped in primary and secondary minerals (Shuman 1985; Mann and Ritchie 1993). Most metals in soils exist in unavailable forms, thus soil conditions have to be altered to elicit phytoextraction, since the phenomenon depends on a relatively abundant source of soluble metal to enable significant metal uptake and translocation to shoots. Metals such as Pb and Cr have their extraction rate limited by their inherently low solubility (Nascimento and Xing 2006).

Owing to the challenges encountered by some natural phytoextraction processes concerning low mobility and bioavailability of some heavy metals in some soil, lower effectiveness of natural organic acids in enhancing phytoextraction, coupled with research results that show that bioavailability of heavy metals in soils decreases above pH 5.5–6 particularly in alkaline soils, synthetic chelating agents capable of inducing desorption of metals from soil particles, solubilizing and complexing heavy metals into the soil solution, making them more bioavailable to plants as well as to promote heavy metal translocation from roots to the harvestable parts of the plants can be utilized as amendments to enhance the process (e.g., Carbonell-Barrachina et al. 1994; Blaylock et al. 1997; Huang et al. 1997; Lasat et al. 1998; Kayser et al. 2000; Pickering et al. 2000; Komárek et al. 2007).

However, it is noteworthy that apart from the criticisms that some of the artificial chelators have some toxic properties themselves, it has been reported that many synthetic chelators capable of inducing phytoextraction might form chemically and microbiologically stable complexes with heavy metals, threatening soil quality (Wu et al. 1999). Furthermore, the amount of heavy metals made soluble by synthetic chelators usually exceeds by far most plant's uptake capacity, especially under field condition, thereby leading to the risk of increased leaching (Ginneken et al. 2007) with serious risk implication concerning groundwater pollution, thus, attempts should be made to minimize this, for instance, by applying the chelator at the time of maximum crop biomass (Salt et al. 1998; Crèman et al. 2001; Gebelen et al. 2002).

27.4 HYPERACCUMULATION AND TOLERANCE TO NOXIOUS SUBSTANCES BY PLANTS

As earlier stated, it has been established through studies that some plants are able to grow on soils contaminated by certain noxious substances without exhibiting symptoms of toxicity, an ability that is termed tolerance. There is therefore, the need to understand the mechanisms through which some species of plant can live unharmed in contaminated environment, noting as well that there are species to species, and even variety to variety differences in tolerance to the harmful effects of noxious substances and their accumulation in different parts of plants.

Tolerance and hyperaccumulation traits have been shown to be genetically determined and genetically independent in plants (Baker et al. 1994; Meerts and Isacker 1997; MacNair et al. 1999). Several studies on the inheritance of metal tolerance have demonstrated that only one or two major genes might control metal tolerance in a completely or partially dominant way (Pollard et al. 2002; Bert et al. 2003), although, researches in this regard concerning hyperaccumulation are not yet conclusive (Cosio 2004). However, depending on plant species, tolerance to metals adjacent to their roots, for instance, may result from two basic strategies: metal exclusion and metal accumulation (Baker 1981; Baker and Walker 1990).

The exclusion strategy, comprising avoidance of metal uptake and restriction of metal transport from the root to shoot (De Vos et al. 1991), is usually used by pseudo-metallophytes. The accumulation strategy which leads to high uptake of metals depends on plants' internal tolerance mechanisms that immobilize, compartmentalize or detoxify metals in the symplasm. This could be through the compartmentation in metabolically less active parts of plants (Wang and Evangelou 1995; Sanità di Toppi and Gabbrielli 1999) and/or production of metal binding compounds (Marschner 1995; Rauser 1995; Küpper et al. 1999). Tolerance to metals and other noxious substances is thought to be based on several mechanisms rather than one alone and they are likely to operate in combination (Baker 1987; MacNair 1997). The lack of a comprehensive understanding of this complex metal homeostatic network in plants remains a major bottleneck in the development of phytoextraction technologies (Hirschi et al. 2000; Krämer 2003; Nascimento and Xing 2006). Compartmentation in the vacuole and chelation in the cytoplasm are among the most significant mechanisms proposed to be related to metal accumulation and the resistance of their damaging effects by plants (Cunningham et al. 1995; Nascimento and Xing 2006). Vacuolar compartmentation appears to play a major role in keeping heavy metals in the form of soluble complexes and/or insoluble precipitates away from the cytoplasm, where important metabolic functions are performed (Wang and Evangelou 1995). Presently, it appears that emphasis is mainly given to internal tolerance mechanisms, particularly vacuolar metal compartmentation and sequestration. Little attention has so far been devoted to other tolerance mechanisms (Cosio 2004). Nevertheless, given that the goal of phytoextraction is to maximize metal accumulation in plant tissues, mechanisms of internal tolerance are likely to be important.

In order to deal with heavy metal stress in plant tissues, chelation and sequestration of metals by particular ligands have also been noted to be mechanisms used by plants. Complexation of metal ions by specific high-affinity ligands reduces the concentration of free metal ions in cells, thereby reducing their phytotoxicity. The two best-characterized metal-binding ligands in plant cells are the phytochelatins (PCs) and metallothioneins (MTs) (Grill et al. 1988; Rauser 1990; Robinson et al. 1993; Zenk 1996; Rea et al. 1998; Cobbett 2000a; Cobbett and Goldsbrough 2002). These ligands are widely distributed in plants and form stable complexes with metals in the cytosol which can be subsequently sequestered into the vacuole (Zenk 1996; Goldsbrough 2000).

Many physiological and genetic studies indicate that PCs and MTs are critical for metal tolerance and accumulation in plants (Howden and Cobbett 1992; Zhu et al. 1999; Schmöger et al. 2000; Inouhe et al. 2000; Hartley-Whitaker et al. 2001; Van Hoof et al. 2001). For instance, methallothioneins-like proteins are implicated in metal homeostasis in cyanobacteria, yeasts, animals and plants (Hamer et al. 1985; Zhou and Goldsbrough 1995; Huckle et al. 1996; Murphy et al. 1997). A number of researchers have introduced MTs from animal sources into plants in a transgenic approach (Yeargan et al. 1992; Hasegawa et al. 1997). These approaches increased plant tolerance, but not accumulation.

On the other hand PCs synthesis seems to be the principal response of plants and many fungi to toxic metal exposure, and apparently also of certain animals (Cobbett 2000b; Clemens et al. 2001; Vatamaniuk et al. 2001). For example, PCs are essential for Cd detoxification in *Arabidopsis thaliana* (Howden et al. 1995). It is believed that PCs and their precursor glutathione could shuttle Cd ions from the cytosol to the vacuoles (Vögeli-Lange and Wagner 1990; Sanità di Toppi and Gabbrielli 1999; Heiss et al. 2003). Studies also showed that other ligands participating in complexing heavy metals in vacuoles may also be metal chaperones, organic acids such as citric, malic and malonic acids or even histidine, nicotianamine and phytates (Krämer et al. 1996; Stephan et al. 1996; Rauser 1999; Salt et al. 1999) among others. Many species of plants known to hyperaccumulate noxious substances have also been noted to be highly tolerant to most of these substances, although, the precise relationship between metal hyperaccumulation and tolerance is still a subject of debate (Nascimento and Xing 2006) with some authors proposing that there is no correlation between these traits (Baker and Walker 1990; Baker et al. 1994), while others suggest that hyperaccumulators possess a high degree of tolerance to substances such as metals (Reeves and Brooks 1983b; Chaney et al. 1997), although, it is noteworthy that the distribution of metals within plant organs and tissues is an indirect indicator of detoxification and tolerance mechanisms employed by plant species (Cosio 2004). Inherent characteristics regarding metal binding in cell walls is considered as one of the possible mechanisms in plant resistance and has been proposed by several authors as an explanation for the differences between species in metal uptake and distribution (e.g., Florjin and Van Beusichem 1993; Cakmak et al. 2000). Accumulator species have evolved specific mechanisms for detoxifying noxious substances, such as heavy metal levels, accumulated in the cells. For instance, tolerance to nickel was achieved by a Ni-accumulating plant through complexation to low molecular weight organic compounds (Lee et al. 1977). Similarly, cadmium has been shown to accumulate in plants where it is detoxified by binding to phytochelatins (Wagner 1984; Steffens 1990; Cobbett and Goldsbrough 1999).

It was also shown that tolerance to high Ni accumulation by *Thlaspi goesingense* was achieved by Ni complexation by histidine, which rendered the metal inactive (Krämer et al. 1996; Krämer et al. 1997). Zinc tolerance by the shoots of *Thlaspi caerulescens* was also suggested to be as a result of sequestration Zn in the vacuole (Lasat et al. 1996; Lasat et al. 1998). Zinc seems to be preferentially sequestrated in vacuoles of epidermal cells in a soluble form (Küpper et al. 1999; Frey et al. 2000) with series of mechanisms proposed to account for Zn inactivation in the vacuole, among which are precipitation as Znphytate (Van Steveninck et al. 1990) and binding to low molecular weight organic acids (Mathys 1977; Tolrà et al. 1996; Salt et al. 1999). In *Arabidopsis halleri* leaves, Zn was found to be predominantly coordinated to malate (Sarret et al. 2002) and accumulated in the mesophyll cells (Küpper et al. 2000; Zhao et al. 2000). Plant may also modify their metabolic activities, and this may be done according to the environmental conditions, plant age and its general physical condition and the vegetation period (Azhar et al. 2006). In addition, sensitive mechanisms maintain intracellular concentration of ions of noxious substances within the physiological range. These characteristics allow transporters to recognize, bind, and mediate the transmembrane transport of specific ions. For example, some transporters mediate the transport of divalent cations but do not recognize mono- or trivalent ions (Lasat 2000), among other mechanisms.

27.5 CONCLUSIONS AND RECOMMENDATIONS

The use of phytoextraction technique to remove noxious substances from contaminated media has been shown not only to be feasible but one that is well known among the various phytoremediation technologies irrespective of the fact that this technique is almost only suitable for the removal of contaminants from diffusely polluted areas contaminated at shallow depths with low to moderate concentration of contaminants as plant growth may not be sustained in heavily polluted medium. Thus, contaminated medium that is conducive to phytoextraction should be of primary importance while continuous search for hyperaccumulating plants having the ability to thrive in toxic environment with little maintenance and other desirable traits are imperative for the success of phytoextraction. Studies have further shown that to be able to use this type of clean-up technology, the plant to be used must either have a natural ability to take up, translocate and hyperaccumulate these substances in the above-ground biomass or they may be assisted through various means in order to be able to perform these tasks. Therefore, the need for studies that can contribute to the optimization of this type of phytoremediation technology are still presently inexhaustible with good prospects for improvement in the removal of other noxious substances apart from heavy metals that phytoextraction is presently well known for. In addition, many of the naturally hyperaccumulating plants tend to grow slowly and usually produce low biomass. Therefore, there is also the need for increased number and intensity in studies focusing on hyperaccumulating plants with high biomass production and accumulation of pollutants of interest in the harvestable parts of the plant through efficient translocation from roots to the above-ground biomass.

Local species are suggested to be of high priority in order to avoid the introduction of inappropriate or invasive plant species that may destroy biodiversity. Presently, no single plant species have been described to possess all the desirable traits for phytoextraction thereby, encouraging series of plant-breeding and genetic-engineering research efforts with the hope that public opinion concerning genetically modified plants for phytoextraction will be different compared to those used for human consumption. The roots, rhizospheric processes and surrounding region have been shown to be very important to the success of phytoextraction, thus, further studies that are aimed at increasing uptake of noxious substances through the roots are necessary. Studies aimed at clarifying if hyperaccumulators exude specific chelators and other organic products including determining their chemical composition and exudation rate in the rhizosphere and how they can enhance phytoextraction are necessary.

Presently, much is still not yet fully understood as it concerns proper identification of root exudates and other natural organic products, how these may act as natural chelators, how they are degraded in the soil and their relationship with micro-organisms in the soil. Outcome from these kinds of studies are expected to contribute to phytoextraction processes and also overcome environmental constraints associated with chemically assisted phytoextraction. Acidification and/or addition of artificial chelating agents to soil may still be relevant for some phytoextraction processes, however adequate researches that will contribute to the efforts at increasing the efficiency of this artificially assisted phytoextraction will be necessary. It is equally important that these studies are also focused on minimizing or eliminating the constraints associated with this method such as claims that many of these chelators are themselves toxic or that they may form complexes that might threaten soil quality or induce the release of noxious substances from soil that usually exceeds by far most plant's uptake capacity, especially under field condition, with serious risk implication concerning soil and groundwater pollution, among others. In addition, tolerance of phytoextracting plants to the targeted pollutants is also very important and necessary for the success of the process. Presently, research gap still exists concerning how this complex homeostatic network is achieved in some plants and how tolerance and hyperaccumulation are controlled and/or influenced by genes. Nevertheless, tolerance is thought to be based on several mechanisms rather than one alone and they are likely to operate in combination. Therefore, more research efforts are required in this regard in order to maximize this technique since the main objective of phytoextraction is to accumulate toxic substance in the above-ground tissues of plants.

In recommending phytoextraction for the cleaning up of contaminated media, it is important not to ignore aspects that have to do with the development of economically feasible techniques for the disposal of metal-enriched plants, although, this may not be of much challenge in the case of phytomining, since accumulated metals are to be recovered. Removal of the whole or above-ground portions of plants after sufficient growth and hyperaccumulation of heavy metals make them to become hazardous material that are to be properly disposed off. This means that research must be focused on finding ways to avoid transfer of metals to other media, the environment in general, and particularly to the food chain. This is particularly important in low income countries such as those in Sub-Sahara Africa.

REFERENCES

Abrams, M.M., C. Shennan, J. Zazoski, and R.G. Burau. 1990. Selenomethionine uptake by wheat seedlings. *Agron. J.* 82: 1127–1130.

Alkorta, I., J. Hernández-Allica, J.M. Becerril, I. Amezaga, I. Albizu, and C. Garbisu. 2004. Recent findings on the phytoremediation of soils contaminated with environmentally toxic heavy metals and metalloids such as zinc, cadmium, lead, and arsenic. *Reviews in Environmental Science and Biotechnology* 3: 71–90.

Arvy, M.P. 1993. Selenate and selenite uptake and translocation in bean plant (*Phaseolus vulgaris*). *Journal of Experimental Botany* 44: 1083–1087.

Ayotamuno, J.M., and R.B. Kogbara. 2007. Determining the tolerance level of *Zea mays* (maize) to a crude oil polluted agricultural soil. *African Journal of Biotechnology* 6: 1332–1337.

Azhar, N., M.Y. Ashraf, M. Hussain, and F. Hussain. 2006. Phytoextraction of lead (Pb) by EDTA application through sunflower (*Helianthus annuus* L.) cultivation: Seedling growth studies. *Pakistan Journal of Botany* 38 (5): 1551–1560.

Baker, A.J. 1999. Metal hyperaccumulator plants: A review of the biological resource for possible exploitation in the phytoremediation of metal-polluted soils. In *Phytoremediation of Contaminated Soil and Water*, eds. Terry, N., and G.S. Bañeulos, 85–107. CRC Press LLC, Boca Raton, FL.

Baker, A.J.M. 1981. Accumulators and excluders strategies in the response of plants to heavy metals. *Journal of Plant Nutrition* 3: 643–654.

Baker, A.J.M. 1987. Metal tolerance. *New Phytologist* 106: 93–111.

Baker, A.J.M., and P.L. Walker. 1990. Ecophysiology of metal uptake by tolerant plants. In *Heavy Metal Tolerance in Plants: Evolutionary Aspects*, ed. Shaw, A.J., 155–178. Boca Raton: CRC Press.

Baker, A.J.M., and R.R. Brooks. 1989. Terrestrial higher plants which hyperaccumulate metal elements: A review of their distribution, ecology, and phytochemistry. *Biorecovery* 1: 81–126.

Baker, A.J.M., R.D. Reeves, and A.S.M. Hajar. 1994. Heavy metal accumulation and tolerance in British populations of the metallophyte *Thlaspi caerulescens* J. & C. Presl (Brassicaceae). *New Phytologist* 127: 61–68.

Baker, A.J.M., R.D. Reeves, and S.P. Mcgrath. 1991. *In situ* decontamination of heavy metal polluted soils using crops of metal-accumulating plants: A feasibility study. In *In situ Bioreclamation*, eds. Hinchee, R.L., and R.F. Olfenbuttel, 600–605. Butterworth-Heinemann, Boston.

Baker, A.J.M., S.P. McGrath, C.M.D. Sidoli, and R.D. Reeves. 1994. The possibility of *in situ* heavy metal decontamination of polluted soils using crops of metal-accumulating plants. *Resour. Conserv. Recycl.* 11: 41–49.

Baumann, A. 1885. Das verhalten von zinksatzen gegen pflanzen und im boden. Landwirtsch. *Vers.-Statn* 31: 1–53.

Bernal, M.P., S.P. McGrath, A.J. Miller, and A.J.M. Baker. 1994. Comparison of the chemical changes in the rhizosphere of the nickel hyperaccumulator *Alyssum murale* with the non-accumulator *Raphanus sativus*. *Plant and Soil* 164: 251–259.

Bert, V., P. Meerts, P. Saumitou-Laprade, P. Salis, W. Gruber, and N. Verbruggen. 2003. Genetic basis of Cd tolerance and hyperaccumulation in *Arabidopsis halleri*. *Plant Soil* 249: 9–48.

Blaylock, M.J., and J.W. Huang. 2000. Phytoextraction of metals. In *Phytoremediation of toxic metals: Using plants to clean-up the environment*, eds. Raskin, I., and B.D. Ensley, 53–70. John Wiley & Sons, Inc., New York.

Blaylock, M.J., D.E. Salt, S. Dushenkov, et al. 1997. Enhanced accumulation of Pb in Indian mustard by soil-applied chelating agents. *Environ. Sci. Technol.* 31 (3): 860–865.

Briggs, G.G., R.H. Bromilow, and A.A. Evans. 1982. Relationships between lipophilicity and root uptake and translocation of non-ionised chemicals by barley. *Pestcide Sci.* 13: 495–504.

Brooks, R.R. 1998. General introduction. In *Plants that Hyperaccumulate Heavy Metals: Their Role in Phytoremediation, Microbiology, Archaeology, Mineral Exploration and Phytomining*, ed. Brooks, R.R., 1–14. CAB International, New York.

Brooks, R.R., J. Lee, R.D. Reeves, and T. Jaffré. 1977. Detection of nickeliferous rocks by analysis of herbarium specimens of indicator plants. *J. Geochem. Explor.* 7: 49–57.

Brooks, R.R., M.F. Chambers, L.J. Nicks, and B.H. Robinson. 1998. Phytomining. *Trends Plant Sci.* 3: 359–362.

Brown, S.L. R.L. Chaney, J.S. Angle, and A.J.M. Baker. 1994. Phytoremediation potential of *Thlaspi caerulescens* and bladder campion for zinc-contaminated and cadmium-contaminated soil. *Journal of Environmental Quality* 23: 1151–1157.

Brown, T.A., and A. Shrift. 1982. Selenium: Toxicity and tolerance in higher plants. *Biol. Rev.* 57: 59–84.

Byers, H.G. 1935. Selenium occurrence in certain soils in the United States, with a discussion of the related topics. *US Dept. Agric. Technol. Bull.* 482: 1–47.

Cakmak, I., R.M. Welch, J.J. Hart, W.A. Norvell, L. Oztürk, and L.V. Kochian. 2000. Uptake and retranslocation of leaf-applied cadmium (^{109}Cd) in diploid, teraploid and hexaploid wheats. *J. Exp. Botany* 51: 221–226.

Cao, A., A. Carucci, T. Lai, P. La Colla, and E. Tamburini. 2007. Effect of biodegradable chelating agents on heavy metals phytoextraction with *Mirabilis jalapa* and on its associated bacteria. *European Journal of Soil Biology* 43: 200–206.

Carbonell-Barrachina, A., F.B. Carbonell, and J.M. Beyeto. 1994. Effect of arsenite on concentrations of micronutrients in tomato plants grown in hydroponic culture. *J. Plant Nutr.* 17: 1887–1903.

Chaney, R.L. 1983. Plant uptake of inorganic waste constituents. In *Land Treatment of Hazardous Waste*, eds. Parr, J.F., P.B. Marsh, and J.M. Kla, 50–76. Noyes Data Crop, Park Ridge, NJ.

Chaney, R.L., M. Malik, Y.M. Li, et al. 1997. Phytoremediation of soil metals. *Current Opinion in Biotechnology* 8: 279–284.

Chaudhry, Q., M. Blom-Zandstra, S. Gupta, and E.J. Joner. 2005. Utilising the synergy between plants and rhizosphere microorganisms to enhance breakdown of organic pollutants in the environment. *Environmental Science and Pollution Research* 12 (1): 34–48.

Chen, Y., X.D. Li, and Z.G. Shen. 2004. Leaching and uptake of heavy metals by ten different species of plants during an EDTA-assisted phytoextraction process. *Chemosphere* 57: 187–196.

Claus, D., H. Dietze, A. Gerth, W. Grossser, and A. Hedner. 2007. Application of agronomic practice improves phytoextraction on a multipolluted site. *Journal of Environmental Engineering and Landscape Management* 15 (4): 208–212.

Clemens, S., J.I. Schroeder, and T. Degenkolb. 2001. *Caenorhabditis elegans* expresses a functional phytochelatin synthase. *Eur. J. Biochem.* 268: 3640–3643.

Clemens, S., M.G. Palmgren, and U. Krämer. 2002. A long way ahead: Understanding and engineering plant metal accumulation. *Trends Plant Sci.* 7: 309–315.

Cobbett, C., and P. Goldsbrough. 2002. Phytochelatins and metallothioneins: Roles in heavy metal detoxification and homeostasis. *Annual Review of Plant Biology* 53: 159–182.

Cobbett, C.S. 2000a. Phytochelatins and their roles in heavy metal detoxification. *Plant Physiology* 123: 825–832.

Cobbett, C.S. 2000b. Phytochelatin biosynthesis and function in heavy metal detoxification. *Curr. Op. Plant Biol.* 3: 211–216.

Cobbett, C.S., and P.B. Goldsbrough. 1999. Mechanisms of metal resistance: Phytochelatins and metallothioneins. In *Phytoremediation of Toxic Metals: Using Plants to Clean Up the Environment*, eds. Raskin, I., and B.D. Ensley, 247–269, John Wiley & Sons Inc, New York.

Connell, D.W., and G.J. Miller. 1984. *Chemistry and Ecotoxicology of Pollution*, 444. John Wiley & Sons, New York.

Cooper, E.M., J.T. Sims, S.D. Cunningham, J.W. Huang, and W.R. Berti. 1999. Chelate-assisted phytoextraction of lead from contaminated soil. *Journal of Environmental Quality* 28: 1709–1719.

Cosio, C. 2004. Phytoextraction of heavy metal by hyperaccumulating and non hyperaccumulating plants: Comparison of cadmium uptake and storage mechanisms in the plants. Thesis for the grade of Docteur ès Sciences presented at the Faculté Environnement Naturel, Architectural et Construit, Institut des Sciences et Technologies de l'Environnement, Section des Sciences et Ingénierie de l'Environnement, École Polytechnique Fédérale de Lausanne, France. Thèse No 2937, 136pp.

Crèman, H., Š. Velikonja–Bolta, D. Vodnik, B. Kos, and D. Leštan. 2001. EDTA enhanced heavy metal phytoextraction: Metal accumulation, leaching and toxicity. *Plant Soil* 235: 105–114.

Cunningham, S.D., and D.W. Ow. 1996. Promises and prospects of phytoremediation. *Plant Physiology* 110: 715–719.

Cunningham, S.D., J.R. Shann, D.E. Crowley, and T.A. Anderson. 1997. Phytoremediation of contaminated water and soil. In *Phytoremediation of Soil and Water Contaminants. ACS Symposium Series 664*, eds. Kruger, E.L., T.A. Anderson, and J.R. Coats, 2–19. American Chemical Society, Washington, DC.

Cunningham, S.D., T.A. Anderson, P. Schwab, and F.C. Hsu. 1996. Phytoremediation of soils contaminated with organic pollutants. *Adv. Agronomy* 56: 55–114.

Cunningham, S.D., W.R. Berti, and J.W. Huang. 1995. Phytoremediation of contaminated soils. *Trends in Biotechnology* 13: 393–397.

De Vos, C.H.R., H. Schat, M.A.M. De Waal, R. Voojs, and W.H.O. Ernst. 1991. Increase resistance to copper induced damage of root cell plasma lemma in copper tolerant *Silene cucubalus*. *Plant Physiology* 82: 523–528.

Dietz, A.C., and J.L. Schnoor. 2001. Advances in phytoremediation. *Environ. Health Perspect.* 109: 163–168.

Dittmer, H.J. 1937. A quantitative study of the roots and root hairs of a winter rye plant (*Secale cereale*). *American Journal of Botany* 24: 417–420.

Dushenkof, S., D. Vasudev, Y. Kapulnik, et al. 1997. Removal of uranium from water using terrestrial plants. *Environ. Sci. Technol.* 31: 3468–3474.

Dushenkov, V., P.B.A. Nanda-Kumar, H. Motto, and I. Raskin. 1995. Rhizofiltration: The use of plants to remove heavy metals from aqueous streams. *Environ. Sci. Tech.* 29: 1239–1245.

Ebbs, S.D., M. Lasat, D. Brady, J. Cornish, R. Gordon, and L.V. Kochian. 1997. Phytoextraction of cadmium and zinc from a contaminated soil. *J. Environ. Qual.* 26: 1424–1430.

Ehsan, M., K. Santamaría-Delgado, A. Vázquez-Alarcón, et al. 2009. Phytostabilization of cadmium contaminated soils by *Lupinus uncinatus* Schldl. *Spanish Journal of Agricultural Research* 7 (2): 390–397.

Elliot, H.A., M.R. Liberali, and C.P. Huang. 1986. Competitive adsorption of heavy metals by soils. *J. Environ. Qual.* 15: 214–219.

Entry, J.A., L.S. Watrud, and M. Reeves. 1999. Accumulation of ^{137}Cs and ^{90}Sr from contaminated soil by three grass species inoculated with mycorrhizal fungi. *Environmental Pollution* 104: 449–457.

Erakhrumen, A.A. 2007. Phytoremediation: An environmentally sound technology for pollution prevention, control and remediation in developing countries. *Educational Research and Review* 2 (7): 151–156.

Erakhrumen, A.A. 2011. Research advances in bioremediation of soils and groundwater using plant-based systems: A case for enlarging and updating information and knowledge in environmental pollution management in developing countries. In *Biomanagement of metal-contaminated soils*, eds. Khan, M.S., A. Zaidi, R. Goel, and J. Musarrat, 143–166, *Environmental Pollution* 20.

Ernst, W. 1998. The origin and ecology of contaminated, stabilized and non-pristine soils. In *Metal-Contaminated Soils: In situ Inactivation and Phytorestoration*, eds. Vangronsveld, J. and S. Cunningham, 17–25. Springer-Verlag and R.G. Landes Company.

Flathman, P.E., and G.R. Lanza. 1998. Phytoremediation: Current views on an emerging green technology. *Journal of Soil Contamination* 7 (4): 415–432.

Florjin, P.J., and M.L. Van Beusichem. 1993. Uptake and distribution of cadmium in maize inbread lines. *Plant Soil* 150: 25–32.

Frey, B., C. Keller, K. Zierold, and R. Schulin. 2000. Distribution of Zn in functionally different leaf epidermal cells of the hyperaccumulator *Thlaspi caerulescens*. *Plant Cell Environ.* 23: 675–687.

Friedland, A.J. 1990. Movement of metals through soils and ecosystems. In *Heavy Metal Tolerance in Plants: Evolutionary Aspects*, ed. Shaw, A.J., 7–19. CRC Press: Boca Raton, FL.

Garcia-Miragaya, J. 1984. Levels, chemical fractions and solubility of lead in roadside soils of Caracas, Venezuela. *J. Soil Sci.* 138: 147–152.

Geebelen, W., J. Vangronsveld, D.C. Adriano, L.C. Van Poucke, and H. Clijsters. 2002. Effects of Pb-EDTA and EDTA on oxidative stress reactions and mineral upake in *Phaseolus vulgaris*. *Physiologia Plantarum* 115 (3): 377–384.

Ginneken, L.V., E. Meers, R. Guisson, et al. 2007. Phytoremediation for heavy metal-contaminated soils combined with bioenergy production. *Journal of Environmental Engineering and Landscape Management* 15 (4): 227–236.

Goldsbrough, P. 2000. Metal tolerance in plants: The role of phytochelatins. In *Phytoremediation of Contaminated Soil and Water*, eds. Terry, N., and G. Bañuelos, 221–233. Boca Raton: Lewis Publishers.

Gonzaga, M.I.S. 2006. Effects of soil and plant on arsenic accumulation by arsenic hyperaccumulator *Pteris vittata* L. A dissertation presented to the Graduate School of the University of Florida in partial fulfilment of the requirements for the degree of Doctor of Philosophy, University of Florida. 146pp.

Gosh, M., and S.P. Singh. 2005. Comparative uptake and phytoextraction study of soil induced chromium by accumulator and high biomass weed species. *Applied Ecology and Environmental Research* 3 (2): 67–69.

Greger, M. 1999. Metal availability and bioconcentration in plants. In *Heavy Metal Stress in Plants. From Molecules to Ecosystem*, eds. Prasad, M.N.V., and J. Hagemeyer, 1–29. Springer-Verlag Berlin-Heidelberg.

Grill, E., E.L. Winnacker, and M.H. Zenk. 1988. Occurrence of heavy metal binding phytochelatins in plants growing in a mining refuse area. *Experientia* 44: 539–540.

Gupta, S.K., T. Herren, K. Wenger, R. Krebs, and T. Hari. 2000. *In situ* gentle remediation measures for heavy metal-polluted soils. In *Phytoremediation of Contaminated Soil and Water*, eds. Terry, N., and G. Bañuelos, 303–322. Boca Raton: Lewis Publishers.

Hamer, D.H., D.J. Thiele, and J.E. Lemontt. 1985. Function and autoregulation of yeast copperthionein. *Science* 228: 685–690.

Hanson, P.J., E.I. Sucoff, and A.H. Markhart III. 1985. Quantifying apoplastic flux through red pine root systems using trisodium, 3-hydroxy-5,8,10-pyrenetrisulfonate. *Plant Physiol.* 77: 21–24.

Hardiman, R.T., B. Jacoby, and A. Banin. 1984. Factors affecting the distribution of cadmium, copper and lead and their effect upon yield and zinc content in bush beans (*Phaseolus vulgaris* L.). *Plant Soil* 81: 17–27.

Hartley-Whitaker, J., G. Ainsworth, R. Vooijs, W. Ten Brookum, H. Schat, and A.A. Meharg, 2001. Phytochelatins are involved in differential arsenate tolerance in *Holcus lanatus*. *Plant Physiology* 126: 299–306.

Hasegawa, I., E. Terada, M. Sunairi, et al. 1997. Genetic improvement of heavy metal tolerance by transfer of the yeast methallothionein gene (CUP1). *Plant Soil* 196: 277–281.

Heaton, A.C.P., C.L. Rugh, N.-J. Wang, and R.B. Meagher. 1998. Phytoremediation of mercury and methylmercury polluted soils using genetically engineered plants. *J. Soil Contam.* 7: 497–509.

Heiss, S., A. Wachter, J. Bogs, C. Cobbett, and T. Rausch. 2003. Phytochelatin synthase (PCS) protein is induced in *Brassica juncea* after prolonged Cd exposure. *J. Exp. Botany* 54: 1833–1839.

Henry, J.R. 2000. An overview of phytoremediation of lead and mercury. NNEMS Report. Washington, DC, pp. 3–9.

Hirschi, K.D., V.D. Korenkov, N.L. Wilganowski, and G.J. Wagner. 2000. Expression of *Arabidopsis* CAX2 in tobacco. Altered metal accumulation and increased manganese tolerance. *Plant Physiology* 124: 125–133.

Howden, R., and C.S. Cobbett. 1992. Cadmium-sensitive mutants of *Arabidopsis thaliana*. *Plant Physiology* 100: 100–107.

Howden, R., P.B. Goldsbrough, C.R. Andersen, and C.S. Cobett. 1995. Cadmium-sensitive *cad1* mutants of *Arabidopsis thaliana* are Phytochelatin deficient. *Plant Physiol.* 107: 1059–1066.

Hsu, F.C., R.L. Marxmiller, and A.Y.S. Yang. 1990. Study of root uptake and xylem translocation of cinmethylin and related compounds in detopped soybean roots using a pressure chamber technique. *Plant Physiol.* 93: 1573–1578.

Huang, J.W.W., J. Chen, W.R. Berti, and S.D. Cunningham. 1997. Phytoremediation of lead-contaminated soils: Role of synthetic chelates in lead phytoextraction. *Environ. Sci. Technol.* 3: 800–805.

Huckle, J.W., A.P. Morby, J.S. Turne, and N.J. Robinson. 1996. Isolation of a prokaryotic metallothionein locus and analysis of transcriptional control by trace metal ion. *Mol. Microbiol.* 7: 177–187.

Inouhe, M., R.O.S. Ito, N. Sasada, H. Tohoyama, and M. Joho. 2000. Azuki bean cells are hypersensitive to cadmium and do not synthesize phytochelatins. *Plant Physiology* 123: 1029–1036.

Kärenlampi, S., H. Schat, J. Vangronsveld, et al. 2000. Genetic engineering in the improvement of plants for phytoremediation of metal polluted soils. *Environ. Pollut.* 107: 225–231.

Kaschl, A., V. Römheld, and Y. Chen. 2002. Cadmium binding by fractions of dissolved organic matter and humic substances from municipal solid waste compost. *Journal of Environmental Quality* 31: 1885–1892.

Kayser, A., K. Wenger, A. Keller, et al. 2000. Enhancement of phytoextraction of Zn, Cd, and Cu from calcareous soil: The use of NTA and sulfur amendments. *Environ. Sci. Technol.* 34: 1778–1783.

Komárek, M., P. Tlustoš, J. Száková, V. Chrastný, and J. Balík. 2007. The role of Fe- and Mn-oxides during EDTA-enhanced phytoextraction of heavy metals. *Plant Soil Environ.* 53 (5): 216–224.

Kos, B., and D. Leštan. 2003. Influence of biodegradable ([S, S]-EDDS) and non degradable (EDTA) chelate and hydrogel modified soil water sorption capacity on Pb phytoextraction and leaching. *Plant Soil* 253: 403–411.

Kos, B., and D. Leštan. 2004. Chelator induced phytoextraction and *in situ* soil washing of Cu. *Environmental Pollution* 132: 333–339.

Krämer, U. 2003. Phytoremediation to phytochelatin—plant trace metal homeostasis. *New Phytologist* 158: 4–6.

Krämer, U., J.D. Cotter-Howells, J.M. Charnock, A.J.M. Baker, and J.A.C. Smith. 1996. Free histidine as a metal chelator in plants that accumulate nickel. *Nature* 379: 635–638.

Krämer, U., R.D. Smith, W. Wenzel, I. Raskin, and D.E. Salt. 1997. The role of metal transport and tolerance in nickel hyperaccumulation by *Thlaspi goesingense* halacsy. *Plant Physiology* 115: 1641–1650.

Krishnamurti, G.S.R., G. Cieslinski, P.M. Huang, and K.C.J. Van Pees. 1997. Kinetics of cadmium release from soils as influenced by organic acids: Implication in cadmium availability. *Journal of Environmental Quality* 26: 271–277.

Küpper, H., E. Lombi, F.J. Zhao, and S.P. McGrath. 2000. Cellular compartmentation of cadmium and zinc in relation to other elements in the hyperaccumulator *Arabidopsis halleri*. *Planta* 212: 75–84.

Küpper, H., F.J. Zhao, and S.P. McGrath. 1999. Cellular compartmentation of zinc in leaves of the hyperaccumulator *Thlaspi caerulescens*. *Plant Physiology* 119: 305–311.

Kumar, P.B.A.N., V. Dushenkov, H. Motto, and I. Raskin. 1995. Phytoextraction: The use of plants to remove heavy metals from soils. *Environmental Science and Technology* 29 (5): 1232–1238.

Lasat, M.M. 2000. Phytoextraction of metals from contaminated soil: A review of plant/soil/metal interaction and assessment of pertinent agronomic issues. *Journal of Hazardous Substance Research* 2: 5–1 to 5–25.

Lasat, M.M., A.J.M. Baker, and L.V. Kochian. 1996. Physiological characterization of root Zn^{2+} absorption and translocation to shoots in Zn hyperaccumulator and nonaccumulator species of *Thlaspi*. *Plant Physiol.* 112: 1715–1722.

Lasat, M.M., A.J.M. Baker, and L.V. Kochian. 1998. Altered Zn compartmentation in the root symplasm and stimulated Zn absorption into the leaf as mechanisms involved in Zn hyperaccumulation in *Thlaspi caerulescens*. *Plant Physiol.* 118: 875–883.

Lee, J., R.D. Reeves, R.R. Brooks, and T. Jaffré. 1977. Isolation and identification of a citratocomplex of nickel from nickel-accumulating plants. *Phytochemistry* 16: 1502–1505.

Li, M.S., Y.P. Luo, and Z.Y. Su. 2007. Heavy metal concentrations in soils and plant accumulation in a restored manganese mineland in Guangxi, South China. *Environmental Pollution* 147: 168–175.

Lin, Q., Y.X. Chen, H.M., Chen, Y.L., Yu, Y.M., Luo, and M.H. Wong. 2003. Chemical behavior of Cd in rice rhizosphere. *Chemosphere* 50: 755–761.

Lombi, E., F.J. Zhao, S.J., Dunham, and S.P. McGrath. 2001. Phytoremediation of heavy metal-contaminated soils: Natural hyperaccumulation versus chemically enhanced phytoextraction. *Journal of Envirnonmental Quality* 30: 1919–1926.

Lombi, E., K.L. Tearall, J.R. Howarth, F.J., Zhao, M.J., Hawkesford, and S.P. McGrath. 2002. Influence of iron status on calcium and zinc uptake by different ecotypes of the hyperaccumulator *Thlaspi caerulescens*. *Plant Physiol.* 128: 1359–1367.

Ma, L.Q., K.M. Komar, C. Tu, W. Zhang, Y. Cai, and E.D. Kennelley. 2001. A fern that hyperaccumulates arsenic. *Nature* 409: 579.

Macek, T., D. Pavlíková, and M. Macková. 2004. Phytoremediation of metals and inorganic pollutants. In *Soil Biology. Vol. 1: Applied Bioremediation and Phytoremediation*, eds. Singh, A., and O.P. Ward, 135–157. Springer Verlag, Berlin, Heidelberg.

MacNair, M.R. 1997. The evolution of plants in metal-contaminated environments. In *Environmental Stress, Adaptation and Evolution*, ed. Loeschcke, R.B.V., 2–24.

MacNair, M.R., V. Bert, S.B. Huitson, P. Saumitou-Laprade, and D. Petit. 1999. Zinc tolerance and hyperaccumulation are genetically independent characters. *Proc. R. Soc. Lond.* B 266: 2175–2179.

Madrid, F., M.S. Liphadzi, and M.B. Kirkham. 2003. Heavy metal displacement in chelate-irrigated soil during phytoremediation. *Journal of Hydrology* 272: 107–119.

Mann, S.S., and G.S.P. Ritchie. 1993. The influence of pH on the forms of cadmium in four west Australian soils. *Australian Journal of Soil Research* 31: 255–270.

Marschner, H. 1995. *Mineral nutrition of higher plants*. 2nd Edition. San Diego: Academic Press, 889 pp.

Martinez, C.E., and H.L. Motto. 2000. Solubility of lead, zinc, and copper added to mineral soils. *Environmental Pollution* 107: 153–158.

Mathys, W. 1977. The role of malate, oxalate, and mustard oil glucosides in the evolution of zinc-resistance in herbage plants. *Physiol. Plant* 40: 130–136.

McCutcheon, S.C., and J.L. Schnoor (eds). 2003. *Phytoremediation: Transformation and Control of Contaminants.* Hoboken, NJ: Wiley-Interscience, Inc.

McGrath, S.P., Z.G. Shen, and F.J. Zhao. 1997. Heavy metal uptake and chemical changes in the rhizosphere of *Thlaspi caerulescens* and *Thlaspi ocholeucum* grown in contaminated soils. *Plant and Soil* 188: 153–159.

McNeil, K.R., and S. Waring. 1992. Vitrification of contaminated soils. In *Contaminated Land Treatment Technologies*, ed. Rees, J.F., 143–159. Society of Chemical Industry. Elsevier Applied Sciences, London.

Meagher, R.B. 2000. Phytoremediation of toxic elemental and organic pollutants. *Current Opinion in Plant Biology* 3: 153–162.

Meerts, P., and N.V. Isacker. 1997. Heavy metal tolerance and accumulation in metallicolous and non-metallicolous populations of *Thlaspi caerulescens* from continental Europe. *Plant Ecology* 133: 221–231.

Mench, M., and E. Martin. 1991. Mobilization of cadmium and other metals from two soils by root exudates of *Zea mays* L., *Nicotiana tabacum* L. and *Nicotiana rustica* L. *Plant and Soil* 132: 187–196.

Moon, G.J., B.F. Clough, C.A. Peterson, and W.G. Allaway. 1986. Apoplastic and symplastic pathways in *Avicennia marina* (Forsk.) Vierh. roots revealed by fluorescent tracer dyes. *Australian Journal of Plant Physiology* 13:637–648.

Murphy, A., J. Zhou, P.B. Goldsbrough, and L. Taiz. 1997. Purification and immunological identification of metallothioneins 1 and 2 from *Arabidopsis thaliana*. *Plant Physiol.* 113: 1293–1301.

Nascimento, C.W.A., and B. Xing. 2006. Phytoextraction: A review on enhanced metal availability and plant accumulation. *Sci. Agric.* (*Piracicaba, Braz.*) 63 (3): 299–311.

Nascimento, C.W.A., D. Amarasiriwardena, and B. Xing. 2006. Comparison of natural organic acids and synthetics chelates at enhancing phytoextraction of metals from a multi-metal contaminated soil. *Environmental Pollution* 140: 114–123.

Newman, L.A., and C.M. Reynolds. 2004. Phytodegradation of organic compounds. *Current Opinion in Biotechnology* 15: 225–230.

Padmavathiamma, P.K., and L.Y. Li. 2007. Phytoremediation technology: Hyper-accumulation metals in plants. *Water Air Soil Pollution* 184: 105–126.

Peijnenburg, W.J.G.M., and T. Jager. 2003. Monitoring approaches to assess bioaccessibility and bioavailability of metals: Matrix issues. *Ecotoxicol. Environ. Saf.* 56: 63–77.

Pickering, I.J., R.C. Prince, M.J. George, R.D. Smith, G.N. George, and D.E. Salt. 2000. Reduction and coordination of arsenic in Indian mustard. *Plant Physiol.* 122: 1171–1177.

Pollard, A., K. Powell, F. Harper, and J. Smith. 2002. The genetic basis of metal hyperaccumulation in plants. *Crit. Rev. Plant Sci.* 21: 539–566.

Prasad, M.N.V., and H.M. Freitas. 2003. Metal hyperaccumulation in plants—biodiversity prospecting for phytoremediation technology. *Electronic Journal of Biotechnology* 6 (3): 285–321.

Quartacci, M.F., A.J.M. Baker, and F. Navari-Izzo. 2005. Nitrilotriacetate- and citric acid-assisted phytoextraction of cadmium by Indian mustard (*Brassica juncea* (L.) Czernj, Brassicaceae). *Chemosphere* 59: 1249–1255.

Raskin, I., R.D. Smith, and D.E. Salt. 1997. Phytoremediation of metals: Using plants to remove pollutants from the environment. *Curr. Opin. Biotechnol.* 8: 221–226.

Rauser, W.E. 1990. Phytochelatins. *Annu. Rev. Biochem.* 59: 61–86.

Rauser, W.E. 1995. Phytochelatins and related peptides—structure, biosynthesis and function. *Plant Physiology* 109: 1141–1149.

Rauser, W.E. 1999. Structure and function of metal chelators produced by plants. *Cell Biochem. Bioph.* 31: 19–48.

Rea, P.A., Z.-S. Li, Y.-P. Lu, Y.M. Drozdowicz, and E. Martinoia. 1998. From vacuolar GS-X pumps to multispecific ABC transporters. *Annu. Rev. Plant Physiol. Plant Mol. Biol.* 49: 727–760.

Reeves, R., and A. Baker. 2000. Metal accumulating plants. In *Phytoremediation of Toxic Metals: Using Plants to Clean Up the Environment*, eds. Raskin, I., and B. Ensley, 193–229. John Wiley & Sons, Inc.

Reeves, R.D., and R.R. Brooks. 1983a. Hyperaccumulation of lead and zinc by two metallophytes from a mining area of central Europe. *Environmental Pollution* Series A31: 277–287.

Reeves, R.D., and R.R. Brooks. 1983b. European species of *Thlaspi* L. (Cruciferae) as indicators of nickel and zinc. *Journal of Geochemical Exploration* 18: 275–283.

Renella, G., L. Landi, and P. Nannipieri. 2004. Degradation of low molecular weight organic acids complexed with heavy metals in soil. *Geoderma* 122: 311–315.

Robinson, B.H., L. Meblanc, D. Petit, R.R. Brooks, J.H. Kirkman, and P.E.H. Gregg. 1998. The potential of *Thlaspi caerulescens* for phytoremediation of contaminated soils. *Plant Soil* 203: 47–56.

Robinson, B.H., S. Green, T. Mills, et al. 2003a. Phytoremediation: Using plants as biopumps to improve degraded environments. *Australian Journal of Soil Research* 41: 599–611.

Robinson, B.H., T.M. Mills, D. Petit, L.E. Fung, S.R. Green, and B.E. Clothier. 2003b. Natural and induced cadmium-accumulation in poplar and willow: Implication for phytoremediation. *Plant Soil* 227: 301–306.

Robinson, N.J., A.M. Tommey, C. Kuske, and P.J. Jackson. 1993. Plant metallothioneins. *Biochem.* 295: 1–10.

Romkens, P., L. Bouwman, J. Japenga, and C. Draaisma. 2002. Potentials and drawbacks of chelate-enhanced phytoremediation of soils. *Environmental Pollution* 116: 109–121.

Rugh, C.L. 2004. Genetically engineered phytoremediation: One man's trash is another man's transgene. *Trends in Biotechnology* 22: 496–468.

Rulkens, W.H., R. Tichy, and J.T.C. Grotenhuis. 1998. Remediation of polluted soil and sediment: Perspectives and failures. *Water Sci. Technol.* 37: 27–35.

Ryan, P.R., E. Delhaize, and D.L. Jones. 2001. Function and mechanism of organic anion exudation from plant roots. *Annual Review of Plant Physiology and Plant Molecular Biology* 5: 527–560.

Salt, D.E., and U. Krämer. 1999. Mechanisms of metal hyperaccumulation in plants. In *Phytoremediation of Toxic Metals: Using Plants to Clean Up the Environment*, eds. Raskin, I., and B. Ensley, 231–246. John Wiley & Sons, Inc., New York.

Salt, D.E., M. Blaylock, N.P.B.A. Kumar, et al. 1995a. Phytoremediation: A novel strategy for the removal of toxic metals from the environment using plants. *Biotechnology* 13: 468–474.

Salt, D.E., R.C. Prince, A.J.M. Baker, I. Raskin, and I.J. Pickering. 1999. Zinc ligands in the metal hyperaccumulator *Thlaspi caerulescens* as determined using x-ray absorption spectroscopy. *Environ Sci Technol.* 33: 713–717.

Salt, D.E., R.C. Prince, I.J., Pickering, and I. Raskin. 1995b. Mechanisms of cadmium mobility and accumulation in Indian mustard. *Plant Physiology* 109: 1427–1433.

Salt, D.E., R.D. Smith, and I. Raskin. 1998. Phytoremediation. *Annual Review Plant Physiology Plant Molecular Biology* 49: 643–668.

Sanità di Toppi, L., and R. Gabbrielli. 1999. Response to cadmium in higher plants. *Environ. Exp. Botany* 41: 105–130.

Sarret, G., P. Saumitou-Laprade, V. Bert, et al. 2002. Forms of zinc accumulated in the hyperaccumulator *Arabidopsis halleri. Plant Physiol.* 130: 1815–1826.

Schmidt, U. 2003. Enhancing phytoextraction: The effect of chemical soil manipulation on mobility, plant accumulation, and leaching of heavy metals. *Journal of Environmental Quality* 32: 1939–1954.

Schmöger, M.E.V., M. Oven, and E. Grill. 2000. Detoxification of arsenic by phytochelatins in plants. *Plant Physiology* 122: 793–801.

Shrift, A., and J.M. Ulrich. 1976. Transport of selenate and selenite into *Astragalus* roots. *Plant Physiol.* 44: 893–896.

Shuman, L.M. 1985. Fractionation method for soil microelements. *Soil Science* 140: 11–22.

Skinner, R.H., and J.W. Radin. 1994. The effect of phosphorus nutrition on water flow through the apoplastic bypass in cotton roots. *Journal of Experimental Botany* 45 (4): 423–428.

Steffens, J.C. 1990. The heavy metal-binding peptides of plants. *Ann. Rev. Plant Physiol. Mol. Biol.* 41: 553–575.

Stephan, U.W., I. Schmidke, V.W. Stephan, and G. Scholz. 1996. The nicotianamine molecule is made-to-measure for complexation of metal micronutrients in plants. *BioMetals* 9: 84–90.

Subramanian, M., D.O. Oliver, and J.V. Shanks. 2006. TNT phytotransformation pathway characteristics in arabidopsis: Role of aromatic hydroxylamines. *Biotechnology Programme* 22: 208–216.

Sun, B., F.J. Zhao, E. Lombi, and S.P. McGrath. 2001. Leaching of heavy metals from contaminated soils using EDTA. *Environmental Pollution* 113: 111–120.

Tandy, S., K. Bossart, R. Mueller, et al. 2004. Extraction of heavy metals from soils using biodegradable chelating agents. *Environ. Sci. Technol.* 38: 937–944.

Tandy, S., R. Schulin, and B. Nowack. 2006. The influence of EDDS on the uptake of heavy metals in hydroponically grown sunflowers. *Chemosphere* 62 (9): 1454–1463.

Tanton, T.W., and S.H. Crowdy. 1972. Water pathways in higher plants: II. Water pathways in boots. *Journal of Experimental Botany* 23 (3): 600–618.

Tester, M., and R.A. Leigh. 2001. Partitioning of nutrient transport processes in roots. *Journal of Experimental Botany* 52 (Supplementary 1): 445–457.

Timofeev-Resovsky, E.A., B.M. Agafonov, and N.V. Timofeev-Resovsky. 1962. Fate of radioisotopes in aquatic environments (In Russian). *Proceedings of Bilogical Institute USSR Academy of Sciences* 22: 49–67.

Tolrà, R.P., C.H. Poschenrieder, and J. Barceló. 1996. Zinc hyperaccumulation in *Thlaspi caerulescens*. II. Influence of organic acids. *J. Plant Nutr*. 19: 1541–1550.

Tomé, F.V., P.B. Rodríguez, and J.C. Lozano. 2008. Elimination of natural uranium and (226)Ra from contaminated waters by rhizofiltration using *Helianthus annuus* L. *The Science of the Total Environment* 393 (2–3): 351–357.

Tu, C., and L.Q. Ma. 2002. Effects of arsenic concentrations and forms on arsenic uptake by the hyperaccumulator ladder brake. *Journal of the Environmental Quality* 31: 641–647.

Van Hoof, N.A.L.M., V.H. Hassinen, H.W.J. Hakvoort, et al. 2001. Enhanced copper tolerance in *Silene vulgaris* (Moench) Garcke populations from copper mines is associated with increased transcript levels of a 2b-type metallothionein gene. *Plant Physiology* 126: 1519–1526.

Van Steveninck, R.F.M, M.E. Van Steveninck, A.J. Wells, and D.R. Fernando. 1990. Zinc tolerance and the binding of zinc as zinc phytate in *Lemna minor*. X-ray microanalytical evidence. *J. Plant Physiol*. 137: 140–146.

Vatamaniuk, O.K., E.A. Bucher, J.T. Ward, and P.A. Rea. 2001. A new pathway for heavy metal detoxification in animals. *J. Biol. Chem*. 276: 20817–20820.

Vazquez, S., A. Agha, A. Granado, et al. 2006. Use of white lupin plant for phytostabilization of Cd and As polluted acid soil. *Water, Air, & Soil Pollution* 177 (1–4): 349–365.

Vögeli-Lange, R., and G.J. Wagner. 1990. Subcellular localization of cadmium and cadmium-binding peptides in tobacco leaves. Implication of a transport function for cadmium-binding peptides. *Plant Physiol*. 92: 1086–1093.

Wagner, G.J. 1984. Characterization of a cadmium-binding complex of cabbage leaves. *Plant Physiol*. 76: 797–805.

Waisel, Y., A. Eshel, and U. Kafkafi. 1996. *Plant Roots: The Hidden Half*. 2nd Edition, Marcel Dekker, Inc., New York. 1002 pp.

Walker, T.S., H.P. Bais, E. Grotewold, and J.M. Vivanco. 2003. Root exudation and rhizosphere biology. *Plant Physiology* 132: 44–51.

Wang, J., and V.P. Evangelou. 1995. Metal tolerance aspects of plant cell wall and vacuole. In *Handbook of Plant and Crop Physiology*, ed. Pessarakli, M., 695–717. Marcel Dekker, Inc., New York.

Wenzel, W.W., R. Unterbrunner, P. Sommer, and P. Sacco. 2003. Chelate assisted phytoextraction using canola (*Brassica napus L.*) in outdoors pot and lysimeter experiments. *Plant and Soil* 249: 83–89.

Widdowson, M.A., S. Shearer, R.G. Andersen and J.T. Novak, 2005. Remediation of polycyclic aromatic hydrocarbon compounds in groundwater using poplar trees. *Environmental Science & Technology* 39 (6): 1598–1605.

Wierzbicka, M. 1987. Lead accumulation and its translocation barriers in roots of *Allium cepa* L.—autoradiographic and ultrastructural studies. *Plant Cell Environ*. 10: 17–26.

Wu, J., F.C. Hsu, and S.D. Cuningham. 1999. Chelate assisted Pb Phytoextraction: Pb availability, uptake, and translocation constraints. *Environmental Science & Technology* 33 (11): 1898–1904.

Wu, L.H., Y.M. Luo, P. Christie, and M.H. Wong. 2003. Effects of EDTA and low molecular weight organic acids on soil solution properties of a heavy metal polluted soil. *Chemosphere* 50: 819–822.

Yang, M., Y. Chang, J. Kim, J. Shin, and M. Lee. 2008. Study of the rhizofiltration by using *Phaseolus vulgaris* var., *Brassica juncea* (L.) Czern., *Helianthus annuus* L. to remove cesium from groundwater. *Geochimica et Cosmochimica Acta* 72 (12): A1056.

Yeargan, R., I.B. Maiti, M.T. Nielsen, A.G. Hunt, and G.J. Wagner. 1992. Tissue partinioning of cadmium in transgenic tobacco. Seedlings and field grown plants expressing mouse metallothionein I gene. *Transgen. Res*. 1: 261–267.

Yeo, A.R., M.E. Yeo, and T.J. Flowers. 1987. The contribution of an apoplastic pathway to sodium uptake by rice roots in saline conditions. *Journal of Experimental Botany* 38 (7): 1141–1153.

Yoon, J., X. Cao, Q. Zhou, and L.Q. Ma. 2006. Accumulation of Pb, Cu, and Zn in native plants growing on a contaminated Florida site. *Science of the Total Environment* 368: 456–464.

Zenk, M.H. 1996. Heavy metal detoxification in higher plants—a review. *Gene* 179: 21–30.

Zhao, F.J., E. Lombi, T. Breedon, and S.P. McGrath. 2000. Zinc Hyperaccumulation and cellular distribution in *Arabidopsis halleri. Plant Cell Environ*. 23: 507–514.

Zhou, F.J., and P.B. Goldsbrough. 1995. Structure, organization and expression of methallothionein gene family in *Arabidopsis. Mol. Gen. Genet*. 248: 318–328.

Zhu, I.L., E.A.H. Pilon-Smits, A.S. Tarun, S.U. Weber, L. Jouanin, and N. Terry. 1999. Cadmium tolerance and accumulation in Indian mustard is enhanced by overexpressing γ-glutamylcysteine synthetase. *Plant Physiology* 121: 1169–1177.

28 Uptake and Metabolism of Pharmaceuticals and Other Emerging Contaminants by Plants

Benoit van Aken, Rouzbeh Tehrani, and Rashid Kaveh

CONTENTS

28.1 INTRODUCTION

Higher plants have been shown to be able to take up efficiently pollutants from soil, sediments, surface water, and groundwater, leading to the concept of phytoremediation. Following the uptake into plant tissues, pollutants can undergo various fates, including biodegradation, immobilization, volatilization, or photolysis (Schnoor et al. 1995; Salt et al. 1998). Numerous bench-scale or greenhouse investigations have proven the efficiency of plants for the treatment of both heavy metal and organic-contaminated soil and groundwater (Newman and Reynolds 2004). However, even though phytoremediation has technically acquired the status of a proven strategy for cleaning up environmental pollution, it has not yet been recognized by the U.S. Environmental Protection Agency (U.S. EPA) as an approved bioremediation technology, and only a few documented cases of full-scale phytoremediation projects have been documented (Schnoor et al. 1995).

Phytoremediation encompasses a range of different processes: pollutants in soil and groundwater can be taken up inside plant tissues (*phytoextraction*) or adsorbed to the roots (*rhizofiltration*); pollutants inside plant tissues can be transformed by plant enzymes (*phytotransformation*) or can volatilize into the atmosphere (*phytovolatilization*); pollutants in soil can be degraded by microbes in the root zone (*rhizosphere bioremediation*) or incorporated to soil material (*phytostabilization*) (Schnoor et al. 1995; Salt et al. 1998; McCutcheon and Schnoor 2003; Van Aken et al. 2010). The *green liver* model describes the transformation of contaminants inside plant tissues following a sequence similar to that which occurs in the liver of mammals (Sandermann 1994).

In the absence of degradation and/or detoxification mechanisms, removal of contaminants by plants *per se* is not beneficial and can even be detrimental to the environment. Contaminants stored or loosely bound inside plant tissues will eventually return to the soil or enter the food chain, resulting in a pollution transfer. For instance, it has been shown that weathering processes could lead to leaching of toxic explosive compounds taken up inside poplar plants (Yoon et al. 2006). These concerns have led to multiple studies showing that, besides single uptake, higher plants can enzymatically metabolize organic contaminants, including pesticides, chlorinated solvents, nitroaromatic explosives, and even highly recalcitrant compounds, such as polyaromatic hydrocarbons (PAHs) and polychlorinated biphenyls (PCBs) (Schnoor et al. 1995; Van Aken et al. 2010). Although the knowledge of biochemical mechanisms and molecular bases of the transformation of pollutants by higher plants is still fragmentary, several enzymes commonly involved in the metabolism of xenobiotic compounds have be identified, including cytochrome P-450 monooxygenases and glutathione *S*-transferases (Sandermann 1994; Coleman et al. 1997; Schroder et al. 2001; Brentner et al. 2008).

In a pioneering study, Kolpin et al. (2002) provided evidence that a range of emerging pollutants contaminate surface water in the United States. Since then, pharmaceuticals and personal care products have been detected in virtually every compartment of the ecosystem. The occurrence of pharmaceuticals and personal care products (PPCP) in water bodies has raised questions about their effects on the environment and human health. Indeed, pharmaceuticals are *biologically active* compounds that have been designed to produce physiological effects on living organisms at very low doses. Although the acute toxicity of pharmaceuticals for higher organisms is generally low, the effects of long-term exposure on aquatic life is usually unknown and the potential impacts on the development of living organisms (including humans) have raised serious environmental and public health concerns.

Pharmaceuticals enter the environment through the effluents from wastewater treatment plants, animal husbandries, and pharmaceutical industry, resulting in widespread contamination of groundwater and surface water, ultimately leading to their detection in water supply and drinking water (Caliman and Gavrilescu 2009). Because wastewater treatment plants are identified as the major points of entry of pharmaceuticals into the environment, biodegradation studies have been conducted mostly on heterotrophic bacteria in activated sludge systems (Onesios et al. 2009).

An increasing body of knowledge has been published about the capability of higher plants for the removal of pharmaceuticals and personal care products from wastewater, soil, and groundwater. As for other conventional pollutants, removal of pharmaceuticals by plants depends largely on the physical and chemical properties of the compound and the selected plant species. Even though many reports have documented the efficient uptake of pharmaceuticals by plants, information about the metabolism of pharmaceuticals by plants is scarce. Especially, very little is known about the metabolic pathways, enzymes, and molecular bases of the transformation of pharmaceutical compounds inside plant tissues. However, understanding the plant metabolism of pharmaceuticals is critical because in the absence of further transformation and detoxification, pharmaceuticals and/or their metabolites taken up by plants will enter the food chain or return to the soil, resulting in a potential hazard for human health. Moreover, the metabolism of xenobiotic pollutants by plants could, in some instance, results in an increase of toxicity, further threatening human health and the environment (Celiz et al. 2009).

Uptake and transformation of pharmaceuticals by higher plants have environmental relevance for two major reasons: first, aquatic plants in constructed wetlands have been shown to be capable to remove pharmaceuticals from wastewater or contaminated treated effluents, second, the practice of irrigation of agricultural fields with reclamation water and land application of municipal sewage sludge may lead to the contamination of edible crops by pharmaceuticals, which may have detrimental effects on human health. Because pharmaceuticals are notoriously not completely removed by conventional wastewater treatments, there is raising interest in using aquatic plants for advanced polishing treatment. Besides the removal of emerging contaminants, higher plants can be used for further removal of nutrients (nitrogen and phosphorous) and/or as feedstock for biofuel production. On the other hand, the practice of irrigation of agricultural fields with reclamation water and land application of municipal sewage sludge have been shown to lead to contamination of edible crops by pharmaceuticals and their potential toxic metabolites, with unpredictable effects on the environment and human health (Daughton and Ternes 1999; Ternes et al. 2004).

28.2 PHYTOREMEDIATION OF ORGANIC POLLUTANTS

28.2.1 The Bioremediation Concept

Living organisms are commonly exposed to natural or xenobiotic toxic compounds. As a consequence, they have developed multiple detoxification mechanisms to prevent harmful effects from exposure to these compounds. Bacteria, more than higher life forms, are extremely versatile organisms, which allows them to constantly develop new metabolic pathways for the degradation of a large range of xenobiotic pollutants (Limbert and Betts 1996). Whilc provided with lower adaptation capabilities, higher organisms, such as plants and mammals, also possess detoxification mechanisms to counteract the harmful effects of toxic contaminants (Sandermann 1994).

Bioremediation exploits the natural capability of living organisms to degrade toxic chemicals. Traditional remediation technologies of polluted sites require soil excavation and transport, prior to treatment by incineration, landfilling or compositing, which is costly, damaging for the environment, and, in many cases, practically infeasible due to the range of the contamination. There is therefore a considerable interest in developing cost-effective alternatives based on microorganisms or plants. Because of its potential for the sustainable mitigation of environmental pollution, bioremediation has been listed among the "top ten biotechnologies for improving human health" (Daar et al. 2002).

28.2.2 Bioremediation by Higher Plants (Phytoremediation)

Phytoremediation is an emerging technology that makes use of plants and associated bacteria for the treatment of soil and groundwater contaminated by toxic pollutants. Remediation by plants involves several plant-mediated mechanisms, such as the enhancement of microbial activity in the root zone and the hydraulic control of contaminant plumes (Limbert and Betts 1996; Ferro et al. 2003). The biodegradation of xenobiotic compounds by plants has been known for a long time. Land farming of wastewaters has been a treatment technology for at least 300 years, and wetland treatment of wastewater and the use of plants to control air pollution have been used for several decades (McCutcheon and Schnoor 2003). Plant-mediated bioremediation of metal-contaminated soil was proposed first in the 1970s (Brooks et al. 1977). Because of their natural capability to absorb minerals, plants efficiently remove heavy metals from polluted soils. However, unlike organic pollutants, elemental metals cannot be "degraded." They typically accumulate in plant tissues, which have to be harvested and safely disposed (Salt et al. 1998).

Although the metabolism of xenobiotics by plants is known for a long time (Castelfranco et al. 1961), the idea that plants can be used to detoxify organic compounds emerged in the 1970s with the discovery of the metabolism of 1,1,1-trichloro-2,2-*bis*-(4′-chlorophenyl) ethane (DDT) and benzo(*a*) pyrene (Cole 1983). Since then, phytoremediation acquired the status of a proven technology for the remediation of soil and groundwater contaminated by a variety of organic compounds, including pesticides, chlorinated solvents, explosives, polyaromatic hydrocarbons (PAHs), and polychlorinated biphenyls (PCBs) (Salt et al. 1998; Schnoor et al. 1995; Van Aken et al. 2010). Phytoremediation has been extensively reviewed in the literature (see for instance Schnoor et al. 1995; Salt et al. 1998; Macek et al. 2000; Meagher 2000; Pilon-Smits 2005; Van Aken and Doty 2009).

28.2.3 Advantages and Limitations of Phytoremediation

Phytoremediation offers several advantages over other remediation strategies: low cost because of the absence of energy-consuming equipment and limited maintenance, no damaging impact on the environment because of the *in situ* nature of the process, and large public acceptance as an attractive green technology. In addition, phytoremediation offers potential beneficial side-effects, such as erosion control, site restoration, carbon sequestration, and feedstock for biofuel production (Doty et al. 2007). As autotrophic organisms, plants use sunlight and carbon dioxide as energy and carbon sources. From an environmental standpoint, plants can be seen as natural, solar-powered, pump-and-treat systems for cleaning up contaminated soils (Eapen et al. 2007).

Even though phytoremediation has been shown to be a promising technology for the treatment of heavy metal and organic-contaminated soil, it also suffers serious limitations that may impair the development of field applications (Schnoor et al. 1995; Salt et al. 1998). First, phytoremediation is limited to shallow contamination of "moderately hydrophobic" compounds susceptible to sorption by/to the roots (Schnoor et al. 1995; Burken and Schnoor 1998). More importantly, remediation by plants is often slow and incomplete. As a corollary to their autotrophic metabolism, plants usually lack the biochemical machinery to achieve full mineralization of many organic pollutants, especially the most recalcitrant, such as PAHs and PCBs (Schnoor et al. 1995). Phytoremediation can therefore lead to non desirable effects such as the accumulation of toxic pollutants and metabolites that may be released to the soil, enter the food chain, or be volatilized into in the atmosphere (Newman et al. 1997; Yoon et al. 2006; Eapen et al. 2007).

28.2.4 Phytoremediation Processes

Phytoremediation encompasses a range of processes beyond direct plant uptake and metabolism, and it is best described as plant-mediated remediation (Schnoor et al. 1995; McCutcheon and Schnoor 2003). While definitions and terminology vary, the different processes involved as part of phytoremediation can be summarized (Figure 28.1). An extensive list of these physiological processes and their multiple names is presented in McCutcheon and Schnoor (2003). The simplified descriptions below are based on the recognition of the following mechanisms: *phytotransformation*, *phytoextraction*, *rhizofiltration*, *phytovolatilization*, *phytophotolysis*, *rhizodegradation*, *phytostabilization*, and *hydrological control*.

(a) *Phytotransformation sensu stricto* means the metabolism of organic pollutants inside plant tissues. In a broader sense, it refers to the uptake, transformation, and/or storage of organic contaminants inside plants (Schnoor et al. 1995; McCutcheon and Schnoor 2003). The process involves first the uptake of pollutants from soil, sediment, surface water, or groundwater by the roots. The potential for uptake by plants depends largely on the hydrophobicity of the chemical, represented by the octanol-water partition coefficient, log K_{ow}. Experimental relationships based on log K_{ow} can be used to predict the sorption of specific chemicals to the roots (root concentration factor, RCF) and the uptake and translocation of chemicals inside plant tissues through the transpiration stream (transpiration stream concentration factor, TSCF). Typically, "moderately hydrophobic" organic compounds (with a log K_{ow} between 0.5 and 4.5) can be significantly taken up by plants (Briggs et al. 1982; Burken and

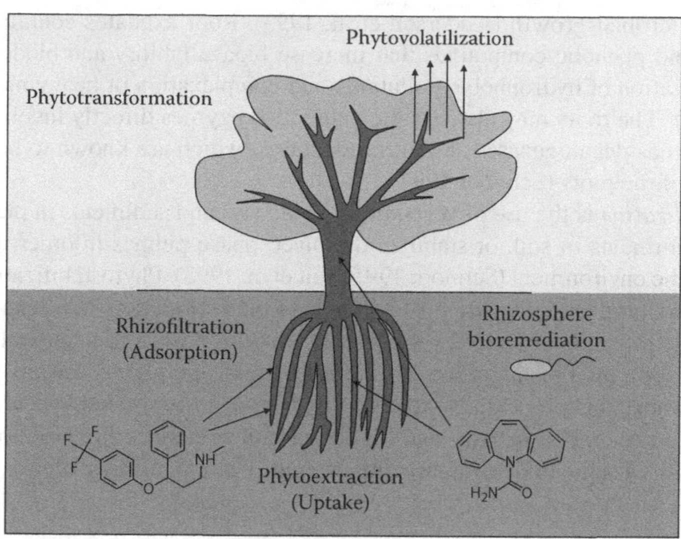

FIGURE 28.1 Phytoremediation involves several processes: Pollutants in soil and groundwater can be taken up inside plant tissues (*phytoextraction*) or adsorbed to the roots (*rhizofiltration*); pollutants inside plant tissues can be transformed by plant enzymes (*phytotransformation*) or can volatilize into the atmosphere (*phytovolatilization*); pollutants in soil can be degraded by microbes in the root zone (*rhizosphere bioremediation*) or incorporated to soil material (*phytostabilization*). (Adapted from Van Aken, B., *Trends in Biotechnology*, 26: 5225–228, 2008.)

Schnoor 1998). Typical applications of phytotransformation include sites contaminated by organic biodegradable contaminants, such as explosives and energetic compounds, petrochemicals and PAHs, chlorinated solvents, pesticides and fertilizers, and landfill leachates (Schnoor et al. 1995; Newman and Reynolds 2004).

(b) *Phytoextraction* refers to the use of plants for the uptake, translocation, and accumulation of contaminants from soil or groundwater inside the roots and aerial plant organs (Chaney et al. 1997). *Rhizofiltration* describes the sorption, concentration, or precipitation of metal contaminants from surface water or groundwater on the plant roots (Salt et al. 1998). Phytoextraction and rhizofiltration typically apply to elemental contaminants (metals) not susceptible to biodegradation. Metals not easily taken up by plants (not suitable for phytoextraction) may be good candidates for rhizofiltration.

(c) *Phytophotolysis* refers to the degradation of contaminants taken up and translocated to leaf tissues by exposure to light. Energetic UV radiation from sunlight has long been recognized as capable to breakdown a range of organic molecules. An indirect effect of sun light is the generation of free radicals and reactive oxygen species (ROS) that can further react with organic molecules. For instance, the light-mediated degradation of explosives in plants has been documented in the literature (Van Aken et al. 2004).

Volatile contaminants taken up inside plant tissues can be volatilized through the leaves through a process known as *phytovolatilization*. Leaves of trees exhibit a very large surface area that can potentially volatilize significant amount of chemicals. Volatile organic compounds, such as solvents and other chlorinated organic chemicals, are therefore susceptible to volatilization into the atmosphere (Burken and Schnoor 1998; Newman et al. 1997).

(d) *Rhizodegradation* (also known as *rhizosphere biodegradation*), refers to the enhancement of microbial activity and biodegradation of pollutants in the root zone (Anderson et al. 1993; Limbert and Betts 1996). Root exudates and root turnover increase soil organic carbon, which is beneficial for both microbial growth and co-metabolic biodegradation. By buffering the pH and biosorption of toxic metals, plants create ecological niches that promote microbial growth (Anderson et al. 1993). Root exudates contain organic acids, alcohols, and phenolic compounds that increase bioavailability and biodegradation, both by solubilization of hydrophobic pollutants and complexation of heavy metals (Chaudhry et al. 2005). The roots may also excrete catabolic enzymes directly involved in biodegradation, such as dehalogenases and nitroreductases, which are known to be active in close proximity of the roots (Schnoor 1997).

(e) *Phytostabilization* is the use of vegetation to hold soil and sediments in place, immobilize toxic contaminants in soil, or stabilize dissolved phase plumes in order to mitigate their impact on the environment (Schnoor 1995; Salt et al. 1998). Phytostabilization is especially useful for contaminants that are not amenable to biodegradation. In a sense, the transpiration of large amounts of water by vigorously growing trees is a stabilization technology, because it prevents the migration of plumes of toxic pollutants toward groundwater or receiving water (this process also refers as *hydraulic control*) (Ferro et al. 2003). Finally, the enhanced microbial activity and higher carbon content in the root zone promote the sequestration of some toxic contaminants by humification, mineralization, and sorption to soil particles.

28.2.5 Metabolism of Xenobiotic Compounds by Plants

Being exposed to a variety of natural allelochemicals and xenobiotic compounds, plants have developed diverse detoxification mechanisms of organic compounds (Singer 2006). Enzymatic reactions performed by plants are highly diverse, and can be summarized according to the *green liver* model (Figure 28.2). Based on the early observations that plants were capable of metabolizing organic xenobiotic pollutants (pesticides), Sandermann (1994) introduced the *green liver* concept, which

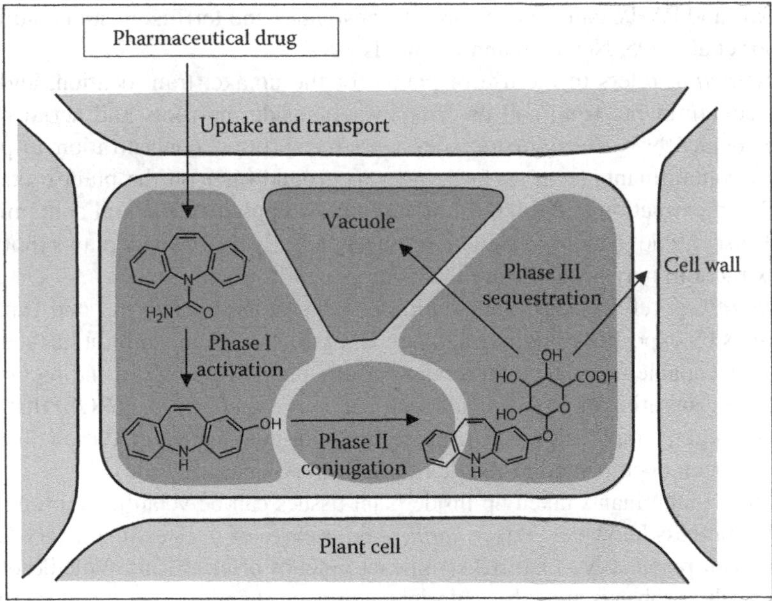

FIGURE 28.2 **(See color insert.)** The three phases of the *green liver* model: Hypothetical pathway representing the metabolism of carbamazepine in plant tissues: phase I: Activation of carbamazepine by hydroxylation and hydrolysis to 2-hydroxyiminostilbene; phase II: Conjugation with a glucose to form an *O*-glucuronide; phase III: Sequestration of the conjugate into the cell wall or in the vacuole. (Adapted from Van Aken, B., *Trends in Biotechnology*, 26: 5225–228, 2008; Celiz, M.D., T.S.O. Jerry, and S. Daina, *Environmental Toxicology and Chemistry*, 28: 122473–2484, 2009.)

typically occurs in three phases. The initial activation (phase I) consists of the oxidation, reduction, or hydrolysis of the xenobiotic compound, which is transformed into a more reactive metabolite. Examples of phase I reactions include oxidation or hydroxylation by cytochrome P-450 monooxygenases and the reduction of nitro groups by nitroreductases. In phase II, the resulting activated product undergoes a transferase-catalyzed conjugation with a molecule of plant origin, forming an adduct less toxic and more soluble than the parent pollutant. Transferases may involve glutathione *S*-transferase or glycosyltransferases. Phase III involves the sequestration of the conjugate, which can be stored in plant organelles such as vacuoles, incorporated into biopolymers such as lignin, or excreted in the case of wetland plants (Cole 1983; Sandermann 1994; Coleman et al. 1997; Schroder et al. 2001). From an environmental viewpoint, phytotransformation must lead to a significant detoxification of organic pollutants. Without further transformation, pollutants and toxic metabolites that accumulate inside plant tissues will sooner or later return to the soil or enter the food chain (Schnoor et al. 1995; Schnoor 1997).

28.3 PHYTOREMEDIATION OF PHARMACEUTICALS

28.3.1 Pharmaceuticals as Environmental Pollutants

Pharmaceuticals used in human and veterinary medicine have contributed to significantly increase the quality of life and life expectancy (Schwab et al. 2005). However, following their use, pharmaceuticals are released into the environment and they are increasingly considered as a new class of emerging contaminants because of their potential impact on wildlife and human health (Kolpin et al. 2002). Because of their designated function, these compounds have specific biological activities and are rather recalcitrant to biological degradation, making them potentially harmful for aquatic organisms. The recent development of analytical methods, such as liquid chromatography–tandem mass spectrometry (LC-MS/MS) and liquid chromatography–time-of-flight mass spectrometry (LC-TOF/MS), has allowed the detection of very low concentrations of pharmaceuticals in environmental samples (Daughton and Ternes 1999; Ternes et al. 2004; Kolpin et al. 2002; Kummerer 2004).

Emerging organic contaminants enter the environment from various anthropogenic activities. Pharmaceuticals and synthetic hormones introduced in the human and animal body for health care or as growth stimulators are only partially metabolized and discharged in significant levels in organic waste. Several thousands of pharmaceutical compounds are used in human and veterinary medicine and hundreds of tons per year are consumed. These compounds are insufficiently reduced in wastewater treatment processes that are not specifically designed for their removal, often resulting in their release into the receiving water. After their discharge into the environment, pharmaceuticals undergo limited biotic or abiotic degradation/transformation, leading to their detection in water supply and drinking water. According to recent reviews, more than 30 pharmaceuticals and other emerging contaminants have been detected in drinking waters around the world (Kleywegt et al. 2011). The presence of these compounds in drinking water has been related primarily to the inability of both wastewater and water treatment facilities to efficiently remove them.

28.3.1.1 Sources of Pharmaceuticals in the Environment

Pharmaceuticals enter the environment primarily through the effluents from wastewater treatment plants, concentrated animal feeding operations, and pharmaceutical industry. Contaminated wastewater generally undergoes treatment that typically fails to completely remove pharmaceuticals from the water, resulting in their discharge with the treated effluent and/or the waste sludge (Ternes et al. 2004). Besides, veterinary pharmaceuticals, including antibiotics and hormones, used in animal husbandries (both for therapeutic purposes and as growth promoters) can be released directly into the environment through non-point sources, such as leakage from storage structures and land application of untreated manure (Kolpin et al. 2002; Kleywgt et al. 2011). Other potential sources of pharmaceuticals found in the environment include wastewater from hospitals, aquaculture effluents, and landfill leachates. As a consequence, pharmaceuticals have been detected in virtually every compartment of the ecosystem, including, surface water, groundwater, soil, and sediments (Daughton 2004).

Following their intake in the body, pharmaceuticals are largely excreted (in their native form or as metabolites) into wastewater. Unused pharmaceuticals are also frequently disposed in wastewater. Municipal sewage is therefore considered is the main route of entrance of human pharmaceuticals into the environment. Conventional wastewater treatment processes are originally designed to remove high concentration of readily biodegradable organic compounds (BOD) found in municipal used water and they are not necessarily well adapted to the removal of low concentrations of potentially toxic and recalcitrant pharmaceuticals. Incomplete removal of pharmaceuticals in wastewater treatment operations has led to their discharge of significant amounts in rivers, lakes, and estuaries (Fent et al. 2006). Many studies about biodegradation of pharmaceuticals have focused on wastewater treatment processes and little is currently known about the environmental fate and transport of pharmaceuticals and hormones. Current research suggests that pharmaceuticals enter the environment and persist to a greater extent than it was first believed (Kolpin et al. 2002).

28.3.1.2 Removal of Pharmaceuticals in Wastewater Treatment

It has been largely demonstrated that pharmaceuticals, although to some extent biodegradable, are not entirely removed by conventional wastewater treatment operations. Two processes contribute to significantly remove pharmaceuticals from wastewater treatment: adsorption to biosolids (municipal wasted sludge) and biodegradation in the activated sludge system. Many pharmaceuticals are hydrophilic and/or in dissociated (ionic) form in wastewater and are poorly removed by adsorption. Acidic pharmaceuticals, such as ibuprofen, are ionic at neutral pH and have little interaction with the negatively charged biosolids (Fent et al. 2006). On the other hand, basic pharmaceuticals, such as fluoroquinolone antibiotics, and hydrophobic compounds, such as 17β-ethynylestradiol, are more susceptible to adsorption to the sludge. Pharmaceuticals in the dissolved phase may undergo biodegradation by microorganisms in the activated sludge (aerobic biodegradation) and for subsequent sludge digestion (anaerobic biodegradation). Reported degradation efficiencies are highly variable between pharmaceuticals, which is expected based on the wide diversity of their chemical structures (Ternes 1998). For instance, the antiepileptic drug, carbamazepine, the X-ray contrast media, diatrizoate, and the anticancer drug, tamoxifen are not efficiently removed by the activated sludge process (less than 10% removal). On the other hand, the non-steroidal antiinflammatory drugs, ibuprofen and acetylsalicylic acid, are highly biodegradable (up to 99% removal). Generally speaking, the biodegradation extent of pharmaceuticals depends on the engineering design and operational parameters, such as hydraulic retention time, solid retention time, and temperature (Fent et al. 2006).

Pharmaceuticals that escape biodegradation in municipal and agricultural wastewater treatments significantly adsorb onto the bacterial biomass and may be discharged into the environment through sewage sludge disposal. Sewage sludge is the solid/semisolid residue left over after the biological treatment of wastewater. Sewage sludge in the United States is classified as either class A or class B. Class A sewage sludge is defined as sewage sludge that has undergone treatment to reduce pathogens as set by U.S. EPA regulations (40 CFR Part 503) and can be applied on agricultural land for irrigation and fertilization purposes (Jones-Lepp and Stevens 2007). Approximately 6.2 million dry metric tons of municipal sewage sludge are produced annually (estimation of 1998) in the United States with about 50% being applied on to land. With new techniques for sludge treatment (e.g., sludge pelletization), this practice is expected to increase in the future (Jones-Lepp and Stevens 2007). Many studies (reviewed in Jones-Lepp and Stevens 2007) have reported the presence of pharmaceuticals in sewage sludge. In a recent U.S. EPA survey, pharmaceuticals and steroids were analyzed in sludge collected from 74 facilities in 35 states. Nine out of 72 pharmaceuticals and 6 out of 25 steroids tested were detected in 95% of the samples. Several compounds were detected in the range of mg kg^{-1} dry solid (e.g., carbamazepine: 9–6,000 μg kg^{-1}, ibuprofen:

100–12,000 μg kg^{-1}, and sulfamethoxazole: 4–650 μg kg^{-1}) (U.S. EPA 2009). Land application of reclamation water and sewage is therefore likely to contribute to contamination of soil and plants by pharmaceuticals.

28.3.1.3 Fate and Occurrence of Pharmaceuticals in the Environment

As a consequence of their incomplete removal from wastewater, pharmaceuticals have been detected in most surface water and groundwater at concentrations ranging from ng L^{-1} to μg L^{-1} (Ternes 1998; Kolpin et al. 2002; Kinney et al. 2006; Barnes et al. 2008; Lapen et al. 2008). To assess the occurrence of organic emergent contaminants in surface water, the U.S. Geological Survey (USGS) conducted in 1999–2000, a study on the concentration of a series of 95 compounds in 139 streams across 30 states (Kolpin et al. 2002). The target compounds included antibiotics, other prescription drugs, nonprescription drugs, steroids, reproductive hormones, personal care products, products of oil use and combustion, and other chemicals. These compounds were selected based on their occurrence in wastewater, quantities consumed, and potential effects on human health and the environment. The results showed that organic emerging contaminants were found in 80% of the streams tested and that 82 out of the 95 compounds under study were detected at least in one stream. Often, mixture of these compounds were observed in streams (with a median of 7 compounds) (Kolpin et al. 2002). However, based on this report, the most frequently detected compounds did not belong to pharmaceuticals, but were natural steroids, biocides, an insect repellent, and caffeine.

In the United States, it is estimated that about 2.4% of wastewater effluent (estimation of 1998) is used for crop irrigation (Kinney et al. 2006). A recent publication has provided evidence that irrigation with reclaimed water from urban wastewater resulted in the contamination of soil with several pharmaceuticals (e.g., erythromycin, carbamazepine, fluoxetine, and diphenhydramine) with typical concentrations ranging from 0.02 to 15 μg kg^{-1} dry soil (Kinney et al. 2006).

Although increasing research has focused on the measurement of pharmaceuticals in wastewater, wastewater treatment plant effluent, surface water, and drinking water, the fate of pharmaceuticals and their metabolites in the aquatic environment is poorly documented. Because of their low volatility, pharmaceuticals are expected to be mainly transported in the aqueous environments. Pharmaceuticals in surface water would then subjected to slow biodegradation by indigenous microorganisms be abiotic transformation (including photolysis), and sorption to organic matter and sediments.

28.3.1.4 Effect of Pharmaceuticals on the Environment and Human Health

Although the occurrence of pharmaceuticals in surface water is today well documented, very little information is available regarding their ecotoxicology and impacts on aquatic and terrestrial wildlife (Fent et al. 2006). Acute toxicity tests have been conducted with some pharmaceuticals, but there still exists a lack of knowledge regarding the chronic effects of these chemicals. Pharmaceuticals, such as analgesics, antibiotics, antiepileptics, β-blockers, blood-lipid regulators, and contraceptives, are *biologically active* compounds that may have effects on living organisms at very low doses. Their unintentional release in the environment may be responsible for altered physiological processes and reproductive functions, increased cancer rates, development of antibiotic resistance, and increased toxicity of chemical mixtures (Kolpin et al. 2002). It is today suspected that micropollutants may have adverse effects on the development without to exhibit recordable toxicity on adults (Boxall 2009; Kumar et al. 2010). Although pharmaceuticals are present in very low concentrations, most of them are hydrophobic (high log K_{ow}) and rather recalcitrant to biodegradation, leading to their bioaccumulation in animal tissues (biomagnification). Human exposure to pharmaceuticals can result from the ingestion of contaminated drinking water or plant products grown on soil irrigated and/or fertilized by contaminated water and sewage sludge (Boxall 2009).

Environmental safety of pharmaceutical products is initially ensured by the extensive testing performed by pharmaceutical companies on animals and humans, as well as post-marketing safety surveillance. Specific regulations for ecological risk assessment of pharmaceuticals are slowly emerging. Since 1995, registration of pharmaceuticals in the European Union (EU) requires ecotoxicological testing. In addition, the European Commission (EC) has recently stated that authorization of human pharmaceuticals must require environmental risk assessment (Fent et al. 2006). In compliance of the U.S. Food and Drug Administration (U.S. FDA) regulation, applicants for the registration of pharmaceutical drugs in the United States are required to provide an environmental assessment report when the expected concentration of the active ingredient in the aquatic environment exceeds one part-per-billion (ppb = $\mu g\ L^{-1}$) (Schwab et al. 2005; Fent et al. 2006).

A special class of emerging contaminants are endocrine disrupting compounds. Endocrine disruptors are chemicals which interfere with one or more functions of the endocrine system. The endocrine system is present in animals and consists of glands and the hormones they produce. Hormones are chemical messengers in the body that guide the development, growth, reproduction, and behavior of animals. Many industrial and environmental chemicals mimic, antagonize or indirectly alter the activity of hormones and they are known as endocrine disruptors. By interfering with the normal functioning of hormonal receptors, endocrine disruptors can perturb the growth and development of exposed organisms. Estrone, 17β-estradiol, and 17α-ethynylestradiol are the principal estrogens that are found in the environment. Although estrone and 17β-estradiol are hormones naturally excreted by mammals, 17α-ethynylestradiol is a major ingredient of birth control pills.

Although pharmaceuticals have been recently detected at low levels in the environment using increasingly sensitive techniques, potential effects of minutes concentrations of pharmaceuticals and their metabolites on human health have received less attention. In a recent report, Schwab et al. (2005) conducted environmental risk assessment of a series of 26 active pharmaceuticals and their metabolites. Predicted no effect concentrations (PNECs) were derived from acceptable daily intakes (ADIs) for human exposure through drinking water and fish consumption. PNECs were then compared with measured environmental concentrations (MECs) from the literature and maximum predicted environmental concentrations (PECs) obtained from the P*h*ATE model. Both MEC-to-PNEC and PEC-to-PNEC ratios for the 26 compounds under study were low and indicated that no significant risk for human health exists from environmental concentrations of pharmaceuticals detected in surface water and drinking water.

28.3.1.5 Environmental Effects of Pharmaceutical Metabolites

Following their introduction in the human or animal body, pharmaceuticals may be partially or totally converted into a range of metabolites. Parent compounds and/or their transformation products are then excreted into the environment where they can be further metabolized by the activities of microorganisms or plants. Most risk assessment investigations have only considered the ecotoxicity of parent pharmaceuticals and little data are available about the environmental hazard posed by the metabolites excreted by animals and humans or formed by microorganisms during wastewater treatment (Celiz et al. 2009). The advance in sensitive extraction and analytical techniques have made the detection of an increasing number of pharmaceutical metabolites possible, which will be required to be included in further environmental risk assessments.

As many toxic compounds that enter the body, pharmaceuticals are metabolized following a typical sequence involving enzymatic transformation into more polar and soluble derivatives, which favors their subsequent elimination or sequestration. Typically, the pharmaceutical molecule is first converted into a more hydrophilic form by cytochrome P-450 enzymes (phase I) (Celiz et al. 2009). These metabolites are therefore susceptible to conjugation with endogenous molecules to form *O*- and *N*-glucuronides, sulfates, or glutathionyl adducts (phase II) that can be easily excreted in urine. Some metabolic products of pharmaceuticals have been shown to exhibit a biological activity and be toxic to living organisms (Celiz et al. 2009). Köhler et al. (2006) observed that a reduction of the concentration of pollutants during the biological treatment resulted in an increase of the ecotoxicity of the treated water, suggesting the transformation of the pollutants into more toxic by-products. Acetylsalicylic acid has been shown to be hydrolyzed to salicylic acid, in turn transformed into gentisic acid, salicyluric acid, and glucuronide conjugates. Salicylic and gentisic acids have been shown to be toxic to embryos of zebrafish and *Daphnia* species, although at levels several orders of magnitude above the concentrations detected in the environment (Celiz et al. 2009). The anticonvulsant drug, carbamazepine, which has been detected in many wastewater effluents and surface water, can be hydroxylated into 2-hydroxycarbamazepine that undergoes hydrolysis to 2-hydroxyiminostilbene and transformation to iminoquinone. The later is believed to be responsible for carbamazepine-mediated idiosyncratic reactions in humans. Fetal malformations in mice have also been attributed to the carbamazepine-10,11-epoxide and subsequent derivatives. These metabolites have been detected in wastewater, treated wastewater, and biosolids (Celiz et al. 2009). Similarly, the nonsteroidal anti-inflammatory drug, diclofenac, is metabolized into acyl glucuronide and *p*-benzoquinone imine derivatives in the liver microsomes, which are responsible for the cytotoxicity of diclofenac in rats and humans. Diclofenac metabolites have been detected in the aqueous environment.

Interestingly, conjugated metabolites of pharmaceuticals have been shown to undergo deconjugation releasing the parent compounds. For instance, the antibiotic sulfamethoxazole is metabolized in the body into acetylsulfamethoxazole, which is inactive. Acetylsulfamethoxazole has been detected in wastewater effluent and has been shown to be hydrolyzed back during the secondary (biological) treatment of wastewater (Ternes et al. 1998; Gobel et al. 2004).

28.3.2 Pharmaceuticals Uptake and Metabolism by Higher Plants

Unlike more traditional pollutants, such as explosives, PCBs, chlorinated solvents, and pesticides, that have been the topic of intense and focused research for decades, pharmaceuticals represent a relatively novel class of contaminants. Because of their diversity of structures and recent occurrence and detection in the environment, these chemicals have not been, for the most part, deeply studied regarding their environmental fate, ecotoxicology, and biodegradation pathways. In this chapter, we focused on two aspects of the uptake of pharmaceuticals by higher plants that are of high environmental relevance: first, the contribution of aquatic plants to the removal of pharmaceuticals from wastewater in constructed wetlands and, second, the absorption and transformation of pharmaceuticals by agricultural edible plants, which can lead to human contamination. A summary of the literature on pharmaceutical uptake and transformation by higher plants is presented in Table 28.1.

The potential for plants to uptake pharmaceuticals is known for several decades because of studies conducted on the effects of antibiotics used to treat agricultural species. In an early publication, Litwack and Pramer (1957) reported the uptake of the antibiotic streptomycin by cells of the alga *Nitella Clavata*. At the same time, Ark and Dekker (1958) showed that the antibiotic GS1 (Pfizer) penetrated pea (*Ascochyta pisi*) and squash (*Fusaruim solani*) seeds after 24 hours of exposure to 50 and 100 mg L^{-1} in solution. The same antibiotic was also shown to be taken up by the roots of barley, bean, and cucumber plants exposed to 200 mg L^{-1} in the hydroponic medium. Following the uptake, the antibiotics were subsequently translocated to the aerial parts of the plants. Peterson and Sinha (1977) studied the uptake, translocation, and persistence of tetracycline antibiotics (chlorotetracycline, doxycycline, oxytetracycline, and tetracycline) in plant species susceptible to mycoplasma infection, including aster (*Callistephus chinensis*), caraway (*Carum carvi*), pot marigold (*Calendula officinalis*), fall dandelion (*Leontodon autumnalis*), peach (*Prunus persica*), poppy (*Papaver somniferum*), and strawberry (*Fragaria ananassa*). Each antibiotic was shown to be taken up from the hydroponic solution after 24 hours of exposure and was subsequently detected into the roots and aerial parts of the plants. Even though the total uptake ranged from 173 to 537 μg g^{-1} plant tissue, the levels of active antibiotics detected inside the plants were rather low, ranging from 3 to 122 μg g^{-1} plant tissue. Although large variations were detected across antibiotic compounds and plants species, the best recoveries were observed with oxytetracycline and tetracycline in poppy plants. Oxytetracycline and tetracycline were shown to persist for 2 weeks in marigold and poppy plants, and four weeks in peach. Chlorotetracycline and doxycycline were less persistent and shown to induce phytotoxicity at the concentration of 100 μg mL^{-1} used for root treatment.

TABLE 28.1
Summary of Selected Publications on the Uptake and Metabolism of Pharmaceuticals by Plants

Pharmaceutical Compound	Plant Species	Process and Toxic Effect	Dosage	Accumulation and Transformation	References
Triclosan (antimicrobial), galaxolide (polycyclic musk)	*Triticum aestivum* (wheat)	Inhibition of chlorophyll accumulation, protein synthesis, peroxidase and superoxide dismutase activities	50,000–250,000 μg L^{-1} in the culture medium	200 to 3,000 μg L^{-1} of the two compounds cause recordable plant response	An et al. (2009)
Ibuprofen, naproxen and diclofenac (antiinflammatory), tonalide (polycyclic musk), bisphenol A (precursor of plastics)	*Phragmites australis* (reed) in constructed wetland	Removal from the medium by adsorption and biodegradation	Ibuprofen: 53 μg L^{-1}, naproxen: 2.2 μg L^{-1}, diclofenac: 0.3 μg L^{-1}, tonalide: 0.1 μg L^{-1}, bisphenol A: 0.3 μg L^{-1}	Removal of ibuprofen: 99%, naproxen and diclofenac: 93%, tonalide: 94%, bisphenol A: 83%	Avila et al. (2010)
Acetaminophen (pain reliever)	*Brassica juncea* (Indian mustard)	Uptake, translocation, and detoxification	~10 mM in the culture medium (1,500,000 μg L^{-1})	Accumulation in roots (1.15 μmol g^{-1}; 174 μg kg^{-1}) and in leaves (0.3 μmol g^{-1}; 45 μg kg^{-1})	Bartha et al. (2010)
Oxytetracycline, norfloxacin (antibiotics)	*Glycine max* (soybean)	Uptake from spiked water	50,000 μg L^{-1} oxytetracycline, 55,000 μg L^{-1} norfloxacin	Accumulation of oxytetracycline: 190,000 μg kg^{-1} dry weight, norfloxacin: 110,000 μg kg^{-1} dry weight (after 2 days of exposure)	Boonsaner and Hawker (2010)
Diazinon, levamisole (ectoparasiticides), amoxicillin, enrofloxacin, florfenicol, oxytetracycline, sulfadiazine, trimethoprim, tylosin (antibiotics), phenylbutazone (antiinflammatory)	*Lactuca sativa* (lettuce), *Daucus carota* (carrot)	Uptake from spiked soil	1,000 μg kg^{-1} soil dry weight	Accumulation of trimethoprim: 6 μg kg^{-1}, levamisole: 170 μg kg^{-1} fresh weight in lettuce leaves, enrofloxacin: 2.8 μg kg^{-1}, diazinon: 13 μg kg^{-1} fresh weight in carrot roots	Boxall et al. (2006)

Norfloxacin (antibiotics)	*Glycine max* (soybean)	Uptake by plants	52,500 μg kg^{-1} dry weight in soil	Accumulation of 2,200 μg kg^{-1} dry weight	Cropp et al. (2010)
Sulfamethazine (antibiotics)	*Zea mays* (maize), *Lactuca sativa* (lettuce), *Solanum tuberosum* (potato)	Uptake by plants	50,000 to 100,000 μg L^{-1} in manure applied on soil	Accumulation in plant tissues from 100 to 1,200 μg kg^{-1} dry weight	Dolliver et al. (2007)
Ibuprofen (antiinflammatory)	*Typha* (cattail)	Uptake from spiked nutrient solution	20 μg L^{-1} in Hoagland nutrient solution	60% removal after 24 h, >95% after 96 h	Dordio et al. (2011)
Chlortetracycline (antibiotics)	*Zea mays* (maize)	Plant metabolism	20,000 μg kg^{-1} spiked in soil	Detection of chlortetracycline glutathionyl derivative, increase of glutathione *S*-transferase (GST) activity	Farkas et al. (2006)
Sulfonamide, flumequine (antibiotics)	*Azolla filicuolides* (water velvet), *Lythrum salicaria* (purple loosestrife), *Lemma minor* (duckweed)	Uptake by hydroponic plants, toxic effects and hormesis, biodegradation in plant tissues	Sulfadimethoxine: 50,000 to 450,000 μgL^{-1}, flumequine: 50 to 1,000 μgL^{-1}		Forni et al. (2001)
Sulfadimethoxine (antibiotics)	*Azolla filiculoides* (water velvet)	Uptake by hydroponic plants, toxic effects and hormesis, biodegradation in plant tissues	50,000 to 450,000 μg L^{-1} in hydroponic medium	Accumulation of 58 to 2,012 μg g^{-1} dry weight	Forni et al. (2002)
Tetracycline, oxytetracycline (antibiotics)	*Myriophyllum* aquaticum (parrot feather), *Pistia stratiotes* (water lettuce)	Uptake and transformation by hydroponic plants	1,000 to 10,000 μg L^{-1} in hydroponic medium		Gujarathi et al. (2005a)
Tetracycline, oxytetracycline (antibiotics)	*Helianthus annuus* (sunflower)	Transformation by hairy root cultures and cell-free root exudates	5,000 μg L^{-1} in hydroponic medium or cell-free root exudates	Uptake of tetracycline: ~50%, oxytetracycline: ~75% after 4 days	Gujarathi et al. (2005b)

(continued)

TABLE 28.1 Continued
Summary of Selected Publications on the Uptake and Metabolism of Pharmaceuticals by Plants

Pharmaceutical Compound	Plant Species	Process and Toxic Effect	Dosage	Accumulation and Transformation	References
Carbamazepine (anticonvulsant), salbutamol (antiasthmatic), sulfamethoxazole, trimethoprim (antibiotics)	*Brassica oleracea* (cabbage), *Brassica rapa* (Wisconsin Fast Plants)	Uptake from spiked nutrient solution	232,500 μg L^{-1} in nutrient solution	Accumulation in cabbage tissues of carbamazepine: 22.73 μg kg^{-1}, salbutamol: 13.33 μg kg^{-1}, sulfamethoxazole: 23.63 μg kg^{-1}, trimethoprim: 13.44 μg kg^{-1} dry weight, accumulation in Wisconsin Fast Plants of carbamazepine: 107.37 μg kg^{-1}, salbutamol: 82.43 μg kg^{-1}, sulfamethoxazole: 440.95 μg kg^{-1}, trimethoprim: 179.60 μg kg^{-1} dry weight	Herklotz et al. (2010)
Acetaminophen (pain reliever)	*Armoracia rusticana* (horseradish)	Uptake by roots	~3.8 μmol g^{-1} (574,000 μg kg^{-1})	Accumulation of ~3.5 μmol g^{-1} (529,000 μg kg^{-1})	Huber et al. (2009)
Bisphenol A, octylphenol, nonylphenol (surfactants), 2,4-dichlorophenol (herbicide), 17β-estradiol (estrogen)	*Portulaca oleracea* (common purslane)	Uptake from water, enzymatic degradation	Bisphenol A: 50 μM (11.4 mg L^{-1}), octylphenol: 25 μM (5.2 mg L^{-1}), onylphenol: 25 μM (5.5 mg L^{-1}), 17β-estradiol: 25 μM (6.8 mg L^{-1})	Bisphenol A: 35% removal from water after 3 hours, 100% after 24 hours, octylphenol and nonylphenol: 100% removal after 24 hours, 2,4-dichlorophenol: 60% removal after 24 hours, 17β-estradiol: 100% removal after 3 days	Imai et al. (2007)
17β-Estradiol (estrogen)	*Arabidopsis thaliana* (thale cress)	Plant genetically modified as biomarker of endocrine disrupting compounds	Detection of 17β-estradiol concentration range of from 0.1 ppt (10^{-12}) to 10 ppb (10^{-9}) using reverse-transcription real-time PCR and 0.1 ppb using GFP		Inui et al. (2005)
Azithromycin, roxithromycin, clindamycin (antibiotics)	*Lactuca sativa* (lettuce), *Spinacia oleracea* (spinach), *Daucus carota* (carrots)	Uptake by roots	Up to 1 μg L^{-1} in irrigation water	Clindamycin detected in spinach roots (<10 ng g^{-1}), lettuce roots (<10 ng g^{-1}), and carrot roots (53 ng g^{-1}), roxithromycin detected in lettuce roots (<10 ng g^{-1}) and carrot roots (115 ng g^{-1})	Jones-Lepp et al. (2010)

17α-Ethynylestradiol (estrogen), triclosan (antimicrobial)	*Phaseolus vulgaris* (bean plant)	Uptake by the roots and leaves	1,000 μg kg^{-1} in sand and soil	Experiments with sand: 17α-ethynylestradiol detected in roots and leaves at up to 1,644,000 and 74,000 μg kg^{-1}, triclosan in roots and leaves detected at up to 2,940,000 and 22,000 μg kg^{-1} dry weight, experiments with soil: 17α-ethynylestradiol detected in roots and leaves at up to 31,000 and 17,000 μg kg^{-1}, triclosan detected in roots and leaves at up to 6,400 and 4,200 μg kg^{-1} dry weight	Karnjanapiboonwong et al. (2011)
Estradiol 17β-glucuronate (estrogen conjugate)	*Secale cereale* (rye), *Hardeum vulgare* (barley)	Vacuolar uptake		Vacuolar uptake involving an ATP-dependent glutathione conjugate pump	Klein et al. (1998, 2000)
Oxytetracycline (antibiotic)	*Medicago sativa* (alfalfa)	Uptake by the roots, decrease of the root biomass at concentrations higher than 0.002 mM (0.92 mg L^{-1}), considerable inhibitory effect at 0.02 mM (9.2 mg L^{-1})	0 to 0.02 mM (0 to 9.2 mg L^{-1})		Kong et al. (2007)
Diclofenac, ibuprofen (antiinflammatory), acetaminophen (pain reliever)	*Armoracia rusticana* (horseradish), *Linum usitatissimum* (common flax) *Lupinus luteolus* (pale yellow lupine), *Hordeum vulgaris* (common barley), *Phragmites australis* (common reed)	Uptake from nutrient solution	Acetaminophen: 15,000 and 30,000 μg L^{-1}, ibuprofen: 20,600 μg L^{-1} and 41,300 μg L^{-1}, diclofenac: 63,700 μg L^{-1}	Lupinus luteolus, 100% removal of acetaminophen, 4-11% of ibuprofen, Hordeum vulgare, 100% removal of acetaminophen, Phragmites australis, 50–60% removal of ibuprofen, limited removal of acetaminophen. Linum usitatissimum, 100% removal of ibuprofen	Kotyza et al. (2010)

(continued)

TABLE 28.1 Continued
Summary of Selected Publications on the Uptake and Metabolism of Pharmaceuticals by Plants

Pharmaceutical Compound	Plant Species	Process and Toxic Effect	Dosage	Accumulation and Transformation	References
Chlortetracycline, tylosin (antibiotics)	*Zea mays* (corn), *Allium cepa* (green onion), *Brassica oleracea* (cabbage)	Uptake from spiked and manured soil	Chlortetracycline: from 587 to 1,587 μg L^{-1}, tylosin: 1,000 μg L^{-1}	Chlortetracycline detected in plant tissues at 2–17 ng g^{-1}, tylosin not detectable	Kumar et al. (2005)
Flumequine (antibiotic)	*Lythrum salicaria* (purple loosestrife)	Uptake from hydroponic solution and phytotoxicity	50 to 100,000 μg L^{-1} in hydroponic medium	Detection in plant tissues at 200 to 64,900 μg g^{-1} dry weight	Migliore et al. (2000)
Endocrine disruptors in wastewater	Constructed wetlands	Endocrine disrupting effect and growth inhibition on fishes			Norris and Burgin (2011)
Chlortetracycline, doxycycline, oxytetracycline, tetracycline (antibiotics)	*Callistephus chinensis* (aster), *Carum carvi* (caraway), *Calendula officinalis* (pot marigold), *Leontodon autumnalis* (fall dandelion), *Prunus persica* (peach), *Papaver somniferum* (poppy), and *Fragaria ananassa* (strawberry)	Uptake, translocation, and accumulation in plant tissues	100,000 μg L^{-1}	Total uptake: 173 to 537 μg g^{-1} plant tissue, active antibiotics uptake 3 to 122 μg g^{-1} plant tissue	Peterson and Sinha (1977)
Fluoxetine (antidepressant)	*Brassica oleracea* (cauliflower)	Uptake and translocation	280 μg L^{-1}	Detected at 490 μg kg^{-1} wet weight in stem, 260 μg kg^{-1} wet weight in leaves	Redshaw et al. (2008)

Fluoxetine (antidepressant), ibuprofen (antiinflammatory), triclosan (antimicrobial)	*Lemna minor, Landoltia punctate* (duckweed)	Uptake and biodegradation in bioreactor containing active plants, chemically inactivated plants, blender-macerated plants, and no-plant controls	5 and 10 μM	80% and 56% removal of fluoxetine with active and chemically inactivated plants after 9 days, 48% and 25% removal of ibuprofen from active and macerated plants after 9 days, significant removal of triclosan in all bioreactors	Reinhold et al. (2010)
Estrone, 17β-estradiol, 17α-ethynylestradiol (estrogens)	*Phragmites australis* (common reed)	Removal from constructed wetland, aerobic biodegradation	Estrone: 0.39 to 10.49 ng L^{-1}, 17β-estradiol: 1.35 to 9.05 ng L^{-1}: 17α-ethynylestradiol: 0.59 to 6.56 ng L^{-1}	68% removal of estrone, 84% removal of 17β-estradiol, 75% removal of 17α-ethynylestradiol	Song et al. (2009)
Carbamazepine (anticonvulsant), ibuprofen (antiinflammatory)	*Lolium perenne* (ryegrass)	Uptake from soil fertilized with pharmaceutical-spiked urine	Carbamazepine: 58 μg L^{-1}, ibuprofen: 844 μg L^{-1} in fertilized soil	Detection of 34% of carbamazepine in aerial parts and 0.3% in roots, ibuprofen not detected in soil or plant tissues after three months	Winker et al. (2010)
Carbamazepine (anticonvulsant), diphenhydramine (antihistamine), fluoxetine (antidepressant), triclosan and triclocarban (antimicrobials)	*Leguminosae* (soybean)	Uptake by the roots and translocation to the leaves	Carbamazepine: 49 μg kg^{-1}, diphenhydramine: 43 μg kg^{-1}, fluoxetine: 40 μg kg^{-1}, triclosan: 13 μg kg^{-1}, triclocarban: 74 μg kg^{-1} in soil	Detection of fluoxetine at 22 μg kg^{-1} and carbamazepine at 153 μg kg^{-1} dry weight in roots, fluoxetine not detected and carbamazepine detected at 216 μg kg^{-1} dry weight in leaves	Wu et al. (2010)
Oxytetracycline (antibiotic)	*Triticum aestivum* (wheat)	Toxic effect on photosynthesis	4,600 to 36,800 μg L^{-1} spiked in soil		Li et al. (2011)

28.3.2.1 Uptake of Pharmaceuticals by Wetland Plants

Constructed wetlands are increasingly considered as sustainable systems for polishing secondary wastewater by the removal of nutrients and emerging contaminants (Song et al. 2009). Constructed wetlands offer several advantages over tertiary or advanced wastewater treatments, including low capital and operating costs, environment friendliness, and the production of feedstock for conversion to biofuel.

Early reports have described the removal of agricultural pharmaceuticals, such as antibiotics and hormones, by aquatic plants in constructed wetland destined to treat agricultural wastewater. Antibiotics are used in agriculture as antimicrobial agents to treat infection and as growth promoters to increase the efficiency of animal production. About 50 million pounds of antibiotics are produced in the United States annually, of which approximately 40% are used as growth promoters (Gujarathi et al. 2005a). It is usually estimated that 30 to 90% of the antibiotics taken by humans and animals are excreted unchanged (Gujarathi et al. 2005a). Migliore et al. (2000) have studied the uptake of the antibiotic flumequine by the aquatic weed, *Lythrum salicaria*. Sterile hydroponic plants were grown on Murashige and Skoog medium and exposed for 30–35 days to concentrations ranging from 50 μg L^{-1} to 100 mg L^{-1}. Hormesis was observed following exposure to concentrations from 50 to 5,000 μg L^{-1} (increase of the number and size of leaves and secondary roots) and high toxicity for exposure to 100 mg L^{-1} (decrease of plant growth). The plants were shown to take up flumequine at levels increasing with the dosage, from 0.2 to 13.3 μg g^{-1} dry weight for exposure at 50 to 5,000 μg L^{-1}. On the other hand, exposure to 100 mg L^{-1} showed a decrease of the plant concentration from 64.9 to 15.7 μg g^{-1} dry weight with the time of exposure (from 10 to 30 days), suggesting biodegradation of flumequine inside plant tissues (Migliore et al. 2000). Forni et al. (2001) showed that aquatic plants were capable to remove antibiotics from the hydroponic medium, making them promising for the treatment of contaminated wastewater in constructed wetlands. The potential use of three macrophytes, *Azolla filiculoides* (water velvet), *Lythrum salicaria* (purple loosestrife), and *Lemna minor* (duckweed), were tested for phytoremediation of two antibiotics used in intensive farming, sulfadimethoxine (sulfonamide) and flumequine (quinolone). Exposure of *Azolla* plants to 50 to 450 mg L^{-1} of sulfadimethoxine in N-free Hoagland mineral medium showed dose-dependent toxic effects on plant growth, nitrogen fixation, and chlorophyll a:chlorophyll b ratio. The plants were reported to degrade the antibiotics at a high rate (although no quantification was provided by the authors). Exposure of *Lythrum salicaria* to flumequine (50 to 1,000 μg L^{-1}) on solid Murashige and Skoog medium showed an increase of the number of leaves and secondary roots at low exposure levels (50 μg L^{-1}) (hormesis). Again, the authors reported a high removal rate of the compound without being more specific about quantification. *Lemna minor* grown on N-free Hoagland mineral medium and exposed to the same concentrations of flumequine showed a decrease in chlorophyll b, which was inversely related to the antibiotic concentration. However, only a mild effect on the plant growth was noticed. In a separate study performed by the same authors, the aquatic fern, *Azolla filiculoides*, was exposed to 50 to 450 mg L^{-1} of sulfadimethoxine, a antibiotic commonly used in intensive farming (Forni et al. 2002). After 5 weeks of exposure, plants were shown to remove significantly the antibiotic from the hydroponic medium (from 88.5% for exposure to 450 mg L^{-1} to 56.3% for exposure to 50 mg L^{-1}). Plant uptake increased with the dosage from 58 to 2,012 μg g^{-1} plant dry weight for exposure to 50 to 450 mg L^{-1}, respectively. Although the plants exposed to the highest concentration survived, the drug was shown to affect the growth rate, nitrogen fixation, and heterocyst frequency. Gujarathi et al. (2005a) studied the biodegradation of tetracycline and oxytetracycline (the most common tetracyclines used in veterinary medicine) by the aquatic plants, *Myriophyllum aquaticum* (parrot feather) and *Pistia stratiotes* (water lettuce). The hydroponic plants, as well as cell-free root exudates, were capable of transforming both antibiotics. Based on kinetic analyses, the authors precluded enzymatic degradation and suggested the involvement of root exudates in the antibiotic biodegradation. The same authors made similar observations studying the transformation of tetracycline and oxytetracycline by hairy root cultures of *Helianthus annuus* (sunflower) (Gujarathi et al. 2005b). Again, transformation of the two antibiotics by cell-free root exudates suggested the implication of root metabolic products. In addition, the decrease of the transformation rates observed in the presence of ascorbic acid (antioxidant) suggested the involvement of reactive oxygen species (ROS).

Using pilot-scale vertical-flow constructed wetland systems vegetated with common reed (*Phragmites australis*), Song et al. (2009) studied the removal of three major endocrine-disrupting compounds found in wastewater, estrone, 17β-estradiol, and 17α-ethynylestradiol. Three parallel constructed wetlands characterized by different sand bed depths (7.5, 30, and 60 cm, respectively) were fed with sewage treatment works effluent from the city of Sendai (Japan). Removal of estrogens was monitored over two years. The concentrations of estrogens in the sewage effluent ranged from 0.39 to 10.49 ng L^{-1} of estrone, from 1.35 to 9.05 ng L^{-1} of 17β-estradiol, and from 0.59 to 6.56 ng L^{-1} of 17α-ethynylestradiol. A significant reduction of estrogens was observed upon treatment in the constructed wetlands with the highest three-month average removal rates observed in the shallowest system (7.5 cm depth): 68% for estrone, 84% for 17β-estradiol, and 75% for 17α-ethynylestradiol. Mean dry-weight estrogen concentrations in the sand bed of the wetlands ranged from 11.3 to 20.7 ng kg^{-1} of estrone, 4.7 to 8.8 ng kg^{-1} of 17β-estradiol, and 9.5 to 16.2 ng kg^{-1} of 17α-ethynylestradiol. The residual concentrations observed in the three wetlands at various depths were not statistically different. Although the plants did not seem to be adversely affected in any of the wetlands, root length and root density were the greatest in the sand bed of the shallowest wetland. Based on these observations, the authors suggested that several factors could explain the higher estrogen removal in the shallowest wetland, including the higher oxygen level favoring microbial aerobic oxidation and the higher root density enhancing microbial activity in the root zone and sorption and uptake by the plants. Using bench-scale flask experiments, Reinhold et al. (2010) investigated the removal of various pharmaceuticals and other emerging contaminants by duckweed species (*Lemna minor* and *Landoltia punctata*) collected from a constructed wetland. The authors studied the removal of atrazine, clofibric acid, fluoxetine, ibuprofen, 2,4-dichlorophenoxyacetic acid, meta-*N*,*N*-diethyl toluamide, picloram, and triclosan dosed at the concentration of 10 μM in the growth medium. Different experimental bioreactors were used that contained active plants, chemically inactivated plants, blender-macerated plants, and no-plant controls. The authors observed that only four out of eight contaminants were significantly removed from the aqueous solution. Ibuprofen was removed by both active and macerated plants (48% and 25% removal after 9 days of exposure, respectively), indicating active plant uptake and microbial degradation of the compound. Fluoxetine was removed from active and chemically inactivated plants (80% and 56% removal after one day of exposure, respectively), indicating that active uptake and sorption on plant material were the major mechanisms of reduction. Triclosan was removed significantly from all the reactors. The slightly higher depletion from active plants as compared with chemically inactivated and macerated plants suggested that the primary mechanism of removal was sorption on the plant material, with minimal active uptake or microbial biodegradation. This study has the merit to show that different mechanisms of removal by plants were at work depending on the chemical nature of the pharmaceuticals. Avila et al. (2010) tested a pilot-scale horizontal subsurface flow constructed wetland system (HSSFCW) for the potential removal of the pharmaceuticals ibuprofen (75 mg L^{-1}) and naproxen (30 mg L^{-1}), as well as other personal care products and estrogenic compounds. Municipal sewer was flowed into an anaerobic upflow sludge bed reactor (HUSB) for primary treatment, followed by two parallel HSSFCW systems (surface area of 0.65 m^2) connected in series to another HSSFCW (surface area of 1.65 m^2). The wetlands were planted with *Phragmites australis* (reed) and the system received a flow of 84 L d^{-1}. The overall removal efficiencies for pharmaceuticals were 98 to 99% for ibuprofen and 99% for naproxen. The removal patterns were different for the two pharmaceuticals depending on their sorption and biodegradation characteristics. About 50% of ibuprofen was removed in the first two wetlands followed by a 99% removal in the second wetland. This could be explained by preferential aerobic biodegradation of ibuprofen as the second wetland showed a significantly higher level of dissolved oxygen. On the other hand, higher removal of naproxen was recorded in the first two wetlands (93%), which may reflect anaerobic biodegradation of this compound under reducing conditions. The authors concluded that these high removal efficiencies were likely to be related to the high temperature of the environment and the high concentration of target compounds in the influent. Recently, Dordio et al. (2011) reported the uptake of the non-steroidal anti-inflammatory (NSAID) drug ibuprofen from modified Hoagland nutrient solution by cattail plants (*Typha* sp.). Sixty percent of the initial 20 μg L^{-1} was removed within 24 h and >99% after 21 days. Exposure to higher ibuprofen concentrations (up to 2.0 mg L^{-1}) did affect *Typha*'s growth, although plant growth rate and photosynthetic pigment measurements approached normal (non-exposed) values after 21 days of exposure. Alteration of antioxidant enzymatic activities, including superoxide dismutase, catalase, and guaiacol peroxidase, in ibuprofen-exposed root and leaf tissues suggested that ibuprofen may exert oxidative stress on *Typha* cells (Dordio et al. 2011). In another recent study, wetland treatment has been shown to result in reduction of the estrogenic activity of wastewater *in vivo*. Norris and Burgin (2011) investigated the endocrine response of mosquito fishes (*Gambusia holbrooki*) resident of wetlands used for the storage of treated sewage effluent and storm water runoff. Fishes in wetlands receiving directly the sewage effluent or runoff exhibited clear morphological traits of endocrine disruption, while fishes in downstream wetlands showed a milder response. Endocrine disruption was determined by observation of the gonopodium of euthanized mature male mosquito fishes under a dissecting microscope. Fishes in the first series of wetlands were also found to be smaller than in the second series.

28.3.2.2 Uptake of Pharmaceuticals by Agricultural Plants

Several studies have been conducted recently that provide evidence that pharmaceuticals present in soil, irrigation water, or organic fertilizers can be taken up by plants, potentially rendering their consumption harmful for human health. In the first study of the kind, Kumar et al. (2005) have conducted greenhouse experiments to determine the potential of edible crop plants to take up the antibiotics, chlortetracycline and tylosin, present in manure-applied soil. The authors reported that corn (*Zea mays*), green onion (*Allium cepa*), and cabbage (*Brassica oleacera*) plants could take up the antibiotics under study. Concentrations reached 2–17 mg kg^{-1} fresh weight and increased with the concentration in soil. Boxall et al. (2006) studied the uptake of 10 veterinary and human antibiotics by lettuce (*Lactuca sativa*) and carrot (*Daucus carota*) plants growing on unmanured soil spiked with 1 mg kg^{-1} of antibiotics. The authors reported a highly variable absorption depending on the compounds and the plant species: concentrations ranged from 6 μg kg^{-1} fresh weight of trimethoprim to 170 μg kg^{-1} of levamisole in lettuce leaves and from 2.8 μg kg^{-1} of enrofloxacin to 13 μg kg^{-1} of diazinon in carrot roots. In a similar study, Dolliver et al. (2007) reported uptake of sulfamethazine by corn, lettuce, and potato (*Solanum tuberosum*), reaching concentrations in plant tissues ranging from 100 to 1,200 μg kg^{-1} dry weight. Sulfamethazine concentrations in plant tissues also increased with the concentration in manure. Redshaw et al. (2008) studied the uptake and translocation of fluoxetine into cauliflower (*Brassica oleracea*) hydroponic cultures. After 12 weeks of exposure, fluoxetine was detected in stem (0.49 μg kg^{-1} fresh weight) and leaf (0.26 μg kg^{-1} fresh weight) tissues.

To determine the potential of food chain contamination by pharmaceuticals in irrigation water, Jones-Lepp et al. (2010) studied the uptake of three antibiotics, azithromycin, roxithromycin, and clindamycin, inside selected crop plants. Greenhouse experiments were conducted with three edible species, lettuce (*Lactuca sativa*), spinach (*Spinacia oleracea*), and carrots (*Daucus carota*), that were irrigated with Colorado River water dosed with increasing concentrations of antibiotics (up to 1,000 ng L^{-1}). Field studies were also conducted in which pepper (*Capsicum annuum*), tomato (*Lycopersicon esculentum*), melon (*Cucumis melo*), lettuce, and watermelon (*Citrullus lanatus*) plants were irrigated with treated wastewater from the city of Tucson (Arizona, U.S.A.) known to contain detectable levels of pharmaceuticals. Greenhouse uptake experiments showed the potential for uptake of the antibiotics under study, albeit at very low levels: trace levels of clindamycin were detected into spinach roots (less than 10 ng g^{-1}), lettuce roots (less than 10 ng g^{-1}), and carrot roots (53 ng g^{-1}), and traces levels of roxithromycin were detected in lettuce roots (less than 10 ng g^{-1})

and carrot roots (115 ng g^{-1}) exposed to 1,000 ng L^{-1} in irrigation water. No antibiotic was detected in plants treated with less than 1,000 ng L^{-1} or in the aerial parts of the plants. Field tests conducted with real wastewater effluent revealed the presence of only one compound (the industrial flavoring agent, *N,N'*-dimethylphenethylamine), while none of the evaluated contaminants were found in crops irrigated with Colorado River water. No detectable levels of the pharmaceuticals under study were recorded in the rhizosphere soil of crops irrigated with Tucson wastewater or Colorado River water. According to the authors, the absence of detection of pharmaceuticals in plants tissues from field studies is attributable to biodegradation and dilution effect in plant tissues.

Soybean plants (*Glycine max*) were shown to remove the antibiotics oxytetracycline and norfloxacin from saline soil collected from shrimp aquaculture facilities (Boonsaner and Hawker 2010). The authors observed a 20% reduction of seed germination in contaminated soil (70 g NaCl kg^{-1} dry weight, 105 mg oxytetracycline kg^{-1}, and 55 mg norfloxacin kg^{-1}) by comparison to non-contaminated, non-saline soil. Soybean plants cultivated in water containing 50 mg oxytetracycline L^{-1} and 55 mg norfloxacin L^{-1} were shown to accumulate up to ~190 mg oxytetracycline kg^{-1} dry weight and ~110 mg norfloxacin kg^{-1} dry weight after 2 days of exposure. Soybean plants growing on saline contaminated soil (70 g NaCl kg^{-1} dry weight, 105 mg oxytetracycline kg^{-1}, and 55 mg norfloxacin kg^{-1}) accumulated a maximum of 44 mg oxytetracycline kg^{-1} dry weight and 2.6 mg norfloxacin kg^{-1} dry weight after 6 days of exposure. In experiments conducted in both contaminated water and soil, these maximum levels decreased the following days, suggesting transformation of the antibiotics inside plant tissues (Boonsaner and Hawker 2010). Hydroponic cabbage (*Brassica oleracea*) and Wisconsin Fast Plants (*Brassica rapa*) were recently shown to uptake carbamazepine, salbutamol, sulfamethoxazole, and trimethoprim spiked into the recirculation nutrient solution (233 mg L^{-1}) (Herklotz et al. 2010). All pharmaceuticals were detected in roots, stems, and leaves at concentrations ranging from 91 μg kg^{-1} fresh weight (trimethoprim) to 138 μg kg^{-1} (sulfamethoxazole) in cabbage roots and from 22 μg kg^{-1} fresh weight (salbutamol) to 118 μg kg^{-1} (sulfamethoxazole) in Wisconsin fast plants. Wu et al. (2010) reported uptake of carbamazepine, diphenhydramine, fluoxetine, triclosan and triclocarban by soybean plants growing on soil, either directly spiked with pharmaceuticals or amended with spiked biosolids. Carbamazepine, triclosan, and triclocarban were detected in root and aerial tissues (including beans), whereas accumulation of diphenhydramine and fluoxetine was more limited. After 60 days of exposure, concentrations ranged from 22 μg fluoxetine kg^{-1} dry weight to 153 μg carbamazepine kg^{-1} in roots and from non-detected for fluoxetine to 216 μg carbamazepine kg^{-1} in leaves. Karnjanapiboonwong et al. (2011) demonstrated recently the potential uptake of the estrogen 17α-ethynylestradiol and biocide triclosan by bean plants (*Phaseolus vulgaris*) grown in sand and soil (dosed at 1 mg kg^{-1}). In sand, 17 α-ethynylestradiol accumulated in roots and leaves at levels up to 1,640 and 74 mg kg^{-1} dry weight, respectively, and triclosan accumulated in roots and leaves at levels up to 2,940 and 22 mg kg^{-1} dry weight, respectively. In soil, 17α-ethynylestradiol accumulated in roots and leaves at levels up to 31 and 17 mg kg^{-1} dry weight, respectively, and triclosan accumulated in roots and leaves at levels up to 6.4 and 4.2 mg kg^{-1} dry weight, respectively. The lower uptake of 17α-ethynylestradiol and triclosan from soil as compared to sand was explained by differential binding to organic carbon in soil. Triclosan, which has a higher K_{ow}, is expected to bind more strongly to soil organic matter and be less available for plant uptake. Kotyza et al. (2010) studied the phytoremediation of selected pharmaceuticals (diclofenac, ibuprofen, and acetaminophen) by cell cultures of *Armoracia rusticana* and *Linum usitatissimum* and hydroponic plants of *Lupinus luteolus, Hordeum vulgaris*, and *Phragmites australis. Lupinus luteolus* (pale yellow lupine) completely removed acetaminophen dosed at 0.1 mM (15 mg L^{-1}) and 0.2 mM (30 mg L^{-1}) from the hydroponic solution after two to four days. *Hordeum vulgare* (common barley) entirely removed acetaminophen after two days, but after eight days of exposure, acetaminophen was released by the roots to the medium. The authors hypothesized that acetaminophen would be stored in the vacuole and then slowly released as a consequence of its toxicity for plant cells. On the other hand, removal of acetaminophen by *Phragmites australis* (common reed) was rather inefficient: after eight days of exposure, initial concentrations of 15 and 30 mg L^{-1} were left almost unchanged (12.6 ± 1.8 and 31.9 ±

4.7 mg L^{-1}, respectively). *Phragmites australis* was shown to decrease the initial concentrations of ibuprofen from 20.6 mg L^{-1} (0.1 mM) and 41.3 mg L^{-1} (0.2 mM) to 8.3 ± 2.0 and 20.9 ± 2.8 mg L^{-1}, respectively. *Lupinus luteolus* was rather inefficient at removing ibuprofen from the hydroponic medium (11% of the initial 0.1 mM and 4% of the initial 0.2 mM), presumably due to its toxicity for the plants. A fresh cell suspension of *Linum usitatissimum* (common flax) was shown to completely remove 0.2 mM of ibuprofen after one day. Diclofenac was not efficiently removed by *Phragmites australis, Hordeum vulgare*, and *Lupinus luteolus* (for instance, *Phragmites australis* reduced an initial concentration of 63.7 mg L^{-1} to 42.7 ± 1.6 mg L^{-1}). Although acetaminophen did not produce recordable toxic effects to *Phragmites australis, Hordeum vulgare*, and *Lupinus luteolus* at the concentrations of 0.1 mM and 0.2 mM, ibuprofen and diclofenac dosed at 0.1 mM and 0.2 mM showed yellowing and dehydration of the shoots after a few days of exposure.

Uptake of three common pharmaceuticals, ibuprofen, carbamazepine, and clofibric acid, from wastewater by *Typha* sp. (cattail) plants was tested using laboratory batch microcosm experiments (Dordio et al. 2010). Polyvinyl chloride (PVC) containers filled with a matrix of light expanded clay aggregates were planted with *Typha* sp. plants collected from streams in Alentejo (Portugal) and fed with wastewater from a local secondary treatment system spiked with a 1 mg L^{-1} of ibuprofen, carbamazepine, and clofibric acid. Unplanted and unfilled vessels were used as controls. Removal efficiencies in planted microcosms, during the winter and summer conditions, were 88.2% and 96.7% for carbamazepine, 81.9% and 96.2% for ibuprofen, and 48.3% and 74.5% for clofibric acid, respectively. Microcosm wetlands with planted beds showed the highest removal efficiencies. Although ibuprofen was efficiently removed from control vessels (90.6% and 73.5% in unplanted beds and 75.2% and 38.2% in wastewater-only vessels during the summer and winter conditions, respectively), removal efficiencies of carbamazepine and clofibric acid were only 12% or lower. In general, the highest removal efficiencies were obtained in planted-bed vessels under summer conditions. The authors observed that the removal kinetics was characterized by a fast initial step (more than 50% removal within 6 h), likely due to adsorption on the expanded clay material, followed by a slower, plant-mediated uptake of the target compounds.

Besides the application of irrigation water and municipal sludge biosolids, fertilization of agricultural lands using human urine constitutes another way of contamination of edible crops by pharmaceuticals. Urine contains high levels of phosphorous and nitrogen and has been used in many countries as agricultural fertilizers. Using greenhouse pot experiments, Winker et al. (2010) investigates the potential for ryegrass (*Lolium perenne*) contamination by application of urine containing pharmaceuticals. Urine from healthy males was spiked with two important pharmaceuticals, carbamazepine and ibuprofen, and applied to soil to initial concentrations of 3.2 and 32 μg kg^{-1} for carbamazepine and 49 and 490 μg kg^{-1} for ibuprofen. After 3 months, no ibuprofen was detected in soil or in plant tissues, presumably because of its fast biodegradation in soil. On the other hand, an average of 53% of the initial carbamazepine (for the high exposure level, 32 μg kg^{-1}) was detected in soil and 34% in plant tissues after three months of exposure. Average concentrations of 277 and 5,592 μg kg^{-1} were detected in roots and aerial parts of the plants, respectively, showing an efficient translocation of the compound, presumably through the transpiration stream. No adverse effect of the pharmaceuticals on plant growth was recorded. A study by Imai et al. (2007) has shown that *Portulaca oleracea* (green purslane), a common garden plant in Japan, had the capability of removing 17β-estradiol dosed at the initial concentration of 25 μM (6.8 mg L^{-1}) in the hydroponic solution, along with other endocrine disrupting compounds, including bisphenol A, 2,4-dichlorophenol, octylphenol, and nonylphenol. However, removal of 17β-estradiol by *Portulaca oleracea* occurred at a slower pace than other target compounds (complete removal of 17β-estradiol was observed after 3 days, while other compounds were removed within 24 hours).

In order to determine the possibility of contamination of edible plants by pharmaceuticals potentially present in irrigation water, cucumber plants (*Cucumis sativus*) were exposed to the antiepileptic drug, carbamazepine (Shenker et al. 2011). Hydroponic cucumber plants were grown for 22 days in medium containing different concentrations of carbamazepine. Plants exposed to concentrations higher than 10,000 μg L^{-1} showed phytotoxic effects, including a 50% decrease of the biomass, a reduction of the number and length of primary and secondary roots, and a reduction of the number and size of mature leaves. Carbamazepine was then dosed at concentrations below 1,000 μg L^{-1} to avoid phytotoxicity. Carbamazepine concentrations in xylem sap and nutrient solution were 65.9 ± 28.4 and 76.1 ± 8.91 μg L^{-1}, respectively, leading the authors to hypothesized that uptake of carbamazepine was not restricted by the root membranes. Carbamazepine in plant tissues was detected mostly in the leaves, while concentration in the roots and stem were markedly lower. Concentration of carbamazepine in leaves was shown to be age-related, with top (youngest) leaves showing a significantly lower concentration (426 μg L^{-1}) than cotyledon leaves (2,354 μg L^{-1}). Based on these observations, the authors suggested that uptake and translocation of carbamazepine occurred primarily through the transpiration (xylem) stream. Greenhouse experiments were conducted with plants growing in loess (sandy) soil irrigated with (1) fresh water spiked with carbamazepine (1.15 μg L^{-1}), (2) reclaimed water spiked with carbamazepine (4.14 μg L^{-1}), and (3) reclaimed water non-spiked with carbamazepine (2.99 μg L^{-1}). In the three different treatments, 76% to 84% of the total carbamazepine uptake was detected in the leaves. Bioaccumulation of carbamazepine in fruits was 11% with spiked fresh water, 18% with non-spiked reclaimed wastewater, and 22% with spiked reclaimed wastewater. Bioaccumulation in roots and stems was between 1% and 3%. Bioaccumulation factors (calculated as the ratio of carbamazepine concentration in the plant to that in the soil solution) were 0.8 to 1 for the fruits and 17.9 to 20 for the leaves. Based on these results, the authors concluded that carbamazepine uptake was dependent of its concentration in soil.

The capability of plants to take up pharmaceuticals has been exploited by genetically engineering plants to be utilized as biomarkers of endocrine disruption compounds. Inui et al. (2005) achieved the *Agrobacterium*-mediated transformation of *Arabidopsis* plants by the introduction of a human estrogen receptor, *LexA-VP16-ER*, and the reporter plasmid, *p*ER8-GFP, expressing a green fluorescent protein (GFP). Upon absorption by the roots, the endocrine disrupting compound binds the *LexA-VP16-ER* transcription factor. The complex endocrine disrupting compound-transcription factor then binds the *LexA* operator region upstream of the *gfp* gene, which induces the expression of GFP. After 7 days of incubation in the presence of the endocrine-disrupting compound, 17β-estradiol, a dose-dependent expression of *gfp* gene was detected by reverse-transcription real-time PCR over a concentration range from 0.1 ppt (10^{-12}) to 10 ppb (10^{-9}). The direct detection of GFP in plant tissues allowed the detection of concentrations of 0.1 ppb. In addition, 11 more endocrine-disrupting compounds were positively detected at ppb and ppm doses by the biomarker plants, including atrazine, estrone, and 4-*t*-octylphenol.

28.3.2.3 Modeling Uptake of Pharmaceuticals by Plants

Various models have been proposed to capture the uptake of xenobiotics from soil and their accumulation inside plant tissues. A classical four-compartment (roots, stems, leaves, and fruits) model for the uptake and biodegradation of non-dissociated organic contaminants by plants has been proposed by Trapp et al. (1994). The model considers the following processes: (1) *uptake* from the soil solution, which is expressed by different partition coefficients; (2) *translocation* with the transpiration stream, which is expressed by the transpiration stream concentration factor (TSCF), (3) *metabolism* inside plant tissues, which is assumed to follow a pseudo-first order kinetics, and (4) *dilution* by plant growth, which is assumed to follow an exponential kinetics. Partition coefficients and TSCF are related to the octanol-water partition coefficient, K_{OW} (Briggs et al. 1982; Burken and Schnoor 1998). Typically, *moderately hydrophobic* pollutants (with a $\log K_{OW}$ between 0.5 and 4.5), such as many pharmaceuticals, are the most susceptible to be efficiently taken up by plants. Due to its relative complexity, this model does not lead to analytic solutions and provides little information about the uptake and translocation mechanisms. As a consequence, simpler one and two-compartment models are frequently used, such as the one-compartment plant model developed by Trapp and Matthies (1995), which has an analytic solution and has been proven to be useful for environmental risk assessment.

A few studies have been conducted to model the uptake of pharmaceuticals from soil and their subsequent accumulation in plant tissues. Cropp et al. (2010) used a simple two-compartment soil/water-plant model to derive analytic expressions for the uptake and accumulation of the antibiotic, norfloxacin, in soybean plants (*Glycine max*). Based on the model, the authors predicted a first order loss rate from the soil/water phase of 0.544 d^{-1} and a maximum concentration in soybean of 52.5 mg kg^{-1} dry weight after 2.79 days of exposure. The predicted concentrations of norfloxacin in soybean plants agreed well with the experimental data, with a reported $R^2 = 0.91$. Based on these findings, the authors comment on other uptake experiments (Dolliver et al. 2007), emphasizing that the highest accumulation level is likely to occur early after the plant exposure to pharmaceuticals, which was not captured by the measurements performed only at the end of the exposure period. Although treating the plant as a single compartment ignores the variation of concentrations within different plant organs and the translocation processes, it provides analytic solutions for the maximum plant concentration and the time when it is likely to occur. Winker et al. (2010) used a three-compartment model (soil, roots, aerial parts) adapted from Trapp et al. (1994) to study the mechanism of entry of carbamazepine and ibuprofen into the roots of exposed ryegrass (*Lolium perenne*) plants. The model assumed no biodegradation in soil and convective transport of carbamazepine with the transpiration stream based on the published transpiration coefficient of ryegrass. The root and Casparian strip were considered as barriers that retard carbamazepine uptake inside plant tissues, which were expressed as transfer coefficients. The different partition and transfer coefficients used in the model were adjusted to obtain optimal fitting with the experimental data. The adjusted model parameters suggest that neither the roots nor the Casparian strip acted as a significant barrier to carbamazepine uptake (transfer coefficients of 1 approximate well the experimental data). The partition coefficient in soil obtained using the model was within the range of values reported in the literature.

28.3.2.4 Metabolism of Pharmaceuticals by Plants

Understanding plant metabolism of pharmaceuticals is critical because in the absence of further transformation and detoxification, pharmaceuticals and/or their metabolites taken up inside plant tissues will enter the food chain or return to the soil, resulting in potential hazard for the environment and human health. However, few studies have been conducted on the metabolism of pharmaceuticals by plants. Farkas et al. (2006) reported transformation of the antibiotic, chlortetracycline, inside root tissues of maize plants (*Zea mays*) growing on spiked soil (20 mg kg^{-1}). Chlortetracycline glutathionyl conjugates (phase II metabolites) were identified by LC-MS/MS in parallel with an increase of glutathione *S*-transferase (GST) activity in the exposed tissues. Huber et al. (2009) showed that incubation of acetaminophen with hairy root cultures of horseradish (*Armoracia rusticana*) resulted in the formation of conjugates with glycosides, glutathione, and cysteine (phase II metabolites). The same authors recently reported that acetaminophen was taken up and metabolized inside the tissues of Indian mustard plants (*Brassica juncea*) (Bartha et al. 2010). Two metabolites were identified in leaf tissues, a glutathionyl and a glycoside conjugate (phase II metabolites). Exposure of plants to acetaminophen resulted in increase of GST activity, which was found to correlate the appearance of acetaminophen conjugates. This led the authors to suggest that, consistently with the *green liver* model, acetaminophen detoxification in plants resembles the mammalian metabolism, including cytochrome P-450-mediated hydroxylation into toxic *N*-acetyl-*p*-benzoquinoneimine (phase I), followed by conjugation with glutathione (phase II).

As an illustration of phase III mechanisms, pharmaceutical compounds, in the form of conjugates with a plant molecule, have been shown to be actively pumped inside the cell vacuoles. Klein et al. (1998, 2000) demonstrated that a conjugate of 17β-estradiol (estradiol 17β-glucuronate) was actively transported inside the vacuole of rye (*Secale cereale*) and barley (*Hordeum vulgare*) cells by the action of an ATP-dependent glutathione conjugate pump. The authors also reported that the presence of natural substrates for these ATPases, namely glutathione conjugates and oxidized glutathione, stimulated up to 7-fold the uptake of estradiol 17β-glucuronate in a concentration-dependent manner.

Although these results are encouraging, very few plant enzymes have been positively identified that are implicated in the metabolism of pharmaceuticals. Similarly, and despite fast advances in high-throughput gene expression and proteomics methods, little information is available regarding the molecular bases of the biodegradation of pharmaceuticals by plants.

28.3.2.5 Phytotoxicity of Pharmaceuticals

Several studies summarized below have recorded various toxic effects toward plants exposed to pharmaceuticals, including acetaminaophen, ibuprofen, and diclofenac (Kotyza et al. 2010), carbamazepine (Shenker et al. 2011), chlorotetracycline and doxycycline (Peterson and Sinha 1977), flumequine (Migliore et al. 2000; Forni et al. 2001), and sulfadimethoxine (Forni et al. 2001). Toxic effects included reduction of the biomass, reduction of the number and size of organs (roots, leaves), inhibition of nitrogen fixation, and alteration of the ratio chlorophyll a:chlorophyll b. In some cases, exposure to pharmaceuticals resulted in a significant increase of the activity of enzymes involved in response to oxidative stress (i.e., peroxidase and catalase) (Farkas et al. 2006; Bartha et al. 2010).

Although adverse effects of pharmaceuticals on plants have been reported, there occurred only at levels far exceeding environmental concentrations. For instance, An et al. (2009) reported the inhibition of elongation of wheat seedlings (*Triticum aestivum*) exposed to triclosan and galaxolide applied in the culture medium at concentrations of 50 to 250 mg L^{-1}. After 21 days of exposure, 0.2–3.0 mg L^{-1} impacted other biochemical indicators including chlorophyll accumulation, protein synthesis, and peroxidase and superoxide dismutase activities. In a recent study, oxytetracycline at levels from 0.01 to 0.08 mmol L^{-1} (4.6 to 36.8 mg L^{-1}) was shown to inhibit biomass growth and photosynthesis of both oxytetracycline-sensitive and tolerant wheat cultivars under hydroponic conditions (Li et al. 2011).

Flumequine dosed at a concentration of 100 mg L^{-1} was shown to exert significant toxic effects (visible after 10 and 30 days) on the growth of *Lythrum salicaria* (aquatic weed), resulting in reduction of the length of roots, hypocotyles, cotyledons, and leaves, and the number of secondary roots and leaves (Migliore et al. 2000). On the other hand, 35 days of exposure to lower concentrations (50 to 5,000 µg L^{-1}) caused alteration of the post-germinative development reflected by hormesis: an increase of the number and size of leaves and roots was observed at concentrations of 50 and 500 µg L^{-1}, and an increase of the size of leaves was observed at the concentration of 5,000 µg L^{-1}. Flumequine is an antibiotic targeting the bacterial DNA-girase, which can also affect the plant DNA-topoisomerase II. Other studies have demonstrated the acute toxicity of antibiotics on plants cells, such as the effect of sulfadimethoxine on the plants of *Hordeum disticum*, *Panicum miliaceum*, *Pisum sativum*, and *Zea mays*, and the weeds of *Amaranthus retroexus*, *Plantago major*, and *Rumex acetosella* (Migliore et al. 2000).

Kong et al. (2007) reported the inhibitory effect of the antibiotic, oxytetracycline, on the growth of *Medicago sativa* (alfalfa). Although concentrations below 0.002 mM (0.92 mg L^{-1}) did not show any effect, a significant decrease of biomass was observed upon exposure to concentrations above 0.002 mM. Exposure to 0.02 M resulted in a 61% decrease of the shoot fresh weight and a 85% decrease of the root fresh weight. In general, oxytetracycline exhibited a stronger inhibitory effect on root growth than shoot growth. The uptake of oxytetracycline by alfalfa was strongly inhibited by the metabolic inhibitor, 2,4-dinitrophenol, and Hg^{2+}, but not by the aquaporin competitors, glycerol and Ag^{+}, suggesting that the inhibition was due to general cellular stress rather than specific effect on aquaporins. This study also showed that the uptake of oxytetracycline was pH-dependent with maximum uptake at pH 7, suggesting the implication of an energy-dependent process.

28.4 CONCLUSIONS AND PERSPECTIVES

As ubiquitous xenobiotic pollutants in wastewater, surface water, groundwater, and water supply, pharmaceuticals have recently raised much concerns about their potential effects on the environment and human health. Even though these chemicals are typically present at very low levels in the environment, the almost inexistence of chronic ecotoxicology data raises questions about the long term effects of pharmaceuticals on living organisms. Indeed, pharmaceuticals have typically been designed to act on specific biological targets and resist fast biodegradation upon introduction in the body. These characteristics make them persistent and active in the environment for long periods of time. Along with many other organic and inorganic compounds, pharmaceuticals are susceptible to be taken up and, to some extent, biodegraded inside plant tissues. This observation has two major practical consequences for phytotechnologies: first, aquatic plants populating constructed and natural wetlands can significantly contribute to the removal of pharmaceuticals from water and wastewater. Aquatic plants in constructed wetland could be used for the combined benefits of pollutant removal (including pharmaceuticals and nutrients) from wastewater and biomass production for conversion into renewable biofuel. On the other hand, agricultural crops may be contaminated by pharmaceuticals through irrigation water, agricultural runoff, and land application of municipal sewage sludge on agricultural fields.

Although a significant number of studies have shown the potential of aquatic plants for the removal pharmaceuticals from wastewater, there are today questions that need to be addressed. First, the single uptake of pharmaceuticals inside plant tissues may lead to accumulation of toxic metabolites and parent compounds inside plant tissues, eventually leading of their release into the environment or transfer to the food chain. Further research is then desirable in order to assess the actual detoxification process that occurs inside plant tissues. Second, research on the phytoremediation of pharmaceuticals by aquatic plants have been conducted using many plant species, many target compounds, and many experimental conditions, making generalization extremely difficult. As a consequence, a large variety of uptake rates and bioaccumulation factors have been reported and *a priori* evaluation of the efficiency of plant species for the removal of pharmaceuticals in constructed wetland is not accurately feasible.

Unlike bacteria and mammals, plants are autotrophic organisms that lack the enzymatic machinery necessary for metabolizing efficiently organic compounds, often resulting in slow and incomplete biodegradation. This has led to various attempts to modify plants genetically by the introduction of bacterial or mammalian genes involved in the breakdown of toxic chemicals. During the last decade, plants have been genetically modified to overcome the inherent limitations of plant detoxification capabilities, following a strategy similar to the development of transgenic crop (Eapen et al. 2007). Bacterial genes encoding enzymes involved in the breakdown of various organic pollutants, such as nitroreductase, cytochrome P-450, and peroxidases have been introduced in higher plants, resulting in significant enhancement of plant tolerance, uptake, and detoxification performances (Van Aken 2009). Similarly, transgenic plants exhibiting enhanced potential for biodegradation of pharmaceuticals would allow more complete detoxification of compounds taken up inside plant tissues and prevent the accumulation and potential release of toxic metabolites. However, this strategy would need a further understanding of the biochemistry and molecular bases of the metabolism of pharmaceuticals by other microorganisms.

ACKNOWLEDGMENTS

The authors are grateful to the Water Environmental Technology (WET) Center at Temple University and the National Institute of Environmental Health Sciences (NIEHS; award number ES05605).

REFERENCES

An, J., Q. Zhou, F. Sun, and L. Zhang. 2009. Ecotoxicological effects of paracetamol on seed germination and seedling development of wheat (*Triticum aestivum* L.). *Journal of Hazardous Materials* 169 (1): 3751–757.

Anderson, T.A., E.A. Guthrie, and B.T. Walton. 1993. Bioremediation in the rhizosphere. *Environmental Science, and Technology* 27 (13): 2630–2636.

Ark, P., and J. Dekker. 1958. Uptake of antibiotic gs1 by seeds and plants. *Phytopathology* 48 (8): 391–391.

Avila, C., A. Pedescoll, V. Matamoros, J. Maria Bayona, and J. Garcia. 2010. Capacity of a horizontal subsurface flow constructed wetland system for the removal of emerging pollutants: An injection experiment. *Chemosphere* 81 (9): 1137–1142.

Barnes, K.K., D.W. Kolpin, E.T. Furlong, et al. 2008. A national reconnaissance of pharmaceuticals and other organic wastewater contaminants in the United States—I) groundwater. *Science of the Total Environment* 402 (2–3): 192-200.

Bartha, B., C. Huber, R. Harpaintner, and P. Schroeder. 2010. Effects of acetaminophen in *Brassica juncea* L. Czern.: investigation of uptake, translocation, detoxification, and the induced defense pathways. *Environmental Science and Pollution Research* 17 (9): 1553–1562.

Boonsaner, M., and D.W. Hawker. 2010. Accumulation of oxytetracycline and norfloxacin from saline soil by soybeans. *Science of the Total Environment* 408 (7): 1731–1737.

Boxall, A.B.A., P. Johnson, E.J. Smith, C.J. Sinclair, E. Stutt, and L.S. Levy. 2006. Uptake of veterinary medicines from soils into plants. *Journal of Agricultural and Food Chemistry* 54 (6): 2288–2297.

Boxall, A.B.A. 2009. Assessing environmental effects of human pharmaceuticals. *Toxicology Letters* 189: S33–S33.

Brentner, L.B., S.T. Mukherji, K.M. Merchie, J.M. Yoon, J.L. Schnoor, and B. Van Aken. 2008. Expression of glutathione *S*-transferases in poplar trees (*Populus trichocarpa*) exposed to 2,4,6-trinitrotoluene (TNT). *Chemosphere* 73 (5): 657–662.

Brian, P., J. Wright, J. Stubbs, and A. Way. 1951. Uptake of antibiotic metabolites of soil microorganisms by plants. *Nature* 167 (42–44): 347–349.

Briggs, G.G., R.H. Bromilow, and A.A. Evans. 1982. Relationships between lipophilicity and root uptake and translocation of non-ionized chemicals by barley. *Pesticide Science* 13 (5): 495–504.

Brooks, R., J. Lee, R. Reeves, and T. Jaffre. 1977. Detection of nickelferous rocks by analysis of herbarium specimens of indicators plants. *Journal of Geochemeical Exploration* 7: 49–57.

Burken, J.G., and J.L. Schnoor. 1998b. Predictive relationships for uptake of organic contaminants by hybrid poplar trees. *Environmental Science, and Technology* 32 (21): 3379–3385.

Caliman, F.A. and M. Gavrilescu. 2009. Pharmaceuticals, personal care products and endocrine disrupting agents in the environment – A review. *CLEAN-Soil, Air, Water* 37: 277–303.

Castelfranco, P., C. Foy, and D. Deutsch. 1961. Non-enzymatic detoxification of 2-chloro-4,6-bis(ehtylamino)-S-triazine (simazine) by extract of *Zea mays*. *Weeds* 9: 580–591.

Celiz, M.D., T.S.O. Jerry, and S. Daina. 2009. Pharmaceutical metabolites in the environment: analytical challenges and ecological risks. *Environmental Toxicology and Chemistry* 28 (12): 2473–2484.

Chaney, R., M. Malik, Y. Li, S. Brown, E. Brewer, J. Angel, and A. Baker. 1997. Phytoremediation of soil metals. *Current Opnion in Biotechnology* 8: 279–284.

Chaudhry, Q., M. Blom-Zandstra, S. Gupta, and E.J. Joner. 2005. Utilising the synergy between plants and rhizosphere microorganisms to enhance breakdown of organic pollutants in the environment. *Environmental Science and Pollution Research* 12 (1): 34–48.

Cole, D.J. 1983. Oxidation of xenobiotics in plants. *Progress in Pest Biochemistry and Technolology* 3: 199–253.

Coleman, J., M. Blake-Kalff, and T. Davies. 1997. Detoxification of xenobiotics by plants: chemical modification and vacuolar compartamentation. *Trends in Plant Science* 2 (4): 144–151.

Cropp, R.A., and D.W. Hawker, M. Boonsaner. 2010. Predicting the accumulation of organic contaminants from soil by plants. *Environmental Contamination and Toxicology* 5 (5): 25–529.

Daar, A.S., H. Thorsteinsdottir, D.K. Martin, A.C. Smith, S. Nast, and P.A. Singer. 2002. Top ten biotechnologies for improving health in developing countries. *Nature Genetics* 32 (2): 229–232.

Daughton, C.G. 2004. Non-regulated water contaminants: emerging research. *Environmental Impact Assessment Review* 24 (7–8): 711–732.

Daughton, C.G., and T.A. Ternes. 1999. Pharmaceuticals and personal care products in the environment: Agents of subtle change? *Environmental Health Perspectives* 107: 907–938.

Dolliver, H., K. Kumar, and S. Gupta. 2007. Sulfamethazine uptake by plants from manure-amended soil. *Journal of Environmental Quality* 36 (4): 1224–1230.

Dordio, A., R. Ferro, D. Teixeira, A.J. Palace, A.P. Pinto, and C.M.B. Dias. 2011. Study on the use of *Typha* spp. for the phytotreatment of water contaminated with ibuprofen. *International Journal of Environmental Analytical Chemistry* 91 (7–8): 654–667.

Dordio, A., A.J. Palace Carvalho, D.M. Teixeira, C.B. Dias, and A.P. Pinto. 2010. Removal of pharmaceuticals in microcosm constructed wetlands using *Typha* spp. and LECA. *Bioresource Technology* 101 (3): 886–892.

Doty, S.L., C.A. James, A.L. Moore, A. Vajzovic et al. 2007. Enhanced phytoremediation of volatile environmental pollutants with transgenic trees. *Proceedings of National Academic of Sciences USA* 104: 16816–16821.

Doty, S.L., M.R. Dosher, G.L. Singleton et al. 2005. Identification of an endophytic rhizobium in stems of *Populus*. *Symbiosis* 39 (1): 27–35.

Eapen, S., S. Singh, and S.F. D'Souza. 2007. Advances in development of transgenic plants for remediation of xenobiotic pollutants. *Biotechnology Advances* 5: 442–451.

Eapen, S., and S.F. D'Souza. 2005. Prospects of genetic engineering of plants for phytoremediation of toxic metals. *Biotechnology Advances* 23: 297–114.

Farkas, M.H., J.O. Berry, and D.S. Aga. 2006. Induction of glutathione transferase activity in plants after tetracycline treatment. *Abstracts of Papers of the American Chemical Society* 231: 9-AGRO.

Fent, K., A. Weston, and D. Caminada. 2006. Ecotoxicology of human pharmaceuticals. *Aquatic Toxicology* 76 (2): 122–159.

Ferro, A., M. Gefell, R. Kjelgren, D. Lipson, N. Zollinger, and S. Jackson. 2003. Maintaining hydraulic control using deep-rooted tree systems. *Advances in Biochemical Engineering/Biotechnology* 78: 125–156.

Forni, C., A. Cascone, S. Cozzolino, and L. Migliore. 2001. Drugs uptake and degradation by aquatic plants as a bioremediation technique. *Minerva Biotecnologica* 13 (2): 151–152.

Forni, C., A. Cascone, M. Fiori, and L. Migliore. 2002. Sulphadimethoxine and *Azolla filiculoides* Lam.: a model for drug remediation. *Water Research* 36 (13): 3398–3403.

Gobel, A., C.S. McArdell, M.J.-F. Suter, and W. Giger. 2004. Trace determination of macrolide and sulfonamide antimicrobials, a human sulfonamide metabolite, and trimethoprim in wastewater using liquid chromatography coupled to electrospray tandem mass spectrometry. *Analytical Chemistry* 76: 4756–4764.

Gujarathi, N.P., B.J. Haney, and J.C. Linden. 2005. Phytoremediation potential of *Myriophyllum aquaticum* and *Pistia stratiotes* to modify antibiotic growth promoters, tetracycline, and oxytetracycline, in aqueous wastewater systems. *International Journal of Phytoremediation* 7: 299–112.

Gujarathi, N.P., B.J. Haney, H.J. Park, S.R. Wickramasinghe, and J.C. Linden. 2005. Hairy roots of *Helianthus annuus*: a model system to study phytoremediation of tetracycline and oxytetracycline. *Biotechnology Progress* 21 (3): 775–780.

Herklotz, P.A., P. Gurung, B.V. Heuvel, and C.A. Kinney. 2010. Uptake of human pharmaceuticals by plants grown under hydroponic conditions. *Chemosphere* 78 (11): 1416–1421.

Huber, C., B. Bartha, R. Harpaintner, and P. Schroeder. 2009. Metabolism of acetaminophen (paracetamol) in plants-two independent pathways result in the formation of a glutathione and a glucose conjugate. *Environmental Science and Pollution Research* 16 (2): 206–213.

Imai, S., A. Shiraishi, K. Gamo, I. Watanabe, H. Okuhata, H. Miyasaka, K. Ikeda, T. Bamba, and K. Hirata. 2007. Removal of phenolic endocrine disruptors by *Portulaca oleracea*. *Journal of Bioscience and Bioengineering* 103 (5): 420–426.

Inui, H., and H. Ohkawa. 2005. Herbicide resistance in transgenic plants with mammalian P450 monooxygenase genes. *Pest Management Science* 61 (3): 286–291.

Jones-Lepp, T.L., and R. Stevens. 2007. Pharmaceuticals and personal care products in biosolids/sewage sludge: the interface between analytical chemistry and regulation. *Analytical and Bioanalytical Chemistry* 387 (4): 1173–1183.

Jones-Lepp, T.L., C.A. Sanchez, T. Moy, and R. Kazemi. 2010. Method development and application to determine potential plant uptake of antibiotics and other drugs in irrigated crop production systems. *Journal of Agricultural and Food Chemistry* 58 (22): 11568–11573.

Karnjanapiboonwong, A., D.A. Chase, J.E. Canas, W.A. Jackson, J.D. Maul, A.N. Morse, and T.A. Anderson. 2011. Uptake of 17 alpha-ethynylestradiol and triclosan in pinto bean, *Phaseolus vulgaris*. *Ecotoxicology and Environmental Safety* 74 (5): 1336–1342.

Kinney, C., E. Furlong, S. Werner, and J. Cahill. 2006. Presence and distribution of wastewater-derived pharmaceuticals in soil irrigated with reclaimed water. *Environmental Toxicology and Chemistry* 25 (2): 317–326.

Klein, M., E. Martinoia, G. Hoffmann-Thoma, and G. Weissenbock. 2000. A membrane-potential dependent ABC-like transporter mediates the vacuolar uptake of rye flavone glucuronides: regulation of glucuronide uptake by glutathione and its conjugates. *Plant Journal* 21 (3): 289–304.

Klein, M., E. Martinoia, and G. Weissenbock. 1998. Directly energized uptake of beta-estradiol 17-(beta-D-glucuronide) in plant vacuoles is strongly stimulated by glutathione conjugates. *Journal of Biological Chemistry* 273 (1): 262–270.

Kleywegt, S., V. Pileggi, P. Yang et al. 2011. Pharmaceuticals, hormones and bisphenol A in untreated source and finished drinking water in Ontario, Canada—and treatment efficiency. *Science of the Total Environment* 409 (8): 1481–1488.

Kohler, A., S. Hellweg, B.I. Escher, and K. Hungerbuhler. 2006. Organic pollutant removal versus toxicity reduction in industrial wastewater treatment: The example of wastewater from fluorescent whitening agent production. *Environmental Science and Technology* 40: 3395–3401.

Kolpin, D., E. Furlong, M. Meyer et al. 2002. Pharmaceuticals, hormones, and other organic wastewater contaminants in US streams, 1999–2000: a national reconnaissance. *Environmental Science, and Technology* 36 (6): 1202–1211.

Kong, W.D., Y.G. Zhu, Y.C. Liang, et al. 2007. Uptake of oxytetracycline and its phytotoxicity to alfalfa (*Medicago sativa* L.). *Environmental Pollution* 147 (1): 187–193.

Kotyza, J., P. Soudek, Z. Kafka, and T. Vanek. 2010. Phytoremediation of pharmaceuticals preliminary study. *International Journal of Phytoremediation* 12 (3): 306–316.

Kumar, A., B. Chang, and I. Xagoraraki. 2010. Human health risk assessment of pharmaceuticals in water: issues and challenges ahead. *International Journal of Environmental Research and Public Health* 7 (11): 3929–3953.

Kumar, K., S.C. Gupta, S.K. Baidoo, Y. Chander, and C.J. Rosen. 2005. Antibiotic uptake by plants from soil fertilized with animal manure. *Journal of Environmental Quality* 34 (6): 2082–2085.

Kummerer, K. 2004. Resistance in the environment. *Journal of Antimicrobial Chemotherapy* 54: 311–320.

Lapen, D.R., E. Topp, C.D. Metcalfe, H. Li, M. Edwards, N. Gottschall, P. Bolton, W. Curnoe, M. Payne, and A. Beck. 2008. Pharmaceutical and personal care products in tile drainage following land application of municipal biosolids. *Science of the Total Environment* 399 (1–3): 50–65.

Li, Z.-J., X.-Y. Xie, S.-Q. Zhang, and Y.-C. Liang. 2011. Wheat growth and photosynthesis as affected by oxytetracycline as a soil contaminant. *Pedosphere* 21 (2): 244–250.

Limbert, E., and W. Betts. 1996. Influence of substrate chemistry and microbial metabolic diversity on the bioremediation of xenobiotic contamination. *The Genetic Engineer and Biotechnologist* 16 (3): 159–180.

Litwack, G., and D. Pramer. 1957. Absorption of antibiotics by plant cells. 3. Kinetics of streptomycin uptake. *Archives of Biochemistry and Biophysics* 68 (2): 396–402.

Macek, T., M. Mackova, and J. Kas. 2000. Exploitation of plants for the removal of organics in environmental remediation. *Biotechnology Advances* 18 (1): 23–34.

McCutcheon, S., and J. Schnoor. 2003. Overview of phytotransformation and control of wastes. In *Phytoremediation. Transformation and Control of Contaminants*, eds. S. McCutcheon, and J. Schnoor, 3–58. John Wiley.

Meagher, R.B. 2000. Phytoremediation of toxic elemental and organic pollutants. *Current Opinion in Plant Biology* 3 (2): 153–162.

Migliore, L., S. Cozzolino, and M. Fiori. 2000. Phytotoxicity to and uptake of flumequine used in intensive aquaculture on the aquatic weed, *Lythrum salicaria* L. *Chemosphere* 40 (7): 741–750.

Newman, L.A., and C.M. Reynolds. 2004. Phytodegradation of organic compounds. *Current Opinion in Biotechnology* 15 (3): 225–230.

Newman, L.A., S.E. Strand, N. Choe, et al. 1997. Uptake and biotransformation of trichloroethylene by hybrid poplars. *Environmental Science, and Technology* 31 (4): 1062–1067.

Norris, A., and S. Burgin. 2011. Apparent rapid loss of endocrine disruptors from wetlands used to store either tertiary treated sewage effluent or stormwater runoff. *Water, Air and Soil Pollution* 219 (1–4): 285–295.

Onesios, K.M., J.T. Yu, and E.J. Bouwer. 2009. Biodegradation and removal of pharmaceuticals and personal care products in treatment systems: a review. *Biodegradation* 20 (4): 441–466.

Peterson, E., and R. Sinha. 1977. Uptake, distribution and persistence of tetracycline antibiotics in various plant species susceptible to mycoplasma-infection. *Journal of Phytopathology* 90 (3): 250–256.

Pilon-Smits, E. 2005. Phytoremediation. *Annual Review of Plant Biology* 56: 15–39.

Redshaw, C.H., V.G. Wootton, and S.J. Rowland. 2008. Uptake of the pharmaceutical fluoxetine hydrochloride from growth medium by Brassicaceae. *Phytochemistry* 69 (13): 2510–2516.

Reinhold, D., S. Vishwanathan, J.J. Park, D. Oh, and F.M. Saunders. 2010. Assessment of plant-driven removal of emerging organic pollutants by duckweed. *Chemosphere* 80 (7): 687–692.

Salt, D., R. Smith, and I. Raskin. 1998. Phytoremediation. *Annual Review of Plant Physiology and Plant Molecular Biology* 49: 643–668.

Sandermann, H. 1994. Higher plant metabolism of xenobiotics: the 'green liver' concept. *Pharmacogenetics* 4: 225–241.

Schnoor, J.L. 1997. Phytoremediation. Technology Evaluation Report TE-98-0. Ground-Water Remediation Technologies Analysis Center, Pittsburgh, PA.

Schnoor, J., L. Licht, S. McCutcheton, N. Wolfe, and L. Carreira 1995. Phytoremediation of organic and nutrient contaminants. *Environ Sci Technol* 29 (7): 318A–323A.

Schroder, P., C. Scheer, and B.J.D. Belford 2001. Metabolism of organic xenobiotics in plants: conjugating enzymes and metabolic end points. *Minerva Biotecnologica* 13 (2): 85–91.

Schwab, B., E. Hayes, J. Fiori, et al. 2005. Human pharmaceuticals in US surface waters: A human health risk assessment. *Regulatory Toxicology and Pharmacology* 42 (3): 296–312.

Shenker, M., D. Harush, J. Ben-Ari, and B. Chefetz. 2011. Uptake of carbamazepine by cucumber plants—a case study related to irrigation with reclaimed wastewater. *Chemosphere* 82 (6): 905–910.

Singer, A. 2006. The chemical ecology of pollutants biodegradation. In *Phytoremediation and Rhizoremediation: Theoretical Background*, eds. M. Mackova, D. Dowling, and T. Macek, 5–21. Springer.

Song, H., K. Nakano, T. Taniguchi, M. Nomura, and O. Nishimura. 2009. Estrogen removal from treated municipal effluent in small-scale constructed wetland with different depth. *Bioresource Technology* 100 (12): 2945–2951.

Ternes, T. 1998. Occurrence of drugs in German sewage treatment plants and rivers. *Water Research* 32 (11): 3245–3260.

Ternes, T., A. Joss, and H. Siegrist. 2004. Scrutinizing pharmaceuticals and personal care products in wastewater treatment. *Environmental Science, and Technology* 38 (20): 392A–399A.

Trapp, S., and M. Matthies. 1995. Generic one-compartment model for uptake of organic-chemicals by foliar vegetation. *Environmental Science, and Technology* 29 (9): 2333–2338.

Trapp, S., C. MCFarlane, and M. Matthies. 1994. Model for uptake of xenobiotics into plants—validation with bromacil experiments. *Environmental Toxicology and Chemistry* 13 (3): 413–422.

U.S. Environmental Protection Agency (USEPA). 2009. *Targeted National Sewage Sludge Survey Sampling and Analysis Technical Report.*

Van Aken, B., P.A. Correa, and J.L. Schnoor. 2010. Phytoremediation of polychlorinated biphenyls: new trends and promises. *Environmental Science, and Technology* 44 (8): 2767–2776.

Van Aken, B. 2009. Transgenic plants for the enhanced phytoremediation of explosives. *Current Opinion in Biotechnology* 20: 1–6.

Van Aken, B. 2008. Transgenic plants for phytoremediation: helping nature to clean up environmental pollution. 26 (5): 225–228.

Van Aken, B., J.M. Yoon, C.L. Just, and J.L. Schnoor. 2004. Metabolism and mineralization of hexahydro-1,3,5-trinitro-1,3,5-triazine inside poplar tissues (*Populus deltoides* x *nigra* DN-34). *Environmental Science, and Technology* 38 (17): 4572–4579.

Winker, M., J. Clemens, M. Reich, H. Gulyas, and R. Otterpohl. 2010. Ryegrass uptake of carbamazepine and ibuprofen applied by urine fertilization. *Science of the Total Environment* 408 (8): 1902–1908.

Wu, C., A.L. Spongberg, J.D. Witter, M. Fang, and K.P. Czajkowski. 2010. Uptake of pharmaceutical and personal care products by soybean plants from soils applied with biosolids and irrigated with contaminated water. *Environmental Science, and Technology* 44 (16): 6157–6161.

Yoon, J.M., B. Van Aken, and J.L. Schnoor. 2006. Leaching of contaminated leaves following uptake and phytoremediation of RDX, HMX, and TNT by poplar. *International Journal of Phytoremediation* 8 (1): 81–94.

Index

Page numbers followed by *f* and *t* indicate figures and tables, respectively.

C

D

F

G

H

I

J

K

L

M

N

O

R

U

V

W

X

Z

9781032340265